INTERNATIONAL ATOMIC MASSES* *(Continued)*

Name	Symbol	Atomic Number	Atomic Mass	Name	Symbol	Atomic Number	Atomic Mass
Xenon	Xe	54	131.29(2)	Zinc	Zn	30	65.39(2)
Ytterbium	Yb	70	173.04(3)	Zirconium	Zr	40	91.224(2)
Yttrium	Y	39	88.90585(2)				

*The 1995 values from the IUPAC World Wide Web page at http://www.chem.qmw.ac.uk/iupac/AtWt/
Numbers in parentheses indicate the uncertainty in the last digit of the atomic masses. Values enclosed in brackets, for example [209], indicate the mass number of the longest-lived isotope of the element. Three such elements (Th, Pa, and U) do have a characteristic terrestrial isotopic composition, and for these elements, an atomic mass is tabulated. For a detailed discussion of the values in this table, see *Pure Appl. Chem.*, **1996**, *68*, 2339.

IMPORTANT PHYSICAL CONSTANTS

Constant	Symbol	Value
Speed of light (*in vacuo*)	c	2.99792×10^{8} ms^{-1}
Planck constant	h	6.62608×10^{-34} J s
Avogadro number	N	6.022137×10^{23} particles mol^{-1}
Faraday constant	F	96485.31 C mol^{-1}
Gas constant	R	$8.31451 \times$ J K^{-1} mol^{-1} 0.0820578 L atm K^{-1} mol^{-1}
Boltzmann constant	k	1.38066×10^{-23} J K^{-1}
Rest mass of the electron	m_c	9.10939×10^{-31} kg
Electronic charge	e	$-1.602177 \times 10^{-19}$ C

ENERGY CONVERSION FACTORS

	Joules	Ergs	Calories	Liter Atmosphere	Electron Volts
1 joule =	1	10^{7}	2.3901×10^{-1}	9.8692×10^{-3}	6.2415×10^{18}
1 erg =	10^{-7}	1	2.3901×10^{-8}	9.8692×10^{-10}	6.2415×10^{11}
1 calorie =	4.1840	4.1840×10^{7}	1	4.1293×10^{-2}	2.6114×10^{19}
1 liter atmosphere =	1.0133×10^{2}	1.0133×10^{9}	24.217	1	6.3242×10^{20}
1 electron volt =	1.6022×10^{-19}	1.6022×10^{-12}	3.8293×10^{-20}	1.5812×10^{-21}	1

PRINCIPLES OF Instrumental Analysis

PRINCIPLES OF Instrumental Analysis

Fifth Edition

Douglas A. Skoog
Stanford University

F. James Holler
University of Kentucky

Timothy A. Nieman
University of Illinois at Urbana-Champaign

BROOKS/COLE
THOMSON LEARNING

Australia • Canada • Mexico • Singapore • Spain
United Kingdom • United States

BROOKS/COLE
THOMSON LEARNING

Chemistry Editor: John Vondeling
Developmental Editors: Marc Sherman, Jennifer Bortel
Project Editor: Frank Messina
Marketing Strategist: Angus McDonald
Production Manager: Charlene Catlett Squibb
Text Designer: Ruth Hoover
Art Director: Lisa Adamitis
Copy Editor: Vince Pecuilis
Cover Designer: Chazz Bjanes
Cover Image: John T. Foley, Department of Physics and Astronomy, Mississippi State University
Cover Printer: Phoenix Color Corp
Compositor: The Clarinda Company
Printer: Phoenix Color Corp

Printed in the United States of America
12 05 04

For more information about our products, contact us at:
Thomson Learning Academic Resource Center
1-800-423-0563

For permission to use material from this text, contact us by:
Phone: 1-800-730-2214 **Fax:** 1-800-730-2215
Web: http//www.thomsonrights.com

Library of Congress Catalog Card Number: 97-67376

PRINCIPLES OF INSTRUMENTAL ANALYSIS,
Fifth Edition

ISBN: 0-03-002078-6

Asia
Thomson Learning
60 Albert Street, #15-01
Albert Complex
Singapore 189969

Australia
Nelson Thomson Learning
102 Dodds Street
South Melbourne, Victoria 3205
Australia

Canada
Nelson Thomson Learning
1120 Birchmount Road
Toronto, Ontario M1K 5G4
Canada

Europe/Middle East/Africa
Thomson Learning
Berkshire House
168-173 High Holborn
London WC1 V7AA
United Kingdom

Latin America
Thomson Learning
Seneca, 53
Colonia Polanco
11560 Mexico D.F.
Mexico

Spain
Paraninfo Thomson Learning
Calle/Magallanes, 25
28015 Madrid, Spain

Preface

Physical scientists, biological scientists, and engineers have an impressive array of powerful and elegant tools for gathering qualitative and quantitative information about the composition and structure of matter. Students of chemistry, biochemistry, physics, geology, health-related sciences, engineering, and environmental sciences must develop an understanding of these tools and how they are used to solve analytical problems. It is to such students that this book is addressed.

We believe that the correct choice and efficient use of modern analytical instruments requires an understanding of the fundamental principles of operation of these measuring devices. Only when the student-analyst attains such understanding can reasonable choices be made among the several alternative means of solving an analytical problem; only then will the student be aware of the pitfalls that accompany physical measurements, and only then will the student be sufficiently attuned to the limitations in sensitivity, precision, and accuracy of instrumental measurement. With these thoughts in mind, it is our objective that this text give the student an introduction to the principles of spectroscopic, electrometric, and chromatographic methods of analysis. With careful and thorough study of this text, the student will discover the kinds of instruments that are currently available as well as their strengths and limitations.

ORGANIZATION OF THE FIFTH EDITION

Users of earlier editions of this text will find that this edition is organized in a somewhat different manner than its predecessors. After a brief introductory chapter that now includes a section on calibration, the book is divided into six major sections:

- Section I comprises four chapters on electronic circuit fundamentals, operational amplifiers, digital electronics and computers, signals, noise, data enhancement techniques, and figures of merit.
- Section II comprises seven chapters devoted to the various branches of atomic spectroscopy including optical absorption, optical emission, and optical fluorescence spectrometry, atomic mass spectrometry, and atomic X-ray spectrometry.
- Section III treats molecular spectroscopy in nine chapters that address optical absorption, emission, luminescence, infrared, Raman, nuclear magnetic resonance, and mass spectrometry. This section concludes with a chapter that presents spectroscopic and microscopic techniques for characterization of surfaces.
- Section IV contains four chapters on electroanalytical chemistry techniques including potentiometry, coulometry, and voltammetry.
- Section V consists of five chapters that discuss various types of chromatography and electrophoresis.
- Section VI consists of three chapters that discuss miscellaneous instrumental approaches to the solution of analytical problems including thermal, radiochemical, and automated methods.

Since the appearance of the first edition of this text in 1971, the field of instrumental analysis has grown to be so large and so diverse that treatment of all of the modern instrumental techniques for chemical analysis is impossible in a one- or even two-semester course. Furthermore, we find that opinions differ among instructors as to which methods should be discussed and which should be omitted. For this reason, we have included in this text far more material than can be covered in a single instrumental analysis course; however, we have organized the material in such a way that instructors can pick and choose the topics to be studied. Thus, as in the fourth edition, introductory chapters on optical spectroscopy, electroanalytical chemistry, and chromatography precede the chapters dealing with specific methods of each type. After the students have mastered these introductory materials, any one of the chapters that follow can be assigned in any order. To assist professors and students in using this book, an **Instructor's Manual** is available, containing answers to all of the Questions and solutions to all of the Problems.

NEW IN THE FIFTH EDITION

- A new chapter on capillary electrophoresis (Chapter 30).
- A new chapter devoted to supercritical fluid chromatography and extraction (Chapter 29).
- A new chapter dealing with atomic mass spectrometry (Chapter 11).
- An addition to the chapter on surface characterization that discusses scanning tunneling and atomic force microscopy (Chapter 21).
- A much expanded and reorganized section dealing with methods of atomization in atomic spectrometry (Chapter 8).
- A description of the piezoelectric effect and the quartz crystal microbalance (Chapter 1).
- An introduction to the data domains concept (Chapter 1).
- A revised and updated discussion of the van Deemter equation (Chapter 26).
- A new section on magnetic resonance imaging (Chapter 19).
- A discussion of some software applications that are useful to the analyst (Chapter 4).
- Revised and updated tables of physicochemical data (Appendix 3 and endpapers).
- A comprehensive list of acronyms and abbreviations that are encountered in analytical chemistry (Appendix 5).

ACKNOWLEDGMENTS

We wish to acknowledge with thanks the significant contributions of the following colleagues who read all or parts of the manuscript in detail and offered numerous helpful suggestions and corrections:

Jennifer Brodbelt, The University of Texas, Austin

Christie G. Enke, University of New Mexico

Peter W. Faguy, University of Louisville

Thomas Gennett, Rochester Institute of Technology

James R. Kincaid, Marquette University

Robert W. Kiser, University of Kentucky

Robert J. Morris, Ball State University

James E. O'Reilly, University of Kentucky

Alexander Scheeline, University of Illinois at Urbana-Champaign

Stanford L. Smith, University of Kentucky

John Walters, St. Olaf College

Kathryn Williams, University of Florida

Steven W. Yates, University of Kentucky

Once again, our team has benefited immeasurably from our association with a superb technical reference librarian, Ms. Maggie Johnson, head of the Chemistry/Physics Library at the University of Kentucky. She has assisted us in performing many tasks in the production of this text, including checking references, performing literature searches, and providing background information for many of the chapters. We are grateful for her competence, enthusiasm, and good humor.

Numerous manufacturers of analytical instrumentation and other products and services related to analytical chemistry have contributed to the writing of this book by providing diagrams, application notes, and photos of their products. We are especially grateful to Perkin-Elmer Corporation, Nicolet Corporation, Bioanalytical Systems, and Mettler Instruments for providing photos for the section openers, as well as to Galactic Industries, Inc. and Mathsoft, Inc. for making their software available to us for executing various tasks related to the production of figures. The authors also are indebted to John T. Foley of the Department of Physics and Astronomy at Mississippi State University for providing the digital image that served as the basis for the cover of this book.

The authors would like to note that certain parts of this text appear in two other books coauthored by two of the present authors and Professor Donald M. West of San Jose State University.[1] The principal overlap occurs in the chapters on electrochemistry and, to a lesser extent, in the chapters on ultraviolet and visible absorp-

[1]D. A. Skoog, D. M. West, and F. J. Holler, *Fundamentals of Analytical Chemistry,* 7th ed., Philadelphia: Saunders College Publishing, 1996; and *Analytical Chemistry: An Introduction,* 6th ed., Philadelphia: Saunders College Publishing, 1994.

tion spectroscopy, atomic spectroscopy, and gas and liquid chromatography. We acknowledge the contributions of Professor West to these chapters and thus to this new edition; we are especially grateful to him for preparing a superb Index to this volume and for contributing many valuable comments and suggestions during the course of the project.

We also want to thank the various persons at Saunders College Publishing for their friendly assistance in completing this project under difficult circumstances. Among these are Project Editor Frank Messina, who wrestled with the many day-to-day details of the production of this book; Associate Editor Marc Sherman; Art Director Lisa Caro; and Production Manager Charlene Catlett Squibb. Finally, we wish to express our thanks to our Acquisitions Editor and Publisher John Vondeling, whose patience and understanding are exceeded only by his talent for motivating authors.

Douglas A. Skoog, Stanford University
F. James Holler, University of Kentucky
Timothy A. Nieman, University of Illinois at Urbana-Champaign
July 1997

Dedication

While this book was in press, our friend, colleague, and coauthor Timothy A. Nieman of the University of Illinois at Urbana-Champaign died after a lengthy illness. His loss is a misfortune for his family, his friends, and future generations of chemistry students who might have benefited from his consummate skill as a teacher, author, research mentor, and chemist. We dedicate the fifth edition of *Principles of Instrumental Analysis* to Tim. We will miss his leadership, wisdom, counsel, and friendship.

Douglas A. Skoog
F. James Holler
August 1997

Contents Overview

Section IV

Electroanalytical Chemistry 563

Section V

Separation Methods 673

Section VI

Miscellaneous Methods 797

Appendices A-1

Contents

Section III

Molecular Spectroscopy 299

Section IV

Electroanalytical Chemistry 563

Section V

Separation Methods 673

Section VI

Miscellaneous Method 797

Appendices

Chapter 1

Introduction

Analytical chemistry deals with methods for determining the chemical composition of samples of matter. A qualitative method yields information about the identity of atomic or molecular species or the functional groups in the sample; a quantitative method, in contrast, provides numerical information as to the relative amount of one or more of these components.

1A CLASSIFICATION OF ANALYTICAL METHODS

Analytical methods are often classified as being either *classical* or *instrumental.* This classification is largely historical with classical methods, sometimes called *wet-chemical methods,* preceding instrumental methods by a century or more.

1A-1 Classical Methods

In the early years of chemistry, most analyses were carried out by separating the components of interest (the *analytes*) in a sample by precipitation, extraction, or distillation. For qualitative analyses, the separated components were then treated with reagents that yielded products that could be recognized by their colors, their boiling or melting points, their solubilities in a series of solvents, their odors, their optical activities, or their refractive indexes. For quantitative analyses, the amount of analyte was determined by *gravimetric* or by *titrimetric* measurements. In gravimetric measurements, the mass of the analyte or some compound produced from the analyte was determined. In titrimetric procedures, the volume or mass of a standard reagent required to react completely with the analyte was measured.

These classical methods for separating and determining analytes still find use in many laboratories. The extent of their general application is, however, decreasing with the passage of time and with the advent of instrumental methods to supplant them.

1A-2 Instrumental Methods

Early in the twentieth century, chemists began to exploit phenomena other than those used for classical methods for solving analytical problems. Thus, measurements of physical properties of analytes—such as conductivity, electrode potential, light absorption or emission, mass-to-charge ratio, and fluorescence—began to be used for quantitative analysis of a variety of inorganic, organic, and biochemical analytes. Furthermore, highly efficient chromatographic and electrophoretic techniques began to replace distillation, extraction, and precipitation for the separation of components of complex mixtures prior to their qualitative or quantitative determination. These newer methods for separating and determining chemical species are known collectively as *instrumental methods of analysis.*

Many of the phenomena that instrumental methods are based on have been known for a century or more. Their application by most chemists, however, was delayed by lack of reliable and simple instrumentation. In fact, the growth of modern instrumental methods of analysis has paralleled the development of the electronics and computer industries.

1B TYPES OF INSTRUMENTAL METHODS

For this discussion, it is useful to consider chemical and physical characteristics that are useful for qualitative or quantitative analysis. Table 1-1 lists most of the characteristic properties that are currently used for instrumental analysis. Most of the characteristics listed in the table require a source of energy to stimulate a measurable response from the analyte. For example, in atomic emission an increase in the temperature of the analyte is required to first produce gaseous analyte atoms and then to excite the atoms to higher energy states. The excited-state atoms then emit characteristic electromagnetic radiation, which is the quantity measured by the instrument. Sources of excitation energy may take the form of a rapid thermal change as in the previous example, electromagnetic radiation from a selected region of the spectrum, application of one of the electrical quantities—voltage, current, or charge—or perhaps subtler forms intrinsic to the analyte itself.

Note that the first six entries in Table 1-1 involve interactions of the analyte with electromagnetic radiation. In the first property, radiant energy is produced by

TABLE 1-1 Chemical and Physical Properties Employed in Instrumental Methods

Characteristic Properties	Instrumental Methods
Emission of radiation	Emission spectroscopy (X-ray, UV, visible, electron, Auger); fluorescence, phosphorescence, and luminescence (X-ray, UV, and visible)
Absorption of radiation	Spectrophotometry and photometry (X-ray, UV, visible, IR); photoacoustic spectroscopy; nuclear magnetic resonance and electron spin resonance spectroscopy
Scattering of radiation	Turbidimetry; nephelometry; Raman spectroscopy
Refraction of radiation	Refractometry; interferometry
Diffraction of radiation	X-Ray and electron diffraction methods
Rotation of radiation	Polarimetry; optical rotary dispersion; circular dichroism
Electrical potential	Potentiometry; chronopotentiometry
Electrical charge	Coulometry
Electrical current	Amperometry; polarography
Electrical resistance	Conductometry
Mass	Gravimetry (quartz crystal microbalance)
Mass-to-charge ratio	Mass spectrometry
Rate of reaction	Kinetic methods
Thermal characteristics	Thermal gravimetry and titrimetry; differential scanning colorimetry; differential thermal analyses; thermal conductometric methods
Radioactivity	Activation and isotope dilution methods

the analyte; the next five properties involve changes in electromagnetic radiation brought about by its interaction with the sample. Four electrical properties then follow. Finally, four miscellaneous properties are grouped together: mass-to-charge ratio, reaction rate, thermal characteristics, and radioactivity.

The second column in Table 1-1 lists the names of instrumental methods that are based upon the various physical and chemical properties. Be aware that it is not always easy to select an optimal method from among available instrumental techniques and their classical counterparts. Some instrumental techniques are more sensitive than classical techniques, but others are not. With certain combinations of elements or compounds, an instrumental method may be more selective; with others, a gravimetric or volumetric approach may suffer less interference. Generalizations on the basis of accuracy, convenience, or expenditure of time are equally difficult to draw. Nor is it necessarily true that instrumental procedures employ more sophisticated or more costly apparatus; indeed, the modern electronic analytical balance used for gravimetric determinations is a more complex and refined instrument than some of those used in the other methods listed in Table 1-1.

As noted earlier, in addition to the numerous methods listed in the second column of Table 1-1, there is a group of instrumental procedures that are used for separation and resolution of closely related compounds. Most of these procedures are based upon chromatography or electrophoresis. One of the characteristics listed in Table 1-1 is ordinarily used to complete the analysis following chromatographic separations. Thus, for example, thermal conductivity, ultraviolet and infrared absorption, refractive index, and electrical conductance have been used for this purpose.

This text deals with the principles, the applications, and the performance characteristics of the instrumental methods listed in Table 1-1 and of chromatographic and electrophoretic separation procedures as well. No space is devoted to the classical methods, the assumption being that the reader will have encountered these techniques in earlier studies.

1C INSTRUMENTS FOR ANALYSIS

An instrument for chemical analysis converts information stored in the physical or chemical characteristics of the analyte to information that may be manipulated and interpreted by a human. Thus, an analytical instrument can be viewed as a communication device between the system under study and the investigator. To retrieve the desired information from the analyte, it is necessary to provide a stimulus, which is usually in the form of electromagnetic, electrical, mechanical, or nuclear energy as illustrated in Figure 1-1. The stimulus elicits a response from the system under study whose nature and magnitude are governed by the fundamental laws of chemistry and physics. The resulting information is contained in the phenomena that result from the interaction of the stimulus with the analyte. A familiar example is the passage of a narrow band of wavelengths of visible light through a sample to measure the extent of its absorption by the analyte. The intensity of the light is determined before and after its interaction with the sample, and the ratio of these intensities provides a measure of the analyte concentration.

Generally, instruments for chemical analysis comprise just a few basic components, some of which are listed in Table 1-2. To understand the relationships among these instrument components and the flow of information from the characteristics of the analyte through the components to the numerical or graphical output produced by the instrument, it is instructive to explore the concept of *data domains.*

1C-1 Data Domains

The measurement process is aided by a wide variety of devices that convert information from one form to another. In order to investigate how instruments function, it is important to understand the way in which information is *encoded,* or transformed from one system of information to another, as a characteristic of *electrical signals*—that is, as voltage, current, charge, or variations in these quantities. The various modes of encoding information electrically are called *data domains.* A classification scheme has been developed based on this

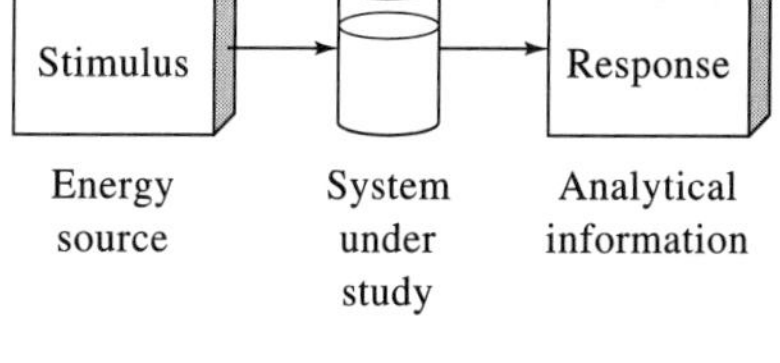

Figure 1-1 Block diagram showing the overall process of an instrumental measurement.

TABLE 1-2 Some Examples of Instrument Components

Instrument	Energy Source (stimulus)	Analytical Information	Input Transducer	Data Domain of Transduced Information	Information Processor	Readout
Photometer	Tungsten lamp, glass filter	Attenuated light beam	Photocell	Electrical current	Meter scale	Current meter
Atomic emission spectrometer	Flame	UV or visible radiation	Photomultiplier tube	Electrical potential	Amplifier, demodulator, monochromator chopper	Chart recorder
Coulometer	DC source	Cell current	Electrodes	Electrical current	Amplifier	Chart recorder
pH meter	Sample/glass electrode	Hydrogen ion activity	Glass-calomel electrodes	Electrical potential	Amplifier, digitizer	Digital unit
X-Ray powder diffractometer	X-Ray tube, sample	Diffracted radiation	Photographic film	Latent image	Chemical developer	Black images on film
Color comparator	Sunlight	Color	Eye	Optic nerve signal	Brain	Visual color response

concept that greatly simplifies the analysis of instrumental systems and promotes understanding of the measurement process.[1] As shown in the data domains map of Figure 1-2, data domains may be broadly classified into *nonelectrical domains* and *electrical domains.*

1C-2 Nonelectrical Domains

The measurement process begins and ends in nonelectrical domains. The physical and chemical characteristics that are of interest in a particular experiment reside in these data domains. Among these characteristics are length, density, chemical composition, intensity of light, pressure, and others listed in the first column of Table 1-1.

It is possible to make a measurement entirely in nonelectrical domains. For instance, the determination of the mass of an object using a mechanical equal-arm balance involves a comparison of the mass of the object, which is placed on one balance pan, with standard masses placed on a second pan. The information repre-

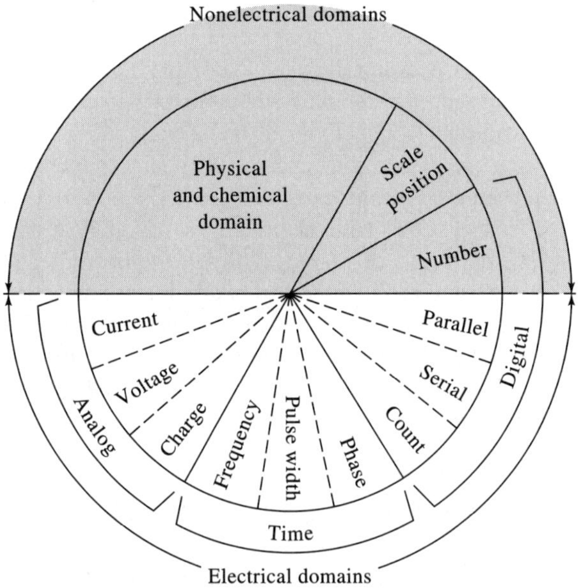

Figure 1-2 Data domains map. The upper (shaded) half of the map comprises nonelectrical domains. The bottom half is made up of electrical domains. Note that the digital domain spans both electrical and nonelectrical domains.

[1] C. G. Enke, *Anal. Chem.*, **1971,** *43,* 69A.

senting the mass of the object in standard units is encoded directly by the experimenter, who provides information processing by summing the masses to arrive at a number. In certain other mechanical balances, the gravitational force on a mass is amplified mechanically by making one of the balance arms longer than the other, thus increasing the resolution of the measurement.

The determination of the linear dimensions of an object with a ruler and the measurement of the volume of a sample of liquid with a graduated cylinder are other examples of measurements carried out exclusively in nonelectrical domains. Such measurements are often associated with classical analytical methods. The advent of inexpensive electronic signal processors, sensitive transducers, and readout devices has led to the development of a host of electronic instruments, which acquire information from nonelectrical domains, process it in electrical domains, and finally present it in nonelectrical domains once again. Electronic devices process information and transform it from one domain to another in ways analogous to the multiplication of mass in mechanical balances with unequal arms. As a consequence of the availability of these electronic devices and their rapid and sophisticated information processing, instruments that rely exclusively on nonelectrical information transfer are rapidly becoming relics of the past. Nonetheless, the information that we seek begins in the properties of the analyte and ends in a number, both of which are nonelectrical domains. The ultimate objective in all measurements is that the final numerical result must be in some manner proportional to the relevant chemical or physical characteristic of the analyte.

1C-3 Electrical Domains

The modes of encoding information as electrical quantities can be subdivided into *analog domains, time domains,* and *digital domains,* as illustrated in the bottom half of the circular map in Figure 1-2. Note that the digital domain spans three electrical domains and one nonelectrical domain because numbers presented on any type of display convey digital information and can also be encoded electrically.

Any measurement process can be represented as a series of *interdomain conversions.* For example, Figure 1-3 illustrates the measurement of the intensity of molecular fluorescence of a sample of tonic water containing a trace of quinine and, in a general way, some of the data-domain conversions that are necessary to arrive at a number expressing the intensity. The intensity of the flu-

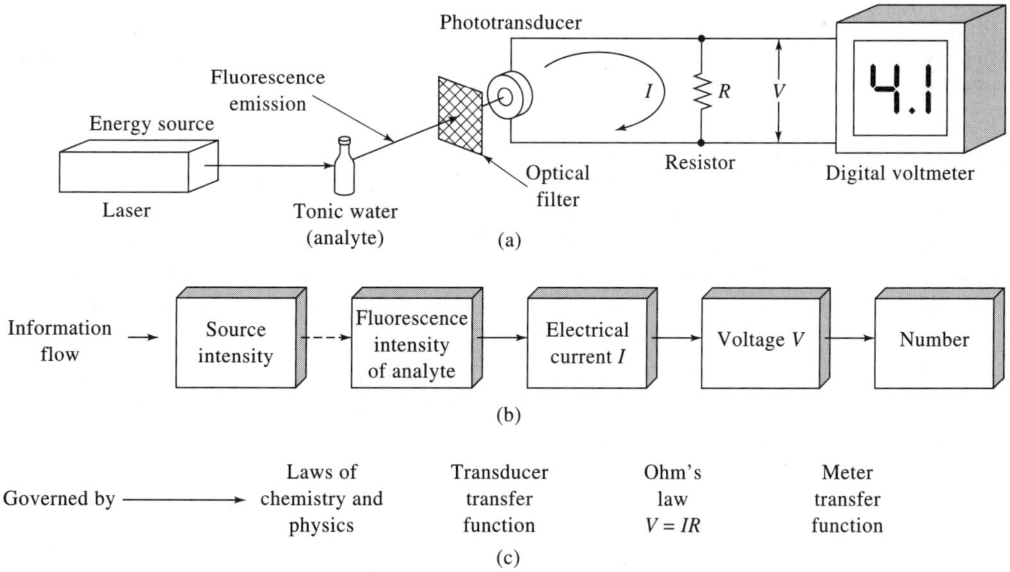

Figure 1-3 A block diagram of a fluorometer showing (a) a general diagram of the instrument, (b) a diagrammatic representation of the flow of information through various data domains in the instrument, and (c) the rules governing the data domain transformations during the measurement process.

orescence is significant in this context because it is proportional to the concentration of the quinine in the tonic water, which is ultimately the information that we desire. The information begins in the solution of tonic water as the concentration of quinine. This information is teased from the sample by applying to it a stimulus in the form of electromagnetic energy from the laser shown in Figure 1-3. The radiation interacts with the quinine molecules in the tonic water to produce fluorescence emission in a region of the spectrum characteristic of quinine and of magnitude proportional to its concentration. Radiation, and thus information, that is unrelated to the concentration of quinine is removed from the beam of light by an optical filter, as shown in Figure 1-3a. The intensity of the fluorescence emission, which is a nonelectrical domain, is encoded into an electrical domain by a special type of device called an *input transducer.* The particular type of transducer used in this experiment is a phototransducer, of which there are numerous types, some of which are discussed in Chapter 7. In this example, the input transducer converts the fluorescence from the tonic water to an electrical current, *I*, proportional to the intensity of the radiation. The mathematical relationship between the electrical output and the input radiant power impinging on its surface is called the *transfer function* of the transducer.

The current from the phototransducer is then passed through a resistor *R*, which according to Ohm's law produces a voltage *V* that is proportional to *I*, which is in turn proportional to the intensity of the fluorescence. Finally, *V* is measured by the digital voltmeter to provide a readout proportional to the concentration of the quinine in the sample.

Voltmeters, alphanumeric displays, electric motors, computer screens, and many other devices that serve to convert data from electrical to nonelectrical domains are called *output transducers.* The digital voltmeter of the fluorometer of Figure 1-3a is a rather complex output transducer that converts the voltage *V* to a number on a liquid crystal display so that it may be read and interpreted by the user of the instrument. We shall consider the detailed nature of the digital voltmeter and various other electrical circuits and signals in Chapters 2 through 4.

Analog Domains

Information in *analog domains* is encoded as the *magnitude* of one of the electrical quantities—voltage, current, charge, or power. These quantities are continuous in both amplitude and time as shown by the typical analog signals of Figure 1-4. Magnitudes of analog quantities can be measured continuously or they can be sampled at specific points in time dictated by the needs of a particular experiment or instrumental method as discussed in Chapter 4. Although the data of Figure 1-4 are recorded as a function of time, any variable such as wavelength, magnetic field strength, or temperature may be the independent variable under appropriate circumstances. The correlation of two analog signals that result from corresponding measured physical or chemical properties is important in a wide variety of instrumental techniques, such as nuclear magnetic resonance spectroscopy, infrared spectroscopy, and differential thermal analysis.

Analog signals are especially susceptible to electrical noise that results from interactions within measurement circuits or from other electrical devices in the vicinity of the measurement system. Such undesirable noise bears no relationship to the information of interest, and methods have been developed to minimize the effects of this unwanted information. Signals, noise, and the optimization of instrumental response are discussed in Chapter 5.

Time Domains

Information is stored in time domains as the time relationship of signal fluctuations, rather than in the amplitudes of the signals. Figure 1-5 illustrates three different time-domain signals recorded as an analog quantity versus time. The horizontal dashed lines represent an arbitrary analog signal threshold that is used to decide whether a signal is HI (above the threshold) or LO (below the threshold). The time relationships between transitions of the signal from HI to LO or from LO to HI contain the information of interest. For instruments that produce periodic signals, the number of cycles of the signal per unit time is the *frequency,* and the time required for each cycle is its *period.* Two examples of instrumental systems that produce information encoded in the frequency domain are Raman spectroscopy and instrumental neutron activation analysis. In these methods, the frequency of arrival of photons at a detector is directly related to the intensity of the emission from the analyte, which is proportional to its concentration.

The time between successive LO to HI transitions is called the *period,* and the time between a LO to HI and a HI to LO transition is called the *pulse width.* De-

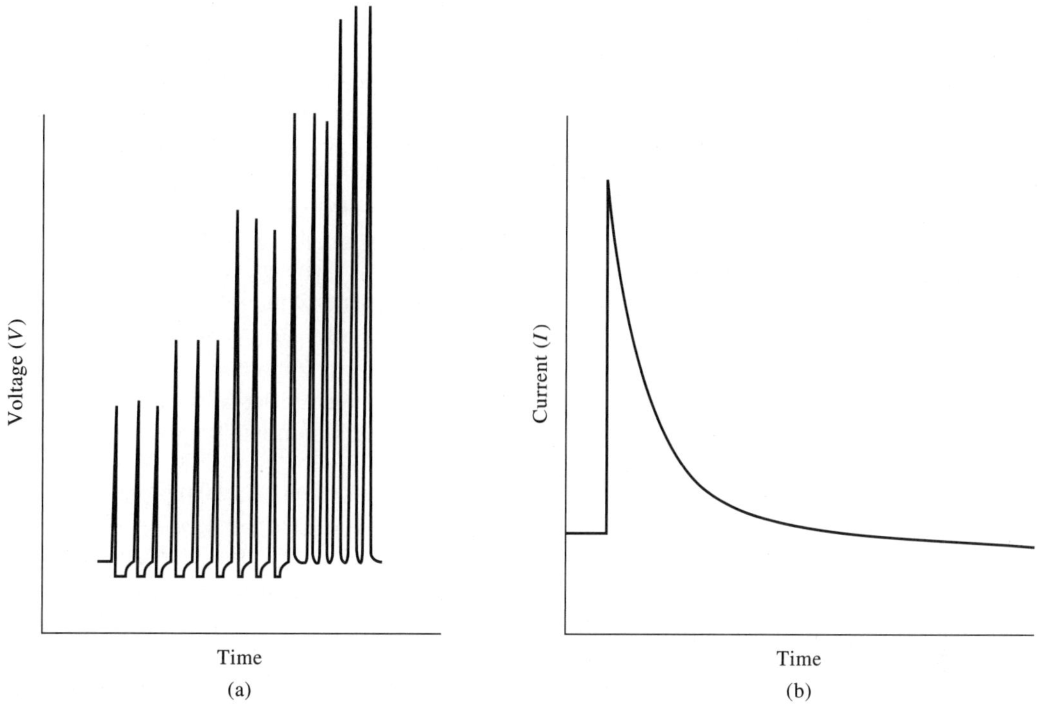

Figure 1-4 Analog signals. (a) Instrument response from the photometric detection system of a flow injection analysis experiment. A stream of reaction mixture containing plugs of red $Fe(SCN)^{2+}$ flows past a monochromatic light source and a phototransducer, which produces a changing voltage as sample concentration changes. (b) The current response of a photomultiplier tube when the light from a pulsed source falls on the photocathode of the device.

vices such as voltage-to-frequency converters and frequency-to-voltage converters may be used to convert time-domain signals to analog-domain signals and vice versa. These and other such *data domain converters* will be discussed in Chapters 3 and 4 as a part of our treatment of electronic devices and will be referred to in other contexts throughout this book.

Digital Domains

Data are encoded in the *digital domain* in a two-level scheme. The information can be represented by the state of a light bulb, a light-emitting diode, a toggle switch, or a logic level signal, to cite but a few examples. The characteristic that these devices share is that each of them must be in one of only two states. For example, lights and switches may be only ON or OFF and logic-level signals may be only HI or LO. The definition of what constitutes ON and OFF for switches and lights is understood, but in the case of electrical signals, as in the case of time domain signals, an arbitrary signal level must be defined that distinguishes between HI and LO. Such a definition may depend on the conditions of an experiment, or it may depend upon the characteristics of the electronic devices in use. For example, the signal represented in Figure 1-5c is a train of pulses from a nuclear detector. The measurement task is to count the pulses during a fixed period of time to obtain a measure of the intensity of radiation. The dashed line represents a signal level that not only is low enough to ensure that no pulses are lost but also is sufficiently high to reject random fluctuations in the signal that are unrelated to the nuclear phenomena of interest. If the signal crosses the threshold fourteen times, as in the case of the signal in Figure 1-5c, then we may be confident that fourteen nuclear events occurred. After the events have been counted, the data are then encoded in the digital domain in the form of the number 14. In

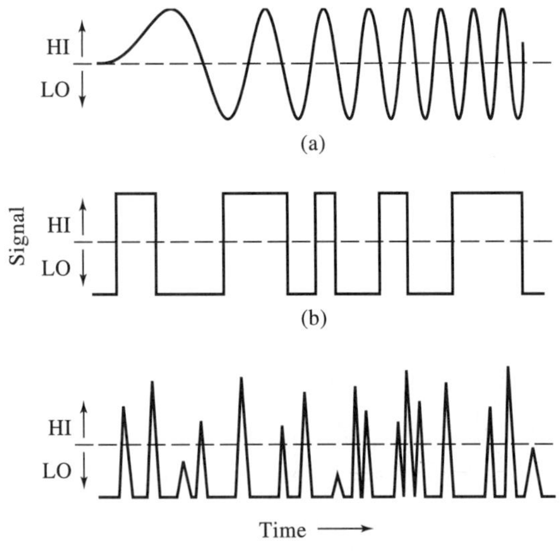

Figure 1-5 Time-domain signals. The horizontal dashed lines represent signal thresholds. When each signal is above the threshold, the signal is HI, and when it is below the threshold, the signal is LO.

Chapter 4, we shall explore the means for making HI-LO electronic decisions and encoding the information in the digital domain.

As suggested by the data domains map of Figure 1-2, the digital domain spans both electrical and nonelectrical domains. In the example just cited, the nuclear events are accumulated by using an electronic counter and are displayed on a digital readout. When the experimenter reads and interprets the display, the number that represents the measured quantity is once again in a nonelectrical domain. Each piece of HI-LO data that represents a nuclear event is a *bit* of information, which is the fundamental unit of information in the digital domain. Bits of information that are transmitted along a single electronic channel or wire may be counted by an observer or by an electronic device that is monitoring the channel; such accumulated data is termed *count digital data,* which appears in the data-domains map of Figure 1-2. For example, the signal in Figure 1-5a corresponds to the number $n = 8$ because there are eight complete cycles in the signal. The signal in the Figure 1-5b corresponds to $n = 5$, and the signal in Figure 1-5c corresponds to $n = 14$. Although effective, this means of transmitting information is not very efficient.

A far more efficient way to encode information is to use binary numbers to represent numeric and alphabetic data. To see how this type of encoding may be accomplished, let us consider the signals in Figure 1-6. The count digital data of the signal in Figure 1-6a represent the number $n = 5$ as before. We monitor the signal and count the number of complete oscillations. The process requires a period of time that is proportional to the number of cycles of the signal, or in this case, five times the length of a single time interval, as indicated in Figure 1-6. Note that the time intervals are numbered consecutively beginning with zero. In a binary encoding scheme, such as the one shown for the signal in Figure 1-6b, we assign a numerical value to each successive interval of time. For example, the zeroth time interval represents $2^0 = 1$, the first time interval represents $2^1 = 2$, the second time interval represents $2^2 = 4$, and so forth, as shown in Figure 1-6. During each time interval, we need only decide whether the signal is HI or LO. If the signal is HI during any given time interval, then the value corresponding to that interval is added to the total. All intervals that are LO contribute zero to the total.

In Figure 1-6b, the signal is HI only in interval 0 and interval 2, so the total value represented is $1 \times 2^0 + 0 \times 2^1 + 1 \times 2^2 = 5$. Thus, in the space of only three time intervals, the number $n = 5$ has been determined. In the count digital example of the signal in Figure 1-6a, five time intervals were required to determine the same number. In this limited example, the binary-coded serial data is nearly twice as efficient as the count serial data. A more dramatic example may be seen in the counting of $n = 10$ oscillations similar to those of the signal in Figure 1-6a. In the same ten time intervals, ten HI-LO bits of information in the serial binary coding scheme enable the representation of the binary numbers from 0 to $2^{10} = 1024$, or 0000000000 to 1111111111. The improvement in efficiency is 1024/10, or about 100-fold. In other words, the count serial scheme requires 1024 time intervals to represent the number 1024, while the binary coding scheme requires only ten time intervals. As a result of the efficiency of binary coding schemes, most digital information is encoded, transferred, processed, and decoded in some form of binary.

Data represented by binary coding on a single transmission line is called *serial-coded binary data,* or simply *serial data.* A common example of serial data transmission is the computer *modem,* which is a device for transmitting data between computers by telephone over a single conductor (and ground).

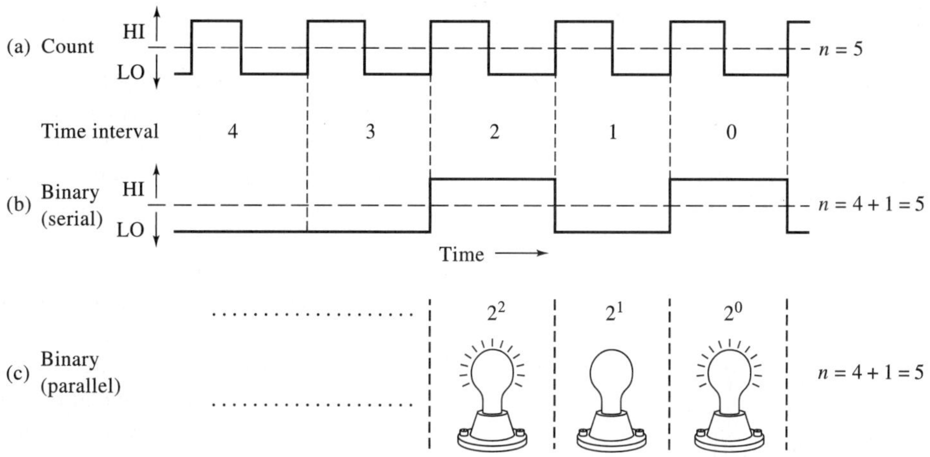

Figure 1-6 Diagram illustrating three types of digital data: (a) count serial data, (b) binary-coded serial data, and (c) parallel binary data. In all three cases, the data represent the number $n = 5$.

A still more efficient method for encoding data in the digital domain is seen in the signal of Figure 1-6c. Here, we use three light bulbs to represent the three binary digits: $2^0 = 1$; $2^1 = 2$; and $2^2 = 4$. However, we could use switches, wires, light-emitting diodes, or any of a host of electronic devices to encode the information. In this scheme, ON = 1 and OFF = 0, so that our number is encoded as shown in Figure 1-6 with the first and third lights ON and the middle light OFF, which represents $4 + 0 + 1 = 5$. This scheme is highly efficient because all of the desired information is presented to us simultaneously, just as all of the digits on the face of the digital voltmeter in Figure 1-3a appear simultaneously. Data presented in this way are referred to as *parallel* digital data. Data are transmitted within analytical instruments and computers by parallel data transmission. Since data usually travel relatively short distances within these devices, it is economical and efficient to use parallel information transfer. This economy of short distances is in contrast to the situation in which data must be transported over long distances from instrument to instrument or from computer to computer. In such instances, communication is carried out serially by using modems or other more sophisticated or faster serial data transmission schemes. We will consider these ideas in somewhat more detail in Chapter 4.

1C-4 Detectors, Transducers, and Sensors

The terms *detector, transducer,* and *sensor* are often used synonymously, but in fact the terms have somewhat different meanings. The most general of the three terms, *detector,* refers to a mechanical, electrical, or chemical device that identifies, records, or indicates a change in one of the variables in its environment, such as pressure, temperature, electrical charge, electromagnetic radiation, nuclear radiation, particulates, or molecules. This term has become a catchall to the extent that entire instruments are often referred to as *detectors.* In the context of instrumental analysis, we shall use the term *detector* in the general sense in which we have just defined it, and we shall use *detection system* to refer to entire assemblies that indicate or record physical or chemical quantities. An example is the UV (ultraviolet) detector often used to indicate and record the presence of eluted analytes in liquid chromatography.

The term *transducer* refers specifically to those devices that convert information in nonelectrical domains to information in electrical domains and the converse. Examples include photodiodes, photomultipliers, and other electronic photodetectors that produce current or voltage proportional to the radiant power of electromagnetic radiation that falls on their surfaces. Other examples include thermistors, strain gauges, and Hall effect magnetic field strength transducers. As suggested previously, the mathematical relationship between the electrical output and the input radiant power, temperature, force, or magnetic field strength is called the *transfer function* of the transducer.

The term *sensor* also has become rather broad, but in this text we shall reserve the term for the class of analytical devices that are capable of monitoring specific chemical species continuously and reversibly. There are numerous examples of sensors throughout this text, including the glass electrode and other ion-selective electrodes, which are treated in Chapter 23, the Clark oxygen electrode, which is described in Chapter 25, and optrodes, or fiber-optic sensors, which appear in Chapter 7. Sensors consist of a transducer coupled with a chemically selective recognition phase. So, for example, optrodes consist of a phototransducer coupled with a fiber optic that is coated on the end opposite the transducer with a substance that responds specifically to a particular physical or chemical characteristic of an analyte.

A sensor that is especially interesting and instructive is the *quartz crystal microbalance,* or QCM. This device is based on the *piezoelectric* characteristics of quartz. When quartz is mechanically deformed, an electrical potential develops across its surface. Furthermore, when a voltage is impressed across the faces of a quartz crystal, the crystal deforms. A crystal connected in an appropriate electrical circuit oscillates at a frequency that is characteristic of the mass and shape of the crystal and that is amazingly constant—provided that the mass of the crystal is constant. This property of some crystalline materials is called the *piezoelectric effect,* and forms the basis for the quartz-crystal microbalance. Moreover, the characteristic constant frequency of the quartz crystal is the basis for modern high-precision clocks, time bases, counters, timers, and frequency meters, which in turn have led to many highly accurate and precise analytical instrumental systems.

If a quartz crystal is coated with a polymer that selectively adsorbs certain molecules, the mass of the crystal increases if the molecules are present, thus decreasing the resonant frequency of the quartz crystal. When the molecules are desorbed from the surface, the crystal returns to its original frequency. The relationship between the change in frequency of the crystal ΔF and the change in mass of the crystal ΔM is given by

$$\Delta F = \frac{CF^2 \Delta M}{A}$$

where M is the mass of the crystal, A is its surface area, F is the frequency of oscillation of the crystal, and C is a proportionality constant. The relationship above indicates that it is possible to measure very small changes in the mass of the crystal if the frequency of the crystal can be measured precisely. As it turns out, it is possible to measure frequency changes of one part in 10^7 quite easily with inexpensive instrumentation. The limit of detection for a piezoelectric sensor of this type is estimated to be about 1 pg, or 10^{-12} g. These sensors have been used to detect a variety of gas-phase analytes including formaldehyde, hydrogen chloride, hydrogen sulfide, and benzene. They have also been proposed as sensors for chemical warfare agents such, as mustard gas and phosgene.

The piezoelectric mass sensor presents an excellent example of a transducer converting a property of the analyte, mass in this case, to a change in an electrical quantity, the resonant frequency of the quartz crystal. This example also illustrates the distinction between a transducer and a sensor. In the quartz-crystal microbalance, the transducer is the quartz crystal, and the selective second phase is the polymeric coating. The combination of the transducer and the selective phase constitute the sensor.

1C-5 Readout Devices

A readout device is a transducer that converts information from an electrical domain to a domain that is understandable by a human observer. Usually, the transduced signal takes the form of the alphanumeric or graphic output of a cathode-ray tube, a series of numbers on a digital display, the position of a pointer on a meter scale, or, occasionally, the blackening of a photographic plate, or a tracing on a recorder paper. In some instances, the readout device may be arranged to give the analyte concentration directly.

1C-6 Microprocessors and Computers in Instruments

Most modern analytical instruments contain or are attached to one or more sophisticated electronic devices and data domain converters, such as operational amplifiers, integrated circuits, analog-to-digital and digital-to-analog converters, counters, microprocessors, and computers. In order to appreciate the power and limitations of such instruments, it is necessary that the scientist develop at least a qualitative understanding of how these devices function and what they can do. Chapters 3 and 4 provide a brief treatment of these important topics.

1D SELECTING AN ANALYTICAL METHOD

It is evident from column 2 of Table 1-1 that the modern chemist has an enormous array of tools for carrying out analyses—so many, in fact, that the choice among them is often difficult. In this section, we describe how such choices are made.

1D-1 Defining the Problem

In order to select an analytical method intelligently, it is essential to define clearly the nature of the analytical problem. Such a definition requires answers to the following questions:

1. What accuracy is required?
2. How much sample is available?
3. What is the concentration range of the analyte?
4. What components of the sample will cause interference?
5. What are the physical and chemical properties of the sample matrix?
6. How many samples are to be analyzed?

The answer to question 1 is of vital importance because it determines how much time and care will be needed for the analysis. The answers to questions 2 and 3 determine how sensitive the method must be and how wide a range of concentrations must be accommodated. The answer to question 4 determines the selectivity required of the method. The answers to question 5 are important because some analytical methods in Table 1-1 are applicable to solutions (usually aqueous) of the analyte. Other methods are more easily applied to gaseous samples, while still other methods are suited to the direct analysis of solids.

The number of samples to be analyzed (question 6) is also an important consideration from the economic standpoint. If this number is large, considerable time and money can be spent on instrumentation, method development, and calibration. Furthermore, if the number is large, a method should be chosen that requires the least operator time per sample. On the other hand, if only a few samples are to be analyzed, a simpler but more time-consuming method that requires little or no preliminary work is often the wiser choice.

With answers to the foregoing six questions, a method can then be chosen, provided that the performance characteristics of the various instruments shown in Table 1-1 are known.

1D-2 Performance Characteristics of Instruments; Figures of Merit

Table 1-3 lists quantitative performance criteria of instruments that can be used to decide whether a given instrumental method is suitable for attacking an analytical problem. These characteristics are expressed in numerical terms that are called *figures of merit.* Figures of merit permit us to narrow the choice of instruments for a given analytical problem to a relatively few. Selection among these few can then be based upon the qualitative performance criteria listed in Table 1-4.

In this section, we define each of the six figures of merit listed in Table 1-3. These figures are then used throughout the remainder of the text in discussing various instruments and instrumental methods.

TABLE 1-3 Numerical Criteria for Selecting Analytical Methods

Criterion	Figure of Merit
1. Precision	Absolute standard deviation, relative standard deviation, coefficient of variation, variance
2. Bias	Absolute systematic error, relative systematic error
3. Sensitivity	Calibration sensitivity, analytical sensitivity
4. Detection limit	Blank plus three times standard deviation of a blank
5. Concentration range	Concentration limit of quantitation (LOQ) to concentration limit of linearity (LOL)
6. Selectivity	Coefficient of selectivity

TABLE 1-4 Other Characteristics to Be Considered in Method Choice

1. Speed
2. Ease and convenience
3. Skill required of operator
4. Cost and availability of equipment
5. Per-sample cost

TABLE 1-5 Figures of Merit for Precision of Analytical Methods

Terms	Definition*
Absolute standard deviation, s	$s = \sqrt{\dfrac{\sum_{i=1}^{N}(x_i - \bar{x})^2}{N-1}}$
Relative standard deviation (RSD)	$\text{RSD} = \dfrac{s}{\bar{x}}$
Standard deviation of the mean, s_m	$s_m = s/\sqrt{N}$
Coefficient of variation, CV	$\text{CV} = \dfrac{s}{\bar{x}} \times 100\%$
Variance	s^2

*x_i = numerical value of the ith measurement.

$\bar{x}$ = mean of N measurements = $\dfrac{\sum_{i=1}^{N} x_i}{N}$

Precision

As we show in Section a1A, Appendix 1, the precision of analytical data is the degree of mutual agreement among data that have been obtained in the same way. Precision provides a measure of the random, or indeterminate, error of an analysis. Figures of merit for precision include *absolute standard deviation, relative standard deviation, coefficient of variation,* and *variance.* These terms are defined in Table 1-5.

Bias

As shown in Section a1A-2, Appendix 1, bias provides a measure of the systematic, or determinate, error of an analytical method. Bias is defined by the equation

$$\text{bias} = \mu - x_t \tag{1-1}$$

where μ is the population mean for the concentration of an analyte in a sample that has a true concentration of x_t. Determining bias involves analyzing one or more standard reference materials whose analyte concentration is known. Sources of such materials are given in references 3 and 4 in Section a1A-2 of Appendix 1. The results from such an analysis will, however, contain both random and systematic errors; but if a sufficient number of analyses are performed, the mean value may be determined with a given level of confidence. As shown in Section a1B-2, Appendix 1, the mean of 20 or 30 replicate analyses can ordinarily be taken as a good estimate of the population mean μ in Equation 1-1. Any difference between this mean and the known value analyte concentration of the standard reference material can be attributed to bias.

If performing 20 replicate analyses on a standard is impractical, the probable presence or absence of bias can be evaluated as shown in Example a1-7 in Appendix 1.

Ordinarily in developing an analytical method, every effort is made to identify the source of bias and eliminate it or correct for it by the use of blanks and by instrument calibration.

Sensitivity

There is general agreement that the sensitivity of an instrument or a method is a measure of its ability to discriminate between small differences in analyte concentration. Two factors limit sensitivity: the slope of the calibration curve and the reproducibility or precision of the measuring device. Of two methods that have equal precision, the one that has the steeper calibration curve will be the more sensitive. A corollary to this statement is that if two methods have calibration curves with equal slopes, the one that exhibits the better precision will be the more sensitive.

The quantitative definition of sensitivity that is accepted by the International Union of Pure and Applied Chemists (IUPAC) is *calibration sensitivity,* which is the slope of the calibration curve at the concentration of interest. Most calibration curves that are used in analytical chemistry are linear and may be represented by the equation

$$S = mc + S_{\text{bl}} \tag{1-2}$$

where S is the measured signal, c is the concentration of the analyte, S_{bl} is the instrumental signal for a blank, and m is the slope of the straight line. The quantity S_{bl} should be the y-intercept of the straight line. With such curves, the calibration sensitivity is independent of the concentration c and is equal to m. The calibration sensitivity as a figure of merit suffers from its failure to take into account the precision of individual measurements.

Mandel and Stiehler[2] recognized the need to include precision in a meaningful mathematical statement of sensitivity and proposed the following definition for *analytical sensitivity*, γ:

$$\gamma = m/s_S \tag{1-3}$$

Here, m is again the slope of the calibration curve, and s_S is the standard deviation of the measurement.

The analytical sensitivity offers the advantage of being relatively insensitive to amplification factors. For example, increasing the gain of an instrument by a factor of five will produce a fivefold increase in m. Ordinarily, however, this increase will be accompanied by a corresponding increase in s_S, thus leaving the analytical sensitivity more or less constant. A second advantage of analytical sensitivity is that it is independent of the measurement units for S.

A disadvantage of analytical sensitivity is that it is often concentration dependent since s_S may vary with concentration.

Detection Limit

The most generally accepted qualitative definition of detection limit is that it is the minimum concentration or mass of analyte that can be detected at a known confidence level. This limit depends upon the ratio of the magnitude of the analytical signal to the size of the statistical fluctuations in the blank signal. That is, unless the analytical signal is larger than the blank by some multiple k of the variation in the blank owing to random errors, it is impossible to detect the analytical signal with certainty. Thus, as the limit of detection is approached, the analytical signal and its standard deviation approach the blank signal S_{bl} and its standard deviation s_{bl}. The minimum distinguishable analytical signal S_m is then taken as the sum of the mean blank signal $\bar{S}_{bl}$ plus a multiple k of the standard deviation of the blank. That is,

$$S_m = \bar{S}_{bl} + ks_{bl} \tag{1-4}$$

Experimentally, S_m can be determined by performing 20 to 30 blank measurements, preferably over an extended period of time. The resulting data are then treated statistically to obtain $\bar{S}_{bl}$ and s_{bl}. Finally, the slope from Equation 1-2 is used to convert S_m to c_m, which is defined as the detection limit. That is, the detection limit is given by

$$c_m = \frac{S_m - \bar{S}_{bl}}{m} \tag{1-5}$$

As pointed out by Ingle,[3] numerous alternatives, based correctly or incorrectly on t and z statistics (Section a1B-2, Appendix 1), have been used to determine a value for k in Equation 1-4. Kaiser[4] argues that a reasonable value for the constant is $k = 3$. He points out that it is wrong to assume a strictly normal distribution of results from blank measurements and that when $k = 3$, the confidence level of detection will be 95% in most cases. He further argues that little is to be gained by using a larger value of k—and thus a greater confidence level. Long and Winefordner,[5] in a discussion of detection limits, also recommend the use of $k = 3$.

EXAMPLE 1-1

A least-squares analysis of calibration data for the determination of lead based upon its flame emission spectrum yielded the equation

$$S = 1.12\, c_{Pb} + 0.312$$

where c_{Pb} is the lead concentration in parts per million and S is a measure of the relative intensity of the lead emission line. The following replicate data were then obtained:

Concn, ppm Pb	No. of Replications	Mean Value of S	s
10.0	10	11.62	0.15
1.00	10	1.12	0.025
0.000	24	0.0296	0.0082

[2] J. Mandel and R. D. Stiehler, *J. Res. Natl. Bur. Std.*, **1964,** *A53,* 155.

[3] J. D. Ingle Jr., *J. Chem. Educ.*, **1970,** *42,* 100.

[4] H. Kaiser, *Anal. Chem.*, **1987,** *42,* 53A.

[5] G. L. Long and J. D. Winefordner, *Anal. Chem.*, **1983,** *55,* 712A.

Calculate **(a)** the calibration sensitivity, **(b)** the analytical sensitivity at 1 and 10 ppm of Pb, and **(c)** the detection limit.

(a) By definition, the calibration sensitivity m is the slope of the straight line. Thus, $m = 1.12$.
(b) At 10 ppm Pb, $\gamma = m/s_S = 1.12/0.15 = 7.5$.
At 1 ppm Pb, $\gamma = 1.12/0.025 = 45$.
(c) Applying Equation 1-4,

$$S_m = 0.0296 + 3 \times 0.0082 = 0.054$$

Substituting into Equation 1-5 gives

$$c_m = \frac{0.054 - 0.0296}{1.12} = 0.022 \text{ ppm Pb.}$$

Dynamic Range

Figure 1-7 illustrates the definition of the *dynamic range* of an analytical method, which extends from the lowest concentration at which quantitative measurements can be made (limit of quantitation, or LOQ) to the concentration at which the calibration curve departs from linearity (limit of linearity, or LOL). The lower limit of quantitative measurements is generally taken to be equal to ten times the standard deviation of repetitive measurements on a blank, or $10s_{bl}$. At this point, the relative standard deviation is about 30% and decreases rapidly as concentrations become larger. At the limit of detection, the relative standard deviation is 100%.

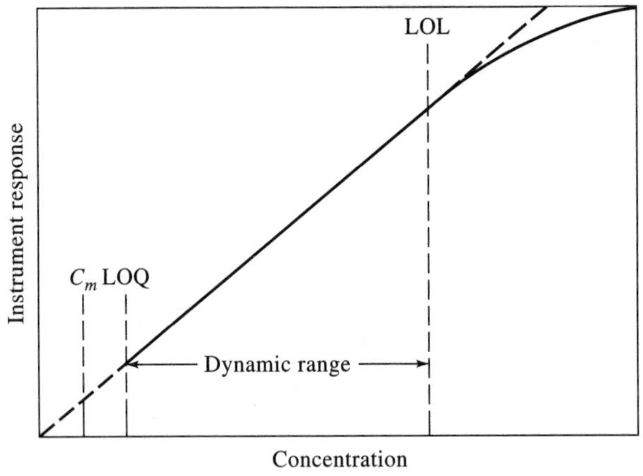

Figure 1-7 Useful range of an analytical method. LOQ = limit of quantitative measurement; LOL = limit of linear response.

To be very useful, an analytical method should have a dynamic range of at least two orders of magnitude. Some methods have applicable concentration ranges of five to six orders of magnitude.

Selectivity

Selectivity of an analytical method refers to the degree to which the method is free from interference by other species contained in the sample matrix. Unfortunately, no analytical method is totally free from interference from other species, and frequently steps must be taken to minimize the effects of these interferences.

Consider, for example, a sample containing an analyte A as well as potential interfering species B and C. If c_A, c_B, and c_C are the concentrations of the three species and m_A, m_B, and m_C are their calibration sensitivities, then the total instrument signal will be given by a modified version of Equation 1-3. That is,

$$S = m_A c_A + m_B c_B + m_C c_C + S_{bl} \qquad (1\text{-}6)$$

Let us now define the selectivity coefficient for A with respect to B as

$$k_{B,A} = m_B/m_A \qquad (1\text{-}7)$$

The selectivity coefficient then gives the relative response of the method to species B as compared with A. A similar coefficient for A with respect to C is

$$k_{C,A} = m_C/m_A \qquad (1\text{-}8)$$

Substituting these relationships into Equation 1-4 leads to

$$S = m_A(c_A + k_{B,A}c_B + k_{C,A}c_C) + S_{bl} \qquad (1\text{-}9)$$

Selectivity coefficients can range from zero (no interference) to values a good deal greater than unity. Note that a coefficient is negative when the interference causes a reduction in the intensity of the output signal of the analyte. For example, if the presence of interferant B causes a reduction in S in Equation 1-7, m_B will carry a negative sign, as will $k_{B,A}$.

Selectivity coefficients are useful figures of merit for describing the selectivity of analytical methods. Unfortunately, they are not widely used except to characterize the performance of ion-selective electrodes (Chapter 23). Example 1-2 illustrates the use of selectivity coefficients when they are available.

EXAMPLE 1-2

The selectivity coefficient for an ion-selective electrode for K^+ with respect to Na^+ is reported to be 0.052. Calculate the relative error in the determination of K^+ in a solution that has a K^+ concentration of 3.00×10^{-3} M if the Na^+ concentration is **(a)** 2.00×10^{-2} M; **(b)** 2.00×10^{-3} M; **(c)** 2.00×10^{-4} M. Assume that S_{bl} for a series of blanks was approximately zero.

(a) Substituting into Equation 1-9 yields

$$S = m_{K^+}(c_{K^+} + k_{Na^+,K^+}c_{Na^+}) + 0$$

$$S/m_{K^+} = 3.00 \times 10^{-3} + 0.052 \times 2.00 \times 10^{-2} = 4.04 \times 10^{-3}$$

If Na^+ were not present

$$S/m_{K^+} = 3.00 \times 10^{-3}$$

The relative error in c_{K^+} will be identical to the relative error in S/m_{K^+} (see Section a1B-5, Appendix 1). Therefore,

$$E_{rel} = \frac{4.04 \times 10^{-3} - 3.00 \times 10^{-3}}{3.00 \times 10^{-3}} \times 100\% = 35\%$$

Proceeding in the same way we find

(b) $E_{rel} = 3.5\%$

(c) $E_{rel} = 0.35\%$

Tue Aug 10. 10.

1E CALIBRATION OF INSTRUMENTAL METHODS

With two exceptions, all types of analytical methods require *calibration,* a process that relates the measured analytical signal to the concentration of analyte.[6] The three most common calibration methods include the preparation and use of a calibration curve, the standard addition method, and the internal standard method.

1E-1 Calibration Curves

To use the calibration curve technique, several standards containing exactly known concentrations of the analyte are introduced into the instrument, and the instrumental response is recorded. Ordinarily, this response is corrected for the instrument output obtained with a blank. Ideally, the blank contains all of the components of the original sample except for the analyte. The resulting data are then plotted to give a graph of corrected instrument response versus analyte concentration.

Figure 1-8 shows a typical *calibration curve* (also called a *working curve* or an *analytical curve*). Plots, such as this, that are linear over a significant concentration range (the *dynamic range*) are often obtained and are desirable because they are less subject to error than are nonlinear curves. Not uncommonly, however, nonlinear plots are observed, which require a larger number of calibration data to establish accurately the relationship between the instrument response and concentration. Usually, an equation is developed for the calibration curve by a least-squares technique (Appendix a1C) so that sample concentrations can be computed directly.

The success of the calibration curve method is critically dependent upon how accurately the analyte concentrations of the standards are known and how closely the matrix[7] of the standards resemble that of the samples to be analyzed. Unfortunately, matching the matrix of complex samples is often difficult or impossible, and matrix effects lead to interference errors. To minimize matrix effects, it is often necessary to separate the analyte from the *interferent* before measuring the instrument response.

1E-2 Standard Addition Methods

Standard addition methods are particularly useful for analyzing complex samples in which the likelihood of matrix effects is substantial. A standard addition method can take several forms.[8] One of the most common forms involves adding one or more increments of a standard solution to sample aliquots of the same size.

[6] The two exceptions are gravimetric and coulometric methods. In both of these cases, the relationship between the quantity measured and the concentration of analyte can be computed from accurately known physical constants.

[7] The term *matrix* refers to the collection of all of the various constituents making up an analytical sample. In addition to the analyte, the sample matrix includes all of the other constituents of the sample, which are sometimes referred to as the *concomitants.*

[8] See M. Bader, *J. Chem. Educ.,* **1980,** *57,* 703.

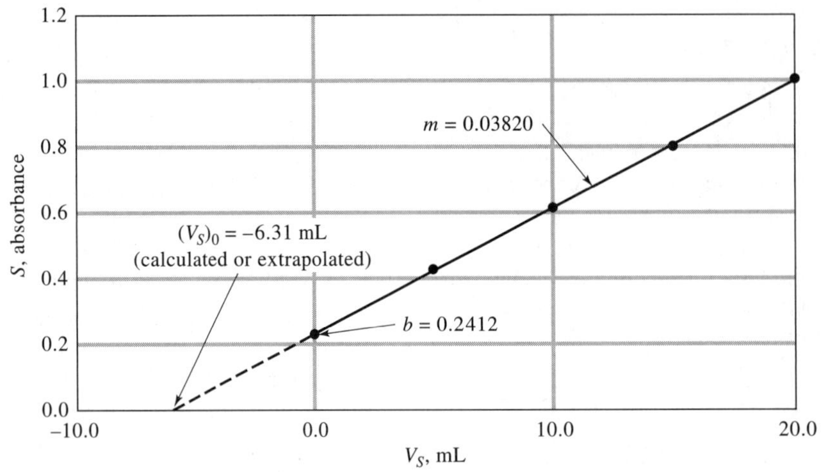

Figure 1-8 Linear calibration plot for the method of standard additions. The concentration of the unknown solution may be calculated from the slope m and the intercept b, or it may be determined by extrapolation as explained in the text.

This process is often called *spiking* the sample. Each solution is then diluted to a fixed volume before measurement. It should be noted that when the amount of sample is limited, standard additions can be carried out by successive introductions of increments of the standard to a single measured volume of the unknown. Measurements are made on the original sample and on the sample plus the standard after each addition. In most versions of the standard addition method, the sample matrix is nearly identical after each addition, the only difference being the concentration of the analyte or, in cases involving the addition of an excess of an analytical reagent, the concentration of the reagent. All other constituents of the reaction mixture should be identical because the standards are prepared in aliquots of the sample.

Assume that several identical aliquots V_x of the unknown solution with a concentration c_x are transferred to volumetric flasks having a volume V_t. To each of these flasks is added a variable volume V_s mL of a standard solution of the analyte having a known concentration c_s. Suitable reagents are then added, and each solution is diluted to volume. Instrumental measurements are then made on each of these solutions to yield an instrument response S. If the instrument response is proportional to concentration, as it must be if the standard addition method is to be applicable, we may write

$$S = \frac{kV_s c_s}{V_t} + \frac{kV_x c_x}{V_t} \tag{1-10}$$

where k is a proportionality constant. A plot of S as a function of V_s is a straight line of the form

$$S = mV_s + b$$

where the slope m and the intercept b are given by

$$m = \frac{kc_s}{V_t}$$

and

$$b = \frac{kV_x c_x}{V_t}$$

Just such a plot is depicted in Figure 1-8.

A least-squares analysis (Section a1C, Appendix 1) can be used to determine m and b; c_x can then be obtained from the ratio of these two quantities and the known values of c_x, V_x, and V_s. Thus,

$$\frac{b}{m} = \frac{kV_x c_x/V_t}{kc_s/V_t} = \frac{V_x c_x}{c_s}$$

or

$$c_x = \frac{bc_s}{mV_x} \tag{1-11}$$

A value for the standard deviation in c_x can then be obtained by assuming that the uncertainties in c_s, V_s, and V_t are negligible with respect to those in m and b. Then, the relative variance of the result $(s_c/c_x)^2$ is assumed to be the sum of the relative variances of m and b. That is,

$$\left(\frac{s_c}{c_x}\right)^2 = \left(\frac{s_m}{m}\right)^2 + \left(\frac{s_b}{b}\right)^2$$

where s_m is the standard deviation of the slope and where s_b is the standard deviation of the intercept. Taking the square root of this equation gives

$$s_c = c_x\sqrt{\left(\frac{s_m}{m}\right)^2 + \left(\frac{s_b}{b}\right)^2} \qquad (1\text{-}12)$$

Alternatively, a manual plot of the data may be constructed, and the linear portion of the plot may be extrapolated to the left of the origin, as shown by the dashed line of Figure 1-8. The difference between the volume of the standard added at the origin (zero) and the value of the volume at the intersection of the straight line with the x-axis, or the x-intercept $(V_x)_0$, is the volume of standard reagent equivalent to the amount of analyte in the sample. In addition, the x-intercept corresponds to zero instrument response, so that we may write

$$S = \frac{kV_sc_s}{V_t} + \frac{kV_xc_x}{V_t} = 0 \qquad (1\text{-}13)$$

By solving Equation 1-13 for c_x, we obtain

$$c_x = -\frac{(V_s)_0c_s}{V_x}$$

EXAMPLE 1-3

Ten-millimeter aliquots of a natural water sample were pipetted into 50.00-mL volumetric flasks. Exactly 0.00, 5.00, 10.00, 15.00, and 20.00 mL of a standard solution containing 11.1 ppm of Fe^{3+} were added to each, followed by an excess of thiocyanate ion to give the red complex $Fe(SCN)^{2+}$. After dilution to volume, the instrument response S for each of the five solutions, measured with a colorimeter, was found to be 0.240, 0.437, 0.621, 0.809, and 1.009, respectively. **(a)** What was the concentration of Fe^{3+} in the water sample? **(b)** Calculate a standard deviation of the slope and of the intercept and the standard deviation for the concentration of Fe^{3+}.

(a) In this problem, $c_s = 11.1$ ppm, $V_x = 10.00$ mL, and $V_t = 50.00$ mL. A plot of the data, shown in Figure 1-8, demonstrates that there is a linear relationship between the instrument response and the iron concentration.

To obtain the equation for the line in Figure 1-8 ($S = mV_s + b$), we follow the procedure illustrated in Example a1-12 in Appendix 1. The result is $m = 0.03820$ and $b = 0.2412$ and thus

$$S = 0.03820\ V_s + 0.2412$$

Substituting into Equation 1-11 gives

$$c_x = \frac{0.2412 \times 11.1}{0.03820 \times 10.00} = 7.01 \text{ ppm } Fe^{3+}$$

This value may be determined by graphical extrapolation as illustrated in the figure as well. The extrapolated value represents the volume of reagent corresponding to zero instrument response, which in this case is −6.31 mL. The unknown concentration of the analyte in the original solution is then calculated as follows:

$$c_x = -\frac{(V_s)_0c_s}{V_x} = \frac{6.31 \text{ mL} \times 11.1 \text{ ppm}}{10.00 \text{ mL}}$$
$$= 7.01 \text{ ppm } Fe^{3+}$$

(b) Equations a1-35 and a1-36 give the standard deviation of the intercept and the slope. That is, $s_b = 3.8 \times 10^{-3}$ and $s_m = 3.1 \times 10^{-4}$.

Substituting into Equation 1-12 gives

$$s_c = 7.01\sqrt{\left(\frac{3.82 \times 10^{-3}}{0.2412}\right)^2 + \left(\frac{3.07 \times 10^{-4}}{0.0382}\right)^2}$$
$$= 0.12 \text{ ppm } Fe^{3+}$$

In the interest of saving time or sample, it is possible to perform a standard addition analysis by using only two increments of sample. Here, a single addition of V_s mL of standard would be added to one of the two samples, and we can write

$$S_1 = \frac{kV_xc_x}{V_t}$$

$$S_2 = \frac{kV_xc_x}{V_t} + \frac{kV_sc_s}{V_t}$$

where S_1 and S_2 are the analytical signals resulting from the diluted sample and the diluted sample plus standard, respectively. Dividing the second equation by the first gives upon rearrangement

$$c_x = \frac{S_1 c_s V_s}{(S_2 - S_1)V_x}$$

1E-3 The Internal Standard Method

An *internal standard* is a substance that is added in a constant amount to all samples, blanks, and calibration standards in an analysis. Alternatively, it may be a major constituent of samples and standards that is present in a large enough amount that its concentration can be assumed to be the same in all cases. Calibration then involves plotting the ratio of the analyte signal to the internal standard signal as a function of the analyte concentration of the standards. This ratio for the samples is then used to obtain their analyte concentrations from a calibration curve.

An internal standard, if properly chosen and used, can compensate for several types of both random and systematic errors. Thus, if the analyte and internal standard signals respond proportionally to random instrumental and method fluctuations, the ratio of these signals is independent of these fluctuations. If the two signals are influenced in the same way by matrix effects, compensation of these effects also occurs. In those instances where the internal standard is a major constituent of samples and standards, compensation for errors that arise in sample preparation, solution, and cleanup may also occur.

A major difficulty in applying the internal standard method is that of finding a suitable substance to serve as the internal standard and of introducing that substance into both samples and standards in a reproducible way. The internal standard should provide a signal that is similar to the analyte signal in most ways but sufficiently different so that the two signals are readily distinguishable by the instrument. The internal standard must be known to be absent from the sample matrix so that the only source of the standard is the added amount. For example, lithium is a good internal standard for the determination of sodium or potassium in blood serum because the chemical behavior of lithium is similar to both analytes, but it does not occur naturally in blood.

As an example, the internal standard method is often used in the determination of trace elements in metals by emission spectroscopy. Thus, in determining parts per million of antimony and tin in lead to be used for the manufacture of storage batteries, the relative intensity of a strong line for each of the minor constituents might be compared with the intensity of a weak line for lead. Ordinarily, these ratios would be less affected by variables that arise in causing the samples to emit radiation. In the development of any new internal standard method, we must verify that changes in concentration of analyte do not affect the signal intensity that results from the internal standard. In order for such a procedure to be successful, a good deal of time and effort would need to be expended in preparing a set of pure lead samples that contains exactly known concentrations of antimony and tin.

1F QUESTIONS AND PROBLEMS

1-1 What is a transducer in an analytical instrument?

1-2 What is the information processor in an instrument for measuring the color of a solution visually?

1-3 What is the detector in a spectrograph in which spectral lines are recorded photographically?

1-4 What is the transducer in a smoke detector?

1-5 What is a data domain?

1-6 What are analog domains? How is information encoded in analog domains?

1-7 List four output transducers and describe how they are used.

1-8 What is a figure of merit?

1-9 The following calibration data were obtained by an instrumental method for the determination of the species X in aqueous solution.

Concn X, C_X ppm	**No. Replications, *N***	**Mean Analytical Signal, *S***	**Standard Deviation, ppm**
0.00	25	0.031	0.0079
2.00	5	0.173	0.0094
6.00	5	0.422	0.0084
10.00	5	0.702	0.0084
14.00	5	0.956	0.0085
18.00	5	1.248	0.0110

(a) Calculate the calibration sensitivity.
(b) Calculate the analytical sensitivity at each concentration.
(c) Calculate the coefficient of variation for the mean for each of the replicate sets.
(d) What is the detection limit for the method?

1-10 A 25.0-mL sample containing Cu^{2+} gave an instrument signal of 23.6 units (corrected for a blank). When exactly 0.500 mL of 0.0287 M $Cu(NO_3)_2$ was added to the solution, the signal increased to 37.9 units. Calculate the molar concentration of Cu^{2+} assuming that the signal was directly proportional to the analyte concentration.

1-11 Exactly 5.00-mL aliquots of a solution containing phenobarbital were measured into 50.00-mL volumetric flasks and made basic with KOH. The following volumes of a standard solution of phenobarbital containing 2.000 μg/mL of phenobarbital were then introduced into each flask and the mixture was diluted to volume: 0.000, 0.500, 1.00, 1.50, and 2.00 mL. A fluorometer reading for each of these solutions was 3.26, 4.80, 6.41, 8.02, and 9.56, respectively.
(a) Plot the data.
(b) Using the plot from (a), calculate the concentration of phenobarbital in the unknown.
(c) Derive a least-squares equation for the data.
(d) Compute the concentration of phenobarbital from the equation in (c).
(e) Calculate a standard deviation for the concentration obtained in (d).

Section I

Measurement Basics

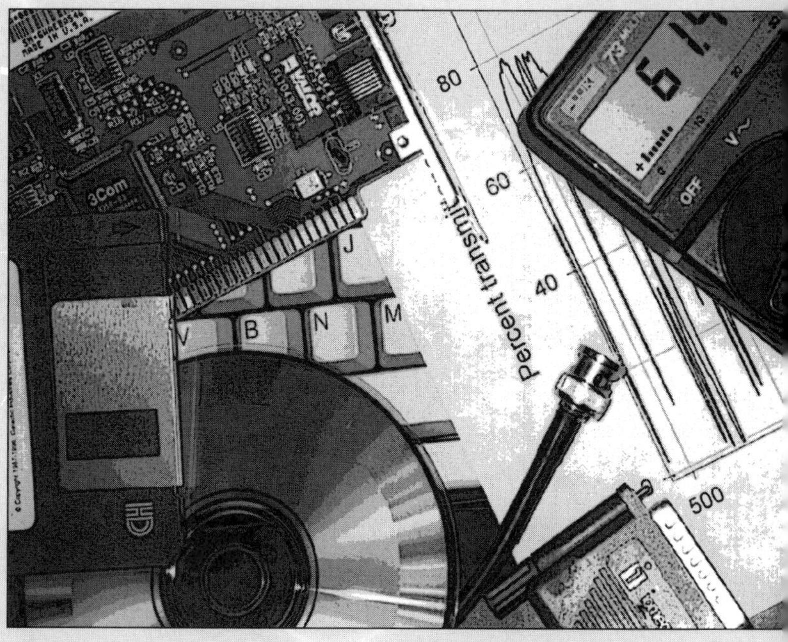

Our lives have been profoundly influenced by the microelectronics revolution. It is not surprising that the field of instrumental analysis has undergone a similar revolution. The montage of computer and electronic components depicted above is symbolic of both the nature and the pace of the changes that are occurring. By the time you read these lines, one or more of the components in this image will have been supplanted by more sophisticated technology or will have become obsolete.

Chapter 1 laid the foundation for the study of instrumental chemical analysis. In the four chapters of Section I, the fundamental concepts of analog electronics, digital electronics, computers, and data manipulation are presented that are essential to understanding how instrumental measurements are accomplished. Chapter 2 provides a brief introduction to the components and principles of operation of basic analog dc and ac circuits. Chapter 3 continues the investigation of the analog domain by presenting the principles of operation and many useful examples of operational amplifier circuits. Digital electronics and the boundary between the analog domain and the digital domain are explored in Chapter 4, as are the nature of computers and their role in instrumental analysis. In Chapter 5, we complete the inquiry of measurement fundamentals by examining the nature of signals and noise, as well as hardware and software methods for increasing the signal-to-noise ratio.

Chapter 2

Electrical Components and Circuits

In Chapter 1, we introduced the concept of data domains and pointed out that modern instruments function by converting data from one domain to another. Most of these conversions are between electrical domains. To understand these conversions, and thus how modern electronic instruments work, some knowledge is required of basic direct current (dc) and alternating current (ac) circuit components. The purpose of this chapter is to survey these topics in preparation for the two following chapters that deal with integrated circuits and microcomputers in instruments for chemical analysis. Armed with this knowledge, you will understand and appreciate the functions of the various measurement systems and methods discussed elsewhere in this text.

2A DIRECT CURRENT CIRCUITS AND MEASUREMENTS

In this section, we consider some basic direct current circuits and how they are used in making current, voltage, and resistance measurements. A general definition of a *circuit* is a closed path that may be followed by an electric current. We begin our discussion of circuits with a survey of four important laws of electricity.

2A-1 Laws of Electricity

Ohm's Law

Ohm's law describes the relationship among potential, resistance, and current in a resistive *series circuit.* In a series circuit, all circuit elements are connected in sequence along a unique path, head to tail, as are the battery and three resistors shown in Figure 2-1. Ohm's law can be written in the form

$$V = IR \qquad (2\text{-}1)$$

where V is the potential difference in volts between two points in a circuit, R is the resistance between the two points in ohms, and I is the resulting current in amperes.[1]

[1] Throughout most of the text, the symbol V will be used to denote electrical potential difference in circuits. In Chapters 22 through 24, however, the electrochemical convention will be followed in which electromotive force is designated as E.

Kirchhoff's Laws

Kirchhoff's current law states that the algebraic sum of currents around any point in a circuit is zero. *Kirchhoff's voltage law* states that the algebraic sum of the voltages around a closed electrical loop is zero.

The applications of Kirchhoff's and Ohm's laws to basic dc circuits are considered in Section 2A-2.

Power Law

The power P in watts dissipated in a resistive element is given by the product of the current in amperes and the potential difference across the resistance in volts:

$$P = IV \tag{2-2}$$

Substituting Ohm's law, gives

$$P = I^2R = V^2/R \tag{2-3}$$

2A-2 **Basic Direct Current Circuits**

In this section we describe two types of basic dc circuits that find widespread use in electrical devices, namely *series resistive circuits* and *parallel resistive circuits,* and analyze their properties with the aid of the laws described in the previous section.

Series Circuits

Figure 2-1 shows a basic series circuit, which consists of a battery, a switch, and three resistors in series. Applying Kirchhoff's current law to point D in this circuit gives

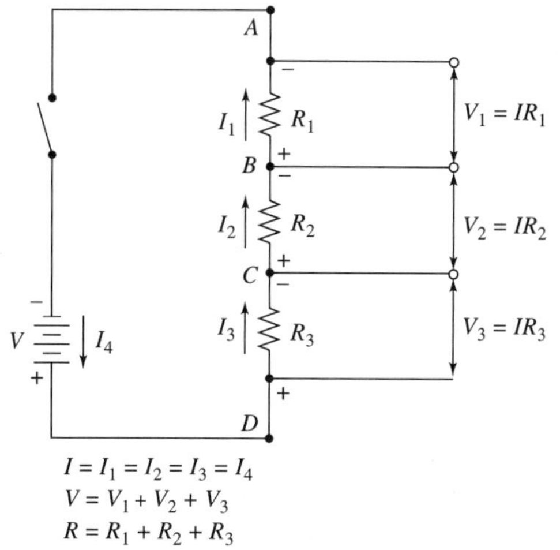

Figure 2-1 Resistors in series; a voltage divider. The current in each resistor is the same in a series circuit.

$$I_4 - I_3 = 0$$

or

$$I_4 = I_3$$

Note that the current out of D must be *opposite* in sign to the input current. Similarly, application of the law to point C gives

$$I_3 = I_2$$

Thus, *the current is the same at all points in a series circuit;* that is,

$$I = I_1 = I_2 = I_3 = I_4 \tag{2-4}$$

Application of Kirchhoff's voltage law to the circuit in Figure 2-1 yields

$$V - V_3 - V_2 - V_1 = 0$$

or

$$V = V_1 + V_2 + V_3 \tag{2-5}$$

Note that point D is positive with respect to point C, which in turn is positive with respect to point B; finally, B is positive with respect to A. The three voltages thus oppose the voltage of the battery and must be given signs that are opposite to V.

Substitution of Ohm's law into Equation 2-5 gives

$$V = I(R_1 + R_2 + R_3) = IR_s \tag{2-6}$$

Note that *the total resistance R_s of a series circuit is equal to the sum of the resistances* of the individual components. That is,

$$R_s = R_1 + R_2 + R_3 \tag{2-7}$$

Applying Ohm's law to the part of the circuit from point B to A gives

$$V_1 = I_1R_1 = IR_1$$

Dividing by Equation 2-6 yields

$$\frac{V_1}{V} = \frac{IR_1}{I(R_1 + R_2 + R_3)}$$

or

$$V_1 = \frac{VR_1}{R_1 + R_2 + R_3} = V\left(\frac{R_1}{R_s}\right) \tag{2-8}$$

In a similar way, we may also write

$$V_2 = VR_2/R_s$$

and

$$V_3 = VR_3/R_s$$

Voltage Dividers

Series resistances are widely used in electrical circuits to provide potentials that are variable functions of an input voltage. Devices of this type are called *voltage dividers.* As shown in Figure 2-2a, one type of voltage divider provides voltages in discrete increments; the second type (Figure 2-2b), called a *potentiometer,*[2] provides a potential that is continuously variable.

In most potentiometers, such as the one shown in Figure 2-2b, the resistance is linear—that is, the resistance between one end, *A,* and any point, *C,* is directly proportional to the length, *AC,* of that portion of the resistor. Then $R_{AC} = kAC$ where AC is expressed in convenient units of length and k is a proportionality constant. Similarly, $R_{AB} = kAB$. Combining these relationships with Equation 2-8 yields

$$V_{AC} = V_{AB}\frac{R_{AC}}{R_{AB}} = V_{AB}\frac{AC}{AB} \quad (2\text{-}9)$$

In commercial potentiometers, R_{AB} is generally a wire-wound resistor formed in a helical coil. A movable contact, called a *wiper,* can be moved from one end of the helix to the other, allowing V_{AC} to be varied continuously from zero to V_{AB}.

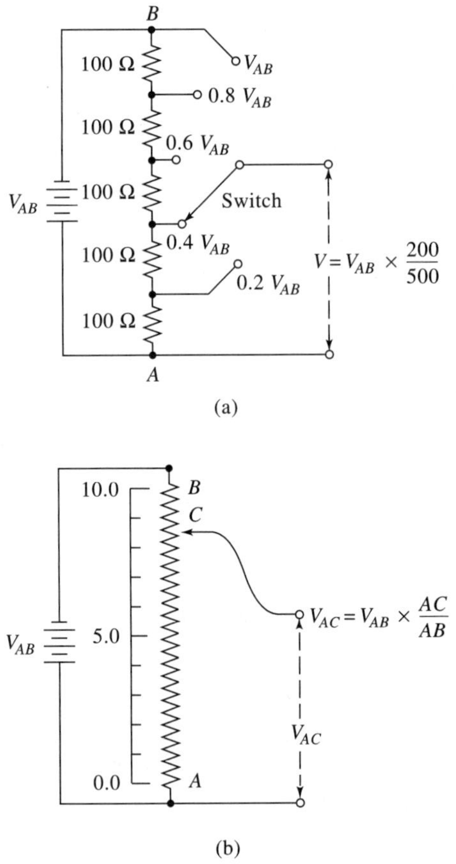

Figure 2-2 Voltage dividers: (a) selector type and (b) continuously variable type (potentiometer).

Parallel Circuits

Figure 2-3 depicts a *parallel* dc circuit. Applying Kirchhoff's current law to point *A* in this figure, we obtain

$$I_1 + I_2 + I_3 - I_t = 0$$

or

$$I_t = I_1 + I_2 + I_3 \quad (2\text{-}10)$$

Applying Kirchhoff's voltage law to this circuit gives three independent equations. Thus, we may write, for the loop that contains the battery and R_1,

$$V - I_1R_1 = 0$$

$$V = I_1R_1$$

For the loop containing V and R_2,

$$V = I_2R_2$$

For the loop containing V and R_3,

$$V = I_3R_3$$

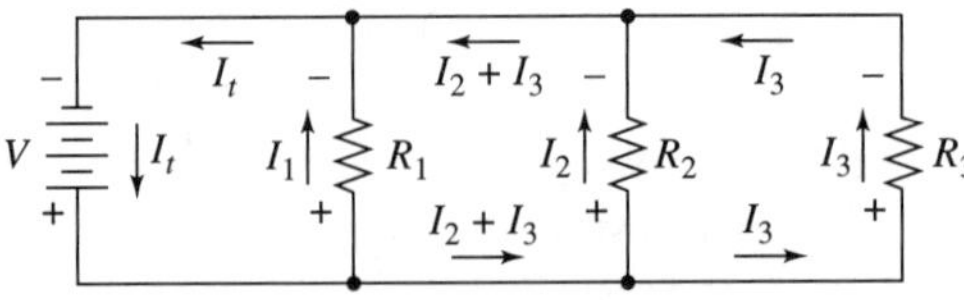

Figure 2-3 Resistors in parallel. The voltage across each resistor is equal to *V,* the battery voltage.

[2] The word *potentiometer* is also used in a different context as the name for a complete instrument that utilizes a linear voltage divider for the accurate measurement of potentials.

We could write additional equations for the loop containing R_1 and R_2 as well as the loop containing R_2 and R_3. However, these equations are not independent of the foregoing three. Substitution of the three independent equations into Equation 2-10 yields

$$I_t = \frac{V}{R_p} = \frac{V}{R_1} + \frac{V}{R_2} + \frac{V}{R_3}$$

By dividing this equation by V, we obtain

$$\frac{1}{R_p} = \frac{1}{R_1} + \frac{1}{R_2} + \frac{1}{R_3} \qquad (2\text{-}11)$$

Since the conductance G of a resistor R is given by $G = 1/R$, we may then write

$$G_p = G_1 + G_2 + G_3 \qquad (2\text{-}12)$$

Equation 2-12 shows that in a parallel circuit, in contrast to a series circuit, it is the conductances G that are additive rather than the resistances.

Current Dividers from Parallel Circuits

Just as series resistances form a voltage divider, parallel resistances create a current divider. The fraction of the total current I_t that is present in R_1 in Figure 2-3 is

$$\frac{I_1}{I_t} = \frac{V/R_1}{V/R_p} = \frac{1/R_1}{1/R_p} = \frac{G_1}{G_p}$$

or

$$I_1 = I_t\frac{R_p}{R_1} = I_t\frac{G_1}{G_p} \qquad (2\text{-}13)$$

An interesting special case occurs when two resistances, R_1 and R_2, form a parallel circuit. The fraction of the current in R_1 is given by

$$\frac{I_1}{I_t} = \frac{G_1}{G_t} = \frac{1/R_1}{1/R_p} = \frac{1/R_1}{1/R_1 + 1/R_2} = \frac{R_2/R_1R_2}{R_2/R_1R_2 + R_1/R_1R_2} = \frac{R_2}{R_1 + R_2}$$

In a similar way, we can show that

$$\frac{I_2}{I_t} = \frac{R_1}{R_1 + R_2}$$

In other words, for two parallel resistors, the fraction of the current in one resistor is just the ratio of the resistance of the second resistor to the sum of the resistances of the two resistors. The equation above is often called the *current-splitting equation.*

EXAMPLE 2-1

For the accompanying circuit, calculate **(a)** the total resistance, **(b)** the current drawn from the battery, **(c)** the current in each of the resistors, and **(d)** the potential difference across each of the resistors.

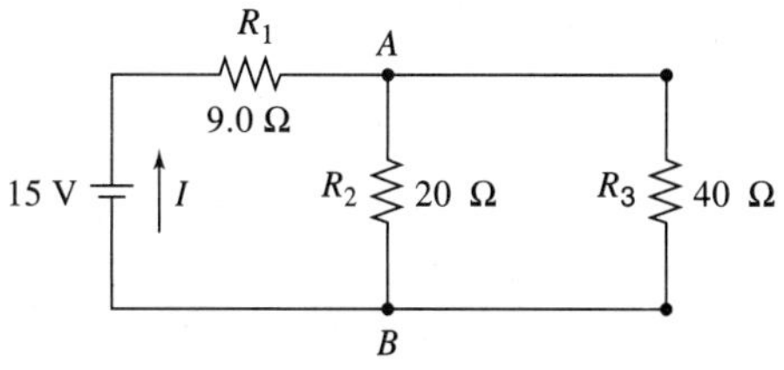

R_2 and R_3 are parallel resistances. Thus, the resistance $R_{2,3}$ between points A and B will be given by Equation 2-11. That is,

$$\frac{1}{R_{2,3}} = \frac{1}{20\ \Omega} + \frac{1}{40\ \Omega}$$

or

$$R_{2,3} = 13.3\ \Omega$$

We can now reduce the original circuit to the following *equivalent circuit.*

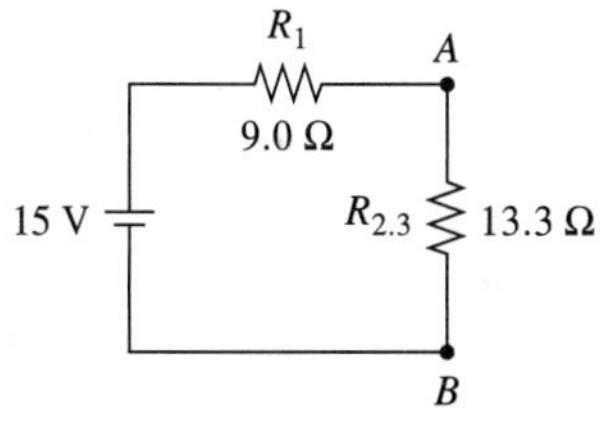

Here, we have the equivalent of two series resistances, and

$$R_s = R_1 + R_{2,3} = 9.0\ \Omega + 13.3\ \Omega = 22.3\ \Omega$$

From Ohm's law, the current I is given by

$$I = 15\ \text{V}/22.3\ \Omega = 0.67\ \text{A}$$

Employing Equation 2-8, the voltage V_1 across R_1 is

$$V = 15\ \text{V} \times 9.0\ \Omega/(9.0\ \Omega + 13.3\ \Omega) = 6.0\ \text{V}$$

Similarly, the voltage across resistors R_2 and R_3 is

$$V_2 = V_3 = V_{2,3} = 15\ \text{V} \times 13.3\ \Omega/22.3\ \Omega$$
$$= 8.95\ \text{V} = 9.0\ \text{V}$$

Note that the sum of the two voltages is 15 V, as required by Kirchhoff's voltage law.

The current R_1 is given by

$$I_1 = I = 0.67 \text{ A}$$

The currents through R_2 and R_3 are found from Ohm's law. Thus,

$$I_2 = 9.0 \text{ V}/20\ \Omega = 0.45 \text{ A}$$

$$I_3 = 9.0 \text{ V}/40\ \Omega = 0.22 \text{ A}$$

Note that the two currents add to give the net current, as required by Kirchhoff's law.

2A-3 DC Current, Voltage, and Resistance Measurements

In this section, we consider (1) how current, potential, and resistance are measured in dc circuits and (2) the uncertainties associated with such measurements.

Digital Voltmeters

Until about thirty years ago, dc electrical measurements were made with a D'Arsonval moving-coil meter, which was invented over a century ago. Now, such meters are largely obsolete, having been replaced by the ubiquitous digital voltmeter (DVM) and the digital multimeter (DMM).

A digital voltmeter usually consists of a single integrated circuit, a power supply that is often a battery, and a liquid-crystal digital display. The heart of the integrated circuit is an *analog-to-digital converter,* which converts the input analog signal to a number that is proportional to the magnitude of the input voltage.[3] A discussion of analog-to-digital converters is given in Section 4C-7. Modern commercial digital voltmeters can be small, are often inexpensive (costing less than \$100), and generally have high input resistances of 10^{10} to $10^{12}\ \Omega$.

Figure 2-4 illustrates how a digital voltmeter, labeled DVM, can be used to measure dc potentials, currents, and resistances. In each schematic, the reading on the voltmeter display is V_M and the internal resistance of the DVM is R_M. The configuration shown in Figure 2-4a is used to determine an unknown potential V_x of a voltage source that has an internal resistance of R_s. The potential displayed by the meter V_M may be somewhat different from the true potential of the source because of the *loading error,* which is discussed in the section that follows. A voltage divider, such as that shown in Figure 2-2a, is usually connected to the input of a digital voltmeter to provide it with several operating ranges.

Digital voltmeters may also be used to measure various ranges of current. The unknown current is passed through one of several small standard resistances built into the meter. The potential difference across this resistance is then measured and is proportional to the current. Figure 2-4b illustrates how an unknown current, I_x, is measured in a circuit consisting of a dc source and a load resistance, R_L. The precision resistors R_{std} in the meter usually range from 0.1 Ω or less to several hundred ohms, thus giving various current ranges. If, for example, $R_{std} = 1.000\ \Omega$ and the DVM reads 1.456 V, then the measured current is 1.456 A. By

[3] An analog signal is one that varies continuously with time and can assume any value within a certain range.

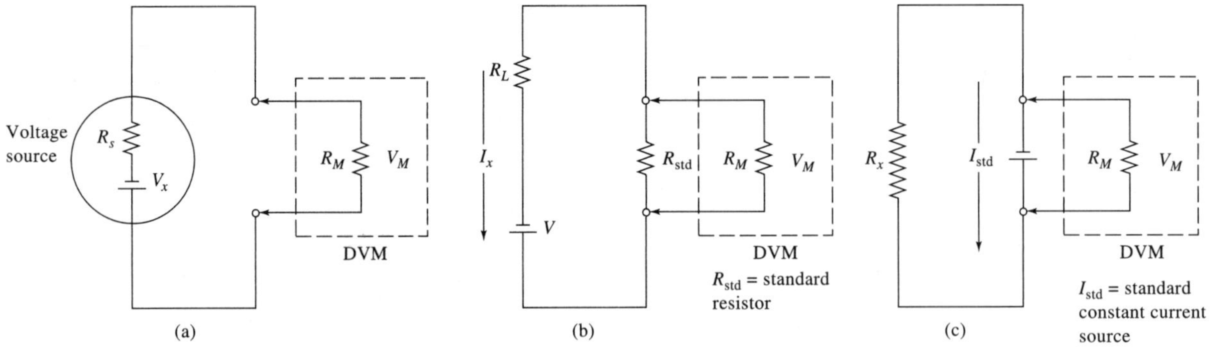

Figure 2-4 Uses of a digital voltmeter. (a) Measurement of the output V_x of a voltage source. (b) Measurement of current I_x through a load resistor R_L. (c) Measurement of resistance R_x of an unknown circuit element.

choosing the standard resistors to be even powers of ten and arranging circuitry to move the display decimal point to match the resistor, the DVM reads the current directly.

Figure 2-4c demonstrates how an unknown resistance R_x is determined with a modern digital voltmeter. For this application, the meter is equipped with a dc source that produces a constant current I_{std} that is directed through the unknown resistance R_x. The DVM indicates the potential difference across R_x when the current I_{std} is passed through the resistor. For example, if the standard current is 0.0100 A, then a DVM reading of 0.945 V yields a measured resistance of 0.945 V/0.0100 A = 94.5 Ω. Once again, we just move the decimal point to obtain a direct readout of resistance.

An instrument that contains circuits for measuring voltages, currents, and resistances is usually called a *digital multimeter* (DMM).

The Loading Error in Potential Measurements

When a meter is used to measure potential, the presence of the meter tends to perturb the circuit in such a way that a *loading error* is introduced. This situation is not peculiar to potential measurements. In fact, it is an example of a fundamental limitation to any physical measurement. That is, the process of measurement inevitably disturbs the system of interest so that the quantity actually measured differs from its value prior to the measurement. This type of error can never be completely eliminated; often, however, it can be reduced to insignificant proportions.

The magnitude of the loading error in potential measurements depends on the ratio of the internal resistance of the meter to the resistance of the circuit under consideration. The percent relative loading error E_r associated with the measured potential V_M in Figure 2-4a is given by

$$E_r = \frac{V_M - V_x}{V_x} \times 100\%$$

where V_x is the true voltage of the source. Applying Equation 2-9 for a voltage divider, we can write

$$V_M = V_x\left(\frac{R_M}{R_M + R_s}\right)$$

When we substitute this equation into the previous one and rearrange the result, we obtain

$$E_r = -\frac{R_s}{R_M + R_s} \times 100\% \qquad (2\text{-}14)$$

TABLE 2-1 Effect of Meter Resistance on the Accuracy of Potential Measurements*

Meter Resistance R_M, Ω	Resistance of Source R_s, Ω	R_M/R_s	Relative Error %
10	20	0.50	−67
50	20	2.5	−29
500	20	25	−3.8
1.0×10^3	20	50	−2.0
1.0×10^4	20	500	−0.20

*See Figure 2-4a.

This equation shows that the relative loading error becomes smaller as the meter resistance R_M becomes larger relative to the source resistance R_s. Table 2-1 illustrates this effect. Digital voltmeters offer the great advantage of having enormous internal resistances of 10^{11} to 10^{12} Ω, and thus loading errors are avoided except in circuits having resistances of greater than about 10^9 Ω. An important example of loading error may occur in the measurement of the voltage of glass pH electrodes, which have resistances of 10^6 to 10^9 Ω or greater. Instruments such as pH meters and pIon meters have very high resistance inputs to guard against loading errors of this kind.

We should note, however, that in a practical DVM used for voltage measurements, the input voltage divider used to provide multiple ranges decreases the internal resistance of the meter to a typical value of 10 MΩ. Thus, voltage measurements on sources with internal resistances greater than 10 kΩ will be in error by 0.1% or more, depending on the exact internal resistance.

The Loading Error in Current Measurements

As shown in Figure 2-4b, in making a current measurement, a small, high-precision standard resistor with a resistance R_{std} is introduced into the circuit. In the absence of this resistance, the current in the circuit would be $I = V/R_L$. With resistor R_{std} in place, it would be $I_M = V/(R_L + R_{std})$. Thus, the loading error is given by

$$E_r = \frac{I_M - I_x}{I_x} \times 100\% = \frac{\dfrac{V}{(R_L + R_{std})} - \dfrac{V}{R_L}}{\dfrac{V}{R_L}} \times 100\%$$

This equation simplifies to

$$E_r = -\frac{R_{std}}{R_L + R_{std}} \times 100\% \qquad (2\text{-}15)$$

Table 2-2 reveals that the loading error in current measurements becomes smaller as the ratio of R_{std} to R_L becomes smaller.

2B ALTERNATING CURRENT CIRCUITS

The electrical output from transducers of analytical signals often fluctuate periodically. These fluctuations can be represented (as in Figure 2-5) by a plot of the instantaneous current or potential as a function of time. The *period* t_p for the signal is the time required for the completion of one cycle.

The reciprocal of the period is the *frequency* f of the signal. That is,

$$f = 1/t_p \qquad (2\text{-}16)$$

The unit of frequency is the hertz, Hz, which is defined as one cycle per second.

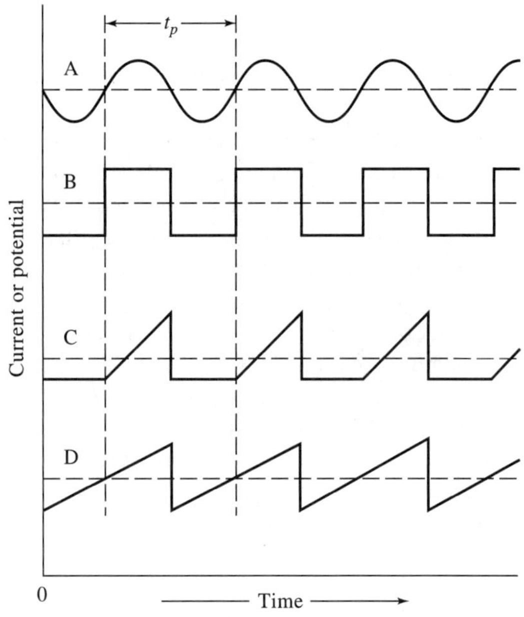

Figure 2-5 Examples of periodic signals: *(A)* sinusoidal, *(B)* square wave, *(C)* ramp, and *(D)* sawtooth.

2B-1 Sinusoidal Currents

The sinusoidal wave (Figure 2-5A) is the most frequently encountered type of periodic electrical signal. A common example is the alternating current produced by rotation of a coil in a magnetic field (as in an electrical generator). Thus, if the instantaneous current or voltage produced by a generator is plotted as a function of time, a sine wave results.

TABLE 2-2 Effect of Resistance of the Standard Resistor, R_{std}, on the Accuracy of Current Measurement*

Circuit Resistance R_L, Ω	Standard Resistance R_{std}, Ω	R_{std}/R_L	Relative Error %
1.0	1.0	1.0	−50
10	1.0	0.10	−9.1
100	1.0	0.010	−0.99
1000	1.0	0.0010	−0.10

*See Figure 2-4b.

A pure sine wave is conveniently represented as a vector of length I_p (or V_p), which is rotating counterclockwise at a constant angular velocity ω. The relationship between the vector representation and the sine-wave plot is shown in Figure 2-6. The vector rotates at a rate of 2π radians in the period t_p; thus, the angular frequency is given by

$$\omega = \frac{2\pi}{t_p} = 2\pi f \qquad (2\text{-}17)$$

If the vector quantity is current or voltage, the instantaneous current i or instantaneous voltage v at time t is given by (see Figure 2-6)[4]

$$i = I_p \sin \omega t = I_p \sin 2\pi f t \qquad (2\text{-}18)$$

or, alternatively,

$$v = V_p \sin \omega t = V_p \sin 2\pi f t \qquad (2\text{-}19)$$

where I_p and V_p, the maximum, or peak, current and voltage, are called the *amplitude, A,* of the sine wave.

[4] It is useful to symbolize the instantaneous value of time-varying current, voltage, or charge with the lower case letters *i, v,* and *q*. On the other hand, capital letters are used for steady current, voltage, or charge or for a specifically defined variable quantity such as a peak voltage current, that is, V_p and I_p.

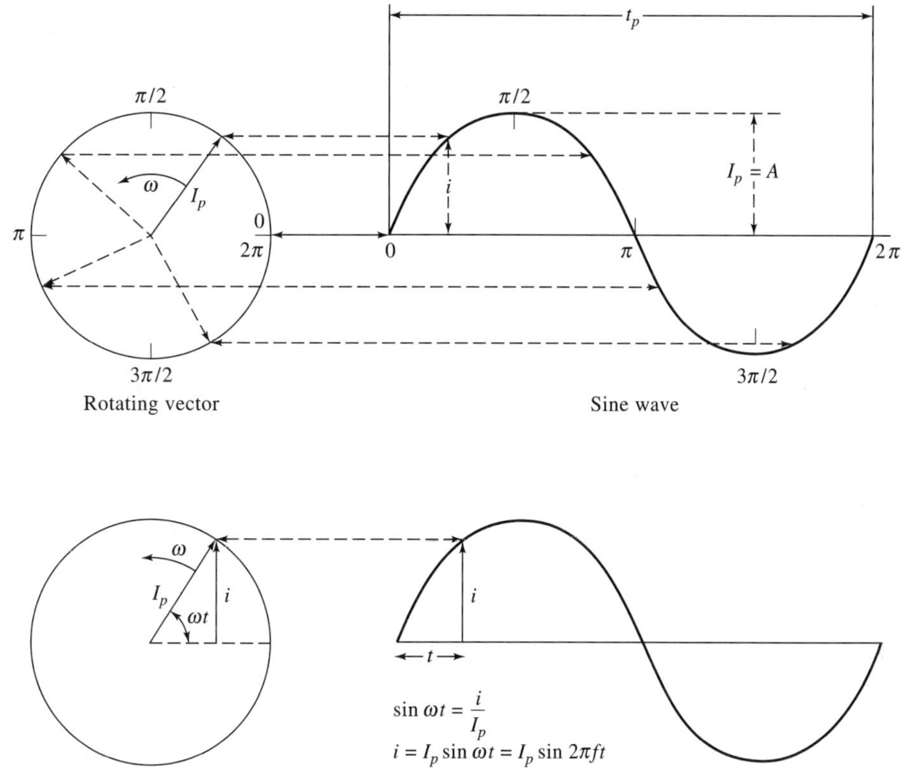

Figure 2-6 Relationship between a sine wave of period t_p and amplitude I_p and the corresponding vector of length I_p rotating at an angular velocity of $\omega = 2\pi f$ radians/second or a frequency of f Hz.

Figure 2-7 shows two sine waves that have different amplitudes. The two waves are also out of *phase* by 90 degrees, or $\pi/2$ radians. The phase difference is called the *phase angle,* which arises when one vector leads or lags a second by this amount. A more generalized equation for a sine wave, then, is

$$i = I_p \sin(\omega t + \phi) = I_p \sin(2\pi ft + \phi) \quad (2\text{-}20)$$

where ϕ is the phase angle from some reference sine wave. An analogous equation can be written in terms of voltage:

$$v = V_p \sin(2\pi ft + \phi) \quad (2\text{-}21)$$

The current or voltage associated with a sinusoidal current can be expressed in several ways. The simplest is the peak amplitude, I_p (or V_p), which is the maximum instantaneous current or voltage during a cycle; the peak-to-peak value, which is $2I_p$, or $2V_p$, is also used occasionally. The *root mean square* or *rms current* in an ac circuit will produce the same heating in a resistor as a direct current of the same magnitude. Thus, the rms current is important in power calculations (Equations 2-2 and 2-3). The rms current is given by

$$I_{rms} = \sqrt{\frac{I_p^2}{2}} = 0.707 I_p \quad \text{and}$$
$$V_{rms} = \sqrt{\frac{V_p^2}{2}} = 0.707 V_p \quad (2\text{-}22)$$

2B-2 Reactance in Electrical Circuits

When the current in an electrical circuit is increased or decreased, energy is required to change the electric and magnetic fields associated with the flow of charge. For example, if the circuit contains a coil of copper wire, or an *inductor,* the coil resists the change in the current as energy is stored in the magnetic field of the inductor. As the current is reversed, the energy is returned to the ac source; and as the second half of the cycle is completed, energy is once again stored in a magnetic field of the

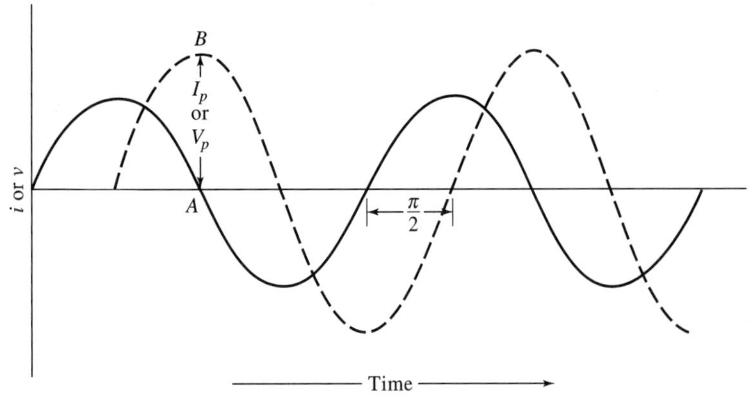

Figure 2-7 Sine waves with different amplitudes (I_p or V_p) and with a phase difference of 90 deg or $\pi/2$ radians.

opposite sense. In a similar way, a *capacitor* in an ac circuit resists changes in voltage. The resistance of inductors to changes in current and the resistance of capacitors to changes in voltage is called *reactance.* As we shall see, reactances in an ac circuit introduce phase shifts in the ac signal. The two types of reactance that characterize capacitors and inductors are *capacitive reactance* and *inductive reactance,* respectively.

Both capacitive reactance and inductive reactance are frequency-dependent quantities. At low frequency when the rate of change in current is low, the effects of reactance in most of the components of a circuit are sufficiently small to be neglected. With rapid changes, on the other hand, circuit elements such as switches, junctions, and resistors may exhibit reactance. When the effects of this type of reactance are undesirable, every effort is made to diminish its magnitude and thus its effects.

Capacitance and inductance are often deliberately introduced into a circuit with *capacitors* and *inductors.* These devices play important roles in such useful functions as converting alternating current to direct current or the converse, discriminating among signals of different frequencies, separating ac and dc signals, differentiating signals, or integrating signals.

In the sections that follow, we shall consider only the properties of capacitors since most modern electronic circuits are based on these devices rather than on inductors.

2B-3 Capacitors and Capacitance: Series *RC* Circuits

A typical capacitor consists of a pair of conductors separated by a thin layer of a *dielectric* substance—that is, by an electrical insulator that contains essentially no mobile, current-carrying, charged species. The simplest capacitor consists of two sheets of metal foil separated by a thin film of a dielectric such as air, oil,

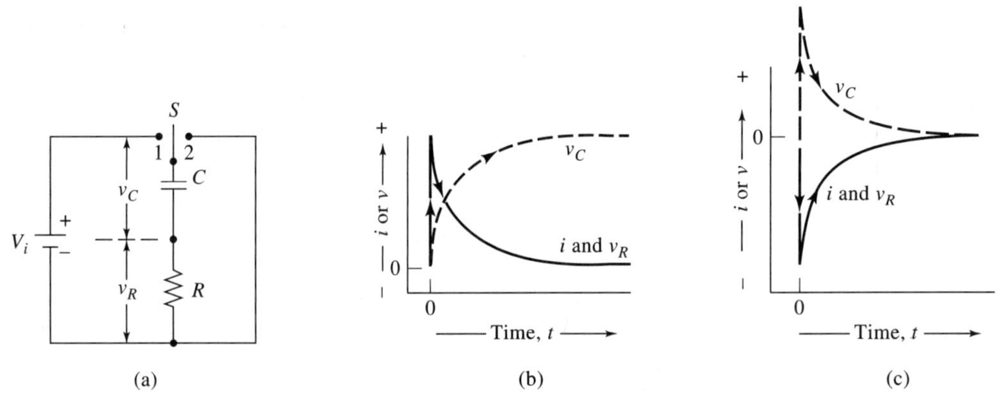

Figure 2-8 (a) A series *RC* circuit. Time response of circuit when switch *S* is (b) in position 1 and (c) in position 2.

plastic, mica, paper, ceramic, or metal oxide. Except for air and mica capacitors, the two layers of foil and the insulator are usually folded or rolled into a compact package and sealed to prevent atmospheric deterioration.

In order to describe the properties of a capacitor, consider the *series RC circuit* shown in Figure 2-8a, which contains a battery V_i, a resistor R, and a capacitor C, in series. The capacitor is symbolized by a pair of parallel lines of identical length.

When the switch S is moved from position 2 to position 1, electrons flow from the negative terminal of the battery through the resistor R into the lower conductor, or *plate,* of the capacitor. Simultaneously, electrons are repelled from the upper plate and flow toward the positive terminal of the battery. This movement constitutes a momentary current, which quickly decays to zero as the potential difference builds up across the plates of the capacitor and eventually reaches the potential of the battery V_i. When the current ceases, the capacitor is said to be *charged.*

If the switch then is moved from position 1 to position 2, electrons will flow from the negatively charged lower plate of the capacitor through the resistor R to the positive upper plate. Again, this movement constitutes a current that decays to zero as the potential difference between the two plates disappears; here, the capacitor is said to be *discharged.*

A useful property of a capacitor is its ability to store an electrical charge for a period of time and then to give up the stored charge when needed. Thus, if S in Figure 2-8a is first held at 1 until C is charged and is then moved to a position *between* 1 and 2, the capacitor will maintain its charge for an extended period. When S is moved to position 2, discharge occurs in the same way as it would if the change from 1 to 2 had been rapid.

The quantity of electricity Q required to charge a capacitor fully depends upon the area of the plates, their shape, the spacing between them, and the dielectric constant for the material that separates them. In addition, the charge Q is directly proportional to the applied voltage. That is,

$$Q = CV \qquad (2\text{-}23)$$

When V is the applied potential in volts and Q is the quantity of charge in coulombs, the proportionality constant C is the *capacitance* of a capacitor in *farads* F. A one-farad capacitor, then, stores one coulomb of charge per applied volt. Most of the capacitors used in electronic circuitry have capacitances in the microfarad (10^{-6} F) to picofarad (10^{-12} F) ranges.

Capacitance is important in ac circuits, particularly because a voltage that varies with time gives rise to a time-varying charge—that is, a *current.* This behavior is seen by differentiating Equation 2-23 to give

$$\frac{dq}{dt} = C\frac{dv_C}{dt} \qquad (2\text{-}24)$$

By definition, the current, i, is the time rate of change of charge; that is, $dq/dt = i$. Thus,

$$i = C\frac{dv_C}{dt} \qquad (2\text{-}25)$$

It is important to note that the current in a capacitor is zero when the voltage is time independent—that is, when the voltage across the capacitor is constant. Furthermore, note that a very large current is required in order to make a rapid change in the voltage across a capacitor. This result imposes a significant limitation on certain electroanalytical methods of analysis, as discussed in Chapter 25.

Rate of Current Change in an *RC* Circuit

The rate at which a capacitor is charged or discharged is finite. Consider, for example, the circuit shown in Figure 2-8a. From Kirchhoff's voltage law, we know that at any instant after the switch is moved to position 1, the sum of the voltage across C and R(v_C and v_R) must equal the input voltage V_i. Thus,

$$V_i = v_C + v_R \qquad (2\text{-}26)$$

Because V_i is constant, the increase in v_C that accompanies the charging of the capacitor must be exactly offset by a decrease in v_R.

Substitution of Equations 2-1 and 2-23 into this equation gives, upon rearrangement,

$$V_i = \frac{q}{C} + iR \qquad (2\text{-}27)$$

In order to determine how the current in an RC circuit changes as a function of time, we differentiate Equation 2-27 with respect to time, remembering that V_i is constant. Thus,

$$\frac{dV_i}{dt} = 0 = \frac{dq/dt}{C} + R\frac{di}{dt} \qquad (2\text{-}28)$$

Here, again, we have used lower case letters to represent instantaneous charge and current.

As noted earlier, $dq/dt = i$. Substituting this expression into Equation 2-28 yields, upon rearrangement,

$$\frac{di}{i} = -\frac{dt}{RC}$$

Integration between the limits of the initial current I_{init} and i gives

$$\int_{I_{init}}^{i} \frac{di}{i} = -\int_{0}^{t} \frac{dt}{RC} \qquad (2\text{-}29)$$

and

$$i = I_{init}\, e^{-t/RC} \qquad (2\text{-}30)$$

This equation shows that the current in the RC circuit decays exponentially with time.

Rate of Voltage Change in an *RC* Circuit

In order to obtain an expression for the instantaneous voltage across the resistor, v_R, we use Ohm's law to substitute $i = v_R/R$ and $I_{init} = V_i/R$ into Equation 2-30 and rearrange to obtain

$$v_R = V_i\, e^{-t/RC} \qquad (2\text{-}31)$$

Substitution of this expression into Equation 2-26 yields upon rearrangement an expression for the instantaneous voltage across the capacitor v_C:

$$v_C = V_i(1 - e^{-t/RC}) \qquad (2\text{-}32)$$

Note that the product RC that appears in the last three equations has the units of time; since $R = v_R/i$ and $C = q/v_C$,

$$RC = \frac{v_R}{i} \times \frac{q}{v_C} \quad \text{and}$$

$$\frac{\cancel{\text{volts}}}{\cancel{\text{coulombs}}/\text{seconds}} \times \frac{\cancel{\text{coulombs}}}{\cancel{\text{volts}}} = \text{seconds}$$

The product RC is called the *time constant* for the circuit and is a measure of the time required for a capacitor to charge or discharge. This dependence of the charging time on RC can be rationalized from the form of Equation 2-32. Because the ratio $-t/RC$ is the exponent in this equation, RC determines the rate of exponential change of the voltage across the capacitor.

The following example illustrates the use of the equations that were just derived.

EXAMPLE 2-2

Values for the components in Figure 2-8a are V_i = 10.0 V, R = 1000 Ω, C = 1.00 μF or 1.00×10^{-6} F. Calculate **(a)** the time constant for the circuit and **(b)** i, v_C, and v_R after two-time constant ($t = 2RC$) have elapsed.

(a) Time constant $= RC = 1000 \times 1.00 \times 10^{-6} = 1.00 \times 10^{-3}$ s or 1.00 ms.

(b) Substituting Ohm's law, $I_{init} = V_i/R$ and t = 2.00 ms into Equation 2-30 reveals

$$i = \frac{V}{R} e^{-t/RC} = \frac{10.0}{1000} e^{-2.00/1.00}$$

$$= 1.35 \times 10^{-3}\ \text{A} \quad \text{or} \quad 1.35\ \text{mA}$$

We find from Equation 2-31 that

$$v_R = 10.0\, e^{-2.00/1.00} = 1.35\ \text{V}$$

And by substituting into Equation 2-26, we find that

$$v_C = V_i - v_R = 10.00 - 1.35$$
$$= 10.0\,(1 - e^{-2.00/1.00}) = 8.65\ \text{V}$$

Phase Relations between Current and Voltage in an *RC* Circuit

Figure 2-8b shows the changes in i, v_R, and v_C that occur during the charging cycle of an RC circuit. These plots are presented in arbitrary units because the shape of the curves is independent of the time constant of the circuit. Note that v_R and i assume their maximum values the instant the switch in Figure 2-8a is moved to 1. At the same instant, on the other hand, the voltage across the capacitor increases rapidly from zero and ultimately approaches a constant value. For practical purposes, a capacitor is considered to be fully charged after five time constants, or $5RC$s, have elapsed. At this point, the current will have decayed to less than 1% of its initial value ($e^{-5RC/RC} = e^{-5} = 0.0067 \approx 0.01$).

When the switch in Figure 2-8a is moved to position 2, the battery is removed from the circuit and the capacitor becomes a source of current. The flow of charge, however, will be in the opposite direction from what it was during charging. Thus,

$$dq/dt = -i$$

The initial potential of the capacitor will be that of the battery. That is,

$$V_C = V_i$$

Using these equations and proceeding as in the earlier derivation, we find that for the discharge cycle

$$i = -\frac{V_C}{R}\,e^{-t/RC} \tag{2-33}$$

$$v_R = -V_C\,e^{-t/RC} \tag{2-34}$$

and because $V_C = 0 = v_C + v_R$ (Equation 2-26)

$$v_C = V_C\,e^{-t/RC} \tag{2-35}$$

Figure 2-8c shows how i, V_R, and v_C change with time.

It is important to note that in each cycle, the change in voltage across the capacitor is *out of phase with and lags behind* that of the current and the potential across the resistor.

2B-4 Response of Series *RC* Circuits to Sinusoidal Inputs

In the sections that follow, the response of series *RC* circuits to a sinusoidal ac voltage signal will be investigated. The input signal v_s is described by Equation 2-19; that is,

$$v_s = V_p \sin\omega t = V_p \sin 2\pi f t \tag{2-36}$$

Phase Changes in Capacitive Circuits

If the switch and battery in the *RC* circuit shown in Figure 2-8a are replaced with a sinusoidal ac source, the capacitor continuously stores and releases charge, thus causing a current that alternates in direction and changes continuously. A phase difference ϕ between the current and voltage is introduced as a consequence of the finite time required to charge and discharge the capacitor (see Figures 2-8b and 2-8c).

We may determine the magnitude of the phase shift by considering a capacitor in an ideal circuit that has no resistance. We first combine Equations 2-18 and 2-25 to give, after rearrangement,

$$C\frac{dv_C}{dt} = I_p \sin 2\pi f t \tag{2-37}$$

At time $t = 0$, $v_C = 0$. Thus, if we rearrange this equation and integrate between times 0 and t, we obtain

$$v_C = \frac{I_p}{C}\int_0^t \sin 2\pi f t\, dt = \frac{I_p}{2\pi f C}(-\cos 2\pi f t)$$

But from trigonometry, $-\cos x = \sin(x - 90)$. Therefore, we may write

$$v_C = \frac{I_p}{2\pi f C}\sin(2\pi f t - 90) \tag{2-38}$$

By comparing Equation 2-38 with Equation 2-21, we see that $I_p/(2\pi f C) = V_p$; therefore, Equation 2-38 can be written in the form

$$v_C = V_p \sin(2\pi f t - 90) \tag{2-39}$$

The instantaneous current, however, is given by Equation 2-18. That is,

$$i = I_p \sin 2\pi f t$$

When we compare the last two equations, we find that the voltage across a pure capacitor that results from a sinusoidal input signal is sinusoidal but lags behind the current by 90 deg (see Figure 2-9). As we shall see later, this lag is smaller than 90 deg in a real circuit that also contains resistance.

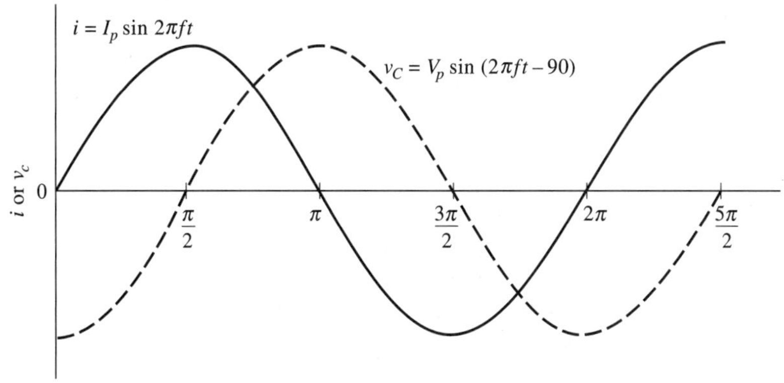

Figure 2-9 Sinusoidal current i and voltage v_c signals in a capacitor.

Capacitive Reactance

Like a resistor, a capacitor during charging impedes the flow of charge, which results in a continuous decrease in the magnitude of the current. This effect results from the limited capacity of the device to hold charge at a given voltage, as given by the expression $Q = CV$. In contrast to a resistor, however, charging a capacitor does not create a permanent loss of energy as heat. Here, the energy stored in the charging process is released to the system during discharge.

Ohm's law can be applied to capacitive ac circuits and takes the form

$$V_p = I_p X_C \tag{2-40}$$

where X_C is the *capacitive reactance,* a property of a capacitor that is analogous to the resistance of a resistor. A comparison of Equations 2-38 and 2-39 shows, however, that

$$V_p = \frac{I_p}{2\pi fC} = \frac{I_p}{\omega C}$$

Thus, the capacitive reactance is given by

$$X_C = \frac{V_p}{I_p} = \frac{I_p}{I_p 2\pi fC} = \frac{1}{2\pi fC} = \frac{1}{\omega C} \tag{2-41}$$

and X_C has the dimensions of ohms.

It should also be noted that in contrast to resistance, capacitive reactance is *frequency dependent* and becomes smaller at higher frequency; at zero frequency X_C becomes extremely large, so that a capacitor acts as an insulator toward a direct current (neglecting the momentary initial charging current).

EXAMPLE 2-3

Calculate the reactance of a 0.020 μF (2.0×10^{-8} F) capacitor at frequencies of **(a)** 3.0 MHz and **(b)** 3.0 kHz.

(a) Substituting 3.0 MHz, or 3×10^6 Hz, into Equation 2-41 yields

$$X_C = \frac{1}{2 \times 3.14 \times 3.0 \times 10^6 \times 2.0 \times 10^{-8}}$$

$$= 2.7\ \Omega$$

(b) At 3.0 kHz, or 3×10^3 Hz,

$$X_C = \frac{1}{2 \times 3.14 \times 3.0 \times 10^3 \times 2.0 \times 10^{-8}}$$

$$= 2700\ \Omega \quad \text{or} \quad 2.7\ \text{k}\Omega$$

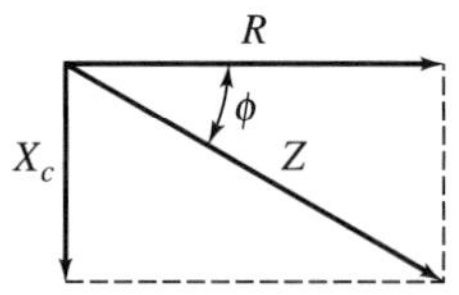

Figure 2-10 Vector diagram for a series *RC* circuit.

Impedance in a Series RC Circuit

The *impedance* Z of an *RC* circuit is made up of two components: the resistance of the resistor and the reactance of the capacitor. Because of the phase shift with the latter, however, the two cannot be combined directly but must be added vectorially, as shown in Figure 2-10. Here the phase angle for R is chosen as zero. As we have shown, the phase angle for a pure capacitive element is -90 deg. Thus the X_C vector is drawn at a right angle to and extends down from the R vector. From the pythagorean theorem, the quantity Z, called the *impedance,* is given by

$$Z = \sqrt{R^2 + X_C^2} \tag{2-42}$$

The phase angle is

$$\phi = \arctan \frac{X_C}{R} \tag{2-43}$$

To show the frequency dependence of the impedance and of the phase angle, we can substitute Equation 2-41 into 2-42 and 2-43, giving

$$Z = \sqrt{R^2 + \left(\frac{1}{2\pi fC}\right)^2} \tag{2-44}$$

and

$$\phi = \arctan \frac{1}{2\pi fRC} \tag{2-45}$$

Note that the extent to which the voltage lags the current in an *RC* circuit, ϕ, is dependent upon the frequency f, the resistance R, and the capacitance C, of the circuit.

Ohm's law for a series *RC* circuit can be written as

$$I_p = \frac{V_p}{Z} = \frac{V_p}{\sqrt{R^2 + \left(\frac{1}{2\pi fC}\right)^2}} \quad (2\text{-}46)$$

or

$$V_p = I_p Z = I_p \sqrt{R^2 + \left(\frac{1}{2\pi fC}\right)^2}$$

EXAMPLE 2-4

A sinusoidal ac source with a peak voltage of 20 V was placed in series with a 1.5×10^4 Ω resistor and a 0.0080 μF capacitor. Calculate the peak current, the phase angle, and the potential difference across each of the components if the frequency of the source was **(a)** 750 Hz and **(b)** 75 kHz.

(a) At 750 Hz, we find by substituting into Equation 2-41

$$X_C = \frac{1}{2\pi fC}$$

$$= \frac{1}{2\pi \times 750 \text{ Hz} \times 8.0 \times 10^{-9} \text{ F}}$$

$$= 2.7 \times 10^4 \ \Omega$$

From Equation 2-42, we find

$$Z = \sqrt{(1.5 \times 10^4)^2 + (2.7 \times 10^4)^2} = 3.0 \times 10^4 \ \Omega$$

Substituting into Equation 2-46 yields

$$I_p = 20 \text{ V}/3.0 \times 10^4 \ \Omega = 6.6 \times 10^{-4} \text{ A}$$

To obtain ϕ, we employ Equation 2-43. Thus,

$$\phi = \arctan \frac{X_C}{R} = \arctan \frac{2.7 \times 10^4 \ \Omega}{1.5 \times 10^4 \ \Omega} = 61 \text{ deg}$$

Application of the equation for a voltage divider gives

$$(V_p)_R = V_p \times \frac{R}{Z} = \frac{20 \text{ V} \times 1.5 \times 10^4 \ \Omega}{3.0 \times 10^4 \ \Omega} = 10.0 \text{ V}$$

$$(V_p)_C = V_p \times \frac{X_C}{Z} = \frac{20 \text{ V} \times 2.7 \times 10^4 \ \Omega}{3.0 \times 10^4 \ \Omega} = 18 \text{ V}$$

where $(V_p)_R$ and $(V_p)_C$ are the peak voltages across the resistor and capacitor, respectively.

(b) Proceeding in a similar way, the following data are obtained for a 75 kHz current.

$X_C = 2.65 \times 10^2 \ \Omega$ $\quad \phi = 1.01$ deg

$Z = 1.50 \times 10^4 \ \Omega$ $\quad (V_p)_R = 20.0$ V

$I_p = 1.33 \times 10^{-3}$ A $\quad (V_p)_C = 0.353$ V

Several important properties of a series *RC* circuit are illustrated by the results obtained in Example 2-4. First, the sum of the peak voltages for the resistor and the capacitor are not equal to the peak voltage of the source. At the lower frequency, for example, the sum is 27.2 V compared with 20.0 V for the source. This apparent anomaly is understandable when it is realized that the peak voltage occurs in the resistor at an earlier time than in the capacitor because of the voltage lag in the latter. At any time, however, the sum of the *instantaneous* voltages across the two elements equals the instantaneous voltage of the source.

A second important point shown by the data in Example 2-4 is that the reactance of the capacitor is two orders of magnitude greater at the lower frequency. As a result, the impedance at the higher frequency is largely associated with the resistor, and the current is significantly greater. Associated with the lowered reactance at the higher frequency is the much smaller voltage of 0.35 V across the capacitor as compared with 17.4 V at the lower frequency.

Finally, the magnitude of the voltage lag in the capacitor is of interest. At the lower frequency, this lag amounted to approximately 60 deg while at the higher frequency it was only about 1 deg.

2B-5 Filters Based on RC Circuits

Series *RC* circuits are often used as filters to attenuate high-frequency signals while passing low-frequency components (a *low-pass filter*) or are alternatively used to reduce low-frequency components while passing the high frequencies (a *high-pass filter*). Figure 2-11 shows how a series *RC* circuit can be arranged to give a high- and a low-pass filter. In each case, the input and output are indicated as the voltages $(V_p)_i$ and $(V_p)_o$.

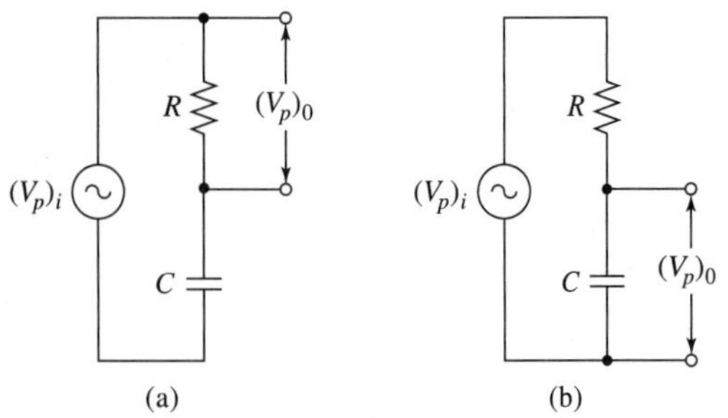

Figure 2-11 Filter circuits: (a) a high-pass filter and (b) a low-pass filter.

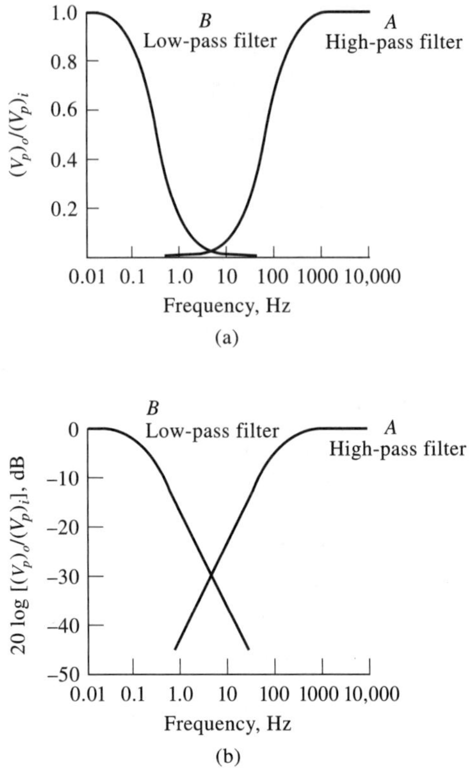

Figure 2-12 (a) Frequency response of high-pass and low-pass filters. (b) Bode diagram for high-pass and low-pass filters. For the high-pass filter, $R = 10\ k\Omega$ and $C = 0.1\ \mu F$. For the low-pass filter, $R = 1\ M\Omega$ and $C = 1\ \mu F$.

High-Pass Filters

In order to use an *RC* circuit as a high-pass filter, the output voltage is taken across the resistor *R* (see Figure 2-11a). The peak current in this circuit can be found by substituting into Equation 2-46. Thus,

$$I_p = \frac{(V_p)_i}{Z} = \frac{(V_p)_i}{\sqrt{R^2 + \left(\frac{1}{2\pi fC}\right)^2}} \tag{2-47}$$

Since the voltage across the resistor is in phase with the current,

$$I_p = \frac{(V_p)_o}{R}$$

The ratio of the peak output to the peak input voltage is obtained by dividing the second equation by the first and rearranging. Thus,

$$\frac{(V_p)_o}{(V_p)_i} = \frac{R}{\sqrt{R^2 + \left(\frac{1}{2\pi fC}\right)^2}} = \frac{R}{Z} \tag{2-48}$$

A plot of this ratio as a function of frequency for a typical high-pass filter is shown as curve *A* in Figure 2-12a. Note that frequencies below 20 Hz have been largely removed from the input signal.

Low-Pass Filters

For the low-pass filter shown in Figure 2-11b, we may write

$$(V_p)_o = I_p X_C$$

Substituting Equation 2-41 gives, upon rearranging,

$$I_p = 2\pi fC(V_p)_o$$

Substituting Equation 2-27 and rearranging yields

$$\frac{(V_p)_o}{(V_p)_i} = \frac{I}{(2\pi fC)\sqrt{R^2 + \left(\frac{I}{2\pi fC}\right)^2}} = \frac{X_C}{Z} \tag{2-49}$$

Curve *B* in Figure 2-12a shows the frequency response of a typical low-pass filter; the data for the plot were obtained with the aid of Equation 2-49. In this case, direct and low-frequency components of the input signal are transferred to the circuit output while high-frequency components are effectively removed.

Figure 2-12b shows *Bode diagrams,* or plots, for the two filters just described. Bode plots are widely encountered in the electronics literature to show the frequency dependence of input/output ratios for various circuits, amplifiers, and filters. The quantity $20\log[(V_p)_o/(V_p)_i]$ gives the gain of an amplifier or a filter in *decibels,* dB.

Low- and high-pass filters are of great importance in the design of electronic circuits.

2B-6 The Response of *RC* Circuits to Pulsed Inputs

When a pulsed input is applied to an *RC* circuit, the voltage across the capacitor and the voltage across the resistor take various forms, depending upon the relationship between the width of the pulse and the time constant for the circuit. These effects are illustrated in Figure 2-13 where the input is a square wave with a pulse width of t_p seconds. The second column shows the variation in capacitor voltage as a function of time, while the third column shows the change in resistor voltage at the same times. In the top set of plots (Figure 2-13a), the time constant of the circuit is much greater than the input pulse width. Under these circumstances, the capacitor can become only partially charged during each pulse. The capacitor then discharges as the input potential returns to zero, and a sawtooth output results. The voltage across the resistor under these circumstances rises instantaneously to a maximum value and then decreases nearly linearly during the pulse lifetime.

The bottom set of graphs (Figure 2-13c) illustrates the two outputs when the time constant of the circuit is much shorter than the pulse width. Here, the charge on the capacitor rises rapidly and approaches full charge near the end of the pulse. As a consequence, the potential across the resistor rapidly decreases to zero after its initial rise. When v_i goes to zero, the capacitor discharges immediately; the output across the resistor peaks in a negative direction and then quickly approaches zero.

These various output wave forms find applications in electronic circuitry. The sharply peaked voltage output shown in Figure 2-13c is particularly important in timing and trigger circuits.

2B-7 AC Current, Voltage, and Impedance Measurements

Alternating current, voltage, and impedance measurements can be carried out with a *digital multimeter.* Digital multimeters are sophisticated instruments that permit the measurement of both ac and dc potentials and currents as well as resistances or impedances over ranges of many orders of magnitude. As shown in Figure 2-14, digital multimeters are built around dc digital

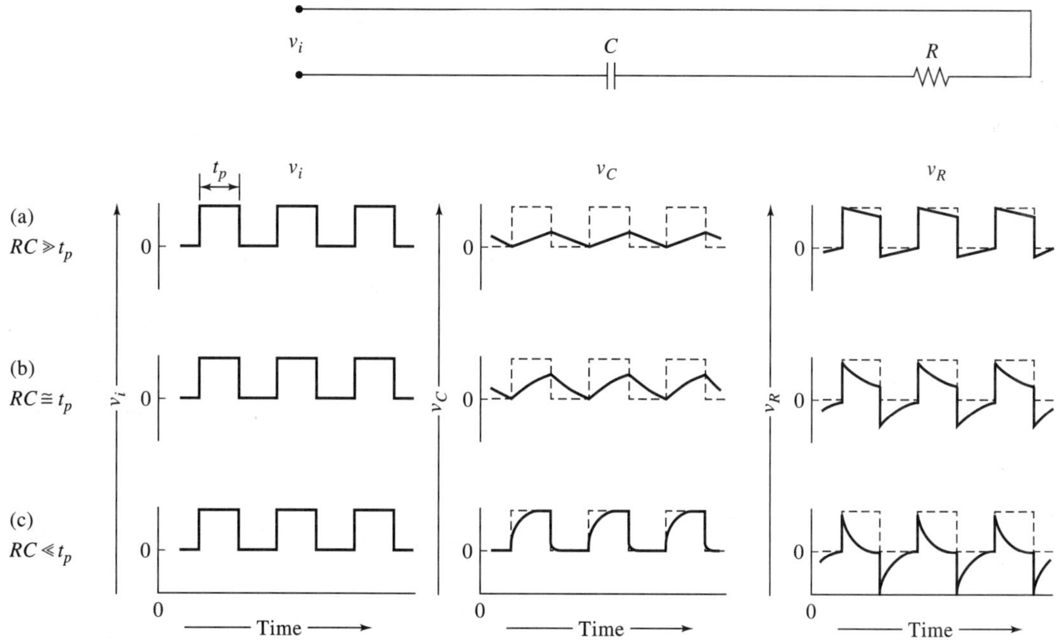

Figure 2-13 Output signals v_r and v_c for pulsed input signal v (a) time constant >> pulse width t_p; (b) time constant = t_p; (c) time constant << t_p.

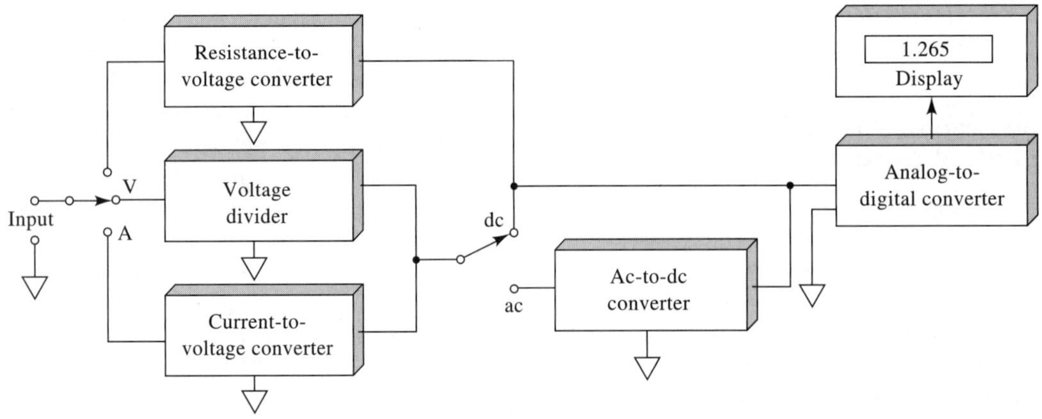

Figure 2-14 Block diagram of a digital multimeter. *(From H. V. Malmstadt, C. G. Enke, and S. R. Crouch,* Electronics and Instrumentation for Scientists. *Menlo Park, CA: Benjamin/Cummings, 1981. With permission.)*

voltmeters, which were discussed in Section 2A-3. In this type of meter, circuits similar to those shown in Figure 2-4 are used with the output from these circuits passing into an ac-to-dc converter before being digitized and displayed.

2C SEMICONDUCTORS AND SEMICONDUCTOR DEVICES

Electronic circuits usually contain one or more nonlinear devices, such as transistors, semiconductor diodes, and vacuum or gas-filled tubes.[5] In contrast to circuit components such as resistors, capacitors, and inductors, the input and output voltages or currents of nonlinear devices are not linearly proportional to one another. As a consequence, nonlinear components can be made to change an electrical signal from ac to dc *(rectification)* or the reverse, to amplify or to attenuate a voltage or current *(amplitude modulation),* or to alter the frequency of an ac signal *(frequency modulation).*

Historically, the vacuum tube was the predominant nonlinear device used in electronic circuitry. In the 1950s, however, tubes were suddenly and essentially completely displaced by *semiconductor-based diodes and transistors,* which have the advantages of low-cost, low-power consumption, small heat generation, long life, and compactness. The era of the individual, or discrete, transistor was remarkably short, however, and electronics is now based largely upon *integrated circuits,* which contain as many as a million transistors, resistors, capacitors, and conductors on a single tiny semiconductor chip. Integrated circuits permit the scientist or engineer to design and construct relatively sophisticated instruments, armed only with their functional properties and input and output characteristics but without detailed knowledge of the internal electronic circuitry of individual chips.

In this section we examine some of the most common components that make up electronic circuits. We then examine a few devices that are important parts of most electronic instruments.

A semiconductor is a crystalline material with a conductivity between that of a conductor and an insulator. There are many types of semiconducting materials, including elemental silicon and germanium; intermetallic compounds, such as silicon carbide and gallium arsenide; and a variety of organic compounds. Two semiconducting materials, which have found widest application for electronic devices, are crystalline silicon and germanium; we shall limit our discussions to these substances.

---

[5] For further information about modern electronic circuits and components, see H. V. Malmstadt, C. G. Enke, and S. R. Crouch, *Microcomputers and Electronic Instrumentation: Making the Right Connections.* Washington, DC: American Chemical Society, 1994; A. J. Diefenderfer and B. E. Holton, *Principles of Electronic Instrumentation,* 3rd ed. Philadelphia: Saunders College Publishing, 1994; J. J. Brophy, *Basic Electronics for Scientists,* 5th ed. New York: McGraw-Hill, 1990; P. Horowitz and W. Hill, *The Art of Electronics,* 2nd ed. New York: Cambridge University Press, 1989.

2C-1 Properties of Silicon and Germanium Semiconductors

Silicon and germanium are Group IV elements and thus have four valence electrons available for bond formation. In a silicon crystal, each of these electrons is localized by combination with an electron from another silicon atom to form a covalent bond. Thus, in principle, there are no free electrons in crystalline silicon, and the material would be expected to be an insulator. In fact, however, sufficient thermal agitation occurs at room temperature to liberate an occasional electron from its bonded state, leaving it free to move through the crystal lattice and thus to conduct electricity. This thermal *excitation* of an electron leaves a positively charged region, termed a *hole,* associated with the silicon atom. The hole, however, like the electron, is mobile and thus also contributes to the electrical conductance of the crystal. The mechanism of hole movement is stepwise; a bound electron from a neighboring silicon atom jumps to the electron-deficient region and thereby leaves a positive hole in its wake. Thus, conduction by a semiconductor involves motion of thermal electrons in one direction and holes in the other.

The conductivity of a silicon or germanium crystal can be greatly enhanced by *doping,* a process whereby a tiny, controlled amount of an impurity is introduced, usually by diffusion, into the heated germanium or silicon crystal. Typically, a silicon or germanium semiconductor is doped with a Group V element, such as arsenic or antimony, or with a Group III element, such as indium or gallium. When an atom of a Group V element replaces a silicon atom in the lattice, one unbound electron is introduced into the structure; only a small amount of thermal energy is then needed to free this electron for conduction. Note that the resulting positive Group V ion does not provide a *mobile* hole since there is little tendency for electrons to move from a covalent silicon bond to this nonbonding position. A semiconductor that has been doped so that it contains nonbonding electrons is termed an *n*-type (negative type) because negatively charged electrons are the *majority carriers* of charge. Holes still exist as in the undoped crystal, which are associated with silicon atoms, but their number is small with respect to the number of electrons; thus, holes represent *minority carriers* in an *n*-type semiconductor.

A *p*-type (positive type) semiconductor is formed when silicon or germanium is doped with a Group III element, which contains only three valence electrons. Here, holes are introduced when electrons from adjoining silicon atoms jump to the vacant orbital associated with the impurity atom. Note that this process imparts a negative charge to the Group III atoms. Movement of the holes from silicon atom to silicon atom, as described earlier, constitutes a current in which the majority carrier is positive. Holes are less mobile than free electrons; thus, the conductivity of a *p*-type semiconductor is inherently less than that of an *n*-type.

2C-2 Semiconductor Diodes

A *diode* is a nonlinear device that has greater conductance in one direction than in another. Useful diodes are manufactured by forming adjacent *n*-type and *p*-type regions within a single germanium or silicon crystal; the interface between these regions is termed a *pn junction.*

Properties of a *pn* Junction

Figure 2-15a is a cross section of one type of *pn* junction, which is formed by diffusing an excess of a *p*-type impurity, such as indium, into a minute silicon chip that has been doped with an *n*-type impurity, such as antimony. A junction of this kind permits movement of holes from the *p* region into the *n* region and movement of electrons in the reverse direction. As holes and electrons diffuse in the opposite direction, a region is created that is depleted of mobile charge carriers and thus has a very high resistance. This region, referred to as the *depletion region,* is depicted in Figure 2-15d. Because there is separation of charge across the depletion region, a potential difference develops across the region that causes a migration of holes and electrons in the opposite direction. The current that results from the diffusion of holes and electrons is balanced by the current produced by migration of the carriers in the electric field, and thus there is no net current. The magnitude of the potential difference across the depletion region depends upon the composition of the materials used in the *pn* junction. For silicon diodes, the potential difference is about 0.6 V, and for germanium, it is about 0.3 V. When a positive potential is applied across a *pn* junction, there is little resistance to current in the direction of the *p*-type to the *n*-type material. On the other hand, the *pn* junction offers a high resistance to the flow of holes in the opposite direction and is thus a *current rectifier.*

Figure 2-15b illustrates the symbol for a diode. The arrow points in the direction of low resistance to positive currents. The triangular portion of the diode symbol

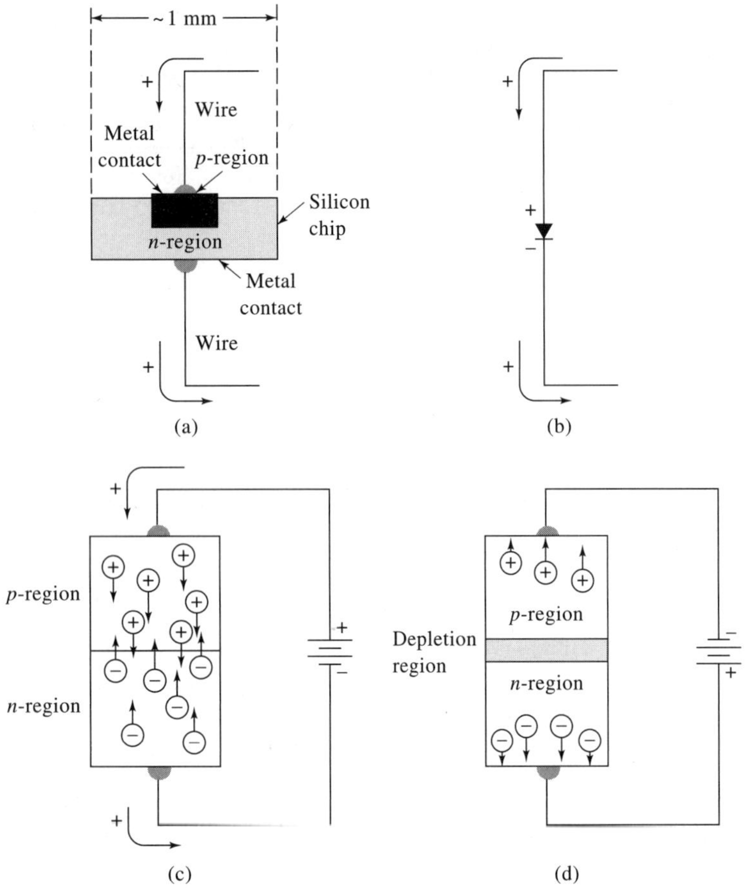

Figure 2-15 A *pn* junction diode. (a) Physical appearance of one type formed by diffusion of a *p*-type impurity into an *n*-type semiconductor, (b) symbol for a diode, (c) current under forward bias, (d) resistance to current under reverse bias.

may be imagined to point in the direction of current in a conducting diode.

Figure 2-15c shows the mechanism of conduction of charge when the *p* region is made positive with respect to the *n* region by application of a potential; this process is called *forward biasing.* Here, the holes in the *p* region and the excess electrons in the *n* region, which are the majority carriers in the two regions, move under the influence of the electric field toward the junction, where they combine and thus annihilate each other. The negative terminal of the battery injects new electrons into the *n* region, which can then continue the conduction process; the positive terminal, on the other hand, extracts electrons from the *p* region, thus creating new holes that are free to migrate toward the *pn* junction.

When the diode is *reverse-biased,* as in Figure 2-15d, the majority carriers in each region drift away from the junction to form the depletion layer, which contains few charges. Only the small concentration of minority carriers present in each region drifts toward the junction and thus creates a current. Consequently, conductance under reverse bias is typically 10^{-6} to 10^{-8} that of conductance under forward bias.

Current-Voltage Curves for Semiconductor Diodes

Figure 2-16 shows the behavior of a typical semiconductor diode under forward and reverse bias. With forward bias, the current increases nearly exponentially with voltage; often currents of several amperes are the result. For a germanium diode under reverse bias, a current on the order of tens of microamperes is observed over a considerable voltage range. Reverse bias current for a silicon diode is on the order of tens of nanoamperes. In this region of the diode characteristic curve, conduction is by the minority carriers. Ordinarily, this reverse current is of no consequence. As the reverse po-

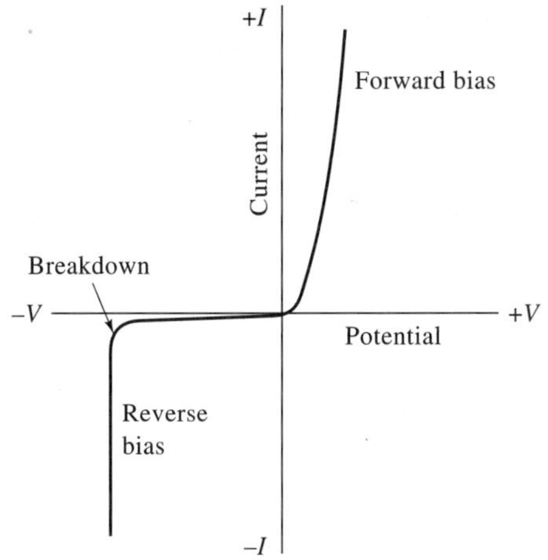

Figure 2-16 Current-voltage characteristics of a silicon semiconductor diode. *(Note that, for the sake of clarity, the small current under reverse bias before breakdown has been greatly exaggerated.)*

tential is increased, however, a *breakdown voltage* is ultimately reached where the reverse current increases abruptly to very high values. Here, holes and electrons, formed by the rupture of covalent bonds of the semiconductor, are accelerated by the field to produce additional electrons and holes by collision. In addition, quantum mechanical tunneling of electrons through the junction layer contributes to the enhanced conductance. This conduction, if sufficiently large, may result in heating and damaging of the diode. The voltage at which the sharp increase in current occurs under reverse bias is called the *Zener breakdown voltage.* By controlling the thickness and type of the junction layer, Zener breakdown voltages that range from a few volts to several hundred volts can be realized. As we shall see, this phenomenon has important practical applications in precision voltage sources.

2C-3 Transistors

The transistor is the basic semiconductor amplifying and switching device. This device performs the same function as the vacuum amplifier tube of yesteryear—that is, it provides an output signal whose magnitude is usually significantly greater than the signal at the input. Several types of transistors are available; two of the most widely used of these, the *bipolar junction transistor* and the *field-effect transistor,* will be described here.

Bipolar Junction Transistors

Bipolar junction transistors (BJT) may be viewed as two back-to-back semiconductor diodes. The *pnp* transistor consists of a very thin *n*-type region sandwiched between two *p*-type regions; the *npn* type has the reverse structure. Bipolar junction transistors are constructed in a variety of ways, two of which are illustrated in Figure 2-17. The symbols for the *pnp* and the *npn* types of transistors are shown on the right in Figure 2-17. In these symbols, the arrow on the emitter lead indicates the direction of positive current. Thus, in the *pnp* type, there is a positive current from the emitter to the base; the reverse is true for the *npn* type.

Electrical Characteristics of a Bipolar Junction Transistor

The discussion that follows will focus upon the behavior of a *pnp*-type bipolar junction transistor. It should be appreciated that the *npn* type acts analogously except for the direction of the current, which is opposite to that of the *pnp* transistor.

When a transistor is to be used in an electronic device, one of its terminals is connected to the input and the second serves as the output; the third terminal is connected to both and is the *common* terminal. Three configurations are thus possible: a common-emitter, a common-collector, and a common-base. The common-emitter configuration has the widest application in amplification and is the one we shall consider in detail.

Figure 2-18 illustrates the current amplification that occurs when a *pnp* transistor is used in the common-emitter mode. Here, a small dc input current I_B, which is to be amplified, is introduced in the emitter-base circuit; this current is labeled as the base current in the figure. As we shall show later, an ac current can also be amplified by superimposing it on I_B. After amplification, the dc component can then be removed by a filter.

The emitter-collector circuit is powered by a dc power supply, such as that described in Section 2D. Typically, the power supply will provide a potential between 9 and 30 V.

Note that, as shown by the breadth of the arrows, the collector or output current I_C is significantly larger than the base input current I_B. Furthermore, the magnitude of the collector current is directly proportional to the input current. That is,

$$I_C = \beta I_B \tag{2-50}$$

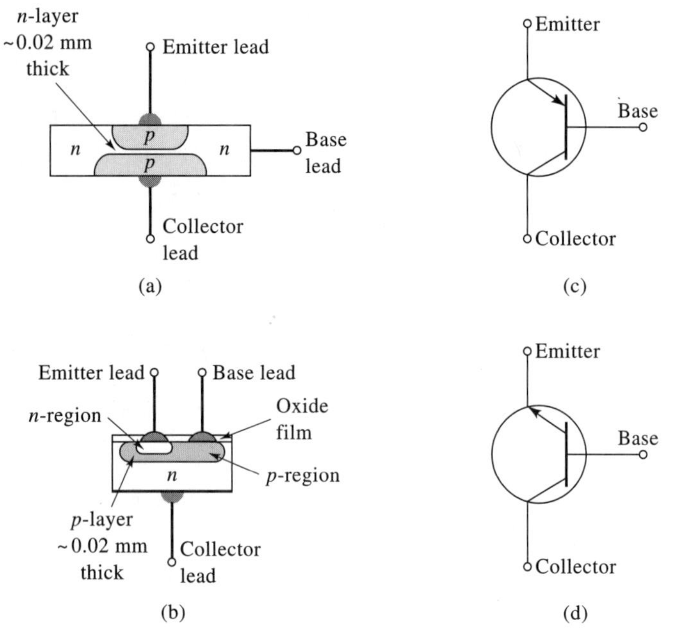

Figure 2-17 Two types of bipolar junction transistors. Construction details are shown in (a) for a ***pnp*** alloy BJT and in (b) for an ***npn*** planar transistor. Symbols for a ***pnp*** and an ***npn*** BJT are shown in (c) and (d), respectively. *(Note that alloy junction transistors may also be fabricated as npn types and planar transistors as pnp.)*

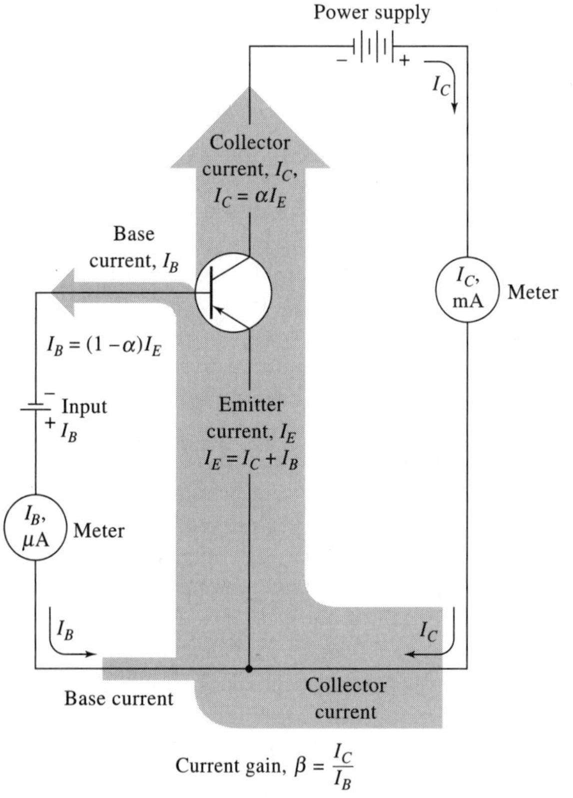

Figure 2-18 Currents in a common-emitter circuit with a transistor. Ordinarily, $\alpha = 0.95$ to 0.995 and $\beta = 20$ to 200.

where the proportionality constant β is the *current gain,* which is a measure of the current amplification that has occurred. Values for β for typical transistors range from 20 to 200.

It is important to note that a BJT requires a current into the base or out of the base in order to initiate conduction between the emitter and the collector. Consequently, circuits built from BJTs draw significant current from their power supplies during operation. We will soon describe another type of transistor, the field-effect transistor, which requires nearly zero current for its operation.

Mechanism of Amplification with a Bipolar Junction Transistor

It should be noted that the emitter-base interface of the transistor shown in Figure 2-18 constitutes a forward-biased *pn* junction similar in behavior to that shown in Figure 2-15c, while the base-collector region is a reverse-biased *pn* junction similar to the circuit shown in Figure 2-15d. Under forward bias, a significant current I_B develops when an input signal of a few tenths of a volt is applied (see Figure 2-16). In contrast, current through the reverse-biased collector-base junction is inhibited by the migration of majority carriers away from the junction, as shown in Figure 2-15d.

In the manufacture of a *pnp* transistor, the *p* region is purposely much more heavily doped than is the *n* region. As a consequence, the concentration of holes in the *p* region is 100-fold or more greater than the concentration of mobile electrons in the *n* layer. Thus, the fraction of the current carried by holes is perhaps 100 times greater than the fraction carried by electrons.

Turning again to Figure 2-18, holes are formed at the *p*-type emitter junction through removal of electrons by the two dc sources, namely, the input signal and the power supply. These holes can then move into the very thin *n*-type base region where some will combine with the electrons from the input source; the base current I_B is the result. The majority of the holes will, however, drift through the narrow base layer and be attracted to the negative collector junction, where they can combine with electrons from the power supply; the collector current I_C is the result.

It is important to appreciate that the magnitude of the collector current is determined by the number of current-carrying holes available in the emitter. This number, however, is a fixed multiple of the number of electrons supplied by the input base current. Thus, when the base current doubles, so also does the collector current. This relationship leads to the current amplification exhibited by a bipolar junction transistor.

Field-Effect Transistors (FET)

Several types of field-effect transistors have been developed and are widely used in integrated circuits. One of these, the insulated-gate field-effect transistor, was the outgrowth of the need to increase the input resistance of amplifiers. Typical insulated-gate field-effect transistors have input impedances that range from 10^9 to 10^{14} Ω. This type of transistor is most commonly referred to as a MOSFET, which is the acronym for *metal oxide semiconductor field-effect transistor.*

Figure 2-19a shows the structural features of an *n-channel* MOSFET. Here, two isolated *n* regions are formed in a *p*-type substrate. Covering both regions is a thin layer of highly insulating silicon dioxide, which may be further covered with a protective layer of silicon nitride. Openings are etched through these layers so that electrical contact can be made to the two *n* regions. Two additional contacts are formed, one to the substrate and the other to the surface of the insulating layer. The contact with the insulating layer is termed the gate because the potential of this contact determines the magnitude of the positive current between the drain and the source. Note that the insulating layer of silicon dioxide between the gate lead and the substrate accounts for the high impedance of a MOSFET.

In the absence of a gate potential, essentially no current develops between drain and source because one of the two *pn* junctions is always reverse-biased regardless of the sign of the potential V_{DS}. MOFSET devices are designed to operate in either an *enhancement* or a *depletion* mode. The former type is shown in Figure 2-19a where current enhancement is brought about by application of a positive potential to the gate. As shown, this positive potential induces a negative substrate *channel* immediately below the layer of silicon

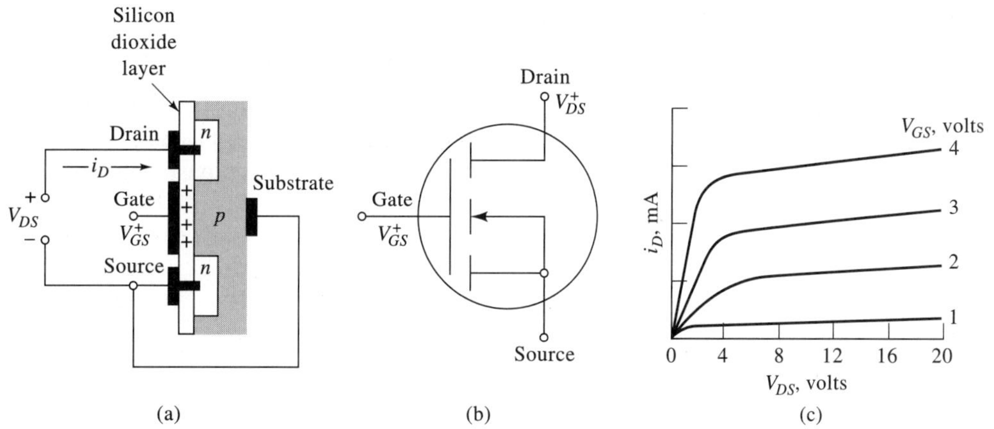

Figure 2-19 An *n*-channel enhancement mode MOSFET. (a) Structure, (b) symbol, (c) performance characteristics.

dioxide that covers the gate electrode. The number of negative charges here, and thus the current, increases as the gate voltage V_{GS} increases. The magnitude of this effect is shown in Figure 2-19c. Also available are *p*-channel enhancement mode MOSFET devices in which the *p* and *n* regions are reversed from that shown in Figure 2-19a.

Depletion mode MOSFET devices are designed to conduct in the *absence* of a gate voltage and to become nonconducting as potential is applied to the gate. An *n*-channel MOSFET of this type is similar in construction to the transistor shown in Figure 2-19a except that the two *n* regions are now connected by a narrow channel of *n*-type semiconductor. Application of a negative voltage at V_{DS} repels electrons out of the channel and thus decreases the conduction through the channel. It is important to note that virtually zero current is required at the gate of a MOSFET device to initiate conduction between the source and drain. This tiny power requirement is in contrast to the rather large power requirements of BJT transistors. The characteristic low power consumption of field-effect devices make them ideal for portable applications requiring battery power.

2D POWER SUPPLIES AND REGULATORS

Generally, laboratory instruments require dc power to operate amplifiers and other reactive components. The most convenient source of electrical power, however, is 110-V ac furnished by public utility companies. As shown in Figure 2-20, laboratory power supply units increase or decrease the potential from the house supply, rectify the current so that it has a single polarity, and finally, smooth the output to give a signal that approximates dc. Most power supplies also contain a voltage regulator that maintains the output voltage at a constant desired level.

2D-1 Transformers

Alternating current is readily increased or decreased in voltage by means of a power transformer such as that shown schematically in Figure 2-21. The varying magnetic field formed around the *primary* coil in this device from the 110-V alternating current induces alternating currents in the *secondary* coils; the potential V_x across each is given by

$$V_x = 115 \times N_2/N_1$$

where N_2 and N_1 are the number of turns in the secondary and primary coils, respectively. Power supplies with multiple taps, as in Figure 2-21, are available commercially; many voltage combinations may be obtained. Thus, a single transformer can serve as a power source for several components of an instrument.

2D-2 Rectifiers and Filters

Figure 2-22 shows three types of rectifiers and their output-signal forms. Each uses semiconductor diodes (see Section 2C-2) to block current in one direction while permitting it in the opposite direction. In order to minimize the current fluctuations shown in Figure 2-14, the output of a rectifier is usually filtered by placing a large capacitance in parallel with the load R_L, as shown in Figure 2-23. The charge and discharge of the capacitor has the effect of decreasing the variations to a relatively small *ripple*. In some applications, an inductor in series and a capacitor in parallel with the load serve as a filter; this type of filter is known as an *L section*. By suitable choice of capacitance and inductance, the peak-to-peak ripple can be reduced to the millivolt range or lower.

2D-3 Voltage Regulators

Often, instrument components require dc voltages that are constant and independent of the current. Voltage regulators serve this purpose. Figure 2-24 illustrates a

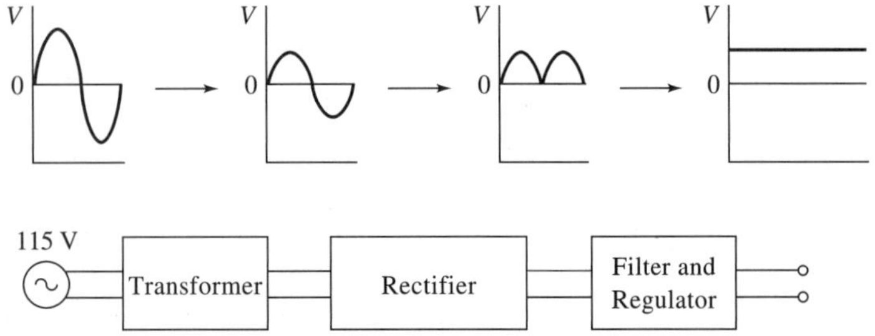

Figure 2-20 Diagram showing the components of a power supply and their effect on a signal.

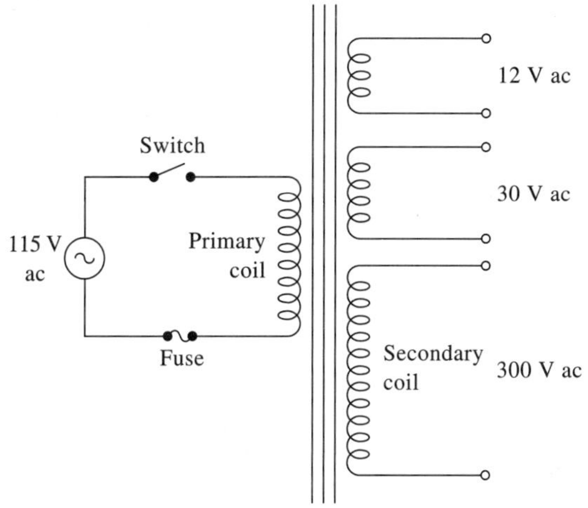

Figure 2-21 Schematic of a typical power transformer with multiple secondary windings.

simple voltage regulator that utilizes a *Zener diode,* a *pn* junction that has been designed to operate under breakdown conditions; note the special symbol for this type of diode. In Figure 2-16 (page 41) it is seen that at a certain reverse bias a semiconductor diode undergoes an abrupt breakdown whereupon the current changes precipitously. For example, under breakdown conditions, a current change of 20 to 30 mA may result from a potential change of 0.1 V or less. Zener diodes with a variety of specified breakdown voltages are available commercially.

For voltage regulators, a Zener diode is chosen such that it always operates under breakdown conditions; that is, the input voltage to be regulated is greater than the breakdown voltage. For the regulator shown in Figure 2-24, an increase in voltage results in an increase in current through the diode. Because of the steepness of the current-voltage curve in the breakdown region (Figure 2-16), however, the voltage drop across the diode, and thus the load, is virtually constant.

Modern integrated-circuit voltage regulators employ the properties of Zener diodes to provide stable reference voltages. These voltages are used in conjunction with feedback circuitry and power transistors to create power supplies regulated to ±10 mV or better. Such regulators have three terminals: input, output, and circuit common. The raw output from a rectified and filtered power supply may be connected to the three-terminal voltage regulator to produce a supply that is stable with respect to temperature fluctuations and that automatically shuts down when the load current exceeds its maximum rating, which is typically one ampere in the most widely used circuits. Integrated-circuit voltage regulators are found in the power supplies of most electronic devices.

Regulators of this type have the disadvantage of dissipating considerable power; so that with the proliferation of computers and other electronic devices, more

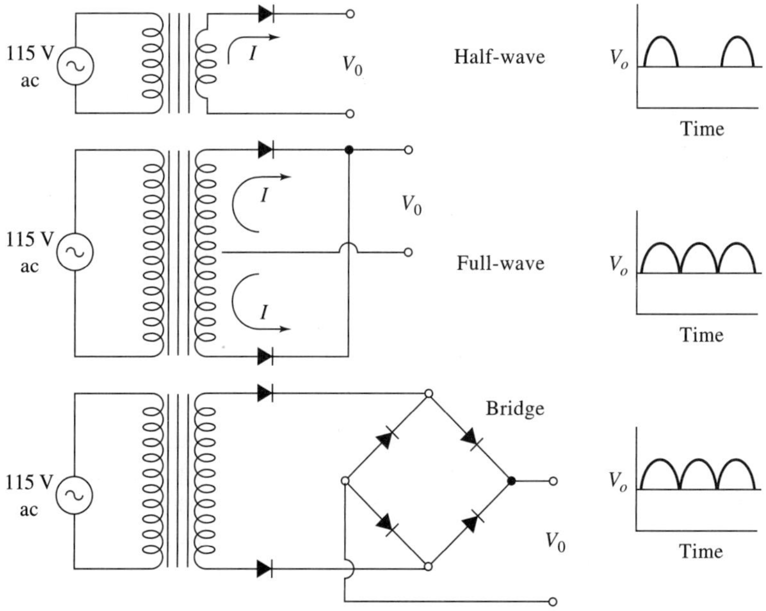

Figure 2-22 Three types of rectifiers.

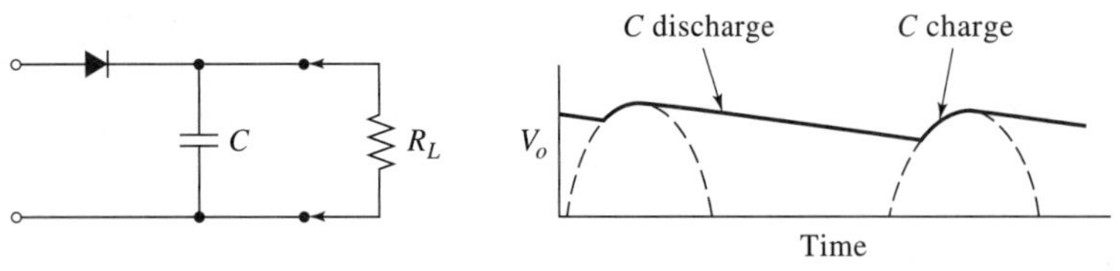

Figure 2-23 Filtering the output from a rectifier.

efficient regulators have become desirable. The solution to this difficulty has been the advent of *switching regulators,* which provide power to the load only when it is needed while maintaining constant voltage. Most computer power supplies contain switching regulators. The details of operation of switching power supplies are beyond the scope of our discussion, but their principles of operation are discussed in the general references given at the beginning of this chapter.

2E READOUT DEVICES

In this section, three common readout devices are described, namely, the cathode-ray tube (CRT), the laboratory recorder, and the alphanumeric display unit.

2E-1 Oscilloscopes

The oscilloscope is a most useful and versatile laboratory instrument that utilizes a cathode-ray tube as a readout device. Both analog and digital oscilloscopes are manufactured. Digital oscilloscopes are used when sophisticated signal processing is required. Analog oscilloscopes are generally simpler than their digital counterparts, are usually portable, are easier to use, and are less expensive, costing as little as $500. We shall confine our discussion to simple analog instruments. The block diagram in Figure 2-25 shows the most important components of such an instrument and the signal pathways for the various components. The actual display is provided by a cathode-ray tube.

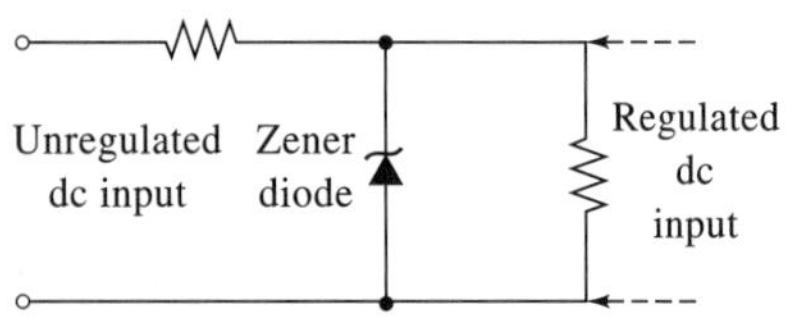

Figure 2-24 A Zener stabilized voltage regulator.

Cathode-Ray Tubes

Figure 2-26 is a block diagram that shows the main components of a cathode-ray tube. Here, the display is formed by the interaction of electrons in a focused beam with a phosphorescent coating on the interior of the large curved surface of the evacuated tube. The electron beam is formed at a heated cathode, which is maintained at ground potential. A multiple anode-focusing array produces a narrow beam of electrons that have been accelerated through a potential of several thousand volts. In the absence of input signals, the beam appears as a small bright dot in the center of the screen.

Horizontal and Vertical Control Plates. Input signals are applied to two sets of plates, one of which deflects the beam horizontally and the other vertically. Thus, it is possible to display an *x*-*y* plot of two related signals on the face of the CRT when the switch in Figure 2-25 is in position 1. Because the screen is phosphorescent, the movement of the dot appears as a lighted continuous trace that fades after a brief period.

The most common way of operating the cathode-ray tube is to cause the dot to sweep periodically at a constant rate across the central horizontal axis of the tube by applying a sawtooth sweep signal to the horizontal deflection plates. The oscilloscope may be operated in this way when the switch in Figure 2-25 is moved to position 2. When operated in this way, the horizontal axis of the display corresponds to time. Application of a periodic signal to the vertical display plates then provides a display of the waveform of the periodic signal. Most analog oscilloscopes have sweep rates that range between 1 μs/cm and 1 ns/cm. Usually the sweep speed can be slowed by factors of 10 until the speed is in the seconds per centimeter range.

The horizontal control section of most oscilloscopes can, if desired, be driven by an external signal rather than by the internal sawtooth signal. In this mode of operation, the oscilloscope becomes an *x*-*y* plotter that displays the functional relationship between two input signals.

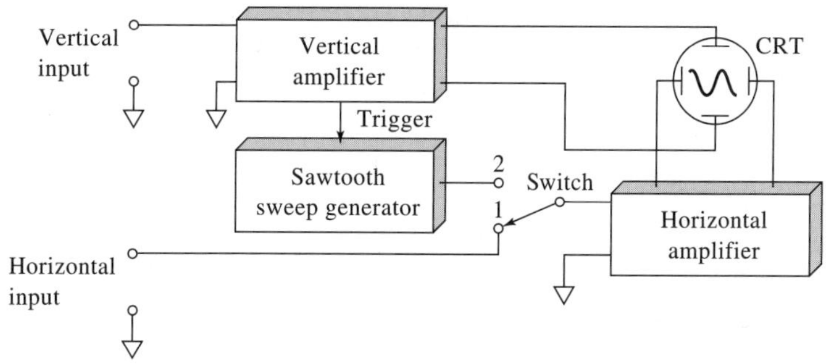

Figure 2-25 Basic analog oscilloscope components.

Trigger Control. In order to display a repetitive signal, such as a sine wave, steadily on the screen, it is essential that each sweep begin at an identical place on the signal profile—for example, at a maximum, a minimum, a zero crossing, or an abrupt change in the signal. Synchronization is usually accomplished by mixing a portion of the test signal with the sweep signal in such a way as to produce a voltage spike for, say, each maximum or some multiple thereof. This spike then serves to trigger the sweep. Thus, the waveform can be observed as a steady image on the screen.

Oscilloscopes are extremely useful in a variety of display and diagnostic applications. They may be used to view the time profile of signals from transducers, to compare the relationships among various repetitive waveforms in analog signal processing circuits, or to reveal high-frequency noise or other signals of interest that cannot be observed by using a DMM or other dc measuring device. The oscilloscope is an essential diagnostic tool in the instrument laboratory.

2E-2 Recorders[6]

The typical laboratory recorder is an example of a *servosystem,* a null device that compares two signals and then makes a mechanical adjustment that reduces their difference to zero; that is, a servosystem continuously seeks the null condition.

In the laboratory recorder, shown schematically in Figure 2-27, the signal to be recorded, V_x, is continu-

[6] For a more extensive discussion of laboratory recorders, see G. W. Ewing, *J. Chem. Educ.,* **1976,** *53,* A361, A407.

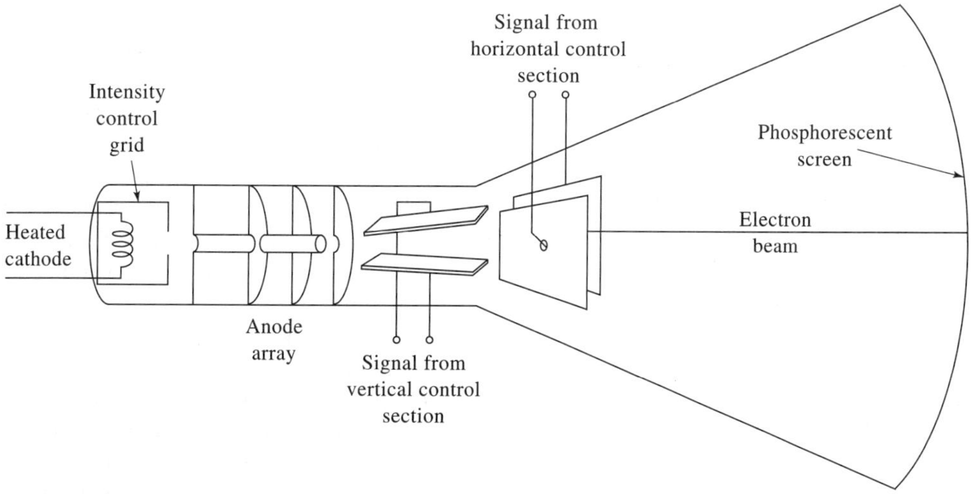

Figure 2-26 Schematic of a cathode-ray tube.

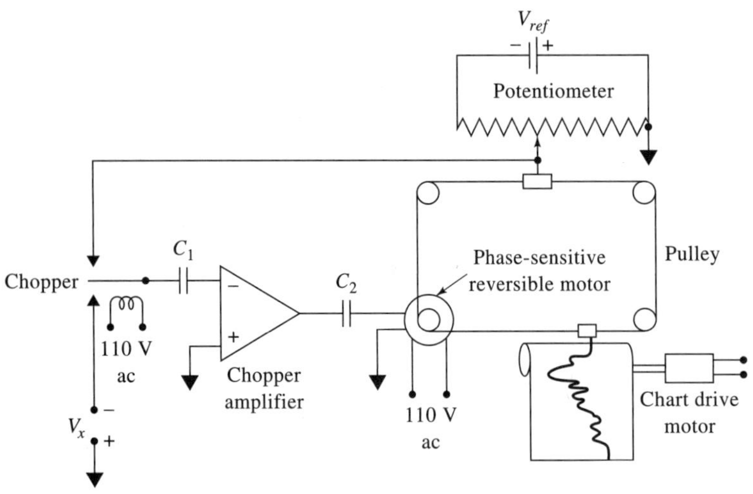

Figure 2-27 Schematic of a self-balancing recording potentiometer.

ously compared with the output from a potentiometer powered by a reference signal, V_{ref}. In most modern recorders, the reference signal is generated by a temperature-compensated Zener diode voltage reference circuit that provides a stable reference potential. Any difference in potential between the potentiometer output and V_x is converted to a 60-cycle ac signal by an electronic chopper; the resulting signal is then amplified sufficiently to activate a small phase-sensitive electrical motor that is mechanically geared or linked (by a pulley arrangement in Figure 2-27) to both a recorder pen and the sliding contact of the potentiometer. The direction of rotation of the motor is such that the potential difference between the potentiometer and V_x is decreased to zero, which causes the motor to stop.

To understand the directional control of the motor, it is important to note that a reversible ac motor has two sets of coils, one of which is fixed (the stator) and the other of which rotates (the rotor). One of these—say, the rotor—is powered from the 110-V power line and thus has a continuously fluctuating magnetic field associated with it. The output from the ac amplifier, on the other hand, is fed to the coils of the stator. The magnetic field induced here interacts with the rotor field and causes the rotor to turn. The direction of motion depends upon the *phase* of the stator current with respect to that of the rotor; the phase of the stator current, however, differs by 180 deg, depending upon whether V_x is greater or smaller than the signal from V_{ref}. Thus, the amplified difference signal can be caused to drive the servomechanism to the null state from either direction.

In most laboratory recorders, the paper is moved at a fixed speed. Thus, a plot of signal intensity as a function of time is obtained. Because the recorder paper is fed from a long roll, or strip, this type of laboratory recorder has come to be called a strip-chart recorder. In *x-y recorders,* the paper is fixed as a single sheet mounted on a flat bed. The paper is traversed by an arm that moves along the *x*-axis. The pen travels along the arm in the *y* direction. The arm drive and the pen drive are connected to the *x* and *y* inputs, respectively, thus permitting both to vary continuously. Often recorders of this type are equipped with two pens, thus allowing the simultaneous plotting of two functions on the *y*-axis. An example of an application of this kind is in chromatography where it is desirable to have a plot of the detector output as a function of time as well as the time integral of this output. Alternatively, a dual-pen recorder might be used to display the outputs of two different detectors that are monitoring the effluent from the same chromatographic column.

A modern laboratory strip-chart recorder has several chart speeds that typically range from 0.1 to 20 cm/min. Most provide a choice of several voltage ranges from 1 mV full scale to several volts. Generally, the precision of these instruments is on the order of a few tenths of a percent of full scale.

Digital recorders and plotters are widely used. Here, the pen may be driven by a stepper motor, which responds to digitized voltage signals by turning some precise fraction of a rotation for each voltage pulse. Modern computer-driven *x-y* plotters often utilize dc servomotors to move either the pen, the paper, or both, to draw graphs of data from analytical instruments. Alternatively, thermal printers similar to those found in adding machines may be used to plot data continuously as well as provide printed numerical records of various attributes of the plotted curves.

2E-3 Alphanumeric Displays

The output from digital equipment is most conveniently displayed in terms of decimal numbers and letters, that is, in *alphanumeric* form. The seven-segment readout device is based on the principle that any alphanumeric character can be represented by lighting an appropriate combination of seven segments as shown in Figure 2-28. Here, for example, a five is formed when segments, *a, f, g, c,* and *d* are lighted; the letter C is observed when segments *a, d, e,* and *f* are displayed. Perhaps the most common method of lighting a seven-segment display is to fashion each segment as a light-emitting diode. A typical LED consists of a *pn* junction shaped as one of the segments and prepared from gallium arsenide, which is doped with phosphorus. Under forward bias, the junction emits red radiation as a consequence of the recombination of minority carriers in the junction region. Each of the seven segments is connected to a decoder logic circuit so that it is activated at the proper time.

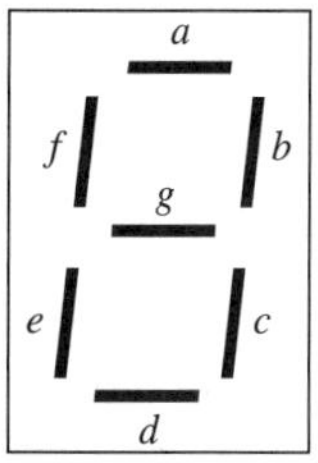

Figure 2-28 A seven-segment display.

Seven-segment liquid-crystal displays, or LCDs, are also widely encountered. Here, a small amount of a liquid crystal is contained in a thin, flat optical cell, the walls of which are coated with a conducting film. Application of an electrical field to a certain region of the cell causes a change in alignment of the molecules in the liquid crystal and a consequent change in its optical appearance.[7] Both LEDs and LCDs find application in many different types of instruments, and each type of electronic readout has its advantages. LCDs are especially useful for battery-operated instruments because they consume very little power, but they are not particularly useful in either very bright or very dim ambient light. On the other hand, LEDs are readable at low ambient light levels as well as in fairly bright light, but they consume considerably more power and thus are not generally used in applications powered by battery.

[7] For discussions of the properties and applications of liquid crystals, see G. H. Brown and P. P. Crooker, *Chem. Eng. News,* **1983,** *Jan. 31, 24;* G. H. Brown, *J. Chem. Educ.,* **1983,** *60,* 900.

2F QUESTIONS AND PROBLEMS

2-1 It was desired to assemble the voltage divider shown below. Two of each of the following resistors were available: 50 Ω, 100 Ω, and 200 Ω.

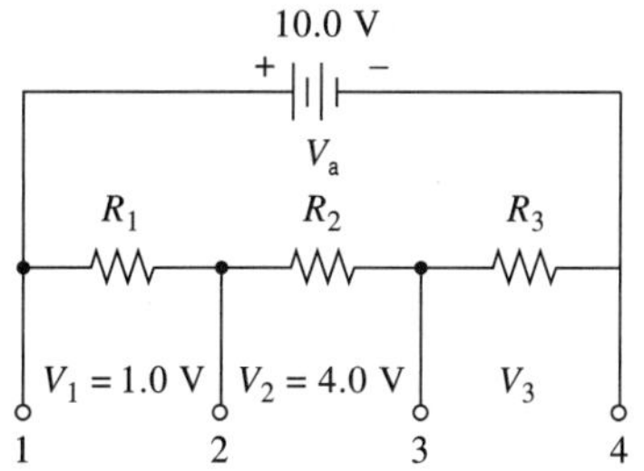

(a) Describe a suitable combination of the resistors that would give the indicated voltages.

(b) What would be the *IR* drop across R_3?

(c) What current would be drawn from the source?
(d) What power is dissipated by the circuit?

2-2 Assume that for a circuit similar to that shown in Problem 2-1, $R_1 = 200\ \Omega$, $R_2 = 500\ \Omega$, $R_3 = 1000\ \Omega$, and $V_b = 15$ V.
(a) Calculate the voltage V_2.
(b) What would be the power loss in resistor R_2?
(c) What fraction of the total power lost by the circuit would be dissipated in resistor R_2?

2-3 For a circuit similar to the one shown in Problem 2-1(a), $R_1 = 1.00$ kΩ, $R_2 = 2.50$ kΩ, $R_3 = 4.00$ kΩ, and $V_B = 12.0$ V. A voltmeter was placed across contacts 2 and 4. Calculate the relative error in the voltage reading if the internal resistance of the voltmeter was (a) 5000 Ω, (b) 50 kΩ, and (c) 500 kΩ.

2-4 A voltmeter was used to measure the potential of a cell with an internal resistance of 750 Ω. What must the internal resistance of the meter be if the relative error in the measurement is to be less than (a) -1.0%, (b) -0.10%?

2-5 For the following circuit, calculate
(a) the potential difference across each of the resistors.
(b) the magnitude of each of the currents shown.
(c) the power dissipated by resistor R_3.
(d) the potential difference between points 3 and 4.

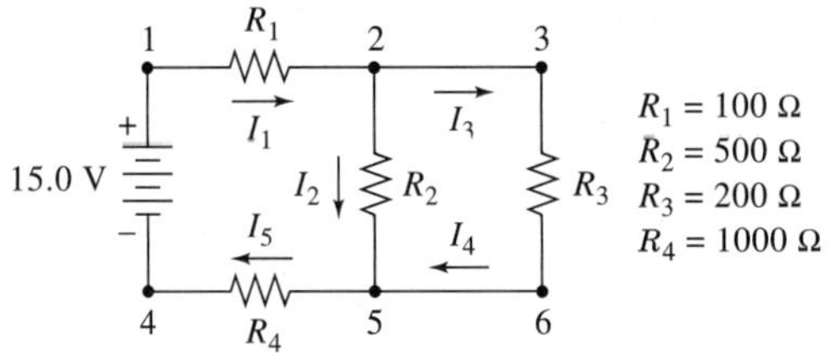

2-6 For the circuit shown below, calculate
(a) the power dissipated between points 1 and 2.
(b) the current drawn from the source.
(c) the potential difference across resistor R_A.
(d) the potential difference across resistor R_D.
(e) the potential difference between points 5 and 4.

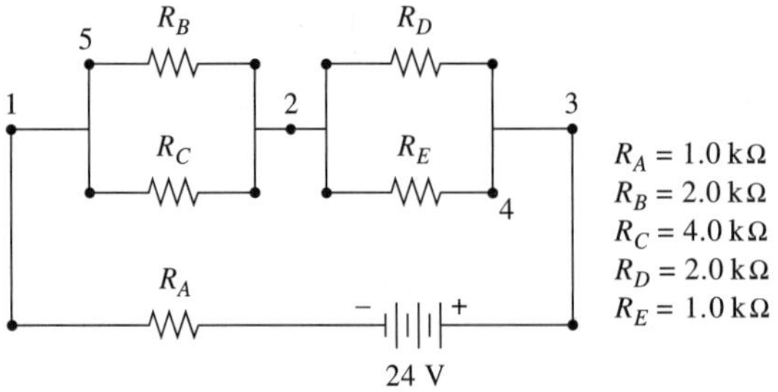

2-7 The circuit that follows is for a laboratory potentiometer for measuring unknown potentials V_x. Assume the resistor AB is a slide wire whose resistance is directly proportional to its length. With the standard Weston cell (1.018 V) in the circuit, a null point was observed when contact C was moved to a position 84.3 cm from point A. When the Weston cell was replaced with an unknown voltage, null was observed at 44.3 cm. Calculate the potential of the unknown.

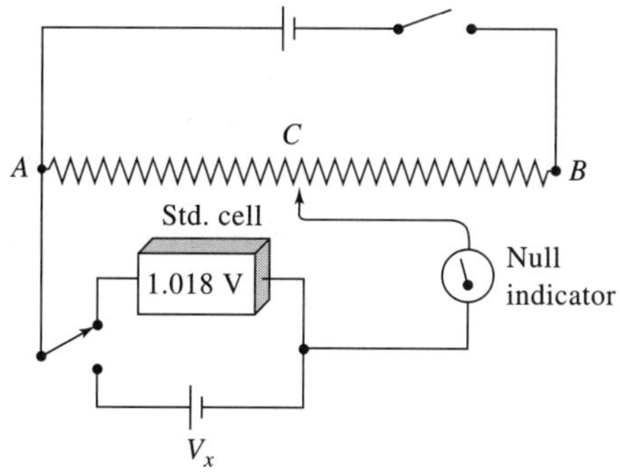

2-8 Show that the data in the fourth column of Table 2-1 are correct.

2-9 Show that the data in the fourth column of Table 2-2 are correct.

2-10 The current in a circuit is to be determined by measuring the potential difference across a precision resistor in series with the circuit.
(a) What should be the resistance of the resistor in ohms if 1.00 V is to correspond to 50 μA?
(b) What must be the resistance of the voltage-measuring device if the error in the current measurement is to be less than 1.0% relative?

2-11 An electrolysis at a nearly constant current can be performed with the following arrangement:

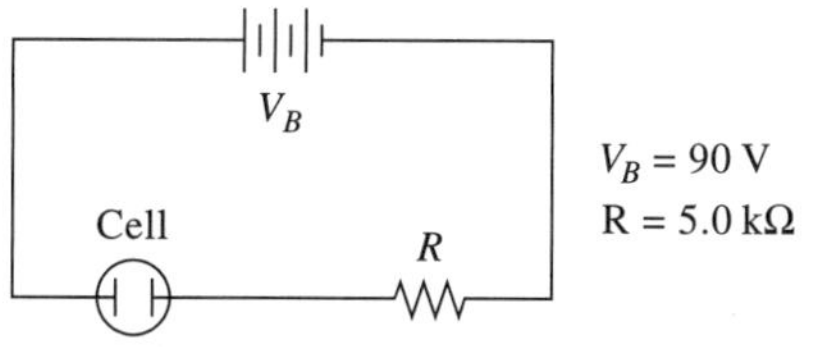

The 90-V source consists of dry cells whose potential can be assumed to remain constant for short periods. During the electrolysis, the resistance of the cell increases from 20 Ω to 40 Ω owing to depletion of ionic species. Calculate the percent change in the current, assuming that the internal resistance of the batteries is zero.

2-12 Repeat the calculations in Problem 2-11, assuming that $V_B = 9.0$ V and $R = 0.50$ kΩ.

2-13 A 24-V dc potential was applied across a resistor and capacitor in series. Calculate the current after 0.00, 0.010, 0.10, 1.0, and 10 s if the resistance was 10 MΩ and the capacitance 0.20 μF.

2-14 How long would it take to discharge a 0.015 μF capacitor to 1% of its full charge through a resistance of (a) 10 MΩ, (b) 1 MΩ, (c) 1 kΩ?

2-15 Calculate time constants for each of the RC circuits described in Problem 2-14.

2-16 A series *RC* circuit consisted of a 25-V dc source, a 50 kΩ resistor, and a 0.035 μF capacitor.
(a) Calculate the time constant for the circuit.
(b) Calculate the current and potential differences across the capacitor and the resistor during a charging cycle; employ as times 0, 1, 2, 3, 4, 5, and 10 ms.
(c) Repeat the calculations in (b) for a discharge cycle.

2-17 Repeat the calculations in Problem 2-16, assuming that the potential was 15 V, the resistance was 20 MΩ, and the capacitance was 0.050 μF. Employ as times 0. 1, 2, 3, 4, 5, and 10 s.

2-18 Calculate the capacitive reactance, the impedance, and the phase angle ϕ for the following series circuits.

Frequency, Hz	*R*, Ω	*C*, μF
(a) 1	20,000	0.033
(b) 10^3	20,000	0.033
(c) 10^6	20,000	0.0033
(d) 1	200	0.0033
(e) 10^3	200	0.0033
(f) 10^6	200	0.0033
(g) 1	2,000	0.33
(h) 10^3	2,000	0.33
(i) 10^6	2,000	0.33

2-19 Derive a frequency response curve for a low-pass *RC* filter in which $R = 2.5 \times 10^3\ \Omega$ and $C = 0.015\ \mu F$. Cover a range of $(V_p)_o/(V_p)_i$ of 0.01 to 0.99.

2-20 Derive a frequency response curve for a high-pass *RC* filter in which $R = 5.0 \times 10^5\ \Omega$ and $C =$ pF (1 pF $= 10^{-12}$ F). Cover a range of $(V_p)_o/(V_p)_i$ of 0.01 to 0.99.

Chapter 3

Operational Amplifiers in Chemical Instrumentation

Most modern analog signal–conditioning circuits owe their success to the class of integrated circuits known as operational amplifiers, which are referred to as op amps or OAs. Operational amplifiers are ubiquitous. Open any instrument or piece of electronic equipment, or scan an instrument schematic, and you will likely find one or more op amps. This fact, coupled with the ease with which relatively complex functions can be accomplished, emphasizes the importance of having a basic understanding of their principles of operation. In this chapter we explore a number of operational amplifier circuits and applications and investigate their properties, advantages, and limitations.[1]

[1] For general information on electronics, computers, and instrumentation, including operational amplifiers and their characteristics, see H. V. Malmstadt, C. G. Enke, and S. R. Crouch, *Microcomputers and Electronic Instrumentation: Making the Right Connections.* Washington, DC: American Chemical Society, 1994; A. J. Diefenderfer and B. E. Holton, *Principles of Electronic Instrumentation,* 3rd ed. Philadelphia: Saunders College Publishing, 1994; J. J. Brophy, *Basic Electronics for Scientists,* 5th ed. New York: McGraw-Hill, 1990; P. Horowitz and W. Hill, *The Art of Electronics,* 2nd ed. New York: Cambridge University Press, 1989.

3A PROPERTIES OF OPERATIONAL AMPLIFIERS

Operational amplifiers derive their name from their original applications in analog computers, where they were used to perform such mathematical operations as summing, multiplying, differentiating, and integrating. Operational amplifiers also find general application in the precise measurement of voltage, current, and resistance, which are measured variables in transducers that are used in chemical instruments. Operational amplifiers also are widely used as constant-current and constant-voltage sources.[2]

3A-1 Symbols for Operational Amplifiers

Figure 3-1 is an equivalent circuit representation of an operational amplifier. In this figure, the input potentials are represented by v_+ and v_-. The input difference voltage v_s is the difference between these two potentials; that is, $v_s = v_- - v_+$. The power supply connections are

[2] For more detailed information about operational amplifiers, see R. Kalvoda, *Operational Amplifiers in Chemical Instrumentation.* New York: Halsted Press, 1975. See also the references in footnote 1.

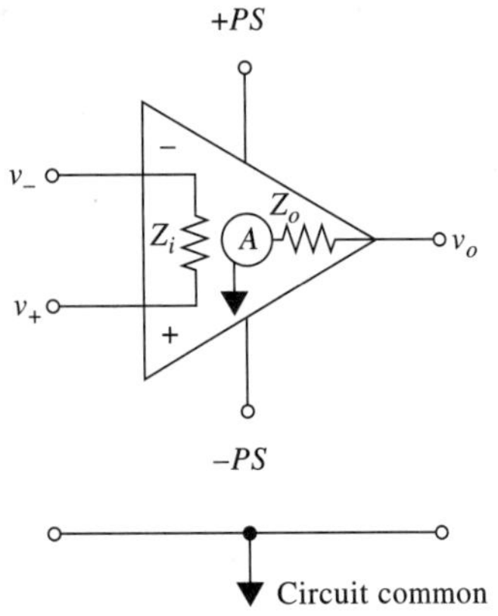

Figure 3-1 Equivalent circuit diagram of an operational amplifier.

labeled $+PS$ and $-PS$ and usually have values of $+15$ and -15 V dc. The so-called *open-loop gain* of the operational amplifier is shown as *A;* thus the output voltage v_o is given by $v_o = -Av_s$. Finally, Z_i and Z_o are the input and output impedance of the operational amplifier. Realize that the input signal may be either ac or dc; the output signal will then be of a corresponding kind.[3] Note that all potentials in operational amplifiers are measured with respect to the *circuit common* shown in Figure 3-1. Circuit common is often referred to somewhat erroneously as *ground.* We shall define these terms carefully and discuss their significance in Section 3A-2.

As shown in Figure 3-2, two alternate versions of Figure 3-1 are commonly used to symbolize operational amplifiers in circuit diagrams. A symbol as complete as that in Figure 3-2a is seldom encountered, and the simplified diagram in Figure 3-2b is used almost exclusively. Here, the power and ground connections are omitted.

3A-2 General Characteristics of Operational Amplifiers

As shown in Figure 3-3, the typical operational amplifier is an analog device that consists of approximately 20 transistors and resistors that have been fabricated on a single integrated circuit chip. Other components, such as capacitors and diodes, may also be integrated into the device. The physical dimensions of an operational amplifier, excluding power supply, are normally on the order of a centimeter or less. Modern operational amplifiers, in addition to being compact, are remarkably reliable and inexpensive. Their cost ranges from a few cents to perhaps one hundred dollars. A wide variety of operational amplifiers is available, each differing in gain, input and output impedance, operating voltage, speed, and maximum power. A typical commercially available operational amplifier comes housed in an 8-pin epoxy or ceramic package as shown in Figure 3-2c.

Operational amplifiers have the following properties: (1) large open-loop gains ($A = 10^4$ to 10^6); (2) high-input impedances ($Z_i \geq 10^6$ M to 10^{13} Ω); (3) low-output impedances ($Z_o \leq 1$ to 10 Ω); and essentially zero output for zero input (ideally < 0.1 mV output). Most operational amplifiers do in fact exhibit a small output voltage with zero input due to circuit characteristics or component instabilities. The *offset voltage* for an operational amplifier is the input voltage required to produce zero output potential. Often operational amplifiers are provided with an *offset trim* adjustment to reduce the offset to a negligible value (see Figure 3-2c).

Circuit Common and Ground Potential

As shown in Figure 3-2a, each of the two input potentials, as well as the output potential, of a typical operational amplifier are measured with respect to *circuit common,* which is symbolized by a triangle (▼). Circuit common is a conductor that provides a common return for all currents to their sources. As a consequence, all voltages in the circuit are measured with respect to the circuit common. Ordinarily electronic equipment is not directly connected to earth ground, which is symbolized by ⏚. Usually, however, the circuit common potential does not differ significantly from the ground potential, but it is important to recognize that circuit common is not necessarily at the same potential as ground. Note that in Figure 3-2b, a circuit common is not shown, but you may assume that there is a circuit common and that all circuit potentials are measured with respect to it.

Inverting and Noninverting Inputs

It is important to realize that in Figure 3-2, the *negative and positive signs indicate the inverting and noninverting inputs* of the amplifier and *do not* imply that they are necessarily to be connected to positive and negative sig-

[3] Throughout this text, we follow the convention of using upper-case *I, V,* and *Q* to represent dc current, voltage, and charge and lower-case *i, v,* and *q* for the corresponding ac quantities. Since the input and output signals of operational amplifiers may be either ac or dc quantities, we use lower-case variables to symbolize these signals unless signals under study are clearly dc quantities.

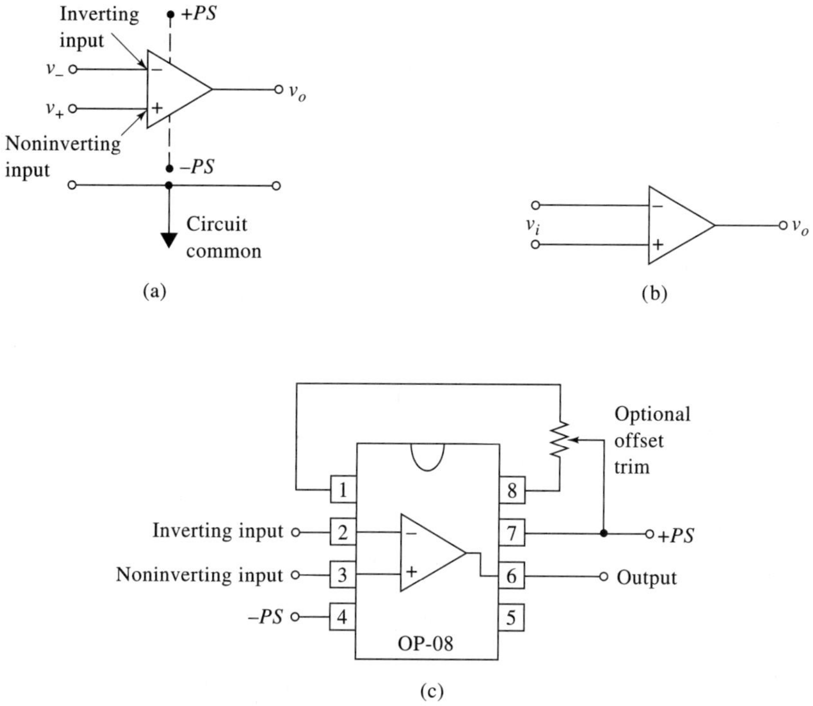

Figure 3-2 Symbols for operational amplifiers. More detail than usual is provided in (a). Note that the two input potentials v_- and v_+ as well as the output potential are measured with respect to the circuit common, which is usually at or near earth ground potential. (b) The usual way of representing an operational amplifier in circuit diagrams. (c) Representation of typical commercial 8-pin operational amplifier.

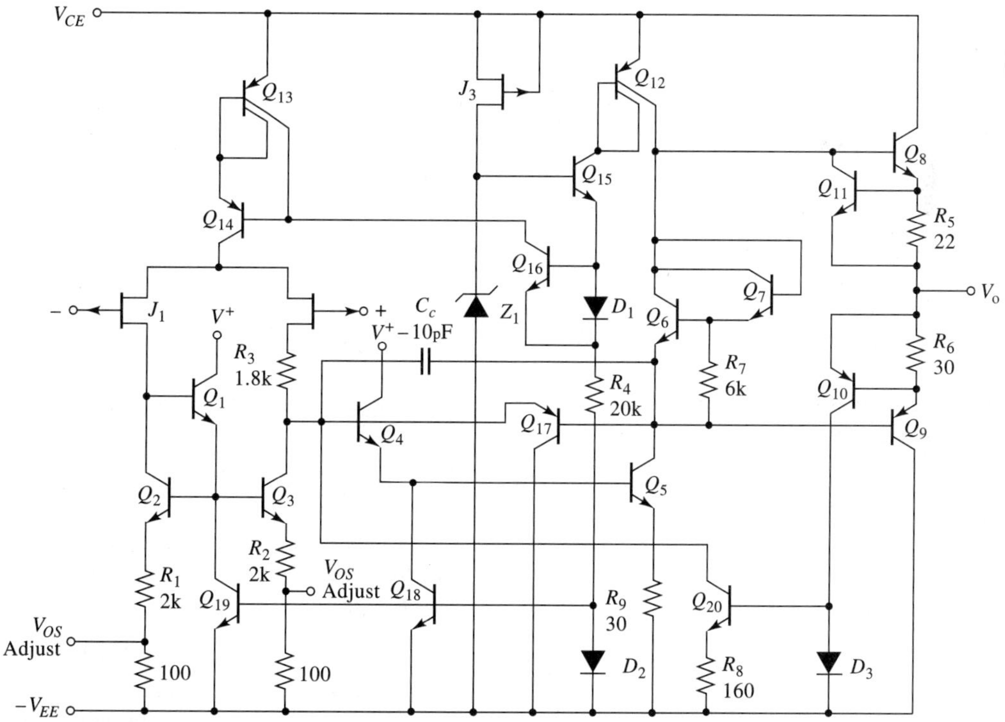

Figure 3-3 Circuit design of a typical operational amplifier. *(Courtesy of National Semiconductor Corporation)*

nals. Either input may be connected to positive or negative signals depending on the application of the circuit. Thus, if a negative voltage is connected to the inverting input, the output of the amplifier is positive with respect to it; if, on the other hand, a positive voltage is connected to the inverting input of the amplifier, a negative output results. An ac signal connected to the inverting input yields an output *that is 180 degrees out of phase* with the signal at the input. The noninverting input of an amplifier, on the other hand, yields an in-phase signal; in the case of a dc signal at the noninverting input, the output will be a dc signal of the same polarity as the input.

3B OPERATIONAL AMPLIFIER CIRCUITS

Operational amplifiers are used in circuit networks that contain various combinations of capacitors, resistors, and other electrical components. Under ideal conditions, the output of the amplifier is determined entirely by the nature of the network and its components and is *independent of the operational amplifier itself.* Thus, it is important to examine some of the many useful operational amplifier networks.

3B-1 Feedback Circuits

It is often desirable to return the output signal or some fraction of the output signal of an operational amplifier to one of the two inputs. When an output signal of an operational amplifier is connected to one of its inputs, the signal is called *feedback.* Figure 3-4 shows an operational amplifier with a feedback loop consisting of the *feedback resistor* R_f that is connected to the input S, which is called the *summing point.* Note that the feedback signal is opposite in sense to the input signal v_i as a result of the characteristics of the inverting input and is therefore called *negative feedback.*

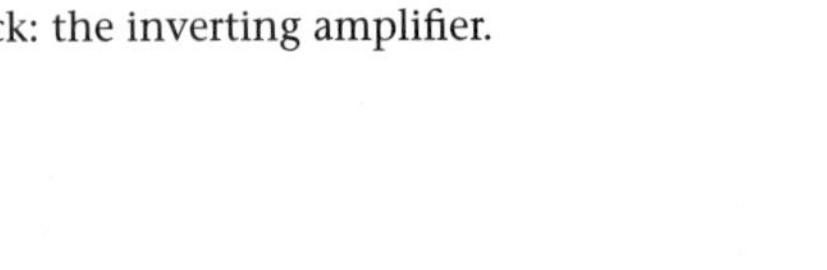

Figure 3-4 An operational amplifier circuit with negative feedback: the inverting amplifier.

The output voltage of the operational amplifier v_o is equal to the difference in voltage between v_- and v_+ multiplied by the open-loop gain of the amplifier, A. That is,

$$v_o = -Av_s = -A(v_- - v_+) \tag{3-1}$$

If we solve this equation for v_-, we find that

$$v_- = v_+ - \frac{v_o}{A} \tag{3-2}$$

Since A is quite large (10^4 to 10^6), Equation 3-2 becomes

$$v_- \approx v_+ \tag{3-3}$$

Because the noninverting input is connected to circuit common, which has a potential of zero volt, the potential at point S must approach the circuit common potential; as a consequence, the summing point is said to be at *virtual common,* or *virtual ground.* In other words, although virtual common is not a true circuit common, it has the characteristics of common.

This result leads to the following generalization that can simplify the analysis of many operational amplifier circuits. An operational amplifier circuit with negative feedback (v_o connected directly or indirectly to v_-), will do what is necessary to satisfy the condition specified in Equation 3-3. Remember, it is the internal circuitry of the operational amplifier, in conjunction with the circuit's ability to draw power from the supply, that makes it possible for the operational amplifier to function in this way.

From Ohm's law, the input current i_i in the circuit shown in Figure 3-4 is given by

$$i_i = \frac{v_i - v_-}{R_i} \tag{3-4}$$

where v_i is the input voltage with respect to common ground.

Similarly, the feedback current, i_f, is given by

$$i_f = \frac{v_- - v_o}{R_f} \tag{3-5}$$

As mentioned earlier, one of the characteristics of an operational amplifier is its very high input resistance. Thus, the current into the operational amplifier i_s is negligible with respect to i_i and i_f, therefore, from Kirchhoff's current law, the latter two currents must be essentially equal. That is $i_i \approx i_f$, and

$$\frac{v_i - v_-}{R_i} = \frac{v_- - v_o}{R_f} \tag{3-6}$$

We then substitute the expression for v_- of Equation 3-2 into Equation 3-6 to obtain

$$\frac{v_i - \left(v_+ - \dfrac{v_o}{A}\right)}{R_i} = \frac{\left(v_+ - \dfrac{v_o}{A}\right) - v_o}{R_f}$$

It is important to appreciate that as a consequence of Equation 3-2, the voltage v_- at the summing point S with respect to the circuit common is necessarily negligible compared with either v_i or v_o. In other words, because A is very large (10^4 to 10^6), v_o/A is extremely small and $v_+ = 0$, the quantity $(v_+ - v_o/A)$ of Equation 3-5 is very nearly equal to zero, and the foregoing equation simplifies to

$$\frac{v_o}{v_i} \approx -\frac{R_f}{R_i} \tag{3-7}$$

Thus, the gain v_o/v_i of a typical operational amplifier circuit with negative feedback depends only upon R_f and R_i and is *independent* of fluctuations in the performance characteristics and gain of the amplifier itself. The operational amplifier circuit of Figure 3-4 is often referred to as an *inverting amplifier* or an *inverter* when $R_i = R_f$ because the sign of the input voltage is inverted.

3B-2 Frequency Response of a Negative Feedback Circuit

The gain of a typical operational amplifier decreases rapidly in response to high-frequency input signals. This frequency dependence arises from small capacitances that develop within the transistors inside the operational amplifier. The frequency response for a typical amplifier is usually given in the form of a *Bode diagram,* such as that shown in Figure 3-5. Here, the solid curve labeled *open-loop gain* represents the behavior of the amplifier *in the absence of the feedback resistor* R_f in Figure 3-4. Note that both the ordinate and abscissa are log scales and that the gain is presented in decibels, or dB, where 1 dB = 20 log(v_o/v_i).

At low input–signal frequencies, $A = 10^5$, or 100 dB; but as the frequency increases above 10 Hz, the open-loop gain rolls off at −20 dB/decade in the same fashion as in a low-pass filter. The frequency range from dc to the frequency at which $A = 1$, or 0 dB, is called the *unit-gain bandwidth,* which for this amplifier is 1 MHz. The point at which $A = 1$ also fixes the *gain bandwidth product* at 1 MHz, which is a constant characteristic of the operational amplifier at all frequencies above 10 Hz. This characteristic permits the calculation of the bandwidth of a given amplifier configuration.

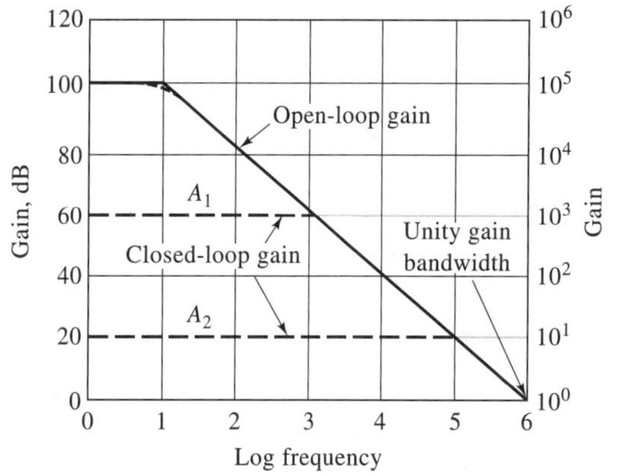

Figure 3-5 A Bode diagram showing the frequency response of a typical operational amplifier. Solid line: operational amplifier open-loop gain without negative feedback; dashed line: inverting amplifier configurations as in Figure 3-4 with $A_1 = 1000$ and $A_2 = 10$.

As an example, consider the amplifier of Figure 3-4 with a signal gain $A_1 = R_f/R_i = 1000$, as indicated by the upper dashed line in Figure 3-5. The bandwidth of the amplifier is then 1 MHz/1000 = 1 kHz, which corresponds to the intersection of the upper dashed line with the open-loop gain curve. A similar amplifier with $A_2 = 10$ then has a bandwidth of 100 kHz as indicated by the lower dashed line, and so on for other values of signal gain that might be chosen for a given op amp. Hence, we see that negative feedback increases amplifier bandwidth, which can be calculated from the signal gain and the unity gain bandwidth of the operational amplifier.

Two other parameters that relate to the speed or bandwidth Δf of an amplifier are illustrated in Figure 3-6. The output response of a voltage follower (which is discussed in the next section) to a step input is characterized by the *rise time* t_r, which is the time required for the output to change from 10% to 90% of the total change. It can be shown that

$$t_r = \frac{1}{3\Delta f} \tag{3-8}$$

For the amplifier with $A_2 = 10$ discussed above, $t_r = 1/(3 \times \text{MHz}) = 0.33\ \mu\text{s}$. From the slope of the change in voltage at the output during the transition

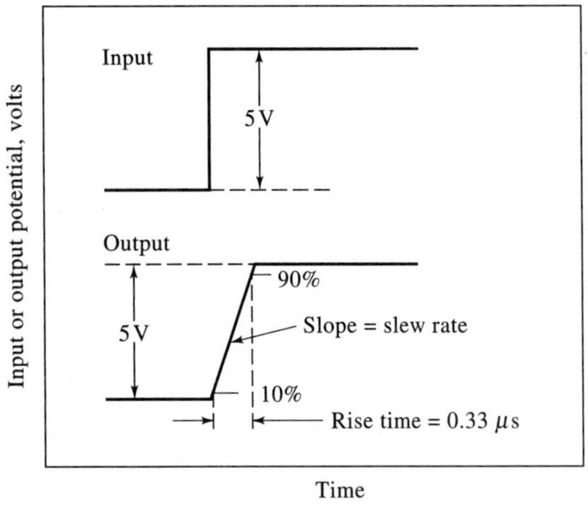

Figure 3-6 Response of an operational amplifier to a rapid step change in input voltage. The slope of the changing portion of the output signal is the slew rate, and the time required for the output to change from 10% to 90% of the total change is the rise time.

from 5 V to 10 V, the *slew rate* can be calculated as follows:

$$\text{slew rate} = \frac{\Delta v}{\Delta t} = \frac{5\ \text{V}}{0.33\ \mu\text{s}} = 17\ \text{V}/\mu\text{s} \qquad (3\text{-}9)$$

The slew rate is the maximum rate of change of the output of an amplifier in response to a step change at the input. Typical values of slew rate are on the order of a few volts per microsecond, but special operational amplifiers may be purchased with slew rates of up to several hundred volts per microsecond.

3B-3 The Voltage-Follower Circuit

Figure 3-7 depicts a *voltage follower*, a circuit in which the input is fed into the *noninverting* input, and a feedback loop is provided from the output to the inverting input. If we invoke the condition that the operational amplifier will do what is necessary to cause v_- to equal v_+ and note that v_- and v_o are connected, we may write the following set of equalities:

$$v_i = v_+ \approx v_- = v_o$$

Thus, although this circuit has a unit voltage gain ($v_o/v_i = 1$), it can have a very large power gain because operational amplifiers have high input impedances but low output impedances. To show the effect of this large difference in impedance, let us define the power gain as P_o/P_i where P_o is the power of the output from the operational amplifier and P_i is the input power. If we then substitute the power law ($P = iv = v^2/R$) and Ohm's law into this definition and recall that v_o and v_i are approximately the same in an operational amplifier, we obtain the expression

$$\text{power gain} = \frac{P_o}{P_i} = \frac{i_o v_o}{i_i v_i} = \frac{v_o^2/Z_o}{v_i^2/Z_i} = \frac{Z_i}{Z_o}$$

where Z_i and Z_o are the input and output impedances of the operational amplifier. This result is important because it means that the voltage follower will draw almost no current from an input; the internal circuitry of the operational amplifier and the power supply, however, can supply large currents at the output of the operational amplifier. As will be shown later, this property is valuable in the measurement of high-impedance sources with low-impedance measuring devices. An important example of such a measurement situation is the monitoring of glass electrodes with a strip-chart recorder.

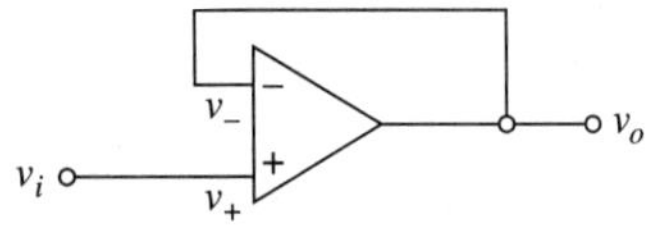

Figure 3-7 Operational amplifier voltage-follower.

3C AMPLIFICATION AND MEASUREMENT OF TRANSDUCER SIGNALS

Operational amplifiers find general application in the amplification and measurement of the electrical signals from transducers. These signals, which are often concentration dependent in analytical instruments, include current, potential, and charge. This section presents basic applications of operational amplifiers to the measurement of each type of signal.

3C-1 Current Measurement

The accurate measurement of small currents is important to such analytical methods as voltammetry, coulometry, photometry, and the ionization detectors

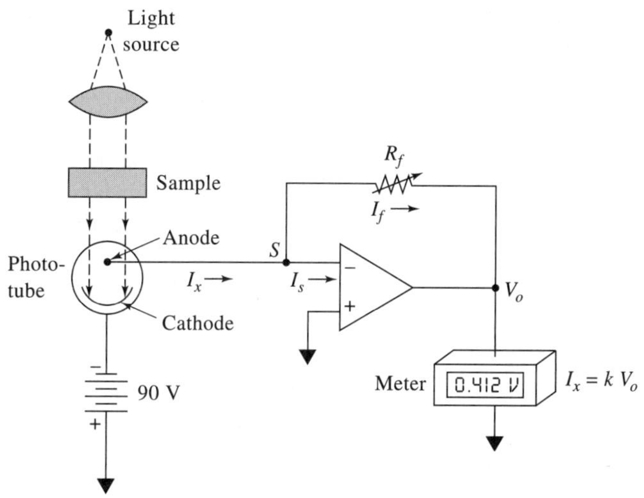

Figure 3-8 Application of an operational amplifier to the measurement of a small photocurrent I_x.

popular in gas chromatography. As pointed out in Chapter 2, an important concern that arises in all physical measurements, including that of current, is whether the measuring process will, in itself, alter significantly the signal being measured, thus leading to a loading error. It is inevitable that any measuring process will perturb a system under study in such a way that the quantity actually measured will differ from its original value before the measurement. We must therefore try to ensure that the perturbation can be kept small. For a current measurement, this consideration requires that the internal resistance of the measuring device be minimized so that it does not alter the current significantly.

A low-resistance current-measuring device is readily obtained by removing the resistor R_i in Figure 3-4 and using the current to be measured as the signal input. An arrangement of this kind is shown in Figure 3-8, where a small direct current I_x is generated by a phototube, a transducer that converts radiant energy such as light into an electrical current. When the cathode of the phototube is maintained at a potential of about −90 V, absorption of radiation by its surface results in the ejection of electrons, which are then accelerated to the wire anode. A current results that is directly proportional to the power of the radiant beam.

If the conclusions reached in the discussion of feedback are applied to this circuit, we may write

$$I_x = I_f + I_s \approx I_f$$

In addition, the point S is at virtual common so that the potential V_o corresponds to the potential difference across the resistor R_f. Therefore, from Ohm's law

$$V_o = -I_f R_f = -I_x R_f$$

and

$$I_x = -V_o/R_f = kV_o$$

Thus, the potential measurement V_o gives the current, provided R_f is known. By making R_f reasonably large, the accurate measurement of small currents is feasible. For example, if R_f is 100 kΩ, a 1 μA current results in an output potential of 0.1 V, a quantity that is easily measured with a high degree of accuracy and precision.

As shown by the following example, an important property of the circuit shown in Figure 3-8 is its low resistance with respect to the current from the transducer. Thus, the meter is driven, not by the transducer, but by the amplified current from the external power supply of the operational amplifier. The result is a minimal measuring error.

EXAMPLE 3-1

Assume that R_f in Figure 3-8 is 200 kΩ, the internal resistance of the phototube is 5.0×10^4 Ω, and the amplifier open-loop gain is 1.0×10^5. Calculate the relative error in the current measurement that results from the presence of the measuring circuit.

Here, the effective resistance of the measuring circuit R_m is the resistance between the summing point S and the circuit common. This resistance is given by Ohm's law. That is,

$$R_m = \frac{v_-}{I_x} = \frac{v_+ - V_o/A}{I_x} = \frac{V_o/A}{I_x}$$

where v_+ is the potential at the noninverting input, v_- is the potential at the inverting input, and A is the open-loop gain. We may also write (Equation 3-5)

$$I_f = \frac{v_- - V_o}{R_f} = \frac{v_+ - V_o/A - V_o}{R_f} = \frac{V_o/A - V_o}{R_f}$$
$$= I_x$$

Combining these two equations yields upon rearrangement

$$R_m = \frac{V_o/A}{(V_o/A - V_o)/R_f}$$

But since $V_o/A - V_o \approx V_o$ in the denominator, the previous equation becomes

$$R_m = \frac{R_f}{A}$$

Substituting numerical values gives

$$R_m = 200 \times 10^3\ \Omega/1.0 \times 10^5 = 2.0\ \Omega$$

Equation 2-15 shows that the relative loading error in a current measurement is given by

$$\text{rel error} = \frac{-R_m}{R_L + R_m}$$

where R_L is the load resistance of the phototube in the absence of the resistance of the measuring device R_m. Thus,

$$\text{rel error} = \frac{-2.0\ \Omega}{5.0 \times 10^4\ \Omega + 2.0\ \Omega}$$

$$= -4.0 \times 10^{-5} \quad \text{or} \quad -0.004\%$$

The instrument shown in Figure 3-8 is called a photometer; it measures the attenuation of a light beam by absorption brought about by a colored analyte in a solution; this parameter is related to the concentration of the species responsible for the absorption. Photometers are described in detail in Section 13D-3.

3C-2 Potential Measurements

Potential measurements are used extensively for the determination of temperature and the concentration of ions in solution. In the former application, the transducer is a thermocouple; in the latter, it is an ion-sensitive electrode.

Equation 2-14 shows that accurate potential measurements require that the resistance of the measuring device be large with respect to the internal resistance of the voltage source to be measured. The need for a high-resistive measuring device becomes particularly acute in the determination of pH with a glass electrode, which typically has an internal resistance on the order of tens to hundreds of megaohms. Because of the high resistance, the basic inverting amplifier shown in Figure 3-4, which has an internal resistance of perhaps $10^5\ \Omega$, is not satisfactory for voltage measurements. On the other hand, an inverting amplifier can be combined with the voltage-follower circuit shown in Figure 3-7 to give a high-impedance voltage-measuring device. An example of such a circuit is shown in Figure 3-9. The first stage consists of a voltage follower, which typically provides an input impedance in excess of $10^{12}\ \Omega$. An inverting amplifier circuit follows, which amplifies the input by R_f/R_i or 20 in this case. An amplifier, such as this, with a resistance of 100 MΩ or more is often called an *electrometer.* Carefully designed, operational amplifier-based electrometers are available commercially.

3C-3 Resistance or Conductance Measurements

Electrolytic cells and temperature-responsive devices, such as thermistors and bolometers, are common examples of transducers whose electrical resistance or con-

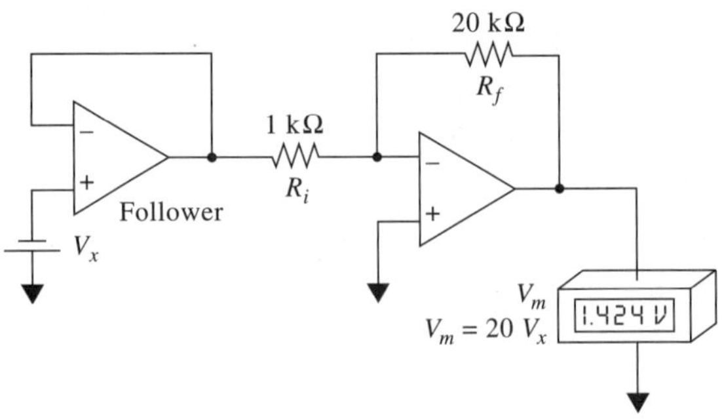

Figure 3-9 A high-impedance circuit for voltage amplification.

ductance varies in response to an analytical signal. These devices are used for conductometric and thermometric titrations, for infrared absorption and emission measurements, and for temperature control in a variety of analytical applications.

The circuit shown in Figure 3-4 provides a convenient means for measurement of the resistance or conductance of a transducer. Here, a constant voltage source is used for V_i and the transducer is substituted for either R_i or R_f in the circuit. The amplified output potential v_o, after being rectified and filtered, is then measured with a suitable meter, potentiometer, or a computerized data acquisition system. Thus, if the transducer is substituted for R_f in Figure 3-4, the output, as can be seen from rearrangement of Equation 3-7, is

$$R_x = -\frac{v_o R_i}{V_i} = k v_o \qquad (3\text{-}10)$$

where R_x is the resistance to be measured and k is a constant that can be calculated if R_i and V_i are known; alternatively, k can be determined by a calibration wherein R_x is replaced by a standard resistor.

If conductance rather than resistance is of interest, the transducer conveniently replaces R_i in the circuit. From Equation 3-7, we find that

$$\frac{1}{R_x} = G_x = -\frac{v_o}{V_i R_f} = k' v_o \qquad (3\text{-}11)$$

where G_x is the desired conductance. Note that in either type of measurement, the value of k, and thus the range of the measured values, can be easily varied by using a variable resistor R_i or R_f.

Figure 3-10 illustrates two basic applications of operational amplifiers for the measurement of conductance or resistance. In (a), the conductance of a cell for a conductometric titration is of interest. Here, an ac input signal v_i of perhaps 5 to 10 V is provided by an ac power supply. The output signal is then rectified, filtered, and measured as a dc potential. The variable resistance R_f provides a means for varying the range of conductances that can be recorded or read. Calibration is provided by switching the standard resistor R_s into the circuit in place of the conductivity cell.

Figure 3-10b illustrates how the circuit in Figure 3-4 can be applied to the measurement of a ratio of resistances or conductances. Here, the absorption of radiant energy by a sample is being compared with that for a reference solution. The two photoconductive transducers, which are devices whose resistances are inversely related to the intensity of light impinging upon their active surfaces, replace R_f and R_i in Figure 3-4. A dc power supply serves as the source of power, and the output potential M, as seen from Equation 3-7, is

$$M = V_0 = -V_i \frac{R_0}{R}$$

Typically, the resistance of a photoconductive cell is inversely proportional to the radiant power P of the radiation striking it. If R and R_0 are matched photoconductors,

$$R = C \cdot \frac{1}{P} \qquad \text{and} \qquad R_0 = C \cdot \frac{1}{P_0}$$

where C is a constant for both photoconductive cells, giving

$$V_0 = M = -V_i \frac{C/P_0}{C/P} = -V_i \frac{P}{P_0} \qquad (3\text{-}12)$$

Thus, the meter reading M is proportional to the ratio of the power of the two beams (P_0/P).

3C-4 Comparison of Transducer Outputs

It is frequently desirable to measure a signal generated by an analyte relative to a reference signal, as in Figure 3-10b. A difference amplifier, such as that shown in Figure 3-11, can also be applied for this purpose. Here, the amplifier is used for a temperature measurement. Note that the two input resistors have equal resistances R_i; similarly, the feedback resistor and the resistor between the noninverting input and ground, which are both labeled R_k, are also alike.

If we apply Ohm's law to the circuit shown in Figure 3-11, we find that (see Equations 3-5 and 3-6)

$$I_1 = \frac{V_1 - v_-}{R_i}$$

and

$$I_f = \frac{v_- - V_o}{R_k}$$

Because the operational amplifier has a high input impedance, I_1 and I_f are approximately equal.

$$I_1 \approx I_f$$

$$\frac{V_1 - v_-}{R_i} = \frac{v_- - V_o}{R_k}$$

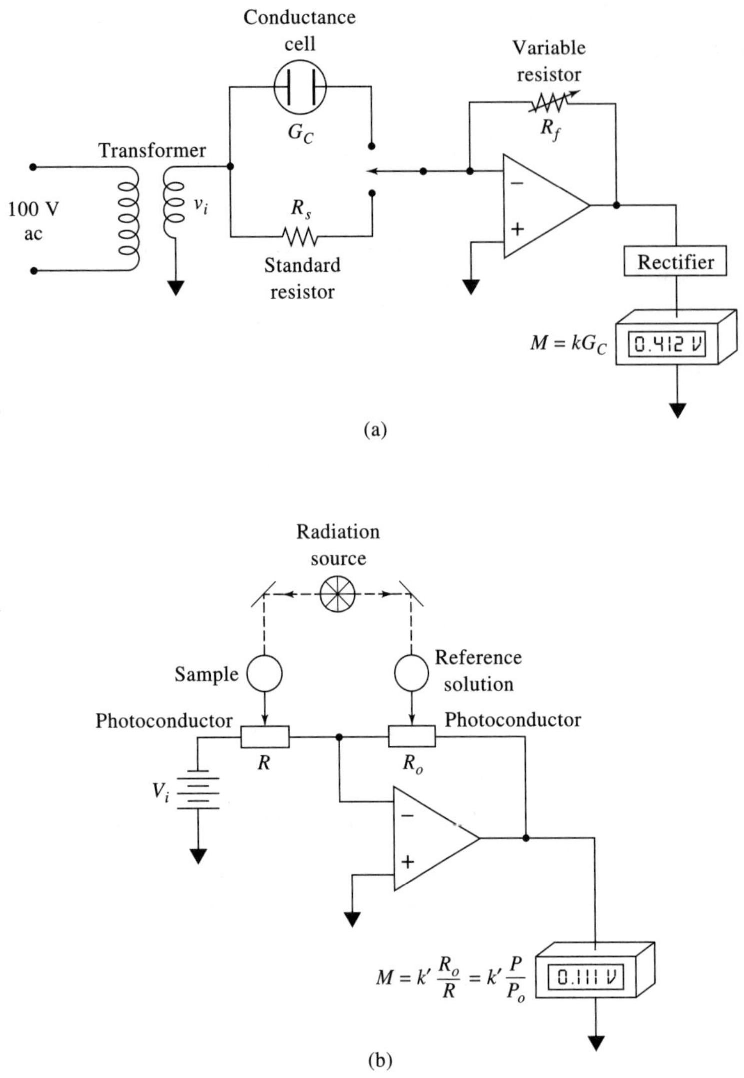

Figure 3-10 Two circuits for transducers whose conductance or resistance is the experimental variable of interest. (a) The output of the cell is a current that is proportional to the conductance of the electrolyte. (b) The ratio of the resistances of the photoconductive cells is proportional to the meter reading.

Solving this equation for v_-, gives

$$v_- = \frac{V_1R_k + V_oR_i}{R_k + R_i} \tag{3-13}$$

The potential v_+ can be written in terms of V_2 via the voltage divider equation, Equation 2-9:

$$v_+ = V_2\left(\frac{R_k}{R_i + R_k}\right) \tag{3-14}$$

Recall that an operational amplifier with a negative feedback loop will do what is necessary to satisfy the equation $V_1 \approx V_2$. When Equations 3-13 and 3-14 are substituted into this relation, we obtain, after rearrangement,

$$V_o = \frac{R_k}{R_i}(V_2 - V_1) \tag{3-15}$$

Thus, it is the difference between the two signals that is amplified. Any extraneous potential *common to the two inputs* shown in Figure 3-10 will be subtracted and will not appear in the output. Thus, any slow drift in the output of the transducers or any 60-cycle currents induced from the laboratory power lines will be eliminated from V_o. This useful property accounts for the widespread use of a differential circuit in the first amplifier stage of many instruments.

An important characteristic of operational amplifier circuits, such as the difference amplifier described above, is the *common mode rejection ratio,* or CMRR. The common mode rejection ratio of a difference amplifier is a measure of how well the amplifier rejects signals that are common to both inputs; it is the ratio of the difference gain A_d to the common mode gain A_{cm}; that is,

$$\text{CMRR} = \frac{A_d}{A_{cm}}$$

Suppose that we apply identical signals to the inputs V_1 and V_2, that $R_k = 1000R_i$, and that $V_o = 0.1V_2$. If the difference amplifier were ideal, V_o should be zero. In real difference amplifiers, some fraction of V_2, which is the signal that is to be rejected, appears at the output. In this case, V_2 is the signal to be rejected, or the common mode signal, so that the common mode gain is $A_{cm} = V_o/V_2 = 0.1$. The difference gain A_d is just the gain of the difference amplifier, which is $A_d = R_k/R_i = 1000$. The common mode rejection ratio for this configuration is then

$$\text{CMRR} = \frac{A_d}{A_{cm}} = 1000/0.1 = 10{,}000$$

The larger the common mode rejection ratio of a difference amplifier, the better it is at rejecting common mode signals.

The transducers shown in Figure 3-11 are a pair of *thermocouple junctions;* one of the transducers is immersed in the sample, and the other transducer is immersed in a reference solution (often an ice bath) held at constant temperature. A temperature-dependent contact potential develops at each of the two junctions formed from wires made of copper and an alloy called constantan (other metal pairs are also used). The potential difference $v_2 - v_1$ is roughly 5 mV per 100°C temperature difference.

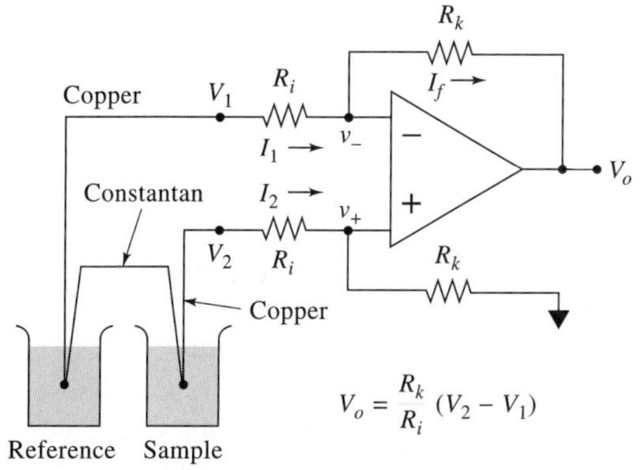

Figure 3-11 An operational amplifier difference amplifier measuring the output voltage of a pair of thermocouples.

3D APPLICATION OF OPERATIONAL AMPLIFIERS TO VOLTAGE AND CURRENT CONTROL

Operational amplifiers are easily used to generate constant-potential or constant-current signals.

3D-1 Constant-Voltage Sources

Several instrumental methods require a dc power source whose potential is precisely known and from which reasonable currents can be obtained without alteration of this potential. A circuit that meets these qualifications is termed a *potentiostat.*

Two potentiostats are illustrated in Figure 3-12. Both employ a standard potential source in a feedback circuit. This source is generally an inexpensive, commercially available, Zener stabilized integrated circuit (see Chapter 2, Section 2D-3) that is capable of producing an output voltage that is constant to a few hundredths of a percent. Such a source will not, however, maintain its voltage when it must deliver a large current.

Recall from our earlier discussions that point S in Figure 3-12a is at virtual ground. For this condition to exist, it is necessary that $V_o = V_{std}$. That is, the current in the load resistance R_L must be such that $I_L R_L = V_{std}$. It is important to appreciate, however, that this current arises from the power source of the operational amplifier and *not from the standard voltage source.* There is essentially no current in the feedback loop because the impedance of the inverting input is huge. Thus, the standard cell controls V_o but provides essentially none of the current through the load.

Figure 3-12b illustrates a modification of the circuit in (a) which permits the output voltage of the potentiostat to be fixed at a level that is a known multiple of the output voltage of the standard potential source.

3D-2 Constant-Current Sources

Constant-current dc sources, called *amperostats,* find application in several analytical instruments. For example, these devices are usually used to maintain a constant current through an electrochemical cell. An amperostat reacts to a change in input power or a change in internal resis-

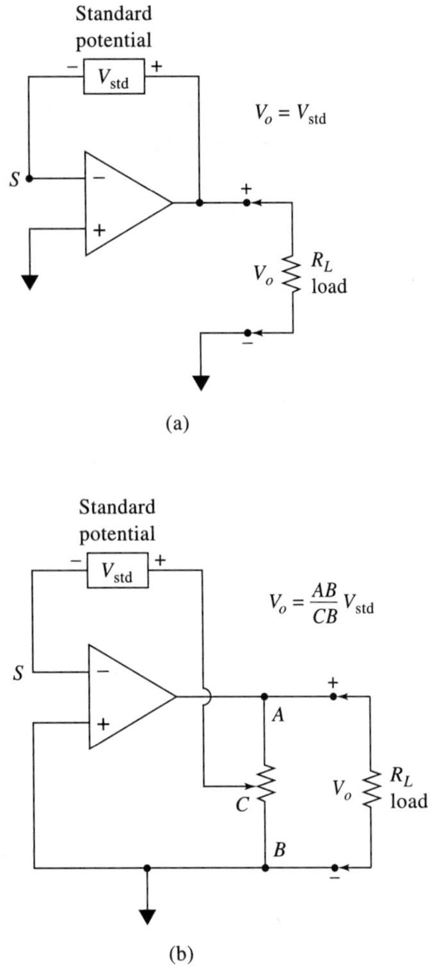

Figure 3-12 Constant-voltage sources.

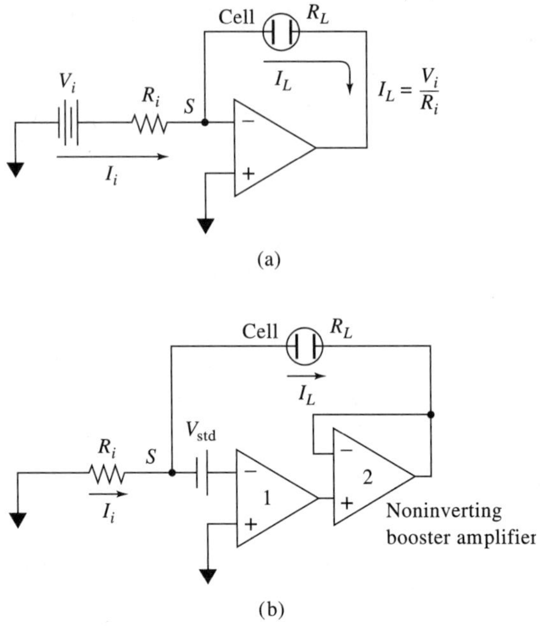

Figure 3-13 Constant-current sources.

tance of the cell by altering its output potential in such a way as to maintain the current at a predetermined level.

Figure 3-13 shows two amperostats. The first requires a voltage input V_i whose potential is constant while it supplies significant current. Recall from our earlier discussion that

$$I_L = I_i = \frac{V_i}{R_i}$$

Thus, the current will be constant and independent of the resistance of the cell, provided that V_i and R_i remain constant.

Figure 3-13b is an amperostat that employs a standard voltage V_{std} to maintain a constant current. Note that operational amplifier 1 has a negative-feedback loop that contains operational amplifier 2. In order to satisfy the condition $v_- = v_+$, the voltage at the summing point S must be equal to $-V_{std}$. Furthermore, we may write that at S,

$$I_i R_i = I_L R_i = -V_{std}$$

Since R_i and V_{std} in this equation are constant, the operational amplifier functions in such a way as to maintain I_L at a constant level that is determined by R_i.

Operational amplifier 2 in Figure 3-12b is simply a voltage follower, which has been inserted into the feedback loop of operational amplifier 1. A voltage follower used in this configuration is often called a noninverting booster amplifier because it can provide the relatively large current that may be required from the amperostat.

3E APPLICATION OF OPERATIONAL AMPLIFIERS TO MATHEMATICAL OPERATIONS

As shown in Figure 3-14, substitution of various circuit elements for R_i and R_f in the circuit shown in Figure 3-4 permits various mathematical operations to be performed on electrical signals as they are generated by an analytical instrument. For example, the output from a chromatographic column usually takes the form of a peak when the electrical signal from a detector is plotted as a function of time. Integration of this peak to find its area is necessary in order to find the analyte concentration. The operational amplifier shown in Figure 3-14c is capable of performing this integration automatically, thus giving a signal that is directly proportional to analyte concentration.

3E-1 Multiplication and Division by a Constant

Figure 3-14a shows how an input signal v_i can be multiplied by a constant whose magnitude is $-R_f/R_i$. The equivalent of division by a constant occurs when this ratio is less than unity.

3E-2 Addition or Subtraction

Figure 3-14b illustrates how an operational amplifier can produce an output signal that is the sum of several input signals. Because the impedance of the amplifier is large and because the output must furnish a sufficient current i_f to keep the summing point S at virtual ground, we may write,

$$i_f \approx i_1 + i_2 + i_3 + i_4 \tag{3-16}$$

But $I_f = -v_o/R_f$, and we may thus write

$$v_o = -R_f\left(\frac{v_1}{R_1} + \frac{v_2}{R_2} + \frac{v_3}{R_3} + \frac{v_4}{R_4}\right) \tag{3-17}$$

If $R_f = R_1 = R_2 = R_3 = R_4$, then the output voltage is the sum of the four inputs but opposite in sign.

$$v_o = -(v_1 + v_2 + v_3 + v_4)$$

To obtain an average of the four signals, let $R_1 = R_2 = R_3 = R_4 = 4R_f$. Substituting into Equation 3-17 gives

$$v_o = -\frac{R}{4}\left(\frac{v_1}{R} + \frac{v_2}{R} + \frac{v_3}{R} + \frac{v_4}{R}\right)$$

and v_o becomes the average of the four inputs. Thus,

$$v_o = \frac{-(v_1 + v_2 + v_3 + v_4)}{4} \tag{3-18}$$

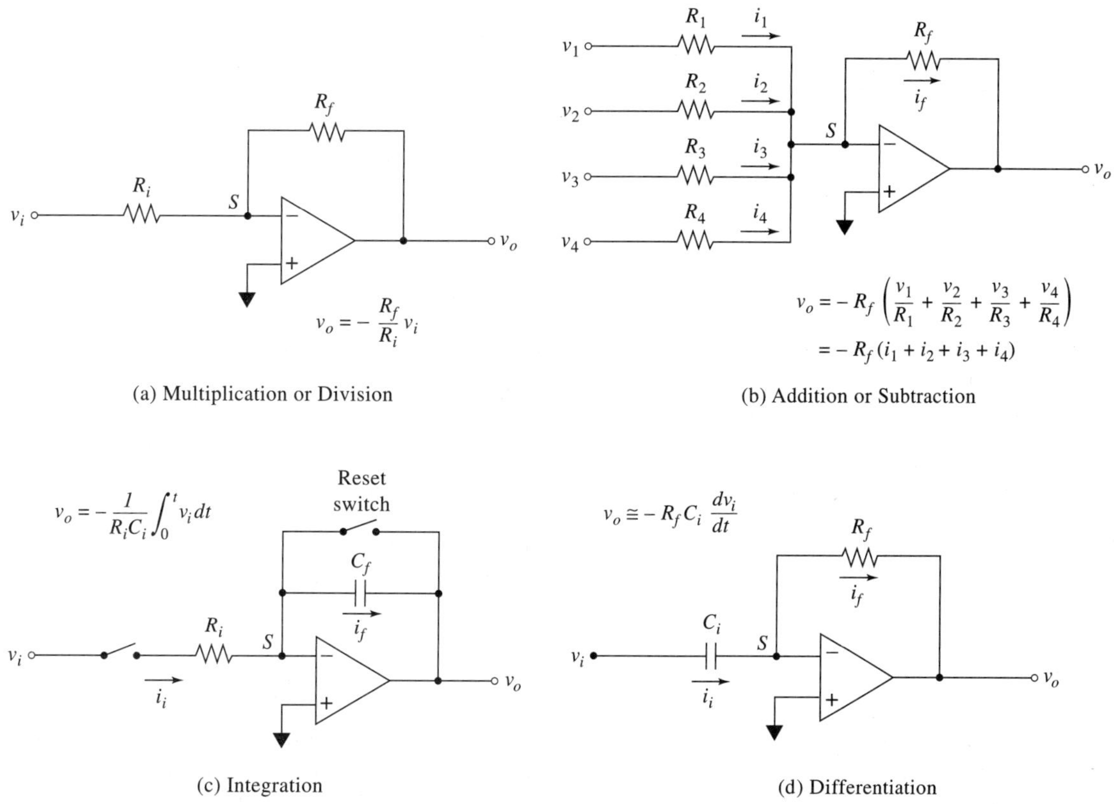

(a) Multiplication or Division

(b) Addition or Subtraction

(c) Integration

(d) Differentiation

Figure 3-14 Mathematical operations with operational amplifiers.

In a similar way, a weighted average can be obtained by varying the ratios of the resistances of the input resistors.

Subtraction can be performed by the circuit in Figure 3-14b by introducing an inverter with $R_i = R_f$ in series with one or more of the resistors, thus changing the sign of one or more of the inputs.

3E-3 Integration

Figure 3-14c illustrates a circuit for integrating a variable input signal v_i with respect to time. When the reset switch is open and the hold switch is closed,

$$i_i = i_f$$

and the capacitor C_f begins to charge. The current in the capacitor i_f is given by Equation 2-25 or

$$i_f = -C\frac{dv_o}{dt}$$

From Ohm's law the current i_i is given by $i_i = v_i/R_i$. Thus, we may write

$$\frac{v_i}{R_i} = -\frac{Cdv_o}{dt}$$

or

$$dv_o = -\frac{v_i}{R_iC}dt \tag{3-19}$$

We then integrate Equation 3-19 to obtain an equation for the output voltage v_o

$$\int_{v_{o1}}^{v_{o2}} dv_o = -\frac{1}{R_iC}\int_{t_1}^{t_2} v_i dt \tag{3-20}$$

or

$$v_{o2} - v_{o1} = -\frac{1}{R_iC}\int_{t_1}^{t_2} v_i dt \tag{3-21}$$

The integral is ordinarily obtained by first opening the hold switch and closing the reset switch to discharge the capacitor, thus making $v_{o1} = 0$ when $t_1 = 0$. Equation 3-21 then simplifies to

$$v_o = -\frac{1}{R_iC}\int_0^t v_i dt \tag{3-22}$$

To begin the integration, the reset switch is opened and the hold switch closed. The integration is stopped at time t by opening the hold switch. The integral over the period of 0 to t is v_o.

3E-4 Differentiation

Figure 3-14d is a basic circuit for differentiation, which is useful when the time rate of change of an experimental quantity is the variable of interest. Note that it differs from the integration circuit only in the respect that the positions of C and R have been reversed. Proceeding as in the previous derivation, we may write

$$C\frac{dv_i}{dt} = -\frac{v_o}{R_f}$$

or

$$v_o = -R_fC\frac{dv_i}{dt} \tag{3-23}$$

The circuit shown in Figure 3-14d is not, in fact, practical for many chemical applications, where the rate of change in the transducer signal is often low. For example, differentiation is a useful way to treat the data from a potentiometric titration; here, the potential change of interest occurs over a period of a second or more ($f \leq 1$ Hz). The input signal will, however, contain extraneous 60-, 120-, and 240-Hz components (see Figure 5-3) that are induced by the ac power supply. In addition, signal fluctuations resulting from incomplete mixing of the reagent and analyte solutions are often encountered. Unfortunately, the output of the circuit in Figure 3-14d has a strong frequency dependence; as a consequence, the output voltage from the extraneous signals often becomes as great or greater than that from the low-frequency transducer signal, even though the magnitude of the former voltage relative to the latter is small.

This problem may be overcome to some extent by introducing a small parallel capacitance, C_f in the feedback circuit and a small series resistor R_i in the input circuit to filter the high-frequency voltages. These added elements are kept small enough so that significant attenuation of the analytical signal does not occur. In general, differentiators are noise-amplifying circuits, while integrators attenuate noise. Thus, analog integrators are more widely used than differentiators. If differentiation of a signal is required, it is often accomplished digitally as will be discussed in Chapter 5.

3E-5 Generation of Logarithms and Antilogarithms

The incorporation of an external transistor into an operational amplifier circuit makes it possible to generate output voltages that are either the logarithm or the an-

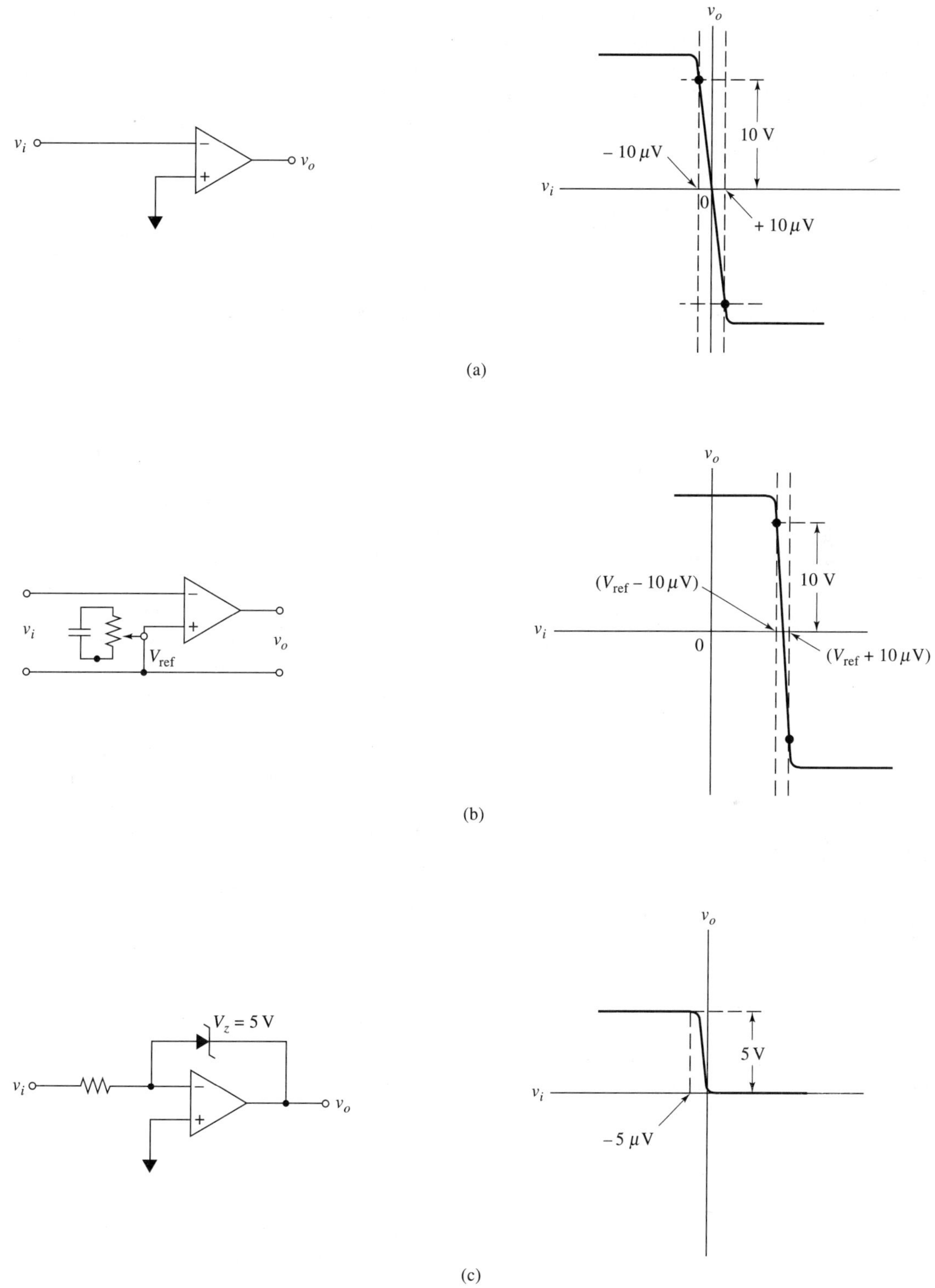

Figure 3-15 Operational amplifier switching circuits and transfer functions.

tilogarithm of the input voltage, depending upon the circuit. Operational amplifier circuits of this kind are, however, highly frequency and temperature dependent and accurate to only a few percent; they are in addition limited to one or two decades of input voltage. Temperature- and frequency-compensated modules for obtaining logarithms and antilogarithms with accuracies of a few tenths of a percent are available commercially for \$20 to \$100. These circuits have been used to produce signals proportional to absorbance in spectrophotometers and find some use in data-compression applications. Currently, logarithms and antilogarithms are usually computed numerically with microcomputers rather than with operational amplifiers.

3F APPLICATION OF OPERATIONAL AMPLIFIERS TO SWITCHING

Another important and widespread application of operational amplifiers is switching. Such circuits are found in a wide variety of applications in which signal levels must be monitored and compared to reference voltages such as sampling circuits, peak detection circuits, analog timers, circuits designed to produce limited signal levels, and circuits at the interface of the boundary between the digital and analog domains. Figures 3-15a and 3-15b show two basic comparator circuits and their output response versus input voltages. In the circuit in (a), the input voltage is compared with the circuit common, while in (b) the comparison is with a reference voltage V_{ref}. The behavior of the first comparator is understood by writing Equation 3-1 and by recognizing that because the noninverting input is connected to common, $v_+ = 0$. We find that

$$v_o = -A\,(v_- - v_+)$$

$$= -Av_- = -Av_{in}$$

As shown by the plot on the right of Figure 3-15a, the equation above applies to a limited region on either side of an input voltage of 0 V. For a typical operational amplifier circuit with an open-loop gain of 10^6, the output voltage v_{out} will be switched from 0 to 10 V when the input voltage v_{in} changes from zero to 10^{-5} V (10 μV). As shown in this plot, the linearity between input and output voltages begins to break down at input voltages greater than about 10 μV or at output voltages greater than about 10 V. Ultimately, the output voltage becomes independent of the input voltage and reaches either the positive or negative output limit. Some operational amplifiers have a tendency to remain at voltage limit, at least temporarily, even after the input voltage returns to zero. To circumvent this problem and to ensure that the operational amplifier stays within the range over which Equation 3-5 applies, a Zener diode (Section 2D-3) with a breakdown voltage less than 10 V is inserted into the feedback loop as shown in Figure 3-15c. In this instance, a change of input voltage of 5 μV causes the circuit to pass from a conducting state to an essentially nonconducting state. Thus, the circuit is an electronic switch that has just two states: one in which the output is 5 V and one in which the output voltage is near zero.

Electronic switches, such as those discussed in this section, have at least two significant advantages relative to their mechanical counterparts; first, they are very sensitive because A is large; and second, they exhibit very rapid responses. Special circuits are available for switching applications called *comparators.* These devices exhibit gains of 10^6 or greater and rise times of 200 ns or less. It is these properties that make them so important for computer interfacing and other switching applications.

3G QUESTIONS AND PROBLEMS

3-1 A low-frequency sine wave voltage is the input to the following circuits. Sketch the anticipated output of each circuit.

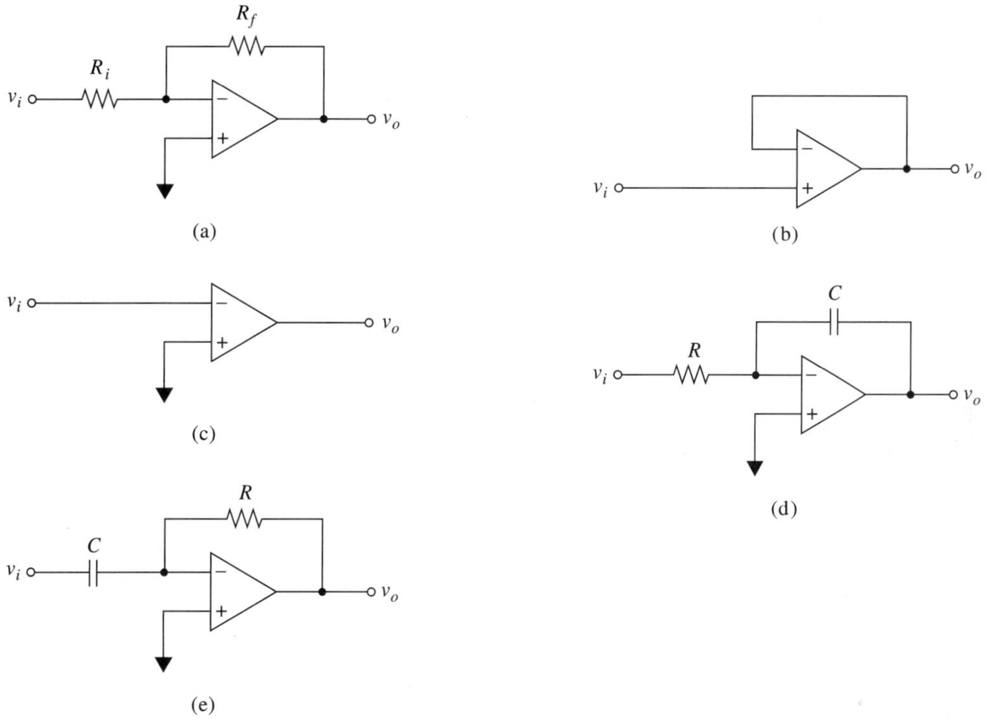

3-2 Assume the following values for the components of the circuit in Figure 3-4: $R_i = 1.00$ kΩ, $R_f = 30.0$ kΩ, $A = 200$, and $V_i = 0.910$ mV dc. Calculate (a) v_o, (b) I_i, (c) I_f.

3-3 Calculate the slew rate and the rise time for an operational amplifier with a 50 MHz bandwidth.

3-4 Determine the relative error that results from using Equation 3-6 to calculate v_o rather than using the more exact expressions if $R_i = 2.00$ MΩ, $R_f = 40.0$ MΩ, and $A = 5 \times 10^4$.

3-5 Design a circuit having an output given by

$$-v_o = 3v_1 + 5v_2 - 6v_3$$

3-6 Design a circuit for calculating the average value of three input voltages multiplied by 1000.

3-7 Design a circuit to perform the following calculation:

$$-y = \frac{1}{10}(5x_1 + 3x_2)$$

3-8 Design a circuit to perform the following calculation:

$$-v_o = 4v_i + 1.00 \times 10^3 I_i$$

3-9 For the following circuit

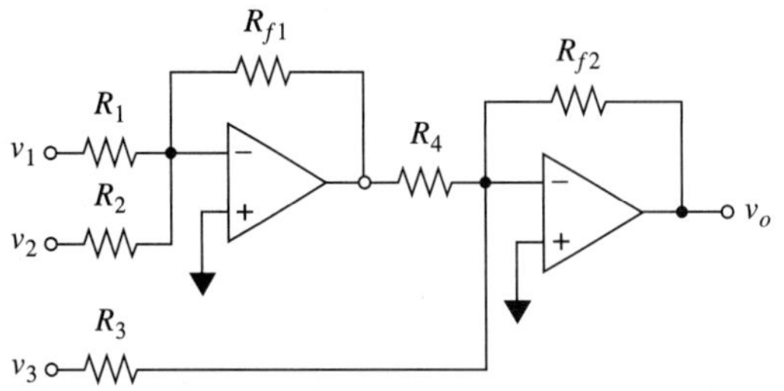

(a) write an expression that gives the output voltage in terms of the three input voltages and the various resistances.
(b) indicate the mathematical operation performed by the circuit when $R_1 = R_{f1} = 200$ kΩ; $R_4 = R_{f2} = 400$ kΩ; $R_2 = 50$ kΩ; $R_3 = 10$ kΩ.

3-10 Show the algebraic relationship between the voltage input and output for the following circuit:

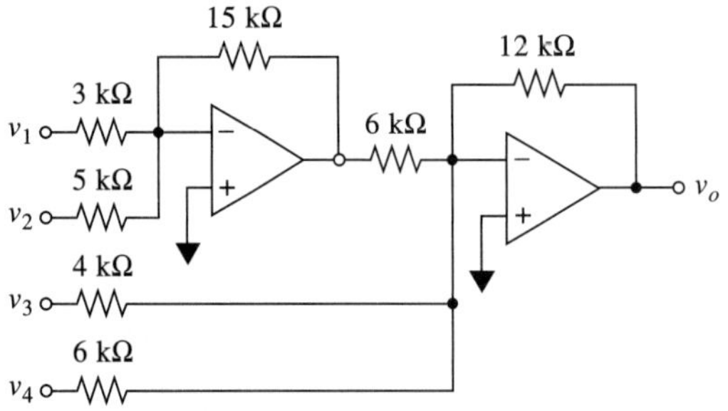

3-11 For the circuit below, sketch the outputs at v_{oA} and v_{oB}, if the input is initially zero, but is switched to a constant positive voltage of time zero.

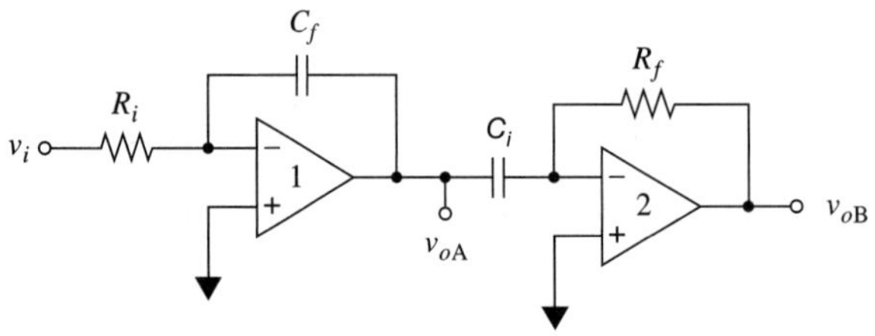

3-12 Derive a relationship between v_{in} and v_{out} for the following circuit:

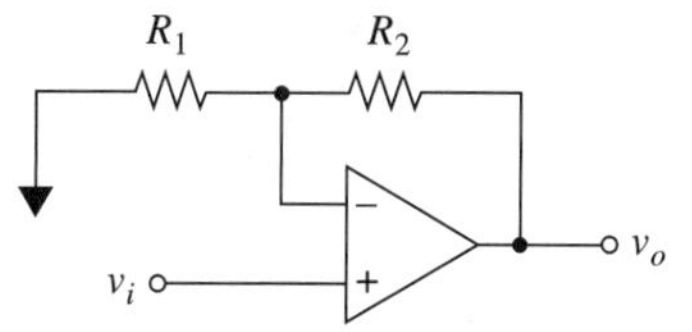

3-13 In the following circuit, R is a variable resistor. Derive an equation that describes v_{out} as a function of v_{in}, and the position of the movable contact of the voltage divider (x). Perform the derivation such that x is zero if there is zero resistance in the feedback loop.

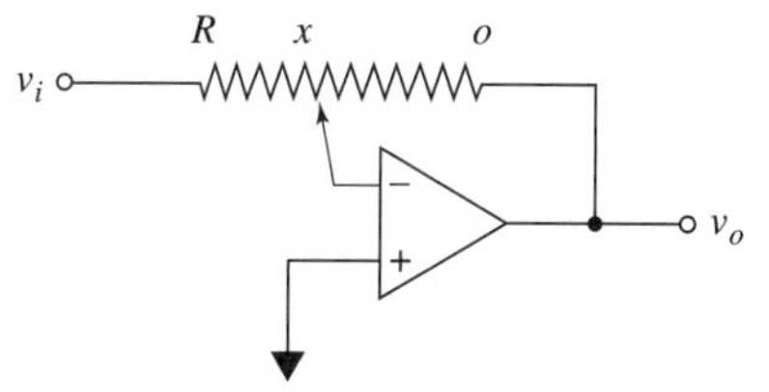

3-14 Derive an expression for the output potential of the following circuit:

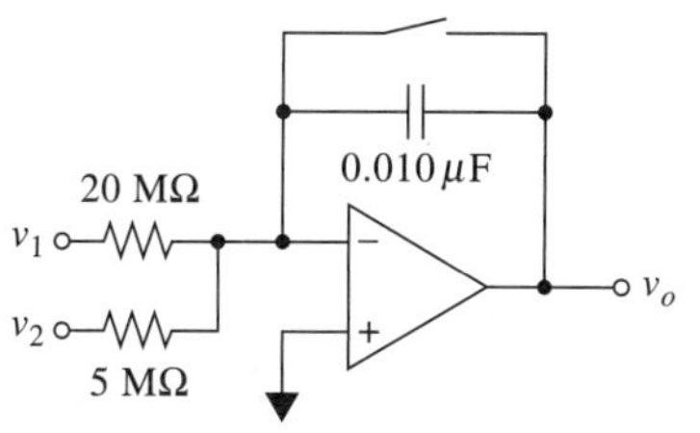

3-15 Show that when the four resistances are equal, the following circuit becomes a subtracting circuit.

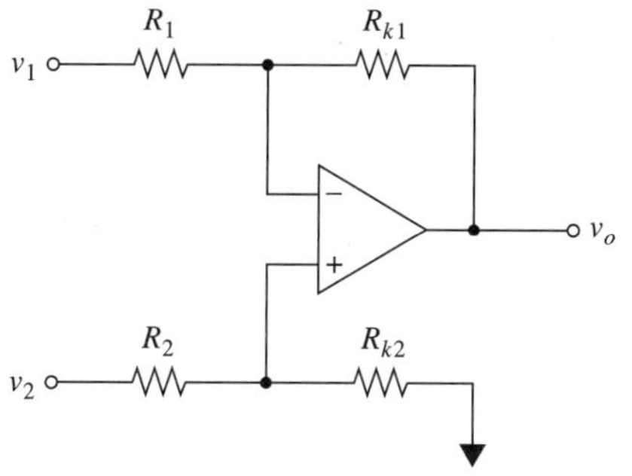

3-16 The linear slide wire AB has a length of 100 cm. Where along its length should contact C be placed in order to provide a potential of exactly 3.00 V? The potential of the Weston cell is 1.0183 V.

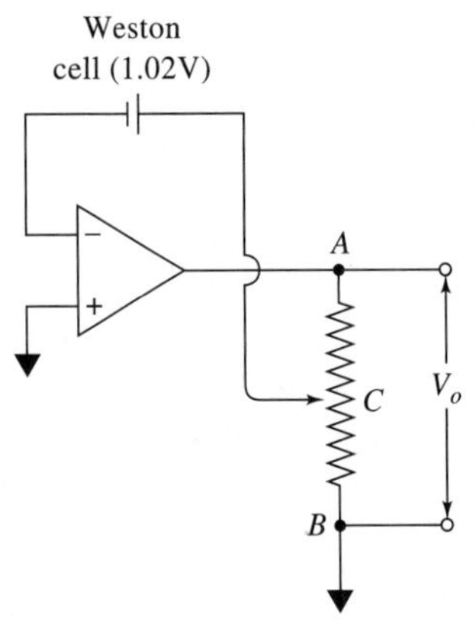

3-17 Design a circuit that will produce the following output:

$$v_o = 4.0 \int_0^t v_1 dt + 5.0 \int_0^t v_2 dt$$

3-18 Design a circuit that will produce the following output:

$$v_o = 2.0 \int_0^t v_1 dt - 6.0(v_2 + v_3)$$

3-19 Plot the output voltage of an integrator after 1, 3, 5, and 7 s after the start of integration if the input resistor is 2.0 MΩ, the feedback capacitor is 0.25 μF, and the input voltage is 4.0 mV.

Chapter 4

Digital Electronics and Microcomputers

The rate of growth of electronics, instrumentation, and microcomputer technology is nearly incomprehensible. Two decades ago, computers were expensive, cumbersome devices rarely seen in chemical laboratories; but in the last decade, the advent of the inexpensive mass-produced microcomputer with its collection of associated peripheral devices has moved computers from the rare to the commonplace. Microcomputers or microprocessors are now found in most laboratory instruments, including even balances and pH meters. Section 4D explains the difference between microprocessors and microcomputers.

The proliferation of automated, computerized instrumentation dictates that chemists develop an understanding of the advantages and limitations of modern electronic devices and computers. Although it is impossible, and perhaps undesirable, for all chemists to attain knowledge of electronics and computers at the design level, the development of high-function integrated circuit modules and data acquisition hardware permits a conceptually straightforward top-down approach to the implementation of electronics and computer technology. In the top-down view, we assume the perspective that an instrument is a collection of functional modules that may be represented as blocks in a schematic diagram such as those depicted in Figure 4–2 or Figure 4–13. Using such an approach, it is possible to accomplish very sophisticated physicochemical measurements by connecting several function modules, integrated circuits, or computers in the correct sequence. It is usually not necessary to have detailed knowledge of the internal design of individual components of an instrumental system. In addition to easing the learning process, the top-down approach aids in diagnosing system malfunctions and in the intelligent application of instrumental systems to the solution of chemical problems.

Digital circuits offer some important advantages over their analog counterparts. For example, digital circuits are less susceptible to environmental noise, and digitally encoded signals can usually be transmitted with a higher degree of signal integrity. Second, digital signals may be transmitted directly to digital computers, which means that software can be used to extract the information from signal outputs of chemical instruments.

This chapter should serve as a springboard for further study and use of modern instrumental systems. Our goals are (1) to provide a brief overview of how digital information can be encoded, (2) to introduce some of the basic components of digital circuits and microcomputers, (3) to describe some of the most common instrument-computer interactions, and (4) to illustrate how computers and software are utilized in an analytical laboratory.[1]

4A ANALOG AND DIGITAL SIGNALS

As described in Chapter 1, chemical data are encoded in *digital, analog,* or *time domains.* An example of a discrete phenomenon in a nonelectrical domain that may be easily converted to the digital domain is the radiant energy produced by the decay of radioactive species. Here, the information consists of a series of pulses of energy that is produced as individual atoms decay. These pulses can be converted to an electrical domain by using an appropriate input transducer, first as analog pulses and then as digital pulses that can be counted. The resulting information can be interpreted and manipulated as an integer number of decays, which is a form of nonelectrical information.

It is important to appreciate that whether a signal resulting from a chemical phenomenon is continuous or discrete may depend upon the intensity of the signal and how it is observed. For example, the yellow radiation produced by heating sodium ions in a flame is often measured with a phototransducer that converts the radiant energy into an analog current, which can vary continuously over a considerable range. However, at low radiation intensity, a properly designed transducer can respond to the individual photons, producing a signal that consists of a series of analog pulses that can be converted to digital pulses and then counted.

Often, in modern instruments, an analog signal, such as shown in Figure 4–1a, is converted to a digital one (Figure 4–1b) by sampling and recording the analog output at regular time intervals. In a later section, we consider how such a conversion is accomplished with an *analog-to-digital converter,* or ADC.

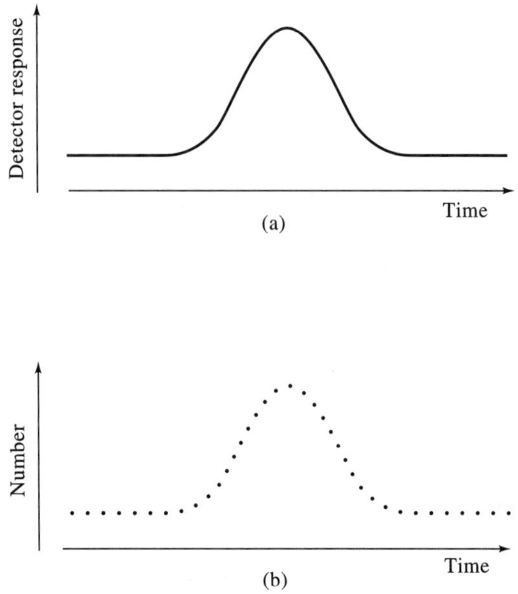

Figure 4-1 Detector response versus time plots for the same signal in (a) the analog domain and (b) the digital domain.

4B COUNTING AND ARITHMETIC WITH BINARY NUMBERS

In a typical digital measurement, a high-speed *electronic counter* is used to count the number of events that occur within a specified set of boundary conditions. Examples of such signals include a number of photons or alpha decay particles emitted by an analyte per second, a number of drops of titrant, or a number of steps of a stepper motor used to deliver reagent from a syringe. Boundary conditions might include a time interval such as one second, which provides the frequency of the signal in hertz, or a given change in an experimental variable such as pH, absorbance, current, or voltage.

Counting such signals electronically requires that they first be converted to digital signal levels to provide a series of pulses of equal voltage compatible with the digital circuitry of the counter. Ultimately these pulses are converted by the counter to a binary number for processing by a computer or to a decimal number for display. Electronic counting is performed in binary-coded-decimal (BCD) numbers or in binary numbers. In both of these coding schemes, only two digits, 0 and 1, are required to represent any number. In electronic counters, the 0 is usually represented by a signal of about 0 V

---

[1] For further information, see H. V. Malmstadt, C. G. Enke, and S. R. Crouch, *Microcomputers and Electronic Instrumentation: Making the Right Connections.* Washington, DC: American Chemical Society, 1994; A. J. Diefenderfer and B. E. Holton, *Principles of Electronic Instrumentation,* 3rd ed. Philadelphia: Saunders College Publishing, 1994; K. L. Ratzlaff, *Introduction to Computer Assisted Experimentation.* New York: Wiley, 1987; S. C. Gates and J. Becker, *Laboratory Automation Using the IBM-PC.* New York: Prentice-Hall, 1989.

and the 1 by a voltage of typically 5 V. It is important to recognize that these voltage levels depend upon current technology and are different for different logic families. For example, many state-of-the-art microprocessors use about 3 V to represent a 1 and use 0 V to signify a 0.

4B-1 The Binary Number System

Each digit in the decimal numbering system represents the coefficient of some power of 10. Thus, the number 3076 can be written as

3 0 7 6

$$6 \times 10^0 = 0006$$
$$7 \times 10^1 = 0070$$
$$0 \times 10^2 = 0000$$
$$3 \times 10^3 = \underline{3000}$$
$$\text{Sum} = 3076$$

Similarly, each digit in the binary system of numbers corresponds to a coefficient of a power of 2.

4B-2 Conversion of Binary and Decimal Numbers

Table 4–1 illustrates the relationship between a few decimal and binary numbers. The examples that follow illustrate methods for conversions between the two systems.

EXAMPLE 4-1

Convert 101011 in the binary system to a decimal number. Binary numbers are expressed in terms of base 2. Thus,

1 0 1 0 1 1

$$1 \times 2^0 = 1$$
$$1 \times 2^1 = 2$$
$$0 \times 2^2 = 0$$
$$1 \times 2^3 = 8$$
$$0 \times 2^4 = 0$$
$$1 \times 2^5 = \underline{32}$$
$$\text{Sum} = 43$$

TABLE 4-1 Relationship between Some Decimal and Binary Numbers

Decimal Number	Binary Representation
0	0
1	1
2	10
3	11
4	100
5	101
6	110
7	111
8	1000
9	1001
10	1010
12	1100
15	1111
16	10000
32	100000
64	1000000

EXAMPLE 4-2

Convert 710 to a binary number.

As a first step, we determine the largest power of 2 that is less than 710. Thus, since $2^{10} = 1024$,

$$2^9 = 512 \quad \text{and} \quad 710 - 512 = 198$$

The process is repeated for 198:

$$2^7 = 128 \quad \text{and} \quad 198 - 128 = 70$$

Continuing, we find that

$$2^6 = 64 \quad \text{and} \quad 70 - 64 = 6$$
$$2^2 = 4 \quad \text{and} \quad 6 - 4 = 2$$
$$2^1 = 2 \quad \text{and} \quad 2 - 2 = 0$$

The binary number is then computed as follows:

1	0	1	1	0	0	0	1	1	0
2^9	–	2^7	2^6	–	–	–	2^2	2^1	–

It is worthwhile noting that in the binary numbering system, the binary digit, or *bit,* lying farthest to the right in a number is termed the *least significant bit, or LSB*; the one on the far left is the *most significant bit, or MSB.*

4B-3 Binary Arithmetic

Arithmetic with binary numbers is similar to, but simpler than, decimal arithmetic. For addition, only four combinations are possible:

$$\begin{array}{r}0\\+0\\\hline 0\end{array}\qquad\begin{array}{r}0\\+1\\\hline 1\end{array}\qquad\begin{array}{r}1\\+0\\\hline 1\end{array}\qquad\begin{array}{r}1\\+1\\\hline 10\end{array}$$

Note that in the last sum, a 1 is carried over to the next higher power of 2. Similarly, for multiplication,

$$\begin{array}{r}0\\\times 0\\\hline 0\end{array}\qquad\begin{array}{r}0\\\times 1\\\hline 0\end{array}\qquad\begin{array}{r}1\\\times 0\\\hline 0\end{array}\qquad\begin{array}{r}1\\\times 1\\\hline 1\end{array}$$

The following example illustrates the use of these operations.

EXAMPLE 4-3

Perform the following calculations with binary arithmetic: **(a)** 7 + 3, **(b)** 19 + 6, **(c)** 7 × 3, and **(d)** 22 × 5.

(a)
$$\begin{array}{r}7\\+3\\\hline 10\end{array}\qquad\begin{array}{r}111\\+11\\\hline 1010\end{array}$$

(b)
$$\begin{array}{r}19\\+6\\\hline 25\end{array}\qquad\begin{array}{r}10011\\+110\\\hline 11001\end{array}$$

(c)
$$\begin{array}{r}7\\\times 3\\\hline 21\end{array}\qquad\begin{array}{r}111\\\times 11\\\hline 111\\111\ \\\hline 10101\end{array}$$

(d)
$$\begin{array}{r}22\\\times 5\\\hline 110\end{array}\qquad\begin{array}{r}10110\\\times 101\\\hline 10110\\00000\ \\10110\ \ \\\hline 1101110\end{array}$$

Note that a carry operation, similar to that in the decimal system, is used. Thus, in **(a)** the sum of the two ones in the right column is equal to 0 plus 1 to carry to the next column. Here, the sum of the three ones is 1 plus 1 to carry to the next column. Finally, this carry combines with the one in the next column to give 0 plus 1 as the most significant digit.

4C BASIC DIGITAL CIRCUIT COMPONENTS

Figure 4-2 is a block diagram of an instrument for counting the number of electrical pulses that are received from a transducer per unit of time. The voltage signal from the transducer first passes into a shaper that removes the small background signals and converts the large signal pulses to rectangular pulses that have the same frequency as the input signal. The resulting signal is then fed into a gate wherein the output from an internal clock provides an exact time interval t during which input pulses are allowed to accumulate in the counter. Finally, the binary-coded-decimal output of the counter is decoded and presented as a decimal number.

4C-1 Signal Shapers

Figure 4-3a shows a circuit of a typical signal shaper. It makes use of a voltage comparator to convert the input signal to the square waveform shown in Figure 4-3c. As shown in Figure 3-15c, the output of a comparator is at one of two voltage levels, +5 V or 0 V; these two levels, often termed *logic states,* have been designated as 1 and 0, respectively, in Figures 4-2 and 4-3. Commercial comparators such as the LM311 are designed so that their outputs have states of 0 V (LO or 0) or 5 V (HI or 1). These logic levels are compatible with most modern digital integrated circuits. When the comparator input voltage v_i is greater than the reference voltage V_{ref}, the output is in the logic state 1. On the other hand, when v_i is less than V_{ref}, the output is in logic state 0. Note that the comparator responds only to signals greater than V_{ref} and ignores the fluctuations in the background signal, provided that any background fluctuations or noise on the signal are small enough that the signal remains below V_{ref}.

4C-2 Binary Counters

Electronic counters employ a series of binary circuits to count electrical pulses. These circuits are basically electronic switches that have only two possible logic states, HI and LO or 1 and 0. Each circuit can then be used to

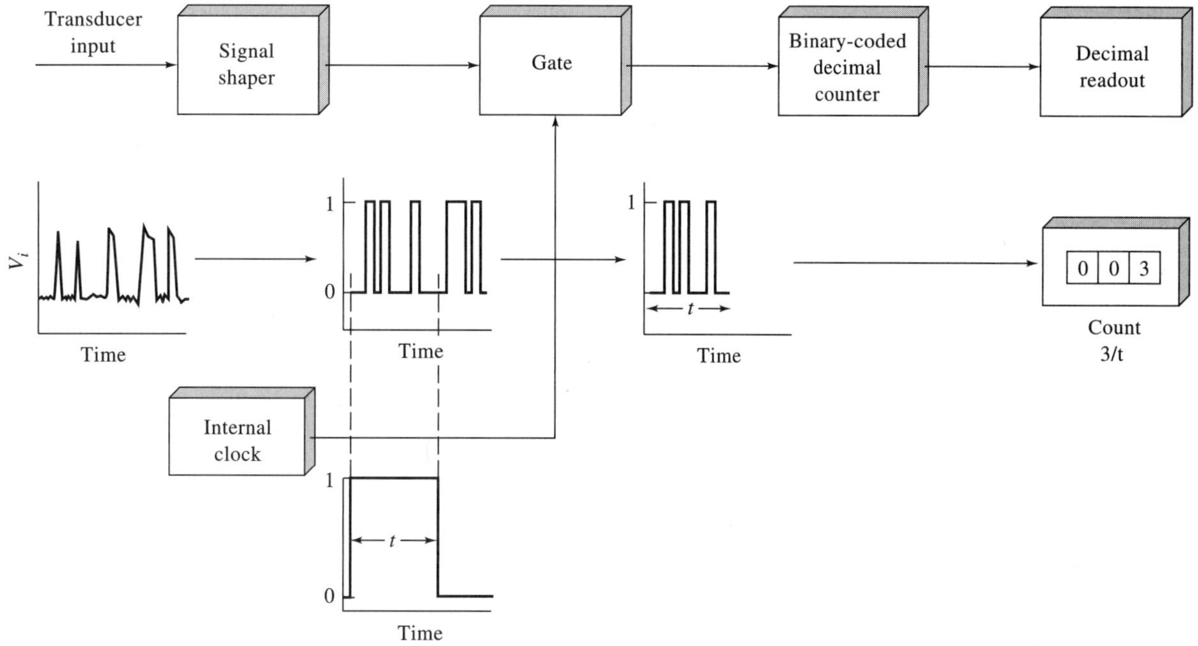

Figure 4-2 A counter for determining voltage pulses per second.

represent one bit of a binary number (or the coefficient of a power of 2). Two circuits can have four possible outputs: 0/0, 0/1, 1/0, and 1/1. We can readily show that three of these circuits have 8 different combinations while four have 16. Thus, n circuits have 2^n distinguishable output combinations. By using a sufficient number of circuit stages, the number of significant bits in a count can be made as large as desired. Thus, seven stages have 128 states, which would provide a count that is accurate to 1 part in 128 or something better than 1% relative.

A convenient circuit for counting is the so-called *JK flip-flop.* This circuit changes its output level whenever the input signal changes from logic state 1 to 0; *no change in output* is associated with an input change of 0 to 1. Flip-flops are integrated circuits that are made up of a suitable combination of diodes, resistors, and transistors.

Figure 4-4 shows how four flip-flops can be arranged to give a counting device that can count from 0 through 15. Additional flip-flops will extend the range to larger numbers. Thus, five flip-flops will provide a range of 0 to 31 and six a range of 0 to 63. Figure 4-5 shows the waveforms of the signals as they appear at flip-flops *A, B, C,* and *D* in the counter shown in Figure 4-4. As shown by the top waveform in Figure 4-5, the input signal *I* is a series of voltage pulses of equal magnitude and frequency, which are first converted in the comparator to the rectangular waveform shown as signal *S.* The gate signal *G* starts the counting as pulse 1 as the output of the comparator goes from 1 to 0. Ultimately, the gate signal terminates the counting after a preset time *t,* which as shown in the figure, corresponds to 12 counts.

Before counting is initiated, all of the flip-flops are cleared; that is, they are brought to their 0 state. At this point all of the light-emitting-diodes, or LEDs, in Figure 4-4 are off. At the start of counting, flip-flop *A* passes from 0 to 1 as a consequence of the comparator signal's changing from 1 to 0 (see first vertical dotted line, Figure 4-5). At this point LED *A* comes on and remains on until signal *S* again shifts from 1 to 0 (see second vertical line). This switching on and off of LED *A* continues until time t_2; note that LED *A* is off at this point.

As can be seen in Figure 4-5, the output of flip-flop *A* is also a square wave having a frequency that is exactly one half that of the input signal *S.* It is apparent from the figure that the response of flip-flop *B* to the signal from flip-flop *A* is exactly analogous to the response of *A* to the signal *S.* Thus, *B* is a square wave that has a frequency just one half that of *A* and one fourth that of

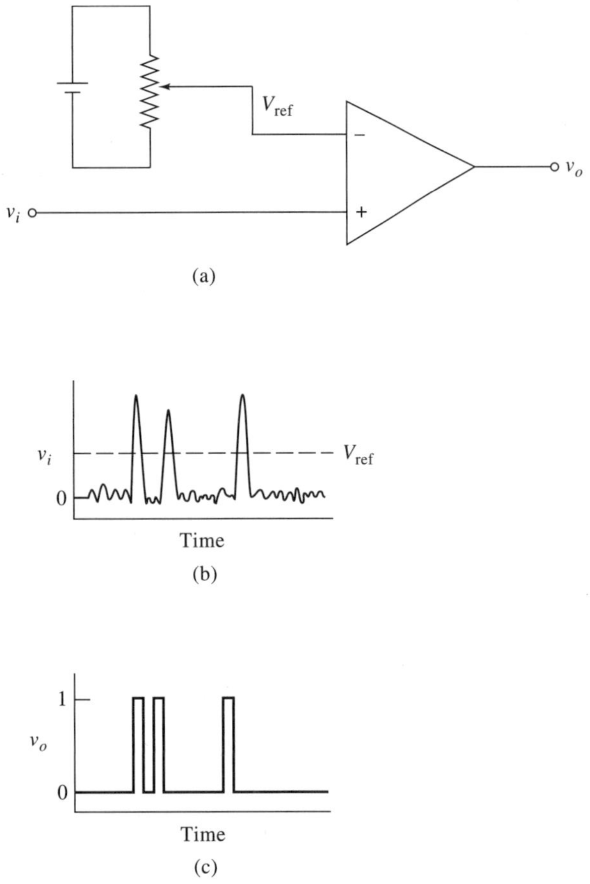

Figure 4-3 A signal shaper: (a) circuit, (b) input signal, (c) output signal.

S. Similarly, the output from binary *C* has a frequency that is one eighth that of *S*, while the frequency of signal *D* is one sixteenth that of *S*. This function is sometimes used to scale the frequency of a signal by a given fraction as described in Section 4C-4.

After time t_2, it is seen that flip-flops *C* and *D* end up in logic state 1, while *A* and *B* are in the 0 condition. These states correspond to the binary number 1100, which is 12 in the decimal system. Figure 4-4 also demonstrates how the binary count could be read out directly by means of a bank of LEDs.

4C-3 Decimal Counting

For most counters, the display is not in the binary form shown by the LEDs in Figure 4-4 but in the more convenient and easily comprehended decimal format. Several systems have been developed for binary to decimal conversion. The most common system is the so-called 8421 system, or the *binary-coded decimal (BCD) system.* Here, each digit in a decimal number is represented by a set of four flip-flops such as the set shown in Figure 4-4. Only 10 of the possible 16 states are then used. The system is so arranged that after a count of nine, the outputs of all of the flip-flops are returned to 0 with the 1 to 0 transition from the *D* binary being fed into the *A* flip-flop of the next set of four. Since a set of four flip-flops is used for one decade of decimal counting, the set is often called a *decade-counting unit* (DCU). Figure 4-6 illustrates how four decade-counting units can yield a decimal number with four significant figures (6395).

4C-4 Scalers

From Figure 4-5, and the accompanying discussion, we see that one output pulse is produced from flip-flop *D* for every 16 input pulses. Thus, by cascading flip-flops it is possible to reduce the number of pulses by a known fraction. Decade-counting units can also be cascaded. Here, each unit reduces the number of counts by a factor of 10. The process of reducing a count by a known fraction is called scaling and becomes important when the frequency of a signal is greater than a counting device can accommodate. In this situation, a *scaler* is introduced between the signal source and the counter.

4C-5 Clocks

Many digital applications require that a highly reproducible and accurately known frequency source be used in conjunction with the measurement of time. Generally, electronic frequency sources are based upon quartz crystals that exhibit the *piezoelectric effect* as described in Section 1C-4. The resonant frequency of a quartz crystal depends upon the mass and dimensions of the crystal. By varying these parameters, electrical output frequencies that range from 10 kHz to 50 MHz or greater can be obtained. Typically, these frequencies are constant to 100 ppm. With special precautions, such as precise temperature control, crystal oscillators can be constructed for time standards that are accurate to 1 part in 10 million.

The use of a series of decade scalers with a quartz oscillator provides a precise clock whose frequency can be varied in steps from perhaps 0.1 Hz to 10 MHz.

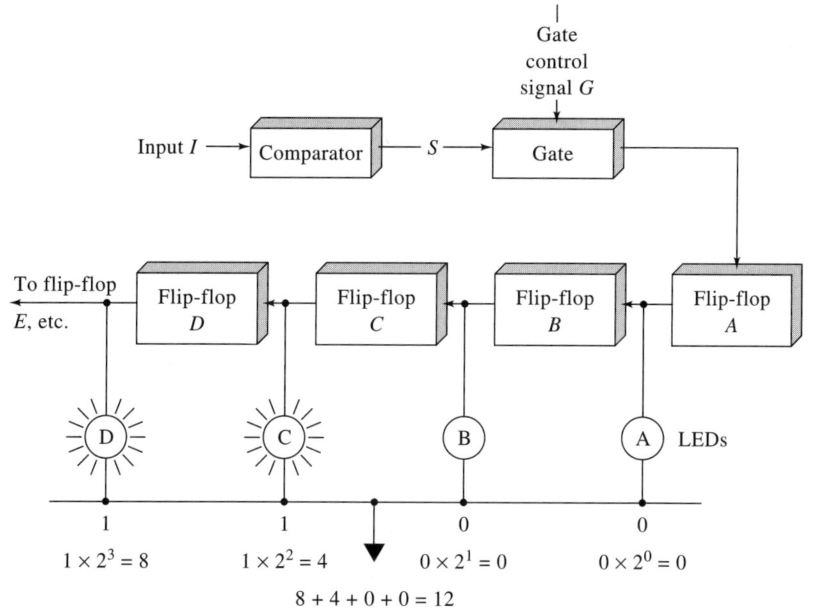

Figure 4-4 A binary counter for numbers 0 through 15. The count shown is binary 1100 or decimal 12.

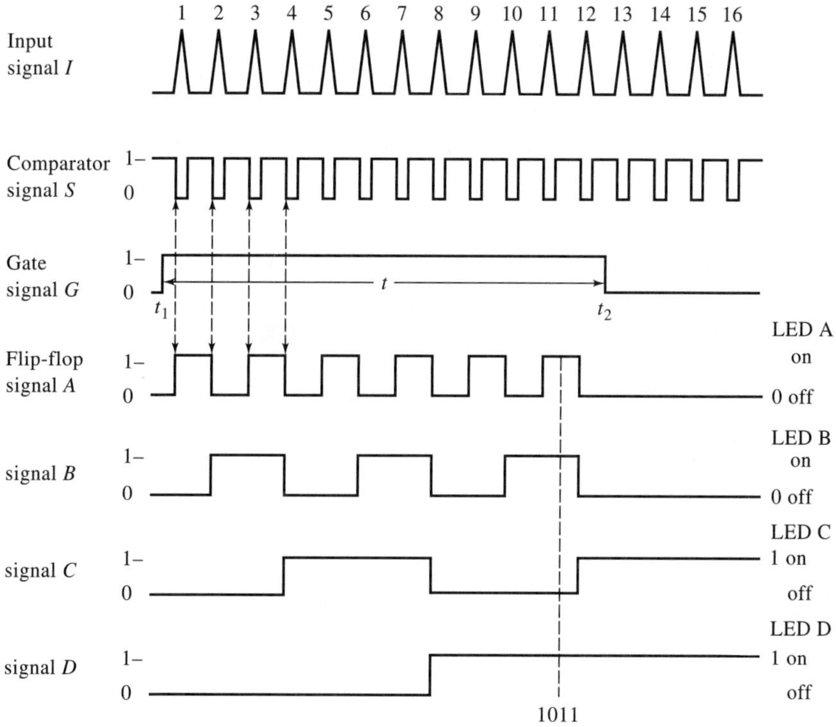

Figure 4-5 Waveforms for the signals at various points in the counter shown in Figure 4-4. Here the count during period t is binary 1011 or decimal 11.

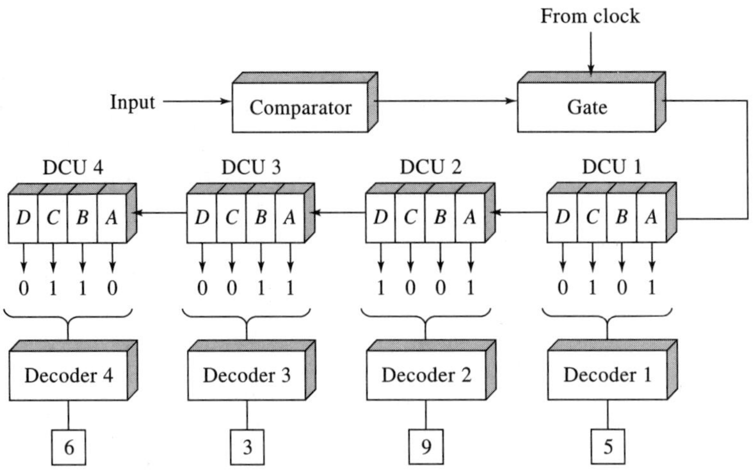

Figure 4-6 A binary-coded decimal counter using four decade-counting units, DCUs.

4C-6 Digital-to-Analog Converters (DAC)

Digital signals are often converted to their analog counterparts for the control of instruments or for display by readout devices such as meters or analog records. Figure 4-7 illustrates the principle of one of the common ways of accomplishing this conversion, which is based upon a *weighted-resistor ladder* network. Note that the circuit is similar to the summing circuit shown in Figure 3-13b, with four resistors weighted in the ratio 8:4:2:1. From the discussion of summing circuits we can show that the output v_{DAC} is given by

$$v_{DAC} = -V_{ref}\left(\frac{D}{1} + \frac{C}{2} + \frac{B}{4} + \frac{A}{8}\right) \qquad (4\text{-}1)$$

where V_{ref} is the voltage associated with logic state 1 and *D, C, B,* and *A* designate the logic states (0 or 1) for a 4-bit binary number in which *A* is the least significant bit and *D* is the most significant one. Table 4-2 shows the analog output from the weighted-resistor ladder shown in Figure 4-7 when *V* is 5 V.

The resolution of a digital-to-analog converter depends upon the number of input bits that the device will accommodate. An *n*-bit device has a resolution of one part in 2^n. Thus, a 10-bit DAC has 2^{10} or 1024 output voltages and, therefore, a resolution of 1 part in 1024, and a 12-bit DAC has a resolution of 1 part in 4096. Note that in our discussion of DACs and ADCs we use the symbol *n* to represent the number of bits of resolution of the device while *N* symbolizes the digital output.

4C-7 Analog-to-Digital Converters (ADC)

The output from most transducers used in analytical instruments is an analog signal. To realize the advantages of digital electronics and computer data processing, it is necessary to convert the analog signal from the analog domain to the digital domain. Figure 4-1 illustrates such a digitization process. Numerous methods are used for this kind of conversion. Two common types of ADC will be described here: the staircase ADC and the successive-approximation ADC.

Staircase Analog-to-Digital Converter

Figure 4-8 depicts a simplified schematic of a device for converting an unknown analog voltage V_i into a digital number *N*. Here, an *n*-bit binary counter, controlled by the signal from a quartz clock, is used to drive an *n*-bit DAC similar to that described in the previous section.

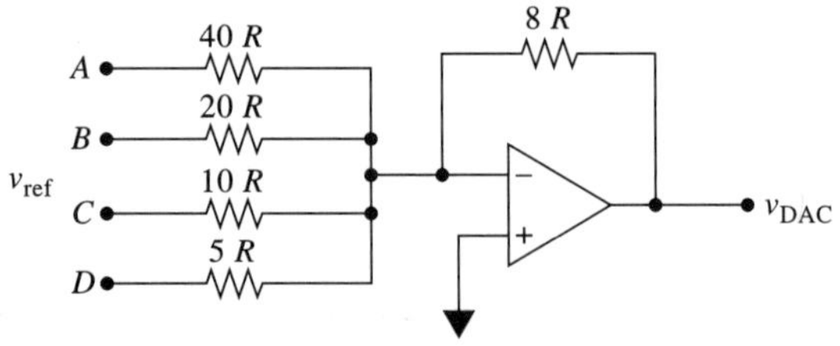

Figure 4-7 A 4-bit digital-to-analog (DAC) converter. Here *A, B, C,* and *D* are +5 V for logic state 1 and are 0 V for logic state 0.

TABLE 4-2 Analog Output from the Digital-to-Analog Converter in Figure 4-7

Binary Number DCBA	Decimal Equivalent	V_a*
0000	0	0 V
0001	1	−1.0 V
0010	2	−2.0 V
0011	3	−3.0 V
0100	4	−4.0 V
0101	5	−5.0 V

*Here, logic state 1 corresponds to +5 V.

The output of the DAC is the *staircase* voltage output v_{DAC} shown in the lower part of the figure. Each step of this signal corresponds to some voltage increment, say 1 mV. The output of the DAC is compared with the unknown input V_i by means of the comparator. When the two voltages become identical within the resolution of the DAC, the comparator shifts from logic state 1 to 0, which in turn stops the counter. The count N then corresponds to the input voltage in units of millivolts. Closing the reset switch sets the counter back to zero in preparation for the conversion of a new voltage, which is begun by opening the reset switch. The input voltage V_i must be held constant during the conversion process to ensure that the digital output corresponds to the desired voltage.

The higher the resolution of the DAC, the more precisely the number will represent v_i. This type of converter clearly illustrates the measurement process. The DAC serves as the reference standard that is compared to v_i by the difference detector (the comparator). The time of conversion is $t_c = Nt_p$, where N is the counter output, which varies with v_i. This conversion time is advantageous if v_i is known to be relatively small most of the time. If v_i is large most of the time, then t_c will be proportionally longer. This type of ADC is often used in nuclear spectroscopy and related fields in which background signals of low intensity are encountered. The oscillator frequency can be as large as 100 MHz when high-speed counters, DACs, and comparators are used.

The operation of the staircase ADC can be made continuous by replacing the simple counter by an up/down counter controlled by the comparator. If v_i increases, the comparator output goes to logic 1 and the counter counts up; and if v_i decreases, the counter counts down. When the DAC output crosses v_i, the counter alternates between N and $N - 1$, a range that is within $\pm\frac{1}{2}$LSB (least-significant bit) of v_i. This type of ADC works well when v_i varies only slowly relative to the conversion time or when a continuous readout is important.

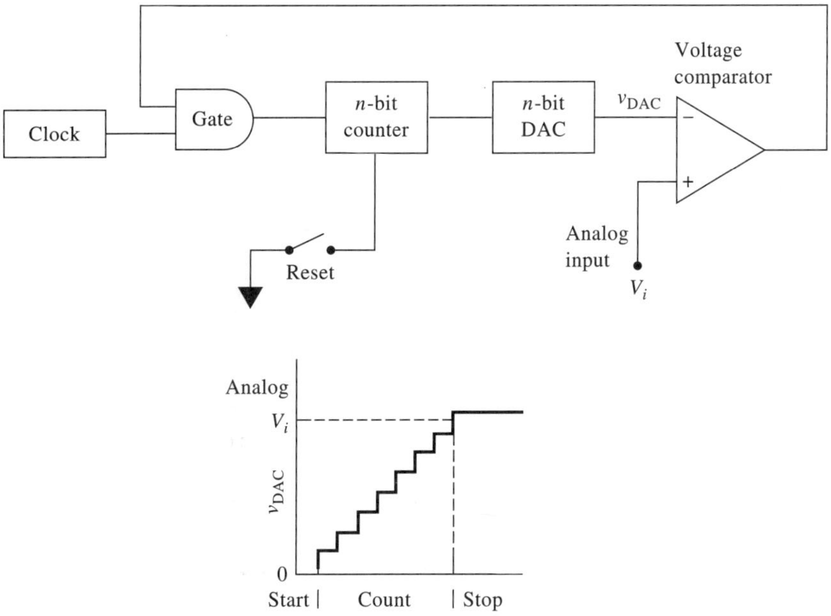

Figure 4-8 A staircase analog-to-digital converter (ADC).

Successive-Approximation Analog-to-Digital Converter

To understand how a successive-approximation ADC works, consider the following question: What is the minimum number of trials necessary to determine, *with certainty,* a number N that lies between 0 and 15? Assume that following each guess you are told whether your number was high or low. The answer is that no more than four guesses are required. To illustrate, suppose that 10 is the target number. Let us first divide the range in half and guess that the unknown number is 7. The number 7 is less than 10, so the upper half of the range is halved again and added to 7 to obtain a second trial number, $N = 7 + 4 = 11$. This number is too large, so 4 is dropped, and half of 4 is added to 7, to yield $N = 9$, which is low. Finally, half of 2 is added to 9 to obtain the target value $N = 10$. The stepwise approach to the final value is shown in the plot of Figure 4-9a. The rules for the successive approximation are as follows:

1. Begin with a guess of one half of full range.
2. If too large, drop guess.
3. If too small, retain guess.
4. Add half of previous increment.
5. Repeat steps 2 through 4 until finished.

Note that n guesses are required to determine a number in the range 0 to 2^{n-1}. For example, 12 guesses are required to determine with certainty a number between 0 and 4095.

The successive-approximation ADC uses exactly the same logic to arrive at a binary or BCD number to represent an unknown voltage V_i as depicted in Figure 4-9b. Here, the 4-bit DAC of Figure 4-7 is used to illustrate how the successive-approximation process can be carried out. Assume that $V_i = 5.1$ V and that all bits are initially set to 0. The first cycle of the oscillator sets the MSB $= 2^3$ to 1, which causes the DAC voltage v_{DAC} to change to 8 V. Because $v_{DAC} > V_i$, the successive-approximation register (SAR) clears the 2^3 bit. The next cycle of the oscillator causes the 2^2 bit to be set to give $v_{DAC} = 4 \text{ V} < V_i$, which causes the comparator output to go to logic 1. The SAR then sets bit $2^2 = 1$, before proceeding to the next cycle, which sets bit $2^1 = 1$, resulting in $v_{DAC} = 6 \text{ V} > V_i$. The comparator output goes to 0, and thus the SAR clears the 2^1 bit. Finally, the SAR sets $2^0 = 1$, to give $v_{DAC} = 5 \text{ V} < V_i$ which keeps bit $2^0 = 1$. The resulting binary number, 0101, represents the input voltage 5 V ± 0.5 V. Note the resolution is $\pm\frac{1}{2}$LSB, which, in this case, is ±0.5 V.

To increase the resolution of the ADC, a DAC of the required resolution must be provided and the SAR must have a correspondingly larger number of bits. Twelve-bit ADCs with input ranges of ±5 V, ±10 V, or 0 to 10 V are typical. Such converters have a fixed conversion time, usually 2 to 8 μs for 12 bits. Successive-approximation converters of this type are widely used

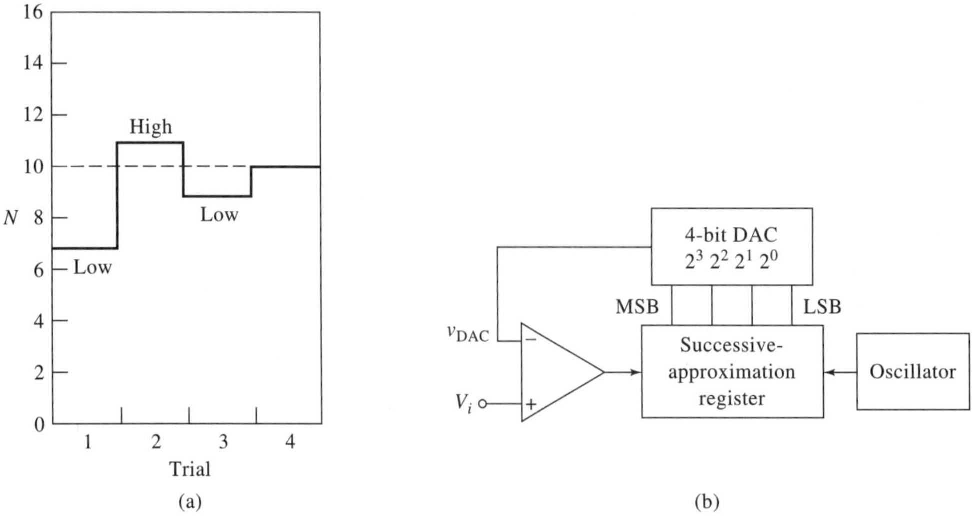

Figure 4-9 Successive-approximation ADC: (a) output of the DAC during the conversion process, (b) block diagram of the ADC.

for computerized timed data acquisition. Because it is important that the voltage to be measured does not vary during the conversion process, a fast sample-and-hold amplifier is almost always used to sample the signal of interest before it is passed to the successive-approximation ADC.

4D MICROPROCESSORS AND MICROCOMPUTERS

A microprocessor is a large-scale integrated circuit made up of hundreds of thousands and even millions of transistors, resistors, diodes, and other circuit elements miniaturized to fit on a single silicon chip with dimensions of a few millimeters on a side. A microprocessor often serves as an arithmetic and logic component, called the *central processing unit* (CPU), of a microcomputer. Microprocessors also find widespread use for the control of the operation of such diverse items as analytical instruments, automobile ignition systems, microwave ovens, cash registers, and electronic games.

Microcomputers consist of one or more microprocessors combined with other circuit components that provide memory, timing, input, and output functions. Microcomputers are finding ever-increasing use for the control of analytical instruments and for processing, storing, and displaying the data derived therefrom. There are at least two reasons for connecting a computer to an analytical instrument. The first is that partial or complete automation of measurements becomes possible. Ordinarily, automation leads to more rapid data acquisition, which shortens the time required for an analysis or increases its precision by providing time for additional replicate measurements to be made. Automation, moreover, frequently provides better and faster control over experimental variables than a human operator can achieve; more precise and accurate data are the result.

A second reason for interfacing computers with instruments is to take advantage of their tremendous computational and data-handling capabilities. These capabilities make possible the routine use of techniques that would ordinarily be impractical because of their excessive computational time requirements. Notable among such applications is the use of Fourier transform calculations, signal averaging, and correlation techniques in spectroscopy to extract small analytical signals from noisy environments.

The interfacing of these devices to instruments is too large and complex a subject to be treated in detail in this text. Thus, the discussion in this chapter is limited to a general summary of computer terminology, the architecture and properties of microprocessors and microcomputers, some useful software packings employed in instrumental applications, and the advantages gained by using these remarkable devices.

4D-1 Computer Terminology

One of the problems that faces the newcomer to the field of computers and computer applications is the bewildering array of new terms and acronyms such as CPU, ALU, PROM, and SSIC. Unfortunately, these terms are not often defined even in elementary presentations. Some of the most important terms and abbreviations are defined here; others will be defined as they appear later in the chapter. A list of the most common of these acronyms is found in Appendix 5.

Bits, Bytes, and Words

Bits are represented in a computer by two electrical states that differ from one another normally by 2 to 5 V. A group of eight bits is called a *byte.* A series of bytes arranged in sequence to represent a piece of data or an instruction is called a *word.* The number of bits (or bytes) per word depends upon the computer; some common sizes include 8, 16, 32, and 64 bits, or 1, 2, 4, and 8 bytes.

Registers

The basic building block of digital computers is the *register,* a physical device that can store a complete byte or a word. A 16-bit binary counter, for example, can serve as a register that is capable of holding a 16-bit word.

Data contained in a register can be manipulated in a number of ways. For example, a register can be *cleared,* a process by which the register is reset to all zeros; the *ones complement* of a register can be taken; that is, every 1 is changed to 0 and every 0 to a 1. The contents of one register can be transferred to another. Moreover, the contents of one register can be added, subtracted, multiplied, or divided by the contents of another. A register in which these processes are performed is often referred to as an *accumulator.* It has been proved that the proper sequence of register operations can solve any computational or informational processing problem no matter how complex provided that there is an *algorithm* for the solution. An algorithm is a detailed statement of the individual steps required to arrive at a solution. One or more algorithms then makes up a *computer program.*

Hardware and Software

Computer *hardware* consists of the physical devices that make up a computer. Examples include disk drives, printers, clocks, memory units, data acquisition modules, and arithmetic logic units. The collection of programs and instructions to the computer, including the disks or tapes for their storage, are the *software*. Hardware and software are equally important to the successful application of computers, and the initial cost of the software may be as great as the cost of the computer. This is especially true of sophisticated software packages designed for special purposes such as data manipulation, curve fitting, or statistical analysis. Over the past few years, the dramatic increase in the availability of fast, high-capacity, low-cost personal microcomputers has produced a corresponding demand for useful and user-friendly software packages. The production and sale of tens or perhaps hundreds of millions of computers worldwide virtually guarantees that a wide variety of software is available at reasonable cost. These market forces have produced lowered costs even for special-purpose scientific software as we shall see in a subsequent section.

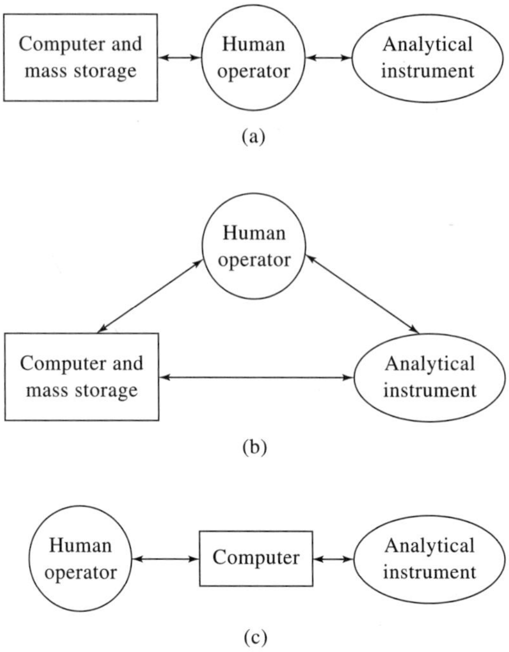

Figure 4-10 Three methods of using computers for analytical measurements: (a) off-line, (b) on-line, (c) in-line.

4D-2 Operational Modes of Computerized Instruments

Figure 4-10 suggests three ways in which computers can be used in conjunction with analytical measurements. In the *off-line* method shown in Figure 4-10a, the data are collected by a human operator and subsequently transferred to the computer for data processing. The *on-line* method of Figure 4-10b differs from the off-line procedure in that direct communication between the instrument and the computer is made possible by means of an electronic *interface* where the signal from the instrument is shaped, digitized, and stored. Here, the computer remains a distinct entity with provision for mass storage of data and instructions for processing these data; off-line operation is also possible with this arrangement.

Most modern instruments are configured as shown in Figure 4-11c. In this *in-line* arrangement, a microcomputer or microprocessor is imbedded in the instrument. Here, the operator communicates with and directs the instrument operation via the computer. The operator does not, however, necessarily program the computer, although often the option is available for doing so. The primary software suite is usually provided with commercial instruments along with a programming language so that the users may program optional modes of data acquisition and manipulation.

In in-line and on-line operations, the data are often transferred to the computer in *real time*, that is, as the data are generated by the instrument. Often, the rate at which data are produced by an instrument is low enough so that only a small fraction of the computer's time is occupied in data acquisition; under this circumstance the periods between data collection can be used for processing the information in various ways. For example, data processing may involve calculation of a concentration, curve smoothing, combining the datum with previously collected and stored data for subsequent averaging, and plotting or printing out the result. *Real-time processing* involves data treatment performed simultaneously with data acquisition. Real-time processing has two major advantages. First, it may reduce significantly the amount of data storage space required, thus making possible the use of a less sophisticated and less expensive computer. Second, if there is sufficient time between the data acquisitions, the processed signal may be used to adjust instrument parameters to improve the quality of subsequent output sig-

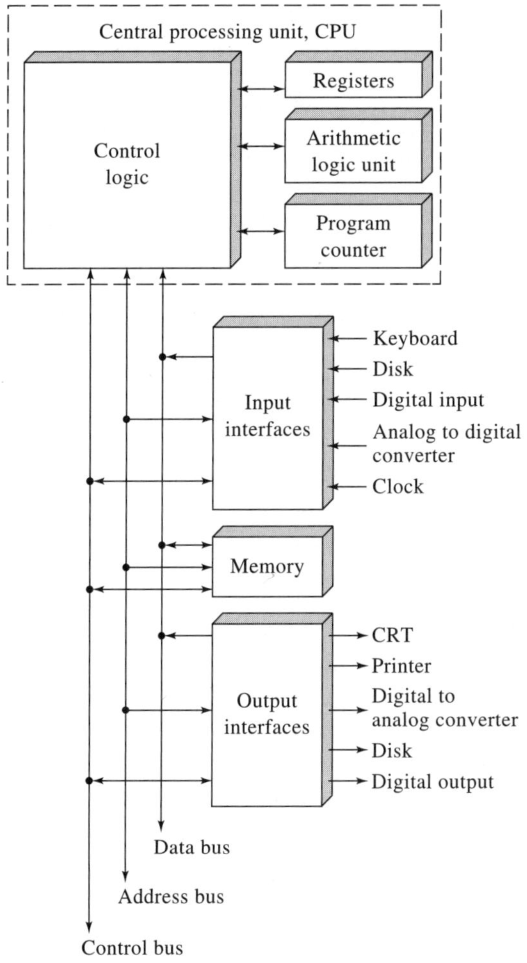

Figure 4-11 Basic components of the digital computer, including peripheral devices.

nals. As the speed and storage capacity of microcomputers has increased, so has the capability for real-time processing.

An example of a real-time processing system is a microprocessor-controlled instrument for automatically performing potentiometric titrations. Usually, such instruments have the storage capacity necessary to store a digitized form of the potential versus reagent volume curve, and all other information that might be required in the process of generating a report about the titration. It is also usually the case that such instruments calculate the first derivative of the potential with respect to volume in real-time and use this information to control the rate at which the titrant is added by a motor-driven syringe. In the early part of the titration, when the rate of potential change is low, the derivative is small, so the titrant is added rapidly. As the equivalence point is approached, the derivative becomes larger, and the computer slows the rate at which the titrant is added. The reverse process occurs beyond the equivalence point.

4E COMPONENTS OF A COMPUTER

Figure 4-11 is a block diagram showing the major hardware components of a computer and its peripheral devices.

4E-1 Central Processing Unit (CPU)

The heart of a computer is the central processing unit, which in the case of a microcomputer is a microprocessor chip. A microprocessor is made up of a control unit and an arithmetic logic unit. The control unit determines the sequence of operations by means of instructions from a program stored in the computer memory. The control unit receives information from the input device, fetches instructions and data from the memory, transmits instructions to the arithmetic logic unit, to the output, and to memory as well.

The arithmetic logic unit, or ALU of a CPU is made up of a series of registers or accumulators in which the intermediate results of binary arithmetic and logic operations are accumulated. A state-of-the-art microprocessor at the time this chapter was written contained three to five million transistors and was capable of executing 100 to 300 million instructions per second.

4E-2 Buses

The various parts of a computer and its memory and peripheral devices are joined by *buses,* each of which is made up of a number of transmission lines. For rapid communication among the various parts of a computer, all of the digital signals making up a word are usually transmitted simultaneously by the parallel lines of the bus. The number of lines in the internal buses of the CPU is then equal to the size of the word processed by the computer. For example, the internal bus for a 16-bit CPU will require 16 parallel transmission lines, each of which transmits one of the 16 bits.

Data are carried into and out of the CPU by a data bus as depicted in Figure 4-11. Both the origin and destination of the signals in the data bus line are specified by the address bus. An address bus with 32 lines can di-

TABLE 4-3 Comparison of Specifications for Three Common Communication Standards

	RS-232	IEEE-488	Ethernet
Configuration	Serial	Parallel	Serial
Distance (m)	30	20	2500
Maximum baud*	19.2 K	10 M	10 M
Cabling	Twisted pair	Shielded bundle	Coaxial

*Baud is a measure of the rate at which information can be transmitted. Baud rate units are bits per second.

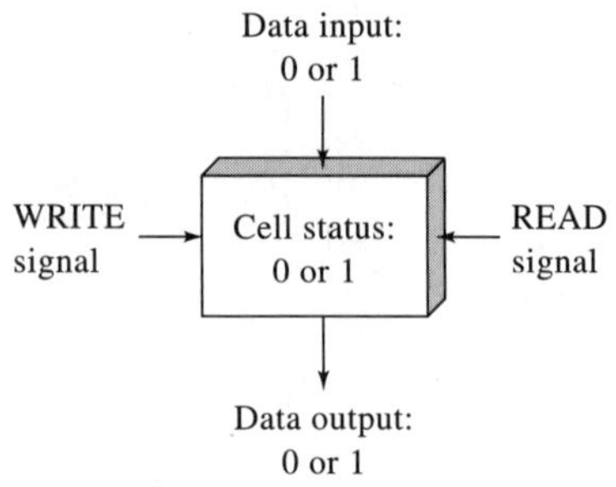

Figure 4-12 An individual computer memory cell for storage of 1 bit.

rectly address 2^{32} or 4,294,967,296 registers or other locations within the computer or 4 gigabytes of memory. The control bus carries control and status information to and from the CPU. These transfers are sequenced by timing signals carried in the bus.

Data must also be transmitted between instrument components or peripheral devices and the CPU; an external bus or communication line is used for this type of data transfer. Table 4-3 summarizes some of the specifications for three popular external communication standards.

4E-3 Memory

In a microcomputer, the memory is a storage area that is directly accessible by the CPU. Because the memory contains both data and program information, the memory must be accessed by the CPU at least once for each program step. The time required to retrieve a piece of information from the memory is called its *access time;* access times are usually in the tens of nanoseconds range.

Memory Chips

The basic unit of a memory chip is a cell, which may have one of two states and which is thus capable of storing one bit of information. Typically, up to a few billion of these cells may be contained on a single silicon memory chip. Figure 4-12 illustrates the functions associated with an individual memory cell. With a READ command from the CPU, the logic state (1 or 0) appears as one of two possible states at the output. A WRITE command allows the 1 or 0 state from the input terminal to displace the datum already present in the cell and to store the new datum in its place.

Individual cells are produced in arrays on memory chips, which in turn are mounted on printed circuit boards that plug directly into the case of the computer. Typically, personal computers have on the order of 8 to 32 megabytes (Mbyte) of memory, but many different configurations are usually available. Because the actual process of addressing and storing information in memory is either established during manufacturing or controlled by the CPU, most chemists have little need for an understanding of the detailed design of memories. However, a little experience with the terminology used when describing memories is often helpful in selecting the proper amount of memory for performing a particular computational task.

Types of Memory

There are two types of memory in most microcomputers: *random access memory* (RAM) and *read only memory* (ROM). The term *random access* is somewhat misleading, because ROM may also be accessed randomly. The term random access means that all locations in the memory are equally accessible and can be reached at about the same speed. Thus, *read/write memory* is a more descriptive term for RAM. Earlier types of semiconductor RAM were *volatile;* that is, the information was not retained unless the memory was refreshed regularly. Many RAM boards now have battery backup power supplies that can prevent the loss of any information if power is lost for eight hours or more. This sort of memory is similar to that found in pocket calculators that retain data and instructions even when turned off.

Read-only memories contain permanent instructions and data that were placed there at the time of their manufacture. These memories are truly static in the sense that they retain their original states for the life of the computer or calculator. The contents of a ROM cannot be altered by reprogramming. A variant of ROM is the *erasable program read only memory* (EPROM or erasable PROM) in which the program contents can be erased by exposure to ultraviolet radiation. After this treatment, the memory can be reprogrammed by means of special equipment. Also available are ROMs that can be reprogrammed by relatively straightforward logic signals. These ROMs are designated EAROMs (electrically alterable ROMs). Bootstrap programs that perform the initialization of a computer system when it is turned on are usually stored in some type of ROM. When system information must be stored when power is off but must occasionally be reprogrammed when new devices are installed in the computer, such data are usually stored in a battery-powered RAM.

In microcomputers and handheld calculator systems, ROM devices are used to store programs needed for performing various mathematical operations, such as obtaining logarithms, exponentials, and trigonometric functions; calculating various statistical quantities such as means, standard deviations, and least-squares parameters; and formatting data in fixed point, scientific notation, or engineering notation.

Bulk Storage Devices

In addition to semiconductor memories, computers are usually equipped with bulk storage devices. Magnetic tapes were for years the primary means for bulk storage, but tapes have now been largely replaced by diskettes and hard disks. Tapes are now used mostly to backup hard disks. Disk storage capacities are constantly increasing. Depending on the type of diskette drive used, capacities in the 1.4 Mbyte to 100 Mbyte range are common. Removable hard drives of several hundred Mbyte capacity as well as a type of fast 100-Mbyte floppy disk called a ZIP® drive are now widely available at reasonable cost. The smallest hard disk drives have capacities in the 300- to 500-Mbyte range, but hard disks capable of storing several gigabytes are now common. The time required to reach a randomly selected location on a disk is the *access time,* and for most hard disks the access time is on the order of 10 to 30 ms. The CD ROM is a particularly attractive storage device for large databases, encyclopedias, and other reference data because of its large storage capacity of about 750 Mbyte. At present, disk drives that are capable of writing CD ROMs are relatively expensive; but as the technology improves, it is expected that CD ROM drives or other similar high-capacity devices for both reading and writing storage media will be widely available and affordable.

4E-4 Input/Output Systems

Input/output devices provide the means by which the user, or attached instruments, communicate with the computer. Familiar input devices include keyboards, magnetic tapes or disks, CD ROMs, and the transduced signals from analytical instruments. Output devices include records, printers, plotters, cathode-ray tubes, and magnetic tapes or disks. It is important to understand that many of these devices provide or use an *analog* signal while, as we have pointed out, the computer can respond only to digital signals. Thus, an important part of the input/output system is an *analog-to-digital converter* for providing data in a form that the computer can use, and a *digital-to-analog converter* for converting the output from the computer to a usable analog signal.

An important piece of hardware in the computerized acquisition, analysis, and output of analytical information is the data acquisition module. These devices may be plugged directly into the bus of a microcomputer and provide a way to acquire data by means of analog-to-digital conversion to feed back data to an experiment by means of ADC, to provide critical timing in the process, and to transfer digital data directly to the computer. Figure 4-13a shows a block function diagram of a typical data acquisition module. It includes an ADC, a DAC, a programmable-gain amplifier, digital input and output lines, memory for temporarily holding collected data, and a precision real-time clock for critical timing of the data acquisition. The entire data acquisition process is carried out by a powerful on-board microprocessor, which receives user-selected instructions from the primary microcomputer to which it is attached. A photo of the data acquisition module is presented in Figure 4-13b.

The main computer carries out data analysis and long-term bulk data storage by using any of a variety of software packages such as National Instruments' LabView® or LabWindows®. The operator controls the overall process through interactions with the main computer. Using microcomputers and data acquisition modules such as these, virtually any instrument or experiment may be interfaced with a computer for efficient analysis and output of instrumental data.

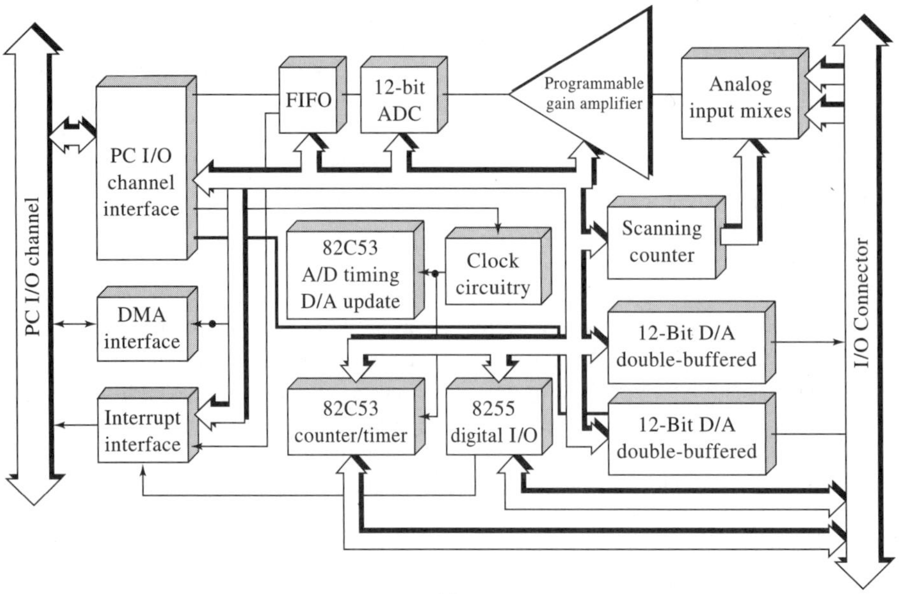

(a)

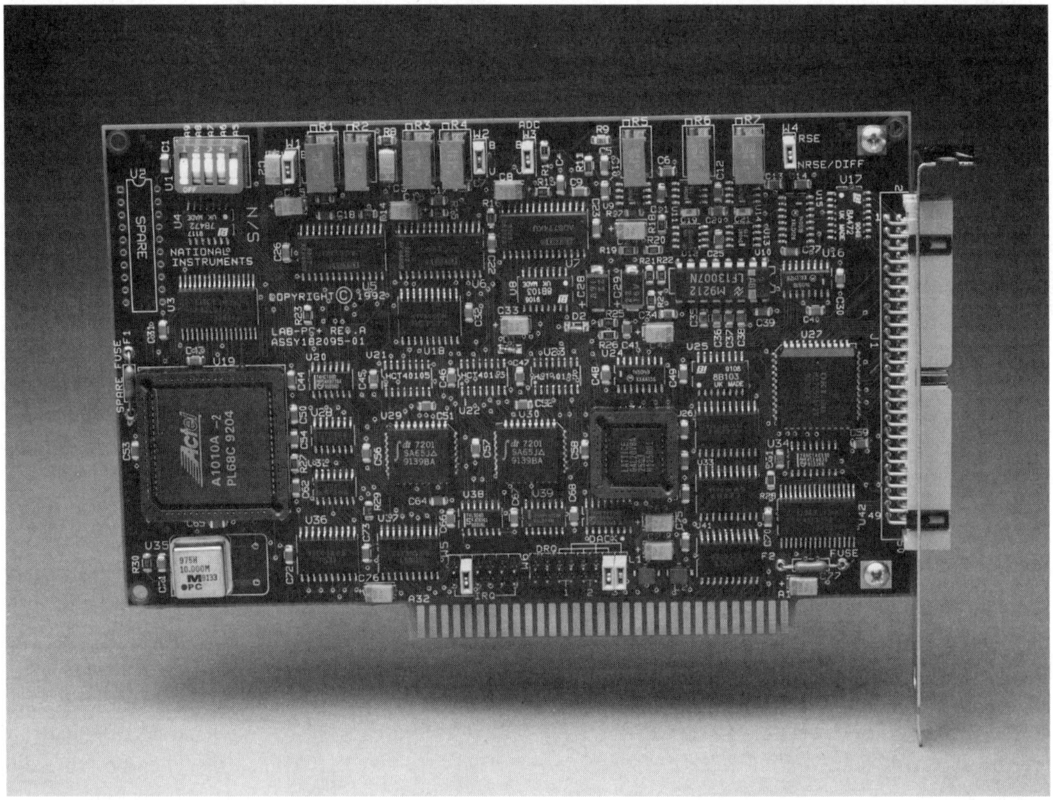

(b)

Figure 4-13 Computer data-acquisition module: (a) block diagram of the module, (b) photo of the printed circuit board containing the module. *(Reprinted with permission of the copyright owner, National Instruments Corporation.)*

4F COMPUTER SOFTWARE[2]

4F-1 Programming

Communication with a computer entails setting an enormous collection of electronic switches to appropriate off or on (0 and 1) states. A program consists of a set of instructions that indicate how these switches are to be set for each step in a program. These instructions must be written in a form to which the computer can respond—that is, a binary *machine code.* Machine coding is tedious and time consuming and is prone to errors. For this reason, *assembly languages* have been developed in which the switch-setting steps are assembled into groups, which can be designated by mnemonics. For example, the mnemonic for subtract might be SUB and might correspond to 101 in machine language. Clearly, it is a good deal easier for the programmer to remember SUB than 101.

Assembly programming, while simpler than machine programming, is still tedious. As a consequence, many high-level languages, such as FORTRAN, BASIC, APL, PASCAL, FORTH, and C, have been developed. These languages have been designed to make communication with the computer relatively straightforward. Here, instructions in the high-level language are translated by a computer program, called a *compiler,* into assembly language and ultimately into machine language, which can then be read and interpreted by the computer. Prior to about 1985, the use of higher-level languages involved a trade-off between available system memory and the time required to program the computer in assembly language because assembly language programming is considerably more efficient with respect to memory than is programming in higher-level languages. However, for more than two decades, the capacity of computer memory has doubled every two years, and the cost per bit of storage has decreased. Thus, the choice of a programming language is largely a matter of taste, of skill, and of the suitability of a given language for a given task. For detailed control of a computer and its hardware, assembly language is essential. For routine data input and output and higher-level functions of hardware, C is the language of choice. For scientific computations, FORTRAN was designed and written nearly 40 years ago and is still the most commonly used language in scientific applications. Vast, well-characterized libraries of scientific FORTRAN subroutines are commonplace.[3] Figure 4-14 illustrates the application of FORTRAN, C, and assembly language to obtaining a sum.

4F-2 High-Level Software Applications

One by-product of the age of personal computers is the wide variety of useful programs accessible to students, teachers, and practicing scientists. Computer languages such as BASIC, FORTRAN, and C, word processors, grammar checkers, spreadsheets, desktop organizers, data-base management systems, Internet browsers, and other productivity tools have appeared in large numbers. The advent of the graphical user interface, or GUI, made popular by Apple Computer in its line of Macintosh computers and perpetuated in Microsoft Windows®, has made highly functional user-friendly software common-

[2] R. E. Dessy, *Anal. Chem.,* **1983,** *55,* 650A, 756A.

[3] W. H. Press, S. A. Teukolsky, William T. Vetterling, and B. P. Flannery, *Numerical Recipes in FORTRAN: The Art of Scientific Computing,* 2nd ed. New York: Cambridge, 1992. These routines are available in versions for FORTRAN 77, FORTRAN 90, C, and Pascal.

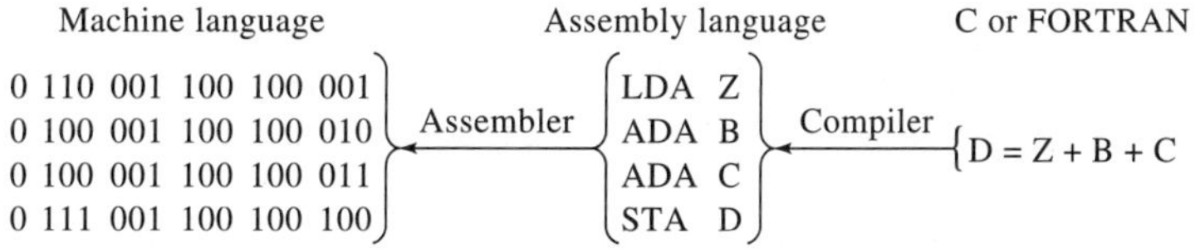

Figure 4-14 Relationships among machine language, assembly language, and the high-level languages FORTRAN and C. (LDA Z = Load the value of Z into Register A; ADA B = add the value of B to the content of Register A; ADA C = add the value of C to the content of Register A; STA D = store the content of Register A as D.)

place. A high-level software application that is particularly useful to chemists is the computer spreadsheet. Microsoft Excel®, Lotus 1-2-3®, and Quattro Pro® are but three of the most prominent examples.

Spreadsheets

Spreadsheets were originally designed as a tool for business, but as they became popular and easy to use, people from all walks of life began to use them to carry out numerical computations in many fields, including science. When software manufacturers realized the diverse nature of their clientele, they began to add highly sophisticated and specialized functions to their spreadsheets. For example, Microsoft Excel® now contains many functions that can be used to save steps when developing complex statistical or engineering analyses. These relationships include basic statistical functions such as mean, standard deviation, median, mode, and histogram; advanced statistical functions such as skew, kurtosis, and various distribution functions; and other advanced functions such as linear and nonlinear least-squares, χ^2 analysis, and analysis of variance. Figure 4-15 shows a Microsoft Excel® spreadsheet for analyzing and plotting spectral data files. Columns A and B of the spreadsheet contain data from the infrared spectrum of 2,2′-bipyridine. The data were collected on a Fourier transform IR spectrometer, stored on a floppy disk, and imported as a text file into Excel® running on a second computer.

Equation Solvers

A tool of special interest to chemists is the equation solver, a program that quickly solves complex mathematical equations such as those found in the study of multiple equilibria. Several equation solvers are available, including TK Solver Plus®, Mathematica®, Maple®, and Mathcad®. All of these programs have been highly rated, and thus the choice of which to use depends upon the tasks at hand and the resources available. Because of its low cost, power, ease of use, and intuitive nature for representing complex mathematical expressions, Mathcad® has become popular in the physical sciences and in engineering for solving a tremendous variety of computation problems from basic statistical analysis to eigen value–eigen vector problems in quantum chemistry.

Figure 4-16 illustrates the application of Mathcad® to statistical analysis of a large set of data that has been acquired by an analytical instrument or typed in by the user. The mean and standard deviation of the data are automatically calculated and displayed in a scatter plot to reveal systematic errors in the data. The mean and standard deviation are plotted as horizontal lines in the graph. Mathcad® application packages for a variety of science and engineering problems and for learning purposes may be purchased or downloaded from the Internet. For more examples of Mathcad® documents and examples relevant to analytical chemistry, see the reference below.[4]

Scientific Packages

A number of software packages have been developed specifically for use in chemistry and related sciences. Programs are available for tasks as diverse as drawing organic molecular structures such as ChemWindows® and ChemDraw®, for performing thermodynamic calculations such as HSC Chemistry for Windows®, for general plotting of scientific data such as Jandel Scientific's SigmaPlot® and Mathsoft's Axum®, for curve fitting such as Jandel Scientific's TableCurve® and PeakFit®, and spectroscopic data analysis and plotting such as Galactic Industries' GRAMS/32®.

To illustrate the utility of this variety of software for data analysis, let us consider GRAMS/32®. The name GRAMS/32® stands for graphic relational array management system, and it also refers to spectrograms and chromatograms. The 32 in the name indicates that it is optimized for operation on 32-bit microcomputers such as the Intel Pentium®. GRAMS/32® is capable of reading, analyzing, and translating data files generated by more than one hundred different chemical instruments and other software packages, including spectrometers, chromatographs, and other instruments. Data files may be translated to and from these file formats, which include ASCII format, comma-delimited format, and various spreadsheet formats as well as standard spectroscopic formats such as JCAMP.

Fourier transformation, data smoothing, digital filtering, deconvolution, curve fitting, and various mathematical operations such as differentiation, integration, interpolation, and conversions between absorbance and transmittance may be easily accomplished with GRAMS/32®. Some of these functions will be discussed in Chapter 5. GRAMS/32® can provide many

[4] F. J. Holler, *Mathcad Applications for Analytical Chemistry,* Philadelphia: Saunders College Publishing, 1994.

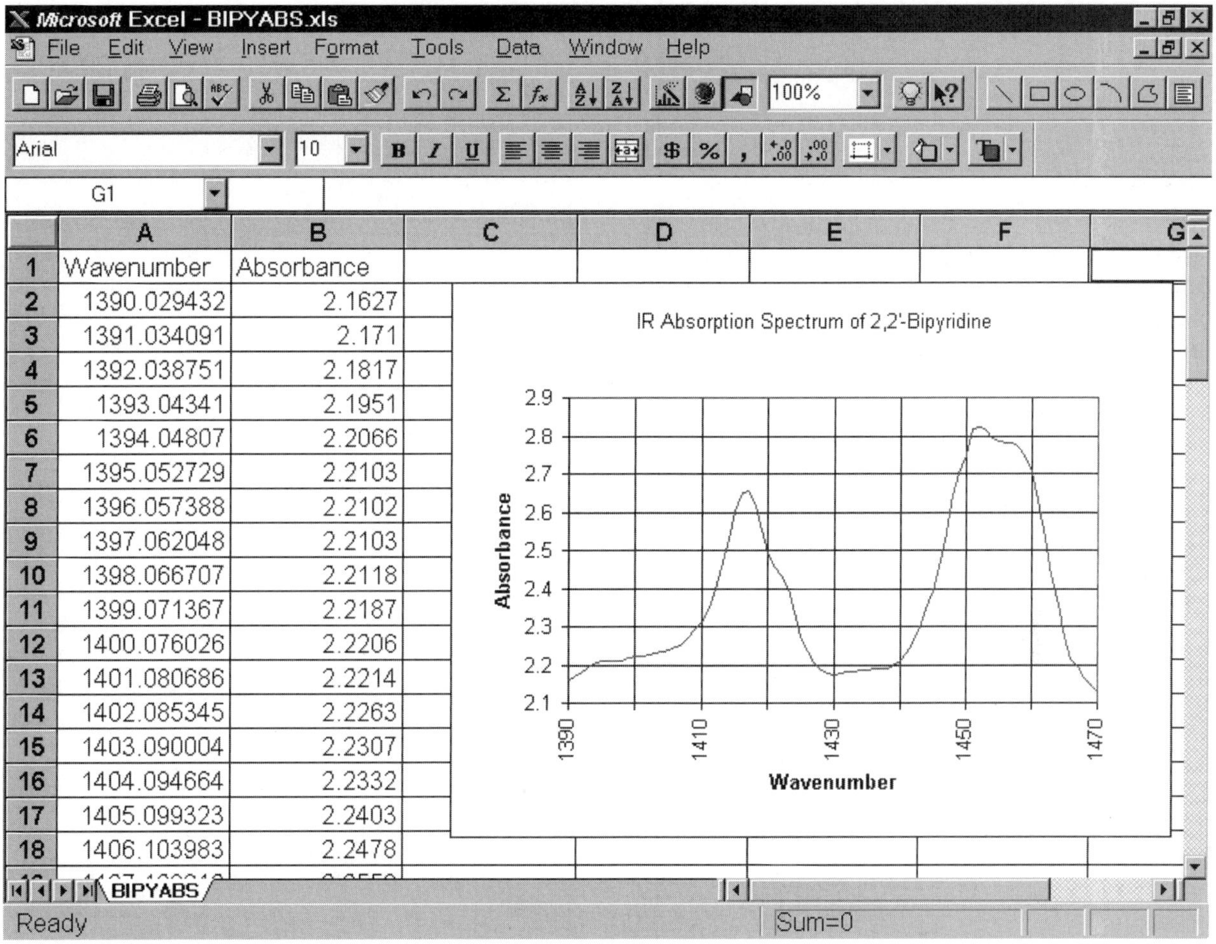

Figure 4-15 Computer screen display of an Excel® spreadsheet containing data from the absorption spectrum of 2,2′-bypyridine.

different types of data plots, including stacked plots, multiple plots, three-dimensional plots, and contour plots for the analysis and interpretation of chemical data. Figure 4-17 depicts a sample plot of data obtained from a UV-visible diode array detector coupled with a chromatograph. The three-dimensional data are plotted three ways: as a contour plot of the entire data array across the bottom half of the screen, as a chromatographic trace taken at 704.84 nm on the upper left of the screen, and as a spectral trace taken at 10.524 minutes into the chromatographic experiment on the upper right of the screen. The crosshairs shown in the contour plot are placed by the analyst to indicate where the spectral and chromatographic traces should be taken. Such software tools are indispensable in finding the spectra of individual components of a mixture of compounds separated by chromatography. GRAMS/32® and other similar packages are rapidly becoming mainstays for efficient analysis in the repertoire of the chemist's tools.

4G APPLICATIONS OF COMPUTERS[5]

Computer interactions with analytical instruments are of two types, *passive* and *active*. In passive applications, the computer does not participate in the control of the experiment but is used only for data handling, processing, storing, file searching, or display. In an active interaction, the output from the computer controls the sequence

[5] A. P. Wade and S. R. Crouch, *Spectroscopy,* **1988,** *3*(10), 24.

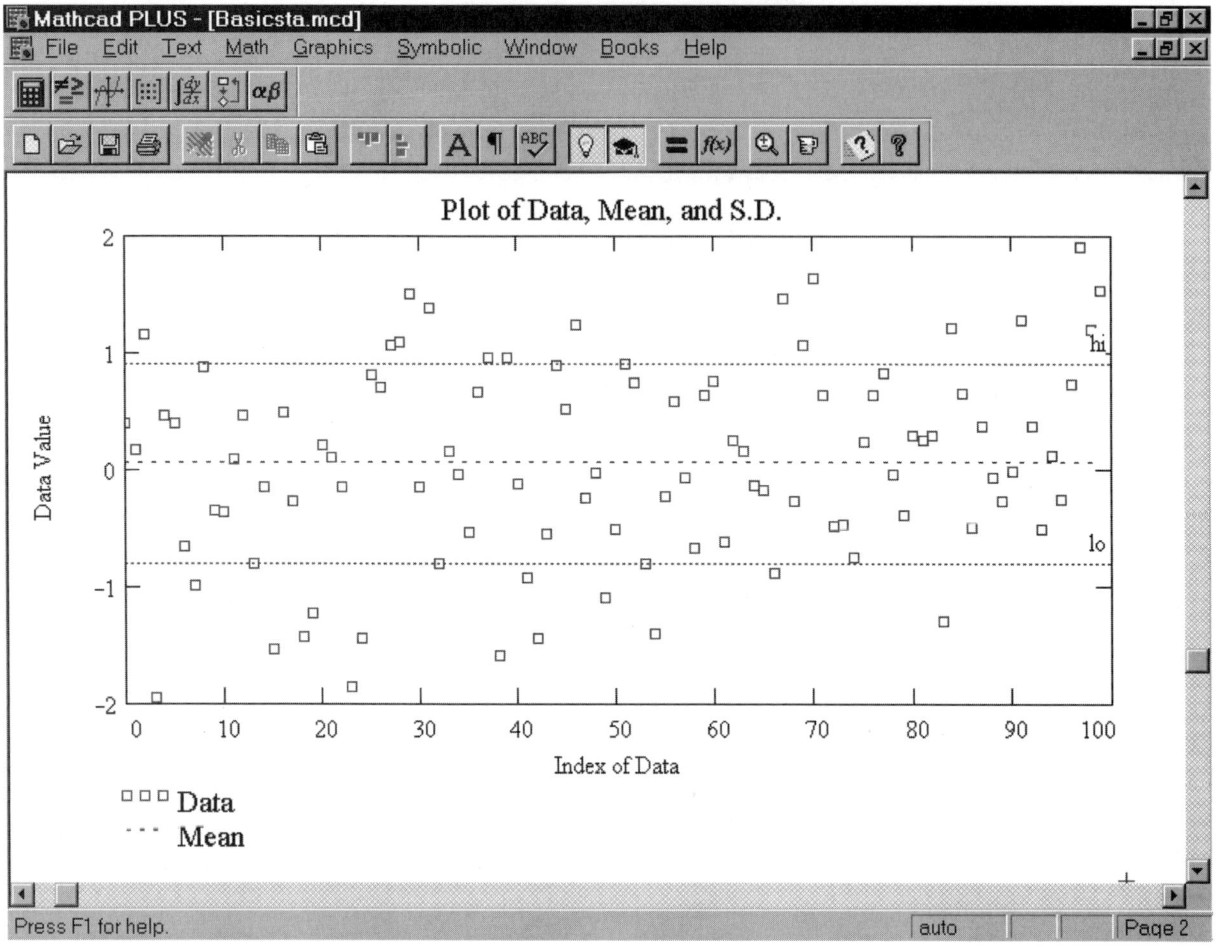

Figure 4-16 Computer screen display of a Mathcad® document for basic statistical analysis of experimental data. *(Software courtesy of Mathsoft, Inc.)*

of steps required for operation of the instrument. For example, in a spectroscopic determination, the computer may choose the proper source, cause this source to be activated and its intensity adjusted to an appropriate level, cause the radiation to pass through the sample and then a blank, control the monochromator so that a proper wavelength is chosen, adjust the detector response, and record the intensity level. In addition, the computer may be programmed to use the data as it is being collected to vary experimental conditions in such a way as to improve the quality of subsequent data. Instruments with computer control are said to be *automated.*

4G-1 Passive Applications

Data processing by a computer may involve relatively simple mathematical operations such as calculation of concentrations, data averaging, least-squares analysis, statistical analysis, and integration to obtain peak areas. More complex calculations may involve the solution of several simultaneous equations, curve fitting, averaging, and Fourier transformations.

Data storage is another important passive function of computers. For example, a powerful tool for the analysis of complex mixtures results when gas chromatography (GC) is linked with mass spectrometry (MS). Gas chromatography separates mixtures on the basis of the time required for the individual components to appear at the end of a suitably packed column. Mass spectrometry permits identification of each component according to the mass of the fragments formed when the compound is bombarded with one of a number of different types of particles such as electrons. Equipment for GC/MS may produce data for as many as 100 spectra in a few minutes,

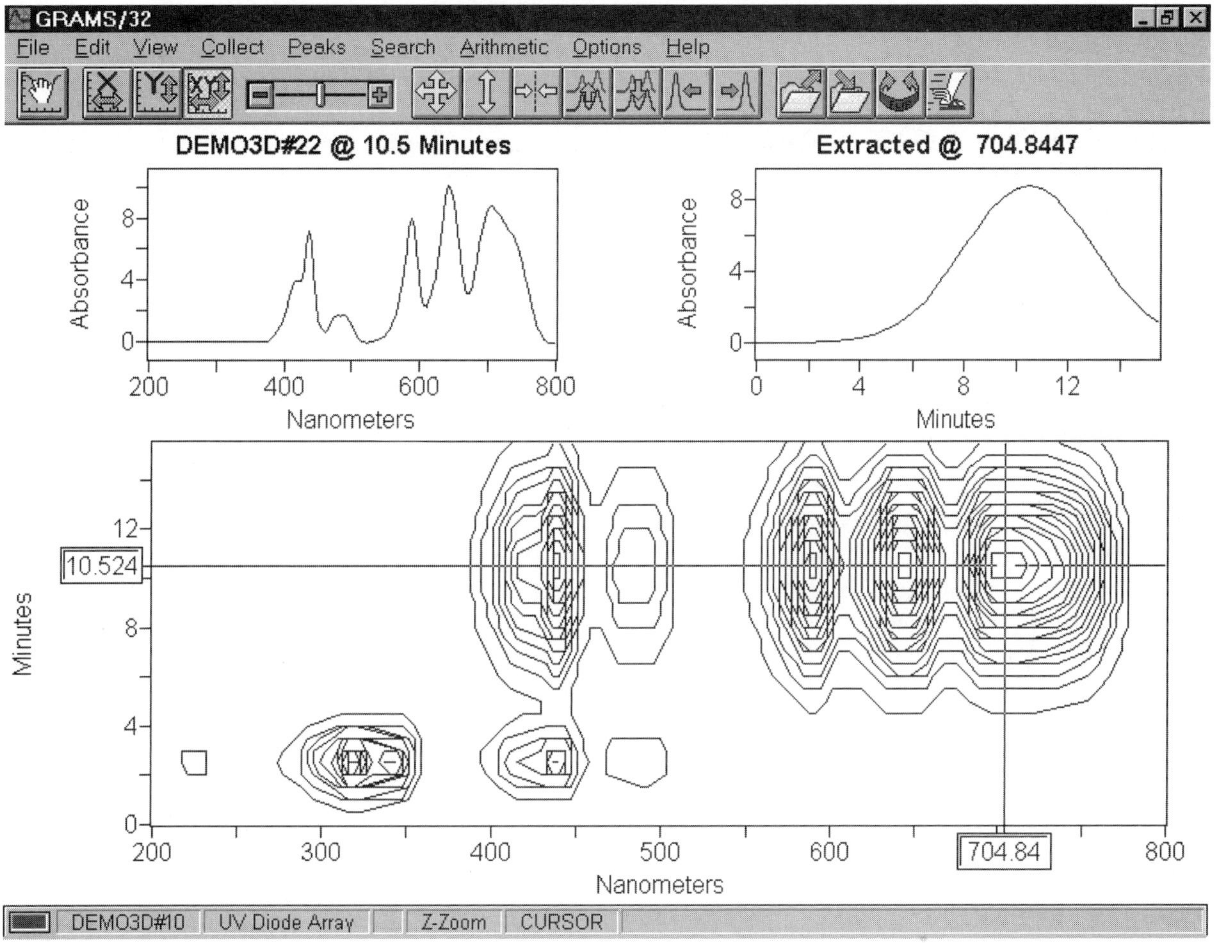

Figure 4-17 Computer screen display of a 2-dimensional spectral chromatogram using GRAMS/32® data-analysis software. *(Software courtesy of Galactic Industries, Inc.)*

with each spectrum being made up of tens to hundreds of peaks. Conversion of these data to an interpretable form (a graph) in real time is often impossible. Thus, the data are usually stored in digital form for subsequent processing and presentation in graphical form.

Identification of a species from its mass spectrum involves a search of files of spectra for pure compounds until a match is found; manually this process is time consuming but can be accomplished quickly by using a computer. Here, the spectra of pure compounds, stored on a hard disk, are searched until spectra are found that are similar to the analyte. Several thousand spectra can be scanned in a minute or less. Such a search usually produces several possible compounds. Further comparison of the spectra by the scientist often makes identification possible.

Another important passive application of the power of computers in GC/MS utilizes the high-speed data fetching and correlating capabilities of the computer. Thus, for example, the computer can be called upon to display on a cathode-ray screen the mass spectrum of any one of the separated components after the component has exited from a gas chromatographic column.

4G-2 Active Applications

In active applications only part of the computer's time is devoted to data collection, while the rest is used for data processing and control. Thus active applications are real-time operations. Most modern instruments contain one or more microprocessors that perform control functions. Examples include adjustment of (1) the slit width and wavelength settings of a monochromator, (2) the temperature of a chromatographic column, (3) the potential applied to an electrode, (4) the rate of addition of a reagent, and (5) the time at which the integration of a peak is to begin. Referring again to the GC/MS instrument considered in the last section, a computer is often used to initiate the collection of mass spectral data each time a compound is sensed at the end of the chromatographic column.

Computer control can be relatively simple, as in the examples just cited, or more complex. For example, the determination of the concentration of elements by atomic emission involves the measurement of the heights of emission peaks, which are found at wavelengths that are characteristic for each element. Here, the computer can cause a monochromator to rapidly sweep a range of wavelengths until a peak is detected. The rate of sweep is then slowed to better determine the exact wavelength at which the maximum output signal is obtained. Intensity measurements are repeated at this point until an average value is obtained that gives a suitable signal-to-noise ratio. The computer then causes the instrument to repeat this operation for each peak of interest in the spectrum. Finally, the computer calculates and prints out the concentrations of the elements present.

Because of its great speed, a computer can often control variables more efficiently than can a human operator. Furthermore, with some experiments, a computer can be programmed to alter the way in which the measurement is being made, according to the nature of the initial data. Here, a feedback loop is used in which the signal output is converted to digital data and fed back through the computer and serves to control and optimize the way in which later measurements are performed.

4H COMPUTER NETWORKS

The connection of two or more computers produces a computer network or simply a *network.* It is possible to significantly increase the efficiency with which information can be transmitted and manipulated, if the communication between computers is software controlled. Networks encompass an enormous number of possible interactions between computers, but they can probably be classified into one or two major classes. First, there is the Internet, which is capable of rapidly transmitting digital representations of an almost unbelievable variety of textual, graphical, audio, and video information throughout the world. The Internet evolved out of a 1960 government project called ARPANET that was developed beginning in the early 1960s at the behest of United States military to provide rapid communication for government and military computer science and engineering researchers. As the personal computer appeared in the early 1980s and users began to realize the power and convenience of network technology, universities and private companies began to connect to the network. When the government dropped support of the network in 1989, a plan was formulated for further commercial development and support of the Internet. The continued development of the personal computer and the technology for connecting to networks, the potential for commercial uses of the network, as well as the popularity of electronic mail and the World Wide Web have led to the massive Internet of today. The Internet is expected to evolve into the so-called Information Superhighway, which will eventually bring Internet technology into virtually every home via high-speed (hundreds of megabits per second) cables or telephone lines. Eventually much of the world's information, including scientific data, journals, and other types of reports, will be available on the Internet.

Parallel with the development of the Internet has been the advent of *Local Area Networks,* or LANs.[6] LANs are found in almost every organization, including scientific laboratories in universities, government agen-

[6] R. E. Dessy, *Anal. Chem.,* **1982,** *54,* 1167A, 1295A.

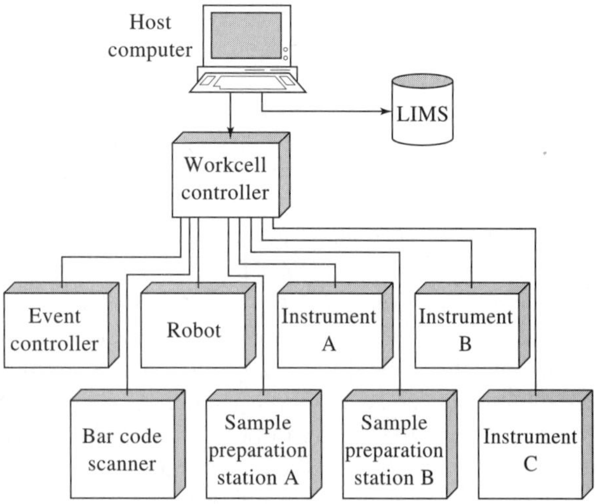

Figure 4-18 Generic hardware configuration of a LAN. *(Reprinted with permission from B. Fowler,* Amer. Lab., ***1988,*** *19(9), 62. Copyright 1988 by International Scientific Communications, Inc.)*

cies, and private companies. The efficient use of LANs in a scientific context requires that they be specifically designed to meet the needs of a particular laboratory. In general LANs provide a common communication network that makes possible the efficient transmission of information between computerized instruments, input/output devices, and a variety of different computers. Figure 4-18 illustrates a generic LAN configuration. Note that although they are not illustrated, all of the building blocks of this system communicate with the work cell controller (a computer responsible for data transmission) through a microprocessor or personal computer. A *Laboratory Information Management System* (LIMS)[7] is included in the upper right corner of this figure. Data handling is one of the major tasks in any laboratory, and a well-designed LIMS will keep track of

[7] R. E. Dessy, *Anal. Chem.*, **1983**, *55*, 70A, 277A; R. Megargle, *Anal. Chem.*, **1989**, *61*, 612A.

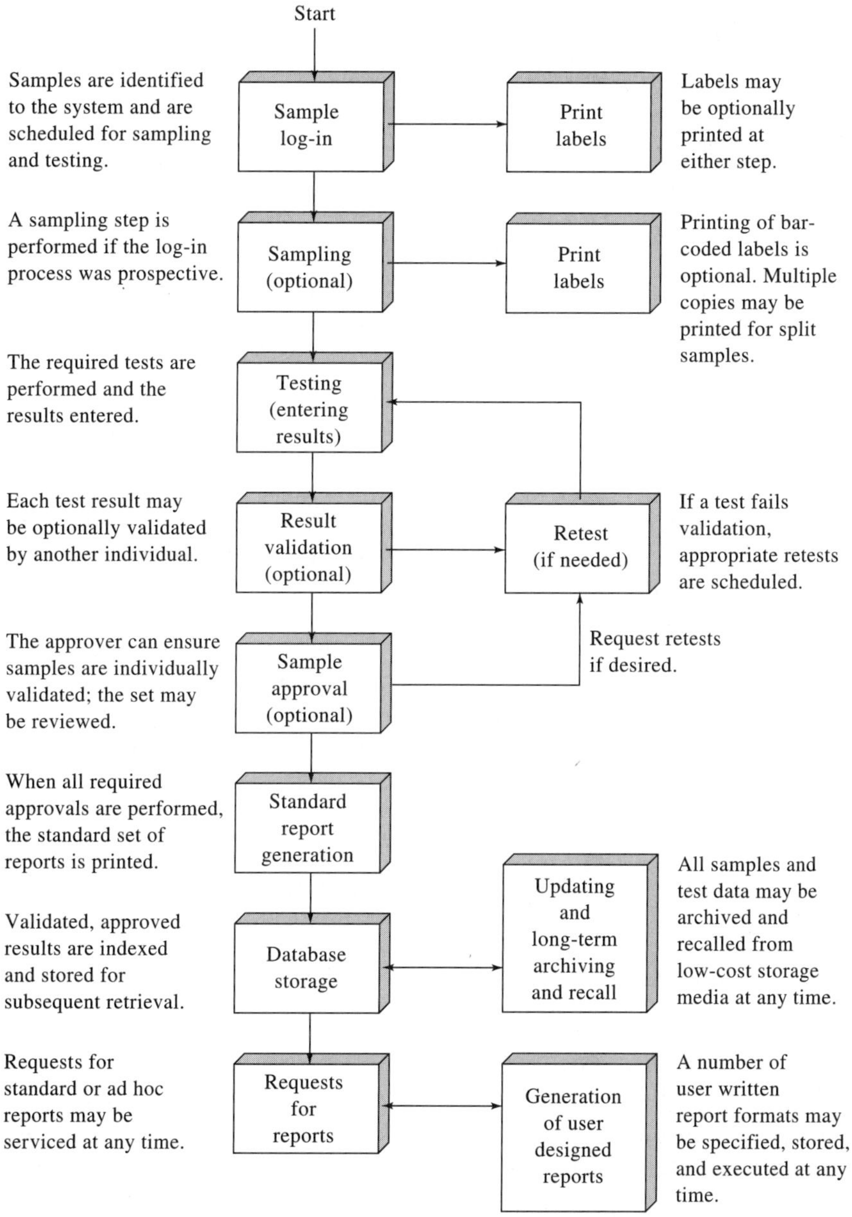

Figure 4-19 LIMS data and sample management overview. *(Reprinted with permission from F. I. Scott,* Amer. Lab., ***1987****, 19(11), 50. Copyright 1987 by International Scientific Communications, Inc.)*

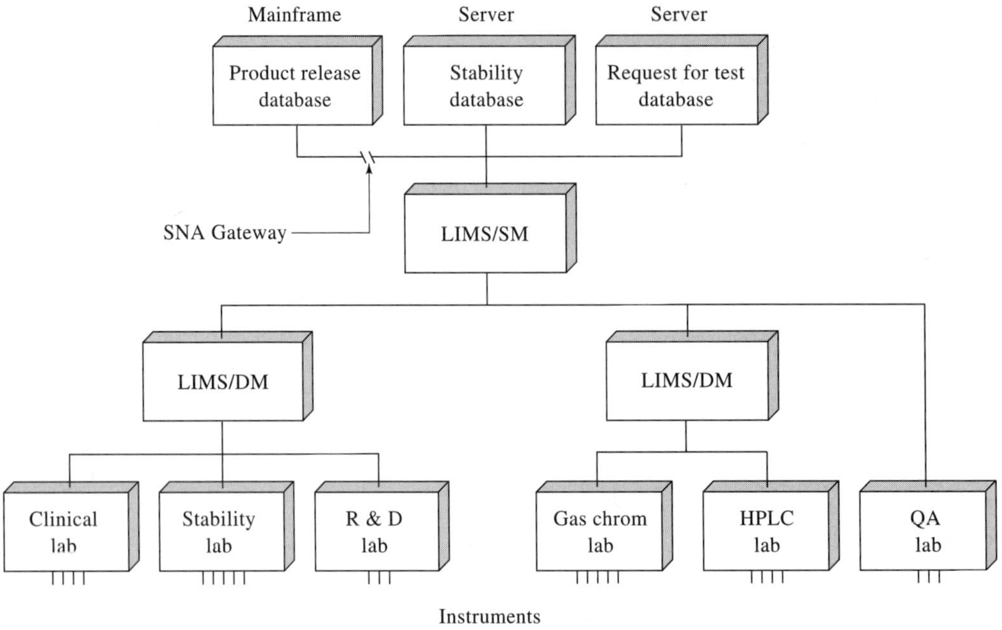

Figure 4-20 Block diagram of an entire automated laboratory system. *(Reprinted with permission from E. L. Copper and E. J. Turkel,* Amer. Lab., ***1988,*** *20(3), 42. Copyright 1988 by International Scientific Communications, Inc.)*

all of the information about all of the samples and projects that have been completed or are in progress. Figure 4-19 summarizes many of the processes that might be controlled by a LIMS in a testing laboratory and provides an overview of some of the options that might be exercised as a sample is processed. Finally, Figure 4-20 is a block diagram of a computer system designed to totally automate an entire laboratory. Note that at the bottom of this figure entire laboratories are designated by boxes; within each of these laboratories a local area network would be used to coordinate activities and communicate with the next level in the hierarchy. In this system we see that two different types of LIMS are used; those designated /DM are standard for data management LIMS, while the /SM designation stands for system/sample management. Essentially the only difference between these coordinating computers, or servers, is the software that controls the communication and the data handling. The SNA Gateway represents a means of connecting this laboratory's cluster of computers with the primary server at the corporate headquarters.

4I QUESTIONS AND PROBLEMS

4-1 Convert each of the following decimal numbers to its binary equivalent.
(a) 24 (b) 79 (c) 136 (d) 581

4-2 Convert each of the following binary numbers into its decimal equivalent.
(a) 101 (b) 10101 (c) 11100010 (d) 1101001001

4-3 Three ADCs (analog-to-digital converters) all have a range of 0 to 10 V. What is the maximum uncertainty in the digitization of a 10-V signal, if the converters have
(a) 8 bits? (b) 12 bits? (c) 16 bits?

4-4 Repeat Problem 4-3 if a 1-V signal is being digitized with the same three converters.

4-5 The maximum percentage error of a voltage processed by an ADC is given by the following equation:

$$\text{max \% error} = (\text{maximum uncertainty/measured voltage}) \times 100\%$$

If the same ADC is used, how do the percentage errors in measured voltages compare, if the measured voltages are 10 V and 1 V?

4-6 Analog-to-digital converters digitize at different rates. What conversion rate is required if a chromatographic peak is to be sampled and digitized 20 times between the first positive deflection from the baseline until the peak returns to the baseline? The total baseline-to-baseline time is: (a) 20 s. (b) 1 s.

4-7 A particular 12-bit ADC has a conversion time of 8 μs. What is the highest frequency that can be accurately recorded with this device, assuming that the *Nyquist* criterion is satisfied (see Section 5C-2)?

4-8 The 200-MHz Pentium processor chip produced by Intel has a 64-bit instruction set, a 36-bit address bus, and is capable of performing more than 200 million machine instructions per second. (a) How many memory locations can be directly accessed by such a device? (b) How many possible instructions can the Pentium execute?

Chapter 5

Signals and Noise

Every analytical measurement is made up of two components. One component, the signal, carries information about the analyte that is of interest to the chemist. The second, called noise[1]*, is made up of extraneous information that is unwanted because it degrades the accuracy and precision of an analysis and also places a lower limit on the amount of analyte that can be detected.*[2] *In this chapter we describe some of the common sources of noise and how their effects can be minimized.*

5A THE SIGNAL-TO-NOISE RATIO

The effect of noise on a signal is shown in Figure 5-1a, which is a strip-chart recording of a tiny direct current of about 10^{-15} A. Figure 5-1b is a theoretical plot of the same current in the absence of noise. The difference between the two plots corresponds to the noise associated with this experiment. Unfortunately, noise-free data, such as that shown in Figure 5-1b can never be realized in the laboratory because some types of noise arise from thermodynamic and quantum effects that are impossible to avoid in a measurement.

In most measurements, the average strength of the noise N is constant and independent of the magnitude of the signal S. Thus, the effect of noise on the relative error of a measurement becomes greater and greater as the quantity being measured decreases in magnitude. For this reason, the *signal-to-noise ratio (S/N)* is a much more useful figure of merit than noise alone for describing the quality of an analytical method or the performance of an instrument.

For a dc signal, such as that shown in Figure 5-1a, the magnitude of the noise is conveniently defined as the standard deviation s of numerous measurements of the signal strength, and the signal is given by the mean $\bar{x}$ of the measurements. Thus, S/N is given by

$$\frac{S}{N} = \frac{\text{mean}}{\text{standard deviation}} = \frac{\bar{x}}{s} \qquad (5\text{-}1)$$

Note that the signal-to-noise ratio $\bar{x}/s$ is the reciprocal of the relative standard deviation, RSD (see Section a1B-1, Appendix 1) of the group of measurements. That is,

$$\frac{S}{N} = \frac{1}{\text{RSD}} \qquad (5\text{-}2)$$

[1] The term ***noise*** is derived from radio engineering where the presence of an unwanted signal is manifested as audio static, or noise. The term is applied now throughout science and engineering to describe the random fluctuations observed whenever replicate measurements are made on signals that are monitored continuously. These fluctuations are random and may thus be described and treated by statistical methods (see Section a1B, Appendix 1).

[2] For a more detailed discussion of noise, see T. Coor, *J. Chem. Educ.*, **1968,** *45,* A533, A583; G. M. Hieftje, *Anal. Chem.*, **1972,** *44*(6), 81A; A. Bezegh and J. Janata, *Anal. Chem.*, **1987,** *59,* 494A; M. E. Green, *J. Chem. Educ.*, **1984,** *61,* 600; H. V. Malmstadt, C. G. Enke, and S. R. Crouch, *Microcomputers and Electronic Instrumentation: Making the Right Connections.* Washington, DC: American Chemical Society, 1994.

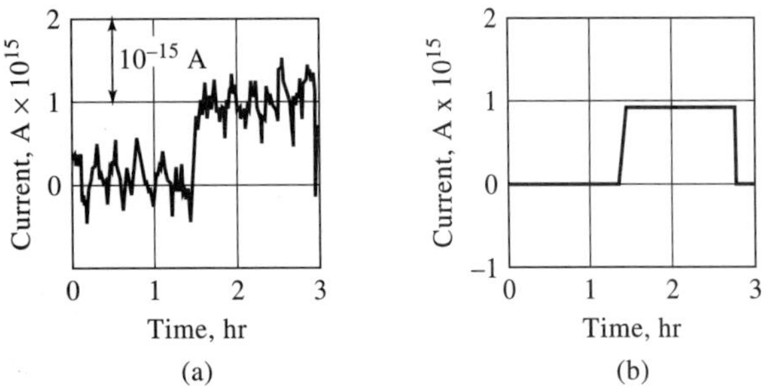

Figure 5-1 Effect of noise on a current measurement: (a) experimental strip-chart recording of a 0.9×10^{-15} A direct current, (b) mean of the fluctuations. *(Adapted from T. Coor,* J. Chem. Educ., ***1968,*** *45, A594. With permission.)*

For a recorded signal such as that shown in Figure 5-1a, the standard deviation can be estimated easily at a 99% confidence level by dividing the difference between the maximum and the minimum signal by five. Here, we assume that the excursions from the mean are random and can thus be treated by the methods of statistics. In Figure a1-5 of Appendix 1, it is seen that 99% of the data under the normal error curve lie within $\pm 2.5\sigma$ of the mean. Thus we can say with 99% certainty that the difference between the maximum and minimum encompasses 5σ. One fifth of the difference is then a good estimate of the standard deviation.

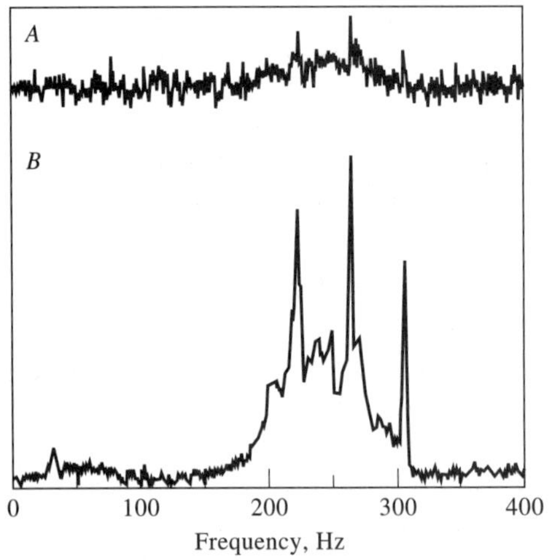

Figure 5-2 Effect of signal-to-noise ratio on the NMR spectrum of progesterone: *A, S/N* = 4.3; *B, S/N* = 43. *(Adapted from R. R. Ernst and W. A. Anderson,* Rev. Sci. Inst., ***1966,*** *37, 101. With permission.)*

As a general rule, it becomes impossible to detect a signal when the signal-to-noise ratio becomes less than about 2 or 3. Figure 5-2 illustrates this rule. The upper plot is a nuclear magnetic resonance spectrum for progesterone with a signal-to-noise ratio of about 4.3. In the lower plot the ratio is 43. At the smaller signal-to-noise ratio, only a few of the several peaks can be recognized with certainty.

5B SOURCES OF NOISE IN INSTRUMENTAL ANALYSES

Chemical analyses are affected by two types of noise: chemical noise and instrumental noise.

5B-1 Chemical Noise

Chemical noise arises from a host of uncontrollable variables that affect the chemistry of the system being analyzed. Examples include undetected variations in temperature or pressure that affect the position of chemical equilibria, fluctuations in relative humidity that cause changes in the moisture content of samples, vibrations that lead to stratification of powdered solids, changes in light intensity that affect photosensitive materials, and laboratory fumes that interact with samples or reagents. Details on the effects of chemical noise appear in later chapters that deal with specific instrumental methods. In this chapter we focus exclusively on instrumental noise.

5B-2 Instrumental Noise

Noise is associated with each component of an instrument—that is, with the source, the input transducer, all signal-processing elements, and the output transducer. Furthermore, the noise from each of these elements may be of several types and may arise from several sources. Thus, the noise that is finally observed is a complex composite that usually cannot be fully characterized. Certain kinds of instrumental noise are recognizable: (1) thermal, or Johnson, noise; (2) shot noise; (3) flicker, or $1/f$, noise; and (4) environmental noise. A consideration of the properties of the four kinds of noise is useful.

Thermal Noise, or Johnson Noise

Thermal noise is caused by the thermal agitation of electrons or other charge carriers in resistors, capacitors, radiation transducers, electrochemical cells, and other resistive elements in an instrument. This agitation of charged particles is random and periodically creates charge inhomogeneities, which in turn create voltage fluctuations that then appear in the readout as noise. It is important to note that thermal noise is present even in the absence of current in a resistive element and disappears only at absolute zero.

The magnitude of thermal noise in a resistive circuit element is readily derived from thermodynamic considerations[3] and is given by

$$\bar{v}_{rms} = \sqrt{4kTR\Delta f} \quad (5\text{-}3)$$

where $\bar{v}_{rms}$ is the root-mean-square noise voltage lying in a frequency bandwidth of Δf Hz, k is the Boltzmann constant (1.38×10^{-23} J/K), T is the temperature in kelvin, and R is the resistance in ohms of the resistive element.

In Section 3B-2 we discussed the relationship between the rise time t_r and the bandwidth Δf of an operational amplifier. These variables are also used to characterize the capability of complete instruments to transduce and transmit information. Recall that

$$\Delta f = \frac{1}{3t_r} \quad (5\text{-}4)$$

The rise time of an instrument is its response time in seconds to an abrupt change in input and normally is taken as the time required for the output to increase from 10% to 90% of its final value. Thus, if the rise time is 0.01 s, the bandwidth is 33 Hz.

Equation 5-3 suggests that thermal noise can be decreased by narrowing the bandwidth. However, as the bandwidth narrows, the instrument becomes slower to respond to a signal change, and more time is required to make a reliable measurement.

EXAMPLE 5-1

What is the effect on thermal noise of decreasing the response time of an instrument from 1 s to 1 μs?

If we assume that the response time is approximately equal to the rise time, we find that the bandwidth has been changed from 1 Hz to 10^6 Hz. According to Equation 5-3, such a change will cause an increase of noise by $(10^6/1)^{1/2}$ or 1000-fold.

[3] For example, see T. Coor, *J. Chem. Educ.*, **1968**, *45*, A534.

As shown by Equation 5-3, thermal noise can also be reduced by lowering the electrical resistance of instrument circuits and by lowering the temperature of instrument components. The thermal noise in transducers is often reduced by cooling. For example, lowering the temperature of a UV-visible photodiode array from a room temperature of 298 K to the temperature of liquid nitrogen of 77 K will halve the thermal noise.

It is important to note that thermal noise, while dependent upon the frequency bandwidth, is independent of frequency itself; thus, it is sometimes termed *white noise* by analogy to white light, which contains all visible frequencies. It is also noteworthy that thermal noise in resistive circuit elements is independent of the physical size of the resistor.

Shot Noise

Shot noise is encountered wherever electrons or other charged particles cross a junction. In typical electronic circuits, these junctions are found at *pn* interfaces; in photocells and vacuum tubes, the junction consists of the evacuated space between the anode and cathode. The currents are comprised of a series of quantized events; that is, the transfer of individual electrons across the junction. These events occur randomly, however, and the rate at which they occur is thus subject to statistical fluctuations, which are described by the equation

$$i_{rms} = \sqrt{2Ie\Delta f} \quad (5\text{-}5)$$

where i_{rms} is the root-mean-square current fluctuation associated with the average direct current, I, e is the charge on the electron of 1.60×10^{-19} C, and Δf is again the bandwidth of frequencies being considered. Like thermal noise, shot noise is white noise.

Equation 5-5 suggests that shot noise in a current measurement can be minimized only by reducing bandwidth.

Flicker Noise

Flicker noise is characterized as having a magnitude that is inversely proportional to the frequency of the signal being observed; it is sometimes termed $1/f$ (one-over-f) noise as a consequence. The causes of flicker noise are not well understood; it is ubiquitous, however, and is recognizable by its frequency dependence. Flicker noise becomes significant at frequencies lower than about 100 Hz. The long-term drift observed in dc amplifiers, meters, and galvanometers is a manifestation of flicker noise. Flicker noise can be reduced significantly by using wire-wound or metallic film resistors rather than the more common carbon-composition type.

Environmental Noise

Environmental noise is a composite of different forms of noise that arise from the surroundings. Figure 5-3 suggests typical sources of environmental noise in a university laboratory.

Much environmental noise occurs because each conductor in an instrument is potentially an antenna capable of picking up electromagnetic radiation and converting it to an electrical signal. There are numerous sources of electromagnetic radiation in the environment, including ac power lines, radio and TV stations, gasoline engine ignition systems, arcing switches, brushes in electrical motors, lightning, and ionospheric disturbances. Note that some of these sources, such as power lines and radio stations, cause noise with relatively narrow frequency bandwidths.

It is also noteworthy that the noise spectrum shown in Figure 5-3 contains a large, continuous noise region at low frequencies. This noise has the properties of flicker noise; its sources are unknown. Superimposed upon the flicker noise are noise peaks associated with yearly and daily temperature fluctuations and other periodic phenomena associated with the use of a laboratory building.

Finally, two quiet-frequency regions in which environmental noise is low are indicated in Figure 5-3: the region extending from about 3 Hz to almost 60 Hz and the region from about 1 kHz to about 500 kHz, or the frequency at which AM radio signals are prevalent. Often, signals are converted to frequencies in these regions in order to reduce noise during signal processing.

5C SIGNAL-TO-NOISE ENHANCEMENT

Many laboratory measurements require only minimal effort to maintain the signal-to-noise ratio at an acceptable level. Examples include the weight determinations made in the course of a chemical synthesis or the color comparison made in determining the chlorine content of the water in a swimming pool. For both examples, the signal is large relative to the noise and the requirements for precision and accuracy are minimal. When the need for sensitivity and accuracy increases, however, the signal-to-noise ratio often becomes the limiting factor in the precision of a measurement.

Both hardware and software methods are available for improving the signal-to-noise ratio of an instrumental method. Hardware noise reduction is accomplished by incorporating into the instrument design components such as filters, choppers, shields, modulators, and synchronous detectors. These devices remove or attenuate the noise without affecting the analytical signal significantly. Software methods are based upon various computer algorithms that permit extraction of signals from noisy data. As a minimum, software methods require sufficient hardware to condition the output signal from the instrument and convert it from analog to digital form. Typically, data are collected by using a computer equipped with a data acquisition module as described in

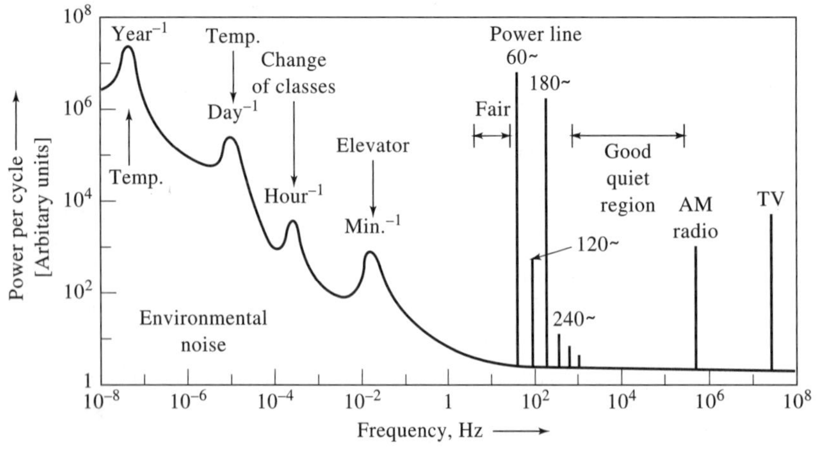

Figure 5-3 Some sources of environmental noise in a university laboratory. Note the frequency dependence and regions where various types of interference occur. *(From T. Coor, J. Chem. Educ., **1968**, 45, A540. With permission.)*

Chapter 4. Signals may then be extracted from noise by using the data acquisition computer or another that is connected to it via a network.

5C-1 Some Hardware Devices for Noise Reduction

This section contains a brief discussion of several hardware devices and techniques used for enhancing the signal-to-noise ratio.

Grounding and Shielding

Noise that arises from environmentally generated electromagnetic radiation can often be substantially reduced by shielding, grounding, and minimizing the lengths of conductors within the instrumental system. Shielding consists of surrounding a circuit, or the most critical of the wires in a circuit, with a conducting material that is attached to earth ground. Electromagnetic radiation is then absorbed by the shield rather than by the enclosed conductors. Noise pickup and possibly amplification of the noise by the instrument circuit may thus be minimized. We should note that the techniques of minimizing noise through grounding and shielding are often more art than science, particularly in instruments that involve both analog and digital circuits. The optimum configuration is often found only after much trial and error. The guidelines given in the references serve as a useful starting point.[4]

Shielding becomes particularly important when the output of a high-resistance transducer, such as the glass electrode, is being amplified. Here, even minuscule randomly induced currents produce relatively large voltage fluctuations in the measured signal.

Difference and Instrumentation Amplifiers

Any noise generated in the transducer circuit is particularly critical because it usually appears in an amplified form in the instrument readout. To attenuate this type of noise, most instruments employ a difference amplifier, such as that shown in Figure 3-11, for the first stage of amplification. Common-mode noise in the transducer circuit generally appears in phase at both the inverting and noninverting inputs of the amplifier and is largely subtracted out by the circuit so that the noise at its output is diminished substantially. For cases in which a difference amplifier is insufficient to remove the noise, an *instrumentation amplifier* such as the one shown in Figure 5-4 is used.

Instrumentation amplifiers are composed of three op amps configured as shown in Figure 5-4. Op amp A and op amp B make up the input stage of the instrumentation amplifier in which the two op amps are cross-coupled through three resistors R_1, R_1/a, and R_1. The second stage of the module is the difference amplifier of op amp C. We will not derive the transfer function of the instrumentation amplifier; however, suffice to say that the overall gain of the circuit is given by

$$v_o = K(2a + 1)(v_2 - v_1) \tag{5-6}$$

Equation 5-6 highlights two advantages of the instrumentation amplifier: (1) the overall gain of the amplifier may be controlled by varying a single resistor R_1/a, and (2) the second difference stage rejects common-mode signals. In addition, op amps A and B are voltage followers with very high input resistance, so the instrumentation amplifier presents a negligible load to its transducer circuit. The combination of the two stages can provide rejection of common-mode noise by a factor of 10^6 or more while amplifying the signal by as much as 1000.

These devices are used often with low-level signals in noisy environments such as in the measurement of signals in biological organisms in which the organism acts as an antenna. Electrocardiographic instrumentation exploits the advantages of instrumentation amplifiers. Another typical application is in computer data ac-

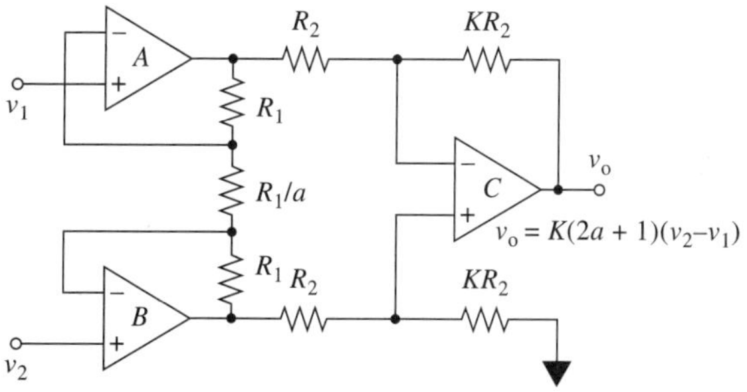

Figure 5-4 An instrumentation amplifier for reducing the effects of noise common to both inputs. The gain of the circuit is controlled by resistors R_1/a and KR_2.

[4] For an excellent discussion of grounding and shielding, see H. V. Malmstadt, C. G. Enke, and S. R. Crouch, *Microcomputers and Electronic Instrumentation: Making the Right Connections,* pp. 401-409. Washington, DC: American Chemical Society, 1994. A venerable but valuable reference is R. Morrison, *Grounding and Shielding Techniques in Instrumentation.* New York: Wiley-Interscience, 1967.

quisition module such as the programmable gain amplifier shown in Figure 4-13. The gain of the instrumentation amplifier is under the control of a computer, which changes resistor R_1/a of Figure 5-4 by using solid-state switches under digital control.

Analog Filtering

One of the most common methods of improving the signal-to-noise ratio in analytical instruments is by use of low-pass analog filters such as that shown in Figure 2-11b. The reason for this widespread application is that many instrument signals are of low frequency with bandwidths extending over a range of only a few hertz. Thus, a low-pass filter characterized by the Bode diagram shown in Figure 2-12b will effectively remove many of the high-frequency components of the signal, including thermal or shot noise. Figure 5-5 illustrates the use of a low-pass *RC* filter for reducing noise from a slowly varying dc signal.

High-pass analog filters such as that shown in Figure 2-11a also find considerable application in analytical instruments where the analyte signal is at relatively high frequency. The high-pass filter reduces the effects of drift and other low-frequency flicker noise.

Narrow-band electronic filters are also available to attenuate noise outside an expected band of signal frequencies. We have pointed out that the magnitude of fundamental noise is directly proportional to the square root of the frequency bandwidth of a signal. Thus, significant noise reduction can be achieved by restricting the input signal to a narrow band of frequencies and using an amplifier that is tuned to this band. It is important to note that the bandpass of the filter must be sufficiently wide to carry all of the signal frequencies.

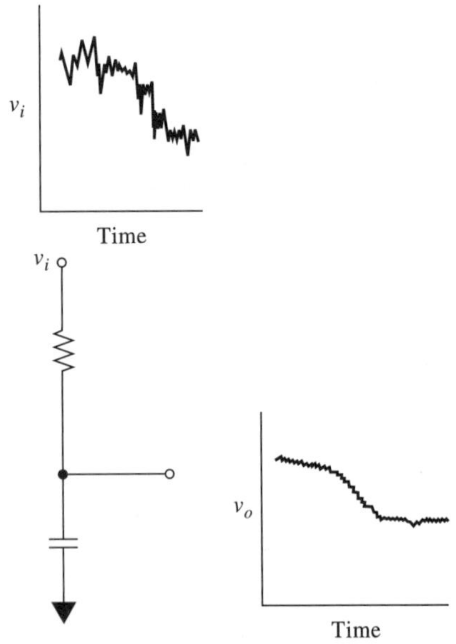

Figure 5-5 Use of a low-pass filter with a large time constant to remove noise from a slowly changing dc voltage.

Modulation

Direct amplification of a low-frequency or dc signal is particularly troublesome when an instrument exhibits amplifier drift and flicker noise. Often, this $1/f$ noise is several times larger than the types of noise that predominate at higher frequencies as is illustrated in the noise power spectrum of Figure 5-3. For this reason, low-frequency or dc signals from transducers are often converted to a higher frequency, where $1/f$ noise is less troublesome. This process is called *modulation.* After amplification, the modulated signal can be freed from amplifier $1/f$ noise by filtering with a high-pass filter; demodulation and filtering with a low-pass filter then produce an amplified dc signal suitable for output.

Figure 5-6 is a schematic showing the flow of information through such a system. The original dc signal shown in power spectrum *A* is modulated to give a narrow-band 400-Hz signal, which is then amplified by a factor of 10^5. As illustrated in power spectrum *B,* shown in the center of the figure, amplification introduces $1/f$ and power-line noise. Much of this noise can be removed with the aid of a suitable high-pass filter such as the one shown in Figure 2-11a. Demodulation of the filtered signal produces the amplified dc signal whose power spectrum is shown in *C.*

Signal Chopping; Chopper Amplifiers

The chopper amplifier provides one means for accomplishing the signal flow shown in Figure 5-6. In this device, the input signal is converted to a square-wave form by an electronic or mechanical chopper. Chopping can be performed either on the physical quantity to be measured or on the electrical signal from the transducer. In general, it is desirable to chop the signal as close to its source as possible because only the noise that arises after chopping is removed by the process.

Atomic absorption spectroscopy provides an example of the use of a mechanical chopper for signal modulation. Noise is a major concern in detecting and measuring signals from atomic absorption sources of the

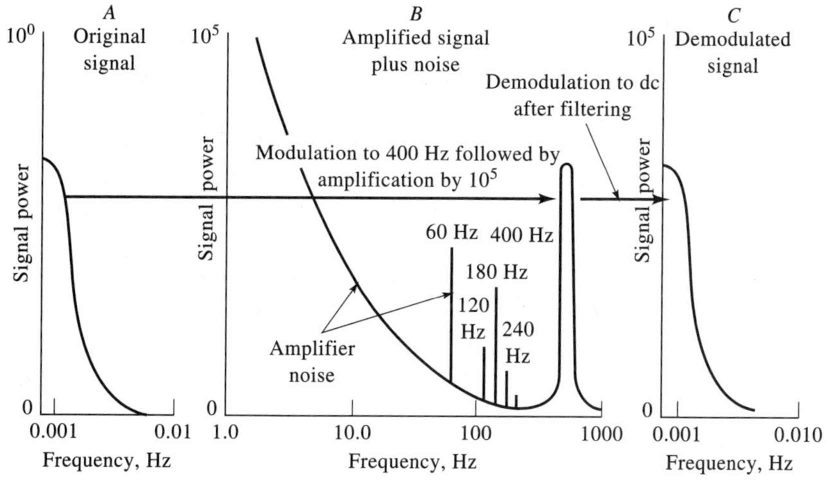

Figure 5-6 Amplification of a dc signal with a chopper amplifier. *(Adapted from T. Coor,* J. Chem. Educ., *1968,* 45, *A540. With permission.)*

low-frequency fluctuations inherent in flames and other atomization devices. In order to minimize these noise problems, light sources in atomic absorption instruments are often chopped by positioning a slotted rotating disk in the beam path as illustrated in Figure 9-13b. The rotation of the chopper produces a radiant signal that fluctuates periodically between zero and some maximum intensity. After interaction with the sample in the flame, the signal is converted by the transducer to a square-wave electrical signal whose frequency depends upon the size of the slots and the rate at which the disk rotates. Noise inherent in flames and other atomization devices is usually of low frequency and can be significantly reduced by the use of a high-pass filter prior to amplification of the transduced electrical signal.

Another example of the use of a chopper is shown in the synchronous modulation and demodulation scheme of Figure 5-7. This device is a chopper ampli-

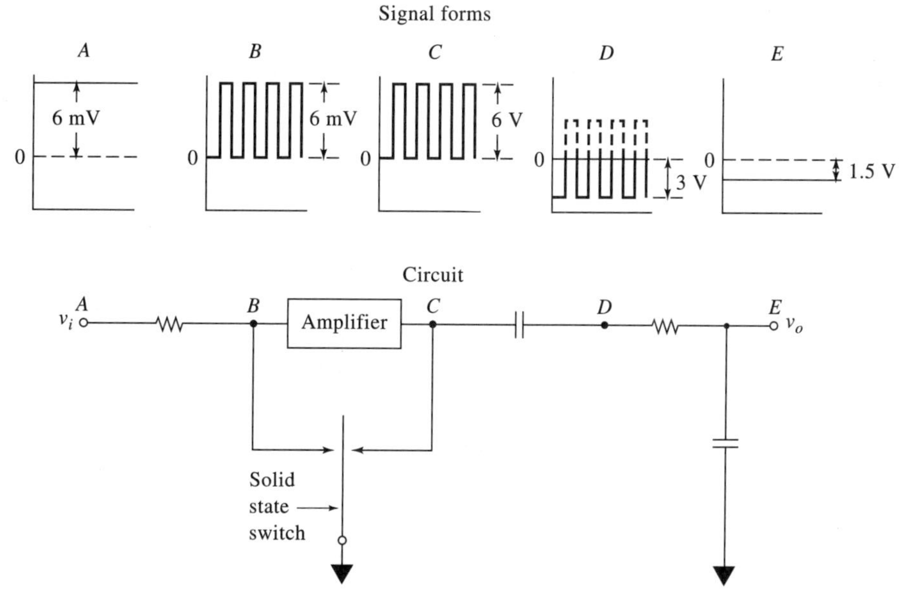

Figure 5-7 A chopper amplifier. The signal forms are idealized waveforms at the various indicated points in the circuit.

fier, which employs an electronic switch that shorts alternatively the input and the output signal of the amplifier to ground. The appearance of the signal at various stages is shown above the circuit diagram. The transducer signal input is a 6-mV dc signal *(A)* in this example. The vibrating switch converts the input to a square-wave signal with an amplitude of 6 mV *(B)*. Amplification by a factor of 1000 produces square wave with an amplitude of 6 V *(C)*, which, however, is shorted to ground periodically; as shown in *(D)*, shorting also reduces the amplitude of the signal to 3 V. Finally, the *RC* filter serves to smooth the signal and produce a 1.5 V-dc output. The synchronous demodulation process has the effect of rejecting the noise generated within the amplifier.

Lock-In Amplifiers[5]

Lock-in amplifiers permit the recovery of signals even when the signal-to-noise ratio is unity or less. Generally, a lock-in amplifier requires a reference signal that has the same frequency and phase as the signal to be amplified. That is, the reference signal must be of the same frequency as the analytical signal and, equally important, must bear a fixed phase relationship to the analytical signal. Figure 5-8a shows a system that utilizes an optical chopper to provide coherent analytical and reference signals. The reference signal is provided by a lamp and can

[5] See T. C. O'Haver, *J. Chem. Educ.*, **1972,** *49,* A131, A211.

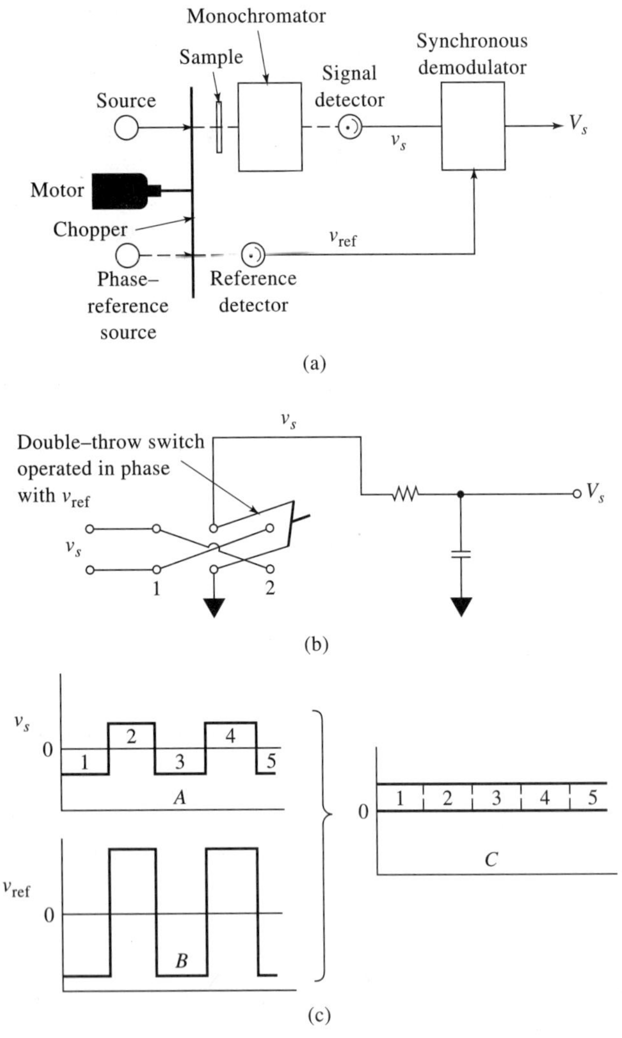

Figure 5-8 A lock-in amplifier system: (a) system for a spectrophotometer, (b) synchronous demodulation (schematic), and (c) signal waveforms.

be quite intense, thus freeing it from potential environmental noise sources. The reference and signal beams are chopped synchronously by the rotating slotted wheel, thus providing signals that are identical in frequency and have a fixed phase angle to one another.

The synchronous demodulator acts in a manner analogous to the double-pole–double-throw switch shown in Figure 5-8b.[6] In this analogy, the reference signal controls the switching so that the polarity of the analytical signal is reversed periodically to provide a rectified dc signal, as shown to the right in Figure 5-8c. Any high-frequency noise is then removed by a low-pass filter.

A lock-in amplifier is generally relatively free of noise because only those signals that are *locked-in* to the reference signal are amplified. All other frequencies are rejected by the system.

5C-2 Software Methods

With the widespread availability of microprocessors and microcomputers, many of the signal-to-noise enhancement devices described in the previous section are being replaced or supplemented by computer programs. Among these programs are routines for various types of averaging, digital filtering, Fourier transformation, smoothing, and correlation techniques. These procedures are applicable to nonperiodic or irregular waveforms, such as an absorption spectrum, to signals having no synchronizing or reference wave, and to periodic signals as well. Some of these relatively common procedures are discussed briefly here.

Ensemble Averaging[7]

In ensemble averaging, successive sets of data stored in memory as *arrays* are collected and summed point by point or in a series of capacitors for hardware averaging. This process is sometimes called *coaddition.* After the collection and summation are complete, the data are averaged by dividing the sum for each point by the number of scans performed. Figure 5-9 illustrates ensemble averaging of an absorption spectrum.

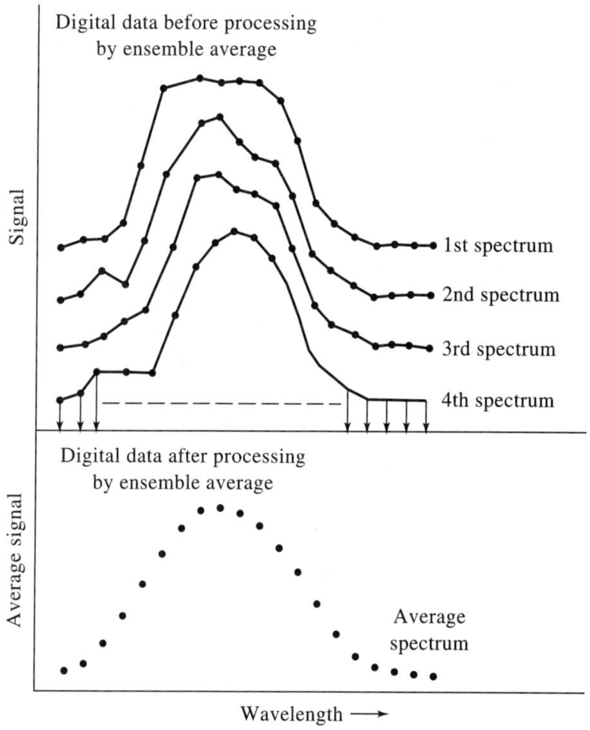

Figure 5-9 Ensemble averaging of a spectrum. *(From D. Binkley and R. Dessy,* J. Chem. Educ., *1979, 6, 150. With permission.)*

To understand why ensemble averaging effectively increases the signal-to-noise ratio of digitally acquired signals let us suppose that we are attempting to measure the magnitude of a dc signal *S*. We make *n* repetitive measurements of *S* and calculate the mean value of the signal from the equation

$$S_x = \frac{\sum_{i=1}^{n} S_i}{n} \tag{5-7}$$

where S_i, $i = 1, 2, 3 \ldots n$ are the individual measurements of the signal, including noise. The noise for each measurement is then $S_x - S_i$. If we square and sum the deviations of the signal from the mean, S_x, and divide by the number of measurements *n*, we obtain the *mean-square noise* given by

$$\text{mean-square noise} = \frac{\sum_{i=1}^{n} (S_x - S_i)^2}{n} \tag{5-8}$$

6 In a state-of-the-art demodulator, the mechanical switch in the figure would be replaced with a solid-state switch because the latter is faster and less noisy.

7 For a more extended description of the various types of signal averaging, see D. Binkley and R. Dessy, *J. Chem. Educ.*, **1979**, *56*, 148; R. L. Rowell, *J. Chem. Educ.*, **1978**, *55*, 148; G. Dulaney, *Anal. Chem.*, **1975**, *47*, 24A.

The mean-square noise is often referred to as the *variance* of the signal, and the *root-mean-square,* or *rms,* noise is its standard deviation, which is given by

$$\text{rms noise} = \sqrt{\frac{\sum_{i=1}^{n}(S_x - S_i)^2}{n}} \tag{5-9}$$

The signal-to-noise ratio for the measurement is then the mean value of the signal divided by its standard deviation, or

$$\frac{S}{N} = \frac{S_x}{\sqrt{\frac{\sum_{i=1}^{n}(S_x - S_i)^2}{n}}} \tag{5-10}$$

If we multiply both the numerator and denominator by *n,* we find that

$$\frac{S}{N} = \frac{nS_x}{\frac{n}{\sqrt{n}}\sqrt{\sum_{i=1}^{n}(S_x - S_i)^2}}$$

$$= \sqrt{n}\frac{S_x}{\sqrt{\sum_{i=1}^{n}(S_x - S_i)^2}} \tag{5-11}$$

The last expression shows that the signal-to-noise ratio is proportional to the square root of the number of data collected to determine the ensemble average. Note that this same signal-to-noise enhancement is realized in boxcar averaging and digital filtering, which are described in subsequent sections.

The improvement in *S/N* that is realized by signal averaging is exploited in many areas of science; nuclear magnetic resonance spectroscopy and Fourier transform infrared spectroscopy are but two of the most prominent examples in chemical instrumentation. We shall consider signal averaging and other aspects of digital data acquisition in more detail in the chapters on those topics.

To realize the advantage of ensemble averaging and still extract all of the information available in a signal waveform, it is necessary to measure points at a frequency that is at least twice as great as the highest frequency component of the waveform. This assertion is a consequence of the *Nyquist sampling theorem,* which states that for bandwidth-limited signals, sampling must occur at a frequency that is at least twice the highest frequency f of the signal of interest. That is, the data acquisition frequency must be at least $2f = 1/(2\,\Delta t)$, where Δt is the time interval between signal samples. For example, if the maximum frequency component in an instrumental signal is 150 Hz, the data must be sampled at a rate of at least 300 samples/s if the signal is to be reproduced accurately.

Sampling frequencies much greater than the Nyquist frequency do not produce significant additional information, and much higher frequencies may actually include undesirable noise. It is customary, however, to sample at a frequency of about ten times the Nyquist frequency to ensure signal integrity. Furthermore, it is very important to sample the waveform reproducibly; that is, the waveform must be sampled beginning at precisely the same point on each successive wave. For example, if the waveform is a visible absorption spectrum, each scan of the spectrum must be synchronized to start at exactly the same wavelength, and each successive sample of the waveform must occur at the same wavelength interval. Generally, synchronization is accomplished by means of a synchronizing pulse, which is derived from the waveform itself or the experimental event that initiated the waveform such as a laser pulse or a pulse of radio-frequency radiation. This pulse then initiates the acquisition of data for each scan of the waveform.

Ensemble averaging can produce dramatic improvements in signal-to-noise ratios as demonstrated by the three NMR spectra in Figure 5-10. Here, only a few of the absorption peaks are discernible in the single scan because their magnitudes are roughly the same as the signal excursions because of random noise. The improvement in the resulting spectrum as a consequence of signal averaging is apparent from the spectra shown in Figure 5-10. Further discussion of the Nyquist sampling theorem and its consequences may be found in the treatment of Fourier transform nuclear magnetic resonance spectroscopy in Chapter 19.

Boxcar Averaging

Boxcar averaging is a digital procedure for smoothing irregularities and enhancing the signal-to-noise ratio in a waveform, the assumption being that these irregularities are the consequence of noise. That is, it is assumed that the analog analytical signal varies only slowly with time and that the average of a small number of adjacent points is a better measure of the signal than any of the

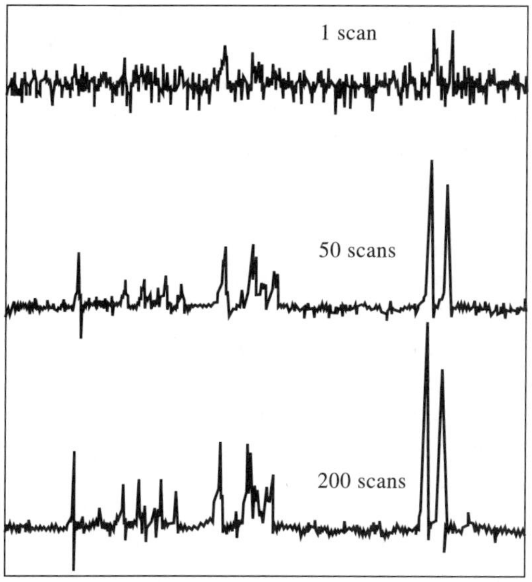

Figure 5-10 Effect of signal averaging. Note that the vertical scale is smaller as the number of scans increases. The signal-to-noise ratio is proportional to $\sqrt{n}$. Random fluctuations in the noise tend to cancel as the number of scans increases, but the signal accumulates; thus, *S/N* increases.

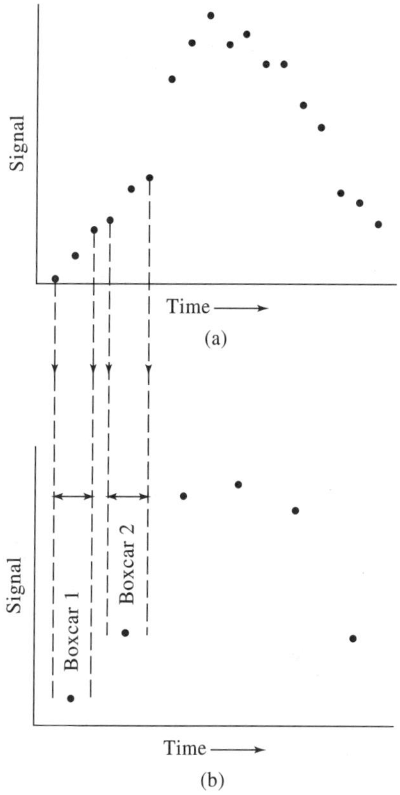

Figure 5-11 Effect of boxcar averaging: (a) original data, (b) data after boxcar averaging. *(Reprinted with permission from G. Dulaney,* Anal. Chem.*, **1975**, 47, 28A. Copyright 1975 American Chemical Society.)*

individual points. Figure 5-11b illustrates the effect of the technique on the data plotted in Figure 5-11a. The first point on the boxcar plot is the mean of points 1, 2, and 3 on the original curve; point 2 is the average of points 4, 5, and 6; and so forth. In practice, 2 to 50 points are averaged to generate a final point. Most often this averaging is performed by a computer in real time, that is, as the data is being collected (in contrast to ensemble averaging, which requires storage of the data for subsequent processing). As Figure 5-11 shows, detail is lost by boxcar averaging, and its utility is limited for complex signals that change rapidly as a function of time. Boxcar averaging is of considerable importance, however, for square-wave or repetitive-pulsed outputs where only the average amplitude is important.

Boxcar averaging can also be accomplished in the analog domain by using a *boxcar integrator.* This device uses a fast electronic switch to sample a repetitive waveform at a programmable time interval from the origin of the waveform. The sampled waveform is connected to an analog integrator to furnish a low-frequency version of the signal at the selected time interval. The instrument may be programmed to scan a very noisy signal waveform from beginning to end. In this way, a profile of the signal is acquired with a signal-to-noise ratio that is selectable by adjusting the time constant of the integrator, the scan speed of the sampling window, and the time window over which the sampling occurs. This window is called the *aperture time.*

Boxcar integrators are often used to sample and measure instrumental waveforms on the picosecond-to-microsecond time scale. Such integrators are particularly useful in connection with pulsed-laser systems in which physical and chemical events occur on such very fast time scales. The output of the integrator may be connected to computer data acquisition systems such as those described in Section 4E-4 for logging of data and subsequent post-experimental analysis and presentation. The advantage of acquisition of signals by means of a boxcar integrator is that the averaging time of the units may be increased to provide *S/N* enhancement. The signal-to-noise ratio is proportional to the square root of the amount of time it takes the integrator to acquire the signal at each time window of the waveform. Such improvement is equivalent to the enhancement realized in digital data acquisition by ensemble averaging.

Digital Filtering

Digital filtering can be accomplished by a number of different well-characterized numerical procedures, including ensemble averaging, which was discussed in the previous section, Fourier transformation, least-squares polynomial smoothing, and correlation. In this section, we shall briefly discuss the Fourier transform procedure and least-squares polynomial smoothing, which is one of the most common of all numerical data enhancement techniques.

In the Fourier transformation, a signal such as the spectrum shown in Figure 5-12a, which is acquired in the time domain, is converted to a frequency-domain signal in which the independent variable is frequency f rather than time as depicted in Figure 5-12b. This transformation, which is discussed in some detail in Section 7I, is accomplished mathematically on a computer by a very fast and efficient algorithm. The frequency-domain signal in b is then multiplied by the frequency response of a digital low-pass filter with upper-cutoff frequency f_0 shown in c, which has the effect of removing all frequency components above f_0 as illustrated in d. The inverse Fourier transform then recovers the filtered time-domain spectrum of Figure 5-12e. The Fourier transformation is used in most modern infrared and NMR spectrometers, as well as in a host of laboratory test instruments and digital oscilloscopes. The procedure is often built in to general-purpose software packages such as Mathcad® and Excel® and is widely available in subroutine packages in a variety of computer languages.

The last and perhaps most widely used digital data enhancement technique that we will discuss is least-squares polynomial data smoothing. In its simplest form, smoothing is very similar to the boxcar averaging scheme of Figure 5-11. Figure 5-13 illustrates how unweighted data smoothing may be accomplished. The 11

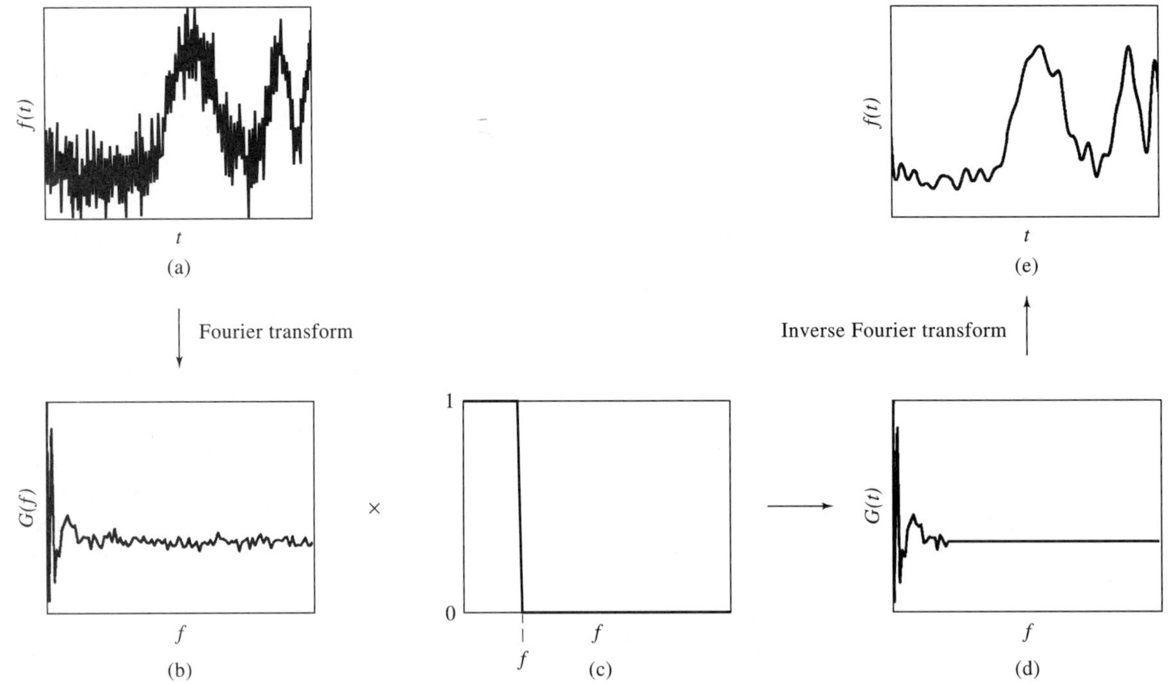

Figure 5-12 Digital filtering with the Fourier transform. (a) noisy spectral peak, (b) the frequency-domain spectrum of a resulting from the Fourier transformation, (c) low-pass digital-filter function, (d) product of (b) times (c). (e) the inverse Fourier transform of (d) with most of the high-frequency noise removed.

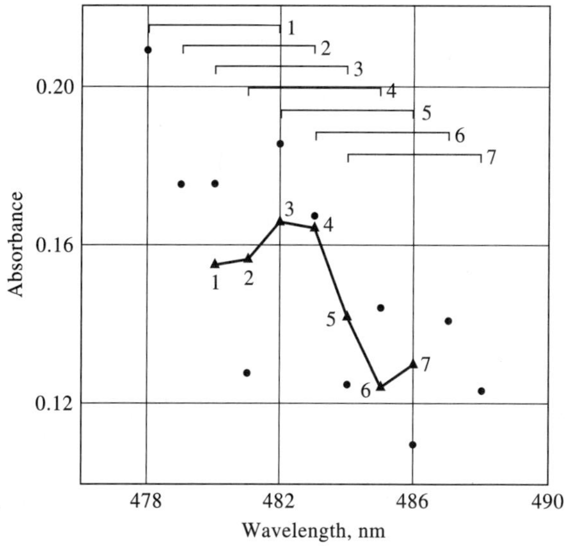

Figure 5-13 The operation of an unweighted moving-window average smoothing function: noisy spectral data (•), smoothed data (▲). See text for a description of the smoothing procedure.

data points represented by filled circles in the plot comprise a section of a noisy absorption spectrum. The first five data encompassed by bracket 1 in the figure are averaged and plotted as data point 3, or triangle 1. The bracket is then moved one point to the right to position 2, points 2 through 6 are averaged, and the average is plotted as triangular point 2. The process is repeated for brackets 3, 4, 5, and so on until all of the points except the last two are averaged to produce a new absorption curve represented by the triangular points and the line connecting them. The new curve is somewhat less noisy than the original data. This procedure is called a 5-point unweighted smooth. In this type of signal-to-noise enhancement, the width of the smoothing function always encompasses an odd number of points, and an even number of points remain unsmoothed at each end of the data set. The number of unsmoothed points is equal to $(n - 1)/2$ where n is the smooth width. For an absorption spectrum consisting of hundreds or perhaps thousands of data, the loss of these data is usually inconsequential.

The signal-to-noise ratio of the data may be enhanced by increasing the width of the smoothing function or by smoothing the data multiple times. Unfortunately, the basic unweighted smoothing function gives too much weight to points that are well removed from the central point. A much better procedure than simply averaging points on a data curve is to perform a least-squares fit of a small section of the curve to a polynomial and take the calculated central point of the fitted polynomial curve as the new smoothed data point. This approach is much superior to the unweighted averaging scheme, but it has the disadvantage of being computationally intense and thus requires considerable computer time. Savitzky and Golay showed that a set of integers could be derived and used as weighting coefficients to carry out the smoothing operation.[8] The use of these weighting coefficients, sometimes referred to as *convolution integers,* turns out to be *exactly equivalent* to fitting the data to a polynomial as just described. The convolution integers for a 5-point quadratic smoothing function are plotted in Figure 5-14a.

The application of the smoothing integers of Figure 5-14a to the data of Figure 5-13 will illustrate the smoothing process. We begin by multiplying the left-most convolution integer, which is -3 in this case, times the absorbance at point 1 in Figure 5-13. The second integer, which is 12, is then multiplied by the second point, and the result is added to the product obtained for the first point. Point 3 is then multiplied by 17, which is the third integer, and the result is summed once again. This process is repeated until each of the five data has been multiplied by its corresponding integer, and the sum of the five results is obtained. Finally the sum of the results is divided by a sixth integer, the so-called *normalization integer,* which is the number 35 in this example of a 5-point quadratic smooth, and the quotient is taken as the new value of the central point of the smoothing interval. The normalization integer also derives from the Savitzky-Golay treatment, as do other similar sets of integers for smoothing, to generate the first and second derivatives of the data. The first derivative convolution integers for a 5-point cubic smooth are plotted in Figure 5-14b, and the second derivative integers for a 5-point quadratic are shown in Figure 5-14c. These sets of integers may be exploited in exactly the same way as the basic smoothing integers to produce the first and second derivatives of the original absorption data. Least-squares polynomial derivative smoothing is used to develop derivative spectra because as noted in our discussion of analog differentiators in Section 3E-4, differentiation is usually a noise-producing process. The effect of derivative smoothing is to minimize the noise generated in the differentiation.

---

[8] See A. Savitzky and M. J. E. Golay, *Anal. Chem.,* **1964,** *36,* 1627.

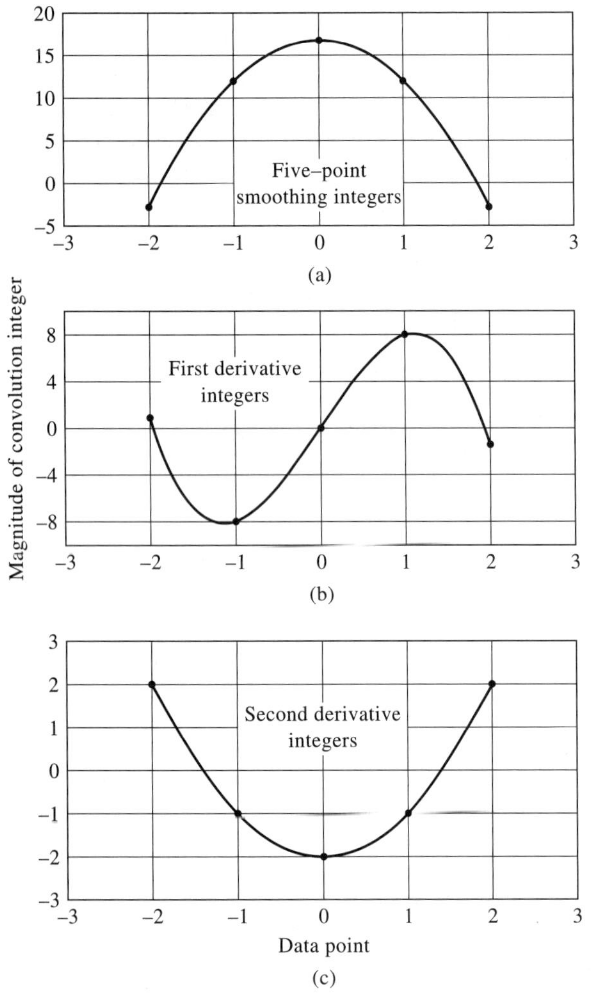

Figure 5-14 Least-squares polynomial smoothing convolution integers: (a) quadratic 5-point integers, (b) first derivative cubic 5-point integers, (c) second derivative quadratic 5-point integers.

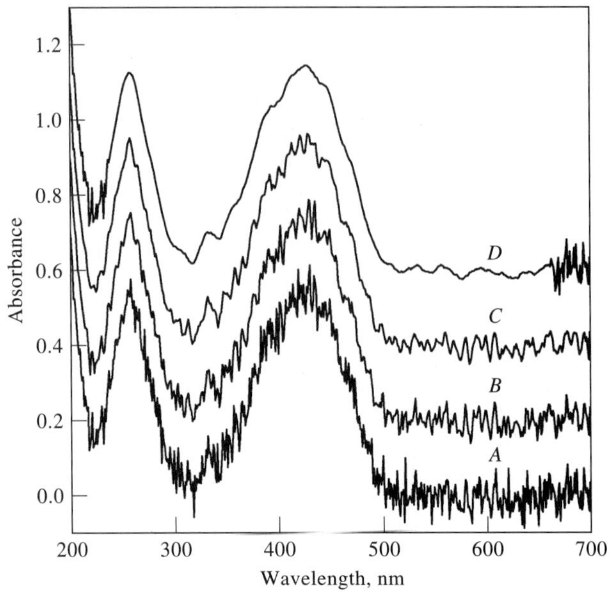

Figure 5-15 Effect of smoothing on a noisy absorption spectrum of tartrazine: *(A)* Raw spectrum, *(B)* quadratic 5-point smooth of the data in *A*, *(C)* fourth-degree 13-point smooth of the same data, *(D)* tenth-degree 77-point smooth of the data.

Because least-squares polynomial smoothing is so widely used to enhance the quality of analytical data, it is important to note the advantages and limitations of the method. The procedure reduces noise and acts as a low-pass filter on data. As with any filtering process, the signal suffers some distortion because of the bandwidth limitation inherent in the process. Users of smoothing must recognize that noise reduction has to be balanced against possible distortion of the signal. The advantage of the procedure is that variables such as the type of smooth, the smooth width, and the number of times that the data are smoothed may be decided after data collection. Furthermore, the smoothing algorithm is computationally trivial and requires minimal computer time. The *S/N* enhancement resulting from smoothing is relatively modest, generally amounting to about a factor of four for spectra that contain peaks with a width of 32 data points and with a smooth width twice that value. However, smoothing produces cleaner data than the original does for human interpretation, and it is widely used for this purpose. When smoothing is applied to data for quantitative analysis, distortion of the data has a minimal effect on the quantitative results because distortion errors tend to cancel when samples and standards are smoothed in the same manner.

The data of Figure 5-15 illustrate the application of least-squares polynomial smoothing to a rather noisy 501-point absorption spectrum of the dye tartrazine shown at the bottom of the figure in curve *A*. Curve *B* is a quadratic 5-point smooth of the data, curve *C* is a fourth-degree 13-point smooth, and curve *D* is a tenth-degree 77-point smooth. Note in curve *D* that 38 points remain unsmoothed at each end of the data set. The effects of the smoothing process are apparent in the progression from curve *A* to curve *D*.

Because of the general utility of smoothing and its wide application, guidelines have been developed for its use, equations are available for calculating the smoothing coefficients, and it has been applied to two-dimensional data such as diode array spectra. See the references in the footnote for further details on the nature of the smoothing process and its implementation.[9]

Correlation Methods

Correlation methods are often used for processing data from analytical instruments. These procedures provide powerful tools for performing such tasks as extracting signals that appear to be hopelessly lost in noise, smoothing noisy data, comparing a spectrum of an analyte with stored spectra of pure compounds, and resolving overlapping or unresolved peaks in spectroscopy and chromatography.[10] Correlation methods are based upon complex mathematical data manipulations that can only be carried out conveniently by means of a computer or by sophisticated analog instrumentation.

Correlation methods will not be discussed in this text. The interested reader should consult the references given in footnote 10.

[9] J. Steinier, Y. Termonia, and J. Deltour, *Anal. Chem.*, **1972**, *44*, 1906; T. A. Nieman and C. G. Enke, *Anal. Chem.*, **1976**, *48*, 705A; H. H. Madden, *Anal. Chem.*, **1978**, *50*, 1383; K. L. Ratzlaff, *Introduction to Computer-Assisted Experimentation*. New York: Wiley, 1987.

[10] For a more detailed discussion of correlation methods, see G. Horlick and G. M. Hieftje, in *Contemporary Topics in Analytical and Clinical Chemistry*, D. M. Hercules, et al., Eds., Vol. 3, pp. 153-216. New York: Plenum Press, 1978. For a briefer discussion, see G. M. Hieftje and G. Horlick, *American Laboratory*, **1981**, *13*(3), 76.

5D QUESTIONS AND PROBLEMS

5-1 What types of noise are frequency dependent? Frequency independent?

5-2 Name the type or types of noise that can be reduced by
(a) decreasing the temperature of a measurement.
(b) decreasing the frequency used for the measurement.
(c) decreasing the bandwidths of the measurement.

5-3 Suggest a frequency range that is well suited for noise minimization. Explain.

5-4 Why is shielding vital in the design of glass electrodes that have an internal resistance of 10^6 ohms or more?

5-5 What type of noise is likely to be reduced by (a) a high-pass filter and (b) a low-pass filter?

5-6 Make a rough estimate of the signal-to-noise ratio for the 0.9×10^{15} A current shown in Figure 5-1a.

5-7 The following data were obtained for repetitive weighings of a 1.004-g standard weight on a top-loading balance:

1.003	1.000	1.001
1.004	1.005	1.006
1.001	0.999	1.007

(a) Assuming the noise is random, calculate the signal-to-noise ratio for the balance.
(b) How many measurements would have to be averaged to measure *S/N* to 500?

5-8 The following data were obtained for a voltage measurement, in mV, on a noisy system: 1.37, 1.84, 1.35, 1.47, 1.10, 1.73, 1.54, 1.08.
(a) Assuming the noise is random, what is the signal-to-noise ratio?
(b) How many measurements would have to be averaged to increase *S/N* to 10?

5-9 Calculate the rms thermal noise associated with a 1.0-MΩ load resistor operated at room temperature if an oscilloscope with a 1-MHz bandwidth is used. If the bandwidth is reduced to 100 Hz, by what factor will the noise be reduced?

5-10 If the spectrum in Figure 5-2a is the result of a single scan and that of Figure 5-2b is the result of ensemble averaging, how many individual spectra were summed to increase *S/N* from 4.3 to 43?

5-11 Calculate the improvement in *S/N* in progressing from the top spectrum to the bottom in Figure 5-10.

5-12 Calculate the improvement in *S/N* in progressing from spectrum *A* to spectrum *D* in Figure 5-15.

5-13 For optimum *S/N* enhancement, spectral data should be smoothed with a smooth width that is no more than twice the width of a spectral peak taken at half its maximum intensity. Estimate from the data of Figure 5-15 (a) the maximum smooth width for the peak on the left side of the spectrum and (b) for the peak on the right.

Section II

Atomic Spectroscopy

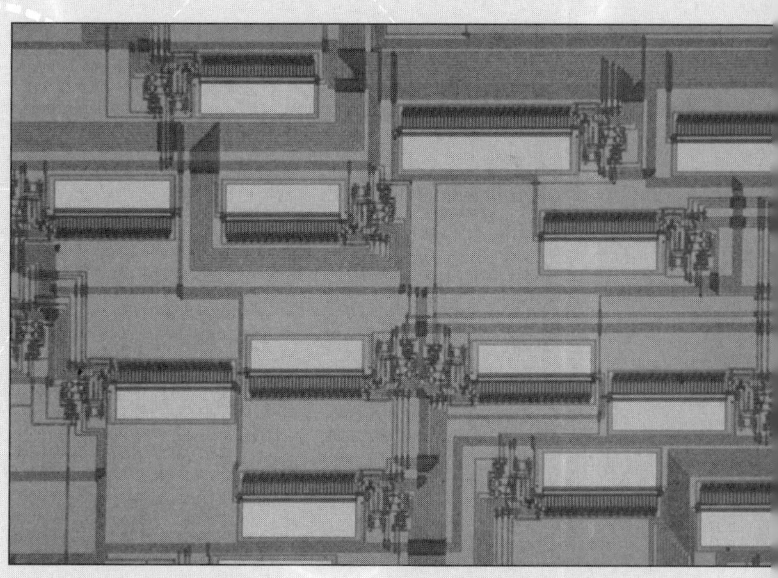

The image shown above is a photomicrograph of a segmented array charge coupled device, which is the transducer used in the Perkin-Elmer Echelle optical emission spectrometer. In this instrument, radiation emitted from the sample is dispersed in two dimensions on the rectangular transducer array. The photosensitive regions of the array, which appear white in the photo, are placed at the precise locations of the surface where emission lines are expected to fall. These devices and instruments are discussed in Scetion 10A-3. The photomicrograph symbolizes the profound influence of microelectronics technology on the science of instrumental measurement. (Photo courtesy of The Perkin-Elmer Corporation, Norwalk, CT.)

Section II comprises the fundamental principles and methods of atomic spectroscopy. While Chapter 6 investigates the nature of light and its interaction with matter, Chapter 7 introduces the optical, electronic, and mechanical components of optical instruments. These two chapters also consider concepts and instrumental components that are useful in our discussion of molecular spectroscopy in Section III. The general nature of atomic spectra and practical aspects of introducing atomic samples into a spectrometer are addressed in Chapter 8. Chapter 9 explores the practice of atomic absorption and atomic fluorscence spectroscopy, while Chapter 10 provides a similar treatment of atomic emission spectroscopy. An introduction to mass spectrometry is presented in Chapter 11, as are descriptions of various instruments and methods of atomic mass spectrometry. Our presentation of atomic spectrometry is completed by a discussion of X-ray spectrometry in Chapter 12.

Chapter 6

An Introduction to Spectrometric Methods

Spectrometric methods are a large group of analytical methods that are based on atomic and molecular spectroscopy. Spectroscopy is a general term for the science that deals with the interactions of various types of radiation with matter. Historically, the interactions of interest were between electromagnetic radiation and matter, but now spectroscopy has been broadened to include interactions between matter and other forms of energy. Examples include acoustic waves and beams of particles such as ions and electrons. Spectrometry and spectrometric methods refer to the measurement of the intensity of radiation with a photoelectric transducer or other type of electronic device.

The most widely used spectrometric methods are based on electromagnetic radiation, which is a type of energy that takes several forms, the most readily recognizable being light and radiant heat. Less obvious manifestations include gamma rays and X-rays as well as ultraviolet, microwave and radio-frequency radiation.

This chapter treats in a general way the interactions of electromagnetic waves with atomic and molecular species. After this introduction, the next five chapters describe the various types of spectrometric methods employed by chemists for the identification and determination of the elements present in various forms of matter. Chapters 13 through 21 then discuss the uses of spectrometry for structural determination of molecular species and describe how these methods are used for their quantitative determination.

6A GENERAL PROPERTIES OF ELECTROMAGNETIC RADIATION

Many of the properties of electromagnetic radiation are conveniently described by means of a classical sinusoidal wave model, which embodies such parameters as wavelength, frequency, velocity, and amplitude. In contrast to other wave phenomena, such as sound, electromagnetic radiation requires no supporting medium for its transmission and thus passes readily through a vacuum.

The wave model fails to account for phenomena associated with the absorption and emission of radiant energy. To understand these processes, it is necessary to invoke a particle model in which electromagnetic radiation is viewed as a stream of discrete particles, or wave packets, of energy called *photons* where energy is proportional to the frequency of the radiation. These dual views of radiation as particles and as waves are not mutually exclusive but, rather, complementary. Indeed, the wave-particle duality is found to apply to the behavior of streams of electrons, protons, and other elementary particles and is completely rationalized by wave mechanics.

6B WAVE PROPERTIES OF ELECTROMAGNETIC RADIATION

For many purposes, electromagnetic radiation is conveniently represented as electric and magnetic fields that undergo in-phase, sinusoidal oscillations at right angles to each other and to the direction of propagation. Figure 6-1a is such a representation of a single ray of plane-polarized electromagnetic radiation. The term *plane polarized* implies that all oscillations of either the electric or the magnetic fields lie within a single plane. Figure 6-1b is a two-dimensional representation of the electric component of the ray in Figure 6-1a. The electric field strength in Figure 6-1 is represented as a vector whose length is proportional to its magnitude. The abscissa of this plot is either time as the radiation passes a fixed point in space or distance when time is held constant. Throughout this chapter and most of the remaining text, only the electric component of radiation will be considered because the electric field is responsible for most of the phenomena that are of interest to us, including transmission, reflection, refraction, and absorption. Note, however, that the magnetic component of electromagnetic radiation is responsible for absorption of radio-frequency waves in nuclear magnetic resonance.

6B-1 Wave Parameters

In Figure 6-1b, the *amplitude A* of the sinusoidal wave is shown as the length of the electric vector at a maximum in the wave. The time in seconds required for the passage of successive maxima or minima through a fixed point in space is called the *period, p,* of the radiation. The *frequency, ν,* is the number of oscillations of the field that occur per second[1] and is equal to $1/p$. Another parameter of interest is the *wavelength,* λ, which is the linear distance between any two equivalent points on successive waves (e.g., successive maxima or minima).[2] Multiplication of the frequency in cycles per second by the wavelength in meters per cy-

[1] The common unit of frequency is the reciprocal second (s^{-1}), or *hertz (Hz),* which corresponds to one cycle per second.

[2] The units commonly used for describing wavelength differ considerably in the various spectral regions. For example, the angstrom unit, Å (10^{-10} m), is convenient for X-ray and short ultraviolet radiation; the nanometer, nm (10^{-9} m), is employed with visible and ultraviolet radiation; the micrometer, mμ (10^{-6} m), is useful for the infrared region. (The *micrometer* was called the *micron* in the early literature; the use of this term is discouraged.)

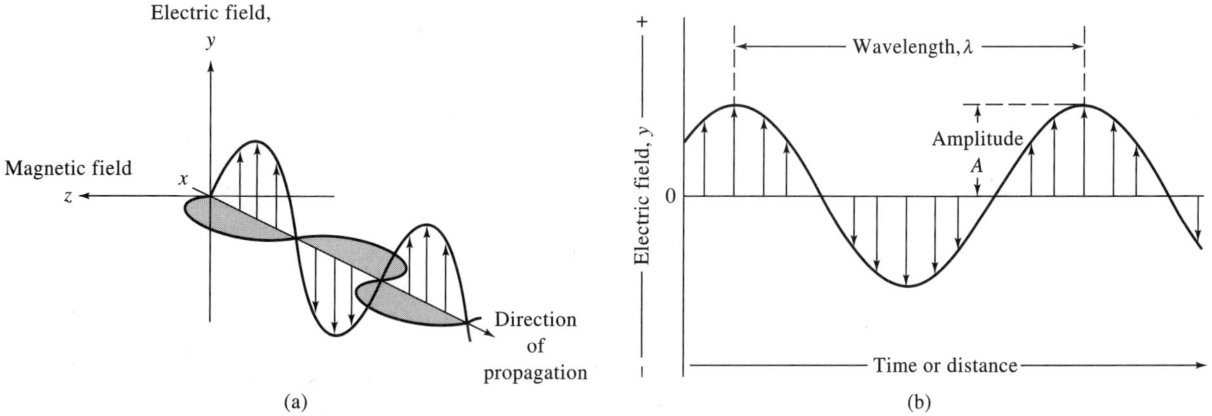

Figure 6-1 Representation of a beam of monochromatic, plane-polarized radiation: (a) electrical and magnetic fields at right angles to one another and direction of propagation, (b) two-dimensional representation of the electric vector.

cle gives the *velocity of propagation* v_i in meters per second:

$$v_i = \nu\lambda_i \tag{6-1}$$

It is important to realize that the frequency of a beam of radiation is determined by the source and *remains invariant.* In contrast, the velocity of radiation depends upon the composition of the medium through which it passes. Thus, Equation 6-1 implies that the wavelength of radiation is also dependent upon the medium. The subscript i in Equation 6-1 emphasizes these dependencies.

In a vacuum, the velocity of radiation is independent of wavelength and is at its maximum. This velocity, given the symbol c, has been determined to be 2.99792×10^8 m/s. It is significant that the velocity of radiation in air differs only slightly from c (about 0.03% less); thus, for either air or vacuum, Equation 6-1 can be written to three significant figures as

$$c = \nu\lambda = 3.00 \times 10^8 \text{ m/s} = 3.00 \times 10^{10} \text{ cm/s} \tag{6-2}$$

In any medium containing matter, propagation of radiation is slowed by the interaction between the electromagnetic field of the radiation and the bound electrons in the matter. Since the radiant frequency is invariant and fixed by the source, the wavelength must decrease as radiation passes from a vacuum to another medium (Equation 6-2). This effect is illustrated in Figure 6-2 for a monochromatic beam of visible radiation.[3]

[3] A *monochromatic* beam is a beam of radiation whose rays have identical wavelengths. A *polychromatic* beam is made up of rays of different wavelengths.

Note that the wavelength shortens nearly 200 nm, or more than 30%, as it passes into glass; a reverse change occurs as the radiation again enters air.

The *wavenumber* $\bar{\nu}$, which is defined as the reciprocal of the wavelength in centimeters, is yet another way of describing electromagnetic radiation. The unit for $\bar{\nu}$ is cm^{-1}. Wavenumber is widely used in infrared spectroscopy. The wavenumber is a useful unit because, in contrast to wavelength, it is directly proportional to the frequency, and thus the energy, of radiation. Thus, we may write

$$\bar{\nu} = k\nu \tag{6-3}$$

where the proportionality constant k depends on the medium and is equal to the reciprocal of the velocity (Equation 6-1).

The *power P* of radiation is the energy of the beam that reaches a given area per second, whereas the *intensity I* is the power per unit solid angle. These quantities are related to the square of the amplitude A (see Figure 6-1). Although it is not strictly correct to do so, power and intensity are often used synonymously.

6B-2 The Electromagnetic Spectrum

As shown in Figure 6-3, the electromagnetic spectrum encompasses an enormous range of wavelengths and frequencies (and thus energies). In fact, the range is so great that a logarithmic scale is required. Figure 6-3 also depicts qualitatively the major spectral regions. The divisions are based on the methods that are required to generate and detect the various kinds of radiation. Several overlaps are evident. Note that the visible portion of the spectrum to which the human eye is sensitive is tiny

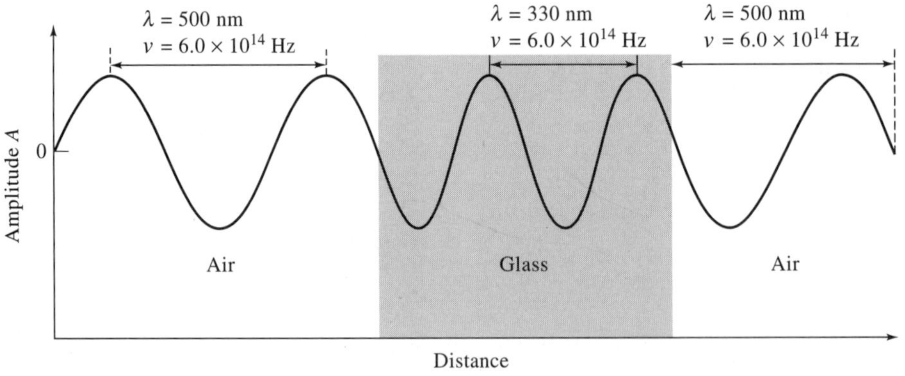

Figure 6-2 Effect of change of medium on a monochromatic beam of radiation.

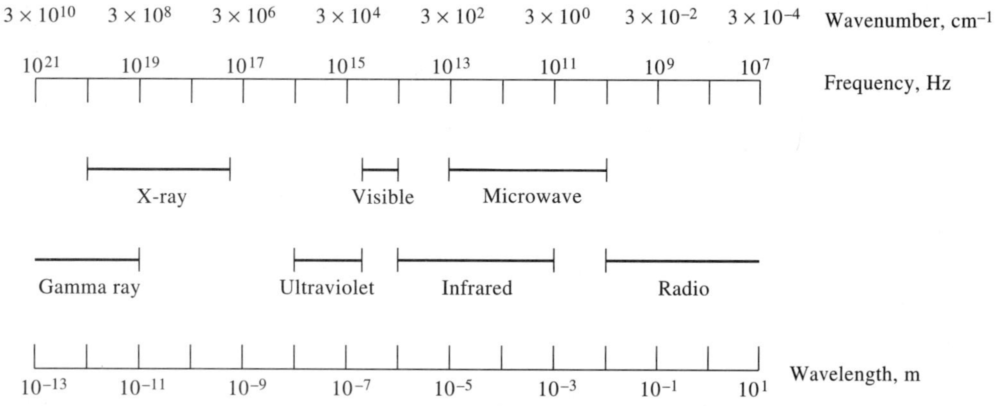

Figure 6-3 Regions of the electromagnetic spectrum.

when compared with other spectral regions. It should also be noted that spectrochemical methods that employ not only visible but also ultraviolet and infrared radiation are often called *optical methods* despite the fact that the human eye is sensitive to neither of the latter two types of radiation. This somewhat ambiguous terminology arises from the many common features of instruments for the three spectral regions and the similarities in the way in which we view the interactions of the three types of radiation with matter.

Table 6-1 lists the wavelength and frequency ranges for the regions of the spectrum that are important

TABLE 6-1 Common Spectroscopic Methods Based on Electromagnetic Radiation

Type Spectroscopy	Usual Wavelength Range*	Usual Wavenumber Range, cm^{-1}	Type of Quantum Transition
Gamma-ray emission	0.005–1.4 Å	—	Nuclear
X-Ray absorption, emission, fluorescence, and diffraction	0.1–100 Å	—	Inner electron
Vacuum ultraviolet absorption	10–180 nm	1×10^6 to 5×10^4	Bonding electrons
Ultraviolet visible absorption, emission, and fluorescence	180–780 nm	5×10^4 to 1.3×10^4	Bonding electrons
Infrared absorption and Raman scattering	0.78–300 μm	1.3×10^4 to 3.3×10^1	Rotation/vibration of molecules
Microwave absorption	0.75–3.75 mm	13–27	Rotation of molecules
Electron spin resonance	3 cm	0.33	Spin of electrons in a magnetic field
Nuclear magnetic resonance	0.6–10 m	1.7×10^{-2} to 1×10^3	Spin of nuclei in a magnetic field

*1 Å = 10^{-10} nm = 10^{-8} cm
1 nm = 10^{-9} m = 10^{-7} cm
1 μm = 10^{-6} m = 10^{-4} cm

for analytical purposes and also gives the names of the various spectroscopic methods associated with each. The last column of the table lists the types of nuclear, atomic, or molecular quantum transitions that serve as the basis for the various spectroscopic techniques.

6B-3 Mathematical Description of a Wave

With time as a variable, the wave in Figure 6-1b can be described by the equation for a sine wave. That is,

$$y = A \sin(\omega t + \phi) \tag{6-4}$$

where y is the *electric field*, A is the amplitude or maximum value for y, t is time, and ϕ is the *phase angle*, a term defined in Section 2B-1, page 28. The angular velocity of the vector ω is related to the frequency of the radiation ν by the equation

$$\omega = 2\pi\nu$$

Substitution of this relationship into Equation 6-4 yields

$$y = A \sin(2\pi\nu t + \phi) \tag{6-5}$$

6B-4 Superposition of Waves

The *principle of superposition* states that when two or more waves traverse the same space, a disturbance occurs that is the sum of the disturbances caused by the individual waves. This principle applies to electromagnetic waves, whereby the disturbances involve an electrical field, as well as to several other types of waves, whereby atoms or molecules are displaced. When n electromagnetic waves differing in frequency, amplitude, and phase angle pass some point in space simultaneously, the principle of superposition and Equation 6-5 permits us to write

$$y = A_1 \sin(2\pi\nu_1 t + \phi_1) + A_2 \sin(2\pi\nu_2 t + \phi_2) + \cdots + A_n \sin(2\pi\nu_n t + \phi_n) \tag{6-6}$$

where y is the resultant field.

The solid line in Figure 6-4a shows the application of Equation 6-6 to two waves of identical frequency but somewhat different amplitude and phase angle. The resultant is a periodic function with the same frequency but larger amplitude than either of the component waves. Figure 6-4b differs from 6-4a in that the phase difference is greater; here, the resultant amplitude is smaller than the amplitudes of the component waves. Clearly, a maximum amplitude occurs when the two waves are completely in phase—a situation that prevails whenever the phase difference between waves $(\phi_1 - \phi_2)$ is 0 deg, 360 deg, or an integer multiple of 360 deg. Under these circumstances, maximum *constructive interference* is said to occur. A maximum *destructive interference* occurs when $(\phi_1 - \phi_2)$ is equal to 180 deg or 180 deg plus an integer multiple of 360 deg. Interference plays an important role in many instrumental methods based on electromagnetic radiation.

Figure 6-5 depicts the superposition of two waves with the identical amplitudes but different frequencies.

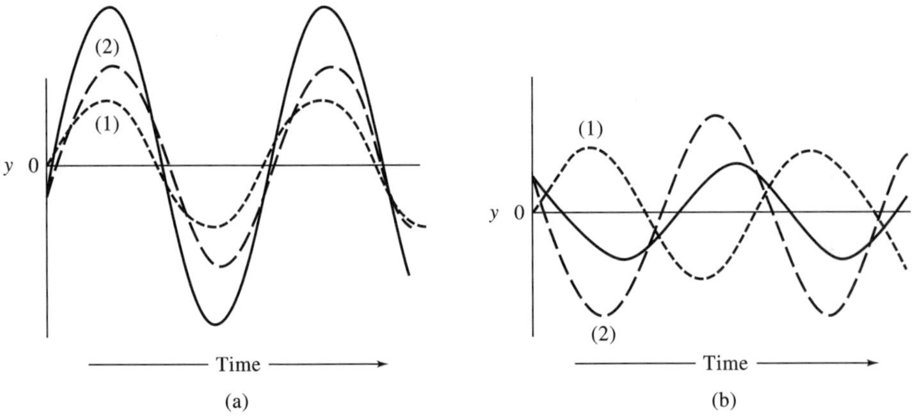

Figure 6-4 Superposition of sinusoidal wave: (a) $A_1 < A_2$, $(\phi_1 - \phi_2) = -20°$, $\nu_1 = \nu_2$; (b) $A_1 < A_2$, $(\phi_1 - \phi_2) = -200°$, $\nu_1 = \nu_2$. In each instance, the solid curve results from the combination of the two dashed curves.

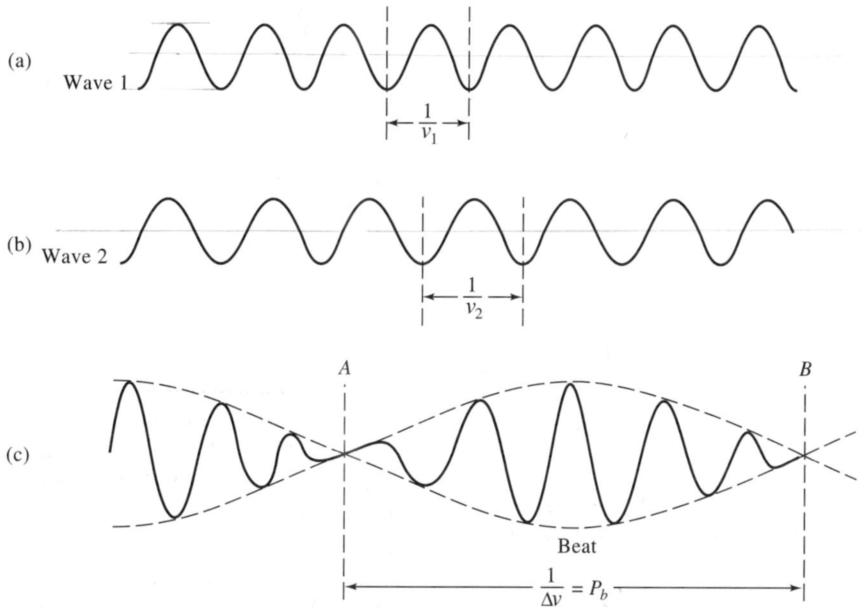

Figure 6-5 Superposition of two waves of different frequencies but identical amplitudes: (a) wave 1 with a period of $1/\nu_1$; (b) wave 2 with a period of $1/\nu_2$ ($\nu_2 = 1.25\ \nu_1$); (c) combined wave pattern. Note that superposition of ν_1 and ν_2 produces a beat pattern with a period of $1/\Delta\nu$ where $\Delta\nu = |\nu_1 - \nu_2|$.

The resulting wave is no longer sinusoidal but does exhibit a periodicity or *beat.* Note that the period of the beat P_b is the reciprocal of the frequency difference $\Delta\nu$ between the two waves. That is,

$$P_b = \frac{1}{\Delta\nu} = \frac{1}{(\nu_2 - \nu_1)} \tag{6-7}$$

An important aspect of superposition is that a complex waveform can be broken down into simple components by a mathematical operation called the *Fourier transformation.* Jean Fourier, an early French mathematician (1768–1830), demonstrated that any periodic function, regardless of complexity, can be described by a sum of simple sine or cosine terms. For example, the square waveform widely encountered in electronics can be described by an equation with the form

$$y = A\left(\sin 2\pi\nu t + \frac{1}{3}\sin 6\pi\nu t + \frac{1}{5}\sin 10\pi\nu t + \cdots + \frac{1}{n}\sin 2n\pi\nu t\right) \tag{6-8}$$

where n takes values of 3, 5, 7, 9, 11, 13, and so forth. A graphical representation of the summation process is shown in Figure 6-6. The solid curve in Figure 6-6a is the sum of three sine waves that differ in amplitude in the ratio of 5:3:1 and in frequency in the ratio of 1:3:5. Note that the resultant approximates the shape of a square wave after including only three terms in Equation 6-8. As shown by the solid line in Figure 6-6b, the resultant more closely approaches a square wave when nine waves are incorporated.

Decomposing a complex waveform into its sine or cosine components is tedious and time consuming when done by hand. Efficient software, however, makes it practical to perform Fourier transformations on a routine basis on a computer. The application of this technique was mentioned in Section 5C-2 and will be considered in the discussion of several types of spectroscopy.

6B-5 Diffraction of Radiation

All types of electromagnetic radiation exhibit *diffraction,* a process in which a parallel beam of radiation is bent as it passes by a sharp barrier or through a narrow opening. Figure 6-7 illustrates the process. Diffraction is a wave property, which can be observed not only for electromagnetic radiation but also for mechanical or

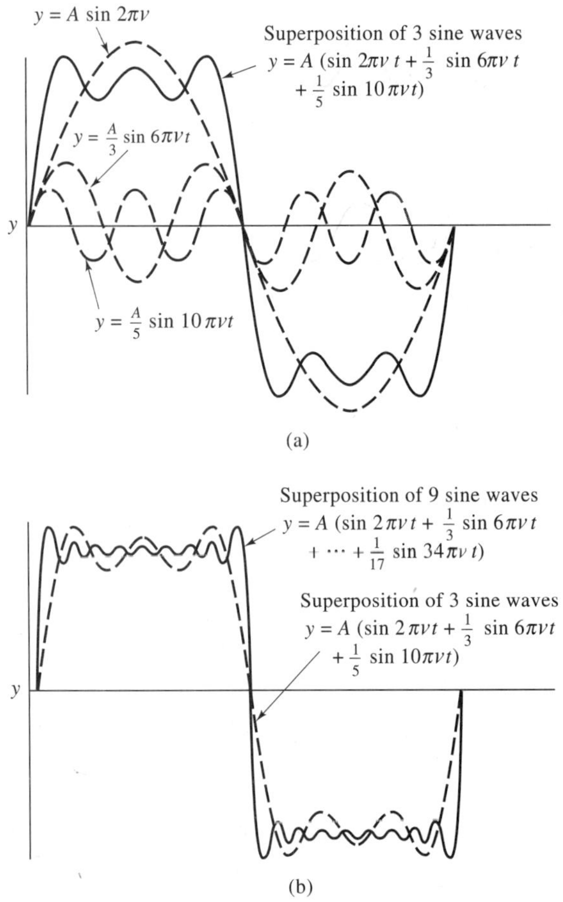

Figure 6-6 Superposition of sine waves to form a square wave: (a) combination of three sine waves; (b) combination of three, as in (a), and nine sine waves.

acoustical waves. For example, diffraction is readily demonstrated in the laboratory by mechanically generating waves of constant frequency in a tank of water and observing the wave crests before and after they pass through a rectangular opening or slit. When the slit is wide relative to the wavelength (Figure 6-7a), diffraction is slight and difficult to detect. On the other hand, when the wavelength and the slit opening are of the same order of magnitude, as in Figure 6-7b, diffraction becomes pronounced. Here, the slit behaves as a new source from which waves radiate in a series of nearly 180-deg arcs. Thus, the direction of the wave front appears to bend as a consequence of passing the two edges of the slit.

Diffraction is a consequence of *interference*. This relationship is most easily understood by considering an experiment, performed first by Thomas Young in 1800, in which the wave nature of light was unambiguously demonstrated. As shown in Figure 6-8a, a parallel beam of light is allowed to pass through a narrow slit *A* (or in Young's experiment, a pinhole) whereupon it is diffracted and illuminates more or less equally two closely spaced slits or pinholes *B* and *C;* the radiation emerging from these slits is then observed on the screen lying in a plane *XY.* If the radiation is monochromatic, a series of dark and light images perpendicular to the plane of the page is observed.

Figure 6-8b is a plot of the intensities of the bands as a function of distance along the length of the screen. If, as in this diagram, the slit widths approach the wavelength of radiation, the band intensities decrease only gradually with increasing distances from the central band. With wider slits, the decrease is much more pronounced.

In Figure 6-8a, the appearance of the central band *E,* which lies in the shadow of the opaque material separating the two slits, is readily explained by noting that the paths from *B* to *E* and *C* to *E* are identical. Thus, constructive interference of the diffracted rays from the two

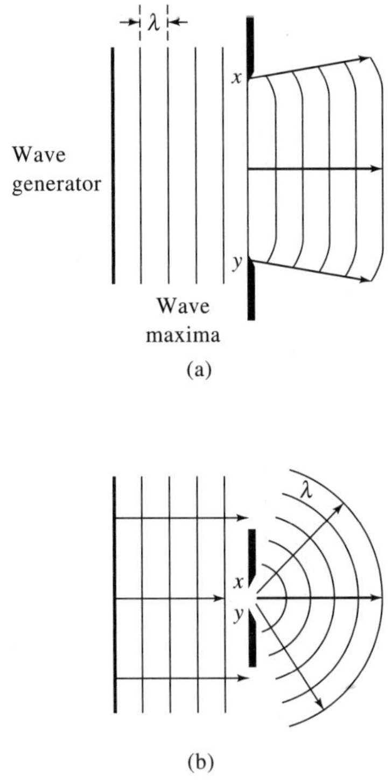

Figure 6-7 Propagation of waves through a slit: (a) $xy >> \lambda$; (b) $xy = \lambda$.

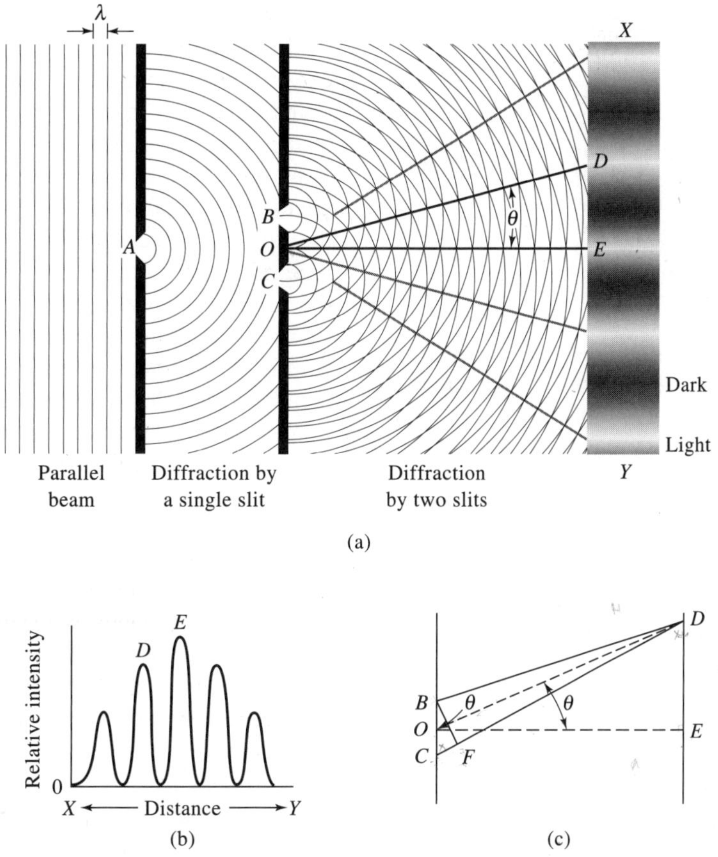

Figure 6-8 Diffraction of monochromatic radiation by slits.

slits occurs, and an intense band is observed. With the aid of Figure 6-8c, the conditions for maximum constructive interference, which result in the other light bands, can be derived. In Figure 6-8c, the angle of diffraction θ is the angle from the normal, formed by the dotted line extending from a point *O*, halfway between the slits, to the point of maximum intensity *D*. The solid lines *BD* and *CD* represent the light paths from the slits *B* and *C* to this point. Ordinarily, the distance $\overline{OE}$ is enormous compared to the distance between the slits $\overline{BC}$. As a consequence, the lines *BD, OD,* and *CD* are, for all practical purposes, parallel. Line *BF* is perpendicular to *CD* and forms the triangle *BCF,* which is, to a close approximation, similar to *DOE;* consequently, the angle *CBF* is equal to the angle of diffraction θ. We may then write

$$\overline{CF} = \overline{BC} \sin \theta$$

Because *BC* is so very small compared to $\overline{OE}$, $\overline{FD}$ closely approximates $\overline{BD}$, and the distance $\overline{CF}$ is a good measure of the difference in path lengths of beams *BD* and *CD.* For the two beams to be in phase at *D,* it is necessary that $\overline{CF}$ correspond to the wavelength of the radiation; that is,

$$\lambda = \overline{CF} = \overline{BC} \sin \theta$$

Reinforcement also occurs when the additional path length corresponds to 2λ, 3λ, and so forth. Thus, a more general expression for the light bands surrounding the central band is

$$\mathbf{n}\lambda = \overline{BC} \sin \theta \qquad (6\text{-}9)$$

where **n** is an integer called the *order* of interference.

The linear displacement $\overline{DE}$ of the diffracted beam along the plane of the screen is a function of the distance $\overline{OE}$ between the screen and the plane of the slits, as well as the spacing between the slits, and is given by

$$\overline{DE} = \overline{OD} \sin \theta$$

Substitution into Equation 6-9 gives

$$n\lambda = \frac{\overline{BC}\,\overline{DE}}{\overline{OD}} = \frac{\overline{BC}\,\overline{DE}}{\overline{OE}} \qquad (6\text{-}10)$$

Equation 6-10 permits the calculation of the wavelength from the three measurable quantities.

EXAMPLE 6-1

Suppose that the screen in Figure 6-8 is 2.00 m from the plane of the slits and that the slit spacing is 0.300 mm. What is the wavelength of radiation if the fourth band is located 15.4 mm from the central band?

Substituting into Equation 6-10 gives

$$4\lambda = \frac{0.300 \text{ mm} \times 15.4 \text{ mm}}{2.00 \text{ m} \times 1000 \text{ mm/m}} = 0.00231 \text{ mm}$$

$$\lambda = 5.78 \times 10^{-4} \text{ mm} = 578 \text{ nm}$$

6B-6 Coherent Radiation

In order to produce a diffraction pattern such as that shown in Figure 6-8a, it is necessary that the electromagnetic waves that travel from slits B and C to any given point on the screen (such as D or E) have sharply defined phase differences that remain *entirely* constant with time; that is, the radiation from slits B and C must be *coherent*. The conditions for coherence are that (1) the two sources of radiation must have identical frequencies (or sets of frequencies) and (2) the phase relationships between the two beams must remain constant with time. The necessity for these requirements can be demonstrated by illuminating the two slits in Figure 6-8a with individual tungsten lamps. Under this circumstance, the well-defined light and dark patterns disappear and are replaced by a more or less uniform illumination of the screen. This behavior is a consequence of the *incoherent* character of filament sources (many other sources of electromagnetic radiation are incoherent as well).

With incoherent sources, light is emitted by individual atoms or molecules, and the resulting beam is the summation of countless individual events, each of which lasts on the order of 10^{-8} s. Thus, a beam of radiation from this type of source is not continuous but instead is composed of a series of *wave trains* that are a few meters in length at most. Because the processes that produce trains are random, the phase differences among the trains must also be variable. A wave train from slit B may arrive at a point on the screen in phase with a wave train from C so that constructive interference occurs; an instant later, the trains may be totally out of phase at the same point, and destructive interference occurs. Thus, the radiation at all points on the screen is governed by the random phase variations among the wave trains; uniform illumination, which represents an average for the trains, is the result.

There are sources that produce electromagnetic radiation in the form of trains with essentially infinite length and constant frequency. Examples include radiofrequency oscillators, microwave sources, and optical lasers. Various mechanical sources, such as a two-pronged vibrating tapper in a water-containing ripple tank, produce a mechanical analog of coherent radiation. When two coherent sources are substituted for slit A in the experiment shown in Figure 6-8a, a regular diffraction pattern is observed.

Diffraction patterns can be obtained from random sources, such as tungsten filaments, provided that an arrangement similar to that shown in Figure 6-8a is employed. Here, the very narrow slit A assures that the radiation reaching B and C emanates from the same small region of the source. Under this circumstance, the various wave trains that exit from slits B and C have a constant set of frequencies and phase relationships to one another and are thus coherent. If the slit at A is widened so that a larger part of the source is sampled, the diffraction pattern becomes less pronounced because the two beams are only partially coherent. If slit A is made sufficiently wide, the incoherence may become great enough to produce only a constant illumination across the screen.

6B-7 Transmission of Radiation

It is observed experimentally that the rate at which radiation is propagated through a transparent substance is less than its velocity in a vacuum and depends upon the kinds and concentrations of atoms, ions, or molecules in the medium. It follows from these observations that the radiation must interact in some way with the matter. Because a frequency change is not observed, however, the interaction *cannot* involve a permanent energy transfer.

The *refractive index* of a medium is one measure of its interaction with radiation and is defined by

$$\eta_i = \frac{c}{v_i} \qquad (6\text{-}11)$$

where η_i is the refractive index at a specified frequency i, v_i is the velocity of the radiation in the medium, and c is its velocity in a vacuum. The refractive index of most liquids lies between 1.3 and 1.8; it is 1.3 to 2.5 or higher for solids.[4]

The interaction involved in transmission can be ascribed to periodic *polarization* of the atomic and molecular species that make up the medium. Polarization in this context means the temporary deformation of the electron clouds associated with atoms or molecules that is brought about by the alternating electromagnetic field of the radiation. Provided that the radiation is not absorbed, the energy required for polarization is only momentarily retained (10^{-14} to 10^{-15} s) by the species and is reemitted without alteration as the substance returns to its original state. Since there is no net energy change in this process, the frequency of the emitted radiation is unchanged, but the rate of its propagation is slowed by the time that is required for retention and reemission to occur. Thus, transmission through a medium can be viewed as a stepwise process that involves polarized atoms, ions, or molecules as intermediates.

Radiation from polarized particles should be emitted in all directions in a medium. If the particles are small, however, it can be shown that destructive interference prevents the propagation of significant amounts in any direction other than that of the original light path. On the other hand, if the medium contains large particles (such as polymer molecules or colloidal particles), this destructive interference is incomplete, and a portion of the beam is scattered in all directions as a consequence of the interaction step. Scattering is considered in a later section of this chapter.

Since the velocity of radiation in matter is wavelength dependent and since c in Equation 6-11 is independent of this parameter, the refractive index of a substance must also change with wavelength. The variation in refractive index of a substance with wavelength or frequency is called its *dispersion.* The dispersion of a typical substance is shown in Figure 6-9. Clearly, the relationship is complex; generally, however, dispersion plots exhibit two types of regions. In the *normal dispersion* region, there is a gradual increase in refractive index with increasing frequency (or decreasing wavelength). Regions of *anomalous dispersion* are frequency ranges in which a sharp change in refractive index is observed. Anomalous dispersion always occurs at frequencies that correspond to the natural harmonic frequency associated with some part of the molecule, atom, or ion of the substance. At such a frequency, permanent energy transfer from the radiation to the substance occurs and *absorption* of the beam is observed. Absorption is discussed in a later section.

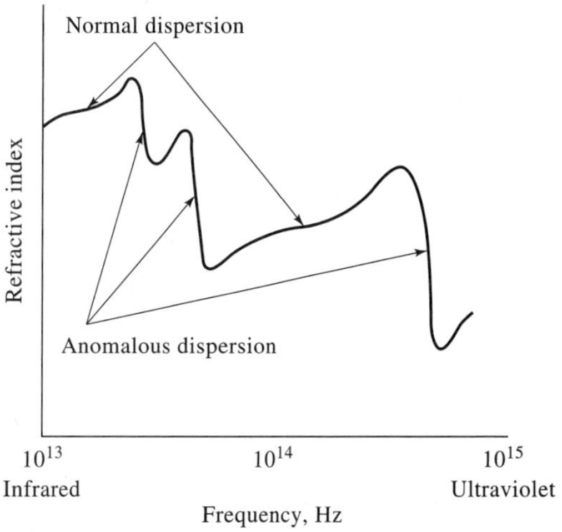

Figure 6-9 A typical dispersion curve.

Dispersion curves are important when choosing materials for the optical components of instruments. A substance that exhibits normal dispersion over the wavelength region of interest is most suitable for the manufacture of lenses, for which a high and relatively constant refractive index is desirable. Chromatic aberrations (formation of colored images) are minimized through the choice of such a material. In contrast, a substance with a refractive index that is not only large but also highly frequency dependent is selected for the fabrication of prisms. The applicable wavelength region for the prism thus approaches the anomalous dispersion region for the material from which it is fabricated.

6B-8 Refraction of Radiation

When radiation passes at an angle through the interface between two transparent media that have different densities, an abrupt change in direction, or *refraction,* of the beam is observed as a consequence of a difference in velocity of the radiation in the two media. When the beam passes from a less dense to a more dense environ-

[4] For a more complete discussion of refractometry, see S. Z. Lewin and N. Bauer, in *Treatise on Analytical Chemistry,* I. M. Kolthoff and P. J. Elving, Eds., Part I, Vol. 6, Chapter 70. New York: Interscience, 1965.

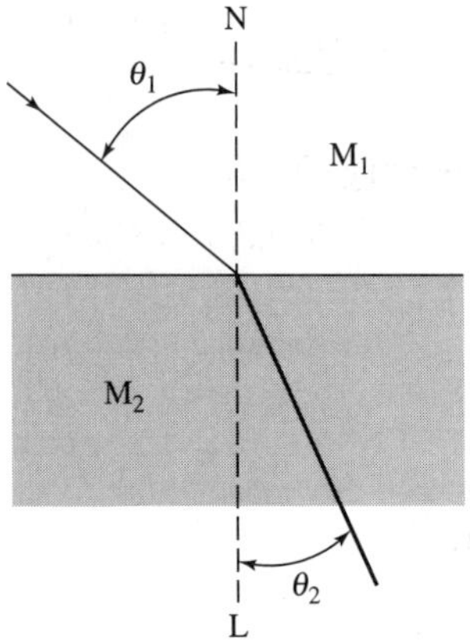

Figure 6-10 Refraction of light in passing from a less dense medium M_1 into a more dense medium M_2, where its velocity is lower.

ment, as in Figure 6-10, the bending is toward the normal to the interface. Bending away from the normal occurs when the beam passes from a more dense to a less dense medium.

The extent of refraction is given by Snell's law:

$$\frac{\sin \theta_1}{\sin \theta_2} = \frac{\eta_2}{\eta_1} = \frac{v_1}{v_2} \qquad (6\text{-}12)$$

If M_1 in Figure 6-10 is a *vacuum*, v_i is equal to c, and η_i is unity (see Equation 6-11); with rearrangement, Equation 6-12 simplifies to

$$(\eta_2)_{vac} = \frac{(\sin \theta_1)_{vac}}{\sin \theta_2} \qquad (6\text{-}13)$$

The refractive indexes of substance M_2 can then be computed from measurements of $(\theta_1)_{vac}$ and θ_2. For convenience, refractive indexes are usually measured and reported with air, rather than a vacuum, as the reference. The refractive index is then

$$(\eta_2)_{air} = \frac{(\sin \theta_1)_{air}}{\sin \theta_2} \qquad (6\text{-}14)$$

Most compilations of refractive indexes provide data in terms of Equation 6-14. Such data are readily converted to refractive indexes with vacuum as a reference by multiplying by the refractive index of air relative to a vacuum. That is,

$$\eta_{vac} = 1.00027\ \eta_{air}$$

This conversion is seldom necessary.

6B-9 Reflection of Radiation

When radiation crosses an interface between media that differ in refractive index, reflection always occurs. The fraction of radiation reflected becomes greater with increasing differences in refractive index. For a beam that enters an interface at right angles, the fraction reflected is given by

$$\frac{I_r}{I_0} = \frac{(\eta_2 - \eta_1)^2}{(\eta_2 + \eta_1)^2} \qquad (6\text{-}15)$$

where I_0 is the intensity of the incident beam and I_r is the reflected intensity; η_1 and η_2 are the refractive indexes of the two media.

EXAMPLE 6-2

Calculate the percent loss of intensity due to reflection of a perpendicular beam of yellow light as it passes through a glass cell that contains water. Assume that for yellow radiation the refractive index of glass is 1.50, of water is 1.33, and of air is 1.00.

The total reflective loss will be the sum of the losses occurring at each of the interfaces. For the first interface (air to glass), we can write

$$\frac{I_{r1}}{I_0} = \frac{(1.50 - 1.00)^2}{(1.50 + 1.00)^2} = 0.040$$

The beam intensity is reduced to $(I_0 - 0.040\ I_0) = 0.960\ I_0$. Reflection loss at the glass-to-water interface is then given by

$$\frac{I_{r2}}{0.960\ I_0} = \frac{(1.50 - 1.33)^2}{(1.50 + 1.33)^2} = 0.0036$$

$$I_{r2} = 0.0035\ I_0$$

The beam intensity is further reduced to $(0.960\ I_0 - 0.0035\ I_0) = 0.957\ I_0$. At the water-to-glass interface

$$\frac{I_{r3}}{0.957\ I_0} = \frac{(1.50 - 1.33)^2}{(1.50 + 1.33)^2} = 0.0036$$

$$I_{r3} = 0.0035\ I_0$$

and the beam intensity becomes $0.953\ I_0$. Finally, the reflection at the second glass-to-air interface will be

$$\frac{I_{r4}}{0.953\ I_0} = \frac{(1.50 - 1.00)^2}{(1.50 + 1.00)^2} = 0.0400$$

$$I_{r4} = 0.038\ I_0$$

The total reflection loss I_{rt} is

$$I_{rt} = 0.040\ I_0 + 0.0035\ I_0 + 0.0035\ I_0 + 0.038\ I_0 = 0.085\ I_0$$

and

$$\frac{I_{rt}}{I_0} = 0.85 \quad \text{or} \quad 8.5\%$$

It will become evident in later chapters that losses such as those shown in Example 6-2 are of considerable significance in various optical instruments.

Reflective losses at a polished glass or quartz surface increase only slightly as the angle of the incident beam increases up to about 60 deg. Beyond this figure, however, the percentage of radiation that is reflected increases rapidly and approaches 100% at 90 deg, or grazing incidence.

6B-10 Scattering of Radiation

As noted earlier, the transmission of radiation in matter can be pictured as a momentary retention of the radiant energy by atoms, ions, or molecules followed by reemission of the radiation in all directions as the particles return to their original state. With atomic or molecular particles that are small relative to the wavelength of the radiation, destructive interference removes most but not all of the reemitted radiation except the radiation that travels in the original direction of the beam; the path of the beam appears to be unaltered as a consequence of the interaction. Careful observation, however, reveals that a very small fraction of the radiation is transmitted at all angles from the original path and that the intensity of this *scattered radiation* increases with particle size.

Rayleigh Scattering

Scattering by molecules or aggregates of molecules with dimensions significantly smaller than the wavelength of the radiation is called *Rayleigh scattering;* its intensity is proportional to the inverse fourth-power of the wavelength, the dimensions of the scattering particles, and the square of the polarizability of the particles. An everyday manifestation of Rayleigh scattering is the blue color of the sky, which results from the greater scattering of the shorter wavelengths of the visible spectrum.

Scattering by Large Molecules

With particles of colloidal dimensions, scattering is sufficiently intense to be seen by the naked eye (the *Tyndall effect*). Measurements of scattered radiation are used to determine the size and shape of polymer molecules and colloidal particles.

Raman Scattering

The Raman scattering effect differs from ordinary scattering in that part of the scattered radiation suffers quantized frequency changes. These changes are the result of vibrational energy level transitions that occur in the molecules as a consequence of the polarization process. Raman spectroscopy is discussed in Chapter 18.

6B-11 Polarization of Radiation

Ordinary radiation consists of a bundle of electromagnetic waves in which the vibrations are equally distributed among a huge number of planes centered along the path of the beam. Viewed end on, a beam of monochromatic radiation can be visualized as an infinite set of electrical vectors that fluctuate in length from zero to a maximum amplitude A. Figure 6-11b depicts an end-on view of these vectors at various times during the passage of one wave of monochromatic radiation through a fixed point in space.

Figure 6-12a shows a few of the vectors depicted in Figure 6-11b at the instant the wave is at its maximum.

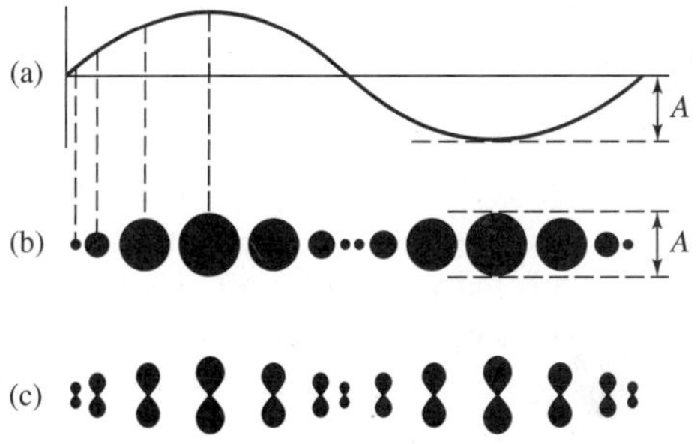

Figure 6-11 Unpolarized and plane-polarized radiation: (a) cross-sectional view of a beam of monochromatic radiation, (b) successive end-on view of the radiation in (a) if it is unpolarized, (c) successive end-on views of the radiation of (a) if it is plane polarized on the vertical axis.

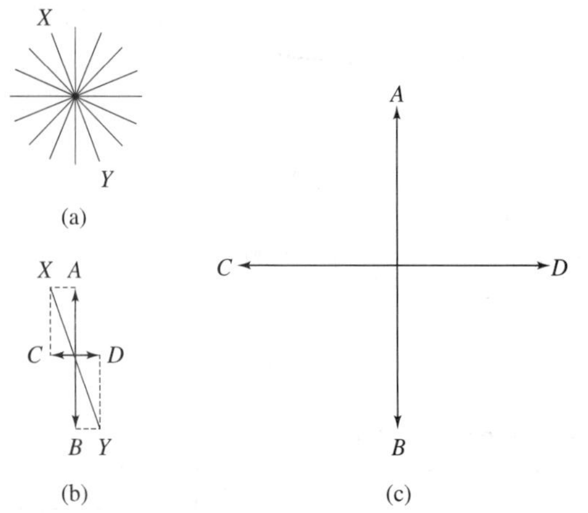

Figure 6-12 (a) A few of the electrical vectors of a beam traveling perpendicular to the page.
(b) The resolution of a vector in plane *XY* into two mutually perpendicular components.
(c) The resultant when all vectors are resolved (not to scale).

The vector in any one plane, say *XY* as depicted in Figure 6-11a, can be resolved into two mutually perpendicular components *AB* and *CD* as shown in Figure 6-12b. If the two components for all of the planes shown in Figure 6-12a are combined, the resultant has the appearance shown in Figure 6-12c. Removal of one of the two resultant planes of vibration in Figure 6-12c produces a beam that is *plane polarized.* The resultant electric vector of a plane-polarized beam then occupies a single plane in space. Figure 6-11c shows an end-on view of a beam of plane-polarized radiation after various time intervals.

Plane-polarized electromagnetic radiation is produced by certain radiant energy sources. For example, the radio waves emanating from an antenna and the microwaves produced by a klystron tube are both plane polarized. Visible and ultraviolet radiation from relaxation of a single excited atom or molecule is also polarized, but the beam from such a source has no net polarization since it is made up of a multitude of individual wave trains produced by an enormous number of individual atomic or molecular events. The plane of polarization of these individual waves is random so that their individual polarizations cancel.

Polarized ultraviolet and visible radiation is produced by passage of radiation through media that selectively absorb, reflect, or refract radiation that vibrates in only one plane.

6C QUANTUM-MECHANICAL PROPERTIES OF RADIATION

When electromagnetic radiation is emitted or absorbed, a permanent transfer of energy from the emitting object or to the absorbing medium occurs. In order to describe these phenomena, it is necessary to treat electromagnetic radiation not as a collection of waves but rather as a stream of discrete particles called *photons* or *quanta.* The need for a particle model for radiation became apparent as a consequence of the discovery of the photoelectric effect in the late nineteenth century.

6C-1 The Photoelectric Effect

The first observation of the photoelectric effect was made in 1887 by Heinrich Hertz, who reported that a spark jumped more readily between two charged spheres when their surfaces were illuminated with light. Between the time of this observation and the theoretical explanation of the photoelectric effect by Einstein in 1905, several important studies of the photoelectric effect were performed with what is now known as a vacuum phototube. Einstein's explanation of the photoelectric effect was both simple and elegant but was far enough ahead of its time that it was not generally accepted until 1916, when Millikan's systematic studies had confirmed Einstein's theoretical conclusions in every detail.

Figure 6-13 is a schematic of vacuum phototube circuit similar to the one used by Millikan to study the

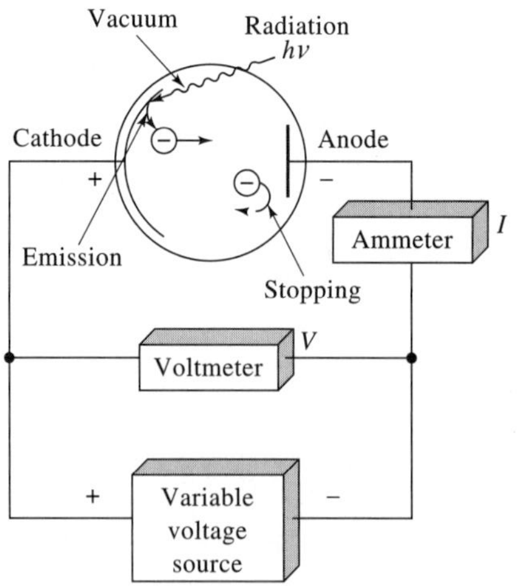

Figure 6-13 Apparatus for studying the photoelectric effect.

photoelectric effect. The surface of the large *photocathode* on the left usually is coated with an alkali metal or one of its compounds, but other metals may be used. When monochromatic radiation impinges on the photocathode, electrons are emitted from its surface with a range of kinetic energies. As long as the voltage V applied between the anode and the cathode is positive, the electrons are swept from left to right through the phototube to produce a current I in the circuit. When the voltage across the phototube is adjusted so that the anode is slightly negative with respect to the cathode, the photoelectrons are repelled by the anode, and the photocurrent decreases as expected. At this point in the experiment, however, some of the electrons have sufficient kinetic energy to overcome the negative potential applied to the anode, and a current is still observed.

This experiment may be repeated for phototubes with different materials coated on the photocathode. In each experiment, the photocurrent is measured as a function of the applied voltage, and the voltage V_0 at which the photocurrent becomes precisely zero is noted. The negative voltage at which the photocurrent is zero is called the *stopping voltage*. It corresponds to the potential at which the most energetic electrons from the cathode are just repelled from the anode. If we multiply the stopping voltage by the charge on the electron, $e = -1.60 \times 10^{-19}$ coulombs, we have a measure of the kinetic energy in joules of the *most energetic* of the emitted electrons. When this experiment is repeated for various frequencies of monochromatic light, the following results are observed:

1. When light of constant frequency is focused on the anode at low applied negative potential, the photocurrent is directly proportional to the intensity of the incident radiation.
2. The magnitude of the stopping voltage depends on the frequency of the radiation impinging on the photocathode.
3. The stopping voltage depends on the chemical composition of the coating on the photocathode.
4. The stopping voltage is *independent of the intensity of the incident radiation.*

These observations suggest that electromagnetic radiation is a form of energy that releases electrons from metallic surfaces and imparts to these electrons sufficient kinetic energy to cause them to travel to a negatively charged electrode. Furthermore, the number of photoelectrons released is proportional to the intensity of the incident beam.

The results of these experiments are shown in the plots of Figure 6-14, in which the maximum kinetic energy, or stopping energy eV_0, of the photoelectrons is plotted against frequency for photocathode surfaces of potassium, sodium, and copper. Other surfaces give plots with identical slopes, h, but different intercepts, ω.

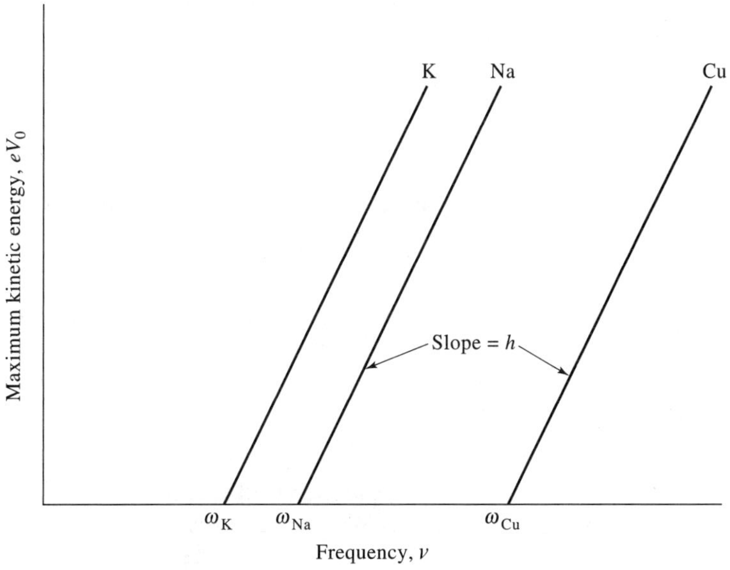

Figure 6-14 Maximum kinetic energy of photoelectrons emitted from three metal surfaces as a function of radiation frequency.

The plots shown in Figure 6-14 are described by the equation

$$eV_0 = h\nu + \omega \qquad (6\text{-}16)$$

In this equation, the slope h is the Planck constant, which is equal to 6.6254×10^{-34} joule second, and the intercept ω is the *work function,* a constant that is characteristic of the surface material. Approximately a decade before Millikan's work that led to Equation 6-16, Einstein had proposed the relationship between frequency ν of light and energy E as embodied by his now famous equation

$$E = h\nu \qquad (6\text{-}17)$$

By substituting Einstein's equation into Equation 6-16 and rearranging, we obtain

$$E = h\nu = eV_0 + \omega \qquad (6\text{-}18)$$

This equation shows that the energy of an incoming photon is equal to the kinetic energy of the ejected photoelectron plus the energy required to eject the photoelectron from the surface being irradiated.

The photoelectric effect cannot be explained by a wave model but requires instead a quantum model in which radiation is viewed as a stream of discrete bundles of energy, or photons. For example, calculations indicate that no single electron could acquire sufficient energy for ejection if the radiation striking the surface is uniformly distributed over the face of the electrode as it is in the wave model; nor could any electron accumulate enough energy rapidly enough to establish the nearly instantaneous currents that are observed. Thus, it is necessary to assume that the energy is not uniformly distributed over the beam front but rather is concentrated in packets, or bundles of energy.

Equation 6-18 can be recast in terms of wavelength by substitution of Equation 6-2. That is,

$$E = h\frac{c}{\lambda} = eV_0 - \omega \qquad (6\text{-}19)$$

Note that although photon energy is directly proportional to frequency, it is a reciprocal function of wavelength.

EXAMPLE 6-3

Calculate the energy of **(a)** a 5.3-Å X-ray photon and **(b)** a 530-nm photon of visible radiation.

$$E = h\nu = \frac{hc}{\lambda}$$

(a) $E = \dfrac{(6.63 \times 10^{-34}\ \text{J} \cdot \text{s}) \times (3.00 \times 10^{8}\ \text{m/s})}{5.30\text{Å} \times (10^{-10}\ \text{m/Å})}$

$= 3.75 \times 10^{-16}\ \text{J}$

The energy of radiation in the X-ray region is commonly expressed in electron volts, the energy acquired by an electron that has been accelerated through a potential of one volt. In the conversion table inside the front cover of this book, we see that $1\ \text{J} = 6.24 \times 10^{18}\ \text{eV}$.

$$E = 3.75 \times 10^{-16}\ \text{J} \times (6.24 \times 10^{18}\ \text{eV/J}) = 2.34 \times 10^{3}\ \text{eV}$$

(b) $E = \dfrac{(6.63 \times 10^{-34}\ \text{J} \cdot \text{s}) \times (3.00 \times 10^{8}\ \text{m/s})}{530\ \text{nm} \times (10^{-9}\ \text{m/nm})}$

$= 3.75 \times 10^{-19}\ \text{J}$

Energy of radiation in the visible region is often expressed in kJ/mol rather than kJ/photon to aid in the discussion of the relationships between the energy of absorbed photons and the energy of chemical bonds.

$$E = 3.75 \times 10^{-19}\ \frac{\text{J}}{\text{photon}} \times \frac{(6.02 \times 10^{23}\ \text{photons})}{\text{mol}} \times 10^{-3}\ \frac{\text{kJ}}{\text{J}}$$

$= 226\ \text{kJ/mol}$

Jan. 05, 05

6C-2 Energy States of Chemical Species

The quantum theory was first proposed in 1900 by Max Planck, a German physicist, to explain the properties of radiation emitted by heated bodies. The theory was later extended to rationalize other types of emission and absorption processes. Two important postulates of quantum theory include:

1. Atoms, ions, and molecules can exist only in certain discrete states, characterized by definite amounts of energy. When a species changes its state, it absorbs

or emits an amount of energy *exactly* equal to the energy difference between the states.

2. When atoms, ions, or molecules absorb or emit radiation in making the transition from one energy state to a second, the frequency ν or the wavelength λ of the radiation is related to the energy difference between the states by the equation

$$E_1 - E_0 = h\nu = \frac{hc}{\lambda} \qquad (6\text{-}20)$$

where E_1 is the energy of the higher state and E_0 the energy of the lower state. The terms c and h are the speed of light and the Planck constant, respectively.

For atoms or ions in the elemental state, the energy of any given state arises from the motion of electrons around the positively charged nucleus. As a consequence the various energy states are called *electronic states.* In addition to having electronic states, molecules also have quantized *vibrational states* that are associated with the energy of interatomic vibrations and quantized *rotational states* that arise from the rotation of molecules around their centers of gravity.

The lowest energy state of an atom or molecule is its *ground state.* Higher energy states are termed *excited states.* Generally at room temperature, chemical species are in their ground state.

6C-3 Emission of Radiation

Electromagnetic radiation is produced when excited particles (atoms, ions, or molecules) relax to lower energy levels by giving up their excess energy as photons. Excitation can be brought about by a variety of means, including (1) bombardment with electrons or other elementary particles, which generally leads to the emission of X-radiation; (2) exposure to an electrical current ac spark or the heat of a flame, an arc, or a furnace, which produces ultraviolet, visible, or infrared radiation; (3) irradiation with a beam of electromagnetic radiation, which produces fluorescent radiation; an exothermic chemical reaction that produces chemiluminescence.

Radiation from an excited source is conveniently characterized by means of an *emission spectrum,* which usually takes the form of a plot of the relative power of the emitted radiation as a function of wavelength or frequency. Figure 6-15 illustrates a typical emission spectrum, which was obtained by aspirating a brine solution into an oxyhydrogen flame. Three types of spectra are evident in the figure: *lines, bands,* and a *continuum.* The line spectrum is made up of a series of sharp, well-defined peaks caused by excitation of individual atoms. The band spectrum consists of several groups of lines so closely spaced that they are not completely resolved. The source of the bands consists of small molecules or radicals. Finally, the continuum portion of the spectrum is responsible for the increase in the background that is evident above about 350 nm. The line and band spectra are superimposed on this continuum. The source of the continuum is described on page 133.

Figure 6-16 is an X-ray emission spectrum produced by bombarding a piece of molybdenum with an energetic stream of electrons. Note that in Figure 6-16, there is also a line spectrum superimposed on a continuum. The source of the continuum is described in Section 12A-1.

Line Spectra

Line spectra in the ultraviolet and visible regions are produced when the radiating species are individual atomic particles that are well separated, in a gas phase. The individual particles in a gas behave independently of one another, and the spectrum consists of a series of sharp lines with widths of about 10^{-4} Å. In Figure 6-15, lines for gas-phase sodium, potassium, strontium, and calcium are identified.

The energy-level diagram in Figure 6-17a shows the source of two of the lines in a typical emission spectrum of an element. The horizontal line labeled E_0 corresponds to the lowest, or ground-state, energy of the atom. The horizontal lines labeled E_1 and E_2 are two higher-energy electronic levels of the species. For example, the single outer electron in the ground state E_0 for a sodium atom is located in the $3s$ orbital. Energy level E_1 then represents the energy of the atom when this electron has been promoted to the $3p$ state by absorption of thermal, electrical, or radiant energy. The promotion is depicted by the shorter wavy arrow on the left in Figure 6-17a. After perhaps 10^{-8} s, the atom returns to the ground state, emitting a photon whose frequency and wavelength are given by Equation 6-20.

$$\nu_1 = (E_1 - E_0)/h$$

$$\lambda_1 = hc/(E_1 - E_0)$$

This emission process is illustrated by the shorter straight arrow on the right in Figure 6-17a.

For the sodium atom, E_2 in Figure 6-17 corresponds to the more energetic $4p$ state; the resulting

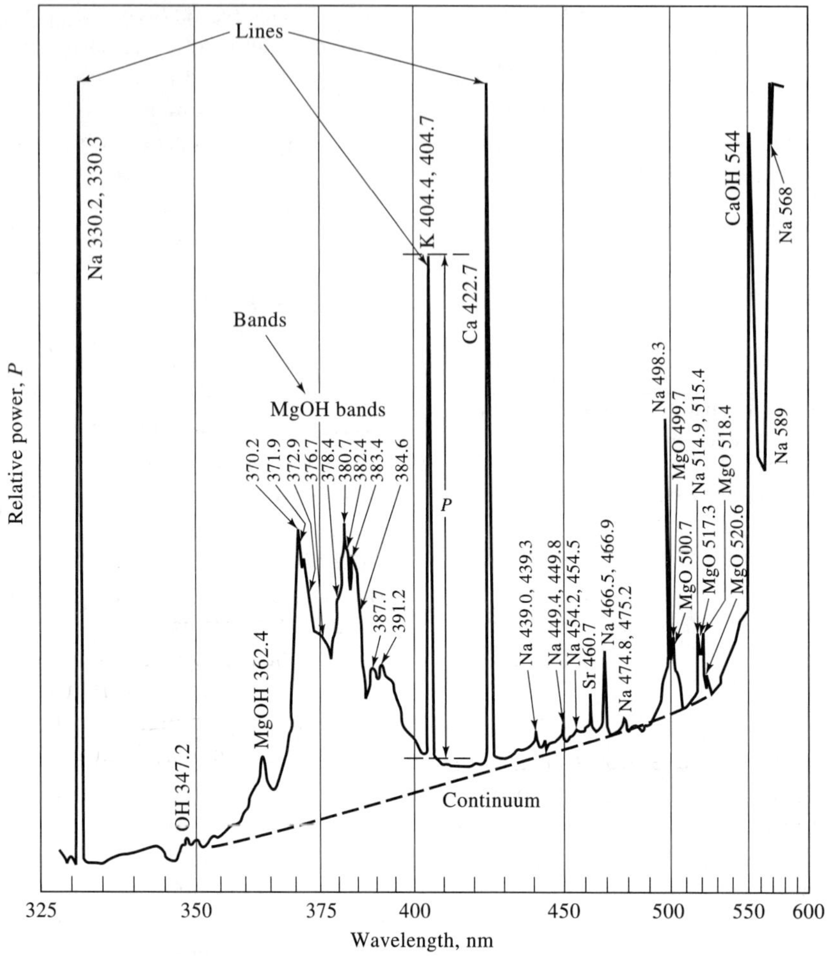

Figure 6-15 Emission spectrum of a brine obtained with an oxyhydrogen flame.
(F. Hermann and C. T. J. Alkemade, Chemical Analysis by Flame Photometry, *2nd ed., p. 484. New York: Interscience, 1963. With permission.)*

emitting radiation λ_2 appears at a shorter wavelength or a higher frequency. The line at about 330 nm in Figure 6-15 results from this transition; the 3*p*-to-3*s* transition provides a line at about 590 nm.

X-Ray line spectra are also produced by electronic transitions. In this case, however, the electrons involved are those in the innermost orbitals. Thus, in contrast to ultraviolet and visible emissions, the X-ray spectrum for an element is independent of its environment. For example, the emission spectrum for molybdenum is the same regardless of whether the sample being excited is molybdenum metal, solid molybdenum sulfide, gaseous molybdenum hexafluoride, or an aqueous solution of an anionic complex of the metal.

Band Spectra

Band spectra are often encountered in spectral sources when gaseous radicals or small molecules are present. For example, in Figure 6-15 bands for OH, MgOH, and MgO are labeled and consist of a series of closely spaced lines that are not fully resolved by the instrument used to obtain the spectrum. Bands arise from numerous quantized vibrational levels that are superimposed on the ground-state electronic energy level of a molecule.

Figure 6-17b is a partial energy-level diagram for a molecule that shows its ground state E_0 and two of its several excited electronic states, E_1 and E_2. A few of the

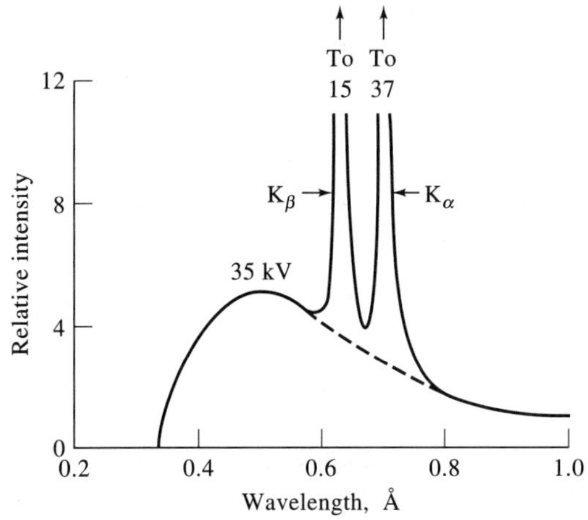

Figure 6-16 X-Ray emission spectrum of molybdenum metal.

many vibrational levels associated with the ground state are also shown. Vibrational levels associated with the two excited states have been omitted because the lifetime of an excited vibrational state is brief compared with that of an electronically excited state (about 10^{-15} s versus 10^{-8} s). A consequence of this tremendous difference in lifetimes is that when an electron is excited to one of the higher vibrational levels of an electronic state, relaxation to the lowest vibrational level of that state occurs before an electronic transition to the ground state can occur. Therefore, the radiation produced by the electrical or thermal excitation of polyatomic species nearly always results from a transition from the *lowest vibrational level of an excited electronic state* to any of the several vibrational levels of the ground state.

The mechanism by which a vibrationally excited species relaxes to the nearest electronic state involves a transfer of its excess energy to other atoms in the system through a series of collisions. As noted, this process takes place at an enormous speed. Relaxation from one electronic state to another can also occur by collisional transfer of energy, but the rate of this process is slow enough that relaxation by photon release is favored.

The energy-level diagram in Figure 6-17b illustrates the mechanism by which two radiation bands that consist of five closely spaced lines are emitted by a molecule excited by thermal or electrical energy. For a real molecule, the number of individual lines is much larger because in addition to the numerous vibrational states, a multitude of rotational states would be superimposed on each. The differences in energy among the rotational levels is perhaps an order of magnitude smaller than that for vibrational states. Thus, a real molecular band would be made up of many more lines than we have shown in Figure 6-17b, and these lines would be much more closely spaced.

Continuum Spectra

As shown in Figure 6-18, truly continuum radiation is produced when solids are heated to incandescence. Thermal radiation of this kind, which is called *blackbody radiation,* is characteristic of the temperature of

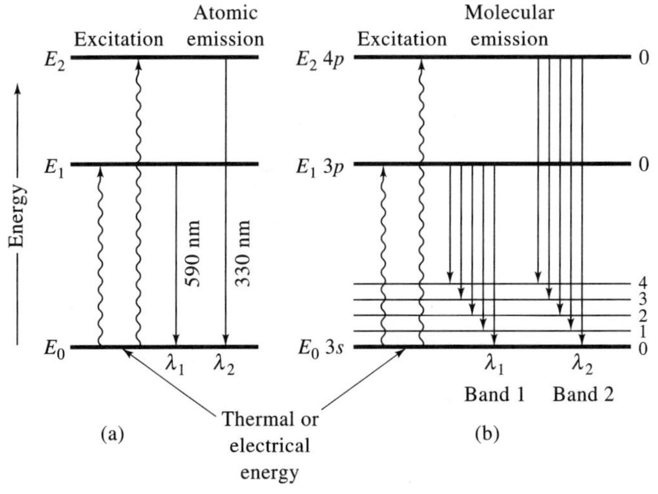

Figure 6-17 Energy-level diagrams for (a) a sodium atom showing the source of a line spectrum and (b) a simple molecule showing the source of a band spectrum.

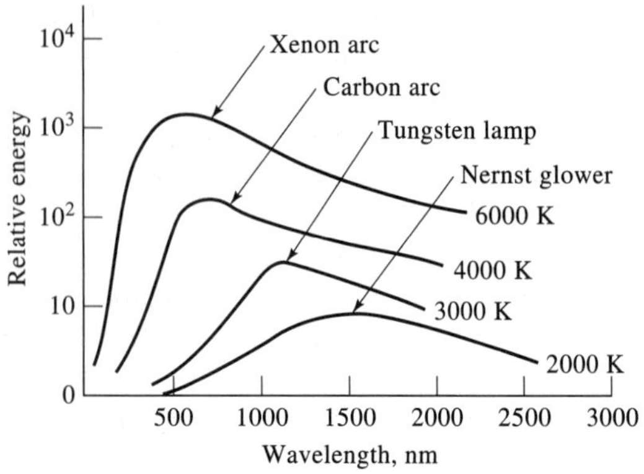

Figure 6-18 Blackbody radiation curves.

the emitting surface rather than the material of which that surface is composed. Blackbody radiation is produced by the innumerable atomic and molecular oscillations excited in the condensed solid by the thermal energy. Note that the energy peaks in Figure 6-18 shift to shorter wavelengths with increasing temperature. It is clear that very high temperatures are needed to cause a thermally excited source to emit a substantial fraction of its energy as ultraviolet radiation.

As noted earlier, part of the continuum background radiation exhibited in the flame spectrum shown in Figure 6-15 is probably thermal emission from incandescent particles in the flame. Note that this background decreases rapidly as the ultraviolet region is approached.

Heated solids are important sources of infrared, visible, and longer-wavelength ultraviolet radiation for analytical instruments.

6C-4 Absorption of Radiation

When radiation passes through a layer of solid, liquid, or gas, certain frequencies may be selectively removed by *absorption,* a process in which electromagnetic energy is transferred to the atoms, ions, or molecules composing the sample. Absorption promotes these particles from their normal room temperature state, or ground state, to one or more higher-energy excited states.

According to quantum theory, atoms, molecules, or ions have only a limited number of discrete, energy levels; for absorption of radiation to occur, the energy of the exciting photon must *exactly* match the energy difference between the ground state and one of the excited states of the absorbing species. Since these energy differences are unique for each species, a study of the frequencies of absorbed radiation provides a means of characterizing the constituents of a sample of matter. For this purpose, a plot of absorbance as a function of wavelength or frequency is experimentally determined (*absorbance,* a measure of the decrease in radiant power, is defined by Equation 6-32 in Section 6D-2). Typical absorption spectra are shown in Figure 6-19.

Examination of the four plots in Figure 6-19 reveals that absorption spectra vary widely in appearance; some are made up of numerous sharp peaks, whereas others consist of smooth continuous curves. In general, the nature of a spectrum is influenced by such variables as the complexity, the physical state, and the environment of the absorbing species. More fundamental, however, are the differences between absorption spectra for atoms and those for molecules.

Atomic Absorption

The passage of polychromatic ultraviolet or visible radiation through a medium that consists of monoatomic particles, such as gaseous mercury or sodium, results in

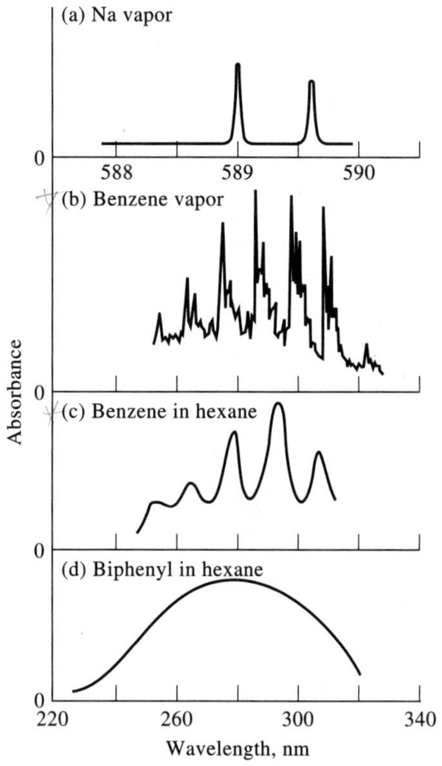

Figure 6-19 Some typical ultraviolet absorption spectra.

the absorption of but a few well-defined frequencies (see Figure 6-19a). The relative simplicity of such spectra is due to the small number of possible energy states for the absorbing particles. Excitation can occur only by an electronic process in which one or more of the electrons of the atom are raised to a higher energy level. For example, sodium vapor exhibits two closely spaced, sharp absorption peaks in the yellow region of the visible spectrum (589.0 and 589.6 nm) as a result of excitation of the 3*s* electron to two 3*p* states that differ only slightly in energy. Several other narrow absorption lines, corresponding to other allowed electronic transitions, are also observed. For example, an ultraviolet peak at about 285 nm results from the excitation of the 3*s* electron in sodium to the excited 5*p* state, a process that requires significantly greater energy than does excitation to the 3*p* state (in fact, the peak at 285 nm is also a doublet; the energy difference between the two peaks is so small, however, that most instruments cannot resolve them).

Ultraviolet and visible radiation has sufficient energy to cause transitions of the outermost or bonding electrons only. X-Ray frequencies, on the other hand, are several orders of magnitude more energetic (see Example 6-3) and are capable of interacting with electrons that are closest to the nuclei of atoms. Absorption peaks that correspond to electronic transitions of these innermost electrons are thus observed in the X-ray region.

Molecular Absorption

Absorption spectra for polyatomic molecules, particularly in the condensed state, are considerably more complex than atomic spectra because the number of energy states of molecules is generally enormous when compared with the number of energy states for isolated atoms. The energy E, associated with the bands of a molecule, is made up of three components. That is,

$$E = E_{\text{electronic}} + E_{\text{vibrational}} + E_{\text{rotational}} \quad (6\text{-}21)$$

where $E_{\text{electronic}}$ describes the electronic energy of the molecule that arises from the energy states of its several bonding electrons. The second term on the right refers to the total energy associated with the multitude of interatomic vibrations that are present in molecular species. Generally, a molecule has many more quantized vibrational energy levels than it does electronic levels. Finally, $E_{\text{rotational}}$ is the energy caused by various rotational motions within a molecule; again the number of rotational states is much larger than the number of vibrational states. Thus, for each electronic energy state of a molecule, several possible vibrational states normally exist; and for each of these vibrational states, in turn, numerous rotational states are possible. As a consequence, the number of possible energy levels for a molecule is normally orders of magnitude greater than the number of possible energy levels for an atomic particle.

Figure 6-20 is a graphical representation of the energy levels associated with a few of the numerous electronic and vibrational states of a molecule. The heavy line labeled E_0 represents the electronic energy of the molecule in its ground state (its state of lowest electronic energy); the lines labeled E_1 and E_2 represent the energies of two excited electronic states. Several of the many vibrational energy levels ($e_0, e_1, \cdots, e_n$) are shown for each of these electronic states.

As can be seen in Figure 6-20, the energy difference between the ground state and an electronically excited state is large relative to the energy differences between vibrational levels in a given electronic state (typically, the two differ by a factor of 10 to 100).

The arrows in Figure 6-20a depict some of the transitions that result from absorption of radiation. Visible radiation causes excitation of an electron from E_0 to any of the n vibrational levels associated with E_1 (only five of the n vibrational levels are shown in Figure 6-20). Potential absorption frequencies are then given by n equations, each with the form

$$\nu_i = \frac{1}{h}(E_1 + e_i' - E_0) \quad (6\text{-}22)$$

where $i = 1, 2, 3, \cdots n$.

Similarly, if the second electronic state has m vibrational levels (four of which are shown), potential absorption frequencies for ultraviolet radiation are given by m equations such as

$$\nu_i = \frac{1}{h}(E_2 + e_i'' - E_0) \quad (6\text{-}23)$$

where $i = 1, 2, 3, \cdots m$.

Finally, as shown in Figure 6-20a, the less energetic near- and mid-infrared radiation can only bring about transitions among the k vibrational levels of the ground state. Here, k potential absorption frequencies are given by k equations, which may be formulated as

$$\nu_i = \frac{1}{h}(e_i - e_0) \quad (6\text{-}24)$$

where $i = 1, 2, 3, \cdots k$.

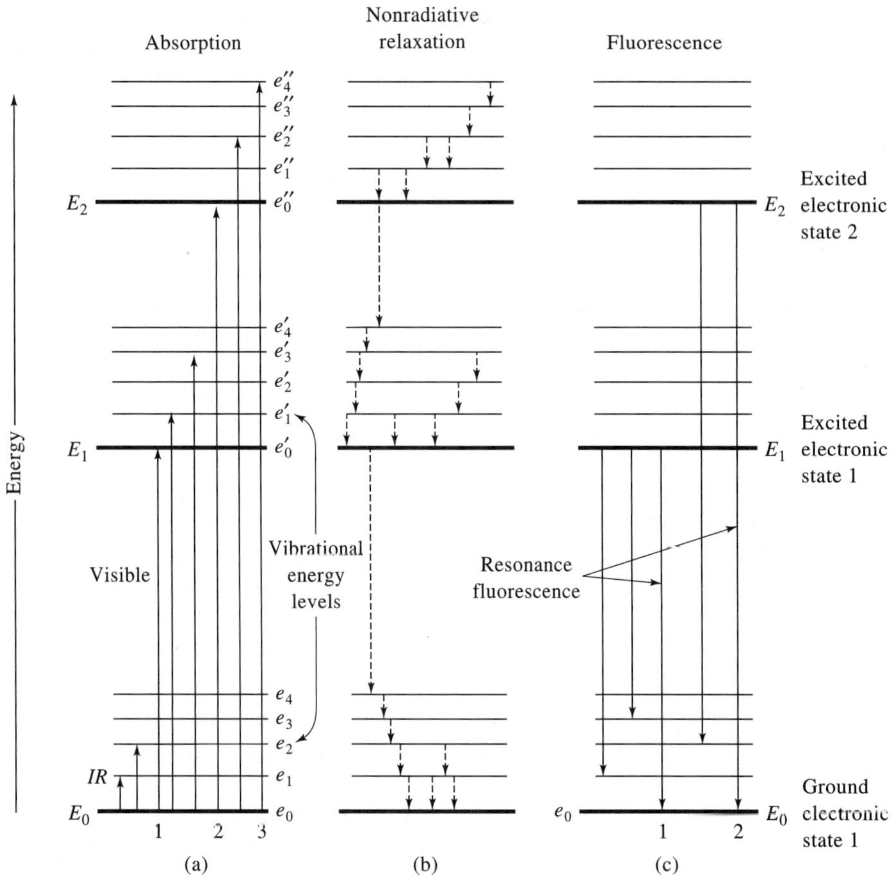

Figure 6-20 Partial energy-level diagrams for a fluorescent organic molecule.

Although they are not shown, several rotational energy levels are associated with each vibrational level in Figure 6-20. The energy difference between the rotational energy levels is small relative to the energy difference between vibrational levels. Transitions between a ground and an excited rotational state is brought about by radiation in the 0.01- to 1-cm range, which includes microwave and longer infrared radiation.

In contrast to atomic absorption spectra, which consist of a series of sharp, well-defined lines, molecular spectra in the ultraviolet and visible regions are ordinarily characterized by absorption regions that often encompass a substantial wavelength range (see Figure 6-19b, c). Molecular absorption also involves electronic transitions. As shown by Equations 6-23 and 6-24, however, several closely spaced absorption lines will be associated with each electronic transition, owing to the existence of numerous vibrational states. Furthermore, as we have mentioned, many rotational energy levels are associated with each vibrational state. As a consequence, the spectrum for a molecule ordinarily consists of a series of closely spaced absorption lines that constitute an *absorption band,* such as those shown for benzene vapor in Figure 6-19b. Unless a high-resolution instrument is employed, the individual peaks may not be detected, and the spectra will appear as broad smooth peaks such as those shown in Figure 6-19c. Finally, in the condensed state, and in the presence of solvent molecules, the individual lines tend to broaden even further to give *continuous spectra* such as that shown in Figure 6-19d. Solvent effects are considered in later chapters.

Pure vibrational absorption is observed in the infrared region, where the energy of radiation is insufficient to cause electronic transitions. Such spectra exhibit narrow, closely spaced absorption peaks that result from transitions among the various vibrational quantum levels (see the transition labeled *IR* at the bottom of Figure 6-20a). Variations in rotational levels may give rise to a series of peaks for each vibrational state; but in liquid and solid samples rotation is often hindered to such an extent that the effects of these small energy differences are not ordinarily detected. Pure rotational spectra for gases can, however, be observed in the microwave region.

Absorption Induced by a Magnetic Field

When electrons of the nuclei of certain elements are subjected to a strong magnetic field, additional quantized energy levels can be observed as a consequence of the magnetic properties of these elementary particles. The differences in energy between the induced states are small, and transitions between the states are brought about only by absorption of long-wavelength (or low-frequency) radiation. With nuclei, radio waves ranging from 30 to 500 MHz ($\lambda = 1000$ to 60 cm) are generally involved; for electrons, microwaves with a frequency of about 9500 MHz ($\lambda = 3$ cm) are absorbed. Absorption by nuclei or by electrons in magnetic fields is studied by *nuclear magnetic resonance* (NMR) and *electron spin resonance* (ESR) techniques, respectively; nuclear magnetic resonance methods are considered in Chapter 19.

6C-5 Relaxation Processes

Ordinarily, the lifetime of an atom or molecule excited by absorption of radiation is brief because there are several *relaxation processes* that permit its return to the ground state.

Nonradiative Relaxation

As shown in Figure 6-20b, *nonradiative relaxation* involves the loss of energy in a series of small steps, the excitation energy being converted to kinetic energy by collision with other molecules. A minute increase in the temperature of the system results.

As shown in Figure 6-20c, relaxation can also occur by emission of fluorescent radiation. Still other relaxation processes are discussed in Chapters 15, 18, and 19.

Fluorescence and Phosphorescence Relaxation

Fluorescence and phosphorescence are analytically important emission processes in which atoms or molecules are excited by absorption of a beam of electromagnetic radiation; radiant emission then occurs as the excited species return to the ground state. Fluorescence occurs more rapidly than phosphorescence and is generally complete after about 10^{-5} s from the time of excitation. Phosphorescence emission takes place over periods longer than 10^{-5} s and may indeed continue for minutes or even hours after irradiation has ceased. Fluorescence and phosphorescence are most easily observed at a 90-deg angle to the excitation beam.

Resonance fluorescence describes the process in which the emitted radiation is identical in frequency to the radiation employed for excitation. The lines labeled 1 and 2 in Figures 6-20a and 6-20c illustrate resonance fluorescence. Here, the species is excited to the energy states E_1 or E_2 by radiation that has an energy of $(E_1 - E_0)$ or $(E_2 - E_0)$. After a brief period, emission of radiation of identical energy occurs, as depicted in Figure 6-20c. Resonance fluorescence is most commonly produced by *atoms* in the gaseous state that do not have vibrational energy states superimposed on electronic energy levels.

Nonresonance fluorescence is brought about by irradiation of *molecules* in solution or in the gaseous state. As shown in Figure 6-20a, absorption of radiation promotes the molecules into any of the several vibrational levels associated with the two excited electronic levels. The lifetimes of these excited vibrational states are, however, only on the order of 10^{-15} s, which is much smaller than the lifetimes of the excited electronic states (10^{-8} s). Therefore, on the average, vibrational relaxation occurs before electronic relaxation. As a consequence, the energy of the emitted radiation is smaller than that of the absorbed by an amount equal to the vibrational excitation energy. For example, for the absorption labeled 3 in Figure 6-20a, the absorbed energy is equal to $(E_2 - E_0 + e_4'' - e_0'')$, whereas the energy of the fluorescent radiation is again given by $(E_2 - E_0)$. Thus, the emitted radiation has a lower frequency or longer wavelength than the radiation that excited the fluorescence. This shift in wavelength to lower frequencies is sometimes called the *Stokes shift.* Both resonance and nonresonance radiation can accompany fluorescence of molecules, although the latter tends to predominate because of the much larger number of vibrationally excited states.

Phosphorescence occurs when an excited molecule relaxes to a metastable excited electronic state (called the *triplet state*), which has an average lifetime of greater than about 10^{-5} s. The nature of this type of excited state is discussed in Chapter 15.

6C-6 The Uncertainty Principle

The *uncertainty principle* was first proposed in 1927 by Werner Heisenberg, who postulated that nature places limits on the precision with which certain pairs of physical measurements can be made. The uncertainty principle, which has important and widespread implications in instrumental analysis, is readily derived from the principle of superposition, which was discussed in Section 6B-4. Applications of this principle will be found in several later chapters that deal with spectroscopic methods.[5]

Let us suppose that we wish to determine the frequency ν_1 of a monochromatic beam of radiation by comparing it with the output of a standard clock, which is an oscillator that produces a light beam that has a precisely known frequency of ν_2. To detect and measure the difference between the known and unknown frequencies, $\Delta\nu = \nu_1 - \nu_2$, we allow the two beams to interfere as in Figure 6-5 and determine the time interval for a beat (*A* to *B* in Figure 6-5). The minimum time Δt required to make this measurement must be equal to or greater than the period of one beat, which as shown in Figure 6-5, is equal to $1/\Delta\nu$. Therefore, the minimum time for a measurement is given by

$$\Delta t \geq 1/\Delta\nu$$

or

$$\Delta t \Delta\nu \geq 1 \qquad (6\text{-}25)$$

Note that to determine $\Delta\nu$ with zero uncertainty, an infinite measurement time is required. If the observation extends over a very short period, the uncertainty will be large.

Let us multiply both sides of Equation 6-25 by the Planck constant to give

$$\Delta t \cdot (h\Delta\nu) = h$$

From Equation 6-17, it is apparent that

$$\Delta E = h\Delta\nu$$

and

$$\boxed{\Delta t \cdot \Delta E = h} \qquad (6\text{-}26)$$

Equation 6-26 is one of several ways of formulating the Heisenberg uncertainty principle. The meaning in words of this equation is as follows. If the energy E of a particle or system of particles—photons, electrons, neutrons, or protons, for example—is measured for an exactly known period of time Δt, then this energy is uncertain by at least $h/\Delta t$. Therefore, the energy of a particle can be known with zero uncertainty only if it is observed for an infinite period. For finite periods, the energy measurement can never be more precise than $h/\Delta t$. The practical consequences of this limitation will appear in several of the chapters that follow.

[5] A general essay on the uncertainty principle, including applications, is given by L. S. Bartell, *J. Chem. Ed.*, **1985**, *62*, 192.

6D QUANTITATIVE ASPECTS OF SPECTROCHEMICAL MEASUREMENTS

As shown in Table 6-2, spectrochemical methods fall into four major categories. All four require the measurement of radiant *power, P,* which is the energy of a beam of radiation that reaches a given area per second. In modern instruments, radiant power is determined with a radiation detector that converts radiant energy into an electrical signal *S*. Generally *S* is a voltage or a current that ideally is directly proportional to radiant power. That is,

$$S = kP \qquad (6\text{-}27)$$

where k is a constant.

Many detectors exhibit a small, constant response, known as a *dark current,* in the absence of radiation. In those cases, the response is described by the relationship

$$S = kP + k_d \qquad (6\text{-}28)$$

where k_d is the dark current, which is generally small and constant at least for short periods of time. Spectrochemical instruments are usually equipped with a compensating circuit that reduces k_d to zero whenever measurements are made. With such instruments, Equation 6-27 then applies.

6D-1 Emission, Luminescence, and Scattering Methods

As shown in column 3 of Table 6-2, in emission, luminescence, and scattering methods, the power of the radiation emitted by an analyte after excitation is ordinarily directly proportional to the analyte concentration c ($P_e = kc$). Combining this equation with Equation 6-28 gives

$$S = k'c \qquad (6\text{-}29)$$

where k' is a constant that can be evaluated by exciting analyte radiation in one or more standards and by measuring *S*. An analogous relationship also applies for luminescence and scattering methods.

TABLE 6-2 Major Classes of Spectrochemical Methods

Class	Radiant Power Measured	Concentration Relationship	Type of Methods
Emission	Emitted, P_e	$P_e = kc$	Atomic emission
Luminescence	Luminescent, P_l	$P_l = kc$	Atomic and molecular fluorescence, phosphorescence, and chemiluminescence
Scattering	Scattered, P_{sc}	$P_{sc} = kc$	Raman scattering, turbidimetry, and nephelometry
Absorption	Incident, P_0, and transmitted, P	$-\log \frac{P}{P_0} = kc$	Atomic and molecular absorption

6D-2 Absorption Methods

As shown in Table 6-2, quantitative absorption methods require two power measurements: one before a beam has passed through the medium that contains the analyte (P_0) and the other after (P). Two terms, which are widely used in absorption spectrometry and are related to the ratio of P_0 and P, are *transmittance* and *absorbance.*

Transmittance

Figure 6-21 depicts a beam of parallel radiation before and after it has passed through a medium that has a thickness of b cm and a concentration c of an absorbing species. As a consequence of interactions between the photons and absorbing atoms or molecules, the power of the beam is attenuated from P_0 to P. The *transmittance* T of the medium is then the fraction of incident radiation transmitted by the medium:

$$T = \frac{P}{P_0} \tag{6-30}$$

Transmittance is often expressed as a percentage or

$$\%T = \frac{P}{P_0} \times 100\% \tag{6-31}$$

Absorbance

The absorbance A of a medium is defined by the equation

$$A = -\log_{10} T = \log \frac{P_0}{P} \tag{6-32}$$

Note that, in contrast to transmittance, the absorbance of a medium increases as attenuation of the beam becomes greater.

Beer's Law

For monochromatic radiation, absorbance is directly proportional to the path length b through the medium and the concentration c of the absorbing species. These relationships are given by

$$A = abc \tag{6-33}$$

where a is a proportionality constant called the *absorptivity.* The magnitude of a will clearly depend upon the units used for b and c. For solutions of an absorbing species, b is often given in terms of centimeters and c in grams per liter. Absorptivity then has units of $\text{L g}^{-1}\,\text{cm}^{-1}$.

When the concentration in Equation 6-32 is expressed in moles per liter and the cell length is in centimeters, the absorptivity is called the *molar absorptivity* and is given the special symbol ϵ. Thus, when b is in centimeters and c is in moles per liter,

$$A = \epsilon bc \tag{6-34}$$

where ϵ has the units $\text{L mol}^{-1}\,\text{cm}^{-1}$.

Equations 6-33 and 6-34 are expressions of *Beer's law,* which serves as the basis for quantitative analyses

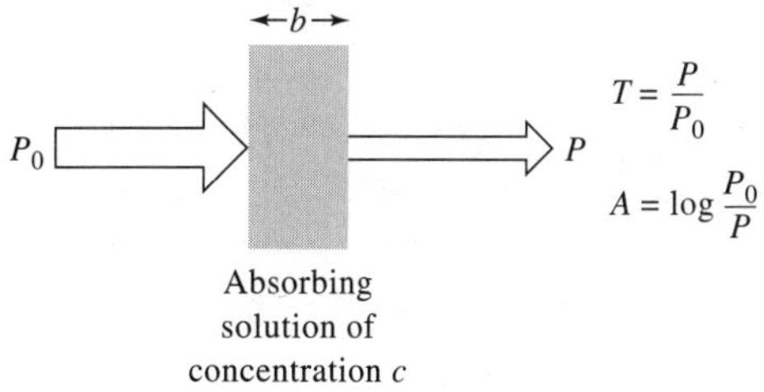

Figure 6-21 Attenuation of a beam of radiation by an absorbing solution.

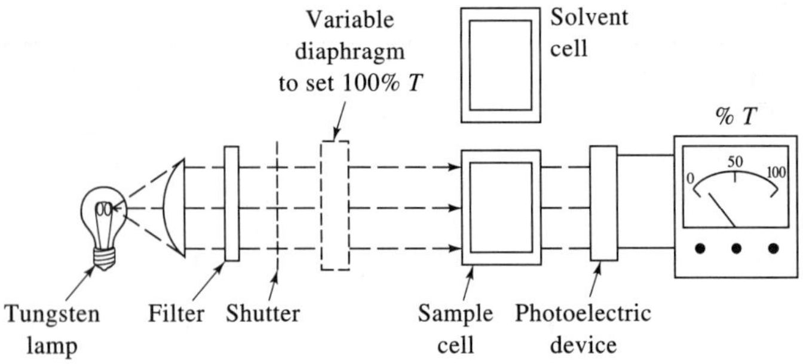

Figure 6-22 Single-beam photometer for absorption measurements in the visible region.

by both atomic and molecular absorption measurements. Certain limitations to the applicability of Beer's law are encountered, and these limitations will be discussed in detail in Section 13B.

Measurement of Transmittance and Absorbance

Figure 6-22 is a schematic of a simple instrument called a photometer used for measuring the transmittance and absorbance of aqueous solutions with a filtered beam of visible radiation. Here, the radiation from a tungsten bulb passes through a colored glass filter that restricts the radiation to a limited band of contiguous wavelengths. The beam then passes through a variable diaphragm that permits adjustment of the power of the radiation that reaches the transparent cell that contains the sample. A shutter can be imposed in front of the diaphragm that completely blocks the beam. With the shutter open, the radiation impinges on a photoelectric device that converts the radiant energy of the beam to a dc current that is detected and measured with a microammeter. The output of the meter S is described by Equation 6-28. Note that the meter has a linear scale that extends from 0 to 100.

In order to make such an instrument direct reading in percent transmittance, two preliminary adjustments are carried out, namely the *0% T,* or *dark current, adjustment,* and the *100% T adjustment.* The 0% T adjustment is made with the detector screened from the source by closing the mechanical shutter. Any small dark current in the detector is nulled electrically until the needle of the detector reads zero.

The 100% T adjustment is made with the shutter open and with the solvent cell in the light path. Usually, the solvent is contained in a cell that is as nearly as possible identical to the cell that contains the sample. The 100% T adjustment with this instrument involves varying the power of the beam by means of the variable diaphragm; in some instruments, the same effect is realized by varying the radiant output of the source electrically. The radiant power that reaches the detector is then varied until the meter reads exactly 100. Effectively, this procedure sets P_0 in Equation 6-31 at 100%. When the solvent is replaced by the cell that contains the sample, the scale then indicates the percent transmittance directly, as shown by the equation

$$\%T = \frac{P}{P_0} \times 100\% = \frac{P}{\cancel{100\%}} \times \cancel{100\%} = P$$

Obviously, an absorbance scale can also be scribed on the readout device. As shown in Figure 6-23, such a scale will be nonlinear unless the output is converted to a logarithmic function by suitable hardware or software.

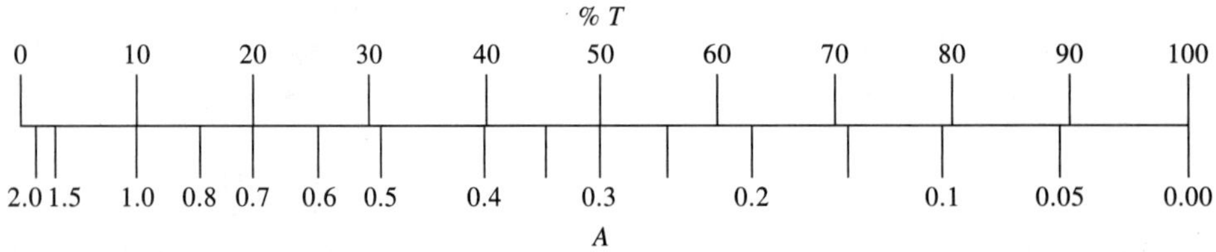

Figure 6-23 Readout for an inexpensive photometer.

6E QUESTIONS AND PROBLEMS

6-1 Define
- (a) coherent radiation.
- (b) dispersion of a transparent substance.
- (c) anomalous dispersion.
- (d) work function of a substance.
- (e) photoelectric effect.
- (f) ground state of a molecule.
- (g) electronic excitation.
- (h) blackbody radiation.
- (i) fluorescence.
- (j) phosphorescence.
- (k) resonance fluorescence.
- (l) photon.
- (m) absorptivity.
- (n) wavenumber.
- (o) relaxation.
- (p) Stokes shift.

6-2 Calculate the frequency in hertz, the energy in joules, and the energy in electron volts of an X-ray photon with a wavelength of 2.70 Å.

6-3 Calculate the frequency in hertz, the wavenumber, the energy in joules, and the energy in kJ/mol associated with the 5.715 μm vibrational absorption band of an aliphatic ketone.

6-4 Calculate the wavelength and the energy in joules associated with an NMR signal at 220 MHz.

6-5 Calculate the velocity, frequency, and wavelength of the sodium D line ($\lambda = 589$ nm) as light from this source passes through a species whose refractive index, η_D, is 1.43.

6-6 When the D line of sodium light impinges an air/diamond interface at an angle of incidence of 30.0 deg, it is found that the angle of refraction is 11.9 deg. What is n_D for diamond?

6-7 What is the wavelength of a photon that has three times as much energy as that of a photon whose wavelength is 500 nm?

6-8 The silver iodide bond energy is approximately 255 kJ/mol (AgI is one of the possible active components in photogray sunglasses). What is the longest wavelength of light that is capable of breaking the bond in silver iodide?

6-9 Cesium is used extensively in photocells and in television cameras because it has the lowest ionization energy of all the stable elements.
- (a) What is the maximum kinetic energy of a photoelectron ejected from cesium by 500 nm light? Note that if the wavelength of the light used to irradiate the cesium surface becomes longer than 660 nm, no photoelectrons are emitted.
- (b) Use the rest mass of the electron to calculate the velocity of the photoelectron in (a).

6-10 The Wien displacement law for blackbody radiators states that the product of temperature in kelvin and the wavelength of maximum emission is a constant $k(k = T \cdot \lambda_{max})$.

Calculate the wavelength of maximum emission for a Globar infrared source operated at 1400 K. Use the data in Figure 6-18 for the Nernst glower for the evaluation of the constant.

6-11 Calculate the wavelength of
(a) the sodium line at 589 nm in an aqueous solution that has a refractive index of 1.35.
(b) the output of a ruby laser at 694.3 when it is passing through a piece of quartz that has a refractive index of 1.55.

6-12 Calculate the reflection loss when a beam of radiant passes through an empty quartz cell assuming the refraction index of quartz is 1.55.

6-13 Explain why the wave model for radiation cannot account for the photoelectric effect.

6-14 Convert the following absorbance data into percent transmittance:
(a) 0.375, (b) 1.325, (c) 0.012.

6-15 Convert the following percent transmittance data into absorbance:
(a) 33.6, (b) 92.1, (c) 1.75.

6-16 Calculate the percent transmittance of solutions with half the absorbance of those in Problem 6-14.

6-17 Calculate the absorbance of solutions with half the percent transmittance of those in Problem 6-15.

6-18 A solution that was 4.14×10^{-3} M in X had a transmittance of 0.126 when measured in a 2.00-cm cell. What concentration of X would be required for the transmittance to be increased by a factor of 3 when a 1.00-cm cell was used?

6-19 A compound had a molar absorptivity of 2.17×10^{3} L cm^{-1} mol^{-1}. What concentration of the compound would be required to produce a solution that has a transmittance of 8.42% in a 2.50-cm cell?

Chapter 7

Components of Optical Instruments

Instruments for ultraviolet and infrared spectroscopy have enough features in common with those designed for the visible region that they are commonly called optical instruments despite the fact that the human eye is sensitive to neither ultraviolet nor infrared wavelengths. In this chapter, we consider the function, the requirements, and the behavior of the components of instruments for optical spectroscopy employing all three types of radiation. Instruments for spectroscopic studies in regions more energetic than the ultraviolet and less energetic than the infrared have characteristics that differ substantially from optical instruments and are considered separately in Chapters 12 and 19.

7A GENERAL DESIGNS OF OPTICAL INSTRUMENTS

Optical spectroscopic methods are based upon six phenomena: (1) absorption, (2) fluorescence, (3) phosphorescence, (4) scattering, (5) emission, and (6) chemiluminescence. While the instruments for measuring each differ somewhat in configuration, most of their basic components are remarkably similar. Furthermore, the required properties of these components are the same regardless of whether they are applied to the ultraviolet, visible, or infrared portion of the spectrum.[1]

Typical spectroscopic instruments contain five components, including: (1) a stable source of radiant energy; (2) a transparent container for holding the sample; (3) a device that isolates a restricted region of the spectrum for measurement[2]; (4) a radiation detector, which converts radiant energy to a usable signal (usually electrical); and (5) a signal processor and readout, which displays the transduced signal on a meter scale, an oscilloscope face, a digital meter, or a recorder chart. Figure 7-1 illustrates the three ways in which these components are configured in order to carry out the six types of spectroscopic measurements mentioned earlier. As can be seen in the figure, components (3), (4), and (5) are arranged in the same way for each type of measurement.

[1] For a more complete discussion of the components of optical instruments, see R. P. Bauman, *Absorption Spectroscopy,* Chapters 2 and 3. New York: Wiley, 1962; E. J. Meehan, in *Treatise on Analytical Chemistry,* P. J. Elving, E. J. Meehan, and I. M. Kolthoff, Eds., Part I, Vol. 7, Chapter 3. New York: Wiley, 1981; J. D. Ingle Jr. and S. R. Crouch, *Spectrochemical Analysis,* Chapters 3 and 4. Englewood Cliffs, NJ: Prentice Hall, 1988.

[2] Fourier transform instruments, which are discussed in Section 7I-3, require no wavelength selecting device but instead use a frequency modulator that provides spectral data in a form that can be interpreted by a mathematical technique called a Fourier transformation.

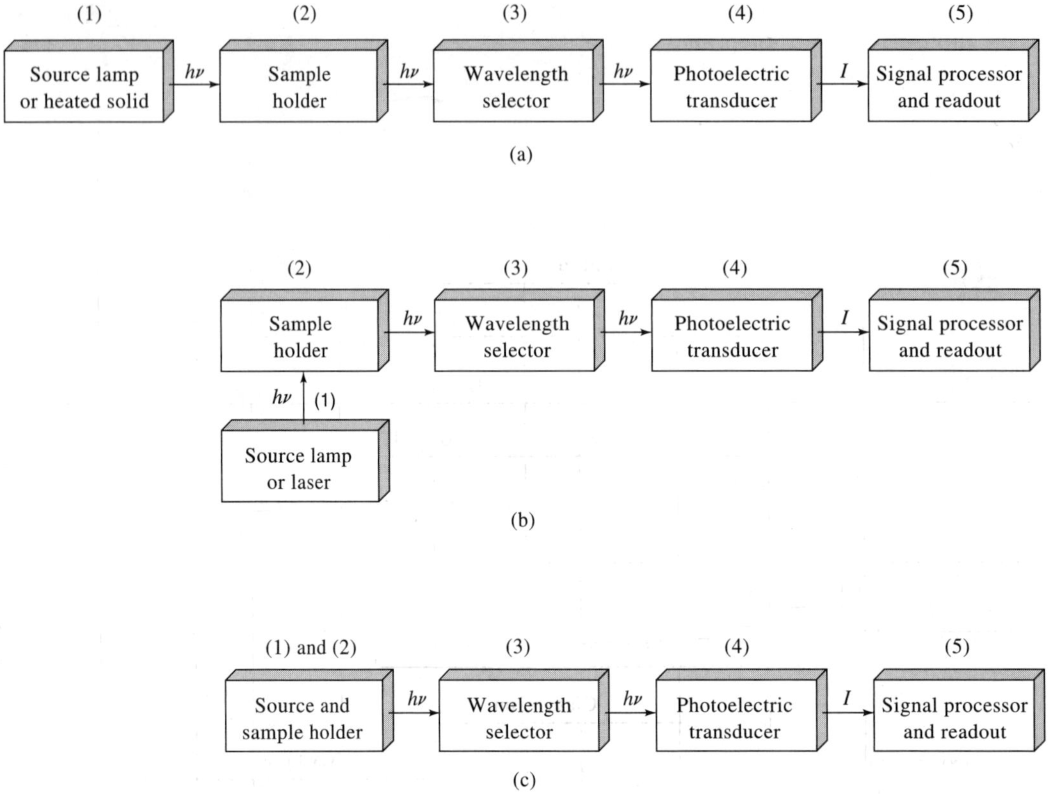

Figure 7-1 Components of various types of instruments for optical spectroscopy: (a) absorption; (b) fluorescence, phosphorescence, and scattering; (c) emission and chemiluminescence.

The first two instrumental configurations, which are used for the measurement of absorption, fluorescence, phosphorescence, and scattering, require an external source of radiant energy. For absorption, the beam from the source passes through the sample directly into the wavelength selector, although in some instruments, the position of the sample and selector is reversed. In the latter three, the source induces the sample, held in a container, to emit characteristic fluorescence, phosphorescence, or scattered radiation, which is usually measured at an angle of 90 deg with respect to the source.

Emission spectroscopy and chemiluminescence spectroscopy differ from the other types in the respect that no external radiation source is required; the sample itself is the emitter (see Figure 7-1c). In emission spectroscopy, the sample container is an arc, a spark, or a flame that both holds the sample and causes it to emit characteristic radiation. In chemiluminescence spectroscopy, the radiation source is a solution of the analyte plus reagents held in a glass sample holder. Emission is brought about by energy released in a chemical reaction in which the analyte takes part directly or indirectly.

Figures 7-2 and 7-3 summarize the optical characteristics of all the components shown in Figure 7-1 with the exception of the signal processor and readout. Note that instrument components differ in detail, depending upon the wavelength region within which they are to be used. Their design also depends on whether the instrument is to be used primarily for qualitative or quantitative analysis and on whether it is to be applied to atomic or molecular spectroscopy. Nevertheless, the general function and performance requirements of each type of component are similar, regardless of wavelength region and application.

J
Feb. 14. 05

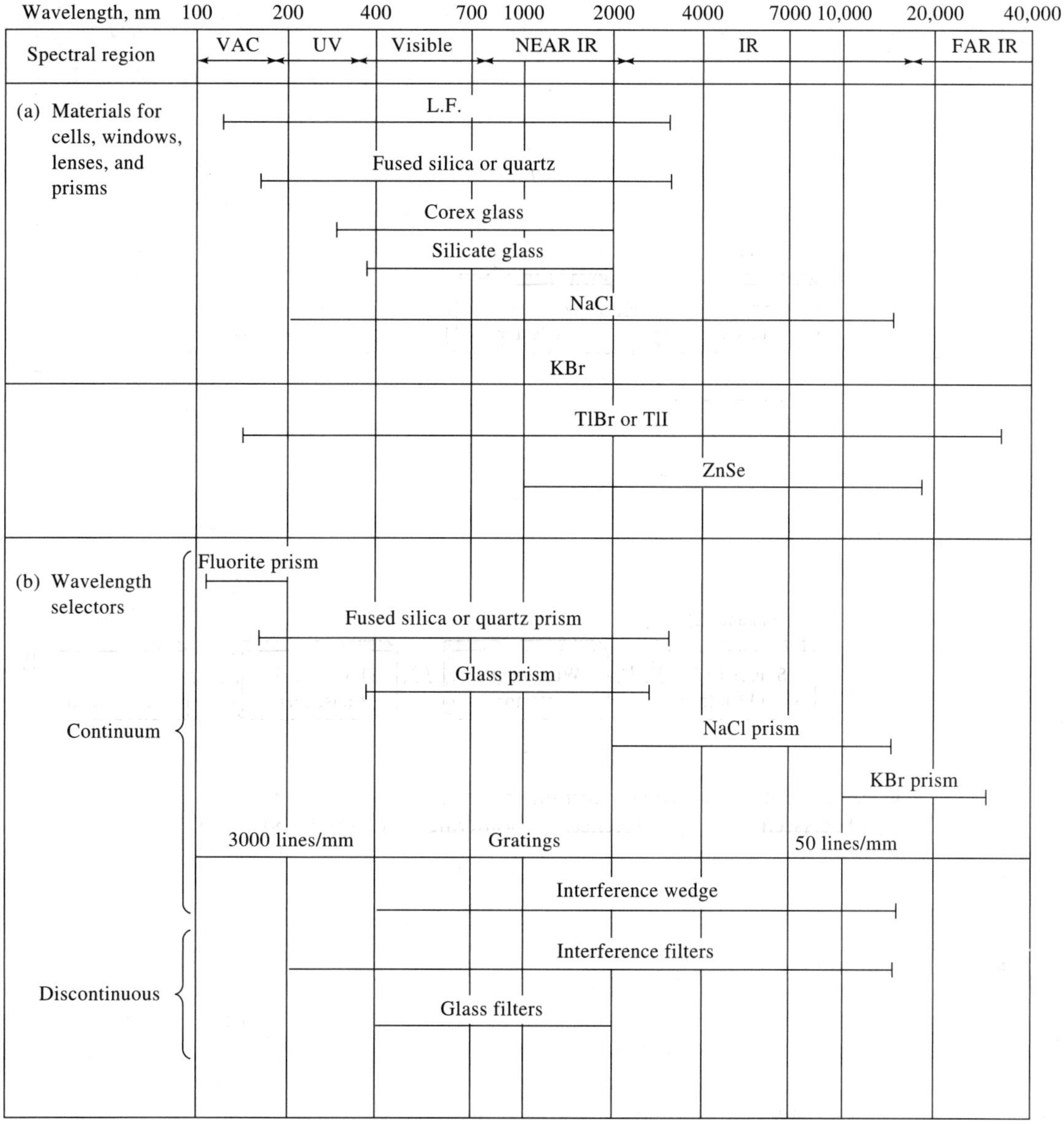

Figure 7-2 (a) Construction materials and (b) wavelength selectors for spectroscopic instruments.

7B SOURCES OF RADIATION

In order to be suitable for spectroscopic studies, a source must generate a beam of radiation with sufficient power for easy detection and measurement. In addition, its output power should be stable for reasonable periods. Typically, the radiant power of a source varies exponentially with the voltage of its power supply. Thus, a regulated power source is often needed to provide the required stability. Alternatively, the problem of source stability is circumvented by double-beam designs in which the ratio of the signal from the sample to that of the source in the absence of sample serves as the analytical variable. In such designs, the intensities of the two beams are measured si-

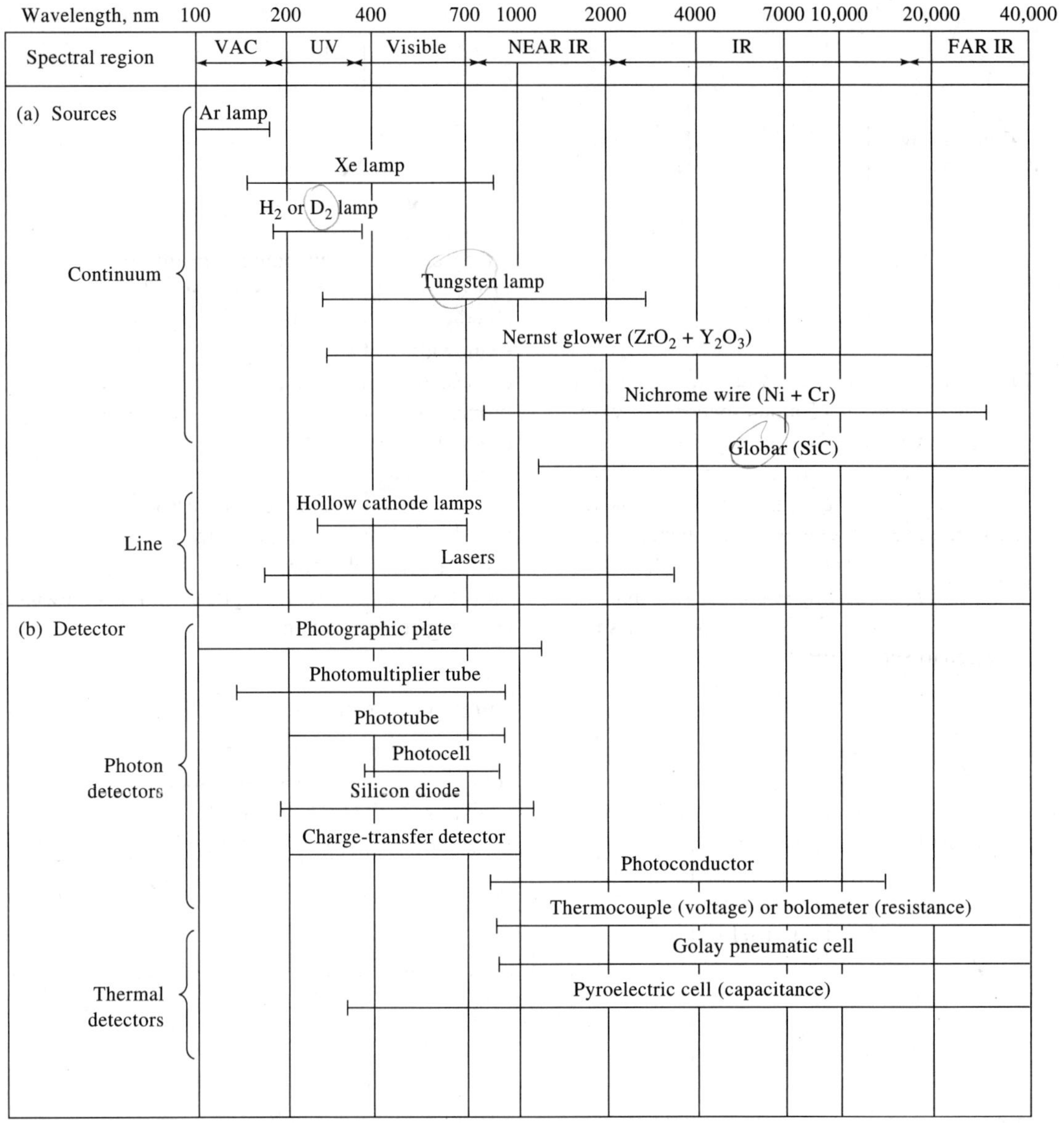

Figure 7-3 (a) Sources and (b) detectors for spectroscopic instruments.

multaneously or nearly simultaneously so that the effect of fluctuations in the source output is largely canceled.

Figure 7-3a lists the most widely used spectroscopic sources. Note that these sources are of two types: *continuum sources,* which emit radiation that changes in intensity only slowly as a function of wavelength, and *line sources,* which emit a limited number of lines or bands of radiation each of which spans a limited range of wavelengths.

7B-1 Continuum Sources

Continuum sources find widespread use in absorption and fluorescence spectroscopy. For the ultraviolet region, the most common source is the deuterium lamp. High-pressure, gas-filled arc lamps that contain argon, xenon, or mercury serve when a particularly intense source is required. For the visible region of the spectrum, the tungsten filament lamp is used almost universally. The common infrared sources are inert solids heated to 1500 to 2000 K, a temperature at which the maximum radiant output occurs at 1.5 to 1.9 μm (see Figure 6-18). Details on the construction and behavior of these various continuum sources will be found in the chapter dealing with specific types of spectroscopic methods.

7B-2 Line Sources

Sources that emit a few discrete lines find wide use in atomic absorption spectroscopy, atomic and molecular fluorescence spectroscopy, and Raman spectroscopy (refractometry and polarimetry also employ line sources). The familiar mercury and sodium vapor lamps provide a relatively few sharp lines in the ultraviolet and visible regions and are used in several spectroscopic instruments. Hollow cathode lamps and electrodeless discharge lamps are the most important line sources for atomic absorption and fluorescence methods. Discussion of such sources is deferred to Section 9B-1.

Jan 8.05 1:26 pm | Feb 14.05 8:14 pm

7B-3 Laser Sources

Lasers are highly useful sources in analytical instrumentation because of their high intensities, their narrow bandwidths, and the coherent nature of their outputs.[3] The first laser was described in 1960. Since that time, chemists have found numerous useful applications for these sources in high-resolution spectroscopy, in kinetic studies of processes with lifetimes in the range of 10^{-9} to 10^{-12} s, in the detection and determination of extremely small concentrations of species in the atmosphere, and in the induction of isotopically selective reactions.[4] In addition, laser sources have become important in several routine analytical methods, including Raman spectroscopy, molecular absorption spectroscopy, emission spectroscopy, and as part of instruments for Fourier transform infrared spectroscopy.

The term *laser* is an acronym for **l**ight **a**mplification by **s**timulated **e**mission of **r**adiation. As a consequence of their light-amplifying characteristics, lasers produce spatially narrow (a few hundredths of a micrometer), extremely intense beams of radiation. The process of stimulated emission, which will be described shortly, produces a beam of highly monochromatic (bandwidths of 0.01 nm or less) and remarkably coherent (Section 6B-6) radiation. Because of these unique properties, lasers have become important sources for use in the ultraviolet, visible, and infrared regions of the spectrum. A limitation of early lasers was that the radiation from a given source was restricted to a relatively few discrete wavelengths or lines. Now, however, dye lasers are available that provide narrow bands of radiation at any chosen wavelength within a somewhat limited range of the source.

8:45pm

Components of Lasers

Figure 7-4 is a schematic representation that shows the components of a typical laser source. The heart of the device is the lasing medium. It may be a solid crystal such as ruby, a semiconductor such as gallium arsenide, a solution of an organic dye, or a gas such as argon or krypton. The lasing material is often activated, or *pumped,* by radiation from an external source so that a few photons of proper energy will trigger the formation of a cascade of photons of the same energy. Pumping can also be accomplished by an electrical current or by an electrical discharge. Thus, gas lasers usually do not have the external radiation source shown in Figure 7-4; instead, the power supply is connected to a pair of electrodes contained in a cell filled with the gas.

A laser normally functions as an oscillator, or a resonator, in the sense that the radiation produced by the lasing action is caused to pass back and forth through the medium numerous times by means of a pair of mirrors as shown in Figure 7-4. Additional photons are generated with each passage, thus leading to enormous amplification. The repeated passage also produces a beam that is highly parallel because nonparallel radiation escapes from the sides of the medium after being reflected a few times (see Figure 7-4). One of the easiest ways to obtain a usable laser beam is to coat one of the mirrors with a sufficiently thin layer of reflecting mate-

---

[3] For a more complete discussion of lasers, see J. Wilson and J. F. B. Hawkes, *Lasers: Principles and Applications.* Englewood Cliffs, NJ: Prentice-Hall, 1987; D. L. Andrews, *Lasers in Chemistry.* New York: Springer-Verlag, 1986; *Laser Spectroscopy and Its Applications,* L. Radziemski, R. Solarz, and J. Paisner, Eds. New York: Marcel Dekker, 1987; *Applications of Lasers,* E. Piepmeier, Ed. New York: Wiley, 1986.

[4] For reviews of some of these applications, see J. C. Wright and M. J. Wirth, *Anal. Chem.,* **1980,** *52,* 988A, 1087A; A. Schawlow, *Science,* **1982,** *217,* 9; E. W. Findsend and M. R. Ondrias, *J. Chem. Educ.,* **1986,** *63,* 479; R. N. Zare, *Science,* **1984,** *226,* 1198; C. P. Christensen, *Science,* **1984,** *224,* 117; J. K. Steehler, *J. Chem. Educ.,* **1990,** *67,* A37.

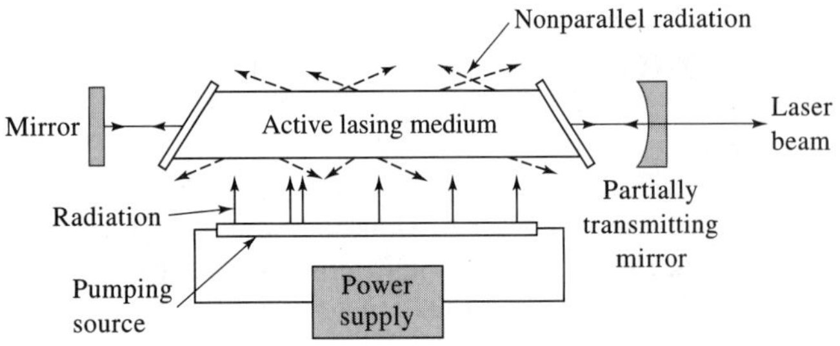

Figure 7-4 Schematic representation of a typical laser source.

rial so that a fraction of the beam is transmitted rather than reflected.

Mechanism of Laser Action

Laser action can be understood by considering the four processes depicted in Figure 7-5, namely: (a) pumping, (b) spontaneous emission (fluorescence), (c) stimulated emission, and (d) absorption. In this figure, we show the behavior of two of the many molecules that make up the lasing medium. Two of the several electronic energy levels of each are shown as having energies E_y and E_x. Note that the higher electronic state for each molecule has several slightly different vibrational energy levels depicted as E_y, E'_y, E''_y, and so forth. We have not shown additional levels for the lower electronic state, although such usually exist.

Pumping. Pumping, which is necessary for laser action, is a process by which the active species of a laser is excited by means of an electrical discharge, passage of an electrical current, or exposure to an intense radiant source. During pumping, several of the higher electronic and vibrational energy levels of the active species are populated. In diagram (1) of Figure 7-5a, one molecule is shown as being promoted to an energy state E''_y; the second is excited to the slightly higher vibrational level E'''_y. The lifetime of excited *vibrational* states is brief, and after 10^{-13} to 10^{-15} s, relaxation to the lowest excited vibrational level [E_y in diagram a(3)] occurs with the production of an undetectable quantity of heat. Some excited electronic states of laser materials have lifetimes considerably longer (often 1 ms or more) than their excited vibrational counterparts; long-lived states are sometimes termed *metastable* as a consequence.

Spontaneous Emission. As was pointed out in the discussion of fluorescence (page 137), a species in an excited electronic state may lose all or part of its excess energy by spontaneous emission of radiation. This process is depicted in the three diagrams shown in Figure 7-5b. Note that the wavelength of the fluorescence radiation is given by the relationship $\lambda = hc/(E_y - E_x)$ where h is the Planck constant and c is the speed of light. It is also important to note that the instant at which emission occurs and the path of the resulting photon vary from excited molecule to excited molecule because spontaneous emission is a random process; thus, as shown in Figure 7-5, the fluorescence radiation produced by one of the species in diagram b(1) differs in direction and phase from that produced by the second species [diagram b(2)]. Spontaneous emission, therefore, yields *incoherent* monochromatic radiation.

Stimulated Emission. Stimulated emission, which is the basis of laser behavior, is depicted in Figure 7-5c. Here, the excited laser species are struck by photons that have precisely the same energies ($E_y - E_x$) as the photons produced by spontaneous emission. Collisions of this type cause the excited species to relax immediately to the lower energy state and to simultaneously emit a photon of exactly the same energy as the photon that stimulated the process. Equally important, the emitted photon *travels* in exactly the same direction and *is precisely in phase* with the photon that caused the emission. Therefore, the stimulated emission is totally *coherent* with the incoming radiation.

Absorption. The absorption process, which competes with stimulated emission, is depicted in Figure 7-5d. Here, two photons with energies exactly equal to ($E_y - E_x$) are absorbed to produce the metastable excited state shown in diagram d(3); note that the state shown in diagram d(3) is identical to that attained in diagram a(3) by pumping.

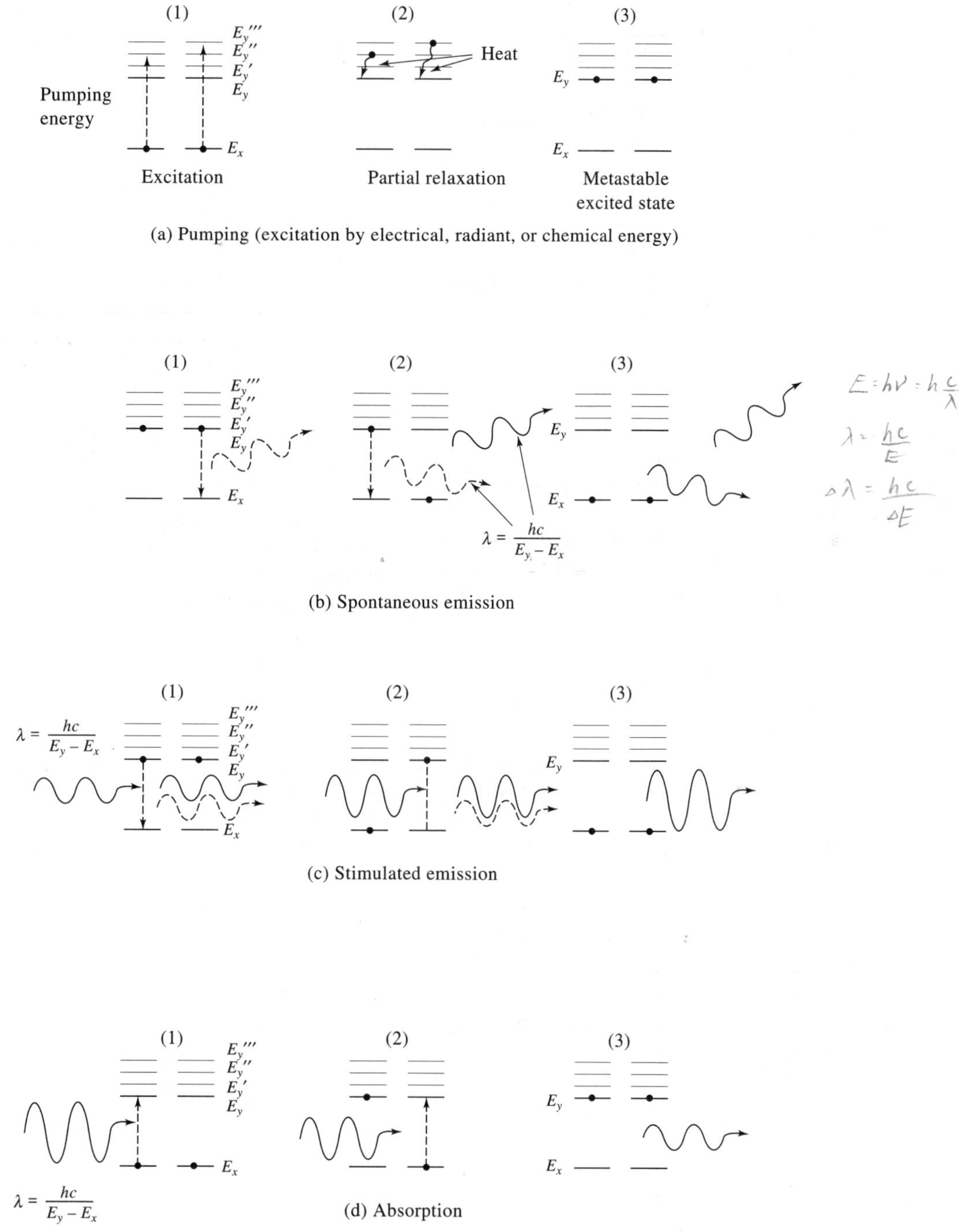

Figure 7-5 Four processes important in laser action: (a) pumping (excitation by electrical, radiant, or chemical energy), (b) spontaneous emission, (c) stimulated emission, and (d) absorption.

Population Inversion and Light Amplification

In order to have light amplification in a laser, it is necessary that the number of photons produced by stimulated emission exceed the number lost by absorption. This condition will prevail only when the number of particles in the higher energy state exceeds the number in the lower; in other words, there must be a *population inversion* from the normal distribution of energy states. Population inversions are brought about by pumping. Figure 7-6 contrasts the effect of incoming radiation on a noninverted population with that of an inverted one. In each case the population is shown as being made up of nine molecules of the lasing medium. In the noninverted system, three molecules are in the excited state and six are in the lower energy level. Three of the incoming photons are absorbed by the medium, thus producing three additional excited molecules. The radiation also, however, stimulates emission of two photons from excited molecules. Thus, the beam is attenuated by one photon. As shown in Figure 7-5b, a net gain in photons is observed in the inverted system because stimulated emission goes on to a greater extent than absorption does.

Three- and Four-Level Laser Systems

Figure 7-7 shows simplified energy diagrams for the two common types of laser systems. In the three-level system, the transition responsible for laser radiation is between an excited state E_y and the ground state E_0; in a four-level system, on the other hand, radiation is generated by a transition from E_y to a state E_x that has a greater energy than the ground state. Furthermore, it is necessary that transitions between E_x and the ground state be rapid. The advantage of the four-level system is that the population inversions necessary for laser action are more readily achieved. To understand this fact, note that at room temperature a large majority of the laser species will be in the ground-state energy level E_0 in both systems. Sufficient energy must thus be provided to convert more than 50% of the lasing species to the E_y level of a three-level system. In contrast, it is only necessary to pump sufficiently to make the number of particles in the E_y energy level exceed the number in E_x of a four-level system. The lifetime of a particle in the E_x state is brief, however, because the transition to E_0 is fast; thus, the number in the E_x state will generally be negligible with respect to the number that has energy E_0 and also (with a modest input of pumping energy) with respect to the number in the E_y state. Therefore, the four-level laser usually achieves a population inversion with a small expenditure of pumping energy.

Some Examples of Useful Lasers[5]

Solid State Lasers. The first successful laser, and one that still finds widespread use, is a three-level device in which a ruby crystal is the active medium. Ruby is pri-

[5] For a review of lasers that are useful in analytical chemistry, see J. C. Wright and M. J. Wirth, *Anal. Chem.*, **1980**, *52*, 1087A.

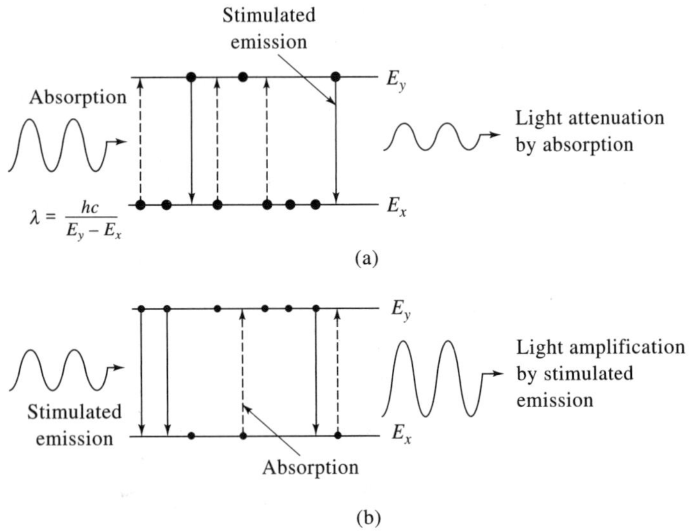

Figure 7-6 Passage of radiation through (a) a noninverted population and (b) an inverted population.

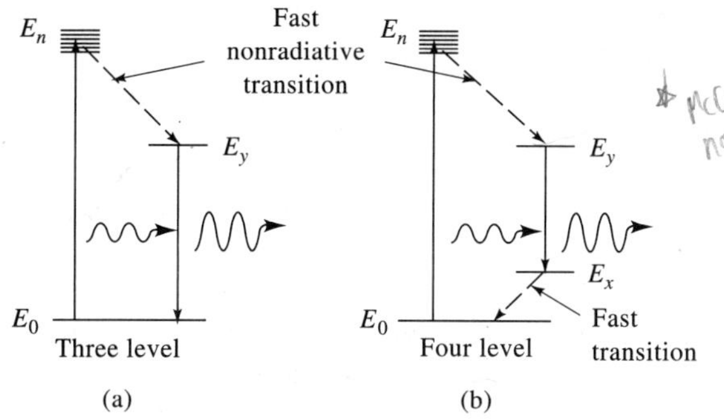

Figure 7-7 Energy level diagrams for two types of laser systems.

marily Al_2O_3 but contains approximately 0.05% chromium(III) distributed among the aluminum(III) lattice sites, which accounts for the red coloration. The chromium(III) ions are the active lasing material. In early lasers, the ruby was machined into a rod about 4 cm in length and 0.5 cm in diameter. A flash tube (often a low-pressure xenon lamp) was coiled around the cylinder to produce intense flashes of light (λ = 694.3 nm). Because the pumping was discontinuous, a pulsed beam was produced. Continuous-wave ruby sources are now available.

The Nd:YAG laser is one of the most widely used solid-state lasers. It consists of neodymium ion in a host crystal of yttrium aluminum garnet. This system offers the advantage of being a four-level laser, which makes it much easier to achieve population inversion than with the ruby laser. The Nd:YAG laser has a very high radiant power output at 1064 nm, which is usually frequency doubled (see page 154) to give an intense line at 532 nm. This radiation is often used for pumping tunable dye lasers.

Gas Lasers. A variety of gas lasers are available commercially. These devices are of four types: (1) neutral atom lasers such as He/Ne; (2) ion lasers in which the active species is Ar^+ or Kr^+; (3) molecular lasers in which the lasing medium is CO_2 or N_2; and (4) eximer lasers. The helium/neon laser is the most widely encountered of all lasers because of its low initial and maintenance costs, its great reliability, and its low power consumption.[6] The most important of its output lines is at 632.8 nm. It is generally operated in the continuous mode rather than a pulsed mode.

*McCurry note

The argon ion laser, which produces intense lines in the green (514.5 nm) and the blue (488.0 nm) regions, is an important example of an ion laser. This laser is a four-level device in which argon ions are formed by an electrical or radio-frequency discharge. The required input energy is high because the argon atoms must first be ionized and then excited from their ground state, with a principal quantum number of 3, to various 4*p* states. Laser activity occurs when the excited ions relax to the 4*s* state. The argon ion laser finds use as a source in fluorescence and Raman spectroscopy because of the high intensity of its lines.

The N_2 laser, which must be operated in the pulsed mode because pumping is carried out with a high potential spark source, provides intense radiation at 337.1 nm. This output has found extensive use for exciting fluorescence in a variety of molecules and for pumping dye lasers. The CO_2 gas laser is used to produce monochromatic infrared radiation at 10.6 μm.

Eximer lasers contain a gaseous mixture of helium, fluorine, and one of the rare gases argon, krypton, or xenon. The rare gas is electronically excited by an electrical current whereupon it reacts with the fluorine to form excited species such as ArF, KrF, or XeF, which are called *eximers* because they are stable only in the excited state. Since the eximer ground state is unstable, rapid dissociation of the compounds occurs as they relax while giving off a photon. Thus, there is a population inversion as long as pumping is carried on. Eximer lasers produce high energy pulses in ultraviolet (351 nm for XeF, 248 nm for KrF, and 193 nm for ArF).

Dye Lasers.[7] Dye lasers have become important radiation sources in analytical chemistry because they are continuously tunable over a range of 20 to 50 nm. The bandwidth of a tunable laser is typically a few hundredths of a nanometer or less. The active materials in dye lasers are solutions of organic compounds capable of fluorescing in the ultraviolet, visible, or infrared regions. Dye lasers are four-level systems. In contrast to the other lasers of this type that we have considered, however, the lower energy level for laser action (E_x in Figure 7-6b) is

[6] For details on the design of commercial He/Ne lasers, see B. Patel, *Photonics Spectra,* **1983** (1), 33.

[7] For further information, see R. B. Green, *J. Chem. Educ.,* **1977,** *54,* A365, A407; M. J. Wirth, *Tunable Laser Systems* in *Lasers in Chemical Analysis,* G. M. Hieftje, J. C. Travis, and F. E. Lytle, Eds. Clifton, NJ: Humana Press, 1981.

not a single energy but a band of energies that arise from the superposition of a large number of closely spaced vibrational and rotational energy states upon the base electronic energy state. Electrons in E_y may then undergo transitions to any of these states, thus producing photons of slightly different energies. Tuning of dye lasers can be readily accomplished by replacing the nontransmitting mirror shown in Figure 7-4 with a monochromator equipped with a reflection grating or a Littrow-type prism (see Figure 7-16) that reflects only a narrow bandwidth of radiation into the laser medium; the peak wavelength can be varied by rotation of the grating or prism. Emission is then stimulated for only part of the fluorescence spectrum, namely, the wavelength selected by the monochromator.

McCreery's note

Semiconductor Diode Lasers.[8] A new and increasingly important source of nearly monochromatic radiation is the laser diode. Laser diodes are products of modern semiconductor technology. Their mechanism of operation may be understood by considering the electrical conduction characteristics of various materials as illustrated in Figure 7-8. A good conductor, such as a metal, consists of a regular arrangement of atoms immersed in a sea of valence electrons. Orbitals on adjacent atoms overlap to form the so-called *valence band*, which is essentially a molecular orbital over the entire metal and which contains the valence electrons of all of the atoms. Empty outer orbitals overlap to form the *conduction band*, which lies at a slightly higher energy than the valence band. The difference in energy between the valence band and the conduction band is the band-gap energy, E_g. Because the band-gap energy is so small in conductors (see Figure 7-8a), electrons in the valence band easily acquire sufficient thermal energy to be promoted to the conduction band, thus providing mobile charge carriers for conduction.

In contrast, insulators have a relatively large band-gap energy, and as a result, electrons in the valence band are unable to acquire enough thermal energy to make the transition to the conduction band, and thus insulators do not conduct electricity (see Figure 7-8c). Semiconductors, such as silicon or germanium, have intermediate band-gap energies, and thus their conduction characteristics are intermediate between conductors and insulators (see Figure 7-8b). We should note that whether a material is a semiconductor or an insulator depends not only on the band-gap energy but also on the temperature of operation and the energy of excitation of the material, which is related to the voltage applied to the material.

When a voltage is impressed across a semiconductor diode in the forward direction (see Section 2C-2),

[8] M. G. D. Bauman, J. C. Wright, A. B. Ellis, T. Kuech, and G. C. Lisensky, *J. Chem. Educ.*, **1992**, *69*, 89; T. Imasaka and N. Ishibashi, *Anal. Chem.*, **1990**, *62*, 363A; R. L. Beyer, *Science*, **1989**, *239*, 742; K. Niemax, A. Zybin, C. Schnürer-Patschan, and H. Groll, *Anal. Chem.*, **1996**, *68*, 351A.

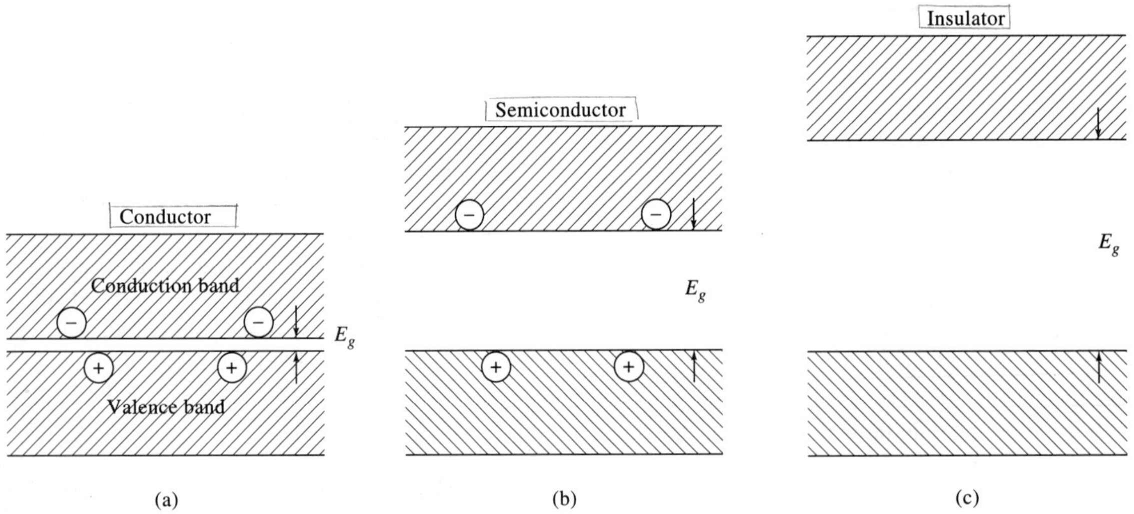

Figure 7-8 Conduction bands and valence bands in three types of materials.

electrons are excited into the conduction band, hole-electron pairs are created, and the diode conducts. Ultimately, some of these electrons relax and go back into the valence band, and energy is released that corresponds to the band-gap energy $E_g = h\nu$. Some of the energy is released in the form of electromagnetic radiation of frequency $\nu = E_g/h$. Diodes that are fabricated to enhance the production of light are called *light-emitting diodes,* or LEDs. Light-emitting diodes are often made of gallium arsenide that is doped with phosphorus, which has a band-gap energy that corresponds to a wavelength of 660 nm. Diodes of this type find wide use as indicators and readouts in electronic instruments. Unfortunately, because of their relatively low intensity and emission wavelengths in the red and infrared, LEDs are of limited utility in spectroscopy.

In recent years semiconductor fabrication techniques have progressed to an extent that permits the construction of highly complex integrated devices such as the distributed-Bragg-reflector (DBR) laser diode shown in Figure 7-9. This device contains a gallium arsenide *pn*-junction diode that produces infrared radiation at about 975 nm. In addition, a stripe of material is fabricated on the chip that acts as a resonant cavity for the radiation so that light amplification can occur within the cavity. An integrated grating provides feedback to the resonant cavity so that the resulting radiation has an extremely narrow bandwidth of about 10^{-5} nm. Laser diodes of this type have achieved continuous power outputs of more than 100 mW with a typical thermal stability of 0.1 nm/°C. Laser diodes may be operated in either a pulsed or continuous (CW) mode, which increases their versatility in a variety of applications. Rapid development of laser diodes has resulted from their utility as light sources for compact-disk players, CD-ROM drives, bar-code scanners, and other familiar optoelectronic devices, and mass production of laser diodes ensures that their cost will continue to be low.

A major impediment to the use of laser diodes in spectroscopic applications has been their limited wavelength range in the red and infrared regions of the spectrum. This disadvantage may be overcome by operating the laser diode in the pulsed mode to achieve sufficient peak power to use nonlinear optics to provide frequency doubling as shown in Figure 7-10. Here, the output of a laser diode is focused in a doubling crystal to provide output in the blue-green region of the spectrum (~490 nm). With proper external optics, frequency-doubled laser diodes can achieve average output powers of 0.5 to 1.0 W with a tunable spectral range of about 30 nm. The advantages of such light sources include compactness, power efficiency, high reliability, and ruggedness. The addition of external optics to the laser diode increases the cost of the devices substantially, but they are competitive with larger, less efficient, and less reliable gas-based lasers.

Recently, it has been reported that gallium nitride laser diodes produce radiation directly in the blue, green, and yellow region of the spectrum.[9] These diodes should prove useful for spectroscopic studies.

The utility of laser diodes for spectroscopic applications has been demonstrated in molecular absorption spectrometry, molecular fluorescence spectrometry, atomic absorption spectrometry, and as light sources for detectors in various chromatographic methods. As practical diode lasers achieve wide commercial availability, they will doubtless appear in increasing numbers as light sources in commercial spectrometric systems.

[9] G. Fasol, *Science*, 1996, *272*, 1751.

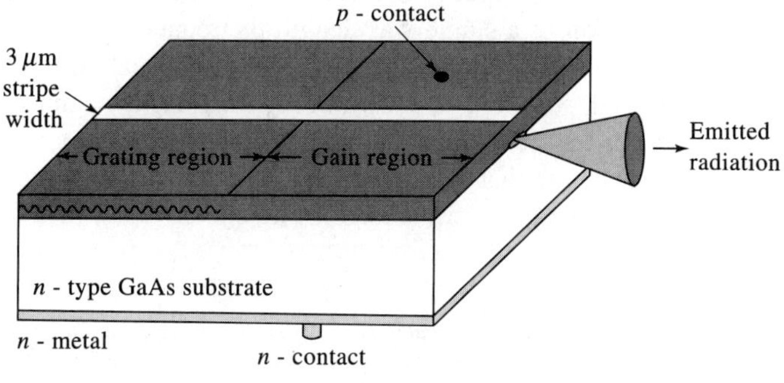

Figure 7-9 A distributed Bragg-reflector laser diode. *(From D. W. Nam and R. G. Waarts,* Laser Focus World, ***1994,*** *30(8), 52. With permission.)*

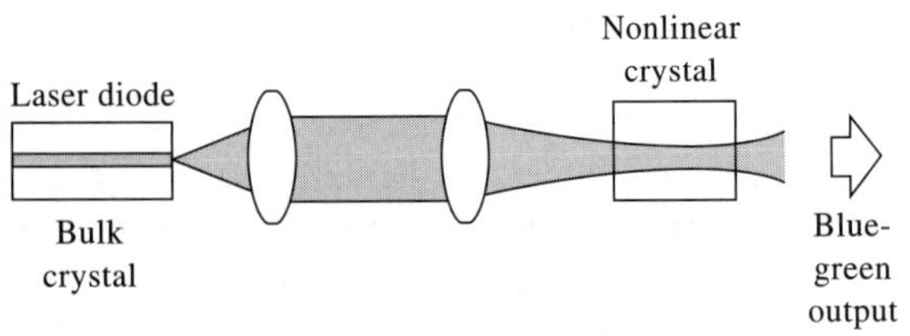

Figure 7-10 A frequency-doubled system for converting 975-nm laser output to 490 nm. *(From D. W. Nam and R. G. Waarts,* Laser Focus World, ***1994,*** 30*(8), 52. With permission.)*

Nonlinear Optical Effects with Lasers

As we have noted in Section 6B-7, when an electromagnetic wave is transmitted through a dielectric[10] medium, the electromagnetic field of the radiation causes momentary distortion, or polarization, of the valence electrons of the molecules that make up the medium. For ordinary radiation the extent of polarization P is directly proportional to the magnitude of the electrical field E of the radiation. Thus we may write

$$P = \alpha E$$

where α is the proportionality constant. Optical phenomena that occur when this situation prevails are said to be *linear.*

At the high radiation intensities encountered with lasers, this relationship breaks down, particularly when E approaches the binding energy of the electrons. Under these circumstances, *nonlinear optical effects* are observed wherein the relationship between polarization and electrical field is given by the equation

$$P = \alpha E + \beta E^2 + \gamma E^3 + \ldots \qquad (7\text{-}1)$$

where the magnitude of the three constants are in the order $\alpha > \beta > \gamma$. At ordinary radiation intensities, only the first term on the right is significant, and the relationship between polarization and field strength is linear. With high-intensity lasers, however, the second term and sometimes even the third term are required to describe the degree of polarization. When only two terms are required, Equation 7-1 can be rewritten in terms of radiation frequency ω and the maximum amplitude of the field strength, E_m. Thus,

$$P = \alpha E_m \sin \omega t + \beta E_m^2 \sin^2 \omega t \qquad (7\text{-}2)$$

Substituting the trigonometric identity $\sin^2 \omega t = \frac{1}{2}(1 - \cos 2\omega t)$ gives

$$P = \alpha E_m \sin \omega t + \frac{\beta E_m^2}{2}(1 - \cos 2\omega t) \qquad (7\text{-}3)$$

The first term in Equation 7-3 is the normal linear term that predominates at low radiation intensities. At sufficiently high intensity, the second-order term becomes significant and results in radiation that has a frequency 2ω that is *double* that of the incident radiation. This frequency-doubling process is now widely used to produce laser frequencies of shorter wavelengths. For example, the 1064-nm near-infrared radiation from a Nd:YAG laser can be frequency doubled to produce a 30% yield of green radiation at 532 nm by passing the radiation through a crystalline material such as potassium dihydrogen phosphate. The 532-nm radiation can then be doubled again to yield ultraviolet radiation at 266 nm by passage through a crystal of ammonium dihydrogen phosphate.

Radiation from laser sources is beginning to find application in several types of nonlinear spectroscopy, most notably in Raman spectroscopy (see Section 18D-3). ✓ Feb 14.05 ✓ Jan 12.08

7C WAVELENGTH SELECTORS

For most spectroscopic analyses, radiation that consists of a limited, narrow, continuous group of wavelengths called a *band* is required.[11] A narrow bandwidth enhances the sensitivity of absorbance measurements, may provide selectivity to both emission and absorption methods, and is frequently a requirement from the standpoint of obtaining a linear relationship between the optical signal and concentration (Equation 6-34). A: εbc Ideally, the output from a wavelength selector would be radiation of a single wavelength or frequency. No real wavelength selector even approaches this ideal; instead, a band, such as that shown in Figure 7-11, is obtained. Here, the percentage of incident radiation of a given wavelength that is transmitted by the selector is plotted as a function of wavelength. The *effective bandwidth,*

[10] Dielectrics are a class of substances that are nonconductors because they contain no free electrons. Generally, dielectrics are optically transparent in contrast to electrically conducting solids, which either absorb radiation or reflect it strongly.

[11] Note that the term *band* in this context has a somewhat different meaning from that used in describing types of spectra in Chapter 6.

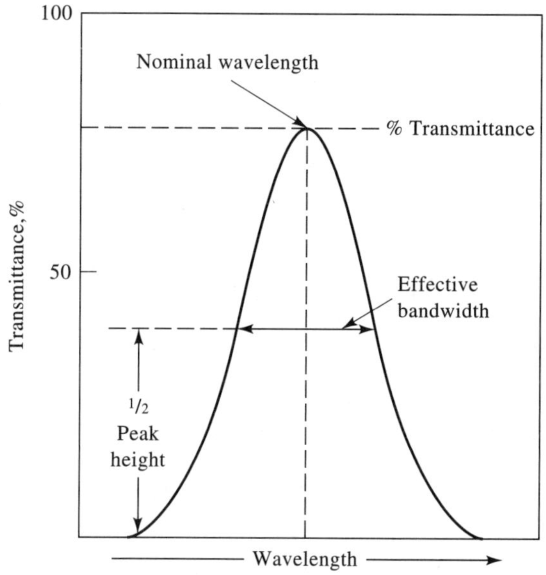

Figure 7-11 Output of a typical wavelength selector.

which is defined in Figure 7-11, is an inverse measure of the quality of the device, a narrower bandwidth representing better performance. Two types of wavelength selectors are encountered, filters and monochromators.

7C-1 Filters

Two types of filters are employed for wavelength selection: *interference filters* and *absorption filters* (the former are sometimes called *Fabry-Perot filters*). Absorption filters are restricted to the visible region of the spectrum; interference filters, on the other hand, are available for the ultraviolet, visible, and well into the infrared region.

Interference Filters

As the name implies, interference filters rely on optical interference to provide narrow bands of radiation. An interference filter consists of a transparent dielectric[12] (frequently calcium fluoride or magnesium fluoride) that occupies the space between two semitransparent metallic films. This array is sandwiched between two plates of glass or other transparent materials (see Figure 7-12a). The thickness of the dielectric layer is carefully controlled and determines the wavelength of the transmitted radiation. When a perpendicular beam of collimated radiation strikes this array, a fraction passes through the first metallic layer while the remainder is reflected. The portion that is passed undergoes a similar partition when it strikes the second metallic film. If the reflected portion from this second interaction is of the proper wavelength, it is partially reflected from the inner side of the first layer in phase with incoming light of the same wavelength. The result is that this particular wavelength is reinforced, while most other wavelengths, being out of phase, suffer destructive interference.

The relationship between the thickness of the dielectric layer t and the transmitted wavelength λ can be found with the aid of Figure 7-12b. For purposes of clarity, the incident beam is shown as arriving at an angle θ from the perpendicular. At point 1, the radiation is partially reflected and partially transmitted to point 1′ where partial reflection and transmission again takes

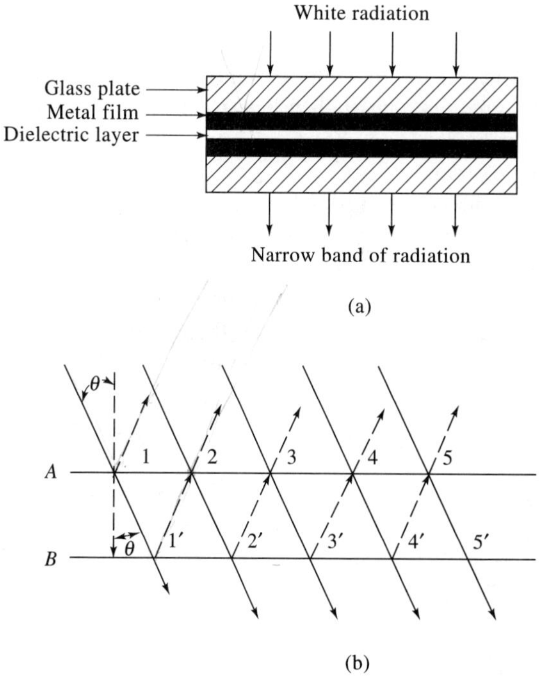

Figure 7-12 (a) Schematic cross section of an interference filter. Note that the drawing is not to scale and that the three central bands are much narrower than shown. (b) Schematic to show the conditions for constructive interference.

[12] A dielectric material is an insulator that contains essentially no current-carrying charged particles. Dielectors are generally transparent in most spectral regions.

place. The same process occurs at 2, 2′, and so forth. For reinforcement to occur at point 2, the distance traveled by the beam reflected at 1′ must be some multiple of its wavelength in the medium λ'. Since the path length between surfaces can be expressed as $t/\cos\theta$, the condition for reinforcement is that

$$\mathbf{n}\lambda' = 2t/\cos\theta$$

where **n** is a small whole number.

In ordinary use, θ approaches zero and $\cos\theta$ approaches unity so that the foregoing equation simplifies to

$$\mathbf{n}\lambda' \approx 2t \qquad (7\text{-}4)$$

where λ' is the wavelength of radiation *in the dielectric* and t is the thickness of the dielectric. The corresponding wavelength in air is given by

$$\lambda = \lambda'\eta$$

where η is the refractive index of the dielectric medium. Thus, the wavelengths of radiation transmitted by the filter are

$$\lambda = \frac{2t\eta}{\mathbf{n}} \qquad (7\text{-}5)$$

The integer **n** is the *order* of interference. The glass layers of the filter are often selected to absorb all but one of the reinforced bands; transmission is thus restricted to a single order.

Figure 7-13 illustrates the performance characteristics of typical interference filters. Ordinarily, filters are characterized, as shown, by the wavelength of their transmittance peaks, the percentage of incident radiation transmitted at the peak (their *percent transmittance,* Equation 6-31), and their effective bandwidths.

Interference filters are available with transmitter peaks throughout the ultraviolet and visible regions and up to about 14 μm in the infrared. Typically, effective bandwidths are about 1.5% of the wavelength at peak transmittance, although this figure is reduced to 0.15% in some narrow-band filters; these have maximum transmittances of 10%.

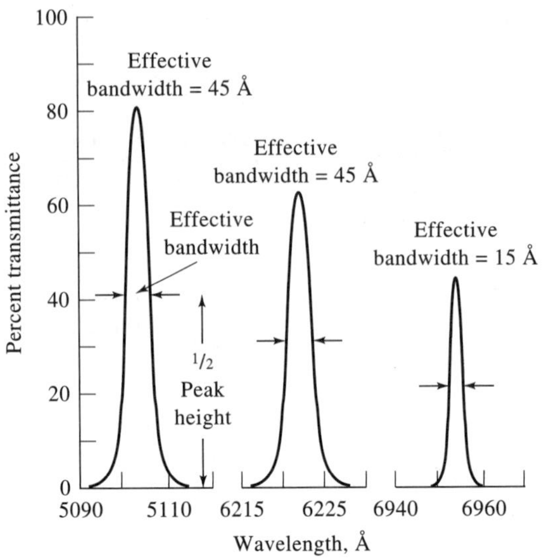

Figure 7-13 Transmission characteristics of typical interference filters.

Interference Wedges

An interference wedge consists of a pair of mirrored, partially transparent plates separated by a wedge-shaped layer of a dielectric material. The length of the plates ranges from about 50 to 200 mm. The radiation transmitted varies continuously in wavelength from one end to the other as the thickness of the wedge varies. By choosing the proper linear position along the wedge, a bandwidth of about 20 nm can be isolated.

Interference wedges are available for the visible region (400 to 700 nm), the near-infrared region (1000 to 2000 nm), and for several parts of the infrared region (2.5 to 14.5 μm). They can serve in place of prisms or gratings in monochromators.

Absorption Filters

Absorption filters, which are generally less expensive than interference filters, have been widely used for band selection in the visible region. These filters function by absorbing certain portions of the spectrum. The most common type consists of colored glass or of a dye suspended in gelatin and sandwiched between glass plates. The former have the advantage of greater thermal stability.

Absorption filters have effective bandwidths that range from perhaps 30 to 250 nm (see Figures 7-14 and 7-15). Filters that provide the narrowest bandwidths also absorb a significant fraction of the desired radiation and may have a transmittance of 10% or less at their band peaks. Glass filters with transmittance maxima throughout the entire visible region are available commercially.

Cut-off filters have transmittances of nearly 100% over a portion of the visible spectrum, but then rapidly decrease to zero transmittance over the remainder. A

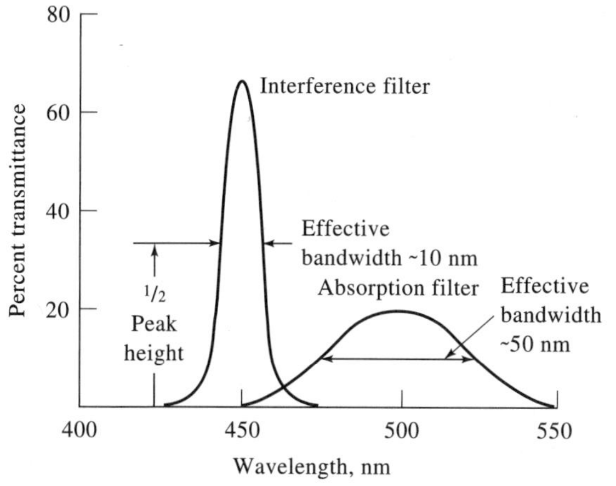

Figure 7-14 Effective bandwidths for two types of filters.

narrow spectral band can be isolated by coupling a cut-off filter with a second filter (see Figure 7-15).

It is apparent from Figure 7-14 that the performance characteristics of absorption filters are significantly inferior to those of interference-type filters. Not only are the bandwidths of absorption filters greater, but for narrow bandwidths the fraction of light transmitted is also less. Nevertheless, absorption filters are totally adequate for many applications.

7C-2 Monochromators

For many spectroscopic methods, it is necessary or desirable to be able to vary the wavelength of radiation continuously over a considerable range. This process is called *scanning* a spectrum. Monochromators are designed for spectral scanning. Monochromators for ultraviolet, visible, and infrared radiation are all similar in mechanical construction in the sense that they employ slits, lenses, mirrors, windows, and gratings or prisms. To be sure, the materials from which these components are fabricated depend upon the wavelength region of intended use (see Figure 7-2).

Components of Monochromators

Figure 7-16 illustrates the optical elements found in all monochromators, which include the following: (1) an entrance slit that provides a rectangular optical image, (2) a collimating lens or mirror that produces a parallel beam of radiation, (3) a prism or a grating that disperses the radiation into its component wavelengths, (4) a focusing element that reforms the image of the entrance slit and focuses it on a planar surface called a *focal plane,* and (5) an exit slit in the focal plane that isolates the desired spectral band. In addition, most monochromators have entrance and exit windows, which are designed to protect the components from dust and corrosive laboratory fumes.

As shown in Figure 7-16, two types of dispersing elements are found in monochromators: reflection gratings and prisms. For purposes of illustration, a beam made up of just two wavelengths, λ_1 and λ_2 ($\lambda_1 > \lambda_2$), is shown. This radiation enters the monochromators via a narrow rectangular opening or *slit,* is collimated, and then strikes the surface of the dispersing element at an angle. For the grating monochromator, angular dispersion of the wavelengths results from diffraction, which occurs at the reflective surface; for the prism, refraction at the two faces results in angular dispersal of the radiation, as shown. In both designs, the dispersed radiation is focused on the focal plane *AB* where it appears as two rectangular images of the entrance slit (one for λ_1 and one for λ_2). By rotating the dispersing element, one band or the other can be focused on the exit slit.

Historically, most monochromators were prism instruments. Currently, however, nearly all commercial monochromators are based upon reflection gratings because they are cheaper to fabricate, provide better wavelength separation for the same size dispersing element, and disperse radiation linearly along the focal plane. As shown in Figure 7-17a, linear dispersion means that the position of a band along the focal plane for a grating varies linearly with its wavelength. For prism instruments, in contrast, shorter wavelengths are

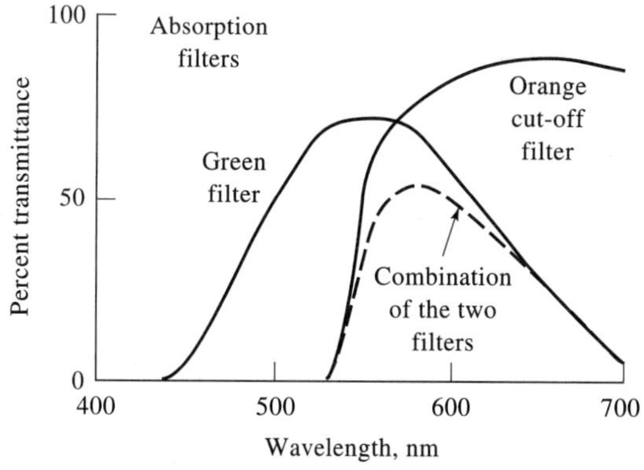

Figure 7-15 Comparison of various types of filters for visible radiation.

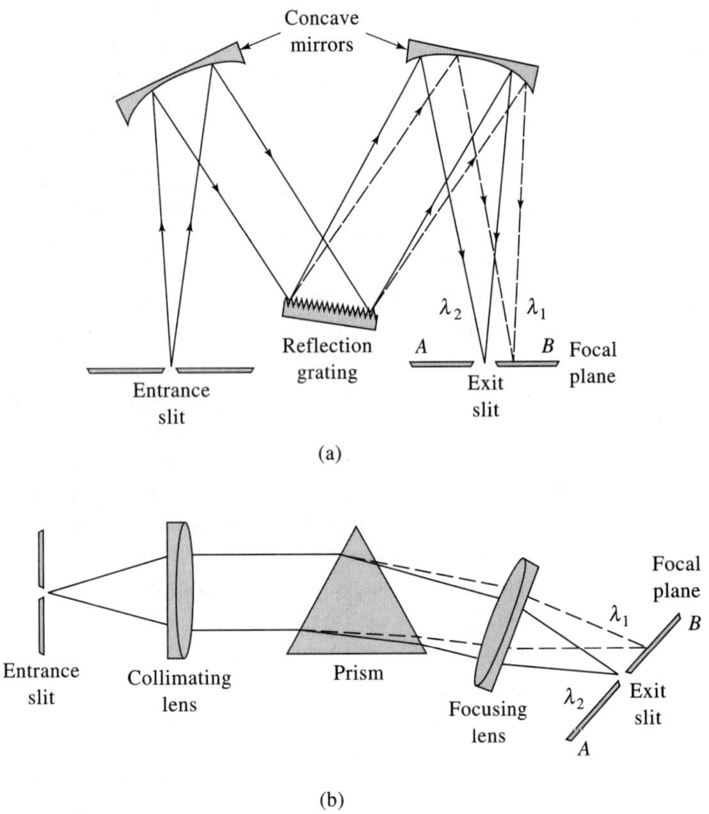

Figure 7-16 Two types of monochromators: (a) Czerney-Turner grating monochromator and (b) Bunsen prism monochromator. (In both instances, $\lambda_1 > \lambda_2$.)

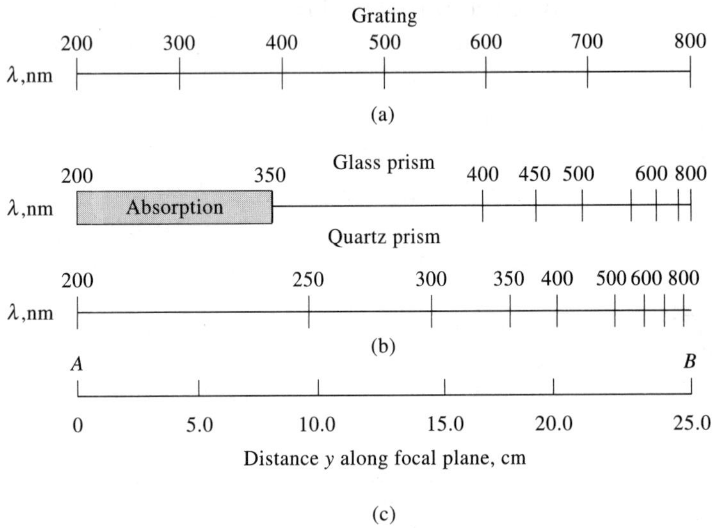

Figure 7-17 Dispersion for three types of monochromators. The points *A* and *B* on the scale in (c) correspond to the points shown in Figure 7-16.

dispersed to a greater degree than are longer ones, which complicates instrument design. The nonlinear dispersion of two types of prism monochromators is illustrated by Figure 7-17b. Because of their more general use, we will largely focus our discussion on grating monochromators.

Prism Monochromators

Prisms can be used to disperse ultraviolet, visible, and infrared radiation. The material used for their construction differs, however, depending upon the wavelength region (see Figure 7-2b).

Figure 7-18 shows the two most common types of prism designs. The first is a 60-deg prism, which is ordinarily fabricated from a single block of material. When crystalline (but not fused) quartz is the construction material, however, the prism is usually formed by cementing two 30-deg prisms together, as shown in Figure 7-18a; one is fabricated from right-handed quartz and the second from left-handed quartz. In this way, the optically active quartz causes no net polarization of the emitted radiation; this type of prism is called a *Cornu prism.* Figure 7-16b shows a *Bunsen monochromator,* which employs a 60-deg prism, likewise often made of quartz.

As shown in Figure 7-18b, the *Littrow prism,* which permits more compact monochromator designs, is a 30-deg prism with a mirrored back. Refraction in this type of prism takes place twice at the same interface so that the performance characteristics are similar to those of a 60-deg prism in a Bunsen mount.

Grating Monochromators

Dispersion of ultraviolet, visible, and infrared radiation can be brought about by directing a polychromatic beam through a *transmission grating* or onto the surface of a *reflection grating;* the latter is by far the more common type. *Replica gratings,* which are used in most monochromators are manufactured from a *master grating.*[13] The latter consists of a hard, optically flat, polished surface upon which have been ruled with a suitably shaped diamond tool a large number of parallel and closely spaced grooves. A magnified cross-sectional view of a few typical grooves is shown in Figure 7-19. A grating for the ultraviolet and visible region will typically contain from 300 to 2000 grooves/mm, with 1200 to 1400 being most common. For the infrared region, 10 to 200 grooves/mm are encountered; for spectrophotometers designed for the most widely used infrared range of 5 to 15 μm, a grating with about 100 grooves/mm is suitable. The construction of good master grating is tedious, time consuming, and expensive because the grooves must be identical in size, exactly parallel, and equally spaced over the length of the grating (3 to 10 cm).

Replica gratings are formed from a master grating by a liquid resin casting process that preserves virtually perfectly the optical accuracy of the original master grating on a clear resin surface. This surface is ordinarily made reflective by a coating of aluminum, or sometimes gold or platinum.

The Echellette Grating. Figure 7-19 is a schematic representation of an *echellette-type* grating, which is grooved or *blazed* such that it has relatively broad faces from which reflection occurs and narrow unused faces. This geometry provides highly efficient diffraction of radiation. Each of the broad faces can be considered to be a point source of radiation; thus interference among the reflected beams 1, 2, and 3 can occur. In order for the interference to be constructive, it is necessary that the path lengths differ by an integral multiple **n** of the wavelength λ of the incident beam.

In Figure 7-19, parallel beams of monochromatic radiation 1 and 2 are shown striking the grating at an incident angle i to the *grating normal.* Maximum constructive interference is shown as occurring at the reflected angle r. It is evident that beam 2 travels a greater distance than beam 1 and the difference in the paths is

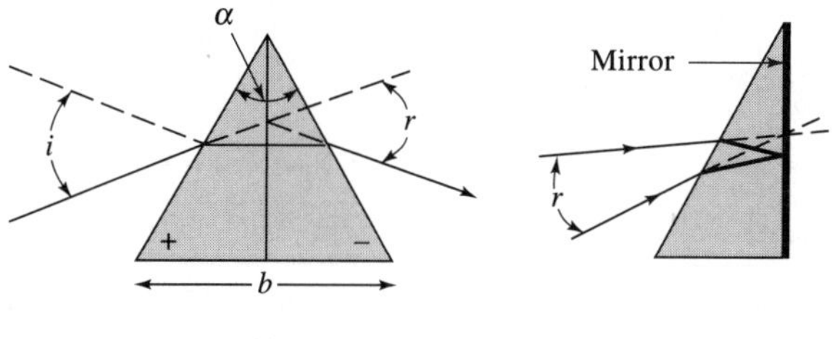

Figure 7-18 Dispersion by a prism: (a) quartz Cornu types and (b) Littrow type.

[13] For an interesting and informative discussion of the manufacture, testing, and performance characteristics of gratings, see *Diffraction Grating Handbook,* Rochester, NY: Bausch and Lomb, Inc. (now Milton Roy Company), 1970. For a historical perspective on the importance of gratings in the advancement of science, see A. G. Ingalls, *Sci. Amer.,* **1952,** *186*(6), 45.

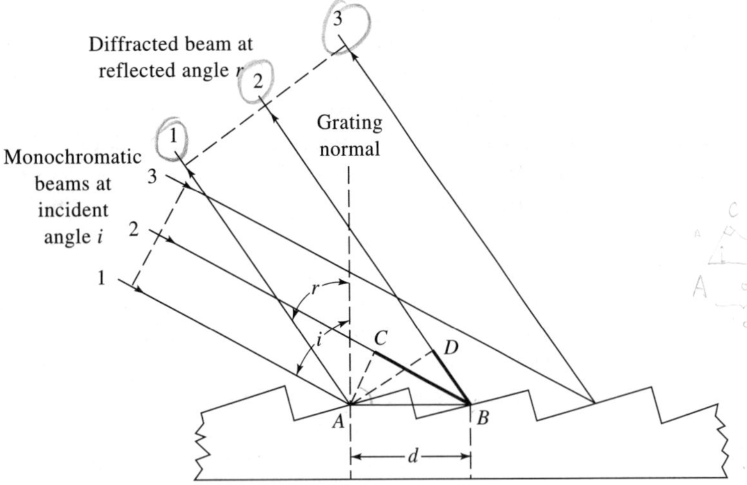

Figure 7-19 Mechanisms of diffraction from an echellette-type grating.

equal to ($\overline{CB} + \overline{BD}$) (shown as a broader line in the figure). For constructive interference to occur, this difference must equal $\mathbf{n}\lambda$. That is,

$$\mathbf{n}\lambda = (\overline{CB} + \overline{BD})$$

where **n**, a small whole number, is called the diffraction *order.* Note, however, that angle *CAB* is equal to angle *i* and that angle *DAB* is identical to angle *r.* Therefore, from simple trigonometry, we may write

$$\overline{CB} = d \sin i$$

where *d* is the spacing between the reflecting surfaces. It is also seen that

$$\overline{BD} = d \sin r$$

Substitution of the last two expressions into the first gives the condition for constructive interference. Thus,

$$\mathbf{n}\lambda = d(\sin i + \sin r) \qquad (7\text{-}6)$$

Equation 7-6 suggests that several values of λ exist for a given diffraction angle *r.* Thus, if a first-order line ($\mathbf{n} = 1$) of 900 nm is found at *r,* second-order (450-nm) and third-order (300-nm) lines also appear at this angle. Ordinarily, the first-order line is the most intense; indeed, it is possible to design gratings that concentrate as much as 90% of the incident intensity in this order. The higher-order lines can generally be removed by filters. For example, glass, which absorbs radiation below 350 nm, eliminates the higher-order spectra associated with first-order radiation in most of the visible region. The example that follows illustrates these points.

EXAMPLE 7-1

An echellette grating that contains 1450 blazes/mm was irradiated with a polychromatic beam at an incident angle 48 deg to the grating normal. Calculate the wavelengths of radiation that would appear at an angle of reflection of +20, +10, and 0 deg (angle *r,* Figure 7-19).

To obtain *d* in Equation 7-6, we write

$$d = \frac{1 \text{ mm}}{1450 \text{ blazes}} \times 10^6 \frac{\text{nm}}{\text{mm}} = 689.7 \frac{\text{nm}}{\text{blaze}}$$

When *r* in Figure 7-19 equals +20 deg,

$$\lambda = \frac{689.7}{\mathbf{n}} \text{ nm} (\sin 48 + \sin 20) = \frac{748.4}{\mathbf{n}} \text{ nm}$$

and the wavelengths for the first-, second-, and third-order reflections are 748, 374, and 249 nm, respectively.

Further calculations of a similar kind yield the following data:

	Wavelength (nm) for		
r, deg	**n = 1**	**n = 2**	**n = 3**
20	748	374	249
10	632	316	211
0	513	256	171

Concave Gratings. Gratings can be formed on a concave surface in much the same way as on a plane surface. A concave grating permits the design of a monochromator without auxiliary collimating and focusing mirrors or lenses because the concave surface both disperses the radiation and focuses it on the exit slit. Such an arrangement is advantageous in terms of cost; in addition, the reduction in number of optical surfaces increases the energy throughput of a monochromator that contains a concave grating.

Holographic Gratings.[14] One of the products from the emergence of laser technology is an optical (rather than mechanical) technique for forming gratings on plane or concave glass surfaces. *Holographic gratings* produced in this way are appearing in ever-increasing numbers in modern optical instruments, even some of the less expensive ones. Holographic gratings, because of their greater perfection with respect to line shape and dimensions, provide spectra that are freer from stray radiation and ghosts (double images).

In the preparation of holographic gratings, the beams from a pair of identical lasers are brought to bear at suitable angles upon a glass surface coated with photoresist. The resulting interference fringes from the two beams sensitize the photoresist so that it can be dissolved away, leaving a grooved structure that can be coated with aluminum or other reflecting substance to produce a reflection grating. The spacing of the grooves can be altered by changing the angle of the two laser beams with respect to one another. Nearly perfect, large (~50 cm) gratings with as many as 6000 lines/mm can be manufactured in this way at a relatively low cost. As with ruled gratings, replica gratings can be cast from a master holographic grating. It has been reported that no optical test exists that can distinguish between a master and replica holographic grating.[15]

Performance Characteristics of Grating Monochromators

resolution

The quality of a monochromator is dependent upon the purity of its radiant output, its ability to resolve adjacent wavelengths, its light-gathering power, and its spectral bandwidth. The last property is discussed in Section 7C-3.

[14] See J. Flamand, A. Grillo, and G. Hayat, *Amer. Lab.*, **1975**, *7* (5), 47; and J. M. Lerner et al., *Proc. Photo-Opt. Instrum. Eng.*, **1980**, *240*, 72, 82.

[15] I. R. Altelmose, *J. Chem. Educ.*, **1986**, *63*, A221.

Spectral Purity. The exit beam of a monochromator is usually contaminated with small amounts of scattered or stray radiation with wavelengths far different from that of the instrument setting. This unwanted radiation can be traced to several sources. Among these sources are reflections of the beam from various optical parts and the monochromator housing; the former arise from mechanical imperfections, particularly in gratings, introduced during manufacture. Scattering by dust particles in the atmosphere or on the surfaces of optical parts also causes stray radiation to reach the exit slit. Generally, the effects of spurious radiation are minimized by introducing baffles in appropriate spots in the monochromator and by coating interior surfaces with flat black paint. In addition, the monochromator is sealed with windows over the slits to prevent entrance of dust and fumes. Despite these precautions, however, some spurious radiation is still emitted; we shall see that its presence can have serious effects on absorption measurements under certain conditions.[16]

Dispersion of Grating Monochromators. The ability of a monochromator to separate different wavelengths is dependent upon its *dispersion.* The *angular dispersion* is given by $dr/d\lambda$ where dr is the change in the angle of reflection or refraction with a change in wavelength $d\lambda$. The angle r is defined in Figures 7-18 and 7-19.

The angular dispersion of a grating can be obtained by differentiating Equation 7-6 while holding i constant. Thus at any given angle of incidence,

$$\frac{dr}{d\lambda} = \frac{\mathbf{n}}{d \cos r} \tag{7-7}$$

The *linear dispersion D* refers to the variation in wavelength as a function of y, the distance along the line AB of the focal planes as shown in Figure 7-16. If F is the focal length of the monochromator, the linear dispersion can be related to the angular dispersion by the relationship

$$D = \frac{dy}{d\lambda} = \frac{F dr}{d\lambda} \tag{7-8}$$

[16] For discussion of the detection, the measurement, and the effects of stray radiation, see W. Kaye, *Anal. Chem.*, **1981**, *53*, 2201; M. R. Sharpe, *Anal. Chem.*, **1984**, *56*, 339A.

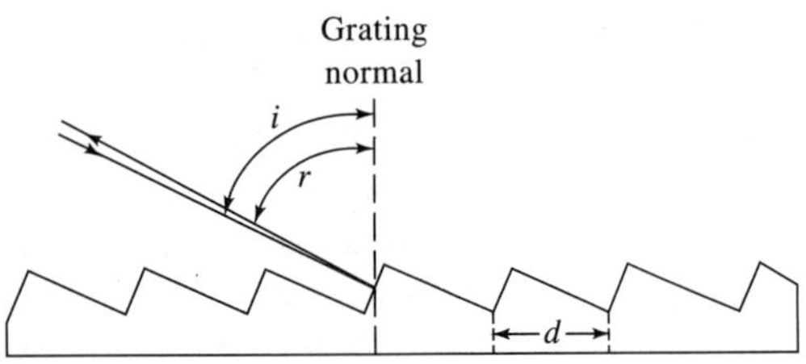

Figure 7-20 Echelle grating: i = angle of incidence; r = angle of reflection; d = groove spacing. In usual practice, $i \approx r = \beta = 63°26'$.

A more useful measure of dispersion is the *reciprocal linear dispersion* D^{-1} where

$$D^{-1} = \frac{d\lambda}{dy} = \frac{1}{F}\frac{d\lambda}{dr} \tag{7-9}$$

The dimensions of D^{-1} are often nm/mm or Å/mm.

Substitution of Equation 7-7 into Equation 7-9 leads to the reciprocal linear dispersion for a grating monochromator:

$$D^{-1} = \frac{d\lambda}{dy} = \frac{d \cos r}{\mathbf{n}F} \tag{7-10}$$

Note that the angular dispersion increases as the distance d between rulings decreases or as the number of lines per millimeter increases. At small angles of diffraction (<20 deg) $\cos r \approx 1$, and Equation 7-10 becomes approximately

$$D^{-1} = \frac{d}{\mathbf{n}F} \tag{7-11}$$

Note that, for all practical purposes, if the angle r is small, the *linear dispersion of a grating monochromator is constant,* a property that greatly simplifies monochromator design.

Resolving Power of Monochromators The *resolving power R* of a monochromator describes the limit of its ability to separate adjacent images that have a slight difference in wavelength. Here, by definition

$$R = \lambda/\Delta\lambda \tag{7-12}$$

where λ is the average wavelength of the two images and $\Delta\lambda$ is their difference. The resolving power of typical bench-top ultraviolet/visible monochromators ranges from 10^3 to 10^4.

It can be shown[17] that the resolving power of a grating is given by the expression

$$R = \frac{\lambda}{\Delta\lambda} = \mathbf{n}N \tag{7-13}$$

when $\mathbf{n}$ is the diffraction order and N is the number of grating blazes illuminated by radiation from the entrance slit. Thus, better resolution is a characteristic of longer gratings, smaller blaze spacings, and higher diffraction orders. This equation applies to both echellette and echelle gratings.

Light-Gathering Power of Monochromators. In order to increase the signal-to-noise ratio of a spectrometer, it is necessary that the radiant energy that reaches the detector must be as large as possible. The *f/number* or *speed* provides a measure of the ability of a monochromator to collect the radiation that emerges from the entrance slit. The *f*/number is defined by the equation

$$f = F/d \tag{7-14}$$

where F is the focal length of the collimating mirror (or lens) and d is its diameter. The light-gathering power of an optical device increases as the inverse square of the *f*/number. Thus, an *f*/2 lens gathers four times more light than an *f*/4 lens. The *f*/numbers for many monochromators lie in the 1 to 10 range.

Echelle Monochromators. Echelle monochromators contain two dispersing elements arranged in series. The first of these elements is a special type of grating called an *echelle grating.* The second, which follows, is usually a low-dispersion prism, or sometimes a grating. The echelle grating, which was first described by G. R. Harrison in 1949, provides higher dispersion and higher resolution than an echellette of the same size.[18] Figure 7-20 shows a cross section of a typical echelle grating. It differs from the echellette grating shown in Figure 7-19 in several respects. First, in order to achieve a high angle of incidence, the blaze angle of an echelle grating is significantly greater than the conventional device, and the short side of the blaze is used rather than the long. Fur-

[17] R. A. Sawyer, *Experimental Spectroscopy,* 2nd ed., p. 130. Englewood Cliffs, NJ: Prentice Hall, 1951.

[18] For a more detailed discussion of the echelle grating, see P. N. Keliher and C. C. Wohlers, *Anal. Chem.,* **1976,** *48,* 333A; D. L. Anderson, A. R. Forster, and M. L. Parsons, *Anal. Chem.,* **1981,** *53,* 770; A. T. Zander and P. N. Keliher, *Appl. Spectrosc.,* **1979,** *33,* 499.

thermore the grating is relatively coarse, having typically 300 or fewer grooves per millimeter for ultraviolet/visible radiation. Note that the angle of refraction r is much higher in the echelle grating than the echellette and approaches the angle of incidence i. That is,

$$r \approx i = \beta$$

Under these circumstances, Equation 7-6 for a grating becomes

$$\mathbf{n}\lambda = 2d \sin \beta \qquad (7\text{-}15)$$

With a normal echellette grating, high dispersion, or low reciprocal dispersion, is obtained by making the groove width d small and the focal length F large. A large focal length reduces light gathering and makes the monochromator large and unwieldy. In contrast, the echelle grating achieves high dispersion by making both the angle β and the order of diffraction **n** large. The reciprocal dispersion for an echelle grating can then be obtained by modifying Equation 7-10 to

$$D^{-1} = \frac{2d \cos \beta}{\mathbf{n}F} \qquad (7\text{-}16)$$

The advantages of the echelle grating are illustrated by the data in Table 7-1, which show the performance characteristics for two typical monochromators, one with a conventional echellette grating and the other with an echelle. Note that for the same focal length the linear dispersion and resolution are an order of magnitude greater for the echelle; the light-gathering power of the echelle is also somewhat superior.

One of the problems encountered with the use of an echelle grating is that the linear dispersion at high orders of refraction is so great that to cover a reasonably broad spectral range it is necessary to use many successive orders. For example, one instrument designed to cover a range of 200 to 800 nm employs diffraction orders 28 to 118 (90 successive orders). Because these orders inevitably overlap, it is essential that a system of cross dispersion, such as that shown in Figure 7-21a, be employed with an echelle grating. Here, the dispersed radiation from the grating is passed through a prism (in some systems, a second grating is used) whose axis is at 90 deg to the grating. The effect of this arrangement is to produce a two-dimensional spectrum as shown schematically in Figure 7-21b. In this figure, the locations of 8 of the 70 orders is indicated by short vertical lines. For any given order, the wavelength dispersion is approximately linear, but, as can be seen, the dispersion lessens at lower orders or higher wavelengths. An actual two-dimensional spectrum from an echelle monochromator consists of a complex series of short vertical lines lying along 50 to 100 horizontal axes, each axis corresponding to one diffraction order. To change wavelength with an echelle monochromator, it is necessary to change the angle of both the grating and the prism.

Recently, several instrument manufacturers have begun to offer echelle-type spectrometers for simultaneous determination of a multitude of elements by atomic emission spectroscopy. The optical designs of two of these instruments are shown in Figures 10-7 and 10-9.

TABLE 7-1 Comparison of Performance Characteristics of a Conventional and Echelle Monochromator*

	Conventional	Echelle
Focal length	0.5 m	0.5 m
Groove density	1200/mm	79/mm
Diffraction angle, β	10°22′	63°26′
Order **n** (at 300 nm)	1	75
Resolution (at 300 nm), $\lambda/\Delta\lambda$	62,400	763,000
Reciprocal linear dispersion, D^{-1}	16 Å/mm	1.5 Å/mm
Light-gathering power, f	f/9.8	f/8.8

*With permission from P. E. Keliher and C. C. Wohlers, *Anal. Chem.*, **1976,** *48,* 334A. Copyright 1976 American Chemical Society.

7C-3 Monochromator Slits

The slits of a monochromator play an important role in determining the monochromator's performance characteristics and quality. Slit jaws are formed by carefully machining two pieces of metal to give sharp edges. Care is taken to assure that the edges of the slit are exactly parallel to one another and that they lie on the same plane. In some monochromators, the openings of the two slits are fixed; more commonly, the spacing can be adjusted with a micrometer mechanism.

The entrance slit (see Figure 7-16) of a monochromator serves as a radiation source; its image is ultimately focused on the focal plane that contains the exit slit. If the radiation source consists of a few discrete

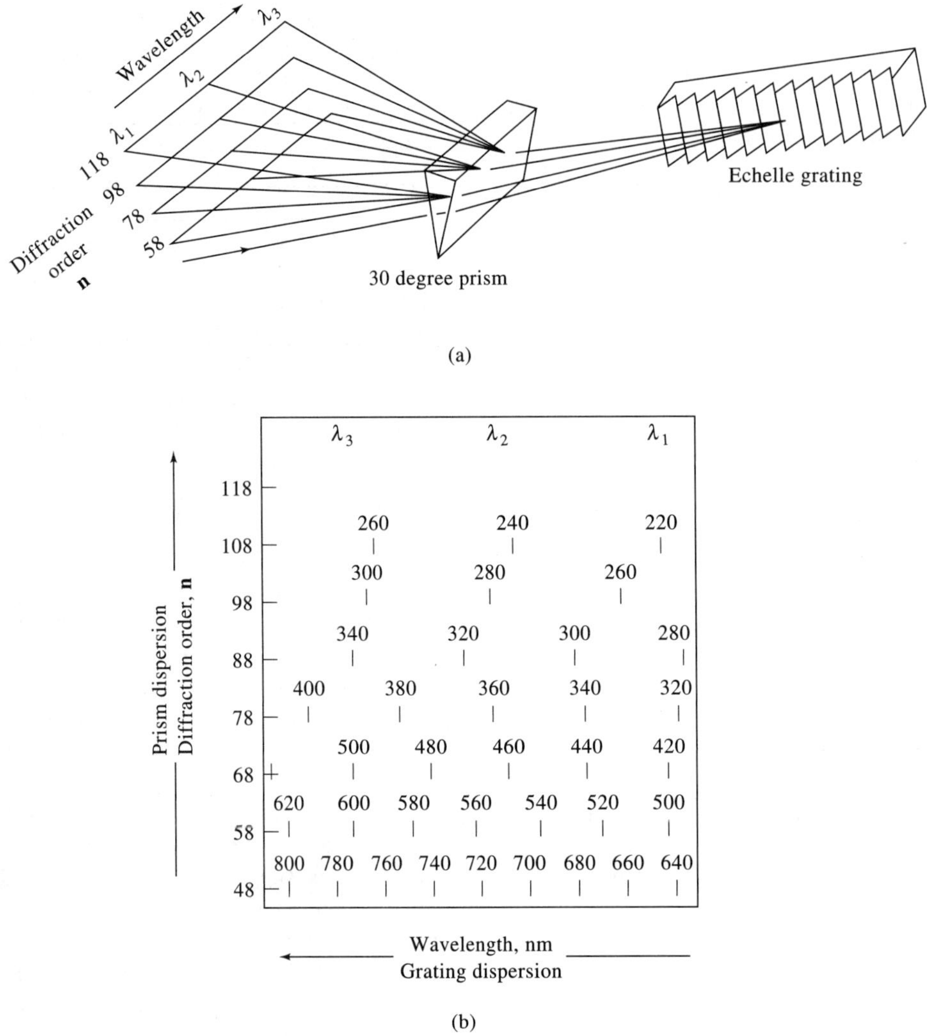

Figure 7-21 An echelle monochromator: (a) arrangement of dispersing elements, and (b) schematic end-on view of the dispersed radiation from the point of view of the transducer.

wavelengths, a series of rectangular images appears on this surface as bright lines, each corresponding to a given wavelength. A particular line can be brought to focus on the exit slit by rotating the dispersing element. If the entrance and exit slits are of the same size (as is usually the case), the image of the entrance slit will in theory just fill the exit-slit opening when the setting of the monochromator corresponds to the wavelength of the radiation. Movement of the monochromator mount in one direction or the other results in a continuous decrease in emitted intensity, zero being reached when the entrance-slit image has been displaced by its full width.

Effect of Slit Width on Resolution

Figure 7-22 illustrates the situation in which monochromatic radiation of wavelength λ_2 strikes the exit slit. Here, the monochromator is set for λ_2 and the two slits are identical in width. The image of the entrance slit just fills the exit slit. Movement of the monochromator to a setting of λ_1 or λ_3 results in the image being moved completely out of the slit. The lower half of Figure 7-22 shows a plot of the radiant power emitted as a function of monochromator setting. Note that the *bandwidth* is defined as the span of monochromator settings (in units of wavelength, or sometimes in units of cm^{-1}) needed

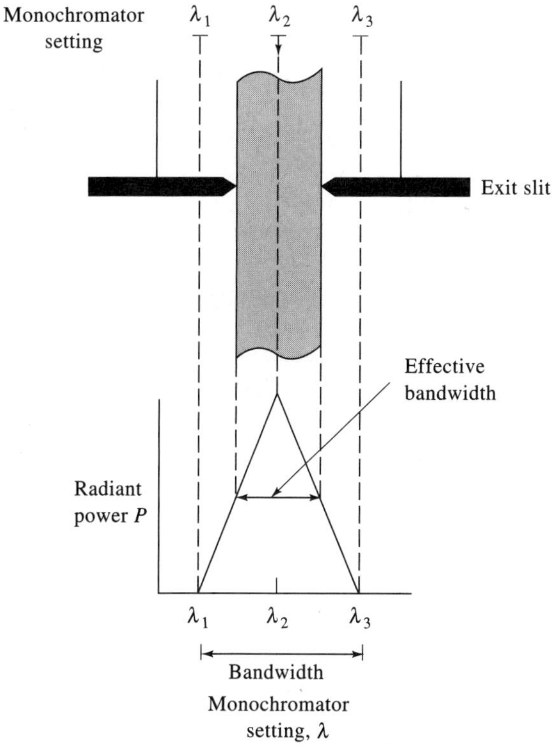

Figure 7-22 Illumination of an exit slit by monochromatic radiation λ_2 at various monochromator settings. Exit and entrance slits are identical.

to move the image of the entrance slit across the exit slit. If polychromatic radiation were employed, it would also represent the span of wavelengths from the exit slit for a given monochromator setting.

The *effective bandwidth,* which is one half the bandwidth when the two slit widths are identical, is seen to be the range of wavelengths that exit the monochromator at a given wavelength setting. The effective bandwidth can be related to the reciprocal linear dispersion by writing Equation 7-8 in the form

$$D^{-1} = \frac{\Delta\lambda}{\Delta y}$$

where $\Delta\lambda$ and Δy are now finite intervals of wavelength and linear distance along the focal plane, respectively. As shown by Figure 7-22, when Δy is equal to the slit width w, $\Delta\lambda$ is the effective bandwidth. That is,

$$\Delta\lambda_{eff} = wD^{-1} \qquad (7\text{-}17)$$

Figure 7-23 illustrates the relationship between the effective bandwidth of an instrument and its ability to resolve spectral peaks. Here, the exit slit of a grating monochromator is illuminated with a beam composed of just three equally spaced lines at wavelengths, λ_1, λ_2, and λ_3; each line is assumed to be of the same intensity. In the top figure, the effective bandwidth of the instrument is exactly equal to the difference in wavelength between λ_1 and λ_2 or λ_2 and λ_3. When the monochromator is set at λ_2, radiation of this wavelength just fills the slit. Movement of the monochromator in either direction diminishes the transmitted intensity of λ_2, but increases the intensity of one of the other lines by an equivalent amount. As shown by the solid line in the plot to the right, no spectral resolution of the three wavelengths is achieved.

In the middle drawing of Figure 7-23, the effective bandwidth of the instrument has been reduced by narrowing the openings of the exit and entrance slits to three quarters that of their original dimensions. The solid line in the plot on the right shows that partial resolution of the three lines results. When the effective bandwidth is decreased to one half the difference in wavelengths of the three beams, complete resolution is achieved, as shown in the bottom drawing. Thus, complete resolution of two lines is feasible only if the slit width is adjusted so that the effective bandwidth of the monochromator is equal to one half the wavelength difference of the lines.

EXAMPLE 7-2

A grating monochromator with a reciprocal linear dispersion of 1.2 nm/mm is to be used to separate the sodium lines at 589.0 and 589.6. In theory, what slit width would be required?

Complete resolution of the two lines requires that

$$\Delta\lambda_{eff} = \frac{1}{2}(589.6 - 589.0) = 0.3 \text{ nm}$$

Substitution into Equation 7-15 after rearrangement gives

$$w = \frac{\Delta\lambda_{eff}}{D^{-1}} = \frac{0.3 \text{ nm}}{1.2 \text{ nm/mm}} = 0.25 \text{ mm}$$

It is important to note that slit widths calculated as in Example 7-2 are theoretical. Imperfections, which are present in most monochromators, are such that slit widths narrower than theoretical are usually required to achieve a desired resolution.

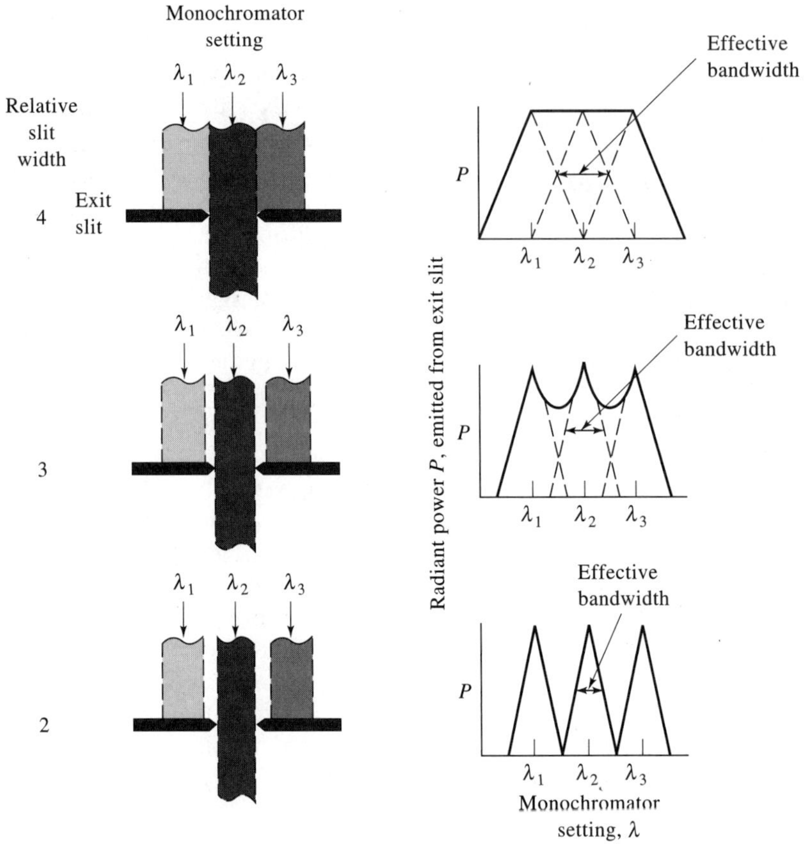

Figure 7-23 The effect of the slit width on spectra. The entrance slit is illuminated with λ_1, λ_2, and λ_3 only. Entrance and exit slits are identical. Plots on the right show changes in emitted power as the setting of monochromator is varied.

Figure 7-24 shows the effect of bandwidth on experimental spectra for benzene vapor. Note the much greater spectral detail realized with the narrowest slit setting and thus the narrowest bandwidth.

Choice of Slit Widths

The effective bandwidth of a monochromator depends upon the dispersion of the grating or prism as well as the width of the entrance and exit slits. Most monochromators are equipped with variable slits so that the effective bandwidth can be changed. The use of minimal slit width is desirable when the resolution of narrow absorption or emission bands is needed. On the other hand, a marked decrease in the available radiant power accompanies a narrowing of slits, and accurate measurement of this power becomes more difficult. Thus, wider slit widths may be used for quantitative analysis rather than for qualitative work, where spectral detail is important.

low resolution
high radiant power Feb 16, 05

7D SAMPLE CONTAINERS

Sample containers are required for all spectroscopic studies except emission spectroscopy. In common with the optical elements of monochromators, the *cells* or *cuvettes* that hold the samples must be made of material that is transparent to radiation in the spectral region of interest. Thus, as shown in Figure 7-2, quartz or fused silica is required for work in the ultraviolet region (below 350 nm); both of these substances are transparent in the visible region and up to about 3 μm in the infrared region as well. Silicate glasses can be employed in the region between 350 and 2000 nm. Plastic containers have also found application in the visible region. Crystalline sodium chloride is the most common substance employed for cell windows in the infrared region; the other infrared transparent materials listed in Figure 7-2 may also be used for this purpose.

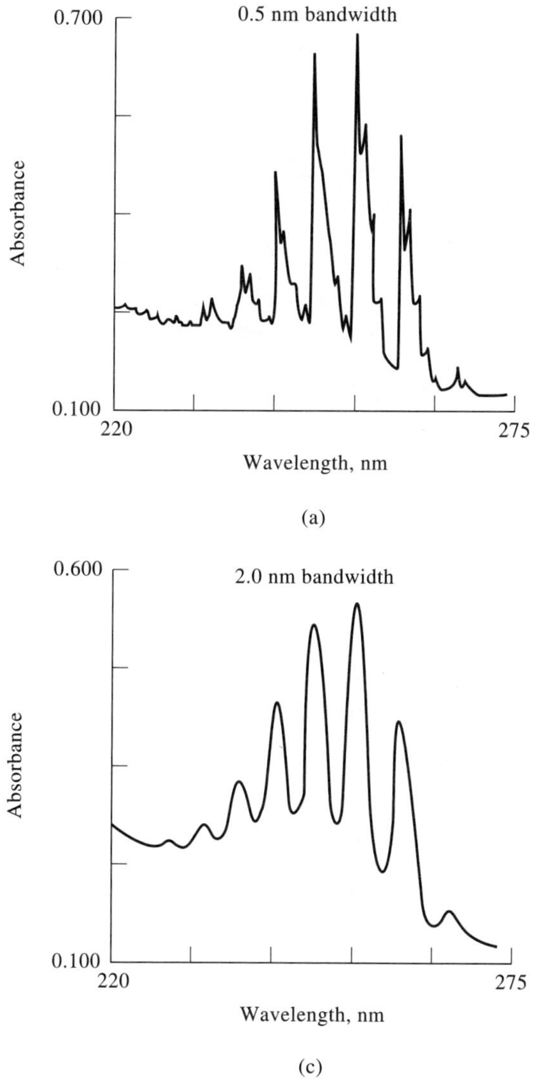

Figure 7-24 Effect of bandwidth on spectral detail for benzene vapor: (a) 0.5 nm; (b) 1.0 nm; (c) 2.0 nm. *(From V. A. Kohler,* Amer. Lab., ***1984*** *(11), 132. Copyright 1984 International Scientific Communications Inc.)*

7E RADIATION TRANSDUCERS

7E-1 Introduction

The detectors for early spectroscopic instruments were the human eye or a photographic plate or film. These detection devices have been largely supplanted by transducers that convert radiant energy into an electrical signal; our discussion will be largely confined to these more modern detectors.

Properties of the Ideal Transducer

The ideal transducer would have a high sensitivity, a high signal-to-noise ratio, and a constant response over a considerable range of wavelengths. In addition, it would exhibit a fast response time and a zero output signal in the absence of illumination. Finally, the electrical signal produced by the transducer would be directly proportional to the radiant power P. That is,

$$S = kP \tag{7-18}$$

where S is the electrical response in terms of current or voltage, and k is the calibration sensitivity (Section 1D-2).

Many real transducers exhibit a small, constant response, known as a *dark current,* in the absence of radiation. For these transducers , the response is described by the relationship

$$S = kP + k_d \tag{7-19}$$

where k_d represents the dark current, which is ordinarily constant over short measurement periods. Instruments with transducers that produce a dark current are usually equipped with a compensating circuit that reduces k_d to zero; Equation 7-18 then applies.

Types of Radiation Transducers[19]

As indicated in Figure 7-3b, two general types of radiation transducers are encountered; one responds to photons, the other to heat. All photon transducers (also called *photoelectric* or quantum detectors) have an active surface, which is capable of absorbing radiation. In some types, the absorbed energy causes emission of electrons and the development of a photocurrent. In others, the radiation promotes electrons into conduction bands; detection here is based on the resulting enhanced conductivity *(photoconduction)*. Photon transducers are used largely for measurement of ultraviolet, visible, and near-infrared radiation. When they are applied to radiation much longer than 3 μm in wavelength, they must be cooled to dry-ice or liquid-nitrogen temperatures to avoid interference from thermal background noise. Photoelectric transducers differ from heat transducers in that the formers' electrical signal results from a series of individual events (absorption of single photons), the probability of which can be described by the use of statistics. In contrast, thermal transducers, which are widely employed for the detection of infrared radiation, respond to the average power of the incident radiation.

As shown in Section 5B-2 the distinction between photon and heat transducers is important because shot noise often limits the behavior of the former while thermal noise frequently limits the latter. As a consequence, the indeterminate errors associated with the two types of transducers are fundamentally different.

Figure 7-25 shows the relative spectral response of the various kinds of transducers that are useful for ultraviolet, visible, and infrared spectroscopy. The ordinate

[19] For a discussion of optical radiation detectors, see E. L. Dereniak and D. G. Crowe, *Optical Radiation Detectors*. New York: Wiley, 1984; F. Grum and R. J. Becherer, *Optical Radiation Measurements*, Vol. 1. New York: Academic Press, 1979; J. D. Ingle Jr., and S. R. Crouch *Spectrochemical Analysis*, pp. 106–117. Englewood Cliffs, NJ: Prentice Hall, 1988.

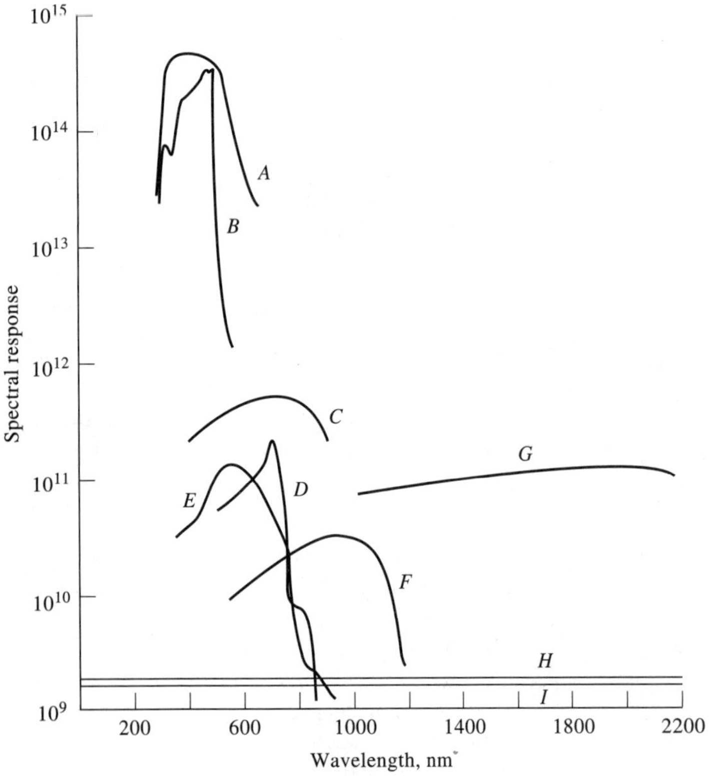

Figure 7-25 Relative response of various types of photoelectric transducers *(A–G)* and heat transducers *H, I*): *A*, photomultiplier tube; *B*, CdS photoconductivity; *C*, GaAs photovoltaic cell; *D*, CdSe photoconductivity cell; *E*, Se/SeO photovoltaic cell; *F*, silicon photodiode; *G*, PbS photoconductivity; *H*, thermocouple; *I*, Golay cell. *(Adapted from P. W. Druse, L. N. McGlauchlin, and R. B. Quistan,* Elements of Infrared Technology, *pp. 424–425. New York: Wiley, 1962. Reprinted by permission of John Wiley & Sons Inc.)*

function is inversely related to the noise of the transducers and directly related to the square root of its surface area. Note that the relative sensitivity of the thermal transducers (curves *H* and *I*) is independent of wavelength but significantly lower than the sensitivity of photoelectric transducers. On the other hand, photon transducers are often far from ideal with respect to constant response versus wavelength.

7E-2 Photon Transducers

Several types of photon transducers are available, including: (1) photovoltaic cells, in which the radiant energy generates a current at the interface of a semiconductor layer and a metal; (2) phototubes, in which radiation causes emission of electrons from a photosensitive solid surface; (3) photomultiplier tubes, which contain a photoemissive surface as well as several additional surfaces that emit a cascade of electrons when struck by electrons from the photosensitive area; (4) photoconductivity transducers in which absorption of radiation by a semiconductor produces electrons and holes, thus leading to enhanced conductivity; (5) silicon photodiodes, in which photons increase the conductance across a reverse-biased *pn* junction; and (6) charge-transfer transducers, in which the charges developed in a silicon crystal as a result of absorption of photons are collected and measured.[20]

Photovoltaic or Barrier-Layer Cells

The photovoltaic cell is a simple device that is used for detecting and measuring radiation in the visible region. The typical cell has a maximum sensitivity at about 550 nm; the response falls off to perhaps 10% of the maximum at 350 and 750 nm (see Figure 7-25*E*). Its range approximates that of the human eye.

The photovoltaic cell consists of a flat copper or iron electrode upon which is deposited a layer of semiconducting material, such as selenium (see Figure 7-26). The outer surface of the semiconductor is coated with a thin transparent metallic film of gold or silver, which serves as the second or collector electrode; the entire array is protected by a transparent envelope. When radiation of sufficient energy reaches the semiconductor, co-

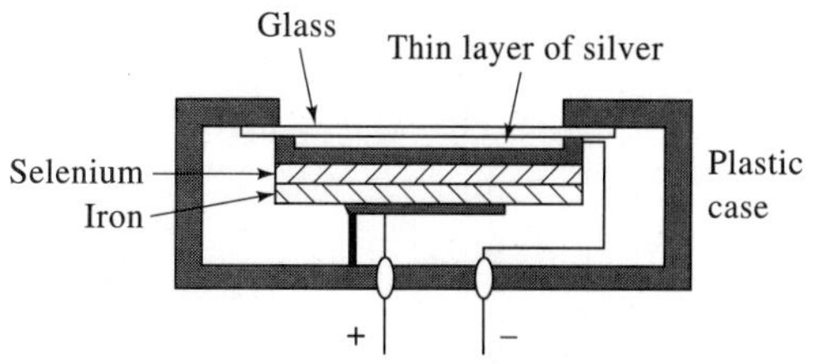

Figure 7-26 Schematic of a typical barrier-layer cell.

valent bonds are broken, with the result that conduction electrons and holes are formed. The electrons then migrate toward the metallic film and the holes toward the base upon which the semiconductor is deposited. The liberated electrons are free to migrate through the external circuit to interact with these holes. The result is an electrical current of a magnitude that is proportional to the number of photons that strike the semiconductor surface. Ordinarily, the currents produced by a photovoltaic cell are large enough to be measured with a microammeter; if the resistance of the external circuit is kept small ($<400\ \Omega$), the photocurrent is directly proportional to the power of the radiation that strikes the cell. Currents on the order of 10 to 100 μA are typical.

The barrier-layer cell constitutes a rugged, low-cost means for measuring radiant power. No external source of electrical energy is required. On the other hand, the low internal resistance of the cell makes the amplification of its output less convenient. Consequently, although the barrier-layer cell provides a readily measured response at high levels of illumination, it suffers from lack of sensitivity at low levels. Another disadvantage of the barrier-type cell is that it exhibits *fatigue* in which its current output decreases gradually during continued illumination; proper circuit design and choice of experimental conditions minimize this effect. Barrier-type cells find use in simple, portable instruments where ruggedness and low cost are important. For routine analyses, these instruments often provide perfectly reliable analytical data.

Vacuum Phototubes[21]

A second type of photoelectric device is the vacuum phototube, which consists of a semicylindrical cathode and a wire anode sealed inside an evacuated transparent

---

[20] For a comparison of the performance characteristics of the three most sensitive and widely used photon transducers, namely photomultipliers, silicon diodes, and charge-transfer devices, see W. E. L. Grossman, *J. Chem. Educ.*, **1989**, *66*, 697.

[21] For a discussion of vacuum phototubes and photomultiplier tubes, see F. E. Lytle, *Anal. Chem.*, **1974**, *46*, 545A.

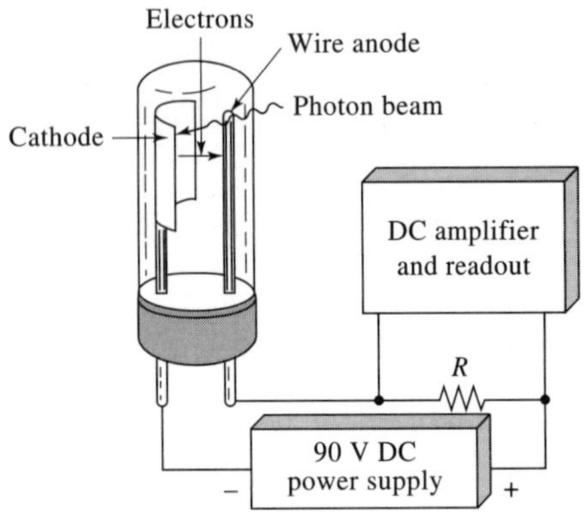

Figure 7-27 A phototube and accessory circuit. The photocurrent induced by the radiation causes a potential drop across *R*, which is then amplified to drive a meter or recorder.

envelope (see Figure 7-27). The concave surface of the electrode supports a layer of photoemissive material (Section 6C-1) that tends to emit electrons when it is irradiated. When a potential is applied across the electrodes, the emitted electrons flow to the wire anode generating a photocurrent that is generally about one tenth as great as that associated with a photovoltaic cell for a given radiant intensity. In contrast, however, amplification is easily accomplished since the phototube has a high electrical resistance.

The number of electrons ejected from a photoemissive surface is directly proportional to the radiant power of the beam that strikes that surface. As the potential applied across the two electrodes of the tube is increased, the fraction of the emitted electrons that reaches the anode rapidly increases; when the saturation potential is achieved, essentially all of the electrons are collected at the anode. The current then becomes independent of potential and directly proportional to the radiant power. Phototubes are usually operated at a potential of about 90 V, which is well within the saturation region.

A variety of photoemissive surfaces are used in commercial phototubes. Typical examples are shown in Figure 7-28. From the user's standpoint, photoemissive surfaces fall into four categories: highly sensitive, red sensitive, ultraviolet sensitive, and flat response. The most sensitive cathodes are bialkali types such as number 117 in Figure 7-28; they are made up of potassium, cesium, and antimony. Red-sensitive materials are multialkali types (for example, Na/K/Cs/Sb), or Ag/O/Cs formulations. The behavior of the latter is shown as S-11 in the figure. Compositions of Ga/In/As extend the red region up to about 1.1 μm. Most formulations are ultraviolet sensitive provided the tube is equipped with transparent windows. Flat responses are obtained with Ga/As compositions such as that labeled 128 in Figure 7-28.

Phototubes frequently produce a small dark current (see Equation 7-19) that results from thermally induced electron emission and natural radioactivity from ^{40}K in the glass housing of the tube.

Photomultiplier Tubes

For the measurement of low radiant power, the *photomultiplier tube* (PMT) offers advantages over the ordinary phototube.[22] Figure 7-29 is a schematic of such a device. The photo cathode surface is similar in composition to the surfaces of the phototubes described in Figure 7-28, and it emits electrons when exposed to radiation. The tube also contains additional electrodes (nine in Figure 7-29) called *dynodes.* Dynode 1 is maintained at a potential 90 V more positive than the cathode, and electrons are accelerated toward it as a consequence. Upon striking the dynode, each photoelectron causes emission of several additional electrons; these, in turn, are accelerated toward dynode 2, which is 90 V more positive than dynode 1. Again, several electrons are emitted for each electron that strikes the surface. By the

[22] For a detailed discussion of the theory and applications of photomultipliers, see R. W. Engstrom, *Photomultiplier Handbook,* Lancaster, PA: RCA Corporation, 1980.

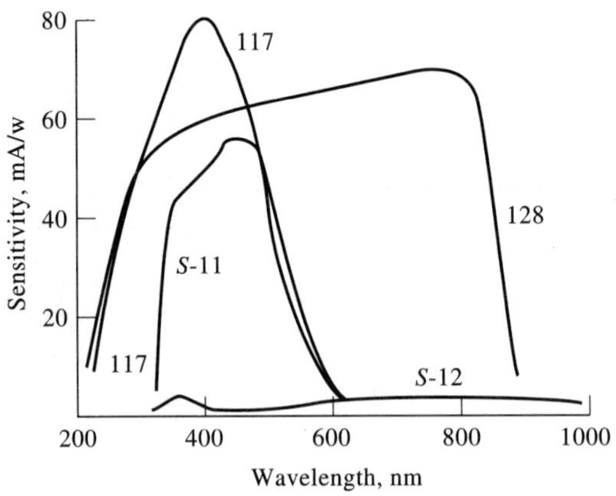

Figure 7-28 Spectral response of some typical photoemissive surfaces. *(From F. E. Lytle,* Anal. Chem., ***1974,*** *46, 546A. Copyright 1974 American Chemical Society.)*

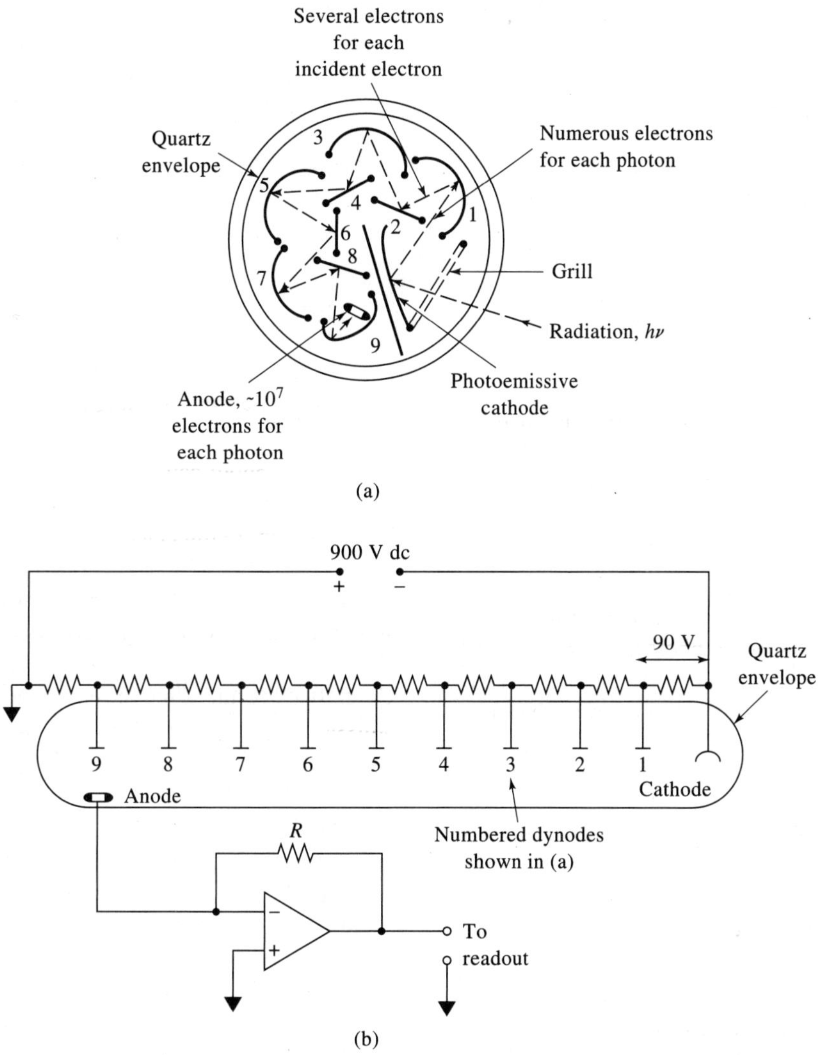

Figure 7-29 Photomultiplier tube: (a) cross-section of the tube and (b) electrical circuit.

time this process has been repeated nine times, 10^6 to 10^7 electrons have been formed for each incident photon; this cascade is finally collected at the anode and the resulting current is then electronically amplified and measured.

As shown by Figure 7-25*A*, photomultipliers are highly sensitive to ultraviolet and visible radiation; in addition, they have extremely fast time responses. Often, the sensitivity of an instrument with a photomultiplier is limited by its dark-current emission. Because thermal emission is the major source of dark-current electrons, the performance of a photomultiplier can be enhanced by cooling. In fact, thermal dark currents can be virtually eliminated by cooling the detector to −30°C. Transducer housings, which can be cooled by circulation of an appropriate coolant, are available commercially.

Photomultiplier tubes are limited to measuring low-power radiation because intense light causes irreversible damage to the photoelectric surface. For this reason, the device is always housed in a light-tight compartment, and care is taken to eliminate the possibility of its being exposed even momentarily to daylight or other strong light. With appropriate external circuitry, photomultiplier tubes can be used to detect the arrival of a single photon at the photocathode.

Silicon Diode Transducers

A silicon diode transducer consists of a reverse-biased *pn* junction formed on a silicon chip. As shown in Figure 7-30, the reverse bias creates a depletion layer that reduces the conductance of the junction to nearly zero. If radiation is allowed to impinge on the chip, however, holes and electrons are formed in the depletion layer and swept through the device to produce a current that is proportional to radiant power.

Silicon diodes are more sensitive than vacuum phototubes but less sensitive than photomultiplier tubes (see Figure 7-25*F*). Photodiodes have spectral ranges from about 190 to 1100 nm.

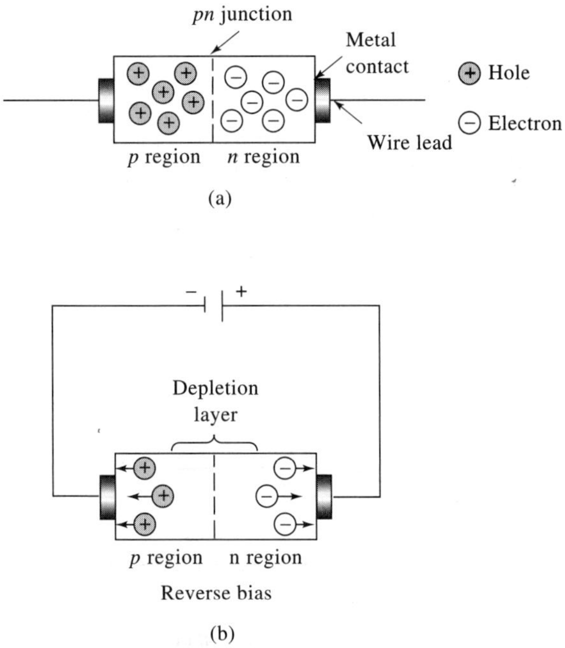

Figure 7-30 (a) Schematic of a silicon diode. (b) Formation of depletion layer, which prevents flow of electricity under reverse bias.

7E-3 Multichannel Photon Transducers[23]

The first multichannel detector used in spectroscopy was a photographic plate or a film strip that was placed along the length of the focal plane of a spectrometer so that all the lines in a spectrum could be recorded simultaneously. Photographic detection is relatively sensitive with some emulsions that respond to as few as 10 to 100 photons. The primary limitation of this type of detector is, however, the time required for developing the image of the spectrum and converting the blackening of the emulsion to radiant intensities.

Modern multichannel transducers consist of an array of small photoelectric-sensitive elements arranged either linearly or in a two-dimensional pattern on a single semiconductor chip. The chip, which is usually silicon and typically has dimensions of a few millimeters on a side, also contains electronic circuitry that makes it possible to determine the electrical output signal from each of the photosensitive elements either sequentially or simultaneously. For spectroscopic studies, a multichannel transducer is generally placed in the focal plane of a spectrometer so that various elements of the dispersed spectrum can be transduced and measured simultaneously.

At the present time three types of multichannel devices are used in commercial spectroscopic instruments: *photodiode arrays* (PDAs), *charge-injection devices* (CIDs), and *charge-coupled devices* (CCDs). Photodiode arrays are one-dimensional transducers in which the photosensitive elements are arranged in a line on the transducer face. In contrast, the individual photosensitive elements of charge-injection and charge-coupled devices are usually formed as two-dimensional arrays. Charge-injection and charge-coupled transducers both function by collecting photogenerated charges in various areas of the transducer surface and then measuring the quantity of charge accumulated in a brief period. In both devices, the measurement is accomplished by transferring the charge from a collection area to a detection area. For this reason, the two types of transducers are sometimes called *charge-transfer devices* (CTDs). These devices have widespread use as image transducers for television applications and in astronomy.

Photodiode Arrays

In a photodiode array, the individual photosensitive elements are small silicon photodiodes each of which consists of a reverse-biased *pn* junction (see previous section). The individual photodiodes are part of a

[23] For a discussion of multichannel photon detectors, see Y. Talmi, *Appl. Spectrosc.*, **1982**, *36*, 1; W. E. Grossman, *J. Chem. Educ.*, **1989**, *66*, 697; J. V. Sweedler, *Crit. Rev. Anal. Chem.*, **1993**, *24*, 59; D. G. Jones, *Anal. Chem.*, **1985**, *57*, 1057A.

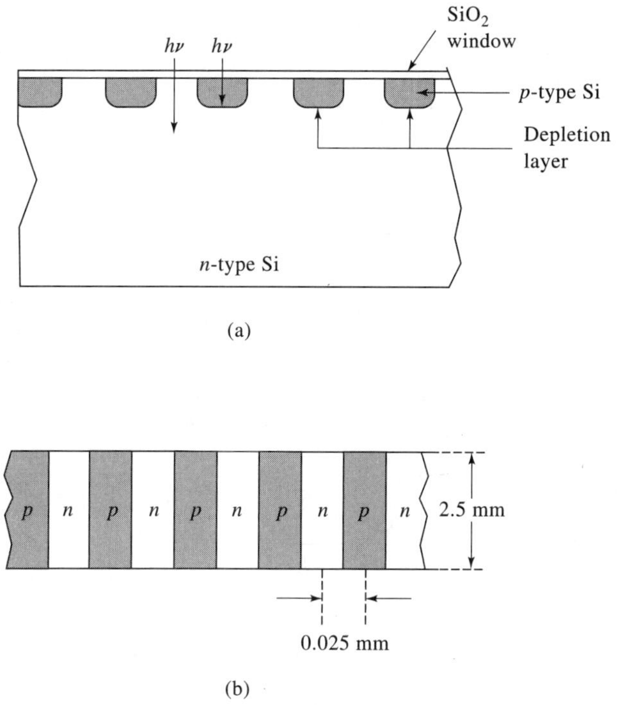

Figure 7-31 A reverse-biased linear diode-array detector: (a) cross section and (b) top view.

large-scale integrated circuit formed on a single silicon chip. Figure 7-31 shows the geometry of the surface region of a few of the transducer elements. Each element consists of a diffused *p*-type bar in an *n*-type silicon substrate to give a surface region that consists of a series of side-by-side elements that have typical dimensions of 2.5 by 0.025 mm (Figure 7-31b). Light that is incident upon these elements creates charges in both the *p* and *n* regions. The positive charges are collected and stored in the *p*-type bars for subsequent integration (the charges formed in the *n*-regions divide themselves proportionally between the two adjacent *p*-regions). The number of transducer elements in a chip ranges from 64 to 4096 with 1024 being perhaps the most widely used.

The integrated circuit that makes up a diode array also contains a storage capacitor and switch for each diode as well as a circuit for sequentially scanning the individual diode-capacitor circuits. Figure 7-32 is a simplified diagram that shows the arrangement of these components. Note that in parallel with each photodiode is a companion 10-pF storage capacitor. Each diode-capacitor pair is sequentially connected to a common output line by the N-bit shift register and the transistor switches. The shift register sequentially closes each of these switches momentarily causing the capacitor to be charged to −5 V, which then creates a reverse bias across the *pn* junction of the detector. Radiation that impinges upon the depletion layer in either the *p* or the *n* region forms charges (electrons and holes) that create a current that partially discharges the capacitor in the circuit. The capacitor charge that is lost in this way is replaced during the next cycle. The resultant charging current is integrated by the preamplifier circuit, which produces a voltage that is proportional to the radiant intensity. After amplification, the analog signal from the preamplifier passes into an analog-to-digital converter and to a microprocessor that controls the readout.

In using a diode-array transducer, the slit width of the spectrometer is usually adjusted so that the image of the entrance slit just fills the surface area of one of the diodes that make up the array. Thus the information obtained is equivalent to that recorded during scanning with a traditional spectrophotometer. With the array, however, information about the entire spectrum is accumulated essentially simultaneously and in discrete elements rather than in a continuous way.

Some of the photoconductor transducers mentioned in the previous section can also be fabricated into linear arrays for use in the infrared region.

Charge-Transfer Devices

Photodiode arrays cannot match the performance of photomultiplier tubes with respect to sensitivity, dynamic range, and signal-to-noise ratio. Thus, their use has been limited to experiments in which the multichannel advantage outweighs their shortcomings. In contrast, performance characteristics of charge-transfer devices appear to approach those of photomultiplier tubes in addition to having the multichannel advantage. As a consequence this type of transducer is now appearing in ever-increasing numbers in modern spectroscopic instruments.[24] A further advantage of charge-transfer devices is that they are two dimensional in the sense that individual transducer elements are arranged in rows and columns. For example, one detector that we describe in the next section consists of 244 rows of transducer ele-

[24] For details on charge-transfer devices, see J. V. Sweedler, *Crit. Rev. Anal. Chem.*, **1993**, *24*, 59; J. V. Sweedler, R. B. Bilhorn, P. M. Epperson, G. R. Sims, and M. B. Denton, *Anal. Chem.*, **1988**, *60*, 282A, 327A; *Charge-Transfer Devices in Spectroscopy*, J. V. Sweedler, K. L. Ratzlaff, and M. B. Denton, Eds. New York: Wiley, 1994.

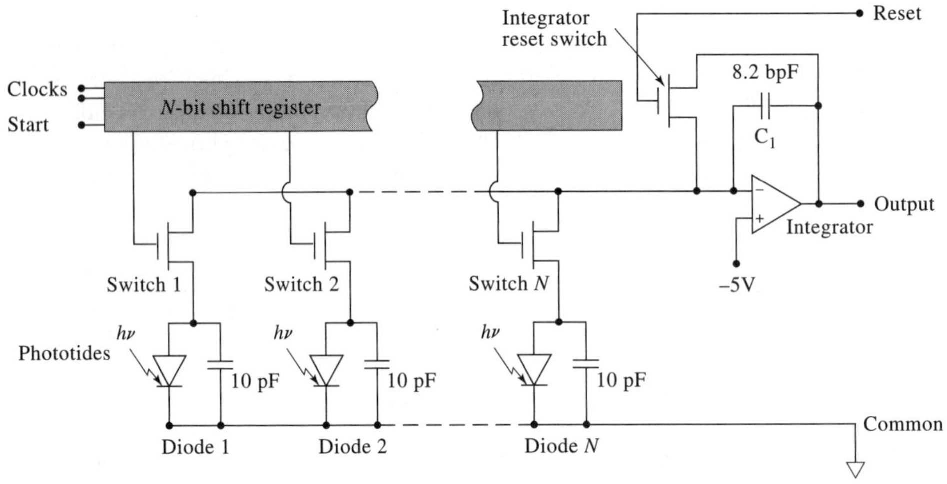

Figure 7-32 Block diagram of a photodiode-array detector chip.

ments, each row composed of 388 detector elements, resulting in a two-dimensional array of 19,672 individual transducers , or *pixels,* on a silicon chip with dimensions of 6.5 mm by 8.7 mm. With this device, it is possible to record an entire two-dimensional spectrum from an echelle spectrometer (Figure 7-21) simultaneously.

Charge-transfer devices operate much like a photographic film in that they integrate signal information as radiation strikes them. Figure 7-33 is a cross-sectional depiction of one of the pixels that composes a charge-transfer array. In this case, the pixel consists of two conductive electrodes that overlie an insulating layer of silica (note that a pixel in some charge-transfer devices is made up of more than two electrodes). This silica layer separates the electrodes from a region of *n*-doped silicon. This assemblage constitutes a metal oxide semiconductor capacitor that stores the charges formed when radiation strikes the doped silicon. When, as shown, a negative charge is applied to the electrodes, a charge inversion region is created under the electrodes, which is energetically favorable for the storage of holes. The mobile holes created by the absorption of photons then migrate and collect in this region. Typically, this region, which is called a *potential well,* is capable of holding as many as 10^5 to 10^6 charges before overflowing into an adjacent pixel. In the figure, one electrode is shown as more negative than the other, making the accumulation of charge under this electrode more favorable. The amount of charge generated during exposure to radiation is measured in either of two ways. In a *charge-injection device,* the voltage change that arises from movement of the charge from the region under one electrode to the region under the other is measured. In a *charge-coupled device,* the charge is moved to a charge-sensing amplifier for measurement.

Charge-Injection Devices. Figure 7-34 is a simplified diagram that shows the steps involved in the collection, storage, and measurement of the charge generated when one pixel of a semiconductor is exposed to photons. To monitor the intensity of the radiation that strikes the sensor element, the potentials applied to the capacitors are cycled as shown in steps (a) through (d) in the figure. In step (a), negative potentials are applied to the two electrodes, which leads to formation of potential wells that collect and store holes formed in the *n* layer by absorption of photons. Because the electrode on the right is at a more negative potential, all the holes are retained under this electrode initially. The magnitude of the charge collected in some brief time interval is deter-

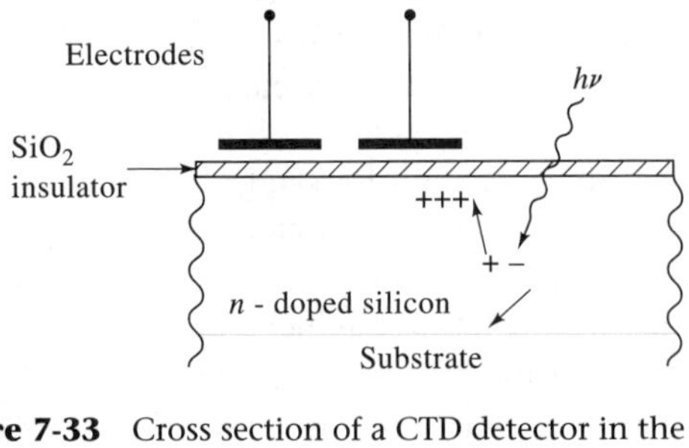

Figure 7-33 Cross section of a CTD detector in the charge integration mode. The positive hole produced by the photon *hν* is collected under the negative electrode.

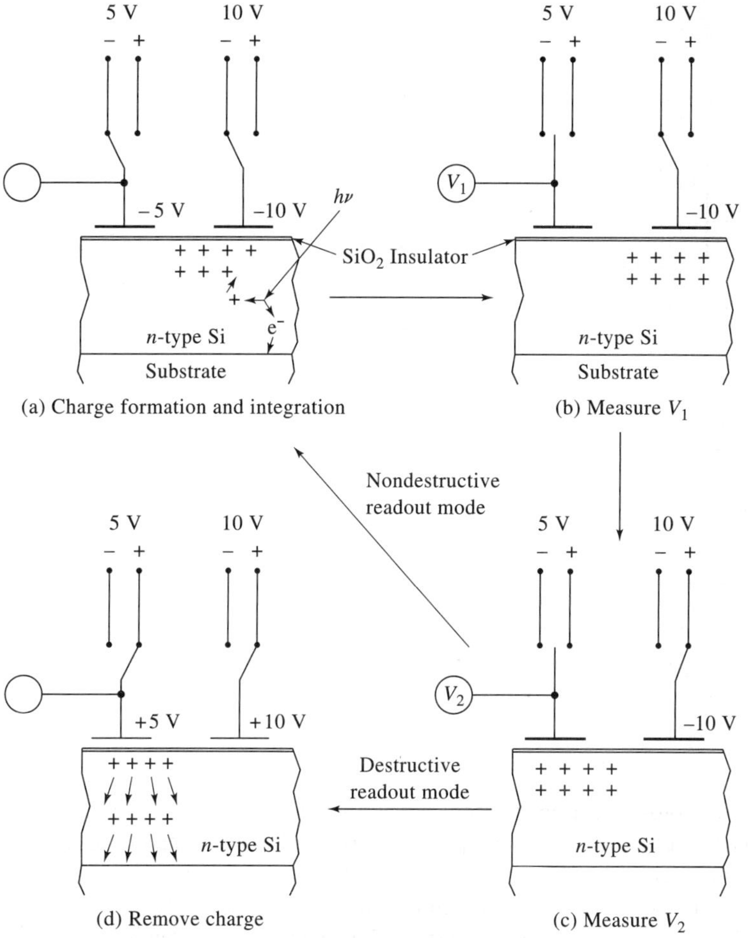

Figure 7-34 Duty cycle of a charge-injection device: (a) production and storage of charge, (b) first charge measurement, (c) second charge measurement after charge transfer, (d) reinjection of charge into the semiconductor.

mined in steps (b) and (c). In (b), the potential of the capacitor on the left (V_1) is determined after removal of its applied potential. In step (c), the holes that have accumulated on the right electrode are transferred to the potential well under the left electrode by switching the potential applied to the former from negative to positive. The new potential of the electrode V_2 is then measured. The magnitude of the accumulated charge is determined from the difference in potential ($V_1 - V_2$). In step (d), the detector is returned to its original state by applying positive potentials to both electrodes, which cause the holes to migrate toward the substrate. As an alternative to step (d), however, the detector can be returned to the condition shown in (a) without the loss of charge that has already accumulated. This process is called the *nondestructive readout mode* (NDRO). A major advantage of charge-injection devices over charge-coupled devices is that successive measurements can be made while integration is taking place.

As was true for the diode-array detector, the chip that contains the array of charge-injection transducer elements also contains appropriate integrated circuits for performing the cycling and measuring steps.

Charge-Coupled Device. Charge-coupled devices are marketed by several manufacturers and come in a variety of shapes and forms. Figure 7-35a illustrates the arrangement of individual detectors in a typical array that is made up of 512 × 320 pixels. Note that in this case the semiconductor is formed from *p*-type silicon, and the capacitor is biased positively so that electrons formed by the absorption of radiation collect in the well below the electrode, whereas holes move away from the *n*-type layer toward the substrate. Note also that each pixel is made up of three electrodes (labeled 1, 2, and 3 in Figure 7-35b) rather than two electrodes as in the charge-injection device. To measure the accumulated charge, a three-phase clock circuit is used to shift the

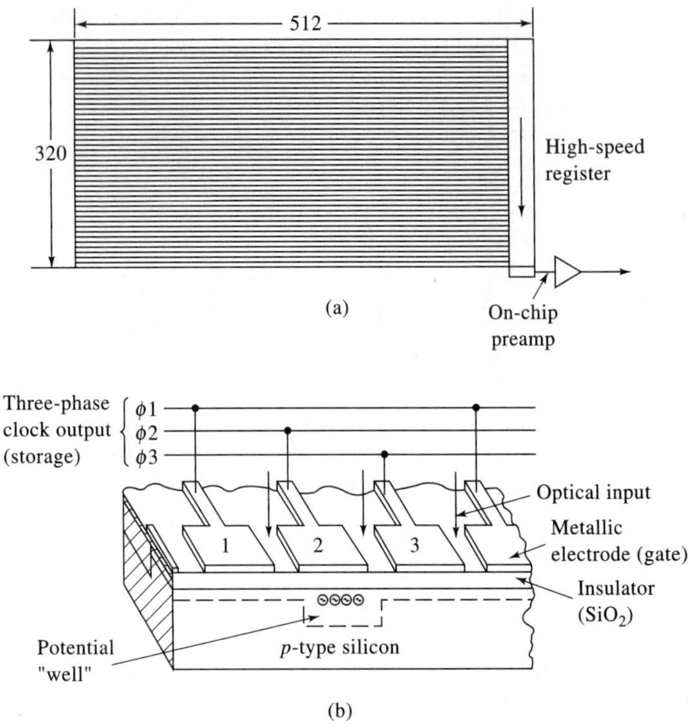

Figure 7-35 A charge-coupled device array: (a) arrangement of 512 × 320 pixels and (b) schematic showing four of the individual detectors.

charge in a stepwise manner to the right to the high-speed shift register shown in Figure 7-35a. The charges are then transferred downward to a preamplifier and then to the readout. Thus, a row-by-row scan of the detector surface is accomplished. In contrast to the charge injection device, the readout in this case neutralizes the accumulated charge. The charge-coupled device offers the advantage of greater sensitivity to low light levels. A disadvantage in some cases, however, is the destructive nature of the readout process.

7E-4 **Photoconductivity Transducers**

The most sensitive transducers for monitoring radiation in the near-infrared region (0.75 to 3 μm) are semiconductors whose resistances decrease when they absorb radiation within this range. The useful range of photoconductors can be extended into the far-infrared region by cooling to suppress noise that arises from thermally induced transitions among closely lying energy levels. This application of photoconductors is important in infrared Fourier transform instrumentation. Crystalline semiconductors are formed from the sulfides, selenides, and stibnides of such metals as lead, cadmium, gallium, and indium. Absorption of radiation by these materials promotes some of their bound electrons into an energy state in which they are free to conduct electricity. The resulting change in conductivity can then be measured with a circuit such as that shown in Figure 3-10a.

Lead sulfide is the most widely used photoconductive material that offers the advantage that it can be used at room temperature. Lead sulfide transducers are sensitive in the region between 0.8 and 3 μm (12,500 to 3300 cm^{-1}). A thin layer of this compound is deposited on glass or quartz plates to form the cell. The entire assembly is then sealed in an evacuated container to protect the semiconductor from reaction with the atmosphere. The sensitivity of cadmium sulfide, cadmium selenide, and lead sulfide transducers is shown by curves *B, D,* and *G* in Figure 7-25.

7E-5 **Thermal Transducers**[25]

The convenient phototransducers just considered are generally not applicable in the infrared because photons in this region lack the energy to cause photoemission of electrons. Thus, thermal transducers or transducers

[25] For a good discussion of optical radiation transducers of all types, including thermal detectors, see E. L. Dereniak and G. D. Crowe, *Optical Radiation Detectors*. New York: Wiley, 1984.

based upon photoconduction (see Section 7E-4) must be employed. Neither of these is as satisfactory as photon transducers.

In thermal transducers, the radiation impinges upon and is absorbed by a small blackbody; the resultant temperature rise is measured. The radiant power level from a typical infrared beam is minute (10^{-7} to 10^{-9} W), so that the heat capacity of the absorbing element must be as small as possible if a detectable temperature change is to be produced. Every effort is made to minimize the size and thickness of the absorbing element and to concentrate the entire infrared beam on its surface. Under the best of circumstances, temperature changes are confined to a few thousandths of a kelvin.

The problem of measuring infrared radiation by thermal means is compounded by thermal noise from the surroundings. For this reason, thermal detectors are housed in a vacuum and are carefully shielded from thermal radiation emitted by other nearby objects. To further minimize the effects of extraneous heat sources, the beam from the source is generally chopped. In this way, the analyte signal, after transduction, has the frequency of the chopper and can be readily separated electronically from extraneous noise, which ordinarily varies only slowly with time.

Thermocouples

In its simplest form, a thermocouple consists of a pair of junctions formed when two pieces of a metal such as copper are fused to each end of a dissimilar metal such as constantan as shown in Figure 3-11. Between the two junctions a potential develops that varies with the *difference* in temperature of the junctions.

The transducer junction for infrared radiation is formed from very fine wires of bismuth and antimony or alternatively by evaporating the metals onto a nonconducting support. In either case, the junction is usually blackened to improve its heat-absorbing capacity and sealed in an evacuated chamber with a window that is transparent to infrared radiation.

The reference junction, which is usually housed in the same chamber as the active junction, is designed to have a relatively large heat capacity and is carefully shielded from the incident radiation. Because the analyte signal is chopped, only the difference in temperature between the two junctions is important; therefore, the reference junction does not need to be maintained at constant temperature. To enhance sensitivity, several thermocouples may be connected in series to fabricate what is called a *thermopile.*

A well-designed thermocouple transducer is capable of responding to temperature differences of 10^{-6} K. This difference corresponds to a potential difference of about 6 to 8 μV/μW. The thermocouple of an infrared detector is a low-impedance device that is usually connected to a high-impedance differential preamplifier, such as the field-effect transistor circuit shown in Figure 7-36. A voltage follower, such as that shown in Figure 3-7, is also used as a signal conditioner in thermocouple detector circuits.

Bolometers

A bolometer is a type of resistance thermometer constructed of strips of metals such as platinum or nickel, or from a semiconductor; the latter devices are often called *thermistors.* These materials exhibit a relatively large change in resistance as a function of temperature. The responsive element is kept small and blackened to absorb radiant heat. Bolometers are not so extensively used as other infrared transducers for the mid-infrared region. However, a germanium bolometer, operated at 1.5 K, is nearly an ideal transducer for radiation in the 5 to 400 cm^{-1} (2000 to 25 μm) range.

Pyroelectric Transducers

Pyroelectric transducers are constructed from single crystalline wafers of *pyroelectric materials,* which are insulators (dielectric materials) with very special thermal and electrical properties. Triglycine sulfate, $(NH_2CH_2COOH)_3 \cdot H_2SO_4$ (usually deuterated or with a fraction of the glycines replaced with alanine), is the most important pyroelectric material used in the construction of infrared transducers.

When an electric field is applied across any dielectric material, electric polarization takes place whose magnitude is a function of the dielectric constant of the material. For most dielectrics, this induced polarization rapidly decays to zero when the external field is removed. Pyroelectric substances, in contrast, retain a strong temperature-dependent polarization after removal of the field. Thus, by sandwiching the pyroelectric crystal between two electrodes (one of which is infrared transparent) a temperature-dependent capacitor is produced. Changing its temperature by irradiating it with infrared radiation alters the charge distribution across the crystal, which creates a measurable current in an external electric circuit that connects the two sides of the capacitor. The magnitude of this current is proportional to the surface area of the crystal and to its rate of change of polarization with temperature. Pyroelectric

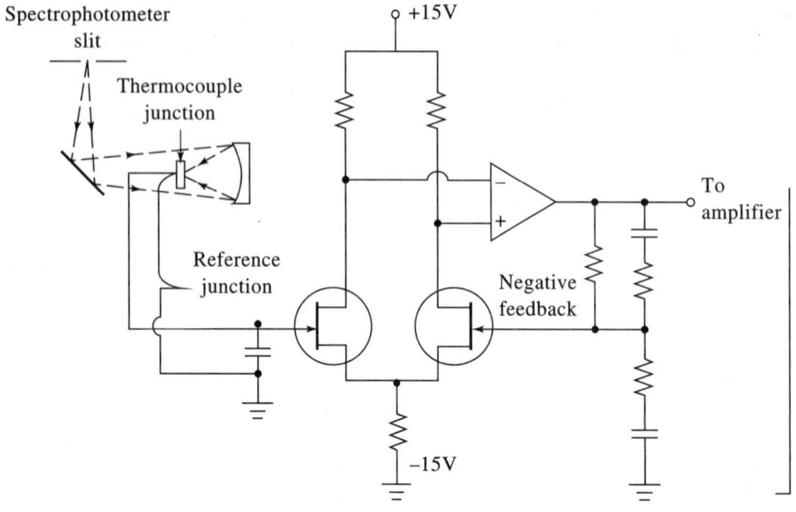

Figure 7-36 Thermocouple and preamplifier. *(Adapted from G. W. Ewing,* J. Chem. Educ., ***1971****, 48, A521. With permission.)*

crystals loose their residual polarization when they are heated to a temperature called the *Curie point.* For triglycine sulfate, the Curie point is 47°C.

Pyroelectric transducers exhibit response times that are fast enough to allow them to track the changes in the time-domain signal from an interferometer. For this reason, most Fourier transform infrared spectrometers employ this type of transducer.

7F SIGNAL PROCESSORS AND READOUTS

The signal processor is ordinarily an electronic device that amplifies the electrical signal from the transducer. In addition, it may alter the signal from dc to ac (or the reverse), change the phase of the signal, and filter it to remove unwanted components. Furthermore, the signal processor may be called upon to perform such mathematical operations on the signal as differentiation, integration, or conversion to a logarithm.

Several types of readout devices are found in modern instruments. Some of these devices include the d'Arsonval meter, digital meters, the scales of potentiometers, recorders, and cathode-ray tubes.

7F-1 Photon Counting

The output from a photomultiplier tube consists of a pulse of electrons for each photon that reaches the detector surface. Ordinarily this analog signal is filtered to remove undesirable fluctuations due to the random appearance of photons at the photocathode and measured as a dc voltage or current. If, however, the intensity of the radiation is too low to provide a satisfactory signal-to-noise ratio, it is possible, and often advantageous, to convert the analog signal to a train of digital pulses that may then be counted as discussed in Section 4C. Here, radiant power is proportional to the number of pulses per unit time rather than to an average current or potential. A measurement of this type is termed *photon counting*.

Counting techniques have been used for many years for measuring the power of X-ray beams and of radiation produced by the decay of radioactive species (these techniques are considered in detail in Chapters 12 and 32). Photon counting has also been applied to ultraviolet and visible radiation.[26] Here, the output of a photomultiplier tube is employed. In the previous section it was indicated that a single photon that strikes the cathode of a photomultiplier ultimately leads to a cascade of 10^6 to 10^7 electrons, which produces a pulse of current that can be amplified and counted.

Generally, the equipment for photon counting is similar to that shown in Figure 4-2 in which a comparator rejects pulses unless they exceed some predetermined minimum voltage. Comparators are useful for this task because dark current and instrument noise are

[26] For a review of photon counting, see H. J. Malmstadt, M. L. Franklin, and G. Horlick, *Anal. Chem.*, 1972, *44* (8), 63A.

often significantly smaller than the signal pulse and are thus not counted; an improved signal-to-noise ratio results.

Photon counting has a number of advantages over analog-signal processing, including improved signal-to-noise ratio, sensitivity to low radiation levels, improved precision for a given measurement time, and lowered sensitivity to photomultiplier tube voltage and temperature fluctuations. The required equipment is, however, more complex and expensive; the technique has thus not been widely applied for routine molecular absorption measurements in the ultraviolet and visible regions. It has, however, become the detection method of choice in fluorescence, chemiluminescence, and Raman spectrometry, where radiant power levels are low.

7G FIBER OPTICS

In the late 1960s, there began to appear on the market analytical instruments that contained fiber optics for transmitting radiation and images from one component of the instrument to another. These useful devices have added a new dimension to optical instrument designs.[27]

7G-1 Properties of Optical Fibers

Optical fibers are fine strands of glass or plastic that are capable of transmitting radiation for distances of several hundred feet or more. The diameter of optical fibers range from 0.05 μm to as large as 0.6 cm. Where images are to be transmitted, bundles of fibers, fused on the ends, are employed. A major application of these fiber bundles has been in medical diagnoses, where their flexibility permits transmission of images of organs through tortuous pathways to the physician. Filter optics are used not only for observation but also for illumination of objects; here, the ability to illuminate without heating is often of considerable importance.

Light transmission in an optical fiber takes place by total internal reflection as shown in Figure 7-37. In order for total internal reflections to occur, it is necessary that the transmitting fiber be coated with a material that has a refractive index that is somewhat smaller than the refractive index of the material from which the fiber is constructed. Thus, a typical glass fiber has a core with a refractive index of about 1.6 and has a glass sheath cladding with a refractive index of approximately 1.5. Typical plastic fibers have a polymethylmethacrylate core with a refractive index of 1.5 and have a polymer coating with a refractive index of 1.4.

A fiber, such as that shown in Figure 7-37, will transmit radiation contained in a limited incident cone with a half angle shown as θ in the figure. The *numerical aperture* of the fiber provides a measure of the magnitude of the so-called *acceptance cone.*

By suitable choice of construction materials, fibers that will transmit ultraviolet, visible, or infrared radiation can be manufactured. Several examples of their application to conventional analytical instruments will be found in the chapters that follow.

7G-2 Fiber-Optic Sensors

Fiber-optic sensors, which are also called *optrodes,* consist of a reagent phase immobilized on the end of a fiber optic.[28] Interaction of the analyte with the reagent creates a change in absorbance, reflectance, fluorescence, or luminescence, which is then transmitted to a detector via the optical fiber. Fiber-optic sensors are

[27] For a review of applications of fiber optics, see I. Chabay, *Anal. Chem.,* **1982,** *54,* 1071A and J. K. Crum, *Anal. Chem.,* **1969,** *41,* 26A.

[28] M. A. Arnold, *Anal. Chem.,* **1992,** *64,* 1015A; R. E. Dessy, *Anal. Chem.,* **1989,** *61,* 1079A; W. R. Seitz, *Anal. Chem.,* **1984,** *56,* 16A; S. Borman, *Anal. Chem.,* **1987,** *59,* 1161A; *Ibid.,* **1986,** *58,* 766A.

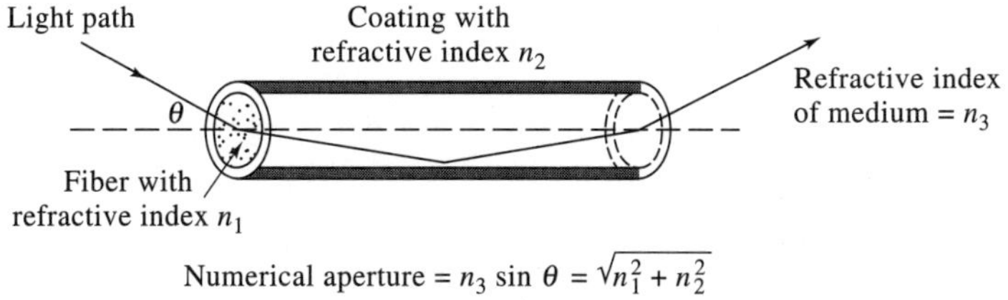

$$\text{Numerical aperture} = n_3 \sin \theta = \sqrt{n_1^2 + n_2^2}$$
$$n_1 > n_2 > n_3$$

Figure 7-37 Schematic of the light path through an optical fiber.

generally simple, inexpensive devices that are easily miniaturized.

7G-3 Fiber Optics for Time Discrimination among Signals

An ingenious application of optical fibers is based on the use of strands of different lengths to vary the time of arrival at a single detector of optical signals from several sources. Time-resolved detection then permits the simultaneous determination of an analyte in several samples by means of a single detection system.[29] Figure 7-38a illustrates how such measurements are accomplished. At the heart of the detection system is a *star coupler*, a bidirectional device for coupling multiple

[29]R.L. Steffen and F.E. Lytle, *Anal. Chim. Acta*, ***1987***, *200*, 491.

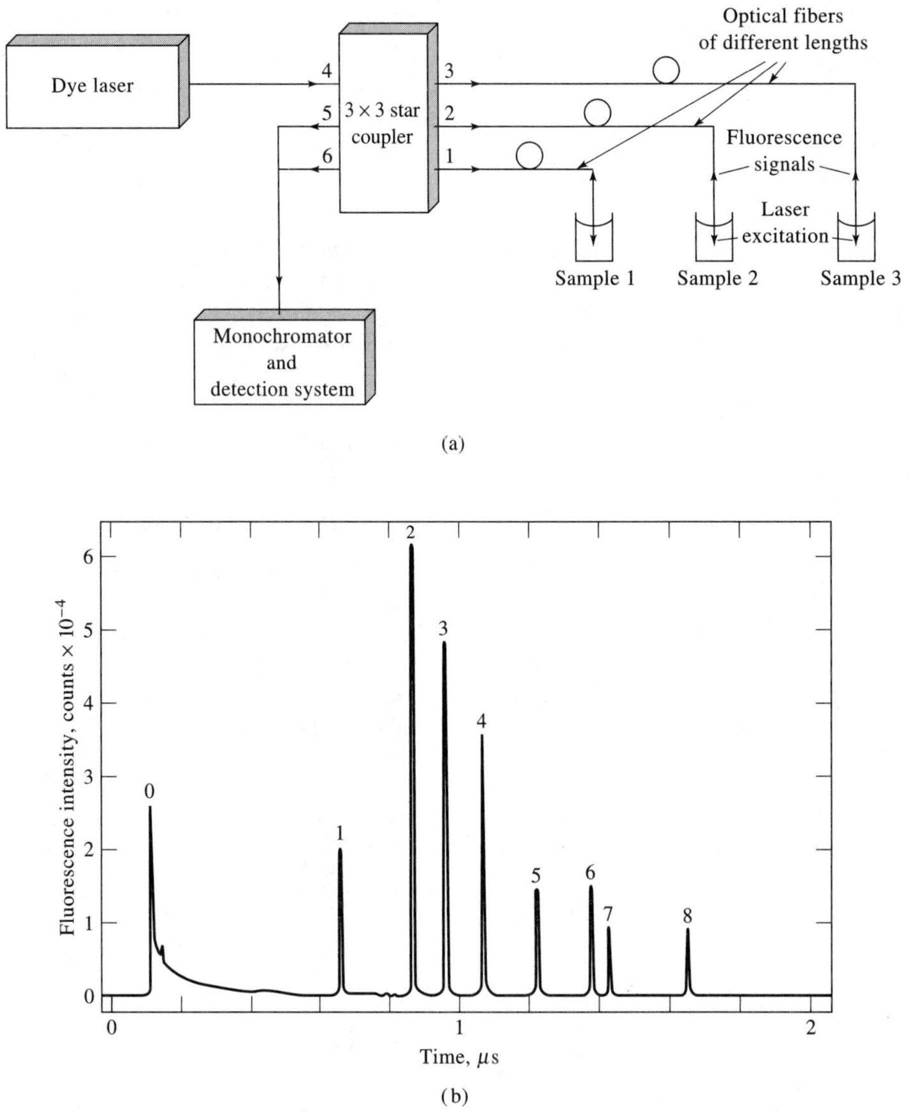

Figure 7-38 Instrument for exciting and separating fluorescence signals based upon the time of arrival of the signals at the transducer: (a) experimental arrangement; (b) output from the instrument. *(Reprinted with permission from R. L.Steffen and F. E. Lytle,* Anal. Chim. Acta, ***1987,*** 200, *491.)*

fiber optic bundles as illustrated in the figure. Star couplers are typically arranged in an $N \times N$ configuration, where N is the number of input ports and the number of output ports so that, as in this example of a 3×3 coupler, three fiber optic bundles may be conveniently coupled to three other bundles. Light entering any one of the six fibers is distributed approximately equally among all of the coupled fibers, including the input fiber. In other words, any light entering the star coupler via a fiber is distributed so that $1/(2N)$ of the incoming light emerges from each of the coupled fibers. In practice, the device is less than 100% efficient because some of the incoming light is lost at the coupling interfaces, but the emerging light is divided approximately equally among the six coupled fiber optic bundles in our example. It is important to recognize that although star couplers are mechanically configured as $N \times N$ devices, all $2N$ of the ports are identical optically and all of the ports are bidirectional.

In the time-correlated fluorescence experiment shown in Figure 7-38a, pulses of radiation from a dye laser pass into port 4 of the star coupler. The device divides the beam into six beams of approximately equal power, three of which are then directed into optical fibers 1, 2, and 3, which differ in length by several meters. About one sixth of the laser excitation passes back through fiber 4 unused. Approximately one third of the excitation emerges from ports 5 and 6 and passes to the detection system to serve as a time marker and power reference. The ends of fibers 1, 2, and 3 are inserted directly into three sample solutions where their output excites the analyte; this in turn produces pulses of fluorescence emission that travel in a reverse direction through the fibers as indicated by the arrowheads in Figure 7-38a. These pulses, however, return to the star coupler at different times. The radiation from sample 1 reaches the star coupler first via fiber 1 and is divided equally among all six of the ports. The fluorescence emission from ports 5 and 6 is then passed into a monochromator and thence to a nanosecond time-resolved detection system. At a subsequent time determined by the difference in the lengths of fiber 1 and fiber 2 and the velocity of light, the radiation from sample 2 arrives at the star coupler, and once again, the combined emission from fibers 5 and 6 is passed onto the detection system. This process is repeated for each of the sample fibers.

A plot of the output from the time-correlated single photon counting detection system is presented in Figure 7-38b for an experimental arrangement similar to that in Figure 7-38a. The system differs from that shown in the figure in that an 8×8 star coupler was used and 8 samples were analyzed. The data are presented as fluorescence intensity versus time of arrival of photons at the transducer. Note that there are nine peaks, the first of which (peak 0) corresponds to the laser excitation pulse mentioned previously. Peak 1 corresponds to a reference solution that was used in place of a sample in position 1 to correct the other peak intensities for fluctuations in the laser pulse power. Peaks 2 to 8 are fluorescence emission from the 7 samples. The magnitude of the peaks is proportional to the concentration of the analyte, 2-(1-naphthyl)-5-phenyloxazole, in the samples. Calibration curves for this detection scheme were linear over three orders of magnitude, and detection limits for the analyte were in the millimolar range and lower.

In the original experiments, optical fibers varying in length from 41 m to 142 m were used. Note that a laser pulse is delayed by roughly 50 ns or 0.05 μs for each 10 m of fiber that it traverses; modern electronic circuitry can easily discriminate among signals on this time scale. Several star coupler configurations were tested to determine the feasiblility of the detection scheme. In one case, a 4×4 coupler was used that permitted monitoring the fluorescence from four samples. By combining a 3×3 coupler with an 8×8 coupler, measurements were made on ten samples nearly simultaneously.

7H TYPES OF OPTICAL INSTRUMENTS

In this section, we define the terminology we will use to describe various types of optical instruments. It is important to realize that the nomenclature proposed here is not agreed upon and used by all scientists; it is simply a common nomenclature and the one that will be encountered throughout this book.

A *spectroscope* is an optical instrument used for the visual identification of atomic emission lines. It consists of a monochromator, such as one of those shown in Figure 7-16, in which the exit slit is replaced by an eyepiece that can be moved along the focal plane. The wavelength of an emission line can then be determined from the angle between the incident and dispersed beam when the line is centered on the eyepiece.

We use the term *colorimeter* to designate an instrument for absorption measurements in which the human eye serves as the detector using one or more color-comparison standards. A *photometer* consists of a source, a filter, and a photoelectric transducer as well as a signal processor and readout. It should be noted that some scientists and instrument manufacturers refer to photometers as colorimeters or photoelectric colorimeters. Filter photometers are commercially available for absorption measurements in the ultraviolet, visible, and infrared regions, as well as emission and fluorescence in the first two wavelength regions. Photometers designed for fluorescence measurements are also called *fluorometers.*

A *spectrograph* is similar in construction to the two monochromators shown in Figure 7-16 except that the slit arrangement is replaced with a large aperture that holds a detector or transducer that is continuously exposed to the entire spectrum of dispersed radiation. Historically, the detector was a photographic film or plate. Currently, however, diode arrays or charge-transfer devices are often used as transducers in spectrographs.

A *spectrometer* is an instrument that provides information about the intensity of radiation as a function of wavelength or frequency. The dispersing modules in some spectrometers are multichannel so that two or more frequencies can be viewed simultaneously. Such instruments are sometimes called *polychromators.* A *spectrophotometer* is a spectrometer equipped with one or more exit slits and photoelectric transducers that permit the determination of the ratio of the power of two beams as a function of wavelength as in absorption spectroscopy. A spectrophotometer for fluorescence analysis is sometimes called a *spectrofluorometer.*

All of the instruments named in this section thus far employ filters or monochromators to isolate a portion of the spectrum for measurement. A *multiplex* instrument, in contrast, obtains spectral information without first dispersing or filtering the radiation to provide wavelengths of interest. The term *multiplex* comes from communication theory, where it is used to describe systems in which many sets of information are transported simultaneously through a single channel. Multiplex analytical instruments then are single-channel devices in which all components of an analytical response are collected *simultaneously.* In order to determine the magnitude of each of these components, it is necessary to modulate the analyte signal in a way that permits subsequent decoding of the response into its components.

Most multiplex analytical instruments depend upon the *Fourier transform* (FT) for signal decoding and are consequently often called Fourier transform spectrometers. Such instruments are by no means confined to optical spectroscopy. Indeed, Fourier transform devices have been described for nuclear magnetic resonance spectrometry, mass, and microwave spectroscopy. Several of these instruments will be described in some detail in subsequent chapters. The section that follows describes the principles on which Fourier transform optical spectrometers are based.

7I PRINCIPLES OF FOURIER TRANSFORM OPTICAL MEASUREMENTS

Fourier transform spectroscopy was first developed by astronomers in the early 1950s in order to study the infrared spectra of distant stars; only by the Fourier technique could the very weak signals from these sources be isolated from environmental noise. The first chemical applications of Fourier transform spectroscopy, which were reported approximately a decade later, were to the energy-starved far-infrared region; by the late 1960s, instruments for chemical studies in both the far-infrared (10 to 400 cm^{-1}) and the ordinary infrared regions were available commercially. Descriptions of Fourier transform instruments for the ultraviolet and visible spectral regions can also be found in the literature, but their adoption has been less widespread.[30]

7I-1 Inherent Advantages of Transform Spectrometry

There are several major advantages to the use of Fourier transform instruments. The first is the *throughput,* or *Jaquinot, advantage,* which is realized because Fourier transform instruments have few optical elements and no slits to attenuate radiation. As a consequence, the power of the radiation that reaches the detector is much greater than that in dispersive instruments, and much greater signal-to-noise ratios are observed.

A second advantage of Fourier transform instruments is their extremely high resolving power and wavelength reproducibility that make possible the analysis of complex spectra in which the sheer number of lines and spectral overlap make the determination of

[30] For more complete discussions of optical Fourier transform spectroscopy, consult the following references: A. G. Marshall and F. R. Verdun, *Fourier Transforms in NMR, Optical, and Mass Spectrometry.* New York: Elsevier, 1990; A. G. Marshall, *Fourier, Hadamard, and Hilbert Transforms in Chemistry.* New York: Plenum Press, 1982; *Transform Techniques in Chemistry,* P. R. Griffiths, Ed. New York: Plenum Press, 1978. For brief reviews, see P. R. Griffiths, *Science,* **1983,** *222,* 297; W. D. Perkins, *J. Chem. Educ.,* **1986,** *63,* A5, A296; L. Glasser, *J. Chem. Educ.,* **1987,** *64,* A228, A260, A306.

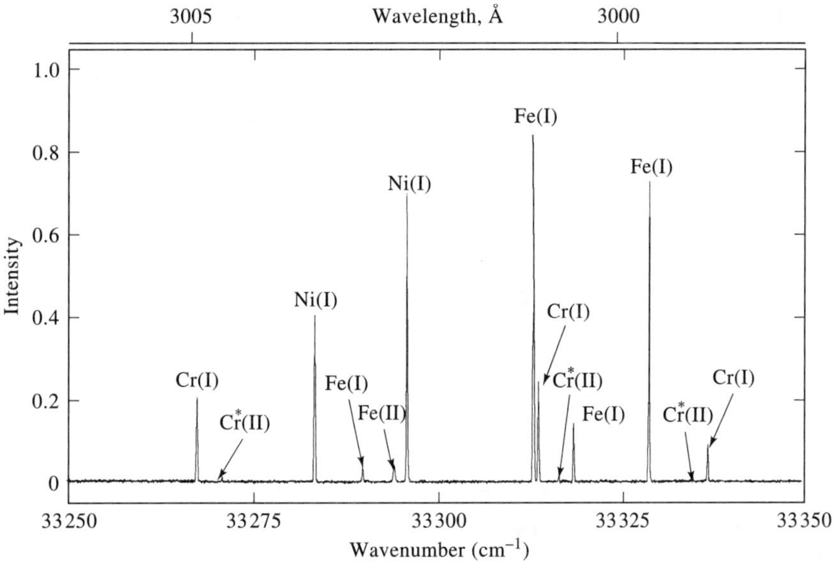

Figure 7-39 An iron emission spectrum illustrating the high resolving power of a Fourier transform emission spectrometer. *(Reprinted with permission from A. P. Thorne,* Anal. Chem., ***199,*** 63, *63A. Copyright 1991 American Chemical Society.)*

individual spectral features difficult. Figure 7-39, which is part of an emission spectrum for a steel, illustrates this advantage. The spectrum, which in Figure 7-39 extends from only 299.85 to 300.75 nm, contains 13 well-separated lines of three elements. The wavelength resolution ($\Delta\lambda/\lambda$) for the closest pair of lines is about 6 ppm.

A third advantage arises because all elements of the source reach the detector simultaneously. This characteristic makes it possible to obtain data for an entire spectrum in one second or less. Let us examine the consequences of this last advantage in further detail.

For purposes of this discussion, it is convenient to think of an experimentally derived spectrum as made up of m individual transmittance measurements at equally spaced frequency or wavelength intervals called *resolution elements.* The quality of the spectrum—that is, the amount of spectral detail—increases as the number of resolution elements becomes larger or as the frequency intervals between measurements become smaller.[31] Thus, in order to increase spectral quality, m must be made larger; clearly, increasing the number of resolution elements must also increase the time required for obtaining a spectrum with a scanning instrument.

Consider, for example, the derivation of an infrared spectrum from 500 to 5000 cm^{-1}. If resolution elements of 3 cm^{-1} were chosen, m would be 1500; if 0.5 s were required for recording the transmittance of each resolution element, 750 s or 12.5 min would be needed to obtain the spectrum. Reducing the width of the resolution element to 1.5 cm^{-1} would be expected to provide significantly greater spectral detail; it would also double the number of resolution elements as well as the time required for their measurement.

For most optical instruments, particularly those designed for the infrared region, decreasing the width of the resolution element has the unfortunate effect of decreasing the signal-to-noise ratio because narrower slits must be used, which lead to weaker source signals that reach the transducer. For infrared detectors, the reduction in signal strength is not, however, accompanied by a corresponding decrease in detector noise. Therefore, a degradation in signal-to-noise ratio results.

On page 108, it was pointed out that marked improvements in signal-to-noise ratios accompany signal averaging. Here, it was shown (Equation 5-11) that the

[31] With a recording spectrophotometer, of course, individual point-by-point measurements are not made; nevertheless, the idea of a resolution element is useful, and the ideas generated from it apply to recording instruments as well.

signal-to-noise ratio S/N for the average of n measurements is given by

$$\frac{S}{N} = \sqrt{n}\frac{S_x}{\sqrt{\sum_{i=1}^{n}(S_x - S_i)^2}} = \frac{S_x}{N_x}\sqrt{n} \quad (7\text{-}20)$$

where S_x and N_x are the averaged signal and noise. The application of signal averaging to conventional spectroscopy is, unfortunately, costly in terms of time. Thus, in the example just considered, 750 s were required to obtain a spectrum of 1500 resolution elements. To improve the signal-to-noise ratio by a factor of 2 would require averaging 4 spectra, which would then require 4×750 s or 50 min.

Fourier transform spectroscopy differs from conventional spectroscopy in that all of the resolution elements for a spectrum are measured *simultaneously,* thus reducing enormously the time required to obtain a spectrum at any chosen signal-to-noise ratio. An entire spectrum of 1500 resolution elements can then be recorded in about the time required to observe just one element by conventional spectroscopy (0.5 s in our earlier example). This large decrease in observation time is often used to markedly enhance the signal-to-noise ratio of Fourier transform measurements. For example, in the 750 s required to derive the spectrum by scanning, 1500 Fourier transform spectra could be recorded and averaged. According to Equation 7-20, the improvement in signal-to-noise ratio would be $\sqrt{1500}$ or about 39. This inherent advantage of Fourier transform spectroscopy was first recognized by P. Fellgett in 1958 and is termed the *Fellgett,* or *multiplex, advantage.* It is worth noting here that for several reasons, the theoretical $\sqrt{n}$ improvement in S/N is seldom entirely realized. Nonetheless, major gains in signal-to-noise ratios are generally observed with the Fourier transform technique.

The multiplex advantage is important enough so that nearly all infrared spectrometers are of the Fourier transform type. Fourier transform instruments are much less common for the ultraviolet, visible, and near-infrared regions, however, because signal-to-noise limitations for spectral measurements with these types of radiation seldom lies in detector noise but instead in shot noise and flicker noise associated with the source. In contrast to detector noise, the magnitudes of both shot and flicker noise increase as the power of the signal increases. Furthermore, the total noise for all of the resolution elements in a Fourier transform measurement tends to be averaged and spread out uniformly over the entire transformed spectrum. Thus, the signal-to-noise ratio for strong peaks in the presence of weak peaks is improved by averaging but degraded for the weaker peaks. For flicker noise, such as is encountered in the background radiation from many spectral sources, degradation of S/N for all peaks is observed. This effect is sometimes termed the *multiplex disadvantage* and is largely responsible for the fact the Fourier transform has not been widely applied for ultraviolet/visible spectroscopy.[32]

7I-2 Time-Domain Spectroscopy

Conventional spectroscopy can be termed *frequency-domain* spectroscopy in that radiant power data are recorded as a function of frequency or the inversely related wavelength. In contrast, *time-domain* spectroscopy, which can be achieved by the Fourier transform, is concerned with changes in radiant power with *time.* Figure 7-40 illustrates the difference.

The plots in Figures 7-40c and 7-40d are conventional spectra of two monochromatic sources with frequencies ν_1 and ν_2 Hz. The curve in Figure 7-40e is the spectrum of a source that contains both frequencies. In each case, some measure of the radiant power, $P(\nu)$ is plotted with respect to the frequency in hertz. The symbol in parentheses is added to emphasize the frequency dependence of the power; time-domain power will be indicated by $P(t)$.

The curves in Figure 7-40a show the time-domain spectra for each of the monochromatic sources. The two have been plotted together in order to make the small frequency difference between them more obvious. Here, the instantaneous power $P(t)$ is plotted as a function of time. The curve in Figure 7-40b is the time-domain spectrum of the source that contains the two frequencies. As is shown by the horizontal arrow, the plot exhibits a periodicity or *beat* as the two waves go in and out of phase.

Examination of Figure 7-41 reveals that the time-domain signal from a source that contains several wavelengths is considerably more complex than those shown in Figure 7-40. Because a large number of wavelengths are involved, a full cycle is not completed in the time period shown. To be sure, a pattern of beats can be ob-

[32] For a further description of this *multiplex disadvantage* in atomic spectroscopy, see A. P. Thorne, *Anal. Chem.,* **1991,** *63,* 62A-63A; A. G. Marshall and F. R. Verdun, *Fourier Transforms in NMR, Optical, and Mass Spectrometry,* Chapter 5. New York: Elsevier, 1990.

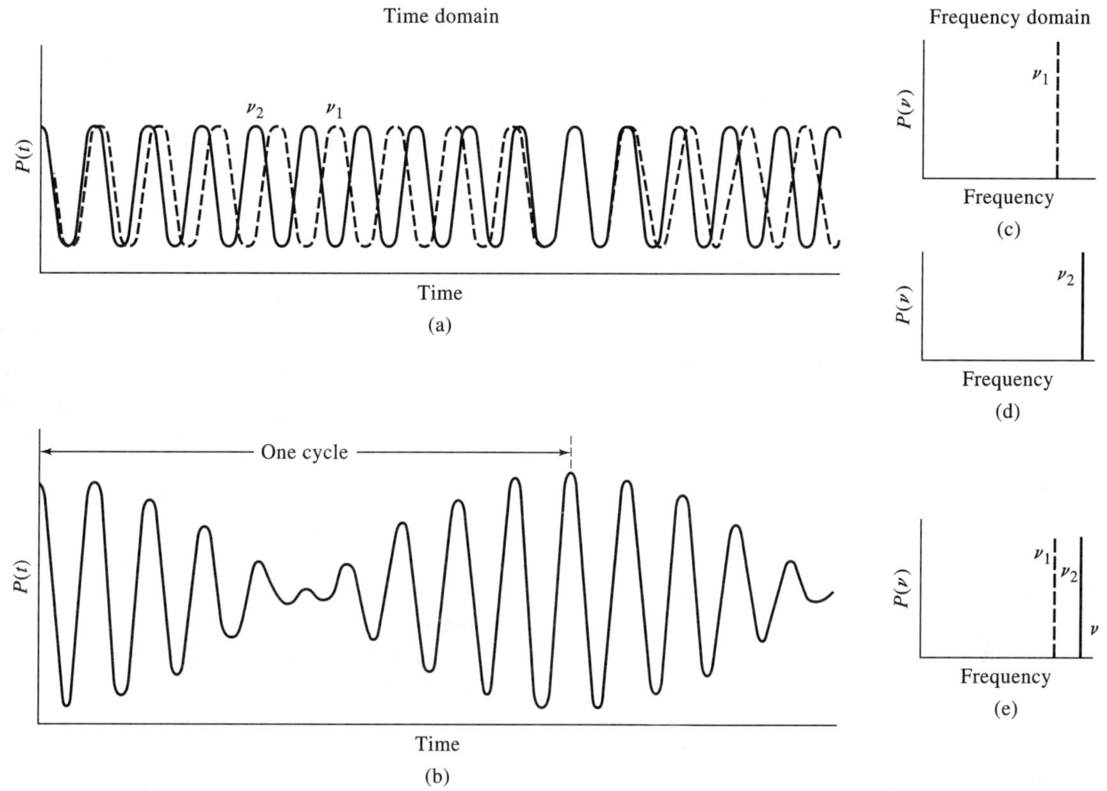

Figure 7-40 Illustrations of (1) time-domain plots (a) and (b) and (2) frequency-domain plots (c), (d), and (e).

served as certain wavelengths pass in and out of phase. In general, the signal power decreases with time as a consequence of the various closely spaced wavelengths becoming more and more out of phase.

It is important to appreciate that a time-domain signal contains the same information as a spectrum does in the frequency domain, and in fact, one can be converted to the other by numerical computations. Thus, Figure 7-40b was derived from Figure 7-40e by means of the equation

$$P(t) = k \cos (2\pi\nu_1 t) + k \cos (2\pi\nu_2 t) \quad (7\text{-}21)$$

where k is a constant and t is the time. The difference in frequency between the two lines was approximately 10% of ν_2.

The interconversion of time- and frequency-domain signals is exceedingly complex and mathematically tedious when more than a few lines are involved; the operation is only practical with a high-speed computer.

7I-3 Obtaining Time-Domain Spectra with a Michelson Interferometer

Time-domain signals, such as those shown in Figures 7-40 and 7-41, cannot be acquired experimentally with radiation of the frequency range that is associated with optical spectroscopy (10^{12} to 10^{15} Hz) because there are no transducers that will respond to power variations at these enormous frequencies. Thus, a typical transducer yields a signal that corresponds to the average power of a high-frequency signal and not to its periodic variation. To obtain time-domain signals requires, therefore, a method of converting (or *modulating*) a high-frequency signal to one of measurable frequency without distorting the time relationships carried in the signal; that is, the frequencies in the modulated signal must be directly proportional to those in the original. Different signal-modulation procedures are employed for the various wavelength regions of the spectrum. The *Michelson interferometer* is used extensively to modulate radiation in the optical region.

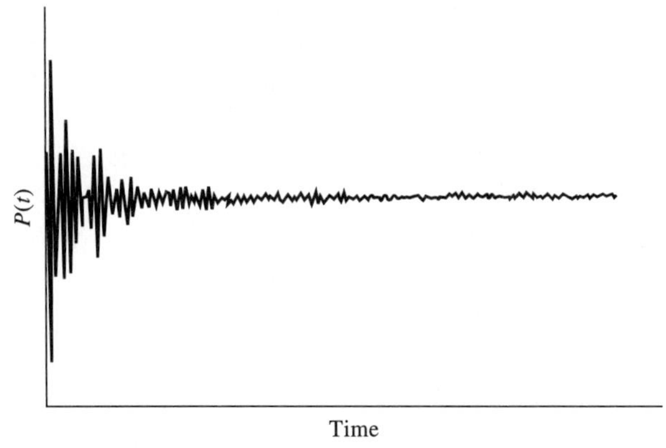

Figure 7-41 Time-domain signal of a source made up of several wavelengths.

The device used for modulating optical radiation is an interferometer similar in design to one first described by Michelson late in the nineteenth century. The Michelson interferometer is a device that splits a beam of radiation into two beams of nearly equal power and then recombines them in such a way that intensity variations of the combined beam can be measured as a function of differences in the lengths of the paths of the two beams. Figure 7-42 is a schematic of such an interferometer as it is used for optical Fourier transform spectroscopy.

As shown in Figure 7-42, a beam of radiation from a source is collimated and impinges on a beam splitter, which transmits approximately half of the radiation and reflects the other half. The resulting twin beams are then reflected from mirrors, one of which is fixed and the other of which is movable. The beams then meet again at the beam splitter, with half of each beam directed toward the sample and detector and the other two halves directed back toward the source. Only the two halves that pass through the sample to the detector are employed for analytical purposes, although the other halves contain the same information about the source.

Horizontal motion of the movable mirror causes the power of the radiation that reaches the detector to fluctuate in a predictable manner. When the two mirrors are equidistant from the splitter (position 0 in Figure 7-42), the two parts of the recombined beam are precisely in phase and the power is a maximum. For a monochromatic source, motion of the movable mirror in either direction by a distance equal to exactly one-quarter wavelength (position *B* or *C* in the figure) changes the path length of the corresponding reflected beam by one-half wavelength (one-quarter wavelength for each direction). Under this circumstance, destructive interference reduces the radiant power of the recombined beams to zero. Further motion to *A* and *D* brings the two halves back in phase so that constructive interference again occurs.

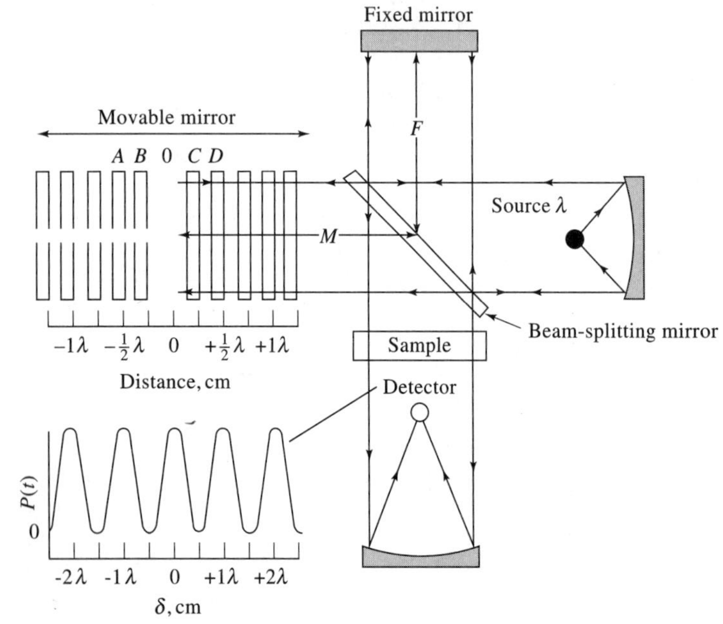

Figure 7-42 Schematic of a Michelson interferometer illuminated by a monochromatic source.

The difference in path lengths for the two beams, $2(M - F)$ in the figure is termed the *retardation* δ. A plot of the output power from the detector versus δ is called an *interferogram;* for monochromatic radiation, the interferogram takes the form of a cosine curve such as that shown in the lower left of Figure 7-42 (cosine rather than sine because the power is always a maximum when δ is zero and the two paths are identical).

The radiation striking the detector after passing through a Michelson interferometer will generally be much lower in frequency than the source frequency. The relationship between the two frequencies is derived by reference to $P(t)$ versus δ plot in Figure 7-42. One cycle of the signal occurs when the mirror moves a distance that corresponds to one half a wavelength ($\lambda/2$). If the mirror is moving at a constant velocity of v_M, and we define τ as time required for the mirror to move $\lambda/2$ cm, we may write

$$v_M\tau = \frac{\lambda}{2} \tag{7-22}$$

The frequency f of the signal at the detector is simply the reciprocal of τ, or

$$f = \frac{1}{\tau} = \frac{v_M}{\lambda/2} = \frac{2v_M}{\lambda} \tag{7-23}$$

We may also relate this frequency to the wavenumber $\overline{\nu}$ of the radiation. Thus,

$$f = 2v_M\overline{\nu} \tag{7-24}$$

The relationship between the *optical frequency* of the radiation and the frequency of the interferogram is obtained by substitution of $\lambda = c/\nu$ into Equation 7-23. Thus,

$$f = \frac{2v_M}{c}\nu \tag{7-25}$$

where ν is the frequency of the radiation and c is the velocity of light (3×10^{10} cm/s). When] v_M is constant, it is evident that the *interferogram frequency f is directly proportional to the optical frequency* ν. Furthermore, the proportionality constant will generally be a very small number. For example, if the mirror is driven at a rate of 1.5 cm/s,

$$\frac{2v_M}{c} = \frac{2 \times 1.5 \text{ cm/s}}{3 \times 10^{10} \text{ cm/s}} = 10^{-10}$$

and

$$f = 10^{-10}\,\nu$$

As shown by the following example, the frequency of visible and infrared radiation is readily modulated into the audio range by a Michelson interferometer.

EXAMPLE 7-3

Calculate the frequency range of a modulated signal from a Michelson interferometer with a mirror velocity of 0.20 cm/s, for visible radiation of 700 nm and infrared radiation of 16 μm (4.3×10^{14} to 1.9×10^{13} Hz).

Employing Equation 7-23, we find

$$f_1 = \frac{2 \times 0.20 \text{ cm/s}}{700 \text{ nm} \times 10^{-7} \text{ cm/nm}} = 5700 \text{ Hz}$$

$$f_2 = \frac{2 \times 0.20 \text{ cm/s}}{16\ \mu\text{m} \times 10^{-4} \text{ cm}/\mu\text{m}} = 250 \text{ Hz}$$

Certain types of visible and infrared transducers are capable of following fluctuations in signal power that fall into the audio-frequency range. Thus, it becomes possible to record a modulated time-domain signal that reflects exactly the appearance of the very high-frequency time-domain signal from a visible or infrared source. Figure 7-43 shows three examples of such time-domain interferograms on the left and their frequency-domain spectra on the right.

Fourier Transformation of Interferograms

The cosine wave of the interferogram shown in Figure 7-43a (and also in Figure 7-42) can be described in theory by the equation

$$P(\delta) = \frac{1}{2}P(\overline{\nu}) \cos 2\pi ft \tag{7-26}$$

where $P(\overline{\nu})$ is the radiant power of the beam that is incident upon the interferometer and $P(\delta)$ is the amplitude or power of the interferogram signal. The parenthetical symbols emphasize that one power is in the frequency domain and the other is in the time domain. In practice, the foregoing equation is modified to take into account the fact that the interferometer ordinarily will not split the source exactly in half and that the detector response and the amplifier behavior are frequency dependent. Thus, it is useful to introduce a new variable $B(\overline{\nu})$ which depends upon $P(\overline{\nu})$ but also takes these factors

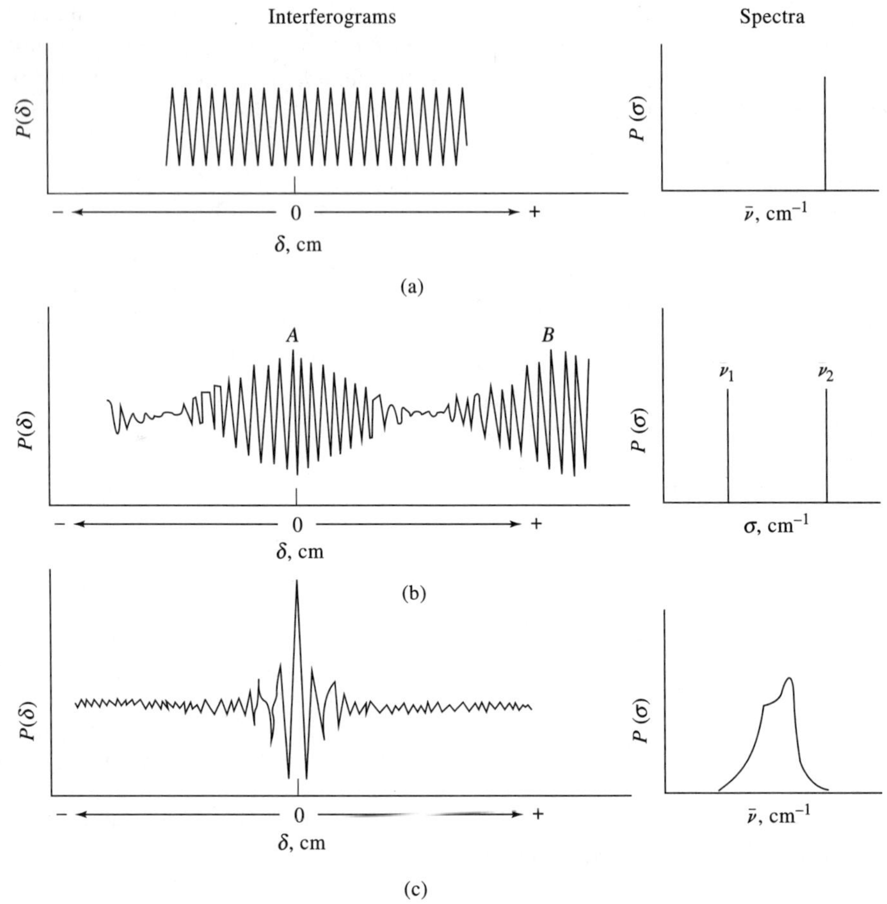

Figure 7-43 Comparison of interferograms and optical spectra.

into account. Therefore, we rewrite the equation in the form

$$P(\delta) = B(\bar{\nu}) \cos 2\pi ft \qquad (7\text{-}27)$$

Substitution of Equation 7-24 into Equation 7-26 leads to

$$P(\delta) = B(\bar{\nu}) \cos 4\pi v_M \bar{\nu} t \qquad (7\text{-}28)$$

But the mirror velocity can be expressed in terms of retardation or

$$v_M = \frac{\delta}{2t}$$

Substitution of this relationship into Equation 7-28 gives

$$P(\delta) = B(\bar{\nu}) \cos 2\pi\delta\bar{\nu}$$

which expresses the magnitude of the interferogram signal as a function of the retardation factor and the wavenumber of the optical input signal.

The interferograms shown in Figure 7-43b can be described by two terms, one for each wavenumber. Thus,

$$P(\delta) = B_1(\bar{\nu}) \cos 2\pi\delta\bar{\nu}_1 + B_2(\bar{\nu}) \cos 2\pi\delta\bar{\nu}_2 \qquad (7\text{-}29)$$

For a continuum source, as in Figure 7-43c, the interferogram can be represented as a sum of an infinite number of cosine terms. That is,

$$P(\delta) = \int_{-\infty}^{+\infty} B(\bar{\nu}) \cos 2\pi\bar{\nu}\delta d\bar{\nu} \qquad (7\text{-}30)$$

The Fourier transform of this integral is

$$B(\bar{\nu}) = \int_{-\infty}^{+\infty} P(\delta) \cos 2\pi\bar{\nu}\delta d\delta \qquad (7\text{-}31)$$

A complete Fourier transformation requires both real (cosine) and imaginary (sine) components; we have presented only the cosine part, which is sufficient for manipulating real/even functions.

Optical Fourier transform spectroscopy consists of recording $P(\delta)$ as a function of δ (Equation 7-30) and then mathematically transforming this relation to one that gives $B(\bar{\nu})$ as a function of $\bar{\nu}$ (the frequency spectrum) as shown by Equation 7-31.

Equations 7-30 and 7-31 cannot be employed as written because they assume that the beam contains radiation from zero to infinite wavenumbers and a mirror drive of infinite length. Furthermore, Fourier transformations with a computer require that the detector output be digitized; that is, the output must be sampled periodically and stored in digital form. Equation 7-31 however, demands that the sampling intervals $d\delta$ be infinitely small; that is, $d\delta \rightarrow 0$. From a practical standpoint, only a finite-sized sampling interval can be summed over a finite retardation range of a few centimeters. These constraints have the effect of limiting the resolution of a Fourier transform instrument and restricting its frequency range.

Resolution

The resolution of a Fourier transform spectrometer can be described in terms of the difference in wavenumber between two lines that can be just separated by the instrument. That is,

$$\Delta\bar{\nu} = \bar{\nu}_1 - \bar{\nu}_2 \tag{7-32}$$

where $\bar{\nu}_1$ and $\bar{\nu}_2$ are wavenumbers for a pair of barely resolvable lines.

It is possible to show that in order to resolve two lines, the time-domain signal must be scanned long enough so that one complete cycle or beat for the two lines is completed; only then will all of the information contained in the spectrum have been recorded. For example, resolution of the two lines $\bar{\nu}_1$ and $\bar{\nu}_2$ in Figure 7-43b requires recording the interferogram from the maximum A at zero retardation to the maximum B where the two waves are again in phase. The maximum at B occurs, however, when $\delta\bar{\nu}_2$ is larger than $\delta\bar{\nu}_1$ by 1 in Equation 7-29. That is, when

$$\delta\bar{\nu}_2 - \delta\bar{\nu}_1 = 1$$

or

$$\bar{\nu}_2 - \bar{\nu}_1 = \frac{1}{\delta}$$

Substitution into Equation 7-32 reveals that the resolution is given by

$$\Delta\bar{\nu} = \bar{\nu}_2 - \bar{\nu}_1 = \frac{1}{\delta} \tag{7-33}$$

This equation means that resolution in wavenumbers will improve in proportion to the reciprocal of the distance that the mirror travels.

EXAMPLE 7-4

What length of mirror drive will provide a resolution of 0.1 cm^{-1}? Substituting into Equation 7-33 gives

$$0.1 = \frac{1}{\delta}$$

$$\delta = 10 \text{ cm}$$

The mirror motion required is one half the retardation, or 5 cm.

Instruments

Details about modern Fourier transform optical spectrometers are found in Section 16C-1. An integral part of these instruments is a sophisticated computer for controlling data acquisition, for storing data, for signal averaging, and for performing the Fourier transformations.

7J QUESTIONS AND PROBLEMS

7-1 Why must the slit width of a prism monochromator be varied to provide constant effective bandwidths whereas a nearly constant slit width may be used with a grating monochromator?

7-2 Why do quantitative and qualitative analyses often require different monochromator slit widths?

7-3 The Wien displacement law states that the wavelength maximum in micrometers for blackbody radiation is given by the relationship

$$\lambda_{max}T = 2.90 \times 10^3$$

where T is the temperature in kelvin. Calculate the wavelength maximum for a blackbody that has been heated to (a) 4000 K, (b) 2000 K, and (c) 1000 K.

7-4 Stefan's law states that the total energy E_T emitted by a blackbody per unit time and per unit area is given by $E_T = \alpha T^4$ where α has a value of 5.69×10^{-8} W M^{-2} K^{-4}. Calculate the total energy output in W/m^2 for each of the blackbodies described in Problem 7-3.

7-5 Relationships described in Problems 7-3 and 7-4 may be of help in solving the following.

(a) Calculate the wavelength of maximum emission of a tungsten filament bulb operated at the usual temperature of 2870 K and at a temperature of 3000 K.
(b) Calculate the total energy output of the bulb in terms of W/cm^2.

7-6 Contrast spontaneous and stimulated emission.

7-7 Describe the advantage of a four-level laser system over a three-level type.

7-8 Define the term *effective bandwidth* of a filter.

7-9 An interference filter is to be constructed for isolation of the CS_2 absorption band at 4.54 μm.

(a) If it is to be based upon first-order interference, what should be the thickness of the dielectric layer (refractive index of 1.34)?
(b) What other wavelengths would be transmitted?

7-10 A 10.0-cm interference wedge is to be built that has a linear dispersion from 400 to 700 nm. Describe details of its construction. Assume that a dielectric with a refractive index of 1.32 is to be employed.

7-11 Why is glass better than fused silica as a prism construction material for a monochromator to be used in the region of 400 to 800 nm?

7-12 For a grating, how many lines per millimeter would be required in order for the first-order diffraction line for $\lambda = 500$ nm to be observed at a reflection angle of 10 deg when the angle of incidence is 60 deg?

7-13 Consider an infrared grating with 72.0 lines per millimeter and 10.0 mm of illuminated area. Calculate the first-order resolution ($\lambda/\Delta\lambda$) of this grating. How far apart (in cm^{-1}) must two lines centered at 1000 cm^{-1} be if they are to be resolved?

7-14 For the grating in Problem 7-13, calculate the wavelengths on the first- and second-order diffraction spectra at reflective angles of (a) 20 deg and (b) 0 deg. Assume the incident angle is 50 deg.

7-15 With the aid of Figures 7-2 and 7-3, suggest instrument components and materials for constructing an instrument that would be well suited for each of the following purposes:

(a) The investigation of the fine structure of absorption bands in the region of 450 to 750 nm.

(b) Obtaining absorption spectra in the far infrared (20 to 50 μm).
(c) A portable device for determining the iron content of natural water based upon the absorption of radiation by the red $Fe(SCN)^{2+}$ complex.
(d) The routine determination of nitrobenzene in air samples based upon its absorption peak at 11.8 μm.
(e) Determining the wavelengths of flame emission lines for metallic elements in the region from 200 to 780 nm.
(f) Spectroscopic studies in the vacuum ultraviolet region.
(g) Spectroscopic studies in the near infrared.

7-16 What is the speed of a lens with a diameter of 4.2 cm and a focal length of 8.2 cm?

7-17 Compare the light-gathering power of the lens described in Problem 7-16 with one that has a diameter of 2.6 cm and a focal length of 8.1 cm.

7-18 A monochromator had a focal length of 1.6 m and a collimating mirror with a diameter of 2.0 cm. The dispersing device was a grating with 1250 lines/mm. For first-order diffraction,
(a) what was the resolving power of the monochromator if a collimated beam illuminated 2.0 cm of the grating?
(b) what is the first- and second-order reciprocal linear dispersion of the monochromator described above?

7-19 A monochromator with a focal length of 0.65 m was equipped with an echellette grating of 2000 blazes per millimeter.
(a) Calculate the reciprocal linear dispersion of the instrument for first-order spectra.
(b) If 3.0 cm of the grating were illuminated, what is the first-order resolving power of the monochromator?
(c) At approximately 560 nm, what minimum wavelength difference could in theory be completely resolved by the instrument?

7-20 Describe the basis for radiation detection with a silicon diode transducer.

7-21 Distinguish among (a) a spectroscope, (b) a spectrograph, and (c) a spectrophotometer.

7-22 A Michelson interferometer had a mirror velocity of 1.25 cm/s. What would be the frequency of the interferogram for (a) UV radiation of 300 nm, (b) visible radiation of 700 nm, (c) infrared radiation of 7.5 μm, and (d) infrared radiation of 20 μm?

7-23 What length of mirror drive in a Michelson interferometer is required to produce a resolution sufficient to separate
(a) infrared peaks at 20.34 and 20.35 μm?
(b) infrared peaks at 2.500 and 2.501 μm?

Chapter 8

An Introduction to Optical Atomic Spectrometry

There are three major types of spectrometric methods for identifying the elements present in samples of matter and determining their concentrations: (1) optical spectrometry, (2) mass spectrometry, and (3) X-ray spectrometry. In optical spectrometry, the elements present in a sample are converted to gaseous atoms or elementary ions by a process called ***atomization.*** *The ultraviolet/visible absorption, emission, or fluorescence of the atomic species in the vapor is then measured. In atomic mass spectrometry, samples are also atomized, but in this case, the gaseous atoms are converted to positive ions (usually singly charged) and separated on the basis of their mass-to-charge ratios. Quantitative data are then obtained by counting the separated ions. In X-ray spectrometry, atomization is not required because X-ray spectra for most elements are largely independent of how they are chemically combined in a sample. Measurements can therefore be based upon the direct measurement of the fluorescence, absorption, or emission spectrum of the sample.*

In this chapter, we first present a theoretical discussion of the sources and properties of optical atomic spectra. We then list the various methods that are used for atomizing samples for elemental analysis. Finally, we describe in some detail the various techniques that are employed for introducing samples into atomizers that are used for optical absorption, emission, and fluorescence spectrometry as well as atomic mass spectrometry.

Chapter 9 is devoted to atomic absorption methods, the most widely used of all of the atomic spectrometric techniques. Chapter 10 deals with several types of atomic emission techniques. This discussion is followed by brief chapters on atomic mass spectrometry and atomic X-ray methods.

8A OPTICAL ATOMIC SPECTRA

In this section, we briefly consider the theoretical basis of optical atomic spectrometry and some of the important characteristics of optical spectra.

8A-1 Energy Level Diagrams

The energy level diagram for the outer electrons of an element provides a convenient method for describing the processes upon which the various methods of atomic spectroscopy are based. The diagram for sodium shown in Figure 8-1a is typical. Note that the energy scale is linear in units of electron volts (eV), with the 3*s* orbital assigned a value of zero. The scale extends to about 5.2 eV, the energy necessary to remove the single 3*s* electron, thus producing a sodium ion.

The energies of several atomic orbitals are indicated on the diagram by horizontal lines. Note that the *p* orbitals are split into two levels that differ but slightly in energy. This difference is rationalized by assuming that an electron spins about its own axis and that the direction of this motion may either be the same as or opposed to its orbital motion. Both the spin and the orbital motions create magnetic fields as a result of the rotation of the charge carried by the electron. The two fields interact in an attractive sense if these two motions are in the opposite direction; a repulsive force is generated when the motions are parallel. As a consequence, the energy of the electron whose spin opposes its orbital motion is slightly smaller than that of the one in which the motions are alike. Similar differences exist in the *d* and *f* orbitals, but their magnitudes are ordinarily so slight as to be undetectable; thus, only a single energy level is indicated for *d* orbitals in Figure 8-1a.

The splitting of higher energy *p, d,* and *f* orbitals into two states is characteristic of all species containing a *single* external electron. Thus, the energy level diagram for the singly charged magnesium ion, shown in Figure 8-1b, has much the same general appearance as that for the uncharged sodium atom. So also do the diagrams for the dipositive aluminum ion and the remainder of the alkali-metal atoms. It is important to note, however, that the energy difference between the 3*p* and 3*s* states is approximately twice as great for the magnesium ion as for the sodium atom because of the larger nuclear charge of the magnesium.

A comparison of Figure 8-1b with Figure 8-2 shows that the energy levels, and thus the spectrum, of an ion are significantly different from that of its parent atom. For atomic magnesium, with two outer electrons,

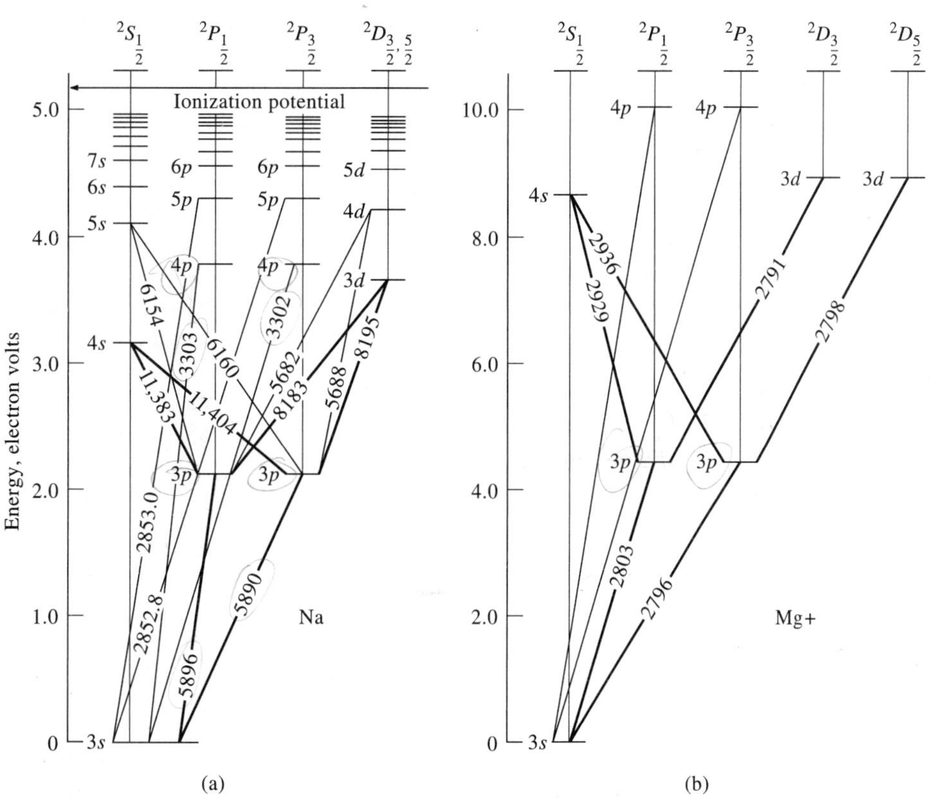

Figure 8-1 Energy level diagrams for (a) atomic sodium and (b) magnesium(I) ion. Note the similarity in pattern of lines but not in actual wavelengths.

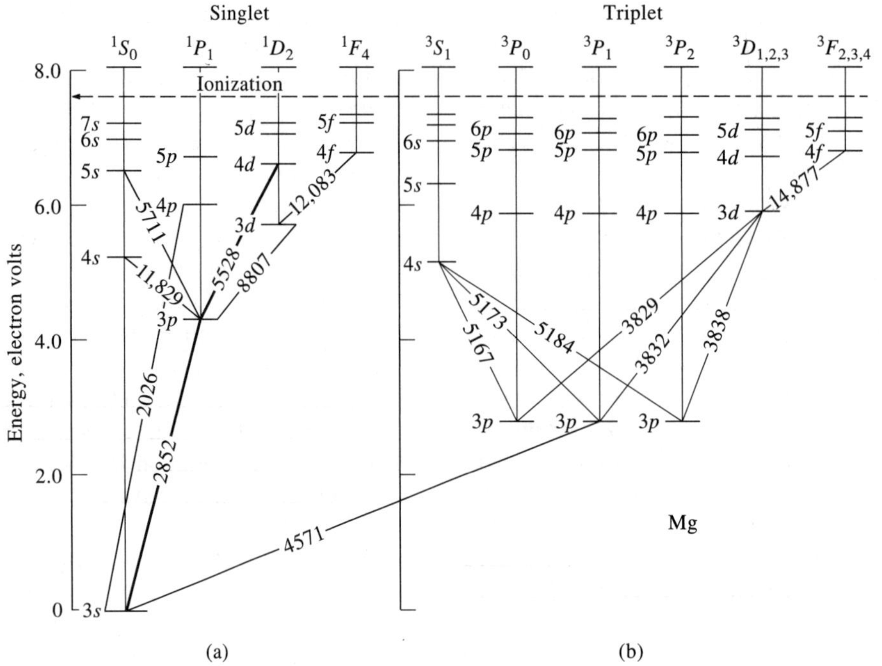

Figure 8-2 Energy level diagram for atomic magnesium. The relative line intensities are indicated very approximately by the width of the lines between states. Note that a singlet/triplet transition is considerably less probable than a singlet to singlet transition.

excited singlet and triplet states with different energies exist. In the excited singlet state, the spins of the two electrons are opposed and said to be paired; in the triplet states, the spins are unpaired or parallel (Section 15A-1). Using arrows to denote the direction of spin, the ground state and the two excited states can be represented as in Figure 8-3. Note that, as is true of molecules, the triplet excited state is of lower energy than the corresponding singlet state.

The *p, d,* and *f* orbitals of the triplet state are split into three levels that differ slightly in energy. These splittings can also be rationalized by taking into account the interaction between the fields associated with the spins of the two outer electrons and the net field that arises from the orbital motions of all the electrons. In the singlet state, the two spins are paired, and their respective magnetic effects cancel; thus, no energy splitting is observed. In the triplet state, however, the two spins are un-

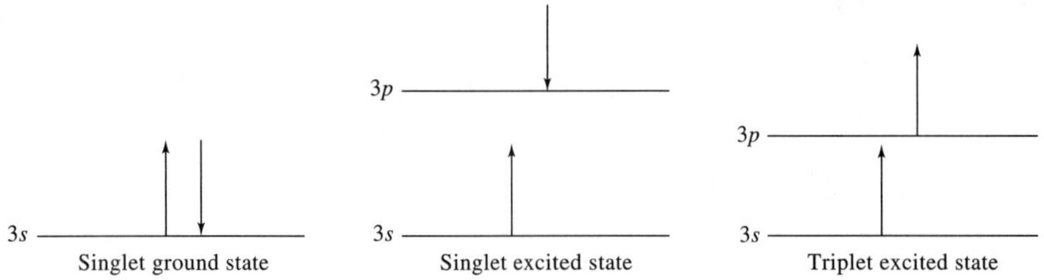

Figure 8-3 Spin orientations in singlet ground and excited states and triplet excited state.

paired (that is, their spin moments lie in the same direction). The effect of the orbital magnetic moment on the magnetic field of the combined spins produces a splitting of the *p* level into a triplet. This behavior is characteristic of all of the alkaline-earth atoms, singly charged aluminum and beryllium ions, and so forth.

As the number of electrons outside the closed shell becomes larger, the energy level diagrams become increasingly complex. Thus, with three outer electrons, a splitting of energy levels into two and four states occurs; with four outer electrons, singlet, triplet, and quintet states exist.

Although correlation of atomic spectra with energy level diagrams for elements such as sodium and magnesium is relatively straightforward and amenable to theoretical interpretation, the same cannot be said for the heavier elements, particularly the transition metals. These species have a larger number of closely spaced energy levels; as a consequence, the number of absorption or emission lines can be enormous. For example, Harvey[1] has listed the number of lines observed in the arc and spark spectra of neutral and singly ionized atoms for a variety of elements. For the alkali metals, this number ranges from 30 for lithium to 645 for cesium; for the alkaline earths, magnesium has 173, calcium 662, and barium 472. Typical of the transition series, on the other hand, are chromium, iron, and cerium with 2277, 4757, and 5755 lines, respectively. Fewer lines are excited in lower-temperature atomizers, such as flames; still, the flame spectra of the transition metals are considerably more complex than the spectra of species with low atomic numbers.

Note that radiation-producing transitions shown in Figures 8-1 and 8-2 are observed only between certain of the energy states. For example, transitions from the 5*s* or the 4*s* to the 3*s* states do not occur, nor do transitions among the various *p* states or the several *d* states. Such transitions are said to be "forbidden," and there are *selection* rules that permit prediction of which transitions are likely to occur and which are not. These rules are beyond the scope of this text.

Atomic Emission Spectra

April 20

At room temperature, essentially all of the atoms of a sample of matter are in the ground state. For example, the single outer electron of a sodium atom occupies the 3*s* orbital under these circumstances. Excitation of this electron to higher orbitals can be brought about by the heat of a flame, a plasma, or an electric arc or spark. The lifetime of the excited atom is brief, however, and its return to the ground state is accompanied by the emission of a photon of radiation. The vertical lines in Figure 8-1a indicate some of the common electronic transitions that follow excitation of sodium atoms; the wavelength of the resulting radiation is also shown. The two lines at 5890 and 5896 Å are the most intense and are responsible for the yellow color that is seen when sodium salts are introduced into a flame.

Figure 8-4 shows a portion of a recorded emission spectrum for sodium. Excitation in this case resulted from spraying a solution of sodium chloride into an oxyhydrogen flame. Note the very large peak at the far right, which is off scale and corresponds to the 3*p* to 3*s* transitions at 5896 and 5890 Å shown in Figure 8-1a. The resolving power of the monochromator used was insufficient to separate the peaks. These two lines are *resonance lines,* which involve transitions between excited electronic state and the ground state. As shown in Figure 8-3, other resonance lines occur at 3302 and 3303 Å as well as at 2853.0 and 2852.8 Å. The much smaller peak at about 5700 Å in Figure 8-4 is in fact two

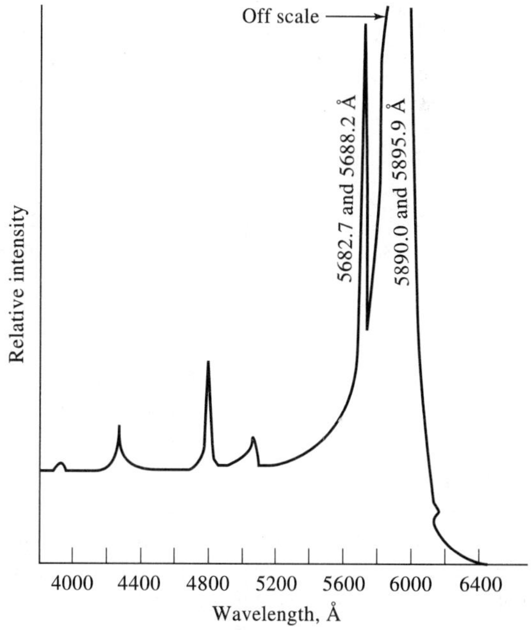

Figure 8-4 A portion of the flame emission spectrum for sodium.

[1] C. E. Harvey, *Spectrochemical Procedures,* Chapter 4. Glendale, CA: Applied Research Laboratories, 1950.

unresolved nonresonance peaks that arise from the two 4*d* to 3*p* transitions also shown in the energy level diagram.

Atomic Absorption Spectra

In a hot gaseous medium, sodium atoms are capable of *absorbing* radiation of wavelengths characteristic of electronic transitions from the 3*s* state to higher excited states. For example, sharp absorption peaks at 5890, 5896, 3302, and 3303 Å are observed experimentally. Referring again to Figure 8-1a, it is apparent that each adjacent pair of these peaks corresponds to transitions from the 3*s* level to the 3*p* and the 4*p* levels, respectively. It should be mentioned that nonresonance absorption due to the 3*p* to 5*s* transition is so weak as to go undetected because the number of sodium atoms in the 3*p* state is generally small at the temperature of a flame. Thus, typically, an atomic absorption spectrum consists predominately of resonance lines, which are the result of transitions from the ground state to upper levels.

Atomic Fluorescence Spectra

Atoms or ions in a flame can be made to fluoresce by irradiation with an intense source containing wavelengths that are absorbed by the element. The fluorescence spectrum is most conveniently measured at an angle of 90 deg to the light path. The observed radiation is most commonly the result of resonance fluorescence involving transitions from excited states returning to the ground state. For example, when magnesium atoms are exposed to an ultraviolet source, radiation of 2852 Å is absorbed as electrons are promoted from the 3*s* to the 3*p* level (see Figure 8-2); the resonance fluorescence emitted at this same wavelength may then be used for analysis. In contrast, when sodium atoms absorb radiation of wavelength 3303 Å, electrons are promoted to the 4*p* state (see Figure 8-1a). A radiationless transition to the two 3*p* states takes place more rapidly than the fluorescence producing transition to the ground state. As a consequence, the observed fluorescence occurs at 5890 and 5896 Å.

Figure 8-5 illustrates yet a third mechanism for atomic fluorescence. Here, some of the thallium atoms, excited in a flame, return to the ground state in two steps, including a fluorescence emission step producing a line at 5350 Å; radiationless deactivation to the ground state quickly follows. Resonance fluorescence at 3776 Å is also observed.

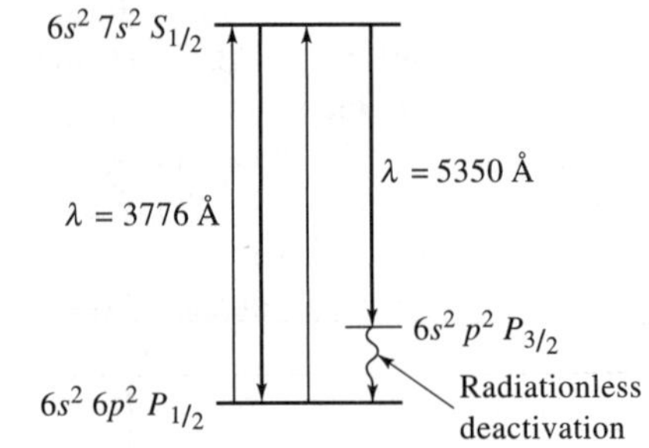

Figure 8-5 Energy level diagram for thallium showing the source of two fluorescence lines.

8A-2 Atomic Line Widths

The widths of atomic lines are of considerable importance in atomic spectroscopy. For example, narrow lines are highly desirable for both absorption and emission work because they reduce the possibility of interference due to overlapping spectra. Furthermore, as will be shown later, line widths are of prime importance in the design of instruments for atomic absorption spectroscopy. For these reasons, it is worthwhile considering some of the variables that influence the width of atomic spectral lines.

As shown in Figurc 8-6, atomic absorption and emission lines generally are found to be made up of a symmetric distribution of wavelengths that centers about a mean wavelength λ_0, which is the wavelength of maximum absorbance for absorbed radiation or maximum intensity for emitted radiation. The energy associated with λ_0 is equal to the exact energy difference be-

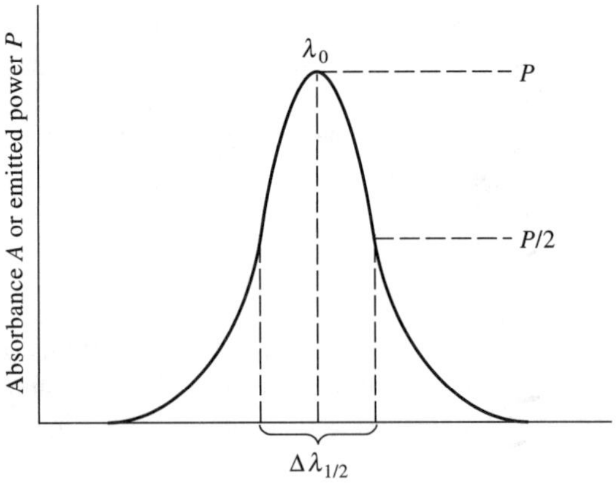

Figure 8-6 Profile of an atomic line showing definition of the effective line width $\Delta\lambda_{1/2}$.

tween the two quantum states responsible for absorption or emission.

An examination of an energy level diagram, such as that shown in Figure 8-1a, suggests that an atomic line contains but a single wavelength λ_0—that is, since a line results from a transition of an electron between two discrete, single-valued energy states, the line width will be zero. Several phenomena however, cause line broadening so that all atomic lines have finite widths as shown in Figure 8-6. Note that the *line width* or *effective line width* $\Delta\lambda_{1/2}$ of an atomic absorption or emission line is defined as its width in wavelength units when measured at one half the maximum signal. This point is chosen because the measurement can be made more accurately at half-peak intensity than at the base.

Line broadening arises from four sources, including: (1) the uncertainty effect, (2) the Doppler effect, (3) pressure effects due to collisions between atoms of the same kind and with foreign atoms, and (4) electric and magnetic field effects. Only the first three of these phenomena are pertinent to this discussion. The magnetic field effect will be discussed in Section 9C-1 in connection with the Zeeman effect.

Line Broadening from the Uncertainty Effect

Spectral lines always have finite widths because the lifetimes of one or both transition states are finite, which leads to uncertainties in the transition times and to line broadening as a consequence of the uncertainty principle (see Section 6C-6). That is, if the lifetimes of two states approached infinity, then the breadth of an atomic line resulting from a transition between the two states approaches zero. While the lifetime of a ground state electron is long, the lifetime of its excited states are generally brief, amounting to 10^{-7} to 10^{-8} s in the typical case. Example 8-1 illustrates how the width of an atomic emission line can be approximated from its mean lifetime and the uncertainty principle.

EXAMPLE 8-1

The mean lifetime of the excited state produced by irradiating mercury vapor with a pulse of 253.7 nm radiation is 2×10^{-8} s. Calculate the approximate value for the width of the fluorescence line produced in this way.

According to the uncertainty principle (Equation 6-25),

$$\Delta\nu \cdot \Delta t > 1$$

Substituting 2×10^{-8} s for Δt gives upon rearranging the uncertainty in $\Delta\nu$ in the frequency of the emitted radiation.

$$\Delta\nu = 1/2 \times 10^{-8} = 5 \times 10^{7}\ \text{s}^{-1}$$

In order to evaluate the relationship between this uncertainty in frequency and the uncertainty in wavelength units, we write Equation 6-2 in the form

$$\nu = c\lambda^{-1}$$

Differentiating with respect to frequency gives

$$d\nu = -1c\lambda^{-2}d\lambda$$

Rearranging and letting $\Delta\nu$ approximate $d\nu$ and $\Delta\lambda_{1/2}$ approximate $d\lambda$, we obtain

$$|\Delta\lambda_{1/2}| = \frac{\lambda^2\Delta\nu}{c} =$$

$$\frac{(253.7 \times 10^{-9}\ \text{m})^2 \times 5 \times 10^{7}\ \text{s}^{-1}}{3 \times 10^{8}\ \text{m/s}} = 1.1 \times 10^{-14}\ \text{m}$$

$$= 1.1 \times 10^{-14}\ \text{m} \times 10^{10}\ \text{Å/m} = 1 \times 10^{-4}\ \text{Å}$$

Line widths due to uncertainty broadening are sometimes termed *natural line widths* and are generally about 10^{-4} Å as shown in Example 8-1.

Doppler Broadening[2]

The wavelength of radiation emitted or absorbed by a rapidly moving atom decreases if the motion is toward a transducer and increases if the atom is receding from the transducer (see Figure 8-7). This phenomenon is known as the *Doppler shift* and is observed not only with electromagnetic radiation but with sound waves as well. For example, the Doppler shift occurs when an automobile blows its horn while it passes a pedestrian. As the auto approaches the observer, each successive sound vibration of the horn is emitted from a distance that is successively closer to the observer. Thus, each sound wave reaches the pedestrian slightly sooner than

[2] For a quantitative treatment of Doppler broadening and pressure broadening, see J. D. Ingle Jr. and S. R. Crouch, *Spectrochemical Analysis*, pp. 210–212. Englewood Cliffs, NJ: Prentice-Hall, 1988.

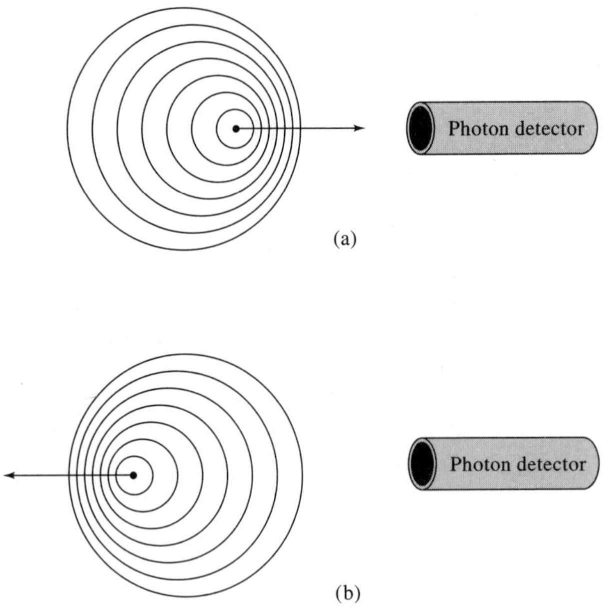

Figure 8-7 Cause of Doppler broadening. (a) Atom moving toward incoming radiation sees wave crests more frequently and thus absorbs radiation that is actually higher in frequency. (b) Atom moving with the direction of radiation encounters wave crests less often and thus absorbs radiation that is actually of lower frequency.

would be expected if the auto were standing still. The result is a highter frequency, or higher pitch, for the horn. When the auto is even with the observer, the waves come directly to the ear of the observer along a line perpendicular to the path of the auto, so there is no shift in the frequency. When the car recedes from the pedestrian, each wave is emitted at a distance that is larger than that for the previous wave; as a result, the frequency is smaller, and this results in a lower pitch.

The magnitude of the Doppler shift increases with the velocity at which the emitting or absorbing species approaches or recedes from the transducer. The Doppler effect is used in astronomy to measure the velocity of recession of galaxies from our position on Earth. The relationship between the so-called *Doppler shift,* or *red shift,* $\Delta\lambda$ and the velocity v of a receding galaxy is embodied in the equation

$$\frac{\Delta\lambda}{\lambda_0} = \frac{v}{c}$$

where λ_0 is the wavelength of an unshifted line of a sample element at rest relative to the observer, and c is the velocity of light. During the 1920s, the famous astronomer Edwin Hubble found a direct proportionality between the *Doppler shifts* of known lines in the spectra of receding galaxies and the distances to the galaxies measured by other means. The slope of the linear plot of galactic recession velocity versus distance from the Earth is the *Hubble constant,* which provides a measure of the age of the known universe as well as basis for the *big bang theory* and other theories of its formation and evolution.

In a collection of atoms in a hot environment, such as a flame, atomic motion occurs in every direction. Individual atoms exhibit a Maxwell-Boltzmann velocity distribution, in which the average velocity of a particular atomic species increases as the square root of the absolute temperature. The maximum Doppler shifts are exhibited by those atoms that are moving with the highest velocities either directly toward or away from the transducer. No shift is associated with atoms that are moving perpendicular to the path to the transducer. Intermediate shifts occur for the remaining atoms that are a function of their speed and direction. Thus, the transducer encounters an approximately symmetrical distribution of wavelengths, with the maximum corresponding to zero Doppler shift. In flames the Doppler effect leads to lines that are about two orders of magnitude greater in breadth than natural line widths.

Pressure Broadening

Pressure, or collisional, broadening arises from collisions of the emitting or absorbing species with other atoms or ions in the heated medium. These collisions cause small changes in ground state energy levels and hence a range of absorbed or emitted wavelengths. In a flame, the collisions are largely between the analyte atoms and the various combustion products of the fuel, which results in broadening that is two or three orders of magnitude greater than the natural line widths. Broadening in hollow cathode lamps and discharge lamps that are used as sources in atomic absorption spectroscopy is largely a consequence of collisions between the emitting atoms and other atoms of the same kind. In high-pressure mercury and xenon lamps, pressure broadening of this type is so extensive that continuum radiation is produced throughout the ultraviolet and visible region.

8A-3 The Effect of Temperature on Atomic Spectra

Temperature exerts a profound effect upon the ratio between the number of excited and unexcited atomic particles in an atomizer. The magnitude of this effect can be derived from the *Boltzmann equation,* which is usually written in the form

$$\frac{N_j}{N_0} = \frac{P_j}{P_0} \exp\left(-\frac{E_j}{kT}\right) \tag{8-1}$$

Here, N_j and N_0 are the number of atoms in an excited state and the ground state, respectively, k is the Boltzmann constant (1.28×10^{-23} J/K), T is the temperature in kelvins, and E_j is the energy difference in joules between the excited state and the ground state. The quantities P_j and P_0 are statistical factors that are determined by the number of states having equal energy at each quantum level.

EXAMPLE 8-2

Calculate the ratio of sodium atoms in the $3p$ excited states to the number in the ground state at 2500 and 2510 K.

In order to calculate E_j in Equation 8-1, we employ an average wavelength of 5893 Å for the two sodium emission lines involving the $3p \rightarrow 3s$ transitions. To obtain the energy in joules, we employ the conversion factors found inside the front cover.

$$\text{wavenumber} = \frac{1}{5893\text{ Å} \times 10^{-8}\text{ cm/Å}} = 1.697 \times 10^{4}\text{ cm}^{-1}$$

$$E_j = \frac{1.697 \times 10^{4}}{\text{cm}} \times 1.986 \times 10^{-23}\text{ J cm} = 3.37 \times 10^{-19}\text{ J}$$

There are two quantum states in the $3s$ level and six in the $3p$. Thus,

$$\frac{P_j}{P_0} = \frac{6}{2} = 3$$

Substituting into Equation 8-1 yields

$$\frac{N_j}{N_0} = 3 \exp\left(\frac{-3.37 \times 10^{-19}\text{ J}}{1.38 \times 10^{-23}\text{ J K}^{-1} \times 2500\text{ K}}\right)$$

or

$$\frac{N_j}{N_0} = 3 \times 5.725 \times 10^{-5} = 1.72 \times 10^{-4}$$

Replacing 2500 with 2510 in the foregoing equations yields

$$\frac{N_j}{N_0} = 1.79 \times 10^{-4}$$

Example 8-2 demonstrates that a temperature fluctuation of only 10 K results in a 4% increase in the number of excited sodium atoms. A corresponding increase in emitted power by the two lines would result. Thus, an analytical method based on the measurement of emission requires close control of atomization temperature.

Absorption and fluorescence methods are theoretically less dependent upon temperature because both measurements are based upon initially *unexcited* atoms rather than thermally excited ones. In the example just considered, only about 0.017% of the sodium atoms were thermally excited at 2500 K. An emission method is based upon this small fraction of the analyte. In contrast, absorption and fluorescence measurements use the 99.98% of the analyte present as unexcited sodium atoms to produce the analytical signals. Note also that while a 10-K temperature change causes a 4% increase in excited atoms, the corresponding *relative* change in percent of unexcited atoms is inconsequential.

Temperature fluctuations actually do exert an indirect influence on atomic absorption and fluorescence measurements in several ways. An increase in temperature usually increases the efficiency of the atomization process and hence the total number of atoms in the vapor. In addition, line broadening and a consequent decrease in peak height occurs because the atomic particles travel at greater rates, which enhances the Doppler effect. Finally, temperature variations influence the degree of ionization of the analyte and thus the concentration of nonionized analyte upon which the analysis is usually based (see page 222). Because of these several effects, reasonable control of the flame temperature is also required for quantitative absorption and fluorescence measurements.

The large ratio of unexcited to excited atoms in atomization media leads to another interesting comparison of the three atomic methods. Because atomic absorption and atomic fluorescence methods are based upon a much larger population of atoms, these two procedures might be expected to be more sensitive than the emission procedure. This apparent advantage is offset in the absorption method, however, by the fact that an absorbance measurement involves evaluation of a difference ($A = \log P_0 - \log P$); when P and P_0 are nearly alike, larger relative errors must be expected in the difference. As a consequence, emission and absorption procedures tend to be complementary in sensitivity, one technique being advantageous for one group of elements and the other for a different group. On the basis of active atom population, atomic fluorescence methods should, at least in theory, be the most sensitive of the three.

8A-4 Band and Continuum Spectra Associated with Atomic Spectra

Generally, in addition to line spectra, both band and continuum radiation are produced when atomic line spectra are generated. For example, Figure 6-15 (page 32) shows the presence of both molecular bands and a continuum, the latter resulting from the thermal radiation from hot particulate matter in the atomization medium. As will be shown later, plasmas, arcs, and sparks also produce both bands and continuum radiation.

Band spectra are often encountered in the determination of elements by atomic absorption and emission spectrometry. For example, when solutions of calcium ion are atomized in a low-temperature flame, molecular absorption and emission bands for CaOH appear in the region of 554 nm (see Figure 8-8), which can be used for the determination of concentrations of the element. More commonly, however, the presence of molecular bands and continuum radiation represents a potential source of interference, which must be dealt with by proper choice of wavelength, by background correction, or by a change in atomization conditions.

8B ATOMIZATION METHODS

In order to obtain both atomic optical and atomic mass spectra, the constituents of a sample must be converted to gaseous atoms or ionized atoms, which can then be determined by emission, absorption, fluorescence, or mass spectral measurements. As noted earlier, the process by which the sample is converted into an atomic vapor is

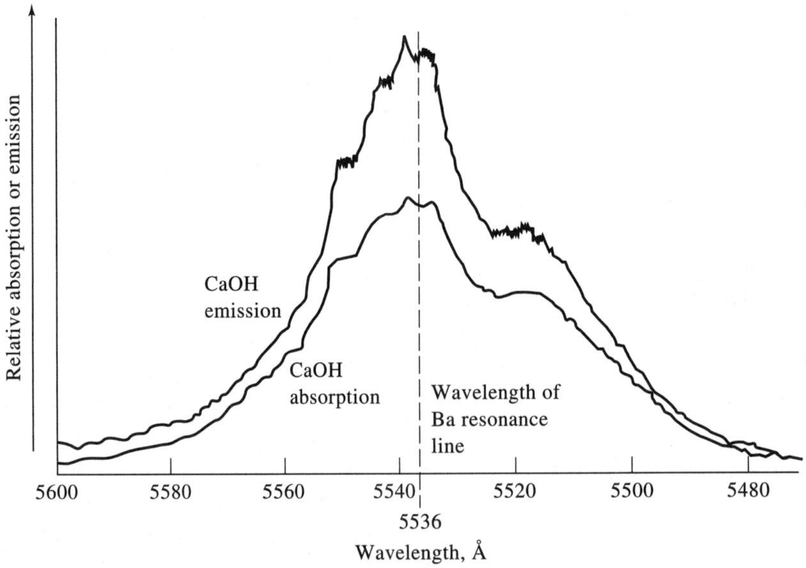

Figure 8-8 Molecular flame and flame absorption spectra for CaOH and Ba. *(Adapted from L. Capacho-Delgado and S. Sprague,* Atomic Absorption Newsletter, ***1965,*** *4, 363. Courtesy of Perkin-Elmer Corporation, Norwalk, CT.)*

called *atomization.* The precision and accuracy of atomic methods are critically dependent upon the atomization step and the method of introduction of the sample into the atomization region. The common types of atomizers are listed in Table 8-1. Detailed descriptions of several of these devices are found in Chapters 9, 10, and 11.

8C SAMPLE INTRODUCTION METHODS

Sample introduction has been termed the "Achilles' Heel" of atomic spectroscopy because in many cases this step limits the accuracy, the precision, and the detection limits of atomic spectrometric measurements.[3] The goal of a sample introduction system in atomic spectrometry is to transfer a reproducible and representative portion of a sample into one of the atomizers listed in Table 8-1 with high efficiency and with no adverse interference effects. The ease with which this goal is realized is strongly dependent on the physical and chemical state of the analyte and the sample matrix. For solid samples of refractory materials, sample introduction is usually a major problem; whereas for solutions and gaseous samples, the introduction step is often trivial. For this reason a majority of atomic spectroscopic studies are performed on solutions.

For the first five atomization sources listed in Table 8-1, samples are most commonly introduced in the form of solutions (usually aqueous) or less often as slurries (a slurry is a suspension of a finely divided powder in a liquid). For samples that are difficult to dissolve, however, several methods have been proposed that permit introduction of samples in the form of solids or finely dispersed powders. Generally, solid introduction techniques are less reproducible and more subject to various errors and as a consequence are not nearly as widely applied as those based upon aqueous solutions. Table 8-2 lists the common methods for sample introduction that are employed in atomic spectroscopy and the type of samples to which each is applicable.

8C-1 Introduction of Solution Samples[4]

Solutions are generally introduced into an atomizer by one of the first three methods listed in Table 8-2. Two of these methods involve *nebulization,* in which the sample is converted to a mist of finely divided droplets (an *aerosol*) by a jet of compressed gas. The flow of gas

[3] R. F. Browner and A. W. Boorn, *Anal. Chem.,* **1984,** *56,* 786A, 875A; *Sample Introduction in Atomic Spectroscopy,* J. Sneddon, Ed. New York: Elsevier, 1990.

[4] For an excellent discussion of liquid introduction methods, see A. G. T. Gustavsson, in *Inductively Coupled Plasmas in Analytical Atomic Spectrometry,* 2nd ed, A. Montaser and D. W. Golightly, Eds., Chapter 15. New York: VCH Publishers, Inc., 1992.

TABLE 8-1 Types of Atomizers Used for Atomic Spectroscopy

Type of Atomizer	Typical Atomization Temperature, °C
Flame	1700–3150
Electrothermal vaporization (ETV)	1200–3000
Inductively coupled argon plasma (ICP)	4000–6000
Direct current argon plasma (DCP)	4000–6000
Microwave-induced argon plasma (MIP)	2000–3000
Glow discharge plasma (GD)	Nonthermal
Electric arc	4000–5000
Electric spark	40,000 (?)

TABLE 8-2 Methods of Sample Introduction in Atomic Spectroscopy

Method	Type of Sample
Pneumatic nebulization	Solution or slurry
Ultrasonic nebulization	Solution
Electrothermal vaporization	Solid, liquid, solution
Hydride generation	Solution of certain elements
Direct insertion	Solid, powder
Laser ablation	Solid, metal
Spark or arc ablation	Conducting solid
Glow discharge sputtering	Conducting solid

then carries the sample into the region where atomization takes place.

Pneumatic Nebulizers

Samples for atomic spectrometric analysis are most commonly dissolved in an aqueous medium and then introduced into the atomizer by means of a *nebulizer* that converts the liquid to a fine mist, or *aerosol.* The most common kind of nebulizer is the concentric-tube pneumatic type, shown in Figure 8-9a, in which the liquid sample is sucked through a capillary tube by a high-pressure stream of gas flowing around the tip of the tube (the Bernoulli effect). This process of liquid transport is called *aspiration.* The high-velocity gas breaks the liquid up into fine droplets of various sizes, which are then carried into the atomizer. Also employed are cross-flow nebulizers, illustrated in Figure 8-9b, in which the high-pressure gas flows across a capillary tip at right angles. Often in this type of nebulizer, the liquid is pumped through the capillary. Figure 8-9c is a schematic of a fritted-disk nebulizer in which the sample solution is pumped onto a fritted surface through which a carrier gas flows. This type of nebulizer produces a much finer aerosol than do the first two. Figure 8-9d shows a Babington nebulizer, which consists of a hollow sphere in which a high-pressure gas is pumped through a small orifice in the sphere surface. Liquid flowing in a thin film over the sphere surface is nebulized by the expanding jet of gas. This type of nebulizer is less subject to clogging and thus has proven useful for samples that have a high salt content or for slurries with a significant particulate content.

Ultrasonic Nebulizers

Several instrument manufacturers also offer ultrasonic nebulizers in which the sample is pumped onto the surface of a piezoelectric crystal that vibrates at a frequency of 20 kHz to several MHz. Such nebulizers produce more dense and more homogeneous aerosols than pneumatic nebulizers do.

Electrothermal Vaporizers

An electrothermal vaporizer (ETV) is an evaporator located in a closed chamber through which an inert gas such as argon flows to carry the vaporized sample into the atomizer. A small liquid or solid sample is placed on a conductor, such as a carbon rod or tantalum filament. An electric current then evaporates the sample rapidly and completely into the argon gas. In contrast to the nebulizer arrangements we have just considered, an electrothermal system produces a discrete signal rather

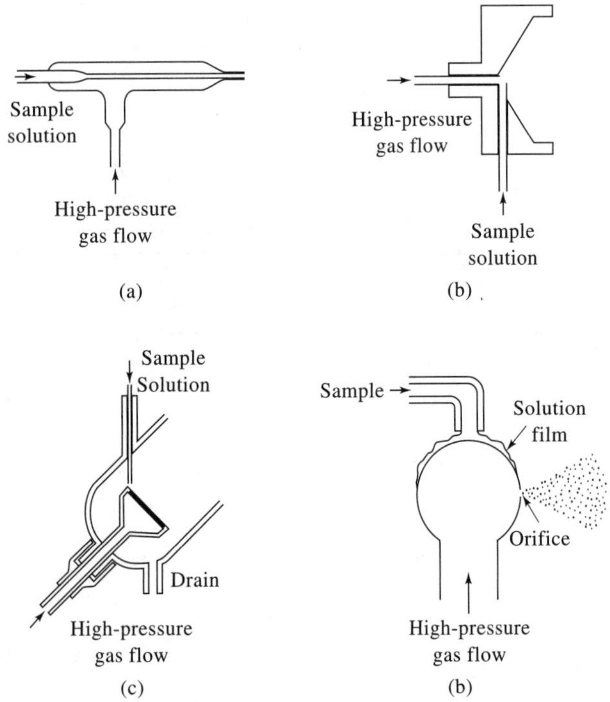

Figure 8-9 Types of pneumatic nebulizers: (a) concentric tube, (b) cross-flow, (c) fritted disk, (d) Babington.

than a continuous one. That is, the signal from the atomized sample increases to a maximum and then decreases to zero as the sample is swept through the atomizer. Quantitative analysis is then obtained from peak heights or areas.

Hydride Generation Techniques

Hydride generation techniques provide a method for introducing samples containing arsenic, antimony, tin, selenium, bismuth, and lead into an atomizer as a gas.[5] Such a procedure enhances the detection limits for these elements by a factor of 10 to 100. Because several of these species are highly toxic, their determination at low concentration levels is of considerable importance. This toxicity also dictates that gases from atomization must be disposed of in a safe and efficient manner.

Rapid generation of volatile hydrides can generally be brought about by addition of an acidified aqueous solution of the sample to a small volume of a 1% aqueous solution of sodium borohydride contained in a glass vessel; a typical reaction is shown by the equation

$$3BH_4^- + 3H^+ + 4H_3AsO_3 \rightarrow 3H_3BO_3 + 4AsH_3\uparrow + 3H_2O$$

The volatile hydride—in this case, arsine—is swept into the atomization chamber by an inert gas. The chamber is usually a silica tube heated to several hundred degrees in a tube furnace or in a flame where decomposition of the hydride takes place leading to formation of atoms of the analyte, the concentration of which is then determined by absorption or emission measurement. The signal is a peak similar to that obtained with electrothermal atomization.

8C-2 Introduction of Solid Samples[6]

The introduction of solids in the form of powders, metals, or particulates into plasma and flame atomizers has the considerable advantage of avoiding the often tedious and time-consuming step of sample decomposition and dissolution. Such procedures, however, often suffer from severe difficulties with calibration, sample conditioning, precision, and accuracy.

A variety of techniques have been proposed during the last two decades for the direct introduction of solids into atomizers, thus avoiding the need to dissolve or decompose the sample. These techniques take a variety of forms, including: (1) direct manual insertion of the solid into the atomization device, (2) electrothermal vaporization of the sample and transfer of the vapor into the atomization region, (3) arc/spark or laser ablation of the solid to produce a vapor that is then swept into the atomizer, (4) slurry nebulization in which the finely divided solid sample is carried into the atomizer as an aerosol consisting of a suspension of the solid in a liquid medium, and (5) sputtering in a glow discharge device. None of these procedures yields as satisfactory results as sample introduction by nebulization of solutions of the sample. Most of these techniques lead to a discrete analytical signal rather than a continuous one.

Direct Sample Insertion

In the direct sample insertion technique, the sample is physically placed into the atomizer. For solids, the sample may be ground into a powder, which is then placed on or into a probe that is inserted directly into the atomizer. With electric arc and spark atomizers, metal samples are frequently introduced as one or both electrodes that are used to form the arc or spark.

Electrothermal Vaporizers

Electrothermal vaporizers, which were described briefly in an earlier section on solutions are also employed for various types of solid samples. Heating is carried out by conductive heating of the sample held on or in a graphite or tantalum rod or boat. An inert gas then carries the vaporized sample into the atomizer.

Arc and Spark Ablation

Electrical discharges of various types are often employed to introduce solid samples into atomizers. Interaction of the discharge with the surface of a solid sample creates a plume that is made up of a particulate and vaporized sample that is then transported into the atomizer by the flow of an inert gas. This process of sample introduction is called *ablation.*

---

[5] For a detailed discussion of these methods, see T. Nakahara, in *Sample Introduction in Atomic Spectroscopy,* J. Sneddon, Ed., Chapter 10. New York: Elsevier, 1990; J. Didina and D. L. Tsalev, *Hydride Generation Atomic Spectroscopy.* New York: Wiley, 1995.

[6] For a description of solid introduction techniques, see C. M. McLeod, M. W. Routh, and M. W. Tikkanen, in *Inductively Coupled Plasmas in Analytical Atomic Spectrometry,* 2nd ed, A. Montaser and D. W. Golightly, Eds., Chapter 16. New York: VCH Publishers, Inc., 1992.

In order for arc and spark ablation to be successful, the sample must be electrically conducting or it must be mixed with a conductor. Ordinarily ablation is carried out in an inert atmosphere such as an argon gas stream. Depending upon the nature of the sample, the resulting analytical signal may be discrete or continuous. Several instrument manufacturers market accessories for electric arc and spark ablation.

Note that arcs and sparks also result in atomization and excitation of atomic emission spectra that are used for analysis. A spark also produces a significant number of ions that can be separated and determined by mass spectrometry.

Laser Ablation

Laser ablation is a relatively new and versatile method of introducing solid samples into atomizers. Similar to arc and spark ablation, sufficient energy in the form of a focused laser beam is directed onto the surface of the solid sample, where ablation takes place to convert the sample into a plume of vapor and particulate matter that is then swept into the atomizer.

Laser ablation is a versatile technique because it is applicable to both conducting and nonconducting solids, inorganic and organic samples, and powder and metallic materials. In addition to bulk analysis, a focused laser beam permits analysis of small areas on the surface of solids. Laser samplers are offered by several instrument makers.

The Glow Discharge Technique[7]

A glow discharge (GD) device is a versatile source that performs both sample introduction and sample atomization simultaneously (see Figure 8-10). Briefly, a glow discharge takes place in a low-pressure atmosphere (1 to 10 torr) of argon gas between a pair of electrodes maintained at a dc potential of 250 to 1000 V. The applied potential causes the argon gas to break down into positively charged argon ions and electrons. The electric field accelerates the argon ions to the cathode surface that contains the sample. Ejection of neutral sample atoms occurs by a process called *sputtering*. The rate of sputtering may be as high as 100 μg/min.

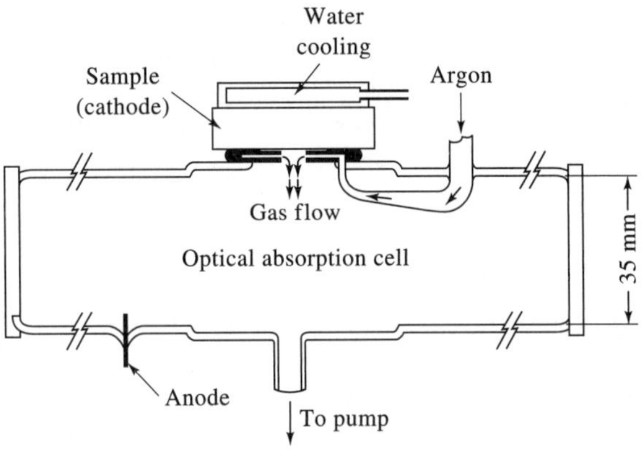

Figure 8-10 A glow discharge atomizer. *(From D. S. Gough, P. Hannaford, and R. M. Lowe,* Anal. Chem., ***1989****, 61, 1653. With permission.)*

The atomic vapor produced in a glow discharge is made up of a mixture of atoms and ions that can be determined by atomic absorption or fluorescence or by mass spectrometry. In addition, a fraction of the atomized species present in the vapor is in an excited state. Relaxation of the excited species produces a low-intensity glow (thus, the name) that can be used for optical emission measurements.

The most important applications of the glow discharge atomizer have been to the analysis of metals and other conducting samples, although with modification, it has also been used with liquid samples and nonconducting materials by mixing them with a conductor such as graphite or pure copper powders.

Glow discharge sources of various kinds are available from several instrument manufacturers.

[7] See R. K. Marcus, T. R. Harville, Y. Mei, and C. R. Shick, *Anal. Chem.*, **1994**, *66*, 902A; W. W. Harrison, C. M. Barshick, J. A. Kingler, P. H. Ratliff, and Y. Mei, *Anal. Chem.*, **1990**, *62*, 943A; R. K. Marcus, *Spectroscopy*, **1992**, *7*(5), 12; *Glow Discharge Spectroscopies*, R. K. Marcus, Ed., New York: Plenum Press, 1993.

8D QUESTION AND PROBLEMS

8-1 Why is the CaOH spectrum in Figure 8-8 so much broader than the barium emission line?

8-2 What is resonance fluorescence?

8-3 Under what conditions can a Stokes shift occur in atomic spectroscopy?

8-4 What are natural line widths for atomic emission and absorption lines?

8-5 It is observed that in a hot flame the emission intensities of the sodium lines at 589.0 and 589.6 nm are greater in a sample solution that contains KCl than when this compound is absent. Suggest an explanation.

8-6 The intensity of a line for atomic Cs is much less in a natural gas flame, which operates at 1800°C, than in a hydrogen/oxygen flame, whose temperature is 2700°C. Explain.

8-7 Name a continuous and a noncontinuous type of atomizer that is used in atomic spectrometry. How do the output signals from a spectrometer differ for each?

8-8 The Doppler effect is one of the sources of the line broadening in atomic absorption spectroscopy. Atoms moving toward the light source are exposed to a higher frequency of radiation than are atoms moving away from the source. The difference in wavelength, $\Delta\lambda$, experienced by an atom moving at speed v (compared to one at rest) is $\Delta\lambda/\lambda = v/c$, where c is the speed of light. Estimate the line width (in Å) of the sodium D line (5893 Å) when the absorbing atoms are at a temperature of (a) 2200 K and (b) 3000 K. The average speed of an atom is given by $v = \sqrt{8kT/\pi m}$, where k is Boltzmann's constant, T is temperature, and m is the mass.

8-9 For Na atoms and Mg^+ ions, compare the ratios of the number of particles in the $3p$ excited state to the number in the ground state in
(a) a natural gas/air flame (2100 K).
(b) a hydrogen/oxygen flame (2900 K).
(c) an inductively coupled plasma source (6000 K).

8-10 In higher-temperature sources, sodium atoms emit a doublet with an average wavelength of 1139 nm. The transition responsible is from the $4s$ to $3p$ state. Calculate the ratio of the number of excited atoms in the $4s$ to the number in the ground $3s$ state in
(a) an acetylene oxygen flame (3000°C).
(b) the hottest part of an inductively coupled plasma source (~9000°C).

8-11 In the concentration range of 500 to 2000 ppm of U, a linear relationship is found between absorbance at 351.5 nm and concentration. At lower concentrations the relationship is nonlinear unless about 2000 ppm of an alkali metal salt is introduced into the sample. Explain.

Chapter 9

Atomic Absorption and Atomic Fluorescence Spectrometry

In this chapter we consider two types of optical atomic spectrometric methods that employ similar techniques for sample introduction and atomization. The first is atomic absorption spectrometry (AAS), which for nearly half a century has been the most widely used method for the determination of single elements in analytical samples. The second is atomic fluorescence spectrometry (AFS), which since the mid-1960s has been studied extensively. In contrast to the absorption method, however, atomic fluorescence has not gained widespread general use for routine elemental analysis. Thus, while several instrument makers currently offer a variety of atomic absorption spectrometers, no atomic fluorescence instruments are marketed commercially at the present time. Because of this difference in usage, we devote the bulk of this chapter to atomic absorption spectrometry[1] and confine our description of atomic fluorescence spectrometry to a brief section at the end.

9A SAMPLE ATOMIZATION TECHNIQUES

In this section we first describe in some detail the two most common methods of sample atomization encountered in atomic absorption and atomic fluorescence spectrometry: flame atomization and electrothermal atomization. We then turn to three rather specialized atomization procedures that are also used in both types of spectrometry.

9A-1 Flame Atomization

In a flame atomizer, a solution of the sample is nebulized by a flow of gaseous oxidant, mixed with a gaseous fuel, and carried into a flame where atomization occurs. As shown in Figure 9-1, a complex set of

---

[1] General references on atomic absorption spectrometry include: L. H. J. Lajunen, *Spectrochemical Analysis by Atomic Absorption and Emission*. Cambridge: Royal Society of Chemistry, 1992; B. Magyar, *Guide-Lines to Planning Atomic Spectrometric Analysis*. New York: Elsevier, 1982; J. D. Ingle Jr. and S. R. Crouch, *Spectrochemical Analysis*, Chapter 10. Englewood Cliffs, NJ: Prentice Hall, 1988; B. Welz, *Atomic Absorption Spectrometry*, 2nd ed. New York: VCH Publishers, 1985. For a history of the development of AAS, see S. R. Koirtyohann, *Anal. Chem.*, **1991**, *63*, 1024A; W. Slavin, *Anal. Chem.*, **1991**, *63*, 1033A

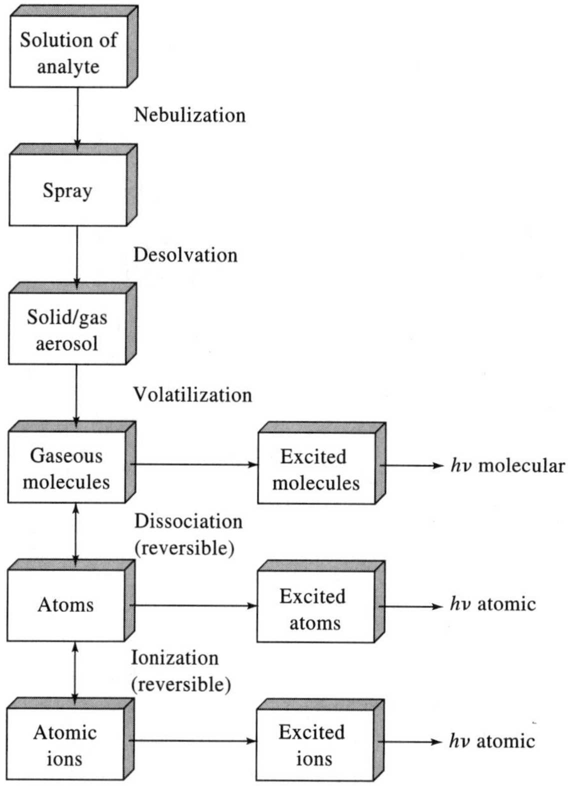

Figure 9-1 Processes occurring during atomization.

interconnected processes then occur in the flame. The first is desolvation, in which the solvent is evaporated to produce a finely divided solid *molecular* aerosol. Dissociation of most of these molecules then leads to an atomic gas. Some of the atoms so formed ionize to give cations and electrons. Undoubtedly other molecules and atoms are also produced in the flame as a result of interactions of the fuel with the oxidant and with the various species in the sample. As indicated in Figure 9-1, a fraction of the molecules, atoms, and ions are also excited by the heat of the flame, thus giving atomic, ionic, and molecular emission spectra. With so many complex processes occurring it is not surprising that atomization is the most critical step in flame spectroscopy and the one that limits the precision of such methods. Because of the critical nature of the atomization step, it is important to understand the characteristics of flames and the variables that affect these characteristics.

Types of Flames

Table 9-1 lists the common fuels and oxidants employed in flame spectroscopy and the approximate range of temperatures realized with each of these mixtures. Note that temperatures of 1700°C to 2400°C are obtained with the various fuels when air serves as the oxidant. At these temperatures, only easily decomposed samples are atomized. For more refractory samples, oxygen or nitrous oxide must be employed as the oxidant. With the common fuels these oxidants produce temperatures of 2500°C to 3100°C.

The burning velocities listed in the fourth column of Table 9-1 are of considerable importance because flames are stable in certain ranges of gas flow rates only. If the gas flow rate does not exceed the burning velocity, the flame propagates itself back into the burner, giving flashback. As the flow rate increases, the flame rises until it reaches a point above the burner where the flow

TABLE 9-1 Properties of Flames

Fuel	Oxidant	Temperatures, °C	Maximum Burning Velocity (cm s^{-1})
Natural gas	Air	1700–1900	39–43
Natural gas	Oxygen	2700–2800	370–390
Hydrogen	Air	2000–2100	300–440
Hydrogen	Oxygen	2550–2700	900–1400
Acetylene	Air	2100–2400	158–266
Acetylene	Oxygen	3050–3150	1100–2480
Acetylene	Nitrous oxide	2600–2800	285

velocity and the burning velocity are equal. This region is where the flame is stable. At higher flow rates, the flame rises and eventually reaches a point where it blows off of the burner. These considerations emphasize the importance of controlling the flow rate of the fuel/oxidant mixture. This flow rate is highly dependent upon the kind of fuel and oxidant being used.

Flame Structure

As shown in Figure 9-2, important regions of a flame include the *primary combustion zone,* the *interzonal region,* and the *secondary combustion zone.* The appearance and relative size of these regions vary considerably with the fuel-to-oxidant ratio as well as with the type of fuel and oxidant. The primary combustion zone in a hydrocarbon flame is recognizable by its blue luminescence arising from the band spectra of C_2, CH, and other radicals. Thermal equilibrium is ordinarily not reached in this region, and it is, therefore, seldom used for flame spectroscopy.

The interzonal area, which is relatively narrow in stoichiometric hydrocarbon flames, may reach several centimeters in height in fuel-rich acetylene/oxygen or acetylene/nitrous oxide sources. The zone is often rich in free atoms and is the most widely used part of the flame for spectroscopy. In the secondary reaction zone, the products of the inner core are converted to stable molecular oxides that are then dispersed into the surroundings.

A flame profile provides useful information about the processes that go on in different parts of a flame; it is a contour plot that reveals regions of the flame that have similar values for a variable of interest. Some of these variables include temperature, chemical composition, absorbance, and radiant or fluorescence intensity.

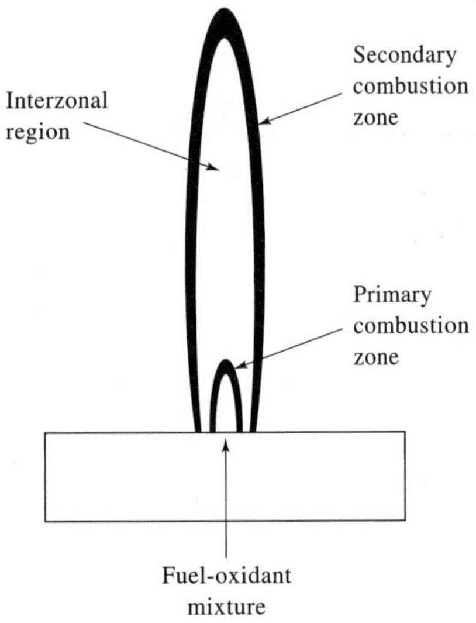

Figure 9-2 Regions in a flame.

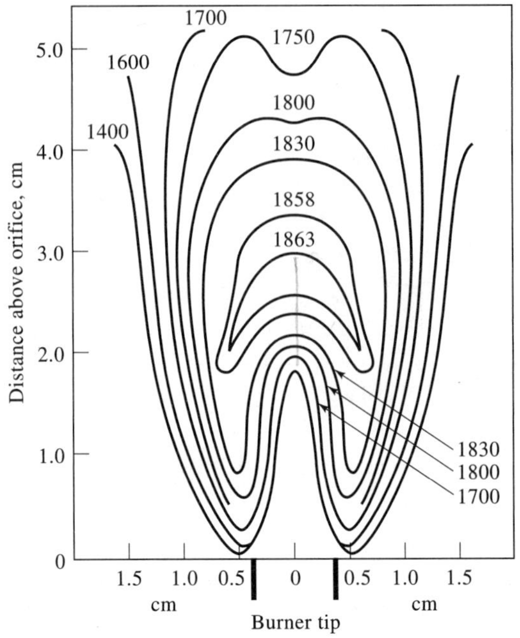

Figure 9-3 Temperature profiles in °C for a natural gas/air flame. *(From B. Lewis and G. vanElbe,* J. Chem. Phys., ***1943,*** 11, *94. With permission.)*

Temperature Profiles. Figure 9-3 is a temperature profile of a typical flame for atomic spectroscopy. The maximum temperature is located in the flame about 1 cm above the primary combustion zone. It is important—particularly for emission methods—to focus the same part of the flame on the entrance slit for all calibrations and analytical measurements.

Flame Absorbance Profiles. Figure 9-4 shows typical absorption profiles for three elements. Magnesium exhibits a maximum in absorbance at about the middle of the flame because of two opposing effects. The initial increase in absorbance as the distance from the base increases results from an increased number of magnesium atoms produced by the longer exposure to the heat of the flame. As the secondary combustion zone is approached, however, appreciable oxidation of the magnesium begins. This process leads to an eventual decrease in absorbance because the oxide particles formed do not absorb at the wavelength used. To obtain maxi-

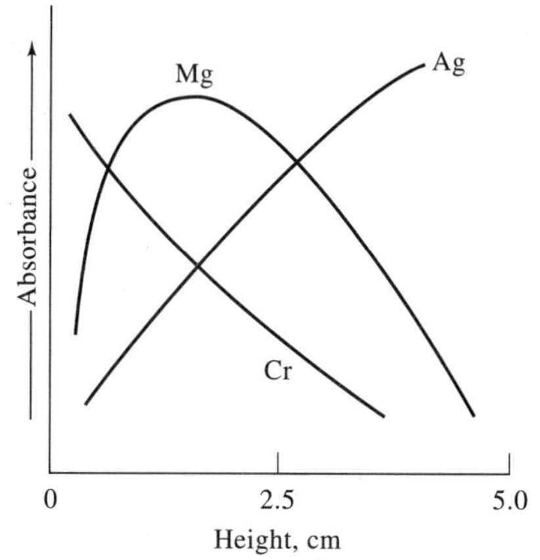

Figure 9-4 Flame absorbance profile for three elements.

mum analytical sensitivity, then, the flame must be moved up and down with respect to the beam until a maximum absorbance is obtained.

The behavior of silver, which is not readily oxidized, is quite different; as shown in Figure 9-4, a continuous increase in the number of atoms, and thus the absorbance, is observed from the base to the periphery of the flame. In contrast, chromium, which forms very stable oxides, shows a continuous decrease in absorbance beginning close to the burner tip; this observation suggests that oxide formation predominates from the start. Clearly, a different portion of the flame should be used for the analysis of each of these elements. The more sophisticated instruments for flame spectroscopy are equipped with monochromators that sample the radiation from a relatively small region of the flame; adjustment of the position of the flame with respect to the entrance slit is thus critical.

Flame Atomizers

Flame atomizers are employed for atomic absorption, fluorescence, and emission spectroscopy. Figure 9-5 is a diagram of a typical commercial laminar flow burner that employs a concentric tube nebulizer, such as that shown in Figure 8-9a. The aerosol, formed by the flow of oxidant, is mixed with fuel and passes a series of baffles that remove all but the finest solution droplets. As a result of the baffles, most of the sample collects in the bottom of the mixing chamber where it drains to a waste container. The aerosol, oxidant, and fuel are then burned in a slotted burner that provides a flame that is usually 5 or 10 cm in length.

Laminar flow burners provide a relatively quiet flame and a long path length. These properties tend to enhance sensitivity and reproducibility. The mixing chamber in this type of burner contains a potentially explosive mixture that can be ignited by flashback if the flow rates are too low. Note that the laminar flow burner in Figure 9-5 is equipped with pressure relief vents for this reason.

Fuel and Oxidant Regulators. An important variable that requires close control in flame spectroscopy is the flow rate of both oxidant and fuel. It is desirable to

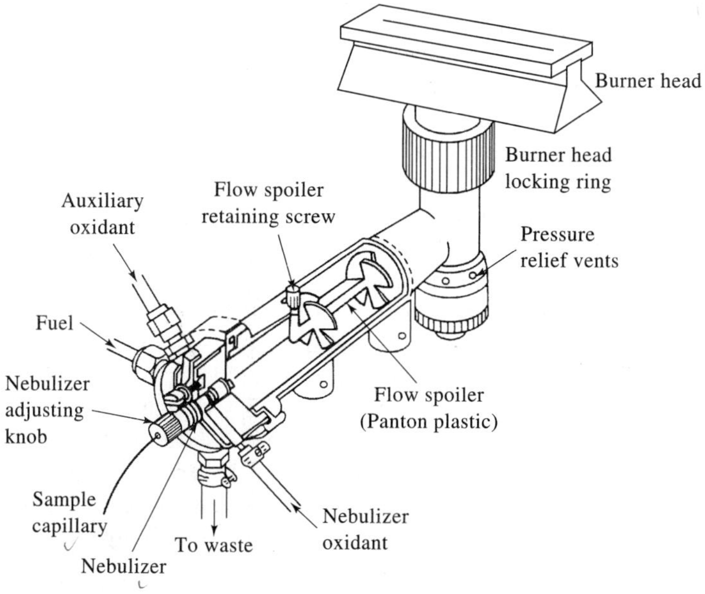

Figure 9-5 A laminar flow burner. *(Courtesy of Perkin-Elmer Corporation, Norwalk, CT.)*

be able to vary each over a considerable range so that optimal atomization conditions can be found experimentally. Fuel and oxidant are ordinarily combined in approximately stoichiometric amounts. For the determination of metals that form stable oxides, however, a flame that contains an excess of fuel may prove more desirable. Flow rates are ordinarily controlled by means of double-diaphragm pressure regulators followed by needle valves in the instrument housing. The most widely used device for measuring flow rates is the rotameter, which consists of a tapered, graduated, transparent tube that is mounted vertically with the smaller end down. A lightweight conical or spherical float is lifted by the gas flow; its vertical position is determined by the flow rate.

Performance Characteristics of Flame Atomizers. In terms of reproducible behavior, flame atomization appears to be superior to all other methods that have been thus far developed for liquid sample introduction for atomic absorption and fluorescence spectrometry. In terms of sampling efficiency and thus sensitivity, however, other atomization methods are markedly better. Two reasons for the lower sampling efficiency of the flame can be cited. First, a large portion of the sample flows down the drain. Second, the residence time of individual atoms in the optical path in the flame is brief ($\approx 10^{-4}$ s).

9A-2 Electrothermal Atomization

Electrothermal atomizers, which first appeared on the market about 1970, generally provide enhanced sensitivity because the entire sample is atomized in a short period, and the average residence time of the atoms in the optical path is a second or more.[2] Electrothermal atomizers are used for atomic absorption and atomic fluorescence measurements, but have not been generally applied for direct production of emission spectra. They are, however, beginning to be used for vaporizing samples for sample introduction in inductively coupled plasma emission spectroscopy.

In electrothermal atomizers, a few microliters of sample are first evaporated at a low temperature and then ashed at a somewhat higher temperature in an electrically heated graphite tube similar to the one in Figure 9-6b or in a graphite cup. After ashing, the current is rapidly increased to several hundred amperes, which causes the temperature to soar to perhaps 2000°C to 3000°C; atomization of the sample occurs in a period of a few milliseconds to seconds. The absorption or fluorescence of the atomized particles is then measured in the region immediately above the heated surface.

Electrothermal Atomizers

Figure 9-6a is a cross-sectional view of a commercial electrothermal atomizer. In this device, atomization occurs in a cylindrical graphite tube that is open at both ends and that has a central hole for introduction of sample by means of a micropipette. The tube is about 5 cm long and has an internal diameter of somewhat less than 1 cm. The interchangeable graphite tube fits snugly into a pair of cylindrical graphite electrical contacts located at the two ends of the tube. These contacts are held in a water-cooled metal housing. Two inert gas streams are provided. The external stream prevents the entrance of outside air and a consequent incineration of the tube. The internal stream flows into the two ends of the tube and out the central sample port. This stream not only excludes air but also serves to carry away vapors generated from the sample matrix during the first two heating stages.

Figure 9-6b illustrates the so-called L'vov platform, which is often used in graphite furnaces such as that shown in Figure 9-6a. The platform is also made of graphite and is located beneath the sample entrance port. The sample is evaporated and ashed on this platform in the usual way. When the tube temperature is raised rapidly, however, atomization is delayed since the sample is no longer directly on the furnace wall. As a consequence, atomization occurs in an environment in which the temperature is not changing so rapidly. More reproducible peaks are obtained as a consequence.

It has been found empirically that some of the sample matrix effects and poor reproducibility associated with graphite furnace atomization can be alleviated by reducing the natural porosity of the graphite tube. During atomization, part of the analyte and matrix apparently diffuse into the tube, which slows the atomization process, thus giving smaller analyte signals. To overcome this effect, most graphite surfaces are coated with a thin layer of pyrolytic carbon, which serves to seal the pores of the graphite tube. Pyrolytic graphite is a type of graphite that has been deposited layer by layer from a

---

[2] For detailed discussions of electrothermal atomizers, see C. W. Fuller, *Electrothermal Atomization for Atomic Absorption Spectroscopy.* London: The Chemical Society, 1978; A. Varma, *CRC Handbook of Furnace Atomic Absorption Spectroscopy.* Boca Raton: CRC Press, 1989. For short reviews, see B. V. L'vov, *Anal. Chem.*, **1991**, *63*, 924A; S. R. Koirtyohann and M. L. Kaiser, *Anal. Chem.*, **1982**, *54*, 1515A; G. R. Carnick and W. Slavin, *Amer. Lab.*, **1988** (11), 88; **1989** (2), 90.

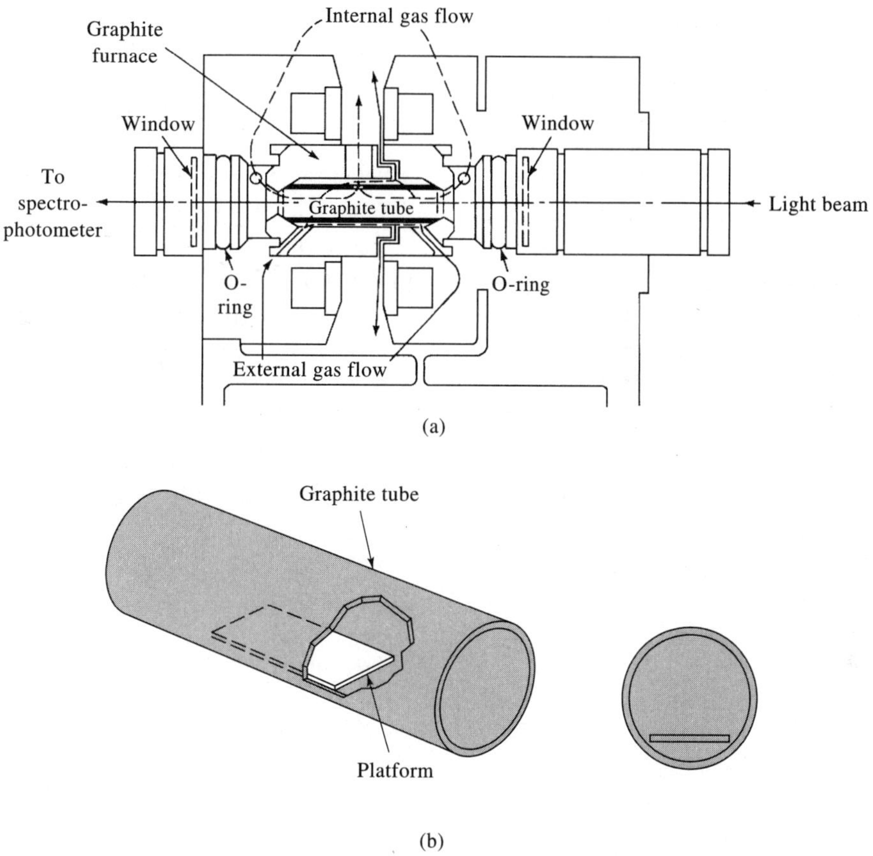

Figure 9-6 (a) Cross-sectional view of a graphite furnace. *(Courtesy of the Perkin-Elmer Corporation, Norwalk, CT.)* (b) The L'vov platform and its position in the graphite furnace. *(Reprinted with permission from W. Slavin,* Anal. Chem., ***1982****, 54, 689A. Copyright 1982 American Chemical Society.)*

highly homogeneous environment. It is formed by passing a mixture of an inert gas and a hydrocarbon such as methane through the tube while it is held at an elevated temperature.

Output Signal

At a wavelength at which absorbance or fluorescence occurs, the transducer output rises to a maximum after a few seconds of ignition followed by a rapid decay back to zero as the atomization products escape into the surroundings. The change is rapid enough (often <1 s) to require a high-speed data acquisition system. Quantitative analyses are usually based on peak height, although peak area has also been used.

Figure 9-7 shows typical output signals from an atomic absorption spectrophotometer equipped with a carbon rod atomizer. The series of four peaks on the right show the absorbance at the wavelength of a lead peak as a function of time when a 2 μL sample of a canned orange juice was atomized. During both drying and ashing, three peaks are produced, which are probably due to molecular evaporation products and particulate ignition products. The three peaks on the left are for lead standards employed for calibration. The sample peak on the far right indicates a lead concentration of about 0.1 μg/mL of juice.

Performance Characteristics of Electrothermal Atomizers

Electrothermal atomizers offer the advantage of unusually high sensitivity for small volumes of sample. Typically, sample volumes between 0.5 and 10 μL are

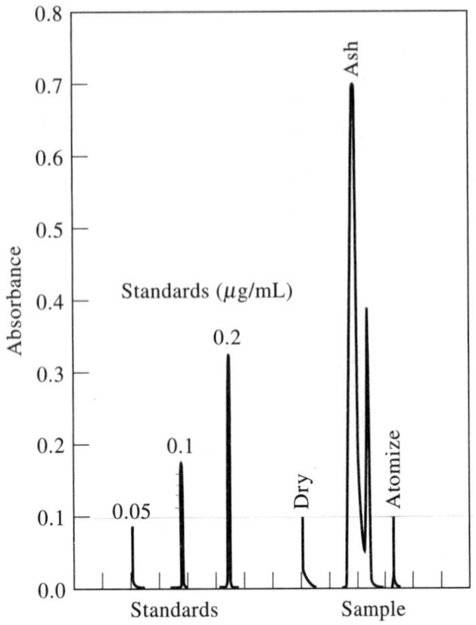

Figure 9-7 Typical output from a spectrophotometer equipped with an electrothermal atomizer. The sample was 2 μL of canned orange juice. The times for drying and ashing are 20 and 60 s, respectively. *(Courtesy of Varian Instrument Division, Palo Alto, CA.)*

used; under these circumstances, absolute detection limits typically lie in the range of 10^{-10} to 10^{-13} g of analyte.[3]

The relative precision of nonflame methods is generally in the range of 5% to 10% compared with the 1% or better that can be expected for flame or plasma atomization. Furthermore, furnace methods are slow—typically requiring several minutes per element. A final disadvantage is that the analytical range is low, being usually less than two orders of magnitude. Consequently, electrothermal atomization is ordinarily applied only when flame or plasma atomization provides inadequate detection limits.

Analysis of Solids with Electrothermal Atomizers

In most methods based upon electrothermal atomizers, samples are introduced as solutions. Several reports, however, have described the use of this type of atomizer for the direct analysis of solid samples. One way of performing such measurements is to weigh the finely ground sample into a graphite boat and insert the boat manually in the furnace. A second way is to prepare a slurry of the powdered sample by ultrasonic agitation in an aqueous medium. The slurry is then pipetted into the furnace for atomization.[4]

9A-3 Specialized Atomization Techniques

By far, the most common sample introduction and atomization technique for atomic absorption analyses are flames or electrothermal vaporizers. Several other atomization methods find occasional use, however. Three of these are described briefly in this section.

Glow Discharge Atomization

As indicated in Section 8C-2 (page 204), a glow discharge device produces an atomized vapor that can be swept into a cell for absorption measurements. Figure 9-8a shows a glow discharge cell that has been marketed since 1987, which can be used as an accessory to most flame absorption spectrometers. It consists of a cylindrical cell about 17 cm in length with a circular hole about 2 cm in diameter cut near the middle of the cylinder. An O-ring surrounds the hole. The sample is pressed against this hole with a torque screw so that it seals the tube. Six fine streams of argon gas from tiny nozzles arranged in a circular pattern above the sample impinge upon the sample surface in a hexagonal pattern. The argon is ionized by a current between an anode supporting the nozzles and the sample, which acts as a cathode. As a result of sputtering, six craters quickly form on the sample surface as shown in Figure 9-8b. The sputtered atoms are drawn by a vacuum to the axis of the cell where they absorb radiation from the spectrometer source.[5]

In order for this technique to be applicable, the sample must be an electrical conductor or must be pelleted with a powdered conductor such as finely ground graphite or copper. Solution samples have also been analyzed by deposition on a graphite, aluminum, or copper cathode. Detection limits with this type of device are reported to be in the low parts per million for solid samples.

---

[3] For a comparison of the performance characteristics of eight modern commercial furnace spectrometers, see *Anal. Chem.*, **1995**, *67*, 51A.

[4] See, for example N. J. Miller-Ihli, *Anal. Chem.*, **1992**, *64*, 964A.

[5] See E. H Piepmeier in *Glow Discharge Spectroscopies*, pp. 69-71, R. K. Marcus, Ed. New York: Plenum Press, 1993.

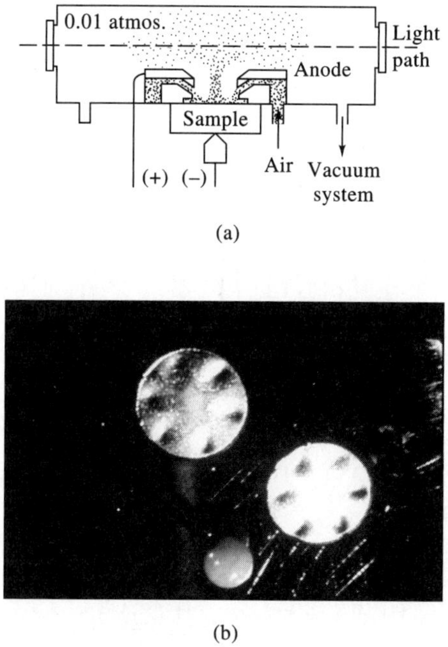

Figure 9-8 (a) Cross section of a cell for glow-discharge atomization of solid samples. (b) Craters formed on sample surface by six jets of ionized argon. *(Courtesy of Leeman Labs, Inc., Bedford, MA.)*

Hydride Atomization

In Section 8C-1 (page 203), methods for introducing solution samples by hydride generation were considered. Atomization of the hydrides requires only that they be heated in a quartz tube as shown in Figure 9-9.

Cold-Vapor Atomization

The cold vapor technique is an atomization method applicable only to the determination of mercury because it is the only metallic element that has an appreciable vapor pressure at ambient temperature.[6] The determination of mercury in various types of samples is of vital importance currently because of the toxicity of various organic mercury compounds and their widespread distribution in the environment.[7] The method of choice for this analysis is cold vaporization followed by atomic absorption spectrophotometry. In performing analyses of this type, mercury is converted to Hg^{2+} by treatment of samples with an oxidizing mixture of nitric and sulfuric acids followed by reduction of the Hg^{2+} to the metal with $SnCl_2$. The elemental mercury is then swept into a long-pass absorption tube by bubbling a stream of inert gas through the mixture from which the element was formed. The analysis is completed by measuring the absorbance at 253.7 nm. Detection limits in the parts-per-billion range are realized. Several manufacturers offer automatic instruments for performing this analysis for a variety of types of samples (see Section 33B-4).

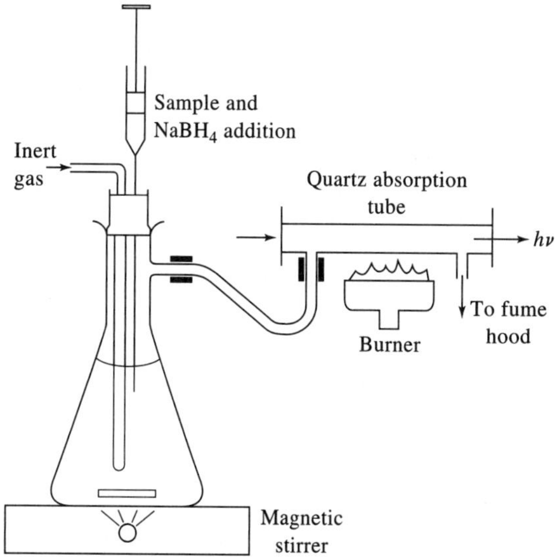

Figure 9-9 A hydride generation and atomization system for atomic absorption spectrometry.

9B ATOMIC ABSORPTION INSTRUMENTATION[8]

Instruments for atomic absorption spectrometry (AAS) are similar in general design to that shown in Figure 7-1a (page 144) and consist of a radiation source, a sample holder, a wavelength selector, a detector, and a

---

[6] See H. J. Lajunen, *Spectrochemical Analysis by Atomic Absorption and Emission*, pp. 117-125. Cambridge: Royal Society of Chemistry, 1992.

[7] See, for example, D. A. Skoog, D. M. West, and F. J. Holler, *Fundamentals of Analytical Chemistry*, 7th ed., pp. 624-631. Philadelphia: Saunders College Publishing, 1996.

[8] Reference books on atomic absorption spectroscopy include: W. Slavin, *Atomic Absorption Spectroscopy*, 2nd ed. New York: Interscience, 1978; B. Welz, *Atomic Absorption Spectroscopy*, 2nd ed. Deerfield Beach, FL: VCH Publishers, 1985; L. H. J. Lajunen, *Spectrochemical Analysis by Atomic Absorption and Emission*. Cambridge, U.K.: Royal Society of Chemistry, 1992; *Atomic Absorption Spectrometry*, J. E. Cantle, Ed. New York: Elsevier, 1982. For short reviews describing the present and future state of atomic spectroscopy, see W. Slavin, *Anal. Chem.*, **1991**, *63*, 1033A; **1986**, *58*, 589A; see also L. de Galan, *Anal. Chem.*, **1986**, *58*, 697A.

signal processor and readout. The sample holder in atomic absorption instruments is the atomizer cell that contains the gaseous atomized sample.

9B-1 Radiation Sources

Analytical methods based on atomic absorption are potentially highly specific because atomic absorption lines are remarkably narrow (0.002 to 0.005 nm) and because electronic transition energies are unique for each element. On the other hand, the limited line widths create a problem not ordinarily encountered in molecular absorption spectroscopy. In Section 13B-2, we show that for there to be a linear relationship between the analytical signal (absorbance) and concentration—that is, for Beer's law as given by Equation 6-34 to be obeyed—it is necessary that the bandwidth of the source must be narrow relative to the width of an absorption peak. Even good-quality monochromators, however, have effective bandwidths that are significantly greater than the width of atomic absorption lines. Consequently, nonlinear calibration curves are inevitable when atomic absorbance measurements are made with an ordinary spectrophotometer equipped with a continuous radiation source. Furthermore, the slopes of calibration curves obtained with such equipment are small because only a small fraction of the radiation from the monochromator slit is absorbed by the sample; poor sensitivities are the consequence.[9]

The problem created by the limited width of atomic absorption peaks has been solved by the use of line sources with bandwidths even narrower than absorption peaks. For example, if the absorbance of the 589.6-nm peak of sodium is to serve as the basis for determining that element, a sodium emission peak at this same wavelength is isolated and used. In this instance, the line is produced by means of a sodium vapor lamp in which sodium atoms are excited by an electrical discharge. The other sodium lines emitted from the source are then removed with filters or with a relatively inexpensive monochromator. Operating conditions for the source are chosen such that Doppler broadening of the emitted lines is less than the broadening of the absorption peak that occurs in the flame or other atomizer.

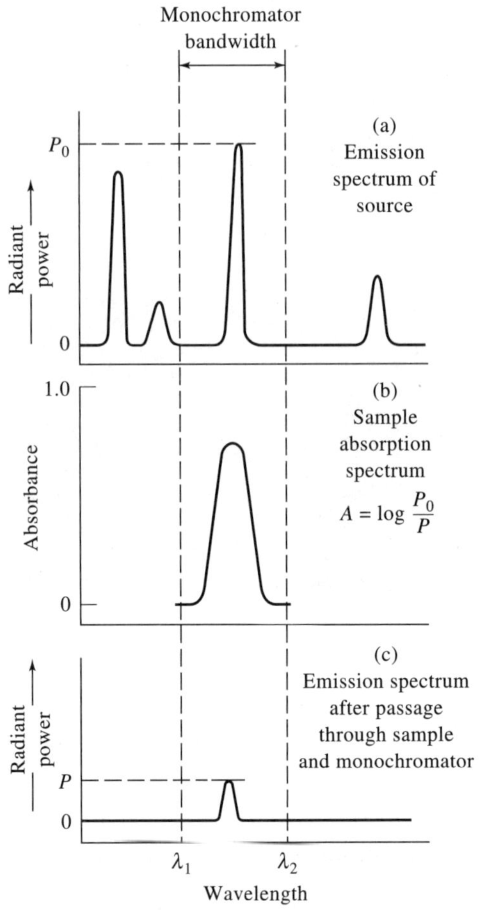

Figure 9-10 Absorption of a resonance line by atoms.

That is, the source temperature is kept below that of the atomizer. Figure 9-10 illustrates the principle of this procedure. Figure 9-10a shows the *emission* spectrum of a typical atomic lamp source, which consists of four narrow lines. With a suitable filter or monochromator, all but one of these lines are removed. Figure 9-10b shows the absorption spectrum for the analyte between wavelengths λ_1 and λ_2. Note that the bandwidth is significantly greater than that of the emission peak. As shown in Figure 9-10c, passage of the line from the source through the flame reduces its intensity from P_0 to P; the absorbance is then given by $\log(P_0/P)$, which is linearly related to the concentration of the analyte in the sample.

A disadvantage of the procedure just described is that a separate lamp source is needed for each element (or sometimes group of elements).

[9] For a paper describing the successful use of a continuum source for atomic absorption spectroscopy, see J. M. Harnly, *Anal. Chem.*, **1986**, *58*, 933A.

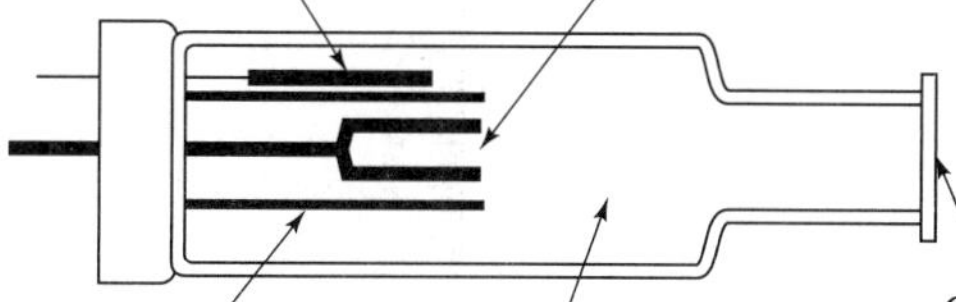

Figure 9-11 Schematic cross section of a hollow cathode lamp.

Hollow Cathode Lamps

The most common source for atomic absorption measurements is the *hollow cathode lamp,* such as the one shown in Figure 9-11.[10] This type of lamp consists of a tungsten anode and a cylindrical cathode sealed in a glass tube that is filled with neon or argon at a pressure of 1 to 5 torr. The cathode is constructed of the metal whose spectrum is desired or serves to support a layer of that metal.

Ionization of the inert gas occurs when a potential on the order of 300 V is applied across the electrodes, which generates a current of about 5 to 15 mA as ions and electrons migrate to the electrodes. If the potential is sufficiently large, the gaseous cations acquire enough kinetic energy to dislodge some of the metal atoms from the cathode surface and produce an atomic cloud in a process called *sputtering.* A portion of the sputtered metal atoms are in excited states and thus emit their characteristic radiation as they return to the ground state. Eventually, the metal atoms diffuse back to the cathode surface or to the glass walls of the tube and are redeposited.

The cylindrical configuration of the cathode tends to concentrate the radiation in a limited region of the metal tube; this design also enhances the probability that redeposition will occur at the cathode rather than on the glass walls.

The efficiency of the hollow cathode lamp depends upon its geometry and the operating potential. High potentials, and thus high currents, lead to greater intensities. This advantage is offset somewhat by an increase in Doppler broadening of the emission lines from the lamp. Furthermore, the greater currents produce an increased number of unexcited atoms in the cloud. The unexcited atoms, in turn, are capable of absorbing the radiation emitted by the excited ones. This *self-absorption* leads to lowered intensities, particularly at the center of the emission band.

A variety of hollow cathode lamps are available commercially. The cathodes of some consist of a mixture of several metals; such lamps permit the determination of more than a single element.

Electrodeless Discharge Lamps

Electrodeless discharge lamps (EDLs) are useful sources of atomic line spectra and provide radiant intensities that are usually one to two orders of magnitude greater than hollow cathode lamps.[11] A typical lamp is constructed from a sealed quartz tube containing a few torr of an inert gas such as argon and a small quantity of the metal (or its salt) whose spectrum is of interest. The lamp contains no electrode but instead is energized by an intense field of radio-frequency or microwave radiation. Ionization of the argon occurs to give ions that are accelerated by the high-frequency component of the field until they gain sufficient energy to excite the atoms of the metal whose spectrum is sought.

Electrodeless discharge lamps are available commercially for 15 or more elements. Their performance does not appear to be as reliable as that of the hollow cathode lamp. Figure 9-12 is a schematic of a commercial electrodeless discharge lamp, which is powered by a 27-MHz radio-frequency source.

Source Modulation

In the typical atomic absorption instrument, it is necessary to eliminate interferences caused by emission of radiation by the flame. Much of this emitted radiation is, of course, removed by the monochromator. Nevertheless, emitted radiation corresponding in wavelength to the monochromator setting is inevitably present in the flame due to excitation and emission by analyte atoms. In order to eliminate the effects of flame emission, it is necessary to modulate the output of the source so that its intensity fluctuates at a constant frequency. The detector then receives two types of signal, an alternating one from the source and a continuous one from

[10] See S. Caroli, *Improved Hollow Cathode Lamps for Atomic Spectroscopy.* New York: Wiley, 1985.

[11] See W. B. Barnett, J. W. Vollmer, and S. M. DeNuzzo, *At. Absorption Newslett.,* **1976,** *15,* 33.

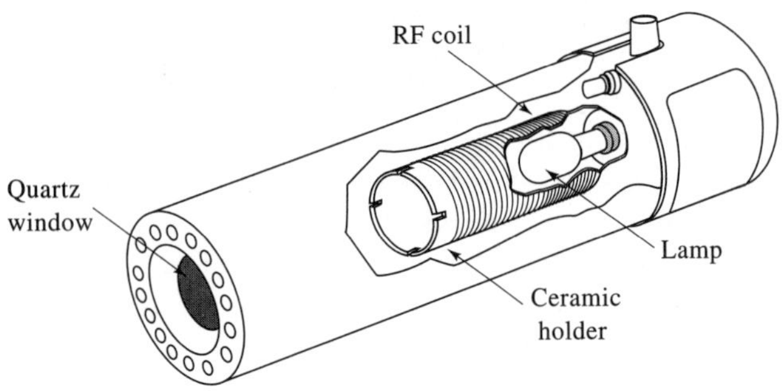

Figure 9-12 Cutaway of electrodeless discharge lamp. *(From W. B. Barnett, J. W. Vollmer, and S. M. DeNuzzo,* At. Absorption Newsletter, ***1976****, 15, 33. With permission.)*

the flame. These signals are converted to the corresponding types of electrical response. A simple high-pass *RC* filter (Section 2B-5) can then be employed to remove the unmodulated dc signal and pass the ac signal for amplification.

A simple and entirely satisfactory way of modulating the emission from the source is to interpose a circular metal disk or chopper in the beam between the source and the flame. Alternate quadrants of this disk are removed to permit passage of light. Rotation of the disk at constant known rate provides a beam that is chopped to the desired frequency. As an alternative, the power supply for the source can be designed for intermittent or ac operation so that the source is switched on and off at the desired constant frequency.

9B-2 Spectrophotometers

Instruments for atomic absorption work are offered by numerous manufacturers; both single- and double-beam designs are available. The range of sophistication and cost (upward from a few thousand dollars) is substantial.

In general, the instrument must be capable of providing a sufficiently narrow bandwidth to isolate the line chosen for the measurement from other lines that may interfere with or diminish the sensitivity of the analysis. A glass filter suffices for some of the alkali metals, which have only a few widely spaced resonance lines in the visible region. An instrument equipped with readily interchangeable interference filters is available commercially. A separate filter and light source is used for each element. Satisfactory results for the analysis of 22 metals are claimed. Most instruments, however, incorporate good-quality ultraviolet/visible monochromators, many of which are capable of achieving a bandwidth on the order of 1 Å.

Most atomic absorption instruments use photomultiplier tubes, which were described in Section 7E-2, as transducers. As pointed out earlier, electronic systems that are capable of discriminating between the modulated signal from the source and the continuous signal from the flame are required. Most instruments currently on the market are equipped with microcomputer systems that are used to control instrument parameters and to control and manipulate data.

Single-Beam Instruments

A typical single-beam instrument, such as that shown in Figure 9-13a, consists of several hollow cathode sources (only one of which is shown), a chopper or a pulsed power supply, an atomizer, and a simple grating spectrophotometer with a photomultiplier transducer. It is used in the way described on page 140. Thus, the dark current is nulled with a shutter in front of the transducer. The 100% T adjustment is then made while a blank is aspirated into the flame, or ignited in a nonflame atomizer. Finally, the transmittance is obtained with the sample replacing the blank.

Double-Beam Instruments

Figure 9-13b is a schematic of a typical double-beam in time instrument. The beam from the hollow cathode source is split by a mirrored chopper, one half passing through the flame and the other half around it. The two beams are then recombined by a half-silvered mirror and passed into a Czerney-Turner grating monochromator; a photomultiplier tube serves as the transducer. The output from the latter is fed to a lock-in amplifier that is synchronized with the chopper drive. The ratio between the reference and sample signal is then amplified and fed to the readout, which may be a digital meter or a signal recorder.

It should be noted that the reference beam in atomic double-beam instruments does not pass through the flame and thus does not correct for loss of radiant power due to absorption or scattering by the flame itself. Methods of correcting for these losses are discussed in the next section.

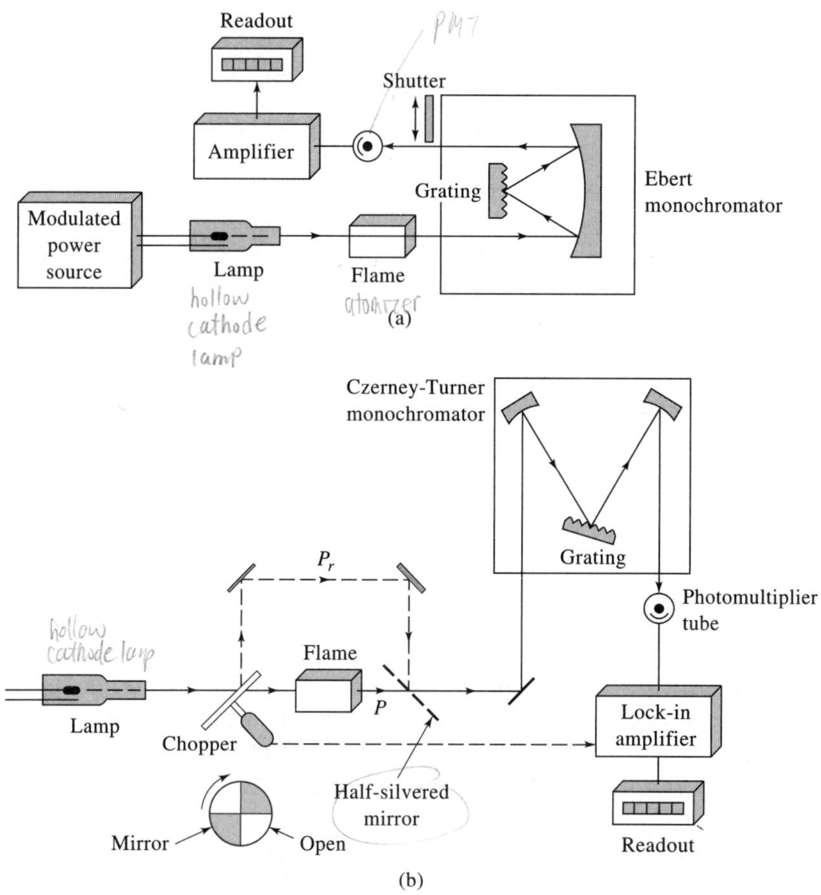

Figure 9-13 Typical flame spectrophotometers: (a) single-beam design and (b) double-beam design.

9C INTERFERENCES IN ATOMIC ABSORPTION SPECTROSCOPY

Interferences of two types are encountered in atomic absorption methods. *Spectral interferences* arise when the absorption or emission of an interfering species either overlaps or lies so close to the analyte absorption or emission that resolution by the monochromator becomes impossible. *Chemical interferences* result from various chemical processes occurring during atomization that alter the absorption characteristics of the analyte.

9C-1 Spectral Interferences

Because the emission lines of hollow cathode sources are so very narrow, interference due to overlapping lines is rare. For such an interference to occur, the separation between the two lines would have to be less than about 0.1 Å. For example, a vanadium line at 3082.11 Å interferes in an analysis based upon the aluminum absorption line at 3082.15 Å. The interference is readily avoided, however, by employing the aluminum line at 3092.7 Å instead.

Spectral interferences also result from the presence of combustion products that exhibit broadband absorption or particulate products that scatter radiation. Both diminish the power of the transmitted beam and lead to positive analytical errors. Where the source of these products is the fuel and oxidant mixture alone, corrections are readily obtained by making absorbance measurements while a blank is aspirated into the flame. Note that this correction must be employed with a double-beam as well as a single-beam instrument because the reference beam of the former does not pass through the flame (see Figure 9-13b).

A much more troublesome problem is encountered when the source of absorption or scattering originates in the sample matrix. In this instance, the power of the transmitted beam P is reduced by the matrix components, but the incident beam power P_r is not; a positive error in absorbance and thus concentration results. An example of a potential matrix interference due to absorption occurs in the determination of barium in alkaline-earth mixtures. As shown by the dashed line in Figure 8-8, the wavelength of the barium line used for atomic absorption analysis appears in the center of a broad absorption band for CaOH; clearly, interference of calcium in a barium analysis is to be expected. In this particular situation, the effect is readily eliminated by substituting nitrous oxide for air as the oxidant, which yields a higher temperature that decomposes the CaOH and eliminates the absorption band.

Spectral interference due to scattering by products of atomization is most often encountered when concentrated solutions containing elements such as Ti, Zr, and W—which form refractory oxides—are aspirated into the flame. Metal oxide particles with diameters greater than the wavelength of light appear to be formed, and scattering of the incident beam results.

Interference due to scattering may also be a problem when the sample contains organic species or when organic solvents are employed to dissolve the sample. Here, incomplete combustion of the organic matrix leaves carbonaceous particles that are capable of scattering light.

Fortunately, with flame atomization, spectral interferences by matrix products are not widely encountered and often can be avoided by variations in the analytical parameters, such as temperature and fuel-to-oxidant ratio. Alternatively, if the source of interference is known, an excess of the interfering substance can be added to both sample and standards. Provided the excess added to the standard sample is large with respect to the concentration from the sample matrix, the contribution from the sample matrix will become insignificant. The added substance is sometimes called a *radiation buffer.*

In the past, the matrix interference problem was severe with electrothermal atomization. Recent developments in platform technology, new high-quality graphite materials, fast photometric instrumentation, and Zeeman background correction are claimed to reduce this type of interference to the level encountered with flames.[12] Several methods that have been developed for correcting for spectral interferences caused by matrix products are discussed in the two references in footnote 13.

The Two-Line Correction Method

The two-line correction procedure employs a line from the source as a reference. This line should lie as close as possible to the analyte line but *must not be absorbed by the analyte.* If these conditions are met, it is assumed that any decrease in power of the reference line from that observed during calibration arises from absorption or scattering by the matrix products of the sample. This decrease in power is then used to correct the absorbance of the analyte line.

The reference line may be from an impurity in the lamp cathode, a neon or argon line from the gas contained in the lamp, or a nonresonant emission line of the element that is being determined. Unfortunately, a suitable reference line is often not available.

The Continuum-Source Correction Method

Figure 9-14 illustrates a second method for background corrections that is widely used. In this technique, a deuterium lamp provides a source of continuous radiation throughout the ultraviolet region. The configuration of the chopper is such that radiation from the continuum source and the hollow cathode lamp are passed alternately through the graphite-tube atomizer. The absorbance of the deuterium radiation is then subtracted from that of the analyte beam. The slit width is kept sufficiently wide so that the fraction of the continuum source that is absorbed by the atoms of the sample is negligible. Therefore, the attenuation of the power of the continuum during passage through the atomized sample reflects only the broadband absorption or scattering by the sample matrix components. A background correction is thus achieved.

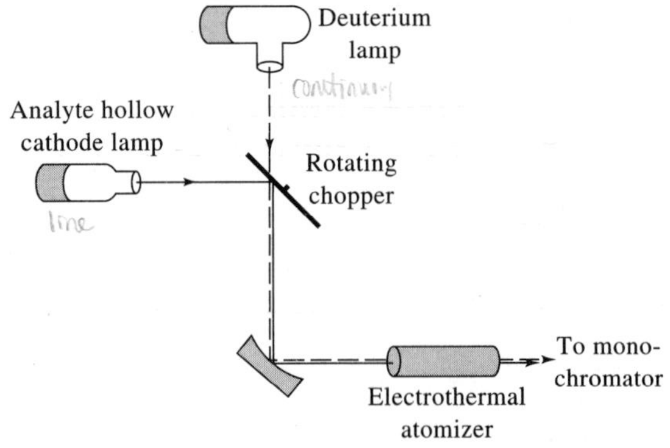

Figure 9-14 Schematic of a continuum-source background correction system. Note that the chopper can be dispensed with by alternately pulsing each lamp.

[12] See W. Slavin, *Anal. Chem.,* **1986,** *58,* 590A.

[13] For a critical discussion of the various methods for background correction, see A. T. Zander, *Amer. Lab.,* 1976, (11), 11; J. J. Sotera and H. L. Kahn, *Amer. Lab.,* **1982,** (11), 100.

Unfortunately, although most instrument manufacturers offer continuum-source background correction systems, the performance of these devices is often less than ideal, which leads to undercorrection in some systems and overcorrection in others. One of the sources of error is the inevitable degradation of signal-to-noise that accompanies the addition of a lamp and chopper. Another is that the hot gaseous media are usually highly inhomogeneous both as to chemical composition and in particulate distribution; thus if the two lamps are not in perfect alignment, an erroneous correction will result that can cause either a positive or a negative error. Finally, the radiant output of the deuterium lamp in the visible region is low enough to preclude the use of this correction procedure for wavelengths greater than about 350 nm.

Background Correction Based on the Zeeman Effect[14]

When an atomic vapor is exposed to a strong magnetic field ($\approx$10 kG) a splitting of electronic energy levels of the atoms takes place that leads to formation of several absorption lines for each electronic transition. These lines are separated from one another by about 0.01 nm, with the sum of the absorbances for the lines being exactly equal to that of the original line from which they were formed. This phenomenon, which is termed the *Zeeman effect,* is general for all atomic spectra. Several splitting patterns arise depending upon the type of electronic transition that is involved in the absorption process. The simplest splitting pattern, which is observed with singlet (page 194) transitions, leads to a central or π line and two equally spaced satellite σ lines. The central line, which is at the original wavelength, has an absorbance that is twice that of each σ line. For more complex transitions, further splitting of the π and σ lines occurs.

Application of the Zeeman effect to atomic absorption instruments is based upon the differing response of the two types of absorption peaks to polarized radiation. The π peak absorbs only that radiation that is plane polarized in a direction parallel to the external magnetic field; the σ peaks, in contrast, absorb only radiation polarized at 90 deg to the field.

Figure 9-15 shows details of an electrothermal atomic absorption instrument, which utilizes the Zeeman effect for background correction. Unpolarized radiation from an ordinary hollow cathode source *A* is passed through a rotating polarizer *B,* which separates the beam into two components that are plane-polarized at 90 deg to one another *C.* These beams pass into a tube-type graphite furnace similar to the one shown in Figure 9-6a. A permanent 11-kG magnet surrounds the furnace and splits the energy levels in such a way as to produce the three absorption peaks shown in *D.* Note that the central peak absorbs only that radiation that is

[14] For a detailed discussion of the application of the Zeeman effect to atomic absorption, see S. D. Brown, *Anal. Chem.,* **1977,** *49* (14), 1269A; F. J. Fernandez, S. A. Myers, and W. Slavin, *Anal. Chem.,* **1980,** *52,* 741.

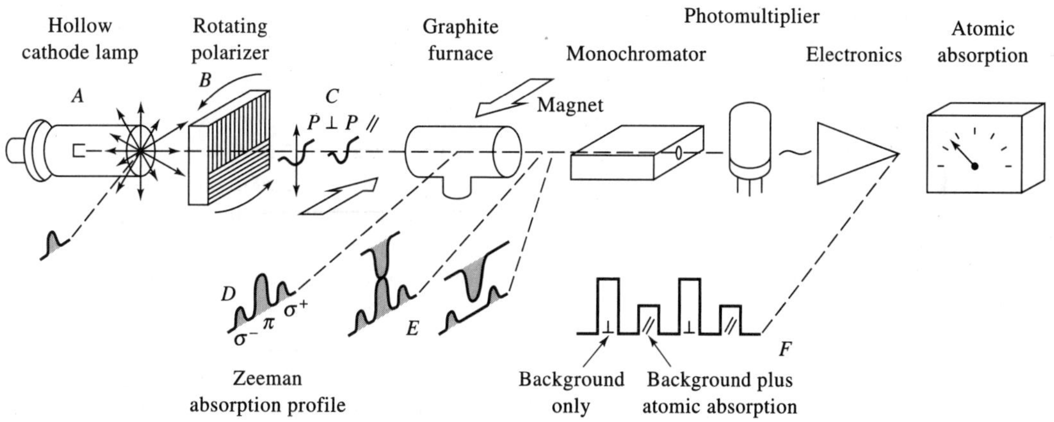

Figure 9-15 Schematic of an electrothermal atomic absorption instrument that provides a background correction based upon the Zeeman effect. *(Courtesy of Hitachi Scientific Instruments, Mountain View, CA.)*

plane polarized with the field. During that part of the cycle when the source radiation is polarized similarly, absorption of radiation by the analyte takes place. During the other half cycle, no analyte absorption can occur. Broad-band molecular absorption and scattering by the matrix products occur during both half cycles, which leads to the cyclical absorbance pattern shown in *F.* The data acquisition system is programmed to subtract the absorbance during the perpendicular half cycle from that for the parallel half cycle, thus giving a background corrected value.

A second type of Zeeman effect instrument has been designed in which a magnet surrounds the hollow cathode source. Here, it is the *emission* spectrum of the source that is split rather than the absorption spectrum of the sample. This instrument configuration provides an analogous correction. To date, most instruments are of the type illustrated in Figure 9-15.

Zeeman effect instruments provide a more accurate correction for background than the methods described earlier. These instruments are particularly useful for electrothermal atomizers and permit the direct determination of elements in samples such as urine and blood. The decomposition of organic material in these samples leads to large background corrections (background $A \geq 1$) and consequent susceptibility to significant error.

Background Correction Based on Source Self-Reversal

A remarkably simple means of background correction is now being marketed and appears to offer most of the advantages of a Zeeman effect instrument.[15] This method, which is sometimes called the *Smith-Hieftje background correction method,* is based upon the self-reversal or self-absorption behavior of radiation emitted from hollow cathode lamps when they are operated at high currents. As was mentioned earlier, high currents produce large concentrations of nonexcited atoms, which are capable of absorbing the radiation produced from the excited species. An additional effect of the high currents is to significantly broaden the emission band of the excited species. The net effect is to produce a band that has a minimum in its center, which corresponds exactly in wavelength to that of the absorption peak (see Figure 9-16).

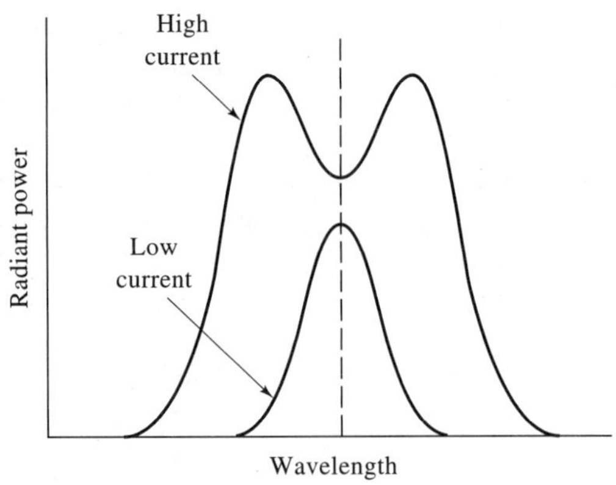

Figure 9-16 Emission line profiles for a hollow-cathode lamp operated at high and low currents.

In order to obtain corrected absorbances, the lamp is programmed to run alternately at low and high currents. The total absorbance is obtained during the low-current operation and the absorbance due to background is provided by measurements during the second part of the cycle when radiation at the absorption peak is at a minimum. The data acquisition system then subtracts the background absorbance from the total to give a corrected value. Recovery of the source to its low-current output takes place in milliseconds when the current is reduced. The measurement cycle can be repeated often enough to give satisfactory signal-to-noise ratios. Equipment for this type of correction is available from commercial sources.

9C-2 Chemical Interferences

Chemical interferences are more common than spectral ones. Their effects can frequently be minimized by a suitable choice of operating conditions.

Both theoretical and experimental evidence suggest that many of the processes occurring in the mantle of a flame are in approximate equilibrium. As a consequence, it becomes possible to regard the burned gases as a solvent medium to which thermodynamic calculations can be applied. The equilibria of principal interest include formation of compounds of low volatility, dissociation reactions, and ionization.

Formation of Compounds of Low Volatility

Perhaps the most common type of interference is by anions that form compounds of low volatility with the analyte and thus reduce the rate at which the analyte is at-

[15] See S. B. Smith Jr. and G. M. Hieftje, *Appl. Spectrosc.,* **1983,** *37,* 419; *Science,* **1983,** *220,* 183.

omized. Low results are the consequence. An example is the decrease in calcium absorbance that is observed with increasing concentrations of sulfate or phosphate. For example, at a fixed calcium concentration, the absorbance falls off nearly linearly with increasing sulfate or phosphate concentrations until the anion-to-calcium ratio is about 0.5; the absorbance then levels off at about 30% to 50% of its original value and becomes independent of anion concentration.

Examples of cation interference have also been recognized. Thus, aluminum is found to cause low results in the determination of magnesium, apparently as a result of the formation of a heat-stable aluminum/magnesium compound (perhaps an oxide).

Interferences due to formation of species of low volatility can often be eliminated or moderated by use of higher temperatures. Alternatively, *releasing agents,* which are cations that react preferentially with the interferant and prevent its interaction with the analyte, can be employed. For example, addition of an excess of strontium or lanthanum ion minimizes the interference of phosphate in the determination of calcium. The same two species have also been employed as releasing agents for the determination of magnesium in the presence of aluminum. In both instances, the strontium or lanthanum replaces the analyte in the compound formed with the interfering species.

Protective agents prevent interference by forming stable but volatile species with the analyte. Three common reagents for this purpose are EDTA, 8-hydroxyquinoline, and APDC, which is the ammonium salt of 1-pyrrolidinecarbodithioic acid. The presence of EDTA has been shown to eliminate the interference of aluminum, silicon, phosphate, and sulfate in the determination of calcium. Similarly, 8-hydroxyquinoline suppresses the interference of aluminum in the determination of calcium and magnesium.

Dissociation Equilibria

In the hot, gaseous environment of a flame or a furnace, numerous dissociation and association reactions lead to conversion of the metallic constituents to the elemental state. It seems probable that at least some of these reactions are reversible and can be treated by the laws of thermodynamics. Thus, it should be possible to formulate equilibrium such as

$$MO \rightleftharpoons M + O$$

$$M(OH)_2 \rightleftharpoons M + 2OH$$

where M is the analyte atom.

In practice, not enough is known about the nature of the chemical reactions in a flame to permit a quantitative treatment such as that for an aqueous solution. Instead, reliance must be placed on empirical observations.

Dissociation reactions involving metal oxides and hydroxides play an important part in determining the nature of the emission or absorption spectra for an element. For example, the alkaline-earth oxides are relatively stable, with dissociation energies in excess of 5 eV. Molecular bands arising from the presence of metal oxides or hydroxides in the flame thus constitute a prominent feature of their spectra (see Figure 8-8). Except at very high temperatures, these bands are more intense than the lines for the atoms or ions. In contrast, the oxides and hydroxides of the alkali metals are much more easily dissociated so that line intensities for these elements are high, even at relatively low temperatures.

It seems probable that dissociation equilibria that involve anions other than oxygen may also influence flame emission and absorption. For example, the line intensity for sodium is markedly decreased by the presence of HCl. A likely explanation is the mass-action effect on the equilibrium

$$NaCl \rightleftharpoons Na + Cl$$

Chlorine atoms formed from the added HCl decrease the atomic sodium concentration and thereby lower the line intensity.

Another example of this type of interference involves the enhancement of the absorption by vanadium when aluminum or titanium is present. The interference is significantly more pronounced in fuel-rich flames than in lean flames. These effects are readily explained by assuming that the three metals interact with such species as O and OH, which are always present in flames. If the oxygen-bearing species are given the general formula Ox, a series of equilibrium reactions can be postulated. Thus,

$$VOx \rightleftharpoons V + Ox$$

$$AlOx \rightleftharpoons Al + Ox$$

$$TiOx \rightleftharpoons Ti + Ox$$

In fuel-rich combustion mixtures, the concentration of Ox is sufficiently small that its concentration is lowered significantly when aluminum or titanium is present in the sample. This decrease causes the first equilibrium to shift to the right with an accompanying increase in metal concentration as well as absorbance. In lean mixtures, on the other hand, the concentration of Ox is apparently high relative to the total concentration of metal atoms. Thus, addition of aluminum or titanium scarcely changes the concentration of Ox. Therefore, the position of the first equilibrium is not disturbed significantly.

Ionization Equilibria

Ionization of atoms and molecules is small in combustion mixtures that involve air as the oxidant, and generally can be neglected. In higher temperatures of flames where oxygen or nitrous oxide serves as the oxidant, however, ionization becomes important, and a significant concentration of free electrons exists as a consequence of the equilibrium

$$M \rightleftharpoons M^+ + e^- \qquad (9\text{-}1)$$

where M represents a neutral atom or molecule and M^+ is its ion. We will focus upon equilibria in which M is a metal atom.

The equilibrium constant K for this reaction takes the form

$$K = \frac{[M^+][e^-]}{[M]} \qquad (9\text{-}2)$$

If no other source of electrons is present in the flame, this equation can be written in the form

$$K = \left(\frac{\alpha^2}{1-\alpha}\right)p$$

where α is the fraction of M that is ionized, and p is the partial pressure of the metal in the gaseous solvent before ionization.

Table 9-2 shows the calculated fraction ionized for several common metals under conditions that approximate those used in flame emission spectroscopy. The temperatures correspond roughly to conditions that exist in air/acetylene and oxygen/acetylene flames, respectively.

It is important to appreciate that treatment of the ionization process as an equilibrium—with free electrons as one of the products—immediately implies that the degree of ionization of a metal will be strongly influenced by the presence of other ionizable metals in the flame. Thus, if the medium contains not only species M but species B as well, and if B ionizes according to the equation

$$B \rightleftharpoons B^+ + e^-$$

TABLE 9-2 Degree of Ionization of Metals at Flame Temperatures*

Element	Ionization Potential, eV	Fraction Ionized at the Indicated Pressure and Temperature: $p = 10^{-4}$ atm, 2000 K	$p = 10^{-4}$ atm, 3500 K	$p = 10^{-6}$ atm, 2000 K	$p = 10^{-6}$ atm, 3500 K
Cs	3.893	0.01	0.86	0.11	>0.99
Rb	4.176	0.004	0.74	0.04	>0.99
K	4.339	0.003	0.66	0.03	0.99
Na	5.138	0.0003	0.26	0.003	0.90
Li	5.390	0.0001	0.18	0.001	0.82
Ba	5.210	0.0006	0.41	0.006	0.95
Sr	5.692	0.0001	0.21	0.001	0.87
Ca	6.111	3×10^{-5}	0.11	0.0003	0.67
Mg	7.644	4×10^{-7}	0.01	4×10^{-6}	0.09

*Data from B. L. Vallee and R. E. Thiers, in *Treatise on Analytical Chemistry,* I. M. Kolthoff and P. J. Elving, Eds., Part I, Vol. 6, p. 3500. New York: Interscience, 1965. Reprinted with permission of John Wiley & Sons, Inc.

then the degree of ionization of M will be decreased by the mass-action effect of the electrons formed from B. Determination of the degree of ionization under these conditions requires a calculation involving the dissociation constant for B and the mass-balance expression

$$[e^-] = [B^+] + [M^+]$$

The presence of atom/ion equilibria in flames has a number of important consequences in flame spectroscopy. For example, the intensity of atomic emission or absorption lines for the alkali metals, particularly potassium, rubidium, and cesium, is affected by temperature in a complex way. Increased temperatures cause an increase in the population of excited atoms, according to the Boltzmann relationship (Equation 8-1). Counteracting this effect, however, is a decrease in concentration of atoms as a result of ionization. Thus, under some circumstances a decrease in emission or absorption may be observed in hotter flames. It is for this reason that lower excitation temperatures are usually specified for the analysis of alkali metals.

The effects of shifts in ionization equilibria can frequently be eliminated by addition of an *ionization suppressor,* which provides a relatively high concentration of electrons to the flame; suppression of ionization of the analyte results. The effect of a suppressor is demonstrated by the calibration curves for strontium shown in Figure 9-17. Note the marked increase in slope of these curves as strontium ionization is repressed by the increasing concentration of potassium ions and electrons. Note also the enhanced sensitivity that results from the use of nitrous oxide instead of air as the oxidant; the higher temperature achieved with nitrous oxide undoubtedly enhances the rate of decomposition and volatilization of the strontium compounds in the plasma.

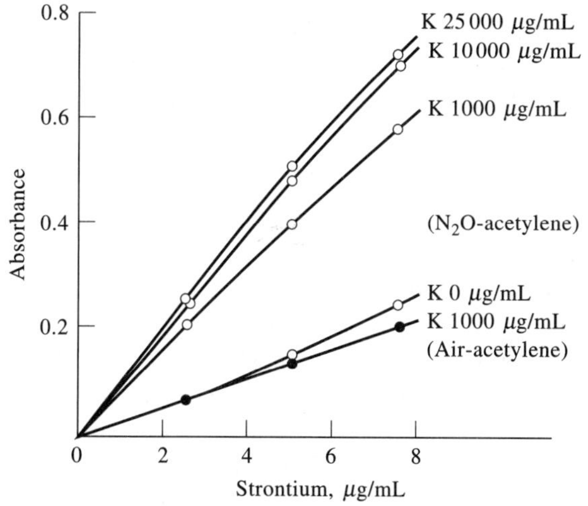

Figure 9-17 Effect of potassium concentration on the calibration curve for strontium. *(Reprinted with permission from J. A. Bowman and J. B. Willis,* Anal. Chem., *1967,* 39, *1220. Copyright 1967 American Chemical Society.)*

April 21, 05

9D ATOMIC ABSORPTION ANALYTICAL TECHNIQUES

This section deals with some of the practical details that must be considered in carrying out an analysis based upon flame or electrothermal atomic absorption.

9D-1 Sample Preparation

A disadvantage of flame spectroscopic methods is the requirement that the sample be introduced into the excitation source in the form of a solution, most commonly an aqueous one. Unfortunately, many materials of interest, such as soils, animal tissues, plants, petroleum products, and minerals are not directly soluble in common solvents, and extensive preliminary treatment is often required to obtain a solution of the analyte in a form ready for atomization. Indeed, the decomposition and solution steps are often more time consuming and introduce more error than the spectroscopic measurement itself.

Decomposition of materials such as those just cited usually require rigorous treatment of the sample at high temperatures with a concomitant potential for loss of the analyte by volatilization or as particulates in a smoke. Furthermore, the reagents used in decomposing a sample often introduce the kinds of chemical and spectral interferences that were discussed earlier. Additionally, the analyte may be present in these reagents as an impurity. In fact, unless considerable care is taken, it is not uncommon in trace analyses to find that reagents are a larger source of the analyte than the samples—a situation that can lead to serious error even with blank corrections.

Some of the common methods used for decomposing and dissolving samples for atomic absorption methods include treatment with hot mineral acids; oxidation with liquid reagents, such as sulfuric, nitric, or perchloric acids (*wet ashing*); combustion in an oxygen bomb or other closed container to avoid loss of analyte; ashing at a high temperature; and high-temperature fusion with reagents such as boric oxide, sodium carbonate, sodium peroxide, or potassium pyrosulfate.[16]

One of the advantages of electrothermal atomization is that some materials can be atomized directly thus avoiding the solution step. For example, liquid samples such as blood, petroleum products, and organic solvents can be pipetted directly into the furnace for ashing and atomization. Solid samples, such as plant leaves, animal tissues, or some inorganic substances can be weighed directly into cup-type atomizers or into tantalum boats for introduction into tube-type furnaces. Calibration is, however, usually difficult and requires standards that approximate the sample in composition.

9D-2 Organic Solvents

Early in the development of atomic absorption spectroscopy it was recognized that enhanced absorption peaks could be obtained from solutions containing low-molecular-weight alcohols, esters, or ketones. The effect of organic solvents is largely attributable to an increased nebulizer efficiency; the lower surface tension of such solutions results in smaller drop sizes and a consequent increase in the amount of sample that reaches the flame. In addition, more rapid solvent evaporation may also contribute to the effect. Leaner fuel/oxidant ratios must be employed with organic solvents in order to offset the presence of the added organic material. Unfortunately, however, the leaner mixture results in lower flame temperatures and an increase in the possibility for chemical interferences.

A most important analytical application of organic solvents to flame spectroscopy is the use of immiscible solvents such as methyl isobutyl ketone to extract chelates of metallic ions. The resulting extract is then nebulized directly into the flame. Here, the sensitivity is increased not only by the enhancement of absorption peaks due to the solvent but also by the fact that for many systems only small volumes of the organic liquid are required to remove metal ions quantitatively from relatively large volumes of aqueous solution. This procedure has the added advantage that at least part of the matrix components are likely to remain in the aqueous solvent; a reduction in interference often results. Common chelating agents include ammonium pyrrolidinedithiocarbamate, diphenylthiocarbazone (dithizone), 8-hydroxyquinoline, and acetylacetone.

[16] R. Bock, *A Handbook of Decomposition Methods in Analytical Chemistry.* New York: Wiley, 1979.

9D-3 Calibration Curves

In theory, atomic absorption should follow Beer's law (Equation 6-34) with absorbance being directly proportional to concentration. In fact, however, departures from linearity are often encountered, and it is foolhardy to perform an atomic absorption analysis without experimentally determining whether or not a linear relationship does exist. Thus periodically, a calibration curve that covers the range of concentrations found in the sample should be prepared. In addition, the number of uncontrolled variables in atomization and absorbance measurements are sufficiently large to warrant measurement of one standard solution each time an analysis is performed (even better is the use of two standards that bracket the analyte concentration). Any deviation of the standard from the original calibration curve can then be used to correct the analytical result.

9D-4 Standard Addition Method

The standard addition method, which was described in Section 1E-2, is widely used in atomic absorption spectroscopy in order to partially or wholly counteract the chemical and spectral interferences introduced by the sample matrix.

9D-5 Applications of Atomic Absorption Spectrometry

Atomic absorption spectrometry is a sensitive means for the quantitative determination of more than 60 metals or metalloid elements. The resonance lines for the nonmetallic elements are generally located below 200 nm, thus preventing their determination by convenient, nonvacuum spectrophotometers.

Detection Limits

Columns two and three of Table 9-3 provide information on detection limits for a number of common elements by flame and electrothermal atomic absorption. For comparison purposes, detection limits for some of the other atomic procedures are also included. Small differences among the quoted values are not significant. Thus, an order of magnitude is probably meaningful, but a factor of 2 or 3 certainly is not.

For many elements, detection limits for atomic absorption spectroscopy with flame atomization lie in the range of 1 to 20 ng/mL or 0.001 to 0.020 ppm; for elec-

TABLE 9-3 Detection Limits (ng/mL)* for Selected Elements†

Element	AAS‡ Flame	AAS§ Electrothermal	AES‡ Flame	AES‡ ICP	AFS‡ Flame
Al	30	0.005	5	2	5
As	100	0.02	0.0005	40	100
Ca	1	0.02	0.1	0.02	0.001
Cd	1	0.0001	800	2	0.01
Cr	3	0.01	4	0.3	4
Cu	2	0.002	10	0.1	1
Fe	5	0.005	30	0.3	8
Hg	500	0.1	0.0004	1	20
Mg	0.1	0.00002	5	0.05	1
Mn	2	0.0002	5	0.06	2
Mo	30	0.005	100	0.2	60
Na	2	0.0002	0.1	0.2	—
Ni	5	0.02	20	0.4	3
Pb	10	0.002	100	2	10
Sn	20	0.1	300	30	50
V	20	0.1	10	0.2	70
Zn	2	0.00005	0.0005	2	0.02

*Nanogram/milliliter = 10^{-3} μg/mL = 10^{-3} ppm.
†AAS = atomic absorption spectroscopy; AES = atomic emission spectroscopy; AFS = atomic fluorescence spectroscopy; ICP = inductively coupled plasma.
‡Reprinted with permission from V. A. Fassel and R. N. Kniseley, *Anal. Chem.*, **1974,** *46,* 1111A. Copyright 1974 American Chemical Society.
§From C. W. Fuller, *Electrothermal Atomization for Atomic Absorption Spectroscopy,* pp. 65-83. London: The Chemical Society, 1977. With permission. The Royal Society of Chemistry, Burlington House: London.

trothermal atomization, the corresponding figures are 0.002 to 0.01 ng/mL or 2×10^{-6} to 1×10^{-5} ppm. In a few cases, detection limits well outside these ranges are encountered.

Accuracy

Under the usual conditions, the relative error associated with a flame absorption analysis is of the order of 1% to 2%. With special precautions, this figure can be lowered to a few tenths of one percent. Errors encountered with electrothermal atomization usually exceed those for flame atomization by a factor of 5 to 10.

9E ATOMIC FLUORESCENCE SPECTROSCOPY

Beginning in about 1964, a significant research effort has been devoted to the development of analytical methods based upon atomic fluorescence.[17] This work has demonstrated clearly that this technique provides a use-

[17] For further information on atomic fluorescence spectroscopy, see C. Veillon, in *Trace Analysis,* J. D. Winefordner, Ed., Chapter 6. New York: Wiley, 1976; N. Omenetto and J. D. Winefordner, *Prog. Anal. At. Spectroscop.,* **1979,** *2,* 1; J. C. Van Loon, *Anal. Chem.,* **1981,** *53,* 332A; D. Butcher et al., *J. Anal. Atom. Spectrom.,* **1988,** *3,* 1059.

March 20. 05

ful and convenient means for the quantitative determination of a reasonably large number of elements. To date, however, the procedure has not found widespread use because of the overwhelming successes of atomic emission and especially atomic absorption methods, which predated atomic fluorescence by more than a decade. As mentioned earlier, these successes have led to the availability of absorption and emission instruments from numerous commercial sources. Currently, no commercial fluorescence instrument is on the market.

The limited use of atomic fluorescence has not arisen so much from any inherent weakness of the procedure but rather because no distinct advantages have been demonstrated relative to the well-established absorption and emission methods. Thus, while fluorescence methods, particularly those based on electrothermal atomization, are somewhat more sensitive for perhaps five to ten elements, the procedure is also less sensitive and appears to have a smaller useful concentration range for several others. Furthermore, dispersive fluorescence instruments appear to be more complex and potentially more expensive to purchase and maintain.[18]

9E-1 Instrumentation

The components of instruments for atomic fluorescence measurements are generally arranged as shown in Figure 7-1b (page 144). The sample container is most commonly a flame but may also be an electrothermal oven, a glow discharge, or an inductively coupled plasma, as described in Section 10A-1.

Sources

A continuum source would be desirable for atomic fluorescence measurements. Unfortunately, however, the output power of most continuum sources over a region as narrow as an atomic absorption line is so low as to restrict the sensitivity of the method severely.

In the early work on atomic fluorescence, conventional hollow cathode lamps often served as excitation sources. In order to enhance the output intensity without destroying the lamp, it was necessary to operate the lamp with short pulses of current that were greater than the lamp could tolerate for continuous operation. The detector, of course, was gated to observe the fluorescent signal only during pulses.

Perhaps the most widely used sources for atomic fluorescence have been the electrodeless discharge lamps (Section 9B-1), which usually produce radiant intensities that exceed those of hollow cathode lamps by an order of magnitude or two. Electrodeless discharge lamps have been operated in both the continuous and pulsed modes. Unfortunately, this type of lamp is not available for many elements.

Lasers, with their high intensities and narrow bandwidths, are obviously the ideal source for atomic fluorescence measurements. Their high cost, however, has discouraged their widespread application to routine atomic fluorescence methods.

Dispersive Instruments

A dispersive system for atomic fluorescence measurements is made up of a modulated source, an atomizer (flame or nonflame), a monochromator or an interference filter system, a detector, and a signal processor and readout. With the exception of the source, most of these components are similar to those discussed in earlier parts of this chapter.

Nondispersive Instruments

In theory, no monochromator or filter should be necessary for atomic fluorescence measurements when an electrodeless discharge lamp or hollow cathode lamp serves as the excitation source because the emitted radiation is, in principle, that of a single element and will thus excite only atoms of that element. A nondispersive system then could be made up of only a source, an atomizer, and a detector. The advantages of such a system are several: (1) simplicity and low-cost instrumentation, (2) ready adaptability to multielement analysis, (3) high-energy throughput and thus high sensitivity, and (4) simultaneous collection of energy from multiple lines, also enhancing sensitivity.

In order to realize these important advantages, it is necessary that the output of the source be free of contaminating lines from other elements; in addition, the atomizer should emit no significant background radiation. The latter consideration may be realized in some instances with electrothermal atomizers but certainly is not with typical flames. To overcome this latter problem, filters, located between the source and detector, have often been used to remove the bulk of the background radiation. Alternatively, solar-blind, photomultiplier detectors, which respond only to radiation below 320 nm in wavelength, have been applied. In this in-

[18] See W. B. Barnett and H. L. Kahn, *Anal. Chem.*, **1972**, *44*, 935.

stance, the analyte must emit lines with wavelengths shorter than 320 nm.

9E-2 Interferences

Interferences encountered in atomic fluorescence spectroscopy appear to be of the same type and of about the same magnitude as those found in atomic absorption spectroscopy.[19]

[19] See J. D. Winefordner and R. C. Elser, *Anal. Chem.*, **1971**, *43* (4), 24A.

9E-3 Applications

Atomic fluorescence methods have been applied to the analysis of metals in such materials as lubricating oils, seawater, biological substances, graphite, and agricultural samples.[20] Detection limits for atomic fluorescence procedures are found in Table 9-3.

[20] For a summary of applications, see J. D. Winefordner, *J. Chem. Educ.*, **1978**, *55*, 72.

9F QUESTIONS AND PROBLEMS

9-1 Define the following terms: (a) releasing agent, (b) protective agent, (c) ionization suppressor, (d) atomization, (e) pressure broadening, (f) hollow cathode lamp, (g) sputtering, (h) self-absorption, (i) spectral interference, (j) chemical interference, (k) radiation buffer, (l) Doppler broadening.

9-2 Describe the effects that are responsible for the three different absorbance profiles in Figure 9-4 and select three additional elements that are expected to have similar profiles.

9-3 Why is an electrothermal atomizer more sensitive than a flame atomizer?

9-4 Describe how a deuterium lamp can be employed to provide a background correction for an atomic absorption spectrum.

9-5 Why is source modulation employed in atomic absorption spectroscopy?

9-6 For the same concentration of nickel, the height of the absorption peak at 352.4 nm was found to be about 30% greater for a solution that contained 50% ethanol than for an aqueous solution. Explain.

9-7 The emission spectrum of a hollow cathode lamp for molybdenum was found to have a sharp peak at 313.3 nm as long as the lamp current was less than 50 mA. At higher currents, however, the peak developed a cup-like crater at its maximum. Explain.

9-8 A chemist attempts to determine strontium with an atomic absorption instrument equipped with a nitrous oxide–acetylene burner, but the sensitivity associated with the 460.7-nm atomic resonance line is not satisfactory. Suggest at least three things that might be tried to increase sensitivity.

9-9 Why is atomic emission more sensitive to flame instability than atomic absorption or fluorescence?

9-10 Figure 9-1 summarizes many of the processes that take place in a laminar flow burner. With specific reference to the analysis of an aqueous $MgCl_2$ solution, describe the processes that are likely to occur.

9-11 Use Equation 7-13 for the resolving power of a grating monochromator to estimate the theoretical minimum size of a diffraction grating that would provide a profile of an atomic absorption line at 500 nm having a line width of 0.002 nm. Assume that the grating is to be used in the first order and that it has been ruled at 2400 groves/mm.

9-12 For the flame shown in Figure 9-3, calculate the relative intensity of the 766.5-nm emission line for potassium at the following heights above the flame (assume no ionization):
(a) 2.0 cm (b) 3.0 cm (c) 4.0 cm (d) 5.0 cm

9-13 In a hydrogen/oxygen flame, an atomic absorption peak for iron was found to decrease in the presence of large concentrations of sulfate ion.
(a) Suggest an explanation for this observation.
(b) Suggest three possible methods for overcoming the potential interference of sulfate in a quantitative determination of iron.

9-14 For Na and Mg^+ atoms, compare the ratios of the number of particles in the 3*p* excited state to the number in the ground state in
(a) a natural gas/air flame (2100 K).
(b) a hydrogen/oxygen flame (2900 K).
(c) an inductively coupled plasma source (6000 K).

9-15 In higher-temperature sources, sodium atoms emit a doublet with an average wavelength of 1139 nm. The transition responsible is from the 4*s* to 3*p* state. Calculate the ratio of the number of excited atoms in the 4*s* to the number in the ground 3*s* state in
(a) an acetylene/oxygen flame (3000°C).
(b) the hottest part of an inductively coupled plasma source (≈9000°C).

9-16 Assume that the peaks shown in Figure 9-7 were obtained for 2 μL aliquots of standards and sample. Calculate the concentration in parts per million of lead in the sample of canned orange juice.

9-17 Suggest sources of the two peaks in Figure 9-7 that appear during the drying and ashing processes.

9-18 In the concentration range of 500 to 2000 ppm of U, a linear relationship is found for absorbance at 351.5 nm and concentration. At lower concentrations the relationship becomes nonlinear unless about 2000 ppm of an alkali metal salt is introduced. Explain.

9-19 What is the purpose of an internal standard in flame emission methods?

9-20 A 5.00-mL sample of blood was treated with trichloroacetic acid to precipitate proteins. After centrifugation, the resulting solution was brought to a pH of 3 and was extracted with two 5-mL portions of methyl isobutyl ketone containing the organic lead complexing agent APCD. The extract was aspirated directly into an air/acetylene flame yielding an absorbance of 0.444 at 283.3 nm. Five-milliliter aliquots of standard solutions containing 0.250 and 0.450 ppm Pb were treated in the same way and yielded absorbances of 0.396 and 0.599. Calculate the ppm Pb in the sample assuming that Beer's law is followed.

9-21 The sodium in a series of cement samples was determined by flame emission spectroscopy. The flame photometer was calibrated with a series of standards that contained 0, 20.0, 40.0, 60.0, and 80.0 μg Na_2O per mL. The instrument readings *R* for these solutions were 3.1, 21.5, 40.9, 57.1, and 77.3.

(a) Plot the data.
(b) Obtain a least-squares line for the data.
(c) Calculate standard deviations for the slope and residuals for the line in (b).
(d) The following data were obtained for replicate 1.000-g samples of cement that were dissolved in HCl and diluted to 100.0 mL after neutralization.

	Emission Reading			
	Blank	**Sample A**	**Sample B**	**Sample C**
Replicate 1	5.1	28.6	40.7	73.1
Replicate 2	4.8	28.2	41.2	72.1
Replicate 3	4.9	28.9	40.2	spilled

Calculate the percent Na_2O in each sample. What are the absolute and relative standard deviation for the average of each determination?

9-22 The chromium in an aqueous sample was determined by pipetting 10.0 mL of the unknown into each of five 50.0-mL volumetric flasks. Various volumes of a standard containing 12.2 ppm Cr were added to the flasks, following which the solutions were diluted to volume.

Unknown, mL	Standard, mL	Absorbance
10.0	0.0	0.201
10.0	10.0	0.292
10.0	20.0	0.378
10.0	30.0	0.467
10.0	40.0	0.554

(a) Plot the data.
(b) Derive an equation for the relationship between absorbance and volume of standard.
(c) Calculate the standard deviation for the slope and the standard deviation about regression in (b).
(d) Calculate the ppm Cr in the sample.
(e) Calculate the standard deviation of the result in (d).

Chapter 10

Atomic Emission Spectrometry

This chapter deals with optical atomic emission spectrometry (AES).[1] Generally, the atomizers listed in Table 8-1 not only convert the components of samples to atoms or simple elementary ions, but in the process excite a fraction of these species to higher electronic states. Rapid relaxation of the excited species is accompanied by the production of ultraviolet and visible line spectra that are useful for qualitative and quantitative elemental analysis. Historically, atomic emission spectroscopy was based upon flame, electric arc, and electric spark atomization and excitation, and these methods still find important applications for analysis of the metallic elements. By now, however, plasma sources, have become the most important and most widely used sources for atomic emission spectrometry. Of the three listed in Table 8-1, the inductively coupled plasma is by far more important than either the direct current plasma or the microwave-induced plasma source.

Plasma, arc, and spark emission spectrometry offer several advantages when compared with the flame and electrothermal absorption methods considered in Chapter 9. Among the advantages is their lower interelement interference, which is a direct consequence of their higher temperatures. Second, good emission spectra are obtained for most elements under a single set of excitation conditions; as a consequence, spectra for dozens of elements can be recorded ***simultaneously.*** *This property is of particular importance for the multielement analysis of very small samples. Flame sources are less satisfactory in this regard because optimum excitation conditions vary widely from element to element; high temperatures are needed for excitation of some elements and low temperatures for others; and finally, the region of the flame that gives rise to optimum line intensities varies from element to element. Another advantage of the more energetic plasma sources is that they permit the determination of low concentrations of elements that tend to form refractory compounds (that is, compounds, such as the oxides of boron, phosphorus, tungsten, uranium, zirconium, and*

[1] For the more extensive treatment of emission spectroscopy, see R. D. Sacks, in *Treatise on Analytical Chemistry*, 2nd ed., P. J. Elving, E. J. Meehan, I. M. Kolthoff, Eds., Part I, Vol. 7, Chapter 6. New York: Wiley, 1981; T. Torok, J. Mika, and E. Gegus, *Emission Spectrochemical Analysis*. New York: Crane Russak, 1978; J. D. Ingle Jr. and S. R. Crouch, *Spectrochemical Analysis*, Chapters 7–9, 11. Englewood Cliffs, NJ: Prentice-Hall, 1988.

niobium that are highly resistant to thermal decomposition). In addition, plasma sources permit the determination of nonmetals such as chlorine, bromine, iodine, and sulfur. Finally, methods based upon plasma sources usually have concentration ranges of several decades in contrast to two or three for the absorption methods described in the previous chapter.

Emission spectra from plasma, arc, and spark sources are often highly complex, being made up of hundreds, or even thousands, of lines. This plethora of lines, while advantageous when qualitative information is sought, enhances the probability of spectral interferences in quantitative analysis. Consequently, emission spectroscopy based upon plasmas, arcs, and sparks requires higher resolution and more expensive optical equipment than is needed for atomic absorption methods with flame or electrothermal sources.

Despite their several advantages, it is unlikely that emission methods based upon high-energy sources will ever completely displace flame and electrothermal atomic absorption procedures. In fact, atomic emission and absorption methods appear to be complementary. Included among the advantages of atomic absorption procedures are simpler and less expensive equipment requirements, lower operating costs, somewhat greater precision (presently, at least), and procedures that require less operator skills to yield satisfactory results.[2]

10A EMISSION SPECTROSCOPY BASED ON PLASMA SOURCES

By definition, a plasma is an electrical conducting gaseous mixture containing a significant concentration of cations and electrons. (The concentrations of the two are such that the net charge approaches zero.) In the argon plasma frequently employed for emission analyses, argon ions and electrons are the principle conducting species, although cations from the sample will also be present in lesser amounts. Argon ions, once formed in a plasma, are capable of absorbing sufficient power from an external source to maintain the temperature at a level at which further ionization sustains the plasma indefinitely; temperatures as great as 10,000 K are encountered. Three types of high-temperature plasmas are encountered: (1) the inductively coupled plasma (ICP), (2) the direct current plasma (DCP), and (3) the microwave induced plasma (MIP).[3] The first two of these sources are marketed by several instrument companies. The microwave induced plasma source is not widely used for general elemental analysis primarily because it is not available from instrument manufacturers. We shall not consider this source further in this chapter. It is noteworthy, however, that one instrument company now offers a microwave plasma source as an element detector for gas chromatography. This instrument is described briefly in Section 27B-4.

10A-1 The Inductively Coupled Plasma Source[4]

Figure 10-1 is a schematic of a typical inductively coupled plasma source called a *torch.* It consists of three concentric quartz tubes through which streams of argon gas flow. Depending upon the torch design, the total rate of argon consumption is 5 to 20 L/min. The diameter of the largest tube is often about 2.5 cm. Surrounding the top of this tube is a water-cooled induction coil that is powered by a radio-frequency generator, which is capable of producing 0.5 to 2 kW of power at about 27 or 41 MHz. Ionization of the flowing argon is initiated by a spark from a Tesla coil. The resulting ions, and their associated electrons, then interact with the fluctuating magnetic field (labeled H in Figure 10-1) produced by the induction coil. This interaction causes the ions and electrons within the coil to flow in the closed annular paths depicted in Figure 10-1; ohmic heating is the consequence of the resistance of the ions and electrons to this movement.

---

[2] For an excellent comparison of the advantages and disadvantages of flames, furnaces, and plasmas as sources for emission spectroscopy, see W. Slavin, *Anal. Chem.*, **1986**, *58*, 589A.

[3] For a comparison of these sources, see R. D. Sacks, in *Treatise on Analytical Chemistry*, 2nd ed., P. J. Elving, E. J. Meehan, and I. M. Kolthoff, Eds., Part I, Vol. 7, pp. 516–526. New York: Wiley, 1981; A. T. Zander, *Anal. Chem.*, **1986**, *58*, 1139A.

[4] For a more complete discussion, see V. A. Fassel, *Science*, **1978**, *202*, 183; V. A. Fassel, *Anal. Chem.*, **1979**, *51*, 1290A; G. A. Meyer, *Anal. Chem.*, **1987**, *59*, 1345A; *Inductively Coupled Plasma Emission Spectroscopy*, Parts 1 and 2, P. W. J. M. Boumans, Ed. New York: Wiley, 1987; A. Varma, *CRC Handbook of Inductively Coupled Plasma Atomic Emission Spectroscopy*. Boca Raton: CRC Press, 1990; *Inductively Coupled Plasma in Analytical Atomic Spectroscopy*, A. Montaser and D. W. Golightly, Eds. New York: VCR Publishers, 1987; *Handbook of Inductively Coupled Plasma Spectroscopy*, M. Thompson and J. N. Walsh, Eds. New York: Chapman & Hall, 1989.

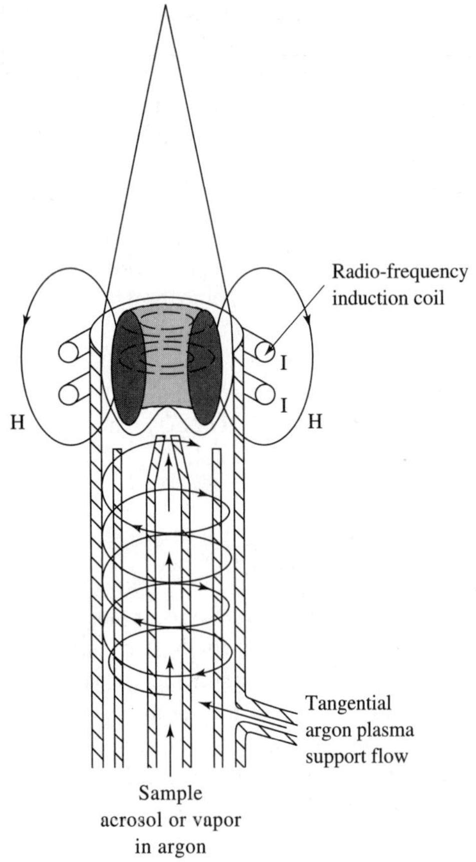

Figure 10-1 A typical inductively coupled plasma. *(From V. A. Fassel,* Science, *1978,* 202, *185. With permission. Copyright 1978 by the American Association for the Advancement of Science.)*

The temperature of the plasma formed in this way is high enough to require thermal isolation of the outer quartz cylinder. This isolation is achieved by flowing argon tangentially around the walls of the tube as indicated by the arrows in Figure 10-1. The tangential flow cools the inside walls of the center tube and centers the plasma radially.[5]

A recent innovation in torch design offered by several manufacturers involves rotating the torch by 90 deg so that it is aligned horizontally with the spectrometer system. The emitted radiation from the center of the plasma is then used for analyses. This arrangement is reported to improve detection limits by a factor of four to ten.

[5] For a description of commercially available torches see, D. Noble, *Anal. Chem.*, **1994**, *66*, 105A.

Note that the argon flow rate through the typical torch is great enough to constitute a significant cost (several thousand dollars annually).

Sample Introduction

Samples are carried into the torch shown in Figure 10-1 by argon flowing at 0.3 to 1.5 L/min through the central quartz tube. Samples are introduced into the argon flow by any of the methods described in Section 8C-1 and listed in Table 8-2. Often, the greatest source of noise in an ICP method resides in the sample introduction step.[6]

The most widely used devices for sample injection are nebulizers, such as those shown in Figure 8-9. Figure 10-2 shows a typical arrangement. Here, the sample is nebulized in a cross-flow nebulizer by a stream of argon, and the resulting finely divided droplets are carried

[6] See R. F. Browner and A. W. Boorn, *Anal. Chem.*, **1984**, *56*, 787A, 875A.

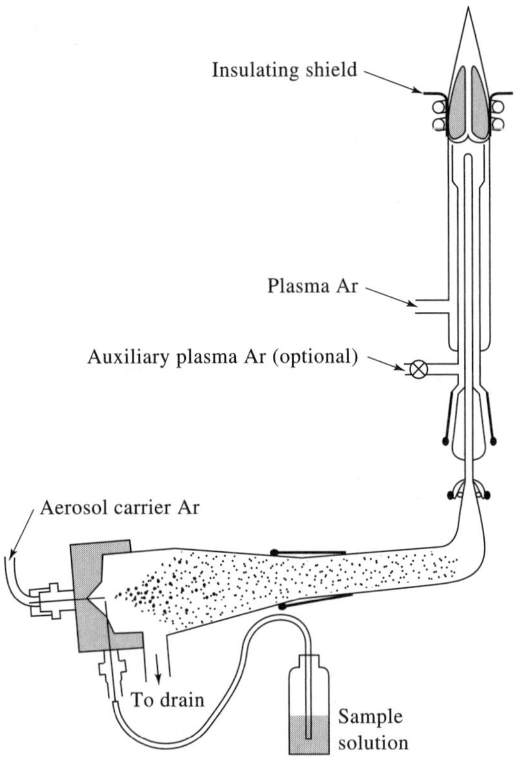

Figure 10-2 A typical nebulizer for sample injection into a plasma source. *(From V. A. Fassel,* Science, *1978,* 202, *186. With permission. Copyright 1978 by the American Association for the Advancement of Science.)*

into the plasma. Aerosols have also been produced from liquids by means of an ultrasonic nebulizer.

Another method of introducing liquid and solid samples into a plasma is by electrothermal vaporization. Here, the sample is vaporized in a furnace similar to that described in Section 8C-1 for electrothermal atomization. In plasma applications, however, the furnace is used for *sample introduction only* rather than for *sample atomization.* Figure 10-3 shows an electrothermal vaporizer offered by one instrument manufacturer in which vapor formation takes place on an open graphite rod. The vapor is then carried into a plasma torch by a flow of argon. The observed signal is a transient peak similar to the peaks obtained in electrothermal atomic absorption. Electrothermal vaporization coupled with a plasma torch offers the microsampling capabilities and low detection limits of electrothermal furnaces while maintaining the wide linear working range, the freedom from interference, and the multielement capabilities of ICP.

The ablation devices for solids described in Section 8C-2 are also marketed by several makers of ICP instruments. With these types of sample introduction systems, the plume of vapor and particulate matter produced by interaction of the sample with an electric arc or spark or with a laser beam are transported by a flow of argon into the torch where further atomization and excitation occur.

Plasma Appearance and Spectra

The typical plasma has a very intense, brilliant white, nontransparent core topped by a flamelike tail. The core, which extends a few millimeters above the tube, is made up of a continuum upon which is superimposed the atomic spectrum for argon. The source of the continuum apparently arises from recombination of argon and other ions with electrons. In the region 10 to 30 mm above the core, the continuum fades, and the plasma is optically transparent. Spectral observations are generally made at a height of 15 to 20 mm above the induction coil. Here, the background radiation is remarkably free of argon lines and is well suited for analysis. Many of the most sensitive analyte lines in this region of the plasma are from ions, such as Ca^+, Ca^{2+}, Cd^+, Cr^{2+}, and Mn^{2+}.

Analyte Atomization and Ionization

Figure 10-4 shows temperatures at various parts of the plasma. By the time the sample atoms have reached the observation point, they will have resided for about 2 ms at temperatures ranging from 4000 to 8000 K. These times and temperatures are roughly two to three times greater than those found in the hottest combustion flames (acetylene/nitrous oxide) used in flame methods. As a consequence, atomization is more com-

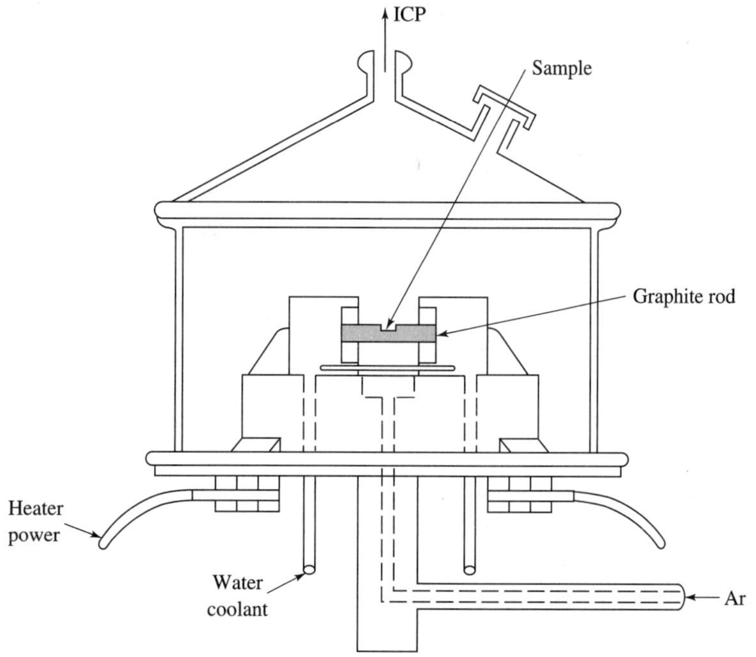

Figure 10-3 Device for electrothermal vaporization.

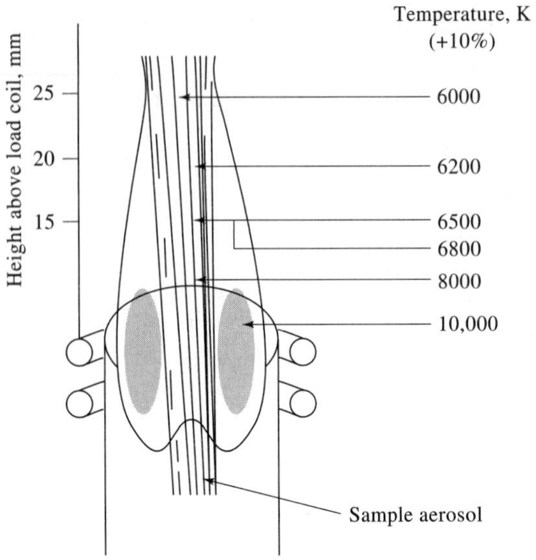

Figure 10-4 Temperatures in a typical inductively coupled plasma source. *(From V. A. Fassel, Science, **1978**, 202, 186. With permission. Copyright 1978 by the American Association for the Advancement of Science.)*

plete, and fewer chemical interference problems arise. Surprisingly, ionization interference effects are small or nonexistent, probably because the electron concentration from ionization of the argon is large compared with that resulting from ionization of sample components.

Several other advantages are associated with the plasma source. First, atomization occurs in a chemically inert environment, which tends to enhance the lifetime of the analyte by preventing oxide formation. In addition, and in contrast to arc, spark, and flame, the temperature cross section of the plasma is relatively uniform; as a consequence, self-absorption and self-reversal effects are not encountered. Thus, linear calibration curves over several orders of magnitude of concentration are usually observed.

10A-2 The Direct Current Plasma Source

Direct current plasma jets were first described in the 1920s and were systematically investigated as sources for emission spectroscopy for several decades. It was not until the 1970s, however, that a source based on this principle was designed that could successfully compete with flame and inductively coupled plasma sources in terms of reproducible behavior.[7]

Figure 10-5 is a schematic of a commercially available dc plasma source that is well suited for excitation of emission spectra for a wide variety of elements. This plasma jet source consists of three electrodes arranged in an inverted Y configuration. A graphite anode is located in each arm of the Y and a tungsten cathode at the inverted base. Argon flows from the two anode blocks toward the cathode. The plasma jet is formed by bringing the cathode momentarily in contact with the anodes. Ionization of the argon occurs and a current develops ($\approx$14 A) that generates additional ions that sustain the current indefinitely. The temperature at the arc core is about 10,000 K and at the viewing region is 5000 K. The sample is aspirated into the area between the two arms of the Y, where it is atomized, excited, and viewed.

Spectra produced from the plasma jet tend to have fewer lines than those produced by the inductively coupled plasma, and the lines present are largely from atoms rather than ions. Sensitivities achieved with the dc jet plasma appear to range from an order of magnitude lower to about the same as those found with the in-

[7] For additional details, see G. W. Johnson, H. E. Taylor, and R. K. Skogerboe, *Anal. Chem.*, **1979**, *51*, 2403; *Spectrochim. Acta*, Part B, **1979**, *34*, 197; R. J. Decker, *Spectrochim. Acta*, Part B, **1980**, *35*, 21.

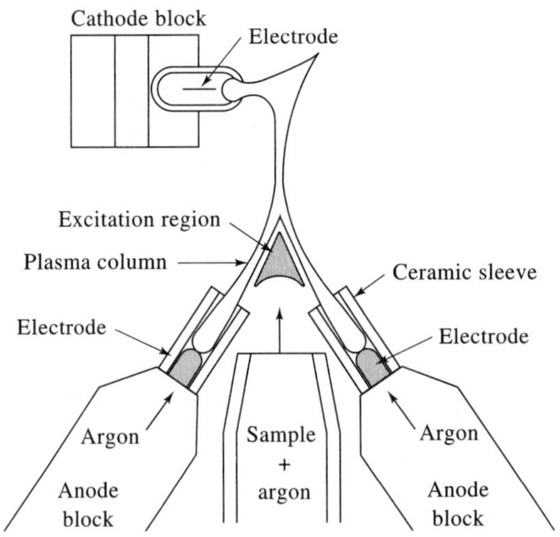

Figure 10-5 A three-electrode dc plasma jet. *(Courtesy of Spectra Metrics, Inc. Haverhill, MA.)*

ductively coupled plasma. Reproducibilities of the two systems are similar. Significantly less argon is required for the dc plasma and the auxiliary power supply is simpler and less expensive. On the other hand, the graphite electrodes must be replaced every few hours, while the inductively coupled plasma source requires little or no maintenance.

10A-3 Plasma Source Spectrometers[8]

Table 10-1 lists the most important properties of the ideal instrument for plasma emission spectroscopy. Needless to say, the ideal spectrometer is not available today partly because some of these properties are mutually exclusive. For example, high resolution requires the use of narrow slits, which usually reduces *S/N* and thus the precision of intensity readings. Nevertheless modern instruments developed during the last decade are beginning to approach many of the ideals listed in Table 10-1.

TABLE 10-1 Desirable Properties of an Emission Spectrometer

1. High resolution (0.01 nm or $\lambda/\Delta\lambda > 100{,}000$)
2. Rapid signal acquisition and recovery
3. Low stray light
4. Wide dynamic range ($>10^6$)
5. Accurate and precise wavelength identification and selection
6. Precise intensity readings (<1% RSD at 500 × the detection limit)
7. High stability with respect to environmental changes
8. Easy background corrections
9. Computerized operation: readout, storage data manipulation, etc.

Plasma emission spectrometers are currently offered by a dozen or more instrument manufacturers.[9] Their designs, performance characteristics, and wavelength ranges vary considerably. Most encompass the entire ultraviolet/visible spectrum from 170 to 800 nm. A few instruments are equipped for vacuum operation, which extends the ultraviolet to 150 to 160 nm. This short wavelength region is important because elements such as phosphorus, sulfur, and carbon have emission lines in this region.

Instruments for emission spectroscopy are of three basic types: *sequential, simultaneous multichannel,* and *Fourier transform.* At the present time, Fourier transform instruments have not been widely used. Sequential instruments are usually programmed to move from the line for one element to that of a second, pausing long enough (a few seconds) at each to obtain a satisfactory signal-to-noise ratio. In contrast, multichannel instruments are designed to measure the intensities of emission lines for a large number of elements (sometimes as many as 50 or 60) *simultaneously,* or nearly so. Where several elements are to be determined, the excitation time with sequential instruments will be significantly greater than with the other two. Thus, these instruments, while simpler, are costly in terms of sample consumption and time.

Both sequential and multichannel emission spectrometers are of two general types, one employing a classical grating monochromator and the other an echelle monochromator, such as that shown in Figure 7-21.

Sequential Instruments

Most sequential instruments incorporate a grating monochromator such as that shown in Figure 7-16a with a photomultiplier detector. Usually the grating is a holographic type having 2400 or 3600 grooves per millimeter. With some instruments of this type, scanning involves rotating the grating with a digitally controlled stepper motor so that different wavelengths are sequentially and precisely focused on the exit slit. In some designs, however, the grating is fixed and the slit and photomultiplier tube are moved along the focal plane or curve. Instruments are also marketed that have two sets of slits and photomultiplier tubes, one for the ultraviolet region and one for the visible. Here, at an appropriate wavelength, the exit beam is switched from one slit to the other by movement of a plane mirror.

[8] For a description of modern emission spectrometers, see A. Strasheim and A. Montaser, in *Inductively Coupled Plasmas in Atomic Spectrometry,* 2nd ed., A. Montaser and D. W. Golightly, Eds., Chapter 3. New York: VCH Publishers, 1992; J. W. Olesik, in *Inductively Coupled Plasma Emission Spectroscopy,* Part I, P. W. J. M. Boumans, Ed. New York: Wiley, 1987; G. A. Meyer, *Anal. Chem.,* **1987,** *59,* 1345A.

[9] For a comparison of commercial instruments, see D. Noble, *Anal. Chem.,* **1994,** *66,* 105A.

Slew-Scan Spectrometers. With complex spectra made up of hundreds of lines, scanning a significant wavelength region takes an inordinate length of time and is thus not practical. To partially overcome this problem, *slew-scan* spectrometers have been developed in which the grating, or the detector and slit, is driven by a two-speed motor. Here, the instrument is rapidly moved or *slewed* to a wavelength near a line. The rate of movement is then quickly slowed so that the instrument sweeps across the line in a series of small (0.01 to 0.001 nm) steps. Computer control of such instruments is essential. With slew scanning, the time spent in wavelength regions devoid of useful data is minimized, but sufficient time is spent at analyte lines to obtain satisfactory signal-to-noise ratios.

Figure 10-6a shows the design of a commercially available slew-scan spectrometer. This instrument has a composite grating system consisting of two gratings mounted one above the other. The second-order spectrum from the grating that contains 2400 lines per millimeter is employed for the wavelength range of 160 to 380 nm, and the first-order spectrum from the grating made up of 1200 lines per millimeter is used for the region from 380 to 850 nm. A baffle falls into place to expose one or the other grating at the appropriate time. This arrangement provides a resolution of 0.008 nm from 160 to 335 nm, 0.018 nm from 335 to 670 nm, and 0.04 nm from 670 to 850 nm. A motor-driven filter wheel, holding a series of interference filters, is located in front of the entrance slit to remove overlapping spectral orders and stray light. In addition two exit slits and two photomultiplier tubes are used for maximum sensitivity in the ultraviolet and the visible regions. Both the filter wheel and the mirrors are under computer control.

A unique feature of this instrument is the galvanometer-type grating drive, an electromagnetic mechanism that can scan from 165 to 800 nm in 20 msec (Figure 10-6b). The drive is controlled by a digital-to-analog converter and a capacitive feedback loop. The mechanism has only one moving part—the center rotor shaft that turns within an electromagnetic coil.

The cost of slew-scan instruments is significantly less than that of the multichannel spectrometers discussed in the next section, and they are considerably more versatile because they are not restricted to a limited number of preselected wavelengths. Generally, however, they are slower and consume more sample than their multichannel simultaneous counterparts.

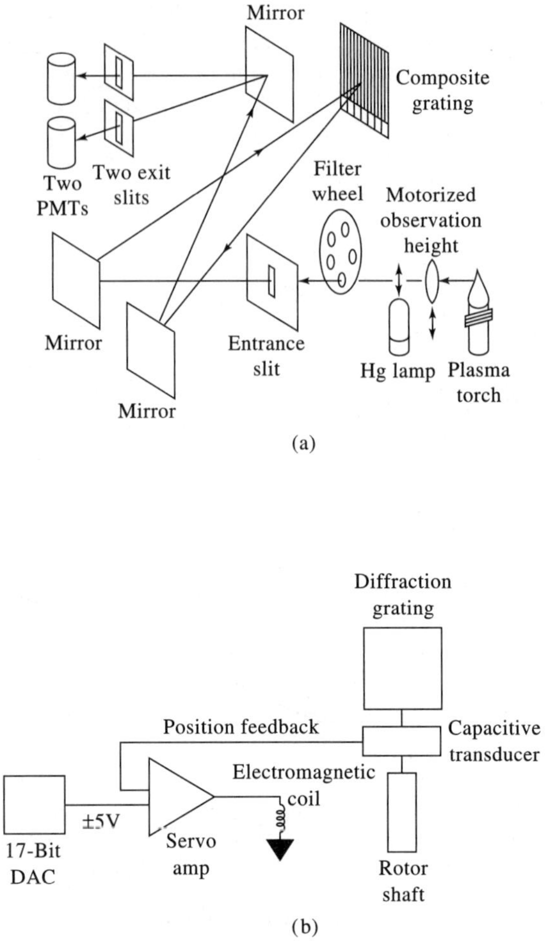

Figure 10-6 (a) Optical diagram of a slew-scan spectrometer for ICP spectrometry. (b) Magnetically driven grating. *(Courtesy of Thermo Jarrell Ash Corp.)*

Scanning Echelle Spectrometers. Figure 10-7 is a schematic of an echelle spectrometer, which first appeared on the market in the 1970s for use with a dc plasma source. It can be operated either as a scanning instrument or as a simultaneous multichannel spectrometer. Scanning is accomplished by moving a photomultiplier tube in both x and y directions in order to scan an aperture plate located on the focal plane of the monochromator. The plate contains as many as 300 photo-etched slits. The time required to move from one slit to another is typically 1 s. The instrument can be operated in a slew-scan mode. This instrument can also be converted to a multichannel polychromator by mounting several small photomul-

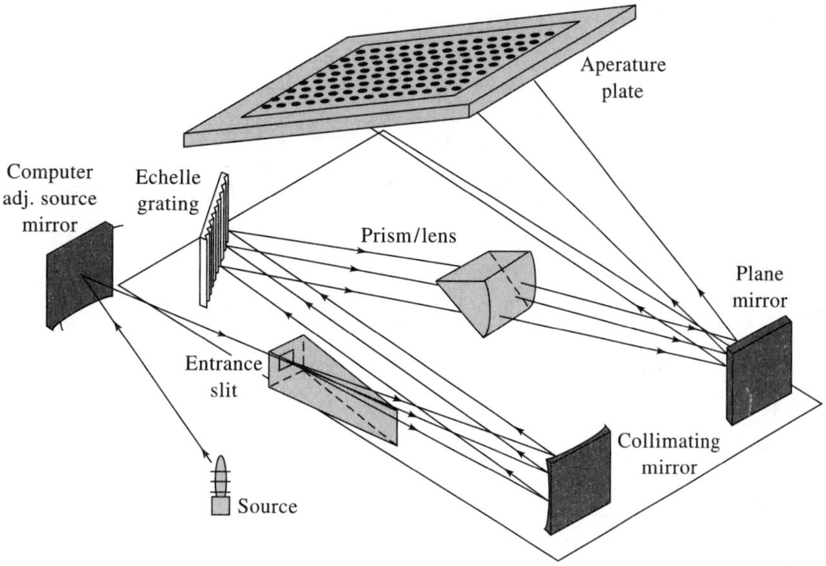

Figure 10-7 Schematic of an echelle polychromator system.

tiplier tubes behind appropriate slits in the aperture plate.

Multichannel Spectrometers

Simultaneous multichannel instruments are of two general types: *polychromators* and *array-based systems.* The former employ a series of photomultiplier tubes for detection, whereas the latter use two dimensional charge-injection devices or charge-coupled devices as detectors.

Polychromators. Several companies offer emission spectrometers in which as many as 60 photomultipliers are located behind fixed slits along the curved focal plane of a concave grating monochromator.[10] The instrument shown schematically in Figure 10-8 is typical. Here the entrance slit, the exit slits, and the grating surface are located along the circumference of a *Rowland circle,* the curvature of which corresponds to the focal curve of the concave grating. Radiation from each of the several fixed slits are reflected by mirrors (two of which are shown) to photomultiplier tubes. The slits are factory configured to transmit lines for elements chosen by the customer. In this instrument, the pattern of lines can, however, be changed relatively inexpensively to accommodate new elements or to delete others. The signals from the several photomultiplier tubes are fed into individual analog integrators; the output voltages are then digitized, converted to con-centrations, stored, and displayed. The entrance slit can be moved tangentially to the Rowland circle by means of a stepper motor. This device permits scanning through peaks and provides information for background corrections.

Spectrometers such as that shown in Figure 10-8 have been used both with plasma and with arc and spark sources. For rapid routine analyses, such instruments are often ideal. For example, in the production of alloys, quantitative determinations for 20 or more elements can be accomplished within five minutes of receipt of a sample; close control over the composition of a final product thus becomes possible.

In addition to speed, photoelectric multichannel spectrometers often offer the advantage of good analytical precision. Under ideal conditions reproducibilities of the order of 1% relative of the amount present have been demonstrated. Several of the newer instruments are provided with a second monochromator that permits spectral scanning, thus adding a versatility that was absent in some earlier instruments.

Multichannel instruments are more expensive (>$80,000) and generally not as versatile as the sequential instruments described in the previous section.

[10] For a review of recent developments in multichannel instruments, see D. D. Nygaard and J. J. Sotera, *Amer. Lab.,* **1988** (7), 30.

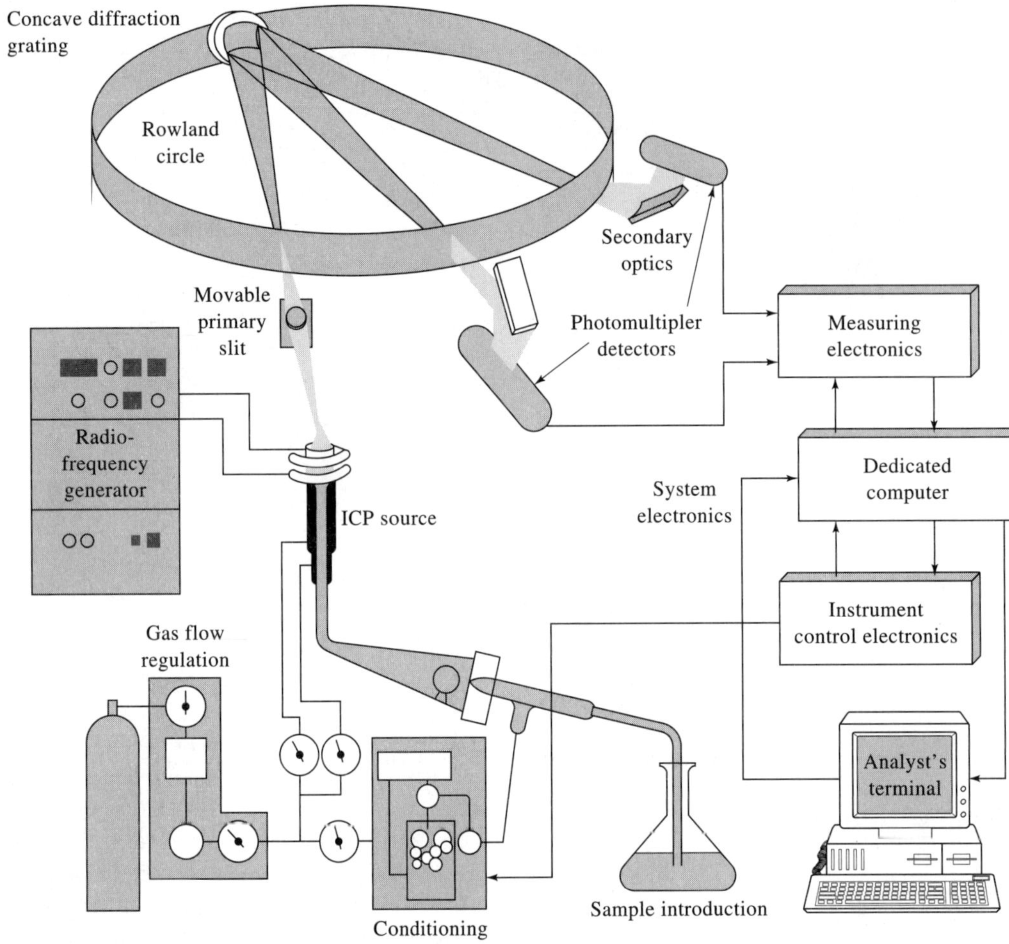

Figure 10-8 Schematic of an ICP polychromator. *(Courtesy of Thermo Jarrell Ash Corp.)*

A Charge-Injection Instrument. In the early 1990s, three American companies began to offer multichannel simultaneous spectrometers based upon echelle spectrometers and two-dimensional charge-transfer detectors.[11] It seems likely that this type of instrument will eventually replace most other types of multichannel emission spectrometers.

Figure 10-9 is an optical diagram of an echelle spectrometer that employs a charge injection detector for simultaneous operation.[12] It uses a calcium fluoride prism to sort the spectral orders that are subsequently formed by the echelle grating (see also Figure 7-21). The transducer is a charge injection device (page 174) with dimensions of 8.7 mm by 6.6 mm and containing 94,672 detector elements. A toroidal camera mirror is used to focus the slit images onto the transducer surface. In order to eliminate dark currents in the transducer elements, the unit is housed in a liquid nitrogen cryostat that maintains a temperature of 135 K.

A set of 39 detector elements, called a *read window,* is used to monitor each spectral line as shown in Figure 10-10a. Normally, as shown by the projected image of one of the windows labeled "examination window," the spectral line is focused on the 9 center elements of the window, whereas the 15 elements on either side of the central set provide background intensity

[11] See J. V. Sweedler, *Crit. Rev. Anal. Chem.,* **1993,** *24,* 59.
[12] See M. J. Pilon, M. B. Denton, R. G. Schleicher, P. M. Moran, and S. B. Smith, *Appl. Spectrosc.* **1990,** *44,* 1613.

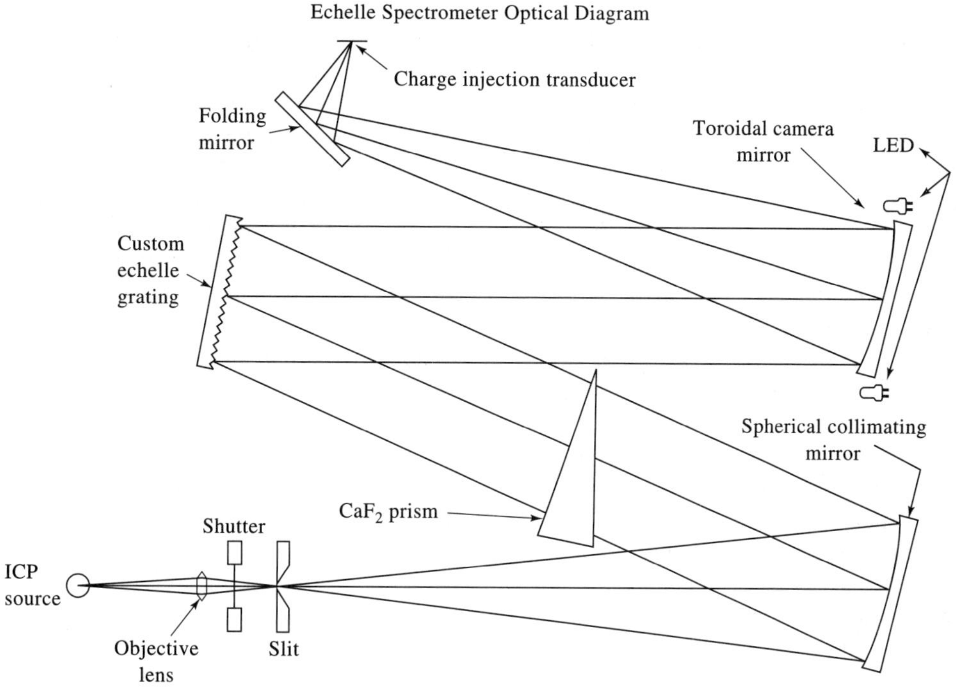

Figure 10-9 Optical diagram of an echelle spectrometer with a charge-injection detector. *(From R. B. Bilhorn and M. B. Denton,* Appl. Spectrosc., ***1990,*** *44, 1615. With permission.)*

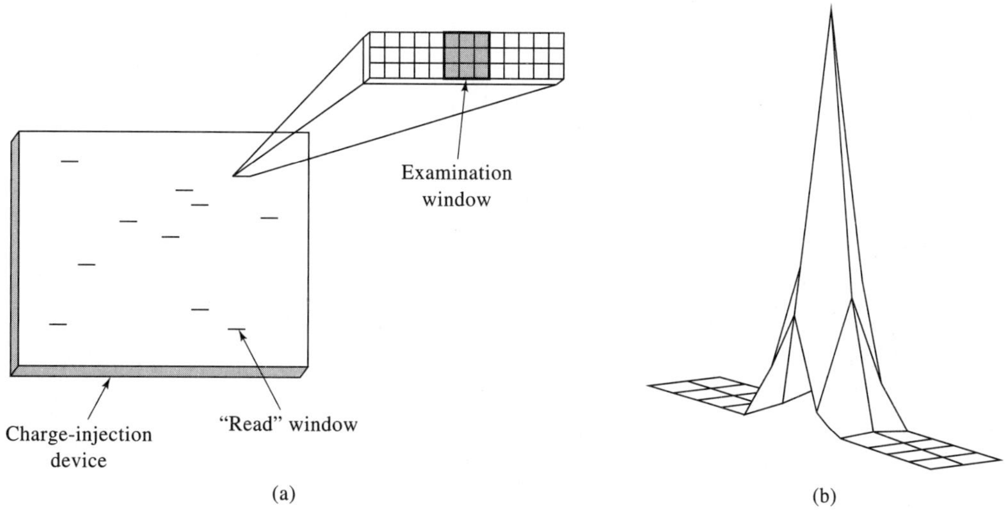

Figure 10-10 (a) Schematic representing the surface of a charge-injection device. The short horizontal lines represent the "read windows." A magnified image of one of the read windows is also shown. The nine central elements form the "examination window" where a line is positioned. (b) Intensity profile for an iron line. All of the radiation from the line falls on the 3 × 3 examination window. *(From R. B. Bilhorn and M. B. Denton,* Appl. Spectrosc., ***1990,*** *44, 1540. With permission.)*

data. Figure 10-10b shows the intensities recorded for the read window for the iron 297.32 nm line. Note that most of the iron radiation falls on the central elements of the window.

One of the useful features of the charge-injection detector, in contrast to the charge-coupled detector, is that the amount of charge accumulated in an element at any instant can be monitored nondestructively; that is, no charge is lost in the measurement process. To make the measurement of line intensity as rapid and efficient as possible, only the charge accumulated in the nine central elements of the window are read initially to determine when sufficient charge has accumulated to provide a satisfactory signal-to-noise ratio. Only then are the remaining elements in the two 15 element sets read in order to correct the observed line intensity for the background radiation. This process goes on simultaneously at the read windows for each element. With an intense line, the time required to accumulate the desired charge is brief. With weak lines, the charge accumulated in a brief period is often used to estimate the integration time required to yield a satisfactory signal-to-noise ratio. In some cases integration times of 100 s or more are required.

Periodic wavelength calibration of the spectrometer just described is maintained through reference to the 253.65 nm mercury line from a small mercury lamp. Data files for line positions for more than 40 elements have been developed. The file for each element contains the wavelengths of up to ten lines and the x and y coordinates of each of these spectral lines with respect to the coordinates for the mercury line. Database recalibration is seldom necessary unless something perturbs the optics of the system significantly. Identification of elements in samples is done by visual inspection with the use of a video monitor and interactive markers. With an ICP excitation source, detection limits that ranged from a few tenths to 10^4 nanogram per milliliter have been reported for most elements. For nonmetals such as phosphorus and arsenic, detection limits were larger by a factor of one hundred.

A Charge-Coupled Instrument. Figure 10-11 is an optical diagram of a recently developed commercial spectrometer with two echelle systems and two charge-

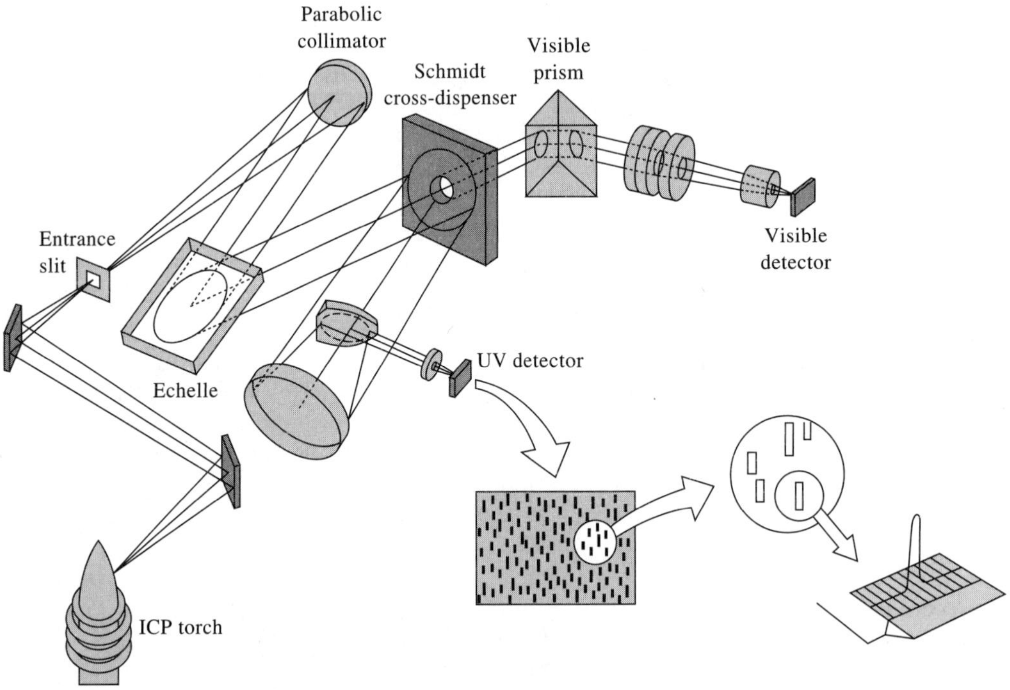

Figure 10-11 An echelle spectrometer with segmented array of charge-coupled devices. *(From T. W. Barnard et al.,* Anal. Chem., ***1993,*** 65, *1232. With permission.)*

coupled detectors, one system having been designed for the 160 to 375 nm region and the other for the 375 to 782 nm range.[13] After entering the spectrometer via a slit, the radiation is dispersed by an echelle grating. The radiation falls on a Schmidt cross-disperser element that separates the orders of the ultraviolet radiation and also separates the ultraviolet and visible optical beams. The Schmidt element consists of a grating ruled on a spherical surface with a hole in the center to pass the visible radiation to a prism where order separation takes place as shown in Figure 7-21. The two dispersed beams are then focused onto the surface of detector elements as shown. Note that a schematic representation of the surface of the ultraviolet detector is shown as an insert in Figure 10-11.

The unique detector systems consist of numerous subarrays, or *array segments,* fabricated on silicon chips with each subarray being custom positioned so that three to four major emission lines for each of 72 elements fall on its surface. Each array segment consists of a *linear* (rather than two-dimensional) charge-coupled transducer that is made up of 20 to 80 pixels. Figure 10-12 is a schematic of one of these array segments, which is made up of the individual photosensitive registers, storage and output registers, and output electronics. Because each array segment can be addressed separately, charge integration times can be varied over a wide enough range to provide dynamic ranges of 10^5 or greater. Although there are only 224 of these array segments in the system, multiple lines fall on many of the subarrays so that several thousand lines can be monitored simultaneously.

[13] For a detailed description of this instrument, see T. W. Barnard et al., *Anal. Chem.,* **1993,** *65,* 1225, 1231.

Fourier Transform Spectrometers

Beginning in the early 1980s, several workers have described applications of Fourier transform instruments to the ultraviolet/visible region of the spectrum by using instruments that are similar in design to the infrared instruments discussed in detail in Section 16C-1.[14] Much of this work has been devoted to the use of such instruments for multielement analyses with inductively coupled plasma sources. The advantages of Fourier transform instruments include their wide wavelength coverage (170 nm to > 1000 nm), speed, high resolution, highly accurate wavelength measurements, large dynamic range, compact size, and large optical throughput. In contrast to Fourier transform infrared instruments, however, ultraviolet/visible instruments of this type often exhibit no multiplex advantage and indeed under some circumstances show a multiplex disadvantage (see page 184). The reason for this difference arises because the performance of infrared instruments is ordinarily limited by detector noise, whereas ultraviolet/visible spectrometers are limited by shot and flicker noise associated with the source.

An ultraviolet/visible Fourier transform instrument first appeared on the market in the mid-1980s.[15] The mechanical tolerances required to achieve good resolution with this type of instrument are very demanding, however. Consequently, these spectrometers are quite expensive and are used largely for research projects rather than for routine analytical applications.

[14] A. P. Thorne, *Anal. Chem.,* **1991,** *63,* 57A; L. M. Faires, *Anal. Chem.* **1986,** *58,* 1023A.

[15] See *Anal. Chem.,* **1985,** *57,* 276A; D. Snook and A. Grillo, *Amer. Lab.,* **1986,** (11), 28; F. L. Boudais and J. Buija, *Amer. Lab.,* **1985,** (2), 31.

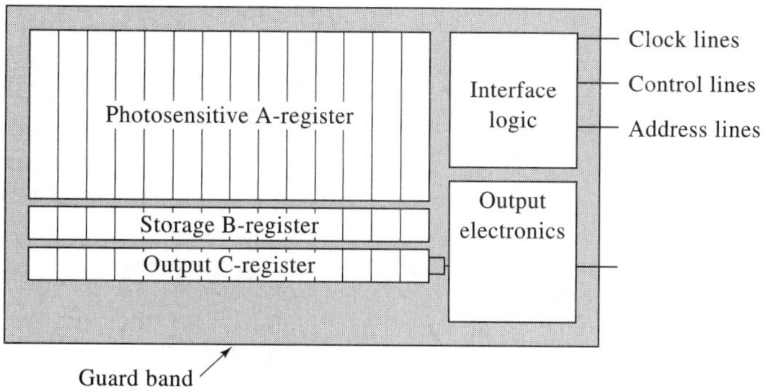

Figure 10-12 Schematic of an array segment showing photodetectors, OCD registers, and readout circuitry. *(From T. W. Marnard et al.,* Anal. Chem., *1993,* 65, *1232. With permission.)*

10A-4 Application of Plasma Sources[16]

Plasma sources are rich in characteristic emission lines, which makes them useful for both qualitative and quantitative elemental analysis. Unquestionably, the inductively coupled plasma and the direct current plasma yield significantly better quantitative analytical data than other emission sources do. The excellence of these results stems from their high stability, low noise, low background, and freedom from interferences when operated under appropriate experimental conditions. The performance of the inductively coupled plasma source is somewhat better than that of the direct current plasma source in terms of detection limits. The latter, however, is less expensive to purchase and operate and is entirely satisfactory for many applications.

Sample Preparation

Inductively coupled plasma emission spectroscopy is used primarily for the qualitative and quantitative analysis of samples that are dissolved or suspended in aqueous or organic liquids. The techniques for preparation of such solutions are similar to those described in Section 9D-1 for flame absorption methods. With plasma emission, however, methods do exist for the direct analysis of solids. These procedures include electrothermal vaporization, laser and spark ablation, and glow discharge, all of which were described in Section 8C-2. Suspensions of solid in solutions can also be handled with the Babington nebulizer, such as that shown in Figure 8-9d.

Elements Determined

In principle, all metallic elements can be determined by plasma emission spectrometry. A vacuum spectrometer is necessary for the determination of boron, phosphorus, nitrogen, sulfur, and carbon because the emission lines for these elements lie at wavelengths below 180 nm where components of the atmosphere absorb. The usefulness for the alkali metals is limited by two difficulties: (1) the compromise operating conditions that can be used to accommodate most other elements are unsuited for the alkalis, and (2) the most prominent lines of Li, K, Rb, and Cs are located at near-infrared wavelengths, which lead to detection problems with many plasma spectrometers that are designed primarily for ultraviolet radiation. Because of problems of this sort, plasma emission spectroscopy is generally limited to the determination of about 60 elements.

Figure 10-13 is a periodic table showing the applicability of inductively coupled plasma emission spectrometry to various elements. The detection limits for the best lines for each of the elements is indicated by the degree of shading. The area of shading indicates the number of lines for each element that yields a detection limit within a factor of three of the best line. The more such lines that are available the greater the chance that a usable line can be found that is free from interference when the matrix yields a line-rich spectrum.

Line Selection

An examination of Figure 10-13 shows that most elements have several prominent lines that can be used for identification and determination purposes. Wavelength data recorded to three decimal places with appropriate intensity information for prominent lines for over 70 elements can be found in several publications.[17] Thus, a suitable line for determination of any element can ordinarily be found. Selection will depend upon a consideration of what other elements are present in the sample and whether there is any likelihood of overlapping lines.

Calibration Curves

Calibration curves for plasma emission spectrometry most often consist of a plot of the output current or voltage of a transducer as a function of analyte concentration. When the range of concentrations is large, log-log plots are employed instead. Figure 10-14 shows typical calibration curves for four trace elements present in steel samples. Often calibration plots are linear as are the central two curves in the figure. Departures from linearity are encountered, however, when large concentration ranges are covered (see the outer two lines in Figure 10-14). A major cause of nonlinearity is self-absorption in which the output signal is reduced by absorption by nonexcited atoms in the medium. Self-

---

[16] For useful discussions of the applications of plasma emission sources, see *Inductively Coupled Plasma Emission Spectroscopy,* Parts I and II, P. W. J. M. Boumans, Ed. New York: Wiley-Interscience, 1987; G. A. Meyer and P. N. Keliher, in *Inductively Coupled Plasmas in Analytical Atomic Spectroscopy,* 2nd ed., Chapter 10, A. Montaser and D. W. Golightly, Eds. New York: VCH Publishers, 1992; M. Thompson and J. N. Walsh, *Handbook of Inductively Coupled Plasma Spectrometry,* 2nd ed. Glasgow: Blackie, 1989.

[17] See R. K. Winge, V. A. Fassel, V. J. Peterson, and M. A. Floyd, *Inductively Coupled Plasma Emission Spectroscopy: An Atlas of Spectral Information.* New York: Elsevier, 1985; P. W. J. M. Boumans, *Line Coincidence Tables for Inductively Coupled Plasma Spectrometry,* 2nd ed. Oxford: Pergamon, 1984; C. C. Wohlers, *ICP Information Newslett.,* **1985,** *10,* 601.

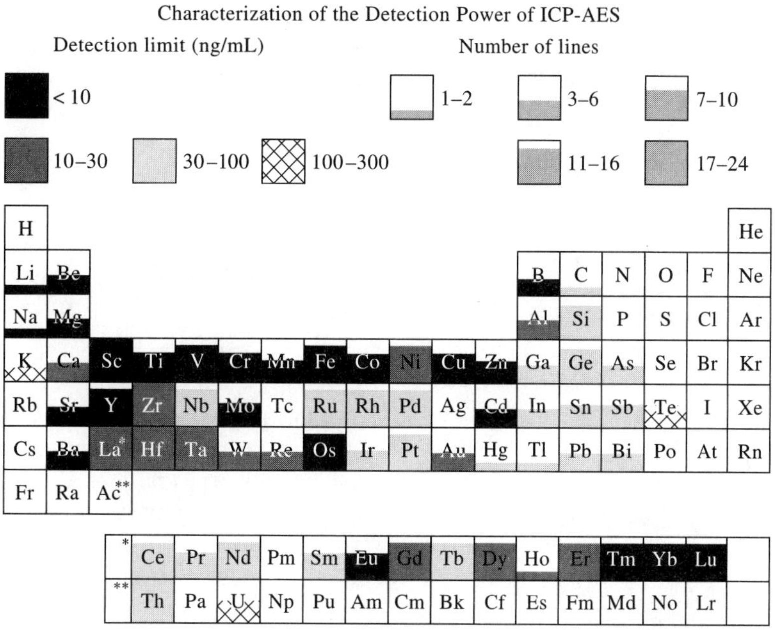

Figure 10-13 Periodic table characterizing the detection power and number of useful emission lines of ICP by employing a pneumatic nebulizer. The degree of shading indicates the range of detection limits for the useful lines. The area of shading indicates the number of useful lines. *(Adapted from* Inductively Coupled Plasma Emission Spectroscopy, *Part 1, p. 143, P. W. J. M. Boumans, Ed. New York: Wiley, 1987. With permission.)*

absorption becomes evident only at high analyte concentrations and causes the calibration curve to bend toward the horizontal axis. None of the plots in Figure 10-14 exhibit self-absorption. Nonlinearity also arises from erroneous background corrections and from nonlinear responses of the detection systems. The nonlinearity in the curves for niobium and thallium at low concentrations are probably a consequence of incorrect background corrections. Note that the departures from linearity are away from the concentration axis.

An internal standard is often used in emission spectrometry. In this case the vertical axis of the calibration curve is the ratio or the log ratio of the detector signal for the analyte to the detector signal for the internal standard. Figure 10-15 shows calibration curves for several elements. In these experiments a fixed amount of yttrium was incorporated in all of the standards, and the relative intensity of the analyte line to that of an yttrium line at 242.2 nm served as the analytical parameter. Note that the curves are all linear and cover a concentration range of nearly three orders of magnitude. Note also that some of the data were obtained by introducing various amounts of the analyte and internal standard into pure

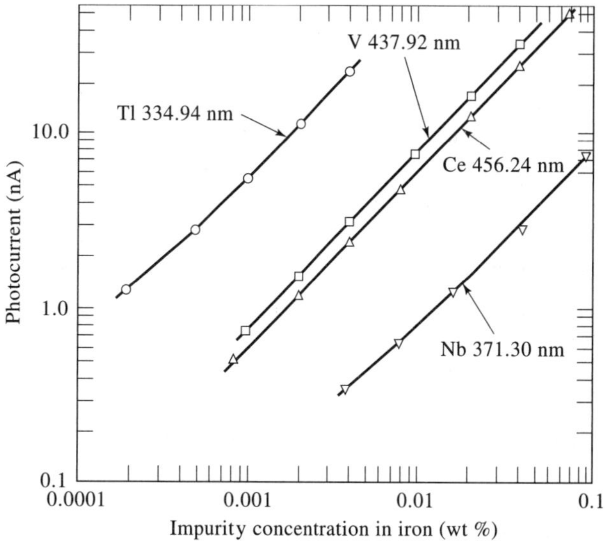

Figure 10-14 Typical calibration curves. *(From V. A. Fassel and R. N. Kniseley,* Anal. Chem., ***1974,*** 46, *1117A. With permission.)*

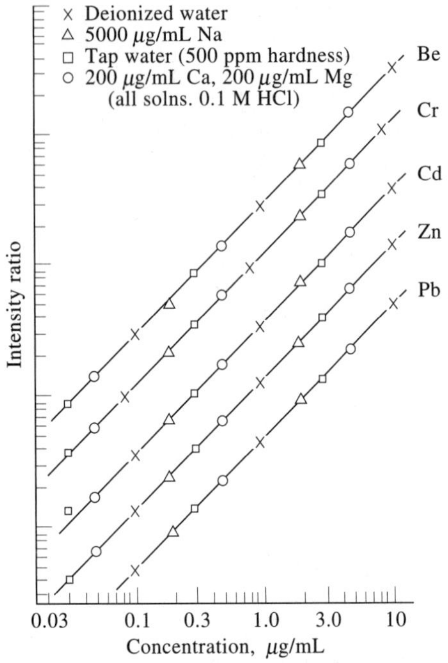

Figure 10-15 Calibration curves with an inductively coupled plasma source. Here, an yttrium line at 242.2 nm served as an internal standard. Notice the lack of interelement interference. *(From V. A. Fassel, Science, **1978,** 202, 187. With permission. Copyright 1978 by the American Association for the Advancement of Science.)*

water. Other data are for solutions that contained relatively high concentrations of different salts, thus demonstrating a freedom from interelement interference.

As in atomic absorption spectroscopy, one or more standards should be run periodically to correct for the effects of instrument drift. The improvement in precision that results from this procedure is illustrated by the data in Table 10-2. Note also the improved precision when higher concentrations of analyte are measured.[18]

Interferences

As was noted earlier, chemical interferences and matrix effects are significantly lower with plasma sources than with other atomizers. At low analyte concentrations, however, the background emission due to recombination of argon ions with electrons is large enough to require careful corrections. For single-channel instruments, this correction is conveniently obtained from measurements on either side of the peak. Many multichannel instruments are equipped with optical elements that permit similar corrections.

Because ICP spectra for many elements are so rich in lines, the possibility of spectral interferences from line overlap is very real. Avoidance of this type of error requires knowledge of all of the components likely to be present in the sample and a careful study of the information in the reference works listed in footnote 16.

Detection Limits

In general, the detection limits with the inductively coupled plasma source appear comparable to or better than other atomic spectral procedures. Table 10-3 compares these limits for several of these methods. Note that more elements can be detected at levels of ten parts-per-billion or less with plasma excitation than with other emission or absorption methods.

10B EMISSION SPECTROSCOPY BASED ON ARC AND SPARK SOURCES

Arc and spark source spectroscopies were the first instrumental methods to become widely used for analysis. These techniques, which began to replace the classical gravimetric and volumetric methods for elemental analysis in the 1920s, were based upon excitation of emission spectra of elements with electric arcs or electric sparks. These spectra permitted the qualitative and quantitative determination of metallic elements in a variety of sample types including metals and alloys, soils, minerals, and rocks.[19] Arc and spark sources still find considerable use in qualitative and semiquantitative analysis but have been largely displaced by plasma methods when quantitative data are required.

In arc and spark sources, sample excitation occurs in the gap between a pair of electrodes. Passage of electricity from the electrodes through the gap provides the necessary energy to atomize the sample and produce atoms or ions in electronic excited states.

---

[18] For a useful discussion of precision in ICP spectrometry, see R. L. Watters Jr., *Amer. Lab.*, **1983,** *15* (3), 16.

[19] For additional details, see R. D. Sacks, in *Treatise on Analytical Chemistry,* 2nd ed., P. J. Elving, E. J. Meehan, and I. M. Kolthoff, Eds., Part I, Vol. 7, Chapter 6. New York: Wiley, 1981; P. W. J. M. Boumans in *Analytical Emission Spectroscopy,* Vol. 1, Part I, E. L. Grove, Ed. New York: Marcel Dekker, 1972; J. D. Ingle Jr. and S. R. Crouch, *Spectrochemical Analysis,* Chapter 9. Englewood Cliffs, NJ: Prentice-Hall, 1988.

TABLE 10-2 Effect of Standardization Frequency on Precision of ICP Data*

Frequency of Recalibration, hr	Relative Standard Deviation, % — Concentration Multiple above Detection Limit 10^1 to 10^2	10^2 to 10^3	10^3 to 10^4	10^4 to 10^5
0.5	3–7	1–3	1–2	1.5–2
2	5–10	2–6	1.5–2.5	2–3
8	8–15	3–10	3–7	4–8

*Data from: R. M. Barnes, in *Applications of Inductively Coupled Plasmas to Emission Spectroscopy,* R. M. Barnes, Ed., p. 16. Philadelphia: The Franklin Institute Press, 1978. With permission.

10B-1 Sample Types and Sample Handling

Currently, arc and spark source methods are largely limited to the elemental analysis of solids because liquid and gaseous samples are so much more conveniently handled by the plasma emission methods that we have discussed in the previous sections.

Metals

If the sample is a metal or an alloy, one or both electrodes can be formed from the sample by milling, by turning, or by casting the molten metal in a mold. Ideally, the electrode will be shaped as a cylindrical rod that is one-eighth inch to one-quarter inch in diameter and tapered at one end. For some samples, it is more convenient to employ the polished, flat surface of a large piece of the metal as one electrode and a tapered cylindrical graphite or metal rod as the other. Regardless of the ultimate shape of the sample, care must be taken to avoid contamination of the surface while it is being formed.

Nonmetallic Solids

For nonmetallic materials, the sample is often supported on an electrode whose emission spectrum will not interfere with the analysis. Carbon is an ideal electrode material for many applications. It can be obtained in a highly pure form, is a good conductor, has good heat resistance, and is readily shaped. Manufacturers offer carbon electrodes in many sizes, shapes, and forms. Frequently, one of the electrodes is a cylinder with a small crater drilled into one end; the finely ground sample is then packed into this cavity. The other electrode is commonly a tapered carbon rod with a slightly rounded tip. This configuration appears to produce the most stable and reproducible arc or spark. Figure 10-16 illustrates some of the common electrode forms.

TABLE 10-3 Comparison of Detection Limits for Several Atomic Spectral Methods*

Method	Number of Elements Detected at Concentrations of <1 ppb	1–10 ppb	11–100 ppb	101–500 ppb	>500 ppb
Inductively coupled plasma emission	9	32	14	6	0
Flame atomic emission	4	12	19	6	19
Flame atomic fluorescence	4	14	16	4	6
Flame atomic absorption	1	14	25	3	14

*Detection limits correspond to a signal that is twice as great as the standard deviation for the background noise. Data abstracted with permission from V. A. Fassel and R. N. Kniseley, *Anal. Chem.*, **1974,** *46*(13), 1111A. Copyright 1974 American Chemical Society.

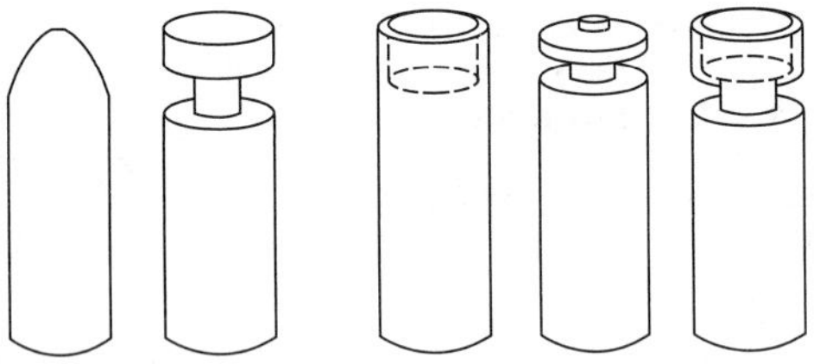

Figure 10-16 Some typical graphite electrode shapes. Narrow necks are to reduce thermal conductivity.

Another common method for atomization of powdered samples is based upon *briquetting* or *pelleting* the sample. Here, the finely ground sample is mixed with a relatively large amount of powdered graphite, copper, or other conducting and compressible substance. The resulting mixture is then compressed at high pressure into the form of an electrode.

10B-2 Instruments for Arc and Spark Source Spectroscopy

Because of their instabilities, it is necessary to integrate the emission signals from arc and spark sources for at least 20 s and often for a minute or more to obtain reproducible analytical data. This requirement makes the use of sequential spectrometers, such as those described in Section 10A-3, impractical for most applications and demands the use of a simultaneous multichannel instrument. Two types of multichannel instruments have been applied to arc and spark spectroscopy: (1) spectrographs, which are considered briefly in the section that follows and (2) multichannel spectrometers, such as those described in Section 10A-3.

Spectrographs

Beginning in the 1930s, a major portion of industrial laboratories throughout the world adopted spark and arc emission spectroscopy for elemental analysis of raw materials, intermediates, and finished products. In the early years, the only instrument available for this type of analysis was the *spectrograph* in which detection and recording of dispersed radiation is accomplished with a photographic film or plate located at the focal plane or curve of a monochromator.

Figure 10-17 is a schematic of an emission spectrograph, which employs a concave grating for dispersing the radiation from a source. Generally, provisions are made that allow vertical movement of the photographic plate or film holder (the camera) so that several spectra can be recorded successively (see Figure 10-18a, b, and c).

A photographic emulsion can serve as the detection system to record, amplify, and integrate the emission from a spectral line. When a photon of radiation is absorbed by a silver halide particle of the emulsion, the radiant energy is converted to a latent image. Treatment of the emulsion with a reducing agent results in the formation of a multitude of silver atoms for each absorbed photon. This process is an example of chemical amplifi-

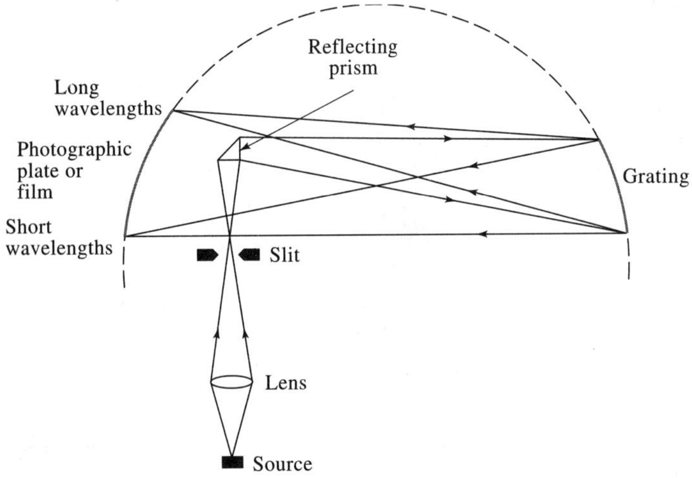

Figure 10-17 The Eagle mounting for a grating spectrograph.

cation. The number of black silver particles, and thus the degree of blackness of any exposed region on an emulsion, is a function of the intensity of the radiation that strikes that region integrated over the time of exposure.

A major disadvantage of spectrographs is the time required to record and analyze spectra. These operations involve exposing the photographic plate or film while samples and standards are being excited, removing the emulsion to a dark room, where developing, fixing, washing, and drying is carried out, and then examining the exposed plate or film in a comparator that permits line identification. This process often consumes hours of operator time. If a quantitative analysis is required, still more time must be expended because the relationship between the relative intensity of the radiation striking a given region of the emulsion and the resulting degree of blackening must be established. This process, called *plate calibration,* must be carried out regularly because the slope and intercept of the resulting calibration curve varies from batch of emulsions to batch of emulsions as well as with the conditions used for development. The calibration curve is also wavelength dependent, so that several regions of the plate or film may have to be calibrated.

Multichannel Photoelectric Spectrometers

Multichannel Photomultiplier Instruments. In the metals industry, there has always been a pressing need for a method for determining the amounts of several metals in samples rapidly enough so that the composition of melts could be adjusted before the melt is poured. This need and others where speed is required led directly in the 1930s to the development and widespread use of photoelectric polychromators, such as that shown in Figure 10-8. These instruments are large and often not very versatile because changing the position of slits and photomultipliers to accommodate a different set of elements is time consuming and with some instruments is not possible in the user's laboratory. On the other hand, once an instrument of this type is set up and calibrated, it is capable of determining 20 or more elements with reasonable accuracy in a few minutes. Thus, the use of these instruments is generally limited to situations where there is a demand for a large number of routine analyses of similar samples. In the metals industries, spark sources were generally employed for excitation of samples and yielded remarkably precise and accurate results, with relative standard deviations of a few percent or even less.

Multichannel Instruments Based on Charge-Transfer Devices. Only recently have there appeared on the market spectrometers that offer the versatility of spectrographs with photographic recording and the speed and precision of spectrometers equipped with multiple photomultipliers. These instruments are the multichannel instruments that employ charge-transfer transducers, such as those illustrated in Figures 10-9 and 10-11. These instruments were originally designed for work plasma sources but now have been applied to spark and arc sources as well.

10B-3 Arc Source Emission Spectroscopy

The usual arc source for a spectrochemical analysis is formed with a pair of graphite or metal electrodes spaced a few millimeters apart. The arc is initially ignited by a low-current spark that causes momentary formation of ions for electrical conduction in the gap; once the arc is struck, thermal ionization maintains the current. Alternatively, the arc can be started by bringing the electrodes together to provide the heat for ionization; the electrodes are then separated to the desired distance.

In the typical arc, currents in the range of 1 to 30 A are used. A dc source usually has an open circuit voltage of about 200 V; ac arc sources operate in a high-voltage range of 2200 to 4400 V or a low-voltage range of 100 to 400 V. With either type, the arc extinguishes at the end of each half cycle. With the high-voltage type, the arc reignites spontaneously; with the low-voltage arc, reignition is brought about by a low-current spark discharge.

Characteristics of Arc Sources

Electricity is carried in an arc by the motion of the electrons and ions formed by thermal ionization; the high temperature that develops is a result of the resistance to this motion by the cations in the arc gap. Thus, the arc temperature depends upon the composition of the plasma, which in turn depends upon the rate of formation of atomic particles from the sample and the electrodes. Typically, the plasma temperature is 4000 to 5000 K.

The spectra produced in the typical arc are rich in intense lines for atoms and contain a lesser number for ionic species ions. Note that in collections of spectral line data, wavelengths are often followed by I, II, and III to indicate the source of the line as that of the neutral atom, the singly charged ion, and the doubly charged ion, respectively.

Cyanogen Spectral Bands. When a carbon or graphite electrode is arced in air, intense bands due to the presence of cyanogen radicals (CN) are emitted, which renders most of the region between 350 to 420 nm useless for elemental analysis. Unfortunately, several elements have their most sensitive lines in this region. To avoid or minimize this interference, arc excitation is usually performed in a controlled gas atmosphere of carbon dioxide, helium, or argon. The CN band emission is not completely eliminated under these conditions, however, unless the electrodes are heated under vacuum to drive off adsorbed nitrogen.

Rates of Emission. It has been shown experimentally that the rates at which various species are volatilized and excited differ widely. The spectra of some species appear early and then disappear as the sample is consumed; the spectra for other species reach their maximum intensities at a later time. Thus, it is necessary to integrate the emission signals for as long as a minute or more. For powder samples the signal for the lines of various elements usually rises rapidly during this period and then decreases to zero as various species are removed by vaporization. Because of this type of behavior, powder samples are often "burned" until they are completely consumed; the integrated output signals then serve as analytical variables.

Applications of Arc Sources

Arc sources are particularly useful for qualitative and semiquantitative analysis of nonmetallic samples, such as soils, plant materials, rocks and minerals. Ordinarily, excitation times and arc currents are adjusted so that complete volatilization of the sample occurs; currents of 5 to 30 A for 20 to 200 s are typical. Commonly, 2 to 50 mg of sample in the form of a powder, small chips, grindings, or filings, often mixed with a weighed amount of graphite, are packed into the cavity of graphite electrodes. Usually the sample-containing electrode is made the anode with a second graphite counter electrode acting as the cathode.

For qualitative and semiquantitative analysis of nonroutine samples, photographic recording is still commonly employed. Several spectra can be obtained on one plate by vertical repositioning of the camera after each excitation. Usually, one or more iron arc spectra are also recorded in order to align the plate or film with a master plate showing the location of characteristic lines.

Figure 10-18 shows a typical portion of a set of arc spectra projected adjacent to a master spectrum in a comparator-densitometer. The wavelength region under examination extends from approximately 3170 to 3310 Å. The upper three spectra are for a sample at three different exposures. The fourth spectrum is that of an iron electrode that was obtained immediately after the three sample exposures. The bottom two spectra are iron spectra on the master plate. After aligning the many iron lines of the master plate with corresponding lines on the sample plates, identification of elements can be carried out. Usually, positive identification of an element requires matching of several lines. Note that the moderately intense lines at about 3174 and 3261 Å suggest that the sample contains tin. Copper, zinc, and perhaps vanadium also appear to be present in significant amounts. These conclusions would need to be confirmed by further observations at other wavelengths, however.

Numerous semiquantitative arc spectrographic methods have been described that provide concentration data reliable to within 30% to 200% of the true

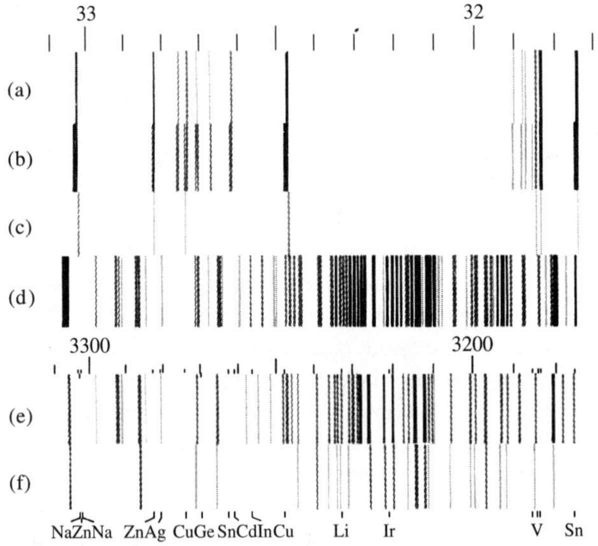

Figure 10-18 Projected spectra by a comparator-densitometer: (a), (b), and (c) spectra of sample at three different exposures; (d) iron spectrum on the sample plate; (e) and (f) iron spectra on the master plate.

amount of an element present in the sample.[20] These methods are useful when the preparation of good standards is not economical. Several such procedures are based upon the total vaporization of a measured quantity (1 to 10 mg) of sample in the arc. The concentration estimate may then be based on a knowledge of the minimum amount of an element required to cause the appearance of each of a series of lines. In other methods, the blackness of elemental lines is measured and compared with the line of an added matrix material or with the background.

Arc excitation is also used for quantitative analysis. The precision obtainable with an arc is, however, generally significantly poorer than that with a spark and much poorer than that with a plasma or flame. Furthermore, emission intensities of solid samples are highly sample dependent. As a consequence matrix matching of standards and samples is usually required for satisfactory results. To partially offset this problem, the internal standard method is always used.

10B-4 Spark Sources and Spark Spectra

A variety of circuits have been developed to produce high-voltage sparks for emission spectroscopy. It has been found that an intermittent spark that always propagates in the same direction gives higher precision and lower drift radiant emission. For this reason, the ac line voltage is often rectified before it is stepped up to 10 to 50 kV in a coil. Solid-state circuitry is used to control both the spark frequency and duration. Typically, four spark discharges take place per half cycle of the 60 Hz line current.

The *average* current with a high-voltage spark is usually significantly less than that of the typical arc, being on the order of a few tenths of an ampere. On the other hand, during the initial phase of the discharge the *instantaneous* current may exceed 1000 A; here, electricity is carried by a narrow streamer that involves but a minuscule fraction of the total space in the spark gap. The temperature within this streamer is estimated to be as great as 40,000 K. Thus, while the average temperature of a spark source is much lower than that of an arc, the *energy* in the small volume of the streamer may be several times greater. As a consequence, ionic spectra are more pronounced in a high-voltage spark than in an arc. As a matter of fact, lines emitted by ions are often termed "spark lines" by spectroscopists.

Applications of Spark Source Spectroscopy[21]

Quantitative spark analyses demand precise control of the many variables involved in sample preparation and excitation, and also in film processing with spectrographs. In addition, quantitative measurements require a set of carefully prepared standards for calibration; these standards should approximate as closely as possible the composition and physical properties of the samples to be analyzed. Generally, spark source analyses are based upon the ratio of intensity of the analyte line to the intensity of an internal standard line (usually a line of a major constituent of the sample). Under ideal conditions, relative standard deviations of a few percent can be realized with spark spectral measurements.

Currently, the primary use of spark source emission spectroscopy is for the identification and analysis of metals and other conducting materials. It is used in preference to plasma methods when dissolving samples is to be avoided because of time considerations. Most commonly detection is carried out with a polychromator equipped with photomultiplier tubes. In fact several modern multichannel instruments are now equipped with interchangeable sources that permit excitation by plasmas, arcs, sparks, glow discharge, and lasers.

An important use of spark source (and to a lesser extent, arc source) spectrometers is in foundry and shop floors, scrap yards, and metal-casting facilities. The instruments for these applications are often mobile and are equipped with a handheld spark or arc gun that the operator can touch to the metal surface to cause excitation. Such equipment is used for the rapid identification of alloy types and for the analysis of melts before casting. One of these instruments is small enough (~30 × 20 × 10 in.) and light enough (25 kg) to be carried from place to place by an operator. It has been designed for sorting, grading, and identifying up to 14 components in steels.

---

[20] For a more complete description of some semiquantitative procedures, see C. E. Harvey, *Spectrochemical Procedures,* Chapter 7. Glendale, CA: Applied Research Laboratories, 1950; T. Torok, J. Mika, and E. Gegus, *Emission Spectrochemical Analysis,* pp. 284–308. New York: Crane Russak, 1978.

[21] See J. P. Walters, *Science,* **1977,** *198,* 787.

10C MISCELLANEOUS SOURCES FOR OPTICAL EMISSION SPECTROSCOPY

In this section we consider three other emission sources that find lesser use than the plasma and arc and spark sources we have just considered.

10C-1 Flame Emission Sources

For many years, flames have been employed to excite emission spectra for various elements, and most modern atomic absorption spectrometers are readily adapted for flame emission work. Flames are not, however, widely used for this purpose because for most single-element determinations, absorption methods provide as good or better results in terms of accuracy, convenience, and detection limits. For multielement analyses, plasma sources are far superior to flames in most regards. For these reasons flame emission spectrometry currently finds little use except for the determination of the alkali metals and occasionally calcium. These elements are excited at the relatively low temperatures of flames to give spectra that are remarkably simple and free of interference from other metallic species. Alkali metal spectra generally consist of a relatively few quite intense lines, many of which are in the visible region and are well suited to quantitative emission measurements.

Because of spectral simplicity, simple filter photometers usually suffice for routine determinations of the alkali and alkaline-earth metals. A low-temperature flame is employed to eliminate excitation of most other metals. As a consequence, the spectra are simple, and interference filters can be used to isolate the desired emission line.

Several instrument manufacturers supply flame photometers designed specifically for the determination

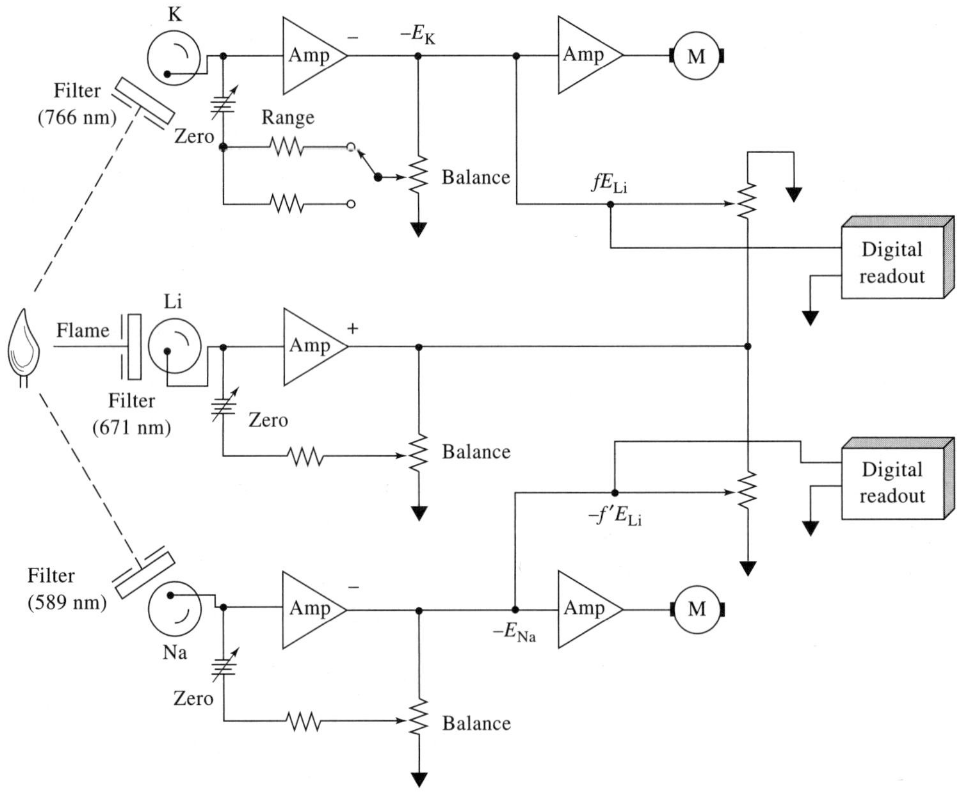

Figure 10-19 A three-channel photometer for monitoring emission by K, Li, and Na. *(From J. D. Ingle Jr. and S. R. Crouch,* Spectrochemical Analysis, *p. 254. Englewood Cliffs, NJ: Prentice-Hall, 1988. With permission.)*

of sodium, potassium, and sometimes calcium in blood serum, urine, and other biological fluids. A typical instrument for determining sodium and potassium is a three-channel device, such as that shown in Figure 10-19. Two of the channels transmit a sodium line and a potassium line, respectively; the third channel is for lithium radiation, an element that is introduced in a fixed amount into all samples and standards as an internal standard. For samples that contain lithium, this third channel can be used for estimating the concentration of that element; in this case relative line intensities rather than the ratio of line intensities serves as the analytical parameter.

In the instrument shown in Figure 10-19, the radiation from the flame is split into three beams of approximately equal power. Each beam then passes into a separate photometric system that consists of an interference filter that transmits an emission line of one of the elements while absorbing those of the other two, a phototube, and an amplifier. The ratios of outputs of the sodium and lithium transducer and the potassium and lithium transducer are then recorded. Many clinical analyzers are provided with automatic dilution systems so that samples and standards are diluted identically with the lithium internal standard solution. A blank containing only the lithium standard is used periodically to zero the instrument. Often a one-point or two-point calibration is performed automatically after a certain number of samples have been analyzed. Automatic instruments of this type can often run as many as 100 samples per hour.

10C-2 Glow Discharge Sources

The glow discharge atomizer described in Section 8C-2 has proven to be a useful source for exciting emission spectra of metals, alloys, and other solid materials. Several instrument manufacturers offer glow discharge sources for this purpose.

10C-3 Laser Microprobe Sources

When a powerful laser pulse is focused on a 5 to 50 μm spot on a sample surface, a small amount of the solid is vaporized regardless of whether or not it is a conductor. The resulting plume is made up of atoms, ions, and molecules. In the microprobe, the contents of the plume are excited by a spark between a pair of small electrodes located immediately above the surface of the sample. The resulting radiation is then focused upon a suitable monochromator/detection system. With this type of source, it has been possible to determine the trace element composition of single blood cells and tiny inclusion areas in alloys.

10D QUESTIONS AND PROBLEMS

10-1 What is an internal standard and why is it used?

10-2 Why are atomic emission methods with an inductively coupled plasma source better suited for multielement analysis than are flame atomic absorption methods?

10-3 Why do ion lines predominate in spark spectra and atom lines in arc and inductively coupled plasma spectra?

10-4 Calculate the theoretical reciprocal linear dispersion of an echelle grating with a focal length of 0.75 m, a groove density of 100 grooves/mm, a diffraction angle of 63°26′ when the diffraction order is (a) 20 and (b) 100.

10-5 Why are arc sources often blanketed with a stream of an inert gas?

10-6 Describe three ways of introducing a sample into an ICP torch.

10-7 What are the relative advantages and disadvantages of ICP torches and dc argon torches?

10-8 Why are ionization interferences less severe in ICP than in flame emission spectroscopy?

10-9 What are some of the advantages of plasma sources compared with flame sources for emission spectrometry?

10-10 Why is the internal standard method often employed in plasma emission spectrometry?

Chapter 11

Atomic Mass Spectrometry

Atomic mass spectrometry is a versatile and widely used tool for identifying the elements present in samples of matter and for determining their concentrations.[1] *Nearly all the elements in the periodic table can be determined by mass spectrometry.*

Atomic mass spectrometry offers a number of advantages over the atomic optical spectrometric methods that we have thus far considered, including: (1) detection limits that are, for many elements, as great as three orders of magnitude better than optical methods; (2) remarkably simple spectra that are usually unique and often easily interpretable; and (3) the ability to measure atomic isotopic ratios. Disadvantages include: (1) instrument costs that are two to three times that of optical atomic instruments, (2) instrument drift that can be as high as 5% to 10% per hour, and (3) certain types of interference effects that will be discussed later.

[1] For a general reference on atomic mass spectrometry, see *Inorganic Mass Spectrometry,* F. Adams, R. Gijbels, and R. Van Grieken, Eds. New York: Wiley, 1988.

11A SOME GENERAL FEATURES OF ATOMIC MASS SPECTROMETRY

An atomic mass spectrometric analysis involves the following steps: (1) atomization, (2) conversion of a substantial fraction of the atoms formed in step 1 to a stream of ions (usually singly charged positive ions), (3) separating the ions formed in step 2 on the basis of their mass-to-charge ratio (m/z), where m is the mass of the ion in atomic mass units and z is its charge, and (4) counting the number of ions of each type or measuring the ion current produced when the ions formed from the sample strike a suitable transducer. Because most of the ions formed in step 2 are singly charged, m/z is usually simply the mass of the ion. Steps 1 and 2 involve the same techniques that were discussed in Section 8C for atomic optical spectroscopy. Steps 3 and 4 are carried out with a mass spectrometer.

11A-1 Atomic Weights in Mass Spectrometry

At the outset, it is worthwhile to point out that the atomic weights in the literature of mass spectrometry, and in this chapter, differ from those used in most other subdisciplines of analytical chemistry because mass spectrometers discriminate among the masses of isotopes, while other analytical instruments generally do not. We will therefore review briefly some terms related to atomic (and molecular) weights.

Atomic and molecular weights are generally expressed in terms of *atomic mass units* (amu), or *daltons* (Da). The atomic mass unit, or dalton, is based upon a relative scale in which the reference is the carbon isotope $^{12}_{6}C$, which is assigned a mass of exactly 12 amu. Thus, *the amu, or Da, is defined as 1/12 of the mass of one neutral* $^{12}_{6}C$ *atom.* This definition makes 1 amu, or 1 Da, of carbon equal to

$$1 \text{ amu} = 1 \text{ Da}$$
$$= \frac{1}{12}\left(\frac{12 \text{ g } ^{12}\text{C/mol } ^{12}\text{C}}{6.0221 \times 10^{23} \text{ atoms } ^{12}\text{C/mol } ^{12}\text{C}}\right)$$
$$= 1.66054 \times 10^{-24} \text{ g/atom } ^{12}\text{C}$$
$$= 1.66054 \times 10^{-27} \text{ kg/atom } ^{12}\text{C}$$

The atomic weight of an isotope such as $^{35}_{17}Cl$ is then related to that of the reference $^{12}_{6}C$ atom by comparing the masses of the two isotopes. Such a comparison reveals that the chlorine-35 isotope has a mass that is 2.91407 times greater than the mass of the carbon isotope. Therefore, the atomic mass of the chlorine isotope is

$$\text{atomic mass } ^{35}_{17}\text{Cl} = 12.0000 \text{ Da} \times 2.91407$$
$$= 34.9688 \text{ Da}$$

Because 1 mol of $^{12}_{6}C$ weighs 12.0000 g, the atomic weight $^{35}_{17}Cl$ is 34.9688 g/mol.[2]

In mass spectrometry, in contrast to most types of chemistry, we are often interested in the *exact mass m* of particular isotopes of an element or the exact mass of compounds containing a particular set of isotopes. Thus, we may need to distinguish between the masses of compounds such as

$$^{12}C^{1}H_4 \qquad m = 12.000 \times 1 + 1.007825 \times 4$$
$$= 16.031 \text{ Da}$$
$$^{13}C^{1}H_4 \qquad m = 13.00335 \times 1 + 1.007825 \times 4$$
$$= 17.035 \text{ Da}$$
$$^{12}C^{1}H_3{}^{2}H_1 \qquad m = 12.000 \times 1 + 1.007825 \times 3 + 2.0140 \times 1$$
$$= 17.037 \text{ Da}$$

Normally, in mass spectrometry, exact masses are quoted to three or four figures to the right of the decimal point in the calculations above, because high-resolution mass spectrometers make measurements at this level of precision.

In other contexts, we shall use the term *nominal mass,* which implies a whole-number precision in a mass measurement. Thus, the nominal masses of the three isomers just cited are 16, 17, and 17 Da, respectively.

The *chemical atomic weight* or the *average atomic weight* (*A*) of an element in nature is given by the equation

$$A = A_1p_1 + A_2p_2 + \cdots + A_np_n = \sum_{i=1}^{n} A_np_n$$

where $A_1, A_2, \cdots A_n$ are the atomic masses in daltons of the *n* isotopes of the element and $p_1, p_2, \cdots p_n$ are the fractional abundances of these isotopes in nature. The chemical atomic weight is, of course, the type of weight of interest to chemists for most purposes. The *average* or *chemical molecular weight* of a compound is then the sum of the chemical atomic weights for the atoms appearing in the formula of the compound. Thus, the chemical molecular weight of CH_4 is $12.01115 + 4 \times 1.00797 = 16.0434$ Da.

11A-2 Mass-to-Charge Ratio

One other term that is used throughout this chapter is the *mass-to-charge ratio* of an atomic or molecular ion. This term is obtained by dividing the atomic or molecular mass of an ion *m* by the number of charges *z* that the ion bears. Thus, for $^{12}C^{1}H_4^{+}$, $m/z = 16.035/1 = 16.035$. For $^{13}C^{1}H_4^{2+}$, $m/z = 17.035/2 = 8.518$. Because most ions in mass spectrometry are singly charged, the term mass-to-charge ratio is often shortened to the more convenient term *mass.* Strictly speaking, this abbreviation is not correct, but it is widely used in the mass spectrometry literature.

11A-3 Types of Atomic Mass Spectrometry

Table 11-1 lists the most important types of atomic mass spectrometry. Historically, thermal ionization mass spectrometry and spark source mass spectrometry

[2] A listing of the isotopes of all of the elements and their atomic masses can be found in several handbooks, such as *Handbook of Chemistry and Physics*. 73rd ed., pp. 11-28 to 11-132. Boca Raton, FL: CRC Press, 1992–1993.

TABLE 11-1 Types of Atomic Mass Spectrometry

Name	Acronym	Atomic Ion Sources	Typical Mass Analyzer
Inductively coupled plasma	ICPMS	High-temperature argon plasma	Quadrupole
Direct current plasma	DCPMS	High-temperature argon plasma	Quadrupole
Microwave-induced plasma	MIPMS	High-temperature argon plasma	Quadrupole
Spark source	SSMS	Radio-frequency electric spark	Double-focusing
Thermal ionization	TIMS	Electrically heated plasma	Double-focusing
Glow discharge	GDMS	Glow-discharge plasma	Double-focusing
Laser microprobe	LMMS	Focused laser beam	Time-of-flight
Secondary ion	SIMS	Accelerated ion bombardment	Double-focusing

were the first mass spectrometric methods developed for qualitative and quantitative elemental analysis, and these types of procedures still find applications, although they are now overshadowed by some of the other methods listed in Table 11-1, particularly inductively coupled plasma mass spectrometry. Note that the first three entries in the table are *hyphenated methods,* which are a combination of two instrumental techniques that produce analytical results that are superior in some way to the results from either of the original individual methods. We shall encounter a number of hyphenated methods in various parts of this book.

Before examining the various methods for atomization and ionization found in the table, we need to describe briefly how mass spectrometers are used to separate and measure ionic species.

11B MASS SPECTROMETERS

A mass spectrometer is an instrument that separates rapidly moving ions on the basis of their mass-to-charge ratios, m/z. Most of the ions we will be discussing are singly charged so that the ratio is simply equal to the mass of the ion. Several types of mass spectrometers are currently available from instrument manufacturers. In this chapter, we describe the three types that have found general use in atomic mass spectrometry: namely, the *quadrupole mass spectrometer,* the *time-of-flight mass spectrometer,* and the *double-focusing mass spectrometer.* Other types of mass spectrometers are considered in Chapter 20, which is devoted to molecular mass spectrometry. The last column in Table 11-1 indicates where each of the three types of atomic mass spectrometry is most commonly applied.

The block diagram in Figure 11-1 shows the principal components of all types of mass spectrometers. The purpose of the inlet system is to introduce a micro amount of sample into the ion source where the components of the sample are converted into gaseous ions by bombardment with electrons, photons, ions, or molecules. Alternatively, ionization is brought about by thermal or electrical energy. The output of the ion source is a stream of positive (most common) or negative

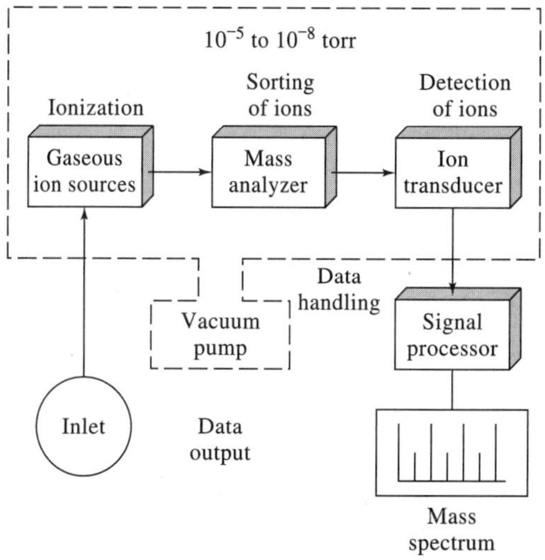

Figure 11-1 Components of a mass spectrometer.

gaseous ions that are then accelerated into the mass analyzer.

The function of the mass analyzer is analogous to the monochromator in an optical spectrometer. In the former, however, dispersion is based upon mass-to-charge ratio of analyte ions rather than upon the wavelength of photons.

Like an optical spectrometer, a mass spectrometer contains a transducer that converts the beam of ions to an electrical signal that can then be processed, stored in the memory of a computer, and displayed or recorded in a number of ways. Unlike most optical spectrometers, mass spectrometers require an elaborate vacuum system to maintain a low pressure in all of the components except the signal processor and readout.

In the sections that follow, we first describe the various transducer systems that find use in mass spectrometers. This is followed by consideration of the three

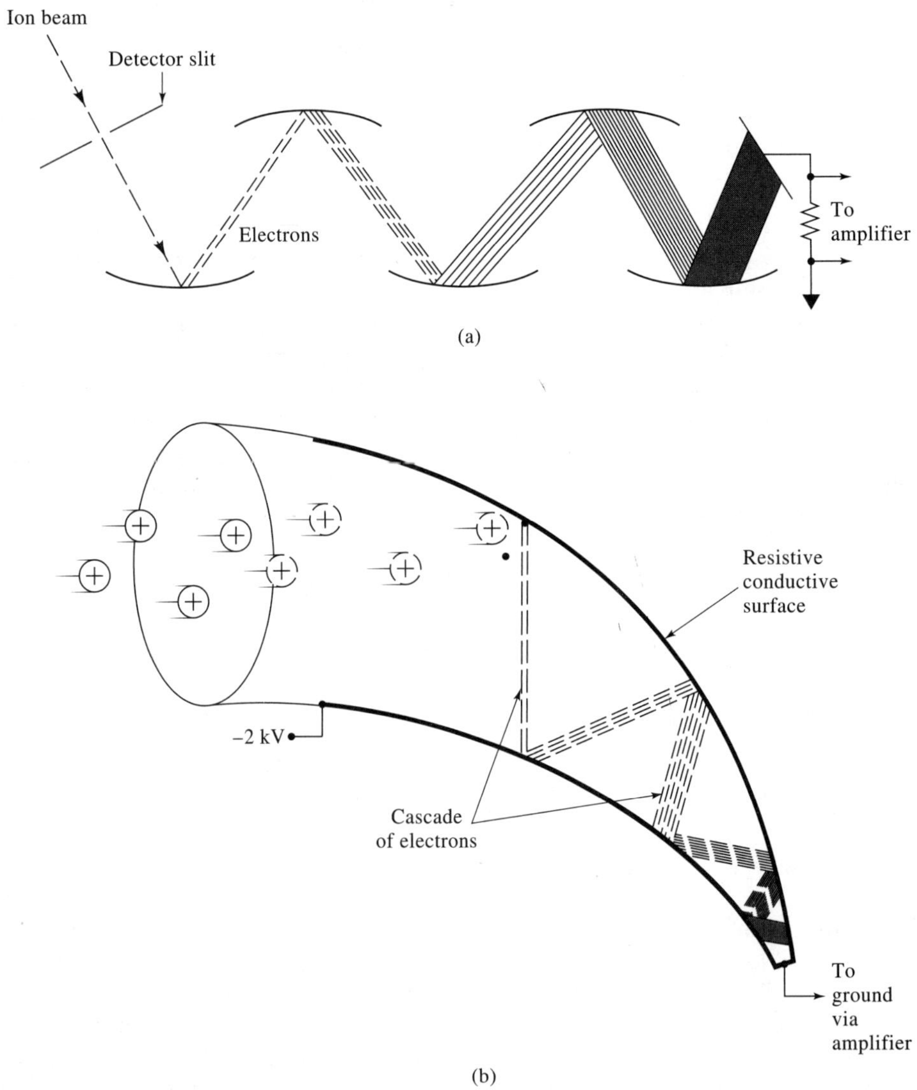

Figure 11-2 (a) Discrete dynode electron multiplier. Dynodes are kept at successively higher potentials via a multistage voltage divider. (b) Continuous dynode electron multiplier. *(Adapted from J. T. Watson,* Introduction to Mass Spectrometry, *p. 247. New York: Raven Press, 1985. With permission.)*

types of mass analyzers that have been used in atomic mass spectrometers. Section 11C contains material on the various ion sources in common usage.

11B-1 Transducers for Mass Spectrometry

Several types of transducers are commercially available for mass spectrometers. The electron multiplier is the transducer of choice for most routine experiments.

Electron Multipliers

Figure 11-2a is a schematic of a discrete-dynode electron multiplier designed for the detection of positive ions. This transducer is very much like the photomultiplier transducer for ultraviolet/visible radiation, with each dynode being held at a successively higher voltage. The cathode and the several dynodes have Cu/Be surfaces from which bursts of electrons are emitted when struck by energetic ions or electrons. Electron multipliers with up to 20 dynodes are available that typically provide a current gain of 10^7.

Figure 11-2b illustrates a continuous-dynode electron multiplier, which is a trumpet-shaped device made of glass that is heavily doped with lead. A potential of 1.8 to 2 kV is impressed across the length of the transducer. Ions that strike the surface near the entrance eject electrons that then skip along the surface, ejecting more electrons with each impact. Transducers of this type typically have current gains of 10^5, but in certain applications gains as high as 10^8 can be achieved.

In general, electron multipliers are rugged and reliable and are capable of providing high-current gains and nanosecond response times. These transducers can be placed directly behind the exit slit of a magnetic sector mass spectrometer, because the ions reaching the transducer usually have enough kinetic energy to eject electrons from the first stage of the device. Electron multipliers can also be used with mass analyzers that utilize low-kinetic-energy ion beams (that is, quadrupoles), but in these applications the ion beam exiting the analyzer is accelerated to several thousand electron volts prior to striking the first stage.

The Faraday Cup

Figure 11-3 is a schematic of a Faraday cup collector. The transducer is aligned so that ions exiting the analyzer strike the collector electrode. This electrode is surrounded by a cage that prevents the escape of reflected ions and ejected secondary electrons. The collector electrode is inclined with respect to the path of the entering ions so that particles striking or leaving the electrode are reflected away from the entrance to the cup. The collector electrode and cage are connected to ground potential through a large resistor. The charge of the positive ions striking the plate is neutralized by a flow of electrons from ground through the resistor. The resulting potential drop across the resistor is amplified by a high-impedance amplifier. The response of this transducer is independent of the energy, the mass, and the chemical nature of the ion. The Faraday cup is inexpensive and simple mechanically and electrically; its main disadvantage is the need for a high-impedance amplifier, which limits the speed at which a spectrum can be scanned. The Faraday cup transducer is also less sensitive than electron multipliers are, because it provides no internal amplification.

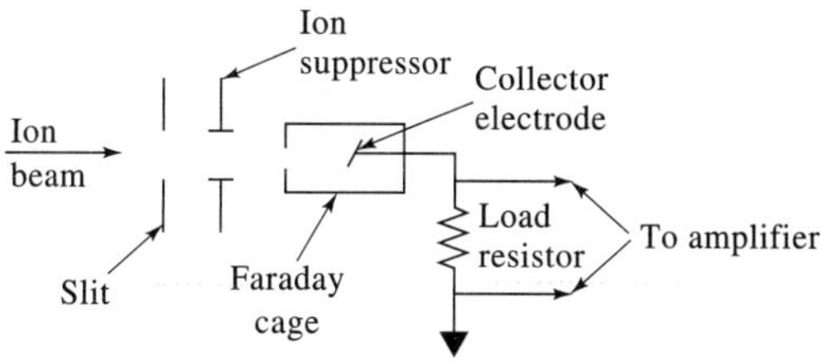

Figure 11-3 Faraday cup detector. The potential on the ion suppressor plates is adjusted to minimize differential response as a function of mass.

Other Types of Detection Systems

Photographic plates coated with a silver bromide emulsion are sensitive to energetic ions. Photographic detection is most frequently encountered in spark source instruments because this type of detection is well suited to the simultaneous observation of a wide range of *m/z* values in instruments that focus ions along a plane (Figure 11-9).

Scintillation-type transducers also find some use. These transducers consist of a crystalline phosphor dispersed on a thin aluminum sheet that is mounted on the window of a photomultiplier tube. When ions (or electrons produced when the ions strike a cathode) impinge upon the phosphor, they produce flashes of light, which are detected by the photomultiplier.

11B-2 Quadrupole Mass Analyzers

The most common type of mass spectrometer used in atomic mass spectroscopy is the quadrupole mass analyzer shown in Figure 11-4. This instrument is more compact, less expensive, and more rugged than most other types of mass spectrometers. It also has the advantage of high scan rates so that an entire mass spectrum can be obtained in less than 100 ms.[3]

The heart of a quadrupole instrument is the four parallel cylindrical rods that serve as electrodes. Opposite rods are connected electrically, one pair being attached to the positive side of a variable dc source and the other pair to the negative terminal. In addition, variable radio-frequency ac potentials, which are 180-deg out of phase, are applied to each pair of rods. In order to obtain a mass spectrum with this device, ions are accelerated into the space between the rods by a potential of 5 to 10 V. Meanwhile, the ac and dc voltages on the rods are increased simultaneously while maintaining their ratio constant. At any given moment, all of the ions except those having a certain *m/z* value strike the rods and are converted to neutral molecules. Thus, only ions having a limited range of *m/z* values reach the transducer. Typically, quadrupole instruments easily resolve ions that differ in mass by one unit.

Note that a quadrupole mass spectrometer is more analogous to an optical variable-band filter photometer than to an optical spectrometer in which a grating simultaneously disperses a spectrum of electromagnetic radiation. Thus, it would perhaps be better to refer to the device as a mass filter rather than a mass spectrometer. However, the latter designation is widely used throughout the literature, and we will follow this practice here and in Chapter 20.

[3] P. E. Miller and M. B. Denton, *Anal. Chem.*, **1986**, *63*, 617.

Ion Trajectories in a Quadrupole

To understand the filtering capability of a quadrupole, we need to consider the effect of the dc and ac potentials on the trajectory of ions as they pass through the channel between the rods.[4] At the outset, let us focus on the pair of positive rods, which are shown in Figure 11-5 as lying in the *xz* plane. In the absence of a dc potential, ions in the channel will tend to converge in the center of the channel during the positive half of the ac cycle and will tend to diverge during the negative half. This behavior is illustrated at points *A* and *B* in the figure. If during the negative half cycle an ion strikes the rod, the positive charge will be neutralized, and the resulting molecule will be carried away. Whether or not a positive ion strikes the rod will depend upon the rate of movement of the ion along the *z* axis, its mass-to-

[4] The discussion of quadrupole mass filters that follows is largely based upon P. E. Miller and M. B. Denton, *J. Chem. Educ.*, **1986**, *63*, 617. See also R. E. Marchand and R. J. Hughes, *Quadrupole Storage Mass Spectrometry*. New York: Wiley, 1989; J. J. Leary and R. L. Schmidt, *J. Chem. Educ.*, **1996**, *73*, 1142.

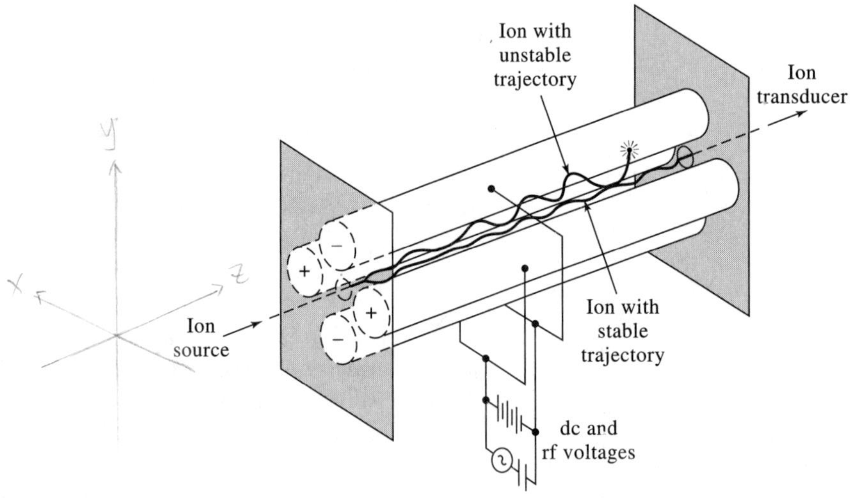

Figure 11-4 A quadrupole mass spectrometer.

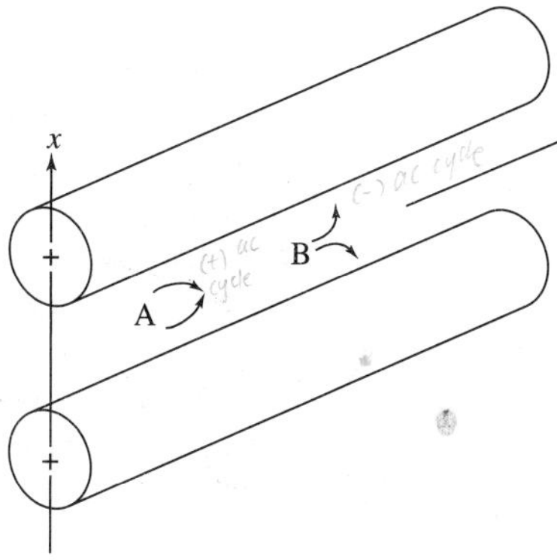

Figure 11-5 Operation of a quadrupole in the *xz* plane. *A:* ions are focused toward *z* axis; *B:* ions are attracted toward *x* rods.

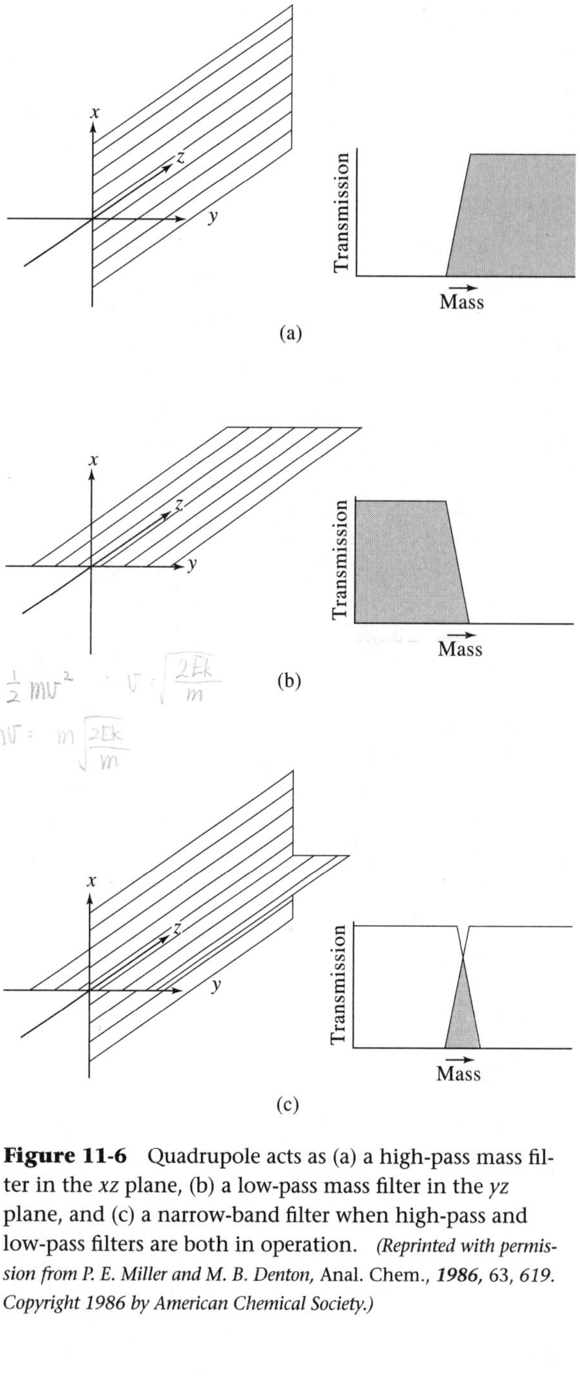

Figure 11-6 Quadrupole acts as (a) a high-pass mass filter in the *xz* plane, (b) a low-pass mass filter in the *yz* plane, and (c) a narrow-band filter when high-pass and low-pass filters are both in operation. *(Reprinted with permission from P. E. Miller and M. B. Denton,* Anal. Chem., ***1986,*** 63, *619. Copyright 1986 by American Chemical Society.)*

charge ratio, and the frequency and magnitude of the ac signal.

Now let us consider the effect of a positive dc potential that is superimposed upon the ac signal. From Newtonian physics, the momentum of ions of equal kinetic energy is directly proportional to the square root of mass. It is therefore more difficult to deflect a heavier ion than to deflect a lighter one. If an ion in the channel is heavy and/or the frequency of the ac potential is large, the ion will not respond significantly to the alternating potential and will be influenced largely by the dc potential. Under these circumstances, the ion will tend to remain in the space between the rods. In contrast, if the ion is light and/or the frequency is low, the ion may collide with the rod and be eliminated during the negative excursion of the ac potential. Thus, as shown in Figure 11-6a, the pair of positive rods form a high-pass mass filter for positive ions traveling in the *xz* plane.

Now let us turn to the pair of rods that are maintained at a negative dc potential. In the absence of the ac potential, all positive ions will tend to be drawn to the rods, where they are annihilated. For the lighter ions, however, this movement may be offset by the excursion of the ac potential. Thus, as shown in Figure 11-6b, the rods in the *yz* plane operate as a low-pass filter.

In order for an ion to travel through the quadrupole to the detector, it must have a stable trajectory in both the *xz* and *yz* planes. Thus, the ion must be sufficiently heavy so that it will not be eliminated by the high-mass filter in the *xz* plane and must be sufficiently light so that it will not be removed by the low-mass filter in the *yz* plane. Therefore, as is shown in Figure 11-6c, the total quadrupole transmits a band of ions that have a limited range of *m/z* values. The center of this band can be varied by adjusting the ac and dc potentials.

Scanning with a Quadrupole Filter

The differential equations that are needed to describe the behavior of ions of different masses in a quadrupole are complex and difficult to treat analytically and are beyond the scope of this text. These equations reveal, however, that the oscillations of charged particles in a quadrupole fall into two categories: (1) those in which the amplitudes of the oscillations are finite and (2) those in which the oscillations grow exponentially and ultimately approach infinity. The variables contained in these equations are mass-to-charge ratio, the dc voltage, the frequency and magnitude of the ac potential, and the distance between the rods. The resolution of a quadrupole is determined by the ratio of the ac to dc potential and becomes a maximum when this ratio is just slightly less than 6. Thus, quadrupole spectrometers are operated at a constant potential ratio of this value.

In order to scan a mass spectrum with a quadrupole instrument, the ac voltage V and the dc voltage U are increased simultaneously from zero to some maximum value while their ratio is maintained at slightly less than 6. The changes in potential during a typical scan are shown in Figure 11-7. The two diverging straight lines show the variation in the two dc potentials as a function of time. The time for a single sweep is a few milliseconds. While the dc potentials are varied from 0 to about ±250 V, the ac signals increase linearly from zero to approximately 1500 V. Note that the ac signals are 180 deg out of phase.

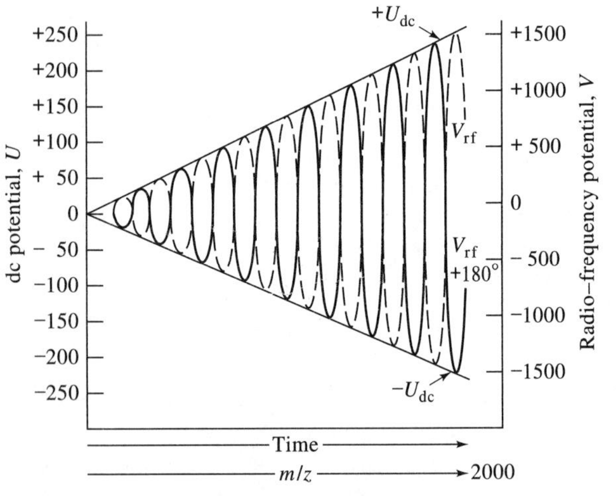

Figure 11-7 Voltage relationships during a mass scan with a quadrupole analyzer.

Quadrupole mass spectrometers with ranges that extend up to 3000 to 4000 m/z are now available from several instrument manufacturers. These instruments easily resolve ions that differ in mass by one unit. Generally, quadrupole instruments are equipped with a circular aperture, rather than a slit (see Figure 11-4), to introduce the sample into the dispersing region. The aperture provides a much greater sample throughput than can be tolerated in magnetic sector instruments, where resolution is inversely related to slit width.

11B-3 Time-of-Flight Mass Analyzers

In time-of-flight (TOF) instruments, positive ions are produced periodically by bombardment of the sample with brief pulses of electrons, secondary ions, or laser-generated photons. These pulses typically have a frequency of 10 to 50 kHz and a lifetime of 0.25 μs. The ions produced in this way are then accelerated by an electric field pulse of 10^3 to 10^4 V that has the same frequency as, but lags behind, the ionization pulse. The accelerated particles pass into a field-free *drift tube* about a meter in length (Figure 11-8). Because all ions entering the tube ideally have the same kinetic energy, their velocities in the tube must vary inversely with their masses, with the lighter particles arriving at the detector earlier than the heavier ones. Typical flight times are 1 to 30 μs.[5]

The transducer in a time-of-flight mass spectrometer is usually an electron multiplier whose output is displayed on the vertical deflection plates of an oscilloscope while the horizontal sweep is synchronized with the accelerator pulses; an essentially instantaneous display of the mass spectrum appears on the oscilloscope screen. Typical flight times are on the order of a microsecond, so digital data acquisition requires extremely fast electronics. Variations in ion energies and initial positions cause peak broadening that generally limits attainable resolution.

From the standpoint of resolution and reproducibility, instruments employing time-of-flight separators are not as satisfactory as those based on magnetic or quadrupole separators. Several advantages partially offset these limitations, however, including simplicity and ruggedness, ease of accessibility to the ion source, virtually unlimited mass range and rapid

[5] W. N. Delgass and R. G. Cook, *Science*, **1987**, *235*, 545.

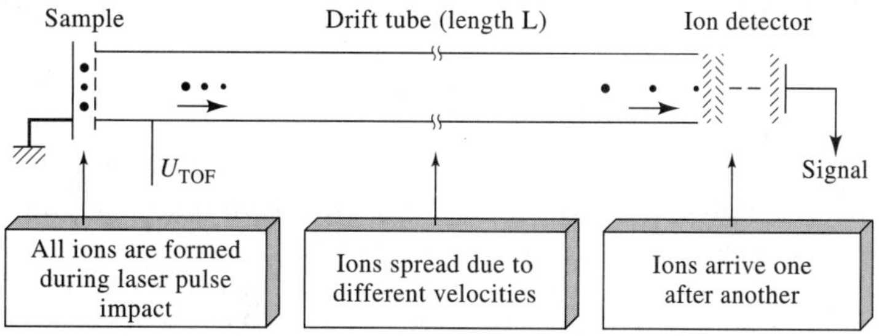

Figure 11-8 Principle of a time-of-flight mass spectrometer. A bunch of ions produced by a laser probe is accelerated into the drift tube where separation occurs. *(From A. H. Verbueken, F. J. Bruynseels, R. Van Grieken, and F. Adams, in* Inorganic Mass Spectrometry, *p. 186. F. Adams, R. Gijbels, and R. Van Grieken, Eds. New York: Wiley, 1988. With permission.)*

data acquisition rate. Several instrument manufacturers now offer time-of-flight instruments, but they are less widely used than are quadrupole mass spectrometers.

11B-4 Double-Focusing Analyzers

As shown in Figure 11-9, a double-focusing mass spectrometer contains two devices for focusing a beam of ions: an *electrostatic analyzer* and a *magnetic sector analyzer.* In this instrument, the ions from a source are accelerated through a slit into a curved electrostatic field that serves to focus a beam of ions having a narrow band of kinetic energies onto a slit that leads to a curved magnetic field. In this field, the lightest ions are deflected the most and the heaviest the least. The dispersed ions then fall on a photographic plate and are thus recorded. A detailed discussion of how the electrostatic and magnetic analyzers function is found in Section 20C-3. As shown in Table 11-1, analyzers of this type find use in several types of atomic mass spectrometry.

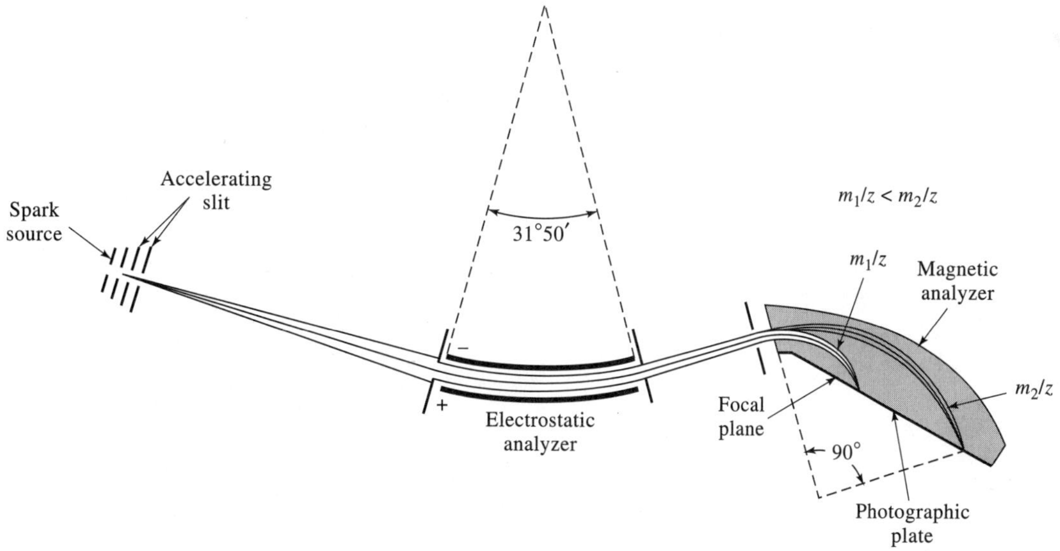

Figure 11-9 Mattacuh-Herzog–type double-focusing mass spectrometer. Resolution > 10^5 has been achieved with instruments based on this design.

11C INDUCTIVELY COUPLED PLASMA MASS SPECTROMETRY

Since the early 1980s, inductively coupled plasma mass spectrometry (ICPMS) has grown to be one of the most important techniques for elemental analysis because of its low detection limits for most elements, its high degree of selectivity, and its reasonably good precision and accuracy.[6] In these applications an ICP torch serves as an atomizer and ionizer. For solutions, sample introduction is accomplished by a conventional or an ultrasonic nebulizer. For solids, one of the other sample introduction techniques discussed in Section 8C-2, such as spark or laser ablation or glow discharge, are employed. Commercial versions of instruments for these various techniques have been on the market since 1983. In these instruments, positive metal ions, produced in a conventional ICP torch, are sampled through a differentially pumped interface linked to a quadrupole mass spectrometer. The spectra produced in this way, which are remarkably simple compared with conventional ICP optical spectra, consist of a simple series of isotope peaks for each element present. These spectra are used for qualitative determination of the elements present in the sample and for the quantitative measurement of their amounts. Usually the latter are based upon calibration curves in which the ratio of the ion count for the analyte to the count for an internal standard is plotted as a function of concentration. Analyses can also be performed by the isotope dilution technique, which is described in detail in Section 32D.

11C-1 Instruments for ICPMS[7]

Figure 11-10 shows schematically the components of a commercial ICPMS system. A critical part of the instrument is the interface that couples the ICP torch, which operates at atmospheric pressure with the mass spectrometer that requires a pressure of less than 10^{-4} torr.

[6] For detailed descriptions of this technique, see K. E. Jarvis, A. L. Gray, and R. S. Houk, *Handbook of Inductively Coupled Plasma Mass Spectrometry.* New York: Chapman and Hall, 1992; *Inductively Coupled Mass Spectrometry,* 2nd ed., A. Montaser and D. W. Golightly, Eds. New York: VCH Publishers, 1992; R. S. Houk, *Anal. Chem.,* **1986,** *58,* 97A; J. W. Olesik, *Anal. Chem.,* **1991,** *63,* 12A; N. P. Vela, L. K. Olson, and J. A. Caruso, *Anal. Chem.,* **1993,** *65,* 585A.

[7] For a description of eleven commercially available instruments, see A. Newman, *Anal. Chem.,* **1996,** *69,* 46A.

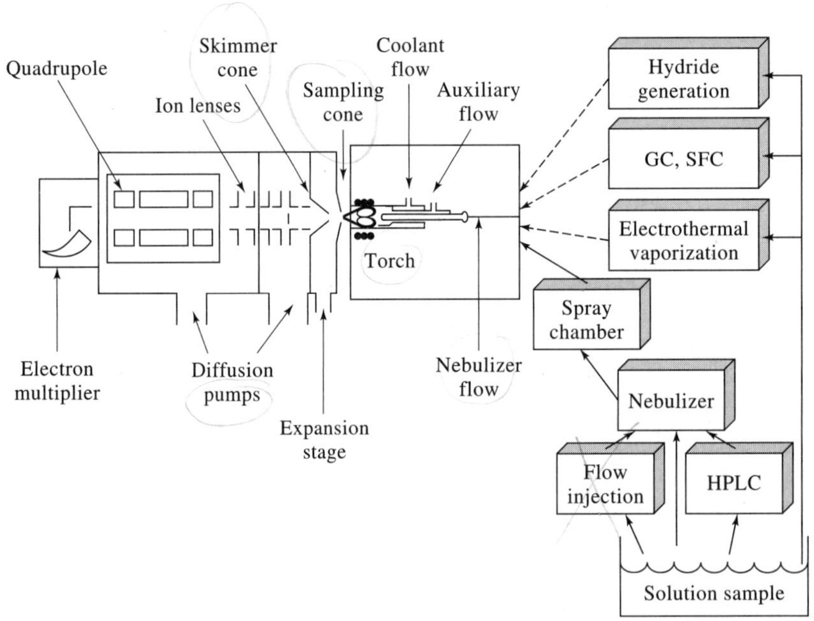

Figure 11-10 Schematic of an ICPMS system. Dotted lines show introduction of gaseous samples; solid lines shown introduction of liquid samples. *(From N. P. Vela, L. K. Olson, and J. A. Caruso,* Anal. Chem., ***1993,*** 65, *587A. With permission.)*

This coupling is accomplished by a differentially pumped interface coupler that consists of a *sampling cone,* which is a water-cooled nickel cone with a small orifice (<1.0 mm) in its center. The hot plasma gas is transmitted through this orifice into a region that is maintained at a pressure of about 1 torr by means of a mechanical pump. In this region, rapid expansion of the gas occurs, which results in its being cooled. A fraction of the gas in this region then passes through a small hole in a second cone called a *skimmer* and into a chamber maintained at the pressure of the mass spectrometer. Here the positive ions are separated from electrons and molecular species by a negative potential, are accelerated, and are focused by a magnetic ion lens onto the entrance orifice of a quadrupole mass analyzer.

Performance specifications for typical commercial atomic mass spectrometers equipped with an ICP torch include a mass range of 3 to 300, the ability to resolve ions differing in m/z by 1, and a dynamic range of 6 orders of magnitude. Over 90% of the elements in the periodic table have been determined with this type of instrument. Measurement times of 10 s per element, with detection limits in the 0.1- to 10-ppb range for most elements, are achieved. Relative standard deviations of 2% to 4% for concentrations in the middle regions of calibration curves are reported.

Recently, at least two instrument manufacturers have coupled laser-sampling systems to ICPMS to produce instruments designed for the elemental analysis of solid samples with minimal sample preparation.[8] In these instruments, pulsed laser beams are focused onto a few square micrometers of a solid, giving power densities of as great as 10^{12} W/cm^2. Such high-intensity radiation rapidly vaporizes most materials, even refractory ones. A flow of argon then carries the vaporized sample into an ICP torch where atomization and ionization take place. The resulting plasma then passes into the mass spectrometer. Instruments of this type appear to be particularly well suited for semiquantitative analysis of samples that are difficult to decompose or dissolve such as geological materials, alloys, glasses, agricultural products, urban particulates, and soils. Figure 11-11 is a mass spectrum of a standard rock sample obtained by laser ablation/ICPMS. To obtain this spectrum, the rock was ground into a fine powder and formed into a small disk under pressure. After ablation with the beam of a laser, the resulting vapor was introduced into an ICP torch by injector gas flow, where ion-

[8] See E. A. Denoyer, K. J. Fredeen, and J. W. Hager, *Anal. Chem.,* **1991**, *63,* 445A.

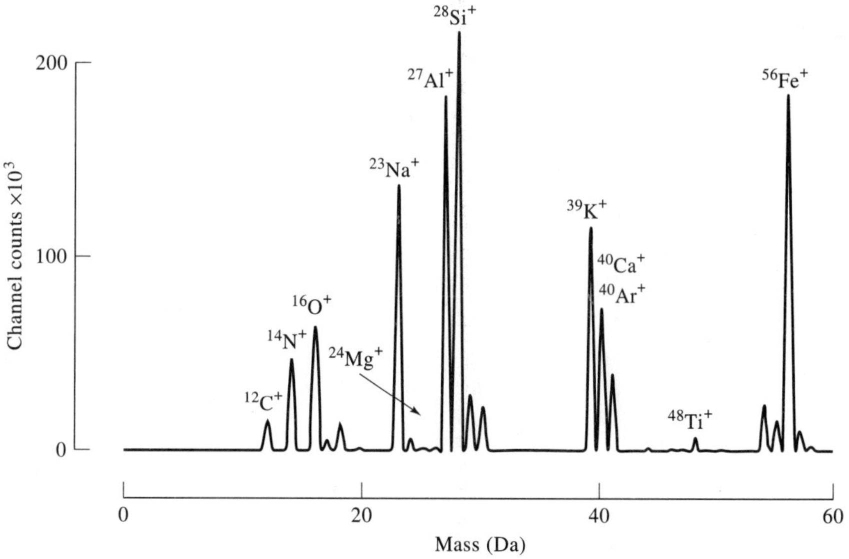

Figure 11-11 Spectrum of a standard rock sample obtained by laser ablation/ICPMS. Major components (%): Na, 5.2; Mg, 0.21; Al, 6.1; Si, 26.3; K, 5.3; Cu, 1.4; Ti, 0.18; Fe, 4.6. *(From* Inorganic Mass Spectrometry, *F. Adams, R. Gijbek, and R. Van Grieken, Eds., p. 297. New York: Wiley, 1988. With permission.)*

ization took place. The gaseous ions were then analyzed in a quadrupole mass spectrometer.

Accessories that permit use of most of the other ion sources listed in Table 11-1 in conjunction with an ICP torch are also available commercially.

11C-2 Atomic Mass Spectra and Interferences

One of the advantages of using mass spectrometric detection with inductively coupled plasmas as opposed to optical detection is that mass spectra are usually much simpler and easier to interpret than corresponding optical spectra. This property is particularly true for those elements such as the rare earths that may exhibit thousands of emission lines. Figure 11-12 illustrates this advantage for a solution containing cerium. Figure 11-12a is the optical emission spectrum of a solution containing 100 ppm cerium; the lower plot is the mass spectrum of a solution having a concentration of 10 ppm of the element. The optical spectrum contains a dozen or more strong cerium lines and hundreds of weak lines for the element, all of which are superimposed upon a complicated background spectrum. The optical back-

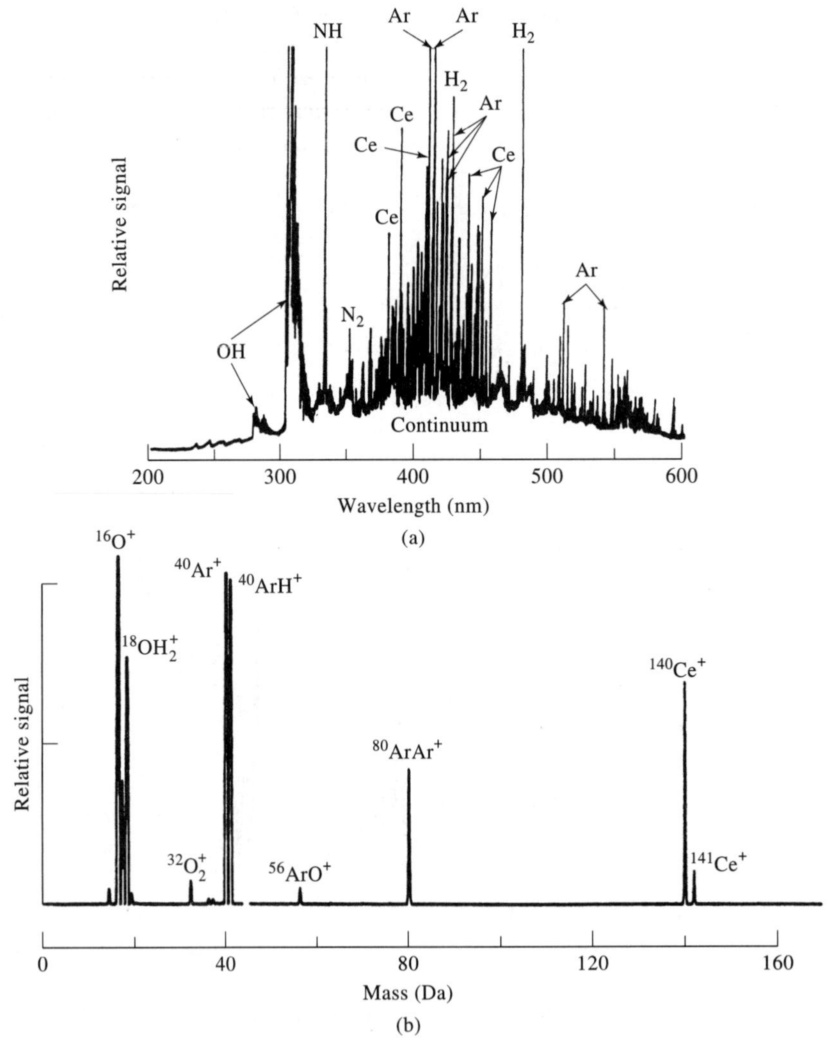

Figure 11-12 Comparison of (a) optical ICP spectrum for 100 ppm cerium and (b) mass ICP spectrum for 10 ppm cerium.

ground is made up of molecular bands from atmospheric contaminants, such as NH, OH, N_2, and H_2, and a continuum from the recombination of argon and other ions with electrons. In contrast, the mass spectrum for cerium, shown in Figure 11-12b, is markedly simpler, consisting of two isotope peaks for $^{140}Ce^+$ and $^{142}Ce^+$ and a small peak for the doubly charged $^{140}Ce^{2+}$, which is located at $m/z = 70$. Note that the background for the spectrum consists of only a few molecular ion species, and all of these occur at *m/z* values of 40 or less.

Spectra such as that in Figure 11-12b led early workers in the field of ICPMS to have hopes of an "interference-free method." Unfortunately, this hope was not realized in further studies, and serious interference problems are sometimes encountered in atomic mass spectrometry just as in optical atomic spectroscopy. Interference effects in atomic mass spectroscopy fall into two broad categories: spectroscopic interferences and nonspectroscopic interferences. The latter effects are similar to the matrix effects encountered in optical emission and absorption methods and generally can be coped with by means of the various methods described in the two previous chapters.

Spectroscopic Interferences

Spectroscopic interferences occur when an ionic species in the plasma has the same *m/z* values as an analyte ion. Such interferences fall into four categories: (1) isobaric ions, (2) polyatomic or adduct ions, (3) doubly charged ions, and (4) refractory oxide ions.[9]

Isobaric Interferences. *Isobaric species* are two elements that have isotopes having substantially the same mass. For atomic mass spectrometry with a quadrupole mass spectrometer, isobaric species are isotopes that differ in mass by less than one unit. With higher resolution instruments, smaller differences can be tolerated.

Most elements in the periodic table have one, two, or even three isotopes that are free from isobaric overlap. An exception is indium, which has two stable isotopes $^{113}In^+$ and $^{115}In^+$. The former overlaps with $^{113}Cd^+$ and the latter with $^{115}Sn^+$. More commonly, an isobaric interference occurs with the most abundant and thus the most sensitive isotope. For example, the very large peak for $^{40}Ar^+$ (see Figure 11-12b) overlaps the peak for the most abundant calcium isotope $^{40}Ca^+$ (97%) making it necessary to use the second most abundant isotope $^{44}Ca^+$ (2.1%). As another example, the most abundant nickel isotope $^{58}Ni^+$ suffers from an isobaric overlap by $^{58}Fe^+$. This interference can be corrected by measuring the peak for $^{56}Fe^+$. From the natural abundance ratio of the $^{56}Fe^+$ isotope to that of $^{58}Fe^+$ isotope, the contribution of Fe to the peak at *m/z* 58 can be obtained and the correction made. Because isobaric overlaps are exactly predictable from abundance tables, corrections for the problem can be carried out with appropriate software. Some current instruments are capable of making such corrections automatically.

Polyatomic Ion Interferences. An interference problem that is usually more serious than that caused by isobaric elements is the problem that arises from polyatomic species formed from interactions between species in the plasma and species in the matrix or the atmosphere. As a consequence, several molecular ions can form and possibly interfere. This type of interference is found largely at *m/z* values of below about 82. The presence of several of these species is shown in Figure 11-12b. Among the potential interferents are $^{40}Ar^{2+}$, $^{40}ArH^+$, $^{16}O_2^+$, $H_2^{16}O^+$, $^{16}OH^+$, $^{14}N^+$, and several others. Some of these can cause serious problems. Examples of serious polyatomic interferences include $^{14}N_2^+$ with $^{28}Si^+$, NOH^+ with $^{31}P^+$, $^{16}O_2^+$ with $^{32}S^+$, $^{40}ArO^+$ with $^{56}Fe^+$, and $^{40}Ar_2^+$ with $^{80}Se^+$. Some of these can be corrected for with a blank. With others, a different analyte isotope may have to be used.

Oxide and Hydroxide Species Interference. The most serious general class of interferences in ICPMS involves oxides and hydroxides formed from the analyte itself, the matrix components, the solvent, and the plasma gases. Some of these oxides are apparent in Figure 11-12b. More serious interferences of this type appear to arise from the oxides and hydroxides of the analytes and of the matrix components. Almost all of these species to some extent form MO^+ and MOH^+ ions, where M represents the analyte or matrix element. The potential then exists for the peaks from these species to overlap the peak of one of the analyte ions. For example, the singly charged oxide ions of the five naturally occurring titanium isotopes have masses of 62, 63, 64, 65, and 66. The peaks for these ions offer potential in-

[9] For comprehensive discussion of interferences in ICPMS, see K. E. Jarvis, A. L. Gray, and R. S. Houk, *Handbook of Inductively Coupled Plasma Mass Spectrometry,* Chapter 5. New York: Blackie, 1992; G. Horlick and Y. Shao, in *Inductively Coupled Plasma in Analytical Atomic Spectrometry,* 2nd ed., pp. 571–596, A. Montaser and D. W. Golightly, Eds. New York: VCH Publishers, 1992.

terferences for the analyte peaks for $^{62}Ni^+$, $^{63}Cu^+$, $^{64}Zn^+$, $^{65}Cu^+$, and $^{66}Zn^+$. As a further example, the potential problems created by the oxide and hydroxide isotopes of calcium in the determination of several metallic species are shown in Table 11-2.

Reduction of oxide and hydroxide formation in plasmas is the focus of much current research. Oxide formation depends upon such experimental variables as injector flow rate, the radio-frequency power, sampler skimmer spacing, sample orifice size, plasma gas composition, oxygen elimination, and solvent removal efficiencies. All of these variables can be manipulated to attack specific oxide and hydroxide overlap problems.

Matrix Effects

In ICPMS, matrix effects become noticeable at concomitant concentrations of greater than about 500 to 1000 μg/mL. Usually these effects cause a reduction in the analyte signal, although under certain experimental conditions signal enhancement is observed. The matrix effect is quite general in that a high concentration of nearly any concomitant element results in a matrix effect. Generally, matrix effects can be minimized by using more dilute solutions, by altering the sample introduction procedure, or by separating out the offending species.

TABLE 11-2 Calcium Oxide and Hydroxide Species and Other Potential Interferences in the Mass Region for Ni Determination

m/z	Element[a]	Interferences
56	Fe(91.66)	^{40}ArO, ^{40}CaO
57	Fe(2.19)	$^{40}ArOH$, $^{40}CaOH$
58	Ni(67.77), Fe(0.33)	^{42}CaO, NaCl
59	Co(100)	^{43}CaO, $^{42}CaOH$
60	Ni(26.16)	$^{43}CaOH$, ^{44}CaO
61	Ni(1.25)	$^{44}CaOH$
62	Ni(3.66)	^{46}CaO, Na_2O, NaK
63	Cu(69.1)	$^{46}CaOH$, $^{40}ArNa$
64	Ni(1.16), Zn(48.89)	$^{32}SO_2$, $^{32}S_2$, ^{48}CaO
65	Cu(30.9)	$^{33}S^{32}S$, $^{33}SO_2$, $^{48}CaOH$

[a]Percentage natural abundance in parentheses. (From M. A. Vaughan and D. M. Templeton, *Appl. Spectrosc.*, **1990,** *44,* 1685. With permission.)

The effects can also be largely eliminated by the use of an appropriate internal standard, that is, by introduction of an internal standard element that has about the same mass and ionization potential as those of the analyte.

11C-3 Applications of ICPMS

ICPMS can be used for qualitative, semiquantitative, and quantitative determination of one or more elements in samples of matter.

Qualitative and Semiquantitative Applications

Because ICPMS is easily adapted to multielement analyses, it is well suited to the rapid characterization and semiquantitative analysis of various types of naturally occurring and man-made complex materials. Generally, detection limits are better than those for optical emission ICP and compete with detection limits for electrothermal atomic absorption spectroscopy.

Generally, atomic mass spectra are a great deal simpler and easier to interpret than optical emission spectra. This property is particularly important for samples that contain rare earth elements and other heavy metals, such as iron, that yield complex emission spectra. Figure 11-12b illustrates this advantage. This spectral simplicity is further illustrated in Figure 11-13, which is the atomic mass spectrum for a mixture of 14 rare earth elements that range in atomic mass from 139 to 175. The optical emission spectrum for such a mixture would be so complex that interpretation would be tedious, time consuming, and perhaps impossible.

Semiquantitative analysis for one or more of the components in a mixture such as the one that yielded Figure 11-13 is possible by measuring the peak ion current or intensity for a solution having a known concentration of element in question. The analyte concentration in the sample is then computed by assuming that the ion current is proportional to concentration. Concentrations calculated by this crude but simple procedure are generally accurate to ±100% relative.

Detection Limits

One of the main attractions of ICPMS lies with the lower detection limits attainable with mass spectrometric detection than with optical detection. These limits in many cases equal and sometimes exceed those that can be realized by electrothermal atomic absorption methods. The ICPMS procedure, of course, offers the great advantages of speed and multielement capability. Figure 11-14 compares detection limits for ICPMS with

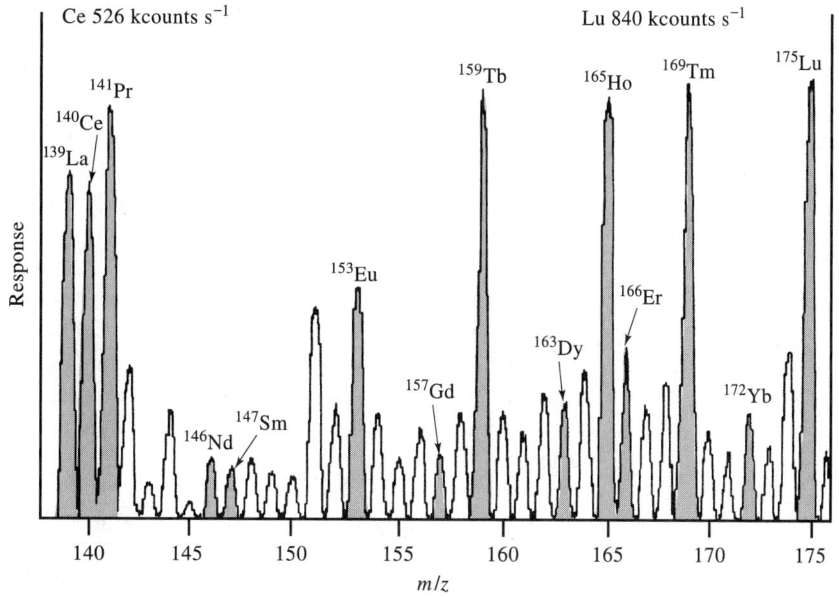

Figure 11-13 ICPMS spectrum for the rare earth elements. Solutions contain 1 μg/mL of each element. *(From K. E. Jarvis,* J. Anal. Atom. Spectrom., ***1989****, 4, 563. With permission.*

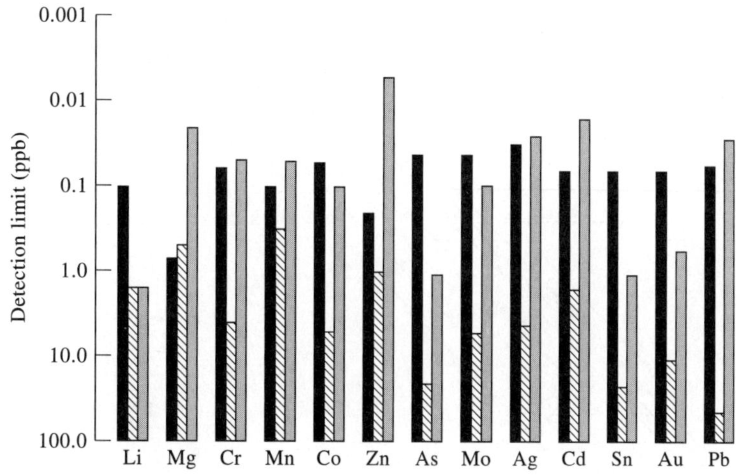

Figure 11-14 Detection limits for selected elements by ICPMS (black bars) compared with those for ICP OES (cross-hatched bars) and ETAAS (gray bars) (inductively coupled plasma optical emission spectrometry and electrothermal atomic absorption spectrometry, respectively), plotted on a logarithmic scale in concentrations of ppb (or μg/L). Because ETAAS detection limits are inherently in mass units (pg), they have been converted to concentration by assuming a 20 μL sample. *(From M. Selby and G. M. Hieftje,* Amer. Lab., ***1987*** *(8), 20. With permission.)*

those for inductively coupled plasma optical emission spectroscopy (ICPOES) and with electrothermal atomic absorption spectroscopy (ETAAS) for selected elements. These data are typical for most other elements in the periodic table. Generally, detection limits with mass spectrometric detection range from 0.02 to 0.7 ppb with the majority of elements in the 0.02- to 0.1-ppb range.

Quantitative Analyses

The most widely used quantitative method of ICPMS uses a set of calibration standards for preparing a calibration curve. Simple aqueous standards are usually adequate if the unknown solutions are sufficiently dilute—less than 2000 μg/mL of total dissolved solids. With higher concentrations of matrix elements, attempts are often made to match the matrix elements in the samples with those in the standards. In order to compensate for instrument drift, instabilities, and matrix effects, an internal standard is usually introduced into the standards and the unknowns. The internal standard is an element that is absent from the samples and that has an atomic mass and ionization potential near those of the analytes. Two elements that are frequently used for internal standards are indium and rhodium. Both lie in the central part of the mass range (115, 113, and 103) and are infrequently found naturally in samples. Generally, log/log plots of ion current, ion count, or intensity ratios for sample and internal standards are linear over several orders of magnitude of concentration. Figure 11-15 shows calibration curves for the determination of several rare earth elements. Note that a linear relationship between peak count and concentration is observed over four orders of concentration.

Table 11-3 illustrates typical results obtained for the quantitative analysis of trace elements in a National

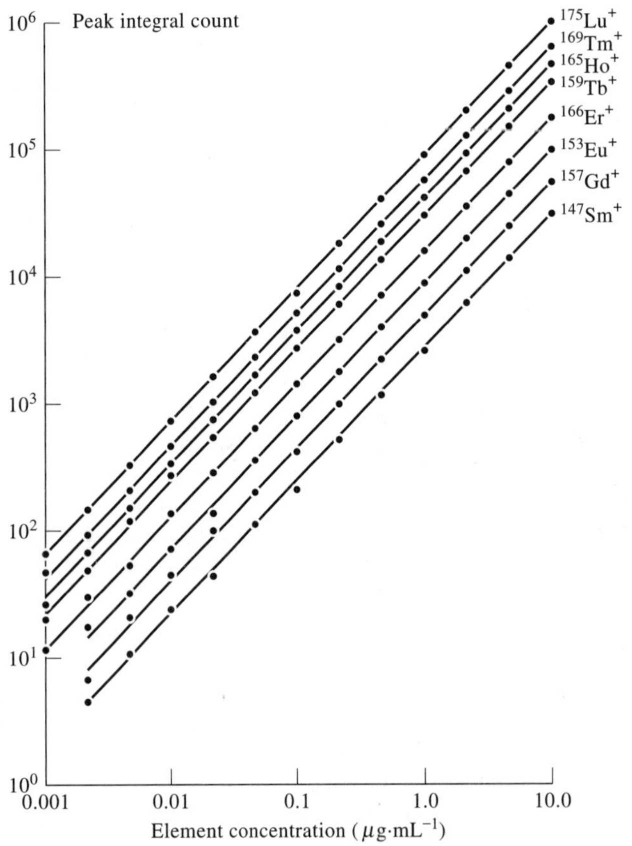

Figure 11-15 ICPMS calibration curves for several rare earth elements.

TABLE 11-3 Quantitative Determination of Trace Elements in a Standard Sample of Water

			ICP-MS[a]	
Element	Ion	NBS[a]	Mean	RSD (%)[b]
Beryllium	$^{9}Be^{+}$	19	21	20
Vanadium	$^{51}V^{+}$	54	52	6
Chromium	$^{52}Cr^{+}$	17	18	12
Manganese	$^{55}Mn^{+}$	32	34	5
Cobalt	$^{59}Co^{+}$	19	21	7
Zinc	$^{66}Zn^{+}$	69	57	11
Arsenic	$^{75}As^{+}$	77	76	5
Strontium	$^{88}Sr^{+}$	243	297	7
Molybdenum	$^{98}Mo^{+}$	97	134	9
Silver	$^{107}Ag^{+}$	2.8	3.5	16
Cadmium	$^{114}Cd^{+}$	10	13	22
Barium	$^{138}Ba^{+}$	47	74	17
Lead	$^{208}Pb^{+}$	27	31	8

[a]Concentration in parts per billion.
[b]Based on 10 determinations.

Bureau of Standards natural water sample. Note that the concentrations are in parts per billion. The 13 analyses were completed in an elapsed time of only 15 min. For most of the elements the results show remarkably good agreement with the standard sample data.

For the most accurate quantitative analyses, an *isotope dilution method* may be employed. The basis of this technique is the measurement of the change in ratio of signal intensities for two isotopes of an element that result from the addition of a known quantity, or a *spike,* of a standard solution that is enriched in one of the isotopes. The principles and mathematical relationships that relate analyte concentration to the quantity of spike added and the change in signal ratio are described in detail in Section 32D, which deals with the use of this technique with radioactive isotopes. The advantages of the isotope dilution method are an improvement in accuracy and its immunity to a wide variety of chemical and physical interferences. Its main limitation is the time required to complete this type of analysis.

Isotope Ratio Measurements

The measurement of isotope ratios is of considerable importance in several fields of science and medicine. For example, archeologists and geologists use such data to establish the age of artifacts and various types of deposits. Chemists and clinicians use isotopically enriched materials as tracers in various types of studies. The outcome of these studies is based upon isotope ratio measurements. Historically, isotope ratio measurements were based upon thermal atomization and ionization in which the sample was decomposed, atomized, and ionized on one or more electrically heated filaments. The ions formed in this way were then carried into a double-focusing mass spectrometer, where isotopic ratios were measured. These measurements were time consuming but quite precise, with relative standard deviations on the order of 0.01%. Now, isotope ratios are often determined by ICPMS, with which measurements can often be completed in five or ten minutes. Precisions in this case, however, are generally poorer by a factor of 10 to 100. For many purposes, reproducibility of this order is perfectly satisfactory.

11D SPARK SOURCE MASS SPECTROMETRY[10]

Spark source atomic mass spectrometry (SSMS) was first introduced in the 1930s as a general tool for multielement and isotope trace analyses. It was not until 1958, however, that the first commercial spark source mass spectrometer appeared on the market. After a period of rapid development in the 1960s, the use of this technique leveled off and then declined with the appearance of ICPMS and some of the other mass spectrometric methods discussed in this chapter. Currently, spark source mass spectrometry still finds application to samples that are not easily dissolved and analyzed by ICP.

In spark source mass spectrometry, the atomic constituents of a sample are converted by a high-potential ($\approx$30 kV), radio-frequency spark to gaseous ions for mass analysis. The spark is housed in a vacuum chamber located immediately adjacent to the mass analyzer. The chamber is equipped with a separate high-speed pumping system that quickly reduces the internal pressure to

[10] See G. Ramendik, J. Verlinden, and R. Gijbels, in *Inorganic Mass Spectrometry,* F. Adams, R. Gijbels, and R. Van Grieken, Eds., Chapter 2. New York: Wiley, 1988.

about 10^{-8} torr after sample changes. Often the sample serves as one or both electrodes. Alternatively, it is mixed with graphite and loaded into a cup-shaped electrode. The gaseous positive ions formed in the spark plasma are accelerated into the analyzer by a dc potential.

A spark source produces ions with a wide range of kinetic energies. Consequently, expensive double-focusing mass spectrometers are required for mass analysis of the ions. The Mattuch-Herzog type shown in Figure 11-9 is generally employed. Modern spark source instruments are designed to utilize both photographic and electrical detection systems. The latter are generally based on electron multipliers, which are described in Section 11B-1. When electron multipliers are used with double-focusing instruments, the spectrum is scanned by varying the magnetic field of the magnetic analyzer. Generally, the resolution of the double-focusing mass spectrometers used with spark sources is orders of magnitude greater than the quadrupole instruments used with ICP. Thus, the number of isobaric interferences is significantly smaller. For example, in the determination of iron in a silicate matrix, as found in samples of rock or glass, there is potential for interference by silicon. Here, the determination is based on the peak for $^{56}Fe^{+}$ (m/z = 55.93494), but a detectable peak for $^{28}Si_2^{+}$ (m/z = 55.95386) is also observed. The resolving power of most double-focusing is sufficient to separate these peaks, whereas with a quadrupole instrument an isobaric interference would result from the silicon dimer.

11D-1 Spectra

Like ICP mass spectra, spark source mass spectra are much simpler than their atomic emission counterparts, consisting of one major peak for each isotope of an element as well as a few weaker lines corresponding to multiply charged ions and ionized oxide and hydroxide species. The presence of these additional ions creates the potential for interference just as in ICPMS.

11D-2 Qualitative Applications

Spark source mass spectrometry is a powerful tool for qualitative and semiquantitative analysis. All elements in the periodic table from ^{7}Li through ^{238}U can be identified in a single excitation. By making multiple exposures, it is possible to determine order of magnitude concentrations for major constituents of a sample as well as for constituents in the parts per billion concentration range. Interpretation of spectra does require skill and experience, however, because of the presence of multiply charged species, polymeric species, and molecular ions.

11D-3 Quantitative Applications

A radio-frequency spark is not a very reproducible source over short periods. As a consequence, it is necessary to integrate the output signals from a spark for periods that range from several seconds to hundreds of seconds if good quantitative data are to be obtained. The photoplate is of course an integrating device; with electrical detection, provision must be made for current integration. In addition to integration, it is common practice to improve reproducibility by using the ratio of the analyte signal to that of an internal standard as the analytical variable. Often one of the major elements of the sample matrix is chosen as the standard; alternatively, a fixed amount of a pure compound is added to each sample and each standard used for calibration. In the latter case, the internal standard substance must be absent from the samples. With these precautions, relative standard deviations that range from a few percent to as much as 20% can be realized.

The advantages of spark source mass spectrometry include its high sensitivity, its applicability to a wide range of sample matrices, the large linear dynamic range of the output signal (often several orders of magnitude), and the wide range of elements that can be detected and measured quantitatively.

11E GLOW-DISCHARGE MASS SPECTROMETRY[11]

As shown in Section 8C-2, a glow-discharge source is used as an atomization device for various types of atomic spectroscopy. In addition to atomizing samples, it also produces a cloud of positive analyte ions from solid samples. This device consists of a simple two-electrode closed system containing argon at a pressure of 0.1 to 10 torr. A potential of 5 to 15 kV from a pulsed dc power supply is applied across the electrodes, causing the formation of positive argon ions, which are then accelerated toward the cathode. The cathode is fabricated from the

[11] See W. W. Harrison, C. M. Barshick, J. A. Klinger, P. H. Ratliff, and Y. Mei, *Anal. Chem.*, **1990**, *62*, 943A; W. W. Harrison, in *Inorganic Mass Spectrometry*, Chapter 3, F. Adams, R. Gijbels, and R. Van Grieken, Eds. New York: Wiley, 1988.

sample, or alternatively, the sample is deposited on an inert metal cathode. Just as in the hollow cathode lamp (Section 9B-1), atoms of the sample are sputtered from the cathode into the region between the two electrodes, where they are converted to positive ions by collision with electrons or positive argon ions. Analyte ions are then drawn into the mass spectrometer by differential pumping. The ions are then filtered in a quadrupole analyzer or dispersed with a magnetic sector for detection and determination.

As mentioned earlier, glow-discharge sources are often used with ICP torches. The glow discharge serves as the atomizer and the ICP torch as the ionizer.

The glow-discharge source appears to be more stable than the spark source and is more economic to purchase and operate. It is particularly useful for the direct analysis of solid samples. Commercial quadrupole and double-focusing mass spectrometers with glow-discharge sources are now on the market.

11F ELEMENTAL SURFACE ANALYSIS BY MASS SPECTROMETRY

In certain areas of science and engineering, the elemental composition of a surface layer of solid that is a few angstrom to a few tens of angstrom units in thickness is of much greater interest than is the bulk elemental composition. Fields where surface composition is of prime importance include heterogeneous catalysis, corrosion and adhesion studies, embrittlement properties, and studies of the behavior and functions of biological membranes.

Two atomic mass spectrometric methods are commonly used to determine the elemental composition of solid surfaces: *secondary ion mass spectrometry* and *laser microprobe mass spectrometry*. These techniques are discussed in detail in Sections 21B-5 and 21B-6.

11G QUESTIONS AND PROBLEMS

11-1 What type of mass spectrometers are used in ICPMS? How do they differ from one another?

11-2 What function does the ICP torch have in ICPMS?

11-3 What are the ordinate and the abscissa of an ordinary atomic mass spectrum?

11-4 Why has ICPMS become an important and widely used analytical method?

11-5 Describe the interface between the ICP torch and the mass spectrometer in an ICPMS instrument.

11-6 How are lasers used as a means of sampling solids for ICPMS?

11-7 What type of interferences are encountered in atomic mass spectrometry?

11-8 Why is an internal standard often employed in quantitative analysis by ICPMS?

11-9 Describe how the isotope dilution technique is used in atomic mass spectrometry.

11-10 Describe how glow-discharge mass spectrometry is carried out. What are the advantages of this technique?

11-11 What is secondary ion-mass spectrometry? What type of information does this method provide?

Chapter 12

Atomic X-Ray Spectrometry

X-Ray spectroscopy, like optical spectroscopy, is based upon measurement of emission, absorption, scattering, fluorescence, and diffraction of electromagnetic radiation. X-Ray fluorescence and X-ray absorption methods are widely used for the qualitative and quantitative determination of all elements in the periodic table having atomic numbers greater than that of sodium.[1] *With special equipment, elements with atomic numbers in the range of 5 to 10 can also be determined.*

12A FUNDAMENTAL PRINCIPLES

X-Rays are short wavelength electromagnetic radiation produced by the deceleration of high-energy electrons or by electronic transitions of electrons in the inner orbitals of atoms. The wavelength range of X-rays is from about 10^{-5} Å to 100 Å; conventional X-ray spectroscopy is, however, largely confined to the region of about 0.1 Å to 25 Å (1 Å = 0.1 nm = 10^{-10} m).

12A-1 Emission of X-Rays

For analytical purposes, X-rays are obtained in four ways: (1) by bombardment of a metal target with a beam of high-energy electrons, (2) by exposure of a substance to a primary beam of X-rays in order to generate a secondary beam of X-ray fluorescence, (3) by use of a radioactive source whose decay process results in X-ray emission,[2] and (4) from a synchrotron radiation source. Only three laboratories in the United States have facilities to produce X-rays from synchrotron radiation.[3] For this reason, we will consider only the first three sources.

---

[1] For a more extensive discussion of the theory and analytical applications of X-rays, see R. Jenkins, *X-Ray Fluorescence Spectrometry.* New York: Wiley, 1988; B. Dziunikowski, *Energy-Dispersive X-Ray Fluorescence Analysis.* New York: Elsevier, 1989; H. A. Liebhafsky, H. G. Pfeiffer, and E. A. Meyers, in *Treatise on Analytical Chemistry,* 2nd ed., P. J. Elving, E. J. Meehan, and I. M. Kolthoff, Eds., Part I, Vol. 8, Chapter 13. New York: Wiley, 1986; R. Jenkins, R. W. Gould, and D. Gedcke, *Quantitative X-Ray Spectrometry,* 2nd ed. New York: Marcel Dekker, 1995.

[2] Electromagnetic radiation produced by radioactive sources is often called *gamma radiation*. Gamma radiation is indistinguishable from X-ray radiation.

[3] The three laboratories are the Brookhaven National Laboratory, the Cornell High-Energy Synchrotron Radiation Laboratory, and the Stanford Synchrotron Radiation Laboratory. The high-intensity and collimated nature of beams from this type of source leads to applications that cannot be accomplished by the other three X-ray sources. For a review of the applications of synchrotron-induced X-rays, see K. W. Jones and B. M. Gordon, *Anal. Chem.*, **1989**, *61*, 341A.

X-Ray sources, like ultraviolet and visible emitters, often produce both continuum and line spectra; both types are of importance in analysis. Continuum radiation is also called *white radiation* or *Bremsstrahlung*. *Bremsstrahlung* means radiation that arises from retardation by particles; such radiation is generally continous in nature.

Continuum Spectra from Electron Beam Sources

In an X-ray tube, electrons produced at a heated cathode are accelerated toward a metal anode (the *target*) by a potential as great as 100 kV; upon collision, part of the energy of the electron beam is converted to X-rays. Under some conditions, only a continuum spectrum such as that shown in Figure 12-1 results; under others, a line spectrum is superimposed upon the continuum (see Figure 12-2).

The continuum X-ray spectrum shown in the two figures is characterized by a well-defined, short-wavelength limit (λ_0), which is dependent upon the accelerating voltage V but independent of the target material. Thus, λ_0 for the spectrum produced with a molybdenum target at 35 kV (Figure 12-2) is identical to λ_0 for a tungsten target at the same voltage (Figure 12-1).

The continuum radiation from an electron beam source results from collisions between the electrons of the beam and the atoms of the target material. At each collision, the electron is decelerated and a photon of X-ray energy is produced. The energy of the photon is equal to the difference in kinetic energies of the electron before and after the collision. Generally, the electrons in a beam are decelerated in a series of collisions; the resulting loss of kinetic energy differs from collision to collision. Thus, the energies of the emitted X-ray photons vary continuously over a considerable range. The maximum photon energy generated corresponds to the instantaneous deceleration of the electron to zero kinetic energy in a single collision. For such an event, we may write

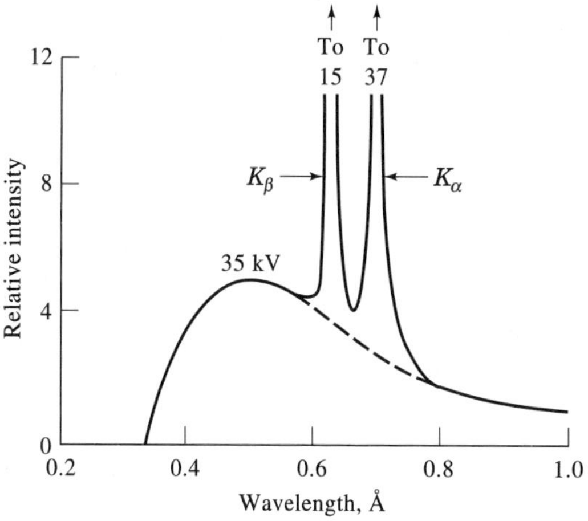

Figure 12-2 Line spectrum for an X-ray with a molybdenum target.

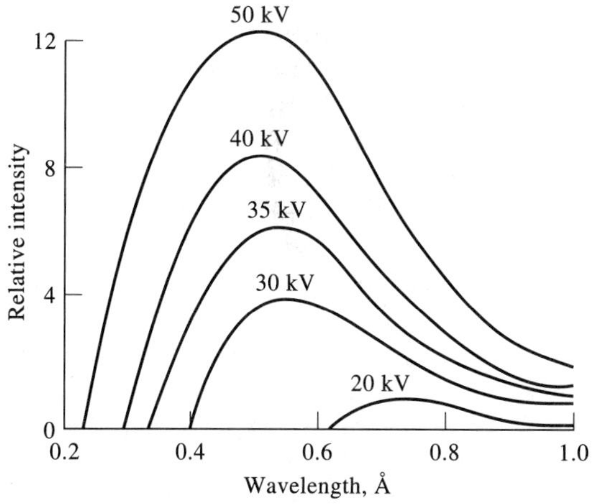

Figure 12-1 Distribution of continuum radiation from an X-ray tube with a tungsten target. The numbers above the curves indicate the accelerating voltages.

$$h\nu_0 = \frac{hc}{\lambda_0} = Ve \qquad (12\text{-}1)$$

where *Ve*, the product of the accelerating voltage and the charge on the electron, is the kinetic energy of all of the electrons in the beam; h is the Planck constant; and c is the velocity of light. The quantity ν_0 is the maximum frequency of radiation that can be produced at voltage V, while λ_0 is the low-wavelength limit for the radiation. This relationship is known as the *Duane-Hunt law*. Upon substituting numerical values for the constants and rearranging, Equation 12-1 becomes

$$\lambda_0 = 12{,}398/V \qquad (12\text{-}2)$$

where λ_0 and V have units of angströms and volts, respectively. It is interesting to note that Equation 12-1 provides a direct means for the highly accurate determination of the Planck constant.

Line Spectra from Electron Beam Sources

As shown in Figure 12-2, bombardment of a molybdenum target produces intense emission lines at about 0.63 and 0.71 Å; an additional simple series of lines occurs in the longer wavelength range of 4 to 6 Å.

The emission behavior of molybdenum is typical of all elements having atomic numbers larger than 23; that is, the X-ray line spectra are remarkably simple when compared with ultraviolet emission and consist of two series of lines. The shorter wavelength group is called the K series, and the other the L series.[4] Elements with atomic numbers smaller than 23 produce only a K series. Table 12-1 presents wavelength data for the emission spectra of a few elements.

A second characteristic of X-ray spectra is that the minimum acceleration voltage required for the excitation of the lines for each element increases with atomic number. Thus, the line spectrum for molybdenum (atomic number = 42) disappears if the excitation voltage drops below 20 kV. As shown in Figure 12-1, bombardment of tungsten (atomic number = 74) produces no lines in the region of 0.1 to 1.0 Å even at 50 kV. Characteristic K lines appear at 0.18 and 0.21 Å, however, if the voltage is raised to 70 kV.

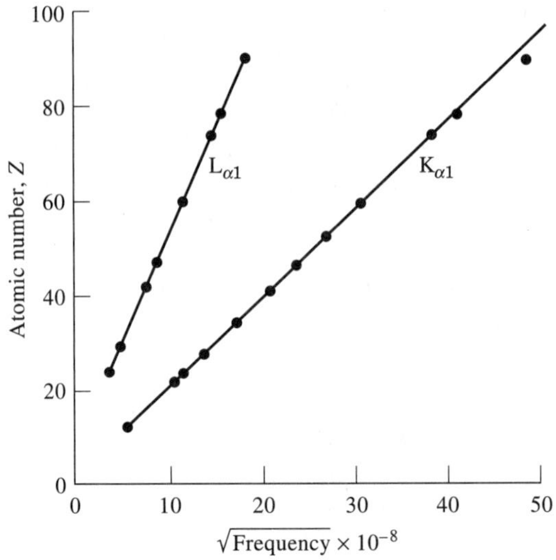

Figure 12-3 Relationship between X-ray emission frequency and atomic number for $K_{\alpha 1}$ and $L_{\alpha 1}$ lines.

Figure 12-3 illustrates the linear relationship between the square root of the frequency for a given (K or L) line and the atomic number of the element responsible for the radiation. This property was first discovered by H. G. J. Moseley in 1914.

X-Ray line spectra result from electronic transitions that involve the innermost atomic orbitals. The short-wavelength K series is produced when the high-energy electrons from the cathode remove electrons from those orbitals nearest to the nucleus of the target atom. The collision results in the formation of excited *ions,* which then emit quanta of X-radiation as electrons from outer orbitals undergo transitions to the vacant orbital. As shown in Figure 12-4, the lines in the K series arise from electronic transitions between higher energy levels and the K shell. The L series of lines results when an electron is lost from the second principal quantum level, either as a consequence of ejection by an electron from the cathode or from the transition of an L electron to the K level that accompanies the production of a quantum of K radiation. It is important to appreciate that the energy scale in

[4] For the heavier elements, additional series of lines (M, N, and so forth) are found at longer wavelengths. Their intensities are low, however, and little use is made of them. The designations K and L arose from the German words *kurtz* and *lang* for short and long wavelengths. The additional alphabetical designations were then added for lines occurring at progressively longer wavelengths.

TABLE 12-1 Wavelengths in Angstrom Units of the More Intense Emission Lines for Some Typical Elements

		K Series		L Series	
Element	Atomic Number	α_1	β_1	α_1	β_1
Na	11	11.909	11.617	—	—
K	19	3.742	3.454	—	—
Cr	24	2.290	2.085	21.714	21.323
Rb	37	0.926	0.829	7.318	7.075
Cs	55	0.401	0.355	2.892	2.683
W	74	0.209	0.184	1.476	1.282
U	92	0.126	0.111	0.911	0.720

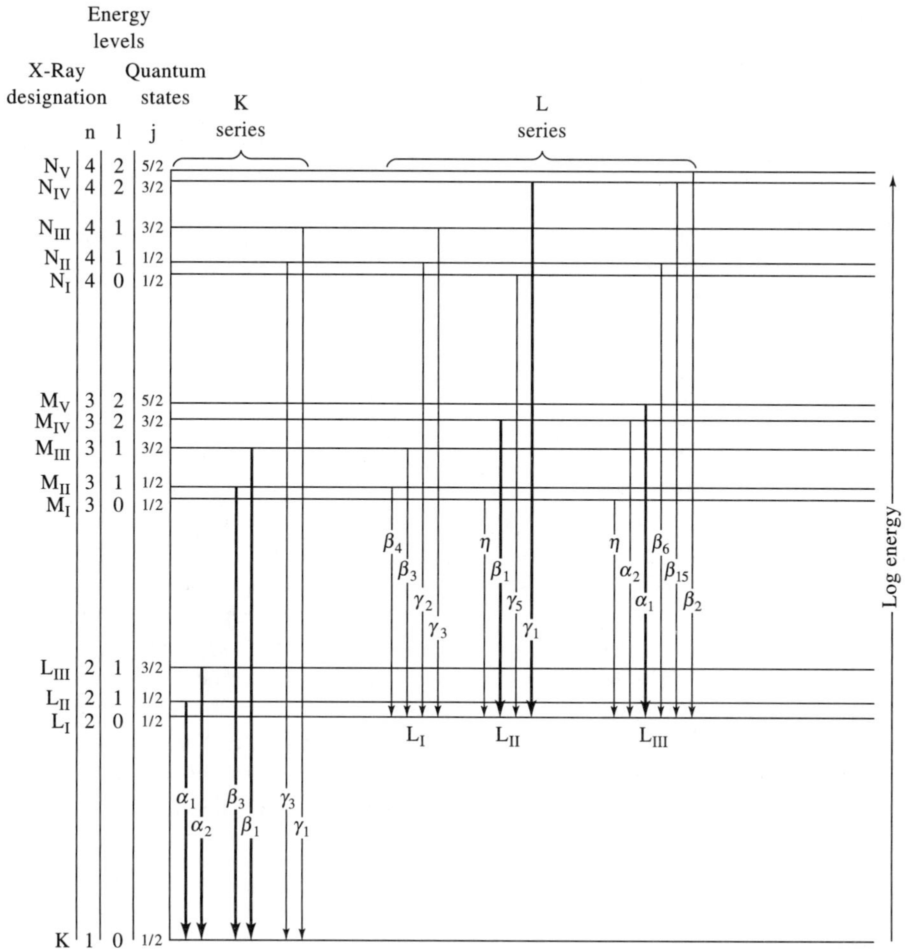

Figure 12-4 Partial energy-level diagram showing common transitions producing X-rays. The most intense lines are indicated by the wider arrows.

Figure 12-4 is logarithmic. Thus, the energy difference between the L and K levels is significantly larger than that between the M and L levels. The K lines therefore appear at shorter wavelengths. It is also important to note that the energy differences between the transitions labeled α_1 and α_2 as well as those between β_1 and β_2 are so small that only single lines are observed in all but the highest resolution spectrometers (see Figure 12-2).

The energy level diagram in Figure 12-4 is applicable to any element with sufficient electrons to permit the number of transitions shown. The differences in energies between the levels increase regularly with atomic number because of the increasing charge on the nucleus; therefore, the radiation for the K series appears at shorter wavelengths for the heavier elements (see Table 12-1). The effect of nuclear charge is also reflected in the increase in minimum voltage required to excite the spectra of these elements.

It is important to note that for all but the lightest elements, the wavelengths of characteristic X-ray lines are independent of the physical and chemical state of the element because the transitions responsible for these lines involve electrons that take no part in bonding. Thus, the position of the K_α lines for molybdenum is the same regardless of whether the target is the pure metal, its sulfide, or its oxide.

Line Spectra from Fluorescent Sources

Another convenient way of producing a line spectrum is to irradiate the element or one of its compounds with the continuum radiation from an X-ray tube. This process is considered further in a later section.

Spectra from Radioactive Sources

X-Radiation is often a product of radioactive decay processes. *Gamma rays,* which are indistinguishable from X-rays, owe their production to intranuclear reactions. Many α and β emission processes (see Section 32A-2) leave a nucleus in an excited state; the nucleus then releases one or more quanta of gamma rays as it returns to its ground state. *Electron capture* or *K capture* also produces X-radiation. This process involves capture of a K electron (less commonly, an L or an M electron) by the nucleus and formation of an element of the next lower atomic number. As a result of K capture, electronic transitions to the vacated orbital occur, and the X-ray line spectrum of the newly formed element is observed. The half-lives of K-capture processes range from a few minutes to several thousand years.

Artificially produced radioactive isotopes provide a very simple source of monoenergetic radiation for certain analytical applications. The best known example is iron-55, which undergoes a K-capture reaction with a half-life of 2.6 years:

$$^{55}\text{Fe} \rightarrow {}^{54}\text{Mn} + h\nu$$

The resulting manganese K_α line at about 2.1 Å has proved to be a useful source for both fluorescence and absorption methods. Table 12-2 lists some additional common radioisotopic sources for X-ray spectroscopy.

12A-2 Absorption Spectra

When a beam of X-rays is passed through a thin layer of matter, its intensity or power is generally diminished as a consequence of absorption and scattering. The effect of scattering for all but the lightest elements is ordinarily small and can be neglected in those wavelength regions where appreciable absorption occurs. As shown in Figure 12-5, the absorption spectrum of an element, like its emission spectrum, is simple and consists of a few well-defined absorption peaks. Here again, the wavelengths of the peaks are characteristic of the element and are largely independent of its chemical state.

A peculiarity of X-ray absorption spectra is the appearance of sharp discontinuities, called *absorption edges,* at wavelengths immediately beyond absorption maxima.

TABLE 12-2 Common Radioisotopic Sources for X-Ray Spectroscopy

Source	Decay Process	Half-Life	Type of Radiation	Energy, keV
$^{3}_{1}$H-Ti[a]	β^-	12.3 years	Continuum	3–10
			Ti-K X-rays	4–5
$^{55}_{26}$Fe	EC[b]	2.7 years	Mn-K X-rays	5.9
$^{57}_{27}$Co	EC	270 days	Fe-K X-rays	6.4
			γ rays	14, 122, 136
$^{109}_{48}$Cd	EC	1.3 years	Ag-K X-rays	22
			γ rays	88
$^{125}_{53}$I	EC	60 days	Te-K X-rays	27
			γ rays	35
$^{147}_{61}$Pm-Al	β^-	2.6 years	Continuum	12–45
$^{210}_{82}$Pb	β^-	22 years	Bi-L X-rays	11
			γ rays	47

[a]Tritium adsorbed on nonradioactive titanium metal.
[b]Electron capture.

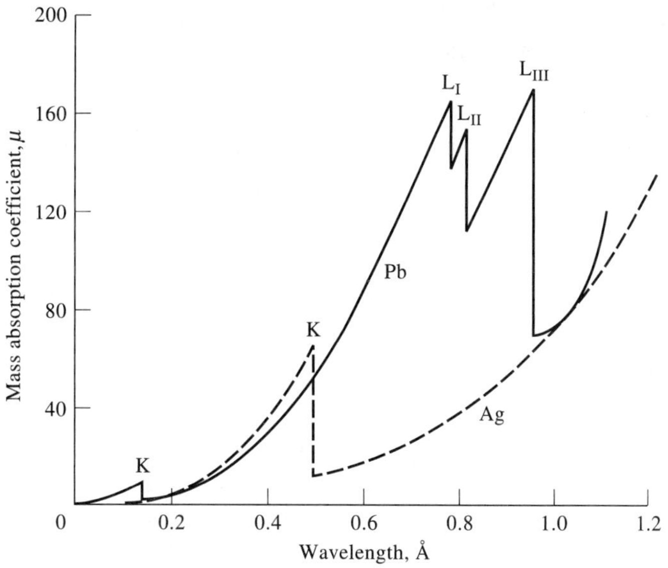

Figure 12-5 X-Ray absorption spectra for lead and silver.

The Absorption Process

Absorption of an X-ray quantum causes ejection of one of the innermost electrons from an atom and the consequent production of an excited ion. In this process, the entire energy $h\nu$ of the radiation is partitioned between the kinetic energy of the electron (the *photoelectron*) and the potential energy of the excited ion. The highest probability for absorption arises when the energy of the quantum is exactly equal to the energy required to remove the electron just to the periphery of the atom (that is, as the kinetic energy of the ejected electron approaches zero).

The absorption spectrum for lead, shown in Figure 12-5, exhibits four peaks, the first occurring at 0.14 Å. The energy of the quantum corresponding to this wavelength exactly matches the energy required to just eject the highest energy K electron of the element; immediately beyond this wavelength, the energy of the radiation is insufficient to bring about removal of a K electron, and an abrupt decrease in absorption occurs. At wavelengths shorter than 0.14 Å, the probability of interaction between the electron and the radiation gradually diminishes; this results in a smooth decrease in absorption. In this region, the kinetic energy of the ejected photoelectron increases continuously with the decrease in wavelength.

The additional peaks at longer wavelengths correspond to the removal of an electron from the L energy levels of lead. Three sets of L levels, differing slightly in energy, exist (see Figure 12-4); three peaks are, therefore, observed. Another set of peaks, arising from ejections of M electrons, are located at still longer wavelengths.

Figure 12-5 also shows the K absorption edge for silver, which occurs at 0.485 Å. The longer wavelength for the silver peak reflects the lower atomic number of the element compared with lead.

The Mass Absorption Coefficient

Beer's law is as applicable to the absorption of X-radiation as to other types of electromagnetic radiation; thus, we may write

$$\ln\frac{P_0}{P} = \mu x$$

where x is the sample thickness in centimeters and P and P_0 are the powers of the transmitted and incident beams. The constant μ is called the *linear absorption coefficient* and is characteristic of the element as well as the number of its atoms in the path of the beam. A more convenient form of Beer's law is

$$\ln\frac{P_0}{P} = \mu_M \rho x \qquad (12\text{-}3)$$

where ρ is the density of the sample and μ_M is the *mass absorption coefficient,* a quantity that is *independent* of the physical and chemical state of the element. Thus, the mass absorption coefficient for bromine has the same value in gaseous HBr as in solid sodium bromate. Note that the mass absorption coefficient carries units of cm^2/g.

Mass absorption coefficients are additive functions of the weight fractions of elements contained in a sample. Thus,

$$\mu_M = W_A\mu_A + W_B\mu_B + W_C\mu_C + \cdots \quad (12\text{-}4)$$

where μ_M is the mass absorption coefficient of a sample containing the weight fractions W_A, W_B, and W_C of elements A, B, and C. The terms μ_A, μ_B, and μ_C are the respective mass absorption coefficients for each of the elements. Tables of mass absorption coefficients for the elements at various wavelengths are found in many handbooks and monographs.[5]

12A-3 X-Ray Fluorescence

The absorption of X-rays produces electronically excited ions that return to their ground state by transitions involving electrons from higher energy levels. Thus, an excited ion with a vacant K shell is produced when lead absorbs radiation of wavelengths shorter than 0.14 Å (Figure 12-5); after a brief period, the ion returns to its ground state via a series of electronic transitions characterized by the emission of X-radiation (fluorescence) of wavelengths identical to those that result from excitation produced by electron bombardment. The wavelengths of the fluorescence lines are always somewhat greater than the wavelength of the corresponding absorption edge, however, because absorption requires a complete removal of the electron (that is, ionization), whereas emission involves transitions of an electron from a higher energy level within the atom. For example, the K absorption edge for silver occurs at 0.485 Å, while the K emission lines for the element have wavelengths at 0.497 and 0.559 Å. When fluorescence is to be excited by radiation from an X-ray tube, the operating voltage must be sufficiently great so that the cutoff wavelength λ_0 (Equation 12-2) is shorter than the absorption edge of the element whose spectrum is to be excited. Thus, to generate the K lines for silver, the tube voltage would need to be (Equation 12-2)

$$V \geq \frac{12398 \text{ V} \cdot \cancel{\text{Å}}}{0.485 \cancel{\text{Å}}} = 25{,}560 \text{ V or } 25.6 \text{ kV}$$

[5] For example, E. P. Bertin, *Principles and Practice of X-Ray Spectrometric Analysis,* 2nd ed., pp. 972–976. New York: Plenum Press, 1975.

12A-4 Diffraction of X-Rays

In common with other types of electromagnetic radiation, interaction between the electric vector of X-radiation and the electrons of the matter through which it passes results in scattering. When X-rays are scattered by the ordered environment in a crystal, interference (both constructive and destructive) takes place among the scattered rays because the distances between the scattering centers are of the same order of magnitude as the wavelength of the radiation. Diffraction is the result.

Bragg's Law

When an X-ray beam strikes a crystal surface at some angle θ, a portion is scattered by the layer of atoms at the surface. The unscattered portion of the beam penetrates to the second layer of atoms where again a fraction is scattered, and the remainder passes on to the third layer (Figure 12-6). The cumulative effect of this scattering from the regularly spaced centers of the crystal is diffraction of the beam in much the same way as visible radiation is diffracted by a reflection grating (Section 7C-2). The requirements for X-ray diffraction are: (1) the spacing between layers of atoms must be roughly the same as the wavelength of the radiation and (2) the scattering centers must be spatially distributed in a highly regular way.

In 1912, W. L. Bragg treated the diffraction of X-rays by crystals as shown in Figure 12-6. Here, a narrow beam of radiation strikes the crystal surface at angle θ; scattering occurs as a consequence of interaction of the radiation with atoms located at *O, P,* and *R*. If the distance

$$AP + PC = \mathbf{n}\lambda$$

where $\mathbf{n}$ is an integer, the scattered radiation will be in phase at *OCD,* and the crystal will appear to reflect the X-radiation. But

$$AP = PC = d \sin \theta \qquad (12\text{-}5)$$

where d is the interplanar distance of the crystal. Thus, we may write that the conditions for constructive interference of the beam at angle θ are

$$\mathbf{n}\lambda = 2d \sin \theta \qquad (12\text{-}6)$$

Equation 12-6 is called the *Bragg equation* and is of fundamental importance. Note that X-rays appear to be

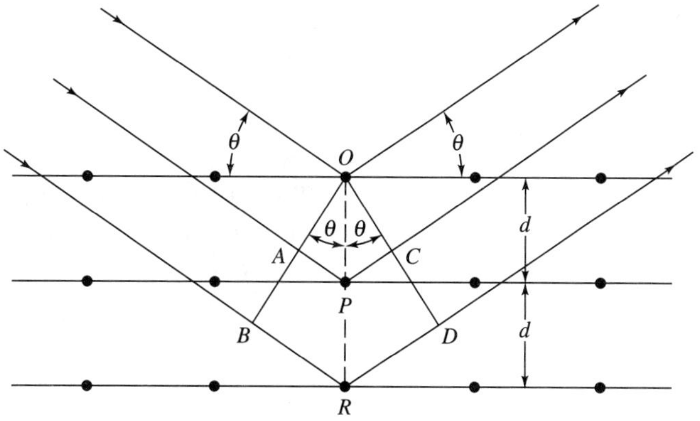

Figure 12-6 Diffraction of X-rays by a crystal.

reflected from the crystal only if the angle of incidence satisfies the condition that

$$\sin \theta = \frac{n\lambda}{2d}$$

At all other angles, destructive interference occurs.

12B INSTRUMENT COMPONENTS

Absorption, emission, fluorescence, and diffraction of X-rays all find applications in analytical chemistry. Instruments for these applications contain components that are analogous in function to the five components of instruments for optical spectroscopic measurement; these components include a source, a device for restricting the wavelength range of incident radiation, a sample holder, a radiation detector or transducer, and a signal processor and readout. These components differ considerably in detail from their optical counterparts. Their functions, however, are the same, and the ways in which they are combined to form instruments are often similar to those shown in Figure 7-1 (page 144).

As with optical instruments, both X-ray photometers and spectrophotometers are encountered, the first using filters and the second using monochromators to restrict radiation from the source. In addition, however, a third method is available for obtaining information about isolated portions of an X-ray spectrum. Here, isolation is achieved electronically with devices that have the power to discriminate between various parts of a spectrum based on the *energy* rather than the *wavelength* of the radiation. Thus, X-ray instruments are often described as *wavelength dispersive instruments* or *energy dispersive instruments,* depending upon the method by which they resolve spectra.

12B-1 Sources

Three types of sources are encountered in X-ray instruments: tubes, radioisotopes, and secondary fluorescent sources.

The X-Ray Tube

The most common source of X-rays for analytical work is the X-ray tube (sometimes called a *Coolidge tube*), which can take a variety of shapes and forms; one design is shown schematically in Figure 12-7. An X-ray source is a highly evacuated tube in which is mounted a tungsten filament cathode and a massive anode. The anode generally consists of a heavy block of copper with a metal target plated on or imbedded in the surface of the copper. Target materials include such metals as tungsten, chromium, copper, molybdenum, rhodium, scandium, silver, iron, and cobalt. Separate circuits are used to heat the filament and to accelerate the electrons to the target. The heater circuit provides the means for controlling the intensity of the emitted X-rays while the accelerating potential determines their energy, or wavelength. For quantitative work, both circuits must be operated with stabilized power supplies that control the current or the potential to 0.1% relative.

The production of X-rays by electron bombardment is a highly inefficient process. Less than one percent of the electrical power is converted to radiant power, the remainder being dissipated as heat. As a consequence, until relatively recently, water cooling of the anodes of X-

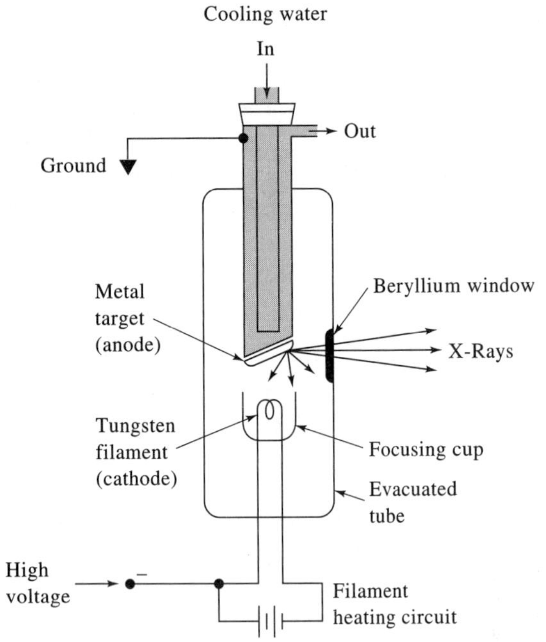

Figure 12-7 Schematic of an X-ray tube.

ray tubes was required. With modern equipment, however, cooling is often unnecessary because tubes can be operated at significantly lower power than formerly. This reduction in power is made possible by the greater sensitivity of modern X-ray transducers.

Radioisotopes

A variety of radioactive substances have been used as sources in X-ray fluorescence and absorption methods (see Table 12-2). Generally, the radioisotope is encapsulated to prevent contamination of the laboratory and shielded to absorb radiation in all but certain directions.

Many of the best radioactive sources provide simple line spectra; others produce a continuum (see Table 12-2). Because of the shape of X-ray absorption curves, a given radioisotope will be suitable for excitation of fluorescence or for absorption studies for a range of elements. For example, a source producing a line in the region between 0.3 and 0.47 Å is suitable for fluorescence or absorption studies involving the K absorption edge for silver (see Figure 12-5). Sensitivity improves as the wavelength of the source line approaches the absorption edge. Iodine-125 with a line at 0.46 Å is ideal for determining silver from this standpoint.

Secondary Fluorescent Sources

In some applications, the fluorescence spectrum of an element that has been excited by radiation from an X-ray tube serves as a source for absorption or fluorescence studies. This arrangement has the advantage of eliminating the continuum emitted by a primary source. For example, an X-ray tube with a tungsten target (Figure 12-1) could be used to excite the K_α and K_β lines of molybdenum (Figure 12-2). The resulting fluorescence spectrum would then be similar to the spectrum in Figure 12-2 except that the continuum would be absent.

12B-2 Filters for X-Rays

In many applications, it is desirable to employ an X-ray beam that is restricted in its wavelength range. As in the visible region, both filters and monochromators are used for this purpose.

Figure 12-8 illustrates a common technique for producing a relatively monochromatic beam by use of a filter. Here, the K_β line and most of the continuum emitted from a molybdenum target is removed by a zirconium filter having a thickness of about 0.01 cm. The pure K_α line is then available for analytical purposes. Several other target-filter combinations of this type have been developed, each of which serves to isolate one of the in-

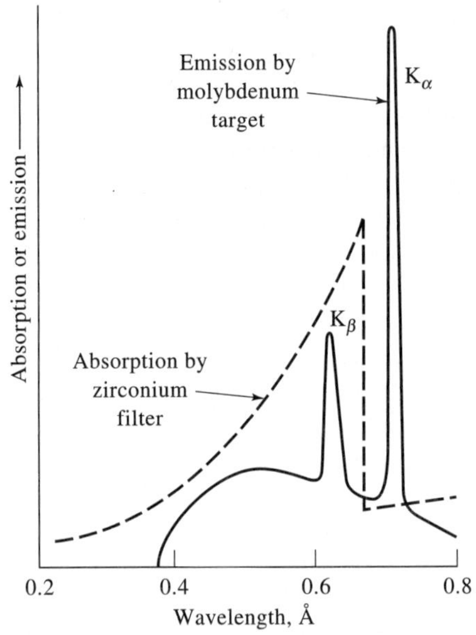

Figure 12-8 Use of a filter to produce monochromatic radiation.

tense lines of a target element. Monochromatic radiation produced in this way is widely used in X-ray diffraction studies. The choice of wavelengths available by this technique is limited by the relatively small number of target-filter combinations that are available.

Filtering the continuum from an X-ray tube is also feasible with thin strips of metal. As with glass filters for visible radiation, relatively broad bands are transmitted with a significant attenuation of the desired wavelengths.

12B-3 X-Ray Monochromators

Figure 12-9 shows the essential components of an X-ray spectrometer. The monochromator consists of a pair of beam collimators, which serve the same purpose as the slits in an optical instrument, and a dispersing element. The latter is a single crystal mounted on a *goniometer* or rotatable table that permits variation and precise determination of the angle θ between the crystal face and the collimated incident beam. From Equation 12-6, it is evident that for any given angular setting of the goniometer, only a few wavelengths are diffracted (λ, $\lambda/2$, $\lambda/3$, . . ., $\lambda/\mathbf{n}$ where $\lambda = 2d \sin \theta$).

In order to produce a spectrum, it is necessary that the exit beam collimator and the detector be mounted on a second table that rotates at twice the rate of the first; that is, as the crystal rotates through an angle θ, the detector must simultaneously move through an angle 2θ. Clearly, the interplanar spacing d for the crystal must be known precisely (Equation 12-6).

The collimators for X-ray monochromators ordinarily consist of a series of closely spaced metal plates or tubes that absorb all but the parallel beams of radiation.

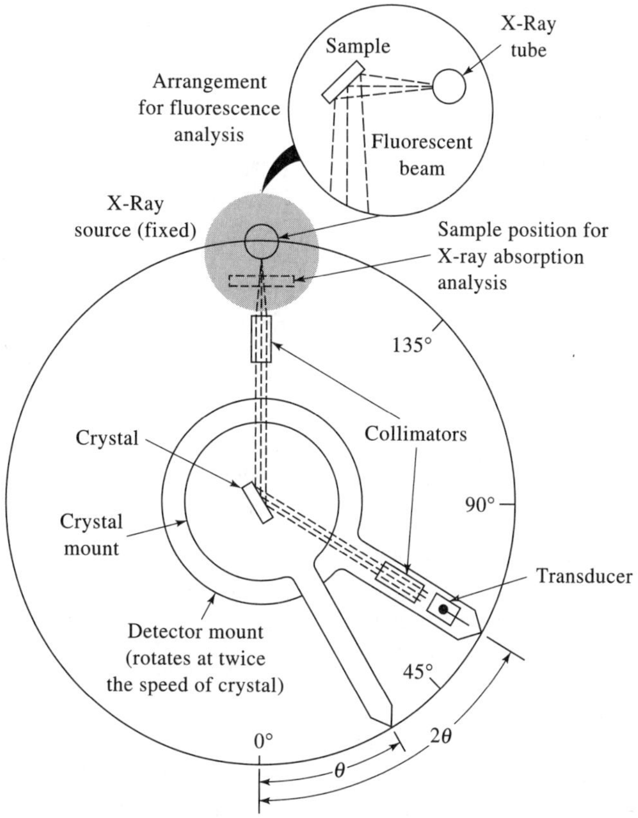

Figure 12-9 An X-ray monochromator and detector. Note that the angle of the detector with respect to the beam (2θ) is twice that of the crystal face. For absorption analysis, the source is an X-ray tube, and the sample is located in the beam as shown. For emission work, the sample becomes a fluorescent source of X-rays as shown in the insert.

X-Radiation longer than about 2 Å is absorbed by constituents of the atmosphere. Therefore, provision is usually made for a continuous flow of helium through the sample compartment and monochromator when longer wavelengths are required. Alternatively, provisions may be made to evacuate these areas by pumping.

The loss of intensity is high in a monochromator equipped with a flat crystal because as much as 99% of the radiation is sufficiently divergent to be absorbed in the collimators. Increased intensities, by as much as a factor of ten, have been realized by using a curved crystal surface that acts not only to diffract but also to focus the divergent beam from the source on the exit collimator.

As illustrated in Table 12-1, most analytically important X-ray lines lie in the region between about 0.1 and 10 Å. A consideration of data in Table 12-3, however, leads to the conclusion that no single crystal satisfactorily disperses radiation over this entire range. As a consequence, an X-ray monochromator must be provided with at least two (and preferably more) interchangeable crystals.

The useful wavelength range for a crystal is determined by its lattice spacing d and the problems associated with detection of the radiation when 2θ approaches zero or 180 deg. When a monochromator is set at angles of 2θ that are much less than 10 deg, the amount of polychromatic radiation scattered from the crystal surface becomes prohibitively high. Generally, values of 2θ greater than about 160 deg cannot be measured because the location of the source unit prohibits positioning of the detector at such an angle (see Figure 12-9). The minimum and maximum values for λ_{max} in Table 12-3 were determined from these limitations.

It will be seen from Table 12-3 that a crystal with a large lattice spacing, such as ammonium dihydrogen phosphate, has a much greater wavelength range than a crystal in which this variable is small. The advantage of large values of d is offset, however, by the consequent lower dispersion. This effect can be seen by differentiation of Equation 12-6, which leads to

$$\frac{d\theta}{d\lambda} = \frac{\mathbf{n}}{2d \cos \theta}$$

Here, $d\theta/d\lambda$, a measure of dispersion, is seen to be inversely proportional to d. Table 12-3 provides dispersion data for the various crystals at their maximum and minimum wavelengths. The low dispersion of ammonium dihydrogen phosphate prohibits its use in the region of short wavelengths; here, a crystal such as topaz or lithium fluoride must be substituted.

TABLE 12-3 Properties of Typical Diffracting Crystals

Crystal	Lattice Spacing d, Å	Wavelength Range[a], Å		Dispersion $d\theta/d\lambda$, deg/Å	
		λ_{max}	λ_{min}	at λ_{max}	at λ_{min}
Topaz	1.356	2.67	0.24	2.12	0.37
LiF	2.014	3.97	0.35	1.43	0.25
NaCl	2.820	5.55	0.49	1.02	0.18
EDDT[b]	4.404	8.67	0.77	0.65	0.11
ADP[c]	5.325	10.50	0.93	0.54	0.09

[a]Based on assumption that the measurable range of 2θ is from 160 deg for λ_{max} to 10 deg for λ_{min}.
[b]Ethylenediamine *d*-tartrate.
[c]Ammonium dihydrogen phosphate.

12B-4 X-Ray Transducers and Signal Processors

Early X-ray equipment employed photographic emulsions for detection and measurement of radiation. For reasons of convenience, speed, and accuracy, however, modern instruments are generally equipped with transducers that convert radiant energy into an electrical signal. Three types of transducers are encountered: gas-filled transducers, scintillation counters, and semiconductor transducers. Before considering how each of these devices functions, it is worthwhile to discuss *photon counting,* a signal-processing method that is often used with X-ray transducers as well as detectors of radiation from radioactive sources (Chapter 32). As was mentioned earlier (Section 7F-1), photon counting is also used in ultraviolet and visible spectroscopy.

Photon Counting

In contrast to the various photoelectric detectors we have thus far considered, X-ray detectors are usually operated as *photon counters.* In this mode of operation, individual pulses of charge are produced as quanta of radiation are absorbed by the transducer and are counted; the power of the beam is then recorded digitally as the number of counts per unit of time. Photon counting requires rapid response times for the transducer and signal processor so that the arrival of individual photons may be accurately transduced and recorded. In addition, the technique is applicable only to beams of relatively low intensity. As the beam intensity increases, photon pulses begin to overlap and only a steady-state current, which represents an average number of pulses per second, can be measured. If the response time of the transducer is long, pulse overlap occurs at relatively low photon intensities. As its response time becomes shorter, the transducer is more capable of detecting individual photons without pulse overlap.

For weak sources of radiation, photon counting generally provides more accurate intensity data than are obtainable by measuring average currents. The improvement can be traced to the fact that signal pulses are generally substantially larger than the pulses arising from background noise in the source, transducer, and associated electronics; separation of the signal from noise can then be achieved with a *pulse-height discriminator,* an electronic device that will be further discussed in a later section.

Photon counting is used in X-ray work because the power of available sources is often low. In addition, photon counting permits spectra to be obtained without the use of a monochromator. This property is considered in the section devoted to energy dispersive systems.

Gas-Filled Transducers

When X-radiation passes through an inert gas such as argon, xenon, or krypton, interactions occur that produce a large number of positive gaseous ions and electrons (ion pairs) for each X-ray quantum. Three types of X-ray transducers, namely, *ionization chambers, proportional counters,* and *Geiger tubes,* are based upon the enhanced conductivity resulting from this phenomenon.

A typical gas-filled transducer is shown schematically in Figure 12-10. Radiation enters the chamber through a transparent window of mica, beryllium, aluminum, or Mylar. Each photon of X-radiation may interact with an atom of argon, causing it to lose one of its outer electrons. This *photoelectron* has a large kinetic energy, which is equal to the difference between the X-ray photon energy and the binding energy of the electron in the argon atom. The photoelectron then loses

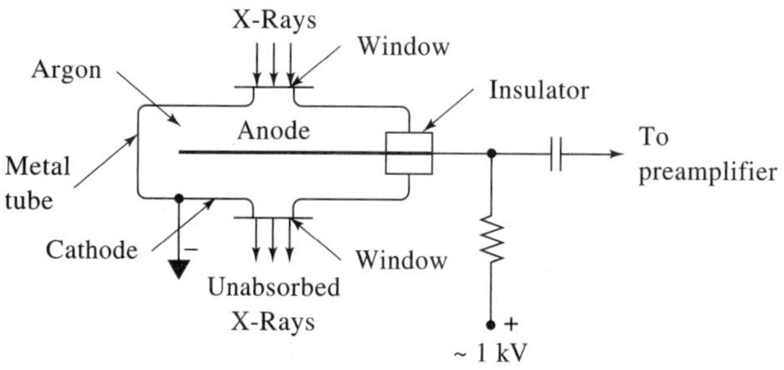

Figure 12-10 Cross section of a gas-filled detector.

this excess kinetic energy by ionizing several hundred additional atoms of the gas. Under the influence of an applied potential, the mobile electrons migrate toward the central wire anode while the slower moving cations are attracted toward the cylindrical metal cathode.

Figure 12-11 shows the effect of applied potential on the number of electrons that reach the anode of a gas-filled transducer for each entering X-ray photon. Three characteristic voltage regions are indicated. At potentials less than V_1, the accelerating force on the ion pairs is low, and the rate at which the positive and negative species separate is insufficient to prevent partial recombination. As a consequence, the number of electrons reaching the anode is smaller than the number produced initially by the incoming radiation.

In the *ionization chamber region* between V_1 and V_2, the number of electrons reaching the anode is reasonably constant and represents the total number formed by a single photon.

In the *proportional counter region* between V_3 and V_4, the number of electrons increases rapidly with applied potential. This increase is the result of secondary ion-pair production caused by collisions between the accelerated electrons and gas molecules; amplification *(gas amplification)* of the ion current results.

In the *Geiger range* V_5 to V_6, amplification of the electrical pulse is enormous but is limited by the positive space charge created as the faster moving electrons migrate away from the slower positive ions. Because of this effect, the number of electrons reaching the anode is independent of the type and energy of incoming radiation and is governed instead by the geometry and gas pressure of the tube.

Figure 12-11 also illustrates that a larger number of electrons is produced by the more energetic 0.6 Å radiation than the longer wavelength 5 Å X-rays. Thus, the

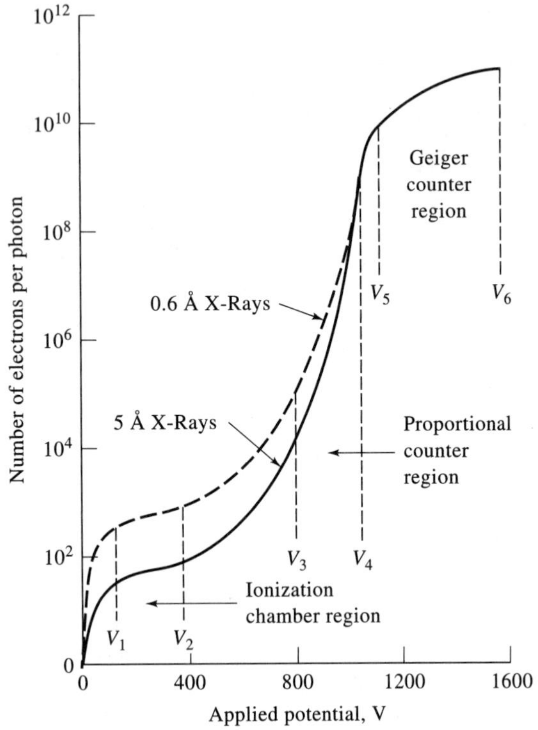

Figure 12-11 Gas amplification for various types of gas-filled detectors.

size of the pulse (the pulse height) is greater for high-frequency X-rays than for low-frequency X-rays.

The Geiger Tube

The Geiger tube is a gas-filled transducer operated in the voltage region between V_5 and V_6 in Figure 12-11; here, gas amplification is greater than 10^9. Each photon produces an avalanche of electrons and cations; the resulting currents are thus large and relatively easy to detect and measure.

The conduction of electricity through a chamber operated in the Geiger region, and in the proportional region as well, is not continuous because the space charge mentioned earlier terminates the flow of electrons to the anode. The net effect is a momentary pulse of current followed by an interval during which the tube does not conduct. Before conduction can again occur, this space charge must be dissipated by migration of the cations to the walls of the chamber. During the *dead time,* when the tube is nonconducting, response to radiation is impossible; the dead time thus represents a lower limit in the response time of the tube. Typically, the dead time of a Geiger tube is in the range from 50 to 200 μs.

Geiger tubes are usually filled with argon; a low concentration of an organic *quenching gas* often alcohol or methane, is also present to minimize the production of secondary electrons when the cations strike the chamber wall. The lifetime of a tube is limited to some 10^8 to 10^9 counts, by which time the quencher has been depleted.

With a Geiger tube, radiation intensity is determined by counting the pulses of current. The device is applicable to all types of nuclear and X-radiation. However, it lacks the large counting range of other detectors because of its relatively long dead time; its use in X-ray spectrometers is limited by this factor. Although quantitative applications of Geiger tube transducers have decreased, transducers of this type are still frequently encountered whenever portability is important.

Proportional Counters

The proportional counter is a gas-filled transducer that is operated in the V_3 to V_4 voltage region in Figure 12-11. Here, the pulse produced by a photon is amplified by a factor of 500 to 10,000, but the number of positive ions produced is small enough so that the dead time is only about 1 μs. In general, the pulses from a proportional counter tube must be amplified before being counted.

The number of electrons per pulse, which is proportional to the *pulse height,* produced in the proportional region depends directly upon the energy, and thus the *frequency,* of the incoming radiation. A proportional counter can be made sensitive to a restricted range of X-ray frequencies with a *pulse-height analyzer,* which counts a pulse only if its amplitude falls within certain limits. A pulse-height analyzer in effect permits electronic sorting of radiation; its function is analogous to that of a monochromator.

Proportional counters have been widely used as detectors in X-ray spectrometers.

Ionization Chambers

Ionization chambers are operated in the voltage range from V_1 to V_2 in Figure 12-11. Here, the currents are small (10^{-13} to 10^{-16} A typically) and relatively independent of applied voltage. Ionization chambers are not employed in X-ray spectrometry because of their lack of sensitivity. They do, however, find application in radiochemical measurements, which are considered in Chapter 32.

Scintillation Counters

The luminescence produced when radiation strikes a phosphor represents one of the oldest methods of detecting radioactivity and X-rays, and one of the newest as well. In its earliest application, the technique involved the manual counting of flashes that resulted when individual photons or radiochemical particles struck a zinc sulfide screen. The tedium of counting individual flashes by eye led Geiger to develop gas-filled transducers, which were not only more convenient and reliable but more responsive to radiation. The advent of the photomultiplier tube (Section 7E-2) and better phosphors, however, has reversed this trend, and scintillation counting has again become one of the important methods for radiation detection.

The most widely used modern scintillation detector consists of a transparent crystal of sodium iodide that has been activated by the introduction of 0.2% thallium iodide. Often, the crystal is shaped as a cylinder that is 3 to 4 in. in each dimension; one of the plane surfaces then faces the cathode of a photomultiplier tube. As the incoming radiation traverses the crystal, its energy is first lost to the scintillator; this energy is subsequently released in the form of photons of fluorescence radiation. Several thousand photons with a wavelength of about 400 nm are produced by each primary particle or photon over a period of about 0.25 μs, which is the *dead time.* The dead time of a scintillation counter is thus significantly smaller than the dead time of a gas-filled detector.

The flashes of light produced in the scintillator crystal are transmitted to the photocathode of a photomultiplier tube and are in turn converted to electrical pulses that can be amplified and counted. An important characteristic of scintillators is that the number of photons produced in each flash is proportional to the energy of the incoming radiation. Thus, incorporation of a pulse-height analyzer to monitor the output of a scintillation counter forms the basis of energy dispersive photometers, to be discussed later.

In addition to sodium iodide crystals, a number of organic scintillators such as stilbene, anthracene, and terphenyl have been used. In crystalline form, these compounds have decay times of 0.01 and 0.1 μs. Organic liquid scintillators have also been developed and are used to advantage because they exhibit less self-absorption of radiation than do solids. An example of a liquid scintillator is a solution of *p*-terphenyl in toluene.

Semiconductor Transducers

Semiconductor transducers have assumed major importance as detectors of X-radiation. These devices are sometimes called *lithium-drifted silicon detectors,* Si(Li), or *lithium-drifted germanium detectors,* Ge(Li).

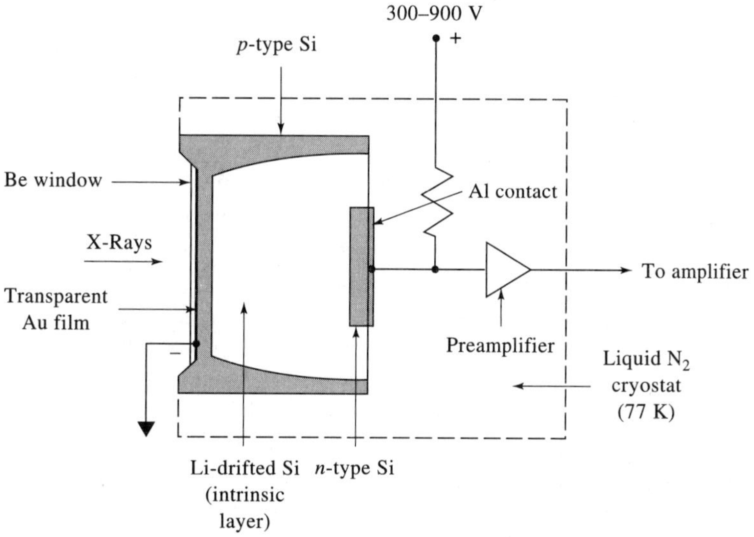

Figure 12-12 Vertical cross section of a lithium-drifted silicon detector for X-rays and radiation from radioactive sources.

Figure 12-12 illustrates one form of a lithium-drifted detector, which is fashioned from a wafer of crystalline silicon. Three layers exist in the crystal, a *p*-type semiconducting layer that faces the X-ray source, a central *intrinsic* zone, and an *n*-type layer. The outer surface of the *p*-type layer is coated with a thin layer of gold for electrical contact; often, it is also covered with a thin beryllium window that is transparent to X-rays. The signal output is taken from an aluminum layer that coats the *n*-type silicon; this output is fed into a preamplifier with an amplification factor of about 10. The preamplifier is frequently a field-effect transistor that is made an integral part of the detector.

A lithium-drifted detector is formed by vapor-depositing lithium on the surface of a *p*-doped silicon crystal. Upon heating to 400° to 500°C, the lithium diffuses into the crystal; because this element easily loses electrons, its presence converts the *p*-type region to an *n*-type region. While still at an elevated temperature, a dc potential is applied across the crystal to cause withdrawal of the electrons from the lithium layer and holes from the *p*-type layer. Current across the *pn* junction requires migration, or drifting, of lithium *ions* into the *p* layer and formation of the intrinsic layer where the lithium ions replace the holes lost by conduction. Upon cooling, this central layer has a high resistance relative to the other layers because the lithium ions in this medium are less mobile than the holes they displaced.

The intrinsic layer of a silicon detector functions in a way that is analogous to argon in the gas-filled transducer. Initially, absorption of a photon results in formation of a highly energetic photoelectron, which then loses its kinetic energy by elevating several thousand electrons in the silicon to the conduction band; a marked increase in conductivity results. When a potential is applied across the crystal, a current pulse accompanies the absorption of each photon. In common with a proportional detector, the size of the pulse is directly proportional to the energy of the absorbed photons. In contrast to the proportional detector, however, secondary amplification of the pulse does not occur.

As shown in Figure 12-12, the detector and preamplifier of a lithium-drifted detector must be thermostatted at the temperature of liquid nitrogen (77 K) to decrease electronic noise to a tolerable level. The original Si(Li) detectors had to be cooled at all times because at room temperature the lithium atoms would diffuse throughout the silicon thereby degrading the performance of the detector. Modern Si(Li) detectors need to be cooled only during use.

Germanium is used in place of silicon to give lithium-drifted detectors that are particularly useful for detection of radiation that is shorter in wavelength than 0.3 Å. These detectors must be cooled at all times. Germanium detectors that do not require lithium drifting have been produced from very pure germanium. These detectors, which are called *intrinsic germanium detectors,* need to be cooled only during use.

Distribution of Pulse-Heights from X-Ray Transducers

To understand the properties of energy dispersive spectrometers, it is important to appreciate that the size of current pulses resulting from absorption of successive X-ray photons of identical energy by the transducer will not be exactly the same. Variations arise because the ejection of photoelectrons and their subsequent generation of conduction electrons are random processes governed by the laws of probability. Thus, a Gaussian distribution of pulse heights around a mean is observed. The breadth of this distribution varies from one type of transducer to another, with semiconductor detectors providing significantly narrower bands of pulses. It is this property that has made lithium-drifted detectors so important for energy dispersive X-ray spectroscopy.

12B-5 Signal Processors

The signal from the preamplifier of an X-ray spectrometer is fed into a linear fast response amplifier whose gain can be varied by a factor up to 10,000. The result is voltage pulses as large as 10 V.

Pulse-Height Selectors

All modern X-ray spectrometers (wavelength dispersive as well as energy dispersive) are equipped with *discriminators* that reject pulses of about 0.5 V or less (after amplification). In this way, transducer and amplifier noise is reduced significantly. In lieu of a discriminator, many instruments are equipped with *pulse-height selectors,* which are electronic circuits that reject not only pulses with heights below some predetermined minimum level but also those above a preset maximum level; that is, they remove all pulses except those that lie within a limited *channel* or *window* of pulse heights.

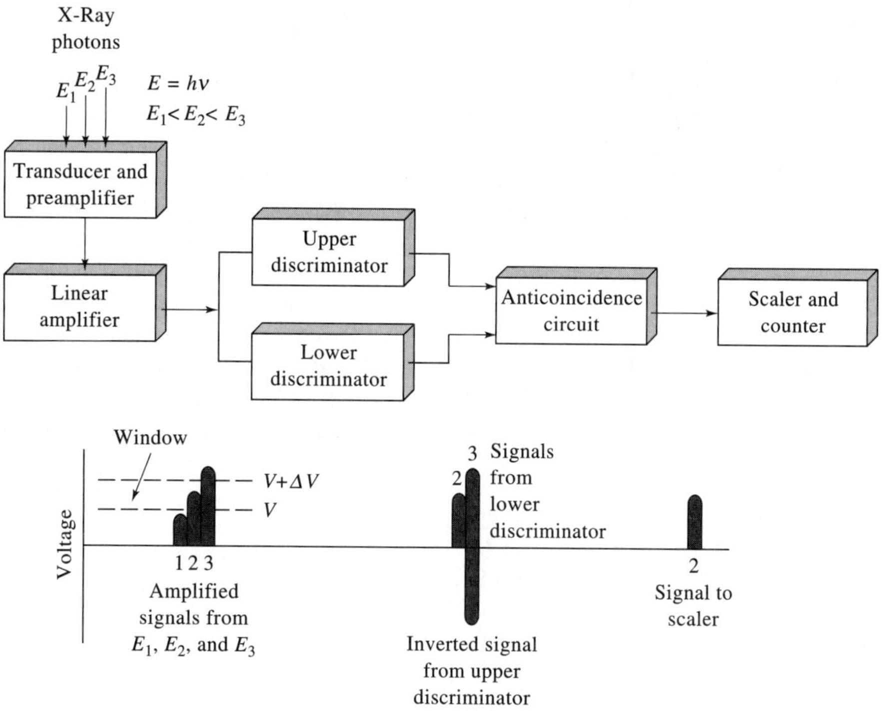

Figure 12-13 Schematic of a signal height selector. The lower discriminator rejects pulses below V only; the upper discriminator rejects pulses below $V + \Delta V$ and inverts the remaining pulse. Lower plot shows the transmitted pulse at the output of the anticoincidence circuit that is counted in this illustration.

Figure 12-13 provides a schematic of a pulse-height selector and its method of operation. Here, the output pulses from the detector and preamplifier are further amplified and appear as voltage signals (in the 10 V range) as shown in the lower part of the figure. These signals are then fed into two discriminator circuits each of which can be set to reject any signal below a certain voltage. As shown in the lower part of Figure 12-13, the lower discriminator rejects signal 1, which is smaller than V, but transmits signals 2 and 3. The upper discriminator, on the other hand, is set to $V + \Delta V$ and thus rejects all but signal 3. In addition, the lower circuit is so arranged that its output signal is reversed in polarity and thus cancels out signal 3 from the upper circuit in the anticoincidence circuit. As a consequence, only signal 2, with a voltage in the range V to $V + \Delta V$, reaches the counter.

Dispersive instruments are often equipped with pulse-height selectors to reject noise and to supplement the monochromator in separating the analyte line from higher order, more energetic radiation that is diffracted at the same crystal setting.

Pulse-Height Analyzers

Pulse-height analyzers consist of one or more pulse-height selectors that are configured in such a way as to provide energy spectra. A single-channel analyzer typically has a voltage range of perhaps 10 V or more with a window of 0.1 to 0.5 V. The window can be manually or automatically adjusted to scan the entire voltage range, thus providing data for an energy dispersion spectrum. Multichannel analyzers typically contain up to a few thousand separate channels, each of which acts as a single channel that corresponds to a different voltage window. The signal from each channel is then accumulated in a memory location of the analyzer corresponding to the energy of the channel thus permitting simultaneous counting and recording of an entire spectrum.

Scalers and Counters

To obtain convenient counting rates, the output from an X-ray transducer is sometimes scaled—that is, the number of pulses is reduced by dividing by some multiple of ten (or occasionally two). A brief description of electronic scalers is found in Section 4C-4. Counting of the scaled pulses is now generally carried out with electronic counters such as those described in Sections 4C-2 and 4C-3.

12C X-RAY FLUORESCENCE METHODS

Although it is feasible to excite an X-ray emission spectrum by incorporating the sample into the target area of an X-ray tube, the inconvenience of this arrangement discourages its application to many types of materials. Instead, excitation is more commonly brought about by irradiation of the sample with a beam of X-rays from an X-ray tube or a radioactive source. Under these circumstances, the elements in the sample are excited by absorption of the primary beam and emit their own characteristic fluorescence X-rays. This procedure is thus properly called an *X-ray fluorescence,* or *emission,* method. X-Ray fluorescence (XRF) is one of the most widely used of all analytical methods for the qualitative identification of elements having atomic numbers greater than that of oxygen (>8); in addition, it is often employed for semiquantitative or quantitative elemental analyses as well.[6] A particular advantage of XRF is that it is, in contrast to most other elemental analysis techniques, nondestructive of the sample.

12C-1 Instruments

Various combinations of the instrument components discussed in the previous section lead to several recognizable types of X-ray fluorescence instruments.[7] The three basic types are *wavelength dispersive, energy dispersive,* and *nondispersive;* the latter two can be further subdivided depending upon whether an X-ray tube or a radioactive substance serves as a radiation source.

Wavelength Dispersive Instruments

Wavelength dispersive instruments always employ tubes as a source because of the large energy losses suffered when an X-ray beam is collimated and dispersed into its component wavelengths. Radioactive sources produce X-ray photons at a rate less than 10^{-4} that of an X-ray tube; the added attenuation by a monochromator would then result in a beam that was difficult or impossible to detect and measure accurately.

Wavelength dispersive instruments are of two types, *single channel,* or *sequential,* and *multichannel,* or *simultaneous.* The spectrometer shown in Figure 12-9 is a sequential instrument that can be readily employed for X-ray fluorescence analysis; here, the X-ray tube and sample are arranged as shown in the circular insert at the top of the figure. Single-channel instruments may be manual or automatic. The former are entirely satisfactory for the quantitative determination of a few elements. In this application, the crystal and transducer are set at the proper angles (θ and 2θ), and counting is continued until sufficient counts have accumulated for precise results. Automatic instruments are much more convenient for qualitative analysis, where an entire spectrum must be scanned. Here, the electric drive for the crystal and detector is synchronized, and the detector output is connected to the data acquisition system.

Most modern single-channel spectrometers are provided with two X-ray sources; typically, one has a chromium target for longer wavelengths and the other a tungsten target for shorter. For wavelengths longer than 2 Å, it is necessary to remove air between the source and detector by pumping or by displacement with a continuous flow of helium. A means must also be provided for convenient interchange of dispersing crystals.

Recording single-channel instruments cost approximately $60,000.

Multichannel dispersive instruments are large, expensive (>$150,000) installations, which permit the simultaneous detection and determination of as many as 24 elements. Here, individual channels consisting of an appropriate crystal and a detector are arranged radially around an X-ray source and sample holder. Ordinarily, the crystals for all or most of the channels are fixed at an appropriate angle for a given analyte line; in some instruments, one or more of the crystals can be moved to permit a spectral scan.

Each transducer in a multichannel instrument is provided with its own amplifier, pulse-height selector,

---

[6] See R. Jenkins, *X-Ray Fluorescence Spectrometry.* New York: Wiley, 1988.

[7] For a discussion of modern X-ray fluorescence instruments, see R. Jenkins, *Anal. Chem.,* **1984,** *56,* 1099A.

scaler, and counter or integrator. These instruments are ordinarily equipped with a computer for instrument control, data processing, and display of analytical results. A determination of 20 or more elements can be completed in a few seconds to a few minutes.

Multichannel instruments are widely used for the determination of several components in materials of industry such as steel, other alloys, cement, ores, and petroleum products. Both multichannel and single-channel instruments are equipped to handle samples in the form of metals, powdered solids, evaporated films, pure liquids, or solutions. Where necessary, the materials are held in a cell with a Mylar or cellophane window.

Energy Dispersive Instruments[8]

As shown in Figure 12-14, an energy dispersive spectrometer consists of a polychromatic source, which may be either an X-ray tube or a radioactive material, a sample holder, a semiconductor detector, and the various electronic components required for energy discrimination.

An obvious advantage of energy dispersive systems is the simplicity and lack of moving parts in the excitation and detection components of the spectrometer. Furthermore, the absence of collimators and a crystal diffractor, as well as the closeness of the detector to the sample, result in a 100-fold or more increase in energy reaching the detector. These features permit the use of weaker sources such as radioactive materials or low-power X-ray tubes, which are cheaper and less likely to cause radiation damage to the sample. Generally energy dispersive instruments cost about one fourth to one fifth the cost of wavelength dispersive systems.

In a multichannel, energy dispersive instrument, all of the emitted X-ray lines are measured simultaneously. Increased sensitivity and improved signal-to-noise ratio result from the Fellgett advantage (see page 184).

The principal disadvantage of energy dispersive systems, when compared with crystal spectrometers, is their lower resolutions at wavelengths longer than about 1 Å. On the other hand, at shorter wavelengths, energy dispersive systems exhibit superior resolution.

[8] See R. Woldseth, *All You Want To Know About XES*. Foster City, CA: Kevex Corporation, 1973.

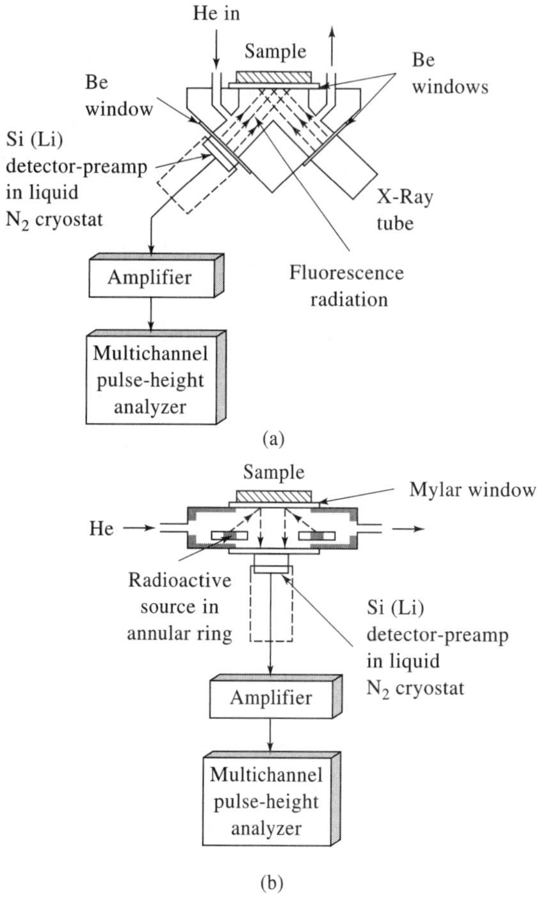

Figure 12-14 Energy dispersive X-ray fluorescence spectrometer. Excitation by X-rays from (a) an X-ray tube and (b) a radioactive substance.

Nondispersive Instruments

Figure 12-15 is a cutaway view of a basic, commercial, nondispersive instrument that has been employed for the routine determination of sulfur and lead in gasoline. For a sulfur determination, the sample is irradiated by X-rays produced by an iron-55 radioactive source; this radiation in turn generates a sulfur fluorescence line at 5.4 Å. The analyte radiation then passes through a pair of adjacent filters and into twin proportional counters. The absorption edge of one of the filters lies just below 5.4 Å, while that of the other is just above it. The difference between the two signals is proportional to the sulfur content of the sample. A sulfur determination with this instrument requires a counting time of about 1 min.

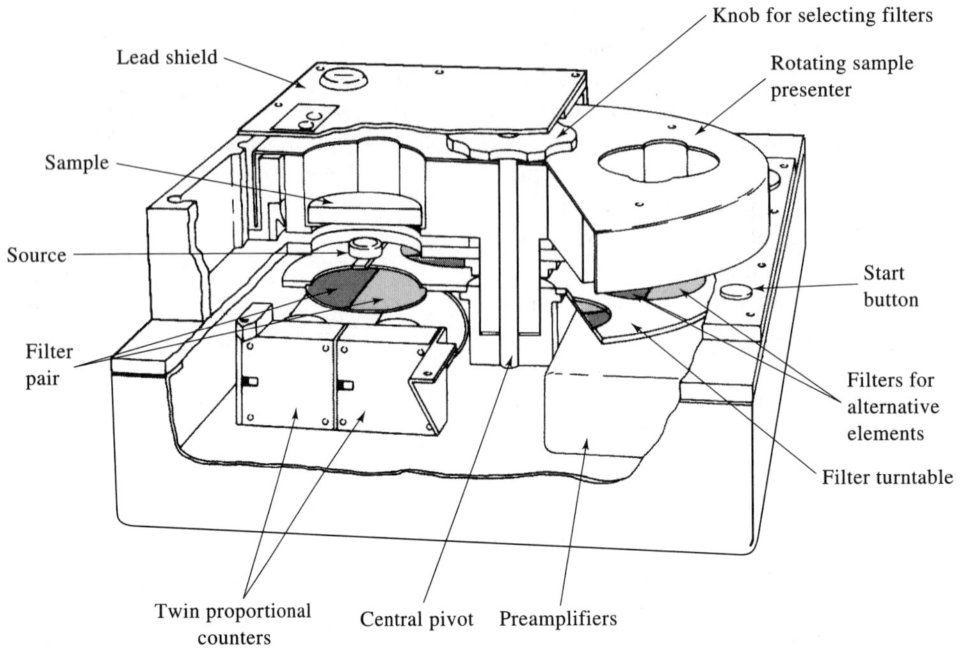

Figure 12-15 Cutaway view of a commercial nondispersive X-ray fluorescence instrument. *(Reprinted with permission from B. J. Price and K. M. Field,* Amer. Lab., ***1974,*** *6(9), 62. Copyright 1974 by International Scientific Communications, Inc.)*

Relative standard deviations of about 1% are obtained for replicate measurements.

12C-2 Qualitative and Semiquantitative Analysis

Figure 12-16 illustrates an interesting qualitative application of the X-ray fluorescence method. Here, the untreated sample, which was excited by radiation from an X-ray tube, was subsequently recovered unchanged. Note that the abscissa for wavelength dispersive instruments is often plotted in terms of the angle 2θ, which can be readily converted to wavelength with knowledge of the crystal spacing of the monochromator (Equation 12-6). Identification of peaks is then accomplished by reference to tables of emission lines of the elements.

Figure 12-17 is a spectrum obtained with an energy dispersive instrument. With such equipment, the abscissa is generally calibrated in channel number or energy in keV. Each dot represents the number of counts accumulated in one of the several hundred channels.

Qualitative information, such as that shown in Figures 12-16 and 12-17, can be converted to semiquantitative data by careful measurement of peak heights. To obtain a rough estimate of concentration, the following relationship is used:

$$P_x = P_s W_x \tag{12-7}$$

where P_x is the relative line intensity measured in terms of number of counts for a fixed period, and W_x is the weight fraction of the desired element in the sample. The factor P_s is the relative intensity of the line that would be observed under identical counting conditions if W_x were unity. The value of P_s is determined with a sample of the pure element or a standard sample of known composition.

The use of Equation 12-7, as outlined in the previous paragraph, carries with it the assumption that the emission from the species of interest is unaffected by the presence of other elements in the sample. We shall see that this assumption may not be justified; as a consequence, a concentration estimate may be in error by a factor of two or more. On the other hand, this uncertainty is significantly smaller than that associated with a semiquantitative analysis by optical emission where order of magnitude errors are not uncommon.

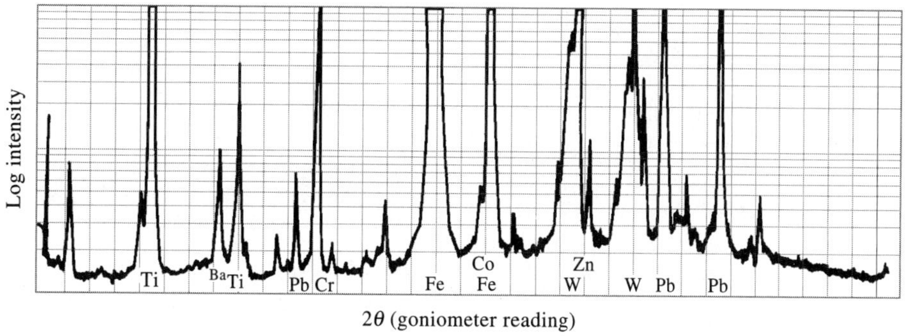

Figure 12-16 X-Ray fluorescence spectrum for a genuine bank note recorded with a wavelength dispersive spectrometer. *(Taken from H. A. Liebhafsky, H. G. Pfeiffer, E. H. Winslow, and P. D. Zeman,* X-Ray Absorption and Emission in Analytical Chemistry, *p. 163. New York: Wiley, 1960. Reprinted by permission of John Wiley & Sons, Inc.)*

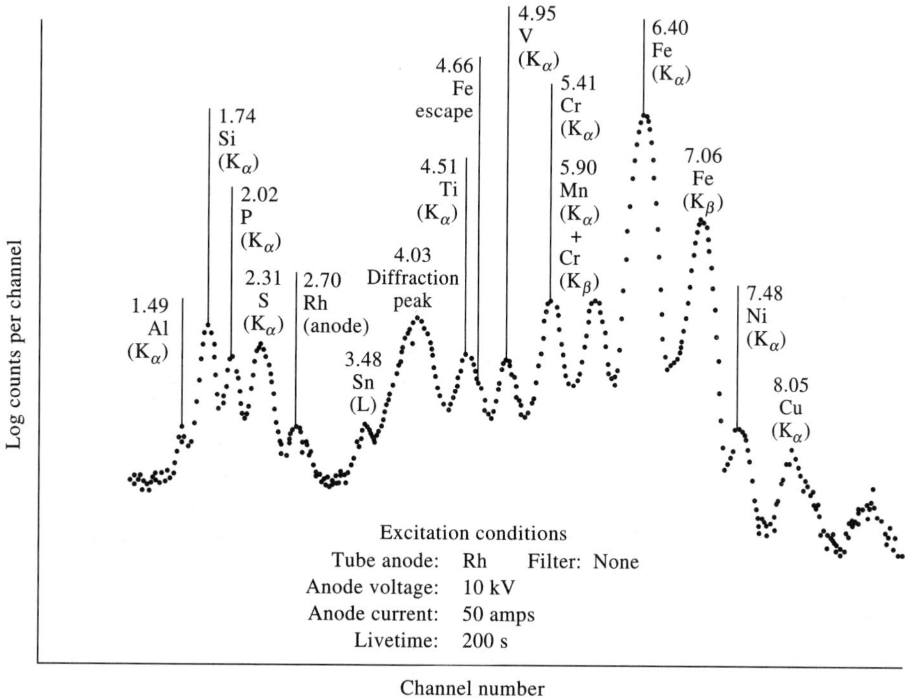

Figure 12-17 Spectrum of an iron sample obtained with an energy dispersive instrument with an Rh anode X-ray tube source. The numbers above the peaks are energies in keV. *(Reprinted with permission from J. A. Cooper,* Amer. Lab., ***1976,*** *8(11), 44. Copyright 1976 by International Scientific Communications, Inc.)*

12C-3 Quantitative Analysis[9]

Modern X-ray fluorescence instruments are capable of producing quantitative analyses of complex materials with precision that equals or exceeds that of the classical wet chemical methods or other instrumental methods. For the accuracy of such analyses to approach this level, however, requires either the availability of calibration standards that closely approach the samples in overall chemical and physical composition or suitable methods for dealing with matrix effects.

Matrix Effects

It is important to realize that the X-rays produced in the fluorescence process are generated not only from atoms at the surface of a sample but also from atoms well below the surface. Thus, a part of both the incident radiation and the resulting fluorescence traverse a significant thickness of sample within which absorption and scattering can occur. The extent to which either beam is attenuated depends upon the mass absorption coefficient of the medium, which in turn is determined by the coefficients of *all* of the elements in the sample. Therefore, while the net intensity of a line reaching the detector in an X-ray fluorescence measurement depends on the concentration of the element producing the line, it is also affected by the concentration and mass absorption coefficients of the matrix elements as well.

Absorption effects by the matrix may cause results calculated by Equation 12-7 to be either high or low. If, for example, the matrix contains a significant amount of an element that absorbs either the incident or the emitted beam more strongly than the element being determined, then W_x will be low, since P_s was evaluated with a standard in which absorption was smaller. On the other hand, if the matrix elements of the sample absorb less than those in the standard, high values for W_x result.

A second matrix effect, called the *enhancement effect,* can also yield results that are greater than expected. This behavior is encountered when the sample contains an element whose characteristic emission spectrum is excited by the incident beam, and this spectrum in turn causes a secondary excitation of the analytical line.

Several techniques have been developed to compensate for absorption and enhancement effects in X-ray fluorescence analyses.

[9] See R. Jenkins, R. W. Gould, and D. Gedcke, *Quantitative X-Ray Spectrometry,* 2nd ed. New York: Marcel Dekker, 1995.

Calibration Against Standards

Here, the relationship between the analytical line intensity and the concentration is determined empirically with a set of standards that closely approximate the samples in overall composition. The assumption is then made that absorption and enhancement effects are identical for both samples and standards, and the empirical data are employed to convert emission data to concentrations. The degree of compensation achieved in this way depends upon the closeness of the match between samples and standards.

Use of Internal Standards

In this procedure, an element is introduced in known and fixed concentration into both the calibration standards and the samples; the added element must be absent in the original sample. The ratio of the intensities between the element being determined and the internal standard serves as the analytical variable. The assumption here is that absorption and enhancement effects are the same for the two lines and that use of intensity ratios compensates for these effects.

Dilution of Samples and Standards

Here, both sample and standards are diluted with a substance that absorbs X-rays only weakly, that is, a substance containing elements with low atomic numbers. Examples of such diluents include water; organic solvents containing carbon, hydrogen, oxygen, and nitrogen only; starch; lithium carbonate; alumina; and boric acid or borate glass. By employing an excess of diluent, matrix effects become essentially constant for the diluted standards and samples, and adequate compensation is achieved. This procedure has proved particularly useful for mineral analyses, where both samples and standards are dissolved in molten borax; after cooling, the fused mass is excited in the usual way.

Some Quantitative Applications of X-Ray Fluorescence

With proper correction for matrix effects, X-ray fluorescence spectrometry is perhaps the most powerful tool available to the chemist for the rapid quantitative determination of all but the lightest elements in complex samples. For example, Baird and Henke[10] have demonstrated that nine elements can be determined in samples of granitic rocks in an elapsed time, including sample preparation, of

[10] A. K. Baird and B. L. Henke, *Anal. Chem.,* **1965,** *37,* 727.

about 12 min. The precision of the method is better than wet chemical analyses and averages 0.08% relative. It is noteworthy that one of the elements reported is oxygen, which ordinarily can be determined by difference only. An excellent overview of X-ray fluorescence analysis of geological materials has also been published.[11]

X-Ray methods also find widespread application for quality control in the manufacture of metals and alloys. Here, the speed of the analysis permits correction of the composition of the alloy during its manufacture.

X-Ray fluorescence methods are readily adapted to liquid samples. Thus, as mentioned earlier, methods have been devised for the direct quantitative determination of lead and bromine in aviation gasoline samples. Similarly, calcium, barium, and zinc have been determined in lubricating oils by excitation of fluorescence in the liquid hydrocarbon samples. The method is also convenient for the direct determination of the pigments in paint samples.

X-Ray fluorescence methods are being widely applied to the analysis of atmosphere pollutants. For example, one procedure for detecting and determining contaminants involves drawing an air sample through a stack consisting of a micropore filter for particulates and three filter paper disks impregnated with orthotolidine, silver nitrate, and sodium hydroxide, respectively. The reagents retain chlorine, sulfides, and sulfur dioxide in that order. The filters then serve as samples for X-ray fluorescence analysis.

Another indication of the versatility of X-ray fluorescence is its use for the quantitative determination of elements heavier than sodium in rocks and soil encountered near the landing sight of the Mars Pathfinder mission.[12] The Pathfinder's microrobot Sojourner was equipped with a sensor head that could be placed flat against a material to be analyzed. The head contained curium 244, which emits α particles that then bombard the sample surface as in Figure 12-14b. The X-ray fluorescence emitted from the sample struck the transducer of an energy dispersive spectrometer in which the radiant power was recorded as a function of energy, transmitted from Mars to Earth, and ultimately analyzed on Earth. Lighter elements were determined by backscattering or proton emission. With these three methods of detection, all of the elements in the periodic table but hydrogen could be determined at concentration levels of a few tenths of a percent.

Advantages and Disadvantages of X-Ray Fluorescence Methods

X-Ray fluorescence offers a number of impressive advantages. The spectra are relatively simple; spectral line interference is thus unlikely. Generally, the X-ray method is nondestructive and can be used for the analysis of paintings, archeological specimens, jewelry, coins, and other valuable objects without harm to the sample. Furthermore, analyses can be performed on samples ranging from a barely visible speck to a massive object. Other advantages include the speed and convenience of the procedure, which permits multielement analyses to be completed in a few minutes. Finally, the accuracy and precision of X-ray fluorescence methods often equal or exceed those of other methods.[13]

X-Ray fluorescence methods are generally not as sensitive as the various optical methods that have been discussed earlier in this text. In the most favorable cases, concentrations of a few parts per million or less can be measured. More commonly, however, the concentration range of the method will be from about 0.01 to 100%. X-Ray fluorescence methods for the lighter elements are inconvenient; difficulties in detection and measurement become progressively worse as atomic numbers become smaller than 23 (vanadium), in part because a competing process, called *Auger emission*, reduces the fluorescence intensity (see Figure 21–7). Present commercial instruments are limited to atomic numbers of 5 (boron) or 6 (carbon). Another disadvantage of the X-ray emission procedure is the high cost of instruments, which ranges from about $5000 for an energy dispersive system with a radioactive source, to well over $500,000 for automated and computerized wavelength dispersive systems.

12D X-RAY ABSORPTION METHODS

In contrast to optical spectroscopy, where absorption methods are of prime importance, X-ray absorption applications are limited when compared with X-ray emission and fluorescence procedures. While absorption measurements can be made relatively free of matrix effects, the required techniques are somewhat cumbersome and time consuming when compared with flu-

[11] J. E. Anzelma and J. R. Lindsay. *J. Chem. Educ.*, **1987**, *64*, A181 and A200.

[12] Burka, Lauren P. "Welcome to the Mars Pathfinder Mission!" 1997. hhtp://mpfwww.jpl.nasa.gov/ (16 July 1997).

[13] For a comparison of X-ray fluorescence and ICP for the analysis of iron oxides, see R. A. Peterson and D. M. Wheeler, *Amer. Lab.*, **1981**, *13* (10), 138.

orescence methods. Thus, most applications are confined to samples in which the effect of the matrix is minimal.

Absorption methods are analogous to optical absorption procedures in which the attenuation of a band or line of X-radiation serves as the analytical variable. Wavelength selection is accomplished with a monochromator such as that shown in Figure 12-9 or by a filter technique similar to that illustrated in Figure 12-8. Alternatively, the monochromatic radiation from a radioactive source is employed.

Because of the breadth of X-ray absorption peaks, direct absorption methods are generally useful only when a single element with a high atomic number is to be determined in a matrix consisting of only lighter elements. Examples of applications of this type are the determination of lead in gasoline and the determination of sulfur or the halogens in hydrocarbons.

12E X-RAY DIFFRACTION METHODS

Since its discovery in 1912 by von Laue, X-ray diffraction has provided a wealth of important information to science and industry. For example, much that is known about the arrangement and the spacing of atoms in crystalline materials has been determined directly from diffraction studies. In addition, such studies have led to a much clearer understanding of the physical properties of metals, polymeric materials, and other solids. X-Ray diffraction is currently of prime importance in elucidating the structures of such complex natural products as steroids, vitamins, and antibiotics. Such applications are beyond the scope of this text.

X-Ray diffraction also provides a convenient and practical means for the qualitative identification of crystalline compounds. The X-ray powder diffraction method is unique in that it is the only analytical method that is capable of providing qualitative and quantitative information about the *compounds* present in a solid sample. For example, the powder method can determine the percent KBr and NaCl in a solid mixture of these two compounds. Other analytical methods reveal only the percent K^+, Na^+, Br^-, and Cl^- in the sample.[14]

[14] For a more detailed discussion of the X-ray powder diffraction method, see R. Jenkins and R. Snyder, *Introduction to X-Ray Powder Diffractometry*. New York: Wiley, 1996.

X-Ray powder methods are based upon the fact that an X-ray diffraction pattern is unique for each crystalline substance. Thus, if an exact match can be found between the pattern of an unknown and an authentic sample, chemical identity can be assumed.

12E-1 Identification of Crystalline Compounds

Sample Preparation

For analytical diffraction studies, the crystalline sample is ground to a fine homogeneous powder. In such a form, the enormous number of small crystallites are oriented in every possible direction; thus, when an X-ray beam traverses the material, a significant number of the particles can be expected to be oriented in such ways as to fulfill the Bragg condition for reflection from every possible interplanar spacing.

Samples may be held in the beam in thin-walled glass, or cellophane capillary tubes. Alternatively, a specimen may be mixed with a suitable noncrystalline binder and molded into an appropriate shape.

Automatic Diffractometers

Diffraction patterns are generally obtained with automated instruments similar in design to that shown in Figure 12-9. Here, the source is an X-ray tube with suitable filters. The powdered sample, however, replaces the single crystal on its mount. In some instances, the sample holder may be rotated in order to increase the randomness of the orientation of the crystals. The diffraction pattern is then obtained by automatic scanning in the same way as for an emission or absorption spectrum. Instruments of this type offer the advantage of high precision for intensity measurements and automated data reduction and report generation.

Photographic Recording

The classical method for recording powder diffraction patterns, and one that still finds use, particularly when the amount of sample is small, is photographic. The most common instrument for this purpose is the *Debye-Scherrer* powder camera, which is shown schematically in Figure 12-18a. Here, the beam from an X-ray tube is filtered to produce a nearly monochromatic beam (often the copper or molybdenum K_α line), which is colli-

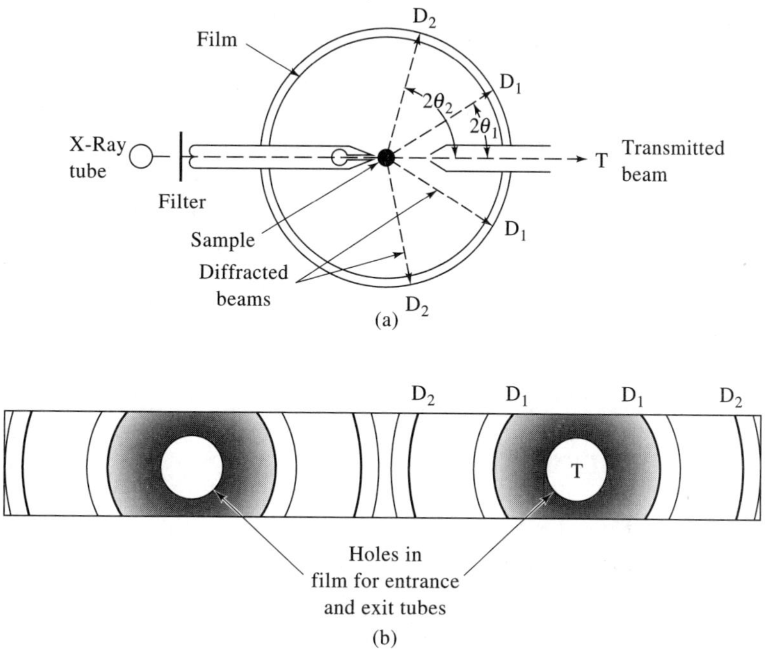

Figure 12-18 Schematic of (a) a Debye-Scherrer powder camera; (b) The film strip after development. D_2, D_1, and T indicate positions of the film in the camera.

mated by passage through a narrow tube. The undiffracted radiation T then passes out of the camera via a narrow exit tube. The camera itself is cylindrical and equipped to hold a strip of film around its inside wall. The inside diameter of the cylinder usually is 5.73 or 11.46 cm, so that each lineal millimeter of film is equivalent to 1.0 or 0.5 deg in θ, respectively. The sample is held in the center of the beam by an adjustable mount.

Figure 12-18b depicts the appearance of the exposed and developed film; each set of lines (D_1, D_2, and so forth) represents diffraction from one set of crystal planes. The Bragg angle θ for each line is easily evaluated from the geometry of the camera.

12E-2 Interpretation of Diffraction Patterns

The identification of a species from its powder diffraction pattern is based upon the position of the lines (in terms of θ or 2θ) and their relative intensities. The diffraction angle 2θ is determined by the spacing between a particular set of planes; with the aid of the Bragg equation, this distance d is calculated from the known wavelength of the source and the measured angle. Line intensities depend upon the number and kind of atomic reflection centers in each set of planes.

Identification of crystals is empirical. A powder diffraction file is maintained by the International Centre for Diffraction Data, Swarthmore, PA. In 1988, this file contained powder diffraction patterns for over 50,000 compounds. Because the file is so large as to make searching difficult and time consuming, the powder data file has been broken down into subfiles that contain listings for inorganics, organics, minerals, metals, alloys, forensic materials, and others. The data in these files is in terms of d spacings and relative line intensities. The entries are arranged in order of the d spacing for the most intense line; entries are withdrawn from this file on the basis of d spacing that lies within a few hundredths of an angstrom of the d spacing of the most intense line for the analyte. Further elimination of possible compounds is accomplished by consideration of the spacing for the second most intense line, then the third, and so forth. Ordinarily, three or four spacings serve to identify the compound unambiguously. Computer search programs are now available to relieve the tedium of the search process.

If the sample contains two or more crystalline compounds, identification becomes more complex. Here, various combinations of the more intense lines are used until a match can be found.

By measuring the intensity of the diffraction lines and comparing with standards, a quantitative analysis of crystalline mixtures is also possible.

12F THE ELECTRON MICROPROBE

An important method for the determination of the elemental composition of surfaces is based upon the *electron microprobe.* In this technique, X-ray emission from the elements on the surface of a sample is stimulated by a narrowly focused beam of electrons. The resulting X-ray emission is detected and analyzed with either a wavelength or an energy dispersive spectrometer. This method is discussed in detail in Section 21B-7.

12G QUESTIONS AND PROBLEMS

12-1 What is the short-wavelength limit of the continuum produced by an X-ray tube having a silver target and operated at 80 kV?

12-2 What minimum tube voltage would be required to excite the K_β and L_β series of lines for (a) U, (b) K, (c) Rb, (d) W?

12-3 The $K_{\alpha 1}$ lines for Ca, Zn, Zr, and Sn occur at 3.36, 1.44, 0.79, and 0.49 Å, respectively. Calculate an approximate wavelength for the K_α lines of (a) V, (b) Ni, (c) Se, (d) Br, (e) Cd, (f) Sb.

12-4 The L_α lines for Ca, Zn, Zr, and Sn are found at 36.3, 11.9, 6.07, and 3.60 Å, respectively. Estimate the wavelengths for the L_α lines for the elements listed in Problem 12-3.

12-5 The mass absorption coefficient for Ni, measured with the Cu K_α line is 49.2 cm^2/g. Calculate the thickness of a nickel foil that was found to transmit 36.1% of the incident power of a beam of Cu K_α radiation. Assume that the density of Ni is 8.9 g/cm^3.

12-6 For Mo K_α radiation (0.711 Å), the mass absorption coefficients for K, I, H, and O are 16.7, 39.2, 0.0, and 1.50 cm^2/g, respectively.

(a) Calculate the mass absorption coefficient for a solution prepared by mixing 8.00 g of KI with 92 g of water.

(b) The density of the solution described in (a) is 1.05 g/cm^3. What fraction of the radiation from a Mo K_α source would be transmitted by a 0.50-cm layer of the solution?

12-7 Aluminum is to be used as windows for a cell for X-ray absorption measurements with the Ag K_α line. The mass absorption coefficient for aluminum at this wavelength is 2.74; its density is 2.70 g/cm^3. What maximum thickness of aluminum foil could be employed to fabricate the windows if no more than 2.0% of the radiation is to be absorbed by them?

12-8 A solution of I_2 in ethanol had a density of 0.794 g/cm^3. A 1.50-cm layer was found to transmit 27.3% of the radiation from a Mo K_α source. Mass absorption coefficients for I, C, H, and O are 39.2, 0.70, 0.00, and 1.50, respectively.

(a) Calculate the percent I_2 present, neglecting absorption by the alcohol.

(b) Correct the results in part (a) for the presence of alcohol.

12-9 Calculate the goniometer setting, in terms of 2θ, required to observe the $K_{\alpha 1}$ lines for Fe (1.76 Å), Se (0.992 Å), and Ag (0.497 Å) when the diffracting crystal is (a) topaz; (b) LiF; (c) NaCl.

12-10 Calculate the goniometer setting, in terms of 2θ, required to observe the $L_{\beta 1}$ lines for Br at 8.126 Å when the diffracting crystal is
(a) ethylenediamine *d*-tartrate.
(b) ammonium dihydrogen phosphate.

12-11 Calculate the minimum tube voltage required to excite the following lines. The numbers in parentheses are the wavelengths in Å for the corresponding absorption edges.
(a) K lines for Ca (3.064)
(b) L_α lines for As (9.370)
(c) L_β lines for U (0.592)
(d) K lines for Mg (0.496)

12-12 Manganese was determined in samples of geological interest via X-ray fluorescence using barium as an internal standard. The fluorescence intensity of isolated lines for each element gave the following data:

	Counts Per Second	
Wt. % Mn	**Ba**	**Mn**
0.00	156	80
0.10	160	106
0.20	159	129
0.30	160	154
0.40	151	167

What is the weight percent manganese in a sample that had a Mn/Ba count ratio of 0.886?

12-13 As discussed on page 293, an important experiment aboard the NASA Mars Pathfinder Sojourner was the APXS, or Alpha Proton X-Ray Spectrometer. The world was captivated by the Sojourner, and the World Wide Web site at http://mpfwww.jpl.nasa.gov/ and its mirror sites were nearly overwhelmed by many millions of hits per hour. For detailed information regarding the instrumentation aboard the Sojourner, please refer to the article at http://mars.sgi.com/mpf/sci_desc.html#APXS, which contains a general description of the instrument package and links to more detailed information including the article by T. Economou at http://astro.uchicago.edu/home/web/papers/economou/pathfinder/ (July 25, 1997). APXS and other data were collected by Pathfinder and transmitted to Earth for further analysis and study. Several open-ended questions regarding the Pathfinder mission and its scientific results are included on the following page.

Figure 12-19 Panoramic Image of Mars from the Pathfinder mission. *(Reproduced from http://mars.sgi.com/default.html, with permission.)*

(a)

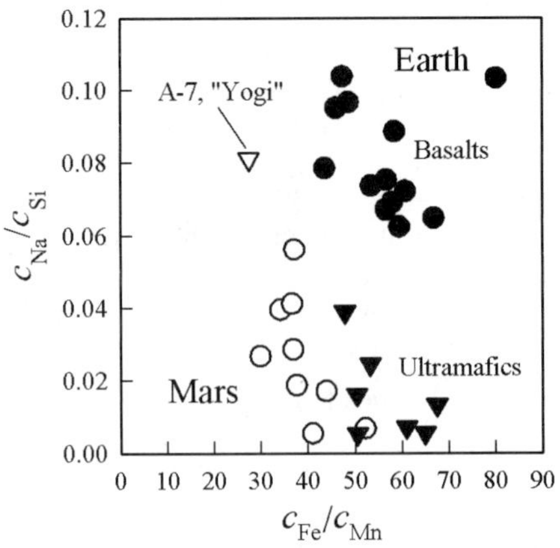

(b)

Figure 12-20 (a) Sojourner deploys the APXS and analyzes "Barnacle Bill". (b) Plot of the concentration ratio Na/Si vs. the concentration ratio Fe/Mn for various terrestrial and Martian rocks. Note that the groupings of the rocks on the plot are characteristic of the different types of rocks.

(a) Consult the article by Economou cited above, and describe the construction and operation of the detector head on the Sojourner rover.

(b) What determines the selectivity and sensitive of the APXS system?

(c) What elements may not be determined by APXS? Why not?

(d) From the table of composition shown below, compute appropriate elemental ratios to classify rock A-3 using Figure 12-20b. Are there other elemental ratios that may be appropriate for such classifications?

(e) Use the data in the table to check the accuracy of the oxygen determination.

(f) Find oxygen data more recent than July 25, 1997 on the Web, and compare to the results in the table. Also find more recent results on K and Mn and compare them to the table. Which results are more accurate and more precise? Explain.

Analysis of Martian Samples by the APX-Spectrometer, First Preliminary Results

Element	A-2, Soil Wt. %	A-4, Soil Wt. %	A-5, Soil Wt. %	A-3, Rock ("Barnacle Bill") Wt. %	A-7, Rock ("Yogi") Wt. %
Fe	15.2	11.2	13.6	9.9	10.7
K	0.5	0.6	0.6	1.2	0.9
Mn	0.4	0.4	0.5	0.7	0.4
Na	3.2	3.8	2.6	3.1	1.9
O	42.5	43.9	43.2	45.0	44.6
Si	21.6	20.2	20.5	25.7	23.8

Section III

Molecular Spectroscopy

The image above shows the optics and the mechanical parts of a Fourier transform infrared spectrometer. This type of spectrometer is discussed in detail in Section 16C-1. (Courtesy of Nicolet Instruments, Inc., Madison, WI.)

Section III begins with an introduction to the theory and practice of ultraviolet, visible, and near-infrared spectrometry in Chapter 13, followed by applications of these regions of the spectrum in Chapter 14. Molecular luminescence spectrometry, including fluorescence, phosphorescence, and chemiluminescence measurement techniques, is described in Chapter 15. Vibrational spectroscopy is discussed in Chapter 16 (theory and instrumentation of infrared spectrometry), Chapters 17 (applications in the near-infrared, mid-infrared, and far-infrared regions of the spectrum), and Chapter 18 (Raman spectroscopy and its applications). Nuclear magnetic resonance spectrometry is investigated in Chapter 19. Molecular mass spectrometry is explored in Chapter 20. Chapter 21 concludes Section III, addressing surface analytical techniques such as electron spectroscopy, secondary-ion mass spectrometry, scanning electron microscopy, and scanning probe microscopy.

Chapter 13

An Introduction to Ultraviolet/Visible Molecular Absorption Spectrometry

This chapter provides introductory material that is primarily applicable to absorption spectroscopy based upon electromagnetic radiation in the wavelength region of 160 to 780 nm. Some of the material, however, is applicable to infrared spectroscopy as well. Absorption measurements based upon ultraviolet and visible radiation find widespread application for the quantitative determination of a large variety of inorganic and organic species.[1]

Molecular absorption spectroscopy is based on the measurement of the transmittance T or the absorbance A of solutions contained in transparent cells having a path length of b cm. Ordinarily, the concentration c of an absorbing analyte is linearly related to absorbance as represented by the equation

$$A = -\log T = \log \frac{P_0}{P} = \epsilon bc \qquad (13\text{-}1)$$

All of the variables in this equation are defined in Table 13-1. This equation is a mathematical representation of Beer's law.

13A MEASUREMENT OF TRANSMITTANCE AND ABSORBANCE

Ordinarily, transmittance and absorbance, as defined in Table 13-1, cannot be measured in the laboratory because the analyte solution must be held in some sort of a transparent container, or cell. As shown in Figure 13-1, reflection occurs at the two air/wall interfaces as well as at the two wall/solution interfaces. The resulting beam attenuation is substantial as we demonstrated in Example 6-2 (page 126), where it was shown that about 8.5% of a beam of yellow light is lost by reflection in passing through a glass cell containing water. In addition, attenuation of a beam may occur as a result of scat-

[1] Some useful references on absorption methods include: E. J. Meehan, in *Treatise on Analytical Chemistry,* 2nd ed., P. J. Elving, E. J. Meehan, and I. M. Kolthoff, Eds., Part I, Vol. 7, Chapters 1–3. New York: Wiley, 1981; R. P. Bauman, *Absorption Spectroscopy.* New York: Wiley, 1962; F. Grum, in *Physical Methods of Chemistry,* A. Weissberger and B. W. Rossiter, Eds., Vol. I, Part III B, Chapter 3. New York: Wiley-Interscience, 1972; H. H. Jaffé and M. Orchin, *Theory and Applications of Ultraviolet Spectroscopy.* New York: Wiley, 1962; G. F. Lothian, *Absorption Spectrophotometry,* 3rd ed. London: Adam Hilger Ltd., 1969; J. D. Ingle Jr. and S. R. Crouch, *Spectrochemical Analysis,* Chapter 13. Englewood Cliffs, NJ: Prentice-Hall, 1988.

TABLE 13-1 Important Terms and Symbols for Measurement of Absorption

Term and Symbol*	Definition	Alternative Name and Symbol
Radiant power P, P_0	Energy of radiation (in ergs) impinging on a 1-cm^2 area of a detector per second	Radiation intensity I, I_0
Absorbance A	$\log \frac{P_0}{P}$	Optical density D; extinction E
Transmittance T	$\frac{P}{P_0}$	Transmission T
Path length of radiation† b	—	l, d
Absorptivity† a	$\frac{A}{bc}$	Extinction coefficient k
Molar absorptivity‡ ϵ	$\frac{A}{bc}$	Molar extinction coefficient

*Terminology recommended by the American Chemical Society (*Anal. Chem.*, **1990,** *62,* 91).
†c may be expressed in g/L or in other specified concentration units; b may be expressed in cm or in other units of length.
‡c is expressed in mol/L; b is expressed in cm.

tering by large molecules and sometimes from absorption by the container walls. To compensate for these effects, the power of the beam transmitted by the analyte solution is usually compared with the power of the beam transmitted by an identical cell containing only solvent. An experimental transmittance and absorbance that closely approximate the true transmittance and absorbance are then obtained with the equations

$$T = \frac{P_{\text{solution}}}{P_{\text{solvent}}} = \frac{P}{P_0} \tag{13-2}$$

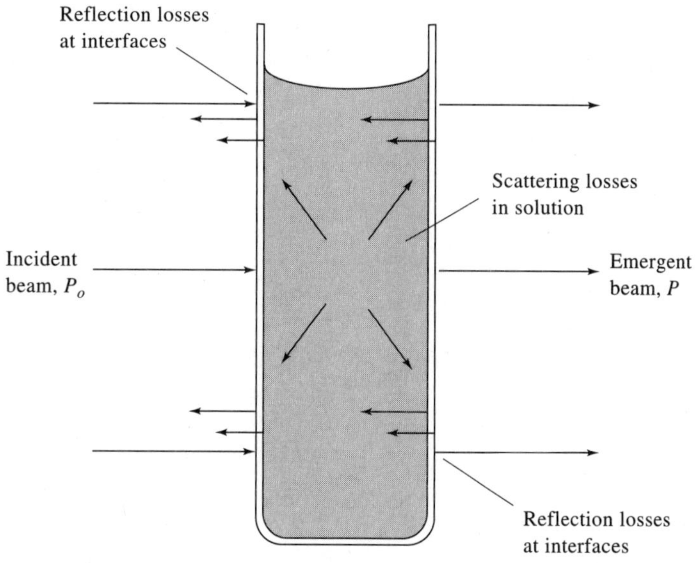

Figure 13-1 Reflection and scattering losses.

$$A = \log \frac{P_{\text{solvent}}}{P_{\text{solution}}} \approx \log \frac{P_0}{P} \tag{13-3}$$

The terms P_0 and P, as used henceforth in this book, refer to the power of radiation after it has passed through cells containing the solvent and the analyte, respectively.

13B BEER'S LAW

Equation 13-1 represents *Beer's law.* This relationship can be rationalized as follows.[2] Consider the block of absorbing matter (solid, liquid, or gas) shown in Figure 13-2. A beam of parallel monochromatic radiation with power P_0 strikes the block perpendicular to a surface; after passing through a length b of the material, which contains n absorbing atoms, ions, or molecules, its power is decreased to P as a result of absorption. Consider now a cross section of the block having an area S and an infinitesimal thickness dx. Within this section there are dn absorbing particles; associated with each particle, we can imagine a surface at which photon capture will occur. That is, if a photon reaches one of these areas by chance, absorption will follow immediately. The total projected area of these capture surfaces within the section is designated as dS; the ratio of the capture area to the total area, then, is dS/S. On a statistical average, this ratio represents the probability for the capture of photons within the section.

The power of the beam entering the section, P_x, is proportional to the number of photons per square centimeter and dP_x represents the quantity absorbed within the section; the fraction absorbed is then $-dP_x/P_x$, and this ratio also equals the average probability for capture. The term is given a minus sign to indicate that P undergoes a decrease. Thus,

$$-\frac{dP_x}{P_x} = \frac{dS}{S} \tag{13-4}$$

Recall, now, that dS is the sum of the capture areas for particles within the section; it must therefore be proportional to the number of particles, or

$$dS = a\,dn \tag{13-5}$$

where dn is the number of particles and a is a proportionality constant, which can be called the *capture cross section.* By combining Equations 13-4 and 13-5 and integrating over the interval between 0 and n, we obtain

$$-\int_{P_0}^{P} \frac{dP_x}{P_x} = \int_0^n \frac{a\,dn}{S}$$

which, on evaluation of the integrals, gives

$$-\ln \frac{P}{P_0} = \frac{an}{S}$$

On converting to base 10 logarithms and inverting the fraction to change the sign, we obtain

$$\log \frac{P_0}{P} = \frac{an}{2.303\,S} \tag{13-6}$$

where n is the total number of particles within the block shown in Figure 13-2. The cross-sectional area S can be expressed in terms of the volume of the block V in cm^3 and its length b in cm. Thus,

$$S = \frac{V}{b}\ \text{cm}^2$$

Substitution of this quantity into Equation 13-6 yields

$$\log \frac{P_0}{P} = \frac{anb}{2.303\,V} \tag{13-7}$$

Note that since n/V has the units of concentration, that is, the number of particles per cubic centimeter, we can then readily convert n/V to moles per liter. Thus, the number of moles is given by

$$\text{number mol} = \frac{n \text{ particles}}{6.02 \times 10^{23} \text{ particles/mol}}$$

and c in mol/L is given by

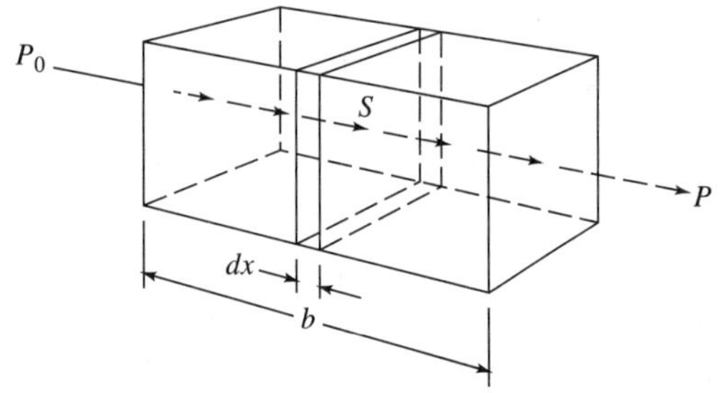

Figure 13-2 Attenuation of radiation with initial power P_0 by a solution containing c moles per liter of absorbing solute and with a path length of b cm. $P < P_0$.

[2] The discussion that follows is based on a paper by F. C. Strong, *Anal. Chem.*, **1952**, *24*, 338.

$$c = \frac{n}{6.02 \times 10^{23}} \text{ mol} \times \frac{1000 \text{ cm}^3\text{/L}}{V \text{ cm}^3}$$

$$= \frac{1000\, n}{6.02 \times 10^{23}\, V} \text{ mol/L}$$

Combining this relationship with Equation 13-7 yields

$$\log \frac{P_0}{P} = \frac{6.02 \times 10^{23}\, abc}{2.303 \times 1000}$$

Finally, the constants in this equation can be collected into a single term ϵ to give

$$\log \frac{P_0}{P} = \epsilon bc = A \qquad (13\text{-}8)$$

which is a formulation of Beer's law.

13B-1 Application of Beer's Law to Mixtures

Beer's law also applies to a medium containing more than one kind of absorbing substance. Provided there is no interaction among the various species, the total absorbance for a multicomponent system is given by

$$A_{\text{total}} = A_1 + A_2 + \cdots + A_n$$
$$= \epsilon_1 bc_1 + \epsilon_2 bc_2 + \cdots + \epsilon_n bc_n \qquad (13\text{-}9)$$

where the subscripts refer to absorbing components $1, 2, \ldots, n$.

13B-2 Limitations to Beer's Law

Few exceptions are found to the generalization that absorbance is linearly related to path length. On the other hand, deviations from the direct proportionality between the measured absorbance and concentration when b is constant are frequently encountered. Some of these deviations are fundamental and represent real limitations of the law. Others occur as a consequence of the manner in which the absorbance measurements are made or as a result of chemical changes associated with concentration changes; the latter two are sometimes known, respectively, as *instrumental deviations* and *chemical deviations.*

Real Limitations to Beer's Law

Beer's law is successful in describing the absorption behavior of media containing relatively low analyte concentrations; in this sense, it is a limiting law. At high concentrations (usually > 0.01 M), the average distance between the molecules responsible for absorption is diminished to the point where each molecule affects the charge distribution of its neighbors. This interaction, in turn, can alter the ability of the molecules to absorb a given wavelength of radiation. Because the extent of interaction depends upon concentration, the occurrence of this phenomenon causes deviations from the linear relationship between absorbance and concentration. A similar effect is sometimes encountered in media containing low absorber concentrations but high concentrations of other species, particularly electrolytes. The close proximity of ions to the absorber alters the molar absorptivity of the latter by electrostatic interactions; the effect is lessened by dilution.

While the effect of molecular interactions is ordinarily not significant at concentrations below 0.01 M, some exceptions occur among certain large organic ions or molecules. For example, the molar absorptivity at 436 nm for the cation of methylene blue in aqueous solutions is reported to increase by 88% as the dye concentration is increased from 10^{-5} to 10^{-2} M; even below 10^{-6} M, strict adherence to Beer's law is not observed.

Deviations from Beer's law also arise because ϵ is dependent upon the refractive index of the medium.[3] Thus, if concentration changes cause significant alterations in the refractive index n of a solution, departures from Beer's law are observed. A correction for this effect can be made by substitution of the quantity $\epsilon n/(n^2 + 2)^2$ for ϵ in Equation 13-8. In general, this correction is never very large and is rarely significant at concentrations less than 0.01 M.

Apparent Chemical Deviations

Apparent deviations from Beer's law arise when an analyte dissociates, associates, or reacts with a solvent to produce a product having a different absorption spectrum from the analyte. A common example of this behavior is found with aqueous solutions of acid/base indicators. For example, the color change associated with a typical indicator HIn arises from shifts in the equilibrium

$$\underset{\text{color 1}}{\text{HIn}} \rightleftharpoons \text{H}^+ + \underset{\text{color 2}}{\text{In}^-}$$

Example 13-1 demonstrates how the shift in this equilibrium with dilution results in deviation from Beer's law.

[3] G. Kortum and M. Seiler, *Angew. Chem.*, **1939**, *52*, 687.

EXAMPLE 13-1

The molar absorptivities at 430 and 570 nm of the weak acid HIn ($K_a = 1.42 \times 10^{-5}$) and its conjugate base In^- were determined by measurements of strongly acidic and strongly basic solutions of the indicator. Under these conditions, essentially all of the indicator was in the HIn and In^- form, respectively. The results were

	ϵ_{430}	ϵ_{570}
HIn	6.30×10^2	7.12×10^3
In^-	2.06×10^4	9.61×10^2

Calculate absorbance data for unbuffered solutions that have total indicator concentrations ranging from 2×10^{-5} to 16×10^{-5} M.

Let us calculate the molar concentrations of [HIn] and [In^-] of the two species in a solution in which the total concentration of indicator is 2.00×10^{-5} M. Here,

$$HIn \rightleftharpoons H^+ + In^-$$

and

$$K_a = 1.42 \times 10^{-5} = \frac{[H^+][In^-]}{[HIn]}$$

From the equation for the dissociation process, we may write

$$[H^+] = [In^-]$$

Furthermore, the sum of the concentrations of the two indicator species must equal the total molar analytical concentration of the indicator. Thus,

$$[In^-] + [HIn] = 2.00 \times 10^{-5}$$

Substitution of these relationships into the expression for K_a gives

$$\frac{[In^-]^2}{2.00 \times 10^{-5} - [In^-]} = 1.42 \times 10^{-5}$$

Rearrangement yields the quadratic expression

$$[In^-]^2 + 1.42 \times 10^{-5}\,[In^-] - 2.84 \times 10^{-10} = 0$$

The positive solution to this equation is

$$[In^-] = 1.12 \times 10^{-5}$$

$$[HIn] = 2.00 \times 10^{-5} - 1.12 \times 10^{-5} = 0.88 \times 10^{-5}$$

We are now able to calculate the absorbance at the two wavelengths. Thus, substituting into Equation 13-9 gives

$$A = \epsilon_{In^-} b[In^-] + \epsilon_{HIn} b[HIn]$$

$$A_{430} = 2.06 \times 10^4 \times 1.00 \times 1.12 \times 10^{-5} + 6.30 \times 10^2 \times 1.00 \times 0.88 \times 10^{-5}$$

$$= 0.236$$

Similarly at 570 nm,

$$A_{570} = 9.61 \times 10^2 \times 1.00 \times 1.12 \times 10^{-5} + 7.12 \times 10^3 \times 1.00 \times 0.88 \times 10^{-5}$$

$$= 0.073$$

Additional data, obtained in the same way, are shown in Table 13-2.

TABLE 13-2 Concentration and Absorbance Data Computed by the Techniques Shown in Example 13-1

C_{HIn}, M	[HIn]	[In^-]	A_{430}	A_{570}
2.00×10^{-5}	0.88×10^{-5}	1.12×10^{-5}	0.236	0.073
4.00×10^{-5}	2.22×10^{-5}	1.78×10^{-5}	0.381	0.175
8.00×10^{-5}	5.27×10^{-5}	2.73×10^{-5}	0.596	0.401
12.00×10^{-5}	8.52×10^{-5}	3.48×10^{-5}	0.771	0.640
16.00×10^{-5}	11.9×10^{-5}	4.11×10^{-5}	0.922	0.887

Figure 13-3 is a plot of the data shown in Table 13-2, which illustrates the kinds of departures from Beer's law that arise when the absorbing system is capable of undergoing dissociation or association. Note that the direction of curvature is opposite at the two wavelengths.

Apparent Instrumental Deviations with Polychromatic Radiation

Strict adherence to Beer's law is observed only with truly monochromatic radiation; this observation is yet another manifestation of the limiting character of the law. Unfortunately, the use of radiation that is restricted to a single wavelength is seldom practical because devices that isolate portions of the output from a continuum source produce a more or less symmetric band of wavelengths around the desired one (see Figures 7-13 and 7-14, for example).

The following derivation shows the effect of polychromatic radiation on Beer's law.

Consider a beam consisting of just two wavelengths λ' and λ''. Assuming that Beer's law applies strictly for each of these individual wavelengths, we may write for radiation λ'

$$A' = \log \frac{P'_0}{P'} = \epsilon' bc$$

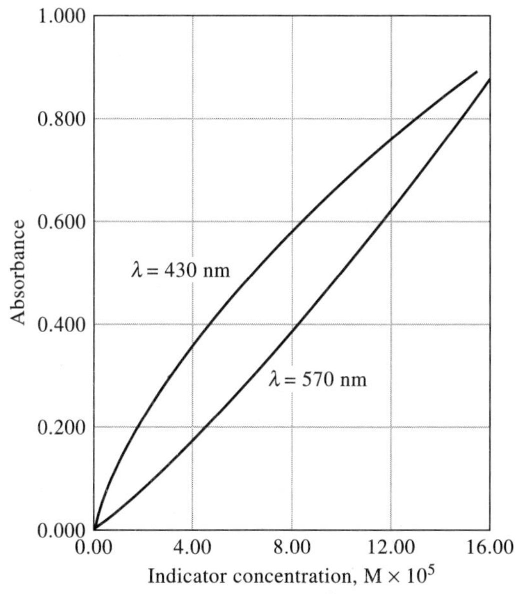

Figure 13-3 Chemical deviations from Beer's law for unbuffered solutions of the indicator HIn. For data, see Example 13-1.

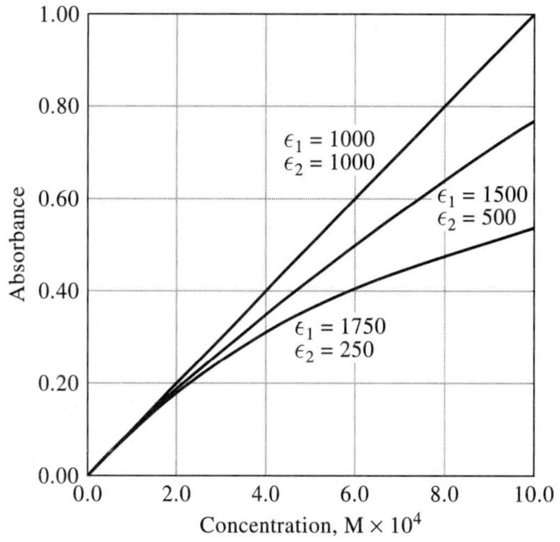

Figure 13-4 Deviations from Beer's law with polychromatic light. The absorber has the indicated molar absorptivities at the two wavelengths λ' and λ''.

or

$$P'_0/P' = 10^{\epsilon' bc}$$

and

$$P' = P'_0 10^{-\epsilon' bc}$$

Similarly, for λ''

$$P'' = P''_0 10^{-\epsilon'' bc}$$

When an absorbance measurement is made with radiation composed of both wavelengths, the power of the beam emerging from the solution is given by $P' + P''$ and that of the beam from the solvent by $P'_0 + P''_0$. Therefore, the measured absorbance A_m is

$$A_m = \log \frac{(P'_0 + P''_0)}{(P' + P'')}$$

Substituting for P' and P'' yields

$$A_m = \log \frac{(P'_0 + P''_0)}{(P'_0 10^{-\epsilon' bc} + P''_0 10^{-\epsilon'' bc})}$$

or

$$A_m = \log(P'_0 + P''_0) - \log(P'_0 10^{-\epsilon' bc} + P''_0 10^{-\epsilon'' bc})$$

Now, when $\epsilon' = \epsilon''$, this equation simplifies to

$$A_m = \epsilon' bc$$

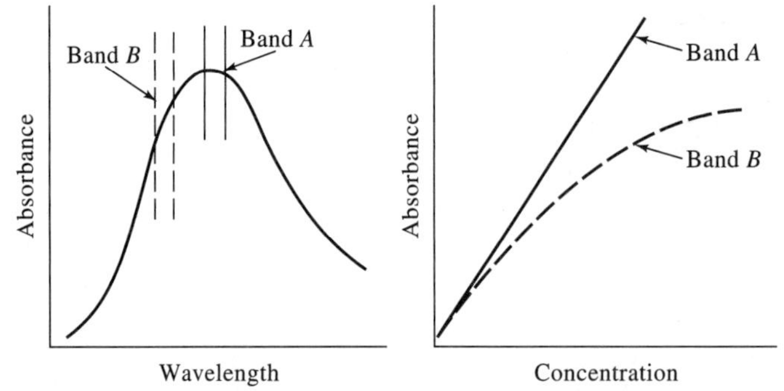

Figure 13-5 The effect of polychromatic radiation upon Beer's law. Band *A* shows little deviation because ϵ does not change greatly throughout the band. Band *B* shows marked deviation because ϵ undergoes significant changes in this region.

and Beer's law is followed. As shown in Figure 13-4, however, the relationship between A_m and concentration is no longer linear when the molar absorptivities differ; moreover, greater departures from linearity can be expected with increasing differences between ϵ' and ϵ''. This derivation can be expanded to include additional wavelengths; the effect remains the same.

It is an experimental fact that deviations from Beer's law resulting from the use of a polychromatic beam are not appreciable, provided the radiation used does not encompass a spectral region in which the absorber exhibits large changes in absorption as a function of wavelength. This observation is illustrated in Figure 13-5.

It is also found experimentally that for absorbance measurements at the maximum of narrow peaks, departures from Beer's law are not significant if the effective bandwidth of the monochromator or filter $\Delta\lambda_{eff}$ (Equation 7-17) is less than 1/10 of the half width of the absorption peak at half height.

Instrumental Deviations in the Presence of Stray Radiation

We show in Chapter 7 that the radiation exiting from a monochromator is ordinarily contaminated with small amounts of scattered or stray radiation, which reaches the exit slit as a result of scattering and reflections from various internal surfaces. Stray radiation often differs greatly in wavelength from that of the principal radiation and, in addition, may not have passed through the sample.

When measurements are made in the presence of stray radiation, the observed absorbance is given by

$$A' = \log \frac{P_0 + P_s}{P + P_s}$$

where P_s is the power of nonabsorbed stray radiation. Figure 13-6 shows a plot of A' versus concentration for various ratios between P_s and P_0. It is noteworthy that at high concentrations and at longer path lengths, stray radiation can also cause significant deviations from the linear relationship between absorbance and path length.[4]

Note also that the instrumental deviations illustrated in Figures 13-5 and 13-6 result in absorbances that are smaller than theoretical. It can be shown that instrumental deviations always lead to negative absorbance errors.[5]

13C THE EFFECTS OF INSTRUMENTAL NOISE ON SPECTROPHOTOMETRIC ANALYSES

The accuracy and precision of spectrophotometric analyses are often limited by the uncertainties or noise associated with the instrument.[6] A general discussion of

[4] For a discussion of the effects of stray radiation, see M. R. Sharpe, *Anal. Chem.*, **1984**, *56*, 339A.

[5] E. J. Meehan, in *Treatise on Analytical Chemistry*, 2nd ed., P. J. Elving, E. J. Meehan, and I. M. Kolthoff, Eds., Part I, Vol. 7, p. 73. New York: Wiley, 1981.

[6] See L. D. Rothman, S. R. Crouch, and J. D. Ingle Jr., *Anal. Chem.*, **1975**, *47*, 1226; J. D. Ingle Jr. and S. R. Crouch, *Anal. Chem.*, **1972**, *44*, 1375; H. L. Pardue, T. E. Hewitt, and M. J. Milano, *Clin. Chem.*, **1974**, *20*, 1028; J. O. Erickson and T. Surles, *Amer. Lab.*, **1976**, *8* (6), 41; *Optimum Parameters for Spectrophotometry*. Palo Alto, CA: Varian Instruments Division, 1977.

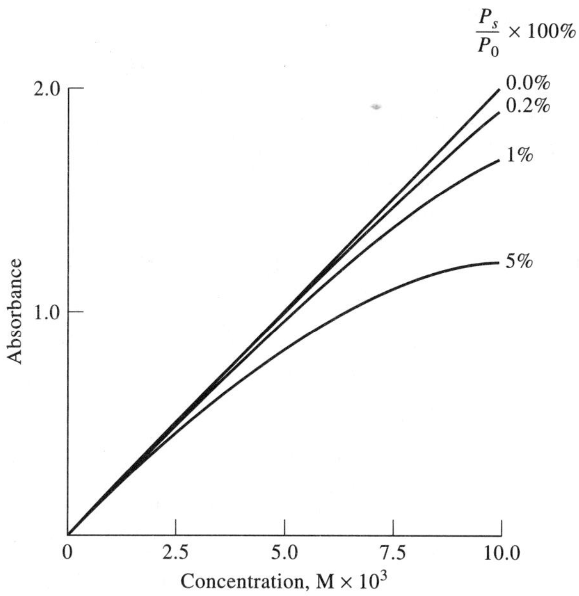

Figure 13-6 Apparent deviation from Beer's law brought about by various amounts of stray radiation.

instrumental noise and signal-to-noise optimization is found in Chapter 5; it may be helpful to review this material before undertaking a detailed study of this section.

13C-1 Instrumental Noise as a Function of Transmittance

As was pointed out earlier, a spectrophotometric measurement entails three steps: a 0% *T* adjustment, a 100% *T* adjustment, and a measurement of % *T* with the sample in the radiation path. The noise associated with each of these steps combines to give a net uncertainty for the final value obtained for *T*. The relationship between the noise encountered in the measurement of *T* and the uncertainty in concentration can be derived by writing Beer's law in the form

$$c = -\frac{1}{\epsilon b}\log T = -\frac{0.434}{\epsilon b}\ln T \qquad (13\text{-}10)$$

In order to relate the standard deviation in concentration σ_c to the standard deviation in transmittance σ_T we proceed as in Section a1B-4, Appendix 1 by taking the partial derivative of this equation with respect to *T*, holding *b* and *c* constant. That is,

$$\frac{\partial c}{\partial T} = -\frac{0.434}{\epsilon b T}$$

Application of Equation a1-28 (Appendix 1) gives

$$\sigma_c^2 = \left(\frac{\partial c}{\partial T}\right)^2 \sigma_T^2 = \left(\frac{-0.434}{\epsilon b T}\right)^2 \sigma_T^2 \qquad (13\text{-}11)$$

Dividing Equation 13-11 by the square of Equation 13-10 gives

$$\left(\frac{\sigma_c}{c}\right)^2 = \left(\frac{\sigma_T}{T \ln T}\right)^2$$

$$\frac{\sigma_c}{c} = \frac{\sigma_T}{T \ln T} = \frac{0.434\,\sigma_T}{T \log T} \qquad (13\text{-}12)$$

For a limited number of measurements, we replace the population standard deviations σ_c and σ_T with the sample standard deviations s_c and s_T (Section a1B-1, Appendix 1) and obtain

$$\frac{s_c}{c} = \frac{0.434\,s_T}{T \log T} \qquad (13\text{-}13)$$

This equation relates the *relative* standard deviation of *c* (s_c/c) to the absolute standard deviation of the transmittance measurement (s_T). Experimentally, s_T can be evaluated by making, say, 20 replicate transmittance measurements ($N = 20$) of the transmittance of a solution in exactly the same way and substituting the data into Equation a1-9, Appendix 1.

From an examination of Equation 13-13 we see that the uncertainty in a photometric concentration measurement varies nonlinearly with the magnitude of the transmittance. The situation is somewhat more complicated than is suggested by Equation 13-13, however, because the uncertainty s_T is, under many circumstances, also *dependent upon T*.

13C-2 Sources of Instrumental Noise

In a detailed theoretical and experimental study, Rothman, Crouch, and Ingle have described several sources of instrumental uncertainties and shown their net effect on the precision of absorbance or transmittance measurements.[7] These uncertainties fall into three categories depending upon how they are affected by the magnitude of the photocurrent and thus *T*. For *Case I*

[7] L. D. Rothman, S. R. Crouch, and J. D. Ingle Jr., *Anal. Chem.*, **1975**, *47*, 1226.

TABLE 13-3 Types and Sources of Uncertainties in Transmittance Measurements

Category	Characterized by[a]	Typical Sources	Likely To Be Important In
Case I	$s_T = k_1$	Limited readout resolution	Inexpensive photometers and spectrophotometers having small transmittance meter scales
		Heat detector Johnson noise	IR and near-IR spectrophotometers and photometers
		Dark current and amplifier noise	Regions where source intensity and detector sensitivity are low
Case II	$s_T = k_2\sqrt{T^2 + T}$	Photon detector shot noise	High-quality UV/visible spectrophotometers
Case III	$s_T = k_3T$	Cell positioning uncertainties	High-quality UV/visible and IR spectrophotometers
		Source flicker	Inexpensive photometers and spectrophotometers

[a] k_1, k_2, and k_3 are constants for a given system.

uncertainties, the precision is independent of T; that is, s_T is equal to a constant k_1. For *Case II uncertainties,* the precision is directly proportional to $\sqrt{T^2 + T}$. Finally, *Case III uncertainties* are directly proportional to T. Table 13-3 summarizes information about the sources of these three types of uncertainty and the kinds of instruments where each is likely to be encountered.

Case I: $s_T = k_1$

Case I uncertainties are often encountered with less expensive ultraviolet and visible spectrophotometers or photometers that are equipped with meters or digital readouts with limited resolution. For example, a typical instrument may be equipped with a meter having a 5- to 7-in. scale which is readable to 0.2% to 0.5% of full scale. Here, the absolute uncertainty in T is the same from one end of the scale to the other. A similar limitation in readout resolution is found in some digital instruments.

Infrared and near-infrared spectrophotometers also exhibit Case I behavior. With these, the limiting random error usually arises from Johnson noise in the thermal detector. Recall (Section 5B-2) that this type of noise is independent of the magnitude of the photocurrent; indeed, fluctuations are observed even in the absence of radiation and, therefore, net current.

Dark current and amplifier noise are usually small compared with other sources of noise in photometric and spectrophotometric instruments and become important only under conditions of low photocurrents where the lamp intensity or the photodetector sensitivity is low. For example, such conditions are often encountered near the wavelength extremes for an instrument.

The precision of concentration data obtained with an instrument that is limited by Case I noise can be obtained directly by substituting an experimentally determined value for $s_T = k_1$ into Equation 13-13. Here, the precision of a particular *concentration* determination depends upon the magnitude of T even though the instrumental precision is independent of T. The third column of Table 13-4 shows data obtained with Equation 13-13 when an absolute standard deviation s_T of ± 0.003 or $\pm 0.3\%$ T was assumed. A plot of the data is shown by curve *A* in Figure 13-7.

An indeterminate uncertainty of 0.3% T is typical of many moderately priced spectrophotometers or photometers. Clearly, concentration errors of 1% to 2% relative are to be expected with such instruments. It is also evident that precision at this level can only be realized if the absorbance of the sample lies between about 0.1 and 1.

TABLE 13-4 Relative Precision of Concentration Measurements as a Function of Transmittance and Absorbance for Three Categories of Instrument Noise

		Relative Standard Deviation in Concentration[a]		
Transmittance, T	Absorbance, A	Case I Noise[b]	Case II Noise[c]	Case III Noise[d]
0.95	0.022	± 6.2	±8.4	±25.3
0.90	0.046	± 3.2	±4.1	±12.3
0.80	0.097	± 1.7	±2.0	± 5.8
0.60	0.222	± 0.98	±0.96	± 2.5
0.40	0.398	± 0.82	±0.61	± 1.4
0.20	0.699	± 0.93	±0.46	± 0.81
0.10	1.00	± 1.3	±0.43	± 0.56
0.032	1.50	± 2.7	±0.50	± 0.38
0.010	2.00	± 6.5	±0.65	± 0.28
0.0032	2.50	±16.3	±0.92	± 0.23
0.0010	3.00	±43.4	±1.4	± 0.19

[a] $(s_c/c) \times 100\%$.
[b] From Equation 13-13 with $s_T = k_1 = \pm 0.0030$
[c] From Equation 13-14 with $k_2 = \pm 0.0030$
[d] From Equation 13-15 with $k_3 = \pm 0.013$

Case II: $s_T = k_2\sqrt{T^2 + T}$

This type of uncertainty often limits the precision of the highest quality instruments. It has its origin in shot noise (Section 5B-2), which must be expected whenever the passage of electricity involves transfer of charge across a junction, such as the movement of electrons from the cathode to the anode of a photomultiplier tube. Here, an electric current results from a series of discrete events (emission of electrons from a cathode), the number of which per unit time is distributed in a random way about a mean value. The magnitude of the current fluctuations is proportional to the square root of current (see Equation 5-5, page 101). The effect of shot noise on s_c is derived by substitution of s_T into Equation 13-13. Rearrangement leads to

$$\frac{s_c}{c} = \frac{0.434\, k_2}{\log T}\sqrt{\frac{1}{T} + 1} \qquad (13\text{-}14)$$

The data in column 4 of Table 13-4 were obtained with the aid of Equation 13-14. Figure 13-7, curve B is

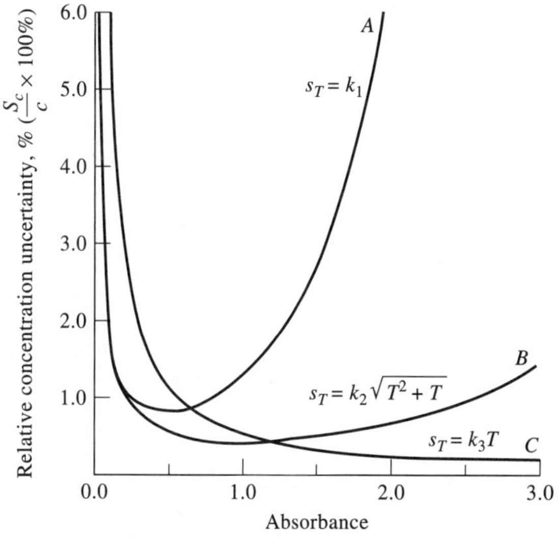

Figure 13-7 Relative concentration uncertainties arising from various categories of instrumental noise: A, Case I; B, Case II; C, Case III. The data are taken from Table 13-4.

a plot of such data. Note the much larger range of absorbances that can be encompassed without serious loss of accuracy when shot noise, rather than Johnson noise, limits the precision. This increased range represents a major advantage of photon-type detectors over thermal types, which are represented by curve *A* in the figure. As with Johnson-noise-limited instruments, shot-limited instruments do not give very reliable concentration data at transmittances greater than 95% (or $A < 0.02$).

Case III: $s_T = k_3 T$

One source of noise of this type is the slow drift in the radiant output of the source; this type of noise can be called *source flicker noise* (Section 5B-2). The effects of fluctuations in the intensity of a source can be minimized by the use of a constant voltage power supply or a split beam arrangement (page 315). With many instruments, source flicker noise does not limit performance.

An important and widely encountered noise source, which is proportional to transmittance, results from failure to position sample and reference cells reproducibly with respect to the beam during replicate transmittance measurements. All cells have minor imperfections. As a consequence, reflection and scattering losses vary as different sections of the cell window are exposed to the beam; small variations in transmittance result. Rothman, Crouch, and Ingle have shown that this uncertainty often is the most common limitation to the accuracy of high-quality ultraviolet/visible spectrophotometers. It is also a serious source of uncertainty in infrared instruments.

One method of reducing the effect of cell positioning with a double-beam instrument is to leave the cells in place during calibration and analysis; new standards and samples are introduced after washing and rinsing the cell in place with a syringe. Care must be taken to avoid touching or jarring the cells during this process.

The effect of uncertainties that are proportional to transmittance on analytical results can be obtained by substitution of $s_T = k_3 T$ into Equation 13-13, which gives

$$\frac{s_c}{c} = -\frac{0.434\, k_3}{\log T} \qquad (13\text{-}15)$$

Column 5 of Table 13-4 contains data obtained from Equation 13-15 when k_3 is assumed to have a value of 0.013, which approximates the value observed in the

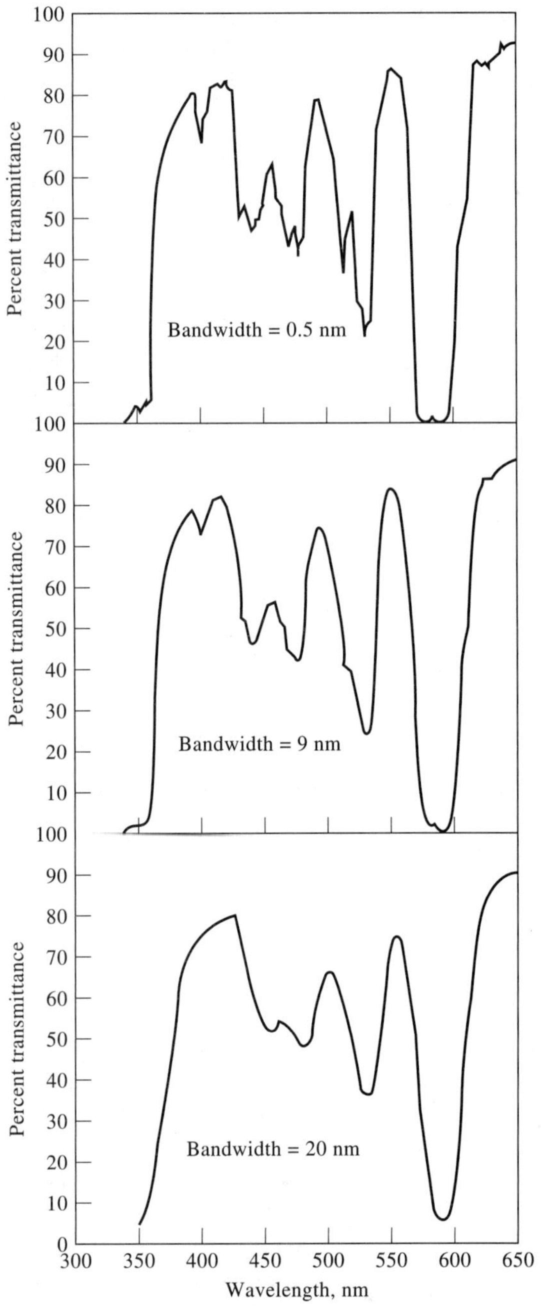

Figure 13-8 Effect of bandwidth on spectral detail. The sample was a didymium glass. *(Spectra provided through the courtesy of Amsco Instrument Company [formerly G. K. Turner Associates, Carpinteria, CA].)*

Rothman, Crouch, and Ingle study. The data are plotted as curve *C* in Figure 13-7.[8]

13C-3 Effect of Slit Width on Absorbance Measurements

As shown in Section 7C-3, narrow slit widths are required to resolve complex spectra.[9] For example, Figure 13-8 illustrates the loss of detail that accompanies the use of wider slits. In this example, the transmittance spectrum of a didymium glass was obtained at slit settings that provided effective bandwidths of 0.5, 9, and 20 nm. The progressive loss of spectral detail is clear. For qualitative studies, such losses often loom important.

Figure 13-9 illustrates a second effect of slit width on spectra made up of narrow peaks. Here, the spectrum of a praseodymium chloride solution was obtained at slit settings of 1.0, 0.5, and 0.1 mm. Note that the peak absorbance values increase significantly (by as much as 70% in one instance) as the slit width decreases. At slit settings less than about 0.14 mm, absorbances were found to become independent of slit width. Careful inspection of Figure 13-8 reveals the same type of effect. In both sets of spectra, the areas under the individual peaks are the same, but wide slit widths result in broader lower peaks.

It is evident from both of these illustrations that quantitative measurement of narrow absorption bands demands the use of narrow slit widths or, alternatively, very reproducible slit-width settings.

Unfortunately, a decrease in slit width is accompanied by a second-order power reduction in the radiant energy; at very narrow settings spectral detail may be lost owing to an increase in the signal-to-noise ratio. The situation becomes particularly serious in spectral regions where the output of the source or the sensitivity of the detector is low. Under such circumstances, noise in either of these components or their associated electronic circuits may result in partial or total loss of spectral fine structure.

In general, it is good practice to narrow slits no more than is necessary for resolution of the spectrum at

[8] To see how curve-fitting may be used to extract k_1, k_2, and k_3 from experimental absorption data, see F. J. Holler, *Mathcad Applications for Analytical Chemistry*, pp. 147–149. Philadelphia: Saunders College Publishing, 1994.

[9] For a discussion of the effects of slit width on spectra, see *Optimum Parameters for Spectrophotometry*. Palo Alto, CA: Varian Instruments Division, 1977; F. C. Strong III, *Anal. Chem.*, **1976**, *48*, 2155; D. D. Gilbert, *J. Chem. Educ.*, **1991**, *68*, A278.

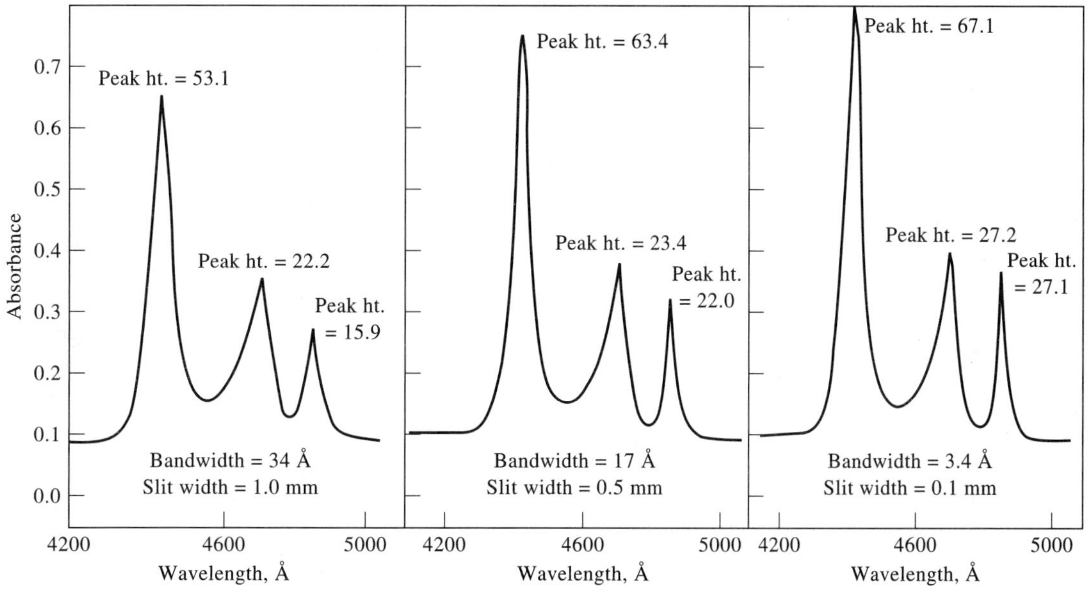

Figure 13-9 Effect of slit width and bandwidth on peak heights. Here, the sample was a solution of praseodymium chloride. *(From* Optimum Spectrophotometer Parameters, Application Report *AR 14-2. Cary Instruments: Monrovia, CA. With permission.)*

hand. With a variable slit spectrophotometer, proper slit adjustment can be determined by acquiring spectra at progressively narrower slits until peak heights become constant. Generally, constant peak heights are observed when the effective bandwidth of the instrument is 0.1 or less of the effective bandwidth of the absorption peak.

13C-4 Effect of Scattered Radiation at Wavelength Extremes of an Instrument

We have already noted that scattered radiation may cause instrumental deviations from Beer's law. When measurements are attempted at the wavelength extremes of an instrument, the effects of stray radiation may be even more serious and on occasion may lead to the appearance of false absorption peaks. For example, consider a spectrophotometer for the visible region equipped with glass optics, a tungsten source, and a photovoltaic cell detector. At wavelengths below about 380 nm, the windows, cells, and prism begin to absorb radiation, thus reducing the energy reaching the transducer. The output of the source falls off rapidly in this region as well; so also does the sensitivity of the photoelectric device. Thus, the total signal for the 100% T adjustment may be as low as 1% to 2% of that in the region between 500 and 650 nm.

The scattered radiation, however, is often made up of wavelengths to which the instrument is highly sensitive. Thus, its effects can be enormously magnified. Indeed, in some instances the output signal produced by the stray radiation may exceed that from the monochromator beam; under these circumstances, the measured absorbance is as much that for the stray radiation as for the radiation to which the instrument is set.

An example of a false peak appearing at the wavelength extremes of a visible-region spectrophotometer is shown in Figure 13-10. The spectrum of a solution of cerium(IV) obtained with an ultraviolet/visible spectrophotometer, sensitive in the region of 200 to 750 nm, is shown by curve *B*. Curve *A* is a spectrum of the same solution obtained with a simple visible spectrophotometer. The apparent maximum shown in curve *A* arises from the instrument responding to stray wavelengths longer than 400 nm, which (as can be seen from the spectra) are not absorbed by the cerium(IV) ions.

This same effect is sometimes observed with ultraviolet/visible instruments when attempts are made to measure absorbances at wavelengths lower than about 200 nm.

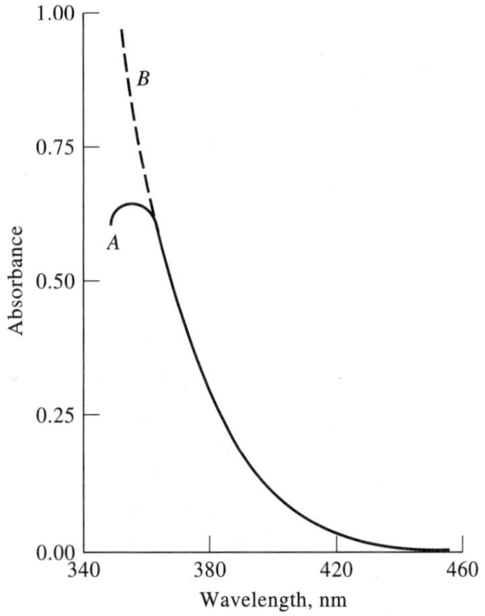

Figure 13-10 Spectra of cerium(IV) obtained with a spectrophotometer having glass optics (*A*) and quartz optics (*B*). The false peak in *A* arises from transmission of stray radiation of longer wavelengths.

13D INSTRUMENTATION

The scientist interested in molecular absorption measurements in the ultraviolet, visible, and near-infrared regions has a hundred or more instrument makes and models to choose from. Some are simple and inexpensive (a few hundred dollars); others are complex, computerized devices costing $30,000 or more. For many applications, the simpler instruments provide information as quickly and as satisfactorily as that obtained by their more sophisticated counterparts. On the other hand, the more complicated instruments have been developed to perform tasks that are difficult, time-consuming, or impossible with the simpler ones.[10]

[10] For an interesting paper discussing commercial instruments for ultraviolet/visible absorption measurements, see R. Jarnutowski, J. R. Ferraro, and D. C. Lankin, *Spectroscopy,* **1992,** *7* (7), 22; I. R. Altemose, *J. Chem. Educ.,* **1986,** *63,* A216, A262.

13D-1 Instrument Components

Instruments for measuring the absorption of ultraviolet, visible, and near-infrared radiation are made up of one or more (1) sources, (2) wavelength selectors, (3) sample containers, (4) radiation transducers, and (5) signal processors and readout devices. The design and performance of components 2, 4, and 5 have been described in considerable detail in Chapter 7 and thus are not discussed further here. We will, however, consider briefly the characteristics of sources and sample containers for the region of 185 to 3000 nm.

Sources

For the purposes of molecular absorption measurements, a continuum source is required whose power does not change sharply over a considerable range of wavelengths.

Deuterium and Hydrogen Lamps. A continuum spectrum in the ultraviolet region is produced by electrical excitation of deuterium or hydrogen at low pressure. The mechanism by which a continuous spectrum is produced involves initial formation of an excited *molecular* species followed by dissociation of the excited molecule to give two atomic species plus an ultraviolet photon. The reactions for deuterium are

$$D_2 + E_e \rightarrow D_2^* \rightarrow D' + D'' + h\nu$$

where E_e is the electrical energy absorbed by the molecule and D_2^* is the excited deuterium molecule. The energetics for the overall process can be represented by the equation

$$E_e = E_{D_2^*} = E_{D'} + E_{D''} + h\nu$$

Here, $E_{D_2^*}$ is the *fixed quantized energy* of D_2^* while $E_{D'}$ and $E_{D''}$ are the *kinetic energies* of the two deuterium atoms. The sum of the latter two can vary continuously from zero to $E_{D_2^*}$. Thus the energy and the frequency of the photon can also vary continuously. That is, when the two kinetic energies are by chance small, $h\nu$ will be large, and conversely. The consequence is a true continuum spectrum from about 160 nm to the beginning of the visible region (see Figure 13-11).

Most modern lamps of this type contain deuterium and are of a low-voltage type in which an arc is formed between a heated, oxide-coated filament and a metal electrode. The heated filament provides electrons to maintain a direct current when about 40 V is applied; a regulated power supply is required for constant intensities.

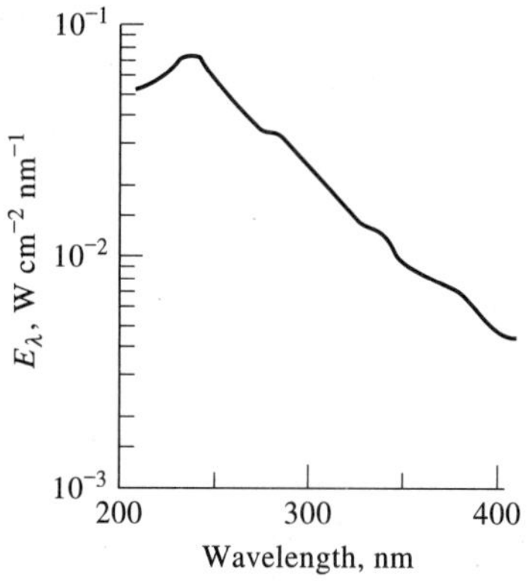

Figure 13-11 Output from a deuterium lamp.

An important feature of deuterium and hydrogen discharge lamps is the shape of the aperture between the two electrodes, which constricts the discharge to a narrow path. As a consequence, an intense ball of radiation about 1 to 1.5 mm in diameter is produced. Deuterium gives a somewhat larger and brighter ball than hydrogen, which accounts for the widespread use of the former.

Both deuterium and hydrogen lamps produce a useful continuum spectrum in the region of 160 to 375 nm. At longer wavelengths (>400 nm), the lamps produce emission lines, which are superimposed on the continuum spectrum. For many applications, these lines represent a nuisance; they can be useful, however, for wavelength calibration of absorption instruments.

Quartz windows must be employed in deuterium and hydrogen lamps since glass absorbs strongly at wavelengths less than about 350 nm.

Tungsten Filament Lamps. The most common source of visible and near-infrared radiation is the tungsten filament lamp. The energy distribution of this source approximates that of a blackbody and is thus temperature dependent. Figure 6-18 (page 134) illustrates the behavior of the tungsten filament lamp at 3000 K. In most absorption instruments, the operating filament temperature is 2870 K; the bulk of the energy is thus emitted in the infrared region. A tungsten filament lamp is useful for the wavelength region between 350 and 2500 nm. The lower limit is imposed by the absorption of radiation by the glass envelope that houses the filament.

In the visible region, the energy output of a tungsten lamp varies approximately as the fourth power of the operating voltage. As a consequence, close voltage control is required for a stable radiation source. Constant voltage transformers or electronic voltage regulators are usually employed to obtain the required stability. As an alternative, the lamp can be operated from a 6-V storage battery, which provides a remarkably stable voltage source if it is maintained in good condition.

Tungsten/halogen lamps contain a small quantity of iodine within a quartz envelope that houses the tungsten filament. Quartz is required because of the high operating temperature (~3500 K) of the lamp. The lifetime of a tungsten/halogen lamp is more than double that of the ordinary lamp. This added life results from the reaction of the iodine with gaseous tungsten that forms by sublimation and ordinarily limits the life of the filament; the product is the volatile WI_2. When molecules of this compound strike the filament, decomposition occurs, which redeposits tungsten. Tungsten/halogen lamps are significantly more efficient and extend the output range well into the ultraviolet. For these reasons, they are found in many modern spectroscopic instruments.

Xenon Arc Lamps. This lamp produces intense radiation by the passage of current through an atmosphere of xenon. The spectrum is continuous over the range between about 200 and 1000 nm, with the peak intensity occurring at about 500 nm (see Figure 6-18, page 134). In some instruments, the lamp is operated intermittently by regular discharges from a capacitor; high intensities are obtained.

Sample Containers

In common with the other optical elements of an absorption instrument, the cells, or *cuvettes,* that hold the sample and solvent must be constructed of a material that passes radiation in the spectral region of interest. Thus, as shown in Figure 7-2a (page 148), quartz or fused silica is required for work in the ultraviolet region (below 350 nm); both of these substances are transparent in the visible region and are transparent to about 3 μm in the infrared region as well. Silicate glasses can be employed in the region between 350 and 2000 nm. Plastic containers have also found application in the visible region.

The best cells have windows that are perfectly normal to the direction of the beam in order to minimize reflection losses. The most common cell length for studies in the ultraviolet and visible regions is 1 cm; matched, calibrated cells of this size are available from several commercial sources. Other path lengths, from 0.1 cm (and shorter) to 10 cm, can also be purchased. Transparent spacers for shortening the path length of 1-cm cells to 0.1 cm are also available.

For reasons of economy, cylindrical cells are sometimes employed in the ultraviolet and visible regions. Special care must be taken to duplicate the position of the cell with respect to the beam; otherwise, variations in path length and reflection losses at the curved surfaces can cause significant errors.

The quality of absorbance data is critically dependent upon the way the matched cells are used and maintained. Fingerprints, grease, or other deposits on the walls alter the transmission characteristics of a cell markedly. Thus, thorough cleaning before and after use is imperative; the surface of the windows must not be touched during the handling. Matched cells should never be dried by heating in an oven or over a flame—such treatment may cause physical damage or a change in path length. The cells should be calibrated against each other regularly with an absorbing solution.

13D-2 Types of Instruments

In this section, we consider four general types of spectroscopic instruments: (1) single-beam, (2) double-beam in space, (3) double-beam in time, and (4) multichannel.

Single-Beam Instruments

Figure 13-12a is a schematic of a single-beam instrument for absorption measurements. It consists of one of the radiation sources described in an earlier section of this chapter, a filter or a monochromator for wavelength selection, matched cells that can be interposed alternately in the radiation beam, one of the transducers described in Section 7E, an amplifier, and a readout device. Normally, a single-beam instrument requires a stabilized voltage supply to avoid errors resulting from changes in the beam intensity during the time required to make the 100% T adjustment and determine % T for the analyte.

Single-beam instruments vary widely in their complexity and performance characteristics. The simplest and least expensive consists of a battery-operated tungsten bulb as the source, a set of glass filters for wavelength selection, test tubes for sample holders, a photovoltaic cell as the transducer, and a small microammeter as the readout device. At the other extreme are sophisticated, computer-controlled instruments with a range of 200 to 1000 nm or more. These spectrophotometers

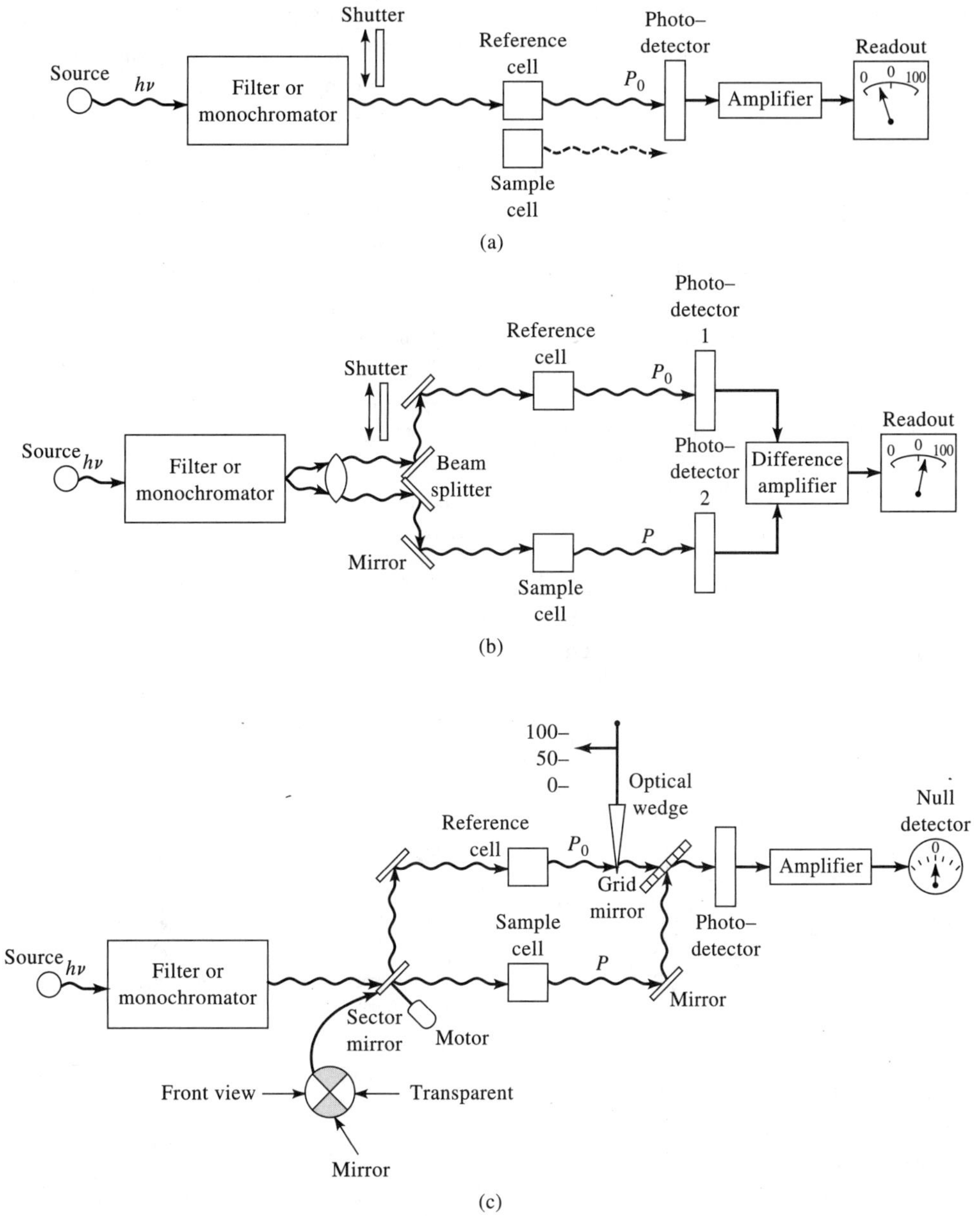

Figure 13-12 Instrument designs for photometers and spectrophotometers: (a) single-beam instrument, (b) double-beam instrument with beams separated in space, (c) double-beam instrument with beams separated in time.

have interchangeable tungsten/deuterium lamp sources, use rectangular silica cells, and are equipped with a high-resolution grating monochromator with variable slits. Photomultiplier tubes are used as transducers, and the output is often digitized and stored so that it can be printed or plotted in several forms.

Double-Beam Instruments

Many modern photometers and spectrophotometers are based upon a double-beam design. Figure 13-12b illustrates a double-beam-in-space instrument in which two beams are formed in space by a V-shaped mirror called a beam splitter. One beam passes through the reference solution to a phototransducer, and the second simultaneously traverses the sample to a second, matched transducer. The two outputs are amplified, and their ratio (or the log of their ratio) is determined electronically and displayed by the readout device. With manual instruments, the measurement is a two-step operation involving first the zero adjustment with a shutter in place between selector and beam splitter. In the second step, the shutter is opened and the transmittance or absorbance is read directly from the meter.

The second type of double-beam instrument is illustrated in Figure 13-12c. Here the beams are separated in time by a rotating sector mirror that directs the entire beam from the monochromator first through the reference cell and then through the sample cell. The pulses of radiation are recombined by another sector mirror, which transmits one pulse and reflects the other to the transducer. As shown by the insert labeled "front view" in Figure 13-12c, the motor-driven sector mirror is made up of pie-shaped segments, half of which are mirrored and half of which are transparent. The mirrored sections are held in place by blackened metal frames that periodically interrupt the beam and prevent its reaching the transducer. The detection circuit is programmed to use these periods to perform the dark-current adjustment.

The instrument shown in Figure 13-12c is a null type, in which the beam passing through the solvent is attenuated until its intensity just matches that of the beam passing through the sample. Attenuation is accomplished in this design with an optical wedge, the transmission of which decreases linearly along its length. Thus, the null point is reached by moving the wedge in the beam until the two electrical pulses are identical as indicated by the null detector. The transmittance (or absorbance) is then read directly from the pointer attached to the wedge.

Double-beam instruments offer the advantage that they compensate for all but the most short-term fluctuations in the radiant output of the source as well as for drift in the transducer and amplifier. They also compensate for wide variations in source intensity with wavelength (see Figure 13-11). Furthermore, the double-beam design lends itself well to the continuous recording of transmittance or absorbance spectra. Consequently, most modern ultraviolet and visible recording instruments are double-beam (usually in time). Scanning infrared spectrophotometers often are based on this design.

Multichannel Instruments

The most recent type of spectrophotometer, which appeared on the market in the early 1980s, is a single beam instrument based upon the diode array transducer described in Section 7E-3. A simplified schematic of such a diode array spectrometer is shown in Figure 13-13. Radiation from a lamp is focused upon the sample or solvent container and then passes into a monochromator with a fixed grating. The dispersed radiation falls on a photodiode array transducer, which, as mentioned earlier, consists of a linear array of several hundred photodiodes that have been formed along the length of a silicon chip. Typically, the chips are 1 to 6 cm in length, and the widths of the individual diodes are 15 to 50 μm (see Figure 13-14). This chip also contains a capacitor and an electronic switch for each diode. A computer-driven shift register sequentially closes each switch momentarily, which causes each capacitor to be charged to −5 V. Radiation impinging on any diode surface causes partial discharge of its capacitor. This lost charge is replaced during the next switching cycle. The resulting charging currents, which are proportional to the radiant power, are amplified, digitized, and stored in computer memory. The entire cycle is completed in a few milliseconds.

The monochromator slit width of a diode array instrument is usually made identical to the width of one of the silicon diodes. Thus, the output of each diode corresponds to the radiation of a different wavelength, and a spectrum is obtained by scanning these outputs sequentially. Since the electronic scanning process is remarkably rapid, data for an entire spectrum are accumulated in 1 s or less.

A diode array instrument is a powerful tool for studies of transient intermediates in moderately fast reactions, for kinetic studies, and for the qualitative and quantitative determination of the components exiting from a liquid chromatographic column or a capillary electrophoresis column. The disadvantages of this type of instrument are its somewhat limited resolution (usually 1 to 2 nm) and its moderately high cost.

13D-3 Some Typical Instruments

In the sections that follow, some typical photometers and spectrophotometers are described. The choice here among the scores of models that are available from instrument manufacturers has not been based upon quality

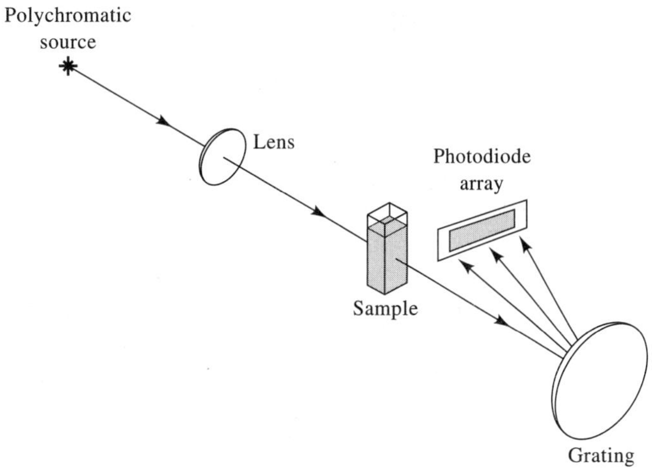

Figure 13-13 Diagram of a multichannel spectrophotometer based on a grating and a photodiode array transducer.

of performance or cost but rather was made to illustrate the wide variety of design variables that are encountered.

Photometers

Photometers provide simple, relatively inexpensive tools for performing absorption analyses. Filter photometers are often more convenient and more rugged and are easier to maintain and use than the more sophisticated spectrophotometers. Furthermore, photometers characteristically have high radiant energy throughputs and thus good signal-to-noise ratios even with relatively simple and inexpensive transducers and circuitry. Where high spectral purity is not important to a method (and often it is not), quantitative analyses can be performed as accurately with

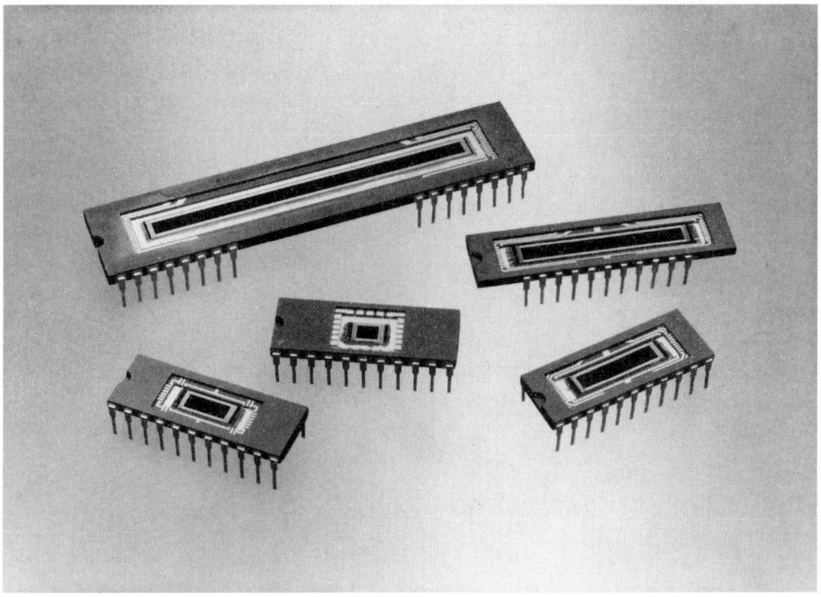

Figure 13-14 Diode arrays of various size. *(Courtesy of EG&G Reticon, Inc.)*

a photometer as with more complex instrumentation.

Visible Photometers. Figure 13-15 presents schematics for two photometers. The upper figure illustrates a single-beam, direct-reading instrument consisting of a tungsten filament lamp, a lens to provide a parallel beam of light, a filter, and a photovoltaic cell. The current produced is indicated with a microammeter, the face of which is ordinarily scribed with a linear scale from 0 to 100. The 0% *T* measurement involves mechanical or electrical adjustment of the meter needle while the shutter interrupts the incident beam. In some instruments, adjustment to obtain a full-scale (100% *T*) response with the solvent in the light path requires changing the voltage applied to the lamp. In others, the aperture size of a diaphragm located in the light path is altered. Since the signal from the photovoltaic cell is linear with respect to the power of the radiation it receives, the scale reading with the sample in the light path will be the percent transmittance, that is, the percent of full scale. A logarithmic scale may be substituted to give the absorbance of the solution directly.

Figure 13-15b is a schematic representation of a double-beam, null-type photometer that is similar to the one shown in Figure 13-12b. Here, the light beam is split by a half-silvered mirror, which transmits about 50% of the radiation striking it and reflects the other 50%. One beam passes through the sample and thence to a photovoltaic cell; the other passes through the solvent to a similar transducer. The electric outputs from the two photovoltaic cells are passed through variable resistors, one of which is calibrated as a transmittance scale in linear units from 0 to 100. A sensitive current

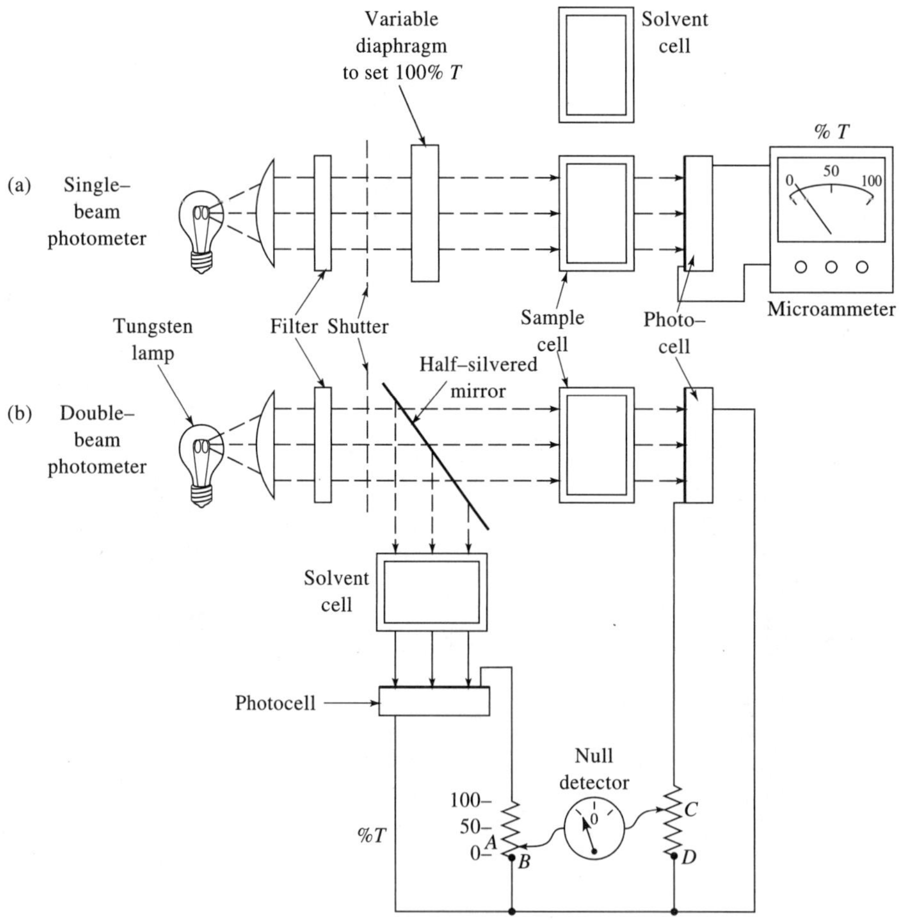

Figure 13-15 A single- and double-beam photometer.

meter, which serves as a null detector, is connected across the two resistors. When the potential difference across *AB* is equal to that across *CD,* no current passes through the null detector; under all other circumstances, a current is indicated. At the outset, the contact on the left is set to 0% *T,* the shutter is closed, and the pointer of the null detector is centered mechanically; the center mark then corresponds to zero current. Next, the solvent is placed in both cells, and contact *A* is set at 100 (the 100% *T* setting); with the shutter open, contact *C* is then adjusted until zero current is indicated. Replacement of the solvent with the sample in one cell results in a decrease in radiant power reaching the working photocell and a corresponding decrease in output potential; a current is then indicated by the galvanometer. The lack of balance is compensated for by moving *A* to a lower value. At balance, the percent transmittance is read directly from the scale.

Probe-Type Photometers. Figure 13-16 is a schematic of an interesting, commercially available, dipping-type photometer, which employs an optical fiber to transmit light from a source to a layer of solution lying between the glass seal at the end of the fiber and a mirror. The reflected radiation from the latter passes to a photodiode detector via a second glass fiber. The photometer uses an amplifier with an electronic chopper that is synchronized with the light source; as a result, the photometer does not respond to extraneous radiation. Six interference filters are provided, which can be interchanged by means of a knob located on the instrument panel. Custom filters are also available. Probe tips are manufactured from stainless steel, Swagelok® Stainless Steel, Pyrex®, and Acid-resistant Lexan® Plastic. Light path lengths that vary from 1 mm to 10 cm are available.

Absorbance is measured by first dipping the probe into the solvent and then into the solution to be measured. The device is particularly useful for photometric titrations (Section 14E).

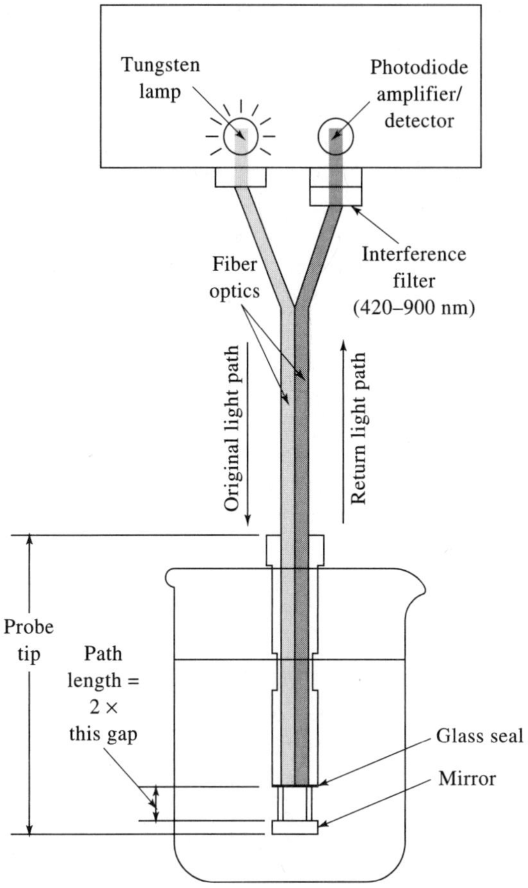

Figure 13-16 Schematic of a probe-type photometer. *(Courtesy of Brinkman Instrument Company, Division of Sybron Corporation, Westbury, NY.)*

Filter Selection. Photometers are generally supplied with several filters, each of which transmits a different portion of the visible spectrum. Selection of the proper filter for a given application is important because the sensitivity of the measurement is directly dependent upon this choice. The color of the light absorbed is the complement of the color of the solution itself. A liquid appears red, for example, because it transmits the red portion of the spectrum but absorbs the green. It is the intensity of green radiation that varies with concentration; a green filter should thus be employed. In general, then, the most suitable filter for a photometric analysis will be the color complement of the solution being analyzed. If several filters possessing the same general hue are available, the one that causes the sample to exhibit the greatest absorbance (or least transmittance) should be used.

Ultraviolet Absorption Photometers

Ultraviolet photometers often serve as detectors in high-performance liquid chromatography. In this application, a mercury vapor lamp usually serves as a source, and the emission line at 254 nm is isolated by filters. This type of detector is described briefly in Section 28C-6.

Ultraviolet photometers are also available for continuously monitoring the concentration of one or more constituents of gas or liquid streams in industrial plants. The instruments are dual channel in space and often employ one of the emission lines of mercury, which has been isolated by a filter system. Typical applications include the determination of low concentrations of phenol in waste water, monitoring the concentration of chlorine, mercury, or aromatics in gases, and the determination of the ratio of hydrogen sulfide to sulfur dioxide in the atmosphere.

Spectrophotometers

Numerous spectrophotometers are available from commercial sources. Some have been designed for the visible region only; others are applicable in the ultraviolet and visible regions. A few have measuring capabilities from the ultraviolet through the near infrared (185 to 3000 nm).

Instruments for the Visible Region. Several spectrophotometers designed to operate within the wavelength range of about 380 to 800 nm are available from commercial sources. These instruments are frequently simple, single-beam grating instruments that are relatively inexpensive (less than $1000 to perhaps $3000), rugged, and readily portable. At least one is battery operated and light and small enough to be handheld. The most common application of these instruments is for quantitative analysis, although several produce surprisingly good absorption spectra as well.

Figure 13-17 shows a simple and inexpensive spectrophotometer, the Spectronic 20. The original version of this instrument first appeared in the market in the mid-1950s, and the modified version shown in the figure is still being manufactured and widely sold. Undoubtedly, more of these instruments are currently in use throughout the world than any other single spectrophotometer model. The instrument owes its popularity, particularly as a teaching tool, to its relatively low cost, its ruggedness, and its satisfactory performance characteristics.

The Spectronic 20 employs a tungsten filament light source, which is operated by a stabilized power supply that provides radiation of constant intensity. After diffraction by a simple reflection grating, the radiation passes through the sample or reference cuvettes to a phototube. The amplified electrical signal from the transducer then powers a meter with a 5½-in. scale calibrated in transmittance and absorbance.

The instrument is equipped with an occluder, which is a vane that automatically falls between the beam and the detector whenever the cuvette is removed from its holder; the 0% *T* adjustment can then be made. As shown in Figure 13-18, the light control device in the Spectronic 20 consists of a V-shaped slot that can be moved in or out of the beam in order to set the meter to 100% *T*.

The wavelength range of the Spectronic 20 is from 340 to 625 nm; an accessory phototube extends this range to 950 nm. Other specifications for the instrument include a bandwidth of 20 nm and a wavelength accuracy of ±2.5 nm. The Spectronic 20 has been upgraded recently to the Spectronic 20+; the newer instrument utilizes a solid-state transducer as well as other enhanced features.

Single-Beam Instruments for the Ultraviolet/Visible Region. Several instrument manufacturers offer nonrecording single-beam instruments which can be used for both ultraviolet and visible measurements. The lower wavelength extremes for these instruments vary from 190 to 210 nm, and the upper from 800 to 1000 nm. All are equipped with interchangeable tungsten and hydrogen or deuterium lamps. Most employ photomultiplier tubes as transducers and gratings for dispersion. Some are equipped with digital readout devices; others employ large meters. The prices for these instruments range from $2000 to $8000.

As might be expected, performance specifications vary considerably among instruments and are related, at least to some degree, to instrument price. Typically, bandwidths vary from 2 to 8 nm; wavelength accuracies of ±0.5 to ±2 nm are reported.

The optical designs for the various grating instruments do not differ greatly from those shown in Figures 13-12a and 13-17. One manufacturer, however, employs a concave rather than a plane grating; a simpler and more compact design results. Instruments equipped with holographic gratings (Section 7C-2) are also beginning to appear on the market.

Single-Beam Computerized Spectrophotometers. One manufacturer is now offering a line of computerized, recording, single-beam spectrophotometers, which operate in the range of 190 to 800 nm (or to 900 nm with an accessory).[11] With these instruments, a

[11] W. Kaye, D. Barber, and R. Marasco, *Anal. Chem.*, **1980,** *52,* 437A; V. A. Kohler and N. Brenner, *Amer. Lab.*, **1981,** *13* (9), 109.

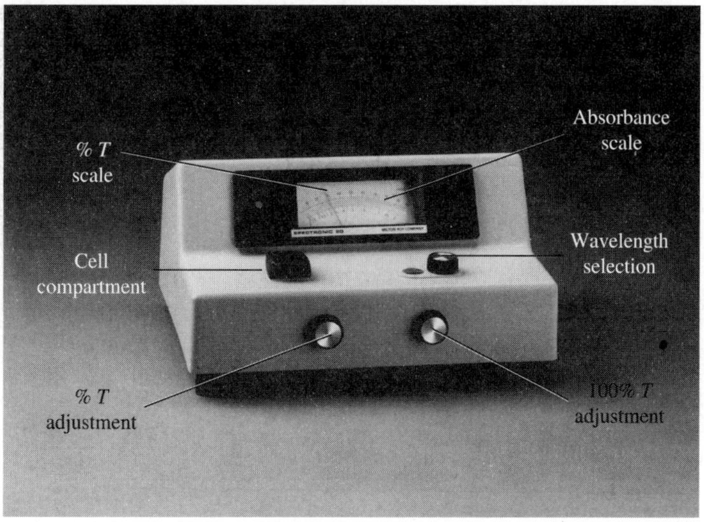

(a)

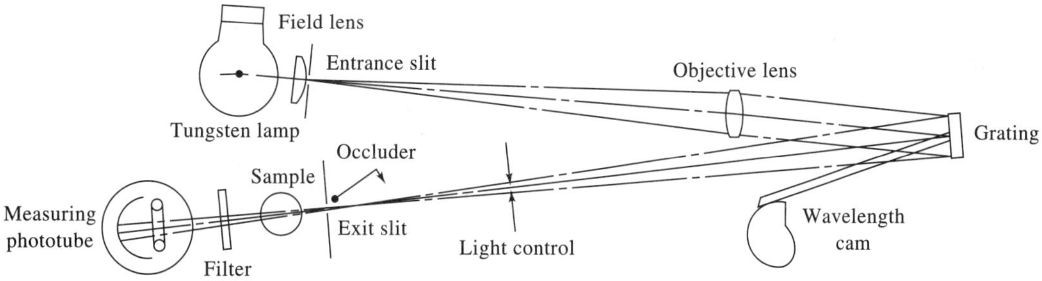

(b)

Figure 13-17 (a) The Spectronic 20 spectrophotometer and (b) its optical diagram. *(Courtesy of Spectronic Instruments, Inc., Rochester, NY.)*

wavelength scan is first performed with the reference solution in the beam path. The resulting transducer output is digitized in real time and stored in the memory of the computer. Samples are then scanned and absorbances calculated with the aid of the stored reference solution data. The complete spectrum is displayed on a cathode-ray tube within 2 s of data acquisition. Scan speeds as great as 1200 nm/min are feasible. The computer associated with the instrument provides several options with regard to data processing and presentation such as log absorbance, transmittance, derivatives, overlaid spectra, repetitive scans, concentration calculations, peak location and height determinations, and kinetic measurements.

As noted earlier, single-beam instruments have the inherent advantages of greater energy throughput, superior signal-to-noise ratios, and less cluttered sample compartments. On the other hand, the process of recording transducer outputs for reference and sample solutions successively for subsequent absorbance or transmittance calculations has heretofore not been very satisfactory because of the drift or flicker noise in sources and transducers. The manufacturer of these new

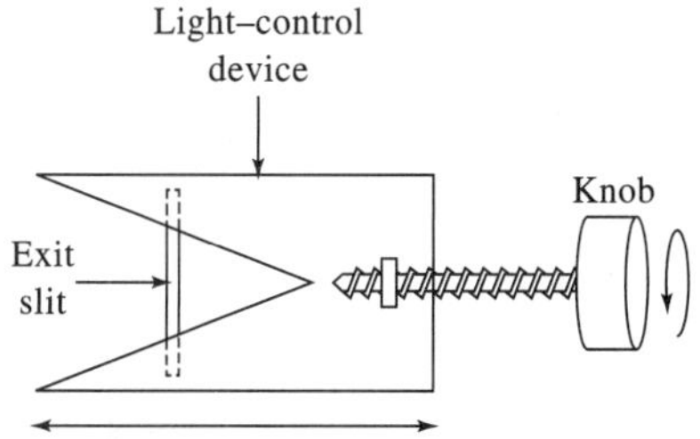

Figure 13-18 End view of the exit slit of the Spectronic 20 spectrophotometer pictured in Figure 13-17b.

single-beam instruments claims to have eliminated these instabilities by means of a fundamentally new source design and a new electronic design that eliminates hysteresis or memory effects in the phototransducer.

The photometric accuracies of the new instruments are reported to be ± 0.005 *A* or $\pm 0.3\%$ *T* with a drift of less than 0.002 *A*/hr. Bandwidths of 0.5, 1, and 2 nm are available by manual interchange of fixed slits.

Double-Beam Instruments. Numerous double-beam spectrophotometers for the ultraviolet/visible region of the spectrum are now available. Generally, these instruments are more expensive than their single-beam counterparts, with the nonrecording variety ranging in cost from about \$4000 to \$15,000.

Figure 13-19 shows construction details of a typical, relatively inexpensive, manual, double-beam ultraviolet/visible spectrophotometer. In this instrument, the radiation is dispersed by a concave grating, which also focuses the beam on a rotating sector mirror. The instrument design is similar to that shown in Figure 13-12c.

The instrument has a wavelength range of 195 to 850 nm, a bandwidth of 4 nm, a photometric accuracy of 0.5% *T*, and a reproducibility of 0.2% *A*; stray radiation is less than 0.1% of P_0 at 240 and 340 nm. This instrument is typical of several similar ones offered by various instrument companies. Such instruments are well suited for quantitative work where acquisition of an entire spectrum is not often required.

Figure 13-20 shows the optics of a more sophisticated, dual-channel (in time), recording spectrophotometer, which employs a 45×45-mm plane grating having 1440 lines/mm. Its range is from 190 to 750 nm with the option of extending this range to 900 nm. Bandwidths of 0.2, 0.5, 1.0, and 3.00 mm can be chosen by exchange of slits. The instrument has a photometric accuracy of ± 0.003 *A*; its stray radiation is less than 0.1% of P_0 at 220 and 340 nm. The performance of this instrument is significantly better than the double-beam instrument described previously; its price is correspondingly higher.

Double-Dispersing Instruments. In order to enhance spectral resolution and achieve a marked reduction in scattered radiation, a number of instruments have been designed with two gratings or prisms serially arranged with an intervening slit; in effect, then, these instruments consist of two monochromators in a series configuration.

The instrument shown in Figure 13-21 achieves the same performance characteristics with a single grating. Note that the radiation passing through entrance slit 1 is dispersed by the grating and travels through exit slit 1 to entrance slit 2. After a second dispersion by the grating, the beam emerges from exit slit 2 where it is split into a sample and a reference beam. The resolution of the instrument is reported to be 0.07 nm, and the stray light

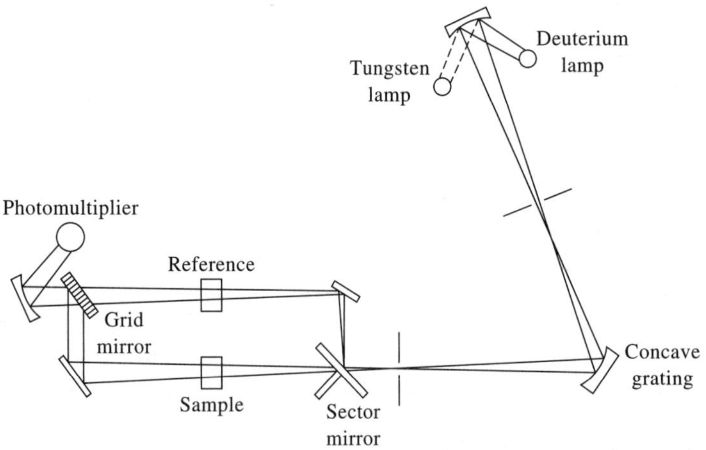

Figure 13-19 Schematic of a typical manual double-beam spectrophotometer for the ultraviolet/visible region.

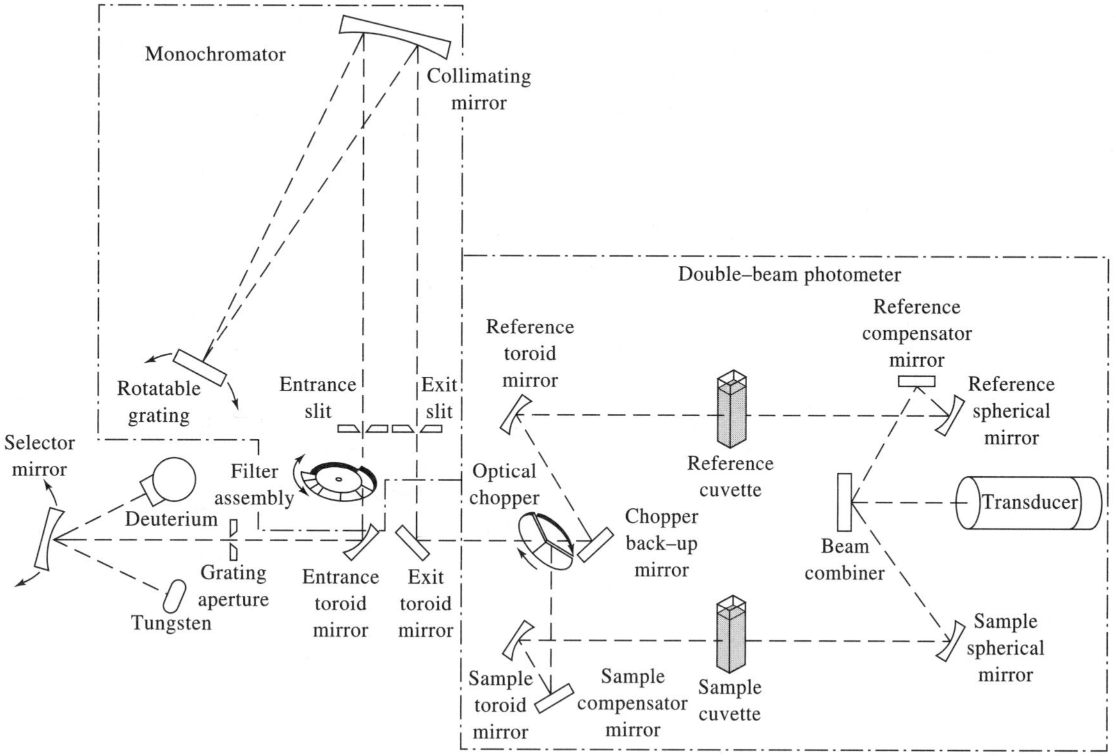

Figure 13-20 A double-beam recording spectrophotometer for the ultraviolet and visible regions; the Perkin-Elmer 57 Series. *(Courtesy of Coleman Instruments Division, Oak Brook, IL.)*

0.0008% from 220 to 800 nm. The wavelength range is 185 to 3125 nm. For wavelengths greater than 800 nm, a lead sulfide photoconductive cell is used with the tungsten source.

Diode Array Instruments. In the mid-1970s, a number of scientific papers and review articles appeared in the literature describing applications of silicon diode arrays (Section 7E-3) as transducers for spectrochemical measurements.[12] With these devices located at the focal plane of a monochromator, a spectrum can be obtained by electronic rather than mechanical scanning; all of the data points needed to define a spectrum can thus be gathered essentially simultaneously. The concept of multichannel instruments is attractive because of the potential speed at which spectra can be acquired as well as their applicability to simultaneous multicomponent determinations. In about 1980, the first electronic multichannel spectrometer designed specifically for molecular absorption measurements in the ultraviolet/visible range became available from a commercial source.[13] By now, several instrument companies offer such instruments.

Figure 13-22 shows an optical diagram of a typical multichannel ultraviolet/visible spectrophotometer. Because of the few optical components, the radiation throughput is much higher than that of traditional spectrophotometers. As a result, a single deuterium lamp can serve as a source for not only the ultraviolet but also the visible region (up to 820 nm). After passing through the solvent or analyte solution, the radiation is focused on an entrance slit and then passes onto the surface of a holographic reflection grating. The transducer is a diode

[12] Y. Talmi, *Appl. Spectrosc.*, **1982**, *36*, 1; Y. Talmi, *Anal. Chem.*, **1975**, *47*, 658A, 697A; Y. Talmi, *Multichannel Image Detectors*, Vol. 1, 1979; Vol. 2, 1983. Washington, DC: American Chemical Society; G. Horlick and E. G. Codding, *Contemp. Top. Anal. Clin. Chem.*, **1977**, *1*, 195.

[13] J. C. Milter, S. A. George, and B. G. Willis, *Science*, **1982**, *218*, 241.

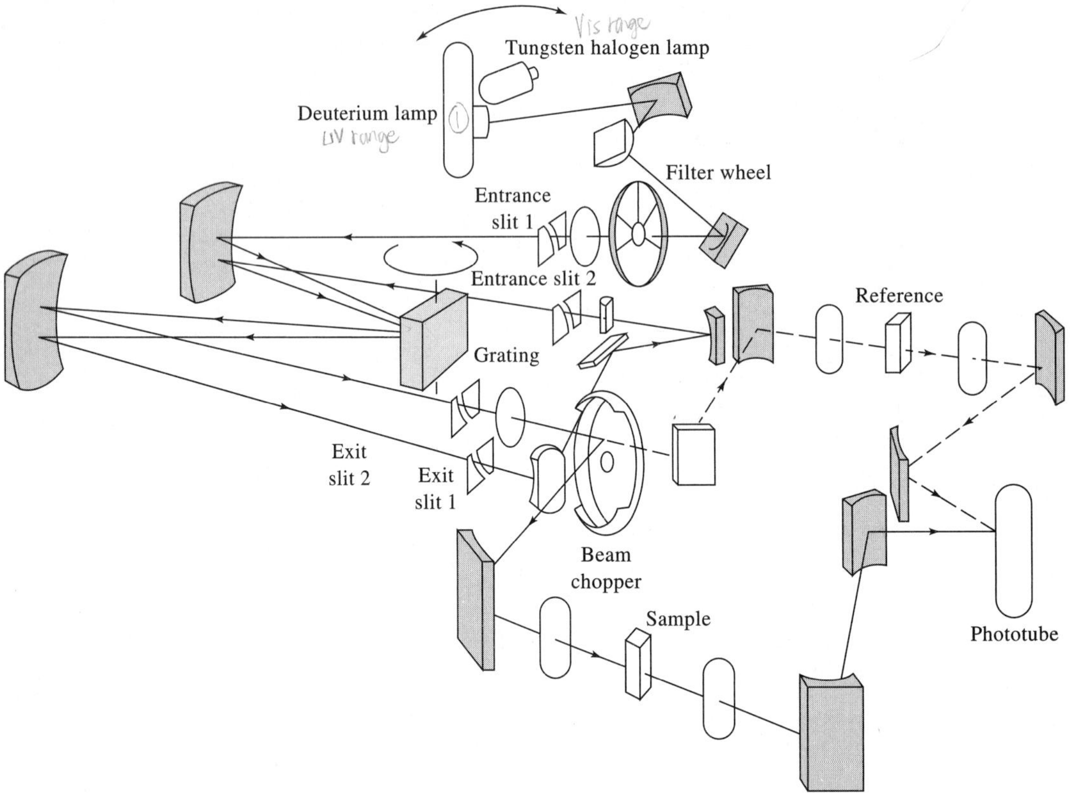

Figure 13-21 A double-dispersing spectrophotometer. *(Courtesy of Varian Instrument Division, Palo Alto, CA.)*

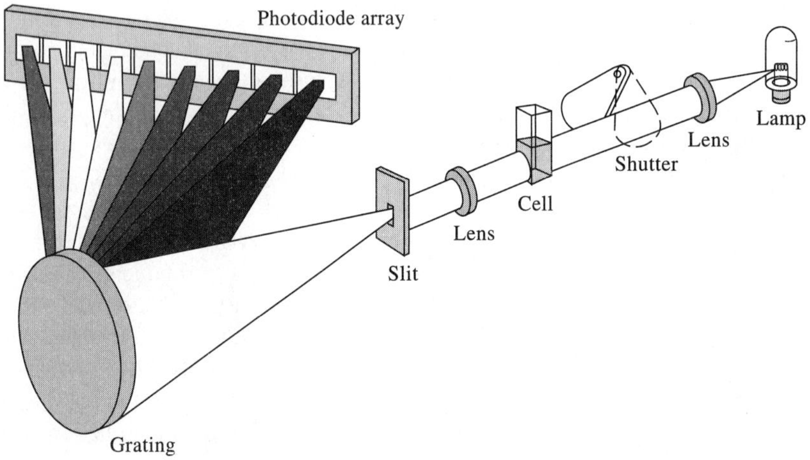

Figure 13-22 A multichannel diode array spectrometer; the HP 8452A. *(Courtesy of Hewlett-Packard Company, Palo Alto, CA.)*

array made up of 328 elements, each having a dimension of 18 by 0.5 mm. The dispersion of the grating and the size of the diode elements are such that a resolution of 2 nm is realized throughout the entire spectral region. Because the system contains no moving parts, the wavelength reproducibility from scan to scan is exceedingly high, and signal averaging can be used to produce marked increases in signal-to-noise ratios.

A single scan from 200 to 820 nm with this instrument requires 0.1 s. In order to improve signal-to-noise ratios, however, spectra are generally scanned for a second or more with the data being accumulated in computer memory. With such short exposure times, photodecomposition of samples is minimized despite the location of the sample between the source and the monochromator. The stability of the source and the electronic system is such that the solvent signal needs to be observed and stored only every five to ten minutes.

The spectrophotometer shown in Figure 13-22 is designed to be interfaced with most personal computer systems. The instrument (without the computer) sells in the \$7000 to \$9000 price range.

13E QUESTIONS AND PROBLEMS

13-1 Express the following absorbances in terms of percent transmittance:

(a) 0.0510. (d) 0.261.
(b) 0.918. (e) 0.485.
(c) 0.379. (f) 0.072.

13-2 Convert the accompanying transmittance data to absorbances:

(a) 25.5% (d) 3.58%.
(b) 0.567. (e) 0.085.
(c) 32.8% (f) 53.8%.

13-3 Calculate the percent transmittance of solutions having twice the absorbance of the solutions in Problem 13-1.

13-4 Calculate the absorbance of solutions having half the percent transmittance of those in Problem 13-2.

13-5 A solution containing 4.48 ppm $KMnO_4$ has a transmittance of 0.309 in a 1.00-cm cell at 520 nm. Calculate the molar absorptivity of $KMnO_4$.

13-6 A solution containing 3.75 mg/100 mL of A (220 g/mol) has a transmittance of 39.6% in a 1.50-cm cell at 480 nm. Calculate the molar absorptivity of A.

13-7 A solution containing the complex formed between Bi(III) and thiourea has a molar absorptivity of 9.32×10^3 L cm^{-1} at 470 nm.

(a) What is the absorbance of a 6.24×10^{-5} M solution of the complex at 470 nm in a 1.00-cm cell?
(b) What is the percent transmittance of the solution described in (a)?
(c) What is the molar concentration of the complex in a solution that has the absorbance described in (a) when measured at 470 nm in a 5.00-cm cell?

13-8 At 580 nm, which is the wavelength of its maximum absorption, the complex $FeSCN^{2+}$ has a molar absorptivity of 7.00×10^3 L cm^{-1} mol^{-1}. Calculate

(a) the absorbance of a 2.50×10^{-5} M solution of the complex at 580 nm in a 1.00-cm cell.
(b) the absorbance of a solution in which the concentration of the complex is twice that in (a).

(c) the transmittance of the solutions described in (a) and (b).

(d) the absorbance of a solution that has half the transmittance of that described in (a).

13-9 A 2.50-mL aliquot of a solution that contains 3.8 ppm iron(III) is treated with an appropriate excess of KSCN and diluted to 50.0 mL. What is the absorbance of the resulting solution at 580 nm in a 2.50-cm cell? See Problem 13-8 for absorptivity data.

13-10 Zinc(II) and the ligand L form a product that absorbs strongly at 600 nm. As long as the molar concentration of L exceeds that of zinc(II) by a factor of 5, the absorbance is dependent only on the cation concentration. Neither zinc(II) nor L absorbs at 600 nm. A solution that is 1.60×10^{-4} M in zinc(II) and 1.00×10^{-3} M in L has an absorbance of 0.464 in a 1.00-cm cell at 600 nm. Calculate

(a) the percentage transmittance of this solution.

(b) the percent transmittance of this solution in a 2.50-cm cell.

(c) the molar absorptivity of the complex.

13-11 The equilibrium constant for the conjugate acid/base pair

$$HIn + H_2O \rightleftharpoons H_3O^+ + In^-$$

is 8.00×10^{-5}. From the additional information

		Molar Absorptivity	
Species	**Absorption Maximum, nm**	**430 nm**	**600 nm**
HIn	430	8.04×10^3	1.23×10^3
In^-	600	0.775×10^3	6.96×10^3

(a) calculate the absorbance at 430 nm and 600 nm for the following indicator concentrations: 3.00×10^{-4} M, 2.00×10^{-4} M, 1.00×10^{-4} M, 0.500×10^{-4} M, and 0.250×10^{-4} M.

(b) plot absorbance as a function of indicator concentration.

13-12 The equilibrium constant for the reaction

$$2CrO_4^{2-} + 2H^+ \rightleftharpoons Cr_2O_7^{2-} + H_2O$$

is 4.1×10^{14}. The molar absorptivities for the two principal species in a solution of $K_2Cr_2O_7$ are

λ	$\epsilon_1(CrO_4^{2-})$	$\epsilon_2(Cr_2O_7^{2-})$
345	1.84×10^3	10.7×10^2
370	4.81×10^3	7.28×10^2
400	1.88×10^3	1.89×10^2

Four solutions were prepared by dissolving 4.00×10^{-4}, 3.00×10^{-4}, 2.00×10^{-4}, and 1.00×10^{-4} moles of $K_2Cr_2O_7$ in water and diluting to 1.00 L with a pH 5.60 buffer. Derive theoretical absorbance values (1.00-cm cells) for each solution and plot the data for (a) 345 nm; (b) 370 nm; (c) 400 nm.

13-13 Describe the differences between the following and list any particular advantages possessed by one over the other.
(a) hydrogen- and deuterium-discharge lamps as sources for ultraviolet radiation
(b) filters and monochromators as wavelength selectors
(c) photovoltaic cells and phototubes as detectors for electromagnetic radiation
(d) phototubes and photomultiplier tubes
(e) photometers and colorimeters
(f) spectrophotometers and photometers
(g) single-beam and double-beam instruments for absorbance measurements
(h) conventional and diode-array spectrophotomers

13-14 A portable photometer with a linear response to radiation registered 73.6 μA with a blank solution in the light path. Replacement of the blank with an absorbing solution yielded a response of 24.9 μA. Calculate
(a) the percent transmittance of the sample solution.
(b) the absorbance of the sample solution.
(c) the transmittance to be expected for a solution in which the concentration of the absorber is one third that of the original sample solution.
(d) the transmittance to be expected for a solution that has twice the concentration of the sample solution.

13-15 A photometer with a linear response to radiation gave a reading of 685 mV with a blank in the light path and 179 mV when the blank was replaced by an absorbing solution. Calculate
(a) the percent transmittance and absorbance of the absorbing solution.
(b) the expected transmittance if the concentration of absorber is one half that of the original solution.
(c) the transmittance to be expected if the light path through the original solution is doubled.

13-16 Why does a deuterium lamp produce a continuum rather than a line spectrum in the ultra-violet?

13-17 What are the differences between a photon detector and heat detector?

13-18 How is the power of infrared radiation measured?

13-19 Why can photomultiplier tubes not be used with infrared radiation?

13-20 Why is iodine sometimes introduced into a tungsten lamp?

13-21 Describe how an absorption photometer and a fluorescence photometer differ from each other.

13-22 Describe the basic design difference between a spectrometer for absorption measurements and one for emission studies.

13-23 What data are needed to describe the performance characteristics of an interference filter?

13-24 Define
(a) dark current.
(b) transducer.
(c) scattered radiation (in a monochromator).

(d) *n*-type semiconductor.
(e) majority carrier.
(f) depletion layer.

13-25 An interference filter is to be constructed for isolation of the CS_2 absorption band at 4.54 μm.
(a) If the determination is to be used upon first-order interference, how thick should the dielectric layer be (refractive index 1.34)?
(b) What other wavelengths will be transmitted?

13-26 Define the term *effective bandwidth of a filter.*

13-27 How many lines per millimeter are required in a grating if the first-order diffraction line at 500 nm is to be observed at a reflection angle of 10 deg when the angle of incidence is 60 deg?

13-28 An infrared grating has 72.0 lines/mm. Calculate the wavelengths of the first- and second-order diffraction spectra at reflection angles of (a) 0 deg and (b) + 15 deg. Assume the incident angle is 50 deg.

13-29 Describe how a spectroscope, a spectrograph, and a spectrophotometer differ from each other.

13-30 Why do quantitative and qualitative analyses often require different monochromator slit widths?

Chapter 14

Applications of Ultraviolet/Visible Molecular Absorption Spectrometry

Absorption measurements based upon ultraviolet and visible radiation find widespread application for the identification and determination of myriad inorganic and organic species.[1] Molecular ultraviolet/visible absorption methods are perhaps the most widely used of all quantitative analysis techniques in chemical and clinical laboratories throughout the world.

14A THE MAGNITUDE OF MOLAR ABSORPTIVITIES

Empirically, molar absorptivities that range from zero up to a maximum on the order of 10^5 are observed in ultraviolet and visible molecular absorption spectrometry.

[1] Some useful references on absorption methods include: E. J. Meehan, in *Treatise on Analytical Chemistry,* 2nd ed., P. J. Elving, E. J. Meehan, and I. M. Kolthoff, Eds., Part I, Vol. 7, Chapters 1–3. New York: Wiley, 1981; R. P. Bauman, *Absorption Spectroscopy.* New York: Wiley, 1962; F. Grum, in *Physical Methods of Chemistry,* A. Weissberger and B. W. Rossiter, Eds., Vol. I, Part III B, Chapter 3. New York: Wiley-Interscience, 1972; H. H. Jaffé and M. Orchin, *Theory and Applications of Ultraviolet Spectroscopy.* New York: Wiley, 1962; G. F. Lothian, *Absorption Spectrophotometry,* 3rd ed. London: Adam Hilger Ltd., 1969; J. D. Ingle Jr. and S. R. Crouch, *Spectrochemical Analysis,* Chapter 13. Englewood Cliffs, NJ: Prentice-Hall, 1988.

For any particular peak, the magnitude of ϵ depends upon the capture cross section (page 302) of the species and the probability for an energy-absorbing transition to occur. The relationship between ϵ and these variables has been shown to be

$$\epsilon = 8.7 \times 10^{19}\, PA$$

where P is the transition probability and A is the cross-section target area in square centimeters.[2] The area for typical organic molecules has been estimated from electron diffraction and X-ray studies to be about 10^{-15} cm^2 while transition probabilities vary from zero to one. For quantum mechanically allowed transitions, values of P range from 0.1 to 1, which leads to strong absorption bands ($\epsilon_{max} = 10^4$ to 10^5). Peaks having molar absorptivities less than about 10^3 are classified as being of low intensity. They result from so-called forbidden transitions, which have probabilities of occurrence that are less than 0.01.

[2] E. A. Braude, *J. Chem. Soc.,* **1950**, 379.

14B ABSORBING SPECIES

The absorption of ultraviolet or visible radiation by an atomic or molecular species M can be considered to be a two-step process, the first of which involves electronic excitation as shown by the equation

$$M + h\nu \rightarrow M^*$$

The product of the reaction between M and the photon $h\nu$ is an electronically excited species symbolized by M^*. The lifetime of the excited species is brief (10^{-8} to 10^{-9} s), its existence being terminated by any of several *relaxation* processes. The most common type of relaxation involves conversion of the excitation energy to heat; that is,

$$M^* \rightarrow M + \text{heat}$$

Relaxation may also occur by decomposition of M^* to form new species; such a process is called a *photochemical reaction.* Alternatively, relaxation may involve reemission of fluorescence or phosphorescence. It is important to note that the lifetime of M^* is usually so very short that its concentration at any instant is ordinarily negligible. Furthermore, the amount of thermal energy evolved by relaxation is usually not detectable. Thus absorption measurements create a minimal disturbance of the system under study except when photochemical decomposition occurs.

The absorption of ultraviolet or visible radiation generally results from excitation of bonding electrons; as a consequence, the wavelengths of absorption peaks can be correlated with the types of bonds in the species under study. Molecular absorption spectroscopy is, therefore, valuable for identifying functional groups in a molecule. More important, however, are the applications of ultraviolet and visible absorption spectroscopy to the quantitative determination of compounds containing absorbing groups.

For purposes of this discussion, it is useful to recognize three types of electronic transitions and to categorize absorbing species on this basis. The three include transitions involving: (1) π, σ, and n electrons, (2) d and f electrons, and (3) charge transfer electrons.

14B-1 Absorbing Species Containing π, σ, and n Electrons

Absorbing species containing π, σ, and n electrons include organic molecules and ions as well as a number of inorganic anions. Our discussion will deal largely with the former, although brief mention will be made of absorption by certain inorganic systems as well.

All organic compounds are capable of absorbing electromagnetic radiation because all contain valence electrons that can be excited to higher energy levels. The excitation energies associated with electrons forming most single bonds are sufficiently high that absorption by them is restricted to the so-called vacuum ultraviolet region ($\lambda < 185$ nm), where components of the atmosphere also absorb strongly. The experimental difficulties associated with the vacuum ultraviolet are significant; as a result, most spectrophotometric investigations of organic compounds have involved the wavelength region greater than 185 nm. Absorption of longer-wavelength ultraviolet and visible radiation is restricted to a limited number of functional groups (called *chromophores*) that contain valence electrons with relatively low excitation energies.

The electronic spectra of organic molecules containing chromophores are usually complex, because the superposition of vibrational transitions on the electronic transitions leads to an intricate combination of overlapping lines; the result is a broad band of absorption that often appears to be continuous. The complex nature of the spectra makes detailed theoretical analysis difficult or impossible. Nevertheless, qualitative or semiquantitative statements concerning the types of electronic transitions responsible for a given absorption spectrum can be deduced from molecular orbital considerations.

Types of Absorbing Electrons[3]

The electrons that contribute to absorption by an organic molecule are: (1) those that participate directly in bond formation between atoms and are thus associated with more than one atom; (2) nonbonding or unshared outer electrons that are largely localized about such atoms as oxygen, the halogens, sulfur, and nitrogen.

Covalent bonding occurs because the electrons forming the bond move in the field about two atomic centers in such a manner as to minimize the repulsive coulombic forces between these centers. The nonlocalized fields between atoms that are occupied by bonding

[3] For further details, see H. H. Jaffé and M. Orchin, *Theory and Applications of Ultraviolet Spectroscopy.* New York: Wiley, 1962; C. N. R. Rao, *Ultra-Violet and Visible Spectroscopy: Chemical Applications,* 3rd ed. London: Butterworths, 1975; R. P. Bauman, *Absorption Spectroscopy,* Chapters 6 and 8. New York: Wiley, 1962; R. M. Silverstein, G. C. Bassler, and T. C. Merrill, *Spectrometric Identification of Organic Compounds,* 5th ed., Chapter 7. New York: Wiley, 1991.

electrons are called *molecular orbitals* and can be considered to result from the overlap of atomic orbitals. When two atomic orbitals combine, a low-energy *bonding molecular orbital* and a high-energy *antibonding molecular orbital* result. The electrons of a molecule occupy the former in the ground state.

The molecular orbitals associated with single bonds in organic molecules are designated as *sigma* (σ) *orbitals,* and the corresponding electrons are σ electrons. As shown in Figure 14-1a, the distribution of charge density of a sigma orbital is rotationally symmetric around the axis of the bond. Here, the average negative charge density arising from the motion of the two electrons around the two positive nuclei is indicated by the degree of shading.

The double bond in an organic molecule contains two types of molecular orbitals: a *sigma* (σ) orbital corresponding to one pair of the bonding electrons and a *pi* (π) *molecular orbital* associated with the other pair. *Pi orbitals* are formed by the parallel overlap of atomic *p* orbitals. Their charge distribution is characterized by a *nodal plane* (a region of low-charge density) along the axis of the bond and a maximum density in regions above and below the plane (see Figure 14-1b).

Also shown, in Figures 14-1c and 14-1d, are the charge-density distributions for antibonding sigma and pi orbitals; these orbitals are designated by σ^* and π^*.

In addition to σ and π electrons, many organic compounds contain nonbonding electrons. These unshared electrons are designated by the symbol *n*. An example showing the three types of electrons in formaldehyde is shown in Figure 14-2.

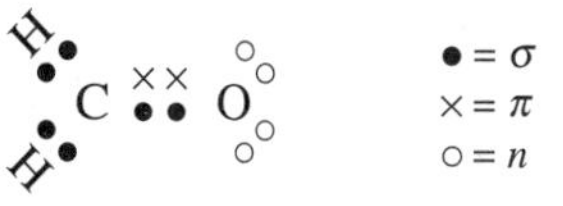

Figure 14-2 Types of molecular orbitals in formaldehyde.

As shown in Figure 14-3, the energies for the various types of molecular orbitals differ significantly. Quite generally, the energy level of a nonbonding electron lies between the energy levels of the bonding and the antibonding π and σ orbitals. Electronic transitions among certain of the energy levels can be brought about by the absorption of radiation. As shown in Figure 14-3, four types of transitions are possible: $\sigma \rightarrow \sigma^*$, $n \rightarrow \sigma^*$, $n \rightarrow \pi^*$, and $\pi \rightarrow \pi^*$.

$\sigma \rightarrow \sigma^*$ Transitions. Here, an electron in a bonding σ orbital of a molecule is excited to the corresponding antibonding orbital by the absorption of radiation. The molecule is then described as being in the σ,σ^* excited state. Relative to other possible transitions, the energy required to induce a $\sigma \rightarrow \sigma^*$ transition is large (see the first arrow in Figure 14-3), corresponding to radiant frequencies in the vacuum ultraviolet region. Methane, for example, which contains only single C—H bonds and can thus undergo only $\sigma \rightarrow \sigma^*$ transitions, exhibits an absorption maximum at 125 nm. Ethane has an absorption peak at 135 nm, which must also arise from the same type of transition, but here electrons of the C—C bond appear to be involved. Because the strength of the C—C bond is less than that of the C—H bond, less en-

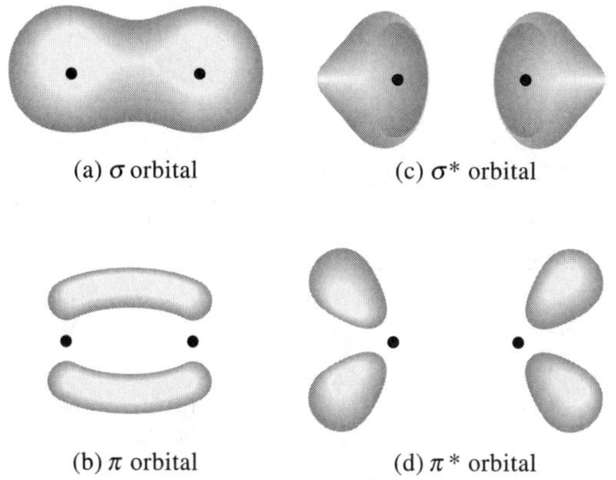

Figure 14-1 Electron distribution in sigma and pi molecular orbitals.

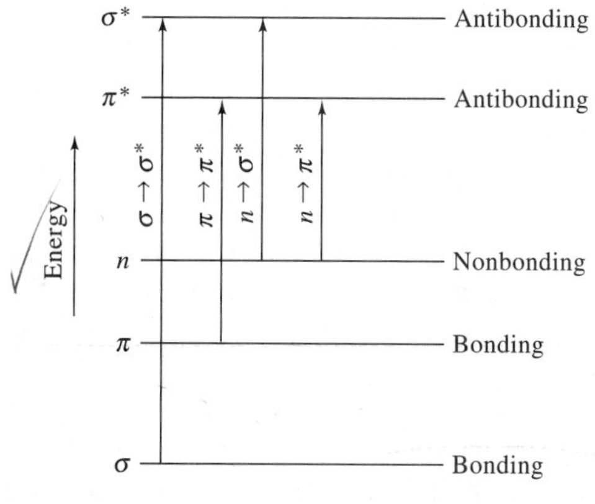

Figure 14-3 Electronic molecular energy levels.

ergy is required for excitation; thus, the absorption peak occurs at a longer wavelength.

Absorption maxima due to $\sigma \rightarrow \sigma^*$ transitions are never observed in the ordinarily accessible ultraviolet region; for this reason, no further discussion will be devoted to this type of absorption.

$n \rightarrow \sigma^*$ Transitions. Saturated compounds containing atoms with unshared electron pairs (nonbonding electrons) are capable of $n \rightarrow \sigma^*$ transitions. In general, these transitions require less energy than the $\sigma \rightarrow \sigma^*$ type and can be brought about by radiation in the region of between 150 and 250 nm, with most absorption peaks appearing below 200 nm. Table 14-1 shows absorption data for some typical $n \rightarrow \sigma^*$ transitions. It will be seen that the energy requirements for such transitions depend primarily upon the kind of bond and to a lesser extent upon the structure of the molecule. The molar absorptivities associated with this type of absorption are low to intermediate in magnitude and usually range between 100 and 3000 L cm^{-1} mol^{-1}.

Absorption maxima for the $n \rightarrow \sigma^*$ transition tend to shift to shorter wavelengths in the presence of polar solvents such as water or ethanol. The number of organic functional groups with $n \rightarrow \sigma^*$ peaks in the readily accessible ultraviolet region is relatively small.

TABLE 14-1 Some Examples of Absorption due to $n \rightarrow \sigma^*$ Transitions[a]

Compound	λ_{max}(nm)	ϵ_{max}
H_2O	167	1480
CH_3OH	184	150
CH_3Cl	173	200
CH_3I	258	365
$(CH_3)_2S$[b]	229	140
$(CH_3)_2O$	184	2520
CH_3NH_2	215	600
$(CH_3)_3N$	227	900

[a]Samples in vapor state.
[b]In ethanol solvent.

$n \rightarrow \pi^*$ and $\pi \rightarrow \pi^*$ Transitions. Most applications of absorption spectroscopy to organic compounds are based upon transitions for n or π electrons to the π^* excited state because the energies required for these processes bring the absorption peaks into an experimentally convenient spectral region (200 to 700 nm). Both transitions require the presence of an unsaturated functional group to provide the π orbitals. Strictly speaking, it is to these unsaturated absorbing centers that the term chromophore applies.

The molar absorptivities for peaks associated with excitation to the n,π^* state are generally low and ordinarily range from 10 to 100 L cm^{-1} mol^{-1}; values for $\pi \rightarrow \pi^*$ transitions, on the other hand, normally fall in the range between 1000 and 10,000. Another characteristic difference between the two types of absorption is the effect exerted by the solvent on the wavelength of the peaks. Peaks associated with $n \rightarrow \pi^*$ transitions are generally shifted to shorter wavelengths (a *hypsochromic* or *blue shift*) with increasing polarity of the solvent. Usually, but not always, the reverse trend (a *bathochromic* or *red shift*) is observed for $\pi \rightarrow \pi^*$ transitions. The hypsochromic effect apparently arises from the increased solvation of the unbonded electron pair, which lowers the energy of the n orbital. The most dramatic effects of this kind (blue shifts of 30 nm or more) are seen with polar hydrolytic solvents, such as water or alcohols, in which hydrogen bond formation between the solvent protons and the nonbonded electron pair is extensive. Here, the energy of the n orbitals is lowered by an amount approximately equal to the energy of the hydrogen bond. When an $n \rightarrow \pi^*$ transition occurs, however, the remaining single n electron cannot maintain the hydrogen bond; thus, the energy of the n,π^* *excited* state is not affected by this type of solvent interaction. A blue shift, also roughly corresponding to the energy of the hydrogen bond, is therefore observed.

A second solvent effect that undoubtedly influences both $\pi \rightarrow \pi^*$ and $n \rightarrow \pi^*$ transitions leads to a bathochromic shift with increased solvent polarity. This effect is small (usually less than 5 nm) and as a result, is completely overshadowed in $n \rightarrow \pi^*$ transitions by the hypsochromic effect just discussed. With the bathochromic shift, attractive polarization forces between the solvent and the absorber tend to lower the energy levels of both the unexcited and the excited states. The effect on the excited state is greater, however, and the energy differences thus become smaller with increased solvent polarity; small bathochromic shifts result.

Organic Chromophores

Table 14-2 lists common organic chromophores and the approximate location of their absorption maxima. These data can serve only as rough guides for the identification of functional groups, since the positions of maxima are also affected by solvent and structural details of the molecule containing the chromophores. Furthermore, the peaks are ordinarily broad because of vibrational effects; the precise determination of the position of a maximum is thus difficult. The spectra shown in Figure 14-4 are typical of those of many organic compounds.

Effect of Conjugation of Chromophores

In the molecular-orbital treatment, π electrons are considered to be further delocalized by conjugation; the orbitals thus involve four (or more) atomic centers. The effect of this delocalization is to lower the energy level of the π^* orbital and give it less antibonding character. Absorption maxima are shifted to longer wavelengths as a consequence.

As seen from the data in Table 14-3, the absorptions of multichromophores in a single organic mole-

TABLE 14-2 Absorption Characteristics of Some Common Chromophores

Chromophore	Example	Solvent	λ_{max}(nm)	ϵ_{max}	Type of Transition
Alkene	$C_6H_{13}CH{=}CH_2$	*n*-Heptane	177	13,000	$\pi \rightarrow \pi^*$
Alkyne	$C_5H_{11}C{\equiv}C{-}CH_3$	*n*-Heptane	178	10,000	$\pi \rightarrow \pi^*$
			196	2,000	—
			225	160	—
Carbonyl	$CH_3\overset{\overset{O}{\|}}{C}CH_3$	*n*-Hexane	186	1,000	$n \rightarrow \sigma^*$
			280	16	$n \rightarrow \pi^*$
	$CH_3\overset{\overset{O}{\|}}{C}H$	*n*-Hexane	180	large	$n \rightarrow \sigma^*$
			293	12	$n \rightarrow \pi^*$
Carboxyl	$CH_3\overset{\overset{O}{\|}}{C}OH$	Ethanol	204	41	$n \rightarrow \pi^*$
Amido	$CH_3\overset{\overset{O}{\|}}{C}NH_2$	Water	214	60	$n \rightarrow \pi^*$
Azo	$CH_3N{=}NCH_3$	Ethanol	339	5	$n \rightarrow \pi^*$
Nitro	CH_3NO_2	Isooctane	280	22	$n \rightarrow \pi^*$
Nitroso	C_4H_9NO	Ethyl ether	300	100	—
			665	20	$n \rightarrow \pi^*$
Nitrate	$C_2H_5ONO_2$	Dioxane	270	12	$n \rightarrow \pi^*$

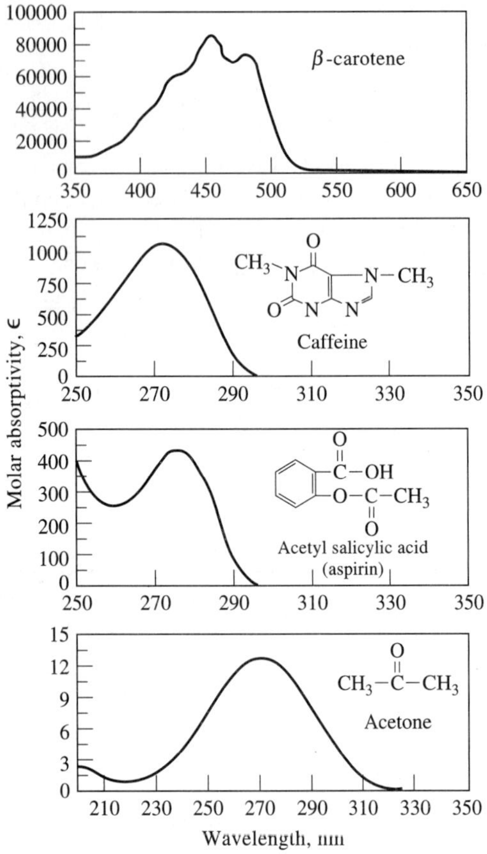

Figure 14-4 Ultraviolet spectra for typical organic compounds.

cule are approximately additive, provided the chromophores are separated from one another by more than one single bond. Conjugation of chromophores, however, has a profound effect on spectral properties. For example, it is seen in Table 14-3 that 1,3-butadiene, $CH_2{=}CHCH{=}CH_2$, has a strong absorption band that is displaced to a longer wavelength by 20 nm compared with the corresponding peak for an unconjugated diene. When three double bonds are conjugated, the bathochromic effect is even larger.

Conjugation between the doubly bonded oxygen of aldehydes, ketones, and carboxylic acids and an olefinic double bond gives rise to similar behavior (see Table 14-3). Analogous effects are also observed when two carbonyl or carboxylate groups are conjugated with one another. For α,β unsaturated aldehydes and ketones, the weak absorption peak due to $n \rightarrow \pi^*$ transitions is shifted to longer wavelengths by 40 nm or more. In addition, a strong absorption peak corresponding to a $\pi \rightarrow \pi^*$ transition appears. This latter peak occurs in the vacuum ultraviolet only if the carbonyl group is not conjugated.

The wavelengths of absorption peaks for conjugated systems are sensitive to the types of groups attached to the doubly bonded atoms. Various empirical rules have been developed for predicting the effect of such substitutions upon absorption maxima and have proved useful for structural determinations.[4]

Absorption by Aromatic Systems

The ultraviolet spectra of aromatic hydrocarbons are characterized by three sets of bands that originate from $\pi \rightarrow \pi^*$ transitions. For example, benzene has a strong absorption peak at 184 nm ($\epsilon_{max} \sim 60{,}000$); a weaker

[4] For a summary of these rules, see R. M. Silverstein, G. C. Bassler, and T. C. Morrill, *Spectrometric Identification of Organic Compounds*, 5th ed., pp. 295–315. New York: Wiley, 1991.

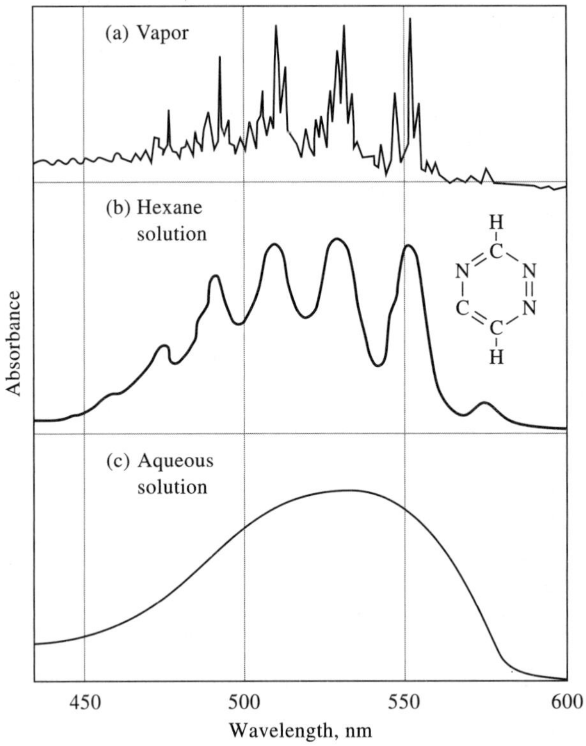

Figure 14-5 Ultraviolet absorption spectra for 1,2,4,5-tetrazine (a) in the vapor phase, (b) in hexane solution, and (c) in aqueous solution. *(From S. F. Mason,* J. Chem. Soc. ***1959****, 1265. With permission.)*

TABLE 14-3 Effect of Multichromophores on Absorption

Compound	Type	λ_{max}(nm)	ϵ_{max}
$CH_3CH_2CH_2CH{=}CH_2$	Olefin	184	~10,000
$CH_2{=}CHCH_2CH_2CH{=}CH_2$	Diolefin (unconjugated)	185	~20,000
$H_2C{=}CHCH{=}CH_2$	Diolefin (conjugated)	217	21,000
$H_2C{=}CHCH{=}CHCH{=}CH_2$	Triolefin (conjugated)	250	—
$CH_3CH_2CH_2CH_2C({=}O)CH_3$	Ketone	282	27
$CH_2{=}CHCH_2CH_2C({=}O)CH_3$	Unsaturated ketone (unconjugated)	278	30
$CH_2{=}CHC({=}O)CH_3$	α,β-Unsaturated ketone (conjugated)	324	24
		219	3,600

band, called the E_2 band, at 204 nm (ϵ_{max} = 7900); and a still weaker peak, termed the B band, at 256 nm (ϵ_{max} = 200). The long-wavelength bands of benzene vapor, 1,2,4,5-tetrazine (see Figure 14-5a), and many other aromatics contain a series of sharp peaks due to the superposition of vibrational transitions on the basic electronic transitions. As shown in Figure 14-5, solvents tend to reduce (or sometimes eliminate) this fine structure as do certain types of substitution.

All three of the characteristic bands for benzene are strongly affected by ring substitution; the effects on the two longer-wavelength bands are of particular interest because they can be studied with ordinary spectrophotometric equipment. Table 14-4 illustrates the effects of some common ring substituents.

By definition, an *auxochrome* is a functional group that does not itself absorb in the ultra-violet region but has the effect of shifting chromophore peaks to longer wavelengths as well as increasing their intensities. It is seen in Table 14-4 that —OH and —NH_2 have an auxochromic effect on the benzene chromophore, particularly with respect to the B band. Auxochromic substituents have at least one pair of *n* electrons capable of interacting with the π electrons of the ring. This interaction apparently has the effect of stabilizing the π* state, thereby lowering its energy; a bathochromic shift results. Note that the auxochromic effect is more pronounced for the phenolate anion than for phenol itself, probably because the anion has an extra pair of unshared electrons to contribute to the interaction. With aniline, on the other hand, the nonbonding electrons are lost by formation of the anilinium cation, and the auxochromic effect disappears as a consequence.

Absorption by Inorganic Anions

A number of inorganic anions exhibit ultraviolet absorption peaks that are a consequence of $n \rightarrow \pi^*$ transitions. Examples include nitrate (313 nm), carbonate (217 nm), nitrite (360 and 280 nm), azide (230 nm), and trithiocarbonate (500 nm) ions.

14B-2 Absorption Involving *d* and *f* Electrons

Most transition-metal ions absorb in the ultraviolet or visible region of the spectrum. For the lanthanide and actinide series, the absorption process results from electronic transitions of 4*f* and 5*f* electrons; for elements of the first and second transition-metal series, the 3*d* and 4*d* electrons are responsible.

TABLE 14-4 Absorption Characteristics of Aromatic Compounds

Compound		E_2 Band		B Band	
		λ_{max}(nm)	ϵ_{max}	λ_{max}(nm)	ϵ_{max}
Benzene	C_6H_6	204	7,900	256	200
Toluene	$C_6H_5CH_3$	207	7,000	261	300
m-Xylene	$C_6H_4(CH_3)_2$	—	—	263	300
Chlorobenzene	C_6H_5Cl	210	7,600	265	240
Phenol	C_6H_5OH	211	6,200	270	1,450
Phenolate ion	$C_6H_5O^-$	235	9,400	287	2,600
Aniline	$C_6H_5NH_2$	230	8,600	280	1,430
Anilinium ion	$C_6H_5NH_3^+$	203	7,500	254	160
Thiophenol	C_6H_5SH	236	10,000	269	700
Naphthalene	$C_{10}H_8$	286	9,300	312	289
Styrene	$C_6H_5CH{=}CH_2$	244	12,000	282	450

McCreary

Absorption by Lanthanide and Actinide Ions

The ions of most lanthanide and actinide elements absorb in the ultraviolet and visible regions. In distinct contrast to the behavior of most inorganic and organic absorbers, their spectra consist of narrow, well-defined, and characteristic absorption peaks, which are little affected by the type of ligand associated with the metal ion. Typical spectra for four lanthanide ions are shown in Figure 14-6.

The transitions responsible for absorption by elements of the lanthanide series appear to involve the various energy levels of 4*f* electrons, while it is the 5*f* electrons of the actinide series that interact with radiation. These inner orbitals are largely screened from external influences by electrons occupying orbitals with higher principal quantum numbers. As a consequence, the bands are narrow and relatively unaffected by the nature of the solvent or the species bonded by the outer electrons.

Absorption by Elements of the First and Second Transition-Metal Series

The ions and complexes of the 18 elements in the first two transition series tend to absorb visible radiation in one if not all of their oxidation states. In contrast to the lanthanide and actinide elements, however, the absorption bands are often broad (Figure 14-7) and are strongly influenced by chemical environmental factors. An example of the environmental effect is found in the pale blue color of the aquo copper(II) ion and the much darker blue of the copper complex with ammonia.

Metals of the transition series are characterized by having five partially occupied *d* orbitals (3*d* in the first series and 4*d* in the second), each capable of accommodating a pair of electrons. The spectral characteristics of transition metals involve electronic transitions among the various energy levels of these *d* orbitals.

Two theories have been advanced to rationalize the colors of transition-metal ions and the profound influence of chemical environment on these colors. The *crystal-field theory,* which will be discussed briefly, is the simpler of the two theories and is adequate for a qualitative understanding. The more complex molecular-orbital treatment, however, provides a better quantitative treatment of the phenomenon.[5]

[5] For a nonmathematical discussion of these theories, see L. E. Orgel, *An Introduction to Transition Metal Chemistry, Ligand-Field Theory.* New York: Wiley, 1960.

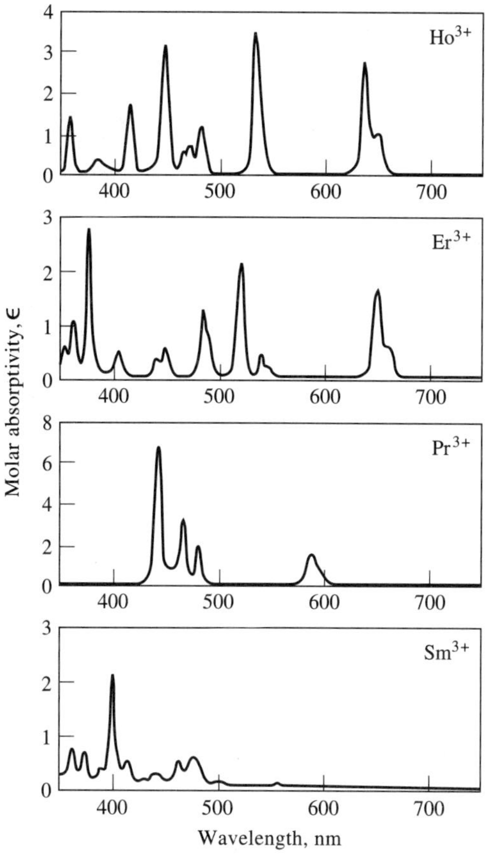

Figure 14-6 Typical absorption spectra for lanthanide ions.

Both theories are based upon the premise that the energies of *d* orbitals of the transition-metal ions in solution are not identical and that absorption involves the transition of electrons from a *d* orbital of lower energy to one of higher energy. In the absence of an external electric or magnetic field (as in the dilute gaseous state), the energies of the five *d* orbitals are identical, and absorption of radiation is not required for an electron to move from one orbital to another. On the other hand, when complex formation occurs in solution between the metal ion and water or some other ligand, splitting of the *d*-orbital energies results. This effect results from the differential forces of electrostatic repulsion between the electron pair of the donor and the electrons in the various *d* orbitals of the central metal ion. In order to understand this splitting of energies, the spatial distribution of electrons in the various *d* orbitals must be considered.

The electron-density distribution of the five *d* orbitals around the nucleus is shown in Figure 14-8. Three of the orbitals, termed d_{xy}, d_{xz}, and d_{yz}, are similar in every regard except for their spatial orientation. Note that these orbitals occupy spaces *between* the three axes; consequently, they have minimum electron densities along the axes and maximum densities on the diagonals between axes. In contrast, the electron densities of the $d_{x^2-y^2}$ and the d_{z^2} orbitals are directed along the axes.

Let us now consider a transition-metal ion that is coordinated to six molecules of water. These ligand molecules can be imagined as being symmetrically distributed around the central atom, one ligand being located at each end of the three axes shown in Figure 14-8; the resulting octahedral structure is the most common orientation for transition-metal complexes. The negative ends of the water dipoles are pointed toward the metal ion, and the electric fields from these dipoles tend to exert a repulsive effect on all of the *d* orbitals, *thus increasing their energy;* the orbitals are then said to have become *destabilized.* The maximum charge density of the d_{z^2} orbital lies along the bonding axis. The negative field of a bonding ligand therefore has a greater effect on this orbital than upon the d_{xy}, d_{xz}, and d_{yz} orbitals, whose charge densities do not coincide with the bonding axes. These latter orbitals will be destabilized equally, inasmuch as they differ from one another only in orientation. The effect of the electric field on the $d_{x^2-y^2}$ orbital is less obvious, but quantum calculations have shown that it is destabilized to the same extent as the d_{z^2} orbital. Thus, the energy-level diagram for the octahedral configuration (Figure 14-9) shows that the energies of all of the *d* orbitals increase in the presence of a ligand field but, in addition, that the *d orbitals are split* into levels differing in energy by Δ. Also shown are energy diagrams for complexes involving four-coordinate bonds. Two configurations are encountered; the *tetrahedral,* in which the four groups are symmetrically distributed in space around the metal ion; and the *square planar,* in which the four ligands lie equidistant from the metal ion in a single plane. Unique *d*-orbital splitting patterns for each configuration can be deduced by arguments similar to those used for the octahedral structure.

The magnitude of Δ (Figure 14-9) depends upon a number of factors, including the charge on the metal ion and the position of the parent element in the periodic table. An important variable attributable to the ligand is

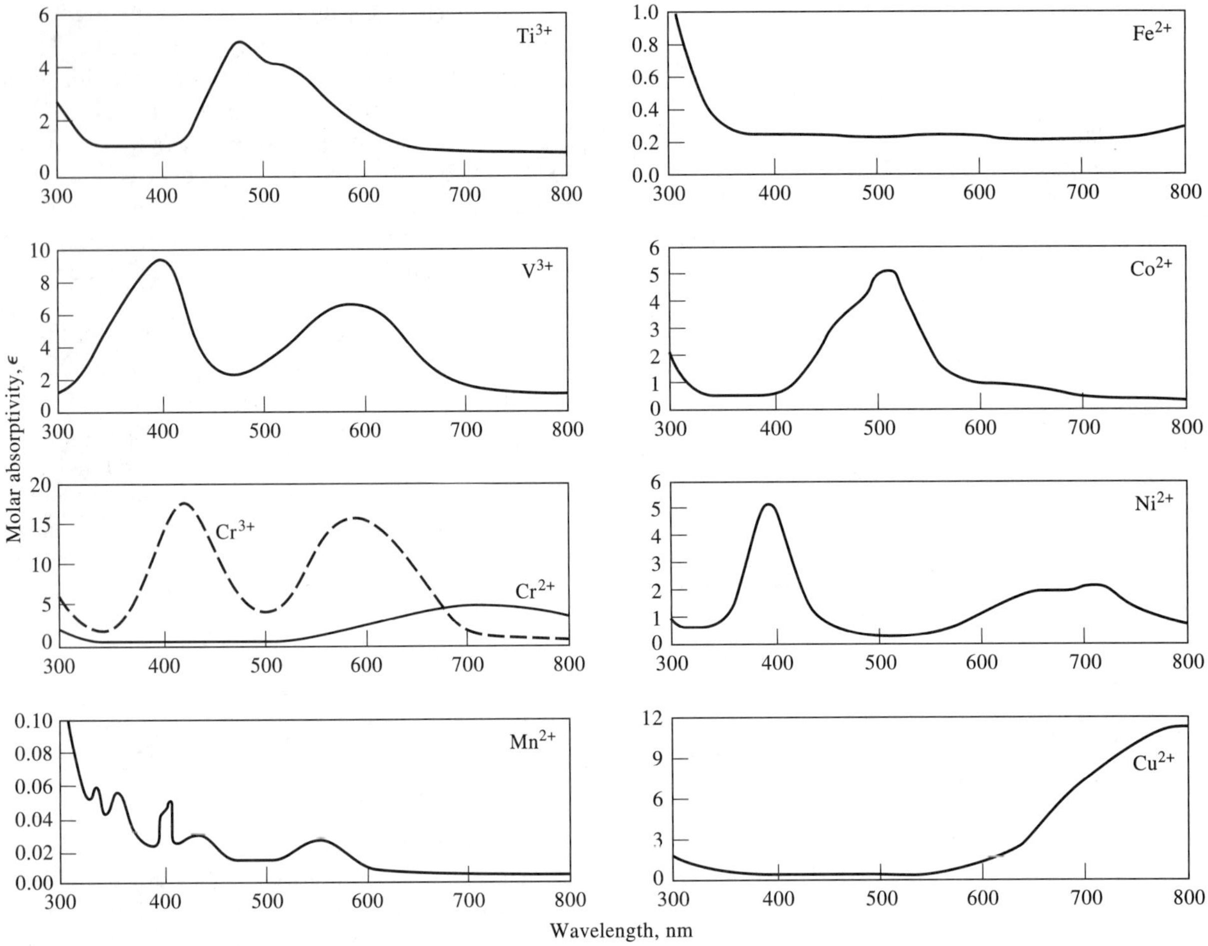

Figure 14-7 Absorption spectra of some transition-metal ions.

the *ligand field strength,* which is a measure of the extent to which a complexing group will split the energies of the *d* electrons; that is, a complexing agent with a high ligand field strength will cause Δ to be large.

It is possible to arrange the common ligands in order of increasing ligand field strengths: $I^- < Br^- < Cl^- < F^- < OH^- < C_2O_4^{2-} \sim H_2O < SCN^- < NH_3$ < ethylenediamine < *o*-phenanthroline < NO_2^- < CN^-. With only minor exceptions, this order applies to all transition-metal ions and permits qualitative predictions about the relative positions of absorption peaks for the various complexes of a given transition-metal ion. Since Δ increases with increasing field strength, the wavelength of the absorption maxima decreases. This effect is demonstrated by the data in Table 14-5.

14B-3 Charge-Transfer Absorption

For analytical purposes, species that exhibit *charge-transfer absorption*[6] are of particular importance because their molar absorptivities are very large (ϵ_{max} > 10,000). Thus, these complexes provide a highly sensitive means for detecting and determining absorbing species. Many inorganic complexes exhibit charge-transfer absorption and are therefore called *charge-transfer complexes.* Common examples of such complexes include the thiocyanate and phenolic complexes of iron(III), the *o*-phenanthroline complex

[6] For a brief discussion of this type of absorption, see C. N. R. Rao, *Ultra-Violet and Visible Spectroscopy: Chemical Applications,* 3rd ed., Chapter 11. London: Butterworths, 1975.

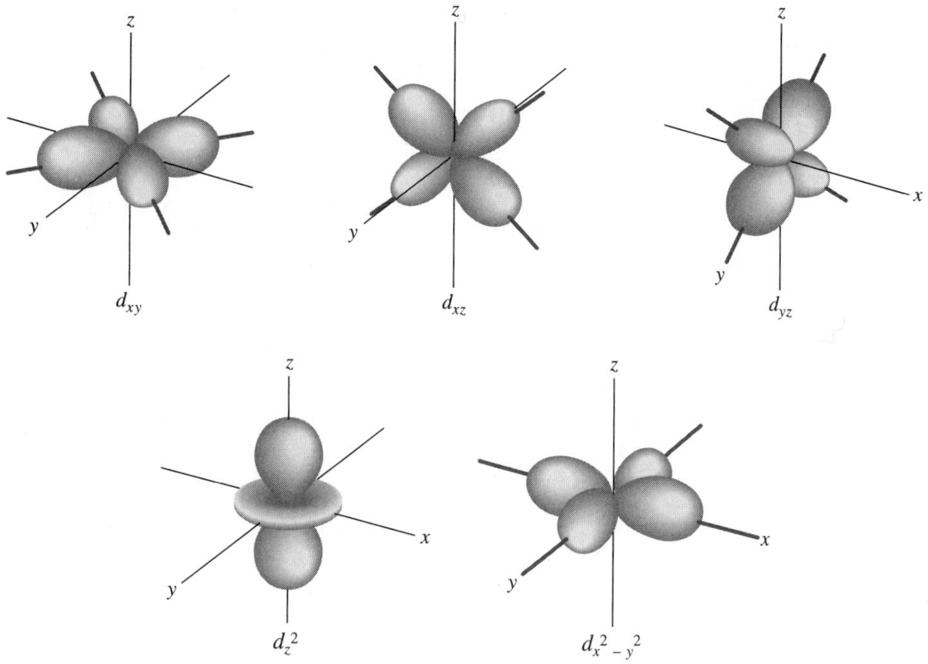

Figure 14-8 Electron-density distribution in the five d orbitals.

of iron(II), the iodide complex of molecular iodine, and the hexacyanoferrate(II)/hexacyanoferrate(III) complex responsible for the color Prussian blue. Figure 14-10 contains spectra for three of these complexes.

In order for a complex to exhibit a charge-transfer spectrum, it is necessary for one of its components to have electron-donor characteristics and for the other component to have electron-acceptor properties. Absorption of radiation then involves transfer of an electron

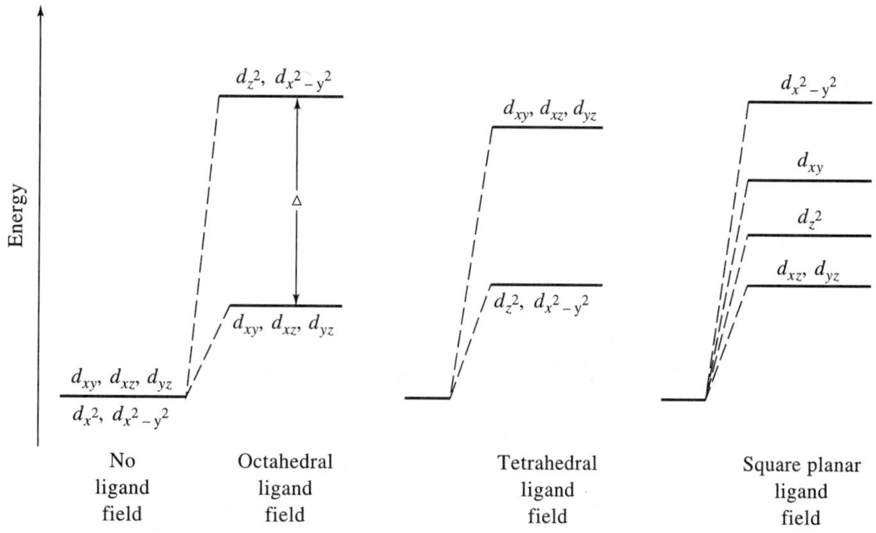

Figure 14-9 Effect of ligand field on d-orbital energies.

TABLE 14-5 Effect of Ligands on Absorption Maxima Associated with $d \rightarrow d$ Transitions

Central Ion	λ_{max}(nm) for the Indicated Ligands				
	Increasing Ligand Field Strength→				
	$6Cl^-$	$6H_2O$	$6NH_3$	3en[a]	$6CN^-$
Cr(III)	736	573	462	456	380
Co(III)	—	538	435	428	294
Co(II)	—	1345	980	909	—
Ni(II)	1370	1279	925	863	—
Cu(II)	—	794	663	610	—

[a]en = ethylenediamine, a bidentate ligand = $NH_2CH_2CH_2NH_2$

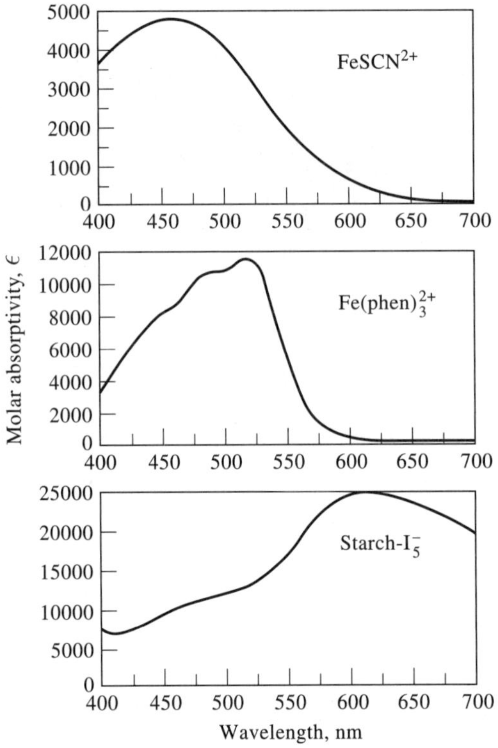

Figure 14-10 Charge-transfer spectra of three species.

from the donor to an orbital that is largely associated with the acceptor. As a consequence, the excited state is the product of a kind of internal oxidation/reduction process. This behavior differs from that of an organic chromophore, where the electron in the excited state is in the *molecular* orbital formed by two or more atoms.

A well-known example of charge-transfer absorption is observed in the iron(III)/thiocyanate complex. Absorption of a photon results in the transfer of an electron from the thiocyanate ion to an orbital associated with the iron(III) ion. The product is thus an excited species consisting of predominantly iron(II) ion and the neutral thiocyanate radical SCN. As with other types of electronic excitation, the electron, under ordinary circumstances, returns to its original state after a brief period. Occasionally, however, dissociation of the excited complex may occur producing photochemical oxidation/reduction products.

In most charge-transfer complexes involving a metal ion, the metal serves as the electron acceptor. An exception is the *o*-phenanthroline complex of iron(II) or copper(I), where the ligand is the acceptor and the metal ion is the donor. Other examples of this type are known.

Organic compounds form many interesting charge-transfer complexes. An example is quinhydrone (a 1:1 complex of quinone and hydroquinone), which exhibits strong absorption in the visible region. Other examples include iodine complexes with amines, aromatics, and sulfides, among others.

14C APPLICATION OF ABSORPTION MEASUREMENT TO QUALITATIVE ANALYSIS

Ultraviolet and visible spectrophotometry have somewhat limited application for qualitative analysis because the number of absorption maxima and minima are relatively few. Thus, unambiguous identification is frequently impossible.

14C-1 Methods of Plotting Spectral Data

Several different types of spectral plots are encountered in qualitative molecular spectroscopy. The ordinate is most commonly percent transmittance, absorbance, log absorbance, or molar absorptivity. The abscissa is usually wavelength or wavenumber, although frequency is occasionally employed. Figure 14-11 shows the effect of analyte concentration on three of the most common types of plots. A comparison of plots (a) and (b) shows that the absorbance plots give a larger difference in

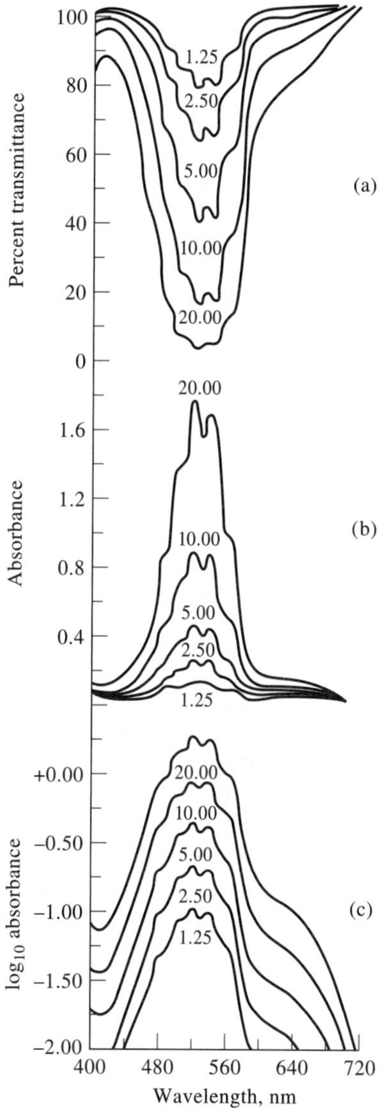

Figure 14-11 Methods for plotting spectral data. The numbers below the curves indicate the concentration in ppm of $KMnO_4$ in the solution; b = 2.00 cm.

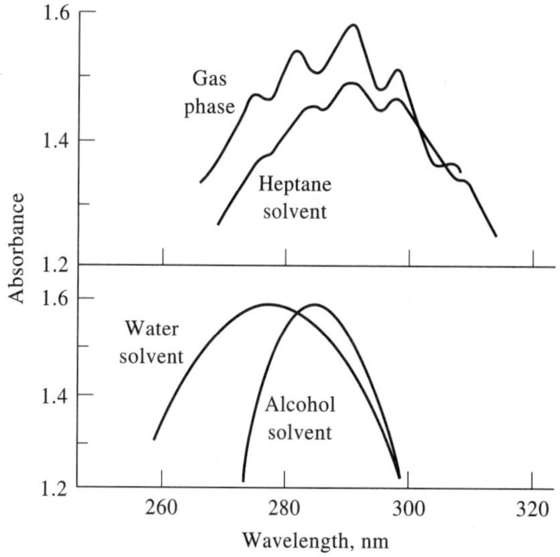

Figure 14-12 Effect of solvent on the absorption spectrum of acetaldehyde.

peak heights as a function of concentration than do transmittance plots. In contrast, the difference between curves is larger at high transmittances. A plot with log absorbance as the ordinate leads to a loss of spectral detail but is convenient for comparing the spectra of solutions of different concentrations because the curves are displaced equally along the vertical axis for equal multiples of concentration.

14C-2 Solvents

In choosing a solvent, consideration must be given not only to its transparency, but also to its possible effects upon the absorbing system. Quite generally, polar solvents such as water, alcohols, esters, and ketones tend to obliterate spectral fine structure arising from vibrational effects; spectra that approach those of the gas phase (see Figure 14-12) are more likely to be observed in nonpolar solvents such as hydrocarbons. In addition, the positions of absorption maxima are influenced by the nature of the solvent. Clearly, the same solvent must be used when comparing absorption spectra for identification purposes.

Table 14-6 lists some common solvents and the approximate wavelength below which they cannot be used because of absorption. These minima depend strongly upon the purity of the solvent.[7] Common solvents for ultraviolet spectrophotometry include water, 95% ethanol, cyclohexane, and 1,4-dioxane. For the visible region, any colorless solvent is suitable.

[7] Most major suppliers of reagent chemicals in the United States offer spectrochemical grades of solvents; these meet or exceed the requirements set forth in *Reagent Chemicals, American Chemical Society Specifications,* 8th ed. Washington, DC: American Chemical Society, 1993. A 1996 supplement to this volume is available as an Adobe Acrobat® PDF file on the World Wide Web at: http://pubs.acs.org/books/reagents/index@.html

14C-3 Detection of Functional Groups

Even though it may not provide the unambiguous identification of an organic compound, an absorption spectrum in the visible and the ultraviolet regions is nevertheless useful for detecting the presence of certain functional groups that act as chromophores.[8] For example, a weak absorption band in the region of 280 to 290 nm, which is displaced toward shorter wavelengths with increased solvent polarity, strongly indicates the presence of the carbonyl group. A weak absorption band at about 260 nm with indications of vibrational fine structure constitutes evidence for the existence of an aromatic ring. Confirmation of the presence of an aromatic amine or a phenolic structure may be obtained by comparing the effects of pH on the spectra of solutions containing the sample with those shown in Table 14-4 for phenol and aniline.

TABLE 14-6 Solvents for the Ultraviolet and the Visible Regions

Solvent	Approximate[a] Transparency Minimum (nm)
Water	190
Ethanol	210
n-Hexane	195
Cyclohexane	210
Benzene	280
Diethyl ether	210
Acetone	330
1,4-Dioxane	220

[a]For 1-cm cells.

14D QUANTITATIVE ANALYSIS BY ABSORPTION MEASUREMENTS

Absorption spectroscopy is one of the most useful and widely used tools available to the chemist for quantitative analysis.[9] Important characteristics of spectrophotometric and photometric methods include: (1) wide applicability to both organic and inorganic systems, (2) typical sensitivities of 10^{-4} to 10^{-5} M (this range can often be extended to 10^{6} to 10^{-7} M by certain modifications)[10], (3) moderate to high selectivity, (4) good accuracy (typically, relative uncertainties of 1% to 3% are encountered, although with special precautions, errors can be reduced to a few tenths of a percent), (5) ease and convenience of data acquisition.

14D-1 Scope

The applications of quantitative, ultraviolet/visible absorption methods not only are numerous, but also touch upon every field in which quantitative chemical information is required. For example, it has been estimated that in the field of health alone, 95% of all quantitative determinations are performed by ultraviolet/visible spectrophotometry and this number represents well over 3,000,000 daily tests carried out in the United States.[11]

The reader can develop a notion of the scope of spectrophotometry by consulting a series of review articles published periodically in *Analytical Chemistry*[12] and from monographs on the subject.[13]

Applications to Absorbing Species

Tables 14-2, 14-3, and 14-4 list many common organic chromophoric groups. Spectrophotometric determination of any organic compound containing one or more

[8] See R. M. Silverstein, G. C. Bassler, and T. C. Morrill, *Spectrometric Identification of Organic Compounds,* 5th ed., Chapter 6. New York: Wiley, 1991.

[9] For a wealth of detailed, practical information on spectrophotometric practices, see *Techniques in Visible and Ultraviolet Spectrometry,* Vol. I, *Standards in Absorption Spectroscopy,* C. Burgess and A. Knowles, Eds. London: Chapman and Hall, 1981; J. R. Edisbury, *Practical Hints on Absorption Spectrometry.* New York: Plenum Press, 1968.

[10] See, for example: T. D. Harris, *Anal. Chem.,* **1982,** *54,* 741A.

[11] R. W. Birke and R. Mavrodineanu, *Accuracy in Analytical Spectrophotometry, NBS Special Publication* 260-81, p. 1. Washington, DC: National Bureau of Standards, 1983.

[12] See J. A. Howell and L. G. Hargis, *Anal. Chem.,* **1996,** *68,* 169R; **1994,** *66,* 445R; **1992,** *64,* 66R; **1990,** *62,* 155R; **1988,** *60,* 131R; **1986,** *58,* 108R; **1984,** *56,* 225R.

[13] For example, see E. B. Sandell and H. Onishi, *Colorimetric Determination of Traces of Metals,* 4th ed. New York: Interscience, 1978; *Colorimetric Determination of Nonmetals,* 2nd ed., D. F. Boltz and J. A. Howell, Eds. New York: Wiley, 1978; Z. Marczenko, *Separation & Spectrophotometric Determination of Elements.* New York: Halsted Press, 1986; M. Pisez and J. Bartos, *Colorimetric and Fluorometric Analysis of Organic Compounds and Drugs.* New York: Marcel Dekker, 1974; F. D. Snell, *Photometric and Fluorometric Method of Analysis,* Parts 1 and 2, "Metals"; Part 3, "Nonmetals." New York: Wiley, 1978–81.

of these groups is potentially feasible; many examples of this type of determination are found in the literature.

A number of inorganic species also absorb and are thus susceptible to direct determination; we have already mentioned the various transition metals. In addition, a number of other species also show characteristic absorption. Examples include nitrite, nitrate, and chromate ions; osmium and ruthenium tetroxides; molecular iodine; and ozone.

Applications to Nonabsorbing Species

Numerous reagents react selectively with nonabsorbing species to yield products that absorb strongly in the ultraviolet or visible regions.[14] The successful application of such reagents to quantitative analysis usually requires that the color-forming reaction be forced to near completion. It should be noted that color-forming reagents are frequently employed as well for the determination of absorbing species, such as transition-metal ions; the molar absorptivity of the product will frequently be orders of magnitude greater than that of the uncombined species.

A host of complexing agents find application for the determination of inorganic species. Typical inorganic reagents include thiocyanate ion for iron, cobalt, and molybdenum; the anion of hydrogen peroxide for titanium, vanadium, and chromium; and iodide ion for bismuth, palladium, and tellurium. Of even more importance are organic chelating agents that form stable, colored complexes with cations. Examples include *o*-phenanthroline for the determination of iron, dimethylglyoxime for nickel, diethyldithiocarbamate for copper, and diphenyldithiocarbazone for lead.

14D-2 Procedural Details

The first steps in a photometric or spectrophotometric analysis involve the establishment of working conditions and the preparation of a calibration curve relating concentration to absorbance.

Selection of Wavelength

Spectrophotometric absorbance measurements are ordinarily made at a wavelength corresponding to an absorption peak because the change in absorbance per unit of concentration is greatest at this point; the maximum sensitivity is thus realized. In addition, the absorption curve is often flat in this region; under these circumstances, good adherence to Beer's law can be expected (Figure 13-5). Finally, the measurements are less sensitive to uncertainties arising from failure to reproduce precisely the wavelength setting of the instrument.

Variables That Influence Absorbance

Common variables that influence the absorption spectrum of a substance include the nature of the solvent, the pH of the solution, the temperature, electrolyte concentrations, and the presence of interfering substances. The effects of these variables must be known; conditions for the analysis must be chosen such that the absorbance will not be materially influenced by small, uncontrolled variations in their magnitudes.

Cleaning and Handling of Cells

Accurate spectrophotometric analysis requires the use of good-quality, matched cells. These should be regularly calibrated against one another to detect differences that can arise from scratches, etching, and wear. Equally important is the use of proper cell cleaning and drying techniques. Erickson and Surles[15] recommend the following cleaning sequence for the outside windows of cells. Prior to measurement, the cell surfaces are cleaned with a lens paper soaked in spectrograde methanol. The paper is held with a hemostat; after wiping, the methanol is allowed to evaporate, leaving the cell surfaces free of contaminants. The authors showed that this method was far superior to the usual procedure of wiping the cell surfaces with a dry lens paper, which apparently leaves lint and films on the surface.

Determination of the Relationship Between Absorbance and Concentration

After deciding upon the conditions for the analysis, it is necessary to prepare a calibration curve from a series of standard solutions that bracket the concentration range expected for the samples. Seldom, if ever, is it safe to assume adherence to Beer's law and use only a single standard to determine the molar absorptivity. The results of an analysis should *never* be based on a literature value for the molar absorptivity.

---

[14] For example, see the first reference in footnote 13.

[15] J. O. Erickson and T. Surles, *Amer. Lab.*, **1976**, *8* (6), 50.

Ideally, calibration standards should approximate the composition of the samples to be analyzed not only with respect to the analyte concentration but also with regard to the concentrations of the other species in the sample matrix in order to minimize the effects of various components of the sample on the measured absorbance. For example, the absorbance of many colored complexes of metal ions is decreased to a varying degree in the presence of sulfate and phosphate ions as a consequence of the tendency of these anions to form colorless complexes with metal ions. The color formation reaction is often less complete as a consequence, and lowered absorbances are the result. The matrix effect of sulfate and phosphate can often be counteracted by introducing into the standards amounts of the two species that approximate the amounts found in the samples. Unfortunately, when complex materials such as soils, minerals, and plant ash are being analyzed, however, preparation of standards that match the samples is often impossible or extremely difficult. When this is the case, the *standard addition method* is often helpful in counteracting matrix effects.

The Standard Addition Method

The standard addition method can take several forms.[16] The one most often chosen for photometric or spectrophotometric analyses, and the one that was discussed in some detail in Section 1E-2, involves adding one or more increments of a standard solution to sample aliquots of the same size. Each solution is then diluted to a fixed volume before measuring its absorbance. Example 1-3 (page 17) illustrates the application of this method to data from the photometric determination of iron(III) based upon the formation of the red thiocyanate complex (see Figure 14-13).

In the interest of saving time or sample, it is possible to perform a standard addition analysis using only two increments of sample. Here, a single addition of V_s mL of standard would be added to one of the two samples. This approach is based upon the equation (see page 18)

$$c_x = \frac{A_1 c_s V_s}{(A_2 - A_1)V_x} \qquad (14\text{-}1)$$

[16] See M. Bader, *J. Chem. Educ.*, **1980**, *57*, 703.

EXAMPLE 14-1

A 2.00-mL urine specimen was treated with reagent to generate a color with phosphate, following which the sample was diluted to 100 mL. To a second 2.00-mL sample was added exactly 5.00 mL of a phosphate solution containing 0.0300 mg phosphate/mL, which was treated in the same way as the original sample. The absorbance of the first solution was 0.428; that of the second was 0.538. Calculate the concentration in milligrams of phosphate per millimeter of the specimen.

Here we substitute into Equation 14-1 and obtain

$$c_x = \frac{0.428 \times 0.0300\ (\text{mg } PO_4^{3-}/\text{mL}) \times 5.00\ \text{mL}}{(0.538 - 0.428) \times 2.00\ \text{mL sample}}$$

$$= 0.292\ \text{mg } PO_4^{3-}/\text{mL sample}$$

Analysis of Mixtures of Absorbing Substances

The total absorbance of a solution at a given wavelength is equal to the sum of the absorbances of the individual components present (Equation 13-9). This relationship makes possible the quantitative determination of the individual constituents of a mixture, even if their spectra overlap. Consider, for example, the spectra of M and N, shown in Figure 14-14. Note that no wavelength exists at which the absorbance of this mixture is due simply to one of the components; thus, determination of either M or N is impossible by a single measurement. However, the absorbances of the mixture at the two wavelengths λ' and λ'' may be expressed as follows:

$$A' = \epsilon'_M b c_M + \epsilon'_N b c_N \qquad (\text{at } \lambda')$$

$$A'' = \epsilon''_M b c_M + \epsilon''_N b c_N \qquad (\text{at } \lambda'')$$

The four molar absorptivities ϵ'_M, ϵ'_N, ϵ''_M, and ϵ''_N can be evaluated from individual standard solutions of M and of N, or better, from the slopes of their Beer's law plots. The absorbances of the mixture, A' and A'', are experimentally measurable, as is b, the cell thickness. Thus, from these two equations, the concentrations of the individual constituents, c_M and c_N, can be calculated. These relationships are valid only if Beer's law is followed and if the two components behave independently of one another. The greatest accuracy in an analysis of this sort is attained by choosing wavelengths at which the differences in molar absorptivities are large.

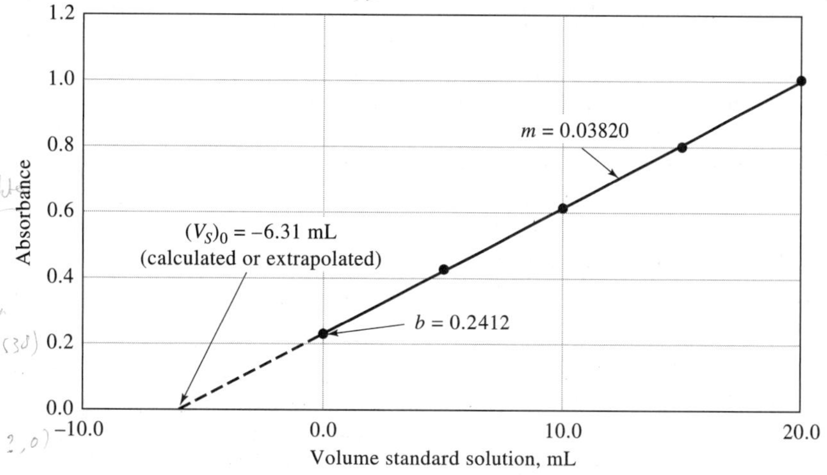

Figure 14-13 Data for standard addition method for the determination of Fe^{3+} as the SCN^- complex.

Mixtures containing more than two absorbing species can be analyzed, in principle at least, if a further absorbance measurement is made for each added component. The uncertainties in the resulting data become greater, however, as the number of measurements increases. Some of the newer computerized spectrophotometers are capable of reducing these uncertainties by overdetermining the system. That is, these instruments use many more data points than unknowns and effectively match the entire spectrum of the unknown as closely as possible by least-squares techniques utilizing the methods of matrix algebra. The spectra for standard solutions of each component are required for the analysis.

14D-3 Derivative and Dual-Wavelength Spectrophotometry[17]

In derivative spectrophotometry, spectra are obtained by plotting the first or a higher order derivative of absorbance or transmittance with respect to wavelength as a function of wavelength. Often these plots reveal spectral detail that is lost in an ordinary spectrum. In addition, concentration measurements of an analyte in the presence of an interference can sometimes be made more easily or more accurately. Unfortunately, the advantages of derivative spectra are at least partially offset by a degradation in signal-to-noise ratio that accompanies obtaining derivatives. In many parts of the ultraviolet and visible regions, however, signal-to-noise ratio is not a serious limiting factor; it is here that deriv-

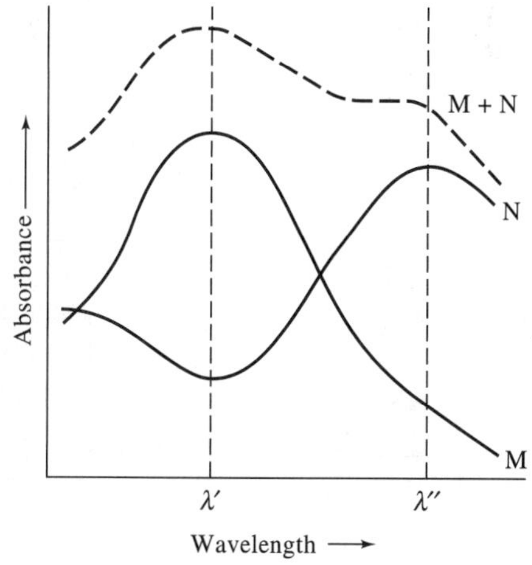

Figure 14-14 Absorption spectrum of a two-component mixture. Optimum wavelengths λ′ and λ″ for the determination of the two components are indicated by the vertical dashed lines.

[17] For additional information, see G. Talsky, *Derivative Spectrophotometry,* New York: VCH Publishers, 1994; T. C. O'Haver, *Anal. Chem.,* **1979,** *51,* 91A; F. Sanchez Rajas, C. Bosch Ojeda, and J. M. Cano Pavon, *Talanta,* **1988,** *35,* 753; J. E. Cahill, *Amer. Lab.,* **1979,** *11* (11), 79; J. E. Cahill and F. G. Padera, *Amer. Lab.,* **1980,** *12* (4), 101; T. J. Porro, *Anal. Chem.,* **1972,** *44* (4), 93A.

ative spectra are of most use. An additional disadvantage of derivative spectrophotometry is that the required equipment is generally more costly.

A variety of methods are used to obtain derivative spectra. For microprocessor-controlled digital spectrophotometers, the differentiation can be performed numerically using derivative least-squares polynomial smoothing, which is discussed in Section 5C-2. With analog instruments, derivatives of spectral data can be obtained by a suitable operational amplifier circuit (see Section 3E-4). A third procedure is accomplished by wavelength modulation.

Wavelength Modulation Devices

Several procedures have been developed for wavelength modulation. In some of these procedures, a wavelength interval of a few nanometers is swept rapidly and repetitively while the spectrum is being scanned in the usual way. The amplitude of the resulting ac signal from the transducer is a good approximation of the derivative of the spectrum with respect to wavelength. The repetitive, rapid sweep has been obtained by any of a number of mechanical means, including oscillation or vibration of a mirror, slit, or dispersing element of a monochromator. A second scheme for wavelength modulation, and one that is offered by several instrument makers, involves the use of dual dispersing systems arranged in such a way that two beams of slightly different wavelengths (typically 1 or 2 nm difference) fall alternatively onto a sample cell and its transducer; no reference beam is used. The ordinate variable is the difference between the alternate signals, which provides a good approximation of the derivative of absorbance as a function of wavelength ($\Delta A/\Delta\lambda$). Figure 14-15 is a schematic of a dual-wavelength instrument. Note that a reference cell is not employed and the sample cell is alternately exposed to radiation from each of the monochromators. Generally, dual-wavelength instruments can also be operated in the single wavelength mode with the radiation passing alternately through a reference and standard cell.

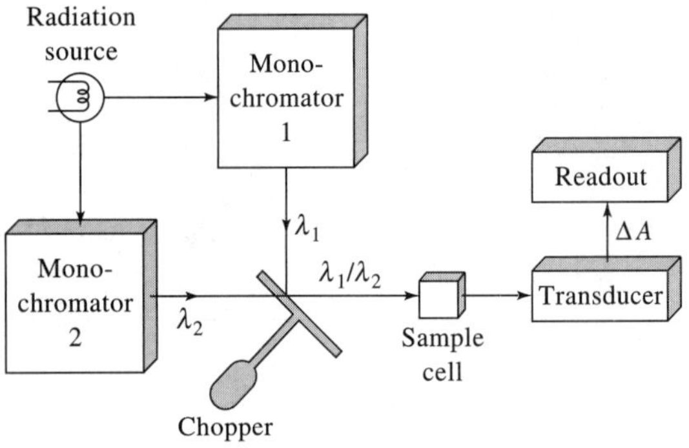

Figure 14-15 Schematic of a dual-wavelength spectrophotometer. The chopper causes λ_1 and λ_2 to pass through the sample alternately.

Applications of Derivative Spectra

Many of the most important applications of derivative spectroscopy in the ultraviolet and visible regions have been for qualitative identification of species in which the enhanced detail of a derivative spectrum makes it possible to distinguish among compounds having overlapping spectra. Figure 14-16 illustrates the way in

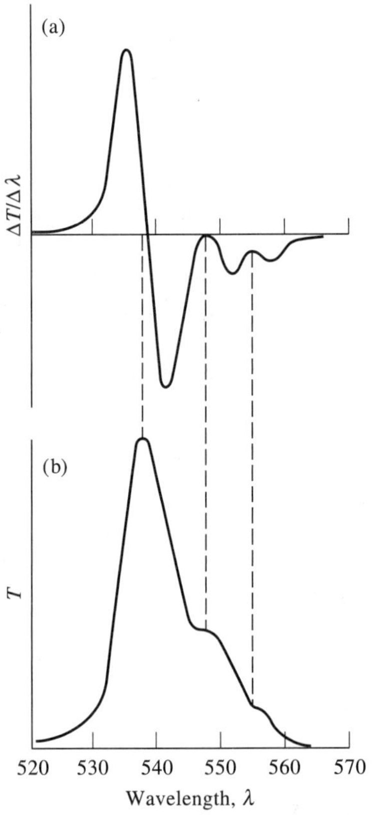

Figure 14-16 Comparison of a derivative spectrum (a) with a standard transmittance spectrum (b).

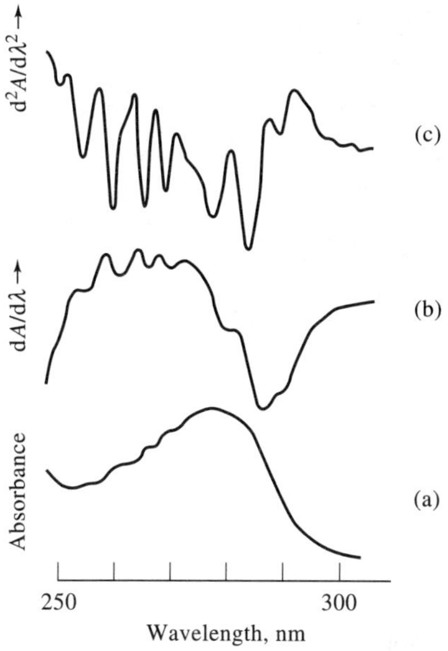

Figure 14-17 Absorption spectra of bovine albumin: (a) ordinary spectrum, (b) first derivative spectrum, (c) second derivative spectrum. *(Reprinted with permission from J. E. Cahill and F. G. Padera,* Amer. Lab., ***1980,*** *12(4), 109. Copyright 1980 by International Scientific Communications, Inc.).*

which a derivative plot can reveal details of a spectrum consisting of three overlapping absorption peaks.

Dual-wavelength spectrophotometry has proven particularly useful for extracting ultraviolet/visible absorption spectra of analytes present in turbid solutions, where light scattering obliterates the details of an absorption spectrum. For example, three amino acids, tryptophan, tyrosine, and phenylalanine, contain aromatic side chains, which exhibit sharp absorption peaks in the 240- to 300-nm range. These sharp peaks are not, however, apparent in spectra of typical protein preparations, such as bovine or egg albumin, because the large protein molecules scatter radiation severely, yielding only a smooth absorption peak such as that shown in Figure 14-17a. As shown in curves (b) and (c), the aromatic fine structure is revealed in first and second derivative spectra.

Dual-wavelength spectrophotometry has also proved useful for determination of an analyte in the presence of a spectral interference. Here, the instrument is operated in the nonscanning mode with absorbances being measured at two wavelengths at which the interference has identical molar absorptivities. In contrast, the analyte must absorb more strongly at one of these wavelengths than the other. The differential absorbance is then directly proportional to the analyte concentration.

14E PHOTOMETRIC TITRATIONS

Photometric or spectrophotometric measurements can be employed to advantage in locating the equivalence point of a titration, provided the analyte, the reagent, or the titration product absorbs radiation.[18] Alternatively, an absorbing indicator can provide the absorbance change necessary for location of equivalence.

14E-1 Titration Curves

A photometric titration curve is a plot of absorbance, corrected for volume changes, as a function of the volume of titrant. If conditions are chosen properly, the curve will consist of two linear regions with differing slopes, one occurring at the beginning of the titration and the other located well beyond the equivalence-point region; the end point is taken as the intersection of the extrapolated linear portions of the curve. Figure 14-18 shows some typical titration curves. Titration of a nonabsorbing species with a colored titrant

[18] For further information concerning this technique, see J. B. Headridge, *Photometric Titrations*. New York: Pergamon Press, 1961; M. A. Leonard, in *Comprehensive Analytical Chemistry,* G. Svehla, Ed., Vol. 8, Chapter 3. New York: Elsevier, 1977.

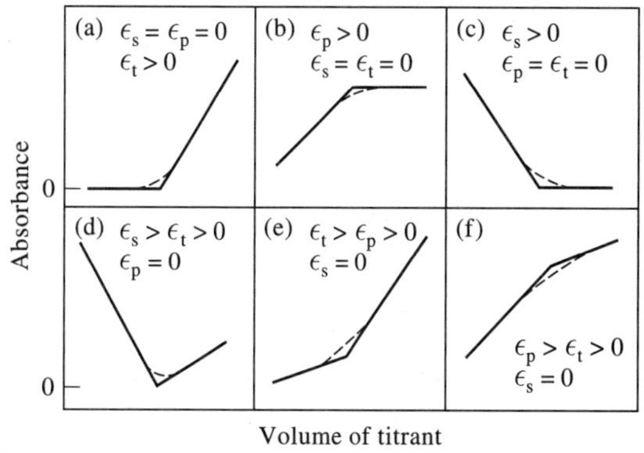

Figure 14-18 Typical photometric titration curves. Molar absorptivities of the substance titrated, the product, and the titrant are given by ϵ_s, ϵ_p, ϵ_t, respectively.

that is decolorized by the reaction produces a horizontal line in the initial stages, followed by a rapid rise in absorbance beyond the equivalence point (Figure 14-18a). The formation of a colored product from colorless reactants, on the other hand, initially produces a linear rise in the absorbance, followed by a region in which the absorbance becomes independent of reagent volume (Figure 14-18b). Depending upon the absorption characteristics of the reactants and the products, the other curve forms shown in Figure 14-18 are also possible.

In order to obtain a satisfactory photometric end point, the absorbing system(s) must obey Beer's law; otherwise, the titration curve will lack the linear regions needed for extrapolation to the end point. Further, it is necessary to correct the absorbance for volume changes; here, the observed values are multiplied by $(V + v)/V$, where V is the original volume of the solution and v is the volume of added titrant. Little error is introduced as a result of the assumption that the volumes V and v are additive.

14E-2 Instrumentation

Photometric titrations are ordinarily performed with a spectrophotometer or a photometer that has been modified to permit insertion of the titration vessel in the light path. Alternatively, a probe-type cell, such as that shown in Figure 13-16 can be employed. After the zero adjustment of the meter scale has been made, radiation is allowed to pass through the solution of the analyte, and the instrument is adjusted by varying the source intensity or the detector sensitivity until a convenient absorbance reading is obtained. Ordinarily, no attempt is made to measure the true absorbance, since relative values are perfectly adequate for the purpose of end-point detection. Data for the titration are then collected without alteration of the instrument setting.

The power of the radiation source and the response of the detector must be reasonably constant during the period required for a photometric titration. Cylindrical containers are ordinarily used, and care must be taken to avoid any movement of the vessel that might alter the length of the radiation path.

Both filter photometers and spectrophotometers have been employed for photometric titrations. The latter are preferred, however, because their narrower bandwidths enhance the probability of adherence to Beer's law.

14E-3 Application of Photometric Titrations

Photometric titrations often provide more accurate results than a direct photometric analysis because the data from several measurements are used to determine the end point. Furthermore, the presence of other absorbing species may not interfere, since only a change in absorbance is being measured.

The photometric end point has the advantage over many other commonly used end points that the experimental data are taken well away from the equivalence-point region. Thus, the titration reactions need not have such favorable equilibrium constants as those required for a titration that depends upon observations near the equivalence point (for example, potentiometric or indicator end points). For the same reason, relatively dilute solutions may be titrated.

The photometric end point has been applied to all types of reactions.[19] Most of the reagents used in oxidation/reduction titrations have characteristic absorption spectra and thus produce photometrically detectable end points. Acid/base indicators have been employed for photometric neutralization titrations. The photometric end point has also been used to great advantage in titrations with EDTA and other complexing agents. Figure 14-19 illustrates the application of this end point to

[19] See, for example, the review: A. L. Underwood, *Advances in Analytical Chemistry and Instrumentation,* C. N. Reilley, Ed., Vol. 3, pp. 31–104. New York: Interscience, 1964.

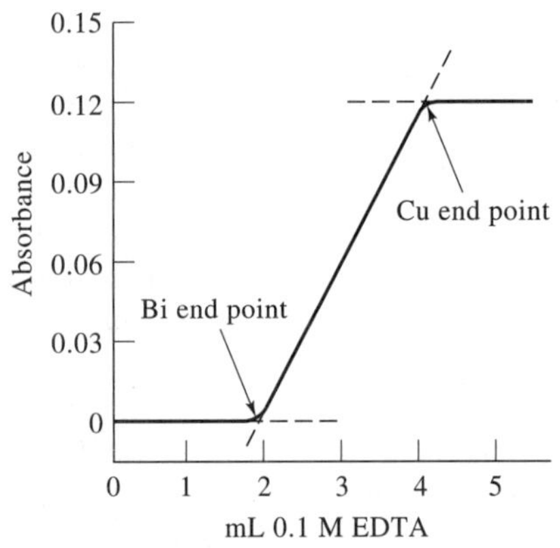

Figure 14-19 Photometric titration curve of 100 mL of a solution that was 2.0×10^{-3} M in Bi^{3+} and Cu^{2+}. Wavelength: 745 nm.

the successive titration of bismuth(III) and copper(II). At 745 nm, neither cation nor the reagent absorbs, nor does the more stable bismuth complex, which is formed in the first part of the titration; the copper complex, however, does absorb. Thus, the solution exhibits no absorbance until essentially all of the bismuth has been titrated. With the first formation of the copper complex, an increase in absorbance occurs. The increase continues until the copper equivalence point is reached. Further reagent additions cause no further absorbance change. Two well-defined end points result.

The photometric end point has also been adapted to precipitation titrations. Here, the suspended solid product has the effect of diminishing the radiant power by scattering; titrations are carried to a condition of constant turbidity.

14F PHOTOACOUSTIC SPECTROSCOPY

Photoacoustic (PAS) or optoacoustic spectroscopy, which was developed in the early 1970s, provides a means for obtaining ultraviolet, visible, and infrared absorption spectra of solids, semisolids, or turbid liquids. Acquisition of spectra for these kinds of samples by ordinary methods is usually difficult at best and often impossible because of light scattering and reflection.

14F-1 The Photoacoustic Effect

Photoacoustic spectroscopy is based upon a light absorption effect that was first investigated in the 1880s by Alexander Graham Bell and others. This effect is observed when a gas in a closed cell is irradiated with a chopped beam of radiation of a wavelength that is absorbed by the gas. The absorbed radiation causes periodic heating of the gas, which in turn results in regular pressure fluctuations within the chamber. If the chopping rate lies in the acoustical frequency range, these pulses of pressure can be detected by a sensitive microphone. The photoacoustic effect has been used since the beginning of the 20th century for the analysis of absorbing gases and has recently taken on new importance for this purpose with the advent of tunable infrared lasers as sources. Of greater importance, however, has been the application of this phenomenon to

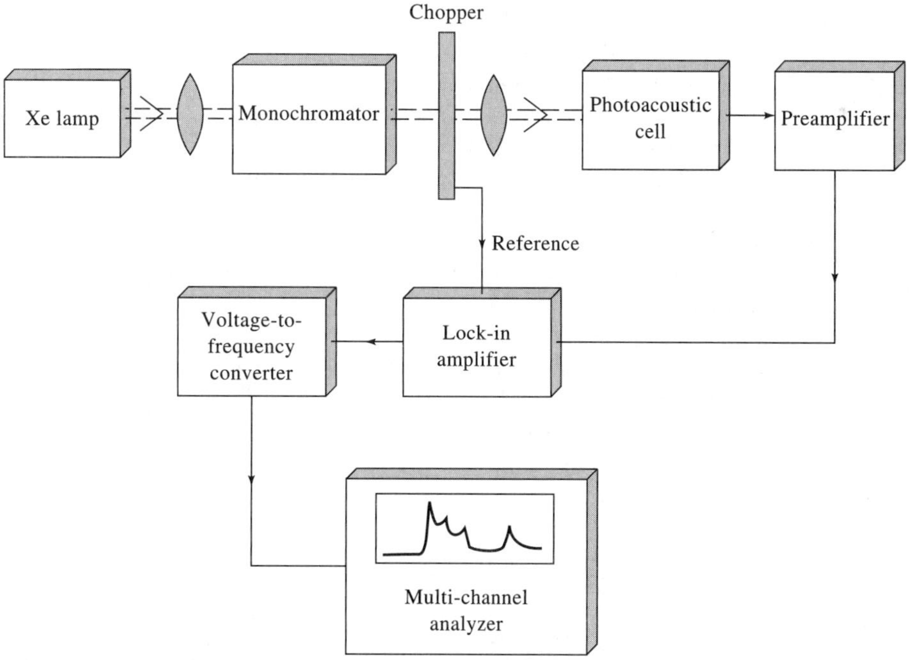

Figure 14-20 Block diagram of a single-beam photoacoustic spectrometer with digital data processing. *(Reprinted with permission from A. Rosencwaig,* Anal. Chem., ***1975,*** 47, *593A. Copyright 1975 American Chemical Society.)*

the derivation of absorption spectra of solids and turbid liquids.[20]

14F-2 Photoacoustic Spectra

In photoacoustic studies of solids, the sample is placed in a closed cell containing air or some other *nonabsorbing* gas and a sensitive microphone. The solid is then irradiated with a chopped beam from a monochromator. The photoacoustic effect is observed *provided the radiation is absorbed by the solid;* the power of the resulting sound is directly related to the extent of absorption. Radiation reflected or scattered by the sample has no effect on the microphone and thus does not interfere. This latter property is perhaps the most important characteristic of the method.

The source of the photoacoustic effect in solids appears to be similar to that in gases. That is, nonradiative relaxation of the absorbing solid causes a periodic heat flow from the solid to the surrounding gas; the resulting pressure fluctuations in the gas are then detected by the microphone.

14F-3 Instruments

Figure 14-20 is a block diagram showing the components of a single-beam photoacoustic spectrometer. In this apparatus, the spectrum from the lamp is first recorded digitally, followed by the spectrum for the sample. The stored lamp data are then used to correct the output from the sample for variations in the lamp output as a function of wavelength. With this technique, it is necessary to assume an absence of drift in the source and detector systems. Double-beam instruments, which largely avoid the drift problem, have also been described. One such instrument is equipped with a pair of matched cells (and transducers), one of which contains the sample and the other a reference material such as finely divided carbon. A commercially available instrument is also based upon the split-beam principle. In this instrument, however, about 8% of the output from the grating monochromator is directed to a pyroelectric transducer, the rest passing into the sample cell. The output from the sample transducer is compared with that from the pyroelectric transducer to produce a spectrum that is corrected for variations in the lamp output as a function of wavelength as well as time.

14F-4 Applications

Figure 14-21 illustrates one application of ultraviolet/visible photoacoustic spectroscopy. Here, photoacoustic spectra of smears of whole blood, blood cells freed of plasma, and hemoglobin extracted from the

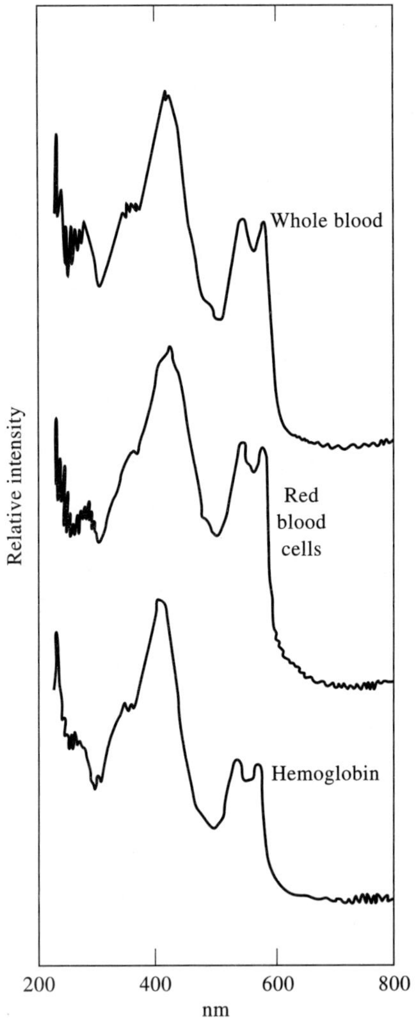

Figure 14-21 Photoacoustic spectra of smears of blood and blood components. *(Reprinted with permission from A. Rosencwaig,* Anal. Chem., *1975,* 47, *596A. Copyright 1975 American Chemical Society.)*

[20] For a review on applications, see A. Rosencwaig, *Anal. Chem.*, **1975**, *47*, 592A; J. W. Lin and L. P. Dubek, *Anal. Chem.*, **1979**, *51*, 1627; J. F. McClelland, *Anal. Chem.*, **1983**, *55*, 89A; D. Betteridge and P. J. Meylor, CRC *Crit. Rev. Anal. Chem.*, **1984**, *14*, 267; A. Rosencwaig, *Photoacoustics and Photoacoustic Spectroscopy.* New York: Wiley, 1980.

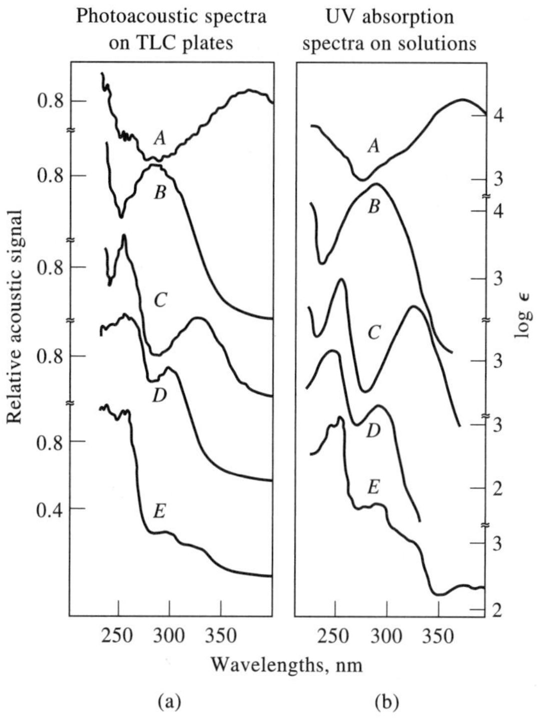

Figure 14-22 Spectra of spots on a thin-layer chromatogram (left) and of solutions of the same compounds (right). The compounds are: *(A) p*-nitroaniline, *(B)* benzylidene acetone, *(C)* salicylaldehyde, *(D)* 1-tetralone, and *(E)* fluorenone. *(Reprinted with permission from A. Rosencwaig,* Anal. Chem., ***1975,*** *47, 600A. Copyright 1975 American Chemical Society.)*

cells are shown. Conventional spectroscopy, even with very dilute solutions of whole blood, does not yield satisfactory spectra because of the strong light-scattering properties of the blood cells, protein, and lipid molecules present. Photoacoustic spectroscopy permits spectroscopic studies of blood without the necessity of a preliminary separation of these large molecules.

Figure 14-22 shows another application of photoacoustic spectroscopy. The five spectra on the left were for five organic compounds that had been separated on thin-layer chromatographic plates (Section 28H). These spectra were taken directly on the thin-layer plates; for comparison, solution spectra are shown on the right. The similarity of the two makes rapid identification of the compound possible.

Other applications of the method have included the study of minerals; semiconductors; natural products, such as marine algae and animal tissue; coatings on surfaces; and catalytic surfaces.

Finally, photoacoustic measurements in the mid-infrared region have proved useful for qualitative identification of components in inorganic and organic solids and in various types of commercial polymers. Usually, Fourier transform techniques are necessary to obtain satisfactory *S/N*. Photoacoustic cells are generally available as accessories for Fourier transform instruments.[21]

[21] For a description of commercially available accessories for photoacoustic FTIR, see D. Noble, *Anal. Chem.,* **1994,** *66,* 757A.

14G QUESTIONS AND PROBLEMS

14-1 A standard solution was put through appropriate dilutions to give the concentrations of iron shown below. The iron(II)/1,10-phenanthroline complex was then developed in 25.0-mL aliquots of these solutions, following which each was diluted to 50.0 mL. The following absorbances were recorded at 510 nm:

Concentration of Fe(II) in the Original Solutions, ppm	Absorbance, *A* (1.00-cm Cells)
2.00	0.164
5.00	0.425
8.00	0.628
12.00	0.951
16.00	1.260
20.00	1.582

(a) Produce a calibration curve from these data.
(b) By the method of least squares (Section a1C, Appendix 1), develop an equation relating absorbance and concentration of iron(II).
(c) Calculate the standard deviation about regression.
(d) Calculate the standard deviation of the slope.

14-2 The method developed in Problem 14-1 was used for the routine determination of iron in 25.0-mL aliquots of ground water. Compute the concentration (as ppm Fe) in samples that yielded the accompanying absorbance data (1.00-cm cell). Estimate standard deviations for the derived concentrations. Repeat the calculations by assuming the absorbance data are means of three measurements.
(a) 0.107 (b) 0.721 (c) 1.538

14-3 Verify the results found in Table 13-2.

14-4 A 25.0-mL aliquot of an aqueous quinine solution was diluted to 50.0 mL and found to have an absorbance of 0.832 at 348 nm when measured in a 2.00-cm cell. A second 25.0-mL aliquot was mixed with 10.00 mL of a solution containing 23.4 ppm of quinine; after dilution to 50.0 mL, this solution had an absorbance of 1.220 (2.00-cm cell). Calculate the concentration in parts per million of quinine in the sample.

14-5 A 5.12-g pesticide sample was decomposed by wet ashing and then diluted to 200.0 mL in a volumetric flask. The analysis was completed by treating aliquots of this solution as indicated.

Volume of Sample Taken, mL	Reagent Volumes Used, mL			Absorbance, *A*, 545 nm (1.00-cm cells)
	3.82 ppm Cu^{2+}	Ligand	H_2O	
50.0	0.00	20.0	30.0	0.512
50.0	4.00	20.0	26.0	0.844

Calculate the percentage of copper in the sample.

14-6 A simultaneous determination for cobalt and nickel can be based upon absorption by their respective 8-hydroxyquinolinol complexes. Molar absorptivities corresponding to their absorption maxima are as follows:

	Molar Absorptivity, ϵ	
	365 nm	700 nm
Co	3529	428.9
Ni	3228	10.2

Calculate the molar concentration of nickel and cobalt in each of the following solutions based upon the accompanying data:

	Absorbance, A (1.00-cm cells)	
Solution	**365 nm**	**700 nm**
(a)	0.598	0.039
(b)	0.902	0.072

14-7 When measured with a 1.00-cm cell, a 8.50×10^{-5} M solution of species A exhibited absorbances of 0.129 and 0.764 at 475 and 700 nm, respectively. A 4.65×10^{-5} M solution of species B gave absorbances of 0.567 and 0.083 under the same circumstances. Calculate the concentrations of A and B in solutions that yielded the following absorbance data in a 1.25-cm cell: (a) 0.502 at 475 nm and 0.912 at 700 nm; (b) 0.675 at 475 nm and 0.696 at 700 nm.

14-8 The acid/base indicator HIn undergoes the following reaction in dilute aqueous solution:

$$\underset{\text{color 1}}{\text{HIn}} \rightleftharpoons \text{H}^+ + \underset{\text{color 2}}{\text{In}^-}$$

The following absorbance data were obtained for a 5.00×10^{-4} M solution of HIn in 0.1 M NaOH and 0.1 M HCl. Measurements were made at a wavelength of 485 nm and 625 nm with 1.00-cm cells.

0.1 M NaOH	$A_{485} = 0.052$	$A_{625} = 0.823$
0.1 M HCl	$A_{485} = 0.454$	$A_{625} = 0.176$

In the NaOH solution, essentially all of the indicator is present as In^-; in the acidic solution, it is essentially all in the form of HIn.

(a) Calculate molar absorptivities for In^- and HIn at 485 and 625 nm.

(b) Calculate the acid dissociation constant for the indicator if a pH 5.00 buffer containing a small amount of the indicator exhibits an absorbance of 0.472 at 485 nm and 0.351 at 625 nm (1.00-cm cells).

(c) What is the pH of a solution containing a small amount of the indicator that exhibits an absorbance of 0.530 at 485 nm and 0.216 at 635 nm (1.00-cm cells)?

(d) A 25.00-mL aliquot of a solution of purified weak organic acid HX required exactly 24.20 mL of a standard solution of a strong base to reach a phenolphthalein end point. When exactly 12.10 mL of the base were added to a second 25.00-mL aliquot of the acid, which contained a small amount of the indicator under consideration, the absorbance was found to be 0.306 at 485 nm and 0.555 at 625 nm (1.00-cm cells). Calculate the pH of the solution and K_a for the weak acid.

(e) What would be the absorbance of a solution at 485 and 625 nm (1.25-cm cells) that was 2.00×10^{-4} M in the indicator and was buffered to a pH of 6.000?

14-9 The absolute error in transmittance for a particular photometer is 0.005 and independent of the magnitude of T. Calculate the percent relative error in concentration that is caused by this source when

(a) $A = 0.585$. (c) $A = 1.800$. (e) $T = 99.25\%$.

(b) $T = 49.6\%$. (d) $T = 0.0592$. (f) $A = 0.0055$.

14-10 Maxima exist at 470 nm in the absorption spectrum for the bismuth(III)/thiourea complex and at 265 nm in the spectrum for the bismuth(III)/EDTA complex. Predict the shape of a curve for the photometric titration of
(a) bismuth(III) with thiourea (tu) at 470 nm.
(b) bismuth(III) with EDTA (H_2Y^{2-}) at 265 nm.
(c) the bismuth(III)/thiourea complex with EDTA at 470 nm. Reaction:

$$Bi(tu)_6^{3+} + H_2Y^{2-} \rightarrow BiY^- + 6tu + 2H^+$$

(d) the reaction in (c) at 265 nm.

14-11 Given the information that

$$Fe^{3+} + Y^{4-} \rightleftharpoons FeY^- \qquad K_f = 1.0 \times 10^{25}$$

$$Cu^{2+} + Y^{4-} \rightleftharpoons CuY^{2-} \qquad K_f = 6.3 \times 10^{18}$$

and the further information that, among the several reactants and products, only CuY^{2-} absorbs at 750 nm, describe how Cu(II) could be used as indicator for the photometric titration of Fe(III) with H_2Y^{2-}. Reaction: $Fe^{3+} + H_2Y^{2-} \rightarrow FeY^- + 2H^+$

14-12 The chelate CuA_2^{2-} exhibits maximum absorption at 480 nm. When the chelating reagent is present in at least a tenfold excess, the absorbance is dependent only upon the analytical concentration of Cu(II) and conforms to Beer's law over a wide range. A solution in which the analytical concentration of Cu^{2+} is 2.30×10^{-4} M and that for A^{2-} is 8.60×10^{-3} M has an absorbance of 0.690 when measured in a 1.00-cm cell at 480 nm. A solution in which the analytical concentrations of Cu^{2+} and A^{2-} are 2.30×10^{-4} M and 5.00×10^{-4} M, respectively, has an absorbance of 0.540 when measured under the same conditions. Use this information to calculate the formation constant K_f for the process

$$Cu^{2+} + 2A^{2-} \rightleftharpoons CuA_2^{2-}$$

14-13 Mixture of the chelating reagent B with Ni(II) gives rise to formation of the highly colored NiB_2^{2+}, solutions of which obey Beer's law over a wide range. Provided the analytical concentration of the chelating reagent exceeds that of Ni(II) by a factor of 5 (or more), the cation exists, within the limits of observation, entirely in the form of the complex. Use the accompanying data to evaluate the formation constant K_f for the process

$$Ni^{2+} + 2B \rightleftharpoons NiB_2^{2+}$$

Analytical Concentration, M		
Ni^{2+}	**B**	**Absorbance, A, 395 nm (1.00-cm cells)**
2.50×10^{-4}	2.20×10^{-1}	0.765
2.50×10^{-4}	1.00×10^{-3}	0.360

Chapter 15

Molecular Luminescence Spectrometry

Three related types of optical methods are considered in this chapter, namely ***molecular fluorescence, phosphorescence,*** *and* ***chemiluminescence.*** *In each, molecules of the analyte are excited to give a species whose emission spectrum provides information for qualitative or quantitative analysis. The methods are known collectively as* ***molecular luminescence*** *procedures.*

Fluorescence and phosphorescence are alike in that excitation is brought about by absorption of photons. As a consequence, the two phenomena are often referred to by the more general term ***photoluminescence.*** *As will be shown later, fluorescence differs from phosphorescence in that the electronic energy transitions responsible for the fluorescence do not involve a change in electron spin. As a consequence, fluorescence is short-lived, with luminescence ceasing almost immediately ($<10^{-5}$ s). In contrast, a change in electron spin accompanies phosphorescence emissions, which causes the radiation to endure for an easily detectable time after termination of irradiation—often several seconds or longer. In most instances, photoluminescence emission, be it fluorescence or phosphorescence, is longer in wavelength than the radiation used for its excitation.*

The third type of luminescence, chemiluminescence, is based upon the emission spectrum of an excited species that is formed in the course of a chemical reaction. In some instances, the excited species is the product of a reaction between the analyte and a suitable reagent (usually a strong oxidant such as ozone or hydrogen peroxide); the result is a spectrum characteristic of the ***oxidation product*** *of the analyte or the reagent rather than the analyte itself. In other instances, the analyte is not directly involved in the chemiluminescence reaction; instead, the inhibiting effect or the catalytic effect of the analyte on a chemiluminescence reaction serves as the analytical parameter.*

Measurement of the intensity of photoluminescence or chemiluminescence permits the quantitative determination of a variety of important inorganic and organic species in trace amounts. At the present time, the number of fluorometric methods is significantly larger than the number of applications of phosphorescence and chemiluminescence procedures.

One of the most attractive features of luminescence methods is their inherent sensitivity, with detection limits often being one to three orders of magnitude smaller than those encountered in absorption spectroscopy. Typical detection limits are in the parts per billion range.

Another advantage of photoluminescence methods is their large linear concentration ranges, which also are often significantly greater than those encountered in absorption methods. Because of their high sensitivity, quantitative luminescence methods are often subject to serious interference effects from sample matrices. For this reason luminescence measurements are often combined with the superb separation techniques of chromatography and electrophoresis. Fluorescence detectors are particularly valuable as detectors for liquid chromatography (Chapter 28) and capillary electrophoresis (Chapter 30) because of their unusual sensitivity.

Generally, luminescence methods are less widely applicable for quantitative analyses than absorption methods because many more species absorb ultraviolet/visible radiation than exhibit photoluminescence upon absorption of radiation in this region of the spectrum.[1]

15A THEORY OF FLUORESCENCE AND PHOSPHORESCENCE

Fluorescence occurs in simple as well as in complex gaseous, liquid, and solid chemical systems. The simplest kind of fluorescence is that exhibited by dilute atomic vapors, which was described in Chapter 9. For example, the 3*s* electrons of vaporized sodium atoms can be excited to the 3*p* state by absorption of radiation of wavelengths 5896 and 5890 Å. After 10^{-5} to 10^{-8} s, the electrons return to the ground state, and in so doing emit radiation of the same two wavelengths in all directions. This type of fluorescence, in which the absorbed radiation is reemitted without a change in frequency, is known as *resonance radiation* or *resonance fluorescence.*

Many molecular species also exhibit resonance fluorescence. Much more often, however, molecular fluorescence (or phosphorescence) bands are found centered at wavelengths that are longer than the resonance line. This shift toward longer wavelengths, or lower energies, is termed the *Stokes shift.*

15A-1 Excited States Producing Fluorescence and Phosphorescence

The characteristics of fluorescence and phosphorescence spectra can be rationalized by means of the simple molecular orbital considerations described in Section 14B-1. However, an understanding of the difference between the two photoluminescence phenomena requires a review of *electron spin* and *single/triplet excited states.*

Electron Spin

The Pauli exclusion principle states that no two electrons in an atom can have the same set of four quantum numbers. This restriction requires that no more than two electrons can occupy an orbital and furthermore the two must have opposed spin states. Under this circumstance, the spins are said to be paired. Because of spin pairing, most molecules exhibit no net magnetic field and are thus said to be *diamagnetic*—that is, they are neither attracted nor repelled by static magnetic fields. In contrast, free radicals, which contain unpaired electrons, have a magnetic moment and consequently are attracted into a magnetic field; free radicals are thus said to be *paramagnetic.*

Singlet/Triplet Excited States

A molecular *electronic* state in which all electron spins are paired is called a *singlet* state, and no splitting of electronic energy levels occurs when the molecule is exposed to a magnetic field. The ground state for a free radical, on the other hand, is a *doublet* state because the odd electron can assume two orientations in a magnetic field, which imparts slightly different energies to the system.

When one of a pair of electrons of a molecule is excited to a higher energy level, either a singlet or a *triplet* state is formed. In the excited singlet state, the spin of the promoted electron is still paired with the ground-state electron; in the triplet state, however, the spins of the two electrons have become unpaired and are thus parallel. These states can be represented as follows where the arrows represent the direction of spin:

---

[1] For further discussion of the theory and applications of fluorescence, phosphorescence, and luminescence, see *Molecular Luminescence Spectroscopy,* S. Schulman, Ed. New York: Wiley, Part 1, 1985; Part 2, 1988, Part 3, 1993; E. L. Wehry, in *Physical Methods of Chemistry,* 2nd ed., Volume VIII, Chapter 3, B. W. Rossiter and R. C. Baetzold, Eds. New York: Wiley, 1993; J. R. Lakowicz, *Principles of Fluorescence Spectroscopy.* New York: Plenum Press, 1983; G. G. Guilbault, *Practical Fluorescence,* 2nd ed. New York: Marcel Dekker, 1990.

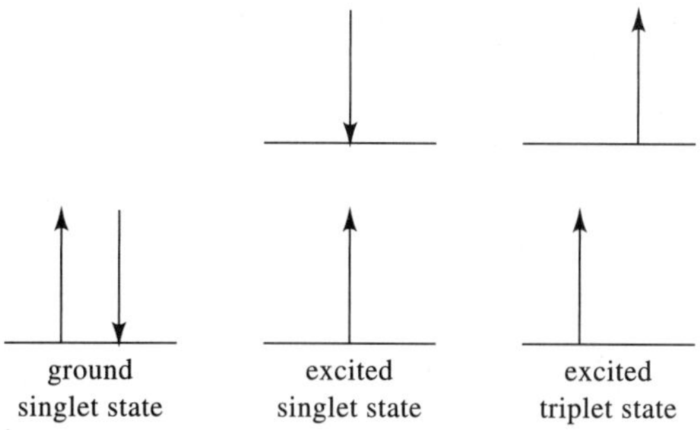

The nomenclature of singlet, doublet, and triplet derives from spectroscopic *multiplicity* considerations, which need not concern us here. Note that the excited triplet state is less energetic than the corresponding excited singlet state.

The properties of a molecule in the excited triplet state differ significantly from those of the excited singlet state. For example, a molecule is paramagnetic in the triplet state and diamagnetic in the singlet. More important, however, is the fact that a singlet/triplet transition (or the reverse), which also involves a change in electronic state, is a significantly less probable event than the corresponding singlet/singlet transition. As a consequence, the average lifetime of an excited triplet state may range from 10^{-4} to several seconds, as compared with an average lifetime of 10^{-5} to 10^{-8} s for an excited singlet state. Furthermore, radiation-induced excitation of a ground-state molecule to an excited triplet state has a low probability of occurring, and absorption peaks due to this process are several orders of magnitude less intense than the analogous singlet/singlet transition. We shall see, however, that an excited triplet state can be populated from an *excited* singlet state of certain molecules; the ultimate consequence of this process is often phosphorescence.

Energy-Level Diagrams for Photoluminescent Molecules

Figure 15-1 is a partial energy-level diagram for a typical photoluminescent molecule. The lowest heavy horizontal line represents the ground-state energy of the molecule, which is normally a singlet state and is labeled S_0. At room temperature, this state represents the energies of essentially all of the molecules in a solution.

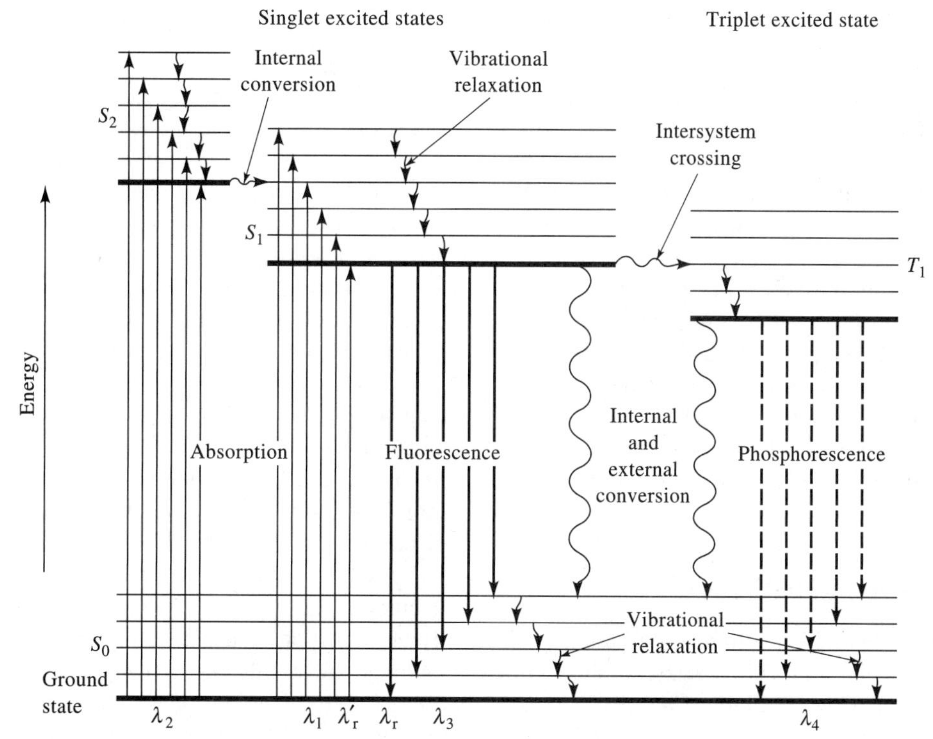

Figure 15-1 Partial energy diagram for a photoluminescent system.

The upper heavy lines are energy levels for the ground vibrational states of three excited electronic states. The two lines on the left represent the first (S_1) and second (S_2) electronic *singlet* states. The one on the right (T_1) represents the energy of the first electronic *triplet* state. As is normally the case, the energy of the first excited triplet state is lower than the energy of the corresponding singlet state.

Numerous vibrational energy levels are associated with each of the four electronic states, as suggested by the lighter horizontal lines.

As shown in Figure 15-1, excitation of this molecule can be brought about by absorption of two bands of radiation, one centered about the wavelength $\lambda_1(S_0 \rightarrow S_1)$ and the second around the shorter wavelength $\lambda_2(S_0 \rightarrow S_2)$. Note that the excitation process results in conversion of the molecule to any of the several excited vibrational states. Note also that direct excitation to the triplet state is not shown because this transition does not occur to any significant extent since this process involves a change in multiplicity, an event which, as we have mentioned, has a low probability of occurrence (a low-probability transition of this type is called *forbidden*).

15A-2 Rates of Absorption and Emission

The rate at which a photon of radiation is absorbed is enormous, the process requiring on the order of 10^{-14} to 10^{-15} s. Fluorescence emission, on the other hand, occurs at a significantly slower rate. Here, the lifetime of the excited state is inversely related to the molar absorptivity of the absorption peak corresponding to the excitation process. Thus, for molar absorptivities in the 10^3 to 10^5 range, lifetimes of excited states are 10^{-7} to 10^{-9} s. For weakly absorbing systems, where the probability of the transition process is smaller, lifetimes may be as long as 10^{-6} to 10^{-5} s. As we have noted, the average rate of a triplet to singlet transition is less than a corresponding singlet to singlet transition. Thus, phosphorescence emission requires times in the range of 10^{-4} to 10 s or more.

15A-3 Deactivation Processes

An excited molecule can return to its ground state by a combination of several mechanistic steps. As shown by the straight vertical arrows in Figure 15-1, two of these steps, fluorescence and phosphorescence, involve the emission of a photon of radiation. The other deactivation steps, indicated by wavy arrows, are radiationless processes. The favored route to the ground state is the one that minimizes the lifetime of the excited state. Thus, if deactivation by fluorescence is rapid with respect to the radiationless processes, such emission is observed. On the other hand, if a radiationless path has a more favorable rate constant, fluorescence is either absent or less intense.

Photoluminescence is limited to a relatively small number of systems incorporating structural and environmental features that cause the rate of radiationless relaxation or deactivation processes to be slowed to a point where the emission reaction can compete kinetically. Information concerning emission processes is sufficiently complete to permit a quantitative accounting of their rates. Understanding of other deactivation routes, however, is rudimentary at best; for these processes, only qualitative statements or speculations about rates and mechanism can be put forth. Nevertheless, the interpretation of photoluminescence requires consideration of these other routes.

Vibrational Relaxation

As shown in Figure 15-1, a molecule may be promoted to any of several vibrational levels during the electronic excitation process. In solution, however, the excess vibrational energy is immediately lost as a consequence of collisions between the molecules of the excited species and those of the solvent; the result is an energy transfer and a minuscule increase in temperature of the solvent. This relaxation process is so efficient that the average lifetime of a *vibrationally* excited molecule is 10^{-12} s or less, a period significantly shorter than the average lifetime of an *electronically* excited state. As a consequence, fluorescence from solution, when it occurs, always involves a transition *from the lowest vibrational level of an excited electronic state.* Several closely spaced peaks are produced, however, since the electron can return *to any one of the vibrational levels of the ground state* (Figure 15-1), whereupon it will rapidly fall to the lowest vibrational level of the ground electronic state by further vibrational relaxation.

A consequence of the efficiency of vibrational relaxation is that the fluorescence band for a given electronic transition is displaced toward lower frequencies or longer wavelengths from the absorption band (the Stokes shift); overlap occurs only for the resonance peak involving transitions between the lowest vibrational level of the ground state and the corresponding level of an excited state. In Figure 15-1, the wavelength of absorbed radiation that produces the resonance peak λ_r is labeled λ_r'.

Internal Conversion

The term *internal conversion* describes intermolecular processes by which a molecule passes to a lower energy *electronic* state without emission of radiation. These processes are neither well defined nor well understood, but it is apparent that they are often highly efficient, because relatively few compounds exhibit fluorescence.

Internal conversion appears to be particularly efficient when two electronic energy levels are sufficiently close for there to be an overlap in vibrational energy levels. This situation is depicted for the two excited singlet states in Figure 15-1. At the overlaps shown, the potential energies of the two excited states are identical; this equality apparently permits an efficient transition. Internal conversion through overlapping vibrational levels is usually more probable than the loss of energy by fluorescence from a higher excited state. Thus, referring again to Figure 15-1, excitation by the band of radiation labeled λ_2 usually produces a fluorescence band centered at wavelength λ_3 to the exclusion of a band that would result from a transition between S_2 and S_0. Here, the excited molecule proceeds from the higher electronic state to the lowest vibrational state of the lower electronic excited state via a series of vibrational relaxations, an internal conversion, and then further relaxations. Under these circumstances, the fluorescence occurs at λ_3 *only,* regardless of whether radiation of wavelength λ_1 or λ_2 was responsible for the excitation. Quinine provides a classical example of this type of behavior (see Problem 15-11); this naturally occurring substance possesses two analytically useful excitation bands, one centered at 250 nm and the other at 350 nm. Regardless of which wavelength is used to excite the molecule, however, the wavelength of maximum emission is 450 nm (see Figure 15-2).

The mechanisms of the internal conversion process $S_1 \rightarrow S_0$ shown in Figure 15-1 are not well understood. The vibrational levels of the ground state may overlap those of the first excited electronic state; under such circumstances, deactivation will occur rapidly by the mechanism just described. This situation prevails with aliphatic compounds, for example, and accounts for the fact that these species seldom fluoresce; in this class of compounds, deactivation by energy transfer through overlapping vibrational levels occurs so rapidly that fluorescence does not have time to occur.

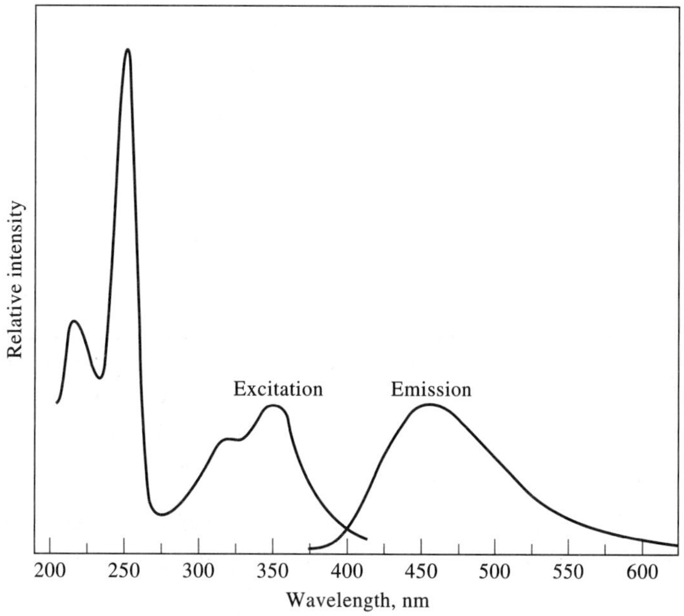

Figure 15-2 Fluorescence excitation and emission spectra for a solution of quinine.

Internal conversion may also result in the phenomenon of *predissociation.* Here, the electron moves from a higher electronic state to an upper vibrational level of a lower electronic state in which the vibrational energy is great enough to cause rupture of a bond. In a large molecule, there is an appreciable probability for the existence of bonds with strengths less than the electronic excitation energy of the chromophores. Rupture of these bonds can occur as a consequence of absorption by the chromophore followed by internal conversion of the electronic energy to vibrational energy associated with the weak bond.

Predissociation should be differentiated from *dissociation,* in which the absorbed radiation excites the electron of a chromophore directly to a sufficiently high vibrational level to cause rupture of the chromophoric bond; no internal conversion is involved. Dissociation processes also compete with the fluorescence process.

External Conversion

Deactivation of an excited electronic state may involve interaction and energy transfer between the excited molecule and the solvent or other solutes. These processes are called collectively *external conversion,* or *collisional quenching.* Evidence for external conversion includes the marked effect upon fluorescence intensity exerted by the solvent; furthermore, those conditions that tend to reduce the number of collisions between particles (low temperature and high viscosity) generally lead to enhanced fluorescence. The details of external conversion processes are not well understood.

Radiationless transitions to the ground state from the lowest excited singlet and triplet states (Figure 15-1) probably involve external conversions, as well as internal conversions.

Intersystem Crossing

Intersystem crossing is a process in which the spin of an excited electron is reversed, and a change in multiplicity of the molecule results. As with internal conversion, the probability of this transition is enhanced if the vibrational levels of the two states overlap. The singlet/triplet transition shown in Figure 15-1 is an example; here, the lowest singlet vibrational state overlaps one of the upper triplet vibrational levels and a change in spin state is thus more probable.

Intersystem crossing is most common in molecules that contain heavy atoms, such as iodine or bromine (the *heavy-atom effect*). Spin/orbital interactions become large in the presence of such atoms, and a change in spin is thus more favorable. The presence of paramagnetic species such as molecular oxygen in solution also enhances intersystem crossing and a consequent decrease in fluorescence.

Phosphorescence

Deactivation of electronic excited states may also involve phosphorescence. After intersystem crossing to an triplet state, further deactivation can occur either by internal or external conversion or by phosphorescence. A triplet→singlet transition is much less probable than a singlet/singlet conversion; thus, as has been noted, the average lifetime of the excited triplet state with respect to emission ranges from 10^{-4} to 10 s or more. Thus, emission from such a transition may persist for some time after irradiation has been discontinued.

External and internal conversions compete so successfully with phosphorescence that this kind of emission is ordinarily observed only at low temperatures, in highly viscous media, or by molecules that are adsorbed on solid surfaces.

15A-4 Variables That Affect Fluorescence and Phosphorescence

Both molecular structure and chemical environment influence whether a substance will or will not luminesce; these factors also determine the intensity of emission when luminescence does occur. The effects of some of these variables are considered briefly in this section.

Quantum Yield

The *quantum yield,* or *quantum efficiency,* for fluorescence or phosphorescence is simply the ratio of the number of molecules that luminesce to the total number of excited molecules. For a highly fluorescent molecule such as fluorescein, the quantum efficiency under some conditions approaches unity. Chemical species that do not fluoresce appreciably have efficiencies that approach zero.

Figure 15-1 and our discussion of deactivation processes suggest that the fluorescence quantum yield ϕ for a compound is determined by the relative rate constants k_x for the processes by which the lowest excited singlet state is deactivated—namely, fluorescence (k_f), intersystem crossing (k_i), external conversion (k_{ec}), internal conversion (k_{ic}), predissociation (k_{pd}), and dissociation (k_d). We may express these relationships by the equation

$$\phi = \frac{k_f}{k_f + k_i + k_{ec} + k_{ic} + k_{pd} + k_d} \qquad (15\text{-}1)$$

where the k terms are the respective rate constants for the several processes enumerated above.

Equation 15-1 permits a qualitative interpretation of many of the structural and environmental factors that influence fluorescence intensity. Those variables that lead to high values for the fluorescence rate constant k_f and low values for the other k terms enhance fluorescence. The magnitude of k_f, the predissociation rate constant k_{pd}, and the dissociation rate constant k_d are mainly dependent upon chemical structure; the remaining constants are strongly influenced by environment and to a somewhat lesser extent by structure.

Transition Types in Fluorescence

It is important to note that fluorescence seldom results from absorption of ultraviolet radiation of wavelengths lower than 250 nm because such radiation is sufficiently energetic to cause deactivation of the excited states by predissociation or dissociation. For example, 200-nm radiation corresponds to about 140 kcal/mol; most organic molecules have at least some bonds that can be ruptured by energies of this magnitude. As a consequence, fluorescence due to $\sigma^* \rightarrow \sigma$ transitions is seldom observed; instead, such emission is confined to the less energetic $\pi^* \rightarrow \pi$ and $\pi^* \rightarrow n$ processes (see Figure 14-3, page 331, for the relative energies associated with these transitions).

As we have noted, an electronically excited molecule ordinarily returns to its *lowest excited state* by a series of rapid vibrational relaxations and internal conversions that produce no emission of radiation. Thus, fluorescence most commonly arises from a transition from the lowest vibrational level of the first excited electronic state to one of the vibrational levels of the electronic ground state. For the majority of fluorescent compounds, then, radiation is produced by either an $n \rightarrow \pi^*$ or a $\pi \rightarrow \pi^*$ transition, depending upon which of these is the less energetic.

Quantum Efficiency and Transition Type

It is observed empirically that fluorescence is more commonly found in compounds in which the lowest energy transition is of a $\pi \rightarrow \pi^*$ type than in compounds in which the lowest energy transition is of the $n \rightarrow \pi^*$ type; that is, the quantum efficiency is greater for $\pi^* \rightarrow \pi$ transitions.

The greater quantum efficiency associated with the $\pi \rightarrow \pi^*$ state can be rationalized in two ways. First, the molar absorptivity of a $\pi \rightarrow \pi^*$ transition is ordinarily 100- to 1000-fold greater than for an $n \rightarrow \pi^*$ process, and this quantity represents a measure of transition probability in either direction. Thus, the inherent lifetime associated with a $\pi \rightarrow \pi^*$ transition is shorter (10^{-7} to 10^{-9} s compared with 10^{-5} to 10^{-7} s for an $n \rightarrow \pi^*$ transition) and k_f in Equation 15-1 is larger.

Thermodynamic considerations suggest that the rate constant for intersystem crossing k_i is smaller for $\pi \rightarrow \pi^*$ excited states because the energy difference between the singlet/triplet states is larger; that is, more energy is required to unpair the electrons of the π^* excited state. As a consequence, overlap of triplet vibrational levels with those of the singlet state is less, and the probability of an intersystem crossing is smaller.

In summary, then, fluorescence is more commonly associated with $\pi \rightarrow \pi^*$ transitions because such transitions exhibit shorter average lifetimes (k_f is larger) and because the deactivation processes that compete with fluorescence are less likely to occur.

Fluorescence and Structure

The most intense and the most useful fluorescence is found in compounds containing aromatic functional groups with low-energy $\pi \rightarrow \pi^*$ transition levels. Compounds containing aliphatic and alicyclic carbonyl structures or highly conjugated double-bond structures may also exhibit fluorescence, but the number of these is small compared with the number in the aromatic systems.

Most unsubstituted aromatic hydrocarbons fluoresce in solution, the quantum efficiency usually increasing with the number of rings and their degree of condensation. The simple heterocyclics, such as pyridine, furan, thiophene, and pyrrole,

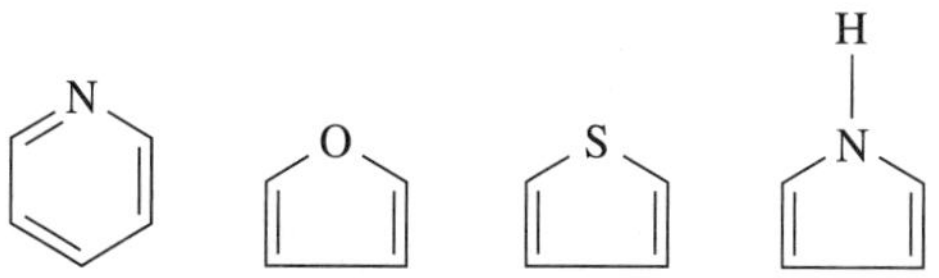

pyridine furan thiophene pyrrole

do not exhibit fluorescence; on the other hand, fused-ring structures ordinarily do. With nitrogen heterocyclics, the lowest-energy electronic transition is believed to involve an $n \rightarrow \pi^*$ system that rapidly converts to the triplet state and prevents fluorescence. Fusion of benzene rings to a heterocyclic nucleus, however, results in an increase in the molar absorptivity of the absorption peak. The lifetime of an excited state is shorter in such structures; fluorescence is thus observed for compounds such as quinoline, isoquinoline, and indole.

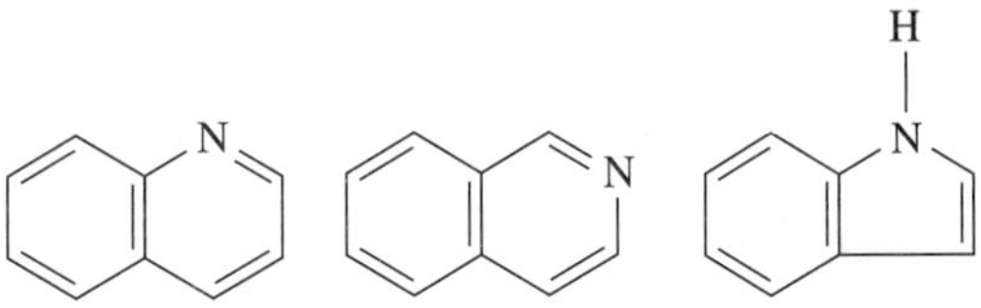

quinoline isoquinoline indole

Substitution on the benzene ring causes shifts in the wavelength of absorption maxima and corresponding changes in the fluorescence peaks. In addition, substitution frequently affects the fluorescence efficiency; some of these effects are illustrated by the data for benzene derivatives in Table 15-1.

The influence of halogen substitution is striking; the decrease in fluorescence with increasing atomic number of the halogen is thought to be due in part to the heavy atom effect (page 360), which increases the probability for intersystem crossing to the triplet state. Predissociation is thought to play an important role in iodobenzene and in nitro derivatives as well; these compounds have easily ruptured bonds that can absorb the excitation energy following internal conversion.

Substitution of a carboxylic acid or carbonyl group on an aromatic ring generally inhibits fluorescence. In these compounds, the energy of the $n \rightarrow \pi^*$ transition is less than in the $\pi \rightarrow \pi^*$ transition; as we have pointed out earlier, the fluorescence yield from the former type of system is ordinarily low.

TABLE 15-1 Effect of Substitution on the Fluorescence of Benzene[a]

Compound	Formula	Wavelength of Fluorescence, nm	Relative Intensity of Fluorescence
Benzene	C_6H_6	270–310	10
Toluene	$C_6H_5CH_3$	270–320	17
Propylbenzene	$C_6H_5C_3H_7$	270–320	17
Fluorobenzene	C_6H_5F	270–320	10
Chlorobenzene	C_6H_5Cl	275–345	7
Bromobenzene	C_6H_5Br	290–380	5
Iodobenzene	C_6H_5I	—	0
Phenol	C_6H_5OH	285–365	18
Phenolate ion	$C_6H_5O^-$	310–400	10
Anisole	$C_6H_5OCH_3$	285–345	20
Aniline	$C_6H_5NH_2$	310–405	20
Anilinium ion	$C_6H_5NH_3^+$	—	0
Benzoic acid	C_6H_5COOH	310–390	3
Benzonitrile	C_6H_5CN	280–360	20
Nitrobenzene	$C_6H_5NO_2$	—	0

[a]In ethanol solution.

Effect of Structural Rigidity

It is found empirically that fluorescence is particularly favored in molecules that possess rigid structures. For example, the quantum efficiencies for fluorene and biphenyl are nearly 1.0 and 0.2, respectively, under similar conditions of measurement. The difference in behavior appears to be largely

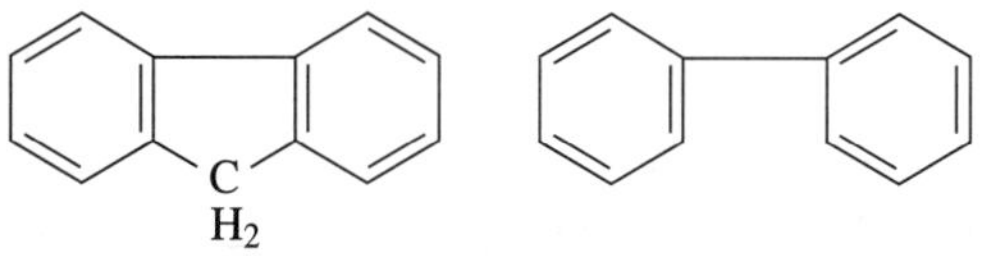

fluorene biphenyl

a result of the increased rigidity furnished by the bridging methylene group in fluorene. Many similar examples can be cited.

The influence of rigidity has also been invoked to account for the increase in fluorescence of certain organic chelating agents when they are complexed with a metal ion. For example, the fluorescence intensity of 8-hydroxyquinoline is much less than that of the zinc complex:

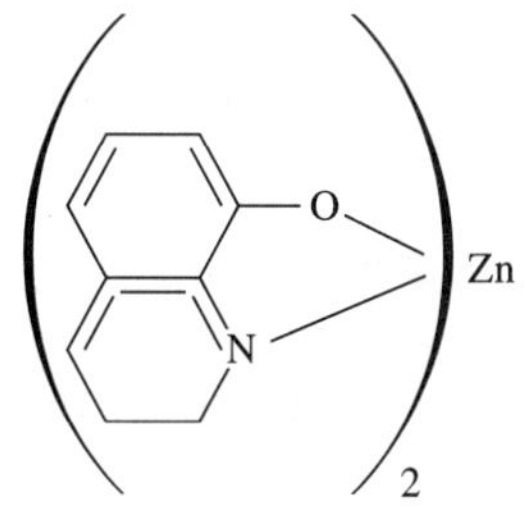

Lack of rigidity in a molecule probably causes an enhanced internal conversion rate (k_{ic} in Equation 15-1) and a consequent increase in the likelihood for radiationless deactivation. One part of a nonrigid molecule can undergo low-frequency vibrations with respect to its other parts; such motions undoubtedly account for some energy loss.

Temperature and Solvent Effects

The quantum efficiency of fluorescence in most molecules decreases with increasing temperature because the increased frequency of collisions at elevated temperatures improves the probability for deactivation by external conversion. A decrease in solvent viscosity also increases the likelihood of external conversion and leads to the same result.

The fluorescence of a molecule is decreased by solvents containing heavy atoms or other solutes with such atoms in their structure; carbon tetrabromide and ethyl iodide are examples. The effect is similar to that which occurs when heavy atoms are substituted into fluorescing compounds; orbital spin interactions result in an increase in the rate of triplet formation and a corresponding decrease in fluorescence. Compounds containing heavy atoms are frequently incorporated into solvents when enhanced phosphorescence is desired.

Effect of pH on Fluorescence

The fluorescence of an aromatic compound with acidic or basic ring substituents is usually pH-dependent. Both the wavelength and the emission intensity are likely to be different for the ionized and nonionized forms of the compound. The data for phenol and aniline shown in Table 15-1 illustrate this effect. The changes in emission of compounds of this type arise from the differing number of resonance species that are associated with the acidic and basic forms of the molecules. For example, aniline has several resonance forms while anilinium has but one. That is,

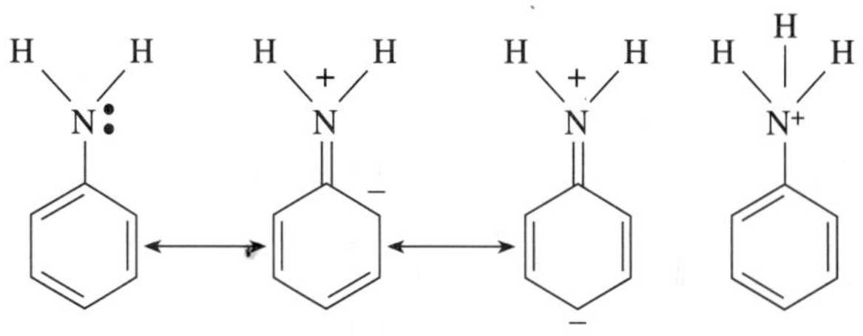

resonance forms of aniline anilinium ion

The additional resonance forms lead to a more stable first excited state; fluorescence in the ultraviolet region is the consequence.

The fluorescence of certain compounds as a function of pH has been used for the detection of end points in acid/base titrations. For example, fluorescence of the phenolic form of 1-naphthol-4-sulfonic acid is not detectable by the eye since it occurs in the ultraviolet region. When the compound is converted to the phenolate ion by the addition of base, however, the emission peak shifts to visible wavelengths where it can readily be seen. It is of interest that this change occurs at a different pH than would be predicted from the acid dissocia-

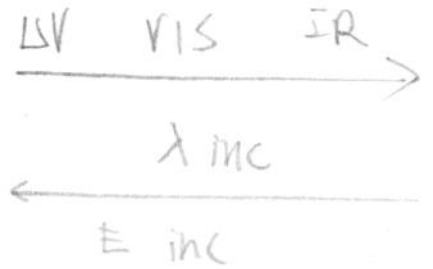

tion constant for the compound. The explanation of this discrepancy is that the acid dissociation constant for the *excited* molecule differs from that for the same species in its ground state. Changes in acid or base dissociation constants with excitation are common and are occasionally as large as four or five orders of magnitude.

These observations suggest that analytical procedures based on fluorescence frequently require close control of pH.

Effect of Dissolved Oxygen

The presence of dissolved oxygen often reduces the intensity of fluorescence in a solution. This effect may be the result of a photochemically induced oxidation of the fluorescing species. More commonly, however, the quenching takes place as a consequence of the paramagnetic properties of molecular oxygen, which promotes intersystem crossing and conversion of excited molecules to the triplet state. Other paramagnetic species also tend to quench fluorescence.

Effect of Concentration on Fluorescence Intensity

The power of fluorescence emission F is proportional to the radiant power of the excitation beam that is absorbed by the system. That is,

$$F = K'(P_0 - P) \tag{15-2}$$

where P_0 is the power of the beam incident upon the solution and P is its power after traversing a length b of the medium. The constant K' depends upon the quantum efficiency of the fluorescence process. In order to relate F to the concentration c of the fluorescing species, we write Beer's law in the form

$$\frac{P}{P_0} = 10^{-\epsilon bc} \tag{15-3}$$

where ϵ is the molar absorptivity of the fluorescing molecules and ϵbc is the absorbance A. By substitution of Equation 15-3 into Equation 15-2, we obtain

$$F = K'P_0(1 - 10^{-\epsilon bc}) \tag{15-4}$$

The exponential term in Equation 15-4 can be expanded as a Maclaurin series to

$$F = K'P_0\left[2.303\epsilon bc - \frac{(2.303\epsilon bc)^2}{2!} + \frac{(2.303\epsilon bc)^3}{3!} \cdots\right] \tag{15-5}$$

Provided $2.303\epsilon bc = A < 0.05$, all of the subsequent terms in the brackets are small with respect to the first; under these conditions, the maximum relative error caused by dropping all but the first term is 0.13%. Thus, we may write

$$F = 2.3\,K'\epsilon bcP_0 \tag{15-6}$$

or at constant P_0,

$$F = Kc \tag{15-7}$$

Thus, a plot of the fluorescence power of a solution versus concentration of the emitting species should be linear at low concentrations. When c becomes great enough so that the absorbance is larger than about 0.05, the higher order terms in Equation 15-5 become important and linearity is lost; F then lies below an extrapolation of the straight-line plot.

Two other factors, also responsible for negative departures from linearity at high concentration, are *self-quenching* and *self-absorption*. The former is the result of collisions between excited molecules. Radiationless transfer of energy occurs, perhaps in a fashion analogous to the transfer to solvent molecules that occurs in an external conversion. Self-quenching can be expected to increase with concentration because of the greater probability of collisions occurring.

Self-absorption occurs when the wavelength of emission overlaps an absorption peak; fluorescence is then decreased as the emission traverses the solution and is reabsorbed by other fluorescent molecules.

The effects of these phenomena are such that a plot of fluorescence versus concentration may exhibit a maximum.

15A-5 Emission and Excitation Spectra

Figure 15-3 shows three types of photoluminescence spectra for phenanthrene. An *excitation* spectrum (E in the figure) is obtained by measuring luminescence intensity at a fixed wavelength while the excitation wavelength is varied. Because the first step in generating fluorescence emission is absorption of radiation to create excited states, an excitation spectrum is essentially identical to an absorbance spectrum taken under the same conditions. Fluorescence and phosphorescence spectra (F and P, respectively), on the other hand, involve excitation at a fixed wavelength while recording the emission intensity as a function of wavelength.

As has been pointed out earlier, photoluminescence usually occurs at wavelengths that are longer than the

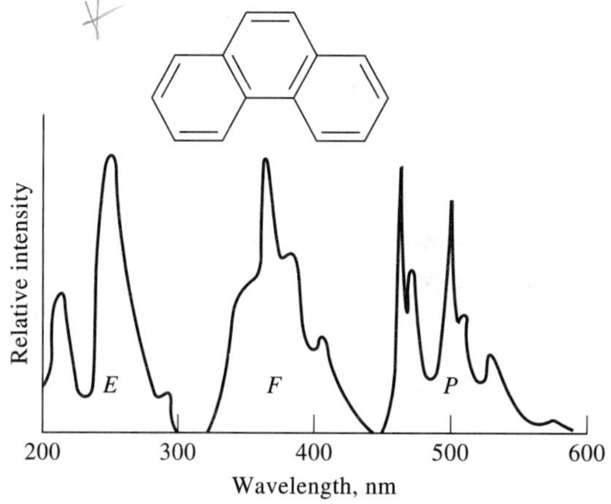

Figure 15-3 Spectra for phenanthrene: *E*, excitation; *F*, fluorescence; *P*, phosphorescence. *(From W. R. Seitz, in* Treatise on Analytical Chemistry, *2nd ed., P. J. Elving, E. J. Meehan, and I. M. Kolthoff, Eds., Part I, Vol. 7, p. 169. New York: Wiley, 1981. Reprinted by permission of John Wiley & Sons, Inc.)*

excitation wavelength. Furthermore, phosphorescence bands are generally found at longer wavelengths than fluorescence bands because the excited triplet state is, in most instances, lower in energy than the corresponding singlet state. In fact, the wavelength difference between the two provides a convenient measure of energy difference between triplet and singlet states.

15B INSTRUMENTS FOR MEASURING FLUORESCENCE AND PHOSPHORESCENCE

The various components of instruments for measuring photoluminescence are similar to those found in ultraviolet/visible photometers or spectrophotometers. Figure 15-4 shows a typical configuration for these components in *fluorometers* and *spectrofluorometers.* Nearly all fluorescence instruments employ double-beam optics as shown in order to compensate for fluctuations in the power of the source. The sample beam first passes through an excitation filter or a monochromator, which transmits radiation that will excite fluorescence but excludes or limits radiation of the wavelength of the fluorescence emission. Fluorescence is propagated from the sample in all directions but is most conveniently observed at right angles to the excitation beam; at other angles, increased scattering from the solution and the cell walls may cause large errors in the measurement of intensity. The emitted radiation reaches a phototransducer after passing through a second filter or monochromator that isolates the fluorescence for measurement.

The reference beam passes through an attenuator that reduces its power to approximately that of the fluorescence radiation (the power reduction is usually by a factor of 100 or more). The signals from the reference and sample photomultiplier tubes are then fed into a difference amplifier whose output is displayed by a meter or recorder. Some fluorescence instruments are of the null type, this state being achieved by optical or electrical attenuators. Newer instruments use an analog divider circuit or digital data acquisition followed by data processing to complete the ratio of the fluorescence emission intensity to the excitation source intensity.

The sophistication, performance characteristics, and costs of fluorometers and spectrofluorometers differ as widely as do the corresponding instruments for absorption measurements. Fluorometers are analogous to absorption photometers in that filters are employed to restrict the wavelengths of the excitation and emission beams. Spectrofluorometers are of two types. The first utilizes a suitable filter to limit the excitation radiation and a grating or prism monochromator to produce the fluorescence spectrum. Several commercial spectrophotometers can be purchased with adapters that permit their use as spectrofluorometers.

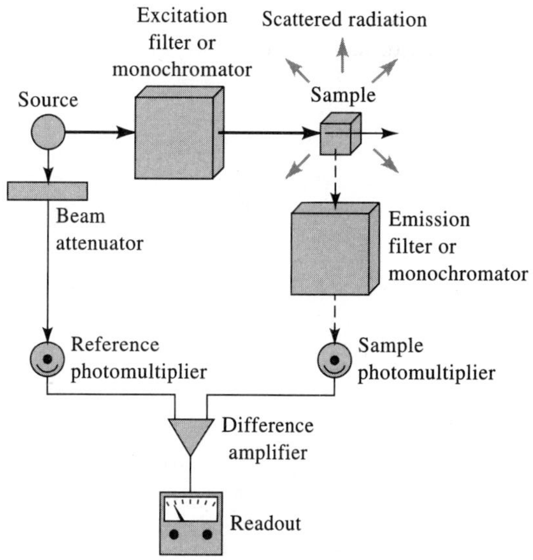

Figure 15-4 Components of a fluorometer or a spectrofluorometer.

True spectrofluorometers are specialized instruments equipped with two monochromators, one of which permits variation in the wavelength of excitation and the other of which allows production of a fluorescence emission spectrum. Figure 15-5a shows an excitation spectrum for anthracene in which the fluorescence emission was measured at a fixed wavelength, while the excitation wavelength was varied. With suitable corrections for variations in source output intensity and detector response as a function of wavelength, an absolute excitation spectrum is obtained that closely resembles an absorption spectrum.

Figure 15-5b is the fluorescence emission spectrum for anthracene; here, the excitation wavelength was held constant while scanning the emission wavelengths. These two spectra bear an approximate mirror image relationship to one another because the vibrational energy differences for the ground and excited electronic states are roughly the same (see Figure 15-1).

The selectivity provided by spectrofluorometers is of prime importance to investigations concerned with the electronic and structural characteristics of molecules and is of value in both qualitative and quantitative analytical work. For concentration measurements, however, the information provided by simpler instruments is often entirely satisfactory. Indeed, relatively inexpensive fluorometers, which have been designed specifically to meet the measurement problems peculiar to fluorescence methods, are frequently as specific and selective as modified absorption spectrophotometers.

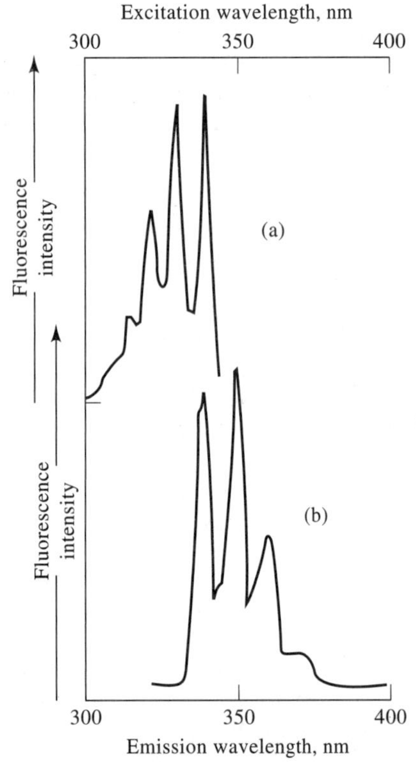

Figure 15-5 Fluorescence spectra for 1 ppm anthracene in alcohol: (a) excitation spectrum; (b) emission spectrum.

The discussion that follows is largely focused on the simpler instruments for fluorescence analysis.

15B-1 Components of Fluorometers and Spectrofluorometers

The components of fluorometers and spectrofluorometers differ only in detail from those of photometers and spectrophotometers; we need to consider only these differences.

Sources

In most applications, a more intense source is needed than the tungsten or hydrogen lamp used for the measurement of absorption. Indeed, as shown by Equation 15-6, the magnitude of the output signal, and thus the sensitivity, is directly proportional to the source power P_0.

Lamps. The most common source for filter fluorometers is a low-pressure mercury vapor lamp equipped with a fused silica window. This source produces useful lines exciting fluorescence at 254, 302, 313, 546, 578, 691, and 773 nm. Individual lines can be isolated with suitable absorption or interference filters. Since fluorescence can be induced in most fluorescing compounds by a variety of wavelengths, at least one of the mercury lines ordinarily proves suitable.

For spectrofluorometers, where a source of continuum radiation is required, a 75- to 450-W high-pressure xenon arc lamp is commonly employed. Such lamps require a large power supply capable of producing direct currents of 5 to 20 A at 15 to 30 V. The spectrum from a xenon arc lamp is a continuum from about 300 to 1300 nm. The spectrum approximates that of a blackbody (see Figure 6-18, page 134). In some instruments, a capacitor is discharged through the lamp at a constant frequency to provide flashes that are regularly spaced in time; higher intensity pulses of light are realized in this way. In addition, the outputs of the transducers are then ac signals, which can be readily amplified and processed.

Lasers. Beginning in the 1970s, various types of lasers were also used as excitation sources for photoluminescence measurements. Of particular interest are tunable dye lasers pumped by a pulsed nitrogen gas laser or Nd:YAG laser.

Most commercial spectrofluorometers utilize lamp sources because they are less expensive and less troublesome to use. Laser sources, however, offer significant advantages in certain instances: for example, (1) when samples are very small, as in microbore chromatography and capillary electrophoresis where the amount of sample is a microliter or less; (2) in remote sensing, as in fluorometric detection of hydroxyl radicals in the atmosphere or of chlorophyll in bodies of water, where the collimated nature of laser beams is vital; or (3) when highly monochromatic excitation is needed to minimize the effects of fluorescing interferences.

Filters and Monochromators

Both interference and absorption filters have been used in fluorometers for wavelength selection of both the excitation beam and the resulting fluorescence radiation. Most spectrofluorometers are equipped with at least one and sometimes two grating monochromators.

Transducers

The typical luminescence signal is of low intensity; large amplifier gains are thus required for its measurement. Photomultiplier tubes are the most common transducers in sensitive fluorescence instruments. Often these are operated in the photon counting mode to give improved signal-to-noise ratios (see page 178). Cooling of transducers is also used sometimes to improve signal-to-noise ratios. Diode-array and charge-transfer detectors have also been proposed for spectrofluorometry.[2] This type of transducer permits the rapid recording of both excitation and emission spectra and is particularly useful in chromatography and electrophoresis.

Cells and Cell Compartments

Both cylindrical and rectangular cells fabricated of glass or silica are employed for fluorescence measurements. Care must be taken in the design of the cell compartment to reduce the amount of scattered radiation reaching the detector. Baffles are often introduced into the compartment for this purpose. Even more than in absorbance measurements, it is important to avoid fingerprints on cells because skin oils often fluoresce.

TF Jan 31.05 12:15 am

15B-2 Instrument Designs

Fluorometers

Filter photometers provide a relatively simple, low-cost way of performing quantitative fluorescence analyses. As noted earlier, either absorption or interference filters are used to limit the wavelengths of the excitation and emitted radiation. Generally, fluorometers are compact, rugged, and easy to use.

Figure 15-6 is a schematic of a typical filter fluorometer that utilizes a mercury lamp for fluorescence excitation and a pair of photomultiplier tubes as transducers. The source beam is split near the source into a reference beam and a sample beam. The reference beam is attenuated by the aperture disk so that its intensity is roughly the same as the fluorescence intensity. Both beams pass through the primary filter with the reference beam then being reflected to the reference photomultiplier tube. The sample beam is focused on the sample by a pair of lenses and causes emission of fluorescence. The emitted radiation passes through a second filter and then is focused on the second photomultiplier tube. The electrical outputs from the two transducers are fed into an analog divider to compute the ratio of the sample to reference intensities, which serves as the analytical variable.

The instrument just described is representative of the dozen or more fluorometers available commercially. Some of these are simpler single-beam instruments. The cost of such fluorometers ranges from a few hundred dollars to about $5000.

Spectrofluorometers

Several instrument manufacturers offer spectrofluorometers capable of providing both excitation and emission spectra. The optical design of one of these, which employs two grating monochromators, is shown in Figure 15-7. Radiation from the first monochromator is split, part passing to a reference photomultiplier and part to the sample. The resulting fluorescence radiation, after dispersion by the second monochromator, is detected by a second photomultiplier.

[2] See Y. Talmi, *Appl. Spectrosc.*, **1982**, *36*, 1; P. M. Epperson, R. D. Jalkian, and M. B. Denton, *Anal. Chem.*, **1989**, *61*, 282.

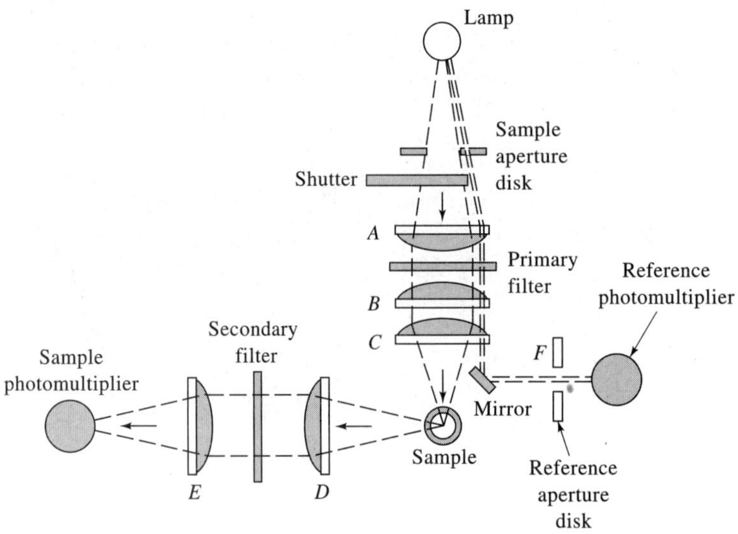

Figure 15-6 A typical fluorometer. *(Courtesy of Farrand Optical Co., Inc.)*

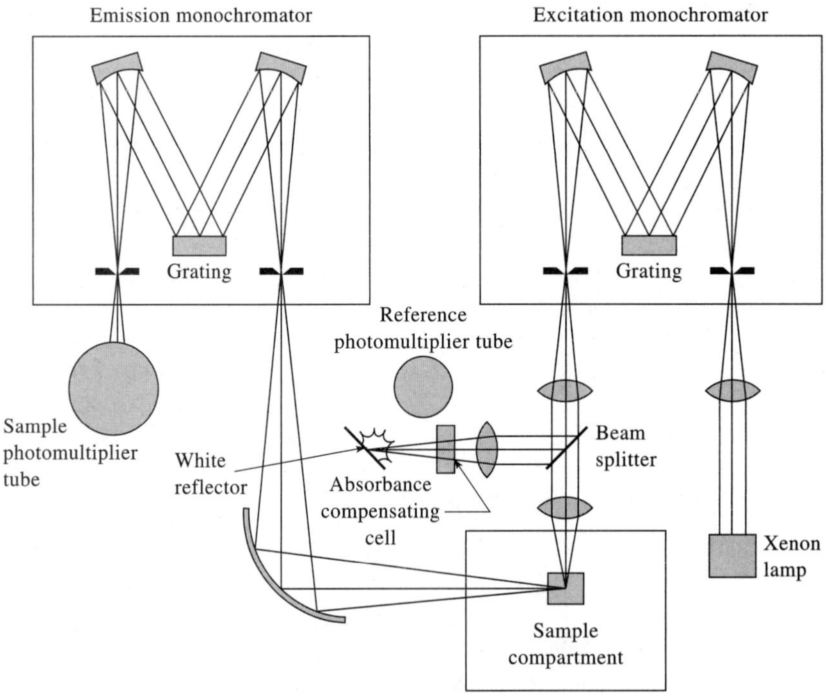

Figure 15-7 A spectrofluorometer. *(Courtesy of SLM Instruments, Inc., Urbana, IL.)*

An instrument such as that shown in Figure 15-7 provides perfectly satisfactory spectra for quantitative analysis. The emission spectra obtained will not, however, necessarily compare well with spectra from other instruments because the output depends not only upon the intensity of fluorescence but also upon the characteristics of the lamp, transducer, and monochromators. All of these instrument characteristics vary with wavelength and differ from instrument to instrument. A number of methods have been developed for obtaining a *corrected* spectrum, which is the true fluorescence spectrum freed from instrumental effects; many of the newer and more sophisticated commercial instruments provide a means for obtaining corrected spectra directly.[3]

Spectrofluorometers Based on Array Transducers

During the past two decades, a number of spectrofluorometers utilizing diode-array and charge-transfer devices have been described that permit fluorescence spectra to be obtained in fractions of a second.[4] The principle of the most sophisticated of these is illustrated in Figure 15-8a. Here, the length of a sample cell is irradiated with an excitation beam that has been dispersed along the *xy* plane by a monochromator that has been rotated 90 deg with respect to its exit slit. The transducer is a two-dimensional charge-coupled device that sees the dispersed excitation radiation in the *xy* plane and the dispersed radiation from the emission mono-

[3] For a summary of correction methods, see N. Wotherspoon, G. K. Oster, and G. Oster, *Physical Methods of Chemistry,* A. Weissberger and B. R. Rossiter, Eds., Vol. 1, Part III B, pp. 460–462 and pp. 473–478. New York: Wiley-Interscience, 1972.

[4] See, for example, Y. Talmi, D. C. Baker, J. R. Jadamec, and W. A. Saner, *Anal. Chem.,* **1978,** *50,* 936A; D. W. Johnson, J. B. Callis, and G. D. Christian, *Anal. Chem.,* **1977,** *49,* 747A; P. M. Epperson, R. D. Jalkian, and M. B. Denton, *Anal. Chem.,* **1989,** *61,* 282.

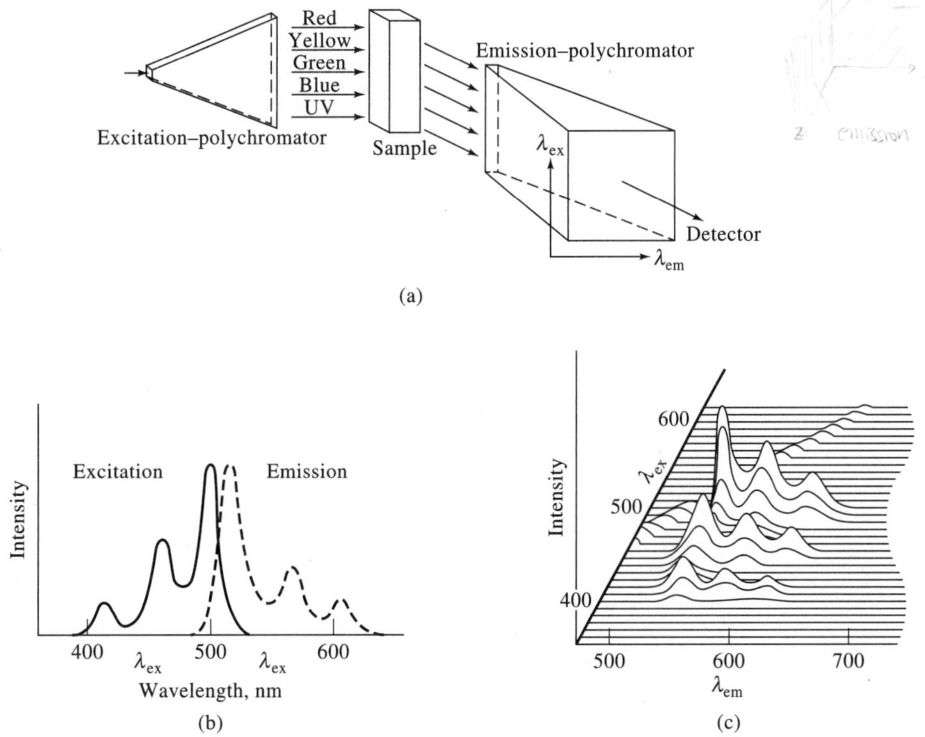

Figure 15-8 (a) Schematic of an optical system for obtaining a total luminescence spectrum with a two-dimensional charge-coupled device. *(With permission from G. W. Suter, A. J. Kallir, and U. P. Wild,* Chimia, ***1983,*** *37, 413.)* (b) Excitation and emission spectra of a hypothetical compound. (c) Total luminescence spectrum of compound in b. *(With permission from D. W. Johnson, J. P. Callis, and G. C. Christian,* Anal. Chem., ***1977,*** *49, 749A.)*

chromator in the *yz* plane. Figure 15-8b shows traditional excitation and emission spectra for a hypothetical molecular species. Figure 15-8c shows the *total luminescence spectrum* for this compound, which is an isometric projection, sometimes called a *stack plot,* of the complete excitation and emission spectra of the compound obtained with the arrangement shown in Figure 15-8a. Total spectra of this type can be obtained in a few seconds or less and are of particular use for analyzing mixtures of fluorescing species. advantage

Jan31.05 12:45am

Fiber Optic Fluorescence Sensors

Fiber optics has been used to demonstrate that several fluorescence analyses can be carried out at various locations well away from a source and a detector. Here, radiation from a laser source travels through an optical fiber and excites fluorescence in sample solutions. Fluorescence emission then travels back through the same fiber to a detector for measurement. The applicability of this type of device has been extended to nonfluorescing analytes by fixing a fluorescing indicator material to the end of the fiber.[5] A discussion of the properties of fiber optics and applications to chemical instrumentation is found in Section 7G.

Phosphorimeters

Instruments that have been used for studying phosphorescence are similar in design to the fluorometers and spectrofluorometers just considered except that two additional components are required.[6] The first is a device that will alternately irradiate the sample and, after a suitable time delay, measure the intensity of phosphorescence. The time delay is required to differentiate between long-lived phosphorescence emission and short-lived fluorescence emission, both of which would originate from the same sample. Both mechanical and electronic devices are used, and many commercial fluorescence instruments have accessories for phosphorescence measurements. An example of one type of mechanical device is shown in Figure 15-9.

Ordinarily, phosphorescence measurements are performed at liquid nitrogen temperature in order to prevent degradation of the output by collisional deactivation. Thus, as shown in Figure 15-9, a Dewar flask with quartz windows is ordinarily a part of a phosphorimeter. At the temperature used, the analyte exists as a solute in a glass or solid solvent. A common solvent for this purpose is a mixture of diethylether, pentane, and ethanol.

[5] For a discussion of fiber optic fluorescence sensors, see O. S. Wolfbeis, in *Molecular Luminescence Spectroscopy,* S. G. Schulman, Ed., Part 2, Chapter 3. New York: Wiley, 1988.

[6] See R. J. Hurtubise, *Anal. Chem.,* **1983,** *55,* 669A; R. J. Hurtubise, *Phosphorimetry: Theory, Instrumentation, and Applications,* Chapter 3. New York: VCH Publishers, 1990.

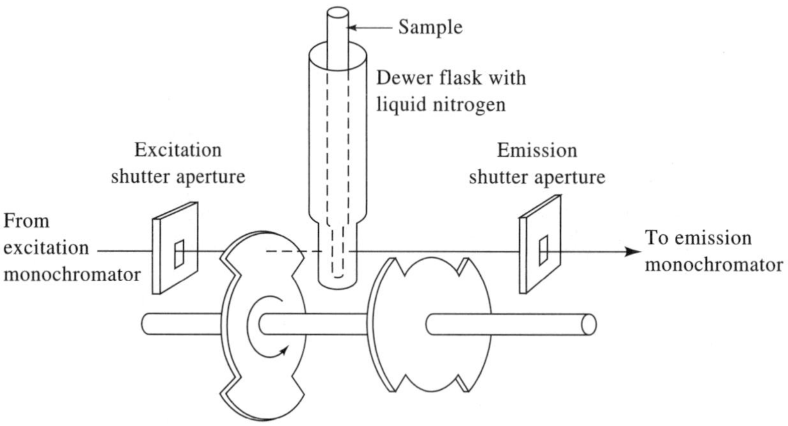

Figure 15-9 Schematic of a device for alternately exciting and observing phosphorescence. *(Reprinted with permission from T. C. O'Haver and J. D. Winefordner,* Anal. Chem., *1966,* 38, *603. Copyright 1966 American Chemical Society.)*

15B-3 Instrument Standardization

Because of variations in source intensity, transducer sensitivity, and other instrumental variables, it is impossible to obtain with a given fluorometer or spectrophotometer exactly the same reading for a solution or a set of solutions from day to day. For this reason it is common practice to standardize an instrument and set it to a reproducible sensitivity level. Standardization is often carried out with a standard solution of a stable fluorophore. The most common reagent for this purpose is a standard solution of quinine sulfate having a concentration of about 10^{-5} M. It is generally excited by radiation at 350 nm and emits radiation of 450 nm. Other compounds have been described for other wavelength regions.

The Perkin-Elmer Corporation offers a set of six fluorescence standards dissolved in a plastic matrix to give stable solid blocks that can be used indefinitely without special storage. With these, the instrument is easily standardized to the wavelength region to be used for the analysis.

15C APPLICATIONS AND PHOTOLUMINESCENCE METHODS

Fluorescence and phosphorescence methods are inherently applicable to lower concentration ranges than are absorbance-based spectrophotometric measurements and are among the most sensitive analytical techniques available to the scientist. The enhanced sensitivity arises from the fact that the concentration-related parameter for fluorometry and phosphorimetry F can be measured independently of the power of the source P_0. In contrast, an absorbance measurement requires evaluation of both P_0 and P, because absorbance, which is proportional to concentration, is dependent upon the ratio between these two quantities. The sensitivity of a fluorometric method can be improved by increasing P_0 or by further amplifying the fluorescence signal. In spectrophotometry, in contrast, an increase in P_0 results in a proportionate change in P and therefore fails to affect A. Thus, fluorometric methods generally have sensitivities that are one to three orders of magnitude better than the corresponding absorbance procedures. On the other hand, the precision and accuracy of photoluminescence methods is usually poorer than spectrophotometric procedures by a factor of perhaps two to five. Generally, phosphorescence methods are less precise than their fluorescence counterparts.

15C-1 Fluorometric Determination of Inorganic Species[7]

Inorganic fluorometric methods are of two types. Direct methods involve the formation of a fluorescing chelate and the measurement of its emission. A second group is based upon the diminution of fluorescence resulting from the quenching action of the substance being determined. The latter technique has been most widely used for anion analysis.

Cations That Form Fluorescing Chelates

Two factors greatly limit the number of transition-metal ions that form fluorescing chelates. First, many of these ions are paramagnetic; this property increases the rate of intersystem crossing to the triplet state. Deactivation by fluorescence is thus unlikely, although phosphorescence may be observed. A second factor is that transition-metal complexes are characterized by many closely spaced energy levels, which enhance the likelihood of deactivation by internal conversion. Nontransition-metal ions are less susceptible to the foregoing deactivation processes; it is for these elements that the principal inorganic applications of fluorometry are to be found. Note that nontransition-metal cations are generally colorless and tend to form chelates that are also without color. Thus, fluorometry often complements spectrophotometry.

Fluorometric Reagents[8]

The most successful fluorometric reagents for cation analyses have aromatic structures with two or more donor functional groups that permit chelate formation

[7] For a review of fluorometric determination of inorganic species, see A. Fernandez-Gutierrez and A. M. De La Pena, in *Molecular Luminescence Spectroscopy,* Part 1, Chapter 4. New York: Wiley, 1985.

[8] For a more detailed discussion of fluorometric reagents, see G. G. Guilbault, in *Comprehensive Analytical Chemistry,* G. Svehla, Ed., Vol. VIII, Chapter 2, pp. 167–178. New York: Elsevier, 1977; P. A. St. Johns, in *Trace Analysis,* J. D. Winefordner, Ed., pp. 263–271. New York: Wiley, 1976.

with the metal ion. The structures of four common reagents follow:

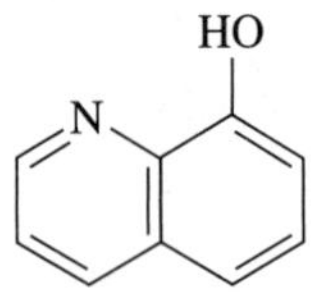

8-hydroxyquinoline
(reagent for Al, Be,
and other metal ions)

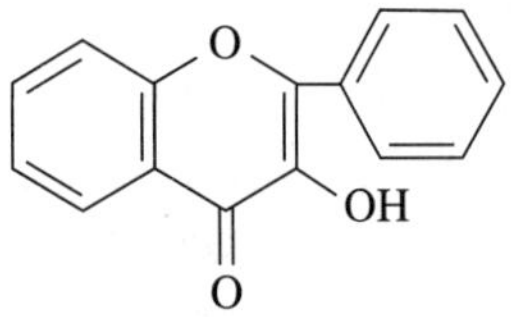

flavanol
(reagent for Zr and Sn)

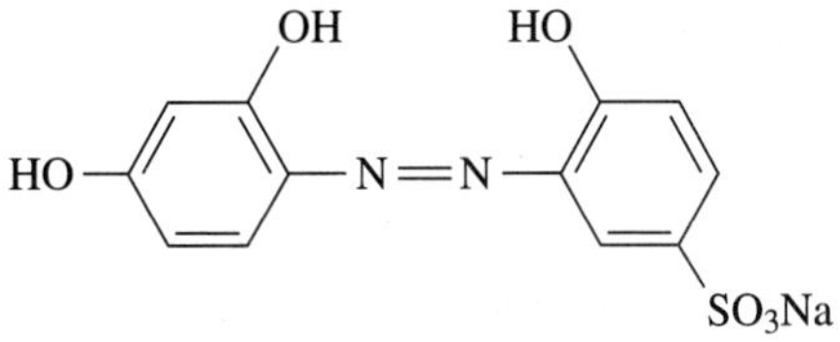

alizarin garnet R
(reagent for Al, F^-)

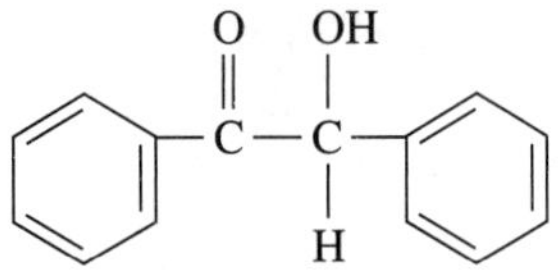

benzoin
(reagent for B, Zn, Ge, and Si)

Selected fluorometric reagents and their applications are presented in Table 15-2.

15C-2 Fluorometric Determination of Organic Species

The number of applications of fluorometric analysis to organic and biochemical species is impressive. For example, Weissler and White have listed methods for the determination of over 200 substances that include a wide variety of organic compounds, enzymes and coenzymes, medicinal agents, plant products, steroids, and vitamins.[9] Without question, the most important applications of fluorometry are in the analyses of food products, pharmaceuticals, clinical samples, and natural products.[10] The sensitivity and selectivity of the method make it a particularly valuable tool in these fields.

[9] A. Weissler and C. E. White, *Handbook of Analytical Chemistry,* L. Meites, Ed., pp. **6**-182 to **6**-196. New York: McGraw-Hill, 1963.

[10] See *Molecular Luminescence Spectroscopy,* S. G. Schulman, Ed., Part 1, Chapters 2, 3, and 5. New York: Wiley, 1985. For a review of bioanalytical applications of fluorescence spectroscopy, see F. V. Bright, *Anal. Chem.,* **1988,** *60,* 1031A.

TABLE 15-2 Selected Fluorometric Methods for Inorganic Species

		Wavelength, nm			
Ion	**Reagent**	**Absorption**	**Fluorescence**	**Sensitivity μg/mL**	**Interference**
Al^{3+}	Alizarin garnet R	470	500	0.007	Be, Co, Cr, Cu, F^-, NO_3^-, Ni, PO_4^{3-}, Th, Zr
F^-	Al complex of Alizarin garnet R (quenching)	470	500	0.001	Be, Co, Cr, Cu, Fe, Ni, PO_4^{3-}, Th, Zr
$B_4O_7^{2-}$	Benzoin	370	450	0.04	Be, Sb
Cd^{2+}	2-(*o*-Hydroxyphenyl)-benzoxazole	365	Blue	2	NH_3
Li^+	8-Hydroxyquinoline	370	580	0.2	Mg
Sn^{4+}	Flavanol	400	470	0.1	F^-, PO_4^{3-}, Zr
Zn^{2+}	Benzoin	—	Green	10	B, Be, Sb, Colored ions

15C-3 Phosphorimetric Methods

Phosphorescence and fluorescence methods tend to be complementary because strongly fluorescing compounds exhibit weak phosphorescence and vice versa.[11] For example, among condensed ring aromatic hydrocarbons, those containing heavier atoms such as halogens or sulfur often phosphoresce strongly; on the other hand, the same compounds in the absence of the heavy atom tend to exhibit fluorescence rather than phosphorescence.

Phosphorimetry has been used for determination of a variety of organic and biochemical species including such substances as nucleic acids, amino acids, pyrine and pyrimidine, enzymes, petroleum hydrocarbons, and pesticides. The method has not, however, found as widespread use as fluorometry perhaps because of the need for low temperatures and the generally poorer precision of phosphorescence measurements. On the other hand, the potentially greater selectivity of phosphorescence procedures is attractive. The reason for this difference in behavior is that efficient phosphorescence requires rapid intersystem crossing to populate the excited triplet state, which in turn reduces the excited singlet concentration and thus the phosphorescence intensity.

During the past two decades, considerable effort has been expended in the development of phosphorimetric methods that can be carried out at room temperature. These efforts have taken two directions. The first is based upon the enhanced phosphorescence that is observed for compounds adsorbed on solid surfaces, such as filter paper. In these applications, a solution of the analyte is dispersed on the solid, and the solvent is evaporated. The phosphorescence of the surface is then measured. The rigid matrix minimizes deactivation of the triplet state by collisional quenching. Collisional quenching has much more of an effect on phosphorescence than on fluorescence because of the much longer lifetime of the triplet state.

The second room temperature method involves solubilizing the analyte in detergent micelles in the presence of heavy metal ions. Apparently the micelles increase the proximity between the heavy metal ions and the phosphor, thus enhancing phosphorescence.[12]

[11] For a review of phosphorimetry, see R. J. Hurtubise, *Anal. Chem.*, **1983**, *55*, 669A; R. J. Hurtubise, *Phosphorimetry, Theory, Instrumentation, and Applications*. New York: VCH Publishers, 1990.

[12] See L. J. Cline Love, M. Skrilec, and J. G. Habarta, *Anal. Chem.*, **1980**, *52*, 754; M. Skrilec and L. J. Cline Love, *Anal. Chem.*, **1980**, *52*, 1559.

15C-4 Application of Fluorometry and Phosphorimetry for Detection in Liquid Chromatography

Photoluminescence measurements provide an important method for detecting and determining components of a sample as they appear at the end of chromatographic or capillary electrophoresis columns. These applications are discussed in Chapters 28 and 30.

15C-5 Lifetime Measurements

The measurement of luminescence lifetimes was initially restricted to phosphorescent systems, where decay times were long enough to permit the easy measurement of emitted intensity as a function of time. Equipment is offered by several instrument manufacturers for studying rates of luminescence decay on the fluorescence time scale (10^{-5} to 10^{-8} s). This equipment employs mode-lock lasers that produce pulses of radiation having widths of 70 to 100 ps for excitation and fast-rise-time photomultiplier tubes for detection. Instruments of this kind provide information that is useful in basic studies of energy transfer and quenching. Furthermore, for analytical work, lifetime measurements enhance the selectivity of luminescence methods because they permit the analysis of mixtures containing two or more luminescent species with different decay rates.

Figure 15-10 shows curves from a typical fluorescence lifetime experiment. Curve *A* gives the output of the source as a function of time, whereas curve *B* shows

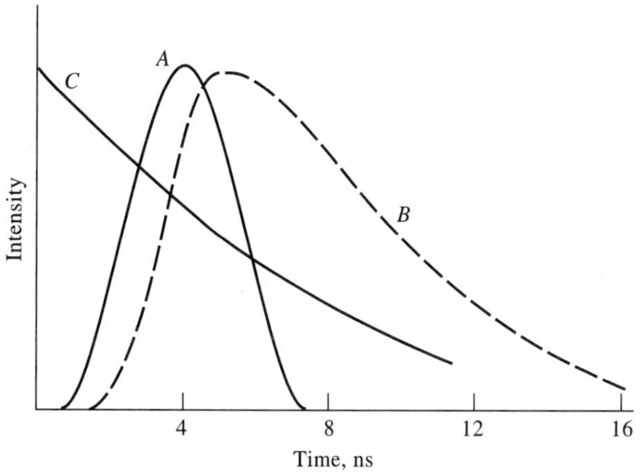

Figure 15-10 Fluorescence lifetime profiles: *A*, excitation pulse; *B*, measured decay curve; *C*, corrected decay curve.

how the observed fluorescence signal decays. Curve *B* is a composite of the decay signal from the source and the decaying emission signal from the analyte. The true fluorescence decay signal *C* is then obtained by deconvolving the contribution of the source to the experimental signal. For details of the various methods for measuring fluorescence lifetimes, the reader is referred to the references in the footnote below.[13]

15D CHEMILUMINESCENCE

The application of chemiluminescence to analytical chemistry is a relatively recent development. The number of chemical reactions that produce chemiluminescence is small, thus limiting the procedure to a relatively small number of species. Nevertheless, some of the compounds that do react to give chemiluminescence are important components of the environment. For these, the high selectivity, the simplicity, and the extreme sensitivity of the method account for its recent growth in usage.[14]

15D-1 The Chemiluminescence Phenomenon

Chemiluminescence is produced when a chemical reaction yields an electronically excited species that emits light as it returns to its ground state or that transfers its energy to another species, which then produces emission. Chemiluminescence reactions are encountered in a number of biological systems, where the process is often termed *bioluminescence.* Examples of species that exhibit bioluminescence include the firefly, the sea pansy and certain jellyfish, bacteria, protozoa, and crustacea. The chemistry of the various natural bioluminescence processes is incompletely understood.

Over a century ago, it was discovered that several relatively simple organic compounds also are capable of exhibiting chemiluminescence. The simplest type of reaction of such compounds to produce chemiluminescence can be formulated as

$$\mathrm{A + B \rightarrow C^* + D}$$

$$\mathrm{C^* \rightarrow C} + h\nu$$

where C* represents the excited state of the species C. Here, the luminescence spectrum is that of the reaction product C. Most chemiluminescence reactions are considerably more complicated than is suggested by the foregoing equations.

For chemiluminescence, the radiant intensity I_{CL} (photons emitted per second) depends upon the rate of the chemical reaction (dC/dt) and the chemiluminescence quantum yield ϕ_{CL} (photons emitted per molecule reacted). The latter term is equal to the product of the excitation quantum yield ϕ_{EX} (excited states per molecule reacted) and the emission quantum yield ϕ_{EM} (photons per excited state). These relationships are described by the equation

$$I_{\mathrm{CL}} = \phi_{\mathrm{CL}} \frac{d\mathrm{C}}{dt} = \phi_{\mathrm{EX}}\, \phi_{\mathrm{EM}} \frac{d\mathrm{C}}{dt}$$

Chemiluminescence systems that are useful in analysis generally have values of ϕ_{CL} of 0.01 to 0.2.

15D-2 Measurement of Chemiluminescence

The instrumentation for chemiluminescence measurements is remarkably simple and may consist of only a suitable reaction vessel and a photomultiplier tube. Generally, no wavelength selection device is necessary since the only source of radiation is the chemical reaction between the analyte and reagent. Several instrument manufacturers offer chemiluminescence photometers.

The typical signal from a chemiluminescence experiment as a function of time rises rapidly to a maximum as mixing of reagent and analyte is complete; then a more or less exponential decay of signal follows (see Figure 15-11). Usually for quantitative analysis, the signal is integrated for a fixed period of time and compared with standards treated in an identical way. Alternatively, peak heights are used. Often a linear relationship between signal and concentration is observed over a concentration range of several orders of magnitude.

---

[13] For references dealing with lifetime measurements, see L. B. McGown and F. V. Bright, *CRC Crit. Rev. Anal. Chem.*, **1987**, *18*, 245; L. B. McGown, *Anal. Chem.*, **1989**, *61*, 839A; G. M. Hieftje and G. R. Haugen, *Anal. Chem.*, **1981**, *53*, 755A; L. B. McGown and F. V. Bright, *Anal. Chem.*, **1984**, *56*, 1401A.

[14] For reviews of analytical applications of chemiluminescence, see M. L. Grayeski, *Anal. Chem.*, **1987**, *59*, 1243A; *Bioluminescence and Chemiluminescence: Instruments and Applications*, K. Van Dyke, Ed. Boca Raton; CRC Press, 1985; A. K. Campbell, *Chemiluminescence: Principles and Applications in Biology and Medicine.* New York: VCH Publishers, 1988.

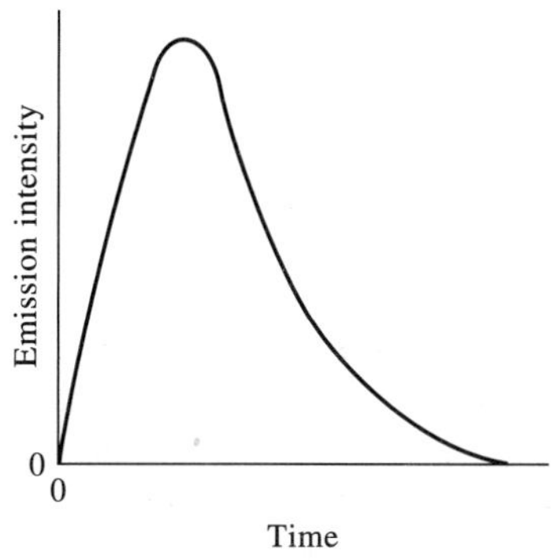

Figure 15-11 Chemiluminescence emission intensity as a function of time after mixing reagents.

15D-3 Analytical Applications of Chemiluminescence[15]

Chemiluminescence methods are generally highly sensitive because low light levels are readily monitored in the absence of noise. Furthermore, radiation attenuation by a filter or a monochromator is avoided. In fact, detection limits are usually determined not by transducer sensitivity but rather by reagent purity. Typical detection limits lie in the parts per billion (or sometimes less) to parts per million range. Typical precisions are difficult to judge from the present literature.

Analysis of Gases

Chemiluminescence methods for determining components of gases originated with the need for highly sensitive means for determining atmospheric pollutants such as ozone, oxides of nitrogen, and sulfur compounds. One of the most widely used of these methods is for the determination of nitrogen monoxide; the reactions are

$$NO + O_3 \rightarrow NO_2^* + O_2$$

$$NO_2^* \rightarrow NO_2 + h\nu \quad (\lambda = 600 \text{ to } 2800 \text{ nm})$$

Ozone from an electrogenerator and the atmospheric sample are drawn continuously into a reaction vessel, where the luminescence radiation is monitored by a photomultiplier tube. A linear response is reported for nitrogen monoxide concentrations of 1 ppb to 10,000 ppm. This procedure has become the predominant one for monitoring the concentration of this important atmospheric constituent from ground level to altitudes as high as 20 km.

The reaction of nitric oxide with ozone has also been applied to the determination of the higher oxides of nitrogen. For example, the nitrogen dioxide content of automobile exhaust gas has been determined by thermal decomposition of the gas at 700°C in a steel tube. The reaction is

$$NO_2 \rightleftharpoons NO + O$$

At least two manufacturers now offer an instrument for determination of nitrogen in solid or liquid materials containing 0.1% to 30% nitrogen. The samples are pyrolyzed in an oxygen atmosphere under conditions whereby the nitrogen is converted quantitatively to nitrogen monoxide; the latter is then measured by the method just described.

Another important chemiluminescence method is used for monitoring atmospheric ozone. In this instance, the determination is based upon the luminescence produced when the analyte reacts with the dye rhodamine-B adsorbed on an activated silica gel surface. This procedure is sensitive to less than 1 ppb ozone; the response is linear up to 400 ppb ozone. Ozone can also be determined in the gas phase based on the chemiluminescence produced when the analyte reacts with ethylene. Both reagents are reported to be specific for ozone.

Still another important gas-phase chemiluminescence method is used for the determination of atmospheric sulfur compounds such as hydrogen sulfide, sulfur dioxide, and mercaptans. Here, the sample is combusted in a hydrogen flame to give a sulfur dimer, which then decomposes with the emission of light. For example with sulfur dioxide the reactions are

$$4H_2 + 2SO_2 \rightleftharpoons S_2^* + 4H_2O$$

$$S_2^* \rightarrow S_2 + h\nu$$

Here, the radiation occurs in the blue with peaks at 384 and 394 nm. Its intensity is proportional to the concentration of the excited sulfur dimer. Similarly combustion of phosphorus compounds in a hydrogen flame gives emission due to HPO* at 526 nm. Linear working curves over four decades of concentration are reported. Both of these flame chemiluminescence techniques have been employed for detection of sulfur and phosphorus species in the effluent from gas chromatographic columns.

[15] For a discussion of various applications of chemiluminescence in chemical analysis, see T. A. Nieman, in *Handbook of Instrumental Techniques for Analytical Chemistry,* F. Settle, Ed., Chapter 27, p. 541. Englewood Cliffs, NJ: Prentice-Hall, 1997.

Analysis for Inorganic Species in the Liquid Phase

Many of the analyses carried out in the liquid phase make use of organic chemiluminescing substances containing the functional group

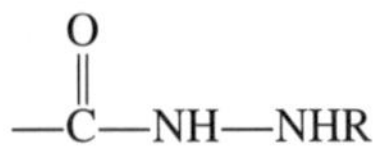

These reagents react with oxygen, hydrogen peroxide, and many other strong oxidizing agents to produce a chemiluminescing oxidation product. Luminol is the most common example of these compounds. Its reaction with strong oxidants, such as oxygen, hydrogen peroxide, hypochlorite ion, and permanganate ion, in the presence of strong base is given below. Often a catalyst is required for this reaction to proceed at a useful rate. The emission produced matches the fluorescence spectrum of the product, 3-aminophthalate anion; the chemiluminescence appears blue and is centered around 425 nm.

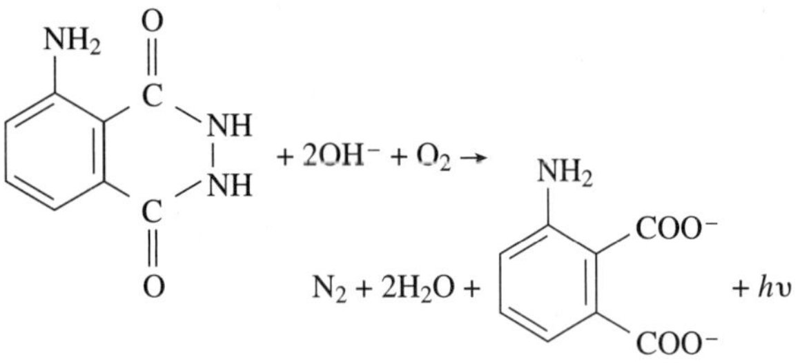

luminol 3-aminophthalate ion

Within certain limits, the chemiluminescence intensity of luminol is directly proportional to the concentration of either the oxidant, the catalyst, or the luminol. Consequently, the reaction provides a sensitive method for determining any one of these species. For example, using hydrogen peroxide as the oxidant, the catalyst Co^{2+} can be estimated at concentrations down to 0.01 nmol/L, Cr^{3+} down to 0.5 nmol/L, and Cu^{2+} down to 1 nmol/L. With a few cations, inhibition of luminescence occurs; in such cases the decrease in intensity permits determination of concentrations.

Analysis for Organic Species

To increase the selectivity of chemiluminescence reactions and to extend chemiluminescence to analytes not directly involved in chemiluminescence reactions, it is common practice to precede a chemiluminescence step by an enzyme reaction for which the desired analyte is the substrate and one of the products is detected by chemiluminescence. This is most commonly done in flow systems with columns of immobilized enzyme. Recently, however, attention has been directed toward biosensor designs using enzymes attached to optical fibers.

Oxidase enzymes that generate H_2O_2 are commonly used in the predetection step. Not only can H_2O_2 be determined with several different chemiluminescence systems, but the necessary oxidant (O_2) is already present in most samples, especially those in aqueous solution. Assuming quantitative conversion by the enzyme, substrates can be determined down to 10 to 100 nM, just as can H_2O_2. Substrates detected this way include glucose, cholesterol, choline, uric acid, amino acids, aldehydes, and lactate. For example:

$$\text{uric acid} + O_2 \xrightarrow{\text{uricase}} \text{allantoin} + H_2O_2$$

The approach can be extended by using sequential enzyme steps to ultimately convert the analyte to an equivalent amount of chemiluminescent product. In this way, sugars other than glucose, glucosides, cholesterol esters, creatinine, and acetylcholine have been determined.[16] For example:

$$\text{sucrose} + H_2O \xrightarrow{\text{invertase}} \alpha\text{-D-glucose} + \text{fructose}$$

$$\alpha\text{-D-glucose} \xrightleftharpoons{\text{mutarotase}} \beta\text{-D-glucose}$$

$$\beta\text{-D-glucose} + O_2 \xrightarrow{\text{glucose oxidase}} \text{gluconic acid} + H_2O_2$$

Luminol plus a peroxidase catalyst appears to be the optimal system for H_2O_2 determination. Peak chemiluminescence intensity is reached in about 100 ms; the solvent is aqueous but is compatible with some organic components; the detection limit is about 0.1 pM, with linearity for three to four decades of concentration.

[16] C. A. K. Swindlehurst and T. A. Nieman, *Anal. Chim. Acta,* **1988,** *205,* 195.

15E QUESTIONS AND PROBLEMS

15-1 Define the following terms: (a) fluorescence, (b) phosphorescence, (c) resonance fluorescence, (d) singlet state, (e) triplet state, (f) vibrational relaxation, (g) internal conversion, (h) external conversion, (i) intersystem crossing, (j) predissociation, (k) dissociation, (l) quantum yield, (m) chemiluminescence.

15-2 Explain the difference between a fluorescence emission spectrum and a fluorescence excitation spectrum. Which more closely resembles an absorption spectrum?

15-3 Why is spectrofluorometry potentially more sensitive than spectrophotometry?

15-4 Which compound below is expected to have the greater quantum yield of fluorescence? Explain.

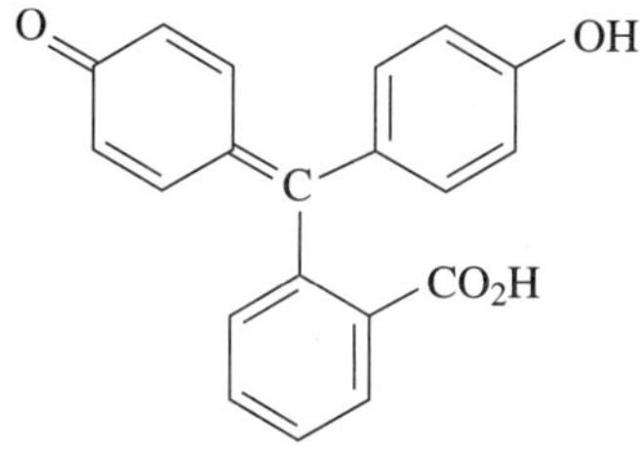

phenolphthalein

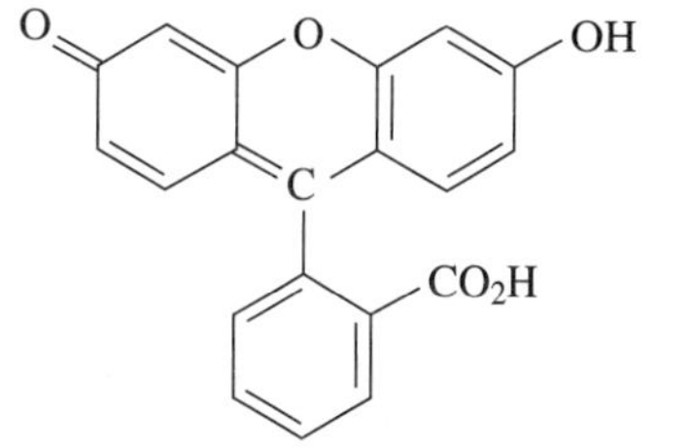

fluorescein

15-5 In which solvent would the fluorescence of naphthalene be expected to be greatest: 1-chloropropane, 1-bromopropane, or 1-iodopropane? Explain.

15-6 The reduced form of nicotinamide adenine dinucleotide (NADH) is an important and highly fluorescent coenzyme. It has an absorption maximum at 340 nm and an emission maximum at 465 nm. Standard solutions of NADH gave the following fluorescence intensities:

Concd NADH (μmol/L)	Relative Intensity, I
0.100	2.24
0.200	4.74
0.300	6.59
0.400	8.98
0.500	10.93
0.600	14.01
0.700	15.49
0.800	18.02

(a) Construct a calibration curve for NADH.
(b) Compute the least-squares regression parameters of a linear equation for the plot in part (a).
(c) Calculate the standard deviation of the slope and the intercept.

(d) An unknown exhibits a relative fluorescence of 12.16. Calculate the concentration of NADH.
(e) Calculate the relative standard deviation for the result in part (d).
(f) Calculate the relative standard deviation for the result in part (d) if the reading of 12.16 was the mean of three measurements.

15-7 The following volumes of a solution containing 1.10 ppm of Zn^{2+} were pipetted into separatory funnels, each containing 5.00 mL of a zinc solution of unknown concentration: 0.00, 5.00, 19.00, and 15.00 mL. Each was extracted with three 5-mL aliquots of CCl_4 containing an excess of 8-hydroxyquinoline. The extracts were then diluted to 25.0 mL and their fluorescence measured with a fluorometer. The results were

Vol Std Zn^{2+}, mL	Fluorometer Reading, *R*
0.000	6.08
5.00	11.24
10.00	15.47
15.00	20.84

(a) Plot the data as fluorometer reading versus volme of standard zinc solution.
(b) Compute by least squares the regression parameters of a linear equation for the plot.
(c) Calculate the standard deviation of the slope and the standard deviation about regression.
(d) Calculate the concentration of zinc in the sample.
(e) Calculate a standard deviation for the result in part (d).

15-8 To four 10.0-mL aliquots of a water sample were added 1.50, 3.00, 4.50 mL of a standard NaF solution containing 10.0 ppb F^-. Exactly 5.00 mL of a solution containing an excess of Al-acid Alizarin Garnet R complex, a strongly fluorescing complex, were added to each, and the solutions were diluted to 50.0 mL. The fluorescence intensity of the four solutions plus a blank were as follows:

mL Sample	mL of Std F^-	Meter Reading, *R*
5.00	0.00	66.3
5.00	1.50	53.7
5.00	3.00	39.0
5.00	4.50	26.4

(a) Explain the chemistry of the analytical method.
(b) Plot the data.
(c) By least squares compute the regression parameters of the linear equation relating the decrease in fluorescence to the volume of standard reagent.
(d) Calculate the standard deviation of the slope and the standard deviation about regression.

(e) Calculate the concentration in ppb of F^- in the sample.
(f) Calculate the standard deviation of the datum in part (e).

15-9 Iron(II) ions catalyze the oxidation of luminol by H_2O_2. The intensity of the resulting chemiluminescence has been shown to increase linearly with iron(II) concentration from 10^{-10} to 10^{-8} M.

Exactly 1.00 mL of water was added to a 2.00-mL aliquot of an Fe(II) solution of unknown concentration, followed by 2.00 mL of a dilute H_2O_2 solution and 1.00 mL of an alkaline solution of luminol. The chemiluminescence signal from the mixture was integrated over a 10.0-s period and found to be 16.1.

To a second 2.00-mL aliquot of the sample was added 1.00 mL of a 4.75×10^{-5} M Fe(II) solution followed by the same volume of H_2O_2 and luminol. The integrated intensity was 29.6. Calculate the Fe(II) molarity of the sample.

15-10 Equations for the chemiluminescence determination of SO_2 are given on page 375. Derive an expression for the relationship between the concentration of SO_2 in a sample, the luminescence intensity, and the equilibrium constant for the first reaction.

15-11 Quinine is one of the best known fluorescent molecules, and the sensitivities of fluorometers are often specified in terms of the detection limit for this molecule. The structure of quinine is given below. Predict the part of the molecule that is most likely to behave as the chromophore and fluorescent center.

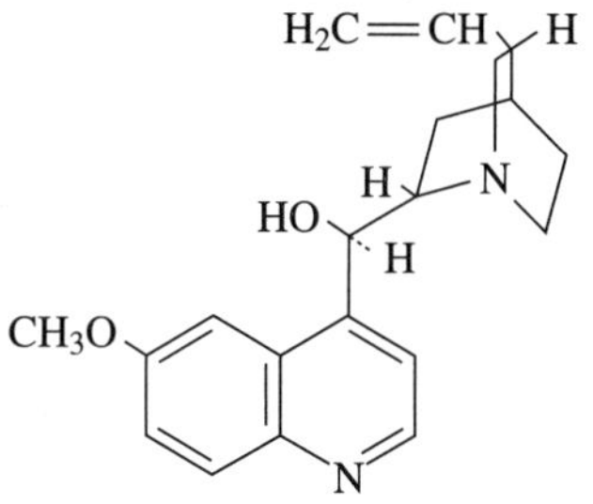

Chapter 16

An Introduction to Infrared Spectrometry

The infrared region of the spectrum encompasses radiation with wavenumbers ranging from about 12,800 to 10 cm^{-1} or wavelengths from 0.78 to 1000 μm.[1] *From the standpoint of both application and instrumentation, the infrared spectrum is conveniently divided into* ***near-***, ***mid-***, *and* ***far-****infrared radiation; rough limits of each are shown in Table 16-1.*

The techniques and the applications of methods based upon the three infrared spectral regions differ considerably. Measurements in the near-infrared region are made with photometers and spectrophotometers similar in design and components to the instruments described in earlier chapters for ultraviolet/visible spectrometry. The most important applications of this spectral region have been to the quantitative analysis of industrial and agricultural materials and for process control. Applications of near-infrared spectrometry are discussed in Section 17D.

Until the early 1980s, instruments for the mid-infrared region were largely of the dispersive type based on diffraction gratings. Since that time, however, there has been a dramatic change in mid-infrared instrumentation so that now the majority of new instruments are of the Fourier transform type. Photometers based upon interference filters also find use for measuring the composition of gases and atmospheric contaminants.

The appearance of relatively inexpensive Fourier transform spectrometers in the last decade has markedly increased the number and type of applications of mid-infrared radiation. This increase has come about as a consequence of an order of magnitude, or more, improvement in signal-to-noise ratios and detection limits that can be realized with interferometric instruments. Before the appearance of these instruments, the mid-infrared spectral region was used largely for qualitative organic analysis and structure determinations based on absorption spectra. Now, in contrast, mid-infrared spectrometry is beginning to be used in addition for quantitative analysis of complex samples by

[1] Until recently, the unit of wavelength μm (that is 10^{-6} m) was called the micron, μ; it is now more properly termed the micrometer, μm.

TABLE 16-1 Infrared Spectral Regions

Region	Wavelength (λ) Range, μm	Wavenumber ($\bar{\nu}$) Range, cm^{-1}	Frequency (ν) Range, Hz
Near	0.78 to 2.5	12,800 to 4000	3.8×10^{14} to 1.2×10^{14}
Middle	2.5 to 50	4000 to 200	1.2×10^{14} to 6.0×10^{12}
Far	50 to 1000	200 to 10	6.0×10^{12} to 3.0×10^{11}
Most used	2.5 to 15	4000 to 670	1.2×10^{14} to 2.0×10^{13}

both absorption and emission spectrometry. Applications of this spectral region are also beginning to appear for microscopic studies of surfaces, analysis of solids by attenuated total reflectance and diffuse reflectance, photoacoustic measurements, and others. Several of these applications are described in Chapter 17.

In the past the far-infrared region of the spectrum, while potentially quite useful, has had limited use because of experimental difficulties. The few sources of this type of radiation are notoriously weak and are further attenuated by the need for order-sorting filters that must be used to prevent radiation of higher grating orders from reaching the detector. Fourier transform spectrometers, with their much higher throughput, largely alleviate this problem and make the far-infrared spectral region much more accessible to chemists. Several applications of far-infrared spectroscopy are described in Section 17E.

In this chapter, we first deal with the mechanisms of the absorption, emission, and reflection of infrared radiation using absorption spectroscopy as the basis for this discussion. We follow this discussion with a description of the components of infrared instruments and how these are arranged in dispersive and nondispersive instruments as well as in Fourier transform spectrometers.[2]

[2] For detailed treatments of infrared spectrometry, see N. B. Colthup, L. H. Daly, and S. E. Wiberley, *Introduction to Infrared and Raman Spectroscopy,* 3rd ed. San Diego: Academic Press, 1990; B. Schrader, *Infrared and Raman Spectroscopy.* New York: VCH Press, 1995; A. L. Smith, in *Treatise on Analytical Chemistry,* 2nd ed, P. J. Elving, E. J. Meehan, and I. M. Kolthoff, Eds., Part I, Vol. 7, Chapter 5. New York: Wiley, 1981.

16A THEORY OF INFRARED ABSORPTION SPECTROMETRY

Infrared absorption, emission, and reflection spectra for molecular species can be rationalized by assuming that all arise from various changes in energy brought about by transitions of molecules from one vibrational or rotational energy state to another. In this section we use molecular absorption to illustrate the nature of these transitions.

16A-1 Introduction

The plot shown in Figure 16-1 is a reproduction of the output of a widely used commercial infrared spectrophotometer. As is ordinarily the case, the ordinate is linear in transmittance. The abscissa in this chart is linear in wavenumbers with units of reciprocal centimeters. Most modern instruments utilize a dedicated microcomputer with versatile software capable of producing a variety of other output formats, such as transmittance versus wavelength and absorbance versus wavenumber or wavelength.

A linear wavenumber scale is usually preferred in infrared spectroscopy because of the direct proportionality between this quantity and both energy and frequency. The frequency of the absorbed radiation is, in turn, the molecular vibrational frequency actually responsible for the absorption process. Frequency, however, is seldom if ever used as the abscissa because of the inconvenient size of the unit; that is, the frequency scale of the plot in Figure 16-1 would extend from 1.2×10^{14} to 2.0×10^{13} Hz. Although a scale in terms of cm^{-1} is often referred to as a frequency scale, keep in mind that this terminology is not strictly correct because the wavenumber is only proportional to frequency.

Finally, note that the horizontal scale of Figure 16-1 changes at 2000 cm^{-1}, with the units at higher

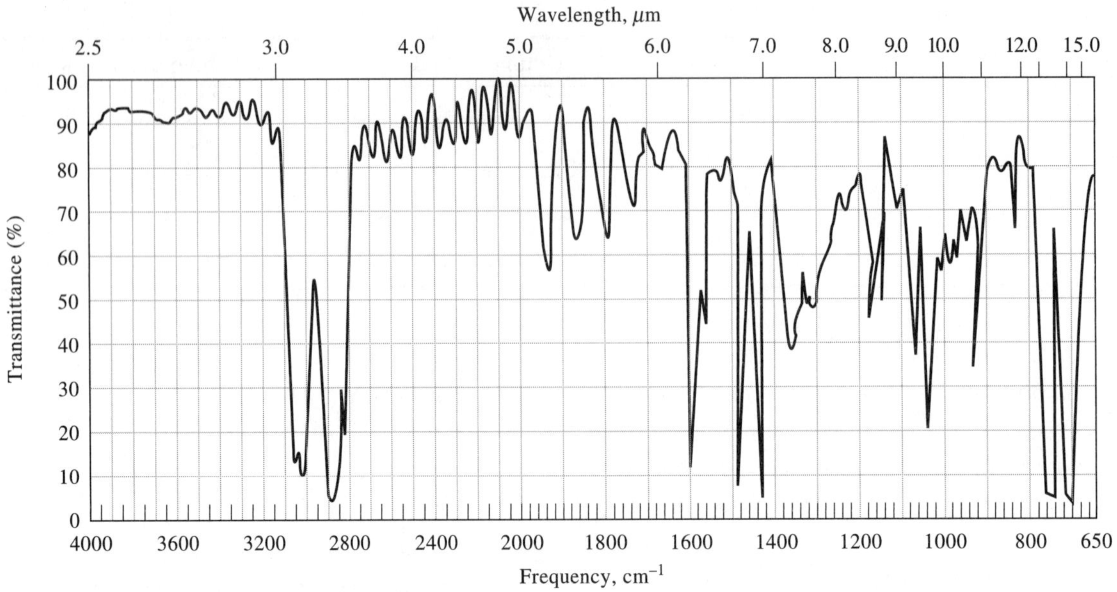

Figure 16-1 Infrared absorption spectrum of a thin polystyrene film recorded with a modern infrared spectrophotometer. Note that the abscissa scale changes at 2000 cm^{-1}.

wavenumbers being represented by half the linear distance of those at lower wavenumbers. This discontinuity is often introduced for convenience, since much useful qualitative infrared detail appears at wavenumbers smaller than 2000 cm^{-1}.

Dipole Changes During Vibrations and Rotations

Infrared radiation is not energetic enough to bring about the kinds of electronic transitions that we have encountered in our discussions of ultraviolet, visible, and X-ray radiation. Absorption of infrared radiation is thus confined largely to molecular species that have small energy differences between various vibrational and rotational states.

In order to absorb infrared radiation, a molecule must undergo a net change in dipole moment as a consequence of its vibrational or rotational motion. Only under these circumstances can the alternating electrical field of the radiation interact with the molecule and cause changes in the amplitude of one of its motions. For example, the charge distribution around a molecule such as hydrogen chloride is not symmetric because the chlorine has a higher electron density than the hydrogen. Thus, hydrogen chloride has a significant dipole moment and is said to be polar. The dipole moment is determined by the magnitude of the charge difference and the distance between the two centers of charge. As a hydrogen chloride molecule vibrates, a regular fluctuation in dipole moment occurs, and a field is established that can interact with the electrical field associated with radiation. If the frequency of the radiation exactly matches a natural vibrational frequency of the molecule, a net transfer of energy takes place that results in a change in the *amplitude* of the molecular vibration; absorption of the radiation is the consequence. Similarly, the rotation of asymmetric molecules around their centers of mass results in a periodic dipole fluctuation that can interact with radiation.

No net change in dipole moment occurs during the vibration or rotation of homonuclear species such as O_2, N_2, or Cl_2; consequently, such compounds cannot absorb in the infrared. With the exception of a few compounds of this type, all other molecular species absorb infrared radiation.

Rotational Transitions

The energy required to cause a change in rotational level is minute and corresponds to radiation of 100 cm^{-1} or less (> 100 μm). Because rotational levels are quantized, absorption by gases in the far-infrared region is characterized by discrete, well-defined lines. In liquids or solids, intramolecular collisions and interactions cause broadening of the lines into a continuum.

Vibrational/Rotational Transitions

Vibrational energy levels are also quantized, and for most molecules the energy differences between quantum states correspond to the mid-infrared region. The infrared spectrum of a gas usually consists of a series of closely spaced lines, because there are several rotational energy states for each vibrational state. On the other hand, rotation is highly restricted in liquids and solids; in such samples, discrete vibrational/rotational lines disappear, leaving only somewhat broadened vibrational peaks.

Types of Molecular Vibrations

The relative positions of atoms in a molecule are not exactly fixed but instead fluctuate continuously as a consequence of a multitude of different types of vibrations and rotations about the bonds in the molecule. For a simple diatomic or triatomic molecule, it is easy to define the number and nature of such vibrations and relate these to energies of absorption. An analysis of this kind becomes difficult if not impossible for molecules made up of several atoms. Not only do large molecules have a large number of vibrating centers, but also interactions among several centers can occur and must be taken into account.

Vibrations fall into the basic categories of *stretching* and *bending.* A stretching vibration involves a continuous change in the interatomic distance along the axis of the bond between two atoms. Bending vibrations are characterized by a change in the angle between two bonds and are of four types: *scissoring, rocking, wagging,* and *twisting.* The various types of vibrations are shown schematically in Figure 16-2.

All of the vibration types shown in Figure 16-2 may be possible in a molecule containing more than two atoms. In addition, interaction or *coupling* of vibrations can occur if the vibrations involve bonds to a single central atom. The result of coupling is a change in the characteristics of the vibrations involved.

In the treatment that follows, we shall first consider isolated vibrations represented by a simple mechanical model called the *harmonic oscillator.* Modifications to the theory of the harmonic oscillator, which are needed to describe a molecular system, will be taken up next. Finally, the effects of vibrational interactions in molecular systems will be discussed.

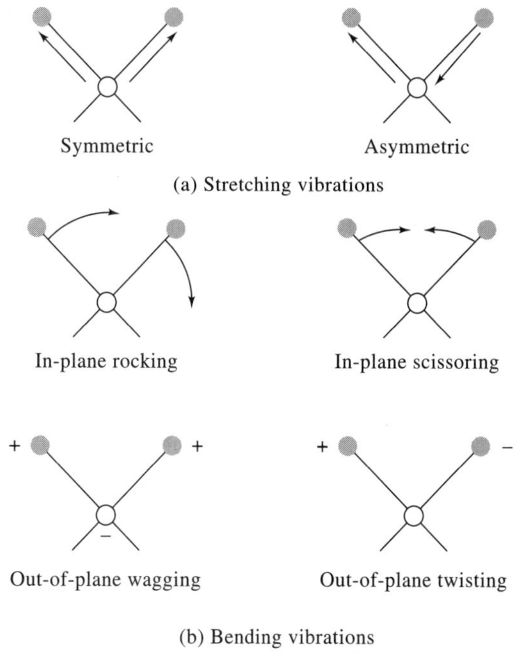

Figure 16-2 Types of molecular vibrations. Note: + indicates motion from the page toward the reader; − indicates motion away from the reader.

16A-2 Mechanical Model of a Stretching Vibration in a Diatomic Molecule

The characteristics of an atomic stretching vibration can be approximated by a mechanical model consisting of two masses connected by a spring. A disturbance of one of these masses along the axis of the spring results in a vibration called a *simple harmonic motion.*

Let us first consider the vibration of a single mass attached to a spring that is hung from an immovable object (see Figure 16-3a). If the mass is displaced a distance y from its equilibrium position by application of a force along the axis of the spring, the restoring force F is proportional to the displacement (Hooke's law). That is,

$$F = -ky \tag{16-1}$$

where k is the *force constant,* which depends upon the stiffness of the spring. The negative sign indicates that F is a restoring force. This means that the direction of the force is opposite to the direction of the displacement. Thus, the force tends to restore the mass to its original position.

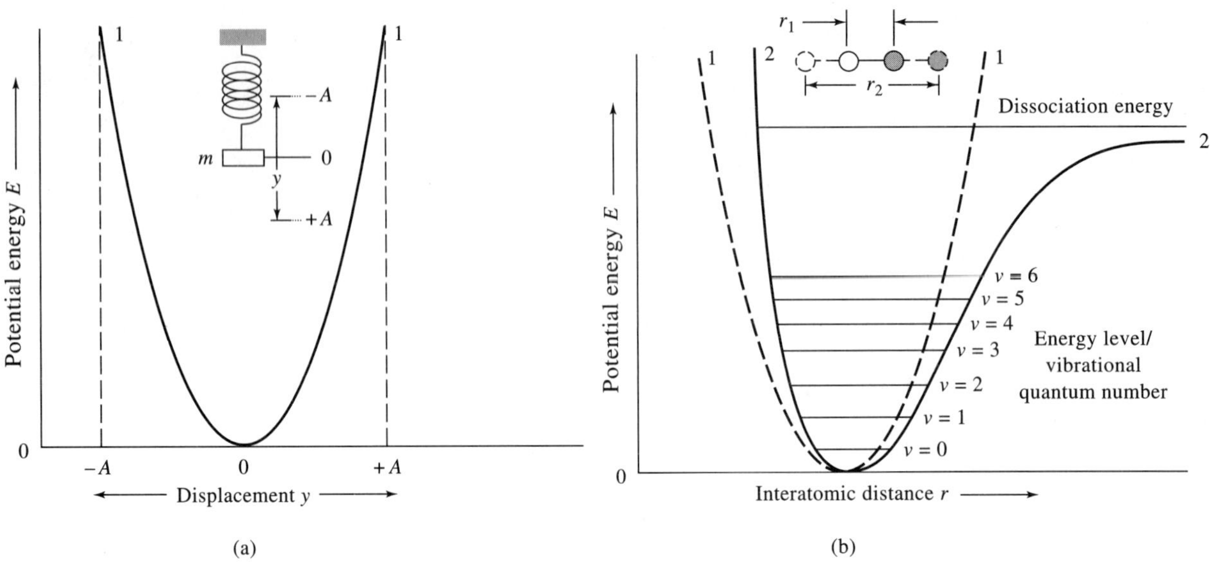

Figure 16-3 Potential energy diagrams. Curve 1, harmonic oscillator. Curve 2, anharmonic oscillator.

Potential Energy of a Harmonic Oscillator

The potential energy E of the mass and spring can be arbitrarily assigned a value of zero when the mass is in its rest or equilibrium position. As the spring is compressed or stretched, however, the potential energy of this system increases by an amount equal to the work required to displace the mass. If, for example, the mass is moved from some position y to $y + dy$, the work and hence the change in potential energy dE is equal to the force F times the distance dy. Thus,

$$dE = -Fdy \tag{16-2}$$

Combining Equations 16-2 and 16-1 yields

$$dE = kydy$$

Integrating between the equilibrium position ($y = 0$) and y gives

$$\int_0^E dE = k\int_0^y ydy$$

$$E = \frac{1}{2}ky^2 \tag{16-3}$$

The potential-energy curve for a simple harmonic oscillation, derived from Equation 16-3, is a parabola as depicted in Figure 16-3a. Notice that the potential energy is a maximum when the spring is stretched or compressed to its maximum amplitude A, and it decreases to zero at the equilibrium position.

Vibrational Frequency

The motion of the mass as a function of time t can be deduced from classical mechanics as follows. Newton's second law states that

$$F = ma$$

where m is the mass and a is its acceleration. But acceleration is the second derivative of distance with respect to time. Thus,

$$a = \frac{d^2y}{dt^2}$$

Substituting these expressions into Equation 16-1 gives

$$m\frac{d^2y}{dt^2} = -ky \tag{16-4}$$

A solution to this equation must be a periodic function such that its second derivative is equal to the original function times $-(k/m)$. A suitable cosine relationship meets this requirement. Thus, the instantaneous displacement of the mass at time t can be written as

$$y = A\cos 2\pi\nu_m t \tag{16-5}$$

where ν_m is the natural vibrational frequency and A is the maximum amplitude of the motion. The second derivative of Equation 16-5 is

$$\frac{d^2y}{dt^2} = -4\,\pi^2\nu_m^2 A\cos 2\pi\nu_m t \tag{16-6}$$

Substitution of Equations 16-5 and 16-6 into Equation 16-4 gives

$$\cancel{A\cos 2\pi\nu_m t} = \frac{4\,\pi^2\nu_m^2 m}{k}\cancel{A\cos 2\pi\nu_m t}$$

The natural frequency of the oscillation is then

$$\nu_m = \frac{1}{2\pi}\sqrt{\frac{k}{m}} \tag{16-7}$$

where ν_m is the *natural frequency* of the mechanical oscillator. While it is dependent upon the force constant of the spring and the mass of the attached body, the natural frequency is *independent* of the energy imparted to the system; changes in energy merely result in a change in the amplitude A of the vibration.

The equation just developed may be modified to describe the behavior of a system consisting of two masses m_1 and m_2 connected by a spring. Here, it is only necessary to substitute the *reduced mass* μ for the single mass m where

$$\mu = \frac{m_1 m_2}{m_1 + m_2} \tag{16-8}$$

Thus, the vibrational frequency for such a system is given by

$$\nu_m = \frac{1}{2\pi}\sqrt{\frac{k}{\mu}} = \frac{1}{2\pi}\sqrt{\frac{k(m_1 + m_2)}{m_1 m_2}} \tag{16-9}$$

Molecular Vibrations

The approximation is ordinarily made that the behavior of a molecular vibration is analogous to the mechanical model just described. Thus, the frequency of the molecular vibration is calculated from Equation 16-9 after substituting the masses of the two atoms for m_1 and m_2 into Equation 16-8 to obtain μ; the quantity k is the force constant for the chemical bond, which is a measure of its stiffness.

16A-3 Quantum Treatment of Vibrations

The equations of classical mechanics, such as we have used thus far, do not completely describe the behavior of particles of atomic dimensions. For example, the quantized nature of molecular vibrational energies, and of other atomic and molecular energies as well, does not appear in these equations. We may, however, invoke the concept of the simple harmonic oscillator to develop the wave equations of quantum mechanics. Solutions of these equations for potential energies have the form

$$E = \left(v + \frac{1}{2}\right)\frac{h}{2\pi}\sqrt{\frac{k}{\mu}} \tag{16-10}$$

where h is the Planck constant, and v is the *vibrational quantum number,* which can take only positive integer values (including zero). Thus, in contrast to ordinary mechanics where vibrators can assume any potential energy, quantum mechanical vibrators can take on only certain discrete energies.

It is interesting to note that the factor $(\sqrt{k/\mu})/2\pi$ appears in both the classical and the quantum equations; by substituting Equation 16-9 into 16-10, we find

$$E = \left(v + \frac{1}{2}\right)h\nu_m \tag{16-11}$$

where ν_m is the vibrational frequency of the classical model.[3]

We now assume that transitions in vibrational energy levels can be brought about by absorption of radiation, provided the energy of the radiation exactly matches the difference in energy levels ΔE between the vibrational quantum states and provided also that the vibration causes a fluctuation in dipole. *This difference is identical between any pair of adjacent levels,* since v in Equations 16-10 and 16-11 can assume only whole numbers; that is,

$$\Delta E = h\nu_m = \frac{h}{2\pi}\sqrt{\frac{k}{\mu}} \tag{16-12}$$

[3] Unfortunately, the generally accepted symbol for the vibrational quantum number v is similar in appearance to the Greek nu (ν) that symbolizes frequency. Thus, care must be exercised continuously to avoid confusing the two in equations such as Equation 16-11.

At room temperature, the majority of molecules are in the ground state ($v = 0$); thus, from Equation 16-11,

$$E_0 = \frac{1}{2}h\nu_m$$

Promotion to the first excited state ($v = 1$) with energy

$$E_1 = \frac{3}{2}h\nu_m$$

requires radiation of energy

$$\left(\frac{3}{2}h\nu_m - \frac{1}{2}h\nu_m\right) = h\nu_m$$

The frequency of radiation ν that will bring about this change is *identical to the classical vibrational frequency of the bond* ν_m. That is,

$$E_{\text{radiation}} = h\nu = \Delta E = h\nu_m = \frac{h}{2\pi}\sqrt{\frac{k}{\mu}}$$

or

$$\nu = \nu_m = \frac{1}{2\pi}\sqrt{\frac{k}{\mu}} \qquad (16\text{-}13)$$

If we wish to express the radiation in wavenumbers, we substitute Equation 6-3 (page 118) and rearrange:

$$\bar{\nu} = \frac{1}{2\pi c}\sqrt{\frac{k}{\mu}} = 5.3 \times 10^{-12}\sqrt{\frac{k}{\mu}} \qquad (16\text{-}14)$$

where $\bar{\nu}$ is the wavenumber of an absorption peak in cm^{-1}, k is the force constant for the bond in newtons per meter (N/m), c is the velocity of light in cm/s, and the reduced mass μ, which is defined by Equation 16-8, has units of kg.[4]

Equation 16-14 and infrared measurements permit the evaluation of the force constants for various types of chemical bonds. Generally, k has been found to lie in the range between 3×10^2 and 8×10^2 N/m for most single bonds, with 5×10^2 serving as a reasonable average value. Double and triple bonds are found by this same means to have force constants of about two and three times this value (1×10^3 and 1.5×10^3, respectively). With these average experimental values, Equation 16-14 can be used to estimate the wavenumber of the fundamental absorption peak, or the absorption peak due to the transition from the ground state to the first excited state, for a variety of bond types. The following example demonstrates such a calculation.

[4] By definition, the newton is $N = kg \cdot m/s^2$ Thus, $\sqrt{k/\mu}$ has units of s^{-1}.

EXAMPLE 16-1

Calculate the approximate wavenumber and wavelength of the fundamental absorption peak due to the stretching vibration of a carbonyl group C═O.

The mass of the carbon atom in kilograms is given by

$$m_1 = \frac{12 \times 10^{-3}\ \text{kg}/\cancel{\text{mol}}}{6.0 \times 10^{23}\ \cancel{\text{atom}}/\cancel{\text{mol}}} \times 1\ \cancel{\text{atom}}$$

$$= 2.0 \times 10^{-26}\ \text{kg}$$

Similarly for oxygen,

$$m_2 = (16 \times 10^{-3})/(6.0 \times 10^{23}) = 2.7 \times 10^{-26}\ \text{kg}$$

and the reduced mass μ is given by (Equation 16-8)

$$\mu = \frac{2.0 \times 10^{-26}\ \text{kg} \times 2.7 \times 10^{-26}\ \cancel{\text{kg}}}{(2.0 + 2.7) \times 10^{-26}\ \cancel{\text{kg}}}$$

$$= 1.1 \times 10^{-26}\ \text{kg}$$

As noted earlier, the force constant for the typical double bond is about 1×10^3 N/m. Substituting this value and μ into Equation 16-14 gives

$$\bar{\nu} = 5.3 \times 10^{-12}\ \text{s/cm}\sqrt{\frac{1 \times 10^3\ \text{N/m}}{1.1 \times 10^{-26}\ \text{kg}}}$$

$$= 1.6 \times 10^3\ \text{cm}^{-1}$$

The carbonyl stretching band is found experimentally to be in the region of 1600 to 1800 cm^{-1} (6.3 to 5.6 μm).

Selection Rules

As given by Equations 16-11 and 16-12, the energy for a transition from energy level 1 to 2 or from level 2 to 3 should be identical to that for the 0 to 1 transition. Furthermore, quantum theory indicates that the only transitions that can take place are those in which the vibrational quantum number changes by unity; that is, the so-called *selection rule* states that $\Delta v = \pm 1$. Since the vibrational levels are equally spaced, only a single absorption peak should be observed for a given molecular vibration.

Anharmonic Oscillator

Thus far, we have considered the classical and quantum mechanical treatments of the harmonic oscillator. The potential energy of such a vibrator changes periodically as the distance between the masses fluctuates (Figure 16-3a). From qualitative considerations, however, this description of a molecular vibration appears imperfect. For example, as the two atoms approach one another, coulombic repulsion between the two nuclei produces a force that acts in the same direction as the restoring force of the bond; thus, the potential energy can be expected to rise more rapidly than the harmonic approximation predicts. At the other extreme of oscillation, a decrease in the restoring force, and thus the potential energy, occurs as the interatomic distance approaches that at which dissociation of the atoms takes place.

In theory, the wave equations of quantum mechanics permit the derivation of more nearly correct potential-energy curves for molecular vibrations. Unfortunately, however, the mathematical complexity of these equations precludes their quantitative application to all but the very simplest of systems. Qualitatively, the curves must take the *anharmonic* form shown as curve 2 in Figure 16-3b. Such curves depart from harmonic behavior by varying degrees, depending upon the nature of the bond and the atoms involved. Note, however, that the harmonic and anharmonic curves are nearly alike at low potential energies. This fact accounts for the success of the approximate methods described.

Anharmonicity leads to deviations of two kinds. At higher quantum numbers, ΔE becomes smaller (see curve 2 in Figure 16-3b), and the selection rule is not rigorously followed; as a result, transitions of $\Delta v = \pm 2$ or ± 3 are observed. Such transitions are responsible for the appearance of *overtone lines* at frequencies approximately two or three times that of the fundamental line; the intensity of overtone absorption is frequently low, and the peaks may not be observed.

Vibrational spectra are further complicated by the fact that two different vibrations in a molecule can interact to give absorption peaks with frequencies that are approximately the sums or differences of their fundamental frequencies. Again, the intensities of combination and difference peaks are generally low.

16A-4 Vibrational Modes

It is ordinarily possible to deduce the number and kinds of vibrations in simple diatomic and triatomic molecules and whether these vibrations will lead to absorption. Complex molecules may contain several types of atoms as well as bonds; for these molecules, the multitude of possible vibrations gives rise to infrared spectra that are difficult, if not impossible, to analyze.

The number of possible vibrations in a polyatomic molecule can be calculated as follows. Three coordinates are needed to locate a point in space; to fix N points requires a set of three coordinates for each point, for a total of $3N$. Each coordinate corresponds to one degree of freedom for one of the atoms in a polyatomic molecule; for this reason, a molecule containing N atoms is said to have *$3N$ degrees of freedom.*

In defining the motion of a molecule, we need to consider: (1) the motion of the entire molecule through space (that is, the translational motion of its center of gravity); (2) the rotational motion of the entire molecule around its center of gravity; and (3) the motion of each of its atoms relative to the other atoms, or in other words, its individual vibrations. Since all atoms in the molecule move in concert through space, definition of translational motion requires three coordinates and thus this common motion requires three of the $3N$ degrees of freedom. Another three degrees of freedom are needed to describe the rotation of the molecule as a whole. The remaining $3N - 6$ degrees of freedom involve interatomic motion and hence represent the number of possible vibrations within the molecule. A linear molecule is a special case since, by definition, all of the atoms lie on a single, straight line. Rotation about the bond axis is not possible, and two degrees of freedom suffice to describe rotational motion. Thus, the number of vibrations for a linear molecule is given by $3N - 5$. Each of the $3N - 6$ or $3N - 5$ vibrations is called a *normal mode.*

For each normal mode of vibration, there is a potential-energy relationship such as that shown by the solid line in Figure 16-3b. The same selection rules discussed earlier apply for each of these relationships. In addition, to the extent that a vibration approximates harmonic behavior, the differences between the energy levels of a given vibration are the same; that is, a single absorption peak should appear for each vibration in which there is a change in dipole.

Four factors tend to produce fewer experimental peaks than would be expected from the theoretical number of normal modes. Fewer peaks are found when (1) the symmetry of the molecules is such that no change in dipole results from a particular vibration; (2) the energies of two or more vibrations are identical or nearly identical; (3) the absorption intensity is so low as to be undetectable by ordinary means; or (4) the vibrational energy is in a wavelength region beyond the range of the instrument.

Occasionally more peaks are found than are expected based upon the number of normal modes. We have already mentioned the occurrence of overtone peaks that occur at two or three times the frequency of a fundamental peak. In addition *combination bands* are sometimes encountered when a photon excites two vibrational modes simultaneously. The frequency of the combination band is approximately the sum or difference of the two fundamental frequencies. This phenomenon occurs when a quantum of energy is absorbed by two bonds rather than one.

16A-5 Vibrational Coupling

The energy of a vibration, and thus the wavelength of its absorption peak, may be influenced by (or coupled with) other vibrators in the molecule. A number of factors influence the extent of such coupling.

1. Strong coupling between stretching vibrations occurs only when there is an atom common to the two vibrations.
2. Interaction between bending vibrations requires a common bond between the vibrating groups.
3. Coupling between a stretching and a bending vibration can occur if the stretching bond forms one side of the angle that varies in the bending vibration.
4. Interaction is greatest when the coupled groups have individual energies that are approximately equal.
5. Little or no interaction is observed between groups separated by two or more bonds.
6. Coupling requires that the vibrations be of the same symmetry species.[5]

As an example of coupling effects, let us consider the infrared spectrum of carbon dioxide. If no coupling occurred between the two C=O bonds, an absorption peak would be expected at the same wavenumber as the peak for the C=O stretching vibration in an aliphatic ketone (about 1700 cm^{-1}, or 6 μm; see Example 16-1). Experimentally, carbon dioxide exhibits two absorption peaks, the one at 2330 cm^{-1} (4.3 μm) and the other at 667 cm^{-1} (15 μm).

Carbon dioxide is a linear molecule and thus has $3 \times 3 - 5 = 4$ normal modes. Two stretching vibrations are possible; furthermore, interaction between the two can occur since the bonds involved are associated with a common carbon atom. As may be seen, one of the coupled vibrations is symmetric and the other is asymmetric.

The symmetric vibration causes no change in dipole, since the two oxygen atoms simultaneously move away from or toward the central carbon atom. Thus, the symmetric vibration is infrared inactive. In the asymmetric vibration, one oxygen moves away from the carbon atom as the carbon atom moves toward the other oxygen. As a consequence, a net change in charge distribution occurs periodically, producing a change in dipole moments, so absorption at 2330 cm^{-1} results.

The remaining two vibrational modes of carbon dioxide involve scissoring, as shown below.

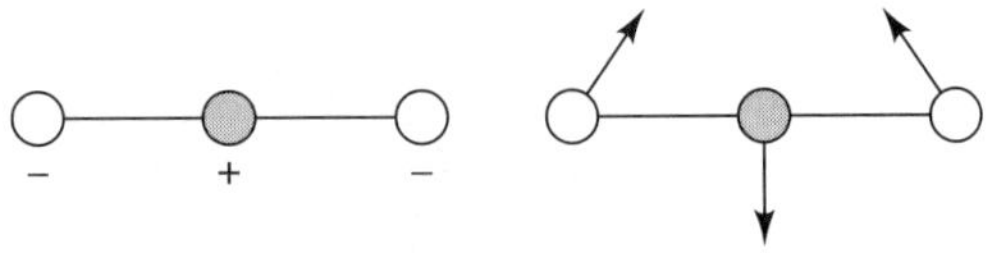

The two bending vibrations are the resolved components at 90 deg to one another of the bending motion in all possible planes around the bond axis. The two vibrations are identical in energy and thus produce a single peak at 667 cm^{-1}. Quantum states that are identical, as these are, are said to be *degenerate*.

It is of interest to compare the spectrum of carbon dioxide with that of a nonlinear, triatomic molecule such as water, sulfur dioxide, or nitrogen dioxide. These molecules have $3 \times 3 - 6 = 3$ vibrational modes which take the following forms:

[5] For a discussion of symmetry operations and symmetry species, see R. P. Bauman, *Absorption Spectroscopy,* Chapter 10. New York: Wiley, 1962; F. A. Cotton, *Chemical Applications of Group Theory.* New York: Wiley, 1971.

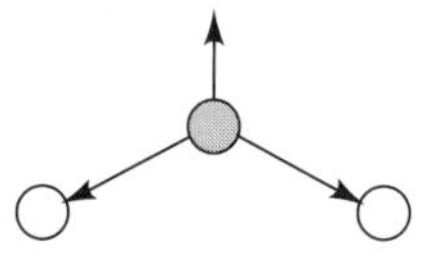

Symmetric stretching

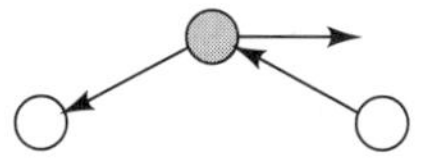

Asymmetric streching

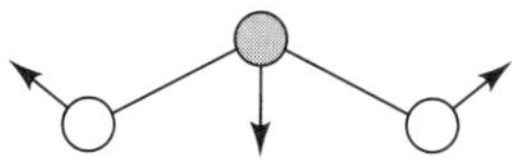

Scissoring

Because the central atom is not in line with the other two, a symmetric stretching vibration will produce a change in dipole and will thus be responsible for infrared absorption. For example, stretching peaks at 3650 and 3760 cm^{-1} (2.74 and 2.66 μm) appear in the infrared spectrum for the symmetric and asymmetric vibrations of the water molecule. There is only one component to the scissoring vibration for this nonlinear molecule because motion in the plane of the molecule constitutes a rotational degree of freedom. For water, the bending vibration causes absorption at 1595 cm^{-1} (6.27 μm). The difference in behavior of linear and nonlinear triatomic molecules with two and three absorption peaks, respectively, illustrates how infrared absorption spectroscopy can sometimes be used to deduce molecular shapes.

Coupling of vibrations is a common phenomenon; as a result, the position of an absorption peak corresponding to a given organic functional group cannot always be specified exactly. For example, the C—O stretching frequency in methanol is 1034 cm^{-1} (9.67 μm), in ethanol it is 1053 cm^{-1} (9.50 μm), and in 2-butanol it is 1105 cm^{-1} (9.05 μm). These variations result from a coupling of the C—O stretching with adjacent C—C stretching or C—H vibrations.

While interaction effects may lead to uncertainties in the identification of functional groups contained in a compound, it is this very effect that provides the unique features of an infrared absorption spectrum that are so important for the positive identification of a specific compound.

16B INFRARED SOURCES AND TRANSDUCERS

Instruments for measuring infrared absorption all require a source of continuous infrared radiation and a sensitive infrared transducer. The desirable characteristics of these instrument components were listed in Sections 7B and 7E. In this section we describe sources and transducers that are found in current-day infrared instruments.

16B-1 Sources

Infrared sources consist of an inert solid that is heated electrically to a temperature between 1500 and 2200 K. Continuum radiation approximating that of a blackbody results (see Figure 6-18, page 134). The maximum radiant intensity at these temperatures occurs between 5000 to 5900 cm^{-1} (2 to 1.7 μm). At longer wavelengths, the intensity falls off smoothly until it is about 1% of the maximum at 670 cm^{-1} (15 μm). On the short wavelength side, the decrease is much more rapid, and a similar reduction in intensity is observed at about 10,000 cm^{-1} (1 μm).

The Nernst Glower

The Nernst glower is composed of rare earth oxides formed into a cylinder having a diameter of 1 to 2 mm and a length of perhaps 20 mm. Platinum leads are sealed to the ends of the cylinder to permit electrical connection to what amounts to a resistive heating element. As current is passed through the device, temperatures between 1200 K and 2200 K result. The Nernst glower has a large negative temperature coefficient of electrical resistance, and it must be heated externally to a dull red heat before the current is large enough to maintain the desired temperature. Because the resistance decreases with increasing temperature, the source circuit must be designed to limit the current; otherwise the glower rapidly becomes so hot that it is destroyed.

Figure 16-4 shows the spectral output of a Nernst glower operated at approximately 2200 K. Note that the curve resembles that of a blackbody; the small peaks and depressions are a function of the chemical composition of the device.

The Globar Source

A Globar is a silicon carbide rod, usually about 50 mm in length and 5 mm in diameter. It also is electrically heated (1300 to 1500 K) and has the advantage of a positive coefficient of resistance. On the other hand, water cooling of the electrical contacts is required to prevent arcing. Spectral energies of the Globar and the Nernst glower are comparable except in the region below 5 μm, where the Globar provides a significantly greater output.

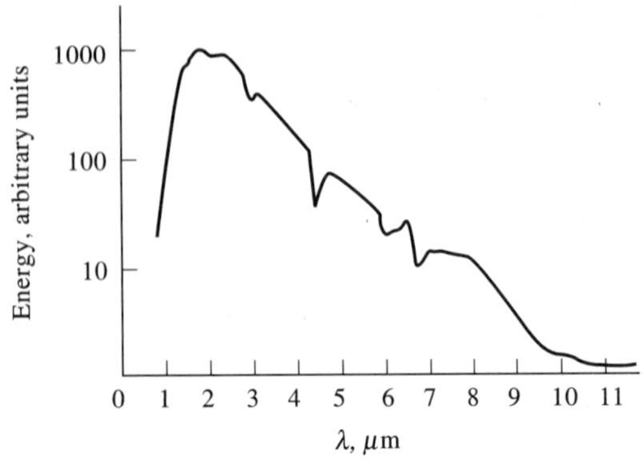

Figure 16-4 Spectral distribution of energy from a Nernst glower operated at approximately 2200 K.

Incandescent Wire Source

A source of somewhat lower intensity but longer life than the Globar or Nernst glower is a tightly wound spiral of nichrome wire heated to about 1100 K by an electrical current. A rhodium wire heater sealed in a ceramic cylinder has similar properties as a source.

The Mercury Arc

For the far-infrared region of the spectrum ($\lambda > 50$ μm), none of the thermal sources just described provides sufficient radiant power for convenient detection. Here, a high-pressure mercury arc is used. This device consists of a quartz-jacketed tube containing mercury vapor at a pressure greater than one atmosphere. Passage of electricity through the vapor forms an internal plasma source that provides continuum radiation in the far-infrared region.

The Tungsten Filament Lamp

An ordinary tungsten filament lamp is a convenient source for the near-infrared region of 4000 to 12,800 cm^{-1} (2.5 to 0.78 μm).

The Carbon Dioxide Laser Source

A tunable carbon dioxide laser is used as an infrared source for monitoring the concentrations of certain atmospheric pollutants and for determining absorbing species in aqueous solutions.[6] A carbon dioxide laser produces a band of radiation in the 900 to 1100 cm^{-1} (11 to 9 μm) range, which consists of about 100 closely spaced discrete lines. As described on page 151, any one of these lines can be chosen by tuning the laser. Although the range of wavelengths available is limited, the 900 to 1100 cm^{-1} region is one particularly rich in absorption bands arising from the interactive stretching modes of CO_2. Thus, this source is useful for quantitative determination of a number of important species such as ammonia, butadiene, benzene, ethanol, nitrogen dioxide, and trichloroethylene. An important property of the laser source is the radiant power available in each line, which is several orders of magnitude greater than that of blackbody sources.

16B-2 Infrared Transducers

Infrared transducers are of three general types: (1) thermal transducers; (2) pyroelectric transducers (a very specialized thermal transducer); (3) photoconducting transducer. The first two are commonly found in photometers and dispersive spectrophotometers. Photoconducting transducers are found in Fourier transform multiplex instruments.

Thermal Transducers

Thermal transducers, whose responses depend upon the heating effect of radiation, are employed for detection of all but the shortest infrared wavelengths. With these devices, the radiation is absorbed by a small blackbody and the resultant temperature rise is measured. The radiant power level from a spectrophotometer beam is minute (10^{-7} to 10^{-9} W), so that the heat capacity of the absorbing element must be as small as possible if a detectable temperature change is to be produced. Every effort is made to minimize the size and thickness of the absorbing element and to concentrate the entire infrared beam on its surface. Under the best of circumstances, temperature changes are confined to a few thousandths of a kelvin.

The problem of measuring infrared radiation by thermal means is compounded by thermal noise from the surroundings. For this reason, thermal transducers are housed in a vacuum and are carefully shielded from thermal radiation emitted by other nearby objects. To further minimize the effects of extraneous heat sources, the beam from the source is always chopped. In this way, the analyte signal, after transduction, has the frequency of the chopper and can be separated electroni-

[6] See H. R. Jones and P. A. Wilks Jr., *Amer. Lab.*, **1982**, *14* (3), 87; L. B. Kreuzer, *Anal. Chem.*, **1974**, *46*, 239A.

cally from extraneous noise signals, which ordinarily vary only slowly with time.

Thermocouples. In its simplest form, a thermocouple consists of a pair of junctions formed when two pieces of a metal such as bismuth are fused to either end of a dissimilar metal such as antimony. A potential develops between the two junctions that varies with their *difference* in temperature.

The transducer junction for infrared radiation is formed from very fine wires or alternatively by evaporating the metals onto a nonconducting support. In either case, the junction is usually blackened (to improve its heat absorbing capacity) and sealed in an evacuated chamber with a window that is transparent to infrared radiation.

The reference junction, which is usually housed in the same chamber as the active junction, is designed to have a relatively large heat capacity and is carefully shielded from the incident radiation. Because the analyte signal is chopped, only the difference in temperature between the two junctions is important; therefore, the reference junction does not need to be maintained at constant temperature. To enhance sensitivity, several thermocouples may be connected in series to give what is called a *thermopile.*

A well-designed thermocouple transducer is capable of responding to temperature differences of 10^{-6} K. This figure corresponds to a potential difference of about 6 to 8 μV/μW. The thermocouple of an infrared transducer is a low-impedance device that is usually connected to a high-impedance preamplifier, such as the field-effect transistor circuit shown in Figure 16-5. A voltage follower operational amplifier, such as that shown in Figure 3-7, is also used as a preamplifier in thermocouple circuits.

Bolometers. A bolometer is a type of resistance thermometer constructed of strips of metals such as platinum or nickel, or from a semiconductor; the latter devices are sometimes called *thermistors.* These materials exhibit a relatively large change in resistance as a function of temperature. The responsive element is kept small and blackened to absorb the radiant heat. Bolometers are not so extensively used as other infrared transducers for the mid-infrared region. However, a germanium bolometer, operated at 1.5 K, is an ideal transducer for radiation in the 5 to 400 cm^{-1} (2000 to 25 mm) range.

Pyroelectric Transducers

Pyroelectric transducers are constructed from single crystalline wafers of *pyroelectric materials,* which are insulators (dielectric materials) with very special thermal and electrical properties. Triglycine sulfate, $(NH_2CH_2COOH)_3 \cdot H_2SO_4$ (usually deuterated or with a fraction of the glycines replaced with alanine), is the

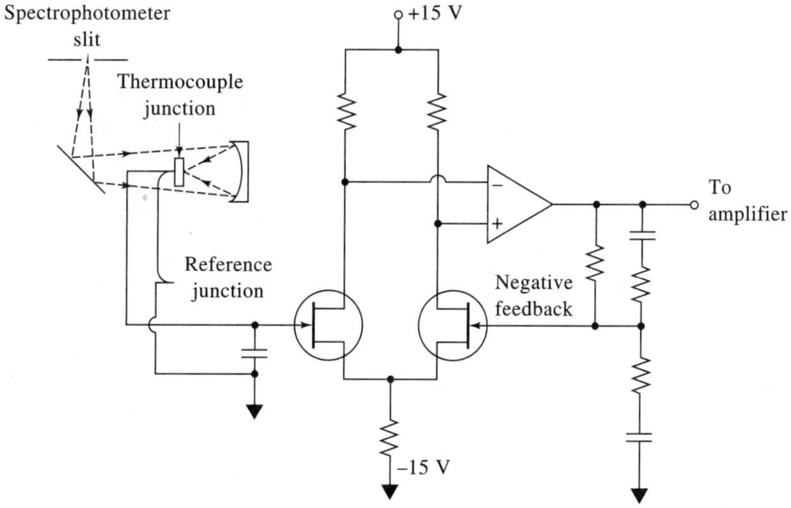

Figure 16-5 Thermocouple and preamplifier. *(Adapted from G. W. Ewing,* J. Chem. Educ., *1971,* 48, *A521. With permission.)*

most important pyroelectric material used in the construction of infrared detection systems.

When an electric field is applied across any dielectric material, electric polarization takes place, with the magnitude of this polarization being a function of the dielectric constant of the material. For most dielectrics, this induced polarization rapidly decays to zero when the external field is removed. Pyroelectric substances, in contrast, retain a strong temperature-dependent polarization after removal of the field. Thus, by sandwiching the pyroelectric crystal between two electrodes, one of which is infrared transparent, a temperature-dependent capacitor is produced. Changing its temperature by irradiating it with infrared radiation alters the charge distribution across the crystal, which can be detected as a current in an external electric circuit connecting the two sides of the capacitor. The magnitude of this current is proportional to the surface area of the crystal and the rate of change of polarization with temperature. Pyroelectric crystals lose their residual polarization when they are heated to a temperature called the *Curie point.* For triglycine sulfate, the Curie point is 47°C.

Pyroelectric transducers exhibit response times that are fast enough to allow them to track the changes in the time domain signal from an interferometer. For this reason, most Fourier transform infrared spectrometers employ this type of transducer.

Photoconducting Transducers

Infrared photoconducting transducers consist of a thin film of a semiconductor material, such as lead sulfide, mercury/cadmium telluride, or indium antimonide, deposited on a nonconducting glass surface and sealed into an evacuated envelope to protect the semiconductor from the atmosphere. Absorption of radiation by these materials promotes nonconducting valence electrons to a higher energy conducting state, thus decreasing the electrical resistance of the semiconductor. Typically, a photoconductor is placed in series with a voltage source and load resistor, and the voltage drop across the load resistor serves as a measure of the power of the beam of radiation.

A lead sulfide photoconductor is the most widely used transducer for the near-infrared region of the spectrum from 10,000 to 333 cm^{-1} (1 to 3 μm). It can be operated at room temperature. For mid- and far-infrared radiation, mercury/cadmium telluride photoconductor transducers are used. They must be cooled with liquid nitrogen (77 K) to minimize thermal noise. The long-wavelength cutoff, and many of the other properties of these transducers, depend upon the mercury telluride/cadmium telluride ratio, which can be varied continuously.

The mercury/cadmium telluride transducer, which offers superior response characteristics to the pyroelectric transducers discussed in the previous section, also finds widespread use in Fourier transform spectrometers, particularly those that are interfaced to gas-chromatographic equipment.

16C INFRARED INSTRUMENTS

Three types of instruments for infrared absorption measurements are available from commercial sources: (1) dispersive grating spectrophotometers that are used primarily for qualitative work; (2) multiplex instruments, employing the Fourier transform (Section 7I), that are suited to both qualitative and quantitative infrared measurements; and (3) nondispersive photometers that have been developed for quantitative determination of a variety of organic species in the atmosphere by absorption, emission, and reflectance spectroscopy.

Until the 1980s, the most widely used instruments for infrared measurements were dispersive spectrophotometers. Now, however, this type of instrument has been largely displaced by Fourier transform spectrometers because of their speed, reliability, and convenience. Only when cost is a major consideration are dispersive instruments still employed.

16C-1 Fourier Transform Spectrometers

The theoretical basis and the inherent advantages of multiplex instruments were discussed in some detail in Section 7I-1, and the reader may find it worthwhile to review that section before proceeding further here. Two types of multiplex instruments have been described for the infrared region. In one, coding is accomplished by splitting the source into two beams whose path lengths can be varied periodically to give interference patterns. The Fourier transform is then used for data processing.[7]

[7] For detailed discussions of Fourier transform infrared spectroscopy, see B. C. Smith, *Fourier Transform Infrared Spectroscopy.* Boca Raton: CRC Press, 1996; P. R. Griffiths and J. A. deHaseth, *Fourier Transform Infrared Spectroscopy.* New York: Wiley, 1986; *Fourier Transform Infrared Spectroscopy,* J. Ferraro and L. Basile, Eds. New York: Academic Press, 1979. For review articles, see W. D. Perkins, *J. Chem. Educ.,* **1986,** *63,* A5, A269; L. Glaser, *J. Chem. Educ.,* **1987,** *64,* A228, A260.

The second is the Hadamard transform spectrometer, which is a dispersive instrument that employs a moving mask at the focal plane of a monochromator for encoding the spectral data. Hadamard transform infrared instruments have not been widely adopted and will, therefore, not be discussed further in this text.[8]

When Fourier transform infrared (FTIR) spectrometers first appeared in the marketplace, they were bulky, expensive (> \$100,000), and required frequent mechanical adjustments. For these reasons, their use was limited to special applications where their unique characteristics (speed, high resolution, sensitivity, and unparalleled wavelength precision and accuracy) were essential. Fourier transform instruments have now been reduced to benchtop size and have become reliable and easy to maintain. Furthermore, the price of simpler models has been reduced to the point where they are competitive with all but the simplest dispersive instruments (\$15,000 to \$20,000). For these reasons, Fourier transform instruments have largely displaced dispersive instruments in the laboratory.[9]

Components of Fourier Transform Instruments

The majority of commercially available Fourier transform infrared instruments are based upon the Michelson interferometer, although other types of optical systems are also encountered. We shall consider the Michelson design only, which is illustrated in Figure 7-42, page 186.[10]

Drive Mechanism. A requirement for satisfactory interferograms (and thus satisfactory spectra) is that the speed of the moving mirror be constant and its position exactly known at any instant. The planarity of the mirror must also remain constant during its entire sweep of 10 cm or more.

In the far-infrared region, where wavelengths range from 50 to 1000 μm (200 to 10 cm^{-1}), displacement of the mirror by a fraction of a wavelength, and accurate measurement of its position, can be accomplished by means of a motor-driven micrometer screw. A more precise and sophisticated mechanism is required for the mid- and near-infrared regions, however. Here, the mirror mount is generally floated on an air bearing held within close-fitting stainless steel sleeves (see Figure 16-6). The mount is driven by a linear drive motor and an electromagnetic coil similar to the voice coil in a loudspeaker; an increasing current in the coil drives the mirror at constant velocity. After reaching its terminus, the mirror is returned rapidly to the starting point for the next sweep by a rapid reversal of the current. The length of travel varies from 1 to about 20 cm; the scan rates range from 0.01 to 10 cm/s.

Two additional features of the mirror system are necessary for successful operation. The first is a means of sampling the interferogram at precisely spaced retardation intervals. The second is a method for determining exactly the zero retardation point in order to permit signal averaging. If this point is not known precisely, the signals from repetitive sweeps would not be in phase; averaging would then tend to degrade rather than improve the signal.

The problem of precise signal sampling and signal averaging can be accomplished by using three interferometers rather than one, with a single mirror mount holding the three movable mirrors. Figure 16-6 is a schematic showing such an arrangement. The components and radiation paths for each of the three interferometer systems are indicated by the subscripts 1, 2, and 3, respectively. System 1 is the infrared system that ultimately provides an interferogram similar to that shown as curve *A* in Figure 16-7. System 2 is a so-called *laser-fringe reference* system, which provides sampling-interval information. It consists of a helium/neon laser S_2, an interferometric system including mirrors MM_2 and M_2, a beam splitter B_2, and a transducer T_2. The output from this system is a cosine wave, as shown in *C* of Figure 16-7. This signal is converted electronically to the square-wave form shown in *D;* sampling begins or terminates at each successive zero crossing. The laser-fringe reference system gives a highly reproducible and regularly spaced sampling interval. In most instruments, the laser signal is also used to control the speed of the mirror-drive system at a constant level.

The third interferometer system, sometimes called the *white-light* system, employs a tungsten source S_3 and transducer T_3 sensitive to visible radiation. Its mirror system is fixed to give a zero retardation that is dis-

8 For a description of the Hadamard transform and Hadamard transform spectrometers, see M. O. Harwit and N. J. A. Sloane, *Hadamard Transform Optics.* New York: Academic Press, 1979; *Transform Techniques in Chemistry,* P. R. Griffiths, Ed. New York: Plenum Press, 1978. For a recent review of analytical applications of the Hadamard transform, see P. J. Treado and M. D. Morris, *Anal. Chem.,* **1989,** *61,* 723A.

9 See S. A. Borman, *Anal. Chem.,* **1983,** *55,* 1054A.

10 The Michelson interferometer was designed and built in 1891 by A. A. Michelson. He was awarded the 1907 Nobel Prize in physics for the invention of interferometry.

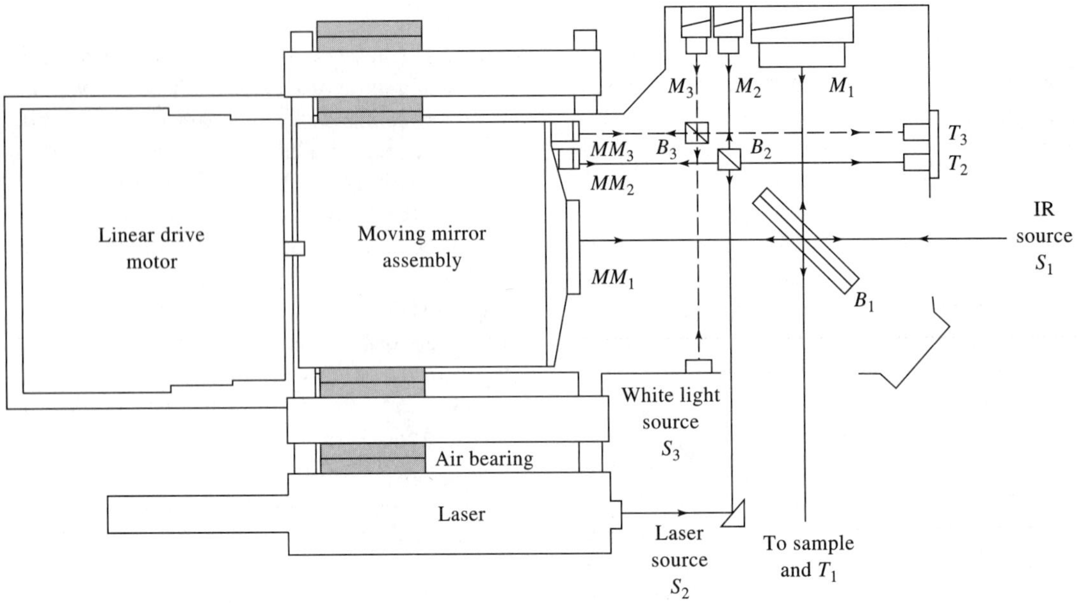

Figure 16-6 Interferometers in an infrared Fourier transform spectrometer. Subscript 1 defines the radiation path in the infrared interferometer; subscripts 2 and 3 refer to the laser and white-light interferometers, respectively. *(Courtesy of Nicolet Analytical Instruments, Madison, WI.)*

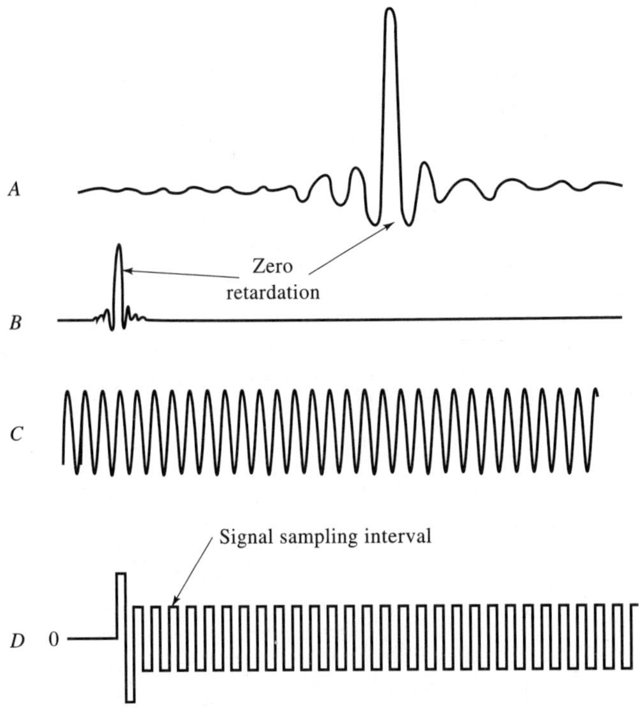

Figure 16-7 Time domain signals for the three interferometers contained in a Fourier transform infrared instrument. Curve *A:* infrared signal; curve *B:* white-light signal; curve *C:* laser-fringe reference signal; curve *D:* square-wave electrical signal formed from the laser signal. *(From P. R. Griffiths,* Chemical Infrared Fourier Transform Spectroscopy, *p. 102. New York: Wiley, 1975. Reprinted by permission of John Wiley & Sons, Inc.)*

placed to the left from that for the analytical signal (see interferogram *B,* Figure 16-7). Because the source is polychromatic, its power at zero retardation is much larger than any signal before and after that point. Thus, this maximum is used to trigger the start of data sampling for each sweep at a highly reproducible point.

The triple interferometer system just described leads to remarkable precision in determining spectral frequencies, which significantly exceeds that realizable with conventional grating instruments. This high reproducibility is particularly important when many scans are to be averaged. Contemporary instruments, such as the one shown in Figure 16-8, are able to achieve this same frequency precision with only a single interferometer. In this instrument, the laser beam is either parallel to or collinear with the infrared beam so that both beams traverse a single interferometer. Furthermore, no white-light source is employed, and the infrared interferogram is used to establish zero retardation. The maximum in the infrared interferogram is an excellent reference because this is the only point at which all wavelengths interfere constructively.

The system shown in Figure 16-6 is capable of providing spectra with a resolution of between 0.1 and 1 cm^{-1}. To obtain resolutions of 0.01 cm^{-1} requires a more sophisticated system for maintaining the alignment of the moving mirror. One mirror alignment system uses three laser-fringe reference systems that are directed at different points on the moving mirror instead of one. Because three points are adequate to define a plane, the use of three lasers significantly increases the accuracy with which the position and orientation of the mirror can be known at any instant.

Beam Splitters. Beam splitters are constructed of transparent materials with refractive indices such that approximately 50% of the radiation is reflected and 50% is transmitted. A widely used material for the far-infrared region is a thin film of Mylar sandwiched between two plates of a low-refractive-index solid. Thin

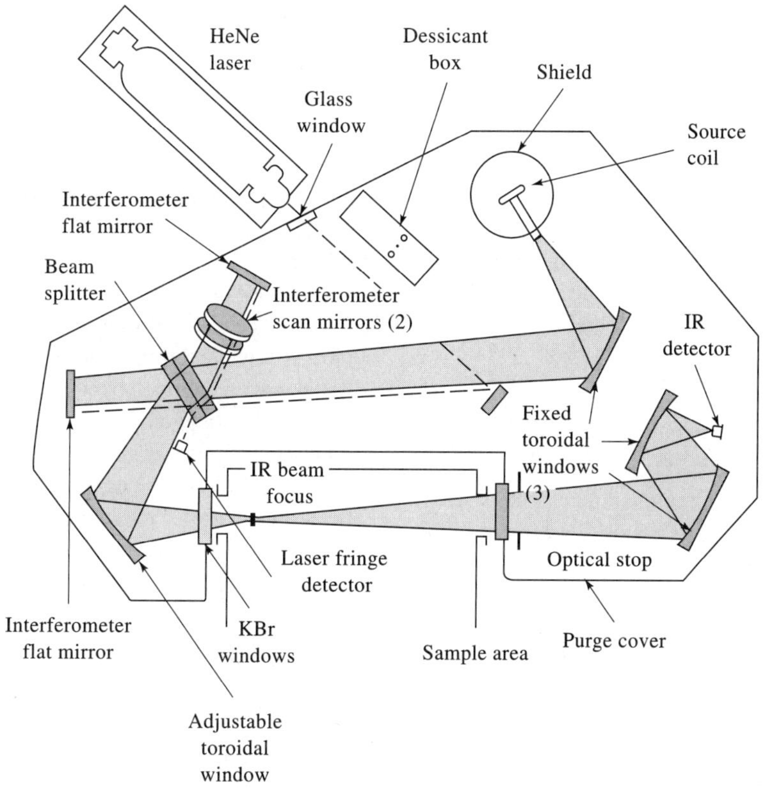

Figure 16-8 A single-beam FTIR spectrometer. *(Courtesy of Perkin-Elmer, Norwalk, CT.)*

films of germanium or silicon deposited on cesium iodide or bromide, sodium chloride, or potassium bromide are satisfactory for the mid-infrared region. A film of iron(III) oxide is deposited on calcium fluoride for work in the near-infrared.

Sources and Transducers. The sources for Fourier transform infrared instruments are similar to those discussed earlier in this chapter. Generally, thermal transducers are not readily adapted to Fourier transform instruments because of their slow response times. Triglycine sulfate pyroelectric transducers are widely used for the mid-infrared region. Where better sensitivity or faster response times are required liquid-nitrogen-cooled mercury/cadmium telluride or indium antimonide photoconductive transducers are employed.

Instrument Designs

Fourier transform infrared spectrometers are usually single-beam instruments. Figure 16-8 shows the optics of a less expensive spectrometer, which sells in the $16,000 to $20,000 range. A typical procedure for determining transmittance or absorbance with this type of instrument is to first obtain a reference interferogram by scanning a reference (usually air) 20 or 30 times, coadding the data, and storing the results in the memory of the instrument computer (usually after transforming it to the spectrum). A sample is then inserted in the radiation path and the process repeated. The ratio of sample and reference spectra data is then computed to give the transmittance at various frequencies. Ordinarily, modern infrared sources and detectors are sufficiently stable so that reference spectra need to be obtained only occasionally.

Performance Characteristics of Commercial Instruments

A number of instrument manufacturers offer several models of Fourier transform infrared instruments. The least expensive of these has a range of 7800 to 350 cm^{-1} (1.3 to 29 μm) with a resolution of 4 cm^{-1}. This performance can be obtained with a scan time as brief as one second. More expensive instruments with interchangeable beam splitters, sources, and transducers offer expanded frequency ranges and higher resolutions. For example, one instrument is reported to produce spectra from the far-infrared (10 cm^{-1} or 1000 μm) through the visible region to 25,000 cm^{-1} or 400 nm. Resolutions for commercial instruments vary from 8 to less than 0.01 cm^{-1}. Several minutes are required to obtain a complete spectrum at the highest resolution.[11]

Advantages of Fourier Transform Spectrometers[12]

Over most of the mid-infrared spectral range, Fourier transform instruments appear to have signal-to-noise ratios that are better than those of a good-quality dispersive instrument by more than an order of magnitude. The enhanced signal-to-noise ratio can, of course, be traded for rapid scanning, with good spectra being attainable in a few seconds in most cases. Interferometric instruments are also characterized by high resolutions (<0.1 cm^{-1}) and highly accurate and reproducible frequency determinations. The latter property is particularly helpful when spectra are to be subtracted for background correction.

A theoretical advantage of Fourier transform instruments is that their optics provide a much larger energy throughput (one or two orders of magnitude) than do dispersive instruments, which are limited in throughput by the necessity of narrow slit widths. The potential gain here, however, may be partially offset by the lower sensitivity of the fast-response detector required for the interferometric measurements. Finally, it should be noted that the interferometer is free from the problem of stray radiation because each infrared frequency is, in effect, chopped at a different frequency.

The areas of chemistry where the extra performance of interferometric instruments have been particularly useful include: (1) very high-resolution work that is encountered with gaseous mixtures having complex spectra resulting from the superposition of vibrational and rotational bands; (2) the study of samples with high absorbances; (3) the study of substances with weak absorption bands (for example, the study of compounds that are chemisorbed on catalyst surfaces); (4) investigations requiring fast scanning such as kinetic studies or detection of chromatographic effluents; (5) collecting infrared data from very small samples; (6) obtaining reflection spectra; (7) infrared emission studies.

---

[11] D. Noble, *Anal. Chem.*, **1995**, *67*, 381A.

[12] For a comparison of performance characteristics of typical dispersive and interferometric instruments, see D. H. Chenery and N. Sheppard, *Appl. Spectrosc.*, **1978**, *32*, 79; W. D. Perkins, *J. Chem. Educ.*, **1987**, *64*, A269.

16C-2 Dispersive Instruments

Dispersive infrared spectrophotometers from commercial sources are generally double-beam, recording instruments, which use reflection gratings for dispersing radiation. As was pointed out in Section 13D-2, the double-beam design is less demanding with respect to the performance of sources and detectors—an important characteristic because of the relatively low intensity of infrared sources, the low sensitivity of infrared transducers, and the consequent need for large signal amplifications.

An additional reason for the general use of double-beam instruments in the infrared region is shown in Figure 16-9. The lower curve reveals that atmospheric water and carbon dioxide absorb in some important spectral regions and can cause serious interference problems. The upper curve shows that the reference beam compensates nearly perfectly for absorption by both compounds. A stable 100% *T* baseline results.

Generally, dispersive infrared spectrophotometers incorporate a low-frequency chopper (5 to 13 cycles per minute) that permits the detector to discriminate between the signal from the source and signals from extraneous radiation, such as infrared emission from various bodies surrounding the transducer. Low chopping rates are demanded by the slow response times of the infrared transducers used in most dispersive instruments. In general, the optical designs of dispersive instruments do not differ greatly from the double-beam ultraviolet/visible spectrophotometers discussed in Chapters 7 and 13 except that the sample and reference compartments are always located between the source and the monochromator in infrared instruments. This arrangement is possible because infrared radiation, in contrast to ultraviolet/visible, is not sufficiently energetic to cause photochemical decomposition of the sample. Placing the sample and reference before the monochromator, however, has the advantage that most scattered radiation, generated within the cell compartment, is effectively removed by the monochromator and thus does not reach the transducer.

Figure 16-10 shows schematically the arrangement of components in a typical infrared spectrophotometer. Like most inexpensive dispersive infrared instruments, it is a null type, in which the power of the reference beam is reduced, or *attenuated,* to match that of the beam passing through the sample. Attenuation is accomplished by imposing a device that removes a continuously variable fraction of the reference beam. The attenuator commonly takes the form of a comb, the teeth of which are tapered so that there is a linear relationship between the lateral movement of the comb and the decrease in power of the beam. Movement of the

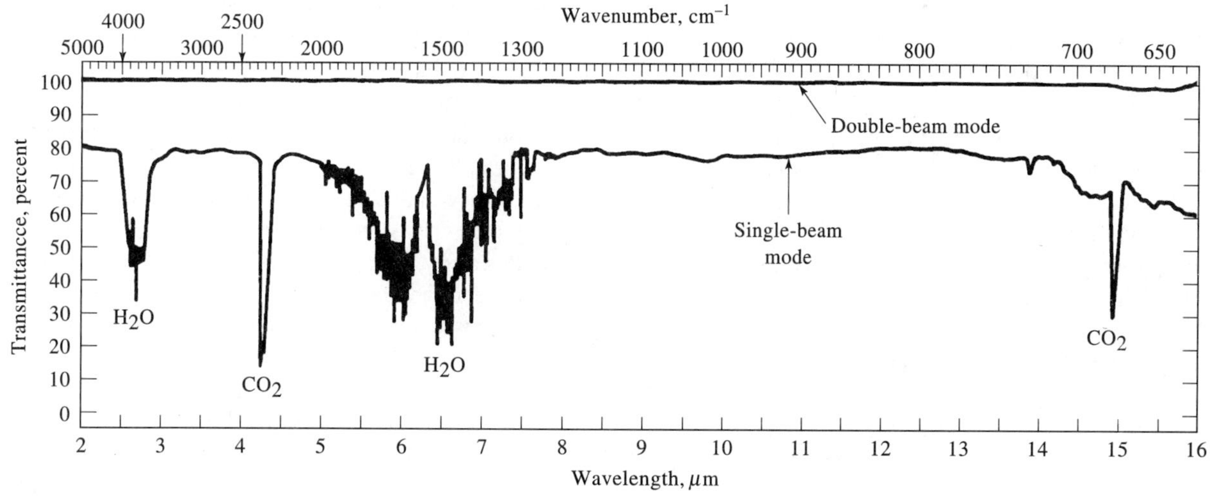

Figure 16-9 Single- and double-beam spectra of atmospheric water vapor and CO_2. In the lower, single-beam trace, the absorption of atmospheric gases is apparent. The top, double-beam trace shows that the reference beam compensates nearly perfectly for this absorption and allows a stable 100% *T* baseline to be obtained. *(From J. D. Ingle Jr. and S. R. Crouch,* Spectrochemical Analysis, *p. 409. Englewood Cliffs, NJ: Prentice-Hall, 1988. With permission.)*

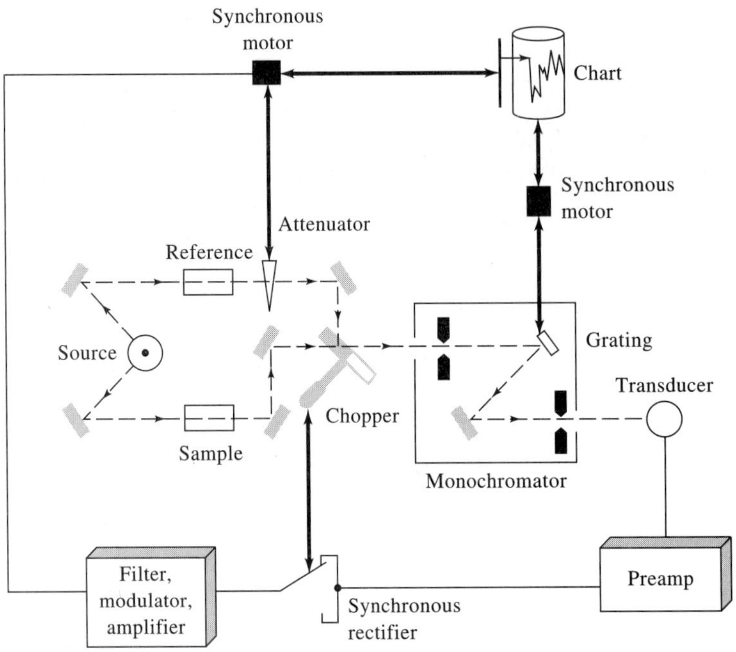

Figure 16-10 Schematic of a double-beam spectrophotometer. Heavy dark line indicates mechanical linkage; light line indicates electrical linkage; dashed line indicates radiation path.

comb occurs when a difference in power of the two beams is sensed by the detection system. This movement is synchronized with the recorder pen so that its position gives a measure of the relative power of the two beams and thus the transmittance of the sample.

Note that three types of systems link the components of the instrument in Figure 16-10: (1) a radiation linkage indicated by dashed lines, (2) a mechanical linkage shown by thick dark lines, (3) an electrical linkage shown by narrow solid lines.

Radiation from the source is split into two beams, half passing into the sample-cell compartment and the other half into the reference area. The reference beam then passes through the attenuator and onto a chopper. The chopper consists of a motor-driven disk that alternately reflects the reference or transmits the sample beam into the monochromator. After dispersion by a prism or grating, the alternating beams fall on the transducer and are converted to an electrical signal. The signal is amplified and passed to the synchronous rectifier, a device that is mechanically or electrically coupled to the chopper to cause the rectifier switch and the beam leaving the chopper to change simultaneously. If the two beams are identical in power, the signal from the rectifier is a constant direct current. If, on the other hand, the two beams differ in power, a fluctuating or ac current is produced, the phase of which is determined by which beam is the more intense. The current from the rectifier is filtered and further amplified to drive a synchronous motor in one direction or the other, depending upon the phase of the input current. The synchronous motor is mechanically linked to both the attenuator and the pen drive of the recorder and causes both to move until a null is achieved. A second synchronous motor drives the chart and varies the wavelength simultaneously. There is frequently a mechanical linkage between the wavelength and slit drives so that the radiant power reaching the detector is kept approximately constant by variations in the slit width.

The reference beam attenuator system, such as the one just described, creates three limitations to the performance of dispersive infrared instruments. First, the response of the attenuator system always lags behind the transmittance changes, particularly in scanning regions where the signal is changing most rapidly. Second, the momentum associated with both the mechanical attenuator and the recorder system may result in the pen drive overshooting the true transmittance. Third, in

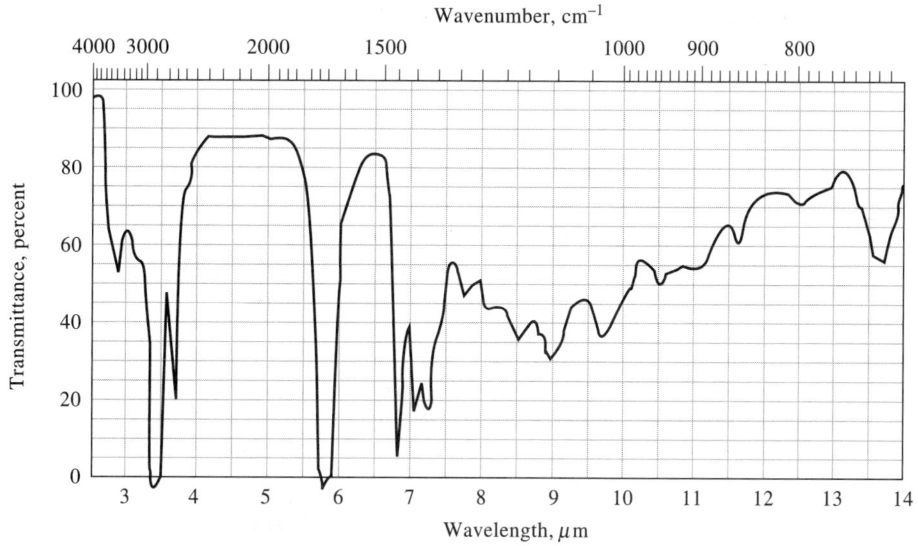

Figure 16-11 Infrared spectrum of *n*-hexanal illustrating overshooting at low %*T*.

regions where the transmittance approaches zero, almost no radiation reaches the transducer, and the exact null position cannot be established accurately. The result is sluggish transducer response and rounded peaks. Figure 16-11 illustrates transmittance overshoot and rounded peaks in regions of low transmittance (1700 and 3000 cm^{-1}).

16C-3 Nondispersive Instruments

A number of simple, rugged instruments have been designed for quantitative infrared analysis. Some are simple filter or nondispersive photometers; others are a type of instrument that employs filter wedges in lieu of a dispersing element to provide entire spectra; and still others use no wavelength selection device at all. Generally, these instruments are less complex, more rugged, easier to maintain, and less expensive than the instruments we have thus far described in this chapter.

Filter Photometers[13]

Figure 16-12 is a schematic of a portable (weight = 18 lb), infrared filter photometer designed for quantitative analysis of various organic substances in the atmosphere. The source is a nichrome-wire-wound ceramic rod; the transducer is a pyroelectric device. A variety of interference filters, which transmit in the range between about 3000 to 750 cm^{-1} (3.3 to 13 μm), are available; each is designed for the determination of a specific compound. The filters are readily interchangeable.

The gaseous sample is introduced into the cell by means of a battery-operated pump. The path length of the cell as shown is 0.5 m; a series of reflecting mirrors (not shown in Figure 16-12) permits increases in cell length to 20 m in increments of 1.5 m. This feature greatly enhances the concentration range of the instrument.

The photometer is reported to be sensitive to a few tenths of a part per million of such substances as acrylonitrile, chlorinated hydrocarbons, carbon monoxide, phosgene, and hydrogen cyanide.

Nonfilter Photometers

Photometers, which employ no wavelength-restricting device, are widely employed to monitor gas streams for a single component.[14] Figure 16-13 shows a typical nondispersive instrument designed to determine carbon

[13] See P. A. Wilks, *Amer. Lab.*, **1994** (12), 44; *Proc. Control Qual.*, **1992**, *3*, 283.

[14] For a description of process infrared measurements, see M. S. Frant and G. LaButti, *Anal. Chem.*, **1980**, *52*, 1331A.

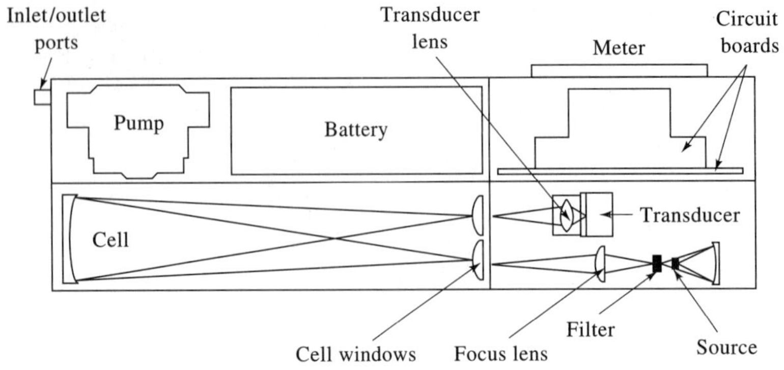

Figure 16-12 A portable, infrared photometer designed for gas analysis. *(Courtesy of The Foxboro Company, Foxboro, MA.)*

monoxide in a gaseous mixture. The reference cell is a sealed container filled with a nonabsorbing gas; as shown in the figure, the sample flows through a second cell that is of similar length. The chopper blade is so arranged that the beams from identical sources are chopped simultaneously at the rate of about five times per second. Selectivity is obtained by filling both compartments of the *sensor cell* with the gas being analyzed, carbon monoxide in this example. The two chambers of the detector are separated by a thin, flexible, metal diaphragm that serves as one plate of a capacitor; the second plate is contained in the sensor compartment on the left.

In the absence of carbon monoxide in the sample cell, the two sensor chambers are heated equally by infrared radiation from the two sources. If the sample contains carbon monoxide, however, the right-hand beam is attenuated somewhat and the corresponding sensor chamber becomes cooler with respect to its reference counterpart; the consequence is a movement of the diaphragm to the right and a change in capacitance of the capacitor. This change in capacitance is sensed by the amplifier system, the output of which drives a servomotor that moves the beam attenuator into the reference beam until the two compartments are again at the same temperature. The instrument thus operates as a null device. The chopper serves to provide an ac signal, which is less sensitive to drift and $1/f$ noise.

The instrument is highly selective because heating of the sensor gas occurs only from that narrow portion of the spectrum that is absorbed by the carbon monoxide in the sample. Clearly, the device can be adapted to the determination of any infrared-absorbing gas.

16C-4 Automated Instruments for Quantitative Analysis

Figure 16-14 is a schematic of a computer-controlled instrument designed specifically for quantitative infrared analyses. The wavelength selector, which consists of three filter wedges (page 156) mounted in the form of a segmented circle, is shown in Figure 16-14b. The motor drive and potentiometric control

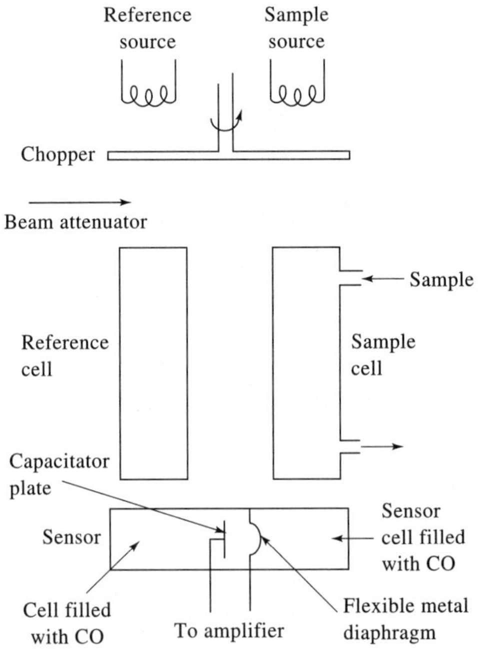

Figure 16-13 A nondispersive infrared photometer for monitoring carbon monoxide.

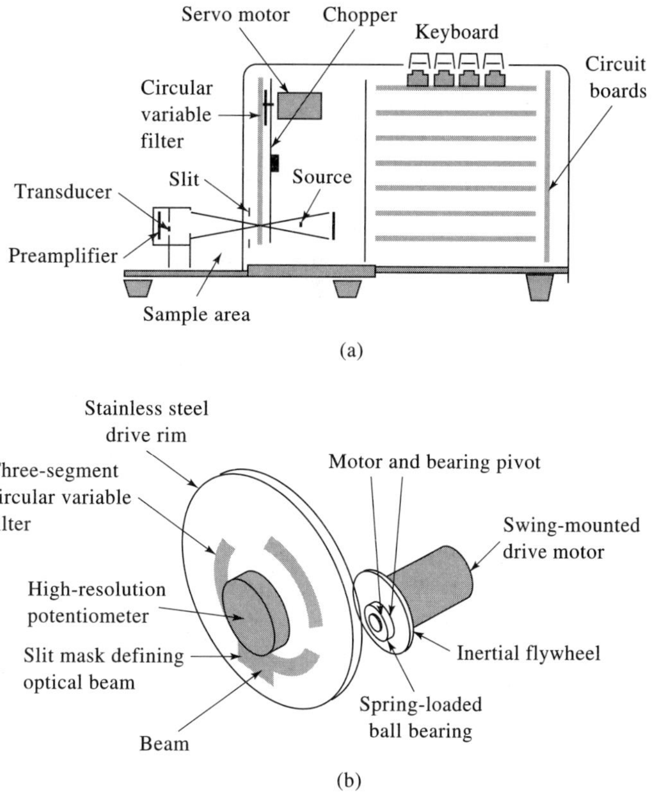

Figure 16-14 An infrared instrument for quantitative analysis. (a) Schematic of the instrument; (b) circular variable filter wheel. *(Courtesy of The Foxboro Company, Foxboro, MA.)*

permit rapid computer-controlled wavelength selection in the region between 4000 and 690 cm^{-1} (or 2.5 to 14.5 μm) with an accuracy of 0.4 cm^{-1}. The source and transducer are similar to those described in the earlier section on filter photometers; note that a beam chopper is used here. The sample area can be readily adapted to solid, liquid, or gaseous samples. The instrument can be programmed to determine the absorbance of a multi-component sample at several wavelengths and then to compute the concentration of each component.

16D QUESTIONS AND PROBLEMS

16-1 The infrared spectrum of CO shows a vibrational absorption peak at 2170 cm^{-1}.
(a) What is the force constant for the CO bond?
(b) At what wavenumber would the corresponding peak for ^{14}CO occur?

16-2 Gaseous HCl exhibits an infrared peak at 2890 cm^{-1} due to the hydrogen/chlorine stretching vibration.
(a) Calculate the force constant for the bond.
(b) Calculate the wavenumber of the absorption peak for DCl assuming the force constant is the same as that calculated in part (a).

16-3 Indicate whether the following vibrations will be active or inactive in the infrared spectrum.

Molecule	Motion
(a) CH_3—CH_3	C—C stretching
(b) CH_3—CCl_3	C—C stretching
(c) SO_2	Symmetric stretching
(d) CH_2═CH_2	C—H stretching: 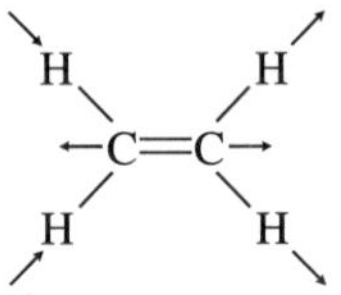
(e) CH_2═CH_2	C—H stretching: 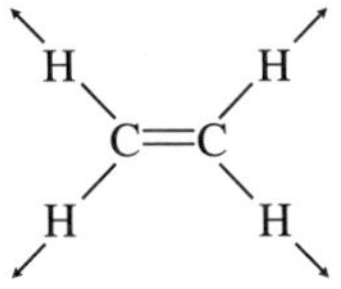
(f) CH_2═CH_2	CH_2 wag: 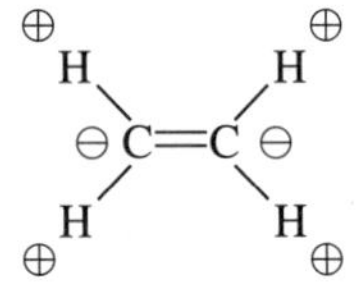
(g) CH_2═CH_2	CH_2 twist: 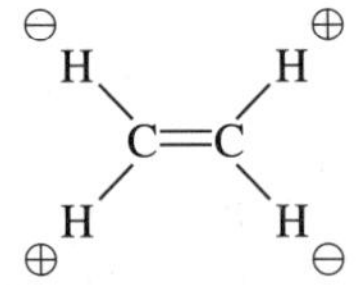

16-4 Calculate the absorption frequency corresponding to the —C—H stretching vibration treating the group as a simple diatomic C—H molecule. Compare the calculated value with the range found in correlation charts (such as the one shown in Figure 17-5). Repeat the calculation for the deuterated bond.

16-5 The wavelength of the fundamental O—H stretching vibration is about 1.4 μm. What is the approximate wavenumber and wavelength of the first overtone peak for the O—H stretch?

16-6 The wavelength of the fundamental N—H stretching vibration is about 1.5 μm. What is the approximate wavenumber and wavelength of the first overtone peak for the N—H stretch?

16-7 Sulfur dioxide is a nonlinear molecule. How many vibrational modes will this compound have? How many absorption peaks would sulfur dioxide be expected to have?

16-8 What are the advantages of a Fourier transform infrared spectrometer compared with a dispersive instrument?

16-9 What length of mirror drive in a Fourier transform spectrometer would be required to provide a resolution of (a) 0.020 cm^{-1} and (b) 2.0 cm^{-1}?

16-10 It was stated that at room temperature (25°C) the majority of molecules are in the ground vibrational energy level ($v = 0$).

(a) Use the Boltzmann equation (Equation 8-1) to calculate the following excited state and ground state population ratios for HCl: $N(v{=}1)/N(v{=}0)$ and $N(v{=}2)/N(v{=}0)$. The fundamental vibrational frequency of HCl occurs at 2885 cm^{-1}.

(b) Use the results of part (a) to predict the intensity of the $v = 1$ to 2 and $v = 2$ to 3 transitions relative to the intensity of $v = 0$ to 1 transition.

16-11 The first FTIR instruments used three different interferometer systems. Briefly, describe how it has been possible to simplify the optical systems in more contemporary instruments.

16-12 In a particular trace analysis via FTIR a set of 16 interferograms were collected. The signal-to-noise ratio associated with a particular spectral peak was approximately 5/1. How many interferograms would have to be collected and averaged if the goal is to obtain a $S/N = 50/1$?

16-13 If a Michelson interferometer has a mirror velocity of 1.00 cm/s, what will be the frequency at the transducer due to light leaving the source at: (a) 1700 cm^{-1}? (b) 1710 cm^{-1}? (c) 1715 cm^{-1}?

Chapter 17

Applications of Infrared Spectrometry

Modern infrared spectrometry is a versatile tool that is applied to the qualitative and quantitative determination of molecular species of all types. As shown in Table 17-1, the applications of infrared spectrometry fall into three major categories based upon the three infrared spectral regions. By far and away, the most widely used region is the mid-infrared that extends from about 670 to 4000 cm^{-1} (2.5 to 14.9 μm). Here, absorption, reflection, and emission spectra are employed for both qualitative and quantitative analysis. The near-infrared region from 4000 to 14,000 cm^{-1} (0.75 to 2.5 μm) also finds considerable use for the routine quantitative determination of certain species, such as water, carbon dioxide, sulfur, low-molecular-weight hydrocarbons, amine nitrogen, and many other simple compounds that are of interest in agriculture and in industry. These determinations are usually based upon diffuse reflectance measurements of untreated solid or liquid samples or transmission studies of gases. The primary use of the far-infrared region has been for the determination of the structures of inorganic and metal-organic species based upon absorption measurements.

In this chapter we first focus on the uses of midinfrared absorption and reflection spectrometry for structural investigations of molecular compounds, particularly organic compounds and species that are of interest in biochemistry. We then examine in less detail some of the other applications listed in Table 17-1.

17A MID-INFRARED ABSORPTION SPECTROMETRY

Mid-infrared absorption and reflection spectrometry are major tools for determining the structure of organic and biochemical species.[1] In this section we examine mid-infrared absorption applications. Section 17B is devoted to mid-infrared reflectance measurements.

[1] For further reading, see N. B. Colthup, L. H. Daly, and S. E. Wiberley, *Introduction to Infrared and Raman Spectroscopy,* 3rd ed. New York: Academic Press, 1990; A. L. Smith, *Applied Infrared Spectroscopy,* New York: Wiley, 1979; R. M. Silverstein, G. C. Bassler, and T. C. Morrill, *Spectrometric Identification of Organic Compounds,* 5th ed., Chapter 3. New York: Wiley, 1991.

TABLE 17-1 Major Applications of Infrared Spectrometry

Spectral Regions	Type Measurement	Type Analysis	Type Samples
Near-infrared	Diffuse reflectance	Quantitative	Solid or liquid materials of commerce
	Absorption	Quantitative	Gaseous mixtures
Mid-infrared	Absorption	Qualitative	Pure solid, liquid, or gaseous compounds
		Quantitative	Complex gaseous, liquid, or solid mixtures
		Chromatographic	Complex gaseous, liquid, or solid mixtures
	Reflectance	Qualitative	Pure solid or liquid compounds
	Emission	Quantitative	Atmospheric samples
Far-infrared	Absorption	Qualitative	Pure inorganic or metal organic species

17A-1 Sample Handling

As we have seen in earlier chapters, ultraviolet and visible molecular spectra are most conveniently obtained from dilute solutions of the analyte. Absorbance measurements in the optimum range are obtained by suitably adjusting either the concentration or the cell length. Unfortunately, this approach is often not applicable for infrared spectroscopy because no good solvents exist that are transparent throughout the region of interest. As a consequence, sample handling is frequently the most difficult and time-consuming part of an infrared spectrometric analysis.[2] In this section we outline some of the common techniques for preparation of samples for infrared absorption measurements.

Gases

The spectrum of a low-boiling liquid or gas can be obtained by permitting the sample to expand into an evacuated cylindrical cell equipped with suitable windows. For this purpose, a variety of cylindrical cells are available with path lengths that range from a few centimeters to 10 m or more. The longer path lengths are obtained in compact cells by providing reflecting internal surfaces, so that the beam makes numerous passes through the sample before exiting from the cell.

[2] See T. J. Porro and S. C. Pattacini, *Spectroscopy,* **1993,** *8*(7), 40; *Ibid., 8*(8), 39; A. L. Smith, *Applied Infrared Spectroscopy,* Chapter 4. New York: Wiley, 1979; *Practical Sampling Techniques for Infrared Analysis,* P. B. Coleman, Ed. Boca Raton, FL: CRC Press, 1993.

Solutions

When feasible, a convenient way of obtaining infrared spectra is on solutions prepared to contain a known concentration of sample, as is generally done in ultraviolet/visible spectrometry. This technique is somewhat limited in its applications, however, by the availability of solvents that are transparent over significant regions in the infrared.

Solvents. Figure 17-1 lists the more common solvents employed for infrared studies of organic compounds. It is apparent that no single solvent is transparent throughout the entire mid-infrared region.

Water and the alcohols are seldom employed, not only because they absorb strongly, but also because they attack alkali-metal halides, the most common materials used for cell windows. For these reasons also, care must be taken to dry the solvents shown in Figure 17-1 before use.

Cells. Because of the tendency for solvents to absorb, infrared cells are ordinarily much narrower (0.01 to 1 mm) than those employed in the ultraviolet and visible regions. Light paths in the infrared range normally require sample concentrations from 0.1% to 10%. The cells are frequently demountable, with Teflon spacers to allow variation in path length (see Figure 17-2). Fixed-path-length cells can be filled or emptied with a hypodermic syringe.

Sodium chloride windows are most commonly employed; even with care, however, their surfaces eventu-

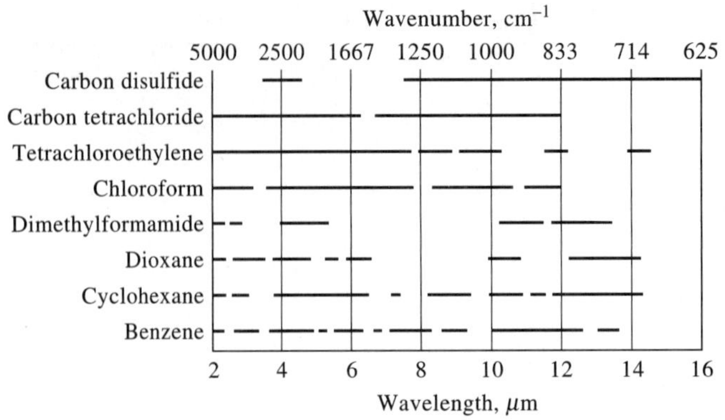

Figure 17-1 Solvents for infrared spectroscopy. Horizontal lines indicate useful regions.

ally become fogged due to absorption of moisture. Polishing with a buffing powder returns them to their original condition.

Figure 17-3 shows how the thickness b of very narrow infrared cells can be determined. The upper figure is a recorder trace obtained by inserting an empty cell in the light path of the sample beam. The reference beam in this case passed unobstructed to the monochromator. The pattern of regular maxima and minima make up an interference fringe. The maxima occur when the radiation reflected off of the two internal surfaces of the cell has traveled a distance that is an integral multiple N of the wavelength of the radiation that was transmitted without reflection. Constructive interference then occurs whenever the wavelength is equal to $2b/N$. That is,

$$\frac{2b}{N} = \lambda \tag{17-1}$$

As shown in Figure 17-3b, the number of interference fringes ΔN between two known wavelengths λ_1 and λ_2 are counted and introduced into the equation

$$\Delta N = \frac{2b}{\lambda_1} - \frac{2b}{\lambda_2}$$
$$= 2b\bar{\nu}_1 - 2b\bar{\nu}_2$$

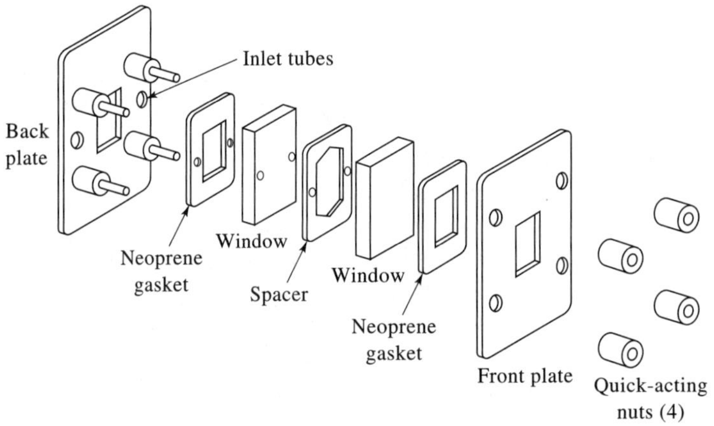

Figure 17-2 Expanded view of a demountable infrared cell for liquid samples. Teflon spacers ranging in thickness from 0.015 to 1 mm are available. *(Courtesy of Perkin-Elmer, Norwalk, CT.)*

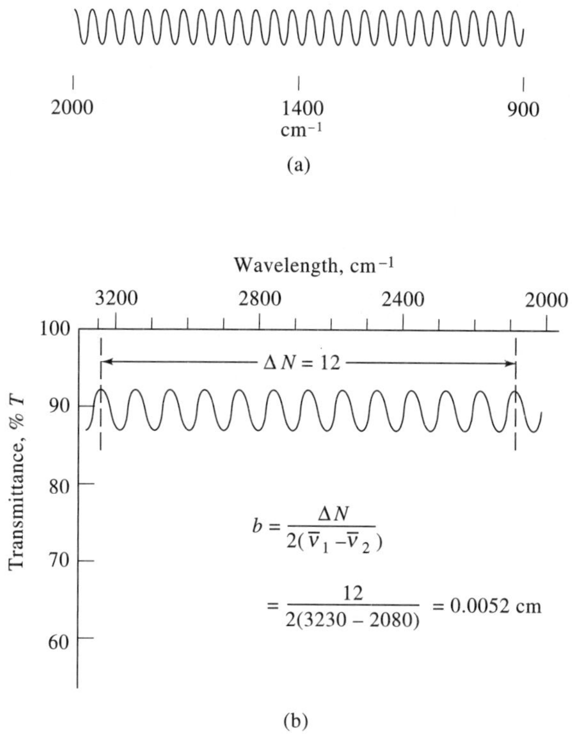

Figure 17-3 (a) Actual recorder tracing with empty cell in sample beam. (b) Illustration of calculation of path length *b*.

or

$$b = \frac{\Delta N}{2(\bar{\nu}_1 - \bar{\nu}_2)} \tag{17-2}$$

It should be noted that interference fringes are ordinarily not seen when a cell is filled with liquid because the refractive index of most liquids approaches that of the window material; reflection is thus reduced (Equation 6-15, page 126). On the other hand, interference can be observed between 2800 and 2000 cm^{-1} in Figure 16-1 (page 382). Here, the sample is a sheet of polystyrene, which has a refractive index considerably different from air; consequently, significant reflection occurs at the two interfaces of the sheet. Equation 17-2 is often used to calculate the thickness of thin polymer films.

Liquids

When the amount of a liquid sample is small or when a suitable solvent is unavailable, it is common practice to obtain spectra on the pure (neat) liquid. Here, only a very thin film has a sufficiently short path length to produce satisfactory spectra. Commonly, a drop of the neat liquid is squeezed between two rock-salt plates to give a layer that has a thickness of 0.01 mm or less. The two plates, held together by capillarity, are then mounted in the beam path. Clearly, such a technique does not give particularly reproducible transmittance data, but the resulting spectra are usually satisfactory for qualitative investigations. Neat liquid samples may contain sufficient water to fog the salt plates and make periodic polishing of the plates necessary.

Solids

Most organic compounds exhibit numerous absorption peaks throughout the mid-infrared region, and finding a solvent that does not have overlapping peaks is often impossible. As a consequence, spectra are often obtained on dispersions of the solid in a liquid or solid matrix. Generally, in these techniques, the solid sample must be ground until its particle size is less than the wavelength of the radiation in order to avoid the effects of scattered radiation.

Pelleting. One of the most popular techniques for handling solid samples has been KBr pelleting (other alkali metal halides have also been used). Halide salts have the property of cold flow in which they have glass-like transparent or translucent properties when sufficient pressure is applied to the finely powdered materials. In using this technique, a milligram or less of the finely ground sample is intimately mixed with about 100 mg of dried potassium bromide powder. Mixing can be carried out with a mortar and pestle, or better, in a small ball mill. The mixture is then pressed in a special die at 10,000 to 15,000 pounds per square inch to yield a transparent disk. Best results are obtained if the disk is formed in a vacuum to eliminate occluded air. The disk is then held in the instrument beam for spectroscopic examination. The resulting spectra frequently exhibit bands at 3450 and 1640 cm^{-1} (2.9 and 6.1 μm) due to absorbed moisture.

With many compounds KBr pelleting produces excellent spectra that appear in many spectral libraries. Being ionic, KBr transmits throughout most of the infrared region with a lower cutoff of about 400 cm^{-1}. Cesium iodide absorbs below 200 cm^{-1} and is sometimes employed because of its greater transparency at low frequencies.

Mulls. Infrared spectra of solids that are not soluble in an infrared-transparent solvent or are not conveniently pelleted in KBr are often obtained by dispersing the analyte in a mineral oil or fluorinated hydrocarbon *mull.* Mulls are formed by grinding 2 to 5 mg of the finely powdered sample (particle size < 2 μm) in the presence of one or two drops of a heavy hydrocarbon oil (Nujol). If hydrocarbon bands are likely to interfere, Fluorolube, a halogenated polymer, can be used instead. In either case, the resulting mull is then examined as a film between flat salt plates.

Other Methods for Solids. The infrared behavior of solids can also be obtained by reflectance techniques and by the photoacoustic method. These spectra are often similar to absorption spectra and provide the same kinds of information. These techniques are discussed in Sections 17B and 17C.

17A-2 Qualitative Analysis

The general use of mid-infrared spectroscopy by bench chemists for identification of organic compounds began in the late 1950s with the appearance in the market of inexpensive and easy-to-use dispersive double-beam recording spectrophotometers that produce spectra in the 5000- to 670-cm^{-1} (2- to 15-μm) range. The appearance of this type of instrument, (as well as nuclear magnetic and mass spectrometers) revolutionized the way

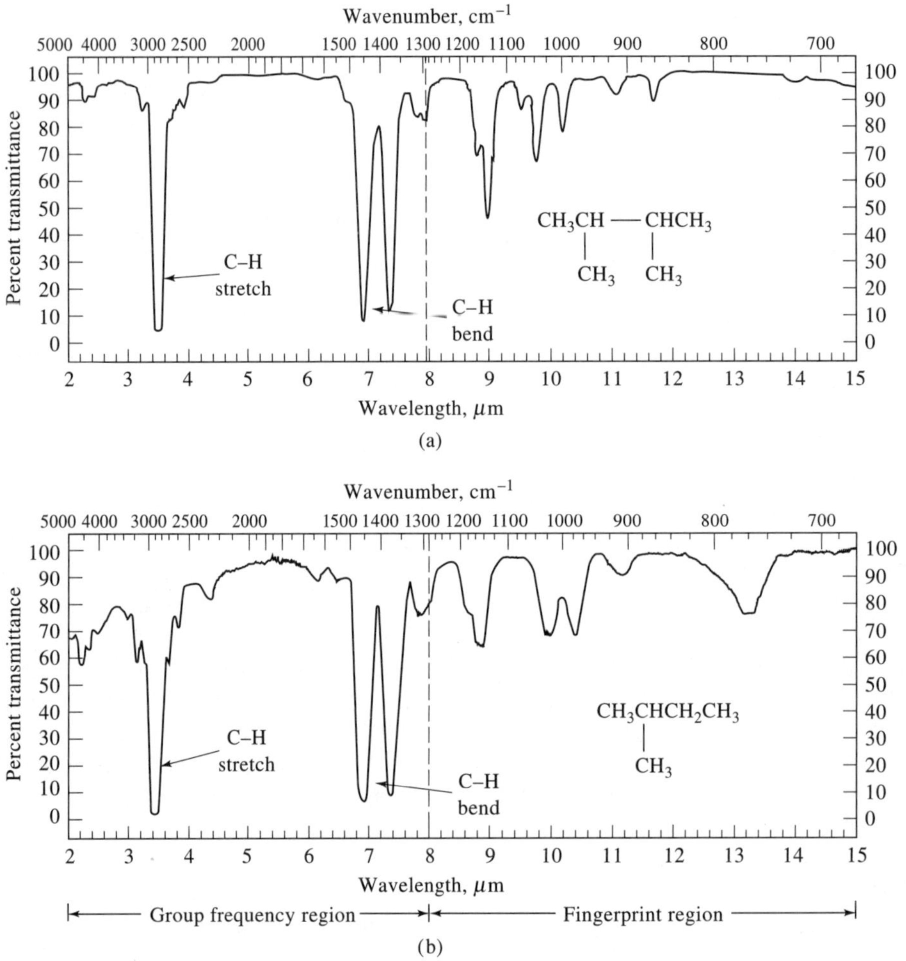

Figure 17-4 Group frequency and fingerprint regions of the mid-infrared spectrum. *(From R. M. Roberts, J. C. Gilbert, L. B. Rodewald, and A. S. Wingrove,* Modern Experimental Organic Chemistry, *4th ed. Philadelphia: Saunders College Publishing, 1985. With permission.)*

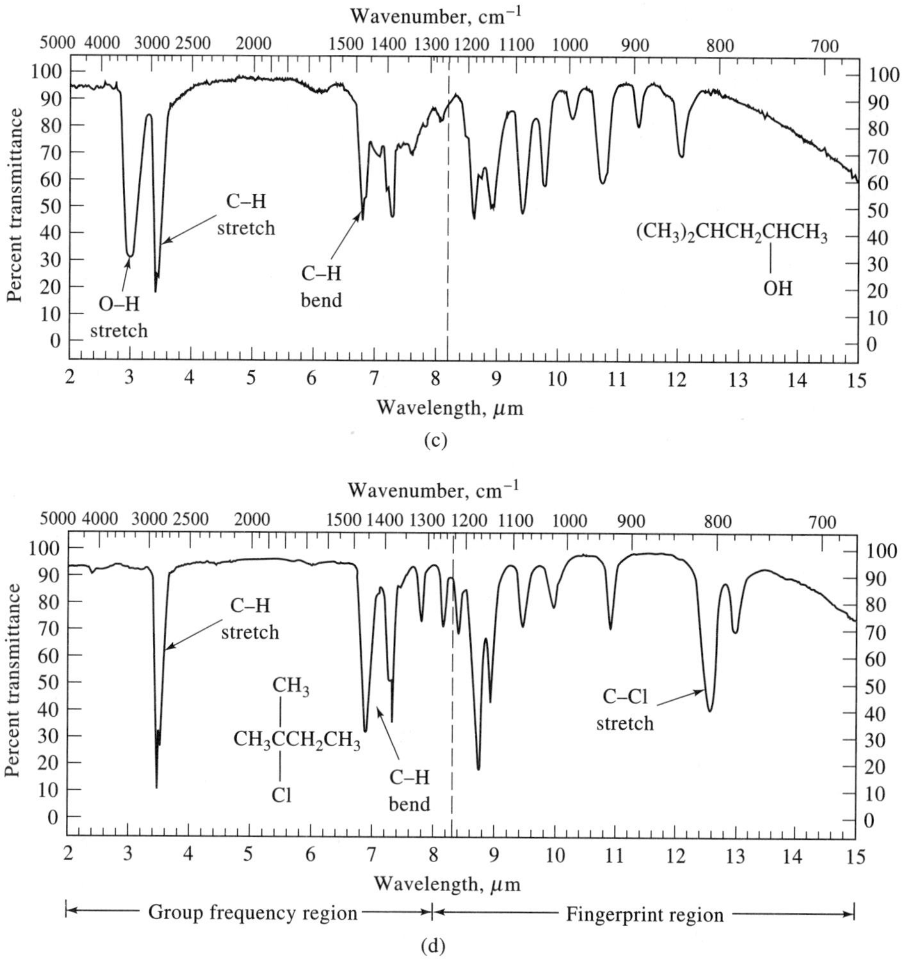

Figure 17-4 *Continued*

chemists went about identifying organic, inorganic, and biological species. Suddenly, the time required to perform a structural determination was reduced by a factor of ten, one hundred, or even one thousand.

Figure 17-4 shows four typical spectra obtained with an inexpensive double-beam dispersive instrument. Identification of an organic compound from a spectrum of this kind is a two-step process. The first step involves determining what functional groups are most likely present by examining the *group frequency region,* which encompasses radiation from about 3600 cm^{-1} to approximately 1200 cm^{-1} (see Figure 17-4). The second step then involves a detailed comparison of the spectrum of the unknown with the spectra of pure compounds that contain all of the functional groups found in the first step. Here, the *fingerprint region,* from 1200 to 600 cm^{-1} (Figure 17-4) is particularly useful because small differences in the structure and constitution of a molecule result in significant changes in the appearance and distribution of absorption peaks in this region. Consequently, a close match between two spectra in the fingerprint region (as well as others) constitutes almost certain evidence for the identity of the compounds yielding the spectra.

Group Frequencies

We have noted that the approximate frequency (or wavenumber) at which an organic functional group, such as C=O, C=C, C—H, C≡C, or O—H absorbs infrared radiation can be calculated from the masses of the atoms and the force constant of the bond between them (Equations 16-13 and 16-14). These frequencies, called *group frequencies,* are seldom totally invariant because of interactions with other vibrations associated with one or both of the atoms composing the group. On the other hand, such interaction effects are ordinarily small; as a result, a range of frequencies can be assigned within which it is highly probable that the absorption peak for a given functional group will be found. Table 17-2 lists group frequencies for several common functional groups. A more detailed presentation of group frequencies is found in the *correlation chart* shown in Figure 17-5.[3] Note that while most group frequencies fall in the range of 3600 to 1250 cm^{-1}, a few fall in the fingerprint region.

[3] Other correlation charts can be found in R. M. Silverstein, G. C. Bassler, and T. C. Morrill, *Spectrometric Identification of Organic Compounds,* 5th ed., pp. 158–163. New York: Wiley, 1991.

TABLE 17-2 Abbreviated Table of Group Frequencies for Organic Groups

Bond	Type of Compound	Frequency Range, cm^{-1}	Intensity
C—H	Alkanes	2850–2970	Strong
		1340–1470	Strong
C—H	Alkenes ($\rangle C{=}C\langle^{H}$)	3010–3095	Medium
		675–995	Strong
C—H	Alkynes (—C≡C—H)	3300	Strong
C—H	Aromatic rings	3010–3100	Medium
		690–900	Strong
O—H	Monomeric alcohols, phenols	3590–3650	Variable
	Hydrogen-bonded alcohols, phenols	3200–3600	Variable, sometimes broad
	Monomeric carboxylic acids	3500–3650	Medium
	Hydrogen-bonded carboxylic acids	2500–2700	Broad
N—H	Amines, amides	3300–3500	Medium
C=C	Alkenes	1610–1680	Variable
C=C	Aromatic rings	1500–1600	Variable
C≡C	Alkynes	2100–2260	Variable
C—N	Amines, amides	1180–1360	Strong
C≡N	Nitriles	2210–2280	Strong
C—O	Alcohols, ethers, carboxylic acids, esters	1050–1300	Strong
C=O	Aldehydes, ketones, carboxylic acids, esters	1690–1760	Strong
NO_2	Nitro compounds	1500–1570	Strong
		1300–1370	Strong

These include the C—O—C stretching vibration at about 1200 cm^{-1} and the C—Cl stretching vibration at 700 to 800 cm^{-1}.

Several group frequencies are identified in the four spectra in Figure 17-4. All four spectra contain a peak at 2900 to 3000 cm^{-1}, which corresponds to a C—H stretching vibration and generally indicates the presence of one or more alkane groups (see Table 17-2). The two peaks at about 1375 cm^{-1} and 1450 cm^{-1} are also characteristic group frequencies for C—H groups and result from bending vibrations in the molecule. Spectrum 17-4c illustrates the group frequency for an O—H stretching vibration at about 3200 cm^{-1} as well as the alkane group frequencies (these peaks are also present in the spectrum for *n*-hexanal shown in Figure 16-11). Finally, the characteristic group frequency for a C—Cl stretching vibration is shown at about 800 cm^{-1} in spectrum 17-4d.

Group frequencies and correlation charts permit intelligent guesses to be made as to what functional groups are likely to be present or absent in a molecule. Ordinarily, it is impossible to identify unambiguously either the sources of all of the peaks in a given spectrum or the exact identity of the molecule. Instead, group frequencies and correlation charts serve as a starting point in the identification process.

The "Fingerprint" Region

Small differences in the structure and constitution of a molecule result in significant changes in the distribution of absorption peaks in this region of the spectrum that extends from about 1200 to 700 cm^{-1} (8 to 14 μm). As a consequence, a close match between two spectra in this fingerprint region (as well as others) constitutes strong evidence for the identity of the compounds yielding the spectra. Most single bonds give rise to absorption bands at these frequencies; because their energies are about the same, strong interaction occurs between neighboring bonds. The absorption bands are thus composites of these various interactions and depend upon the overall skeletal structure of the molecule. Exact interpretation of spectra in this region is seldom possible because of the complexity of the spectra; on the other hand, it is this complexity that leads to uniqueness and the consequent usefulness of the region for final identification purposes.

Figures 17-4a and 17-4b illustrate the unique character of infrared spectra, particularly in the fingerprint region. The two molecules differ by just one methyl group, yet the two spectra differ dramatically in appearance in the fingerprint region.

As shown in Figure 17-5, a number of inorganic groups such as sulfate, phosphate, nitrate, and carbonate absorb in the fingerprint region (<1200 cm^{-1} or >8.3 μm).

Limitations to the Use of Correlation Charts

The unambiguous establishment of the identity or the structure of a compound is seldom possible from correlation charts alone. Uncertainties frequently arise from overlapping group frequencies, spectral variations as a function of the physical state of the sample (that is, whether it is a solution, a mull, in a pelleted form, and so forth), and instrumental limitations.

In employing group frequencies, it is essential that the entire spectrum, rather than a small isolated portion, be considered and interrelated. Interpretation based on one part of the spectrum should be confirmed or rejected by study of other regions.

To summarize, then, correlation charts serve only as a guide for further and more careful study. Several excellent monographs describe the absorption characteristics of functional groups in detail.[4] A study of these characteristics, as well as the other physical properties of the sample, may permit unambiguous identification. Infrared spectroscopy, when used in conjunction with other methods such as mass spectroscopy, nuclear magnetic resonance, and elemental analysis, usually makes possible the positive identification of a species.

Collections of Spectra

As just noted, correlation charts seldom suffice for the positive identification of an organic compound from its infrared spectrum. There are available, however, several catalogs of infrared spectra that assist in qualitative identification by providing comparison spectra for a large number of pure compounds. Manually searching large catalogs of spectra is slow and tedious. For this reason, computer-based search systems have become widely used in the last few years.

---

[4] N. B. Colthup, L. N. Daly, and S. E. Wilberley, *Introduction to Infrared and Raman Spectroscopy,* 3rd ed. New York: Academic Press, 1990; H. A. Szymanski, *Theory and Practice of Infrared Spectroscopy.* New York: Plenum Press, 1964; R. M. Silverstein, G. C. Bassler, and T. C. Morrill, *Spectrometric Identification of Organic Compounds,* 5th ed., Chapter 3. New York: Wiley, 1991.

Figure 17-5 Correlation chart. *(From N. B. Colthup,* J. Optical Soc. Am., *1950,* 40, *397. With permission.)*

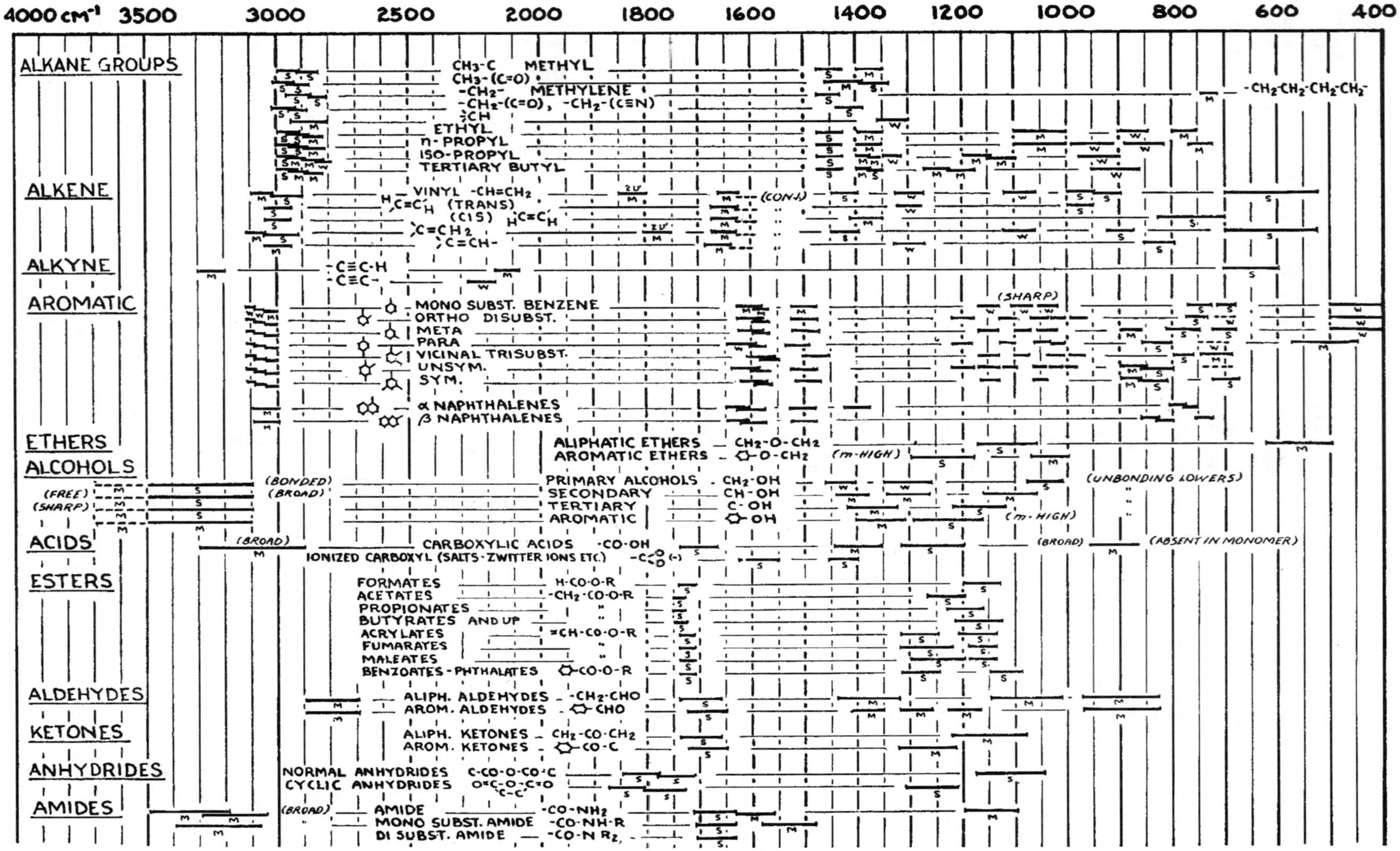

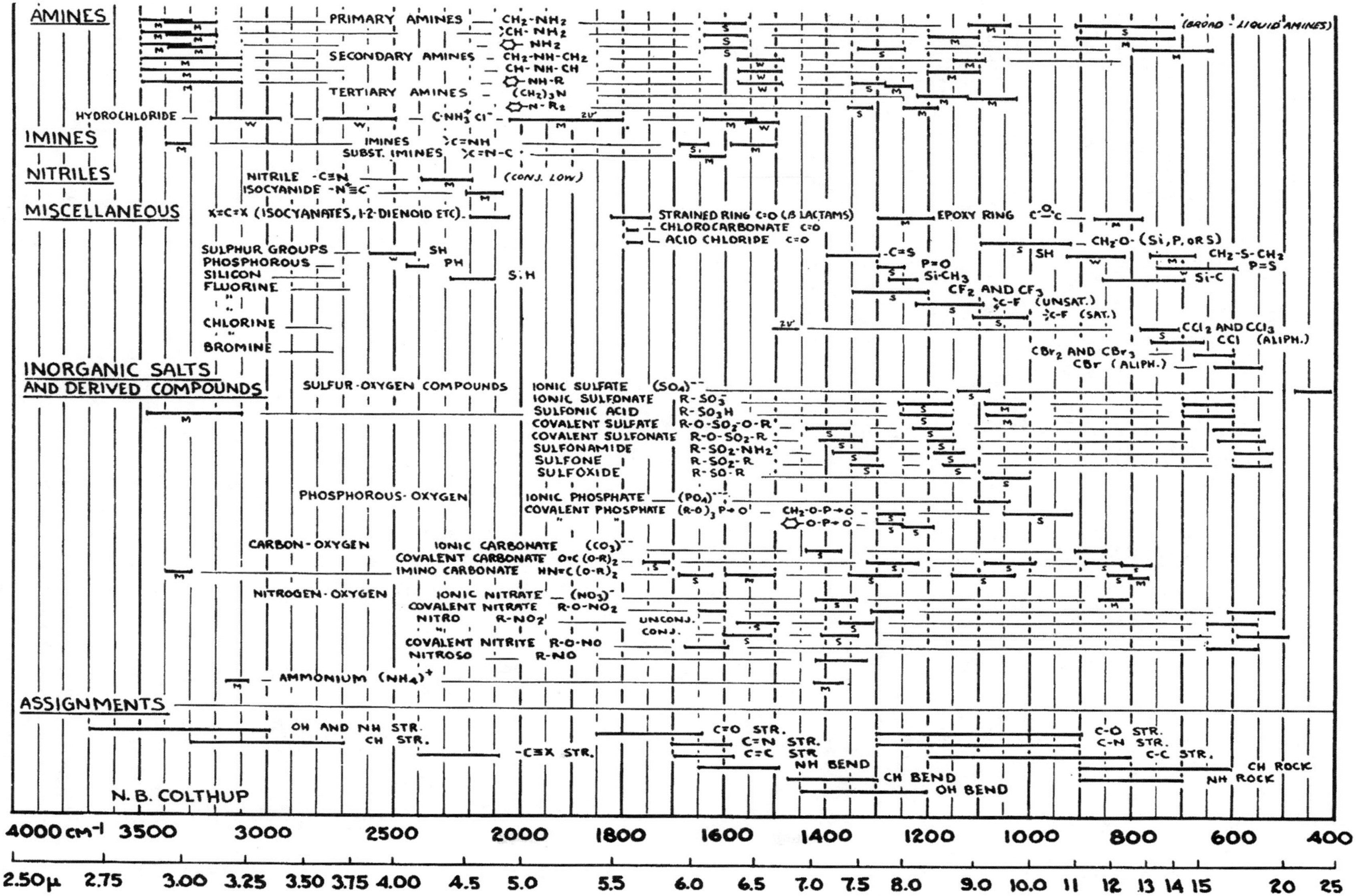
AMINES
PRIMARY AMINES
CH_2-NH_2
$>CH-NH_2$
$\bigcirc-NH_2$
SECONDARY AMINES
$CH_2-NH-CH_2$
$CH-NH-CH$
$\bigcirc-NH-R$
TERTIARY AMINES
$(CH_2)_3N$
$\bigcirc-N-R_2$
HYDROCHLORIDE
$C-NH_3^+Cl^-$
(BROAD - LIQUID AMINES)
IMINES
IMINES $>C=NH$
SUBST. IMINES $>C=N-C$
NITRILES
NITRILE $-C\equiv N$
ISOCYANIDE $-N\equiv C$
(CONJ. LOW)
MISCELLANEOUS
X=C=X (ISOCYANATES, 1-2-DIENOID ETC).
STRAINED RING C=O (β LACTAMS)
CHLOROCARBONATE C=O
ACID CHLORIDE C=O
EPOXY RING
SULPHUR GROUPS
SH
PHOSPHOROUS
PH
SILICON
SiH
FLUORINE
CHLORINE
BROMINE
-C=S
P=O
$Si-CH_3$
CF_2 AND CF_3
C-F (UNSAT.)
C-F (SAT.)
SH
CH_2-O- (Si, P, or S)
CH_2-S-CH_2
P=S
Si-C
CCl_2 AND CCl_3
CCl (ALIPH.)
CBr_2 AND CBr_3
CBr (ALIPH.)
INORGANIC SALTS AND DERIVED COMPOUNDS
SULFUR-OXYGEN COMPOUNDS
IONIC SULFATE $(SO_4)^{--}$
IONIC SULFONATE $R-SO_3^-$
SULFONIC ACID $R-SO_3H$
COVALENT SULFATE $R-O-SO_2-O-R'$
COVALENT SULFONATE $R-O-SO_2-R$
SULFONAMIDE $R-SO_2-NH_2$
SULFONE $R-SO_2-R$
SULFOXIDE $R-SO-R$
PHOSPHOROUS-OXYGEN
IONIC PHOSPHATE $(PO_4)^{---}$
COVALENT PHOSPHATE $(R-O)_3P\to O$
$CH_2-O-P\to O$
$\bigcirc-O-P\to O$
CARBON-OXYGEN
IONIC CARBONATE $(CO_3)^{--}$
COVALENT CARBONATE $O=C(O-R)_2$
IMINO CARBONATE $HN=C(O-R)_2$
NITROGEN-OXYGEN
IONIC NITRATE $(NO_3)^-$
COVALENT NITRATE $R-O-NO_2$
NITRO $R-NO_2$
UNCONJ.
CONJ.
COVALENT NITRITE R-O-NO
NITROSO R-NO
AMMONIUM $(NH_4)^+$
ASSIGNMENTS
OH AND NH STR.
CH STR.
$-C\equiv X$ STR.
C=O STR.
C=N STR.
C=C STR.
NH BEND
CH BEND
OH BEND
C-O STR.
C-N STR.
C-C STR.
CH ROCK
NH ROCK
N. B. COLTHUP
4000 cm⁻¹ 3500 3000 2500 2000 1800 1600 1400 1200 1000 800 600 400
2.50μ 2.75 3.00 3.25 3.50 3.75 4.00 4.5 5.0 5.5 6.0 6.5 7.0 7.5 8.0 9.0 10.0 11 12 13 14 15 20 25

Computer Search Systems[5]

Virtually all infrared instrument manufacturers now offer computer search systems to assist the chemist in identifying compounds from stored infrared spectral data. The position and relative magnitudes of peaks in the spectrum of the analyte are determined and stored in memory to give a peak profile, which can then be compared with profiles of pure compounds stored on high-density magnetic disks or laser disks. The computer then matches profiles and prints a list of compounds having spectra similar to that of the analyte. Usually the spectrum of the analyte and that of each potential match can then be shown simultaneously on the computer display for comparison (see Figure 17-6). The memory banks of these instruments are capable of storing profiles for as many as a hundred thousand pure compounds.

In 1980, the Sadtler Standard Infrared Collection and the Sadtler Commercial Infrared Collection became available as software packages. Currently this library contains over 130,000 spectra. Included are 9,200 vapor-phase spectra, 59,000 condensed-phase spectra of pure compounds, and 53,000 spectra of commercial products, such as monomers, polymers, surfactants, adhesives, inorganics, plasticisers, pharmaceuticals, drugs of abuse and many others. Several manufacturers of Fourier transform instruments have now incorporated these packages into their instrument computers, thus creating instantly available infrared libraries of over 100,000 compounds.

The Sadtler algorithm consists of a search system in which the spectrum of the unknown compound is first coded according to the location of its strongest absorption peak; then each additional strong band ($\% T <$ 60%) in 10 regions 200 cm^{-1} wide from 4000 to 2000 cm^{-1} are coded by their location. Finally, the strong bands in 17 regions 100 cm^{-1} wide from 2100 to 400 cm^{-1} are coded in a similar way.[6] The compounds in the library are coded in this same way. The data are or-

[5] See W. O. George and H. Willis, *Computer Methods in UV, Visible, and IR Spectroscopy.* Boca Raton, FL: CRC Press, 1990; W. A. Warr, *Anal. Chem.*, **1993**, *65*, 1045A, 1087A; G. W. Small, *Anal. Chem.*, **1987**, *59*, 535A.

[6] For a fuller description of this system, see R. H. Shaps and J. F. Sprouse, *Ind. Res. and Dev.*, **1981**, *23*, 168.

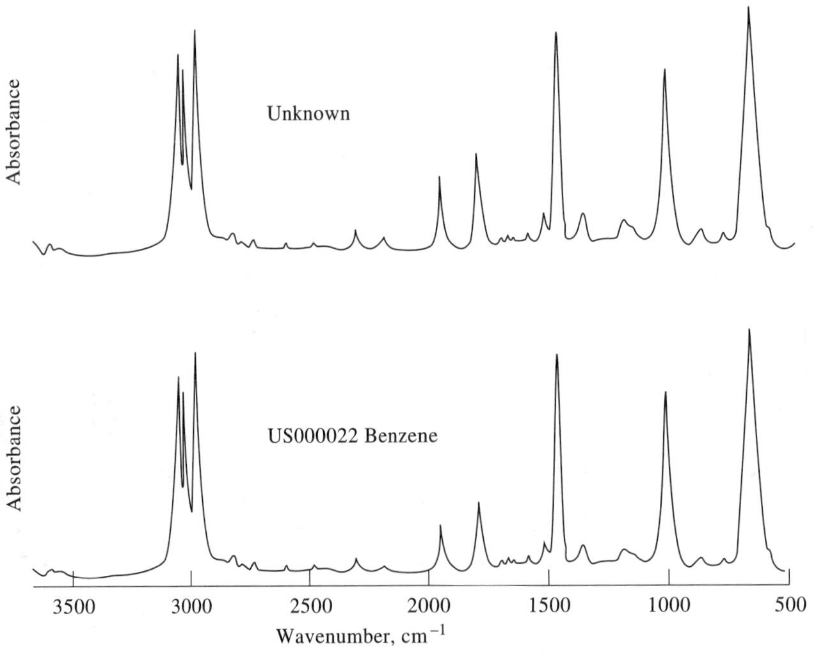

Figure 17-6 Plot of an unknown spectrum and the best match from a search report. *(Courtesy of Sadtler Research Laboratories, Bio-Rad Division, Philadelphia, PA.)*

ganized by the location of the strongest band with only those compounds having the same strongest band being considered in any sample identification. This procedure is rapid and produces a list of potential matches within a short period. For example, approximately 40 s are required to search a base library of 25,000 compounds by this procedure.

17A-3 Quantitative Applications

Quantitative infrared absorption methods differ somewhat from ultraviolet/visible molecular spectroscopic methods because of the greater complexity of the spectra, the narrowness of the absorption bands, and the instrumental limitations of infrared instruments. Quantitative data obtained with dispersive infrared instruments are generally significantly inferior in quality to data obtained with ultraviolet/visible spectrophotometers. The precision and accuracy of measurements with Fourier transform instruments is distinctly better than those with dispersive instruments. Meticulous attention to detail is, however, essential for obtaining good-quality results.[7]

Deviations from Beer's Law

With infrared radiation, instrumental deviations from Beer's law are more common than with ultraviolet and visible wavelengths because infrared absorption bands are relatively narrow. Furthermore, with dispersive instruments, the low intensity of sources and low sensitivities of detectors in this region require the use of relatively wide monochromator slit widths; thus, the bandwidths employed are frequently of the same order of magnitude as the widths of absorption peaks. We have pointed out (Section 13B-2) that this combination of circumstances usually leads to a nonlinear relationship between absorbance and concentration. Calibration curves, determined empirically, are therefore often required for quantitative work.

Absorbance Measurement

Matched absorption cells for solvent and solution are ordinarily employed in the ultraviolet and visible regions, and the measured absorbance is then found from the relation

$$A = \log \frac{P_{\text{solvent}}}{P_{\text{solution}}}$$

The use of the solvent in a matched cell as a reference absorber has the advantage of largely canceling out the effects of radiation losses due to reflection at the various interfaces, scattering and absorption by the solvent, and absorption by the container windows. This technique is seldom practical for measurements in the infrared region because of the difficulty in obtaining cells whose transmission characteristics are identical. Most infrared cells have very short path lengths, which are difficult to duplicate exactly. In addition, the cell windows are readily attacked by contaminants in the atmosphere and the solvent; thus, their transmission characteristics change continually with use. For these reasons, a reference absorber is often dispensed with entirely in infrared work, and the intensity of the radiation passing through the sample is simply compared with that of the unobstructed beam. In either case, the resulting transmittance is ordinarily less than 100%, even in regions of the spectrum where no absorption by the sample occurs (see Figure 17-4).

For quantitative work, it is necessary to correct for the scattering and absorption by the solvent and the cell. Two methods are employed. In the so-called *cell-in/cell-out* procedure, spectra of the pure solvent and the analyte solution are obtained successively with respect to the unobstructed reference beam. The same cell is used for both measurements. The transmittance of each solution versus the reference beam is then determined at an absorption maximum of the analyte. These transmittances can be written as

$$T_0 = P_0/P_r$$

and

$$T_s = P/P_r$$

where P_r is the power of the unobstructed beam and T_0 and T_s are the transmittances of the solvent and analyte solution, respectively, against this reference. If P_r remains constant during the two measurements, then the transmittance of the sample with respect to the solvent can be obtained by division of the two equations. That is,

$$T = T_s/T_0 = P/P_0$$

[7] For discussions of quantitative analysis with Fourier transform instruments see P. R. Griffiths and J. A. deHasech, *Fourier Transform Infrared Spectroscopy,* Chapter 10. New York: Wiley, 1986; T. Hirschfeld, in *Fourier Transform Infrared Spectroscopy,* J. Ferraro and L. Basile, Eds., Vol. 2, pp. 193–239. New York: Academic Press, 1979.

An alternative way of obtaining P_0 and T is the *baseline* method, in which the solvent transmittance is assumed to be constant or at least to change linearly between the shoulders of the absorption peak. This technique is demonstrated in Figure 17-7.

Typical Applications

With the exception of homonuclear molecules, all organic and inorganic molecular species absorb in the infrared region; thus, infrared spectrophotometry offers the potential for determining an unusually large number of substances. Moreover, the uniqueness of an infrared spectrum leads to a degree of specificity that is matched or exceeded by relatively few other analytical methods. This specificity has found particular application to analysis of mixtures of closely related organic compounds. Two examples that typify these applications follow.

Analysis of a Mixture of Aromatic Hydrocarbons. A typical application of quantitative infrared spectroscopy involves the resolution of C_8H_{10} isomers in a mixture that includes *o*-xylene, *m*-xylene, *p*-xylene, and ethylbenzene. The infrared absorption spectra of the individual components in the 12 to 15 μm range is shown in Figure 17-8; cyclohexane is the solvent. Useful absorption peaks for determination of the individual compounds occur at 13.47, 13.01, 12.58, and 14.36 μm, respectively. Unfortunately, however, the absorbance of a mixture at any one of these wavelengths is not entirely determined by the concentration of just one component because of overlapping absorption bands. Thus, molar absorptivities for each of the four compounds must be determined at the four wavelengths. Then, four simultaneous equations can be written that permit the calculation of the concentration of each species from four absorbance measurements (see page 344). Such calculations are most easily performed with a computer.

When the relationship between absorbance and concentration is nonlinear (as frequently occurs in the infrared region), the algebraic manipulations associated with an analysis of several components having overlapping absorption peaks are considerably more complex.[8]

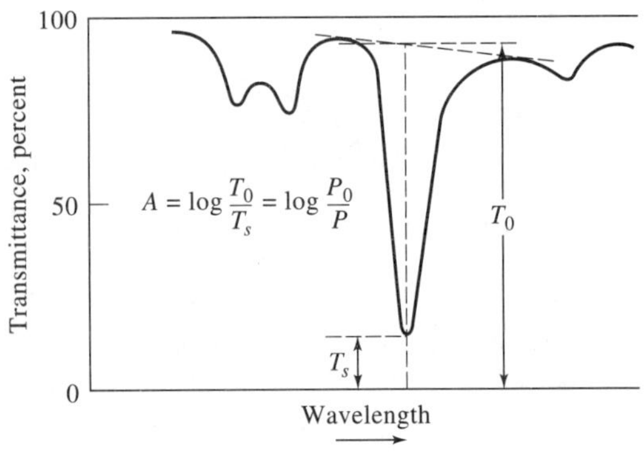

Figure 17-7 Baseline method for determination of absorbance.

TABLE 17-3 An Example of Infrared Analysis of Air Contaminants*

Contaminants	Concn, ppm	Found, ppm	Relative Error, %
Carbon monoxide	50	49.1	1.8
Methylethyl ketone	100	98.3	1.7
Methyl alcohol	100	99.0	1.0
Ethylene oxide	50	49.9	0.2
Chloroform	100	99.5	0.5

*Courtesy of The Foxboro Company, Foxboro, MA.

Determination of Air Contaminants. The recent proliferation of government regulations with respect to atmospheric contaminants has demanded the development of sensitive, rapid, and highly specific methods for a variety of chemical compounds. Infrared absorption procedures appear to meet this need better than any other single analytical tool.

Table 17-3 demonstrates the potential of infrared spectroscopy for the analysis of mixtures of gases. The standard sample of air containing five species in known concentration was analyzed with the computerized instrument shown in Figure 16-14; a 20-m gas cell was employed. The data were printed out within a minute or two after sample injection.

Table 17-4 shows potential applications of infrared filter photometers (such as that shown in Figure 16-12) for the quantitative determination of various chemicals in the atmosphere for the purpose of assuring compli-

[8] See C. L. Lin, et al., *Appl. Spectrosc.*, **1979**, *33*, 481, 487; D. M. Haaland and R. G. Easterling, *Ibid.*, **1980**, *34*, 539; **1982**, *36*, 665.

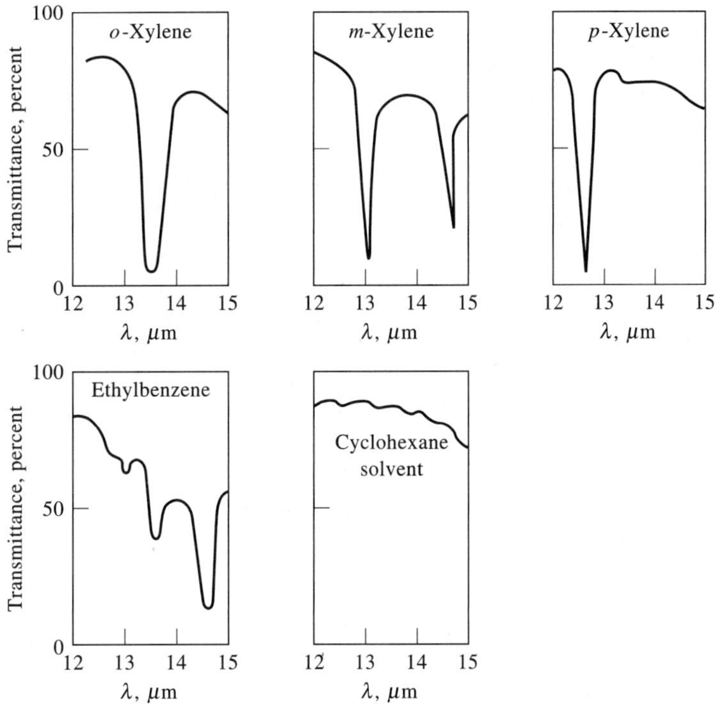

Figure 17-8 Spectra of C_8H_{10} isomers in cyclohexane.

TABLE 17-4 Examples of Infrared Vapor Analysis for OSHA Compliance*

Compound	Allowable Exposure, ppm†	Wavelength, μm	Minimum Detectable Concentration, ppm‡
Carbon disulfide	4	4.54	0.5
Chloroprene	10	11.4	4
Diborane	0.1	3.9	0.05
Ethylenediamine	10	13.0	0.4
Hydrogen cyanide	4.7§	3.04	0.4
Methyl mercaptan	0.5	3.38	0.4
Nitrobenzene	1	11.8	0.2
Pyridine	5	14.2	0.2
Sulfur dioxide	2	8.6	0.5
Vinyl chloride	1	10.9	0.3

*Courtesy of The Foxboro Company, Foxboro, MA 02035.
†1992–1993 OSHA exposure limits for 8-hr weighted average.
‡For 20.25-m cell.
§Short-term exposure limit: 15-min time-weighted average that shall not be exceeded at any time during the work day.

ance with the Occupational Safety and Health Administration (OSHA) regulations.

Of the more than 400 chemicals for which maximum tolerable limits have been set by OSHA more than half appear to have absorption characteristics suitable for determination by means of infrared filter photometers or spectrophotometers. Obviously, among all of these absorbing compounds, peak overlaps are to be expected; yet the method should provide a moderately high degree of selectivity.

Disadvantages and Limitations to Quantitative Infrared Methods

Several disadvantages attend the application of infrared methods to quantitative analysis. Among these are the frequent nonadherence to Beer's law and the complexity of spectra; the latter enhances the probability of the overlap of absorption peaks. In addition, the narrowness of peaks and the effects of stray radiation make absorbance measurements critically dependent upon the slit width and the wavelength setting. Finally, the narrow cells required for many analyses are inconvenient to use and may lead to significant analytical uncertainties. For these reasons, the analytical errors associated with a quantitative infrared analysis often cannot be reduced to the level associated with ultraviolet and visible methods, even with considerable care and effort.

17B MID-INFRARED REFLECTION SPECTROMETRY[9]

Infrared reflection spectrometry has found a number of applications, particularly for dealing with solid samples that are difficult to handle, such as polymer films and fibers, foods, rubbers, agriculture products, and many others. Mid-infrared reflection spectra, while not identical to the corresponding absorption spectra, are similar in general appearance and provide the same information as do their absorption counterparts. Reflectance spectra can be used for both qualitative and quantitative analysis. Most instrument manufacturers now offer adapters that fit into the cell compartments of infrared absorption instruments and make it possible to obtain reflection spectra readily.

17B-1 Types of Reflection

Reflection of radiation is of four types: *specular reflection, diffuse reflection, internal reflection,* and *attenuated total reflection* (ATR). Specular reflection is encountered when the reflecting medium is a smooth polished surface. Here the angle of reflection is identical to the incident angle of the radiation. If the surface is made up of an infrared absorbent, the relative intensity of reflection is less for wavelengths that are absorbed than for wavelengths that are not. Thus, a plot of reflectance R, which is the fraction of the incident radiant energy that is reflected, versus wavelength or wavenumber provides a spectrum for a compound that is similar in general appearance to a transmission spectrum for the species. Specular reflection spectra find some use for examining and characterizing the smooth surfaces of solids and coated solids but is not as widely used as diffuse and total reflection spectra. We will, therefore, focus on the latter two.

17B-2 Diffuse Reflectance Spectrometry

Diffuse reflectance infrared Fourier transform spectrometry (DRIFTS) is an effective way of obtaining infrared spectra directly on powdered samples with a minimum of sample preparation.[10] In addition to saving time in sample preparation, it permits conventional infrared spectral data to be gathered on samples that are not appreciably altered from their original state. The widespread use of diffuse reflectance measurements had to await the general availability of Fourier transform instruments in the mid-1970s because the intensity of radiation reflected from powders is too low to be measured at medium resolution and adequate signal-to-noise ratios with dispersive instruments.

Diffuse reflection is a complex process that occurs when a beam of radiation strikes the surface of a finely divided powder. With this type of sample, specular reflection occurs at each plane surface. However, since there are many of these surfaces and they are randomly oriented, radiation is reflected in all directions. Typi-

[9] See G. Kortum, *Reflectance Spectroscopy.* New York: Springer, 1969; N. J. Harrick, *Internal Reflection Spectroscopy.* New York: Interscience, 1967.

[10] For further information, see M. P. Fuller and P. R. Griffiths, *Anal. Chem.,* **1978,** *50,* 1906; M. P. Fuller and P. R. Griffiths, *Appl. Spectrosc.,* **1980,** *34,* 533.

cally, the intensity of the reflected radiation is roughly independent of the viewing angle.

A number of models have been developed to describe in quantitative terms the intensity of diffuse reflected radiation. The most widely used of these models was developed by Kubelka and Munk.[11] Fuller and Griffiths in their discussion of this model show that the relative reflectance intensity for a powder $f(R'_\infty)$ is given by[12]

$$f(R'_\infty) = \frac{(1-R'_\infty)^2}{2R'_\infty} = \frac{k}{s}$$

where R'_∞ is the ratio of the reflected intensity of the sample to that of a nonabsorbing standard, such as finely ground potassium chloride. The quantity k is the molar absorption coefficient of the analyte, and s is a scattering coefficient. For a dilute sample, k is related to the molar absorptivity ϵ and the molar concentration of the analyte c by the relationship

$$k = 2.303\ \epsilon c$$

Reflectance spectra then consist of a plot of $f(R'_\infty)$ versus wavenumber (see Figure 17-10b).

[11] P. Kubelka, *J. Opt. Soc.*, **1948**, *38*, 448.
[12] M. P. Fuller and P. R. Griffiths, *Anal. Chem.*, **1978**, *50*, 1906.

Instrumentation

Currently, most manufacturers of FTIR instruments offer adapters that fit in cell compartments and permit diffuse reflectance measurements. Figure 17-9 illustrates one type of adapter. The collimated beam from the interferometer is reflected by two plane mirrors onto the off-axis paraboloid P_1 that focuses the radiation onto a small container S holding the powdered sample. Diffuse reflected radiation from the sample, whose surface is about 4 mm in diameter, is collected by the ellipsoidal mirror E and focused by the second paraboloid P_2 onto the detector D. To obtain a spectrum with a single-beam instrument, the signal for the sample is first stored. A reference signal with a good reflector, such as finely ground potassium chloride, in place of the sample is then recorded. These signals are ratioed to give the reflectance.

Comparison of Absorption and Reflection Spectra

Figure 17-10 compares the conventional infrared absorption spectrum for carbazole obtained by means of a KBr pellet with the diffuse reflectance spectrum of a 5% finely ground mixture of carbazole in potassium chloride. Note that the peak locations are the same in the two spectra but that the relative peak heights differ considerably. The differences are typical, with minor peaks generally appearing larger in reflection spectra.

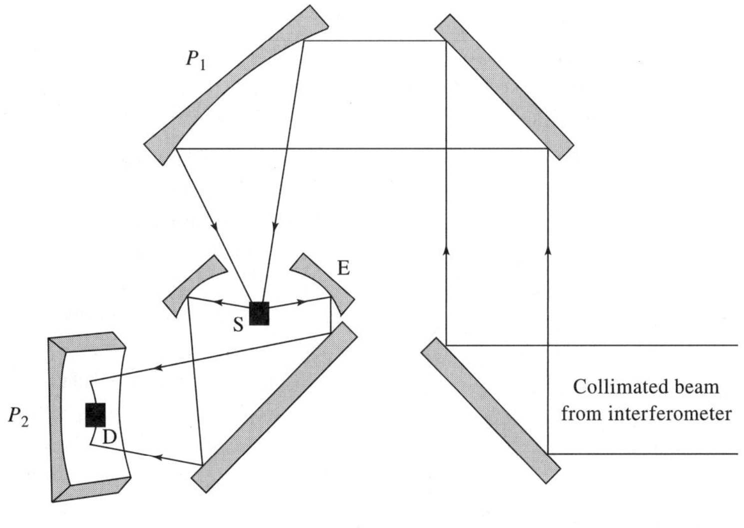

Figure 17-9 A diffuse reflectance attachment for a Fourier transform spectrometer. *(From M. P. Fuller and P. R. Griffiths,* Anal. Chem., ***1978,*** 50, *1907. With permission.)*

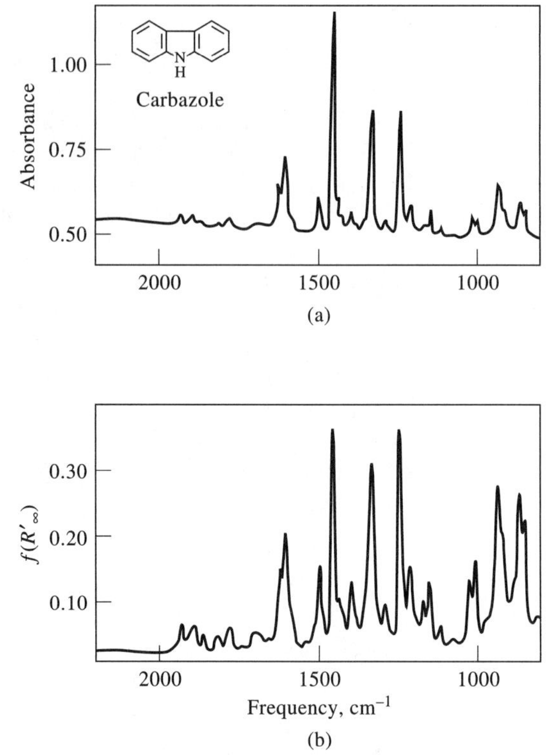

Figure 17-10 Comparison of the absorption spectrum (a) for carbazole with its diffuse reflectance spectrum (b).

17B-3 Attenuated Total Reflectance Spectrometry[13]

Internal-reflection spectroscopy is a technique for obtaining infrared spectra of samples that are difficult to deal with, such as solids of limited solubility, films, threads, pastes, adhesives, and powders.

Principles of the Method

When a beam of radiation passes from a more dense to a less dense medium, reflection occurs. The fraction of the incident beam that is reflected increases as the angle of incidence becomes larger; beyond a certain critical angle, reflection is complete. It has been shown both theoretically and experimentally that during the reflection process the beam acts as if, in fact, it penetrates a small distance into the less dense medium before reflection occurs.[14] The depth of penetration, which varies from a fraction of a wavelength up to several wavelengths, depends upon the wavelength of the incident radiation, the index of refraction of the two materials, and the angle of the incident beam with respect to the interface. The penetrating radiation is called the *evanescent wave*. If the less dense medium absorbs the evanescent radiation, attenuation of the beam occurs at wavelengths of absorption bands. This phenomenon is known as *attenuated total reflectance* (ATR).

Instrumentation

Figure 17-11 shows an apparatus for attenuated total reflectance measurements. As can be seen from the upper figure, the sample (here, a solid) is placed on opposite sides of a transparent crystalline material of high refractive index; a mixed crystal of thallium bromide/thallium iodide is frequently employed, as are plates of germanium and zinc selenide. By proper adjustment of the incident angle, the radiation undergoes multiple internal reflections before passing from the crystal to the detector. Absorption and attenuation take place at each of these reflections.

Figure 17-11b is an optical diagram of a commercially available adapter that will fit into the cell area of most infrared spectrometers and will permit attenuated total reflectance measurements. Note that an incident angle of 30, 45, or 60 deg can be chosen. Cells for liquid samples are also available.

Attenuated Total Reflectance Spectra

Attenuated total reflectance spectra are similar but not identical to ordinary absorption spectra. In general, while the same peaks are observed, their relative intensities differ. The absorbances, while dependent upon the angle of incidence, are independent of sample thickness, since the radiation penetrates only a few micrometers into the sample.

One of the major advantages of attenuated total reflectance spectroscopy is that absorption spectra are readily obtainable on a wide variety of sample types with a minimum of preparation. Threads, yarns, fabrics,

[13] See G. Kortum, *Reflectance Spectroscopy*. New York: Springer, 1969; N. J. Harrick, *Internal Reflection Spectroscopy*. New York: Interscience, 1967; P. A. Wilks, *Amer. Lab.*, **1980** (6), 92.

[14] J. Fahrenfort, *Spectrochem. Acta*, **1961**, *17*, 698.

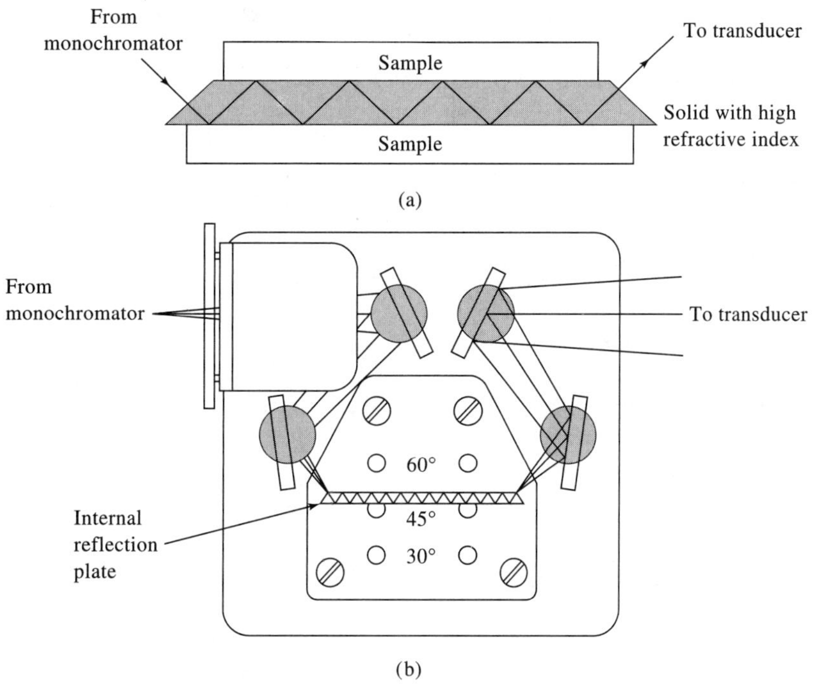

Figure 17-11 Attenuated total reflectance apparatus. (a) Sample mounted on reflection plate; (b) internal reflection adapter. *(Courtesy of The Foxboro Company, Foxboro, MA.)*

and fibers can be studied by pressing the samples against the dense crystal. Pastes, powders, or suspensions can be handled in a similar way. Water solutions can also be accommodated provided the crystal is not water soluble. For liquid samples, the ATR crystal can be dipped into the liquid. Attenuated total reflectance spectroscopy has been applied to many substances such as polymers, rubbers, and other solids. It is of interest to note that the resulting spectra are free from the interference fringes mentioned in the previous section.

17C PHOTOACOUSTIC INFRARED SPECTROSCOPY

In Section 14F, which deals briefly with the principles of photoacoustic measurements, it was noted that this technique has been used to obtain spectra for qualitative studies in the mid-infrared region.[15] As with ultraviolet and visible spectral studies, the technique is profitably applied to those solid and liquid samples that are difficult to handle by ordinary techniques because of their tendency to scatter radiation. In addition, the method has been used for detecting the components of mixtures separated by thin-layer and high-performance liquid chromatography. Most of this work has been carried out with Fourier transform instruments because of their better signal-to-noise characteristics. Most manufacturers offer photoacoustic cells as accessories for FTIR instruments.

Photoacoustic infrared spectroscopy has also been used for monitoring the concentrations of gaseous pollutants in the atmosphere.[16] Here, a tunable carbon dioxide laser source is used in conjunction with a photoacoustic cell. A system of this kind has been designed to analyze a mixture of 10 gases with a sensitivity of 1 ppb and a cycle time of 5 min.

---

[15] For a review of applications of this technique, see J. F. McClelland, *Anal. Chem.*, **1983**, *55*, 89A.

[16] See, for example, L. B. Kreuzer, *Anal. Chem.*, **1974**, *46*, 239A.

17D NEAR-INFRARED SPECTROSCOPY[17]

The near-infrared (NIR) region of the spectrum extends from the upper wavelength end of the visible region at about 770 nm to 2500 nm (13,000 to 4000 cm^{-1}).[18] Absorption bands in this region are overtones or combinations (page 382) of fundamental stretching vibrational bands that occur in the region of 3000 to 1700 cm^{-1}. The bonds involved are usually C—H, N—H, and O—H. Because the bands are overtones or combinations, their molar absorptivities are low and detection limits are on the order of 0.1%.

In contrast to mid-infrared spectrometry, the most important uses of near-infrared radiation are for the routine *quantitative* determination of species, such as water, proteins, low-molecular-weight hydrocarbons, and fats in products of the agricultural, food, petroleum and chemical industries. Both diffuse reflection and transmission measurements are used, although diffuse reflectance is by far the more widely used.

[17] For a general reference on NIR, see *Handbook of Near-Infrared Analysis,* D. R. Burns and E. W. Ciurczak, Eds. New York: Marcel Dekker, 1992. For a review of applications and modern instruments for NIR, see W. F. McClure, *Anal. Chem.,* **1994,** *66,* 43A.

[18] In the literature of near-infrared spectrometry, the abscissa is usually in wavelength in nanometers or micrometers in contrast to mid-infrared spectra in which the abscissa is wavenumber in units of cm^{-1}.

17D-1 Instrumentation

Instrumentation for the near-infrared region is similar to that used for ultraviolet/visible absorption spectroscopy. Tungsten-halogen lamps with quartz windows serve as sources; cells for absorption measurements are usually quartz or fused silica that are transparent up to about 3000 nm. Cell lengths vary from 0.1 to 10 cm. Detectors are generally lead sulfide photoconductors. Several commercial spectrophotometers are designed to operate from 180 to 2500 nm and can thus be used to obtain near-infrared spectra.

A host of photometers and spectrophotometers that have been designed specifically for the near-infrared region are available commercially.[19] The variety of instruments is enormous. The most sophisticated are dual-beam grating, diode-array, or Fourier-transform instruments. Simpler instruments based upon one or more interference filters are also available (see Figure 17-14).

Several solvents are used for near-infrared studies. Some of these are listed in Figure 17-12. Note that only carbon tetrachloride and carbon disulfide are transparent throughout the entire near-infrared region.

[19] For a listing of sources of near-infrared equipment, see W. F. McClure, *Anal. Chem.,* **1994,** *66,* 46A; D. Noble, *Anal. Chem.,* **1995,** *67,* 735A.

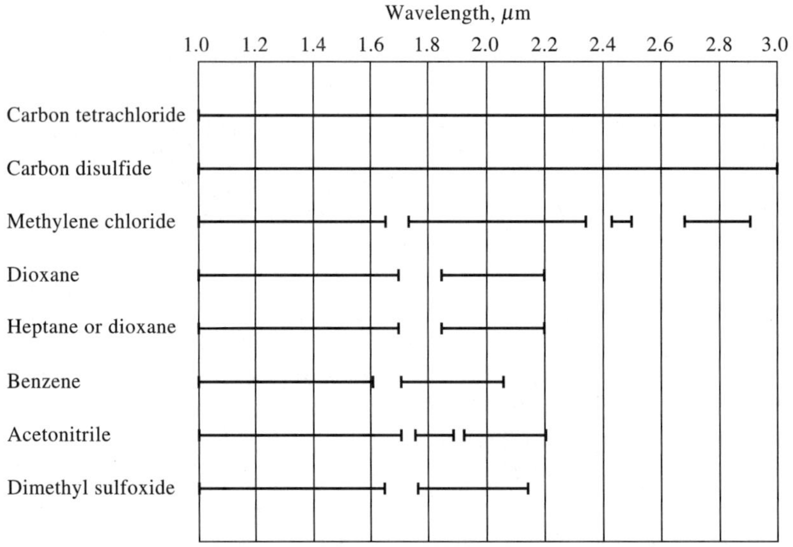

Figure 17-12 Some useful solvents for near-infrared spectroscopy. Solid lines indicate satisfactory transparency for use with 1-cm cells.

17D-2 Applications of Near-Infrared Absorption Spectrometry

In contrast to mid-infrared spectroscopy, near-infrared absorption spectra are less useful for identification and more useful for quantitative analysis of compounds containing functional groups that are made up of hydrogen bonded to carbon, nitrogen, and oxygen. Such compounds can often be determined with accuracies and precisions equivalent to ultraviolet/visible spectroscopy, rather than mid-infrared spectroscopy. Some applications include the determination of water in a variety of samples, including glycerol, hydrazine, organic films, and fuming nitric acid; the quantitative determination of phenols, alcohols, organic acids, and hydroperoxides based upon the first overtone of the O—H stretching vibration that absorbs at about 7100 cm^{-1} (1.4 μm); the determination of esters, ketones, and carboxylic acids based upon their absorption in the region of 3300 to 3600 cm^{-1} (2.8 to 3.0 μm). The absorption in this last case is the first overtone of the carbonyl stretching vibration.

Near-infrared spectrophotometry is also a valuable tool for identification and determination of primary and secondary amines in the presence of tertiary amines in mixtures. The analyses are generally carried out in carbon tetrachloride solutions in 10-cm cells. Primary amines are determined directly by measurement of the absorbance of a combination N—H stretching band at about 5000 cm^{-1} (2.0 μm); neither secondary nor tertiary amines absorb in this region. Primary and secondary amines have several overlapping absorption bands in the 3300 to 10,000 cm^{-1} (1 to 3 μm) region due to various N—H stretching vibrations and their overtones, whereas tertiary amines can have no such bands. Thus, one of these bands gives the secondary amine concentration after correction for the absorption by the primary amine.

17D-3 Applications of Near-Infrared Reflectance Spectrometry

Near-infrared reflectance spectroscopy has become a most important tool for the routine quantitative determination of constituents in finely ground solids. The most widespread use of this technique has been for the determination of protein, moisture, starch, oil, lipids, and cellulose in agricultural products such as grains and oilseeds.[20] For example, Canada sells all its wheat based on guaranteed protein content, and the Canadian Grain Commission formerly ran over 600,000 Kjeldahl protein determinations annually as a consequence. Now, it is estimated that 80% to 90% of all Canadian grain is analyzed for protein by near-infrared reflectance spectroscopy at a savings of over $500,000 a year in analytical costs.[21]

In near-infrared reflectance spectroscopy, the finely ground solid sample is irradiated with one or more narrow bands of radiation ranging in wavelength from 1 to 2.5 μm or 10,000 to 4000 cm^{-1}. *Diffuse reflectance* occurs in which the radiation penetrates the surface layer of the particles, excites vibrational modes of the analyte molecule, and is then scattered in all directions. A reflectance spectrum is thus produced that is dependent upon the composition of the sample. A typical reflectance spectrum for a sample of wheat is shown in Figure 17-13. The ordinate in this case is the logarithm of the reciprocal of reflectance R, $\log(1/R) = -\log R$, where R is the ratio of the intensity of radiation reflected from the sample to the reflectance from a standard reflector, such as finely ground barium sulfate or magnesium oxide. The reflectance band at 1940 nm is a water peak that is used for moisture determinations. The peak at about 2100 nm is, in fact, two overlapping peaks, one for starch and the other for protein. By making measurements at two wavelengths in this region, the

[20] See C. Watson, *Anal. Chem.*, **1977**, *49*, 835A; D. L. Wetzel, *Anal. Chem.*, **1983**, *55*, 1165A.

[21] S. A. Borman, *Anal. Chem.*, **1984**, *56*, 933A.

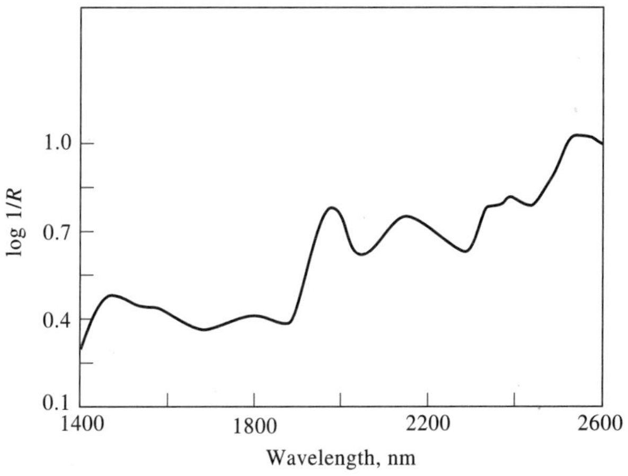

Figure 17-13 Diffuse-reflectance spectrum of a sample of wheat.

concentrations of each of these components can be determined.

Instruments for diffuse reflectance measurements are available from commercial sources. Some of these employ several interference filters to provide narrow bands of radiation. Others are equipped with grating monochromators. Ordinarily, reflectance measurements are made at two or more wavelengths for each analyte species that is being determined. Figure 17-14 is a schematic of a typical filter-type instrument. The interior walls of the integrating sphere surrounding the sample are coated with a nearly perfect diffuse reflecting material such as barium sulfate. Radiation reflected from the sample eventually reaches the detectors after multiple reflections off the walls of the sphere. The tilting mirror permits measurement of the ratio of the intensity of the radiation reflected from the sample and that reflected off the wall.

Calibration of an instrument, such as that shown in Figure 17-14, is often a major undertaking. First, 30 or more samples of the material being analyzed that contain the range of analyte concentrations likely to be encountered must be acquired. For example, for determining protein in wheat, 30 to 50 wheat samples containing 10% to 20% protein are required. Each is then carefully analyzed chemically by the standard Kjeldahl procedure for determining protein. For the reflectance measurements, samples are ground to a reproducible particle size and their reflectance measured at two or more wavelengths. Considerable time and effort has to be expended in determining optimal wavelengths to be used for the analysis. From this study equations are developed and tested that relate the measured reflectances to percent protein. Details of such a calibration procedure are discussed by Honigs and others.[22]

The great advantage of near-infrared reflectance methods is their speed and the simplicity of sample preparation. Once method development has been completed, analysis of solid samples for several species can be completed in a few minutes. Accuracies and precisions of 1% to 2% relative are regularly reported.

17E FAR-INFRARED SPECTROSCOPY

The far-infrared region is particularly useful for inorganic studies because absorption due to stretching and bending vibrations of bonds between metal atoms and

[22] D. E. Honigs, G. M. Hieftje, H. L. Mark, and T. Hirschfeld, *Anal. Chem.*, **1985**, *57*, 2299; D. E. Honigs, G. M. Hieftje, and T. Hirschfeld, *Appl. Spectrosc.*, **1984**, *38*, 1984.

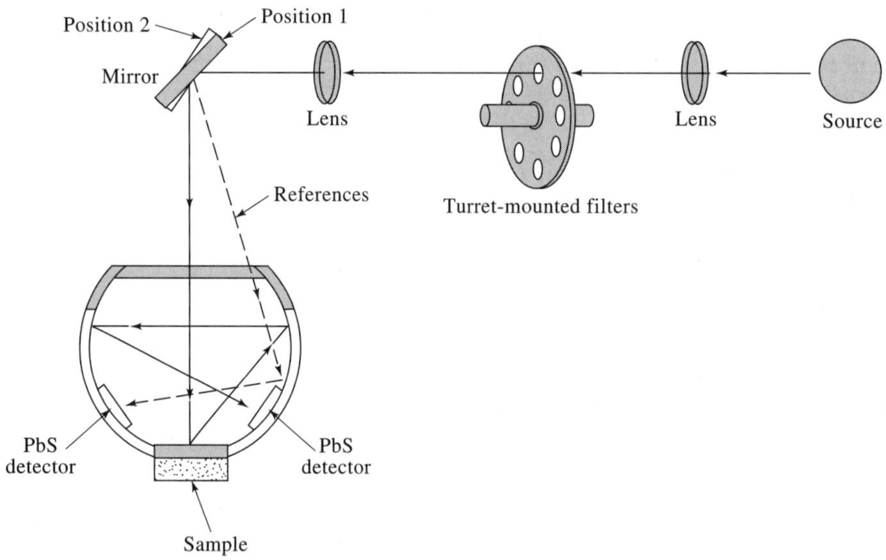

Figure 17-14 A diffuse-reflectance photometer. Mirror position 1: reflectance of sample gives I. Mirror position 2 gives reflectance of the standard reflectance area (I_0). Reflectance $= R = I/I_0$. *(From D. L. Wetzel,* Anal. Chem., ***1983****, 55, 1174A. With permission.)*

both inorganic and organic ligands generally occur at frequencies lower than 650 cm^{-1} (>15 μm). For example, heavy-metal iodides generally absorb in the region below 100 cm^{-1}, while the bromides and chlorides have bands at higher frequencies. Absorption frequencies for metal-organic bonds are ordinarily dependent upon both the metal atom and the organic portion of the species.

Far-infrared studies of inorganic solids have also provided useful information about lattice energies of crystals and transition energies of semiconducting materials.

Molecules composed only of light atoms absorb in the far-infrared if they have skeletal bending modes that involve more than two atoms other than hydrogen. Important examples are substituted benzene derivatives, which generally show several absorption peaks. The spectra are frequently quite specific and useful for identifying a particular compound; to be sure, characteristic group frequencies also exist in the far-infrared region.

Pure rotational absorption by gases is observed in the far-infrared region, provided the molecules have permanent dipole moments. Examples include H_2O, O_3, HCl, and AsH_3. Absorption by water is troublesome; elimination of its interference requires evacuation or at least purging of the spectrometer.

Fourier transform spectrometers are particularly useful for far-infrared studies. The energy advantage of the interferometric system over a dispersive one generally results in a significant improvement in spectral quality. In addition, the application of gratings to this wavelength region is complicated by the overlapping of several orders of diffracted radiation.

17F INFRARED EMISSION SPECTROSCOPY

Upon being heated, molecules that absorb infrared radiation are also capable of emitting characteristic infrared wavelengths. The principal deterrent to the analytical application of this phenomenon has been the poor signal-to-noise characteristics of the infrared emission signal, particularly when the sample is at a temperature only slightly higher than its surroundings. With the interferometric method, interesting and useful applications are now appearing in the literature.

An early example of the application of infrared emission spectroscopy is found in a paper that describes the use of a Fourier transform spectrometer for the identification of microgram quantities of pesticides.[23] Samples were prepared by dissolving them in a suitable solvent and evaporating on a NaCl or KBr plate. The plate was then heated electrically near the spectrometer entrance. Pesticides such as DDT, malathion, and dieldrin were identified in amounts as low as 1 to 10 μg.

Equally interesting has been the use of the interferometric technique for the remote detection of components emitted from industrial stacks. In one of these applications, an interferometer was mounted on an 8-inch reflecting telescope.[24] With the telescope focused on the plume from an industrial plant, CO_2 and SO_2 were readily detected at a distance of several hundred feet. Similarly, the U.S. Environmental Protection Agency's remote optical sensing of emissions (ROSE) system uses a 30-cm telescope to collect radiation from distant industrial plumes and direct it into the interferometer of a Fourier transform infrared instrument.[25]

17G INFRARED MICROSPECTROMETRY

Infrared microspectrometry is used for obtaining absorption or reflection spectra of species in samples having physical dimensions in the 10 to 500 μm range.[26] Infrared microscopes, which were first introduced by several instrument manufacturers in the 1980s, generally consist of two microscopes, one an ordinary optical microscope and the other an infrared device with reflection optics that reduce the size of the infrared beam to about that of the sample. The optical microscope is used to visually locate the sample particle or spot to be studied with the infrared beam. The infrared source is an ordinary Fourier transform spectrometer and not a grating instrument because of the much greater sensitivity of the former. The transducer is usually a liquid-nitrogen-cooled MCT (mercury/cadmium/telluride) photoconductive device that is more sensitive than other types of infrared transducers.

---

[23] I. Coleman and M. J. D. Low, *Spectrochim. Acta*, **1966**, *22*, 1293.

[24] M. J. D. Low and F. K. Clancy, *Env. Sci. Technol.*, **1967**, *1*, 73.

[25] W. F. Herget, in *Fourier Transform Infrared Spectroscopy; Applications to Chemical Systems*, J. R. Ferraro and L. J. Basile, Eds., Vol. 2, pp. 111–127. New York: Academic Press, 1979.

[26] For monographs on infrared microspectroscopy, see *Practical Guide To Infrared Microspectroscopy*, H. J. Humecki, Ed. New York: Marcel Dekker, 1995; *Infrared Microscopy*, R. G. Messerschmidt and M. A. Harthcock, Eds. New York: Marcel Dekker, 1988. For shorter reviews, see J. E. Caton, G. E. Pacey, and J. F. O'Keefe, *Anal. Chem.*, **1986**, *58*, 465A; J. A. Caton and A. J. Sommer, *Anal. Chem.*, **1992**, *64*, 931A.

The use of infrared microscopy is still in its infancy but appears to be a technique that will find increasing applications in the future. Some of its current applications include: identification of polymer contaminants, imperfections in polymer films, and individual layers of laminated polymer sheets; identification of tiny samples of fibers, paint, and explosives in criminalistics; characterization of single fibers in the textile industry; and identification of contaminants on electronic components.

17H QUESTIONS AND PROBLEMS

17-1 Cyclohexanone exhibits its strongest infrared peak at 5.86 μm, and at this wavelength a linear relationship exists between absorbance and concentration.
(a) Identify the part of the molecule responsible for the absorbance at this wavelength.
(b) Suggest a solvent that would be suitable for a quantitative analysis of cyclohexanone at this wavelength.
(c) A solution of cyclohexanone (2.0 mg/mL) in the solvent selected in part (b) exhibits an absorbance of 0.40, in a cell with a path length of 0.025 mm. What is the detection limit for this compound under these conditions, if the noise associated with the spectrum of the solvent is 0.001 absorbance units?

17-2 The spectrum in Figure 17-15 was obtained for a liquid with an empirical formula of C_3H_6O. Identify the compound.

17-3 The spectrum in Figure 17-16 is that of a high-boiling liquid having an empirical formula $C_9H_{10}O$. Identify the compound as closely as possible.

17-4 The spectrum in Figure 17-17 is for an acrid-smelling liquid that boils at 50°C and has a molecular weight of about 56. What is the compound? What impurity is clearly present?

17-5 The spectrum in Figure 17-18 is that of a nitrogen containing substance that boils at 97°C. What is the compound?

17-6 Why are quantitative analytical methods based upon near-infrared radiation likely to be more precise and accurate than methods based upon mid-infrared radiation?

17-7 Why do mid-infrared recorded spectra often have regions of negative transmittance readings?

17-8 Why do infrared spectra seldom show regions at which the transmittance is 100%?

17-9 An empty cell showed 12 interference peaks in the wavelength range of 6.0 to 12.2 μm. Calculate the path length of the cell.

17-10 An empty cell exhibited 9.5 interference peaks in the region of 1250 to 1480 cm^{-1}. What was the path length of the cell?

17-11 Estimate the thickness of the polystyrene film that yielded the spectrum shown in Figure 16-1.

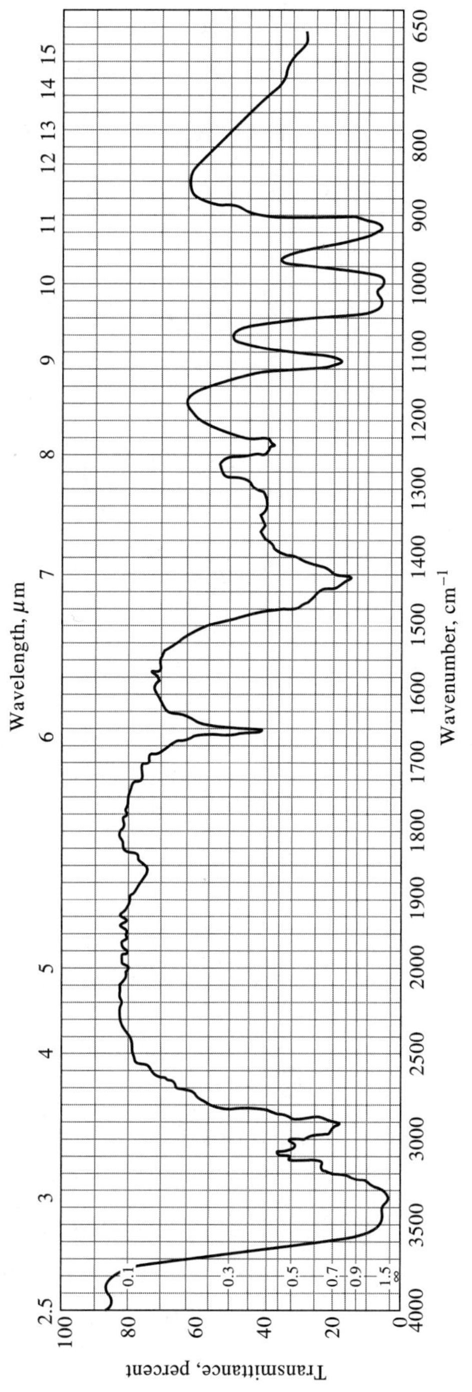

Figure 17-15 See Problem 17-2. *(Spectrum courtesy of Thermodynamics Research Center Data Project, Texas A & M University, College Station, Texas.)*

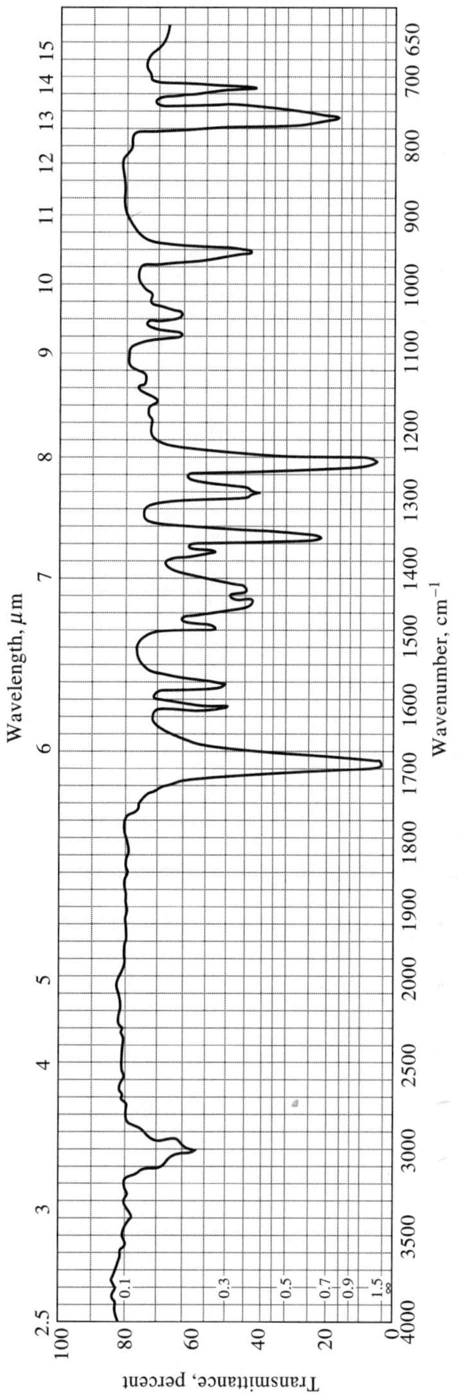

Figure 17-16 See Problem 17-3. *(Spectrum courtesy of Thermodynamics Research Center Data Project, Texas A & M University, College Station, Texas.)*

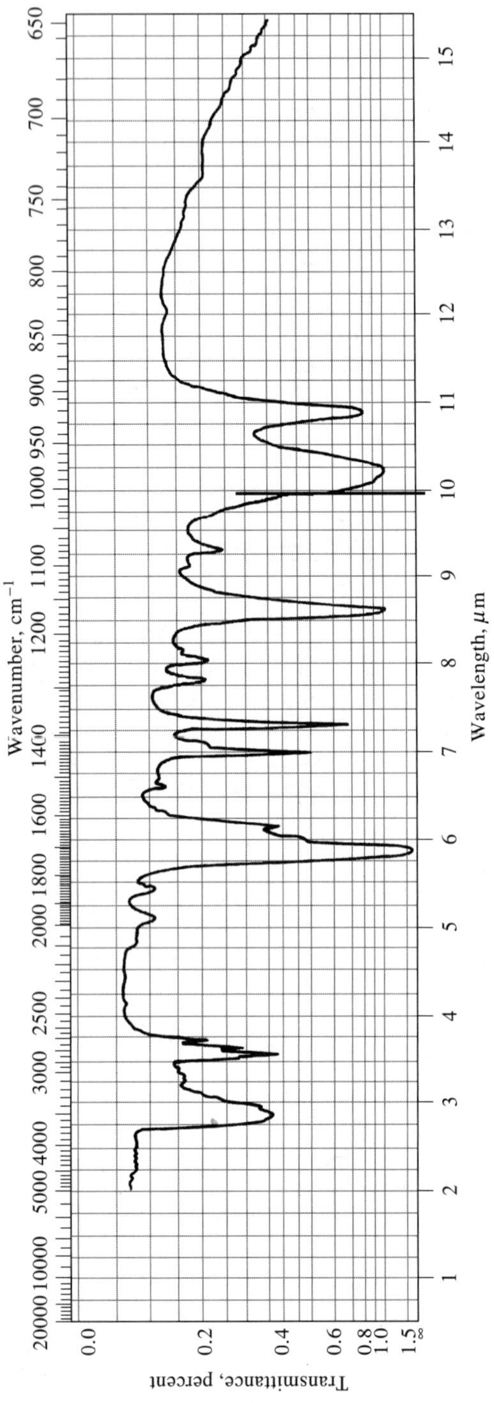

Figure 17-17 See Problem 17-4. *(Spectrum courtesy of Thermodynamics Research Center Data Project, Texas A & M University, College Station, Texas.)*

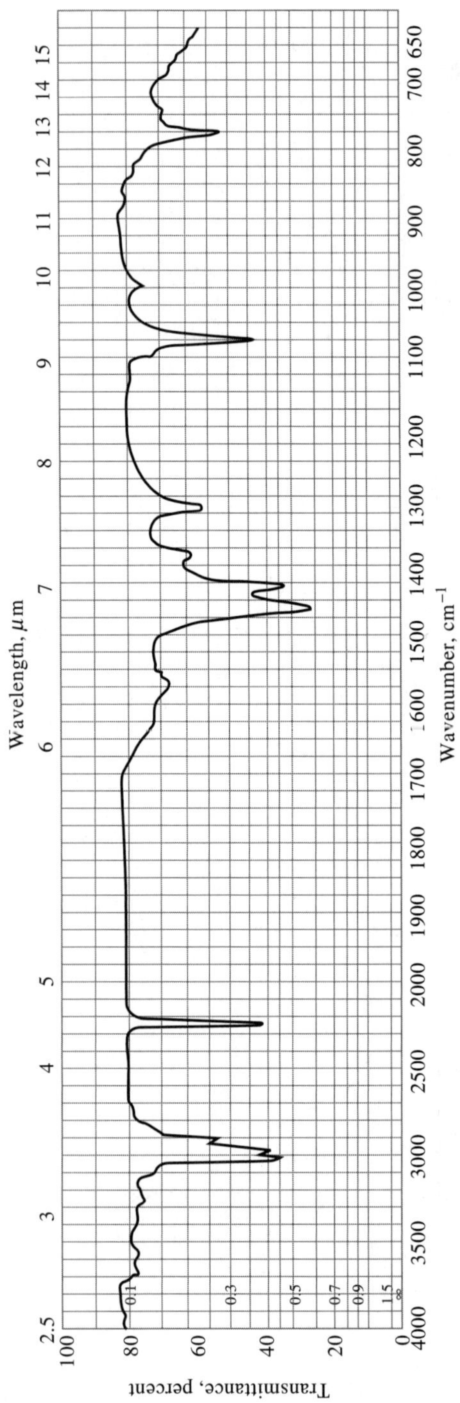

Figure 17-18 See Problem 17-5. *(Spectrum courtesy of Thermodynamics Research Center Data Project, Texas A & M University, College Station, Texas.)*

Chapter 18

Raman Spectroscopy

When radiation passes through a transparent medium, the species present scatter a fraction of the beam in all directions (Section 6B-10). In 1928, the Indian physicist C. V. Raman discovered that the visible wavelength of a small fraction of the radiation scattered by certain molecules differs from that of the incident beam and furthermore that the shifts in wavelength depend upon the chemical structure of the molecules responsible for the scattering. He was awarded the 1931 Nobel Prize in physics for this discovery and for his systematic exploration of the phenomenon.[1]

The theory of Raman scattering, which now is well understood, shows that the phenomenon results from the same type of quantized vibrational changes that are associated with infrared absorption. Thus, the ***difference*** *in wavelength between the incident and scattered visible radiation corresponds to wavelengths in the mid-infrared region. Indeed, the Raman scattering spectrum and infrared absorption spectrum for a given species often resemble one another quite closely. There are, however, enough differences between the kinds of groups that are infrared active and Raman active to make the techniques complementary rather than competitive. For some problems, the infrared method is the superior tool; for others, the Raman procedure offers more useful spectra.*

An important advantage of Raman spectra over infrared lies in the fact that water does not cause interference; indeed, Raman spectra can be obtained from aqueous solutions. In addition, glass or quartz cells can be employed, thus avoiding the inconvenience of working with sodium chloride or other atmospherically unstable window materials. Despite these advantages, Raman spectroscopy was not widely used by chemists for structural studies until lasers became available in the 1960s, which made spectra a good deal easier to obtain. A second deterrent to the general use of Raman spectroscopy was interference by fluorescence of the sample or impurities in the sample. This problem has

[1] For more complete discussions of the theory and practice of Raman spectroscopy, see J. R. Ferraro and K. Nakamoto, *Introductory Raman Spectroscopy.* New York: Academic Press, 1994; *Modern Techniques in Raman Spectroscopy,* J. J. Laserna, Ed. New York: Wiley, 1996; D. P. Strommen and K. Nakamoto, *Laboratory Raman Spectroscopy.* New York: Wiley, 1984; *Analytical Raman Spectroscopy,* J. G. Grasselli and B. J. Bulkin, Eds. New York: Wiley, 1991; P. Hendra, G. Jones, G. Warnes, *Fourier Transform Raman Spectroscopy.* London: Ellis Harwood, 1991.

by now been largely overcome by the use of near-infrared laser sources.

18A THEORY OF RAMAN SPECTROSCOPY

Raman spectra are acquired by irradiating a sample with a powerful laser source of visible or near-infrared monochromatic radiation. During irradiation, the spectrum of the scattered radiation is measured at some angle (often 90 deg) with a suitable spectrometer. At the very most, the intensities of Raman lines are 0.001% of the intensity of the source; as a consequence, their detection and measurement are somewhat more difficult than are infrared spectra. An exception to this statement may be found in resonance Raman spectroscopy, which is considerably more sensitive than normal Raman spectroscopy. Resonance Raman spectroscopy is described in Section 18D-1.

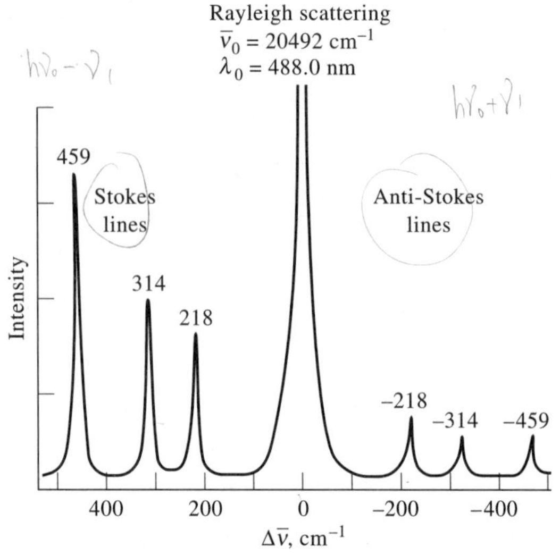

Figure 18-1 Raman spectrum for CCl_4 excited by laser radiation of λ_0 = 488 nm and $\bar{\nu}_0$ = 20,492 cm^{-1}. The number above the peaks is the Raman shift, $\Delta\bar{\nu} = (\bar{\nu}_s - \bar{\mu}_0)$ cm^{-1}. *(Reprinted with permission from D. P. Strommen and K. Nakamato,* Amer. Lab., ***1981****, 43, (10), 72. Copyright 1981 by International Scientific Communications, Inc.)*

18A-1 Excitation of Raman Spectra

Figure 18-1 depicts a portion of a Raman spectrum that was obtained by irradiating a sample of carbon tetrachloride with an intense beam of an argon ion laser having a wavelength of 488.0 nm (20492 cm^{-1}). The emitted radiation is of three types: *Stokes* scattering, *anti-Stokes* scattering, and *Rayleigh* scattering. The last, whose wavelength is exactly that of the excitation source, is significantly more intense than either of the other two types.

As is usually the case for Raman spectra, the abscissa of Figure 18-1 is the wavenumber shift $\Delta\bar{\nu}$, which is defined as the difference in wavenumbers (cm^{-1}) between the observed radiation and that of the source. Note that three Raman peaks are found on both sides of the Rayleigh peak and that the pattern of shifts on each side is identical. That is, Stokes lines are found at wavenumbers that are 218, 314, and 459 cm^{-1} *smaller* than the Rayleigh peak while anti-Stokes peaks occur at 218, 314, and 459 cm^{-1} *greater* than the wavenumber of the source. It should also be noted that additional lines can be found at ±762 and ±790 cm^{-1} as well. Quite generally, anti-Stokes lines are appreciably less intense than the corresponding Stokes lines. For this reason, only the Stokes part of a spectrum is generally used. Furthermore, the abscissa of the plot is often labeled simply frequency in cm^{-1} rather than wavenumber shift $\Delta\bar{\nu}$. It is noteworthy that fluorescence may interfere seriously with the observation of Stokes shifts but not with anti-Stokes. With fluorescing samples, anti-Stokes signals may, therefore, be more useful despite their lower intensities.

It is important to appreciate that the magnitude of Raman shifts are *independent of the wavelength of excitation.* Thus, shift patterns identical to those shown in Figure 18-1 would be observed for carbon tetrachloride regardless of whether excitation was carried out with a krypton ion laser (488.0), a helium/neon laser (632.8 nm), or a Nd:YAG laser (1064 nm).

Superficially, the appearance of Raman spectral lines at lower energies (longer wavelengths) is analogous to the Stokes shifts found in a fluorescence experiment (page 137); for this reason, Raman shifts to longer wavelengths are called *Stokes shifts.* We shall see, however, that Raman and fluorescence spectra arise from fundamentally different processes; thus, the application of the same terminology to both fluorescent and Raman spectra is perhaps unfortunate.

18A-2 Mechanism of Raman and Rayleigh Scattering

In Raman spectroscopy, spectral excitation is normally carried out by radiation having a wavelength that is well away from any absorption peaks of the analyte. The

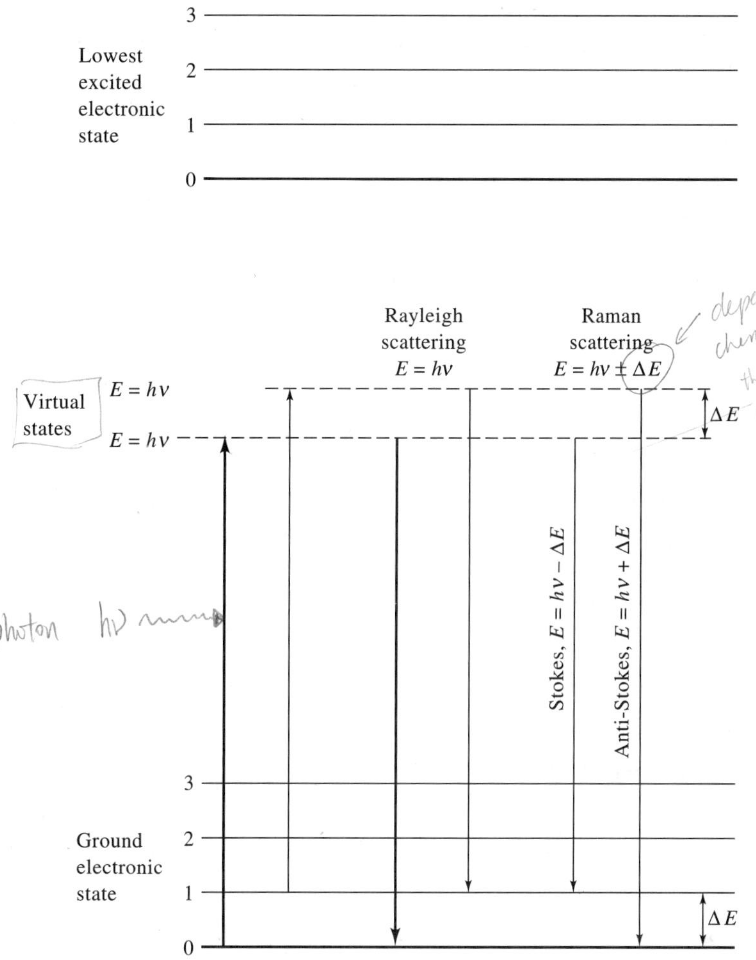

Figure 18-2 Origin of Rayleigh and Raman scattering.

energy-level diagram in Figure 18-2 provides a qualitative picture of the sources of Raman and Rayleigh scattering. The heavy arrow on the far left depicts the energy change in the molecule when it interacts with a photon from the source. The increase in energy is equal to the energy of the photon $h\nu$. It is important to appreciate that the process shown is *not quantized;* thus, depending upon the frequency of the radiation from the source, the energy of the molecule can assume any of an infinite number of values, or *virtual states,* between the ground state and the first electronic excited state shown in the upper part of the diagram. The second and narrower arrow shows the type of change that would occur if the molecule encountered by the photon happened to be in the first vibrational level of the electronic ground state. At room temperature, the fraction of the molecules in this state is small. Thus, as indicated by the width of the arrows, the probability of this process occurring is much smaller.

The middle set of arrows depicts the changes that produce Rayleigh scattering. Again the more probable change is shown by the wider arrow. Note that no energy is lost in Rayleigh scattering. As a consequence, the collisions between the photon and the molecule are said to be *elastic.*

Finally, the energy changes that produce Stokes and anti-Stokes emission are depicted on the right. The two differ from the Rayleigh radiation by frequencies corresponding to $\pm\Delta E$, the energy of the first vibrational level of the ground state. Note that if the bond were infrared active, the energy of its absorption would also be ΔE. Thus, the Raman *frequency shift* and the infrared absorption *peak frequency* are identical.

Note also that the relative populations of the two upper energy states are such that Stokes emission is much favored over anti-Stokes. In addition, Rayleigh scattering has a considerably higher probability of occurring than Raman because the most probable event is the energy transfer to molecules in the ground state and reemission by the return of these molecules to the ground state. Finally, it should be noted that the ratio of anti-Stokes to Stokes intensities will increase with temperature because a larger fraction of the molecules will be in the first vibrationally excited state under these circumstances.

18A-3 Wave Model of Raman and Rayleigh Scattering

Let us assume that a beam of radiation having a frequency ν_{ex} is incident upon a solution of an analyte. The electric field E of this radiation can be described by the equation

$$E = E_0 \cos(2\pi\nu_{ex}t) \qquad (18\text{-}1)$$

where E_0 is the amplitude of the wave. When the electric field of the radiation interacts with an electron cloud of an analyte bond, it induces a dipole moment m in the bond that is given by

$$m = \alpha E = \alpha E_0 \cos(2\pi\nu_{ex}t) \qquad (18\text{-}2)$$

where α is a proportionality constant called the *polarizability* of the bond. This constant is a measure of the deformability of the bond in an electric field.

In order to be Raman active, the polarizability α of a bond must vary as a function of the distance between nuclei according to the equation

$$\alpha = \alpha_0 + (r - r_{eq})\left(\frac{\partial\alpha}{\partial r}\right) \qquad (18\text{-}3)$$

where α_0 is the polarizability of the bond at the equilibrium internuclear distance r_{eq} and r is the internuclear separation at any instant. The change in internuclear separation varies with the frequency of the vibration ν_v as given by

$$r - r_{eq} = r_m \cos(2\pi\nu_v t) \qquad (18\text{-}4)$$

where r_m is the maximum internuclear separation relative to the equilibrium position. Substituting Equation 18-4 into 18-3 gives

$$\alpha = \alpha_0 + \left(\frac{\partial\alpha}{\partial r}\right) r_m \cos(2\pi\nu_v t) \qquad (18\text{-}5)$$

We can then obtain an expression for the induced dipole moment m by substituting Equation 18-5 into Equation 18-2. Thus,

$$m = \alpha_0 E_0 \cos(2\pi\nu_{ex}t) + E_0 r_m \left(\frac{\partial\alpha}{\partial r}\right) \cos(2\pi\nu_v t)\cos(2\pi\nu_{ex}t) \qquad (18\text{-}6)$$

Recall from trigonometry that

$$\cos x \cos y = [\cos(x + y) + \cos(x - y)]/2$$

Applying this identity to Equation 18-6 gives

$$m = \alpha_0 E_0 \cos(2\pi\nu_{ex}t) + \frac{E_0}{2} r_m \left(\frac{\partial\alpha}{\partial r}\right) \cos[2\pi(\nu_{ex} - \nu_v)t] + \frac{E_0}{2} r_m \left(\frac{\partial\alpha}{\partial r}\right) \cos[2\pi(\nu_{ex} + \nu_v)t] \qquad (18\text{-}7)$$

The first term in this equation represents Rayleigh scattering, which occurs at the excitation frequency ν_{ex}. The second and third terms in Equation 18-7 correspond respectively to the Stokes and anti-Stokes frequencies of $\nu_{ex} - \nu_v$ and $\nu_{ex} + \nu_v$. Here, the excitation frequency has been modulated by the vibrational frequency of the bond. It is important to note that Raman scattering *requires* that the polarizability of a bond varies as a function of distance—that is, $\partial\alpha/\partial r$ in Equation 18-7 must be greater than zero for Raman lines to appear.

We have noted, for a given bond, the energy *shifts* observed in a Raman experiment should be identical to the *energies* of its infrared absorption bands, provided that the vibrational modes involved are active toward both infrared absorption and Raman scattering. Figure 18-3 illustrates the similarity of the two types of spectra; it is seen that several peaks with identical $\bar{\nu}$ and $\Delta\bar{\nu}$ values exist for the two compounds. It is also noteworthy, however, that the relative size of the corresponding peaks is frequently quite different; moreover, certain peaks that occur in one spectrum are absent in the other.

The differences between a Raman spectrum and an infrared spectrum are not surprising when it is considered that the basic mechanisms, although dependent upon the same vibrational modes, arise from processes that are mechanistically different. Infrared absorption requires that a vibrational mode of the molecule have a change in dipole moment or charge distribution associated with it. Only then can radiation of the same frequency interact with the molecule and promote it to an

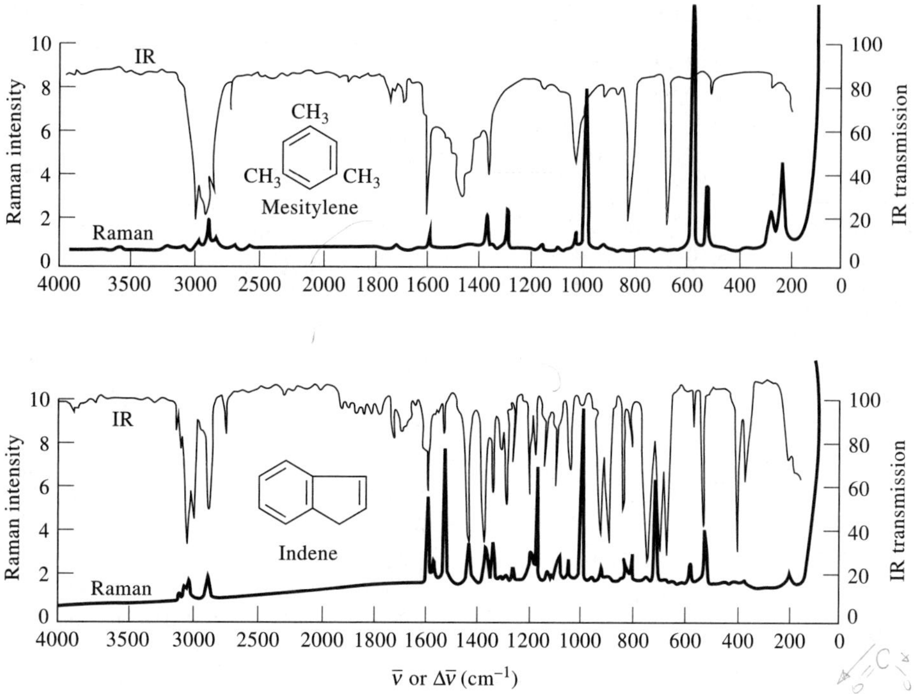

Figure 18-3 Comparison of Raman and infrared spectra. *(Courtesy Perkin-Elmer Corp., Norwalk, CT.)*

excited vibrational state. In contrast, scattering involves a momentary distortion of the electrons distributed around a bond in a molecule, followed by reemission of the radiation as the bond returns to its normal state. In its distorted form, the molecule is temporarily polarized; that is, it develops momentarily an induced dipole that disappears upon relaxation and reemission. Because of this fundamental difference in mechanism, the Raman activity of a given vibrational mode may differ markedly from its infrared activity. For example, a homonuclear molecule such as nitrogen, chlorine, or hydrogen has no dipole moment either in its equilibrium position or when a stretching vibration causes a change in the distance between the two nuclei. Thus, absorption of radiation of the vibration frequency cannot occur. On the other hand, the polarizability of the bond between the two atoms of such a molecule varies periodically in phase with the stretching vibrations, reaching a maximum at the greatest separation and a minimum at the closest approach. A Raman shift corresponding in frequency to that of the vibrational mode results.

It is of interest to compare the infrared and the Raman activities of coupled vibrational modes such as those described earlier (page 388) for the carbon dioxide molecule. In the symmetric mode, no change in the dipole moment occurs as the two oxygen atoms move away from or toward the central carbon atom; thus, this mode is infrared inactive. The polarizability, however, fluctuates in phase with the vibration since distortion of bonds becomes easier as they lengthen and more difficult as they shorten. Raman activity is associated with this mode.

In contrast, the dipole moment of carbon dioxide fluctuates in phase with the anti-symmetric vibrational mode. Thus, an infrared absorption peak arises from this mode. On the other hand, as the polarizability of one of the bonds increases as it lengthens, the polarizability of the other decreases resulting in no net change in the polarizability. Thus, the asymmetric stretching vibration is Raman inactive.

Often, as in the foregoing examples, parts of Raman and infrared spectra are complementary, each being associated with a different set of vibrational modes within a molecule. Other vibrational modes may be both Raman and infrared active. For example, all of the vibrational modes of sulfur dioxide yield both Raman and infrared peaks. The size of the peaks differ, however, because the probability for the transitions are different for the two mechanisms.

18A-4 Intensity of Normal Raman Peaks

The intensity or power of a normal Raman peak depends in a complex way upon the polarizability of the molecule, the intensity of the source, and the concentration of the active group, as well as other factors. In the absence of absorption, the power of Raman emission increases with the fourth power of the frequency of the source; however, advantage can seldom be taken of this relationship because of the likelihood that ultraviolet irradiation will cause photodecomposition.

Raman intensities are usually directly proportional to the concentration of the active species. In this regard, Raman spectroscopy more closely resembles fluorescence than absorption, in which the concentration-intensity relationship is logarithmic.

18A-5 Raman Depolarization Ratios[2]

Raman measurements provide, in addition to intensity and frequency information, one additional variable that is sometimes useful in determining the structure of molecules, namely, the *depolarization ratio.* Here, it is important to carefully distinguish between the terms *polarizability* and *polarization*. The former term describes a *molecular* property having to do with the deformability of a bond. Polarization, in contrast, is a property of a beam of radiation and describes the plane in which the radiation vibrates.

When Raman spectra are excited by plane-polarized radiation, as they are when a laser source is used, the scattered radiation is found to be polarized to various degrees depending upon the type of vibration responsible for the scattering. The nature of this effect is illustrated in Figure 18-4 where radiation from a laser source is shown as being polarized in the yz plane. Part of the resulting scattered radiation is shown as being polarized parallel to the original beam, that is, in the xz plane; the intensity of this radiation is symbolized by the subscript $\parallel$. The remainder of the scattered beam is polarized in the xy plane, which is perpendicular to the polarization of the original beam; the intensity of this perpendicularly polarized radiation is shown by the subscript $\perp$. The depolarization ratio p is defined as

$$p = \frac{I_\perp}{I_\parallel} \qquad (18\text{-}8)$$

Experimentally, the depolarization ratio may be obtained by inserting a Polaroid sheet or other polarizer between the sample and the monochromator. Spectra

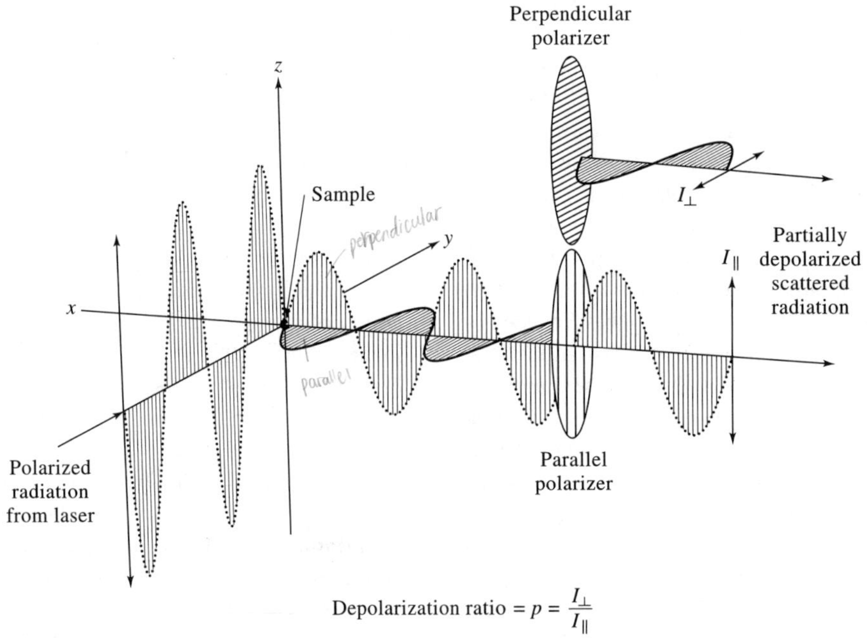

Figure 18-4 Depolarization resulting from Raman scattering.

[2] D.P. Strommen, *J. Chem. Educ.*, **1992,** *69,* 803.

are then obtained with the axis of the sheet oriented parallel with first the *xz* and then the *xy* plane shown in Figure 18-4.

The depolarization ratio is dependent upon the symmetry of the vibrations responsible for scattering. For example, the peak for carbon tetrachloride at 459 cm^{-1} (Figure 18-1) arises from a totally symmetric breathing vibration involving the simultaneous movement of the four tetrahedrally arranged chlorine atoms toward and away from the central carbon atom. The depolarization ratio is 0.005, indicating minimal depolarization; the 459-cm^{-1} line is thus said to be *polarized*. In contrast, the carbon tetrachloride peaks at 218 and 314 cm^{-1}, which arise from nonsymmetric vibrations, have depolarization ratios of about 0.75. From scattering theory it is possible to demonstrate that the maximum depolarization for nonsymmetric vibrations is 6/7 while for symmetric vibrations, the ratio is always less than this number. The depolarization ratio is thus useful in correlating Raman lines with modes of vibration.

18B INSTRUMENTATION

Instrumentation for modern Raman spectroscopy consists of three components: a laser source, a sample-illumination system, and a suitable spectrometer. The performance requirements for these components are more stringent than for the molecular spectrometers we have already described, however, because of the inherent weakness of the Raman scattering signal compared with the signal produced by the Rayleigh scattering.

18B-1 Sources

The sources used in modern Raman spectrometry are nearly always lasers because their high intensity is necessary to produce Raman scattering of sufficient intensity to be measured with a reasonable signal-to-noise ratio. Five of the most common lasers used for Raman spectroscopy are listed in Table 18-1. Because the intensity of Raman scattering varies as the fourth power of the frequency, argon and krypton ion sources that emit in the blue and green region of the spectrum have an advantage over the other sources shown in the table. For example, the argon line at 488 nm provides Raman lines that are nearly three times as intense as those excited by the helium/neon source, given the same input power.

The last two sources in the table, which emit near-infrared radiation, are finding more and more use as excitation sources. Near-infrared sources have two major advantages over shorter wavelength lasers. The first is that they can be operated at much higher power (up to 50 W) without causing photodecomposition of the sample. The second is that they are not energetic enough to populate a significant number of fluorescence-producing excited electronic energy states in most molecules. Consequently, fluorescence is generally much less intense or nonexistent with these lasers. The Nd/YAG line at 1064 nm is particularly effective in eliminating fluorescence. The two lines of the diode array laser at 782 and 830 nm also markedly reduce fluorescence in most cases.

TABLE 18-1 Some Common Laser Sources for Raman Spectroscopy

Type Source	Wavelength, nm
Argon ion	488.0 or 514.5
Krypton ion	530.9 or 647.1
Helium/Neon	632.8
Diode Laser	782 or 830
Nd/YAG	1064

Figure 18-5 illustrates an example where the Nd/YAG source completely eliminates background fluorescence. The upper curve was obtained with conventional Raman equipment using the 514.5-nm line from an argon-ion laser for excitation. The sample was anthracene, and most of the recorded signal arises from the fluorescence of that compound. The lower curve is for the same sample recorded with a Fourier transform spectrometer equipped with a Nd/YAG laser that emitted at 1064 nm. Note the total absence of fluorescence background signal.

A variety of other laser sources are available; undoubtedly new and improved sources will appear in the future. There is a need for several sources because one must be chosen that is not absorbed by the sample (except with resonance Raman measurements) or the solvent.

18B-2 Sample Illumination System

Sample handling for Raman spectroscopic measurements is simpler than for infrared spectroscopy because glass can be used for windows, lenses, and other optical components instead of the more fragile and atmospherically less stable crystalline halides. In addition, the

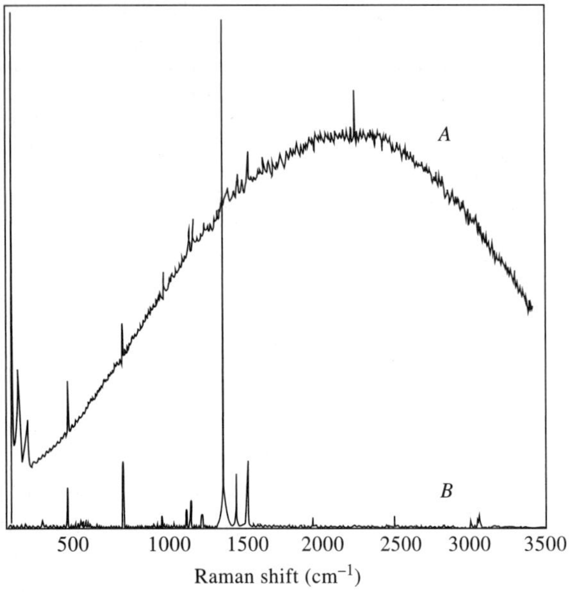

Figure 18-5 Spectra of anthracene. *A:* Conventional instrument, 5145 Å excitation. *B:* FT instrument, 1.064 μm excitation. *(From B. Chase,* Anal. Chem., *1987,* 59, *888A. With permission. Copyright 1987 American Chemical Society.)*

laser source is easily focused on a small sample area and the emitted radiation efficiently focused on a slit. Consequently, very small samples can be investigated. In fact, a common sample holder for nonabsorbing liquid samples is an ordinary glass melting-point capillary.

Liquid Samples

Figure 18-6 shows two of many systems for handling liquids. Each contains a narrow band-pass interference filter that transmits a single line from the laser source and rejects (>99.9%) of the other lines. The size of the tube in Figure 18-6b has been enlarged to show details of the reflection of the Raman radiation off the walls; in fact, the holder is a 1-mm o.d. glass capillary that is about 5 cm long.

A major advantage of sample handling in Raman spectroscopy compared with infrared arises because water is a weak Raman scatterer but a strong absorber of infrared radiation. Thus, aqueous solutions can be studied by Raman spectroscopy but not by infrared. This advantage is particularly important for biological and inorganic systems and in studies dealing with water pollution problems.

Solid Samples

Raman spectra of solid samples are often acquired by filling a small cavity with the sample after it has been ground to a fine powder. Polymers can usually be examined directly with no sample pretreatment. For unusually weak scatterers, such as dilute gas samples, the cell is sometimes placed between the mirrors of the laser source; enhanced excitation power results. Another means for producing Raman spectra from gas samples is the multipass gas cell. A cylindrical glass tube fitted with mirrors at both ends is filled with the sample gas. The excitation laser beam is passed through a small window in one of the mirrors and subsequently passes through the gaseous sample numerous times. The resulting Raman scattering perpendicular to the sample tube and to the excitation laser beam is then focused on the entrance slit of the spectrometer by a large lens.

Fiber-Optic Sampling

One of the significant advantages of Raman spectrometry is that it is based upon visible or near-infrared radiation that can be transmitted for a considerable distance (as much as 100 m or more) through optical fibers. Figure 18-7a is a schematic of typical instrumentation for sampling with a fiber-optic probe. Here, a microscope objective lens is used to focus the laser excitation beam onto one end of a bundle of input fibers, which carries the radiation to a fiber-optic probe that is immersed in the sample. As shown in Figure 18-7b, the probe consists of the input fibers surrounded by several collection fibers that transport the scattered radiation to the monochromator. At the entrance to the spectrometer the collection fibers are arranged linearly as in Figure 18-7c so that they illuminate the entire slit.

Figure 18-8 illustrates how fiber-optic probes can be used to obtain spectra directly under relatively severe sampling conditions. Here, the probe encased in a protective glass sheath was inserted directly into a sample of molten polyethylene.

18B-3 Raman Spectrometers

Until the early 1980s, Raman spectrometers were similar in design and used the same type of components as the classical ultraviolet/visible dispersing instruments described in Section 13D-3. Most employed double grating systems to minimize the spurious radiation reaching the transducer. Photomultipliers served as transducers. Now, however, most Raman spectrometers

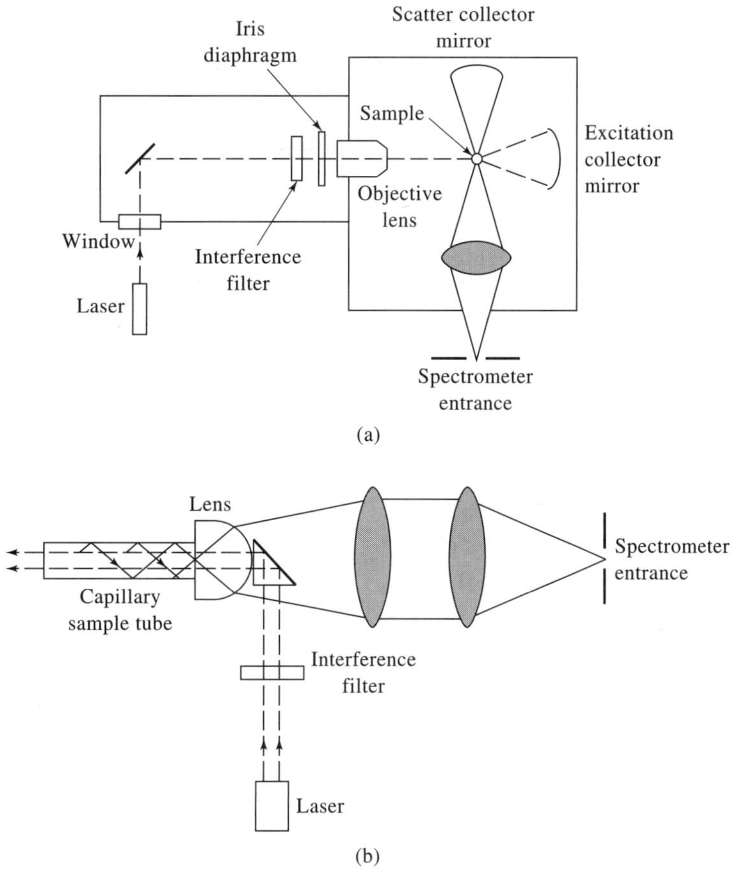

Figure 18-6 Two sample excitation systems.

being marketed are either Fourier transform instruments equipped with cooled germanium transducers or multichannel instruments based upon charge-coupled devices. These transducers, in contrast to photomultiplier tubes, are sensitive to radiation at 782 nm produced by diode lasers, which provide Raman excitation of many compounds without significant fluorescence. Charge-coupled devices are unfortunately not sensitive to the 1064 nm radiation from a Nd/YAG laser.

Currently, there is a controversy in the literature as to which is the better type of instrument for Raman measurements on samples that fluoresce: Fourier transform instruments or dispersing instruments based upon charge-coupled detectors. At this point, it is not clear which of these types of instrument will be the most widely adopted in the future.

Figure 18-9 is a schematic of an instrument for Fourier-transform Raman measurements. The interferometer is the same type as is used in infrared instruments. The transducer is a liquid nitrogen–cooled germanium photoconductor. Because the intensity of the Rayleigh line is several orders of magnitude greater than that of the Raman line, holographic interference filters called *notch filters* or a monochromator are usually used in the instrument to limit the radiation reaching the transducer to wavelengths longer than that of the source. With this arrangement, only the Stokes portion of the spectrum is used. Some Fourier transform instruments utilize filters designed to remove only the wavelength of the laser source. Regardless of the filter system used, the sharpness of the filter's transmittance profile limits the lowest frequencies that can be observed.

Figure 18-10 is a schematic of a typical Raman dispersing spectrometer with a charge-coupled detector. The source is a diode laser and filter system that yields

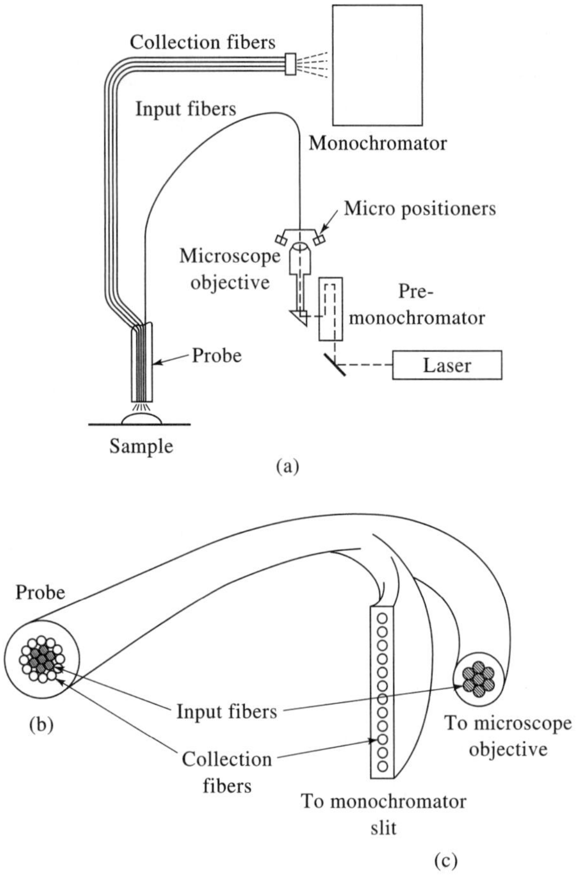

Figure 18-7 (a) Schematic of a system for obtaining Raman spectra with a fiber-optic probe; (b) end view of the probe; (c) end view of the collection fibers at the entrance slit of the monochromator. The blackened circle represents the input fiber, and the hatched circles the collection fibers. *(Adapted from R. L. McCreery, M. Fleischmann, and P. Hendra,* Anal. Chem., ***1983,*** *55, 146. With permission.)*

radiation of 783 nm. This beam is focused on the end of the excitation fiber by means of a lens and transmitted to a fiber probe that consists of the excitation fiber surrounded by 19 collection fibers. The latter then carry the Raman emission to the slit of the monochromator, where it passes through a filter that rejects the Rayleigh scattered radiation. A diffraction grating containing 600 grooves/mm disperses the radiation and reflects it onto a charge-coupled device that is made up of 258 by 1152 pixels. Each of the 258 vertical pixels is summed before being read out.

Figure 18-11 shows some typical applications of a dispersing spectrometer with a fiber-optic probe. To obtain the upper two spectra, the probe was immersed directly in the neat samples. The lower two spectra were obtained by placing the probe approximately 3 mm above the solid samples. Note that the rhodamine 6G is strongly fluorescent, making it extremely difficult to obtain a Raman spectrum with visible radiation. By using 783-nm radiation from a diode laser, this problem is avoided.

18C APPLICATIONS OF RAMAN SPECTROSCOPY

Raman spectroscopy has been applied to the qualitative and quantitative analysis of inorganic, organic, and biological systems.[3]

18C-1 Raman Spectra of Inorganic Species[4]

The Raman technique is often superior to infrared for spectroscopy investigating inorganic systems because aqueous solutions can be employed. In addition, the vibrational energies of metal-ligand bonds are generally in the range of 100 to 700 cm^{-1}, a region of the infrared that is experimentally difficult to study. These vibrations are frequently Raman active, however, and peaks with $\Delta\bar{\nu}$ values in this range are readily observed. Raman studies are potentially useful sources of information concerning the composition, structure, and stability of coordination compounds. For example, numerous halogen and halogenoid complexes produce Raman spectra and thus are amenable to investigation by this means. Metal-oxygen bonds are also Raman active. Spectra for such species as VO_4^{3-}, $Al(OH)_4^-$, $Si(OH)_6^{2-}$, and $Sn(OH)_6^{2-}$ have been obtained; Raman studies have been useful in determining the probable structures of such species. For example, in perchloric acid solutions, vanadium(IV) appears to be present as $VO^{2+}(aq)$ rather than as $V(OH)_2^{2+}(aq)$. Studies of boric acid solutions show that the anion formed by acid dissociation is the tetrahedral $B(OH)_4^-$ rather than $H_2BO_3^-$. Dissoci-

---

[3] See *Analytical Raman Spectroscopy,* J. G. Grasselli and B. J. Bulkin, Eds. New York: Wiley, 1991.

[4] See K. Nakamoto, *Infrared and Raman Spectra of Inorganic and Coordination Compounds,* 5th ed. New York: Wiley, 1996.

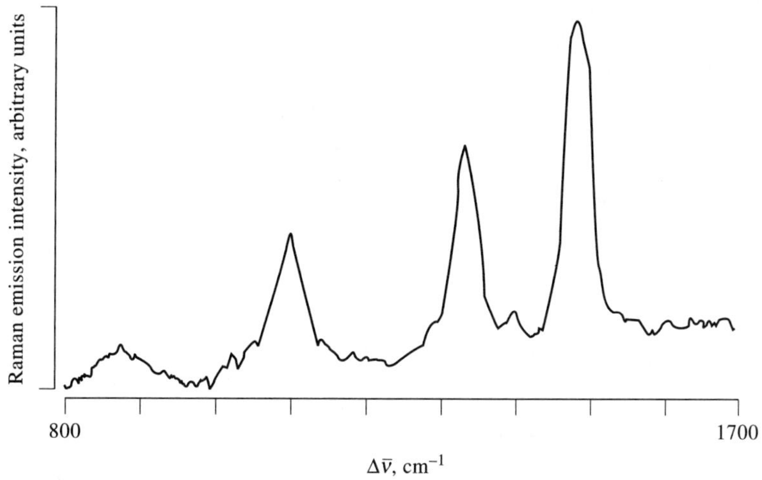

Figure 18-8 Spectrum of molten polyethylene obtained using fiber-optic probe inserted into the polymer melt. *(From R. L. McCreery, M. Fleischmann, and P. Hendra,* Anal. Chem., ***1983,*** *55, 146. With permission.)*

ation constants for strong acids such as H_2SO_4, HNO_3, H_2SeO_4, and H_5IO_6 have been calculated from Raman measurements. It seems probable that the future will see even wider use of Raman spectroscopy for theoretical verification and structural studies of inorganic systems.

18C-2 Raman Spectra of Organic Species

Raman spectra are similar to infrared spectra in that they have regions that are useful for functional group detection and fingerprint regions that permit the identification of specific compounds. Daimay et al. have pub-

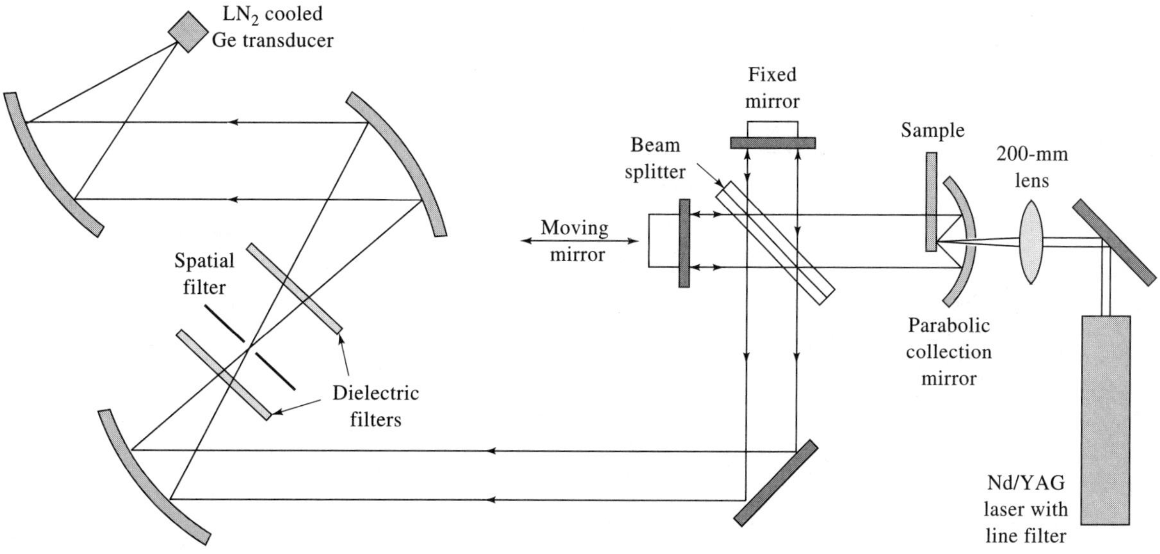

Figure 18-9 Optical diagram of an FT-Raman spectrometer (LN_2 = liquid nitrogen). *(From B. Chase,* Anal. Chem., ***1987,*** *59, 884A. With permission.)*

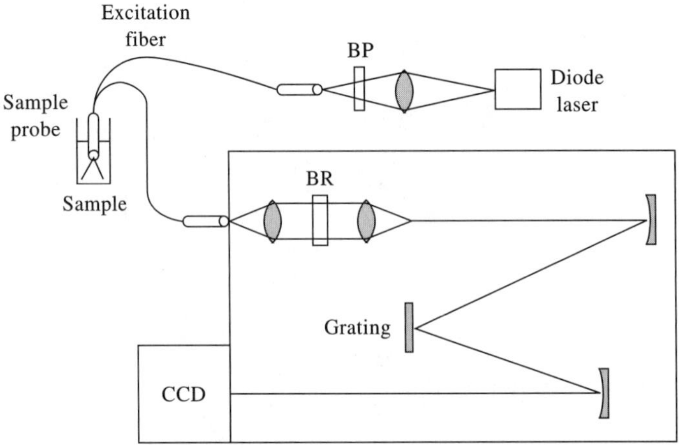

Figure 18-10 Multichannel dispersive Raman spectrometer with a charge-coupled device (CCD). BP is an interference band-pass filter; BR is a Rayleigh band-rejection filter. *(Adapted from C. D. Newman, G. C. Bret, and R. L. McCreery,* Appl. Spectros., ***1992,*** *46, 263. With permission.)*

lished a comprehensive treatment of Raman functional group frequencies.[5]

Raman spectra yield more information about certain types of organic compounds than do their infrared counterparts. For example, the double-bond stretching vibration for olefins results in weak and sometimes undetected infrared absorption. On the other hand, the Raman band (which like the infrared band, occurs at about 1600 cm^{-1}) is intense, and its position is sensitive to the

[5] L. Daimay, N.B. Colthup, W.G. Fately, and J.G. Grasselli, *The Handbook of Infrared & Raman Characteristic Frequencies of Organic Compounds.* New York: Wiley-Interscience, 1971.

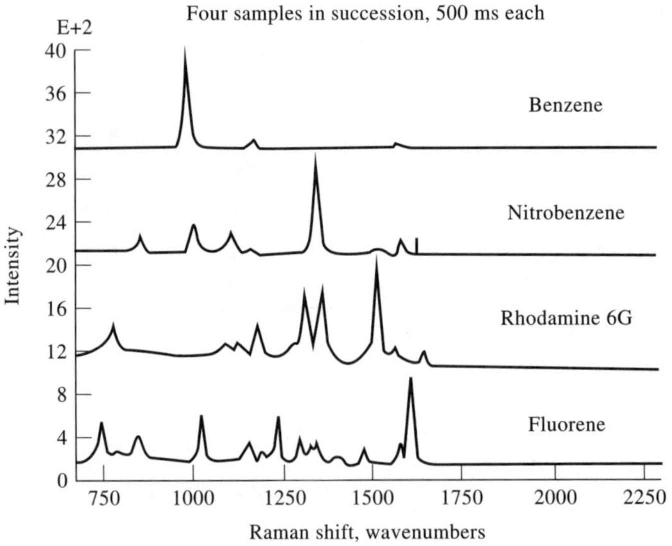

Figure 18-11 Four Raman spectra obtained with the apparatus shown in Figure 18-10. *(Adapted from C. D. Newman, G. C. Bret, and R. L. McCreery,* Appl. Spectros., ***1992,*** *46, 263. With permission.)*

nature of substituents as well as to their geometry. Thus, Raman studies are likely to yield useful information about the olefinic functional group that may not be revealed by infrared spectra. This statement applies to cycloparaffin derivatives as well; these compounds have a characteristic Raman peak in the region of 700 to 1200 cm^{-1}. This peak has been attributed to a breathing vibration in which the nuclei move in and out symmetrically with respect to the center of the ring. The position of the peak decreases continuously from 1190 cm^{-1} for cyclopropane to 700 cm^{-1} for cyclooctane; Raman spectroscopy thus appears to be an excellent diagnostic tool for the estimation of ring size in paraffins. The infrared peak associated with this vibration is weak or nonexistent.

18C-3 Biological Applications of Raman Spectroscopy

Raman spectroscopy has been applied widely for the study of biological systems.[6] The advantages of this technique include the small sample requirement, the minimal sensitivity toward interference by water, the spectral detail, and the conformational and environmental sensitivity.

18C-4 Quantitative Applications

Raman spectra tend to be less cluttered with peaks than infrared spectra. As a consequence, peak overlap in mixtures is less likely, and quantitative measurements are simpler. In addition, Raman sampling devices are not subject to attack by moisture, and small amounts of water in a sample do not interfere. Despite these advantages, Raman spectroscopy has not yet been exploited widely for quantitative analysis. This lack of use has been due largely to the rather high cost of Raman spectrometers relative to that of absorption instrumentation. This barrier is likely to become less important as suitable low-cost laser diodes proliferate.

Because laser beams can be precisely focused, it becomes possible to perform quantitative analyses on very small samples. For this work, instruments called *laser microprobes* are employed. Laser microprobes have been used to determine analytes in single bacterial cells, components in individual particles of smoke and fly ash, and species in microscopic inclusions in minerals.

18D OTHER TYPES OF RAMAN SPECTROSCOPY

With the development of tunable lasers, several new Raman spectroscopic methods were developed in the early 1970s. A brief discussion of the applications of some of these techniques follows.

18D-1 Resonance Raman Spectroscopy[7]

Resonance Raman scattering refers to a phenomenon in which Raman line intensities are greatly enhanced by excitation with wavelengths that closely approach that of an *electronic* absorption peak of an analyte. Under this circumstance, the magnitudes of Raman peaks associated with the most symmetric vibrations are enhanced by a factor of 10^2 to 10^6. As a consequence, resonance Raman spectra have been obtained at analyte concentrations as low as 10^{-8} M. This level of sensitivity is in contrast to normal Raman studies, which are ordinarily limited to concentrations greater than 0.1 M. Because resonance enhancement is restricted to the Raman bands associated with the chromophore, resonance Raman spectra usually consist of only a few lines. Furthermore, resonance Raman spectra may be obtained quite selectively because excitation can be targeted for specific absorption bands.

Figure 18-12a illustrates the energy changes responsible for resonance Raman scattering. This figure differs from the energy diagram for normal Raman scattering (Figure 18-2) in that the electron is promoted into an excited electronic state followed by an immediate relaxation to a vibrational level of the electronic ground state. As shown in the figure, resonance Raman scattering differs from fluorescence (Figure 18-12b) in that relaxation to the ground state is *not* preceded by prior relaxation to the lowest vibrational level of the excited electronic state. The time scales for the two phenomena

[6] For reviews of biological applications, see P. R. Carey, *Biochemical Applications of Raman and Resonance Raman Spectroscopy.* New York: Academic Press, 1982; A. T. Tu, *Raman Spectroscopy in Biology.* New York: Wiley, 1982; I. W. Levin and E. N. Lewis, *Anal. Chem.*, **1990**, *62*, 1101A; P. V. Huong, in *Analytical Raman Spectroscopy*, J. G. Grasselli and B. J. Bulkin, Eds., Chapter 11. New York: Wiley, 1991.

[7] For brief reviews of this topic, see M. D. Morris and D. J. Wallan, *Anal. Chem.*, **1979**, *51*, 182A; D. P. Strommen and K. Nakamoto, *J. Chem. Educ.*, **1977**, *54*, 474; S. A. Asher, *Anal. Chem.*, **1993**, *65*, 59A, 201A.

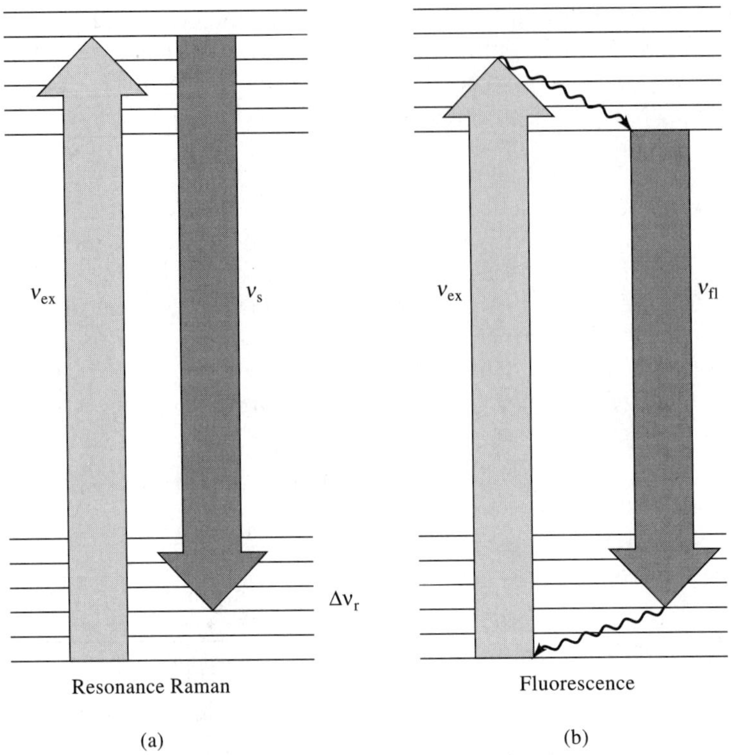

Figure 18-12 Energy diagram for (a) resonance Raman scattering and (b) fluorescence emission. Radiationless relaxation shown as wavy arrows. *(Reprinted with permission from M. D. Morris and D. J. Wallin,* Anal. Chem., ***1979,*** 51, *185A. Copyright 1979 American Chemical Society.)*

are also quite different with Raman relaxation occurring in less than 10^{-14} s compared with the 10^{-6} to 10^{-8} s for fluorescence emission.

Line intensities in a resonance Raman experiment increase rapidly as the excitation wavelength approaches the wavelength of the electronic absorption peak. Thus, to achieve the greatest signal enhancement for a broad range of absorption maxima, a tunable laser is required. With intense laser radiation, sample decomposition can become a major problem because electronic absorption peaks often occur in the ultraviolet region. Specifically, intense absorption causes local heating in the sample, which leads to decomposition of the analyte. To circumvent this problem, it is common practice to circulate the sample past the focused beam of the laser. Circulation is normally accomplished in one of two ways: by pumping a solution or liquid through a capillary mounted in the sample position, or by rotating a cylindrical cell containing the sample through the laser beam. Thus, only a small fraction of the sample is irradiated at any instant, and heating and sample decomposition are minimized.

Perhaps the most important application of resonance Raman spectroscopy has been to the study of biological molecules under physiologically significant conditions; that is, in the presence of water and at low to moderate concentration levels. As an example, the technique has been used to determine the oxidation state and spin of iron atoms in hemoglobin and cytochrome-*c*. In these molecules, the resonance Raman bands are due solely to vibrational modes of the tetrapyrrole chromophore. None of the other bands associated with the protein is enhanced, and at the concentrations normally used these bands do not interfere as a consequence.

A major limitation to resonance Raman spectroscopy is interference by fluorescence either by the analyte itself or by other species present in the sample.

18D-2 Surface-Enhanced Raman Spectroscopy (SERS)[8]

Surface enhanced Raman spectroscopy involves obtaining Raman spectra in the usual way on samples that are adsorbed on the surface of colloidal metal particles (usually silver, gold, or copper) or on roughened surfaces of pieces of these metals. For reasons that are not fully understood, the Raman lines of the adsorbed molecule are often enhanced by a factor of 10^3 to 10^6. When surface enhancement is combined with the resonance enhancement technique discussed in the previous section, the net increase in signal intensity is roughly the product of the intensity produced by each of the techniques. Consequently, detection limits in the 10^{-9} to 10^{-12} M range have been observed.

Several sample-handling techniques are employed for surface-enhanced spectroscopy. In one technique, colloidal silver or gold particles are suspended in a dilute solution (usually aqueous) of the sample. The solution is then held or flowed through a narrow glass tube while it is excited by a laser beam. In another method, a thin film of colloidal metal particles is deposited on a glass slide and a drop or two of the sample solution spotted on the film. The Raman spectrum is then obtained in the usual manner. Alternatively, the sample may be deposited on a roughened metal electrode, which is then removed from the solution and exposed to the laser excitation source.

[8] For a review of SERS, see R. L. Garrell, *Anal. Chem.* **1989**, *61*, 401A.

18D-3 Nonlinear Raman Spectroscopy[9]

In Section 7B-3, we pointed out that many lasers are intense enough to produce significant amounts of nonlinear radiation. Throughout the 1970s and 1980s, many Raman techniques were developed that depend upon polarization induced by second-order and higher field strengths. These techniques are termed nonlinear Raman methods. Included in these methods are: *stimulated Raman scattering, hyper Raman effect, stimulated Raman gain, inverse Raman spectroscopy, coherent anti-Stokes Raman spectroscopy,* and *coherent Stokes Raman spectroscopy.* The most widely used of these methods is coherent anti-Stokes Raman spectroscopy, or CARS.[10]

Nonlinear techniques have been used to overcome some of the drawbacks of conventional Raman spectroscopy, namely, its low efficiency, its limitation to the visible and near-ultraviolet regions, and its susceptibility to interference from fluorescence. A major disadvantage of nonlinear methods is that they tend to be analyte specific and often require several different tunable lasers to be applicable to diverse species. To date, none of the nonlinear methods has found widespread applications among nonspecialists. For this reason we will not consider them further.

[9] For brief reviews, see S. A. Borman, *Anal. Chem.*, **1982**, *54*, 1021A; M. D. Morris and D. J. Wallin, *Anal. Chem.*, **1979**, *51*, 182A. For more detailed discussions see *Chemical Applications of Nonlinear Raman Spectroscopy*, A. B. Harvey, Ed. New York: Academic Press, 1981; G. L. Eesely, *Coherent Raman Spectroscopy.* Elmsford, NY: Pergamon Press, 1981.

[10] For reviews of CARS, see A. B. Harvey, *Anal. Chem.*, **1978**, *50*, 905A; J. F. Verdieck, R. J. Hall, and A. C. Eckbreth, *J. Chem. Educ.*, **1982**, *59*, 495.

18E QUESTIONS AND PROBLEMS

18-1 Why does the ratio of anti-Stokes to Stokes intensities increase with sample temperature?

18-2 What is a virtual state?

18-3 At what wavelengths in nanometers would the Stokes and anti-Stokes Raman lines for carbon tetrachloride ($\bar{\nu}$ = 218, 314, 459, 762, and 790 cm^{-1}) appear if the source was
(a) a helium/neon laser (632.8 nm)?
(b) an argon-ion laser (488.0 nm)?

18-4 Assume the excitation sources in Problem 18-3 have the same power. (a) Compare the relative intensities of the CCl_4 Raman lines when each of the two excitation sources is used. (b) If the intensities were recorded with a typical monochromator/photomultiplier system, why would the measured intensity ratios differ from the ratio calculated in part (a)?

18-5 Under what circumstances would a helium/neon laser be preferable to an argon-ion laser as a Raman source?

18-6 For vibrational states, the Boltzmann equation can be written as

$$\frac{N_1}{N_0} = e^{-\Delta E/kT}$$

where N_0 and N_1 are the populations of the lower and higher energy states, respectively, ΔE is the energy difference between the states, k is the Boltzmann constant, and T is the temperature in kelvins.

For temperatures of 20° and 40°C, calculate the ratios of the intensities of the anti-Stokes and Stokes lines for CCl_4 at (a) 218 cm^{-1}; (b) 459 cm^{-1}; (c) 790 cm^{-1}.

18-7 The following Raman data were obtained for $CHCl_3$ with the polarizer of the spectrometer set (1) parallel to the plane of polarization of the laser and (2) at 90 deg to the plane of the source.

Calculate the depolarization ratio and indicate which Raman peaks are polarized.

		Relative Intensities	
	$\Delta\bar{\nu}$, cm^{-1}	(1) $I_{\parallel}$	(2) $I_{\perp}$
(a)	760	0.60	0.46
(b)	660	8.4	0.1
(c)	357	7.9	0.6
(d)	258	4.2	3.2

Chapter 19

Nuclear Magnetic Resonance Spectroscopy

job: build + design an electrochemical NMR probe for use in a high-resolution NMR spectrometer as part of the work in designing materials for lithium ion batteries.

Nuclear magnetic resonance (NMR) spectroscopy is based on the measurement of absorption of electromagnetic radiation in the radio-frequency region of roughly 4 to 900 MHz. In contrast to ultraviolet, visible, and infrared absorption, nuclei of atoms rather than outer electrons are involved in the absorption process. Furthermore, in order to cause nuclei to develop the energy states required for absorption to occur, it is necessary to place the analyte in an intense magnetic field.

Nuclear magnetic resonance spectroscopy is one of the most powerful tools available to the chemist and biochemist for elucidating the structure of chemical species. The technique is also useful for the quantitative determination of absorbing species.[1]

The theoretical basis for NMR spectroscopy was proposed by W. Pauli in 1924 who suggested that certain atomic nuclei should have the properties of spin and magnetic moment and that, as a consequence, exposure to a magnetic field would lead to splitting of their energy levels. During the next decade, these postulates were verified experimentally. It was not until 1946, however, that Bloch at Stanford and Purcell at Harvard, working independently, demonstrated that nuclei absorb electromagnetic radiation in a strong magnetic field as a consequence of the energy level splitting that is induced by the magnetic field. The two physicists shared the 1952 Nobel Prize for their work.

In the first five years following the discovery of NMR, chemists became aware that molecular environment influences the absorption of radio-frequency (RF) radiation by a nucleus in a magnetic field and that this effect can be correlated with molecular structure. In 1953 the first high-resolution NMR spectrometer designed for chemical structural studies was marketed by Varian Associates. Since then, the growth of NMR spectroscopy has been explosive, and the technique has

[1] The following references are recommended for additional study: E. D. Becker, *High Resolution NMR,* 2nd ed. New York: Academic Press, 1980; F. A. Bovey, *Nuclear Magnetic Resonance Spectroscopy,* 2nd ed. New York: Academic Press, 1988; R. H. Cox and D. E. Leyden, in *Treatise on Analytical Chemistry,* 2nd ed., P. J. Elving, M. M. Bursey, and I. M. Kolthoff, Eds., Part I, Vol. 10. New York: Wiley, 1983; R. J. Abraham, J. Fisher, and P. Loftus, *Introduction to NMR Spectroscopy.* New York: Wiley, 1988; J. W. Akitt, *NMR and Chemistry,* 2nd ed. New York: Chapman and Hall, 1992; J. Wennel and J. Klinowski, *Fundamentals of Nuclear Magnetic Resonance.* New York: Wiley, 1993.

had profound effects on the development of organic, inorganic, and biochemistry. It is unlikely that there has ever been as short a delay between scientific discovery and its widespread acceptance and application.[2]

Two general types of NMR spectrometers are currently in use, ***continuous-wave*** *(CW) and* ***pulsed,*** *or* ***Fourier transform*** *(FT NMR) spectrometers. All early studies were carried out with continuous-wave instruments. In about 1970, however, Fourier transform spectrometers became available commercially, and now this type of instrument dominates the market. In both types of instruments, the sample is positioned in a powerful magnetic field that has a strength of several tesla.*[3] *Continuous-wave spectrometers are similar in principle to optical absorption instruments in that an absorption signal is monitored as the frequency of the source is slowly scanned. In some instruments, the frequency of the source is held constant while the strength of the field is scanned. In pulsed instruments, the sample is irradiated with periodic pulses of RF energy that are directed through the sample at right angles to the magnetic field. These excitation pulses elicit a time-domain signal that decays in the interval between pulses. This signal is then converted to a frequency domain signal by using a Fourier transformation to give a spectrum that is similar to that obtained by using a continuous-wave instrument.*

Nearly all NMR instruments produced at the present time are of the Fourier transform type, and the use of continuous-wave instruments is largely limited to special routine applications, such as the determination of the extent of hydrogenation in petroleum process streams and the determination of water in oils, food products, and agricultural materials. In spite of this predominance of pulsed instruments in the marketplace, we find it convenient to base our initial development of NMR theory on continuous-wave experiments and move from there to a discussion of pulsed Fourier transform measurements.

[2] For interesting discussions of the history of NMR, see E. D. Becker, *Anal. Chem.*, **1993**, *65*, 295A; D. C. Lankin, R. R. Ferraro, and R. Jarnutowski, *Spectroscopy*, **1992**, *7*(8), 18.

[3] The SI symbol for magnetic fields is *B;* an older convention, however, which is still widely used, employed the symbol *H* instead. The derived unit for describing the field strength is the tesla (T), which is defined as $1\ T = 1\ kg\ s^{-2}\ A^{-1}$. Another unit that was popular in the past and still is frequently encountered is the gauss (G). The relationship between the two units is $10^4\ G = 1\ T$.

19A THEORY OF NUCLEAR MAGNETIC RESONANCE

In common with optical spectroscopy, both classical mechanics and quantum mechanics are useful in explaining the NMR phenomenon. The two treatments yield identical relationships. Quantum mechanics, however, provides a useful relationship between absorption frequencies and nuclear energy states, while classical mechanics yields a clear physical picture of the absorption process and how it is measured.

In this section, we first provide a quantum description of NMR that is applicable to both continuous-wave and pulsed NMR measurements. Then, we take a classical approach to NMR and show how it provides a useful picture of continuous-wave NMR. Finally, we complete this section with a discussion of Fourier transform measurements based again on a classical picture.

19A-1 Quantum Description of NMR

To account for the properties of certain nuclei, we must assume that they rotate about an axis and thus have the property of *spin.* Nuclei with spin have angular momentum p. Furthermore, the maximum observable component of this angular momentum is quantized and must be an integral or a half-integral multiple of $h/2\pi$, where h is the Planck constant. The maximum number of spin components or values for p for a particular nucleus is its *spin quantum number I.* The nucleus will then have $2I + 1$ discrete states. The component of angular momentum for these states in any chosen direction will have values of $I, I - 1, I - 2, \ldots, -I$. In the absence of an external field, the various states have *identical* energies.

The four nuclei that have been of greatest use to organic chemists and biochemists are ^1H, ^{13}C, ^{19}F, and ^{31}P, and they are the only four that we shall consider. The spin quantum number for these nuclei is 1/2. Thus, each nucleus has two spin states corresponding to $I = +1/2$ and $I = -1/2$. Heavier nuclei have spin numbers that range from zero, which implies that they have no net spin component, to at least 9/2.

TABLE 19-1 Magnetic Properties of Four Important Nuclei Having Spin Quantum Numbers of 1/2

Nucleus	Magnetogyric Ratio (radian T^{-1} s^{-1})	Isotopic Abundance, %	Relative Sensitivity[a]	Absorption Frequency, MHz[b]
^{1}H	2.6752×10^{8}	99.98	1.00	200.00
^{13}C	6.7283×10^{7}	1.11	0.016	50.30
^{19}F	2.5181×10^{8}	100.00	0.83	188.25
^{31}P	1.0841×10^{8}	100.00	0.066	81.05

[a]At constant field for equal number of nuclei.
[b]At a field strength of 4.69 T.

A spinning, charged nucleus creates a magnetic field that is analogous to the field produced when electricity flows through a coil of wire. The resulting magnetic moment μ is oriented along the axis of spin and is proportional to the angular momentum p. Thus,

$$\mu = \gamma p \qquad (19\text{-}1)$$

where the proportionality constant γ is the *magnetogyric ratio,* which has a different value for each type of nucleus. As we shall see, the magnetogyric ratio is also a factor in the proportionality constant in the relationship between the frequency of the absorbed energy and the magnetic field strength shown in Equation 19-5. Magnetogyric ratios for the four elements we will be dealing with are found in the second column of Table 19-1. The SI unit for these constants is radian · tesla^{-1} · second^{-1}.

The relationship between nuclear spin and magnetic moment leads to a set of observable magnetic quantum states m given by

$$m = I, I-1, I-2, \cdots, -I \qquad (19\text{-}2)$$

Thus, the nuclei that we will consider have two magnetic quantum numbers, $m = +1/2$ and $m = -1/2$. Note that the rules for determining nuclear quantum numbers are similar to those for electronic quantum numbers.

Energy Levels in a Magnetic Field

As shown in Figure 19-1, when a nucleus with a spin quantum number of one half is brought into an external magnetic field B_0, its magnetic moment becomes ori-

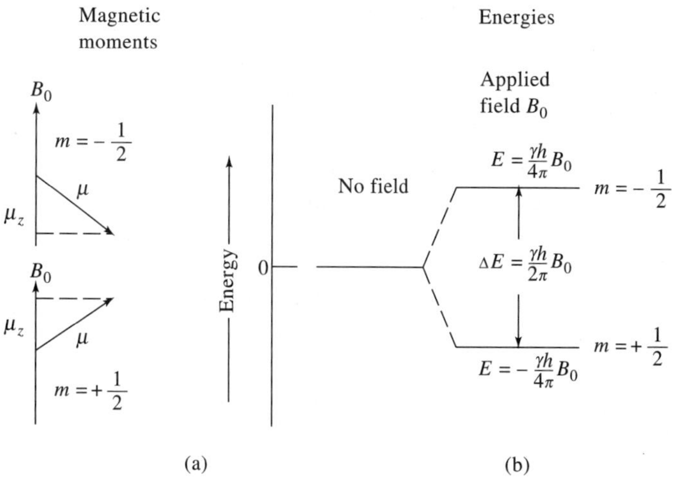

Figure 19-1 Magnetic moments and energy levels for a nucleus with a spin quantum number of ±1/2.

ented in one of two directions with respect to the field depending upon its magnetic quantum state. The potential energy E of a nucleus in these two orientations, or quantum states, is given by

$$E = -\frac{\gamma m h}{2\pi} B_0 \qquad (19\text{-}3)$$

The energy for the lower energy $m = +1/2$ state (see Figure 19-1) is given by

$$E_{+1/2} = -\frac{\gamma h}{4\pi} B_0$$

For the $m = -1/2$ state the energy is

$$E_{-1/2} = \frac{\gamma h}{4\pi} B_0$$

Thus, the difference in energy ΔE between the two states is given by

$$\Delta E = \frac{\gamma h}{4\pi} B_0 - \left(-\frac{\gamma h}{4\pi} B_0\right) = \frac{\gamma h}{2\pi} B_0 \qquad (19\text{-}4)$$

As in other types of spectroscopy, transitions between energy states can be brought about by absorption or emission of electromagnetic radiation of a frequency ν_0 that corresponds in energy to ΔE. Thus, by substituting the Planck relationship $\Delta E = h\nu_0$ into Equation 19-4, we obtain the frequency of the radiation required to bring about the transition

$$\nu_0 = \frac{\gamma B_0}{2\pi} \qquad (19\text{-}5)$$

As we suggested in a previous paragraph, the frequency of a magnetic transition is proportional to the applied field strength B_0 with a proportionality constant of $\gamma/2\pi$.

EXAMPLE 19-1

Many modern proton NMR instruments employ a magnet that provides a field strength of 4.69 T. At what frequency would the hydrogen nucleus absorb in such a field?

Substituting the magnetogyric ratio for the proton (Table 19-1) into Equation 19-5, we find

$$\nu_0 = \frac{(2.68 \times 10^8\ \mathrm{T^{-1}\ s^{-1}})(4.69\ \mathrm{T})}{2\pi}$$

$$= 2.00 \times 10^8\ \mathrm{s^{-1}} = 200\ \mathrm{MHz}$$

The foregoing example reveals that radio-frequency radiation of approximately 200 MHz is required to bring about a change in alignment of the magnetic moment of the proton from a direction that parallels the field to a direction that opposes it.

Distribution of Particles between Magnetic Quantum States

In the absence of a magnetic field, the energies of the magnetic quantum states of a nucleus are identical. Consequently, a large collection of protons contains an identical number of nuclei with magnetic quantum numbers of $+1/2$ and $-1/2$. When placed in a magnetic field, however, the nuclei tend to orient themselves so that the lower energy state ($m = +1/2$) predominates. It is instructive to calculate the extent of this predominance in a typical NMR experiment. For this purpose, the Boltzmann equation (page 199) can be written in the form

$$\frac{N_j}{N_0} = \exp\left(\frac{-\Delta E}{kT}\right) \qquad (19\text{-}6)$$

where N_j is the number of protons in the higher energy state ($m = -1/2$), N_0 is the number in the lower state ($m = +1/2$), k is the Boltzmann constant (1.38×10^{-23} J K^{-1}), T is the absolute temperature, and ΔE is defined by Equation 19-4.

Substituting Equation 19-4 into 19-6 gives

$$\frac{N_j}{N_0} = \exp\left(\frac{-\gamma h B_0}{2\pi kT}\right) \qquad (19\text{-}7)$$

EXAMPLE 19-2

Calculate the relative number of protons in the higher and lower magnetic states when a sample is placed in a 4.69 T field at 20°C.

Substituting numerical values into Equation 19-7 gives

$$\frac{N_j}{N_0} = e^{\left(\frac{-(2.68 \times 10^8\ \mathrm{T^{-1}\ s^{-1}})(6.63 \times 10^{-34}\ \mathrm{J \cdot s})(4.69\ \mathrm{T})}{2\pi(1.38 \times 10^{-23}\ \mathrm{J\ K^{-1}})\ (293\ \mathrm{K})}\right)}$$

$$\frac{N_j}{N_0} = e^{-3.28 \times 10^{-5}} = 0.999967$$

or

$$\frac{N_0}{N_j} = 1.000033$$

Thus, for exactly 10^6 protons in higher energy states there will be

$$N_0 = 10^6/0.999967 = 1{,}000{,}033$$

in the lower energy state. This figure corresponds to a 33-ppm excess.

Example 19-2 demonstrates that the success of the NMR measurement depends upon a remarkably small, ~33 ppm, excess of lower energy protons. If the numbers of nuclei in the two states were identical, however, we would observe no net absorption because the number of particles excited by the radiation would exactly equal the number producing *induced* emission.

If we expand the right side of Equation 19-7 as a Maclaurin series, and truncate the series after the second term, we obtain the important result that

$$\frac{N_j}{N_0} = 1 - \frac{\nu h B_0}{2\pi k T} \qquad (19\text{-}8)$$

Equation 19-8 demonstrates that the relative number of excess low-energy nuclei is linearly related to the magnetic field strength. Thus, the intensity of an NMR signal increases linearly as the field strength increases. This dependence of signal sensitivity on magnetic field strength has led manufacturers to produce magnets with field strengths as large as 14 T.

19A-2 Classical Description of NMR

To understand the absorption process, and in particular the measurement of absorption, a classical picture of the behavior of a charged particle in a magnetic field is helpful.

Precession of Nuclei in a Field

Let us first consider the behavior of a nonrotating magnetic body, such as a compass needle, in an external magnetic field. If momentarily displaced from alignment with the field, the needle will swing in a plane about its pivot as a consequence of the force exerted by the field on its two ends; in the absence of friction, the ends of the needle will fluctuate back and forth indefinitely about the axis of the field. A quite different motion occurs, however, if the magnet is spinning rapidly around its north-south axis. Because of the gyroscopic effect, the force applied by the field to the axis of rotation causes movement not in the plane of the force but perpendicular to this plane; the axis of the rotating particle, therefore, moves in a circular path. That is, the rotational axis of the rotating particle *precesses* around the vector representing the applied magnetic field. This motion, illustrated in Figure 19-2, is similar to the motion of a gyroscope when it is displaced from the vertical by application of a lateral force. The angular velocity of this motion ω_0, in radians per second, is given by

$$\omega_0 = \gamma B_0 \qquad (19\text{-}9)$$

The angular velocity can be converted to the frequency of precession ν_0, known as the *Larmor frequency,* by dividing by 2π. Thus,

$$\nu = \frac{\gamma B_0}{2\pi} \qquad (19\text{-}10)$$

A comparison of Equation 19-10 with Equation 19-5 reveals that the Larmor frequency is identical to the frequency of absorbed radiation derived from quantum mechanical considerations.

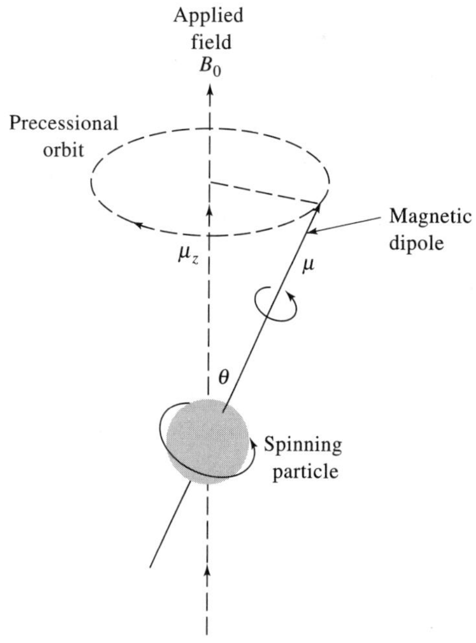

Figure 19-2 Precession of a rotating particle in a magnetic field.

Absorption in Continuous-Wave Experiments

The potential energy E of the precessing charged particle shown in Figure 19-2 is given by

$$E = -\mu_z B_0 = -\mu B_0 \cos \theta \qquad (19\text{-}11)$$

where θ is the angle between the magnetic field vector and the spin axis of the particle, μ is the magnetic moment of the particle, and μ_z is the component of μ in the direction of the magnetic field. Thus, when radio-frequency energy is absorbed by a nucleus, its angle of precession θ must change. Hence, we imagine for a nucleus having a spin quantum number of 1/2 that absorption involves a flipping of the magnetic moment that is oriented in the field direction to a state in which the moment is in the opposite direction. The process is pictured in Figure 19-3. In order for the magnetic dipole to flip, there must be a magnetic force at right angles to the fixed field that moves in a circular path in phase with the precessing dipole. The magnetic moment of *circularly polarized radiation* of a suitable frequency has these necessary properties; that is, the magnetic vector of such radiation has a circular component, as represented by the dashed circle in Figure 19-3.[4] If the rotational frequency of the magnetic vector of the radiation is the same as the precessional frequency of a nucleus, absorption and flipping can occur. As shown in the next paragraph, circularly polarized radiation of suitable frequency can be produced by a radio-frequency oscillator coil.

The radiation produced by the coil of an RF oscillator, which serves as the source in NMR instruments, is plane polarized. Plane-polarized radiation, however, consists of d and l circularly polarized radiation. As shown in the lower part of Figure 19-4b, the vector of the d component rotates in a clockwise manner as the radiation approaches the observer; the vector of the l component rotates in the opposite sense. As shown in the lower part of the figure, addition of the two vectors leads to a vector sum that vibrates in a single plane (Figure 19-4a).

Thus, electromagnetic radiation from an oscillator coil oriented at 90 deg to the direction of the fixed magnetic field introduces circularly polarized radiation into the sample volume in the proper plane for absorption by sample nuclei. Only that magnetic component of the excitation radiation that rotates in the precessional direction is absorbed.

Relaxation Processes in NMR

When a nucleus is exposed to radiation of a suitable frequency, absorption occurs because of the slight excess of lower energy state nuclei that are present in the strong magnetic field. This excess is small as is indicated by the result of Example 19-2, so there is always danger that the absorption process will equalize the number of nuclei in the two states and cause the absorption signal to decrease and to approach zero. When this occurs the spin system is said to be *saturated.* In order to avoid saturation, the rate of relaxation of excited nuclei to their lower energy state must be as great or greater than the rate at which they absorb the radio-frequency energy. One apparent relaxation path is the emission of radiation of a frequency corresponding to

[4] It is important to note here that in contrast to optical spectroscopy, where it is the electrical field of electromagnetic radiation that interacts with absorbing species, in NMR spectroscopy, it is the *magnetic field* of the radiation that excites absorbing species.

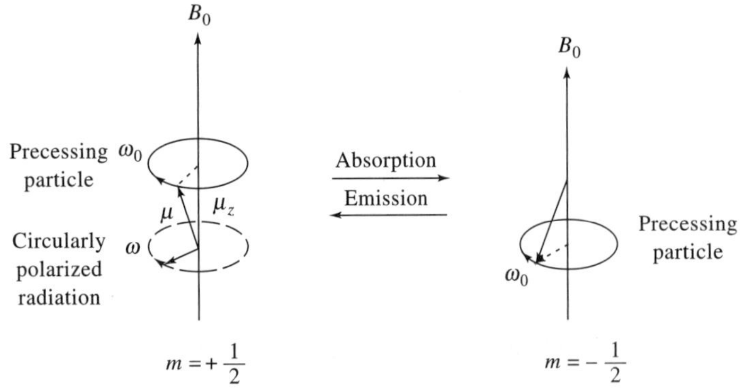

Figure 19-3 Model for the absorption of radiation by a precessing particle.

the energy difference between the states, which results in fluorescence. Radiation theory, however, shows that the probability of spontaneous reemission of photons varies as the cube of the frequency and that at radio frequencies this process does not occur to a significant extent. In NMR studies, then, nonradiative relaxation processes are of prime importance.

To reduce saturation and produce a readily detectable absorption signal, relaxation should occur as rapidly as possible; that is, the lifetime of the excited state should be small. A second factor—the inverse relationship between the lifetime of an excited state and the width of its absorption line—negates the advantage of very short lifetimes. Thus, when relaxation rates are high, or the lifetimes low, line broadening prevents high-resolution measurements. These two opposing factors cause the optimum half-life for an excited species to range from about 0.1 to 10 s.

There are two important types of relaxation processes in NMR spectroscopy: (1) *spin-lattice,* or *longitudinal, relaxation* and (2) *spin-spin,* or *transverse, relaxation.*

Spin-Lattice Relaxation. The absorbing nuclei in an NMR experiment are part of the larger collection of atoms that constitute the sample. The entire collection is termed the *lattice,* regardless of whether the sample is a solid, a liquid, or a gas. In the latter two states particularly, the various nuclei comprising the lattice are in violent vibrational and rotational motion, which creates a complex field about each magnetic nucleus. The resulting lattice field thus contains a continuum of magnetic components, at least some of which must correspond in frequency and phase with the precessional frequency of the magnetic nucleus of interest. These vibrationally and rotationally developed components interact with and convert nuclei from a higher to a lower spin state; the absorbed energy then simply increases the amplitude of the thermal vibrations or rotations. This change produces a minuscule temperature rise in the sample.

Spin-lattice relaxation is a first-order exponential decay characterized by a relaxation time T_1, which is a measure of the average lifetime of the nuclei in the higher energy state. In addition to depending upon the magnetogyric ratio of the absorbing nuclei, T_1 is strongly affected by the mobility of the lattice. In crystalline solids and viscous liquids, where mobilities are low, T_1 is large. As the mobility increases (at higher temperatures, for example), the vibrational and rotational frequencies increase, thus enhancing the probability of a magnetic fluctuation of the proper magnitude for a relaxation transition; T_1 becomes shorter as a consequence. At very high mobilities, on the other hand, the fluctuation frequencies are further increased and spread over such a broad range that the probability of a suitable frequency for a spin-lattice transition again decreases. The result is a minimum in the relationship between T_1 and lattice mobility.

Spin-Spin Relaxation. Several other effects tend to diminish relaxation times and thereby broaden NMR lines. These effects are normally lumped together and

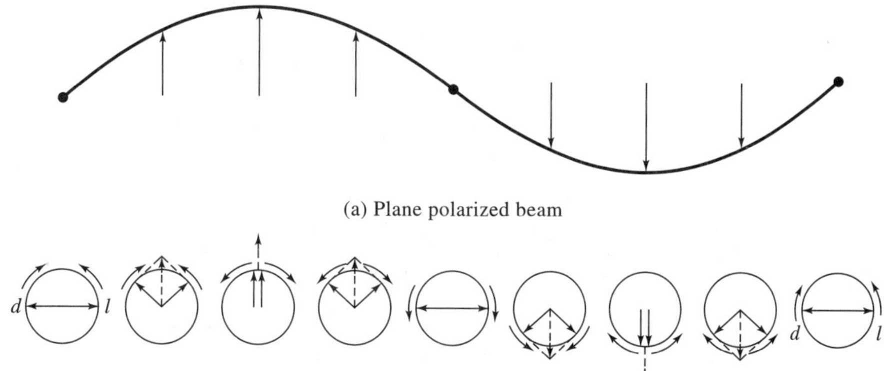

(a) Plane polarized beam

(b) *d, l* circularly polarized beams

Figure 19-4 Equivalency of a plane-polarized beam to two (*d, l*) circularly polarized beams of radiation.

described by a transverse, or spin-spin, relaxation time T_2. Values for T_2 are generally so small for crystalline solids or viscous liquids (as low as 10^{-4} s) as to preclude the use of samples of these kinds for high-resolution spectra unless special techniques are employed. These techniques are described briefly in a later section dealing with ^{13}C NMR studies of solids.

When two neighboring nuclei of the same kind have identical precession rates, but are in different magnetic quantum states, the magnetic fields of each can interact to cause an interchange of states. That is, a nucleus in the lower spin state is excited while the excited nucleus relaxes to the lower energy state. No net change in the relative spin state population, and thus no decrease in saturation, results, but the average lifetime of a particular excited nucleus is shortened. Line broadening is the result.

Two other causes of line broadening should be noted. Both arise if B_0 in Equation 19-10 differs slightly from nucleus to nucleus; under these circumstances, a band of frequencies, rather than a single frequency, is absorbed. One cause for such a variation in the static field is the presence in the sample of other magnetic nuclei whose spins create local fields that may act to enhance or diminish the external field acting on the nucleus of interest. In a mobile lattice, these local fields tend to cancel because the nuclei causing them are in rapid and random motion. In a solid or a viscous liquid, however, the local fields may persist long enough to produce a range of field strengths and thus a range of absorption frequencies. Variations in the static field also result from small inhomogeneities in the field source itself. This effect can be largely offset by rapidly spinning the entire sample in the magnetic field.

19A-3 Fourier Transform NMR[5]

In pulsed NMR measurements, nuclei in a strong magnetic field are subjected periodically to very brief pulses of intense radio-frequency radiation as shown in Figure 19-5. The waveform in part (a) of the figure illustrates the pulse train, pulse width, and time interval between pulses. The expanded view of one of the pulses shows that each pulse is actually a packet of RF radiation. The waveforms are intended to be illustrative and are not drawn to scale. The packet of radiation consists of many

[5] For more complete discussions of Fourier transform NMR, see R. W. King and K. R. Williams, *J. Chem. Educ.*, **1989**, *66*, A213, A243; **1990**, *67*, A93, A125; A. E. Derome, *Modern NMR Techniques for Chemistry Research*. New York: Pergamon Press, 1987; J. W. Akitt, *NMR and Chemistry*, 3rd ed. New York: Chapman and Hall, 1992; R. J. Abraham, J. Fisher, and P. Loftus. *Introduction to NMR Spectroscopy*. New York: Wiley, 1988.

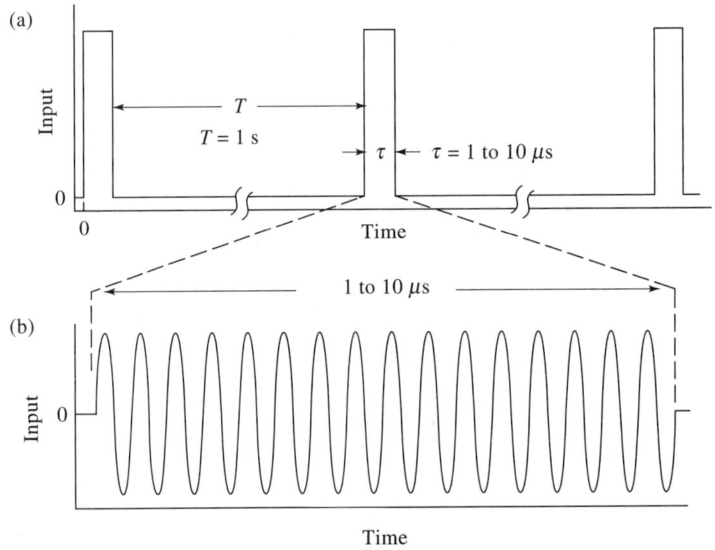

Figure 19-5 Typical input signal for pulsed NMR: (a) pulse sequence; (b) expanded view of radio-frequency pulse, typically at a frequency of several hundred MHz. The time axis is not drawn to scale.

more cycles than are depicted. The length of the pulses τ is usually less than 10 μs, and the frequency of the radiation is on the order of 10^2 to 10^3 MHz. The interval between pulses T is typically one to several seconds. During T, a time-domain, radio-frequency signal, called the *free induction decay* (FID) *signal* is emitted by the excited nuclei as they relax. Free induction decay can be detected with a radio receiver coil that is perpendicular to the static magnetic field. As a matter of fact, a single coil is frequently used to both pulse the sample and detect the decay signal. The FID signal is digitized and stored in a computer for data processing. Ordinarily, the time-domain decay signals from numerous successive pulses are added to improve the signal-to-noise ratio as described in Section 5C-2. The resulting summed data are then converted to a frequency-domain signal by a Fourier transformation, and finally, digital filtering may be applied to the data to further increase the signal-to-noise ratio. The resulting frequency-domain output is similar to the spectrum produced by a scanning continuous-wave experiment.

In order to describe the events that occur in a pulsed NMR experiment, it is helpful to use a set of Cartesian coordinates with the magnetic field pointing along the z axis as shown in Figure 19-6a. The narrow arrows are the magnetic moment vectors of a few of the nuclei in the lower energy ($m = +1/2$) state. The orientations of these vectors around the z axis are random, and they are all rotating at the Larmor frequency ω_0. These excess nuclei impart a stationary net magnetic moment M aligned along the z axis as shown by the broad arrow.

It is helpful in the discussion that follows to imagine that the coordinates in Figure 19-6 are rotating

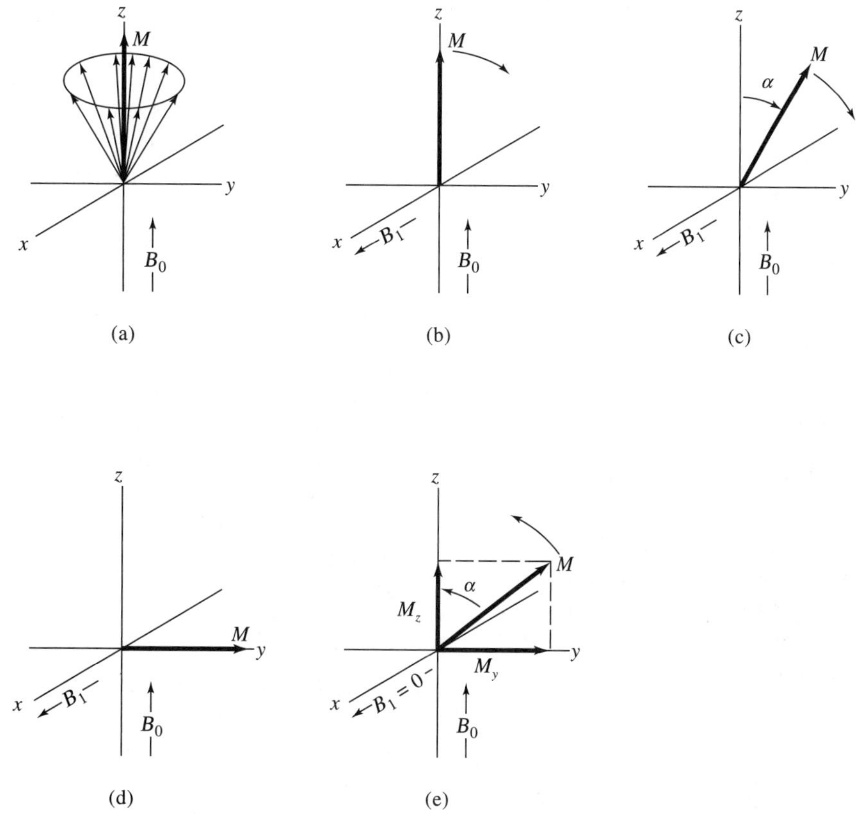

Figure 19-6 Behavior of magnetic moments of nuclei in a rotating field of reference 90-deg pulse experiment: (a) magnetic vectors of excess lower energy nuclei just before pulse; (b), (c), (d) rotation of the sample magnetization vector M during lifetime of the pulse; (e) relaxation after termination of the pulse.

around the z axis at exactly the Larmor frequency. With such a *rotating frame of reference,* the individual magnetic moment vectors in Figure 19-6a become fixed in space at the orientation shown in the figure. Unless otherwise noted, the remaining parts of this figure are discussed in terms of this *rotating frame of reference* rather than a *static,* or *laboratory frame of reference.*

Pulsed Excitation

Figure 19-6b shows the position of the net magnetic moment at the instant the pulse of radio-frequency radiation, traveling along the x axis, strikes the sample. The *magnetic* field of the incident electromagnetic radiation is given the symbol B_1. In the rotating frame, B_1 and the sample magnetization vector M are both static, one along the x axis and the other at right angles to it. A study of basic physics reveals that with each pulse, M experiences a torque that tips it off the z axis. As shown in Figures in 19-6c and 19-6d, this torque rotates the sample magnetic moment M around the x axis in the y-z plane.[6] The extent of rotation depends upon the length of the pulse τ as given by the equation

$$\alpha = \gamma B_1 \tau \qquad (19\text{-}12)$$

where α is the angle of rotation in radians. For many Fourier transform experiments, a pulse length is chosen so that α is 90 deg or $\pi/2$ radians as shown in Figure 19-6d. Typically, the time required to achieve this angle is 1 to 10 μs. Once the pulse is terminated, nuclei begin to relax and return to their equilibrium position as shown in Figure 19-6e. As discussed in the previous section, relaxation takes place by two independent mechanisms: spin-lattice and spin-spin interactions. After several seconds, as a result of these interactions, the nuclei return to their original state as depicted in Figure 19-6a.

When a nucleus returns to its equilibrium state after being tipped by a pulse of RF radiation as shown in Figure 19-6e, there is a decrease in the magnetic moment M_y along the y axis and an increase in the magnetic moment M_z along the z axis. Figure 19-7 provides a more detailed picture of the mechanisms of the two relaxation processes as viewed now in the stationary frame of reference. In spin-lattice relaxation, the magnetization along the z axis increases until it returns to its original value as shown in Figure 19-6a. In spin-spin relaxation, nuclei exchange spin energy with one another so that some now precess faster than the Larmor frequency, while others travel more slowly. The result is that the spins begin to fan out in the x-y plane as shown on the right-hand side of Figure 19-7. Ultimately, this leads to a decrease to zero of the magnetic moment along the y axis. There can be no residual magnetic component in the x-y plane by the time relaxation is complete along the z axis, which means that $T_2 \leq T_1$.

[6] An insight into why NMR is a "resonance" technique can be gained from Figure 19-6. Resonance is a condition in which energy is transferred in such a way that a small periodic perturbation produces a large change in some parameter of the system being perturbed. NMR is a resonance technique because the small periodic perturbation B_1 produces a large change in the orientation of the sample magnetization vector M (Figure 19-6d). In most experiments, B_1 is two or more orders of magnitude smaller than B_0.

Free Induction Decay (FID)

Let us turn again to Figure 19-6d and consider the situation when the signal B_1 goes to zero at the end of the pulse. Now, however, it is useful to picture what is happening in the static laboratory frame of reference rather than the rotating frame. If the coordinates are fixed, the magnetic moment M must now rotate in a clockwise direction around the z axis at the Larmor frequency. This motion gives rise to a radio-frequency signal that can be detected by a coil along the x axis. As mentioned earlier, it can be detected with the same coil that is used to produce the original pulse. As relaxation proceeds, this signal decreases exponentially and approaches zero as the magnetic vector approaches the z axis. This time-domain signal is the free induction decay mentioned earlier; it is ultimately converted to a frequency-domain signal by the Fourier transformation.

Figure 19-8 illustrates the free induction decay that is observed for ^{13}C nuclei when they are excited by a pulse of RF having a frequency that is *exactly* the same as the Larmor frequency of the nuclei. The signal is produced by the four ^{13}C nuclei in dioxane, which behave identically in the magnetic field. The FID in Figure 19-8a is an exponential curve that approaches zero after a few tenths of a second. The apparent noise superimposed on the decay pattern is an experimental artifact caused by spinning sidebands and for the purposes of our discussion can be disregarded. Figure 19-8b is the Fourier transform of the curve in Figure 19-8a, which shows on the left the single ^{13}C absorption peak for dioxane. When the irradiation frequency ν differs from the Larmor frequency $\omega_0/2\pi$ by a small amount, as it usually will, the exponential decay is modulated by a sine wave of frequency $|\nu - \omega_0/2\pi|$. This effect is

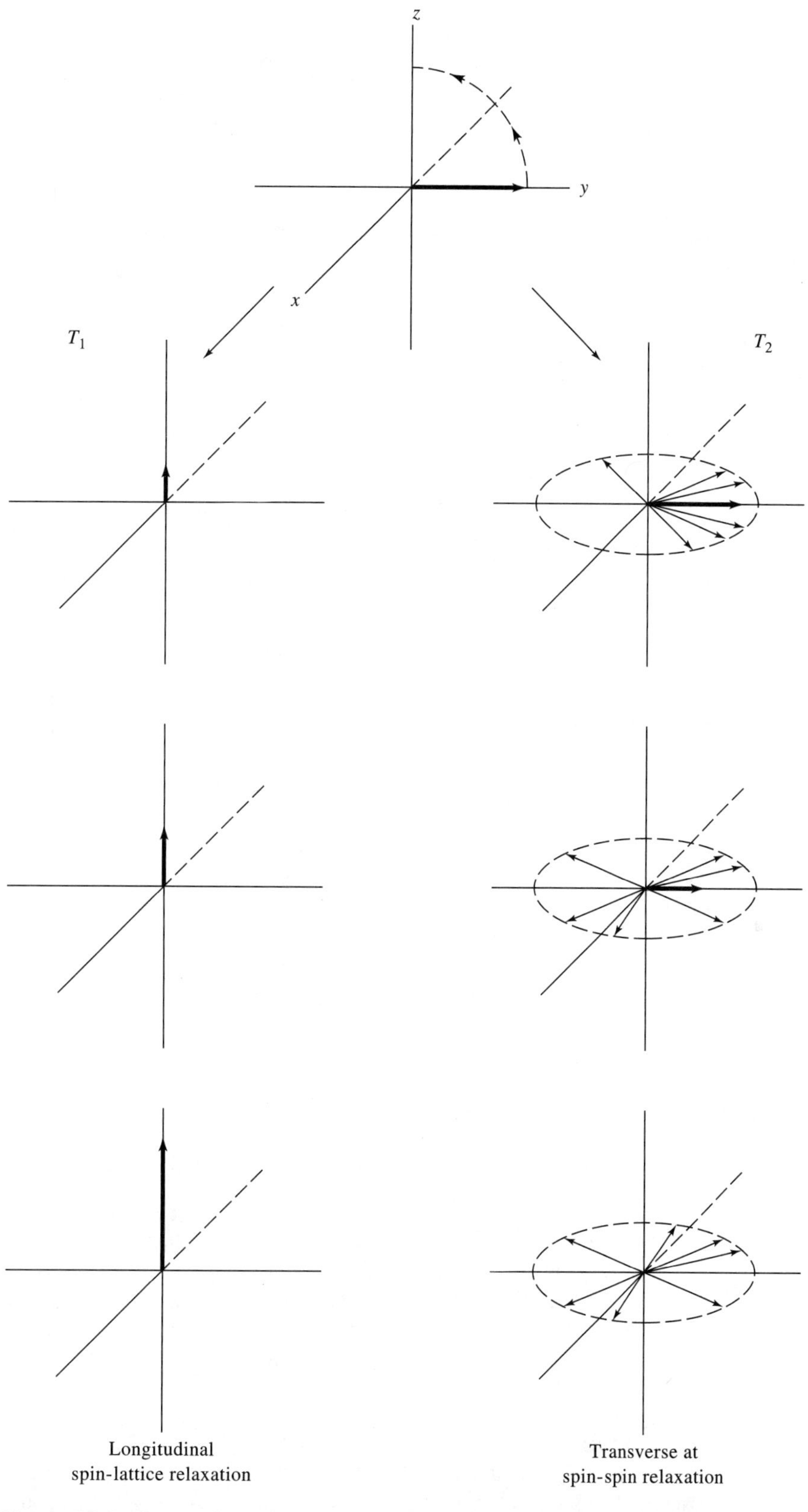

Figure 19-7 Two nuclear relaxation processes. Longitudinal relaxation takes place in the *x-y* plane; transverse in the *x-y* plane. *(Courtesy of Professor Stanford L. Smith, University of Kentucky, Lexington, KY.)*

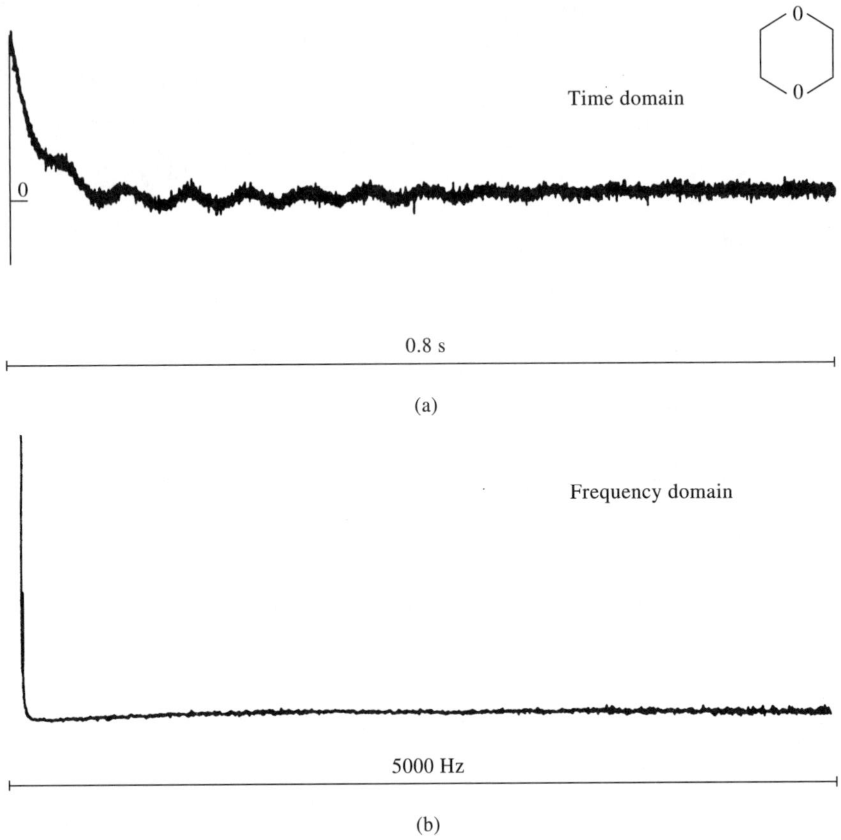

Figure 19-8 (a) FID signal for ^{13}C in dioxane when pulse frequency is identical to Larmor frequency; (b) Fourier transform of (a). *(From R. J. Abraham, J. Fisher, and P. Loftus,* Introduction to NMR Spectroscopy, *p. 89. New York: Wiley, 1988. Reprinted by permission of John Wiley & Sons, Inc.)*

shown in Figure 19-9 in which the difference in the two frequencies is 50 Hz.

When magnetically different nuclei are present, the FID develops a distinct beat pattern such as that in Figure 19-10a, which is the spectrum for the ^{13}C nuclei in cyclohexene. This compound contains three pairs of magnetically different carbon atoms: the pair of olefinic carbons, the pair of aliphatic carbons adjacent to the olefinic pair, and the pair that is directly opposite to the olefinic group. The lines in Figure 19-10b that differ by 62 Hz arise from the two pairs of aliphatic carbon atoms. The pair of olefinic carbons is responsible for the single peak on the left. With compounds having several absorption lines, the FID becomes very complex. In every case, however, the time-domain decay signal contains all of the information required to produce an absorption spectrum in the frequency domain using the Fourier transformation.

19A-4 Types of NMR Spectra

There are several types of NMR spectra, which depend upon the kind of instrument used, the type of nucleus involved, the physical state of the sample, the environment of the analyte nucleus, and the purpose of the data collection. Most NMR spectra can, however, be categorized as either *wide line* or *high resolution.*

Wide-Line Spectra

Wide-line spectra are those in which the bandwidth of the source of the lines is large enough so that the fine

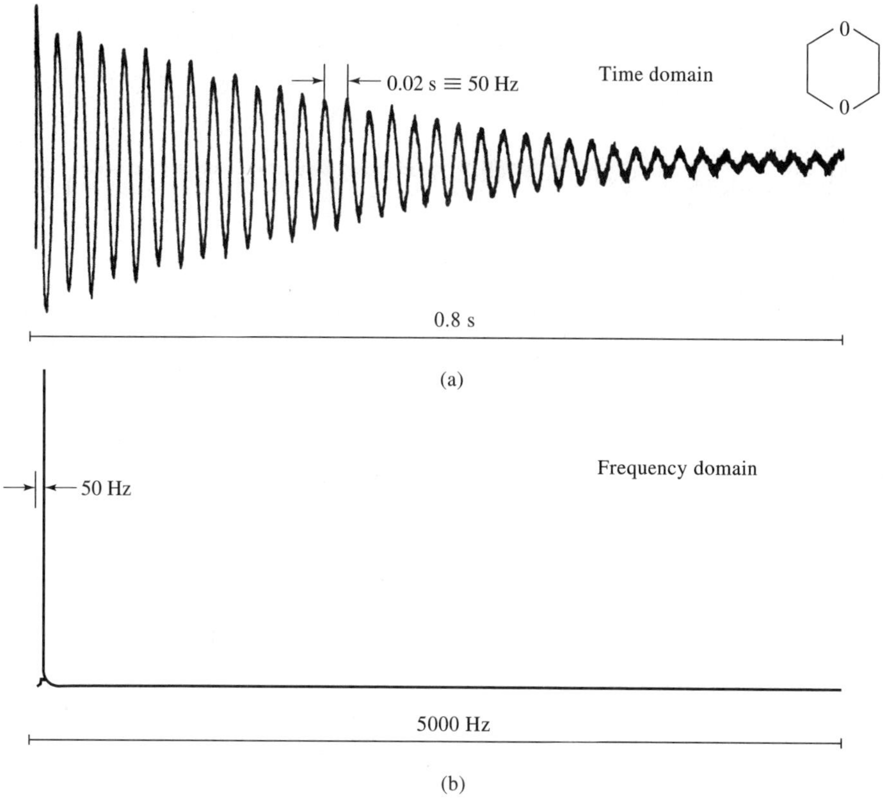

Figure 19-9 (a) FID signal for ^{13}C in dioxane when pulse frequency differs from Larmor frequency by 50 Hz; (b) Fourier transform of (a). *(From R. J. Abraham, J. Fisher, and P. Loftus,* Introduction to NMR Spectroscopy, *p. 90. New York: Wiley, 1988. Reprinted by permission of John Wiley & Sons, Inc.)*

structure due to chemical environment is obscured. Figure 19-11 is a wide-line spectrum for a mixture of several isotopes. A single peak is associated with each species. Wide-line spectra are useful for the quantitative determination of isotopes and for studies of the physical environment of the absorbing species. Wide-line spectra are usually obtained at relatively low magnetic field strength.

High-Resolution Spectra

Most NMR spectra are *high resolution* and are collected by instruments capable of differentiating between very small frequency differences of 0.01 ppm or less. For a given isotope such spectra usually exhibit several peaks that result from differences in their chemical environment. Figure 19-12 illustrates two high-resolution spectra for the protons in ethanol. In the upper spectrum, three peaks are observed arising from absorption by the CH_3, CH_2, and OH protons. As shown in the higher resolution spectrum of Figure 19-12b, two of the three peaks can be resolved into additional peaks. The discussions that follow deal exclusively with high-resolution spectra.

19B ENVIRONMENTAL EFFECTS ON NMR SPECTRA

The frequency of RF radiation that is absorbed by a given nucleus is strongly affected by its chemical environment—that is, by nearby electrons and nuclei. As a consequence, even simple molecules provide a wealth

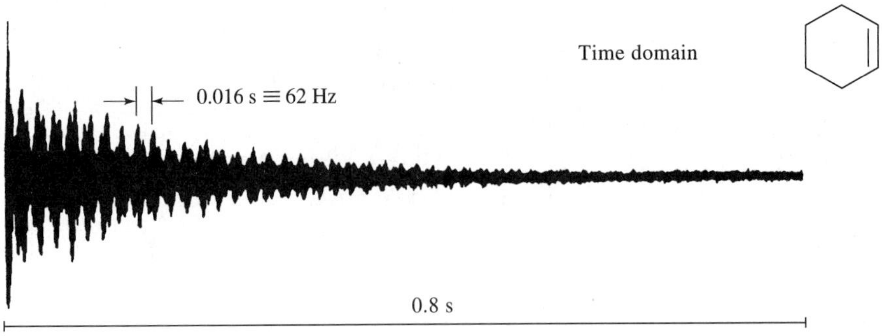

(a)

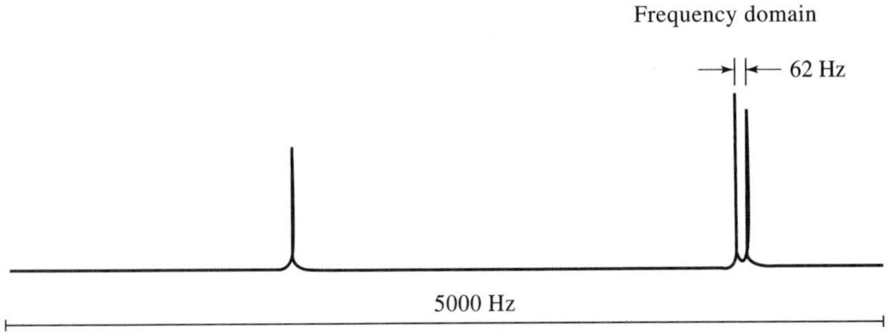

(b)

Figure 19-10 (a) FID signal for ^{13}C in cyclohexane; (b) Fourier transform of (a). *(From R. J. Abraham, J. Fisher and P. Loftus,* Introduction to NMR Spectroscopy, *p. 91. New York: Wiley, 1988. Reprinted by permission of John Wiley & Sons, Inc.)*

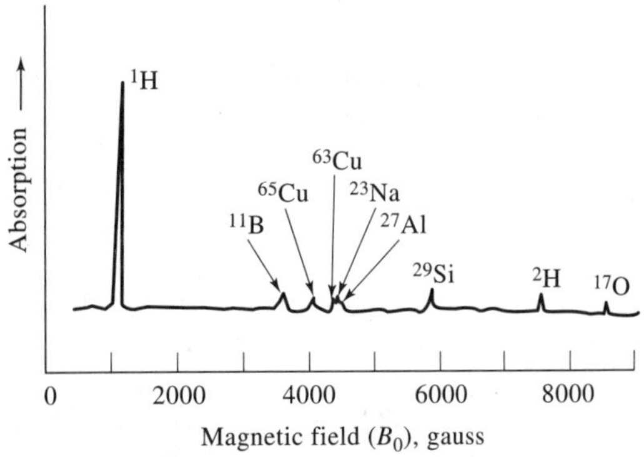

Figure 19-11 A low-resolution NMR spectrum of water in a glass container. Frequency = 5 MHz.

of spectral information that can serve to elucidate their chemical structure. The discussion that follows emphasizes proton spectra because ^{1}H is the isotope that has been studied most widely. It is convenient that most of the concepts of this discussion apply to the spectra of other isotopes as well.

19B-1 Types of Environmental Effects

The spectra for ethyl alcohol, shown in Figure 19-12, illustrate two types of environmental effects. The curve in Figure 19-12a, obtained with a lower resolution instrument, shows three proton peaks with areas in the ratio 1:2:3 (left to right). On the basis of this ratio, it appears logical to attribute the peaks to the hydroxyl, the methylene, and the methyl protons, respectively. Other evidence confirms this conclusion; for example, if the hydrogen atom of the hydroxyl group is replaced by deuterium, the first peak disappears from this part of the

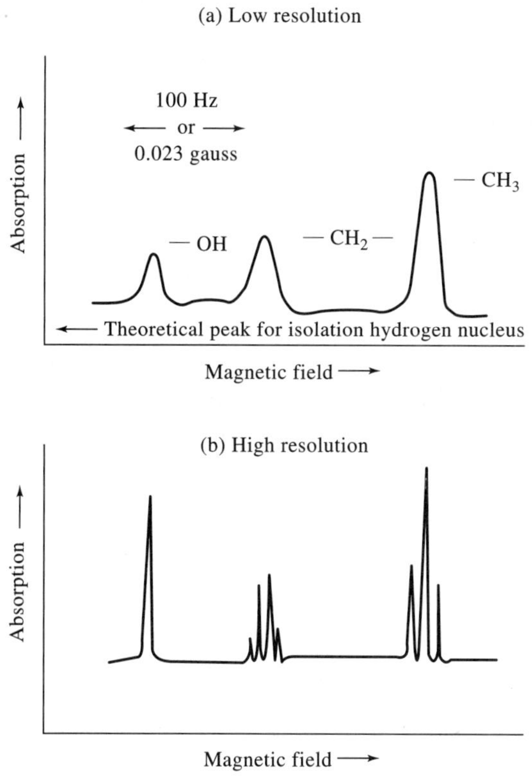

Figure 19-12 NMR spectra of ethanol at a frequency of 60 MHz. Resolution: (a) ~$1/10^6$; (b) ~$1/10^7$.

spectrum. Thus, small differences occur in the absorption frequency of the proton; such differences depend upon the group to which the hydrogen atom is bonded. This effect is called the *chemical shift.*

The higher resolution spectrum of ethanol, shown in Figure 19-12b, reveals that two of the three proton peaks are split into additional peaks. This secondary environmental effect, which is superimposed upon the chemical shift, has a different cause; it is termed *spin-spin splitting.*

Both the chemical shift and spin-spin splitting are important in structural analysis. Experimentally, the two are easily distinguished, because the peak separations resulting from a chemical shift are directly proportional to the field strength or to the oscillator frequency. Thus, if the spectrum in Figure 19-12a is acquired at 100 MHz rather than at 60 MHz, the horizontal distance between any set of peaks is increased by 100/60 as illustrated in Figure 19-13. In contrast, the distance between the fine-structure peaks within a group is not altered by this frequency change.

Origin of the Chemical Shift

The chemical shift is caused by small magnetic fields that are generated by electrons as they circulate around nuclei. These fields usually oppose the applied field. As a consequence, the nuclei are exposed to an effective field that is usually somewhat smaller than the external field. The magnitude of the field developed internally is directly proportional to the applied external field, so that we may write

$$B_0 = B_{appl} - \sigma B_{appl} = B_{appl}(1 - \sigma) \qquad (19\text{-}13)$$

where B_{appl} is the magnitude of the applied field and B_0 *is the magnitude of the resultant field, which determines the resonance behavior of the nucleus.* The quantity σ is the *screening constant,* which is determined by the electron density and its spatial distribution around the nucleus; electron density depends upon the structure of the compound containing the nucleus. Substituting Equation 19-5 into Equation 19-13 gives the resonance condition in terms of frequency. That is,

$$\nu_0 = \frac{\gamma}{2\pi} B_0(1 - \sigma) = k(1 - \sigma) \qquad (19\text{-}14)$$

where $k = \gamma B_0/2\pi$.

The screening constant for protons in a methyl group is larger than the corresponding constant for methylene protons, and it is even smaller for the proton in an —OH group. For an isolated hydrogen nucleus, the screening constant is zero. Thus, in order to bring any of the protons in ethanol into resonance at a given excitation oscillator frequency ν, it is necessary to employ a field B_{appl} that is greater than B_0 (Equation 19-13), the resonance value for the isolated proton. Alternatively, if the applied field is held constant, the oscillator frequency must be increased in order to bring about the resonance condition. Since σ differs for protons in various functional groups, the required applied field differs from group to group. This effect is shown in the ethanol spectrum of Figure 19-12a, in which the hydroxyl proton appears at the lowest applied field, the methylene protons next, and finally the methyl protons. Notice that all of these peaks occur at an applied field greater than the theoretical one for the isolated hydrogen nucleus, which lies far to the left of the spectra of Figure 19-12a. Note also that if the strength of the applied magnetic field is held constant at a level necessary to excite the methyl proton, the frequency must be increased to bring the methylene protons into resonance.

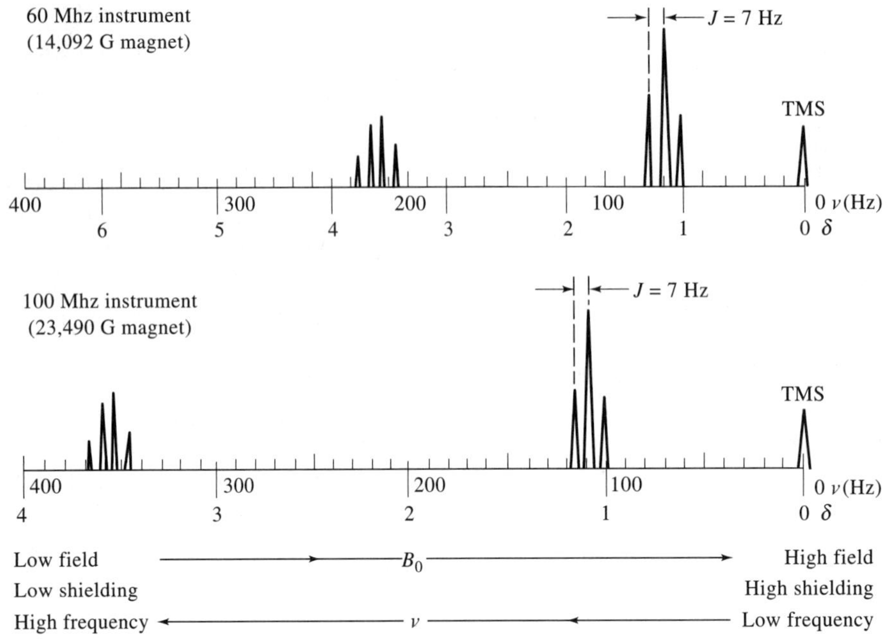

Figure 19-13 Abscissa scales for NMR spectra.

Origin of Spin-Spin Splitting

The splitting of chemical shift peaks occurs as the magnetic moment of a nucleus interacts with the magnetic moments of immediately adjacent nuclei. The magnetic field created by a spinning nucleus affects the distribution of electrons in its bonds to other nuclei. This change in electron distribution then produces changes in the magnetic field of adjacent nuclei and causes splitting of energy levels and hence multiple transitions. This magnetic coupling of nuclei that is transmitted by bonding electrons is often referred to as a *polarization interaction.* Thus, the fine structure of the methylene peak shown in Figure 19-12b can be attributed to the effect of the spins of the adjacent methyl protons. Conversely, the splitting of the methyl peak into three smaller peaks is caused by the adjacent methylene protons. These effects are independent of the applied field and are superimposed on the effects of the chemical shift. Spin-spin splitting is discussed in greater detail in Section 19B-3.

Abscissa Scales for NMR Spectra

The determination of the absolute magnetic field strength with the accuracy required for high-resolution NMR measurements is difficult or impossible. On the other hand, as will be shown in Section 19C, it is entirely feasible to determine the magnitude of a *change* in field strength to within a tenth milligauss or better. Thus, it is expedient to report the position of resonance absorption peaks relative to the resonance peak for an internal standard substance that can be measured during the experiment. The use of an internal standard is also advantageous in that chemical shifts can be reported in terms that are independent of the oscillator frequency.

The internal standard that is used depends upon the nucleus being studied and the solvent system. The compound most generally used for proton studies is tetramethylsilane (TMS), $(CH_3)_4Si$. All of the protons in this compound are identical; and for reasons to be considered later, the screening constant for TMS is larger than for most other protons. Thus, the compound provides a single sharp peak at a high applied field that is nearly always isolated from the peaks of interest in a spectrum. In addition, TMS is inert, readily soluble in most organic liquids, and easily removed from samples by distillation (b.p. = 27°C). Unfortunately, TMS is not water soluble; in aqueous media, the sodium salt of 2,2-dimethyl-2-silapentane-5-sulfonic acid (DSS), $(CH_3)_3SiCH_2CH_2CH_2SO_3Na$, is normally used in its place. The methyl protons of this compound produce a peak at virtually the same place in the spectrum as that of TMS. The methylene protons of DSS give a series of small peaks that may interfere, however. For this reason, most DSS now on the market has the methylene groups deuterated, which eliminates these undesirable peaks.

To express the chemical shift for a sample nucleus relative to TMS in quantitative terms when measurements are made at a constant field strength B_0, we apply Equation 19-14 to the sample and the TMS resonances to obtain

$$\nu_s = k(1 - \sigma_s) \qquad (19\text{-}15)$$

$$\nu_r = k(1 - \sigma_r) \qquad (19\text{-}16)$$

where the subscripts r and s refer to the TMS reference and the analyte sample, respectively. Subtracting the first equation from the second gives

$$\nu_r - \nu_s = k(\sigma_s - \sigma_r) \qquad (19\text{-}17)$$

Dividing this equation by Equation 19-15 in order to eliminate k, gives

$$\frac{\nu_r - \nu_s}{\nu_r} = \frac{\sigma_r - \sigma_s}{1 - \delta_r}$$

Generally, δ_r is much smaller than 1. Thus, this equation simplifies to

$$\frac{\nu_r - \nu_s}{\nu_r} = \sigma_r - \sigma_s \qquad (19\text{-}18)$$

We then define the *chemical shift parameter* δ as

$$\delta = (\sigma_{ref} - \sigma_{sple}) \times 10^6 \text{ ppm} \qquad (19\text{-}19)$$

The quantity δ is dimensionless and expresses the relative shift in parts per million. A distinct advantage of this approach is that for a given peak, δ is the same regardless of whether a 200- or a 500-MHz instrument is employed. Most proton peaks lie in the δ range of 1 to 13. For other nuclei, the range of chemical shifts is greater because of the associated $2p$ electrons. For example, the chemical shift for ^{13}C in various functional groups typically lies in the range 0 to 220 ppm but may be as large as 400 ppm or more. For ^{19}F, the range of chemical shifts may be as large as 800 ppm, while for ^{31}P it is 300 ppm or more.

Generally, NMR plots have scales that are linear in δ, and historically the data were plotted with the field increasing from left to right (see Figure 19-13). Thus, if TMS is used as the reference, its peak appears on the far right-hand side of the plot, since σ for TMS is quite large. As shown, the zero value for the δ scale corresponds to the TMS peak, and the value of δ increases from right to left. Refer again to Figure 19-13 and note that the various peaks appear at the same δ value in spite of the fact that the two spectra were obtained with instruments having markedly different fixed fields.

Spin-spin splitting is generally reported in units of hertz. It can be seen in Figure 19-13 that the spin-spin splitting in frequency units (J) is the same for the 60-MHz and the 100-MHz instruments. Note, however, that the chemical shift *in frequency units* is enhanced with the higher frequency instrument.

19B-2 Theory of the Chemical Shift

As noted earlier, chemical shifts arise from the secondary magnetic fields produced by the circulation of electrons in the molecule. These so-called local *diamagnetic currents*[7] are induced by the fixed magnetic field and result in secondary fields that may either decrease or enhance the field to which a given proton responds. The effects are complex, and we consider only the major aspects of the phenomenon here. More complete treatments can be found in the several reference works listed under footnote 1 at the beginning of this chapter.

Under the influence of the magnetic field, electrons bonding the proton tend to precess around the nucleus in a plane perpendicular to the magnetic field (see Figure 19-14). A consequence of this motion is the development of a secondary field, which opposes the primary field; the behavior here is analogous to the passage of electrons through a wire loop. The nucleus then experiences a resultant field that is smaller, so the nucleus is said to be *shielded* from the full effect of the primary field. As a consequence, the external field must be increased to cause nuclear resonance.

The shielding experienced by a given nucleus is directly related to the electron density surrounding it. Thus, in the absence of the other influences, shielding would be expected to decrease with increasing electronegativity of adjacent groups. This effect is illustrated by the δ values for the protons in the methyl

[7] The intensity of magnetization induced in a *diamagnetic* substance is smaller than that produced in a vacuum with the same field. Diamagnetism is the result of motion induced in bonding electrons by the applied field; this motion, termed a *diamagnetic current,* creates a secondary field that opposes the applied field. *Paramagnetism* and the resulting *paramagnetic currents* operate in just the opposite sense.

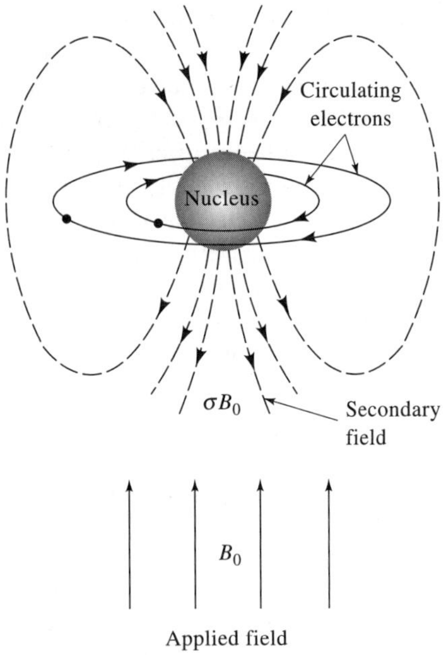

Figure 19-14 Diamagnetic shielding of a nucleus.

halides, CH_3X, which lie in the order I (2.16), Br (2.68), Cl (3.05), and F (4.26). In this example, the least electronegative halogen, iodine, is also the least effective at withdrawing electrons from the methyl protons. Thus, the electrons of iodine provide the smallest shielding effect. The position of the proton peaks in TMS is also explained by this model, since silicon is relatively electropositive.

Effect of Magnetic Anisotropy

An examination of the spectra of compounds containing double or triple bonds reveals that local diamagnetic effects are not sufficient to explain the position of certain proton peaks. Consider, for example, the irregular change in δ values for protons in the following hydrocarbons, arranged in order of increasing acidity, or increased electronegativity of the groups to which the protons are bonded: $CH_3—CH_3$ ($\delta = 0.9$), $CH_2{=}CH_2$ ($\delta = 5.8$), and $HC{\equiv}CH$ ($\delta = 2.9$). Furthermore, the aldehydic proton RCHO ($\delta \sim 10$) and the protons in benzene ($\delta \sim 7.3$) appear considerably farther downfield than is expected on the basis of the electronegativity of the attached groups.

The effects of multiple bonds on the chemical shift can be explained by taking into account the anisotropic magnetic properties of these compounds. For example, the magnetic susceptibilities[8] of crystalline aromatic compounds have been found to differ appreciably from one another, depending upon the orientation of the ring with respect to the applied field. This anisotropy is readily understood from the model shown in Figure 19-15. In this model, the plane of the ring is perpendicular to the magnetic field. In this position, the field can induce a flow of the π electrons around the ring to create a so-called *ring current.* A ring current is similar to a current in a wire loop; that is, a secondary field is induced that acts in opposition to the applied field. This induced field, however, exerts a magnetic effect on the protons attached to the ring in the direction of the field as shown in Figure 19-15. Thus, the aromatic protons require a lower external field to bring them into resonance. This effect is either absent or self-canceling in other orientations of the ring.

An analogous model operates for ethylenic or carbonyl double bonds. In such cases, we may imagine π electrons circulating in a plane along the bond axis when the molecule is oriented with the field as presented in Figure 19-16a. Again, the secondary field produced acts upon the proton to reinforce the applied field. Thus, deshielding shifts the peak to larger values of δ. With an aldehyde, this effect combines with the deshielding brought about by the electronegative nature of the carbonyl group, and a very large value of δ results.

[8] The magnetic susceptibility of a substance can be thought of as the extent to which it is susceptible to induced magnetization by an external field.

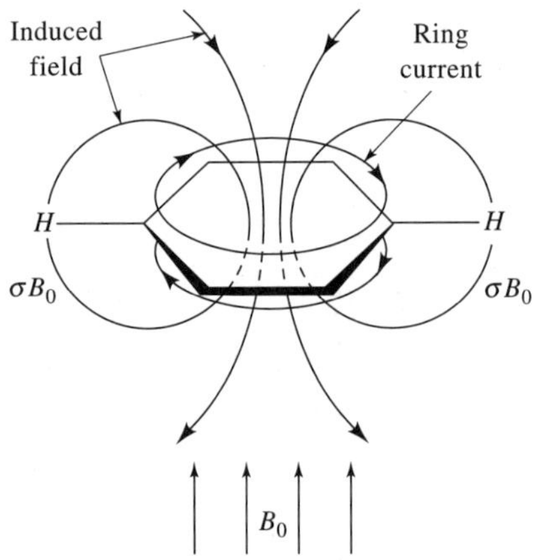

Figure 19-15 Deshielding of aromatic protons brought about by ring current.

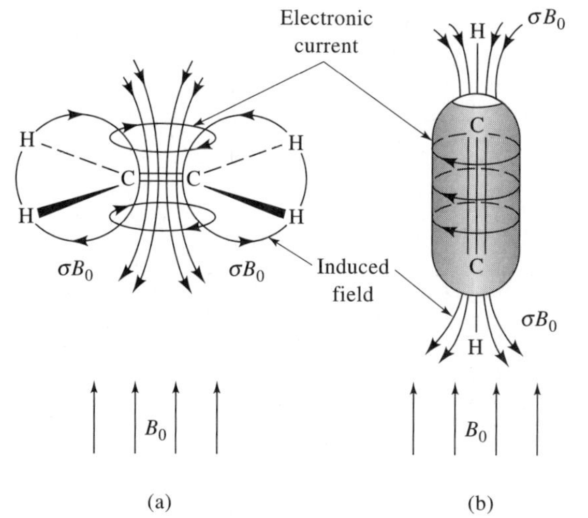

Figure 19-16 Deshielding of ethylene and shielding of acetylene brought about by electronic currents.

In an acetylenic bond, the symmetrical distribution of π electrons about the bond axis permits electrons to circulate around the bond. In contrast, such circulation is prohibited by the nodal plane in the electron distribution of a double bond. From Figure 19-16b, we see that in this orientation the protons are shielded. This effect is apparently large enough to offset the deshielding resulting from the acidity of the protons and from the electronic currents at perpendicular orientations of the bond.

Correlation of Chemical Shift with Structure

The chemical shift is used to identify functional groups and to aid in determining structural arrangements of groups. These applications are based upon empirical correlations between structure and shift. A number of correlation charts and tables[9] have been published, two of which are shown in Figure 19-17 and Table 19-2. Keep in mind that the exact values for δ may depend upon the nature of the solvent as well as upon the concentration of solute. These effects are particularly pronounced for protons involved in hydrogen bonding. An excellent example of this effect is the proton of the alcoholic functional group.

[9] R. M. Silverstein, G. C. Bassler, and T. C. Morrill, *Spectrometric Identification of Organic Compounds,* 5th ed., Chapter 4. New York: Wiley, 1991; L. M. Jackman and S. Sternhell, *Nuclear Magnetic Resonance Spectroscopy,* 2nd ed. New York: Pergamon Press, 1969.

19B-3 Spin-Spin Splitting

As may be seen in Figure 19-12, the absorption bands for the methyl and methylene protons in ethanol consist of several narrow peaks that can be routinely separated with a high-resolution instrument. Careful examination of these peaks shows that the spacing for the three components of the methylene band is identical to that for the four peaks of the methylene band. This spacing in hertz is called the *coupling constant* for the interaction and is given the symbol *J*. Moreover, the areas of the peaks in a multiplet approximate an integer ratio to one another. Thus, for the methyl triplet, the ratio of areas is 1:2:1; for the quartet of methylene peaks, it is 1:3:3:1.

Origin

The foregoing observations may be attributed to the effect that the spins of one set of nuclei exert upon the resonance behavior of another. That is to say, there is a small interaction or coupling between the two adjacent groups of protons. The results of detailed theoretical calculations are consistent with the concept that coupling takes place via interactions between the nuclei and the bonding electrons rather than through free space. For our purposes a detailed discussion of the details of the mechanism is unnecessary.

Let us first consider the effect of the methylene protons in ethanol on the resonance of the methyl protons. Recall that the ratio of protons in the two possible spin states is very nearly unity, even in a strong magnetic field. We then imagine that the two methylene protons in the molecule can have four possible combinations of spin states and that in an entire sample the number of each of these combinations will be approximately equal. If we represent the spin orientation of each nucleus with a small arrow, the four states are as follows:

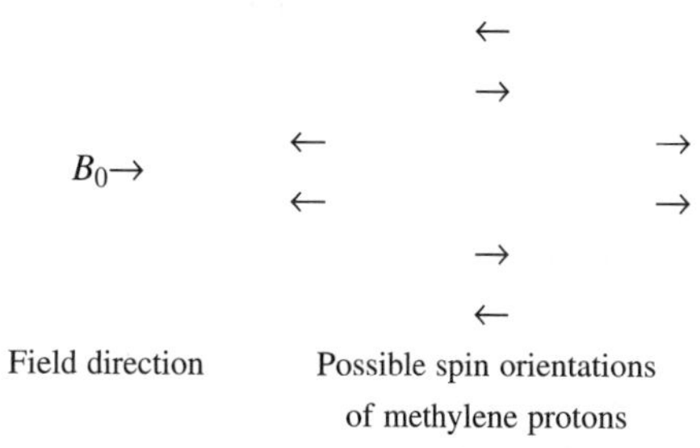

In the first combination on the left, spins of the two methylene protons are paired and aligned against the field, while in the second combination on the right, the

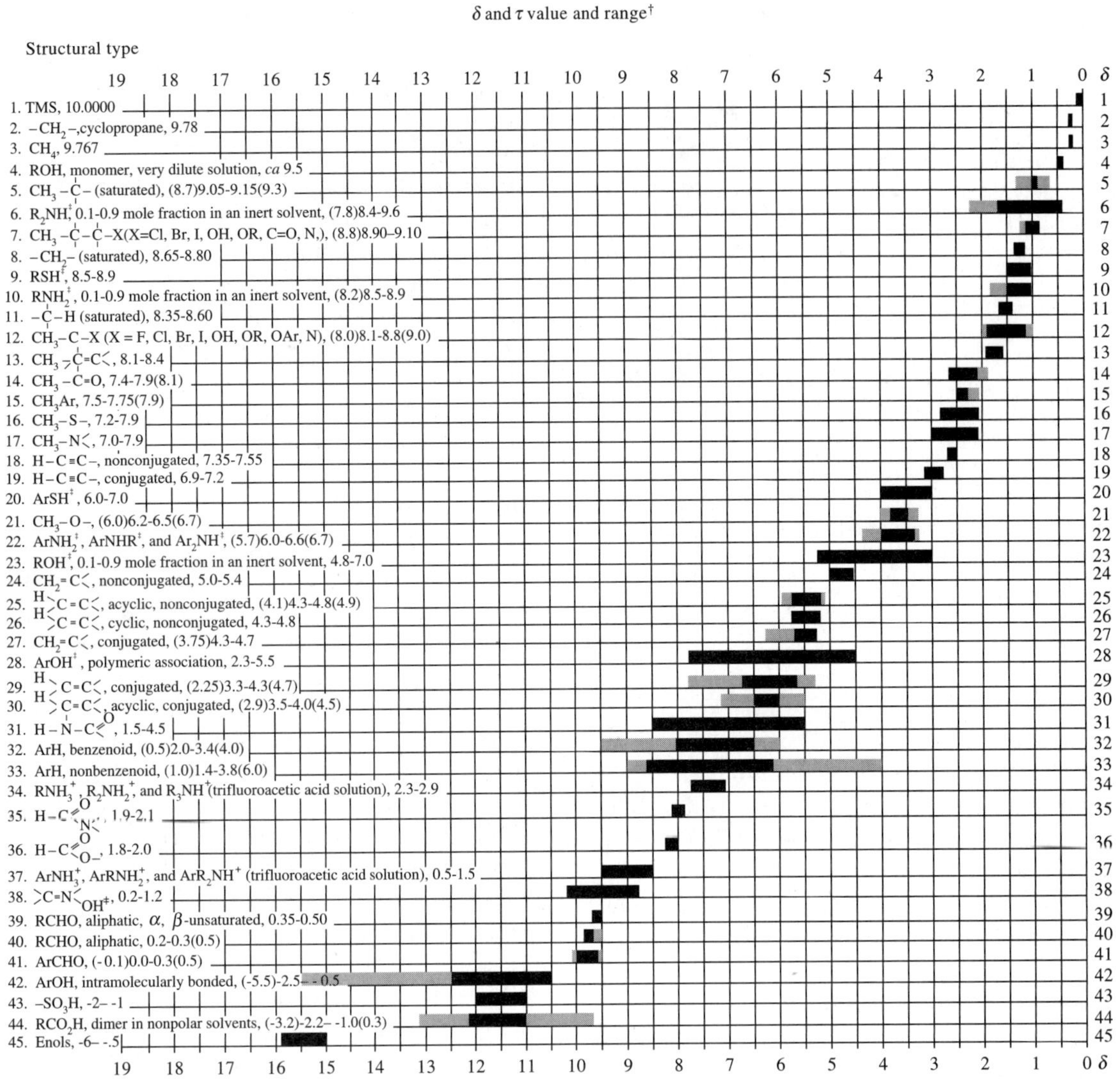

†Normally, absorptions for the functional groups indicated will be found within the range shown. Occasionally, a functional group will absorb outside this range. Approximate limits for this are indicated by absorption values in parentheses and by shading in the figure.
‡The absorption positions of these groups are concentration-dependent and are shifted to higher τ values in more dilute solutions.

Figure 19-17 Absorption positions of protons in various structural environments. *(Table taken from J. R. Dyer,* Applications of Absorption Spectroscopy by Organic Compounds, *p. 85. Englewood Cliffs, NJ: Prentice-Hall, 1965. With permission.)*

paired spins are reversed. There are also two combinations in which the spins are opposed to one another shown in the center. The magnetic effect that is transmitted to the methyl protons on the adjacent carbon atom is determined by the instantaneous spin combinations in the methylene group. If the spins are paired and opposed to the external field, the effective applied field on the methyl protons is slightly decreased. Thus, a somewhat higher field is needed to bring them into resonance, and an upfield shift results. Spins paired and aligned with the field result in a downfield shift. Neither of the combinations of opposed spin has an effect on the resonance of the methyl protons. Thus, splitting into three peaks results. The area under the middle peak is twice that of either of the other two, since two spin combinations are involved.

TABLE 19-2 Approximate Chemical Shifts for Certain Methyl, Methylene, and Methine Protons

	δ, ppm		
Structure	$M = CH_3$	$M = CH_2$	$M = CH$
Aliphatic α substituents			
M—Cl	3.0	3.5	4.0
M—Br	2.7	3.4	4.1
$M—NO_2$	4.3	4.4	4.6
M—OH (or OR)	3.2	3.4	3.6
M—O—ϕ	3.8	4.0	4.6
M—OC(=O)R	3.6	4.1	5.0
M—C=C	1.6	1.9	—
M—C≡C	1.7	2.2	2.8
M—C(=O)H	2.2	2.4	—
M—C(=O)R	2.1	2.4	2.6
M—C(=O)ϕ	2.4	2.7	3.4
M—C(=O)OR	2.2	2.2	2.5
M—ϕ	2.2	2.6	2.8
Aliphatic β substituents			
M—C—Cl	1.5	1.8	2.0
M—C—Br	1.8	1.8	1.9
$M—C—NO_2$	1.6	2.1	2.5
M—C—OH (or OR)	1.2	1.5	1.8
M—C—OC(=O)R	1.3	1.6	1.8
M—C—C(=O)H	1.1	1.7	—
M—C—C(=O)R	1.1	1.6	2.0
M—C—C(=O)OR	1.1	1.7	1.9
M—C—ϕ	1.1	1.6	1.8

Let us now consider the effect of the three methyl protons upon the methylene peak. Possible spin combinations for the methyl protons are as follows:

	←	←	←	→	←	→	→	→
$B_0 \rightarrow$	←	←	→	←	→	←	→	→
	←	→	←	←	→	→	←	→

In this instance, we have eight possible spin combinations; however, among these are two groups containing three combinations that have equivalent magnetic effects. The methylene peak is thus split into four peaks having areas in the ratio 1:3:3:1. These two examples of the adjacent methyl and methylene groups in ethanol suggest the general rule that the number of peaks in a split band in a first-order spectrum is equal to the number n of magnetically equivalent protons[10] on adjacent atoms plus one. The number of such peaks is referred to as the *multiplicity.*

The interpretation of spin-spin splitting patterns is relatively straightforward for *first-order* spectra. First-order spectra are those in which the chemical shift between interacting groups of nuclei is large with respect to their coupling constant J. Rigorous first-order behavior requires that J/δ be smaller than 0.05. Frequently, however, analysis of spectra by first-order techniques can be accomplished down to a value of $\Delta\nu/J$ of somewhat greater than 0.1. The ethanol spectrum shown in Figure 19-13 is an example of a pure first-order spectrum with J for the methyl and methylene peaks being 7 Hz and the separation between the centers of the two multiplets being about 140 Hz.

Interpretation of second-order NMR spectra is relatively complex and will not be dealt with in this text. Note, however, that because δ increases with increases in the magnetic field while J does not, spectra obtained with an instrument having a high magnetic field are much more readily interpreted than those produced by a spectrometer with a weaker magnet.

[10] Magnetically equivalent protons are those that have identical chemical shifts and identical coupling constants.

Rules Governing the Interpretation of First-Order Spectra

The following rules govern the appearance of first-order spectra.

1. Equivalent nuclei do not interact with one another to give multiple absorption peaks. The three protons in the methyl groups in ethanol give rise to splitting of the adjacent methylene protons only and not to splitting among themselves.
2. Coupling constants decrease significantly with separation of groups, and coupling is seldom observed at distances greater than four bond lengths.
3. The multiplicity of a band is determined by the number n of magnetically equivalent protons on the neighboring atoms and is given by the quantity $n + 1$. Thus, the multiplicity for the methylene band in ethanol is determined by the number of protons in the adjacent methyl groups and is equal to $3 + 1 = 4$.
4. If the protons on atom B are affected by protons on atoms A and C that are nonequivalent, the multiplicity of B is equal to $(n_A + 1)(n_C + 1)$, where n_A and n_C are the number of equivalent protons on A and C, respectively.
5. The approximate relative areas of a multiplet are symmetric around the midpoint of the band and are proportional to the coefficients of the terms in the expansion $(x + 1)^n$. The application of this rule is demonstrated in Table 19-3 and in the examples that follow.
6. The coupling constant is independent of the applied field; thus, multiplets are readily distinguished from closely spaced chemical shift peaks by running spectra at two different field strengths.

EXAMPLE 19-3

For each of the following compounds, calculate the number of multiplets for each band and their relative areas.

(a) $ClCH_2CH_2CH_2Cl$. The multiplicity of the band associated with the four equivalent protons on the two ends of the molecule is determined by the number of protons on the central carbon; thus, the multiplicity is $2 + 1 = 3$ and the areas have the ratio 1:2:1. The multiplicity of two central methylene protons is determined by the four equivalent protons at the ends and is 4 + 1 = 5. Expansion of $(x + 1)^4$ gives the following coefficients (Table 19-3), which are proportional to the areas of the peaks 1:4:6:4:1.

(b) $CH_3CHBrCH_3$. The band for the six methyl protons is made up of 1 + 1 = 2 peaks having relative areas of 1:1; the proton on the central carbon atom has a multiplicity of 6 + 1 = 7. These peaks have areas (Table 19-3) in the ratio of 1:6:15:20:15:6:1.

(c) $CH_3CH_2OCH_3$. The methyl protons on the right are separated from the other protons by more than three bonds so that only a single peak is observed for them. The protons of the central methylene groups have a multiplicity of 3 + 1 = 4 and a ratio of 1:3:3:1. The methyl protons on the left have a multiplicity of 2 + 1 = 3 and an area ratio of 1:2:1.

The foregoing examples are relatively straightforward because all of the protons influencing the multiplicity of any single peak are magnetically equivalent. A more complex splitting pattern results when a set of protons is influenced by two or more nonequivalent protons. As an example, consider the spectrum of 1-iodopropane, $CH_3CH_2CH_2I$. If we label the three carbon atoms (a), (b), and (c) from left to right, the chemical shift bands are found at $\delta_{(a)} = 1.02$, $\delta_{(b)} = 1.86$, and $\delta_{(c)} = 3.17$. The band at $\delta_{(a)} = 1.02$ will be split by the two methylene protons on (b) into 2 + 1 = 3 peaks having relative areas of 1:2:1. A similar splitting of the band $\delta_{(c)} = 3.17$ will also be observed. The experimental coupling constants for the two shifts are $J_{(ab)} = 7.3$ and $J_{(bc)} = 6.8$. The band for the methylene protons (b) is influenced by two groups of protons that are not magnetically equivalent, as is evident from the difference between $J_{(ab)}$ and $J_{(bc)}$. Thus, invoking rule 4, the number of peaks is (3 + 1)(2 + 1) = 12. In cases such as this, derivation of a splitting pattern, as shown in Figure 19-18, is helpful. In this example, the effect of the (a) proton is first shown and leads to four peaks of relative areas 1:3:3:1 spaced at 7.3 Hz. Each of these is then split into three new peaks spaced at 6.8 Hz, having relative areas of 1:2:1. The same final pattern is produced if the original band is first split into a triplet. At a very high resolution, the spectrum for 1-iodopropane exhibits a series of peaks that approximates the series shown at the bottom of Figure 19-18. At resolution sufficiently low that the instrument does not detect the difference between $J_{(ab)}$ and $J_{(bc)}$, only six peaks are observed with relative areas of 1:5:10:10:5:1.

Second-Order Spectra

Coupling constants are usually smaller than 20 Hz, while chemical shifts may be as high as several thousand Hz. Therefore, the splitting behavior described by the rules in the previous section is common. However,

TABLE 19-3 **Relative Intensities of First-Order Multiplets ($I = 1/2$)**

Number of Equivalent Protons, n	Multiplicity, $(n + 1)$	Relative Peak Areas														
0	1								1							
1	2							1		1						
2	3						1		2		1					
3	4					1		3		3		1				
4	5				1		4		6		4		1			
5	6			1		5		10		10		5		1		
6	7		1		6		15		20		15		6		1	
7	8	1		7		21		35		35		21		7		1

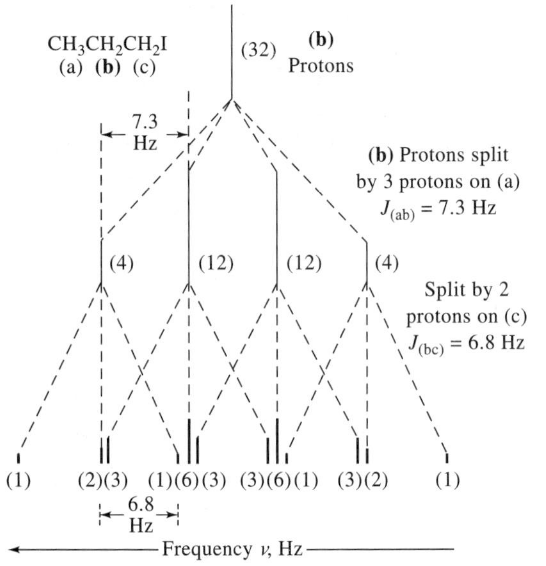

Figure 19-18 Splitting pattern for methylene (b) protons in $CH_3CH_2CH_2I$. Figures in parentheses are relative areas under peaks.

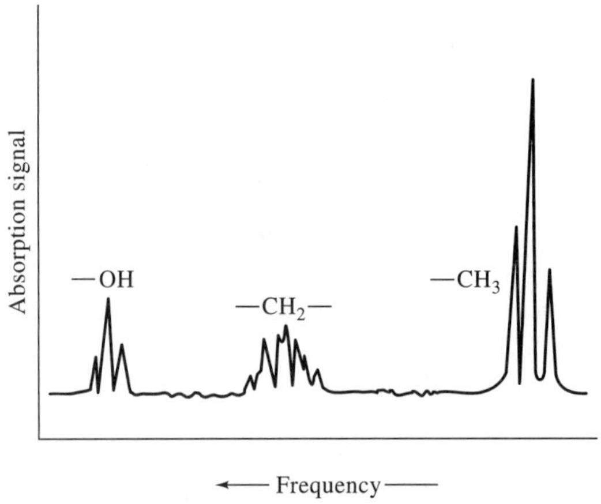

Figure 19-19 Spectrum of highly purified ethanol showing additional splitting of OH and CH_2 peaks (compare with Figure 19-12).

when J/δ becomes greater than about 0.1 to 0.15, these rules no longer apply. Generally, as δ approaches J, the peaks on the inner side of two multiplets tend to be enhanced at the expense of the peaks on the outer side, and the symmetry of each multiplet is thus destroyed, as noted earlier. In addition, more, and sometimes many more, lines appear, so that the spacing between lines no longer has anything to do with the magnitude of the coupling constants. Analysis of a spectrum under these circumstances is difficult.

Effect of Chemical Exchange on Spectra

We now turn again to the NMR spectrum of ethanol (Figure 19-12) and consider why the OH proton appears as a singlet rather than a triplet. The methylene protons and the OH proton are separated by only three bonds, so coupling should increase the multiplicity of both OH and the methylene peaks. As shown in Figure 19-19, we actually observe the expected multiplicity in the NMR spectrum of a highly purified sample of the alcohol. Examine the triplet OH peaks and the eight methylene peaks in this spectrum. If we add a trace of acid or base to the pure sample, the spectrum reverts to the form shown in Figure 19-12.

The exchange of OH protons among alcohol molecules is known to be catalyzed by both acids and bases, as well as by the impurities that commonly occur in alcohol. It is thus plausible to associate the decoupling observed in the presence of these catalysts to an exchange process. If exchange is rapid, each OH group will have several protons associated with it during any brief period; within this interval, all of the OH protons will experience the effects of the three spin arrangements of the methylene protons. Thus, the magnetic effects on the alcoholic proton are averaged, and a single sharp peak is observed. Spin decoupling always occurs when the exchange frequency is greater than the separation frequency between the interacting components.

Chemical exchange can affect not only spin-spin spectra but also chemical shift spectra. Purified alcohol/water mixtures have two well-defined and easily separated OH proton peaks. When acid or base is added to the mixture, however, the two peaks coalesce to form a single sharp line. In this instance, the catalyst enhances the rate of proton exchange between the alcohol and the water and thus averages the shielding effect. A single sharp line is obtained when the exchange rate is significantly greater than the separation frequency between the alcohol and water lines. On the other hand, if the exchange frequency is about the same as this frequency difference, shielding is only partially averaged and a broad line results. The correlation of line breadth with exchange rates has provided a direct means for investigating the kinetics of such processes and represents an important application of the NMR experiment. Such studies are often accomplished by acquiring spectra at several different temperatures and determining the temperature at which multiplets coalesce to a simple band. Armed with this information, the activation energy and other thermodynamic and kinetic parameters for the exchange process may be computed, and the mechanisms of exchange processes may be investigated.[11]

19B-4 Double Resonance Techniques

Double resonance experiments include a group of techniques in which a sample is simultaneously irradiated with two or more signals of different radio frequency. Among these methods are *spin decoupling, nuclear Overhauser effect, spin tickling,* and *internuclear double resonance.* These procedures aid in the interpretation of complex NMR spectra and enhance the information that can be obtained from them.[12] Only the first of these techniques, spin decoupling, will be described here. The nuclear Overhauser effect is discussed briefly in Section 19E-1.

Figure 19-20 illustrates the spectral simplification that accompanies *homonuclear spin decoupling,* which is decoupling carried out between similar nuclei. Spectrum *B* shows the absorption associated with the four protons on the pyridine ring of nicotine. Spectrum *C* was obtained by examining the same portion of the spectrum while simultaneously irradiating the sample with a second radio-frequency signal having a frequency of about 8.6 ppm, which corresponds to a chemical shift, centered on the absorption peaks of protons (c) and (d). The strength of the second signal is suffi-

[11] R. S. Drago, *Physical Methods for Chemists,* 2nd ed. Chapter 8, p. 290. Philadelphia: Saunders College Publishing, 1992.

[12] For a more detailed discussion of double resonance methods, see any of the monographs listed in footnote 1.

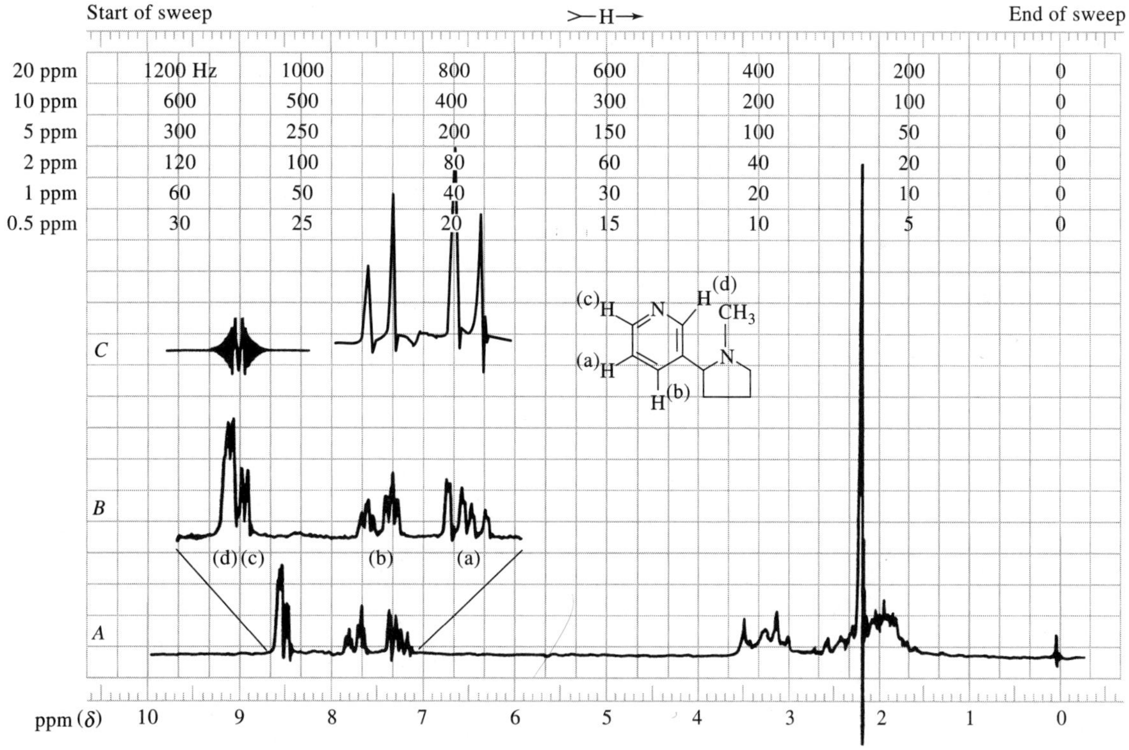

Figure 19-20 Effect of spin decoupling on the NMR spectrum of nicotine dissolved in $CDCl_3$. Curve ***A***, the entire spectrum. Curve ***B***, expanded spectrum for the four protons on the pyridine ring. Curve ***C***, spectrum for protons (a) and (b) when decoupled from (d) and (c) by irradiation with a second beam that has a frequency corresponding to about 8.6 ppm. *(Courtesy of Varian Instrument Division, Palo Alto, CA 94303.)*

cient to cause saturation of the signal for these protons. The consequence is a decoupling of the interaction between the (c) and (d) protons and protons (a) and (b). Here, the complex absorption spectrum for (a) and (b) collapses to two doublet peaks that arise from coupling between these protons. Similarly, the spectrum for (c) and (d) is simplified by decoupling with a signal of frequency corresponding to the peaks for protons (a) or (b).

Heteronuclear decoupling, which is decoupling of the interaction between dissimilar nuclei, is easily accomplished with modern NMR instrumentation. The most important example is encountered in ^{13}C NMR where the technique is used to simplify spectra by decoupling protons (Section 19E-1).

19C NMR SPECTROMETERS

Instrument manufacturers market two general types of NMR spectrometers: *wide-line spectrometers* and *high-resolution spectrometers.*[13] Wide-line instruments have magnets with strengths of a few tenths of a tesla and are considerably simpler and less expensive than are high-resolution instruments. High-resolution instruments are equipped with magnets with strengths that range from 1.4 to 14 T, which corresponds to proton frequencies of 60 to 1000 MHz. Before about 1970, high-resolution NMR spectrometers were all of the continuous-wave type that used permanent magnets or electromagnets to supply the magnetic field. This type of instrument has been largely replaced by Fourier transform spectrometers equipped with superconducting solenoids to provide the magnetic field. Computers are an integral part of modern NMR instruments; they digitize and store the signal, perform Fourier transformation of the FID to provide the frequency-domain signal, and deliver many other data-treatment and instrument-control functions. A major reason why Fourier transform instruments have become so popular is that they permit efficient signal averaging and thus yield greatly enhanced sensitivity (see Sections 5C-2 and 16C-1). As a consequence of this greater sensitivity, routine applications of NMR to naturally occurring carbon-13, to protons in microgram quantities, and to other nuclei, such as fluorine, phosphorus, and silicon, have become widespread.

[13] For a description of commercially available high resolution NMR spectrometers, see D. Noble, *Anal. Chem.*, **1995**, *67*, 559A.

Fourier transform instruments are significantly more expensive than continuous-wave instruments, costing at a minimum $100,000 and as much as $1,000,000 or more. Despite this cost differential, Fourier transform spectrometers now dominate the market to the exclusion of continuous-wave instruments. Thus, we will largely confine our discussion to Fourier transform instruments.

19C-1 Components of Fourier Transform Spectrometers

Figure 19-21 is a simplified block diagram showing the instrument components of a typical Fourier transform NMR spectrometer. The central component of the instrument is a highly stable magnet in which the sample is placed. The sample is surrounded by a transmitter/receiver coil.

Radio-frequency radiation is produced by a crystal controlled frequency synthesizer having an output frequency of ν_c. This signal passes into a pulser switch and power amplifier, which creates an intense and reproducible pulse of radio-frequency current in the transmitter coil. The resulting RF radiation impinges on the sample, contained inside the coil. The length, amplitude, shape, and phase of the pulse are selected by the operator, entered into the console, and controlled by the computer. In Figure 19-21, a 5-μs pulse is shown. The resulting FID signal is picked up by the same coil, which now serves as a receiver. The signal is then amplified and transmitted to a phase-sensitive detector. The detector circuitry produces the difference between the nuclear signals ν_n and the crystal oscillator output ν_c, which leads to the low-frequency, time-domain signal shown on the right of the figure. This signal is digitized and collected in the memory of the computer for analysis by a Fourier transform program and other data analysis software. The output from this program is plotted, giving a frequency-domain spectrum.

19C-2 Magnets

The heart of both continuous-wave and Fourier transform NMR instruments is the magnet. The sensitivity and resolution of both types of spectrometers are critically dependent upon the strength and quality of their magnets (see Example 19-2 and Figure 19-13). Since both sensitivity and resolution increase with increases in field strength, it is advantageous to operate at the highest possible field strength. In addition, the field must be highly homogeneous and reproducible. These requirements ensure that the magnet is by far the most expensive component of an NMR spectrometer.

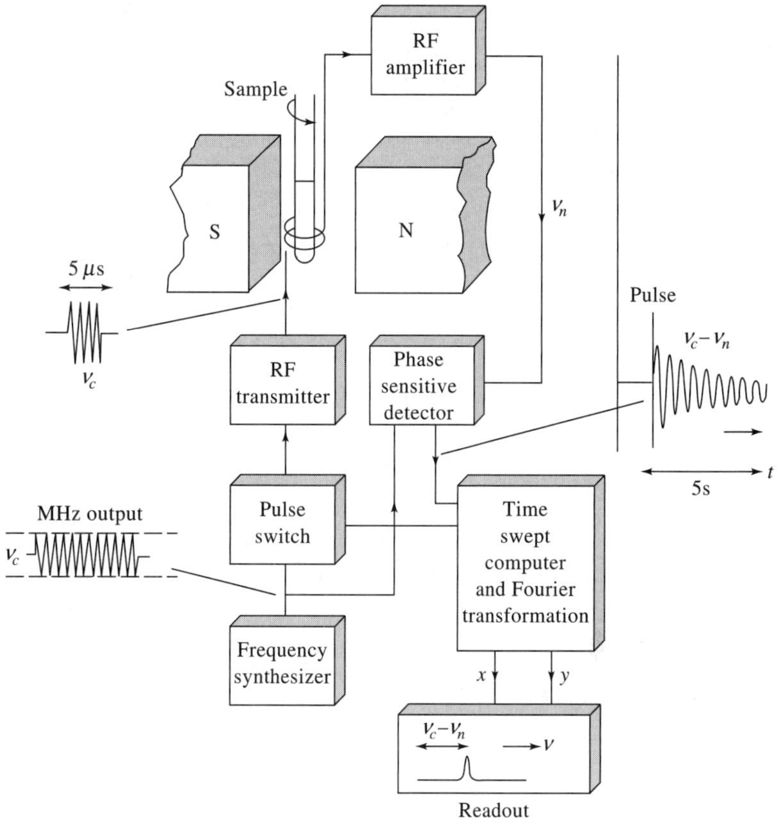

Figure 19-21 Block diagram of a Fourier transform NMR spectrometer. *(Adapted from J. W. Akitt,* NMR and Chemistry, *2nd ed., p. 14. London: Chapman and Hall, 1983. With permission.)*

Spectrometer magnets of three types have in the past been used in NMR spectrometers: permanent magnets, conventional electromagnets, and superconducting solenoids. At present, conventional electromagnets are seldom incorporated into NMR instruments. Permanent magnets with field strengths of 0.7, 1.4, and 2.1 T have been used in the past in commercial continuous-wave instruments; corresponding oscillator frequencies for proton studies are 30, 60, and 90 MHz. Permanent magnets are highly temperature-sensitive and require extensive thermostating and shielding as a consequence. Because of field drift problems, permanent magnets are not ideal for extended periods of data accumulation, as are often employed in Fourier transform experiments.

Superconducting magnets are used in most modern high-resolution instruments. These magnets attain fields as large as 21 T, corresponding to a proton frequency of 900 MHz. In order to superconduct, the solenoid, wound from superconducting niobium/tin or niobium/titanium wire, is bathed in liquid helium at a temperature of 4 K. The helium dewar is held in an outer liquid nitrogen dewar. Most superconducting magnet systems must be filled with liquid nitrogen every 10 days and with liquid helium every 80 to 130 days. The advantages of superconducting solenoids in addition to their high field strengths are their high stability, low operating cost, and their simplicity and small size compared with an electromagnet.

The performance specifications for a spectrometer magnet are stringent. The field produced must be homogeneous to a few parts per billion within the sample area and must be stable to a similar degree for the length of time required to acquire the sample data. Unfortunately, the inherent stability of most magnets is considerably lower than this figure, with variations as large as one part in 10^7 being observed over a period of 1 hr. Several measures are employed in modern NMR instruments to compensate for both drift and field inhomogeneity.

Locking the Magnetic Field

In order to offset the effect of field fluctuations, a *field/frequency lock system* is employed in commercial NMR instruments. In these systems, a reference nucleus is continuously irradiated and monitored at a frequency corresponding to its resonance maximum at the rated field strength of the magnet. Changes in the intensity of the reference absorption signal control a feedback circuit, whose output is fed into coils in the magnetic gap in such a way as to correct for the drift. Recall that for a given type of nucleus, the ratio between the field strength and resonance frequencies is a constant *regardless of the nucleus involved* (Equation 19-5). Thus, the drift correction for the reference signal is applicable to the signals for all nuclei in the sample area. In modern spectrometers, the reference signal is provided by the deuterium in the solvent, and a second transmitter coil tuned to the frequency for deuterium monitors the reference. Most modern superconducting magnets are sufficiently stable so that spectra can be obtained in an unlocked mode for a period of 1 to 20 min.

Shimming

Shim coils are pairs of wire loops through which carefully controlled currents are passed, producing small magnetic fields that compensate for inhomogeneities in the primary magnetic field. In older instruments several potentiometers were used to manually adjust the currents in the various pairs of coils. Because the coils interact, shimming was a tedious and sometimes frustrating experience. In contemporary instruments, the shim controls are under computer control with several algorithms available for optimizing field homogeneity. Ordinarily, shimming must be carried out each time a new sample is introduced into the spectrometer. Shim coils are not shown in the simplified diagram in Figure 19-21.

Sample Spinning

The effects of field inhomogeneities are also counteracted by spinning the sample along its longitudinal axis. Spinning is accomplished by means of a small plastic turbine that slips over the sample tube. A stream of air drives the turbine at a rate of 20 to 50 revolutions per second. If this frequency is much greater than the frequency spread caused by magnetic inhomogeneities, the nuclei experience an averaged environment that causes apparent frequency dispersions to collapse toward zero. A minor disadvantage of spinning is that the magnetic field is modulated at the spinning frequency, which may lead to *sidebands*, or *spinning sidebands*, on each side of absorption peaks.

11:25 am May 9.05

19C-3 The Sample Probe

A key component of an NMR spectrometer is the sample probe, which serves multiple functions. It holds the sample in a fixed position in the magnetic field, contains an air turbine to spin the sample, and houses the coil or coils that permit excitation and detection of the NMR signal. In addition, the probe ordinarily contains two other transmitter coils, one for locking and the other for decoupling experiments that are discussed in Section 19E-1. Finally, most probes have variable temperature capability. The usual NMR sample cell consists of a 5-mm o.d. glass tube containing about 0.4 mL of liquid. Microtubes for smaller sample volumes and larger tubes for special needs are also available.

Transmitter/Receiver Coils

Early NMR instruments contained separate coils mounted at right angles to one another for producing the excitation pulse and detecting the generated NMR signal. These *crossed-coil detectors* have been largely supplanted by the single-coil probe as shown in Figure 19-21 because this design is simpler and more efficient than the crossed-coil device.

The Pulse Generator. Radio-frequency generators and frequency synthesizers produce a signal of essentially a single frequency. To generate Fourier transform spectra, however, the sample must be irradiated with a range of frequencies sufficiently great to excite nuclei with different resonance frequencies. Fortunately, a sufficiently short pulse of radiation, such as that shown in Figure 19-5, provides a relatively broad band of frequencies centered on the frequency of the oscillator. The frequency range of this band is about $1/(4\,\tau)$ Hz, where τ is the length in seconds of each pulse. Thus, a 1-μs pulse from a 100-MHz oscillator produces a frequency range of 100 MHz ± 125 kHz. This production of a band of frequencies from a narrow pulse can be understood by reference to Figure 6-6 (page 122), in which it is shown that a rectangular waveform is synthesized from a series of sine or cosine waves differing from one another by small frequency increments. Conversely, Fourier analysis of a square waveform reveals that it consists of a broad range of frequency components. In fact, the narrower the square wave is, the broader is the range of component frequencies. If the waveform could be infinitesimally narrow, it would in principle consist of *all* frequencies. Thus, a narrow pulse generated by rapidly switching a radio-frequency oscillator from off to on and then on to off will consist of a band of frequencies capable of exciting all nuclei of interest whose resonances occur in the vicinity of the frequency of the oscillator.

As shown in Figure 19-21, the typical pulse generator consists of three parts: a frequency synthesizer, a gate to switch the pulse on and off at appropriate times, and a power amplifier to amplify the pulse to about 50 to 100 W.

The Receiver System. Voltages generated by the current in the detector coil are in the nanovolt to microvolt range and thus must be amplified to a range of about 0 to 10 V before the signal can be further processed and digitized. The first stage of amplification generally takes place in a preamplifier, which is mounted in the probe so that it is as close to the receiver coil as possible to minimize the effects of noise from other parts of the instrument. Further amplification then is carried out in an external RF amplifier as shown in Figure 19-21.

19C-4 The Detector and Data-Processing System

In the detector system shown in Figure 19-21, the high-frequency radio signal is first converted to an audio-frequency signal, which is much easier to digitize. The signal from the radio-frequency amplifier can be thought of as being made up of two components: a *carrier signal*, which has the frequency of the oscillator that is used to produce it, and a superimposed NMR signal from the analyte. The analyte signal differs in frequency from that of the carrier by a few parts per million. For example, the chemical shifts in a proton spectrum typically encompass a range of 10 ppm. Thus, the proton NMR data generated by a 200-MHz spectrometer would lie in the frequency range of 200,000,000 to 200,002,000 Hz. To digitize such high-frequency signals and extract the small differences attributed to the sample in the digital domain is not practical; therefore, the carrier frequency ν_c is subtracted electronically from the analyte signal frequency ν_n in the analog domain. In our example, this process yields to a difference signal $(\nu_n - \nu_c)$ that lies in the audio-frequency range of 0 to 2000 Hz. This process is identical to the way in which audio signals are separated from a radio-frequency carrier signal in home radios.

Sampling the Audio Signal

The sinusoidal audio signal obtained after subtracting the carrier frequency is then digitized by sampling the signal voltage periodically and then converting the voltage to a digital form with an analog-to-digital converter. In order to accurately represent a sine or a cosine wave digitally, it is necessary, according to the *Nyquist theorem,* to sample the signal at least twice during each cycle. If sampling is done at a frequency that is less than twice the signal frequency, *folding* or *aliasing* of the signal occurs. The effect of folding is illustrated in Figure 19-22a, which shows, as the dashed curve, a 1600-

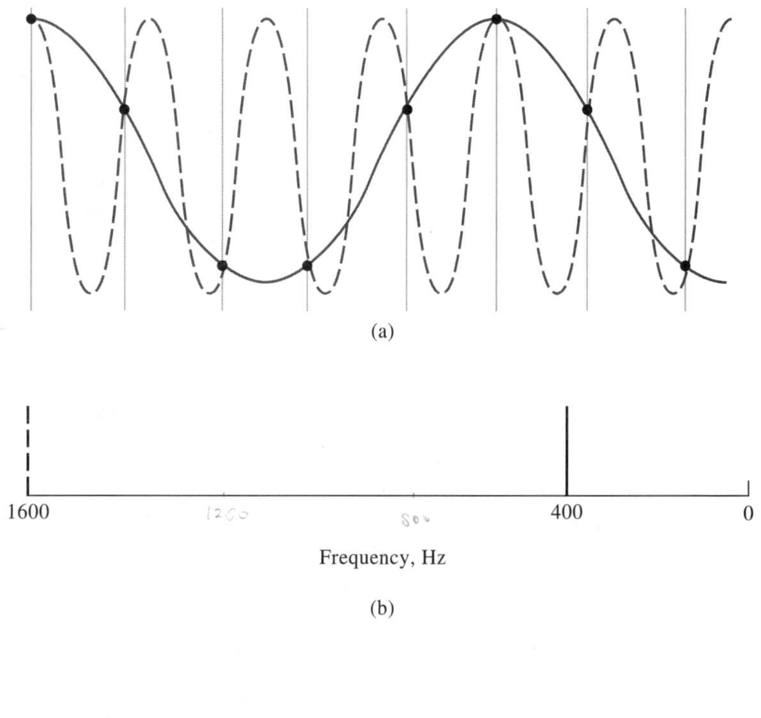

Figure 19-22 Folding of a spectral line brought about by sampling a 1600-Hz signal, which is the dashed curve in (a), at a frequency that is less than the Nyquist frequency of 3200 Hz and that is sampled at a frequency of 2000 samples/second as shown by the dots; the solid line is the apparent signal, which is a cosine wave having a frequency of 400 Hz. (b) frequency domain spectrum of the undersampled dashed signal in (a) showing the resulting folded line at 400 Hz. *(Adapted from D. Shaw,* Fourier Transform NMR Spectroscopy, *2nd ed., p. 159. New York: Elsevier, 1987. With permission, Elsevier Science Publishers.)*

Hz cosine signal that is sampled at a rate of 2000 data per second. The solid dots represent the times at which the computer sampled and digitized the data. This sampling rate is less than the Nyquist frequency, which is 2×1600 Hz $= 3200$ Hz. The effect of this inadequate sampling rate is demonstrated by the solid curve in Figure 19-22a, which is a 400-Hz cosine wave. Thus, as shown in Figure 19-23b, the line at 1600 Hz is absent and the apparent signal is a folded line that appears at 400 Hz in the frequency-domain spectrum.

Modern NMR spectrometers use quadrature phase-sensitive detectors that make possible the determination of positive and negative differences between the carrier frequency and NMR frequencies. Thus, as shown in Figure 19-23c, the carrier signal is set at a frequency that lies in the middle of a spectrum. In this way, folding is avoided because quadrature detectors are able to sense the sign of the frequency difference.

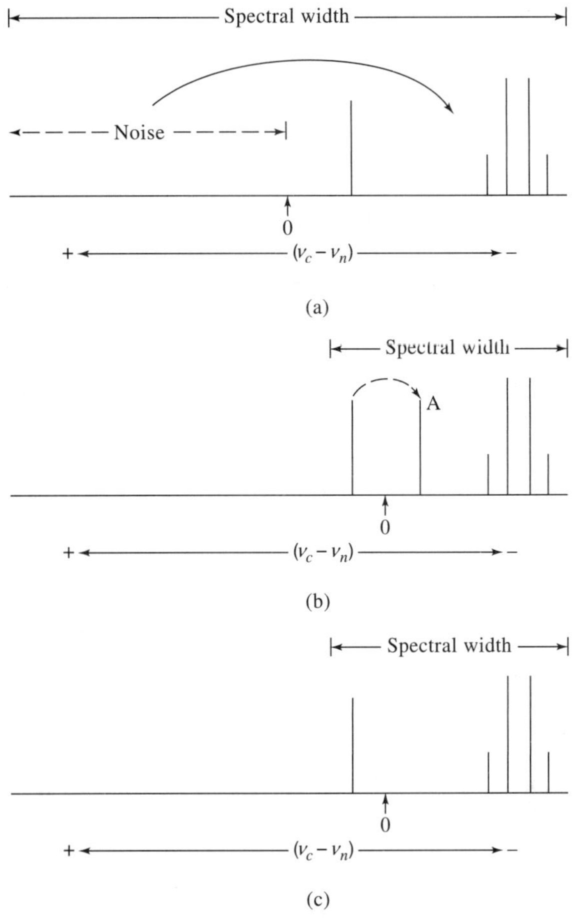

Figure 19-23 (a) Single-detector spectrum with carrier frequency ν_n located at one end of the spectrum; (b) Single-detector spectrum with carrier frequency in center of the spectrum; higher frequency line is aliased and appears at lower frequency. (c) Quadrature-detector spectrum with carrier frequency located in the center of spectrum.

Quadrature Detection Systems

Although the theory of quadrature detection is beyond the scope of this text, we shall briefly discuss the functions and interactions of the basic components of a quadrature phase-sensitive detector as shown in the block diagram in Figure 19-24. In such a system, the incoming NMR signal is split and fed into two identical detectors. In one detector, the signal is treated in the same way as in a single-channel detector, with the NMR signal being subtracted from the carrier frequency to give an audio-frequency signal for digitization. In the second detector, the phase of the carrier signal is shifted by 90 deg before subtraction. As a result, the two audio signals are identical in all respects except phase, so that the signal without phase change is a cosine signal, while the phase-changed signal is a sine wave. These two signals are then digitized, transformed separately into frequency-domain signals in the usual way, and then combined to produce the spectrum. Parts *A* and *B* of Figure 19–25 show the transformed outputs from each channel of the detector in Figure 19-24 when the input signal is from a single NMR line. Each output is made up of two peaks because the frequency-domain input signal makes no distinction between positive and negative frequency differences. Note, however, that when the input is a cosine signal, the signs of the two peaks are the same, whereas for a sine input, they are opposite in sign. Therefore, when the two signals are added, the folded signal disappears leaving only the correct peak depicted in Figure 19-25C.

Signal Integrators

All modern NMR recorders are equipped with electronic or digital integrators to provide areas under absorption peaks. Usually, the integral data appear as step functions superimposed on the NMR spectrum as illustrated in Figure 19-26. Generally, the area data are reproducible to a few percent relative; but with careful selection of experimental variables and data acquisition parameters, integration precision of 1% to 2% is possible.

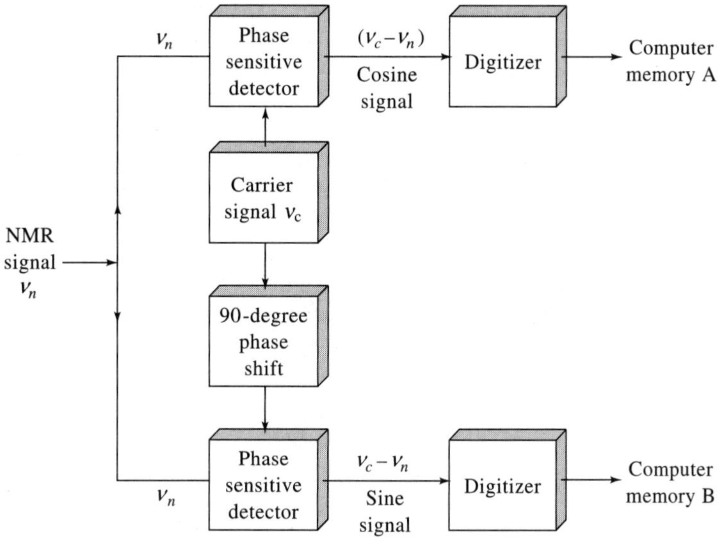

Figure 19-24 Block diagram of a quadrature detector system.

19C-5 Sample Handling

Until recently, high-resolution NMR studies have been restricted to samples that could be converted to a nonviscous liquid state. Most often, a 2% to 15% solution of the sample is used, although pure liquid samples can also be examined if their viscosities are sufficiently low.

The best solvents for proton NMR spectroscopy contain no protons; from this standpoint, carbon tetrachloride is ideal. The low solubility of many compounds in carbon tetrachloride limits its general utility

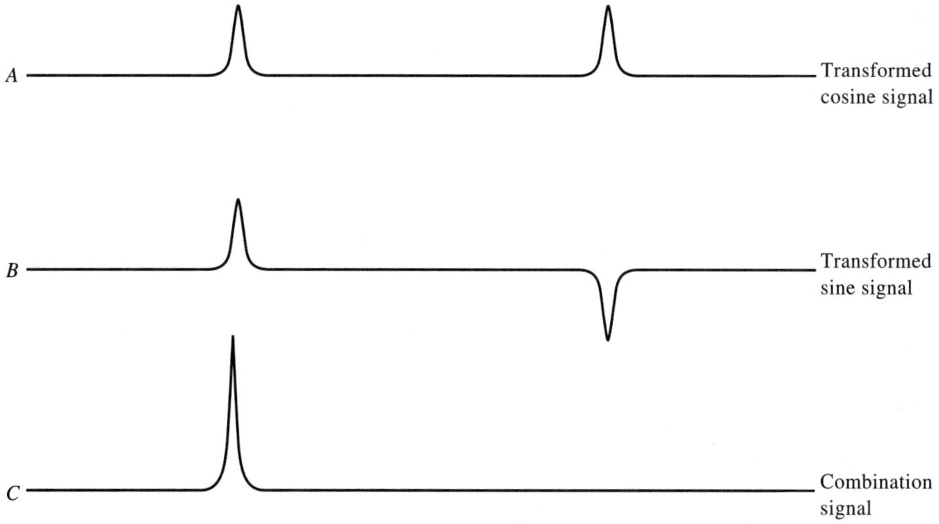

Figure 19-25 Fourier transform of (*A*) a cosine and (*B*) a sine time-domain signal from the quadrature detector shown in Figure 19-24. (*C*) Combination of *A* and *B* giving an NMR spectrum. *(Adapted from A. E. Derome,* Modern NMR Techniques for Chemistry Research, *p. 79. Oxford: Pergamon Press, 1987. With permission.)*

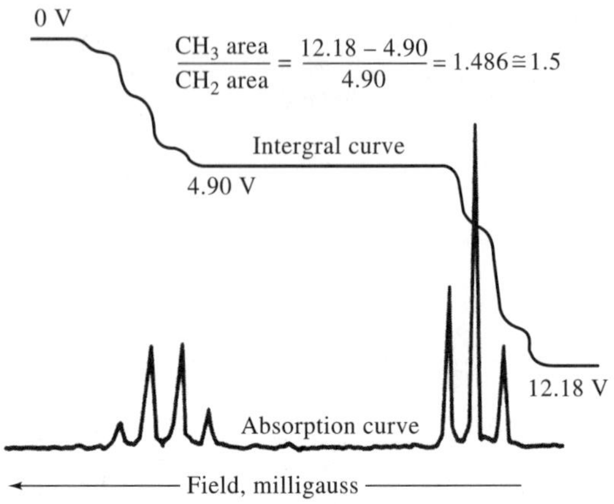

Figure 19-26 Absorption and integral curve for a dilute ethylbenzene solution (aliphatic region). *(Courtesy of Varian Instrument Division. Palo, Alto, CA.)*

as a solvent for NMR experiments, however, and a variety of deuterated solvents are used instead. Deuterated chloroform ($CDCl_3$) and deuterated benzene (C_6D_6) are common solvents for this purpose.

It is now possible to collect high-resolution spectra for solid samples routinely. Techniques have been developed and are being applied in increasing numbers to obtain ^{13}C spectra of polymers, fossil fuels, and other high-molecular-weight substances. There is a brief discussion of the modifications necessary to produce useful solid NMR spectra in Section 19E.

19D APPLICATIONS OF PROTON NMR

Unquestionably, the most important chemical applications of proton NMR spectroscopy have been to the identification and structural elucidation of organic, metal-organic, and biochemical molecules. In addition, however, the method is often useful for quantitative determination of absorbing species.

19D-1 Identification of Compounds

An NMR spectrum, like an infrared spectrum, seldom suffices by itself for the identification of an organic compound. However, in conjunction with other observations such as mass, infrared, and ultraviolet spectra, as well as elemental analysis, NMR is a powerful and indispensable tool for the characterization of pure compounds. The examples that follow give some idea of the kinds of information that can be extracted from NMR spectra.

19D-2 Application of NMR to Quantitative Analysis[14]

A unique aspect of NMR spectra is the direct proportionality between peak areas and the number of nuclei responsible for the peak. As a consequence, a quantitative determination of a specific compound does not require pure samples for calibration. Thus, if an identifiable peak for one of the constituents of a sample does not overlap the peaks of the other constituents, the area of this peak can be used to establish the concentration of the species directly, provided only that the signal area per proton is known. This latter quantity can be obtained conveniently from a known concentration of an internal standard. For example, if the solvent present in a known amount were benzene, cyclohexane, or water, the areas of the single-proton peak for these compounds could be used to give the desired information. Of course, the peak of the internal standard should not overlap with any of the sample peaks. Organic silicon derivatives are uniquely attractive for calibration purposes because of the high upfield location of their proton peaks.

The widespread use of NMR spectroscopy for quantitative work has been inhibited by the cost of the instruments. In addition, the probability that resonance peaks will overlap becomes greater as the complexity of the sample increases. Often, too, analyses that are possible by the NMR method can be accomplished just as conveniently by other techniques.

Analysis of Multicomponent Mixtures

Methods for the analysis of many multicomponent mixtures have been reported. For example, Hollis[15] has described a method for the determination of aspirin, phenacetin, and caffeine in commercial analgesic preparations. The procedure requires about 20 min, and the relative errors are in the range of 1% to 3%. Chamberlain[16] describes a procedure for the rapid determination of benzene, heptane, ethylene glycol, and water in mixtures. A wide range of these mixtures was analyzed with a precision of 0.5%.

---

[14] For a recent treatment of quantitative proton and ^{13}C NMR, see *Analytical NMR*, L. D. Fields and S. Sternhell, Eds. New York: Wiley, 1989. For a description of the use of NMR in trace analysis, see D. L. Rabenstein and T. T. Nakashima, in *Trace Analysis*, G. D. Christian and J. B. Callis, Eds., Chapter 4. New York: Wiley, 1986.

[15] D. P. Hollis, *Anal. Chem.*, **1963**, *35*, 1682.

[16] N. F. Chamberlain, in *Treatise on Analytical Chemistry*, I. M. Kolthoff and P. J. Elving, Eds., Part I, Vol. 4, p. 1932. New York: Interscience, 1963.

EXAMPLE 19-4

The NMR spectrum shown in Figure 19-27 is for an organic compound having the empirical formula $C_5H_{10}O_2$. Identify the compound.

The spectrum suggests the presence of four types of protons. From the integration of the peaks and the empirical formula, we deduce that these four types are populated by 3, 2, 2, and 3 protons, respectively. The single peak at $\delta = 3.6$ must be due to an isolated methyl group. Inspection of Figure 19-17 and Table 19-3 suggests that the peak may result from the functional group $CH_3OC(=O)$—. The empirical formula and the 2:2:3 distribution of the remaining protons indicate the presence of an *n*-propyl group as well. The structure $CH_3OC(=O)CH_2CH_2CH_3$ is consistent with all of these observations. In addition, the positions and the splitting patterns of the three remaining peaks are entirely compatible with this hypothesis. The triplet at $\delta = 0.9$ is typical of a methyl group adjacent to a methylene. From Table 19-3, the two protons of the methylene adjacent to the carboxylate peak should yield the observed triplet peak at about $\delta = 2.2$. The other methylene group would be expected to produce a pattern of $3 \times 4 = 12$ peaks at about $\delta = 1.7$. Only six are observed, presumably because the resolution of the instrument is insufficient to produce the fine structure of the band.

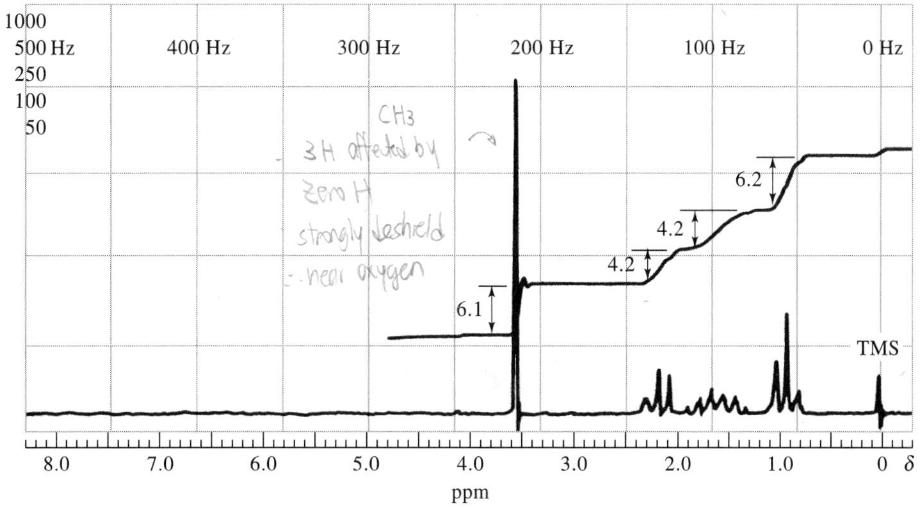

Figure 19-27 NMR spectrum and peak integral curve for the organic compound $C_5H_{10}O_2$ in CCl_4. *(From R. M. Silverstein, G. C. Bassler, and T. C. Morrill,* Spectrometric Identification of Organic Compounds, *3rd ed., p. 296. New York: Wiley, 1974. Reprinted by permission of John Wiley & Sons, Inc.)*

Quantitative Organic Functional Group Analysis

One of the useful applications of NMR has been in the determination of functional groups, such as hydroxyl groups in alcohols and phenols, aldehydes, carboxylic acids, olefins, acetylenic hydrogens, amines, and amides.[17] Relative errors in the 1% to 5% range are reported.

Elemental Analysis

NMR spectroscopy can be used to determine the total concentration of a given NMR-active nucleus in a sample. For example, Jungnickel and Forbes[18] have investigated the integrated NMR intensities of the proton peaks for numerous organic compounds and have concluded that it is possible to make accurate quantitative determinations of total hydrogen in organic mixtures.

[17] See R. H. Cox and D. E. Leyden, in *Treatise on Analytical Chemistry,* 2nd ed., P. J. Elving, M. M. Bursey, and I. M. Kolthoff, Eds., Part I, Vol. 10, pp. 127–136. New York: Wiley, 1983.

[18] J. L. Jungnickel and J. W. Forbes, *Anal. Chem.,* **1963,** *35,* 938.

EXAMPLE 19-5

The spectra shown in Figure 19-28 are for colorless, isomeric, liquids containing only carbon and hydrogen. Identify the two compounds.

The single peak at about $\delta = 7.2$ in the upper figure suggests an aromatic structure; the relative area of this peak corresponds to 5 protons. This information suggests that we may have a monosubstituted derivative of benzene. The seven peaks for the single proton appearing at $\delta = 2.9$ and the six-proton doublet at $\delta = 1.2$ can only be explained by the structure

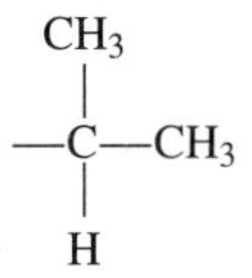

Thus, we conclude that this compound is cumene.

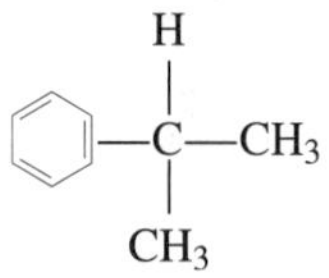

The isomeric compound has an aromatic peak at $\delta = 6.8$. Its relative area suggests a trisubstituted benzene, which can only mean that the compound is $C_6H_3(CH_3)_3$. The relative peak areas confirm this diagnosis.

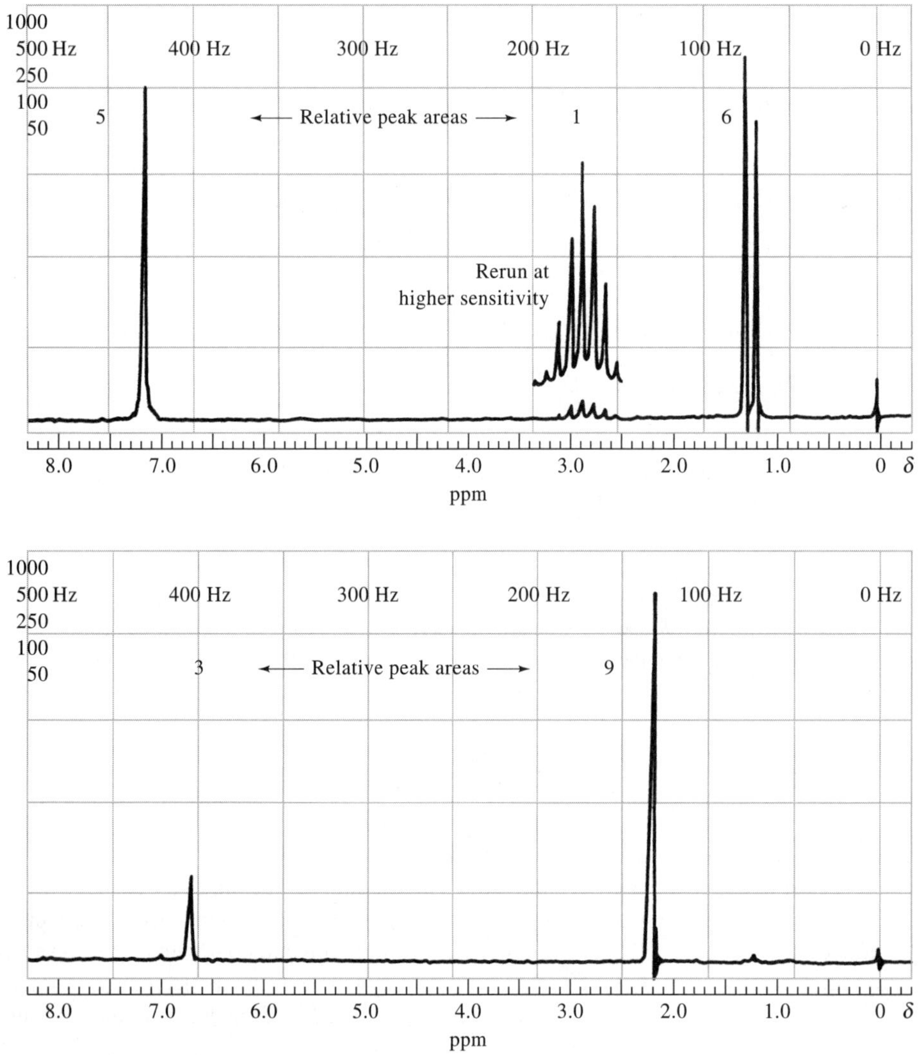

Figure 19-28 NMR spectra for two organic isomers in $CDCl_3$. *(Courtesy of Varian Instrument Division, Palo Alto, CA.)*

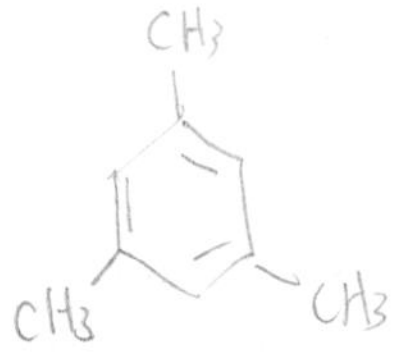

EXAMPLE 19-6

The spectrum shown in Figure 19-29 is for an organic compound having a molecular weight of 72 and containing carbon, hydrogen, and oxygen only. Identify the compound.

The triplet peak at $\delta = 9.8$ appears (Figure 19-17) to be that of an aliphatic aldehyde, RCHO. If this hypothesis is valid, R has a molecular weight of 43, which corresponds to a C_3H_7 fragment. The triplet at $\delta = 9.8$ requires that there be a methylene group adjacent to the carbonyl. Thus, the compound appears to be *n*-butyraldehyde, $CH_3CH_2CH_2CHO$.

The triplet peak at $\delta = 0.97$ appears to be that of the terminal methyl. We expect the protons on the adjacent methylene to show a complicated splitting pattern of $4 \times 3 = 12$ peaks, and the grouping of peaks around $\delta = 1.7$ is compatible with this prediction.

Finally, the peak for the protons on the methylene group adjacent to the carbonyl should appear as a sextet downfield from the other methylene proton peaks. The group at $\delta = 2.4$ is consistent with our prediction.

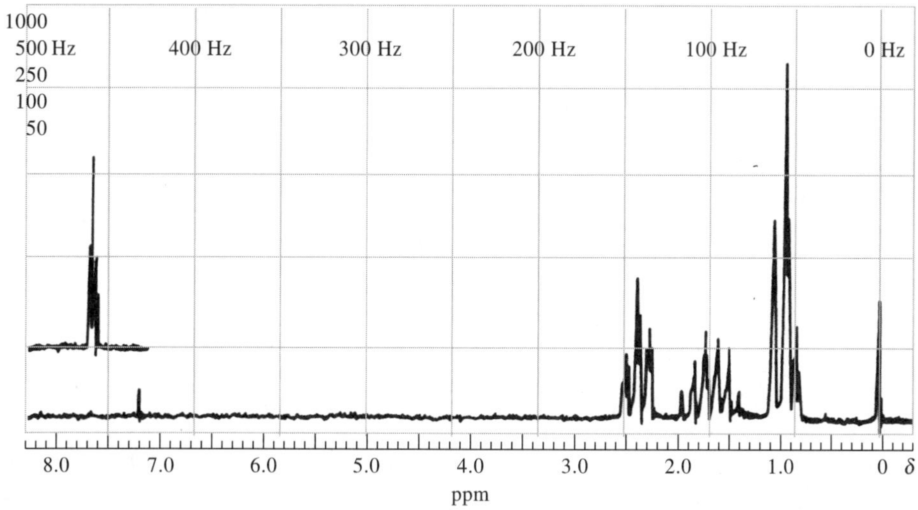

Figure 19-29 NMR spectrum of a pure organic compound containing C, H, and O only. *(Courtesy of Varian Instrument Division, Palo Alto, CA.)*

Paulsen and Cooke[19] have shown that the resonance of fluorine-19 can be used for the quantitative analysis of that element in an organic compound—an analysis that is difficult to carry out by classical methods. For quantitative work, a low-resolution or wide-line spectrometer may be used.

19E CARBON-13 NMR

Carbon-13 NMR was first studied in 1957, but its widespread use did not begin until the early 1970s. The reason for this delay was the time required for the development of instruments sensitive enough to detect the weak NMR signals from the ^{13}C nucleus. This low signal strength results from the low natural isotopic abundance of the isotope (1.1%) and the small magnetogyric ratio, which is about 0.25 that of the proton. These factors combine to make ^{13}C NMR about 6000 times less sensitive than proton NMR.

The most important developments in NMR signal enhancement that have led directly to the explosive growth of ^{13}C magnetic resonance spectroscopy include higher field strength magnets and Fourier transform instruments. Without these developments, ^{13}C NMR would be restricted to the study of highly soluble low-

[19] P. J. Paulsen and W. D. Cooke, *Anal. Chem.*, **1964**, *36*, 1721.

molecular-weight solids, neat liquids, and isotopically enriched compounds.[20]

Carbon-13 NMR has several advantages over proton NMR in terms of its power to elucidate organic and biochemical structures. First, ^{13}C NMR provides information about the backbone of molecules rather than about the periphery. In addition, the chemical shift range for ^{13}C for most organic compounds is about 200 ppm, compared with 10 to 15 ppm for the proton. As a result, there is less overlap of peaks in ^{13}C spectra than in proton spectra. For example, it is often possible to observe individual resonance peaks for each carbon atom in compounds ranging in molecular weight from 200 to 400. In addition, homonuclear, spin-spin coupling between carbon atoms is not observed because in natural abundance samples the probability of two ^{13}C atoms occurring adjacent to each other is small. Furthermore, heteronuclear spin coupling between ^{13}C and ^{12}C does not occur because the spin quantum number ^{12}C is zero. Finally, there are a number of excellent methods for decoupling the interaction between ^{13}C atoms and protons. With decoupling, the spectrum for a particular type of carbon generally exhibits only a single line.

19E-1 Proton Decoupling

Three primary types of proton decoupling experiments are used in ^{13}C NMR, *broad-band decoupling, off-resonance decoupling,* and *pulsed,* or *gated, decoupling.*

Broad-Band Decoupling

Broad-band decoupling is a type of heteronuclear decoupling in which spin-spin splitting of ^{13}C lines by ^{1}H nuclei is avoided by irradiating the sample with a broad-band radio-frequency signal that encompasses the entire proton spectral region, while the ^{13}C spectrum is obtained in the usual way. Ordinarily, the proton signal is produced by a second coil located in the sample probe. The effect of broad-band decoupling is demonstrated in Figure 19-30.

[20] For a thorough discussion of carbon-13 NMR spectroscopy, see A. Lombardo and G. C. Levy, in *Treatise on Analytical Chemistry,* 2nd ed., P. J. Elving, M. M. Bursey, and I. M. Kolthoff, Eds., Part I, Vol. 10, Chapter 2. New York: Wiley, 1983; E. Breitmaier and W. Voelters, *Carbon-13 NMR Spectroscopy,* 3rd ed. New York: VCH Publishers, 1987; R. J. Abraham, J. Fisher, and P. Loftus, *Introduction to NMR Spectroscopy.* New York: Wiley, 1988.

Off-Resonance Decoupling

Although broad-band decoupling considerably simplifies most ^{13}C spectra, it also removes spin-spin splitting information that may be of importance in structural assignments. In the past, this limitation was sometimes rectified by substituting off-resonance decoupling.

In this technique, the decoupling frequency is set at 1000 to 2000 Hz above the proton spectral region, which leads to a partially decoupled spectrum in which all but the largest spin-spin shifts are absent. Under this circumstance, primary carbon nuclei (bearing three protons) yield a quartet of peaks, secondary carbons give three peaks, tertiary carbon nuclei appear as doublets, and quaternary carbons exhibit a single peak. Figure 19-31 demonstrates the utility of this technique for identifying the source of peaks in a ^{13}C spectrum.

Pulsed Decoupling

Modern NMR spectrometers provide decoupling information through the application of complex pulsing schemes, which yield higher *S/N* more rapidly than off-resonance decoupling. These techniques are beyond the scope of this chapter.[21]

Nuclear Overhauser Enhancement

Under conditions of broad-band decoupling, it is found that the areas of ^{13}C peaks are enhanced by a factor that is significantly greater than would be expected from the collapse of the multiple structures into single lines. This phenomenon is a manifestation of *nuclear Overhauser enhancement* (NOE), which is a general effect encountered in decoupling experiments. This effect arises from direct magnetic coupling between a decoupled proton and a neighboring ^{13}C nucleus that results in an increase in the population of the lower energy state of the ^{13}C nucleus over that predicted from the Boltzmann relation. The ^{13}C signal is enhanced by as much as a factor of three as a result. While NOE does increase the sensitivity of ^{13}C measurements, it has the disadvantage that the proportionality between peak areas and number of nuclei may be lost. The theory of NOE, which is based upon dipole-dipole interactions, is beyond the scope of this text.[22]

[21] See J. K. M. Sanders and B. K. Hunter, *Modern NMR Spectroscopy: A Guide for Chemists.* Chapter 3. New York: Oxford, 1988.

[22] See D. Shaw, *Fourier Transform Spectroscopy,* 2nd ed., p. 233. New York: Elsevier, 1984.

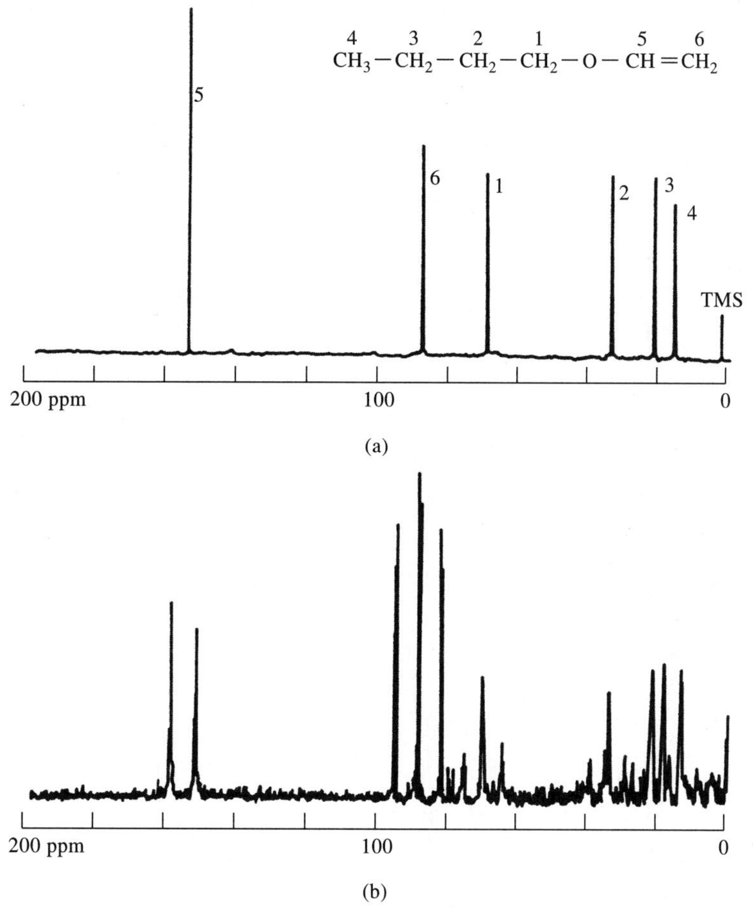

Figure 19-30 Carbon-13 NMR spectra for *n*-butyl vinyl ether obtained at 25.2 MHz: (a) proton decoupled spectrum; (b) spectrum showing effect of coupling between ^{13}C atom and attached protons. *(From R. J. Abraham, J. Fisher, and P. Loftus,* Introduction to NMR Spectroscopy, *p. 104. New York: Wiley, 1988. With permission.)*

19E-2 Application of ^{13}C NMR To Structure Determination

As with proton NMR, the most important and widespread applications of ^{13}C NMR are for the determination of structures of organic and biochemical species. Such determinations are based largely upon chemical shifts, with spin-spin data playing a lesser role than in proton NMR. Figure 19-32 shows some of the chemical shifts that are observed for ^{13}C in various chemical environments. As with proton spectra, these shifts are relative to tetramethylsilane, with δ values ranging from 0 to 200 ppm. In general, environmental effects are analogous to those for the proton, which were described in Section 19B-2. In contrast to proton spectra, however, the effect of substituents on ^{13}C shifts is not limited to the nearest atom. For example, substitution of chlorine on the C_1 carbon in *n*-pentane results in a chemical shift for that carbon of 31 ppm. When chlorine substitution is on the C_2 carbon, the shift for the C_1 carbon is 10 ppm; similarly substitutions on the C_3, C_4, and C_5 carbons result in shifts of 5.3, 0.5, and 0.1 ppm, respectively.

Application of ^{13}C NMR to Solid Samples[23]

As was noted earlier, NMR spectra for solids have in the past not been very useful for structural studies because of

[23] See B. F. Chmelka and A. Pines, *Science,* **1989,** *246,* 71; G. E. Maciel, *Science,* **1984,** *226,* 282; B. C. Gerstein, *Anal. Chem.,* **1983,** *55,* 781A, 899A; M. Mehring, *High Resolution NMR Spectroscopy in Solids,* 2nd ed. New York: Springer-Verlag, 1983.

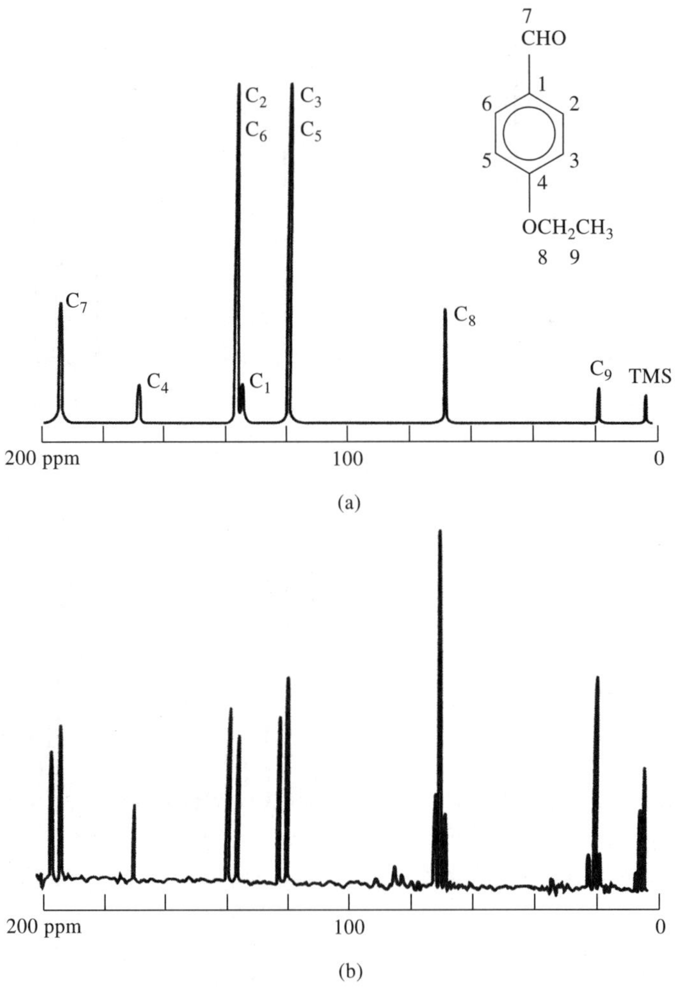

Figure 19-31 Comparison of (a) broad-band and (b) off-resonance decoupling in ^{13}C spectra of *p*-ethoxybenzaldehyde. *(From R. J. Abraham, J. Fisher, and P. Loftus,* Introduction to NMR Spectroscopy, *p. 106. New York: Wiley, 1988. With permission.)*

line broadening, which eliminates or obscures the characteristic sharp individual peaks of NMR. Much of this broadening is attributable to two causes: static dipolar interactions between ^{13}C and ^{1}H and from anisotropy in ^{13}C-shielding tensors. In isotropic liquids, these effects are averaged to zero because of the rapid and random motion of molecules. In solids, heteronuclear dipolar interactions between magnetic nuclei, such as ^{13}C and protons, result in characteristic dipolar line splittings, which depend upon the angle between C—H bonds and the external field. In an amorphous solid, there is a large number of fixed orientations of these bonds, and therefore a large number of splittings can occur. The broad absorption bands in this instance are made up of the numerous lines arising from these individual dipolar interactions. It is possible to remove dipolar splitting from a ^{13}C spectrum by irradiating the sample at proton frequencies while the spectrum is being obtained. This procedure, called *dipolar decoupling,* is similar to spin decoupling, which was described earlier for liquids, except that a much higher power level is required.

A second type of line broadening for solids is caused by chemical shift anisotropy, which was discussed in Section 19B-2. The broadening produced here results from changes in the chemical shift with the orientation of the molecule or part of the molecule with re-

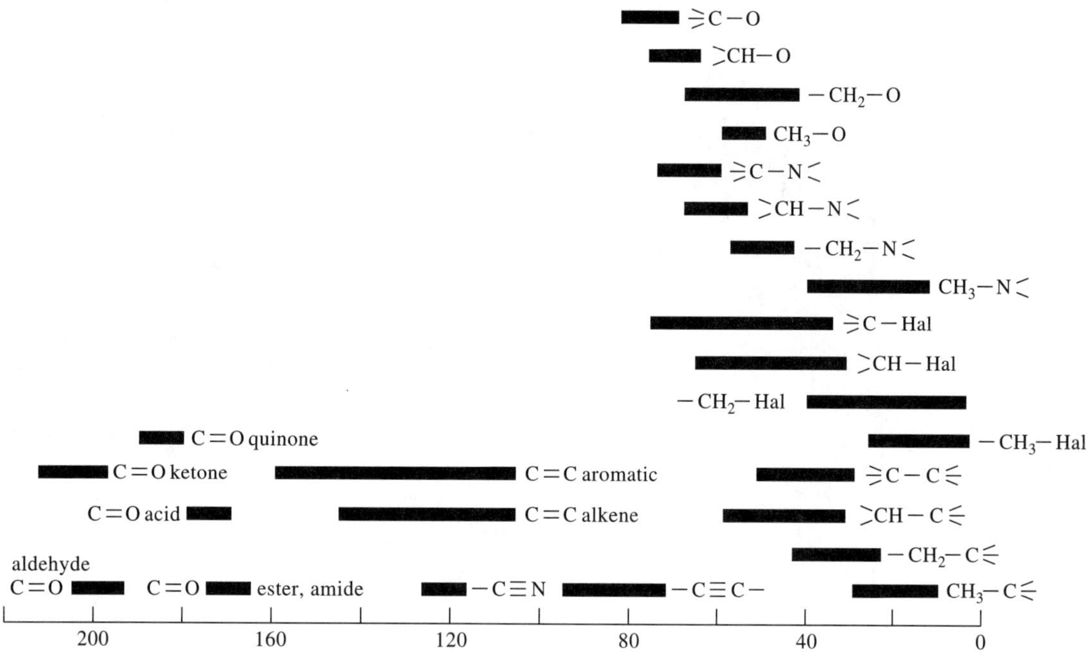

Figure 19-32 Chemical shifts for ^{13}C. *(From D. E. Leyden and R. H. Cox,* Analytical Applications of NMR, *p. 196. New York: Wiley, 1977. Reprinted by permission of John Wiley & Sons, Inc.)*

spect to an external magnetic field. From NMR theory, it is known that the chemical shift $\Delta\delta$ brought about by magnetic anisotropy is given by the equation

$$\Delta\delta = \Delta\chi(3\cos^2\theta - 1)/R^3 \qquad (19\text{-}20)$$

where θ is the angle between the double bond and the applied field, $\Delta\chi$ is the difference in magnetic susceptibilities between the parallel and perpendicular orientation of the double bond, and R is the distance between the anisotropic functional group and the nucleus. When θ is exactly 54.7 deg, $\Delta\delta$, as defined by Equation 19-20, is zero. Experimentally, line broadening due to chemical shift anisotropy is eliminated by *magic angle spinning* (MAS), which involves rotating solid samples rapidly at a frequency greater than 2 kHz in a special sample holder that is maintained at an angle of 57.4 deg with respect to the applied field. In effect, the solid then acts like a liquid being rotated in the field.

One further limitation in ^{13}C Fourier transform NMR of solids is the long spin-lattice relaxation time for excited ^{13}C nuclei. The rate at which the sample can be pulsed is dependent upon the relaxation rate. That is, after each excitation pulse, enough time must elapse for the nuclei to return to the equilibrium ground state. Unfortunately, spin-lattice relaxation times for ^{13}C nuclei in solids are often several minutes; thus, signal averaging sufficient to give a good spectrum would often require several hours or even days.

The problem caused by slow spin-lattice relaxation times is overcome by *cross polarization,* a complicated pulsed technique that causes the Larmor frequencies of proton nuclei and ^{13}C nuclei to become identical—that is, $\gamma_C B_{1C} = \gamma_H B_{1H}$. Under these conditions, the magnetic fields of the precessing proton nuclei interact with the fields of the ^{13}C nuclei, causing the latter to relax.

Instruments are now available commercially that incorporate dipolar decoupling, magic angle spinning, and cross polarization, thus making possible the acquisition of high-resolution ^{13}C spectra from solids. Figure 19-33, which shows spectra for crystalline adamantane collected under various conditions, illustrates the power of these instruments.

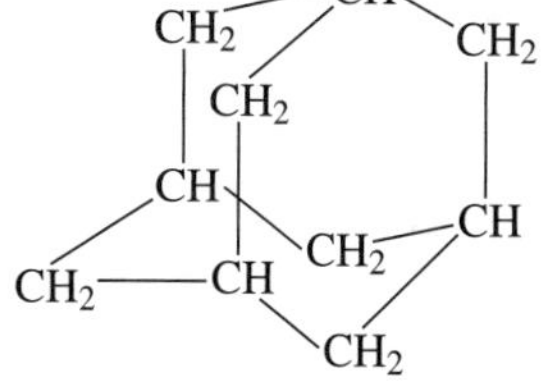

adamantane

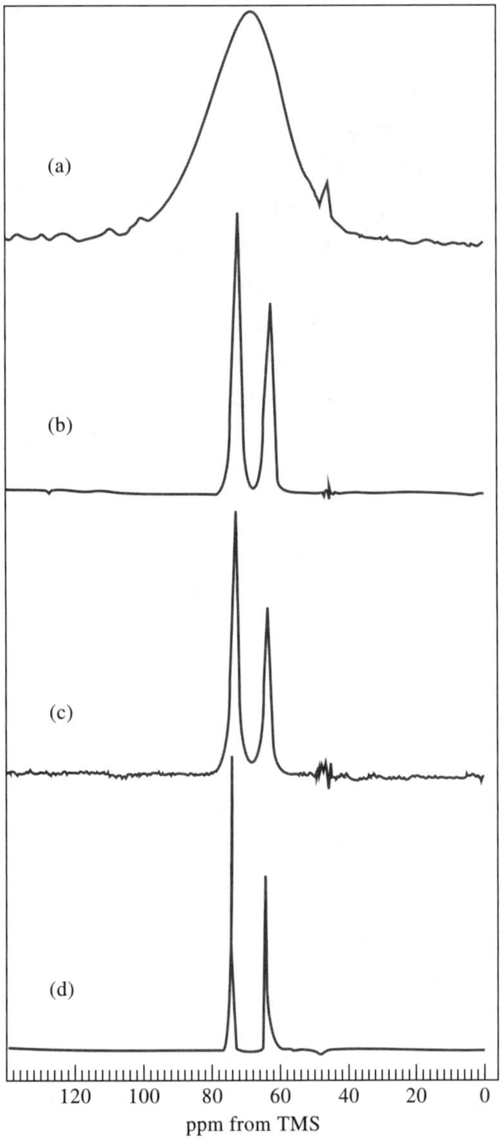

Figure 19-33 Carbon-13 spectra of crystalline adamantane: (a) nonspinning and with no proton decoupling; (b) nonspinning but with dipolar decoupling and cross polarization; (c) with magic angle spinning but without dipolar decoupling or cross polarization; (d) with magic angle spinning, dipolar decoupling, and cross polarization. *(From F. A. Bovey,* Nuclear Magnetic Resonance Spectroscopy, *2nd ed., p. 415. New York: Academic Press, 1988. With permission.)*

19F APPLICATION OF NMR TO OTHER NUCLEI

More than 200 isotopes have magnetic moments and thus, in principle, can be studied by NMR. Among the most widely studied nuclei are ^{31}P, ^{15}N, ^{19}F, ^{2}D, ^{11}B, ^{23}Na, ^{15}N, ^{29}Si, ^{109}Ag, ^{199}Hg, ^{113}Cd, and ^{207}Pb. The first three of these are particularly important in the fields of organic chemistry, biochemistry, and biology.

19F-1 Phosphorus-31

Phosphorus-31, with spin number 1/2, exhibits sharp NMR peaks with chemical shifts extending over a range of 700 ppm. The resonance frequency of ^{31}P at 4.7 T is 81.0 MHz. Numerous investigations, particularly in the biochemical field, have been based upon ^{31}P resonance. An example is shown in Figure 19-34. The species under study is adenosine triphosphate (ATP), a triply charged anion that plays a vital role in carbohydrate metabolism and in energy storage and release in the body.

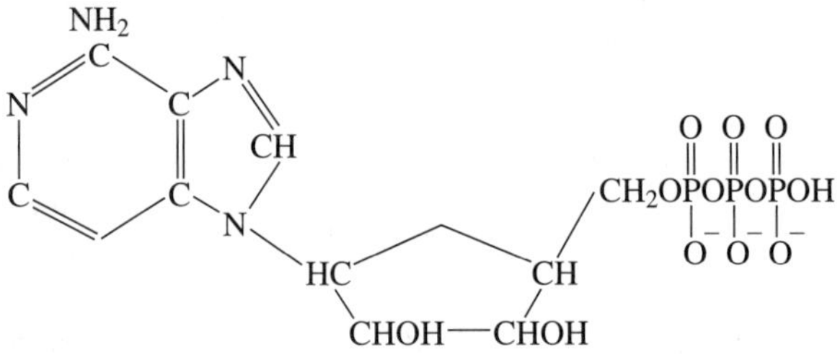

adenosine triphosphate (ATP)

The bottom spectrum, which is for ATP in an aqueous environment, is made up of three sets of peaks corresponding to the three phosphorus atoms. The triplet undoubtedly arises from the central phosphorus, which is coupled to the other two phosphorus atoms. The doublet at about 14 ppm shows some poorly defined indications of proton coupling and thus probably arises from the phosphorus that is adjacent to the methylene group.

Magnesium ion is known to play a part in the metabolic role of ATP, and the upper six spectra in Figure 19-34 suggest that complex formation between the anionic phosphorus and the cation takes place to cause the phosphorus chemical shifts to move downfield as the magnesium ion concentration is increased.

19F-2 Fluorine-19

Fluorine-19 has a spin quantum number of 1/2 and a magnetogyric ratio close to that of ^{1}H. Thus, the resonance frequency of fluorine at 188 MHz is only slightly

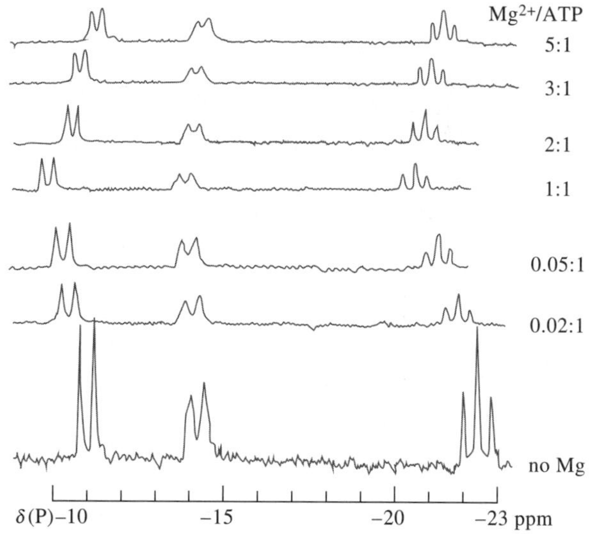

Figure 19-34 Fourier transform phosphorus-31 NMR spectra for ATP solution containing magnesium ions. The ratios on the right are moles of Mg^{2+}/moles ATP. *(From J. W. Akitt,* NMR and Chemistry, *2nd ed., p. 245. London: Chapman and Hall, 1983. With permission.)*

lower than that of the proton at 200 MHz when both are measured at 4.69 T.

Fluorine absorption is also sensitive to the environment, and the resulting chemical shifts extend over a range of about 300 ppm. In addition, the solvent plays a much more important role in determining fluorine peak positions than with proton peaks.

Empirical correlations of the fluorine shift with structure are relatively sparse when compared with information concerning proton behavior. It seems probable, however, that the future will see further developments in this field, particularly for structural investigation of organic fluorine compounds.

Proton, ^{19}F, and ^{31}P spectra for the inorganic species PHF_2 are shown in Figure 19-35. In each spectrum, spin-spin splitting assignments can easily be made based upon the discussion in Section 19B-3. Using a modern multinuclear instrument, three of these spectra could be obtained at a single magnetic field strength.

19G TWO-DIMENSIONAL FOURIER TRANSFORM NMR

Two-dimensional NMR, or 2D NMR, comprises a relatively new set of multipulse techniques that make it possible to unravel complex spectra.[24] In two-dimen-

[24] For brief reviews of 2D NMR, see K. R. Williams and R. W. King, *Anal. Chem.*, **1990**, *67*, A125; A. Bax and L. Lerner, *Science*, **1986**, *232*, 960. For a detailed treatment, see R. R. Ernst, G. Bodenhausen, and A. Wokaun, *Principles of Nuclear Magnetic Resonance in One and Two Dimensions*. Oxford: Oxford University Press, 1987.

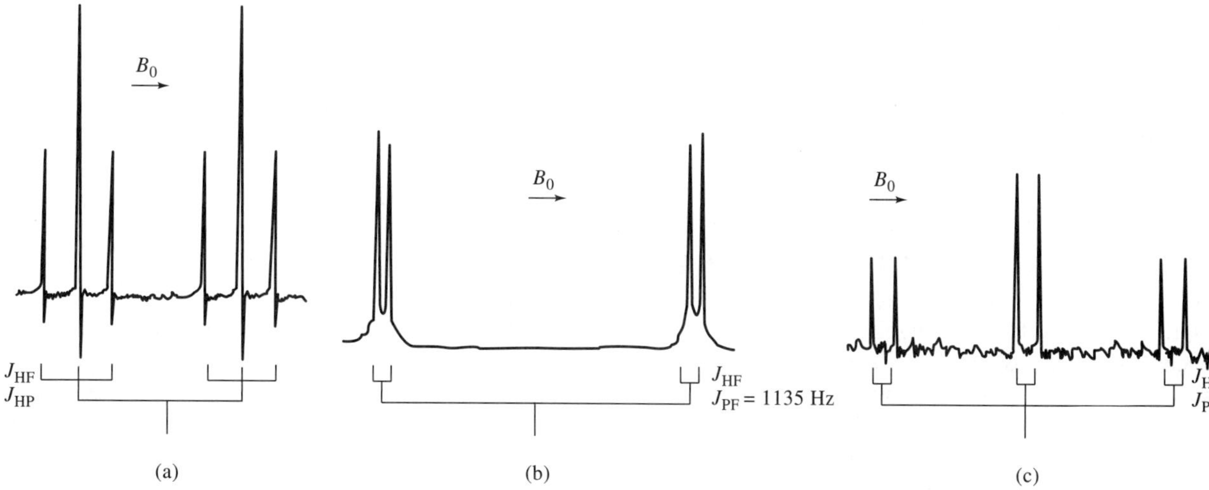

Figure 19-35 Spectra of liquid PHF_2 at −20°C: (a) ^1H spectrum at 60 MHz; (b) ^{19}F spectrum at 94.1 MHz; (c) ^{31}P spectrum at 40.4 MHz. *(From R. J. Myers,* Molecular Magnetism and Molecular Resonance Spectroscopy. *Englewood Cliffs, NJ: Prentice-Hall, 1973. With permission.)*

sional methods, data are acquired as a function of time t_2 just as in ordinary Fourier transform NMR. Prior to obtaining this FID signal, however, the system is perturbed by a pulse for a period t_1. Fourier transformation of the FID as a function of t_2 for a fixed t_1 yields a spectrum similar to that obtained in an ordinary pulse experiment. This process is then repeated for various values of t_1, thus giving a two-dimensional spectrum in terms of two frequency variables ν_1 and ν_2 or sometimes the chemical shift parameters δ_1 and δ_2. The nature and timing of the pulses that have been used in 2D NMR vary widely, and in some cases more than two repetitive pulses are used. Thus, the number of types of two-dimensional experiments that have appeared in the literature is large—perhaps as large as 100 or more. We will describe one such experiment to illustrate the power of the method.

Figure 19-36a is a ^{13}C 2D NMR spectrum for 1,3-butanediol. Figure 19-36b is the ordinary one-dimensional spectrum for the compound. The two-dimensional spectrum is obtained as follows: With the proton broad-band decoupler turned off, a 90-deg pulse is applied to the sample. After a time t_1, the decoupler is turned on and another pulse is applied, and the resulting FID is digitized and transformed. After equilibrium has been reestablished, this process is repeated for other values of t_1, which leads to a series of spectra that are plotted horizontally in the figure. That is, the projections along the δ_1 axis are the spectra that would be obtained without decoupling. The projection along the δ_2 axis is the same as the completely decoupled carbon-13 spectrum. It is obvious that the spectrum is made up of a quartet, two triplets, and a doublet, whose source is apparent from a consideration of the number of protons bonded to each of the four ^{13}C atoms in the molecule. This information is not obvious in the one-dimensional spectrum.

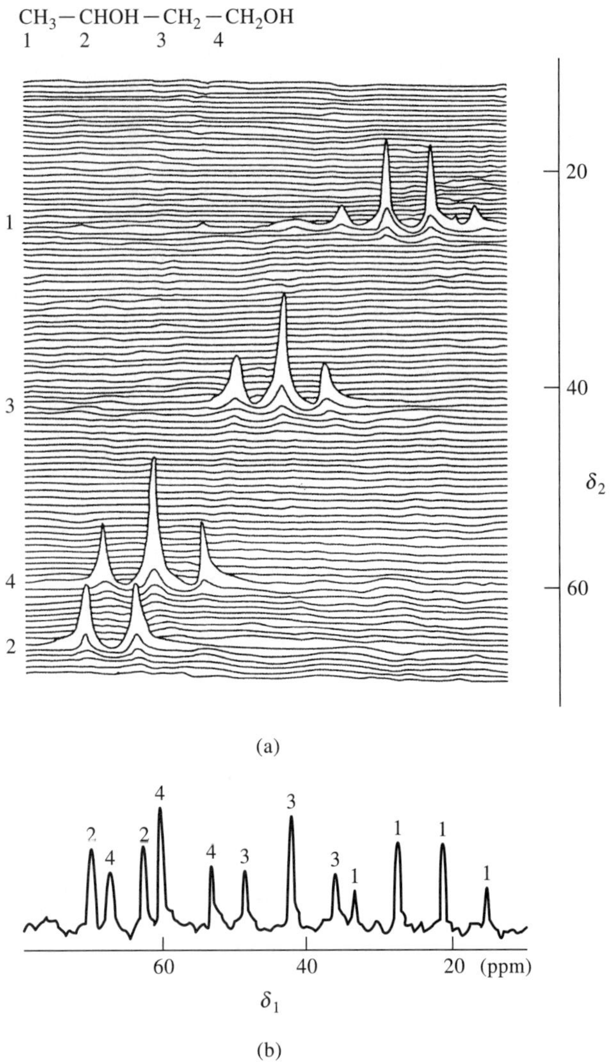

Figure 19-36 Illustration of the use of the two-dimensional spectrum (a) to identify the ^{13}C peaks in a one-dimensional spectrum (b). *(Adapted with permission from S. Borman,* Anal. Chem., ***1982***, 54, *1129A. Copyright 1982 American Chemical Society.)*

19H MAGNETIC RESONANCE IMAGING

During the past two decades, NMR technology has been applied increasingly to fields other than chemistry, such as biology, engineering, industrial quality control, and medicine. One of the most prominent NMR applications is that of magnetic resonance imaging, or MRI. In MRI, data from pulsed RF excitation of solid or semisolid objects are subjected to Fourier transformation and converted to three dimensional images of the interior of the objects. The primary advantage of MRI is that images of objects are formed noninvasively. This means that there is little or no potential of radiation injury or other damage to human or animal subjects as might be encountered with X-ray CAT scans or other similar methods. In the final paragraphs of this chapter, we shall briefly describe the principles of magnetic resonance imaging and discuss some of its applications. For detailed nonmathematical discussions of MRI theory, consult the sources presented in the footnote below.[25]

The fundamental concept of MRI is depicted in Figure 19-37. In these experiments, the magnetic field strength is deliberately varied throughout the subject under study to give a profile such as that shown at the top of the figure. This linear variation, or field gradient, in B is created by auxiliary coils in the magnet bore that are under the control of the computer of the magnetic resonance instrument. Protons in different locations in the subject experience different magnetic field strengths B_1 and B_2 for two of the regions shown in the figure. Equation 19-10, the Larmor equation, suggests that these nuclei exhibit different resonance frequencies, ν_1 and ν_2 in this instance. Thus, for example, if a magnetic field gradient of 1×10^{-5} tesla/cm is applied along the bore, or the z-axis, of an MRI magnet, a resonance frequency range of $(2.68 \times 10^8 \text{ radians s}^{-1}\text{ T}^{-1})(1 \times 10^{-5}\text{ T/cm})/(2\pi \text{ radians}) = 425$ Hz results. In other words, protons 1 cm apart along the field gradient in the subject have resonance frequencies that differ by 425 Hz. Thus, by changing the center frequency of the NMR probe pulse in increments of 425 Hz, it is possible to probe successive 1-cm positions in the direction of the magnetic field gradient. Each consecutive RF pulse produces an FID signal that encodes the concentration of protons at each 1-cm position along the direction of the field gradient. When the FIDs are subjected to Fourier transformation, concentration information is produced as indicated by the heights of the peaks at the bottom of Figure 19-37. In practice, the position of the slice along the z-axis may be changed by adding a dc offset to the auxiliary coils as shown

[25] S. L. Smith, *Anal. Chem.*, **1985**, *57*, 595A; H. J. Smith and F. N. Ranallo, *A Non-Mathematical Approach to Basic MRI.* Madison, WI: Medical Physics Publishing, 1989.

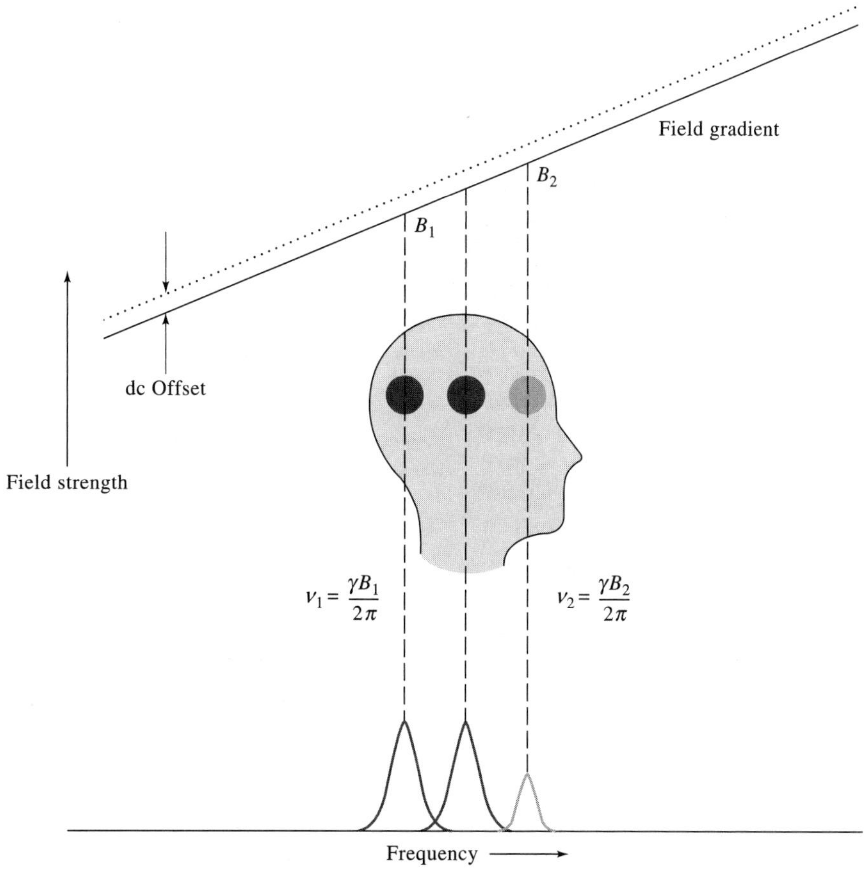

Figure 19-37 Fundamental concept of MRI.

by the dotted field gradient in Figure 19-37. Each slice is probed by changing the pulse width of the RF pulse to tune its frequency to the precession frequency of the protons in the slice. The middle peak in Figure 19-37 corresponds to the center circle representing the signal from protons in the center position of the head of the subject. If the applied field gradient is made larger, the thickness of the slice that is examined is made smaller.

The position selection process described in the previous paragraph provides spatial information in one dimension along the z axis by flipping the spins of the protons *within* the selected slice. Information within each of the slices along the z axis is acquired in a slightly different way as shown schematically in Figure 19-38. The chosen slice is represented in the figure as a two-dimensional grid lying in the x-y slice with rows 1, 2, and 3 and columns a, b, and c. For simplicity, the grid is shown as a 3 × 3 array of elements, but in practice the slice might have 128 × 128, 256 × 256, or 512 × 512 elements, with each element, or *pixel,* having an area of 1 mm × 1 mm in the sample. In such three-dimensional imaging experiments, the z axis field gradient is selected to give a slice thickness of 1 mm, then it is turned off. At this point in the measurement sequence, all protons in the slice have their spins flipped. A second gradient is applied to the subject along a direction perpendicular to the z axis by using a second set of y axis coils oriented at a right angle to the axis of the magnet bore. Nuclei at different positions along the y axis precess at different frequencies depending on their position in the field gradient. When this second y axis field gradient is turned off, all nuclei at a distance corresponding to row 1 on the y axis of Figure 19-38

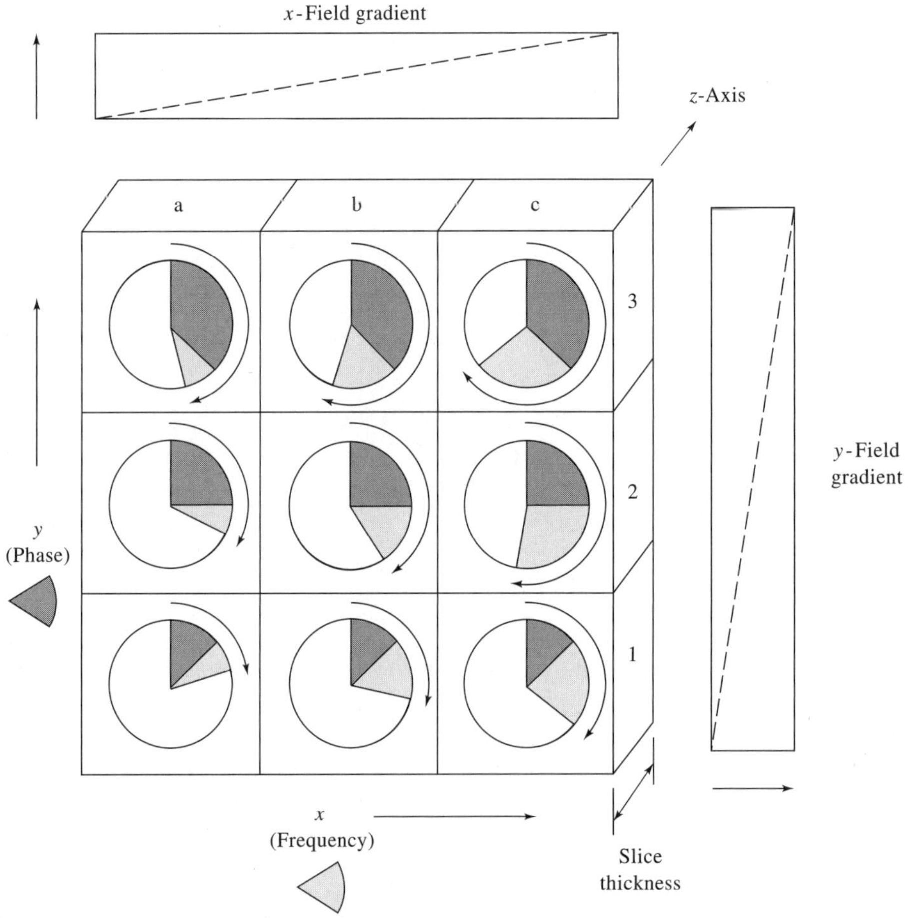

Figure 19-38 Acquisition of information within slices along the z-axis.

have precessed through a fixed phase angle whose magnitude is proportional to the size of the dark gray, pie-shaped angular sections of the circles in row 1. Thus, the distance along the y axis *is encoded in the phase angle* of the precessing protons represented by the dark gray pie slices.

As soon as the y axis gradient is terminated, a third set of magnet coils that is perpendicular to both the z axis and the y axis coils is then activated. This action produces a field gradient in the third dimension along the x axis. Because the nuclei in column (b) have experienced a stronger magnetic field than those in column (a), they precess at a higher frequency whose magnitude is represented by the light gray pie slices. Similarly, the nuclei in column (c) have the largest frequency in the 3×3 pixel array of Figure 19-38. The result is that the positions of the nuclei in the x dimension *are encoded by their precession frequencies.* The total precession angle, which is indicated by the circular arrows in the figure, is unique for each element of the grid, and thus the position of each is encoded in the total angle. The receiver coils are turned on during the application of this last frequency-encoding phase, and a FID is recorded. The complete cycle of (1) slice selection, (2) phase encoding in the y dimension, (3) frequency encoding in the x dimension, and (4) recording the FID requires only a few hundred milliseconds at most. The next step in collecting the data is to turn on the z axis coils to once again flip the spins of the nuclei in the slice and repeat the sequence with a somewhat larger y axis gradient to phase encode row 2. This step is followed by the application of the x axis gradient to frequency encode columns (a), (b), and (c) and the acquisition of the FID for row 2. Finally, the entire measurement sequence is repeated for row 3, and a corresponding FID is acquired. The result of the process for the entire 3×3 slice is a set of 3 FIDs representing the time-domain resonance signal S, each of which is a function of the x and y field gradients G_x and G_y, respectively. The time-domain signal $S(G_x, G_y)$, which contains information related to the total precession angle, is subjected to a two-dimensional Fourier transformation. The result is a frequency-domain signal $S(\omega_x, \omega_y)$ whose frequencies in the x and y dimensions are directly proportional to the distances d_x and d_y. That is,

$$S(G_x, G_y) \rightarrow S(\omega_x, \omega_y) \rightarrow S(d_x, d_y)$$

The two-dimensional distance information in $S(d_x, d_y)$ is at last combined for all of the slices in the z dimension to provide a three-dimensional array of data. Each element of the array contains an intensity that is proportional to the concentration of protons in a volume element, or *voxel,* corresponding to each set of coordinates x, y, and z. Note that with the exception of the magnet, which must have a rather large bore and large static field strength of 0.5 to 4.7 T, instruments for MRI and high-resolution NMR are identical in function. Precisely the same pulse-sequencing techniques and data-enhancement procedures are used in both types of instruments. Although the foregoing discussion is considerably simplified, the basic sequencing and encoding schemes are realistic. Clever and timely application of various RF pulse sequences and magnetic field gradients, appropriate Fourier transformations, and data analysis and reconstruction software routines, produce three-dimensional images. Structures inside subjects may be reconstructed from the three-dimensional data arrays as illustrated in the group of four MRI images of Figure 19-39. The images are four true three-dimensional brain reconstructions calculated by using a single set of MRI data from a young female patient with Rasmussen's encephalitis. These images show dramatic shrinkage in the front right portion of the brain characteristic of this rare illness. The ability to generate high resolution (millimeter) images from true three-dimensional data is a unique strength of the MRI technique.[26]

A second MR image, presented in Figure 19-40, demonstrates functional magnetic resonance image (fMRI) mapping of visual confrontation naming in the left hemisphere of the brain. The light portion of the map demonstrates statistical differences in signal intensity due to changes in cerebral blood oxygenation from activation of brain regions engaged in a naming task compared to a control task. The task undertaken by the subject in this case was the naming of standardized line drawings of a series of seven subjects. The map shows activation of the *angular gyrus,* which is the large light-patch in the upper right region of the brain in this picture, and the so-called *area 37,* which is the smaller light gray region in the lower right region of the picture. Both regions of the brain are known to be involved in language processing. Mapping of brain regions involved in specific types of human brain processing and execution is a current intensive area of research. Advantages of fMRI compared with other techniques are the noninvasive na-

[26] C. D. Smith, Memory Disorders Clinic, Sanders-Brown Research Center on Aging, University of Kentucky Medical Center, unpublished results.

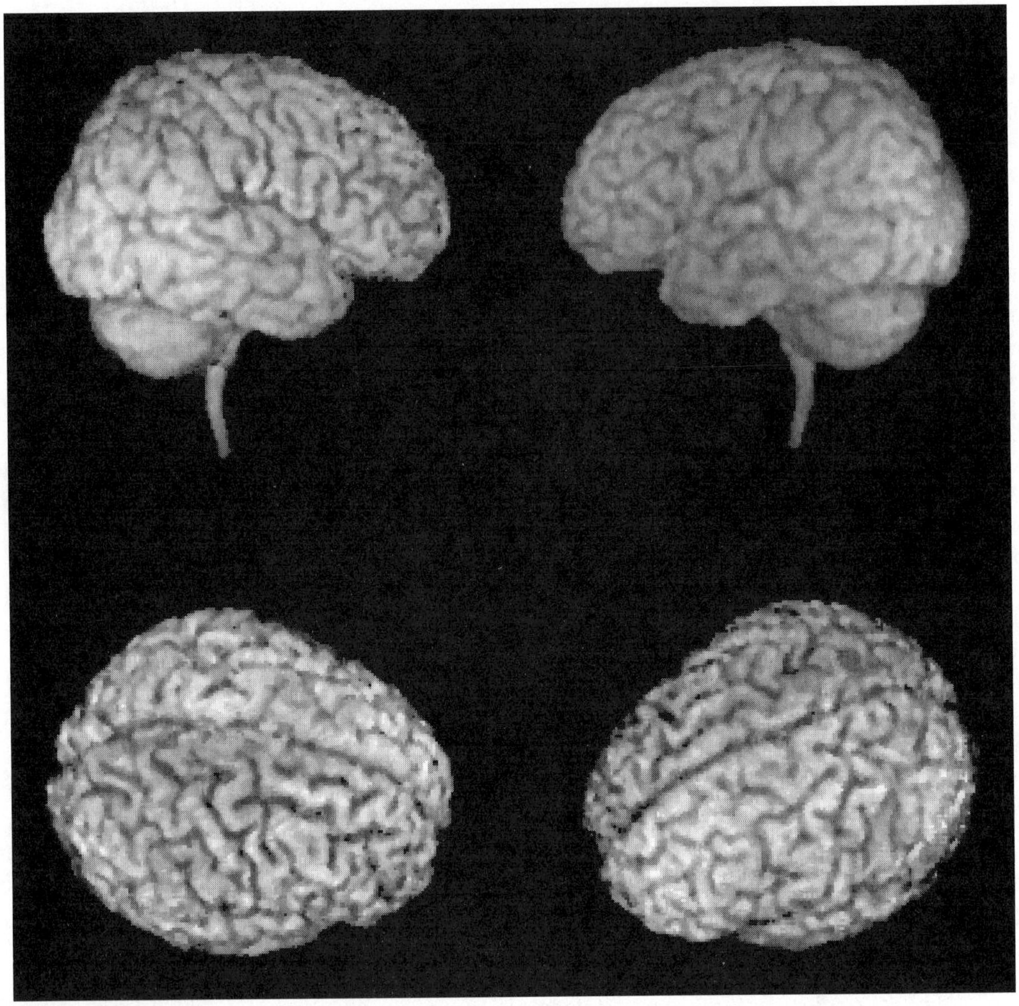

Figure 19-39 Structures inside subjects may be reconstructed from the three-dimensional data NMR arrays. *(Image courtesy of C. D. Smith, Memory Disorders Clinic, Sanders-Brown Research Center on Aging, University of Kentucky Medical Center. With permission.)*

ture of the technique, reproducibility, speed of data acquisition, and high intrinsic spatial resolution.[27]

MRI has become a mainstay in the arsenal of medical diagnostic tools, but its application in other areas has been slow in coming. This circumstance probably results from the very high cost of MRI installations, which may be up to one million dollars for a research-grade instrument and up to three million dollars for a medical diagnostic unit. Much of the cost of these units is in site preparation, magnetic shielding, and the essential high-stability, large-bore magnet. As an example of the effects of the high cost of the instrumentation, note that a single medical MRI session costs the patient or insurance company about a thousand dollars. The inherent advantages of MRI have begun to overshadow the cost factor in recent years, and imaging has been used in other areas of science and commerce, principally in the food industry, for noninvasive examination of a variety of foodstuffs. MRI has been used for such diverse tasks as the exploration of the fat and muscle distribution in meats; the determination of NMR relaxation parameters in cured pork; the determination of water and lipid content in sunflower oil and in water/meat emulsions; the study of heat transfer, composition, and structure in

[27] C. D. Smith, A. H. Andersen, Q. Chen, L. X. Blonder, J. E. Kirsch, M. J. Avison, *Neuro Report,* **1996,** *7,* 2.

foods such as cookies during processing; the measurement of transport phenomena in food processing; the investigation of fat crystallization; the visualization of the hydration of foods; the examination of the internal structure and chemistry of polymers; and the analysis of particle filtration in sandstone cores. As the sophistication and ease of application of data enhancement procedures, exotic pulse sequencing, and data acquisition protocols continue to evolve, especially for the determination of nuclei other than hydrogen, MRI will doubtless become an indispensable tool in the noninvasive investigation of materials.

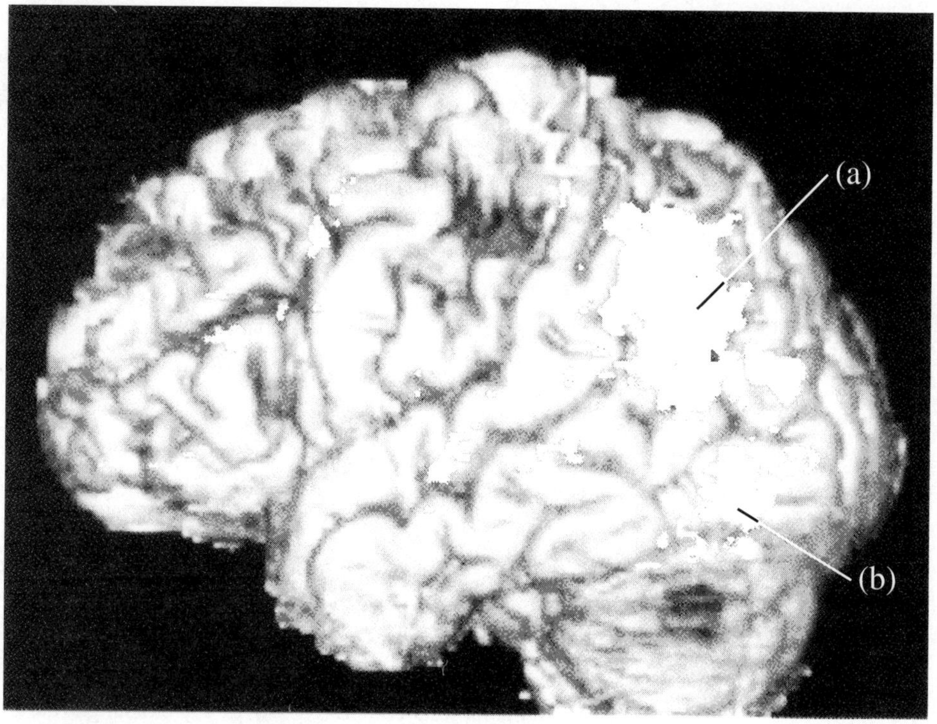

Figure 19-40 fMRI image showing brain activity in the left hemisphere resulting from naming tasks. *(From C. D. Smith, A. H. Andersen, Q. Chen, L. X. Blonder, J. E. Kirsch, M. J. Avison, Neuro Report, 1996, 7, 2. With permission.)*

19I QUESTIONS AND PROBLEMS

19-1 Explain the difference in the way a continuous-wave and a Fourier transform NMR experiment is performed.

19-2 What are the advantages of a Fourier transform NMR measurement over a continuous-wave measurement? What are the disadvantages?

19-3 In NMR spectroscopy, what are the advantages of using a magnet with as great a field strength as possible?

19-4 How can spin-spin splitting lines be differentiated from chemical shift lines?

19-5 Define
(a) magnetic anisotropy.
(b) the screening constant.
(c) the chemical shift parameter.
(d) continuous-wave NMR measurements.
(e) Larmor frequency.
(f) coupling constants.
(g) first-order NMR spectra.

19-6 A nucleus has a spin quantum number of 5/2. How many magnetic energy states does this nucleus have? What is the magnetic quantum number of each?

19-7 What is the absorption frequency in a 2.4-T magnetic field of (a) ^{1}H, (b) ^{13}C, (c) ^{19}F, (d) ^{31}P?

19-8 Why is $^{13}C/^{13}C$ spin-spin splitting not observed in ordinary organic compounds?

19-9 Calculate the relative number of ^{13}C nuclei in the higher and lower magnetic states at 25°C in 2.4-T magnetic field.

19-10 What is the difference between longitudinal and transverse relaxation?

19-11 Explain the source of an FID signal in FT NMR.

19-12 What is a rotating frame of reference?

19-13 How will ΔE for an isolated ^{13}C nucleus compare with that of a ^{1}H nucleus?

19-14 Calculate the resonance frequency of each of the following nuclei in a 5.0-T magnetic field: (a) ^{19}F and (b) ^{31}P.

19-15 What is the ratio of the number of nuclei in the upper magnetic energy state to the number in the lower energy state of ^{13}C in a 220-MHz instrument if the temperature is 300 K?

19-16 Briefly compare the ^{1}H and ^{31}P NMR spectra of methyl phosphorous acid $P(OCH_3)_3$ at 1.4 T. There is a weak spin-spin coupling between phosphorus and hydrogen nuclei in the compound.

19-17 In the room temperature ^{1}H spectrum of methanol, no spin-spin coupling is observed, but when a methanol sample is cooled to −40°C, the exchange rate of the hydroxyl proton slows sufficiently so that splitting is observed. Sketch spectra for methanol at the two temperatures.

19-18 Use the following coupling constant data to predict the ^{1}H and the ^{19}F spectra of the following:

Species	*J* (Hz)
(a) F—C≡C—H	21
(b) CF_3—CH_3	12.8
(c) $(CH_3)_3$—CF	20.4

19-19 Predict the appearance of the high-resolution ^{13}C spectrum of (proton decoupled)
(a) methyl formate.
(b) acetaldehyde.
(c) acetone.

19-20 Repeat Question 19-19 when the protons are not decoupled.

19-21 Predict the appearance of the high-resolution proton NMR spectrum of propionic acid.

19-22 Predict the appearance of the high-resolution proton NMR spectrum of
(a) acetaldehyde.
(b) acetic acid.
(c) ethyl nitrite.

19-23 Predict the appearance of the high-resolution proton NMR spectrum of
(a) acetone.
(b) methyl ethyl ketone.
(c) methyl *i*-propyl ketone.

19-24 Predict the appearance of the high-resolution proton NMR spectrum of
(a) cyclohexane.
(b) 1,2-dimethoxyethane, $CH_3OCH_2CH_2OCH_3$.
(c) diethylether.

19-25 Predict the appearance of the high-resolution proton NMR spectrum of
(a) toluene.
(b) ethyl benzene.
(c) *i*-butane.

19-26 The proton spectrum in Figure 19-41 is for an organic compound containing a single atom of bromine. Identify the compound.

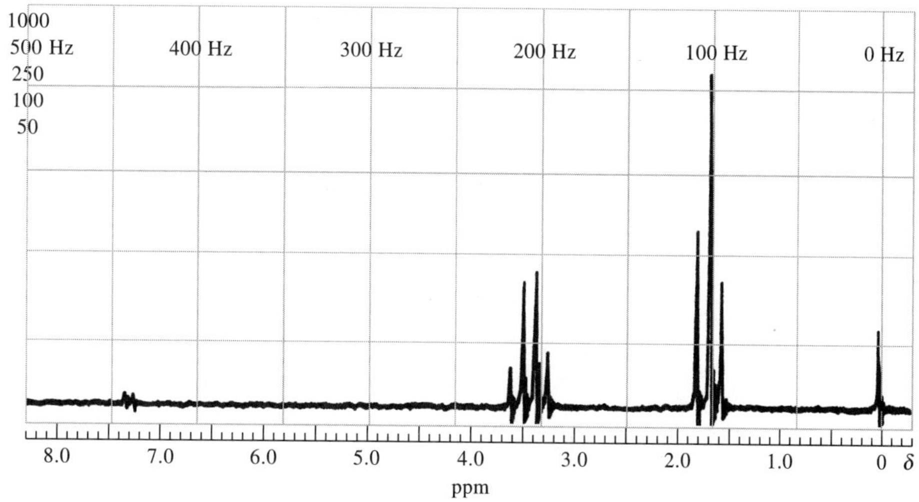

Figure 19-41 *(Courtesy of Varian Instrument Division, Palo Alto, CA.)*

19-27 The proton spectrum in Figure 19-42 is for a compound having an empirical formula $C_4H_7BrO_2$. Identify the compound.

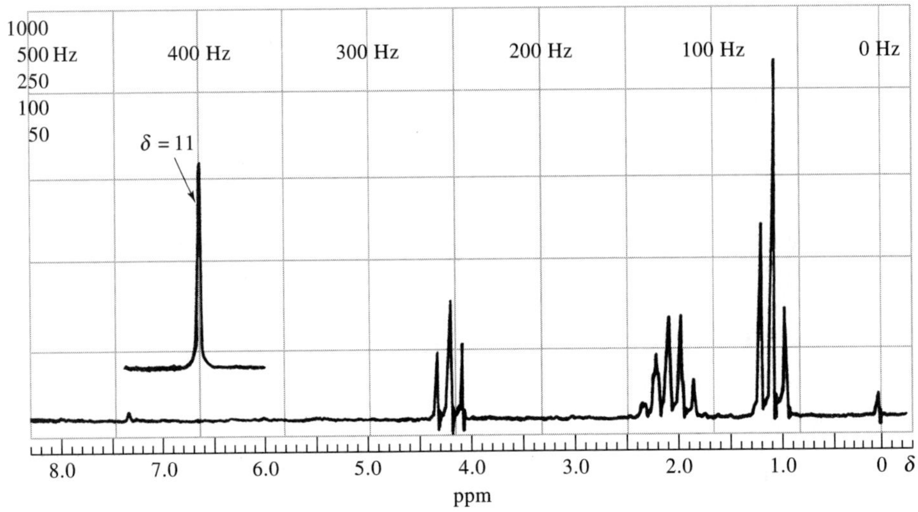

Figure 19-42 *(Courtesy of Varian Instrument Division, Palo Alto, CA.)*

19-28 The proton spectrum in Figure 19-43 is for a compound of empirical formula C_4H_8O. Identify the compound.

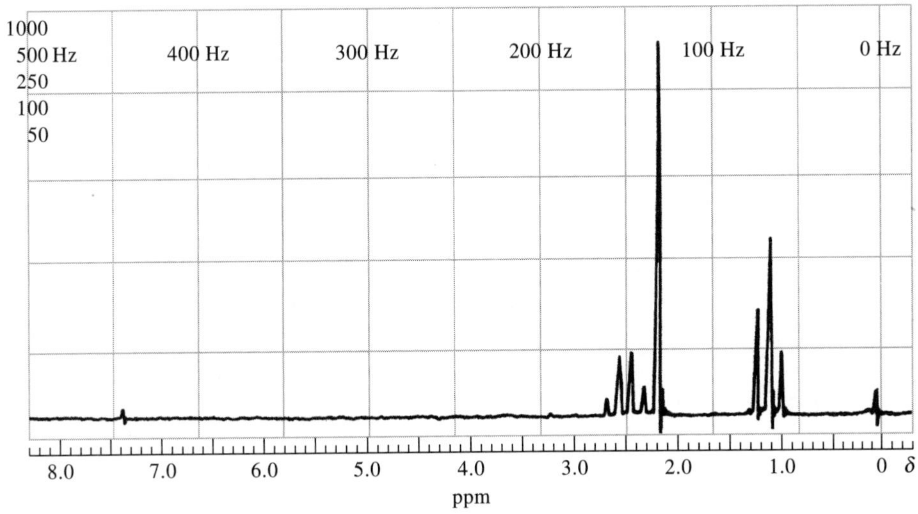

Figure 19-43 *(Courtesy of Varian Instrument Division, Palo Alto, CA.)*

19-29 The proton spectrum in Figure 19-44 is for a compound having an empirical formula $C_4H_8O_2$. Identify the compound.

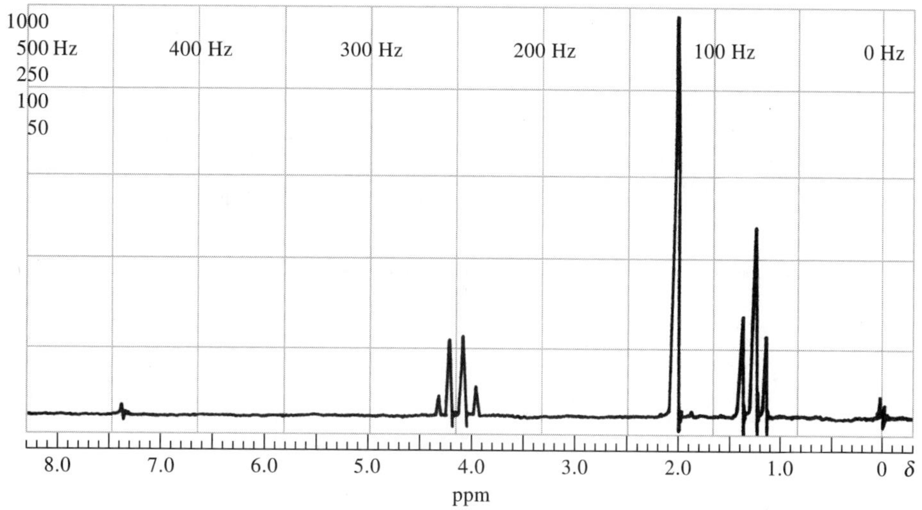

Figure 19-44 *(Courtesy of Varian Instrument Division, Palo Alto, CA.)*

19-30 The proton spectra in Figures 19-45a and 19-45b are for compounds with empirical formulas C_7H_{10}. Identify the compounds.

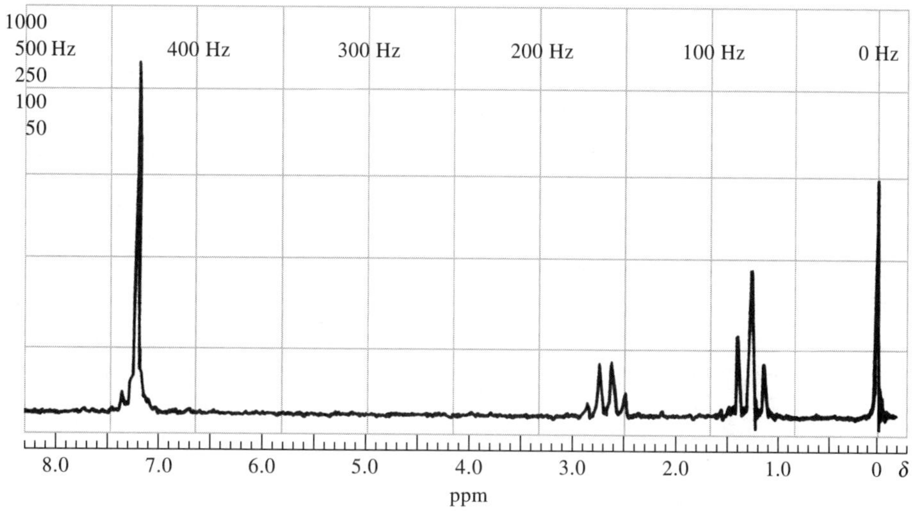

(a)

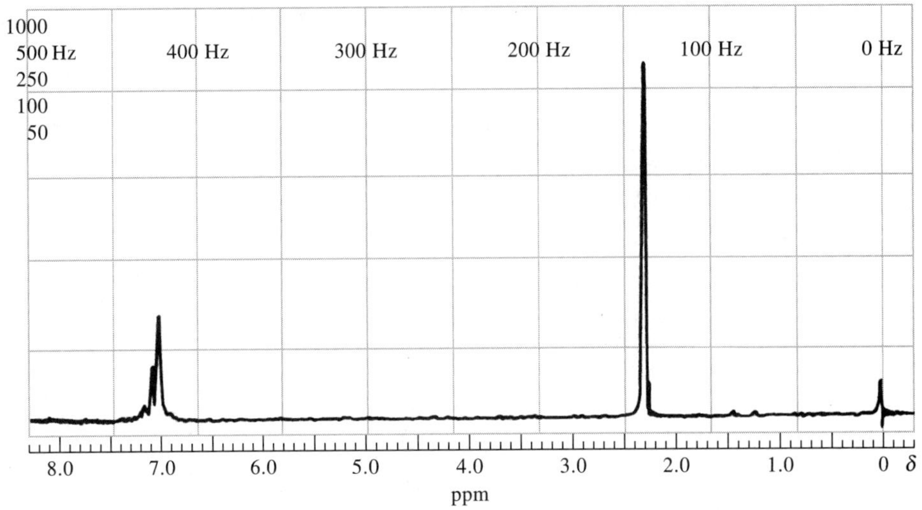

(b)

Figure 19-45 *(Courtesy of Varian Instrument Division, Palo Alto, CA.)*

19-31 From the proton spectrum in Figure 19-46, deduce the structure of this hydrocarbon.

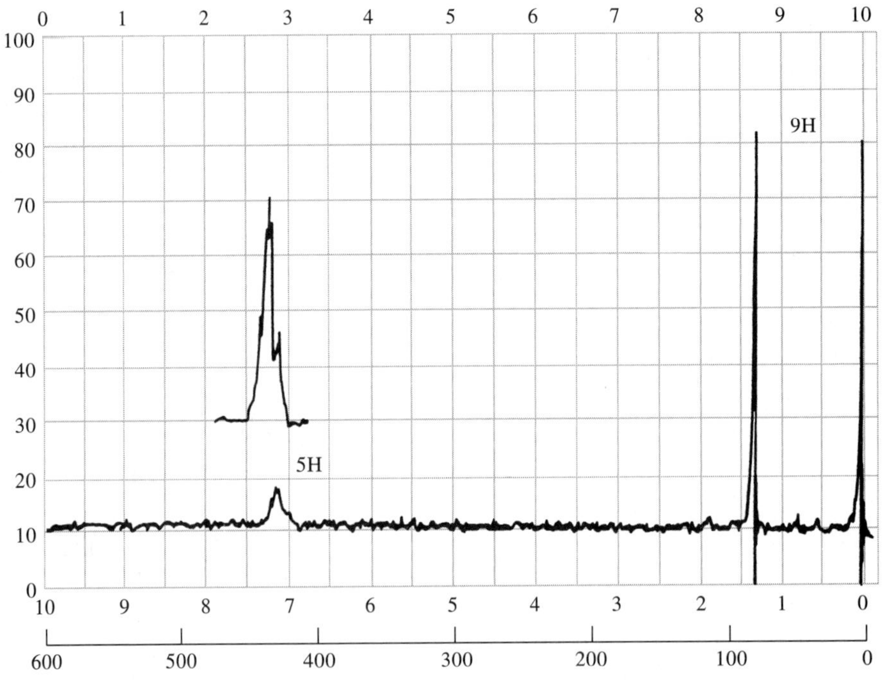

Figure 19-46 *(From C. J. Pouchert,* The Aldrich Library of NMR Spectra, *2nd ed. Milwaukee, WI: The Aldrich Chemical Company. With permission.)*

19-32 From the proton spectrum given in Figure 19-47, determine the structure of this compound, which is a commonly used pain killer; its empirical formula is $C_{10}H_{13}NO_2$.

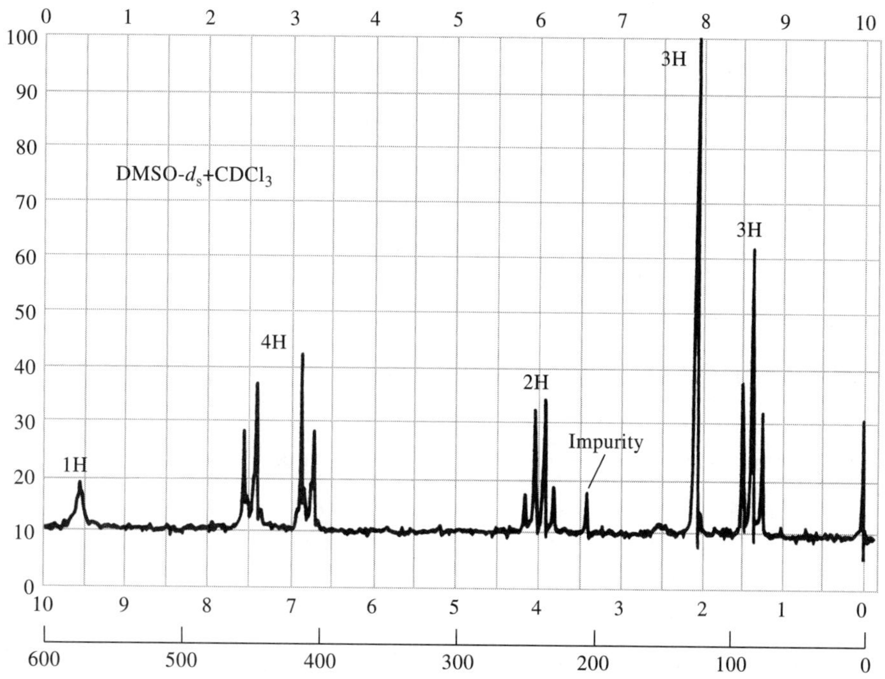

Figure 19-47 *(From C. J. Pouchert,* The Aldrich Library of NMR Spectra, *2nd ed. Milwaukee, WI: The Aldrich Chemical Company. With permission.)*

19-33 What is a frequency lock system in an NMR spectrometer? Describe the two types of lock systems.

19-34 What are shims in an NMR spectrometer, and what are their purpose?

19-35 Why are liquid samples spun while being examined in an NMR spectrometer?

19-36 Explain how a band of frequencies is obtained from an oscillator, which is essentially a monochromatic source of radio-frequency radiation. How could a band sufficiently broad to cover the entire ^{13}C spectrum (200 ppm) be obtained?

19-37 Describe sources of folded spectral lines.

19-38 In NMR spectroscopy how are signals converted to the audio-frequency range?

19-39 Explain the principles of quadrature detectors for FT NMR.

19-40 Describe the differences between off-resonance and broad-band proton decoupling in ^{13}C NMR spectroscopy.

19-41 What is the nuclear Overhauser effect, and what is its source?

19-42 What are the sources of band broadening in ^{13}C spectra of solids? How are lines narrowed so that high-resolution spectra can be obtained?

Chapter

20

Molecular Mass Spectrometry

Mass spectrometry is perhaps the most widely applicable of all of the analytical tools available to the scientist in the sense that the technique is capable of providing information about (1) the elemental composition of samples of matter; (2) the structures of inorganic, organic, and biological molecules; (3) the qualitative and quantitative composition of complex mixtures; (4) the structure and composition of solid surfaces; and (5) isotopic ratios of atoms in samples.

We have already discussed in Chapter 11 how mass spectrometry is used by chemists for the identification and quantitative determination of one or more elements in a sample of matter. This chapter is devoted to describing how mass spectrometry is used to obtain the information of the types listed in (2) and (3) in the previous paragraph. Chapter 21 describes how mass spectrometry is employed for elucidating the structure and composition of surfaces. Finally, in Section 32D, the use of isotopic ratios that have been determined by mass spectrometry is discussed.

The first general application of molecular mass spectrometry for routine chemical analysis occurred in the early 1940s when the techniques began to be adopted by the petroleum industry for the quantitative analysis of hydrocarbon mixtures produced in catalytic crackers. Before this time, analyses of mixtures of this type, which often contained as many as nine hydrocarbon components, was carried out by fractional distillation followed by refractive index measurements of the separated components. Typically 200 hours or more of operator time was required to complete an analysis. It was found that similar information could be obtained in a few hours or less with a mass spectrometer. This improved efficiency led to the appearance and rapid improvement of commercial mass spectrometers. Beginning in the 1950s, these commercial instruments began to be adapted by chemists for the identification and structural elucidation of a wide variety of organic compounds. This use of the mass spectrometer combined with the invention of nuclear magnetic resonance and the development of infrared spectrometry revolutionized the way organic chemists went about identifying and determining the structure of molecules. This application of mass spectrometry is still of prime importance.

Applications of molecular mass spectrometry underwent dramatic changes in the decade of the 1980s, brought about by the development of methods of producing ions from nonvolatile or thermally labile molecules, such as those frequently encountered by biochemists and biologists. Since about 1990, there has been an explosive growth in the area of biological mass spectrometry as a consequence of these new ionization methods. Now, mass spectrometry is being applied to the determination of the structure of polypeptides, proteins, and other high-molecular-weight biopolymers.

In this chapter, we will first describe briefly the nature of molecular mass spectra and define some terms used in molecular mass spectrometry. We next consider the various techniques that are used to form ions from analyte molecules in mass spectrometers and the types of spectra produced by these techniques. We then describe in some detail the various types of mass spectrometers that are used in molecular mass spectrometry (other than the quadrupole and time-of-flight instruments, which received detailed treatment in Section 11B-2). Finally, we describe several of the current applications of molecular mass spectrometry.[1]

20A MOLECULAR MASS SPECTRA

Figure 20-1 illustrates the way in which mass spectral data are usually presented. The analyte was ethyl benzene, which has a nominal molecular mass of 106 dalton. To obtain this spectrum, ethyl benzene vapor was bombarded with a stream of electrons that led to the loss of an electron by the analyte and formation of the molecular ion M^+ as shown by the reaction

$$C_6H_5CH_2CH_3 + e^- \rightarrow C_6H_5CH_2CH_3^{\bullet +} + 2e^- \quad (20\text{-}1)$$

The charged species $C_6H_5CH_2CH_3^{\bullet +}$ is the *molecular ion.* As indicated by the dot, the molecular ion is a *radical ion* that has the same molecular weight as the molecule.

The collision between energetic electrons and analyte molecules usually imparts enough energy to the molecules to leave them in an excited state. Relaxation then often occurs by fragmentation of part of the molecular ions to produce ions of lower masses. For example, a major product in the case of ethyl benzene is $C_6H_5CH_2^+$, which results from the loss of a CH_3 group. Other smaller positively charged fragments are also formed in lesser amounts.

The positive ions produced on electron impact are attracted through the slit of a mass spectrometer where they are sorted according to their mass-to-charge ratios and displayed in the form of a mass spectrum. The plot shown in Figure 20-1 demonstrates the way in which mass spectral data are often treated for presentation. Note that the plot is in the form of a bar graph that relates the relative intensity of mass peaks to their mass-to-charge ratio. Note also that in each spectrum, the largest peak, termed the *base peak,* has been arbitrarily assigned a value of 100. The heights of the remaining peaks have then been computed as a percentage of the base-peak height. Modern mass spectrometers are programmed to recognize the base peak and normalize the remaining peaks relative to that peak.

20B ION SOURCES

The starting point for a mass spectrometric analysis is the formation of gaseous analyte ions, and the scope and the utility of a mass spectrometric method is dictated by the ionization process. The appearance of mass spectra for a given molecular species is highly dependent upon the method used for ion formation. Table 20-1 lists many of the ion sources that have been employed in molecular mass spectrometry.[2] Note that these methods fall into two major categories: *gas-phase sources* and *desorption sources.* In the former, which includes the first three sources in the table, the sample is first vaporized and then ionized. In the latter, the sample

[1] For monographs dealing with molecular mass spectrometry, see F. A. White and G. M. Wood, *Mass Spectrometry: Applications in Science and Engineering.* New York: Wiley-Interscience, 1986; J. T. Watson, *Introduction to Mass Spectrometry,* 3rd ed. New York: Lippincott-Raven, 1997; H. E. Duckworth, R. C. Barber, and V. S. Venkatasubramanian, *Mass Spectroscopy,* 2nd ed. Cambridge, UK: Cambridge University Press, 1986; M. E. Rose and A. W. Johnstone, *Mass Spectrometry for Chemists and Biochemists.* New York: Cambridge University Press, 1982; G. A. Eadon, in *Treatise on Analytical Chemistry,* 2nd ed., J. D. Winefordner, M. M. Bursey, and I. M. Kolthoff, Eds., Part I, Vol. 11, Chapter 1. New York: Wiley, 1989.

[2] For a review of modern ion sources, see R. P. Lattimer and H. R. Schulten, *Anal. Chem.,* **1989,** *61,* 1201A; K. L. Busch and R. G. Cooks, *Science,* **1982,** *218,* 247; E. R. Grant and R. G. Cooks, *Science,* **1990,** *250,* 61; R. J. Cotter, *Anal. Chem.,* **1988,** *60,* 781A; J. B. Fenn et al., *Science,* **1989,** *246,* 64.

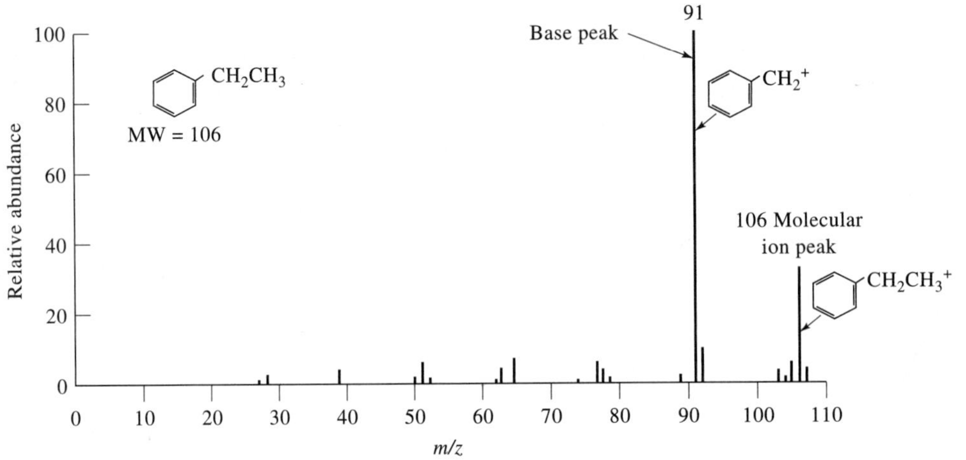

Figure 20-1 Mass spectrum of ethyl benzene.

in a solid or liquid state is converted directly into gaseous ions. An advantage of desorption sources is that they are applicable to nonvolatile and thermally unstable samples. Currently, commercial mass spectrometers are equipped with accessories that permit use of several of these sources interchangeably.

Gas-phase sources are generally restricted to thermally stable compounds that have boiling points less than about 500°C. In most cases, this requirement limits gaseous sources to compounds with molecular weights less than roughly 10^3 dalton. Desorption sources, which do not require volatilization of analyte molecules, are applicable to analytes having molecular weights as large as 10^5 dalton.

Ion sources are also classified as being *hard sources* or *soft sources.* Hard sources impart sufficient energy to analyte molecules so that they are left in a highly excited energy state. Relaxation then involves rupture of bonds, producing fragment ions that have mass-to-charge ratios less than that of the molecular

TABLE 20-1 Ion Sources for Molecular Mass Spectrometry

Basic Type	Name and Acronym	Ionizing Agent
Gas phase	Electron impact (EI)	Energetic electrons
	Chemical ionization (CI)	Reagent gaseous ions
	Field ionization (FI)	High-potential electrode
Desorption	Field desorption (FD)	High-potential electrode
	Electrospray ionization (ESI)	High electrical field
	Matrix-assisted desorption/ionization (MALDI)	Laser beam
	Plasma desorption (PD)	Fission fragments from ^{252}Cf
	Fast atom bombardment (FAB)	Energetic atomic beam
	Secondary ion mass spectrometry (SIMS)	Energetic beam of ions
	Thermospray ionization (TS)	High temperature

ion. Soft sources cause little fragmentation. Consequently, the resulting mass spectrum often consists of the molecular ion peak and only a few, if any other, peaks. Figure 20-2 illustrates the difference in appearance of a hard source and a soft source spectrum.

Both hard and soft source spectra are useful for analysis. The many peaks in a hard source spectrum provide useful information about the kinds of functional groups and thus structural information about analytes. Soft source spectra are useful because they supply accurate information about the molecular weight of the analyte molecule or molecules.

20B-1 The Electron-Impact Source

Historically, ions for mass analysis were produced by electron impact. Here, the sample is brought to a temperature high enough to produce a molecular vapor,

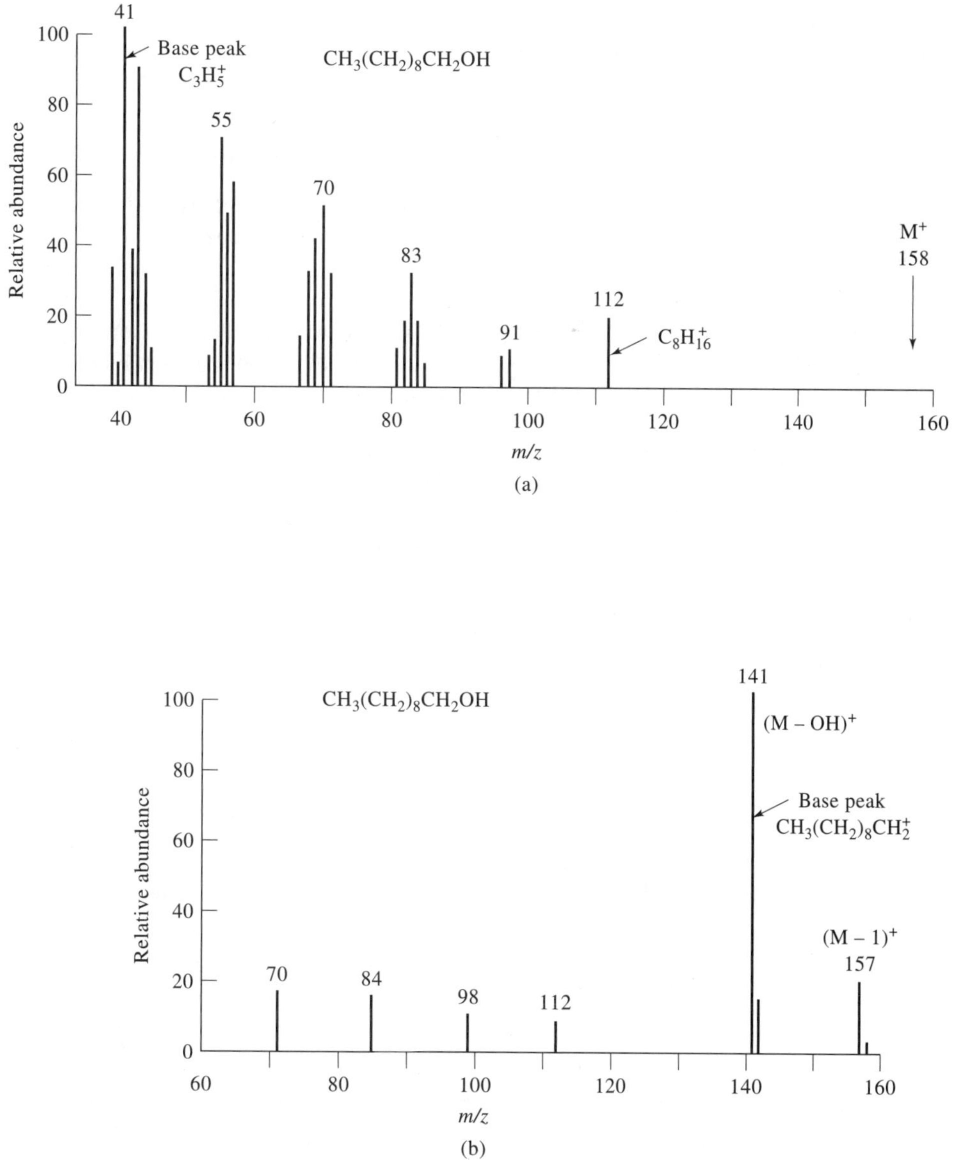

Figure 20-2 Mass spectrum of 1-decanol from (a) a hard source and (b) a soft source.

which is then ionized by bombarding the resulting molecules with a beam of energetic electrons. Despite certain disadvantages, this technique is still of major importance and is the one upon which many libraries of mass spectral data are based.

Figure 20-3 is a schematic of a simple electron-impact ion source. Electrons are emitted from a heated tungsten or rhenium filament and accelerated by a potential of approximately 70 V that is impressed between the filament and the anode. As shown in the figure, the paths of the electrons and molecules are at right angles and intersect at the center of the source, where collision and ionization occur. The primary product is singly charged positive ions formed when the energetic electrons approach molecules closely enough to cause them to lose electrons by electrostatic repulsion. Electron-impact ionization is not very efficient, and only about one molecule in a million undergoes the primary reaction

$$M + e^- \rightarrow M^{\bullet +} + 2e^- \qquad (20\text{-}2)$$

Here M represents the analyte molecule, and $M^{\bullet +}$ is its molecular ion. The positive ions produced on electron impact are attracted through the slit in the first accelerating plate by a small potential difference (typically 5 V) that is applied between this plate and the repellers shown in Figure 20-3. With magnetic sector instruments, high potentials (10^3 to 10^4 V) are applied to the accelerator plates, which give the ions their final velocities before they enter the mass analyzer. Commercial electron impact sources are more complex than that shown in Figure 20-3 and may use additional electrostatic or magnetic fields to manipulate the electron and/or ion beam.

EXAMPLE 20-1

(a) Calculate the kinetic energy that a singly charged ion ($z = 1$) will acquire if it is accelerated through a potential of 10^3 V in an electron-impact source. **(b)** Does the kinetic energy of the ion depend upon its mass? **(c)** Does the velocity of the ion depend upon its mass?

(a) The kinetic energy (KE) added to the ion is due to the accelerating potential V and is given by the equation

$$KE = qV = zeV$$

where e is the electronic charge (1.6×10^{-19} coulombs). Thus, for $z = 1$

$$KE = 1 \times 1.6 \times 10^{-19}\ C \times 10^3\ V = 1.6 \times 10^{-16}\ J$$

(b) The kinetic energy that an ion acquires in the source is independent of its mass and depends only upon its charge and the accelerating potential.

(c) The translational component of the kinetic energy of an ion is a function of the ion mass m and its velocity v as given by the equation

$$KE = (1/2)mv^2 \text{ or } v = (2\ KE/m)^{1/2}$$

Thus, if all ions acquire the same amount of kinetic energy, those ions with largest mass must have the smallest velocity.

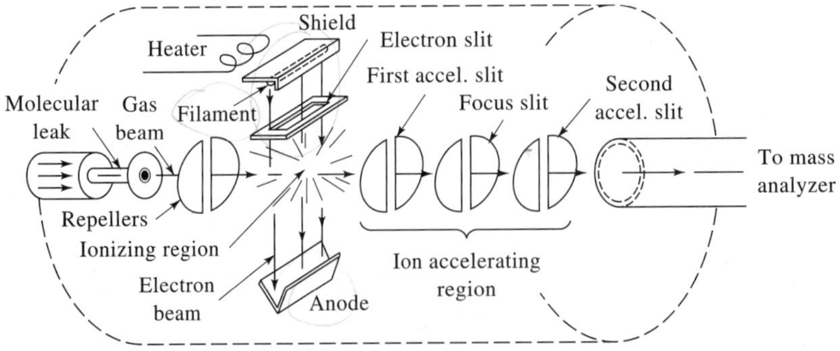

Figure 20-3 An electron-impact ion source. *(From R. M. Silverstein, G. C. Bassler, and T. C. Morrill,* Spectrometric Identification of Organic Compounds, *5th ed., p. 4. New York: Wiley, 1991. Reprinted by permission of John Wiley & Sons, Inc.)*

Electron-Impact Spectra

In order to form a significant number of gaseous ions at a reproducible rate, it is necessary that electrons from the filament in the source be accelerated by a potential of greater than about 50 V. The low mass and high kinetic energy of the resulting electrons cause little increase in the translational energy of impacted molecules. Instead, the molecules are left in highly excited vibrational and rotational states. Subsequent relaxation then usually takes place by extensive fragmentation, giving a large number of positive ions of various masses that are less than (and occasionally greater than) that of the molecular ion. These lower mass ions are called *daughter ions*. Table 20-2 shows some typical fragmentation reactions that follow electron-impact formation of a parent ion from a hypothetical molecule ABCD.

EXAMPLE 20-2

(a) Calculate the energy (in J/mol) that electrons acquire as a result of being accelerated through a potential of 70 V. **(b)** How does this energy compare to that of a typical chemical bond?

(a) The kinetic energy KE of an individual electron is equal to the product of the charge on the electron e times the potential V through which it has been accelerated. Multiplying the kinetic energy of a single electron by Avogadro's number, N, gives the energy per mole:

$$\begin{aligned}\text{KE} &= eVN\\ &= (1.60 \times 10^{-19}\ \text{C/e}^-)(70\ \text{V})N\\ &= (1.12 \times 10^{-17}\ \text{CV/e}^-)(6.02 \times 10^{23}\ \text{e}^-/\text{mol})\\ &= 6.7 \times 10^{6}\ \text{J/mol} \quad \text{or} \quad 6.7 \times 10^{3}\ \text{kJ/mol}\end{aligned}$$

(b) Typical bond energies fall in the 200- to 600-kJ/mol range. Therefore, an electron that has been accelerated through 70 V generally has considerably more energy than that required to break a chemical bond.

The complex mass spectra that result from electron-impact ionization are useful for compound identification. On the other hand, with certain types of molecules, fragmentation is so pervasive that no molecular ion persists; information of prime importance in determining the molecular weight of the analyte is therefore lost. Figure 20-4 shows typical electron-impact spectra for two simple organic molecules: methylene chloride and 1-pentanol. Note that in each of the spectra, the base peak corresponds to a fragment of the molecule, which has a mass significantly less than the molecular mass of the original compound. For methylene chloride, the base peak occurs at mass 49, which corresponds to the loss of one Cl atom. For 1-pentanol, the base peak is found at an *m/z* of 44, which is that of the daughter CH_2CHOH^+. More often than not, the base peaks in electron-impact spectra arise from fragments such as these rather than from the molecular ion.

In each case the molecular ion peak occurs at a mass corresponding to the molecular weight of the ana-

TABLE 20-2 Some Typical Reactions in an Electron-Impact Source

Molecular ion formation	$ABCD + e^- \rightarrow ABCD^{\bullet+} + 2e^-$
Fragmentation	$ABCD^{\bullet+} \rightarrow A^+ + BCD^\bullet$
	$\rightarrow A^\bullet + BCD^+ \rightarrow BC^+ + D$
	$\rightarrow CD^\bullet + AB^+ \rightarrow B + A^+$; $\rightarrow A + B^+$
	$\rightarrow AB^\bullet + CD^+ \rightarrow D + C^+$; $\rightarrow C + D^+$
Rearrangement followed by fragmentation	$ABCD^{\bullet+} \rightarrow ADBC^{\bullet+} \rightarrow BC^\bullet + AD^+$; $\rightarrow AD^\bullet + BC^+$
Collision followed by fragmentation	$ABCD^{\bullet+} + ABCD \rightarrow (ABCD)^{\bullet 2+} \rightarrow BCD^\bullet + ABCDA^+$

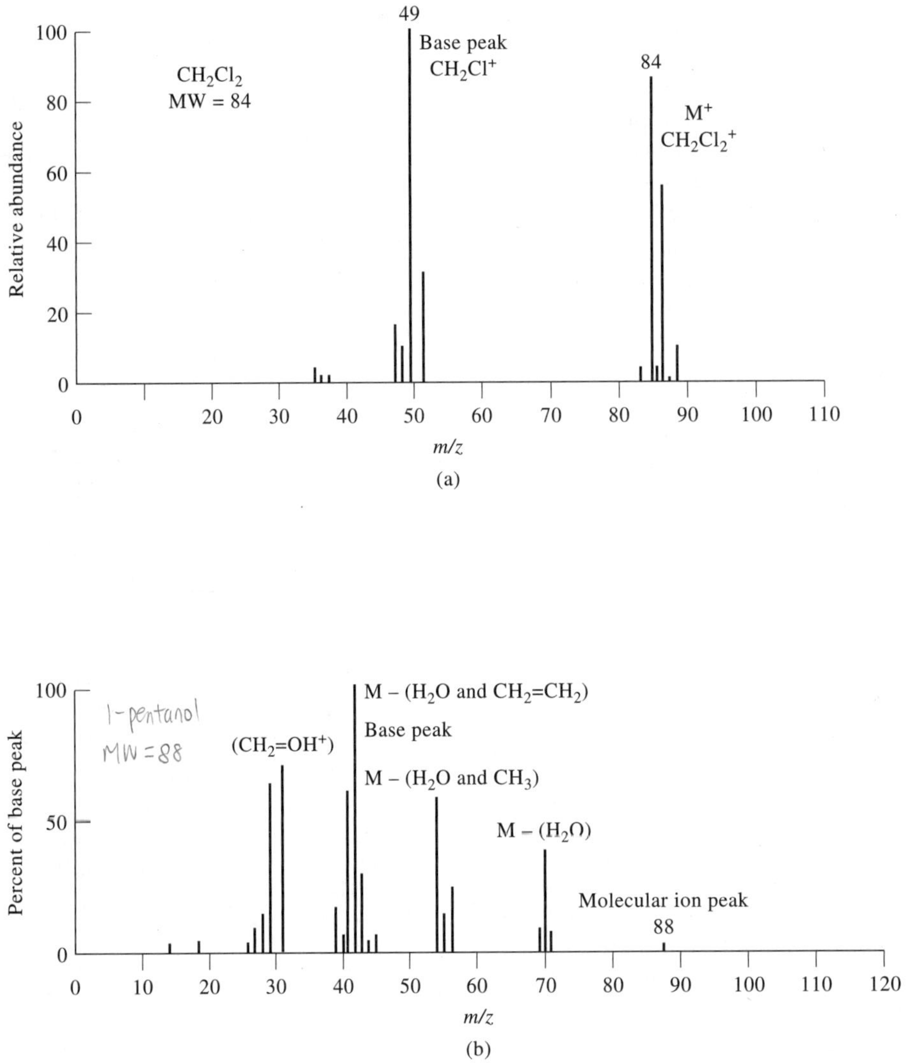

Figure 20-4 Electron-impact mass spectra of (a) methylene chloride and (b) 1-pentanol.

lyte. Thus, molecular ion peaks are observed at 84 for methylene chloride and 88 for 1-pentanol. The molecular ion peak is, of course, of prime importance in structural determinations, because its mass provides the molecular weight of the unknown. Unfortunately, it is not always possible to identify the molecular ion peak. Indeed, with electron-impact ionization, certain molecules yield no molecular ion peak (see Figure 20-2a).

Isotope Peaks

It is of interest to note in the spectra shown in Figures 20-1 and 20-4 that peaks occur at masses that are greater than that of the molecular ion. These peaks are attributable to ions having the same chemical formula, but different isotopic compositions. For example, for methylene chloride, the more important isotopic species are ${}^{12}C^{1}H_2{}^{35}Cl_2$ (m = 84), ${}^{13}C^{1}H_2{}^{35}Cl_2$ (m = 85), ${}^{12}C^{1}H_2{}^{35}Cl^{37}Cl$ (m = 86), ${}^{13}C^{1}H_2{}^{35}Cl^{37}Cl$ (m = 87), ${}^{12}C^{1}H_2{}^{37}Cl_2$ (m = 88). Peaks for each of these species can be seen in Figure 20-4a. The size of the various peaks depends upon the relative natural abundance of the isotopes. Table 20-3 lists the most common isotopes for atoms that occur widely in organic compounds. Note that fluorine, phosphorus, iodine, and sodium occur only as single isotopes.

The small peak for ethyl benzene at mass 107 in Figure 20-1 is due to the presence of ${}^{13}C$ in some of the

TABLE 20-3 Natural Abundance of Isotopes of Some Common Elements

Element[a]	Most Abundant Isotope	Abundance of Other Isotopes Relative to 100 Parts of the Most Abundant[b]	
Hydrogen	^{1}H	^{2}H	0.015
Carbon	^{12}C	^{13}C	1.08
Nitrogen	^{14}N	^{15}N	0.37
Oxygen	^{16}O	^{17}O	0.04
		^{18}O	0.20
Sulfur	^{32}S	^{33}S	0.80
		^{34}S	4.40
Chlorine	^{35}Cl	^{37}Cl	32.5
Bromine	^{79}Br	^{81}Br	98.0
Silicon	^{28}Si	^{29}Si	5.1
		^{30}Si	3.4

[a]Fluorine (^{19}F), phosphorus (^{31}P), sodium (^{23}Na), and iodine (^{127}I) have no additional naturally occurring isotopes.
[b]The numerical entries indicate the average number of isotopic atoms present for each 100 atoms of the most abundant isotope; thus, for every 100 ^{12}C atoms there will be an average of 1.08 ^{13}C atoms.

molecules. The intensities of peaks due to incorporation of two or more ^{13}C atoms in ethyl benzene can be predicted with good precision but are so small as to be undetectable because of the low probability of there being more than one ^{13}C atom in a small molecule. As will be shown in Section 20D-1, isotope peaks sometimes provide a useful means for determining the formula for a compound.

Collision Product Peaks

Ion/molecule collisions, such as that shown by the last equation in Table 20-2, can produce peaks at higher mass numbers than that of the molecular ion. At ordinary sample pressures, however, the only important reaction of this type is one in which the collision transfers a hydrogen atom to the ion to give a protonated molecular ion; an enhanced $(M + 1)^+$ peak results. This transfer is a second-order reaction, and the amount of product depends strongly upon the reactant concentration. Consequently, the height of an $(M + 1)^+$ peak due to this reaction increases much more rapidly with increases in sample pressure than do the heights of other peaks; thus, detection of this reaction is usually possible.

Advantages and Disadvantages of Electron-Impact Sources

Electron-impact sources are convenient to use and produce high ion currents, thus giving good sensitivities. The extensive fragmentation and consequent large number of peaks is also an advantage because it often makes unambiguous identification of analytes possible. This fragmentation can also be a disadvantage, however, when it results in the disappearance of the molecular ion peak so that the molecular weight of analytes cannot be established. Another limitation of the electron-impact source is the need to volatilize the sample, which may result in thermal degradation of some analytes before ionization can occur. The effects of thermal decomposition can sometimes be minimized by carrying out the volatilization from a heated probe that is located close to the entrance slit of the spectrometer. At the lower pressure of the source area, volatilization occurs at a lower temperature. Furthermore, less time is allowed for thermal decomposition to take place. As was mentioned earlier, electron-impact sources are only applicable to analytes having molecular weights smaller than about 10^3 daltons.

20B-2 Chemical Ionization Sources and Spectra

Most modern mass spectrometers are designed so that electron-impact ionization and chemical ionization can be carried out interchangeably. In chemical ionization, gaseous atoms of the sample (from either a batch inlet or a heated probe) are ionized by collision with ions produced by electron bombardment of an excess of a reagent gas. Usually positive ions are used, but negative ion chemical ionization[3] is occasionally used with analytes that contain very electronegative atoms. Chemical ionization is probably the second most common procedure for producing ions for mass spectrometry.[4]

In order to carry out chemical ionization experiments, it is necessary to modify the electron beam ionization area shown in Figure 20-3 by adding vacuum pump capacity and by reducing the width of the slit to the mass analyzer. These measures allow a reagent pressure of about 1 torr to be maintained in the ionization area while maintaining the pressure in the analyzer below 10^{-5} torr. With these changes, a gaseous reagent is introduced into the ionization region in an amount such that the concentration ratio of reagent to sample is 10^3 to 10^4. Because of this large concentration difference, the electron beam reacts nearly exclusively with reagent molecules.

One of the most common reagents is methane, which reacts with high-energy electrons to give several ions such as CH_4^+, CH_3^+, and CH_2^+. The first two predominate and represent about 90% of the reaction products. These ions react rapidly with additional methane molecules as follows:

$$CH_4^+ + CH_4 \rightarrow CH_5^+ + CH_3$$

$$CH_3^+ + CH_4 \rightarrow C_2H_5^+ + H_2$$

Generally, collisions between the sample molecule MH and CH_5^+ or $C_2H_5^+$ are highly reactive and involve proton or hydride transfer. For example,

$$CH_5^+ + MH \rightarrow MH_2^+ + CH_4 \qquad \text{proton transfer}$$

$$C_2H_5^+ + MH \rightarrow MH_2^+ + C_2H_4 \qquad \text{proton transfer}$$

$$C_2H_5^+ + MH \rightarrow M^+ + C_2H_6 \qquad \text{hydride transfer}$$

Note that proton transfer reactions give the $(M + 1)^+$ ion whereas the hydride transfer produces an ion with a mass one less than the analyte, or the $(M - 1)^+$ ion. With some compounds, an $(M + 29)^+$ peak is also produced from transfer of a $C_2H_5^+$ ion to the analyte. A variety of other reagents, including propane, isobutane, and ammonia, are used for chemical ionization. Each produces a somewhat different spectrum with a given analyte.

Figure 20-2 contrasts the chemical ionization and electron-impact spectra for 1-decanol. The electron-impact spectrum (Figure 20-2a) shows evidence for rapid and extensive fragmentation of the molecular ion. Thus, no detectable peaks are observed above mass 112, which corresponds to the ion $C_8H_{16}^+$.

The base peak is provided by the ion $C_3H_5^+$ at mass 41; other peaks for various C_3 species are grouped around the base peak. A similar series of peaks, found at 14, 28, and 42 mass units greater, corresponds to ions with one, two, and three additional CH_2 groups.

Relative to the electron-impact spectrum, the chemical ionization spectrum shown in Figure 20-2b is simple indeed, consisting of the $(M - 1)^+$ peak, a base peak corresponding to a molecular ion that has lost an OH group, and a series of peaks differing from one another by 14 mass units. As in the electron-impact spectrum, these peaks arise from ions formed by cleavage of adjacent carbon-carbon bonds. As we have just noted, chemical ionization spectra generally contain well-defined $(M + 1)^+$ or $(M - 1)^+$ peaks resulting from the addition or abstraction of a proton in the presence of the reagent ion.

20B-3 Field Ionization Sources and Spectra[5]

In *field ionization* sources, ions are formed under the influence of a large electric field (10^8 V/cm). Such fields are produced by applying high voltages (10 to 20 kV) to specially formed emitters consisting of numerous fine tips having diameters of less than 1 μm. The emitter often takes the form of a fine tungsten wire (≈10-μm diameter) on which microscopic carbon dendrites, or whiskers, have been grown by the pyrolysis of benzonitrile in a high electric field. The result of this treatment

[3] R. C. Dougherty, *Anal. Chem.*, **1981**, *53*, 625A.

[4] For a more detailed discussion of chemical ionization, see B. Munson, *Anal. Chem.*, **1977**, *49*, 772A; A. Harrison, *Chemical Ionization Mass Spectrometry.* Boca Raton, FL CRC Press, 1983.

[5] For reviews of these types of sources, see R. P. Lattimer and H. R. Schulten, *Anal. Chem.*, **1989**, *61*, 1201A; T. Komori, T. Kawasaki, and H. R. Schulten, *Mass Spectrom. Rev.*, **1985**, *4*, 155.

is a growth of many hundreds of carbon microtips projecting from the surface of the wire (Figure 20-5).

Field ionization emitters are mounted 0.5 to 2 mm from the cathode, which often also serves as a slit. The gaseous sample from a batch inlet system is allowed to diffuse into the high-field area around the microtips of the anode. The electric field is concentrated at the emitter tips, and ionization occurs via a quantum mechanical tunneling mechanism in which electrons from the analyte are extracted by the microtips of the anode. Little vibrational or rotational energy is imparted to the analyte; thus, little fragmentation occurs.

Figure 20-6 shows spectra for glutamic acid obtained by (a) electron-impact ionization and (b) field ionization. In the electron-impact spectrum, the parent ion peak at 147 is not detectable. The highest observable peak (mass 129) is due to the loss of water by the molecular ion. The base peak at mass 84 arises from a loss of water plus a —COOH group. Numerous other fragments are also found at lower masses. In contrast,

Figure 20-5 Photomicrograph of a carbon microneedle emitter. *(Courtesy of R. P. Lattimer, BF Goodrich Research and Development Center.)*

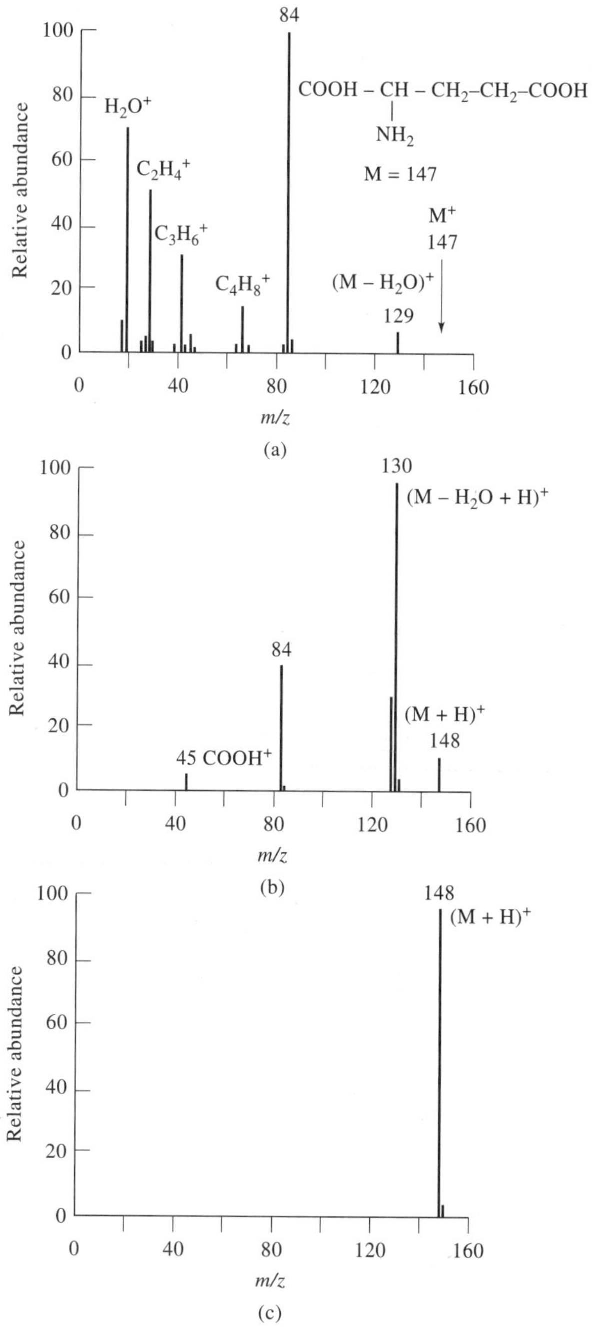

Figure 20-6 Mass spectra for glutamic acid: (a) electron-impact ionization, (b) field ionization, and (c) field desorption. *(From H. D. Beckey, A. Heindrich, and H. U. Winkler,* Int. J. Mass Spec. Ion Phys., ***1970****, 3, App. 11. With permission.)*

the field ionization spectrum is relatively simple, with an easily distinguished $(M + 1)^+$ peak at mass 148.

A limitation to field ionization is its sensitivity, which is at least an order of magnitude less than that of electron-impact sources; maximum currents are on the order of 10^{-11} A.

20B-4 Desorption Sources[6]

The ionization methods we have discussed thus far require that the ionizing agents act on gaseous samples. Such methods are not applicable to nonvolatile or thermally unstable samples. A number of *desorption ionization* methods have been developed in the last two decades for dealing with this type of sample (see Table 20-1). As a consequence, mass spectra for thermally delicate biochemical species and species having molecular weights of greater than 100,000 Da have now been reported.

Desorption methods dispense with volatilization followed by ionization of the gaseous analyte molecules. Instead, energy in various forms is introduced into the solid or liquid sample in such a way as to cause direct formation of gaseous ions. As a consequence, spectra are greatly simplified and often consist of only the molecular ion or the protonated molecular ion. In most cases the exact mechanism of how ions are formed without fragmentation is not understood. We describe several of these methods in this section.

Field Desorption Sources[7]

In field desorption, a multitipped emitter similar to that described in Section 20B-3 is used. In this case, the electrode is mounted on a probe that can be removed from the sample compartment and coated with a solution of the sample. After the probe is reinserted into the sample compartment, ionization again takes place by the application of a high potential to this electrode. With some samples it is necessary to heat the emitter by passing a current through the wire. As a consequence, thermal degradation may occur before ionization is complete.

Figure 20-6c is a field desorption spectrum for glutamic acid. It is even simpler than the spectrum from field ionization and consists of only the protonated molecular ion peak at mass 148 and an isotope peak at mass 149.

Matrix-Assisted Laser Desorption/Ionization

Matrix-assisted laser desorption/ionization spectrometry (MALDI), is a new ionization method, which shows great promise for obtaining accurate molecular weight information about polar biopolymers ranging in a molecular mass from a few thousand to several hundred thousand dalton. The method was first described nearly simultaneously in 1988 by two research groups, one German and the other Japanese.[8]

In the technique developed by the German group, an aqueous/alcohol solution of the sample was mixed with a large excess of a radiation-absorbing matrix material, such as one of those listed in Table 20-4. The resulting solution was evaporated on the surface of a metallic probe that was used for introduction of the sample into the mass spectrometer. The solid mixture was then exposed to a pulsed laser beam, which resulted in sublimation of the analyte as ions were drawn into a time-of-flight spectrometer for mass analysis.[9] The timing of the laser pulses was such that the entire spectrum could be recorded during the period between pulses. The mass spectrum shown in Figure 20-7 was obtained in this way. The analyte was a monoclonal antibody from a mouse having a molecular mass of approximately 150,000 Da. The matrix material was nicotinic acid, which absorbs strongly at 266 nm, the output wavelength of the laser. The analyte exhibited little absorbance at this wavelength. Note that the spectrum is characterized by very low background noise and a complete absence of fragmentation of the large analyte ion. Note also that multiply charged ions having +2 and +3 charges are present as well as peaks for two dimer and one trimer species.

The mechanism of the process described in the previous paragraph is not completely understood nor are all the properties required of a compound to make it suitable as the matrix. Empirical observations show that the matrix compound must absorb the laser radia-

---

[6] For a general discussion of desorption methods, see K. L. Busch and R. G. Cooks, *Science*, **1982**, *218*, 247.

[7] For reviews of this type of source, see L. Prakai, *Field Desorption Mass Spectrometry*. New York: Marcel Dekker, 1989; R. P. Lattimer and H. R. Schulten, *Anal. Chem.*, **1989**, *61*, 1201A; T. Kormori, T. Kawasaki, and H. R. Schulten, *Mass Spectrom. Rev.*, **1985**, *4*, 255.

[8] See M. Karas and F. Hillenkamp, *Anal. Chem.*, **1988**, *60*, 2299; K. Tanaka, H. Waki, Y. Ido, S. Akita, Y. Yoshidda, and T. Yoshidda, *Rapid Commun. Mass Spectro.*, **1988**, *2*, 151. See also F. Hillenkamp, M. Karas, and B. T. Chait, *Anal. Chem.*, **1991**, *63*, 1193A.

[9] Instruments for MALDI are now available from commercial sources. See D. Noble, *Anal. Chem.*, **1995**, *67*, 497A.

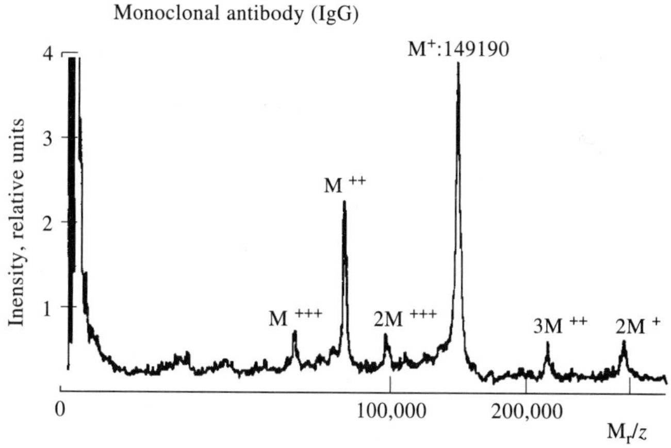

Figure 20-7 MALDI spectrum from a nicotinic acid matrix irradiated with a 266-nm laser beam. *(From M. Karas and U. Bahr,* Trends Anal. Chem., ***1990,*** *9, 323.)*

tion strongly and must be soluble enough in the sample solvent so that it can be present in large excess in the solid mixture deposited on the probe. (The analyte should not absorb the laser radiation to any significant extent because fragmentation will occur to the exclusion of molecular ion formation.) Only a few compounds have been found that are suitable as matrices for biopolymers. Several of these are shown in Table 20-4.

TABLE 20-4 Matrices Most Frequently Used for MALDI Together with the Usable Wavelengths*

Matrix	Wavelength (nm)
Nicotinic acid	266, 220–290
Benzoic acid derivatives:	
2,5-Dihydroxybenzoic acid	266, 337, 355
Vanillic acid	266
2-Amino-benzoic acid	266, 337, 355
Pyrazine-carboxylic acid	266
3-Aminopyrazine-2-carboxylic acid	337
Cinnamic acid derivatives:	
Ferulic acid	266, 377, 355
Sinapinic acid	266, 337, 355
Caffeic acid	266, 337, 355
3-Nitrobenzylalcohol	266

*From M. Karas and U. Bahr, *Trends Anal. Chem.*, **1990**, *9*, 322.

Electrospray Ionization[10]

Electrospray ionization/mass spectrometry (ESI/MS), which was first described in 1984, has now become one of the most important techniques for analyzing biomolecules, such as polypeptides, proteins, and oligonucleotides having molecular weights of 100,000 Da or more.[11] In addition, this method is beginning to find applications to the characterization of inorganic species and synthetic polymers.[12]

Electrospray ionization takes place under atmospheric pressures and temperatures in an apparatus such as that shown in Figure 20-8. A solution of the sample is pumped through a stainless steel capillary needle at a rate of a few microliters per minute. The needle is maintained at several kilovolts with respect to a cylindrical electrode that surrounds the needle. The resulting charged spray of fine droplets then passes through a desolvating capillary, where evaporation of the solvent and attachment of charge to the analyte molecules take place. As the droplets become smaller as a consequence of evaporation of the solvent, their charge density becomes greater and desorption of ions into the ambient gas occurs.

An interesting and useful feature of the electrospray process is that little fragmentation of large and

[10] R. D. Smith, J. A. Loo, C. G. Edmonds, C. J. Baringa, and H. R. Udseth, *Anal. Chem.,* **1990,** *62,* 882; J. B. Fenn, M. Mann, C. K. Meng, S. F. Wong, and C. M. Whitehouse, *Science,* **1989,** *246,* 64; P. Kebarle and L. Tang, *Anal. Chem.,* **1993,** *65,* 972A; S. A. Hofstadler, R. Bakhtiar, and R. D. Smith, *J. Chem. Educ.,* **1996,** *73,* A82.

[11] R. Bakhtiar, S. A. Hofstadler, and R. D. Smith, *J. Chem. Educ.,* **1996,** *73,* A118.

[12] C. E. C. A. Hop and R. Bakhtiar, *J. Chem. Educ.,* **1996,** *73,* A162.

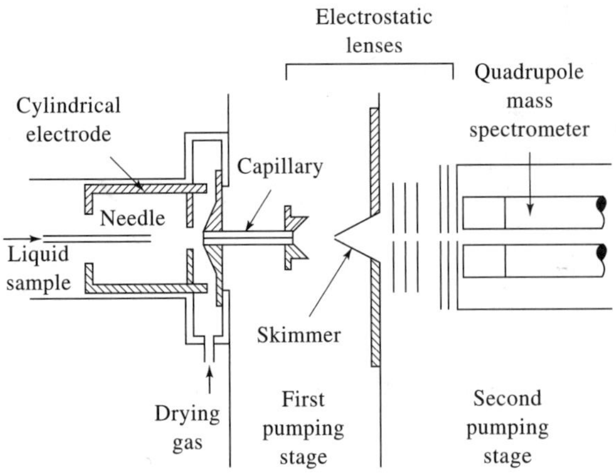

Figure 20-8 Apparatus for electrospray ionization. *(From J. B. Fenn et al.,* Science, *1989, 246, 65.)*

thermally fragile biomolecules occurs. Furthermore, the ions formed are multiply charged so that their *m/z* values are small enough to make them detectable with a quadrupole instrument with a range of 1500 or less.[13] This important property is demonstrated by the mass spectra of four proteins of varying molecular weight (M_r) shown in Figure 20-9. In these spectra, adjacent peaks are for analyte ions that differ by one charge. A striking feature of the spectra for proteins, such as those in the figure, is that the average charge state increases in approximately linear fashion with molecular weight. The charge state corresponding to each peak can be determined from peak distribution, thus making it possi-

[13] For commercially available spectrometers designed specifically for electrospray mass spectrometry, see L. Voress, *Anal. Chem.,* **1994,** *66,* 481A.

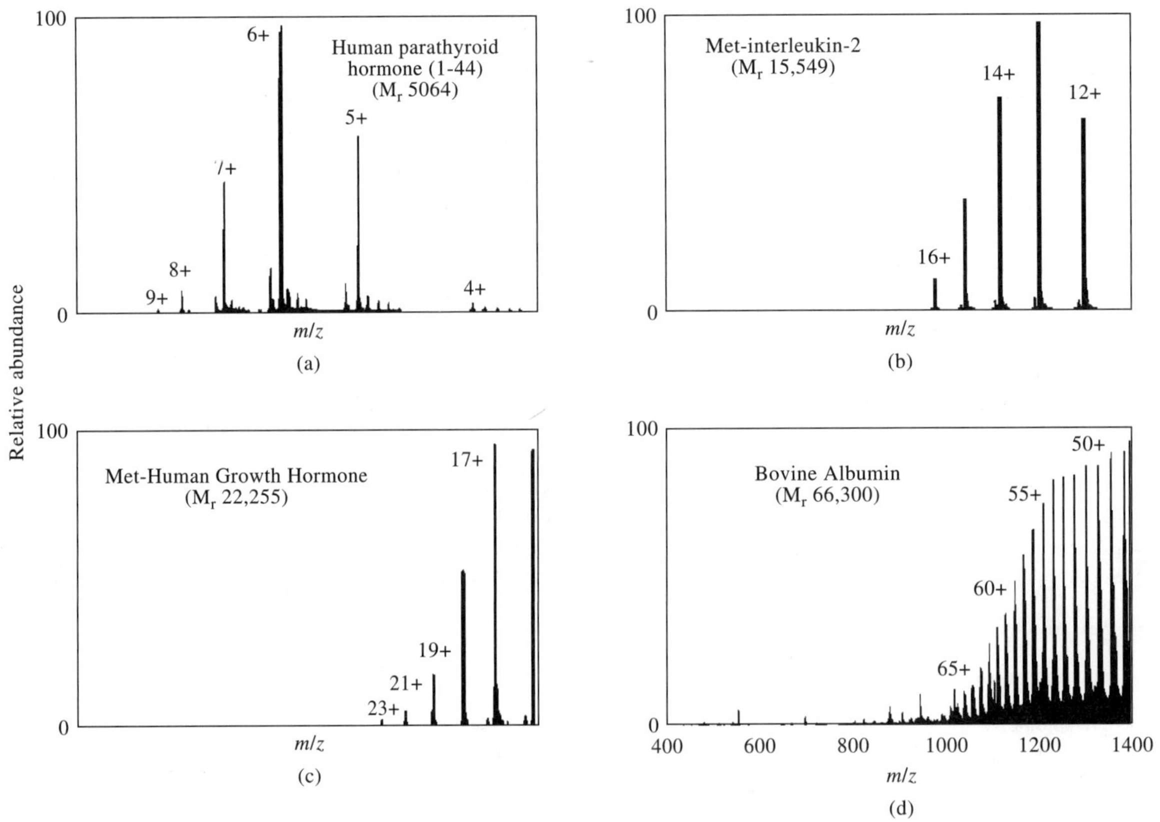

Figure 20-9 Typical electrospray mass spectra of proteins and peptides. The numbers above the peaks represent the molecular charge associated with each peak. *(From R. D. Smith et al.,* Anal. Chem., *1990, 62, 887.)*

ble to determine the molecular weight of a protein from spectra such as those shown in Figure 20-9.[14]

An important characteristic of electrospray ionization is that it is readily adapted to direct sample introduction from high-performance liquid chromatography and capillary electrophoresis columns. These applications are described in Chapters 28 and 30.

Fast Atom Bombardment Sources[15]

Fast atom bombardment (FAB) sources have assumed a major role in the production of ions for mass spectrometric studies of polar high-molecular-weight species. With this type of source, samples in a condensed state, often in a glycerol solution matrix, are ionized by bombardment with energetic (several keV) xenon or argon *atoms*. Both positive and negative analyte ions are sputtered from the surface of the sample in a desorption process. This treatment provides very rapid sample heating, which reduces sample fragmentation. The liquid matrix helps to reduce the lattice energy, which must be overcome to desorb an ion from a condensed phase, and provides a means of "healing" the damage induced by bombardment.

A beam of fast atoms is obtained by passing accelerated argon or xenon ions from an ion source, or gun, through a chamber containing argon or xenon atoms at a pressure of about 10^{-5} torr. The high-velocity ions undergo a resonant electron-exchange reaction with the atoms without substantial loss of translational energy. Thus, a beam of energetic *atoms* is formed. The lower-energy ions from the exchange are readily removed by an electrostatic deflector. Fast atom guns are now available from commercial sources, and most older spectrometers can be adapted to their uses. Newer spectrometers offer accessories permitting this kind of sample ionization.

Fast atom bombardment of organic or biochemical compounds usually produces significant amounts of molecular ions (as well as ion fragments) even for high-molecular-weight and thermally unstable samples. For example, with fast atom bombardment, molecular weights over 10,000 have been determined, and detailed structural information has been obtained for compounds with molecular weights on the order of 3000.

[14] See H. Drossman et al., *Anal. Chem.*, **1990**, *62*, 900.

[15] See K. L. Rinehart Jr., *Science*, **1982**, *218*, 254; M. Barber, R. S. Bordoli, G. J. Elliott, R. D. Sedgwick, and A. N. Taylor, *Anal. Chem.*, **1982**, *54*, 645A; K. Biemann, *Anal. Chem.*, **1986**, *58*, 1288A.

Other Desorption Methods

An examination of Table 20-1 reveals that several other desorption methods are available. Generally these produce spectra similar in character to the desorption methods we have just described.

20C MASS SPECTROMETERS

Several types of instruments are currently used for molecular mass spectrometric measurements. Two of these, the quadrupole spectrometer and the time-of-flight spectrometer, have been described in some detail in Sections 11B-2 and 11B-3. In this section, we consider other common types of mass spectrometers.

20C-1 General Description of Instrument Components

The block diagram in Figure 20-10 shows the major components of mass spectrometers. The purpose of the inlet system is to introduce a very small amount of sample (a micromole or less) into the mass spectrometer, where its components are converted to gaseous ions. Often the inlet system contains a means for volatilizing solid or liquid samples.

Ion sources of mass spectrometers, which were discussed in some detail in the previous section, convert the components of a sample into ions. In many cases the inlet system and the ion source are combined into a single component. In either case, the output is a stream of positive or negative ions (more commonly positive) that are then accelerated into the mass analyzer.

The function of the mass analyzer is analogous to that of the grating in an optical spectrometer. In the former, however, dispersion is based upon the mass-to-charge ratios of the analyte ions rather than upon the wavelength of photons. Mass spectrometers fall into several categories, depending upon the nature of the mass analyzer.

Like an optical spectrometer, a mass spectrometer contains a transducer (for ions) that converts the beam of ions into an electrical signal that can then be processed, stored in the memory of a computer, and displayed or recorded in a variety of ways. The characteristics of several types of transducers employed in mass spectrometers were described in Section 11B-1.

A characteristic feature of mass spectrometers, which is not shared by optical instruments, is the re-

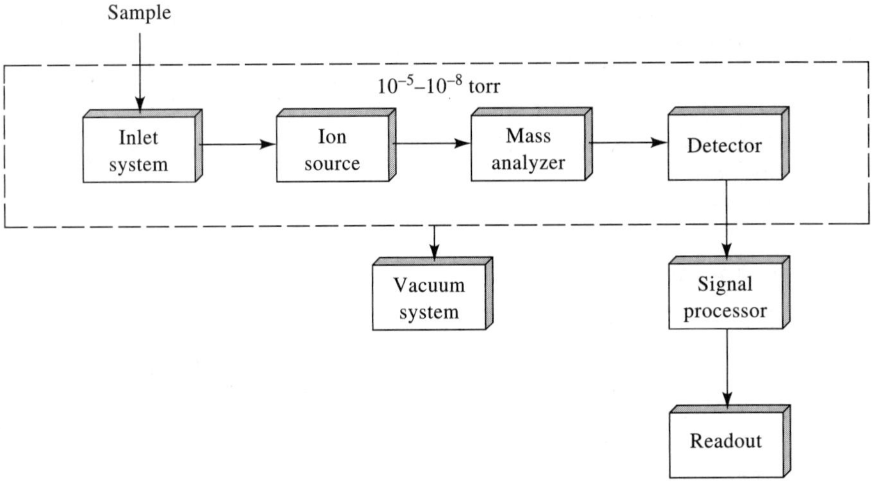

Figure 20-10 Components of a mass spectrometer.

quirement of an elaborate vacuum system to create low pressures (10^{-4} to 10^{-8} torr) in all of the instrument components except the signal processor and readout. The need for a high vacuum arises because charged particles, including electrons, interact with components of the atmosphere and are annihilated as a consequence.

In the sections that follow, we first describe inlet systems that are common to all types of mass spectrometer. We then define *resolution,* which describes the ability of a mass spectrometer to differentiate between ions with differing masses. Finally, several types of mass analyzers that lead to different categories of mass spectrometers are described.

20C-2 Sample Inlet Systems

The purpose of the inlet system is to permit introduction of a representative sample into the ion source with minimal loss of vacuum. Most modern mass spectrometers are equipped with several types of inlets to accommodate various kinds of samples; these include batch inlets, direct probe inlets, chromatographic inlets, and capillary electrophoretic inlets.

Batch Inlet Systems

The classical (and simplest) inlet system is the batch type, in which the sample is volatilized externally and then allowed to leak into the evacuated ionization region. Figure 20-11a is a schematic of a typical system that is applicable to gaseous and liquid samples having boiling points up to about 500°C. For gaseous samples, a small measured volume of gas is trapped between the two valves enclosing the metering area and is then expanded into the reservoir flask. For liquids, a small quantity of sample is introduced into a reservoir, usually with a microliter syringe. In either case, the vacuum system is used to achieve a sample pressure of 10^{-4} to 10^{-5} torr. For samples with boiling points greater than 150°C, the reservoir and tubing must be maintained at an elevated temperature by means of an oven and heating tapes. The maximum temperature of the oven is about 350°C. This maximum limits the system to liquids with boiling points below about 500°C. The sample, which is now in the gas phase, is leaked into the ionization area of the spectrometer via a metal or glass diaphragm containing one or more pinholes. The inlet system is often lined with glass to avoid losses of polar analytes by adsorption.

The Direct Probe Inlet

Solids and nonvolatile liquids can be introduced into the ionization region by means of a sample holder, or probe, which is inserted through a vacuum lock (see Figure 20-11b). The lock system is designed to limit the volume of air that must be pumped from the system after insertion of the probe into the ionization region. Probes are also used when the quantity of sample is limited, because much less sample is wasted than with the batch system. Thus, mass spectra can often be obtained with as little as a few nanograms of sample.

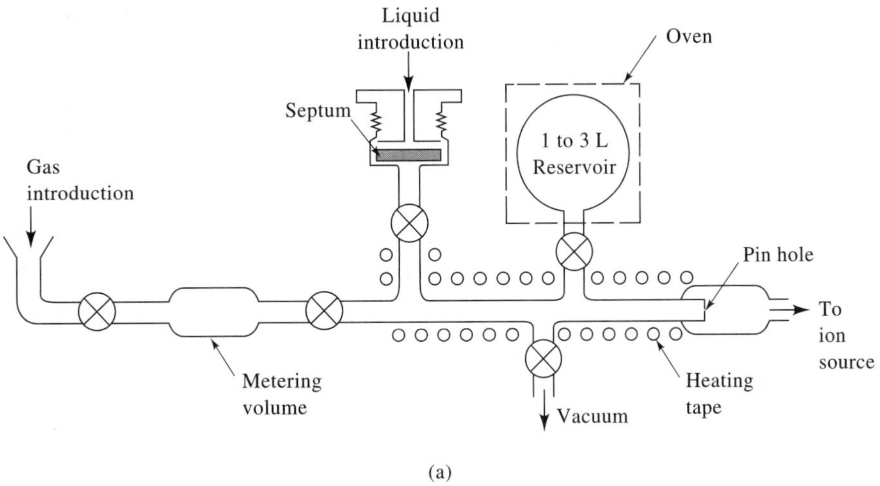

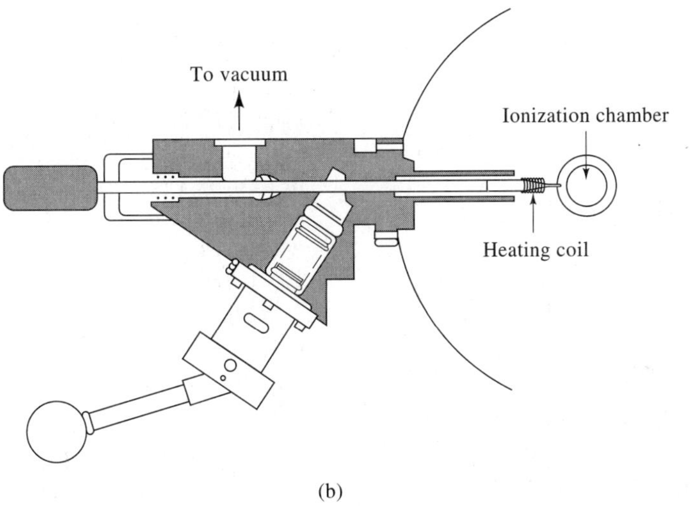

Figure 20-11 Schematic of (a) an external sample introduction system—note that the various parts are not to scale—and (b) a sample probe for inserting a sample directly into the ion source. *(From G. A. Eadon, in* Treatise on Analytical Chemistry, *2nd ed., J. D. Winefordner, M. M. Bursey, and I. M. Kolthoff, Eds., Part I. Vol. 11, p. 9. New York: Wiley, 1989. Reprinted by permission of John Wiley & Sons, Inc.)*

With a probe, the sample is generally held on the surface of a glass or aluminum capillary tube, a fine wire, or a small cup. The probe is positioned within a few millimeters of the ionization source and the slit leading to the spectrometer. Usually, provision is made for both cooling and heating the sample on the probe.

The low pressure in the ionization area and the proximity of the sample to the ionization source often make it possible to obtain spectra of thermally unstable compounds before major decomposition has time to occur. The low pressure also leads to elevated concentrations of relatively nonvolatile compounds in the ionization area. Thus, the probe permits the study of such nonvolatile materials as carbohydrates, steroids, metal-organic species, and low-molecular-weight polymeric substances. The principal sample requirement is attainment of a partial pressure of at least 10^{-8} torr before the onset of decomposition.

Chromatographic and Capillary Electrophoretic Inlet Systems

Mass spectrometers are often coupled with gas or high-performance liquid chromatographic systems or capillary electrophoresis columns to permit the separation and determination of the components of complex mixtures. Linking a chromatographic or electrophoretic column to a mass spectrometer requires the use of specialized inlet systems, some of which are described in Sections 27D-3, 28C-6, and 30B-4.

20C-3 Mass Analyzers

Several devices are available for separating ions with different mass-to-charge ratios. Ideally, the mass analyzer should be capable of distinguishing between minute mass differences. In addition, the analyzer should allow passage of a sufficient number of ions to yield readily measurable ion currents. As with an optical monochromator, to which the analyzer is analogous, these two properties are not entirely compatible, and design compromises must always be made.

Resolution of Mass Spectrometers

The capability of a mass spectrometer to differentiate between masses is usually stated in terms of its *resolution, R,* which is defined as

$$R = m/\Delta m \qquad (20\text{-}3)$$

where Δm is the mass difference between two adjacent peaks that are just resolved and m is the nominal mass of the first peak (the mean mass of the two peaks is sometimes used instead). Two peaks are considered to be separated if the height of the valley between them is no more than a given fraction of their height (often 10%). Thus, a spectrometer with a resolution of 4000 would resolve peaks occurring at m/z values of 400.0 and 400.1 (or 40.00 and 40.01).

The resolution needed in a mass spectrometer depends greatly upon its application. For example, discrimination among ions of the same nominal mass such as $C_2H_4^+$, CH_2N^+, N_2^+, and CO^+ (all ions of nominal mass 28 Da but exact masses of 28.0313, 28.0187, 28.0061, and 27.9949 Da, respectively) requires an instrument with a resolution of several thousand. On the other hand, low-molecular-weight ions differing by a unit of mass or more NH_3^+ ($m = 17$) and CH_4^+ ($m = 16$), for example, can be distinguished with an instrument having a resolution smaller than 50. Commercial spectrometers are available with resolutions ranging from about 500 to 500,000.

EXAMPLE 20-3

What resolution is needed to separate the first two ions mentioned in the previous paragraph? Here,

$$\Delta m = 28.0313 - 28.0187 = 0.0126$$

Substituting into Equation 20-1 gives

$$R = m/\Delta m = 28.025/0.0126 = 2.22 \times 10^3$$

where 28.025 is the mean mass for the two species.

Magnetic Sector Analyzers

Magnetic sector analyzers employ a permanent magnet or an electromagnet to cause the beam from the ion source to travel in a circular path, most commonly of 180, 90, or 60 deg. Figure 20-12 shows a 90-deg sector instrument in which ions formed by electron impact are accelerated through slit B into the metal analyzer tube, which is maintained at an internal pressure of about 10^{-7} torr. Ions of different mass can be scanned across the exit slit by varying the field strength of the magnet or the accelerating potential between slits A and B. The ions passing through the exit slit fall on a collector electrode, resulting in an ion current that is amplified and recorded.

The translational, or kinetic, energy KE of an ion of mass m bearing a charge z upon exiting slit B is given by

$$\text{KE} = zeV = \frac{1}{2}mv^2 \qquad (20\text{-}4)$$

where V is the voltage between A and B, v is the velocity of the ion after acceleration, and e is the electronic charge ($e = 1.60 \times 10^{-19}$ C). Note that all ions having the same number of charges z are assumed to have the same kinetic energy after acceleration regardless of their mass. This assumption is only approximately true, because before acceleration, the ions possess a statistical distribution of velocities (speeds and directions), which will be reflected in a similar distribution for the accelerated ion. The limitations of this assumption are discussed in the next section, on double-focusing instruments. Because all ions leaving the slit have approximately the same kinetic en-

Figure 20-12 Schematic of a magnetic sector spectrometer.

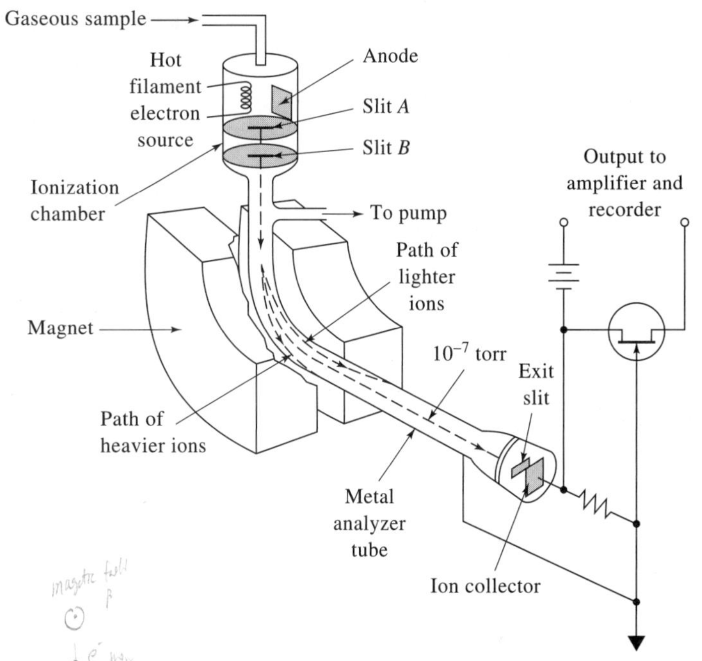

ergy, the heavier ions must travel through the magnetic sector at lower velocities.

The path in the sector described by ions of a given mass and charge represents a balance between two forces acting upon them. The magnetic force F_M is given by the relationship

$$F_M = Bzev \tag{20-5}$$

where B is the magnetic field strength. The balancing centripetal force F_c, is given by

$$F_c = \frac{mv^2}{r} \tag{20-6}$$

where r is the radius of curvature of the magnetic sector. In order for an ion to traverse the circular path to the collector, F_M and F_c must be equal. Thus, equating Equations 20-5 and 20-6 leads to

$$Bzev = \frac{mv^2}{r} \tag{20-7}$$

which rearranges to

$$v = \frac{Bzer}{m} \tag{20-8}$$

Substituting Equation 20-8 into 20-4 gives, after rearranging,

$$\boxed{\frac{m}{z} = \frac{B^2r^2e}{2V}} \tag{20-9}$$

Equation 20-9 reveals that mass spectra can be acquired by varying one of three variables (B, V, or r) while holding the other two constant. Most modern sector mass spectrometers contain an electromagnet in which ions are sorted by holding V and r constant while varying the current in the magnet and thus B. In sector spectrometers that use photographic recording, B and V are constant, and r is the variable (see Figure 11-9).

EXAMPLE 20-4

What accelerating potential will be required to direct a singly charged water molecule through the exit slit of a magnetic mass spectrometer if the magnet has a field strength of 0.240 T (tesla) and the radius of curvature of the ion through the magnetic field is 12.7 cm?

First, we convert all experimental variables into SI units. Thus,

$$\text{charge per ion } ez = 1.60 \times 10^{-19}\ \text{C} \times 1$$

$$\text{radius } r = 0.127\ \text{m}$$

$$\text{mass } m = \frac{18.02 \text{ g H}_2\text{O}^+/\text{mol}}{6.02 \times 10^{23} \text{ H}_2\text{O}^+/\text{mol}} \times 10^{-3} \frac{\text{kg}}{\text{g}}$$

$$= 2.99 \times 10^{-26} \text{ kg H}_2\text{O}^+$$

$$\text{magnetic field } B = 0.240 \text{ T} = 0.240 \text{ W/m}^2$$

We then substitute into Equation 20-9 and solve for the accelerating potential V:

$$V = \frac{B^2 r^2 ez}{2\text{m}}$$

$$= \frac{[0.240 \text{ W/m}^2]^2 [0.127 \text{ m}]^2 [1.60 \times 10^{-19} \text{ C}]}{2 \times 2.99 \times 10^{-26} \text{ kg}}$$

$$= 2.49 \times 10^3 \frac{\text{W}^2\text{C}}{\text{m}^2 \text{ kg}} = 2.49 \times 10^3 \text{ V}$$

It is not obvious that $W^2C/(m^2 \text{ kg})$ and volts are equivalent, but they are.

Double-Focusing Spectrometers

The magnetic sector instruments discussed in the previous section are sometimes called *single-focusing* spectrometers. This terminology is used because a collection of ions exiting the source with the same mass-to-charge ratio but with small diverging directional distribution will be acted upon by the magnetic field in such a way that a converging directional distribution is produced as the ions leave the field. The ability of a magnetic field to bring ions with different directional orientations to focus means that the distribution of translational energies of ions leaving the source is the factor most responsible for limiting the resolution of magnetic sector instruments ($R \leq 2000$).

The translational energy distribution of ions leaving a source arises from the Boltzmann distribution of energies of the molecules from which the ions are formed and from field inhomogeneities in the source. The spread of kinetic energies causes a broadening of the beam reaching the transducer and thus a loss of resolution. In order to measure atomic and molecular masses with a precision of a few parts per million, it is necessary to design instruments that correct for both the directional distribution and energy distribution of ions leaving the source. The term *double focusing* is applied to mass spectrometers in which the directional aberrations and the energy aberrations of a population of ions are simultaneously minimized. Double focusing is usually achieved by the use of carefully selected combinations of electrostatic and magnetic fields. In the double-focusing instrument, shown schematically in Figure 20-13, the ion beam is first passed through an electrostatic analyzer (ESA) consisting of two smooth curved metallic plates across which a dc potential is applied. This potential has the effect of limiting the kinetic energy of the ions reaching the magnetic sector to a closely defined range. Ions with energies greater than average strike the upper side of the ESA slit and are lost to ground. Ions with energies less than average strike the lower side of the ESA slit and are thus removed.

Directional focusing in the magnetic sector occurs along the focal plane labeled *d* in Figure 20-13, energy focusing takes place along the plane labeled *e*. Thus, only ions of one m/z are double focused at the intersection of *d* and *e* for any given accelerating voltage and magnetic field strength. Therefore, the collector slit is located at this locus of double focus.

A wide variety of double-focusing mass spectrometers is available commercially. The most sophisticated of these are capable of resolution in the 10^5 range. More compact double-focusing instruments can also be purchased (for considerably less money). A typical instrument of this type will have a 6-in. electrostatic sector and a 4-in. 90-deg magnetic deflector. Resolutions of about 2500 are common with such instruments. Often they are employed as detection systems for chromatographic columns.

The spectrometer shown in Figure 20-13 is based on the so-called *Nier-Johnson* design. Another double-focusing design, which employs *Mattauch-Herzog* geometry, is shown in Figure 11-9. The geometry of this type of instrument is unique in that the energy and direction focal planes coincide; for this reason, the Mattauch-Herzog design often uses a photographic plate for recording the spectrum. The photographic plate is located along the focal plane, where all of the ions are in focus, regardless of mass-to-charge ratio.

Quadrupole Mass Spectrometers

Quadrupole mass spectrometers are usually less expensive and more rugged than their magnetic sector counterparts. They are generally considerably more compact than magnetic sector instruments and are commonly found in commercial benchtop mass spectrometers. They also offer the advantage of low scan times (that is, $<$100 ms), which is particularly useful for real-time scanning of chromatographic peaks. Quadrupole ana-

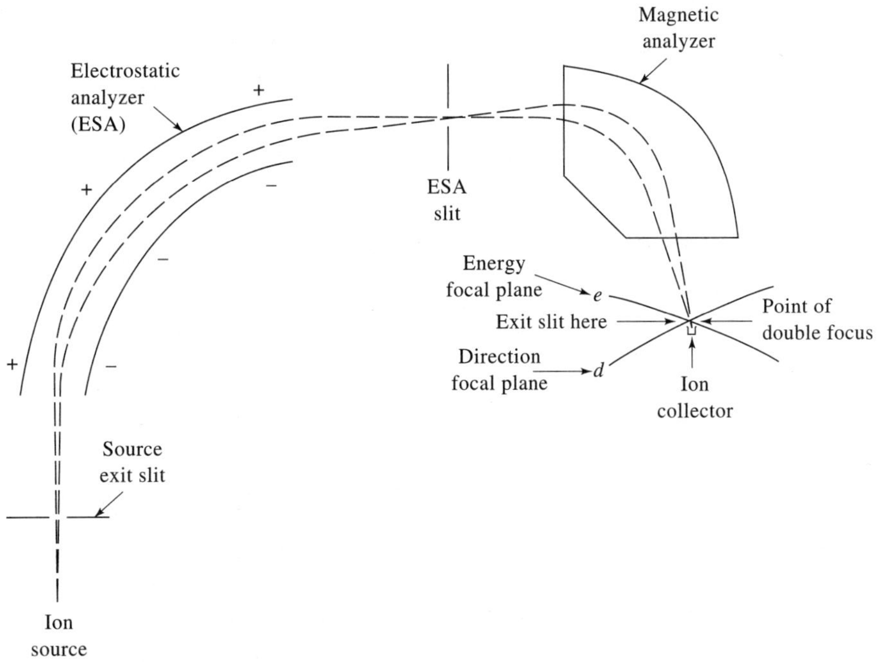

Figure 20-13 Nier-Johnson design of a double-focusing mass spectrometer.

lyzers are by far the most common mass analyzers in use today. A detailed discussion of quadrupole mass spectrometers is found in Section 11B-2.

Time-of-Flight Mass Analyzers

As shown in Section 11B-3, in time-of-flight (TOF) instruments, positive ions are produced periodically by bombardment of the sample with brief pulses of electrons, secondary ions, or laser-generated photons. The ions produced in this way are then accelerated into a field-free drift tube by an electric field pulse of 10^3 to 10^4 V (see Figure 20-14). Separation of ions on the basis of mass occurs during the transit of the ions to the detector located at the end of the tube. Because all ions entering the tube have the same kinetic energy, their velocities in the tube vary inversely with their masses (Equation 20-4), with the lighter particles arriving at the detector earlier than the heavier ones. Typical flight times are 1 to 30 μs.

Time-of-flight instruments offer several advantages over other types of mass spectrometers, including simplicity and ruggedness, ease of accessibility of the ion source, and virtually unlimited mass range. They suffer, however from limited resolution and sensitivity. Several instrument manufacturers now offer time-of-flight instruments, but these are less widely used than are magnetic sector and quadrupole mass spectrometers.

Ion Trap Analyzers

An ion trap is a device in which gaseous anions or cations can be formed and confined for extended periods by electric and/or magnetic fields. Several types of ion traps have been developed,[16] and two are currently used in commercial mass spectrometers. One of these traps is the ion cyclotron resonance trap discussed in Section 20C-4. In this section, we discuss a simpler type of ion trap that was first developed as a detector for gas chromatography (GC/MS) but is now used to obtain mass spectra of a variety of analytes.

Figure 20-15 is a cross-sectional view of a simple commercially available ion trap.[17] It consists of a cen-

[16] For a discussion of the various types of ion traps, see J. Allison and R. M. Stepnowski, *Anal. Chem.*, **1987**, *59*, 1072A; R. G. Cooks, S. A. McLuckey, and R. E. Kaiser, *Chem. & Eng. News*, **1991**, March 25; S. A. McLuckey, G. J. Van Berkel, D. E. Goeringer, A. L. Gish, *Anal. Chem.*, **1994**, *66*, 689A, 737A.

[17] For a detailed description of this device, see G. C. Stafford, P. E. Kelly, J. E. P. Syka, W. E. Reynolds, and J. F. J. Todd, *Int. J. Mass Spectrom. Ion Proc.*, **1984**, *60*, 85.

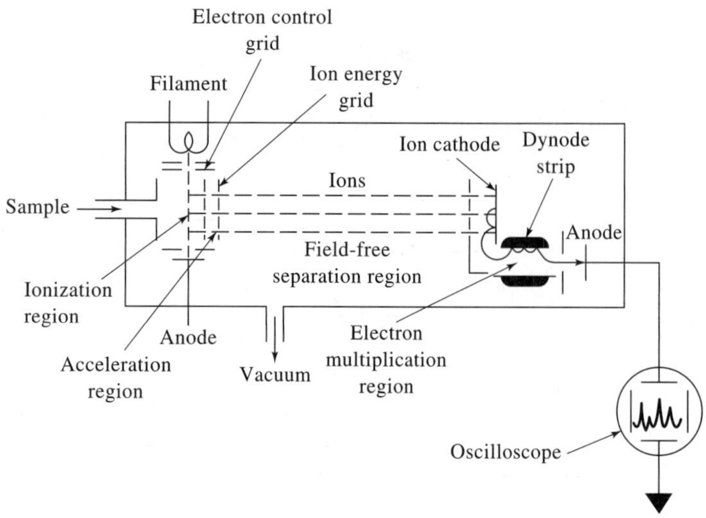

Figure 20-14 Schematic of a time-of-flight mass spectrometer.

tral doughnut-shaped ring electrode and a pair of end-cap electrodes. A variable radio-frequency voltage is applied to the ring electrode while the two end-cap electrodes are grounded. Ions with an appropriate *m/z* value circulate in a stable orbit within the cavity surrounded by the ring. As the radio-frequency voltage is increased, the orbits of heavier ions become stabilized, while those for lighter ions become destabilized, causing them to collide with the wall of the ring electrode.

In operating this device as a mass spectrometer, a burst of analyte ions from an electron-impact or chemical ionization source is admitted through a grid in the upper end cap. The radio-frequency voltage is then scanned, and the trapped ions, as they become destabilized, leave the ring electrode cavity via openings in the lower end cap. The emitted ions then pass into a transducer.

Ion trap spectrometers are rugged, compact, and less costly than sector or quadrupole instruments. A current commercial version of this device, which is illustrated in Figure 27-15, is capable of resolving ions that differ in mass by one unit in the 500 to 1000 Da mass range.

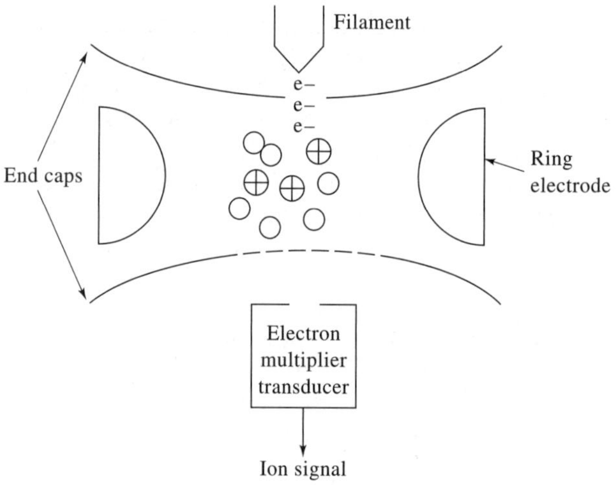

Figure 20-15 Ion trap mass spectrometer. *(Adapted from J. T. Watson,* Introduction to Mass Spectrometry, *p. 89. Philadelphia: Lippincott-Raven Press, 1997. With permission.)*

20C-4 Fourier Transform (FT) Instruments[18]

As was true with infrared and nuclear magnetic resonance instruments, Fourier transform mass spectrometers provide improved signal-to-noise ratios, greater speed, and higher sensitivity and resolution. Commercial Fourier transform mass spectrometers appeared on the market in the early 1980s and are now offered by several manufacturers.

The heart of a Fourier transform instrument is an *ion trap* within which ions can circulate in well-defined orbits for extended periods. Such cavities are constructed to take advantage of the *ion cyclotron resonance phenomenon.*

The Ion Cyclotron Resonance (ICR) Phenomenon[19]

When a gaseous ion drifts into or is formed in a strong magnetic field, its motion becomes circular in a plane that is perpendicular to the direction of the field. The angular frequency of this motion is called the *cyclotron frequency,* ω_c. Equation 20-8 can be rearranged and solved for v/r, which is the cyclotron frequency in radians per second.

$$\omega_c = \frac{v}{r} = \frac{zeB}{m} \qquad (20\text{-}10)$$

Note that in a fixed field, the cyclotron frequency depends only upon the inverse of the m/z value. Increases in the velocity of an ion will be accompanied by a corresponding increase in the radius of rotation of the ion.

An ion trapped in a circular path in a magnetic field is capable of absorbing energy from an ac electric field, provided the frequency of the field matches the cyclotron frequency. The absorbed energy then increases the velocity of the ion (and thus its radius of travel) without disturbing ω_c. This effect is illustrated in Figure 20-16. Here, the original path of an ion trapped in a magnetic field is depicted by the inner solid circle. Brief application of ac voltage creates a fluctuating field between the plates that interacts with the ion, provided the frequency of the source is close to the cyclotron frequency of the ion. Under this circumstance, the velocity of the ion increases continuously as does the radius of its path (see dashed line). When the ac electrical signal is terminated, the radius of the path of the ion again becomes constant, as is suggested by the outer solid circle in the figure.

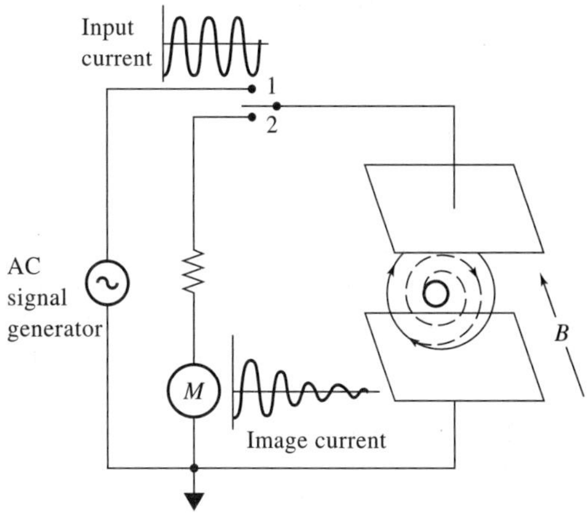

Figure 20-16 Path of an ion in a strong magnetic field. Inner solid line represents the original circular path of the ion. Dashed line shows spiral path when switch is moved briefly to position 1. Outer solid line is new circular path when switch is again opened.

When the region between the plates in Figure 20-16 contains an ensemble of ions of the same mass-to-charge ratio, application of the ac signal having the cyclotron resonance frequency sets all of the particles into coherent motion that is in phase with the field. Ions of different cyclotron frequency, that is, those with different mass-to-charge ratios, are unaffected by the ac field.

Measurement of the ICR Signal

The coherent circular motion of resonant ions creates a so-called *image current* that can be conveniently observed after termination of the frequency sweep signal. Thus, if the switch in Figure 20-16 is moved from position 1 to position 2, a current is observed that decreases exponentially with time. This image current is a capacitor current induced by the circular movement

[18] For greater details on Fourier transform mass spectrometry, see M. L. Gross and D. L. Rempel, *Science,* **1984,** *226,* 261; A. G. Marshall and P. B. Grosshans, *Anal. Chem.,* **1991,** *63,* 215A; M. Johnson, *Spectroscopy,* **1987,** *2*(2), 14 and **1987,** *2*(3), 14; C. D. Hanson, E. L. Kerley, and D. H. Russell, in *Treatise on Analytical Chemistry,* 2nd ed., J. D. Winefordner, M. M. Bursey, and I. M. Kolthoff, Eds., Part I, Vol. 11, Chapter 2. New York: Wiley, 1989.

[19] For detailed discussion, see T. A. Lehman and M. M. Bursey, *Ion Cyclotron Resonance Spectroscopy.* New York, Wiley, 1976; M. L. Gross and C. L. Wilkins, *Anal. Chem.,* **1981,** *43,* 65A.

of a packet of ions with the same mass-to-charge ratios. For example, as a packet of positive ions approaches the upper plate in Figure 20-16, electrons are attracted from circuit common to this plate, causing a momentary current. As the packet continues around toward the other plate, the direction of external electron flow is reversed. The magnitude of the resulting alternating current depends upon the number of ions in the packet. The frequency of the current is characteristic of the mass-to-charge value of the ions in the packet. This current is employed in ion cyclotron spectrometers to measure the concentration of ions that are brought into resonance at various applied signal frequencies.

The induced image current just described decays over a period of a few tenths of a second to several seconds as the coherent character of the circulating packet of ions is lost. Collisions between ions provide the mechanism by which the coherently circulating ions lose energy and the ions return to a condition of thermal equilibrium. This decay of the image current provides a time-domain signal that is similar to the FID signal encountered in FT-NMR experiments (see Section 19A-3).

Fourier Transform Spectrometers

Fourier transform mass spectrometers are generally equipped with a *trapped ion analyzer cell* such as that shown in Figure 20-17. Gaseous sample molecules are ionized in the center of the cell by electrons that are accelerated from the filament through the cell to a collector plate. A pulsed voltage applied at the grid serves as a gate to switch the electron beam on and off periodically. The ions are held in the cell by a 1 to 5 V potential applied to the trap plate. The ions are accelerated by a radio-frequency signal applied to the transmitter plate as shown. The receiver plate is connected to a preamplifier that amplifies the image current. This approach for confining ions is highly efficient, and storage times of up to several minutes have been observed. The dimensions of the cell are not critical but are usually a few centimeters on a side.

The basis of the Fourier transform measurement is illustrated in Figure 20-18. Ions are first generated by a brief electron beam pulse (not shown) and stored in the trapped ion cell. After a brief delay, the trapped ions are subjected to a short radio-frequency pulse that increases linearly in frequency during its lifetime. (Figure 20-18a shows a pulse of 5 ms, during which time the frequency increases

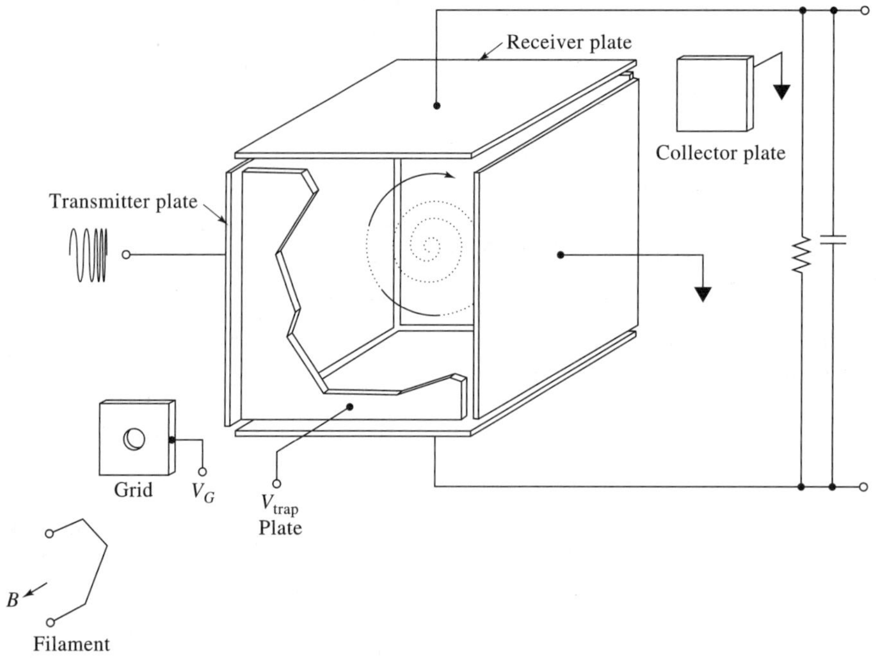

Figure 20-17 A trapped ion analyzer cell. *(Reprinted from E. B. Ledford Jr., R. L. White, S. Ghaderi, and C. L. Wilkins,* Anal. Chem., ***1980****, 52, 1091. Copyright 1980 American Chemical Society.)*

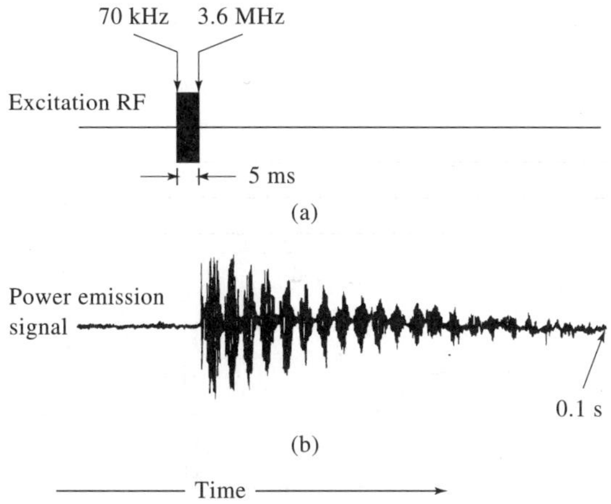

Figure 20-18 Schematic showing the timing of (a) the radio-frequency signal and (b) the transient image signal (lower). *(Reprinted with permission from R. T. McIver Jr.,* Amer. Lab., ***1980,*** *12(11), 26. Copyright 1980 by International Scientific Communications, Inc.)*

linearly from 0.070 to 3.6 MHz.) After the frequency sweep is discontinued, the image current, induced by the various ion packets, is amplified, digitized, and stored in memory. The time-domain decay signal, shown in Figure 20-18b, is then transformed to yield a frequency-domain signal that can be converted to a mass-domain signal via Equation 20-10. Figure 20-19 illustrates the relationship between a time-domain spectrum, its frequency-domain counterpart, and the resulting mass spectrum.

Fourier transform spectrometers are expensive instruments (>$400,000). One commercial model employs a superconducting magnet with a nominal field of 1.9 tesla. Resolution in Fourier transform mass spectrometry is limited by the precision of the frequency measurement rather than slits or field measurements. Because frequency measurements can be made with high precision, extremely high resolution is possible (in excess of 10^6).

20C-5 Computerized Mass Spectrometers

Microcomputers and microprocessors are an integral part of modern mass spectrometers.[20] A characteristic of a mass spectrum is the wealth of structural data that it

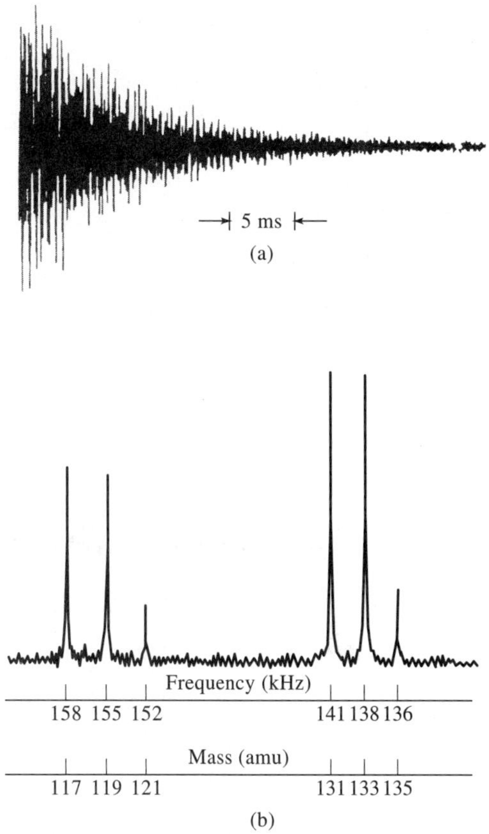

Figure 20-19 Time-domain (a) and (b) frequency- or mass-domain spectrum for 1,1,1,2-tetrachloroethane. *(Reprinted with permission from E. B. Ledford Jr. et al.,* Anal. Chem., ***1980,*** *52, 466. Copyright 1980 American Chemical Society.)*

provides. For example, a molecule with a molecular weight of 500 may be fragmented by an electron beam into 100 or more different ions, each of which leads to a discrete spectral peak. For a structural determination, the heights, and mass-to-charge ratios of each peak must be determined, stored, and ultimately displayed. Because the amount of information is so large, it is essential that acquisition and processing be rapid; the computer is ideally suited for these tasks. Moreover, in order for mass spectral data to be useful, several instrumental variables must be closely controlled or monitored during data collection. Computers and microprocessors are much more efficient than a human operator in exercising such controls.

Figure 20-20 is a block diagram of the computerized control and data-acquisition system of a quadrupole mass spectrometer. This figure shows two features

[20] For a detailed discussion of computerized mass spectrometry, see J. R. Chapman, *Computers in Mass Spectrometry.* New York: Academic Press, 1978.

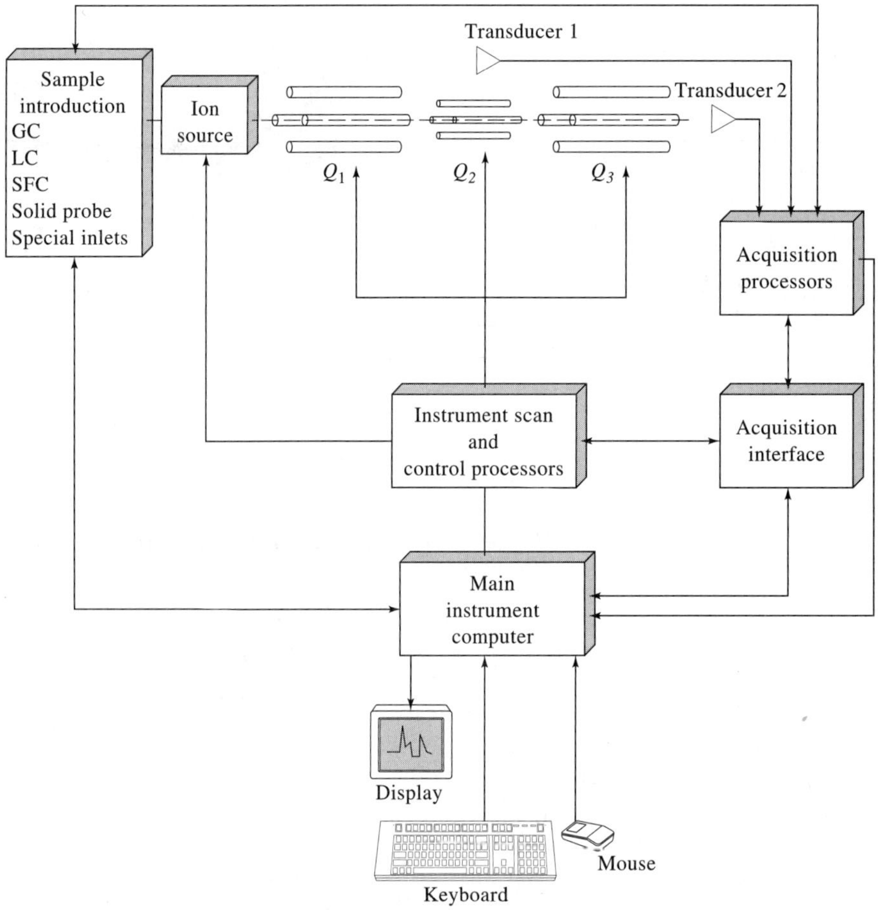

Figure 20-20 Instrument control and data-processing architecture of the VG instrument TRIO-3 triple quadrupole MS/MS. *(Courtesy of VG Instruments, Inc., Stamford, CT.)*

that will be encountered in any modern instrument. The first is a computer that serves as the main instrument controller. The operator communicates via a keyboard with the spectrometer by selecting operating parameters and conditions via easy-to-use interactive software. The computer also controls the programs responsible for data manipulations and output. The second feature common to almost all instruments is a set of microprocessors (often as many as six) that are responsible for specific aspects of instrument control and/or the transmission of information between the computer and spectrometer.

The interface between a mass spectrometer and a computer usually has provisions for digitizing the amplified ion-current signal plus several other signals that are used for control of instrumental variables. Examples of the latter are source temperature, accelerating voltage, scan rate, and magnetic field strength or quadrupole voltages.

The digitized ion-current signal ordinarily requires considerable processing before it is ready for display. First, the peaks must be normalized, a process whereby the height of each peak relative to some reference peak is calculated. Most often the *base peak,* which is the largest peak in a spectrum, serves as the reference and is arbitrarily given a peak height of 100 (sometimes 1000). The *m/z* value for each peak must also be determined. This assignment is frequently based on the time of the peak's appearance and the scan rate. Data are acquired as intensity versus time during a carefully controlled scan of the magnetic and/or electric fields. Conversion from time to *m/z* requires careful periodic

Figure 20-21 A computer display of mass-spectral data. The compound was isolated from a blood serum extract by chromatography. The spectrum showed it to be the barbiturate, pentobarbital. The instrument was a DuPont Model 21-094 computerized mass spectrometer. *(Courtesy of DuPont Instrument System, Wilmington, DE.)*

BARBITURATE

BACKGR 0 BASE 0 SUBTRT 0
IGNORE 0, 0, 0, 0
% F.S. 100 SEQUEN 191
BASE 10012 *2** 0

36	3	61	2	83	30	108	4	133	3	166	1
37	3	62	1	84	13	109	14	134	2	167	2
38	12	63	3	85	38	110	9	135	2	168	3
39	139	64	1	86	8	111	10	136	2	169	5
40	38	65	13	87	9	112	73	137	6	179	1
41	364	66	13	88	1	113	21	138	7	181	4
42	83	67	58	91	5	114	18	139	6	183	7
43	378	68	33	92	3	115	2	140	20	185	2
44	84	69	120	93	4	116	1	141	826	191	1
45	6	70	67	94	21	117	1	142	71	193	1
50	5	71	89	95	17	119	2	143	11	195	2
51	11	72	5	96	20	120	1	144	1	197	65
52	14	73	9	97	55	121	2	151	1	198	10
53	64	74	8	98	127	122	4	152	1	199	2
54	31	75	3	99	9	123	5	153	6	204	1
55	162	77	10	100	3	124	7	155	133	207	6
56	30	78	5	101	3	125	4	156	1000	208	2
57	17	79	12	103	1	126	12	157	273	209	1
58	5	80	22	105	2	128	14	158	29	227	6
59	1	81	18	106	3	129	12	159	3	228	1
60	5	82	13	107	2	130	2	165	1		

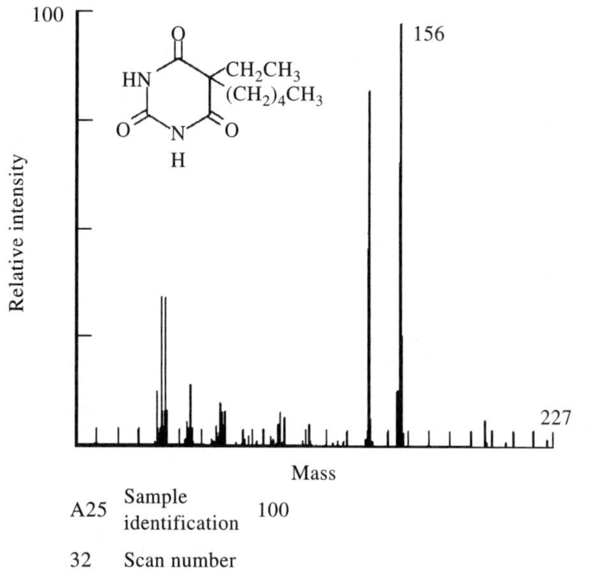

calibration; for this purpose, perfluorotri-*n*-butylamine (PFTBA) or perfluorokerosene is often used as a standard. For high-resolution work, the standard may be admitted with the sample. The computer is programmed to then recognize and use the peaks of the standard as references for mass assignments. For low-resolution instruments, the calibration must generally be obtained separately from the sample, because of the likelihood of peak overlaps.

With most systems the computer stores all spectra and related information on a disk. In routine applications, bar graphs of the normalized spectra can be sent directly to a printer/plotter. However, in many cases, the spectroscopist will use data-reduction software to extract specific information prior to producing a printed copy of a spectrum. Figure 20-21 is an example of the printout from a computerized mass spectrometer. The odd columns in the table list m/z values in increasing order. The even columns contain the corresponding ion currents normalized to the largest peak found at mass 156. The current for this ion is assigned the number of 1000, and all other peaks are relative to this one. Thus, the height of the peak at mass 141 is 82.6% of the base peak.

As is true with infrared and nuclear magnetic resonance spectroscopy, large libraries of mass spectra (>150,000 entries) are available in computer-compatible formats. Most commercial mass spectrometer computer systems have the ability to rapidly search all or part of such files for spectra that match or closely match the spectrum of an analyte.

20D APPLICATIONS OF MOLECULAR MASS SPECTROMETRY

The applications of molecular mass spectrometry are so numerous and widespread that describing them adequately in a brief space is not possible. Table 20-5 lists some of these applications to provide some idea of the ubiquity of mass spectrometry. In this section we describe a few of the most widely used and important of these applications.

20D-1 Identification of Pure Compounds[21]

The mass spectrum of a pure compound provides several kinds of data that are useful for its identification. The first is the molecular weight of the compound, and the second is its molecular formula. In addition, study of fragmentation patterns revealed by the mass spectrum often provides information about the presence or absence of various functional groups. Finally, the actual identity of a compound can often be established by comparing its mass spectrum with those of known compounds until a close match is realized.

Molecular Weights from Mass Spectra

For compounds that can be ionized to give a molecular ion or a protonated or a deprotonated molecular ion by one of the methods described earlier, the mass spectrometer is an unsurpassed tool for the determination of molecular weight. This determination, of course, requires the identification of the molecular ion peak, or in

[21] F. W. McLafferty, *Interpretation of Mass Spectra,* 3rd ed. Mill Valley, CA: University Science Books, 1980; R. N. Silverstein, G. C. Bassler, and T. C. Morrill, *Spectrometric Identification of Organic Compounds,* 5th ed. New York: Wiley, 1991.

TABLE 20-5 Applications of Molecular Mass Spectrometry

1. Elucidation of the structure of organic and biological molecules
2. Determination of the molecular weight of peptides, proteins, and oligonucleotides
3. Identification of components in thin-layer and paper chromatograms
4. Determination of amino acid sequences in sample of polypeptides and proteins
5. Detection and identification of species separated by chromatography and capillary electrophoresis
6. Identification of drugs of abuse and metabolites of drugs of abuse in blood, urine, and saliva
7. Monitoring gases in patient's breath during surgery
8. Testing for the presence of drugs in blood in thoroughbred racing horses and in Olympic athletes
9. Dating archaeological specimens
10. Analyses of aerosol particles
11. Determination of pesticide residues in food
12. Monitoring volatile organic species in water supplies

some cases, the $(M + 1)^+$ or the $(M - 1)^+$ peak. The location of the peak on the abscissa then gives the molecular weight with an accuracy that cannot be realized easily by any other method.

A mass spectrometric molecular weight determination requires the sure knowledge of the identity of the molecular ion peak. Caution is therefore always advisable, particularly with electron-impact sources, where the molecular ion peak may be absent or small relative to impurity peaks. When there is doubt, additional spectra by chemical, field, and desorption ionization are particularly useful.

Molecular Formulas from Exact Molecular Weights

Molecular formulas can be determined from the mass spectrum of a compound, provided the molecular ion peak can be identified and its *exact* mass determined. This application, however, requires a high-resolution instrument capable of detecting mass differences of a few thousandths of a mass unit. Consider, for example, the mass-to-charge ratios of the molecular ions of the following compounds: purine, $C_5H_4N_4$ (m = 120.044); benzamidine, $C_7H_8N_2$ (m = 120.069); ethyltoluene, C_9H_{12} (m = 120.096); and acetophenone, C_8H_8O (m = 120.058). If the measured mass of the molecular ion peak is 120.070 (±0.005), then all but $C_7H_8N_2$ are excluded as possible formulas. Note that the precision in this example is about 40 ppm; precisions on the order of a few parts per million are routinely achievable with high-resolution, double-focusing instruments. Tables that list all reasonable combinations of C, H, N, and O by molecular weight to the third or fourth decimal place have been compiled.[22] A small portion of such a compilation is shown in the fifth column of Table 20-6.

Molecular Formulas from Isotopic Ratios

The data from a low-resolution instrument that can only discriminate between ions differing in mass by whole mass numbers can also yield useful information about the formula of a compound, provided only that the molecular ion peak is sufficiently intense that its height and the heights of the $(M + 1)^+$ and $(M + 2)^+$ isotope peaks can be determined accurately. The following example illustrates this type of analysis.

EXAMPLE 20-5

Calculate the ratios of the $(M + 1)^+$ to M^+ peak heights for the following two compounds: dinitrobenzene, $C_6H_4N_2O_4$ (m = 168), and an olefin, $C_{12}H_{24}$ (m = 168).

From Table 20-4, we see that for every 100 ^{12}C atoms there are 1.08 ^{13}C atoms. Because there are six carbon atoms in nitrobenzene, we would expect there to be 6.48 = 6 × 1.08 molecules of nitrobenzene having one ^{13}C atom for every 100 molecules having none. Thus, from this effect alone the $(M + 1)^+$ peak will be 6.48% of the M^+ peak. The isotopes of the other elements also contribute to this peak; we may tabulate their effects as follows:

$C_6H_4N_2O_4$	
^{13}C	6 × 1.08 = 6.48%
^{2}H	4 × 0.015 = 0.060%
^{15}N	2 × 0.37 = 0.74%
^{17}O	4 × 0.04 = 0.16%
	$(M + 1)^+/M^+$ = 7.44%

$C_{12}H_{24}$	
^{13}C	12 × 1.08 = 12.96%
^{2}H	24 × 0.015 = 0.36%
	$(M + 1)^+/M^+$ = 13.32%

Thus, if the heights of the M^+ and $(M + 1)^+$ peaks can be measured, it is possible to discriminate between these two compounds that have identical whole-number molecular weights.

The use of relative isotope peak heights for the determination of molecular formulas is greatly expedited by the tables referred to in footnote 21; a small part of such a table is shown in Table 20-6. In the latter, a listing of all reasonable combinations of C, H, O, and N is

[22] J. H. Beynon and A. E. Williams, *Mass and Abundance Tables for Use in Mass Spectrometry.* New York: Elsevier, 1963.

TABLE 20-6 Isotopic Abundance Percentages and Molecular Weights for Various Combinations of Carbon, Hydrogen, Oxygen, and Nitrogen[a]

		Abundance, % M Peak Height		
	Formula	M + 1	M + 2	Molecular Weight
M = 83	C_2HN_3O	3.36	0.24	83.0120
	$C_2H_3N_4$	3.74	0.06	83.0359
	C_3HNO_2	3.72	0.45	83.0007
	$C_3H_3N_2O$	4.09	0.27	83.0246
	$C_3H_5N_3$	4.47	0.08	83.0484
	$C_4H_3O_2$	4.45	0.48	83.0133
	C_4H_5NO	4.82	0.29	83.0371
	$C_4H_7N_2$	5.20	0.11	83.0610
	C_5H_7O	5.55	0.33	83.0497
	C_5H_9N	5.93	0.15	83.0736
	C_6H_{11}	6.66	0.19	83.0861
M = 84	CN_4O	2.65	0.23	84.0073
	$C_2N_2O_2$	3.00	0.43	83.9960
	$C_2H_2N_3O$	3.38	0.24	84.0198
	$C_2H_4N_4$	3.75	0.06	84.0437
	C_3O_3	3.36	0.64	83.9847
	$C_3H_2NO_2$	3.73	0.45	84.0085
	$C_3H_4N_2O$	4.11	0.27	84.0324
	$C_3H_6N_3$	4.48	0.08	84.0563
	$C_4H_4O_2$	4.46	0.48	84.0211
	C_4H_6NO	4.84	0.29	84.0449
	$C_4H_8N_2$	5.21	0.11	84.0688
	C_5H_8O	5.57	0.33	84.0575
	$C_5H_{10}N$	5.94	0.15	84.0814
	C_6H_{12}	6.68	0.19	84.0939
	C_7	7.56	0.25	84.0000

[a]Taken from R. M. Silverstein, G. C. Bassler, and T. C. Morrill, *Spectrometric Identification of Organic Compounds,* 4th ed., p. 49. New York: Wiley, 1981.

given for mass numbers 83 and 84 (the original tables extend to mass number 500). Tabulated are the heights of the $(M + 1)^+$ and $(M + 2)^+$ peaks reported as percentages of the height of the M^+ peak. If a reasonably accurate experimental determination of these percentages can be made, a likely formula can be deduced. For example, a molecular ion peak at mass 84 with $(M + 1)^+$ and $(M + 2)^+$ values of 5.6 and 0.3% of M^+ suggests a compound having the formula C_5H_8O.

The isotopic ratio is particularly useful for the detection and estimation of the number of sulfur, chlorine, and bromine atoms in a molecule because of the large contribution they make to the $(M + 2)^+$ (see Table 20-3). For example, an $(M + 2)^+$ that is about 65% of the M^+ peak is strong evidence for a molecule containing two chlorine atoms; an $(M + 2)^+$ peak of 4%, on the other hand, suggests the presence of one atom of sulfur.

Structural Information From Fragmentation Patterns

Systematic studies of fragmentation patterns for pure substances have led to rational guidelines to predict fragmentation mechanisms and a series of general rules that are helpful in interpreting spectra.[23] It is seldom possible (or desirable) to account for all of the peaks in the spectrum. Instead, characteristic patterns of fragmentation are sought. For example, the spectrum in Figure 20-22 is characterized by clusters of peaks differing in mass by 14. Such a pattern is typical of straight-chain paraffins, in which cleavage of adjacent carbon-carbon bonds results in the loss of successive CH_2 groups having this mass. Quite generally, the most stable hydrocarbon fragments contain three or four carbon atoms, and the corresponding peaks are thus the largest.

Alcohols usually have a very weak or nonexistent molecular ion peak but often lose water to give a strong peak at $(M - 18)^+$. Cleavage of the C—C bond next to an oxygen is also common, and primary alcohols always have a strong peak at mass 31 due to the CH_2OH^+. Extensive compilations of generalizations concerning the use of mass spectral data for the identification of organic compounds are available, and the interested reader should consult the references in footnote 21.

Compound Identification from Comparison Spectra

Generally, after determining the molecular weight of the analyte and studying its isotopic distribution and fragmentation patterns, the experienced mass spectroscopist is able to narrow the possible structures down to a handful. When reference compounds are available, final identification is then based upon a comparison of the mass spectrum of the unknown with spectra for authentic samples of the suspected compounds. The procedure is based upon the assumptions that (1) mass fragmentation patterns are unique and (2) experimental conditions can be sufficiently controlled to produce re-

[23] For example, see R. M. Silverstein, G. C. Bassler, and T. C. Morrill, *Spectrophotometric Identification of Organic Compounds,* 5th ed., pp. 13–16. New York: Wiley, 1991.

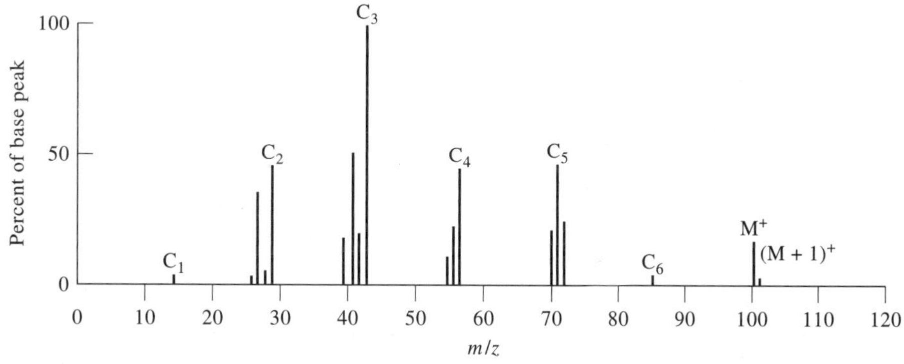

Figure 20-22 Electron-impact spectrum of *n*-heptanal. The peaks labeled C_6, C_5, . . . C_1 correspond to the successive losses of a CH_2 group.

producible spectra. The first assumption often is not valid for spectra of stereo- and geometric isomers and occasionally is not valid for certain types of closely related compounds. The probability that different compounds will yield the same spectrum becomes markedly smaller as the number of spectral peaks increases. For this reason, electron-impact ionization is the method of choice for spectral comparison.

Unfortunately, heights of mass spectral peaks are strongly dependent upon such variables as the energy of the electron beam, the location of the sample with respect to the beam, the sample pressure and temperature, and the general geometry of the mass spectrometer. As a consequence, significant variations in relative abundance are observed for spectra acquired in different laboratories and from different instruments. Nevertheless, it has proven possible in a remarkably large number of cases to identify unknowns from library spectra obtained with a variety of instruments and operating conditions. Generally, however, it is desirable to confirm the identity of a compound by comparison of its spectrum to the spectrum of an authentic compound obtained with the same instrument under identical conditions.

Computerized Library Search Systems. Although libraries of mass spectral data are available in text form,[24] most modern mass spectrometers are equipped with highly efficient computerized library search systems. Two types of libraries are currently encountered: large comprehensive ones and small specific ones. The largest commercially available mass spectral library (>150,000 spectra) is marketed by John Wiley and Sons.[25] A unique feature of this compilation is that it is available on compact disks and can be searched on a personal computer with Cornell University's matching and interpretative software (PBM-STIRS).[26] Small libraries usually contain a few hundred to a few thousand spectra for application to a limited area, such as pesticide residues, drugs, or forensics. Small libraries are often part of the equipment packages offered by instrument manufacturers, and it is almost always possible for the instrument user to generate a library or to add to an existing library.

For large numbers of spectra, such as are obtained when a mass spectrometer is coupled with a chromatograph for identifying components of a mixture, the instrument's computer system can be instructed to perform a library search on all, or any subset, of the mass spectra associated with a particular sample. The results are reported to the user, and if desired, the reference spectra can be displayed on a monitor or plotted for visual comparison.

20D-2 Analysis of Mixtures by Hyphenated Mass Spectral Methods

While ordinary mass spectrometry is a powerful tool for the identification of pure compounds, its usefulness for analysis of all but the simplest mixtures is limited because of the immense number of fragments of differing *m/z* values produced in the typical case. Interpretation of the resulting complex spectrum is often impossible. For this reason, chemists have developed methods in which mass spectrometers are coupled with various efficient separation devices. When two or more analytical techniques or instruments are combined to form a new, more efficient device, the resulting methodology is often termed a *hyphenated method.*

Chromatography/Mass Spectrometry

Gas chromatography/mass spectrometry (GC/MS) has become one of the most powerful tools available to the chemists for the analysis of complex organic and biochemical mixtures. In this application, spectra are collected for compounds as they exit from a chromatographic column. These spectra are then stored in a computer for subsequent processing. Mass spectrometry has also been coupled with liquid chromatography (LC/MS) for the analysis of samples that contain nonvolatile constituents. A major problem that had to be overcome in the development of both of these hyphenated methods is that the sample in the chromatographic column is highly diluted by the gas or liquid carrying it through the column. Thus, methods had to be developed for removing the diluent before introducing the sample into the mass spectrometer. Instruments and applications of GC/MS and LC/MS are described in Sections 27D-3 and 28C-6.

---

[24] F. W. McLafferty and D. A. Stauffer, *The Wiley/NBS Registry of Mass Spectral Data,* 7 vols. New York: Wiley, 1989.

[25] F. W. McLafferty, *Registry of Mass Spectral Data,* 5th ed. New York: Wiley, 1989. On hard disk, tape, and CD-ROM.

[26] G. M. Pesyna, R. Venkataraghavan, H. E. Daryringer, and F. W. McLafferty, *Anal. Chem.,* **1976,** *48,* 1362.

Capillary Electrophoresis/Mass Spectrometry

The first report on coupling capillary electrophoresis with mass spectrometry was published in 1987[27]. Since then, it has become obvious that this hyphenated method will become a powerful and important tool in the analysis of large biopolymers, such as proteins, polypeptides, and DNA species. In most of the applications reported to date, the capillary effluent is passed directly into an electrospray ionization device, and the products then enter a quadrupole mass filter for analysis. Continuous flow fast atom bombardment has also been used for ionization in some applications. Capillary electrophoresis/mass spectrometry is discussed in more detail in Section 30B-4.

Tandem Mass Spectrometry[28]

Another important hyphenated technique involves coupling one mass spectrometer to a second. In this method, the first spectrometer serves to isolate the molecular ions of various components of a mixture. These ions are then introduced one at a time into a second mass spectrometer, where they are fragmented to give a series of mass spectra, one for each molecular ion produced in the first spectrometer. This technique is called *tandem mass spectrometry* (often abbreviated as MS/MS).

The first spectrometer in a tandem instrument is ordinarily equipped with a soft ionization source (often a chemical ionization source) so that its output is largely molecular ions or protonated molecular ions. These ions then pass into the ion source for the second spectrometer. Most often this second ion source consists of a field-free collision chamber through which helium is pumped. Collisions between the fast-moving parent ions and helium atoms cause further fragmentation of the former to give numerous daughter ions. The spectrum of these daughter ions is then scanned by the second spectrometer. In this type of application, the first spectrometer serves the same function as the chromatographic column in GC/MS or LC/MS in that it provides pure ionic species one by one for identification in the second spectrometer.

To illustrate the power of this type of MS/MS, consider a hypothetical mixture of the isomers ABCD and BCDA, and other molecules such as UKL and UMN. To discriminate between the two isomers, a soft ionization source is used to produce predominantly singly charged molecular ions. The first spectrometer is then set to transmit ions with an m/z value corresponding to $ABCD^+$ and $BCDA^+$ (their m/z values are identical). Thus, the molecular ions of the isomers are separated from other components of the mixture. In the ionization chamber of the second spectrometer, collision fragmentation takes place to produce daughter ions such as AB^+, CD^+, BC^+, DA^+, and so forth. Because these fragments have unique mass-to-charge ratios, the identification of the two original isomers is possible in the second analyzer. The first spectrometer can then be set to transmit $IJKL^+$ or $IJMN^+$ ions, which in turn yield U^+, JKL^+, MN^+, JMN^+, and other characteristic ions that are also identified by the second analyzer.

Figure 20-23 illustrates a practical application of this technique, which is sometimes called *daughter-ion MS/MS*. The analyte consists of two very different compounds that have identical whole-number masses of 278. To discriminate between the two, the first spec-

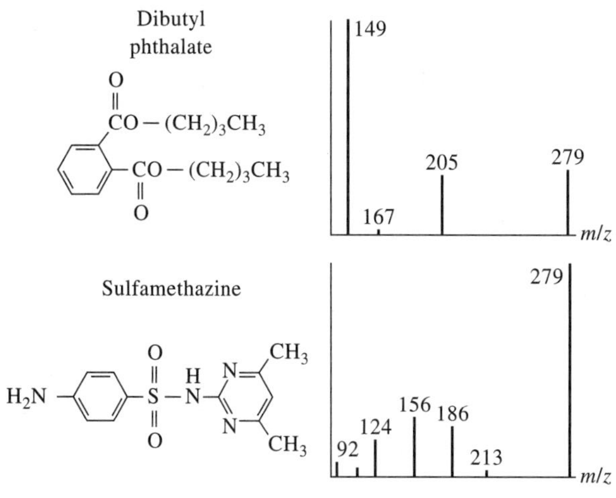

Figure 20-23 Daughter-ion spectra for dibutylphthalate and sulfamethazine obtained after the protonated parent-ion peaks at 279 dalton were isolated by the first spectrometer of an MS/MS instrument. *(Reprinted from K. L. Busch and G. C. DiDonato,* Amer. Lab., *1986, 18(8), 17. Copyright 1986 by International Scientific Communications, Inc.)*

---

[27] J. A. Olivares, N. T. Nguyen, N. T. Yonker, R. D. Smith, *Anal. Chem.*, **1987**, *59*, 1230. See also R. D. Smith, J. A. Olivares, N. T. Nguyen, and H. R. Hudseth, *Anal. Chem.*, **1988**, *60*, 436.

[28] J. V. Johnson and R. A. Yost, *Anal. Chem.*, **1985**, *57*, 758A; R. G. Cooks and G. L. Glish, *Chem. Eng. News*, **Nov. 30, 1981**, 40; *Tandem Mass Spectrometry*, F. W. McLafferty, Ed. New York: Wiley, 1983; K. L. Busch, G. L. Glish, and S. A. McLuckey, *Mass Spectrometry/Mass Spectrometry: Techniques and Applications of Tandem Mass Spectrometry.* New York: VCH Publishers, 1988.

trometer was set to the mass of the protonated parent ions (279). The two quite different daughter-ion spectra shown were obtained after further ionization by collision and passage through the second spectrometer.

In another type of tandem mass spectrometry, termed selected *parent-ion MS/MS,* the first spectrometer is scanned while the second spectrometer is set to the mass of one of the daughter ions. Closely related compounds usually give several of the same daughter ions, so that this method of operation provides a measure of identity and concentration of the *members of a class* of closely related compounds. For example, the hypothetical mixture described earlier contained the related compounds IJKL and IJMN. To identify species containing the IJ group, the second spectrometer would be fixed on the mass corresponding to the IJ^+ ion while the mass spectrum of the parent ions ($ABCD^+$, $BCDA^+$, $IJKL^+$, and $IJMN^+$) was scanned. Ion currents in the second spectrometer would be observed only when the latter two ions were exiting the first spectrometer.

A practical example of parent-ion MS/MS involves the determination of alkylphenols ($HOC_6H_4CH_2R$) in solvent-refined coal. In this experiment, the second spectrometer is set at a *m/z* value of 107, which corresponds to the ion $HOC_6H_4CH_2^+$; the sample is then scanned with the first spectrometer. All of the alkylphenols in the samples yield an ion of mass 107 regardless of the nature of R. Thus, it is possible to measure this class of compounds in a complex sample.

Instruments For Tandem Mass Spectrometry. Tandem mass spectrometers are made up of various combinations of magnetic sectors, electrostatic sectors, and quadrupole filter separators.[29] In describing tandem instruments, these mass separators are designated by B, E, and Q, respectively. For example, a BE instrument consists of a magnetic sector followed by an electrostatic sector; an EBEB instrument is made up of two double-focusing mass spectrometers, each made up of an electrostatic and a magnetic sector. Currently, the most widely used tandem mass spectrometer has the configuration QQQ, and this configuration is illustrated in Figures 20-20 and 20-24. Here, the sample is introduced into one of the soft ionization sources, such as a chemical ionization type. The ions are then accelerated into the first-stage, or parent-ion, separator, which is an ordi-

[29] For a description of tandem mass spectrometers available commercially, see D. Noble, *Anal. Chem.*, **1995**, *67*, 265A. The cost of these instruments range from $300,000 to $1,000,000.

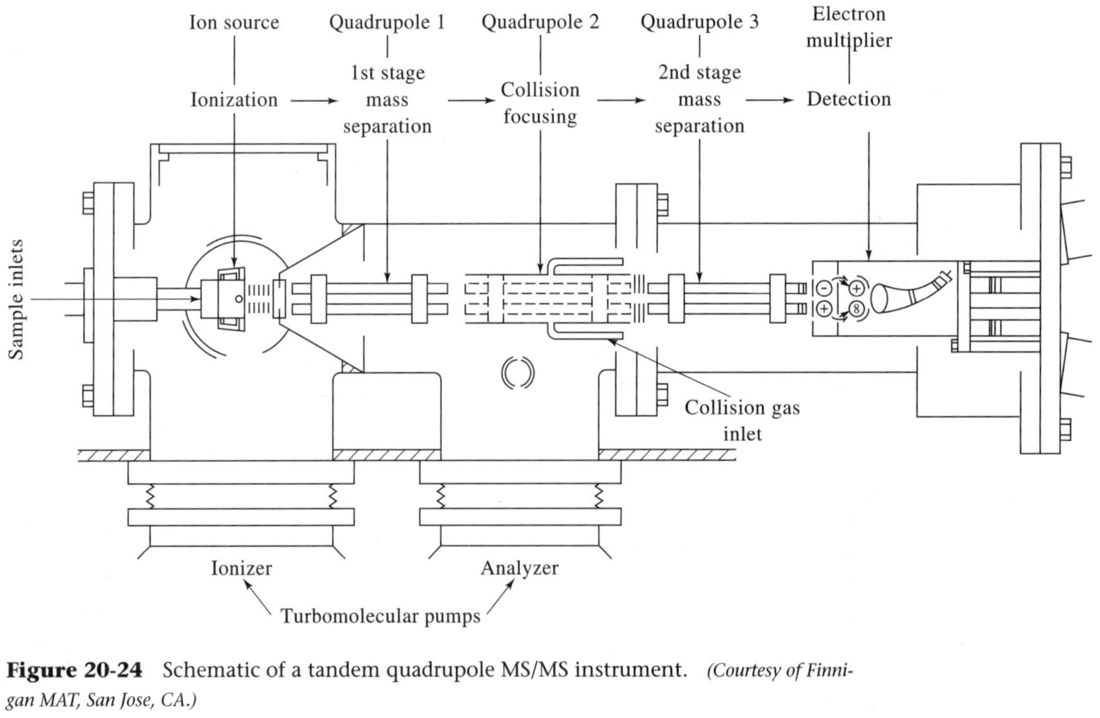

Figure 20-24 Schematic of a tandem quadrupole MS/MS instrument. *(Courtesy of Finnigan MAT, San Jose, CA.)*

nary quadrupole filter. The fast-moving separated ions then pass into quadrupole 2, which is a collision chamber where further ionization of the parent ions from quadrupole 1 occurs. This quadrupole is operated in the radio-frequency-only mode—that is, no dc potential is applied across the rods.[30] This mode provides a highly efficient way of focusing scattered ions but does not act as a mass filter. Helium is pumped into this chamber, giving a pressure of 10^{-3} to 10^{-4} torr. Further ionization occurs here as a consequence of collision of the rapidly moving parent ions with the resident helium atoms. The resulting daughter ions pass into quadrupole 3, where they are scanned and recorded in the usual way.

Applications of Tandem Mass Spectrometry. Dramatic progress in the analysis of complex organic and biological mixtures began when the mass spectrometer was first combined with gas chromatography and subsequently with liquid chromatography. Tandem mass spectrometry appears to offer the same advantages as GC/MS and LC/MS but is significantly faster. Whereas separations on a chromatographic column are achieved in a time scale of a few minutes to hours, equally satisfactory separations in tandem mass spectrometers are complete in milliseconds. In addition, the chromatographic techniques require dilution of the sample with large excesses of a mobile phase and subsequent removal of the mobile phase, which greatly enhances the probability of introduction of interferences. Consequently, tandem mass spectrometry is potentially more sensitive than either of the hyphenated chromatographic techniques because the chemical noise associated with its use is generally smaller. A current disadvantage of tandem mass spectrometry with respect to the other two chromatographic procedures is the greater cost of the equipment required; this gap appears to be lessening as tandem mass spectrometers gain wider use.

To date, tandem mass spectrometry has been applied to the qualitative and quantitative determination of the components of a wide variety of complex materials encountered in nature and industry. Some examples include the identification and determination of drug metabolites, insect pheromones, alkaloids in plants, trace contaminants in air, polymer sequences, petrochemicals, polychlorinated biphenyls, prostaglandins, diesel exhausts, and odors in air. Tandem mass spectrometry appears to be a technique that will find wider and wider application among scientists and engineers.

[30] The absence of a dc potential in the second quadrupole in Figure 20-24 is sometimes indicated by describing the instrument as QqQ instead of QQQ.

20E QUANTITATIVE APPLICATIONS OF MASS SPECTROMETRY

Applications of mass spectrometry for quantitative analyses fall into two categories. The first involves the quantitative determination of molecular species or types of molecular species in organic, biological, and occasionally inorganic samples. The second involves the determination of the concentration of elements in inorganic and, less commonly, organic and biological samples. In the first type of analysis, all of the ionization sources listed in Table 20-1 are used. Mass spectroscopic elemental analyses, which are discussed in detail in Chapter 11, are currently based largely upon inductively coupled plasma sources although radio-frequency spark, laser, thermal, secondary ion, and glow discharge sources have also found use.

20E-1 Quantitative Determination of Molecular Species

Mass spectrometry has been widely applied to the quantitative determination of one or more components of complex organic (and sometimes inorganic) systems such as those encountered in the petroleum and pharmaceutical industries and in studies of environmental problems.[31] Currently, such analyses are usually performed by passage of the sample through a chromatographic or capillary electrophoretic column and into the spectrometer. With the spectrometer set at a suitable *m/z* value, the ion current is then recorded as a function of time. This technique is termed *selected ion monitoring*. In some instances, currents at three or four *m/z* values are monitored in a cyclic manner by rapid switching from one peak to another. The plot of the data consists of a series of peaks with each appearing at a time that is characteristic of one of the several components of the sample that yields ions of the chosen value or values for *m/z*. Generally, the areas under the peaks are directly proportional to the component concentrations and thus

[31] For a monograph devoted to quantitative mass spectrometry, see B. J. Millard, *Quantitative Mass Spectrometry*. London: Heyden, 1978.

serve as the analytical parameter. In this type of procedure, the mass spectrometer simply serves as a sophisticated selective detector for quantitative *chromatographic* or *electrophoretic analyses.* Further details on quantitative gas and liquid chromatography are given in Sections 27D-3 and 28C-6. The use of a mass spectrometer as a detector in capillary electrophoresis is described in Section 30B-4.

In the second type of quantitative mass spectrometry for molecular species, analyte concentrations are obtained directly from the heights of the mass spectral peaks. For simple mixtures, it is sometimes possible to find peaks at unique *m/z* values for each component. Under these circumstances, calibration curves of peak heights versus concentration can be prepared and used for analysis of unknowns. More accurate results can ordinarily be realized, however, by incorporating a fixed amount of an internal standard substance in both samples and calibration standards. The ratio of the peak intensity of the analyte species to that of the internal standard is then plotted as a function of analyte concentration. The internal standard tends to reduce uncertainties arising in sample preparation and introduction. With the small samples needed for mass spectrometry, these uncertainties are often a major source of indeterminate error. Internal standards are also used in GC/MS and LC/MS; here, the ratio of peak areas serves as the analytical variable.

A convenient type of internal standard is a stable, isotopically labeled analog of the analyte. Usually, labeling involves preparation of samples of the analyte in which one or more atoms of deuterium, carbon-13, or nitrogen-15 have been incorporated. It is then assumed that during the analysis, the labeled molecules behave in the same way as do the unlabeled ones. The mass spectrometer easily distinguishes between the two.

Another type of internal standard is a homolog of the analyte that yields a reasonably intense ion peak for a fragment that is chemically similar to the analyte fragment being measured.

With low-resolution instruments, it is seldom possible to locate peaks that are unique to each component of a mixture. In this situation, it is still possible to complete an analysis by collecting intensity data at a number of *m/z* values that equal or exceed the number of sample components. Simultaneous equations are then developed that relate the intensity of each *m/z* value to the contribution made by each component to this intensity. Solving these equations then provides the desired quantitative information. The procedure here is analogous to the method described in Section 14D-2 for the spectrophotometric determination of the components of mixtures of species having overlapping absorption spectra in the ultraviolet/visible region.

Precision and Accuracy

The precision of quantitative mass spectral measurements by the procedure just described appears to range between 2% and 10% relative. The analytical accuracy varies considerably depending upon the complexity of the mixture being analyzed and the nature of its components. For gaseous hydrocarbon mixtures containing 5 to 10 components, absolute errors of 0.2 to 0.8 mole percent appear to be typical.

Applications

The early literature dealing with the direct quantitative applications of mass spectrometry is so extensive as to make a summary difficult. The listing of typical applications assembled by Melpolder and Brown[32] demonstrates clearly the versatility of the method. For example, some of the mixtures that can be analyzed without sample heating include natural gas; C_3–C_5 hydrocarbons; C_6–C_8 saturated hydrocarbons; C_1–C_4 chlorides and iodides; fluorocarbons, thiophenes, atmospheric pollutants, exhaust gases; and many others. By employing higher temperatures, successful analytical methods have been reported for C_{16}–C_{27} alcohols, aromatic acids and esters, steroids, fluorinated polyphenyls, aliphatic amides, halogenated aromatic derivatives, and aromatic nitriles.

Mass spectrometry has also been used for the characterization and analysis of high-molecular-weight polymeric materials. In this application, the sample is first pyrolyzed; the volatile products are then admitted into the spectrometer for examination. Alternatively, heating can be performed on the probe of a direct inlet system. Some polymers yield essentially a single fragment: for example, isoprene from natural rubber, styrene from polystyrene, ethylene from polyethylene, and $CF_2{=}CFCl$ from Kel–F. Other polymers yield two or more products, which depend in amount and kind upon the pyrolysis temperature. Studies of temperature effects can provide information regarding the stabilities of the various bonds, as well as the approximate molecular weight distribution.

[32] See F. W. Melpolder and R. A. Brown, in *Treatise on Analytical Chemistry,* I. M. Kolthoff and P. J. Elving, Eds., Part I, Vol. 4, p. 2047. New York: Interscience, 1963.

20F QUESTIONS AND PROBLEMS

20-1 How do gaseous and desorption sources differ? What are the advantages of each?

20-2 How do the spectra for electron-impact, field ionization, and chemical ionization sources differ from one another?

20-3 Describe gaseous field ionization and desorption field ionization sources.

20-4 The following figure is a simplified diagram of a commercially available electron-impact source.
(a) What potential must be applied between the filament and target so that electrons interacting with molecules at the point marked *SS* (sample source) will have 70 eV of kinetic energy?
(b) What will happen to a molecule that diffuses toward the filament, and is ionized at point *P?*

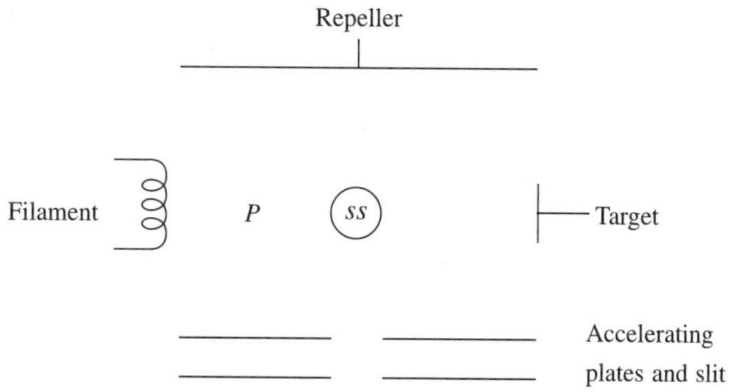

20-5 When a magnetic sector instrument was operated with an accelerating voltage of 3.00×10^3 V, a field of 0.126 T was required to focus the CH_4^+ on the detector.
(a) What range of field strengths would be required to scan the mass range between 16 and 250, for singly charged ions, if the accelerating voltage is held constant?
(b) What range of accelerating voltages would be required to scan the mass range between 16 and 250, for singly charged ions, if the field strength is held constant?

20-6 Calculate the accelerating voltage that would be required to direct singly charged ions of mass 10,000 through an instrument that is identical to the one described in Example 20-2 (page 503).

20-7 The ion-accelerating voltage in a particular quadrupole mass spectrometer is 5.00 V. How long will it take a singly charged benzene ion to travel the length of the rod assembly, a distance of 15.0 cm?

20-8 On page 258 a qualitative discussion was provided that described how a positive ion would behave in the *xz* plane (positive dc potential plane) of a quadrupole mass filter. Construct a similar argument for the behavior of positive ions in the *yz* plane (negative dc potential plane).

20-9 Why do double-focusing mass spectrometers give narrower peaks and higher resolutions?

20-10 Calculate the resolution required to resolve peaks for
(a) CH_2N (MW = 28.0187) and N_2^+ (MW = 28.0061)
(b) $C_2H_4^+$ (MW = 28.0313) and CO^+ (MW = 27.9949)

(c) $C_3H_7N_3^+$ (MW = 85.0641) and $C_5H_9O^+$ (MW = 85.0653)
(d) $^{116}Sn^+$ (AtW = 115.90219) and $^{232}Th^{2+}$ (AtW = 232.03800)

20-11 Calculate the ratio of the $(M + 2)^+$ to M^+ peak heights for (a) $C_{10}H_6Br_2$, (b) C_3H_7ClBr, (c) $C_6H_4Cl_2$.

20-12 (a) Draw an operational amplifier circuit (see Chapter 3) that would be compatible with the Faraday cup detector illustrated in Figure 11-4; this circuit should provide a voltage gain of 100.
(b) Draw a second operational amplifier circuit that monitors current rather than voltage; this circuit should provide a voltage output that is the same as that in the circuits in part (a).

20-13 In a magnetic sector (single-focusing) mass spectrometer, it might be reasonable under some circumstances to monitor one *m/z* value, to then monitor a second *m/z*, and to repeat this pattern in a cyclic manner. Rapidly switching between two accelerating voltages while keeping all other conditions constant is called *peak matching*.
(a) Derive a general expression that relates the ratio of the accelerating voltages to the ratio of the corresponding *m/z* values.
(b) Use this equation to calculate *m/z* of an unknown peak, if *m/z* of the compound that is being used as a standard is 69.00 and the ratio of $V_{unknown}/V_{standard}$ is 0.965035.
(c) Based upon your answer in part (b), and the assumption that the unknown is an organic compound that has a mass of 143, draw some conclusions about your answer in part (b), and about the compound.

20-14 If one wishes to measure the approximate mass of an ion without using a standard, it can be accomplished via the following variant of the peak-matching technique described in Problem 20-13. The peak-matching technique is used to alternately cause the P^+ ion and the $(P + 1)^+$ ions to reach the detector. It is assumed that the difference in mass between P^+ and $(P + 1)^+$ is due to a single ^{13}C.
(a) Assume that all ions are singly charged; derive a relationship that relates the ratio of the accelerating voltage $V(P + 1)/V(P)$ to the mass of P^+.
(b) If $V(P + 1)/V(P) = 0.987753$, calculate the mass of the P^+ ion.

20-15 Identify the ions responsible for the peaks in the mass spectrum shown in Figure 20-19b.

20-16 Identify the ions responsible for the four peaks having greater masses than the M^+ peak in Figure 20-4a.

Chapter 21

Surface Characterization by Spectroscopy and Microscopy

The surface of a solid in contact with a liquid or gaseous phase usually differs substantially from the interior of the solid both in chemical composition and physical properties. Characterization of these surface properties is often of vital importance in a number of fields, including heterogeneous catalysis, semiconductor thin-film technology, corrosion and adhesion mechanisms, activity of metal surfaces, embrittlement properties, and studies of the behavior and functions of biological membranes. This chapter deals with the characterization of solid surfaces by spectroscopic and microscopic methods.

21A INTRODUCTION TO THE STUDY OF SURFACES

Before considering how surfaces are characterized, we first need to define what constitutes the surface of a solid that is in contact with a gaseous or liquid second phase.

21A-1 Definition of a Solid Surface

We shall consider a surface to be the boundary layer between a solid and a vacuum, a gas, or a liquid. Generally we think of a surface as a part of the solid that differs in composition from the average composition of the bulk of the solid. By this definition, the surface comprises not only the top layer of atoms or molecules of a solid but also a transition layer with a nonuniform composition that varies continuously from that of the outer layer to that of the bulk. Thus, a surface may be several or even several tens of atomic layers deep. Ordinarily, however, the difference in composition of the surface layer does not significantly affect the measured overall average composition of the bulk because the surface layer is generally only a tiny fraction of the total solid. From a practical standpoint, it appears best to adopt an operational definition of a surface as that volume of the solid that a specific measurement technique samples. This definition recognizes the fact that if several surface techniques are employed, the chemist may in fact be sampling different surfaces and may obtain different, albeit useful, results as a consequence.

21A-2 Types of Surface Measurements

During the last century, a wide variety of methods have been developed for characterizing surfaces. The classical methods, which are still of importance, provide much useful information about the *physical* nature of surfaces but less about their chemical nature. These methods involve obtaining optical and electron microscopic images of surfaces as well as measurements of adsorption isotherms, surface areas, surface roughness, pore sizes, and reflectivity. Beginning in the 1950s, spectroscopic surface methods began to appear that provided information about the chemical nature of surfaces.

This chapter is divided into two major parts. The first (Section 21B) is concerned with spectroscopic methods for identifying the chemical species making up the surface of solids and determining their concentrations. The next two sections (Section 21C and Section 21D) describe modern microscopic methods for imaging surfaces and determining their morphology, or physical, features.

21B SPECTROSCOPIC SURFACE METHODS

Generally, the chemical composition of the surface of a solid differs, often significantly, from the interior or bulk of the solid. Thus far in this text, we have focused on analytical methods that provide information about bulk composition of solids only. In certain areas of science and engineering, however, the chemical composition of a surface layer of a solid is of much greater importance than is the bulk composition of the material.

Spectroscopic surface methods provide both qualitative and quantitative *chemical* information about the composition of a surface layer of a solid that is a few angstrom units to a few tens of angstrom units in thickness. In this section we describe five of the most widely used of these spectroscopic techniques.[1]

21B-1 General Technique in Surface Spectroscopy

Figure 21-1 illustrates the general way in which a spectroscopic examination of a surface is performed. Here, the solid sample is irradiated with a *primary beam* made up of photons, electrons, ions, or neutral molecules. Impact of this beam on a surface results in formation of a *secondary beam* also consisting of photons, electrons, molecules, or ions from the solid surface. Note that the type of particle making up the primary beam is not necessarily the same as the particle of the secondary beam. The secondary beam, which results from scattering, sputtering, or emission, is then studied by a variety of spectroscopic methods.

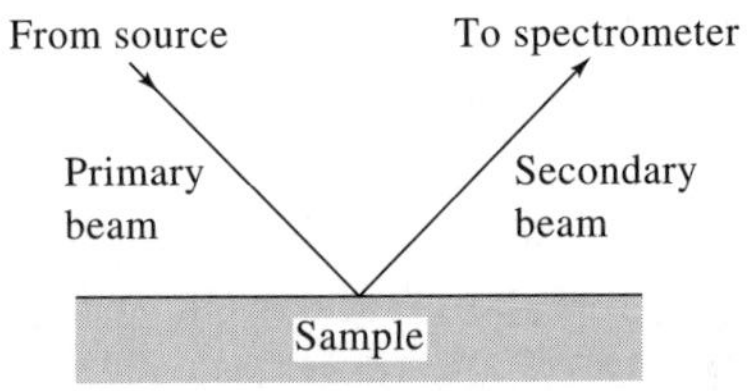

Figure 21-1 General scheme for surface spectrometry. Beam may be photons, electrons, or ions.

The most effective surface methods are those in which the primary beam, the secondary beam, or both is made up of either electrons, ions, or molecules and not photons because this limitation assures that the measurements are restricted to the surface of a sample and not to its bulk. For example, the maximum penetration depth of a beam of 1-keV electrons or ions is approximately 25 Å, whereas the penetration depth of a photon beam of the same energy is about 10^4 Å. Thus, for methods that involve two beams of photons, such as X-ray fluorescence, resonance Raman, or infrared reflection spectroscopy, considerable precaution must be taken to limit the measurements to a surface layer. Although studying surfaces by these methods is possible, avoidance of interference from the bulk is often troublesome. We shall not consider methods that are based upon two beams of photons.

Surface Spectroscopic Methods

Hercules and Hercules have listed about two dozen surface spectroscopic techniques that have been developed over the last four decades and have classified these techniques based on the nature of the primary and secondary beams.[2] Six of the more widely used of these spectroscopic methods, and the ones that we will describe in this section, are listed in Table 21-1.[3]

[1] For a description of several surface spectroscopic techniques, see *Spectroscopy of Surfaces,* R. G. H. Clark and R. E. Hester, Eds. New York: Wiley, 1988.

[2] D. M. Hercules and S. H. Hercules, *J. Chem. Educ.,* **1984,** *61,* 403.

[3] For a review of several of the most important of these methods, see D. M. Hercules, *Anal. Chem.,* **1986,** *58,* 1177A; D. M. Hercules and S. H. Hercules, *J. Chem. Educ.,* **1984,** *61,* 403, 483, 592.

TABLE 21-1 Some Common Types of Spectroscopic Methods for Analysis of Surfaces

Method and Common Acronym	Primary Beam	Secondary Beam
X-Ray Photoelectron Spectroscopy (XPS), or Electron Spectroscopy for Chemical Analysis (ESCA)	X-Ray photons	Electrons
Auger Electron Spectroscopy (AES)	Electrons, or X-ray photons	Electrons
Ultraviolet Photoelectron Spectroscopy (UPS)	UV photons	Electrons
Secondary Ion Mass Spectrometry (SIMS)	Ions	Ions
Laser Microprobe Mass Spectrometry (LMMS)	Photons	Ions
Electron Microprobe (EM)	Electrons	X-Ray photons

Sampling Surfaces

Regardless of the type of spectroscopic surface method being used, three types of sampling methods are employed. The first involves focusing the primary beam on a single small area of the sample and observing the secondary beam. Often the spot is chosen visually with an optical microscope. The second method involves mapping the surface, in which a region of the surface is scanned by moving the primary beam across the surface in a *raster pattern* of measured increments and observing changes in the secondary beam that result. The mapping may be linear or two dimensional. The third technique is known as *depth profiling*. Here a beam of ions from an ion gun is used to etch a hole in the surface by sputtering. During this process a finer primary beam is used to produce a secondary beam from the center of the hole, which provides the analytical data on the surface composition as a function of depth.

Surface Contamination

A problem that is frequently encountered in surface analyses is contamination of the surface being examined by adsorption of components of the atmosphere, such as oxygen, water, or carbon dioxide. Even in a vacuum this type of contamination occurs in a relatively short time. For example at a pressure of 10^{-6} torr, it has been shown that a clean surface will become covered by a monolayer of gas molecules in just 3 s. At 10^{-8} torr coverage occurs in about 1 hr. At 10^{-10} torr, 10 hr is required.[4] As a consequence of adsorption problems, provisions must often be made to clean the sample surface, usually in the chamber used for irradiating the sample. Cleaning may involve baking the sample at a high temperature; sputtering the sample with a beam of inert gas ions from an electron gun; mechanical scraping or polishing the sample surface with an abrasive; ultrasonic washing of the sample in various solvents; and bathing the sample in a reducing atmosphere to remove oxides.

In addition to atmospheric contamination, it should be realized that the primary beam itself can alter the surface as a measurement progresses. Damage caused by the primary beam depends upon the momentum of the primary beam particles; so, of the beams listed in Table 21-1, ions are the most damaging and photons the least.

21B-2 Electron Spectroscopy

The first three methods listed in Table 21-1 are based upon analysis of emitted electrons produced by various incident beams. Here, the signal from the analyte consists of a beam of electrons rather than photons. The spectrometric measurements then consist of the determination of the power of this beam as a function of the energy (or frequency $h\nu$) of the electrons. This type of spectroscopy is termed *electron spectroscopy.*

Although the basic principles of electron spectroscopy were well understood a century ago, the widespread application of this technique to chemical problems did not occur until relatively recently. An important factor that inhibited studies in the field was the lack of engineering technology necessary for performing high-resolution spectral measurements of electrons having energies varying from a few tenths to sev-

[4] D. M. Hercules and S. H. Hercules, *J. Chem. Educ.*, **1984**, *61*, 403.

eral thousand electron volts. By the late 1960s, this technology had developed, and commercial electron spectrometers began to appear in the marketplace. With their appearance, an explosive growth in the number of publications devoted to electron spectroscopy occurred.[5]

For the study of surfaces, three types of electron spectroscopy are encountered. The most common type, which is based upon irradiation of the sample surface with monochromatic X-radiation, is called *X-ray photoelectron spectroscopy* (XPS). It is also termed *electron spectroscopy for chemical analysis* (ESCA). Much of the material in this chapter is devoted to XPS. The second type of electron spectroscopy is called *Auger* (pronounced oh-ZHAY) *electron spectroscopy* (AES). Most commonly, Auger spectra are excited by a beam of electrons, although X-rays are also used. Auger spectroscopy is discussed in Section 21B-4. The third type of electron spectroscopy is *ultraviolet photoelectron spectroscopy* (UPS). Here, a monochromatic beam of ultraviolet radiation causes ejection of electrons from the analyte. This type of electron spectroscopy is not as common as the other two, and we shall not discuss it further.

Electron spectroscopy is a powerful tool for the identification of all of the elements in the periodic table with the exception of hydrogen and helium. More important, the method permits determination of the oxidation state of an element and the type of species to which it is bonded. Finally, the technique provides useful information about the electronic structure of molecules.

Electron spectroscopy has been successfully applied to gases and solids and more recently to solutions and liquids. Because of the poor penetrating power of electrons, however, these methods provide information about solids that is restricted largely to a surface layer that is a few atomic layers thick (20 to 50 Å). Usually, the composition of such surface layers is significantly different from the average composition of the entire sample. Indeed, the most important and valuable current applications of electron spectroscopy are to the qualitative analysis of the surfaces of solids, such as metals, alloys, semiconductors, and heterogeneous catalysts. Quantitative analysis by electron spectroscopy finds somewhat limited applications.

21B-3 X-Ray Photoelectron Spectroscopy

It is important to emphasize the fundamental difference between electron spectroscopy (both XPS and AES) and the other types of spectroscopy we have thus far encountered. In electron spectroscopy, the kinetic energy of emitted electrons is recorded. The spectrum thus consists of a plot of the number of emitted electrons, or the power of the electron beam, as a function of the energy (or the frequency or wavelength) of the emitted electrons (see Figure 21-2).

Principles of XPS

The use of XPS was pioneered by the Swedish physicist K. Siegbahn, who subsequently received the 1981 Nobel Prize for his work.[6] Siegbahn chose to call the technique electron spectroscopy for chemical analysis (ESCA) because, in contrast to the other two electron spectroscopies, XPS provides not only information about the atomic composition of a sample but also information about the structure and oxidation state of the compounds being examined. Figure 21-3 is a schematic representation of the physical process involved in XPS. Here, the three lower lines labeled, E_b, E'_b, and E''_b represent energies of the inner shell K and L electrons of an atom. The upper three lines represent some of the energy levels of the outer shell or valence electrons. As shown in the illustration, one of the photons of a monochromatic X-ray beam of known energy $h\nu$ displaces an electron e^- from a K orbital E_b. The reaction can be represented as

$$A + h\nu \rightarrow A^{+*} + e^- \qquad (21\text{-}1)$$

where A can be an atom, a molecule, or an ion and A^{+*} is an electronically excited ion with a positive charge one greater than that of A.

---

[5] References include *Electron Spectroscopy: Theory, Techniques, and Applications,* 4 vols, C. R. Brundle and A. D. Baker, Eds. New York: Academic Press, 1977–81; P. K. Ghosh, *Introduction to Photoelectron Spectroscopy.* New York: Wiley, 1983; *Practical Surface Analysis by Auger and X-Ray Photoelectron Spectroscopy,* D. Briggs and M. P. Seah, Eds. New York: Wiley, 1983; G. Margaritondo and J. E. Rowe, in *Treatise on Analytical Chemistry,* 2nd ed., P. J. Elving, E. J. Meehan, and I. M. Kolthoff, Eds., Part I, Vol. 8, Chapter 17. New York: Wiley, 1986.

[6] For monographs on XPS, see K. Siegbahn, et al., *ESCA: Atomic, Molecular and Solid State Structure by Means of Electron Spectroscopy.* Upsala: Olmquist and Wiksells, 1967, and *XPS Applied to Free Molecules.* Amsterdam: North-Holland Publishing Co., 1969; P. K. Ghosh, *Introduction to Photoelectron Spectroscopy.* New York: Wiley, 1983; T. L. Barr, *Modern XPS.* Boca Raton: CRC Press, 1994. For a brief description of the history of XPS, see K. Siegbahn, *Science,* **1981,** *217,* 111.

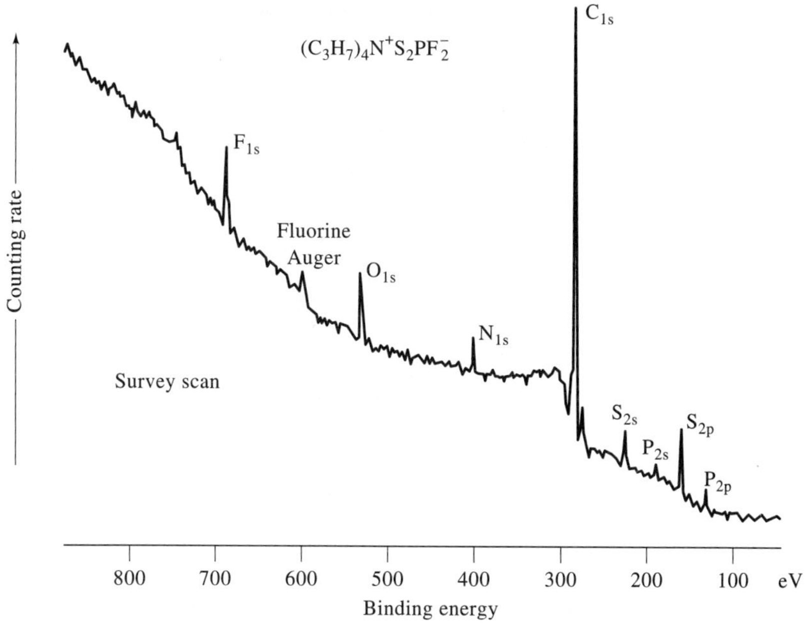

Figure 21-2 X-Ray photoelectron spectrum of tetrapropylammonium-difluoridethiophosphate. The peaks are labeled according to the element and orbital from which the emitted electrons originate.

The kinetic energy of the emitted electron E_k is measured in an electron spectrometer. The *binding energy* of the electron E_b can then be calculated by means of the equation

$$E_b = h\nu - E_k - w \qquad (21\text{-}2)$$

In this equation, w is the so-called *work function* of the spectrometer, a factor that corrects for the electrostatic environment in which the electron is formed and measured. Various methods are available to determine a value for w. The binding energy of an electron is characteristic of the atom and orbital from which the electron was emitted.

Figure 21-2 shows a low-resolution or survey XPS spectrum consisting of a plot of electron counting rate as a function of binding energy E_b. The analyte consisted of an organic compound made up of six elements. With the exception of hydrogen, well-separated peaks for each of the elements can be observed. In addition, a peak for oxygen is present, suggesting that some surface oxidation of the compound had occurred. Note that, as expected, the binding energies for 1*s* electrons increase with atomic number because of the increased positive charge of the nucleus. Note also that more than one peak for a given element can be observed; thus peaks for both 2*s* and 2*p* electrons for sulfur and phosphorus can be seen. The large background count arises because associated with each characteristic peak is a tail due to ejected electrons that have lost part of their energy by inelastic collisions within the solid sample. These electrons have less kinetic energy than their nonscattered counterparts and will thus appear at lower kinetic energies or higher binding energies (Equation 21-2). It is evident from Figure 21-2 that XPS provides a means of qualitative identification of the elements present on the surface of solids.

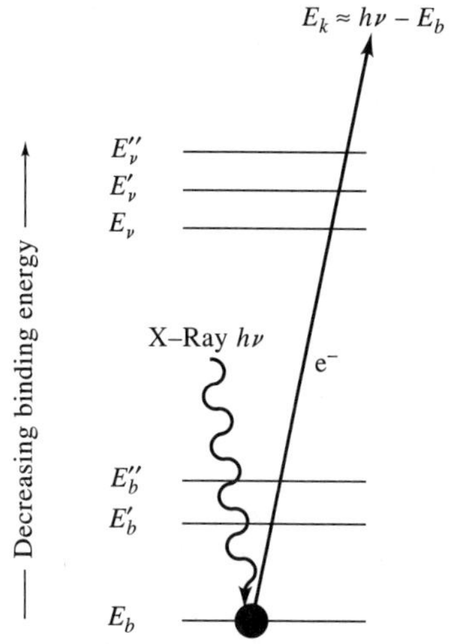

Figure 21-3 Schematic representation of the ESCA process. The incident beam is monoenergetic X-ray photons. The emitted beam is made up of electrons.

Instrumentation

Instruments for electron spectroscopy are offered by perhaps a dozen instrument manufacturers. These products differ considerably in types of components, configurations, and costs. Some are designed for a single type of application, such as XPS, while others can be adapted to Auger and ultraviolet photoelectron spectroscopy as well by purchase of suitable accessories. All are expensive ($300,000 to $900,000).[7]

Electron spectrometers are made up of components whose functions are analogous to those encountered in optical spectroscopic instruments. These components include: (1) a source; (2) a sample holder; (3) an analyzer, which has the same function as a monochromator; (4) a detector; (5) a signal processor and readout. Figure 21-4 shows a typical arrangement of these components. Electron spectrometers generally require elaborate vacuum systems to reduce the pressure in all of the components to as low as 10^{-5} to 10^{-10} torr.[8]

[7] For a useful review of recent developments in electron spectrometers, see J. A. Gardella Jr., *Anal. Chem.*, **1989**, *61*, 589A.

[8] Specifications for several representative commercial instruments are given in D. Noble, *Anal. Chem.*, **1995**, *67*, 675A.

Sources. The simplest X-ray sources for XPS spectrometers are X-ray tubes equipped with magnesium or aluminum targets and suitable filters. The K_α lines for these two elements have considerably narrower bandwidths (0.8 to 0.9 eV) than those encountered with higher atomic number targets; narrow bands are desirable since they lead to enhanced resolution.

Relatively sophisticated XPS instruments, such as that shown in Figure 21-4, employ a crystal monochromator (Section 12B-3) to provide an X-ray beam having a bandwidth of about 0.3 eV. Monochromators eliminate Brehmsstrahlung background, thus improving *S/N* ratios. They also allow much smaller spots on a surface to be examined.

Sample Holders. Solid samples are mounted in a fixed position as close to the photon or electron source and the entrance slit of the spectrometer as possible (see Figure 21-4). In order to avoid attenuation of the electron beam, the sample compartment must be evacuated to a pressure of 10^{-5} torr or smaller. Often, however, much better vacuums (10^{-9} to 10^{-10} torr) are required to avoid contamination of the sample surface by substances such as oxygen or water that react with or are adsorbed on the surface.

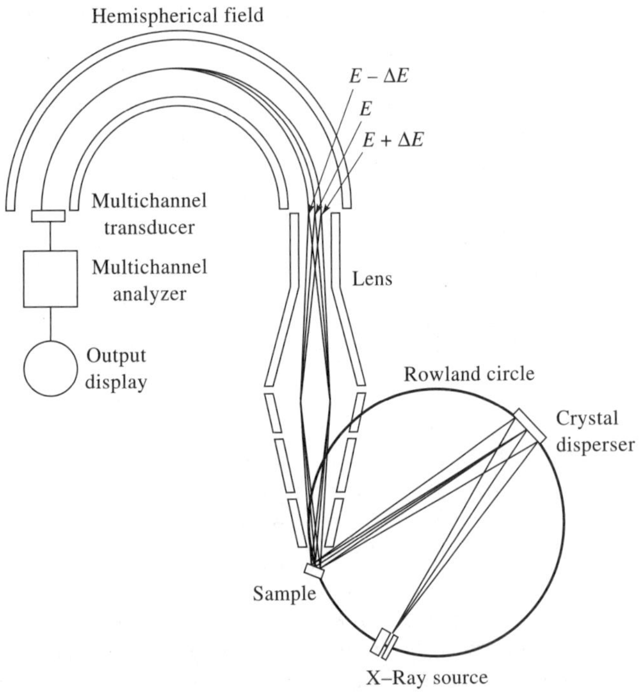

Figure 21-4 Principle of a modern ESCA instrument (HP5950A) using an X-ray monochromatic source and a hemispherical field spectrometer.

Gas samples are leaked into the sample area through a slit of such a size as to provide a pressure of perhaps 10^{-2} torr. Higher pressures lead to excessive attenuation of the electron beam due to inelastic collisions; on the other hand, if the sample pressure is too low, weakened signals are obtained.

Analyzers. Most electron spectrometers are of the hemispherical type in which the electron beam is deflected by an electrostatic magnetic field in such a way that the electrons travel in a curved path (see Figure 21-4). The radius of curvature is dependent upon the kinetic energy of the electron and the magnitude of the field. By varying the field, electrons of various kinetic energies can be focused on the detector.

Typically, pressures in the analyzer of an electron spectrometer are maintained at or below 10^{-5} torr.

Transducers. Most modern electron spectrometers are based upon solid-state, channel electron multipliers, which consist of tubes of glass that have been doped with lead or vanadium. When a potential of several kilovolts is applied across these materials, a cascade or pulse of 10^6 to 10^8 electrons is produced for each incident electron. The pulses are then counted electronically. Several manufacturers are now offering two-dimensional multichannel electron transducers that are analogous in construction and application to the multichannel photon detectors described in Section 7E-3. Here, all of the resolution elements of an electron spectrum are monitored simultaneously and the data stored in a computer for subsequent display. The advantages of such a system are similar to those realized with multichannel photon transducers.

Applications of XPS

Electron spectroscopy for chemical analysis provides qualitative and quantitative information about the elemental composition of matter, particularly of solid surfaces. It also often provides useful structural information.[9]

Qualitative Analysis. A low-resolution, wide-scan XPS spectrum (often called a survey spectrum) such as that shown in Figure 21-2, serves as the basis for the determination of the elemental composition of samples. With a magnesium or aluminum K_α source, all elements except hydrogen and helium emit core electrons having characteristic binding energies. Typically, a survey spectrum encompasses a kinetic energy range of 250 to 1500 eV, which corresponds to binding energies of about 0 to 1250 eV. Every element in the periodic table has one or more energy levels that will result in the appearance of peaks in this region. In most instances, the peaks are well resolved and lead to unambiguous identification provided the element is present in concentrations greater than about 0.1%. Occasionally peak overlap is encountered such as O1*s*/Sb3*d* or Al2*s*, 2*p*/Cu3*s*, 3*p*. Usually, problems due to spectral overlap can be resolved by investigating other spectral regions for additional peaks. Often peaks resulting from Auger electrons are found in XPS spectra (see for example, the peak at about 610 eV in Figure 21-2). Such peaks are readily identified by comparing spectra produced by two X-ray sources (usually magnesium and aluminum K_α). Auger peaks remain unchanged on the kinetic energy scale while photoelectric peaks are displaced. The reason for the behavior of Auger electrons will become apparent in Section 21B-4.

Chemical Shifts and Oxidation States. When one of the peaks of a survey spectrum is examined under conditions of higher energy resolution, the position of the maximum is found to depend to a small degree upon the chemical environment of the atom responsible for the peak. That is, variations in the number of valence electrons, and the type of bonds they form, influence the binding energies of core electrons. The effect of number of valence electrons and thus the oxidation state is demonstrated by the data for several elements shown in Table 21-2. Note that in each case, binding energies increase as the oxidation state becomes more positive. This *chemical shift* can be explained by assuming that the attraction of the nucleus for a core electron is diminished by the presence of outer electrons. When one of these electrons is removed, the effective charge sensed for the core electron is increased; thus an increase in binding energy results.

One of the most important applications of XPS has been the identification of oxidation states of elements contained in various kinds of inorganic compounds.

Chemical Shifts and Structure. Figure 21-5 illustrates the effect of structure on the position of peaks for an element. Each peak corresponds to the 1*s* electron of

[9] For reviews of applications of XPS (and AES as well) see, N. H. Turner and J. A. Schreifels, *Anal. Chem.*, **1996**, *68*, 309R; **1994**, *66*, 163R; **1992**, *64*, 302R; **1990**, *62*, 113R; **1988**, *60*, 377R.

TABLE 21-2 Chemical Shifts as a Function of Oxidation State[a]

Element[b]	Oxidation State									
	−2	−1	0	+1	+2	+3	+4	+5	+6	+7
Nitrogen (1*s*)	—	*0[c]	—	+4.5[d]	—	+5.1	—	+8.0	—	—
Sulfur (1*s*)	−2.0	—	*0	—	—	—	+4.5	—	+5.8	—
Chlorine (2*p*)	—	*0	—	—	—	+3.8	—	+7.1	—	+9.5
Copper (1*s*)	—	—	*0	+0.7	+4.4	—	—	—	—	—
Iodine (4*s*)	—	*0	—	—	—	—	—	+5.3	—	+6.5
Europium (3*d*)	—	—	—	—	*0	+9.6	—	—	—	—

[a]All shifts are in electron volts measured relative to the oxidation states indicated by (*). (Reprinted with permission from D. M. Hercules, *Anal. Chem.*, **1970,** *42,* 28A. Copyright 1970 American Chemical Society.)
[b]Type of electrons given in parentheses.
[c]Arbitrary zero for measurement, end nitrogen in NaN_3.
[d]*Middle nitrogen in* NaN_3.

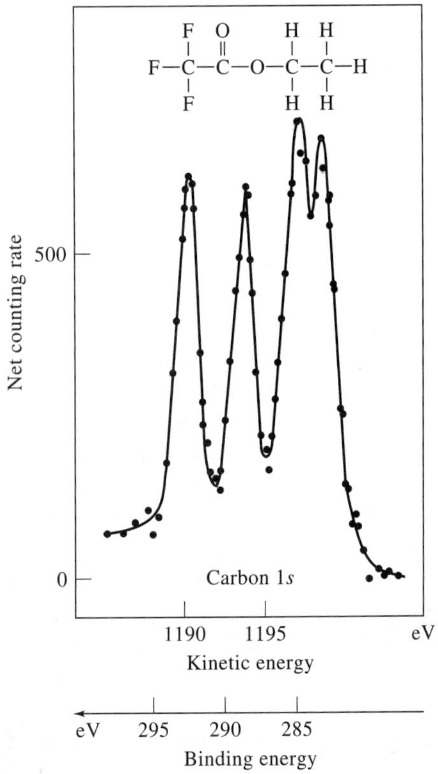

Figure 21-5 Carbon 1*s* X-ray photoelectron spectrum for ethyl trifluoroacetate. *(From K. Siegbahn et al.,* ESCA: Atomic, Molecular, and Solid-State Studies by Means of Electron Spectroscopy, *p. 21. Upsala: Almquist and Wiksells, 1967. With permission.)*

the carbon atom located directly above it in the structural formula. Here, the shift in binding energies can be rationalized by taking into account the effect of the various functional groups on the effective nuclear charge experienced by the 1*s* core electron. For example, of all of the attached groups, fluorine atoms have the greatest ability to withdraw electron density from the carbon atom. The effective nuclear charge felt by the carbon 1*s* electron is therefore a maximum, as is the binding energy.

Figure 21-6 indicates the position of peaks for sulfur in its several oxidation states and in various types of organic compounds. The data in the top row clearly demonstrate the effect of oxidation state. Note also in the last four rows of the chart that XPS discriminates between two sulfur atoms contained in a single ion or molecule. Thus, two peaks are observed for thiosulfate ion ($S_2O_3^{2-}$), suggesting different oxidation states for the two sulfur atoms contained therein.

XPS spectra provide not only qualitative information about types of atoms present in a compound but also the relative number of each type. Thus, the nitrogen 1*s* spectrum for sodium azide ($Na^+N_3^-$) is made up of two peaks having relative areas in the ratio of 2:1 corresponding to the two end nitrogens and the center nitrogen, respectively.

It is worthwhile pointing out again that the photoelectrons produced in XPS are incapable of passing through more than perhaps 10 to 50 Å of a solid. Thus, the most important applications of electron spec-

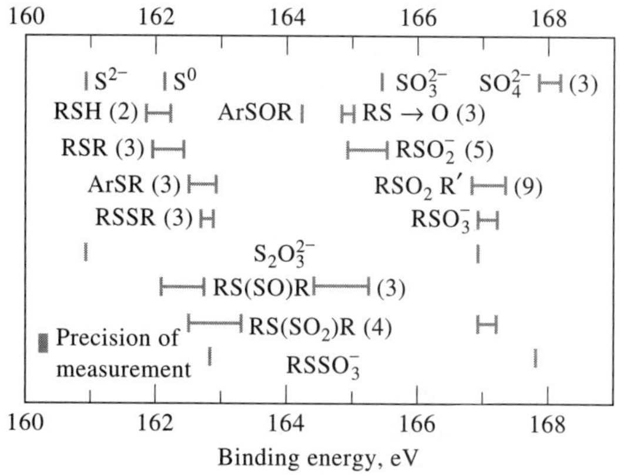

Figure 21-6 Correlation chart for sulfur 2*s* electron binding energies. The numbers in parentheses indicate the number of compounds examined. *(Reprinted with permission from D. M. Hercules,* Anal. Chem., *1970, 42(1), 35A. Copyright 1970 American Chemical Society.)*

troscopy, like X-ray microprobe spectroscopy, are for the accumulation of information about surfaces. Examples of some of its uses include identification of active sites and poisons on catalytic surfaces, determination of surface contaminants on semiconductors, analysis of the composition of human skin, and study of oxide surface layers on metals and alloys.

It is also evident that the method has a substantial potential in the elucidation of chemical structure (see Figures 21-5 and 21-6); the information obtained appears comparable to that from NMR or infrared spectroscopy. A noteworthy attribute is the ability of XPS to distinguish among oxidation states of an element.

It is of interest to note that the information obtained by XPS must also be present in the absorption edge of an X-ray absorption spectrum for a compound. Most X-ray spectrometers, however, do not have sufficient resolution to permit ready extraction of the structural information.

Quantitative Applications. Although several authors have reported using XPS for quantitative determination of the elemental composition of various inorganic and organic materials, the method has not enjoyed widespread application for this purpose.[10] Both peak intensities and peak areas have been used as the analytical parameter, with the relationship between these quantities and concentration being established empirically. Often, internal standards have been recommended. Relative precisions of 3% to 10% have been claimed. For the analysis of solids and liquids, it is necessary to assume that the surface composition of the sample is the same as its bulk composition. For many applications this assumption can lead to significant errors.

21B-4 Auger Electron Spectroscopy[11]

In contrast to XPS, Auger electron spectroscopy is based upon a two-step process in which the first step involves formation of an electronically excited ion A^{+*} by exposing the analyte to a beam of electrons, or sometimes X-rays. With X-rays, the reaction shown by Equation 21-1 occurs; for an electron beam, the excitation reaction can be written

$$A + e_i^- \rightarrow A^{+*} + e_i'^- + e_A^- \qquad (21\text{-}3)$$

where e_i^- represents an incident electron from the source, $e_i'^-$ represents the same electron after it has interacted with A and has thus lost some of its energy, and e_A^- represents an electron that is ejected from one of the inner orbitals of A.

As shown in Figures 21-7a and 21-7b, relaxation of the excited ion A^{+*} can occur in either of two ways; these include

$$A^{+*} \rightarrow A^{++} + e_A^- \qquad (21\text{-}4)$$

or

$$A^{+*} \rightarrow A^{+} + h\nu_f \qquad (21\text{-}5)$$

Here, e_A^- corresponds to an Auger electron while $h\nu_f$ represents a fluorescence photon.

The relaxation process described by Equation 21-5 will be recognized as X-ray fluorescence, which was described in the previous chapter. Note that the energy of the fluorescent radiation $h\nu_f$ is independent of the excitation energy. Thus, polychromatic radiation can be used for the excitation step. Auger emission, shown by Equation 21-4 is a process in which the energy given up in relaxation results in the ejection of an electron (an Auger electron e_A^-) with a kinetic energy E_k. Note that the energy of the Auger electron is *independent* of

[10] For a review of quantitative applications of XPS and Auger spectroscopy, see K. W. Nebesny, B. L. Maschhoff, and N. R. Armstrong, *Anal. Chem.*, **1989**, *61*, 469A.

[11] See M. Thompson, M. D. Baker, A. Christie, and J. F. Tyson, *Auger Electron Spectroscopy.* New York: Wiley, 1985.

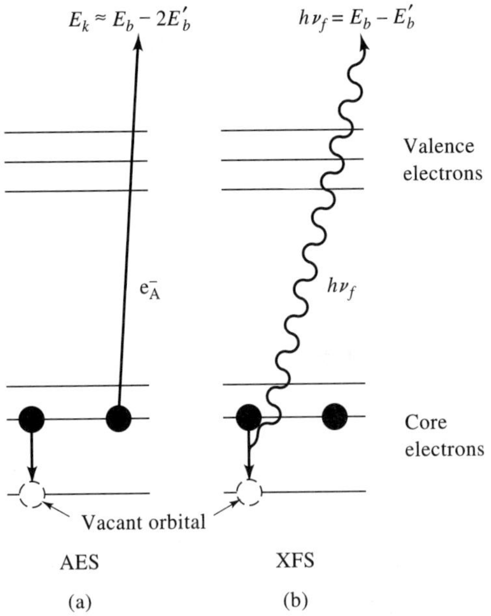

Figure 21-7 Schematic representation of the source of (a) Auger electron emission and (b) X-ray fluorescence that competes with Auger emission.

the energy of the photon or electron that originally created the vacancy in energy level E_b. Thus, as is true in fluorescence spectroscopy, a monoenergetic source is *not* required for excitation. Note that it is this independence of Auger peaks from the input energy that makes it possible for the chemist to differentiate between the Auger peak in a spectrum and the XPS peaks.

The kinetic energy of the Auger electron is the difference between the energy released in relaxation of the excited ion ($E_b - E_b'$) and the energy required to remove the second electron from its orbit (E_b'). Thus,

$$E_k = (E_b - E_b') - E_b' = E_b - 2E_b' \qquad (21\text{-}6)$$

Auger emissions are described in terms of the type of orbital transitions involved in the production of the electron. For example, a KLL Auger transition involves an initial removal of a K electron followed by a transition of an L electron to the K orbital with the simultaneous ejection of a second L electron. Other common transitions are LMM and MNN.

Like XPS spectra, Auger spectra consist of a few characteristic peaks lying in the region of 20 to 1000 eV. Figure 21-8 shows typical Auger spectra obtained for two samples of a 70% copper/30% nickel alloy. Note that the derivative of the counting rate as a function of the kinetic energy of the electron $dN(E)/dE$ serves as the ordinate. Derivative spectra are standard for Auger spectroscopy in order to enhance the small peaks and to repress the effect of the large, but slowly changing, scattered electron background radiation. Also note that the peaks are well separated, making qualitative identification quite straightforward.

Auger electron emission and X-ray fluorescence (Figure 21-7) are competitive processes, and their relative rates depend upon the atomic number of the element involved. High atomic numbers favor fluorescence, while Auger emission predominates with atoms of low atomic numbers. As a consequence, X-ray fluorescence is not a very sensitive means for detecting elements with atomic numbers smaller than about 10.

Auger and X-ray photoelectron spectroscopy provide similar information about the composition of matter. The methods tend to be complementary rather than competitive, however, with Auger spectroscopy being more reliable and efficient for certain applications and XPS for others. As mentioned earlier, most instrument manufacturers recognize the complementary nature of Auger and XPS by making provisions for both kinds of measurements with a single instrument.

The particular strengths of Auger spectroscopy are its sensitivity for atoms of low atomic number, its minimal matrix effects, and above all its high spatial resolution, which permits detailed examination of solid surfaces. The high spatial resolution arises because the primary beam is made up of electrons that can be more tightly focused on a surface than can X-rays. To date, Auger spectroscopy has not been extensively used to provide the kind of structural and oxidation state information that was described for XPS. Quantitative analysis by the procedure is difficult.

Instrumentation

The instrumentation for AES is similar to that for XPS except that the source is usually an *electron gun* rather than an X-ray tube. A schematic of the most common type of electron gun is found in Figure 21-9. This source consists of a heated tungsten filament, which is usually about 0.1 mm in diameter and bent into the shape of a hairpin with a V-shaped tip. The cathodic filament is maintained at a potential of 1 to 50 kV with respect to the anode contained in the gun. Surrounding the filament is a grid cap, or *Wehnelt cylinder,* which is biased negatively with respect to the filament. The effect of the electric field in the gun is to cause the emitted electrons to converge on a tiny spot called the *crossover* that has a diameter d_0.

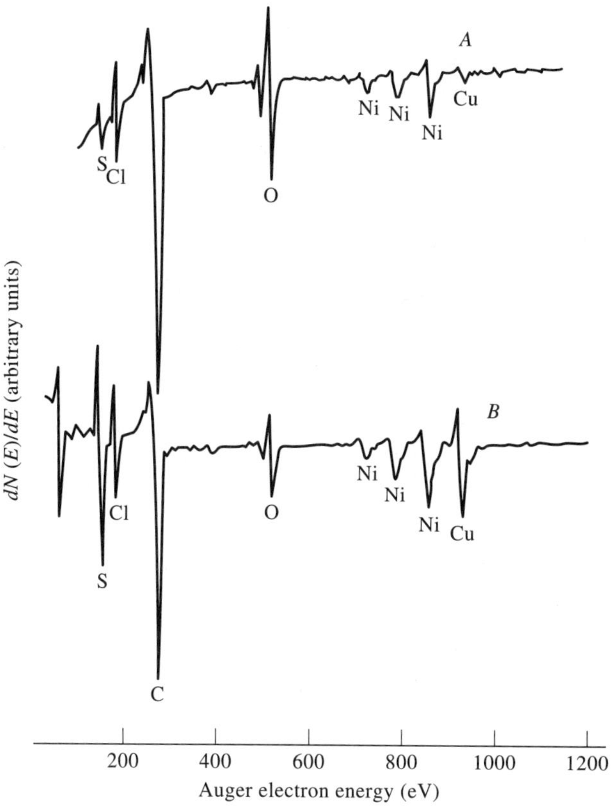

Figure 21-8 Auger electron spectra for a 70% Cu/30% Ni alloy. *A*, passivated by anodic oxidation; *B*, not passivated. *(Adapted from G. E. McGuire et al., J. Electrochem. Soc., **1978**, 125, 1802. Reprinted by permission of the publisher, The Electrochemical Society, Inc.)*

Cathodes constructed in the form of lanthanum hexaboride (LaB_6) rods are also used in electron guns when a source of greater brightness is desired. This type of source is expensive and requires a better vacuum system to prevent oxide formation, which causes the efficiency of the source to deteriorate rapidly. A third type of source is based upon a process called *field emission.* Here, the source is a tungsten or carbon cathode shaped so that it has a very sharp tip (100 nm or less). When this type of cathode is held at a high potential, the electric field at the tip is so intense ($> 10^7$ V/cm) that electrons are produced by a "tunneling process" in which no thermal energy is required to free the electrons from the potential barrier that normally prevents their emission. Field emission sources provide a beam of electrons that have a crossover diameter of only 10 nm compared with 10 μm for LaB_6 rods and 50 μm for tungsten hairpins. The disadvantages of this type of source is its fragility and the fact that it also requires a better vacuum than does an ordinary filament source.

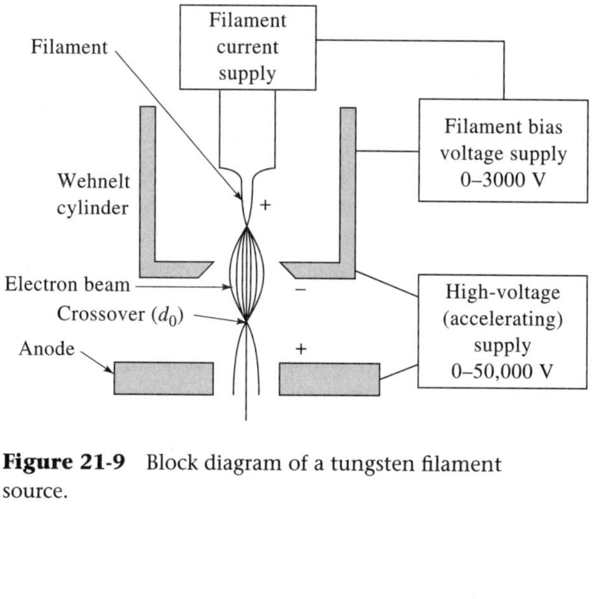

Figure 21-9 Block diagram of a tungsten filament source.

Electron guns produce a beam of electrons with energies of 1 to 10 keV, which can be focused on the surface of a sample for Auger electron studies. One of the special advantages of Auger spectroscopy is its capability for very high spatial-resolution scanning of solid surfaces. Normally electron beams with diameters ranging from 500 to 5 μm are used for this purpose. Guns producing beams of approximately 5 μm are called

Auger microprobes and are employed for scanning solid surfaces in order to detect and determine the elemental composition of inhomogeneities.

Applications of Auger Electron Spectroscopy

Qualitative Analysis of Solid Surfaces. Typically, an Auger spectrum is obtained by bombarding a small area of a surface (diameter, 5 to 500 μm) with a beam of electrons from a gun. A derivative electron spectrum, such as that shown in Figure 21-8, is then obtained with an analyzer. An advantage of Auger spectroscopy for surface studies is that the low-energy Auger electrons (20 to 1000 eV) are able to penetrate only a few atomic layers (3 to 20 Å) of solid. Thus, while the electrons from the electron guns penetrate to a considerably greater depth below the sample surface, only those Auger electrons from the first four or five atomic layers escape to reach the analyzer. Consequently, an Auger spectrum is likely to reflect the true *surface* composition of solids.

The two Auger spectra in Figure 21-8 are for samples of a 70% copper/30% nickel alloy, which is often used for structures where saltwater corrosion resistance is required. Corrosion resistance of this alloy is markedly enhanced by preliminary anodic oxidation in a strong solution of chloride. Figure 21-8A is the spectrum of an alloy surface that has been *passivated* in this way. Spectrum *B* is for another sample of the alloy in which the anodic oxidation potential was not great enough to cause significant passivation. The two spectra clearly reveal the chemical differences between the two samples that account for the greater corrosion resistance of the former. First, the copper-to-nickel ratio in the surface layer of the nonpassivated sample is approximately that for the bulk, whereas in the passivated material the nickel peaks completely overshadow the copper peak. Furthermore, the oxygen-to-nickel ratio in the passivated sample approaches that for pure anodized nickel, which also has a high corrosion resistance. Thus, the resistance toward corrosion of the alloy appears to result from the creation of a surface that is largely nickel oxide. The advantage of the alloy over pure nickel is its significantly lower cost.

Depth Profiling of Surfaces. Depth profiling involves the determination of the elemental composition of a surface as it is being etched away (sputtered) by a beam of argon ions. Either XPS or Auger spectroscopy can be used for elemental detection, although the latter is the more common. Figure 21-10 shows schematically how the process is carried out with a highly focused Auger microprobe; the diameter of the microprobe beam is about 5 μm. The microprobe and etching beams are operated simultaneously, with the intensity of one or more of the resulting Auger peaks being recorded as a function of time. Since the etching rate is related to time, a depth profile of elemental composition is obtained. Such information is of vital importance in a variety of studies such as corrosion chemistry, catalyst behavior, and properties of semiconductor junctions.

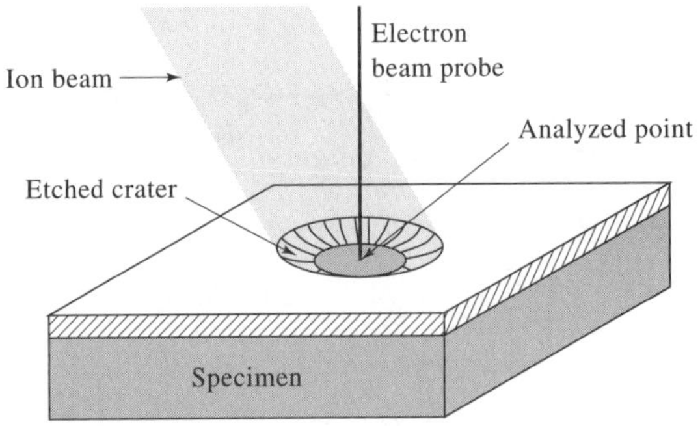

Figure 21-10 Schematic representation of the simultaneous use of ion sputter etching and Auger spectroscopy for determining depth profiles. *(Courtesy of Physical Electronic Industries, Inc., Eden Prairie, MN.*

Figure 21-11 gives a depth profile for the copper/nickel alloy described in the previous section (Figure 21-8). Here, the ratio of the peak intensities for copper versus nickel are recorded as a function of sputtering time. Curve *A* is the profile for the sample that had been passivated by anodic oxidation. With this sample, the copper/nickel ratio is essentially zero for the first ten minutes of sputtering, which corresponds to a depth of about 500 Å. The ratio then rises and approaches that for a sample of alloy that had been chemically etched so that its surface is approximately that of the bulk sample (curve *C*). The profile for the nonpassivated sample (curve *B*) resembles that of the chemically etched sample, although some evidence is seen for a thin nickel oxide coating.

Line Scanning. Line scans are used to characterize the surface composition of solids as a function of distance along a straight line of 100 μm or more. For this purpose, an Auger microprobe is used that produces a beam that can be moved across a surface in a reproducible way. Figure 21-12 shows Auger line scans

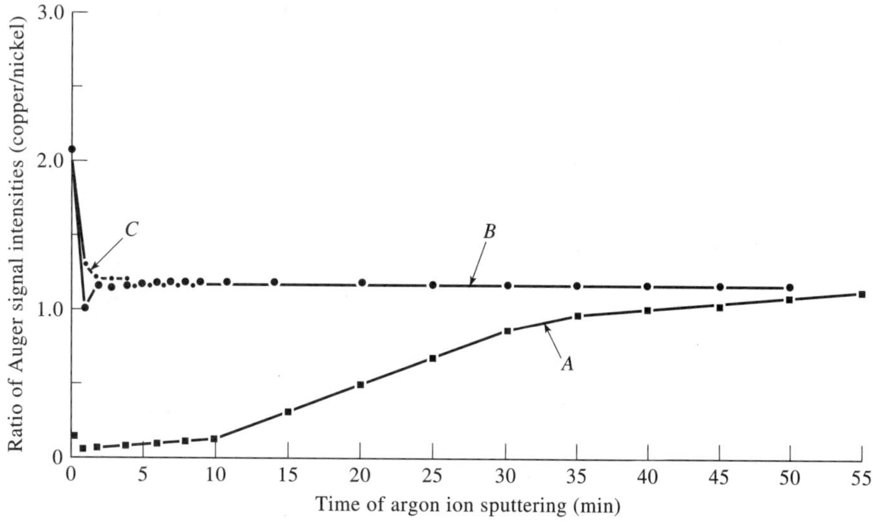

Figure 21-11 Auger sputtering profiles for the copper/nickel alloys shown in Figure 21- 8; *A*, passivated sample; *B*, nonpassivated sample; *C*, chemically etched sample representing the bulk material. *(Adapted from G.E. McGuire et al.,* J. Electrochem. Soc., ***1978,*** *125, 1802. Reprinted by permission of the publisher, The Electrochemical Society, Inc.)*

along the surface of semiconductor device. In the upper figure, the relative peak amplitude of an oxygen peak is recorded as a function of distance along a line; the lower figure is the same scan produced when the analyzer was set to a peak for gold.

21B-5 Secondary-Ion Mass Spectrometry

Secondary-ion mass spectrometry (SIMS) is the most highly developed of the mass spectrometric surface methods, with several manufacturers offering instruments for this technique. SIMS has proven useful for determining both the atomic and the molecular composition of solid surfaces.[12]

Two types of instruments are encountered: *secondary-ion mass analyzers* and *microprobe analyzers.* Both are based upon bombarding the surface of the sample with a beam of 5- to 20-keV ions such as Ar^+, Cs^+, N_2^+, or O_2^+. The ion beam is formed in an *ion gun* in which the gaseous atoms or molecules are ionized by an electron impact source. The positive ions are then accelerated by applying a high dc potential. The impact of these primary ions causes the surface layer of atoms of the sample to be stripped (sputtered) off, largely as neutral atoms. A small fraction, however, forms as positive (or negative) secondary ions that are drawn into a spectrometer for mass analysis.

In secondary-ion mass analyzers, which serve for general surface analysis and for depth profiling, the primary ion beam diameter ranges from 0.3 to 5 mm. Double-focusing, single-focusing, time-of-flight, and quadrupole spectrometers are used for mass determination. These spectrometers yield qualitative and quantitative information about all of the isotopes (hydrogen through uranium) present on a surface. Sensitivities of 10^{-15} g or better are typical. By monitoring peaks for one or a few isotopes, as a function of time, concentration profiles can be obtained with a depth resolution of 50 to 100 Å.

Ion microprobe analyzers are more sophisticated (and more expensive) instruments that are based upon a focused beam of primary ions that has a diameter of 1 to 2 μm. This beam can be moved across a surface for about 300 μm in both the *x* and *y* directions. A microscope is provided to permit visual adjustment of the beam position. Mass analysis is performed with a double-focusing spectrometer. In some instruments, the primary ion beam passes through an additional low-resolu-

[12] A. Benninghoven, F. G. Rudenauer, and H. W. Werner, *Secondary Ion Mass Spectrometry: Basic Concepts, Instrumental Aspects, and Applications and Trends.* New York: Wiley, 1987; W. H. Christie, *Anal. Chem.,* **1981,** *53,* 1240A.

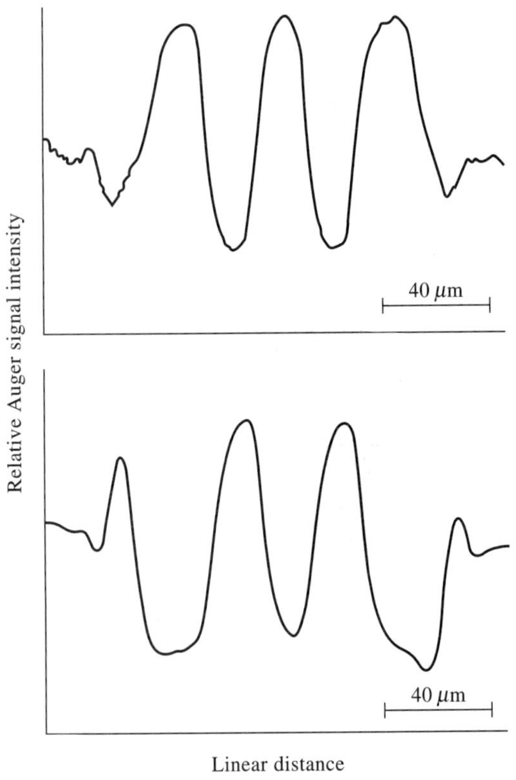

Figure 21-12 Auger line scans for oxygen *(top)* and gold *(bottom)* obtained for the surface of a semiconductor device. *(Courtesy of Physical Electronics Industries, Inc. Eden Prairie, MN.)*

tion mass spectrometer so that only a single type of primary ion bombards the sample. The ion microprobe version of SIMS permits detailed studies of solid surfaces.

21B-6 Laser-Microprobe Mass Spectrometry

Commercial laser-microprobe mass spectrometers are now available for the study of solid surfaces. Ionization and volatilization are accomplished with a pulsed, neodymium-YAG laser, which, after frequency quadrupling, produces a 0.5-μm spot of 266-nm radiation. The power density of the radiation within this spot is 10^{10} to 10^{11} W/cm^2. The power of the beam can be attenuated to 1% by means of a 25-step optical filter. Collinear with the ionization beam is the beam from a second, low-power, He-Ne laser ($\lambda = 633$ nm), which serves as illumination so that the area to be analyzed can be chosen visually. The instrument has an unusually high sensitivity (down to 10^{-20} g), is applicable to both inorganic and organic (including biological) samples, has a spatial resolution of about 1 μm, and produces data at a rapid rate. Some typical applications of this instrument include determination of Na/K concentration ratios in frog nerve fiber, determination of the calcium distribution in retinas, classification of asbestos and coal mine dusts, determination of fluorine distributions in dental hard tissue, analysis of amino acids, and study of polymer surfaces.[13]

21B-7 The Electron Microprobe

With the electron microprobe method, X-ray emission is stimulated on the surface of the sample by a narrow, focused beam of electrons. The resulting X-ray emission is detected and analyzed with either a wavelength or an energy dispersive spectrometer.[14]

Instruments

Figure 21-13 is a schematic of an electron microprobe system. The instrument employs three integrated beams of radiation, namely, electron, light, and X-ray. In addition, a vacuum system is required that provides a pressure of less than 10^{-5} torr and a wavelength- or an energy-dispersive X-ray spectrometer (a wavelength-dispersive system is shown in Figure 21-13). The electron beam is produced by a heated tungsten cathode and an accelerating anode (not shown). Two electromagnetic lenses focus the beam on the specimen; the diameter of the beam is between 0.1 and 1 μm. An associated optical microscope is used to locate the area to be bombarded. Finally, the fluorescent X-rays produced by the electron beam are collimated, dispersed by a single crystal, and detected by a gas-filled transducer. Considerable design effort is required to arrange the three systems spatially so that they do not interfere with one another.

---

[13] For further details, see L. Van Vaeck and R. Gijbels, *Fresenius, J. Anal. Chem.*, **1990**, *337*, 743, 755; E. Denoyer, R. Van Grieken, F. Adams, and D. F. S. Natusch, *Anal. Chem.*, **1982**, *54*, 26A, 280A; R. J. Cotter, *Anal. Chem.*, **1984**, *56*, 485A.

[14] For a detailed discussion of this method, see L. S. Birks, *Electron Probe Microanalysis*, 2nd ed. New York: Wiley-Interscience, 1971; K. F. J. Heinrich, *Electron Beam X-Ray Microanalysis*. New York: Van Nostrand, 1981. For a recent review of the use of the scanning electron microbe for elemental analysis of surfaces see D. E. Newbury et al., *Anal. Chem.*, **1990**, *62*, 1159A, 1245A.

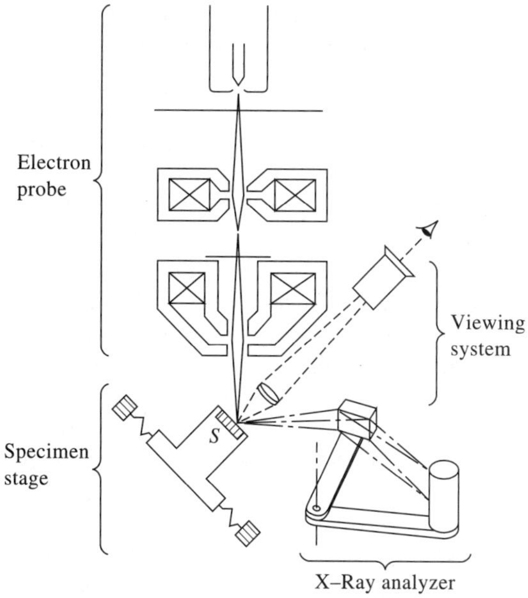

Figure 21-13 Schematic view of an electron-microprobe instrument. *(From D. B. Wittry, in* Treatise on Analytical Chemistry, *I. M. Kolthoff and P. J. Elving, Eds., Vol. 5 Part I, p. 3178. New York: Interscience, 1964. Reprinted by permission of John Wiley & Sons, Inc.)*

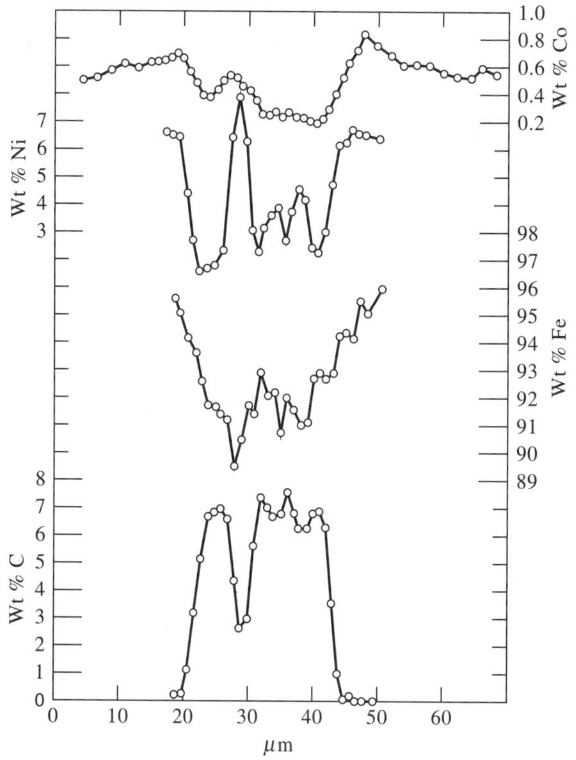

Figure 21-14 Scanning electron microprobe output across the surface of an α-cohenite particle in a lunar rock.

In addition to the foregoing components, the specimen stage is provided with a mechanism whereby the sample can be moved in two mutually perpendicular directions and rotated as well, thus permitting scanning of the surface.

Applications

The electron microprobe provides a wealth of information about the physical and chemical nature of surfaces. It has had important applications to phase studies in metallurgy and ceramics, the investigation of grain boundaries in alloys, the measurement of diffusion rates of impurities in semiconductors, the determination of occluded species in crystals, and the study of the active sites of heterogeneous catalysts. In all of these applications, both qualitative and quantitative information about surfaces is obtained.

Figure 21-14 illustrates the use of the electron microprobe for the analysis of an α-cohenite (Fe_3C) particle in a lunar rock. The data were obtained by a linear scan of the particle observed visually on the surface and by measurement of the intensity of the characteristic emission line for each of four elements.

21C SCANNING ELECTRON MICROSCOPY

In many fields of chemistry, material science, geology, and biology, detailed knowledge of the physical nature of the surfaces of solids is of great importance. The classical method for obtaining such information was by optical microscopy, which is still an important technique for characterizing surfaces. The resolution of optical microscopy is, however, limited by diffraction effects to about the wavelength of light. Currently, surface information at considerably higher resolution is obtained by three techniques, *scanning electron microscopy* (SEM), *scanning tunneling microscopy* (STM), and *atomic force microscopy* (AFM). The latter two methods are sometimes referred collectively as *scanning probe microscopy* (SPM). This section deals with scanning electron microscopy. The two sections that follow are devoted to the two methods for scanning probe microscopy.

In obtaining an image by each of these techniques, the surface of a solid sample is swept in a *raster pattern* with a finely focused beam of electrons or with a suitable probe. A raster is a scanning pattern similar to that used in a cathode-ray tube or in a television set in which an electron beam is (1) swept across a surface in a straight line (the x direction), (2) returned to its starting position, and (3) shifted downward (the y direction) by a standard increment. This process is repeated until a desired area of the surface has been scanned. During this scanning process, a signal is received above the surface (the z direction) and stored in a computer system, where it is ultimately converted to an image.

21C-1 The Scanning Electron Microscope (SEM)[15]

In a scanning electron microscope, the surface of a solid sample is scanned in a raster pattern with a beam of energetic electrons. Several types of signals are produced from a surface in this process, including backscattered, secondary, and Auger electrons; X-ray fluorescence photons; and other photons of various energies. All of these signals have been used for surface studies, but the two most common are (1) backscattered and secondary electrons, which serve as the basis of scanning electron microscopy, and (2) X-ray emission, which is used in electron microprobe analysis.

Instrumentation

Figure 21-15 is a schematic of a combined instrument that is both a scanning electron microscope and a scanning electron microprobe. Note that a common electron gun source and electron focusing system is used but that the electron microscope employs an electron detector, whereas the microprobe uses an X-ray detector, such as those described in Section 12B-4.

Electron Optics. The magnetic condenser and objective lens system shown in Figure 21-15 serve to reduce the image to a final spot size on the sample of 5 to 200 nm. The condenser lens system, which may consist of one or more lenses, is responsible for the throughput of the electron beam reaching the objective lens; the objective lens is responsible for the size of the electron beam impinging on the surface of the sample. Typically, an individual lens is cylindrically symmetrical and between 10 and 15 cm in height. The details of the optics of magnetic lens systems is beyond the scope of this text.

Scanning with a SEM is accomplished by the two pairs of electromagnetic coils located within the objective lens (see Figure 21-15); one pair deflects the beam in the x direction across the sample, and the other pair deflects it in the y direction. Scanning is controlled by applying an electrical signal to one pair of scan coils, such that the electron beam strikes the sample to one side of the center axis of the lens system. By varying the electrical signal to this pair of coils (that is, the x coils) as a function of time, the electron beam is moved in a straight line across the sample and then returned to its original position. After completion of the line scan, the other set of coils (y coils in this case) is used to deflect the beam slightly, and the deflection of the beam using the x coils is repeated. Thus, by rapidly moving the beam, the entire sample surface can be irradiated with the electron beam. The signals to the scan coils can be either analog or digital. Digital scanning has the advantage that it offers very reproducible movement and location of the electron beam. The signal from the sample can be encoded and stored in digital form along with digital representations of the x and y positions of the beam.

The signals that are used to drive the electron beam in the x and y directions are also used to drive the horizontal and vertical scans of a cathode-ray tube (CRT). The image of the sample is produced by using the output of a detector to control the intensity of the spot on the CRT. Thus, this method of scanning produces a map of the sample in which there is a one-to-one correlation between the signal produced at a particular location on the sample surface and a corresponding point on the CRT display. The magnification (M) achievable in the SEM image is given by

$$M = W/w \qquad (21\text{-}7)$$

where W is the width of the CRT display and w is the width of a single line scan across the sample. Because W is a constant, increased magnification is achieved by decreasing w. The inverse relationship between magnification and the width of the scan across the sample implies that a beam of electrons that has been focused to an infinitely small point could provide infinite magnification. A variety of other factors, however, limit the magnification that is achievable to a range from about 10× to 100,000×.

---

[15] Monographs dealing with SEM include Goldstein et al., *Scanning Electron Microscopy and X-Ray Microanalysis*. New York: Plenum Press, 1981; O. C. Wells et al., *Scanning Electron Microscopy*. New York: McGraw-Hill, 1974.

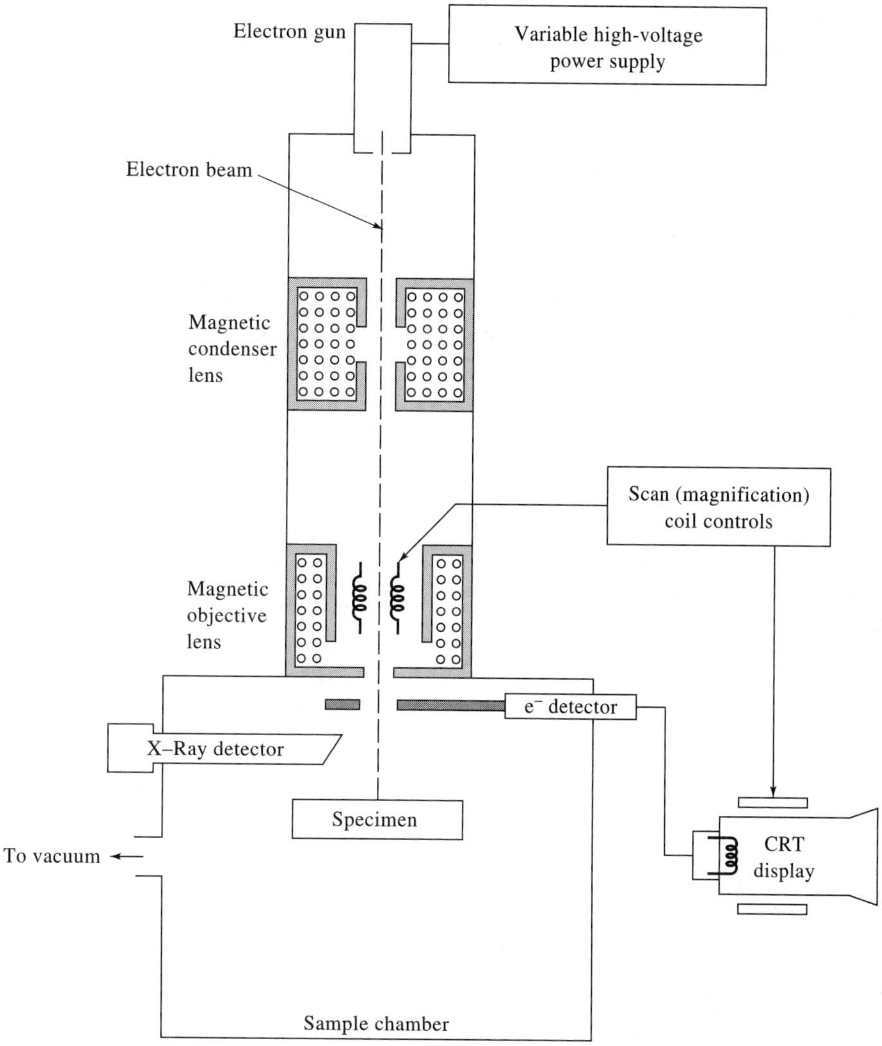

Figure 21-15 Schematic of an SEM.

Sample and Sampler Holder. Sample chambers are designed for rapid changing of samples. Large-capacity vacuum pumps are used to hasten the switch from ambient pressure to 10^{-4} torr or less. The sample holder, or stage, in most instruments is capable of holding samples many centimeters on an edge. Furthermore, the stage can be moved in the *x, y,* and *z* directions, and it can be rotated about each axis. As a consequence, the surfaces of most samples can be viewed from almost any perspective.

Samples that conduct electricity are easiest to study, because the unimpeded flow of electrons to ground minimizes artifacts associated with the buildup of charge. In addition, samples that are good conductors of electricity are usually also good conductors of heat, which minimizes the likelihood of their thermal degradation. Unfortunately, most biological specimens and most mineral samples do not conduct. A variety of techniques have been developed for obtaining SEM images of nonconducting samples, but the most common approaches involve coating the surface of the sample with a thin metallic film produced by sputtering or by vacuum evaporation. Regardless of the method of producing a conductive coating, a delicate balance must be struck between the thinnest uniform coating achievable and an excessively thick coating that obscures surface details.

Transducers. The most common type of transducer for electrons in scanning electron microscopes are scintillation devices that function in an analogous way to the X-ray scintillation transducers described in Section 12B-4. Here, the detector consists of a doped glass or plastic target that emits a cascade of visible photons when struck by an electron. The photons are conducted by a light pipe to a photomultiplier tube that is housed outside the high-vacuum region of the instrument. Typical gains with scintillation transducers are 10^5 to 10^6.

Semiconductor transducers, which consist of flat wafers of semiconductor material, are also used in electron microscopy. When a high-energy electron strikes the detector, electron-hole pairs are produced that result in increased conductivity. Current gains with typical semiconductor transducers are 10^3 to 10^4, but this type of device is small enough so that it can be placed immediately adjacent to the sample, which leads to high collection efficiency. Furthermore, these devices are easy to use and are less expensive than scintillation transducers. In many instances, these advantages more than offset the lower gain of the semiconductor transducer.

The X-rays produced in the scanning electron microprobe mode are usually detected and measured with energy-dispersive systems such as those discussed in Sections 12B-4 and 12B-5. Wavelength-dispersive systems, such as those discussed in Section 12B-3 have also been used in electron microprobe analyses.

The Interaction of Electron Beams With Solids

The versatility of the scanning electron microscope and microprobe for the study of solids arises from the wide variety of signals that are generated when the electron beam interacts with the solid. We shall consider just three of these signals: backscattered electrons, secondary electrons, and X-ray emission. The interactions of a solid with an electron beam can be divided into two categories: *elastic* interactions, which affect the trajectories of the electrons in the beam without altering their energies significantly, and *inelastic* interactions, which result in transfer of part or all of the energy of the electrons to the solid. The excited solid then emits secondary electrons, Auger electrons, X-rays, and sometimes longer wavelength photons.

Elastic Scattering. When an electron collides elastically with an atom, the direction of the electron changes, but the speed of the electron is virtually unaffected, so that the kinetic energy of the electron remains essentially constant. The angle of deflection for any given collision is random and can vary from 0 to 180 deg. Figure 21-16 is a computer simulation of the random behavior of 5 electrons and 100 electrons when they enter a solid normal to the surface. The energy of the beam is assumed to be 20 keV, which is typical. Note that such a beam penetrates to a depth of 1.5 μm or more. Some of the electrons eventually lose energy by inelastic collisions and remain in the solid; the majority, however, undergo numerous collisions and as a result eventually exit from the surface as backscattered electrons. It is important to note that the beam of backscattered electrons has a much larger diameter than the incident beam—that is, for a 5-nm incident beam, the backscattered beam may have a diameter of

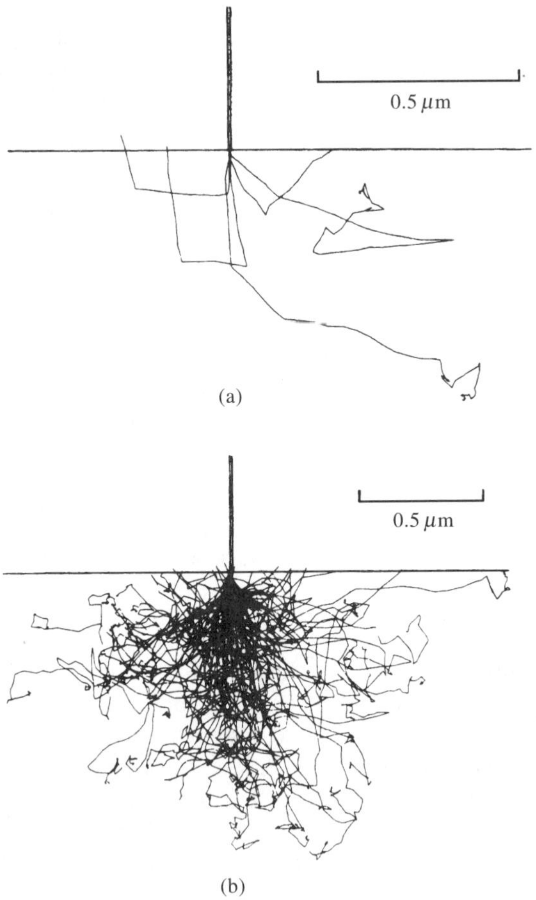

Figure 21-16 Simulation of electron trajectories showing the scattering volume of 20-keV electrons in an iron sample. (a) 5 electrons; (b) 100 electrons. *(From J. I. Goldstein et al.,* Scanning Electron Microscopy and X-Ray Microanalysis, *p.62. New York Plenum Press, 1981. With permission)*

several micrometers. The diameter of the backscattered beam is one of the factors limiting the resolution of an electron microscope.

Secondary Electron Production. It is observed that when the surface of a solid is bombarded with an electron beam having an energy of several keV that electrons having energies of 50 keV or less are emitted from the surface along with the backscattered electrons. The number of these *secondary electrons* is generally one half to one fifth or less of the number of backscattered electrons. Secondary electrons are produced as a result of interactions between the energetic beam electrons and weakly bound conduction electrons in the solid, which leads to ejection of the conduction band electrons with a few electron volts of energy. Secondary electrons are produced from a depth of only 50 to 500 Å and exit in a beam that is slightly larger in diameter than the incident beam. Secondary electrons can be prevented from reaching the transducer by applying a small negative bias to the transducer housing.

X-Ray Emission. A third product of electron bombardment of a solid is X-ray photons. The mechanism of their formation is the same as that described in Section 12A-1. Both characteristic line spectra and an X-ray continuum are produced and emitted from the surface of the sample. This radiation serves as a basis for the electron microprobe.

Applications

Scanning electron microscopy provides morphologic and topographic information about the surfaces of solids that is usually necessary in understanding the behavior of surfaces.[16] Thus, an electron microscopic examination is often the first step in the study of the surface properties of a solid. Figure 21-17 contains several electron micrographs that illustrate the kind of information derived by this method.

21D SCANNING PROBE MICROSCOPES

Scanning probe microscopes (SPMs) are capable of resolving details of surfaces down to the atomic level. The first example of this type of microscope, the scanning tunneling microscope, was described in 1982. Only four years later in 1986, its inventors, G. Binnig and H. Roher, were awarded the Nobel Prize in Physics for their work. Currently the primary use of scanning probe microscopes is for measuring surface topography of samples.

Unlike optical and electron microscopes, scanning probe microscopes reveal details not only on the lateral x and y axis of a sample but also on the z axis, which is perpendicular to the surface. Typically the resolution of SPMs is about 20 Å in the x and y directions, but with ideal samples and the best instruments as low as 1 Å. Resolution in the z dimension is generally better than 1 Å. For comparison the resolution of typical electron microscope is about 50 Å.

We shall discuss the two types of scanning probe microscopes that are currently the most widely used and available from several commercial sources: the *scanning tunneling microscope* (STM) and the *atomic force microscope* (AFM). Both are based upon scanning the surface of the sample in an x/y raster pattern with a very sharp tip that moves up and down along the z axis as the surface topography changes. This movement is measured and translated by a computer into an image of the surface topography. This image often shows details on an atomic size scale.[17]

21D-1 The Scanning Tunneling Microscope

The Binnig/Roher microscope for which they received the Nobel Prize was a scanning tunneling microscope. This device was found to be capable of resolving features on an atomic scale on the surface of a conducting solid surface.[18] By now, these instruments are available from several instrument manufacturers and are used on a routine basis in hundreds of laboratories throughout the world. Their main disadvantage is the requirement that the surface being examined must conduct electricity. The atomic force microscope, discussed in the next section, does not suffer from this limitation.

---

[16] See D. E. Newbury et al., *Anal. Chem.*, **1990**, *62*, 1159A, 1245A.

[17] For a general reference on scanning probe microscopy, see R. Weisendanger, *Scanning Probe Microanalysis and Spectroscopy.* New York: Cambridge University Press, 1994.

[18] G. Binnig, H. Roher, C. Gerber, and E. Weibel, *Phys. Rev. Lett.*, **1982**, *49*, 57.

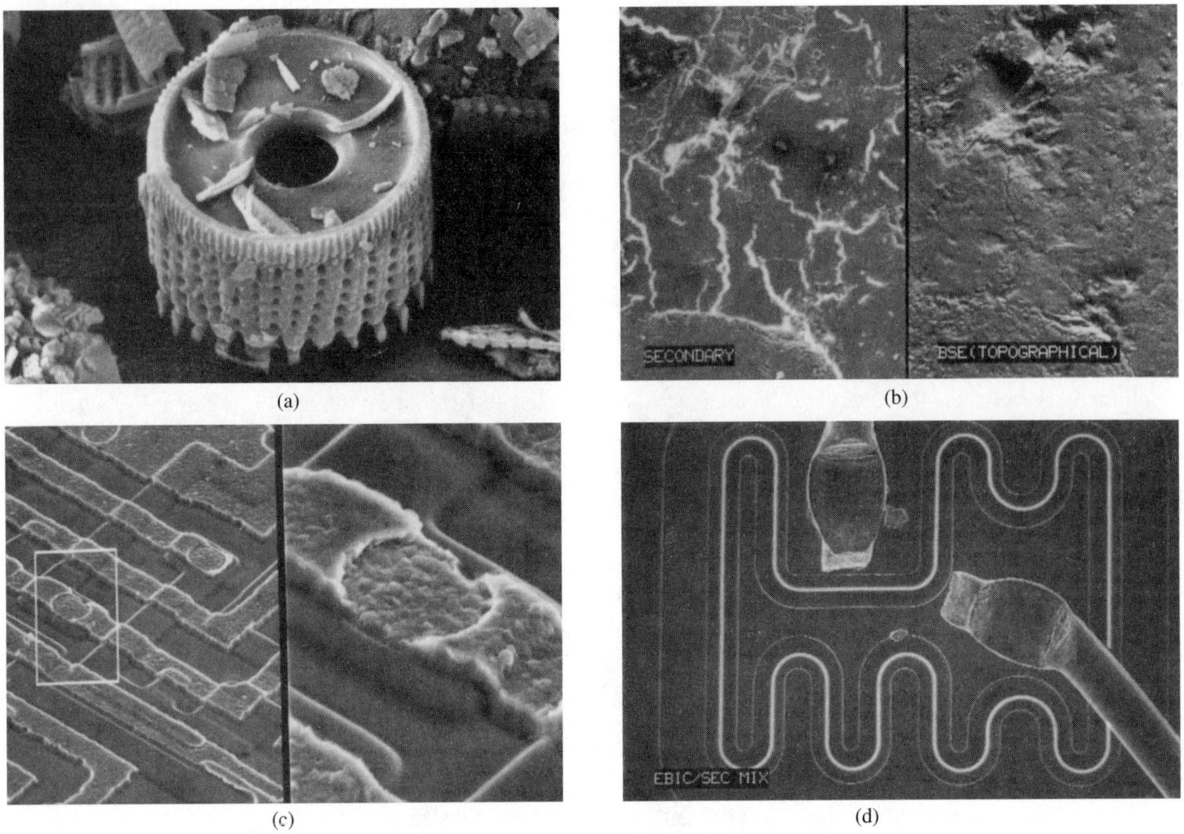

Figure 21-17 A representative set of photomicrographs with magnification and accelerating voltage specified for each image. *(Courtesy of AMRAY, Bedford, MA.)* (a) Image of a diatom (magnification = 5000 ×; accelerating voltage = 10.0 kV. (b) Two views of the same cement sample: Left side is the secondary electron image; right side is the backscattered electron image (magnification = 150 ×; accelerating voltage = 20.0 kV. (c) Left side is an image of part of an integrated circuit; right side is an enlargement of the highlighted rectangle on the left (magnification = 850 × *left*, 3340 × *right*; accelerating voltage = 10.0 kV). (d) Secondary electron beam image of part of a circuit component. The light line winding through the image is due to an induced current called the electron beam induced current (EBIC). This current is induced in the device by the electron beam from the source (magnification = 320 ×; accelerating voltage = 10.0 kV).

Principle of Scanning Tunneling Microscopes

In a scanning tunneling microscope, the surface of a sample is scanned in a raster pattern by a very fine metallic tip. As shown schematically in Figure 21-18a, the tip is maintained at a constant distance d above the surface throughout the scan. The up and down motion of the tip then reflects the topography of the surface. The round balls in the figure represent individual carbon atoms in a sample of pyrolytic graphite. For clarity, the tip is shown as a rounded cone, and the path of the tip during a scan in the x direction is shown by the dashed line. In order to maintain the tip at a constant distance from the sample, a *tunneling current* between the tip and the sample surface is monitored and held at a constant level. The tunneling current is generated by the voltage V that is applied between the tip and the sample as shown.

A tunneling current is a current that passes through a medium that contains no electrons; examples of such media include a vacuum, a nonpolar liquid, or even an aqueous electrolyte solution. The mechanism of current flow through an insulating medium can be rationalized by quantum mechanics but will not be dealt with here.

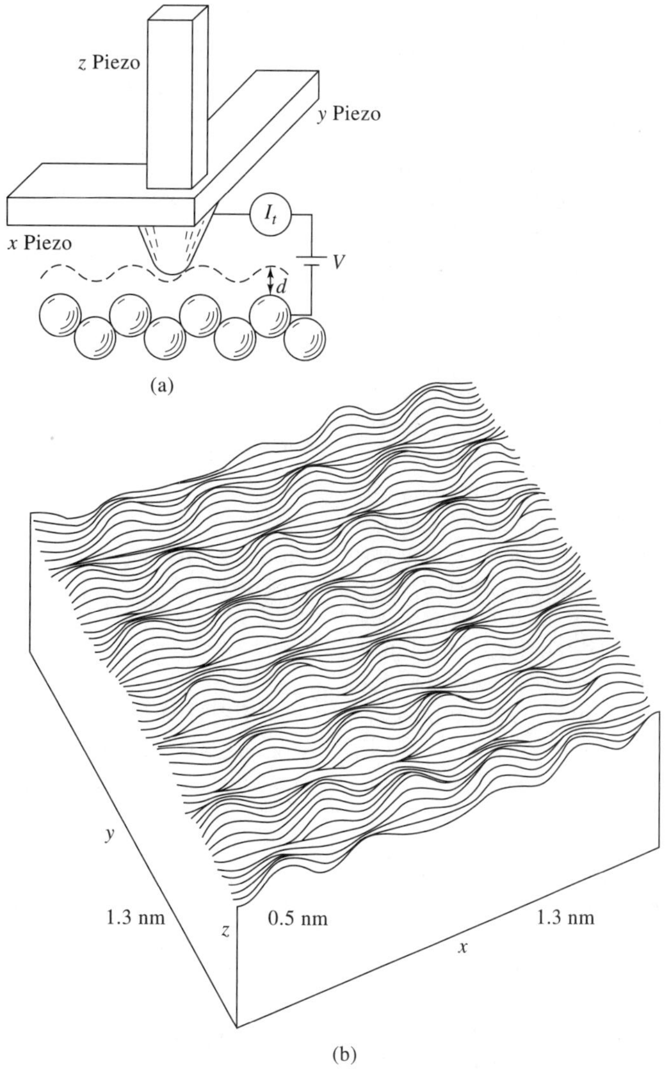

Figure 21-18 (a) Schematic view of an STM tip scanning a sample of pyrolytic carbon in the x direction. Dashed line is the path of the tip over the individual carbon atoms shown as regularly arranged circles. (b) Contour map of the sample surface. *(From P.K. Hansma et al.,* Science, ***1988,*** 242, *209. With permission.)*

Tunneling currents become significant when two conductors are within a few nanometers of one another and when one of the conductors is in the form of a sharp tip. In the case of a scanning tunneling microscopic tip, the magnitude of the tunneling current I_t is given approximately by

$$I_t = Ve^{-Cd} \tag{21-8}$$

where V is the bias voltage between the conductors, C is a constant that is characteristic of the composition of the conductors, and d is the spacing between the lowest atom on the tip and the highest atom on the sample. In a tunneling microscope, the tunneling current is held constant by a feedback mechanism that moves the tip up and down so that d remains constant. Tip movement is controlled by a piezoelectric transducer.

Equation 21-8 reveals that the tunneling current decreases exponentially with the separation between the tip and the sample. This rapid decrease in current with distance causes the tunneling current to be significant only for very small tip-sample separation and is responsible for the high resolution achieved in the *z* direction.

Sample Scanners[19]

In early scanning tunneling microscopes, three-dimensional motion of the tip was controlled by three piezoelectric transducers arranged in an orthogonal pattern as shown in Figure 21-18a. The length of each piezoelectric transducer can be varied by application of a dc potential along its length, thus making it possible to move the tip in a three-dimensional pattern (see Section 1C-4 for a discussion of the composition and properties of piezoelectric transducers). Depending upon the composition of the piezoelectric ceramic material and the dimensions of the transducer, the degree of expansion or contraction can be made as small as 1 nm for every volt applied, thus providing a remarkably sensitive way of controlling the position of a tip attached to its end.

Modern scanning microscopes no longer use the tripod design shown in Figure 21-18a, but instead are based upon a hollow-tube piezoelectric device shown in Figure 21-19. The outer surface of the tube, which is typically between 12 and 24 mm in length and 6 and 12 mm in diameter, is coated with a thin layer of metal. This conducting layer is divided into four equal segments by vertical strips containing no metal coating. Application of potentials to opposite strips of metal give rise to bending of the tube in the *x* and *y* directions as indicated. Likewise application of a potential along the interior axis of the cylinder causes lengthening or shortening of the tube in the *z* direction. A tip located in the center of one end of the cylinder can then be made to move in three dimensions by application of suitable potentials.

Scanning microscopes typically employ scanners with lateral scan ranges of tens of angstroms to over 100 μm. Height differences from the sub-angstrom range to perhaps 10 μm are encountered. The maximum scan size is determined by the length, diameter, and wall thickness of the cylinder as well as the strain coefficient of the ceramic material from which it is fabricated.

[19] For description of sample scanners, see R. S. Howland, in *Atomic Force/Scanning Tunneling Microscopy,* S. H. Cohen, M. T. Bray, and M. L. Lightbody, Eds., pp. 347–358. New York: Plenum Press, 1994; S. Park and R. C. Barrett in *Scanning Tunneling Microscopy,* J. A. Stroscio and W. J. Kaiser, Eds., pp. 51–58. New York: Academic Press, 1993.

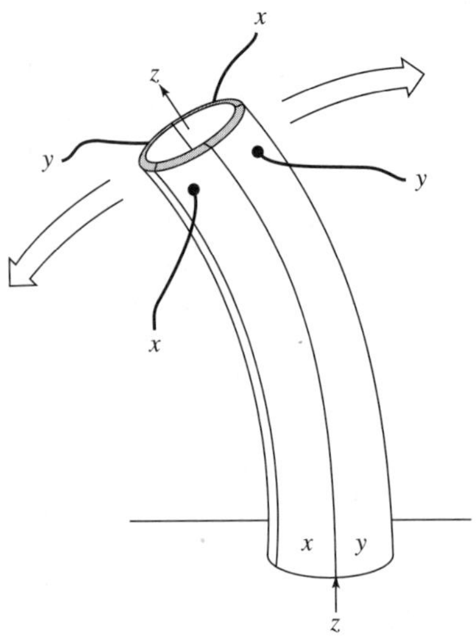

Figure 21-19 Piezoelectric scanner of the segmented tube design.

Computer Interface

Computer control is an essential part of all scanning tunneling microscopes. Most commercial STMs use software and digital-to-analog converters to generate the *x/y* raster scan. The voltages applied to the *x, y,* and *z* piezoelectric elements are fed into a computer where they are processed and converted into contour maps such as that shown in Figure 21-18b. With more sophisticated instruments the images may take the form of gray-level images or pseudocolored elevation maps.

The dashed line in Figure 21-18a shows the path of the tip as it is scanned in the *x* direction over the surface of a sample of highly oriented pyrolytic graphite, the individual carbon atoms of which are represented as round balls. After a scan in the *x* direction is completed, the tip is returned to its original position and then moved down one line by applying a slightly different potential to the *y* piezoelectric transducer. This process is repeated until a plot of the entire sample is obtained. Such a plot is shown in Figure 21-18b, which consists of a series of contour lines that show clearly the position of the electron cloud of each carbon atom on the surface of the sample.

Because the output signal from the detector is so sensitive to the distance between the sample and the tip, differences in distances along a given contour in the figure are revealed to 1/100 of an atomic dimension. The lateral resolution along a given contour depends upon the radius of curvature of the tip. When this radius is that of a single atom, as it usually is, atomic resolution is achieved.

Tips[20]

The tunneling tip is a crucial component of the STM. The best images are obtained when tunneling is limited to a single metal atom at the end of the tip. Fortunately, with a little care, it is possible to construct this type of tip by cutting platinum/iridium wires or by electrochemical etching of tungsten metal. The reason a single-atom tip is not as difficult to prepare as might be expected is the exponential increase in tunneling current with decreasing gap (Equation 21-8). Thus, typically the tunneling current increases by a factor of ten when the gap distances decreases by 1 Å. That is, if there is one atom at the apex of the tip that is closer to the surface of the sample by 1 Å than all the other atoms on the tip, nearly all of the current will flow through that atom to the sample, and atomic resolution is achieved.

21D-2 The Atomic Force Microscope

The atomic force microscope, which was invented in 1986[21], permits resolution of individual atoms on both conducting and *insulating* surfaces. In this procedure, a flexible force-sensing cantilever stylus is scanned in a raster pattern over the surface of the sample. The force acting between the cantilever and the sample surface causes minute deflections of the cantilever, which are detected by optical means. As in STM, the motion of the tip, or sometimes the sample, is achieved with a piezoelectric tube. During a scan, the force on the tip is held constant by the up-and-down motion of the tip, which then provides the topographic information. The advantage of the atomic force microscope is that it is applicable to nonconducting samples.

Figure 21-20 shows schematically the most common method of detecting the deflection of the cantilever holding the tip. A laser beam is reflected off a spot on

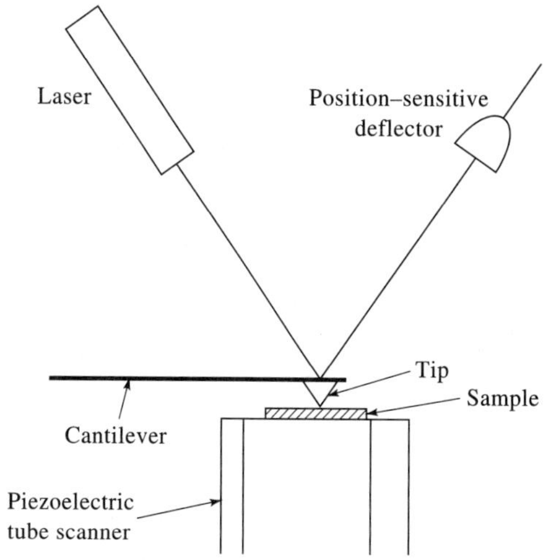

Figure 21-20 Side view of an optical beam deflection detector. Typically, the system is sensitive to 0.01 mm as the tip scans the sample surface.

the cantilever to a segmented photodiode that detects the motion of the probe. The output from the photodiode then controls the force applied to the tip so that it remains constant. In other words, the optical control system is analogous to the tunneling-current control system in the STM.

Figure 21-21 shows a common design of an AFM. The movement system is a tubular piezoelectric device that moves the sample in an x, y, and z direction under the tip. The signal from the laser beam detector is then fed back into the sample piezoelectric transducer, which causes the sample to move up and down to maintain a constant force between the tip and the sample.

The Tip and Cantilever

The performance of an atomic force microscope is critically dependent upon the physical characteristics of the cantilever and tip. In early AFMs, cantilevers were cut from metal foil and tips were made from crushed diamond particles. The tips were painstakingly glued manually to the cantilevers. Currently, this crude method has been replaced by semiconductor mass production methods in which integral cantilever/tip assemblies are produced by etching single chips of silicon, silicon oxide, or silicon nitride. The cantilevers and tips are remarkably small and delicate as suggested by Figure 21-22. Typically, the cantilevers are a few tens of micrometers in length, less than ten micrometers in width,

[20] For a description of how tips are prepared, see S. Park and R. C. Barrett, in *Scanning Tunneling Microscopy*, J. A. Stroscio and W. J. Kaiser, Eds., pp. 51–53. New York: Academic Press, 1993.

[21] G. Binnig, C. F. Quate, and C. Gerber, *Phys. Rev. Lett.*, **1986**, *56*, 930. For a review article on AFM, see D. R. Louder and B. A. Parkinson, *Anal. Chem.*, **1995**, *67*, 297A. For descriptions of several commercial AFM instruments, see A. Newman, *Anal. Chem.*, **1996**, *68*, 267A.

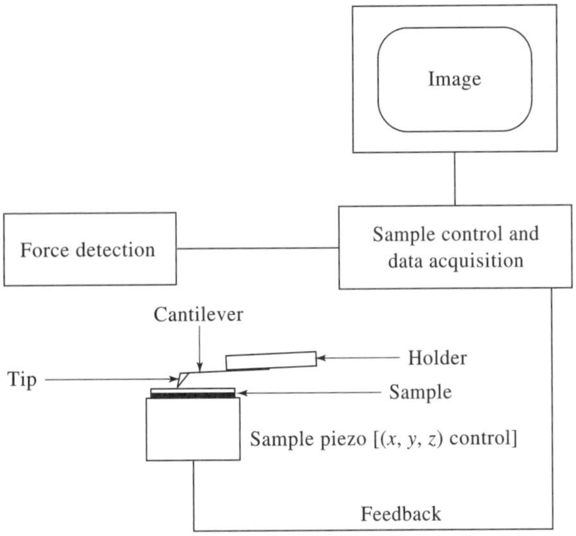

Figure 21-21 Typical design of an atomic force microscope. *(From D. R. Louder and B. A. Parkinson,* Anal. Chem., ***1995****, 67 298A. With permission.)*

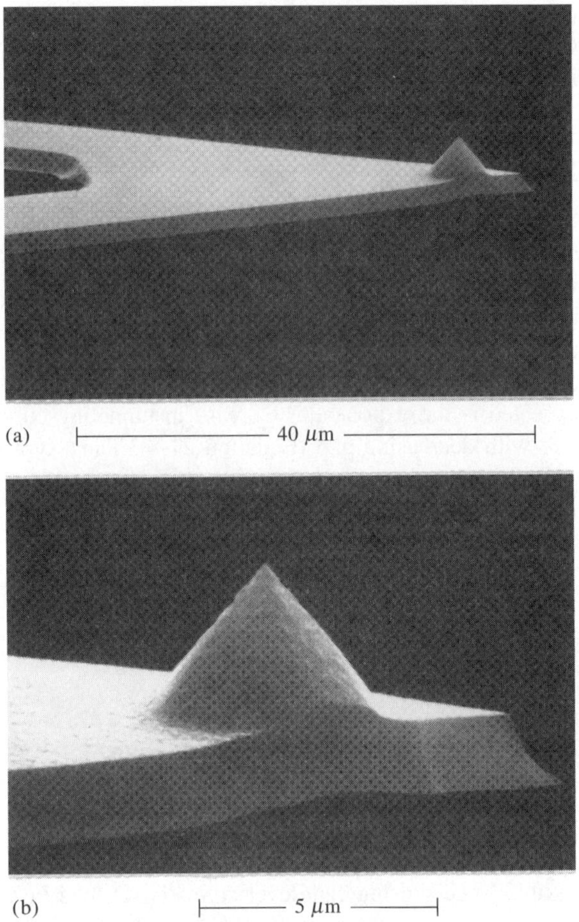

Figure 21-22 Micrograph of (a) an SiO_2 cantilever and tip and (b) a SiO_2 tip. *(From T. R. Albrecht, S. Akamini, T. E. Carver, and C. F. Quate,* J. Vac. Sci. Technol. A, ***1990****, 62, 3392. With permission.)*

and about one micrometer in thickness. The pyramid- or cone-shaped tips are a few micrometers in height and width at the base.

Tapping Mode Operation

One of the disadvantages of the contact mode scanning as described in the previous paragraphs is that the tip is in constant contact with the surface of the sample, and the downward force of the tip may not be low enough to avoid damage to the sample surface and distortion of the image as a consequence. This problem is particularly bothersome with softer materials, such as biological samples, polymers, and even some seemingly hard materials, such as silicon wafers.

It has been found that the problem of surface damage can be largely overcome by a process in which the tip contacts the surface for only a brief time periodically and then is removed from the surface.[22] In this *tapping mode* operation, the cantilever is oscillated at a frequency of a few hundred kilohertz. The oscillation is driven by a constant driving force and the amplitude is monitored continuously. The cantilever is positioned so that the tip touches the surface only at the bottom of each oscillation cycle. This technique has been used successfully to image a wide variety of materials that have been difficult or impossible to image by the ordinary constant-contact mode.

Some Typical Applications of Scanning Probe Microscopes

Scanning probe microscopes have allowed scientists and engineers to see surface structures with unprecedented resolution. As a consequence, they have found widespread use in a number of fields. For example, in the semiconductor field, they have been used for the characterization of silicon surfaces and defects on these surfaces, as well as imaging of magnetic domains on magnetic materials; in biotechnology, imaging of such

[22] See Y. E. Strausser and M. G. Heaton, *Amer. Lab.*, **1994** (4), 20.

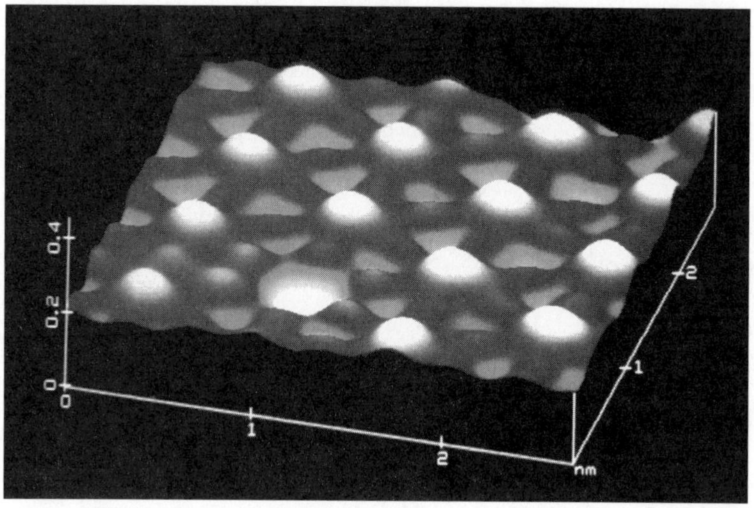

Figure 21-23 STM scan of iodine atoms in a 3-nm × 3-nm array adsorbed on platinum. Note the missing iodine atom in the bottom center of the image. *(Courtesy of B. Schardt and Digital Instruments, Inc., with permission.)*

materials as DNA, chromatin, protein/enzyme interactions, membrane viruses, and so on. An advantage of AFM is that it permits imaging of biological samples under water under conditions wherein less distortion of the image is encountered. For softer samples, distortion often arises because a microdrop of water forms at the tip/surface interface. Capillary forces from this drop often exceed the normal force between the tip and sample and obscure surface details. Surprisingly, if the sample is in water, there is no distortion because with water above the tip as well as below, capillary forces in the up and down directions cancel.

An interesting example of the potential of scanning tunneling microscopy is illustrated in Figure 21-23. The

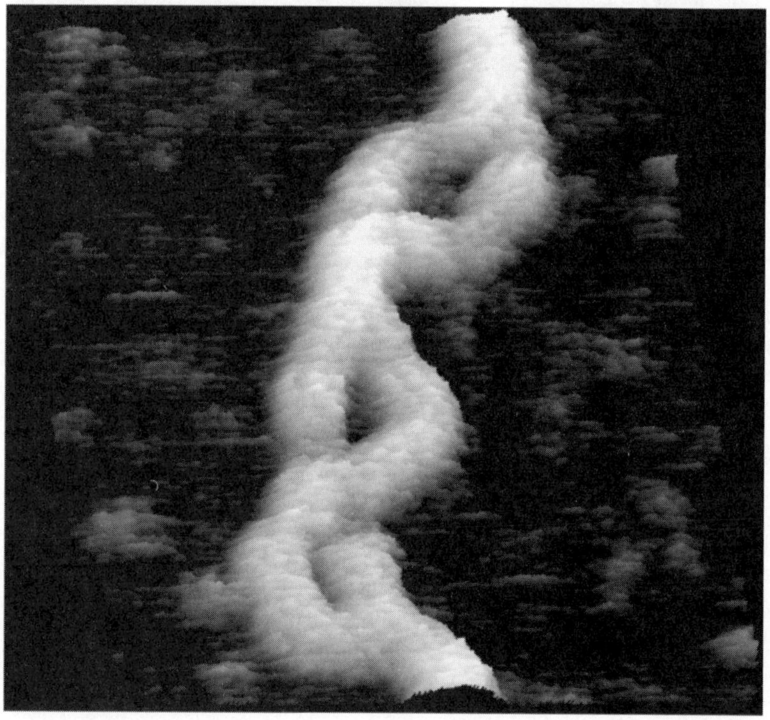

Figure 21-24 Two double-stranded DNA molecules imaged on mica, showing the ability of STM to resolve overlapping molecules. *(Courtesy of W. B. Stine, University of California, San Diego. With permission.)*

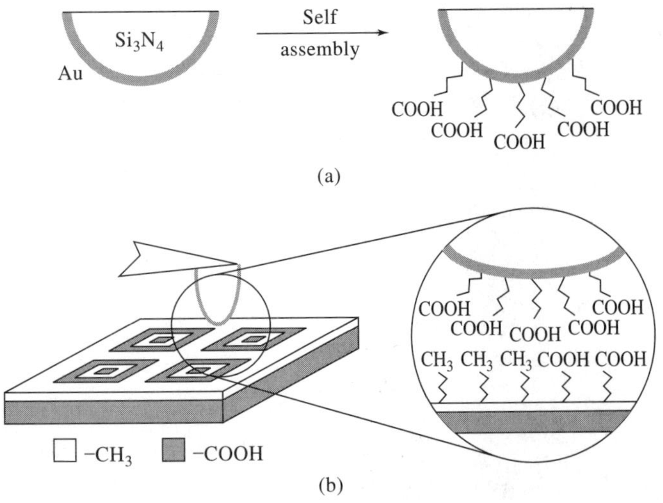

Figure 21-25 Detection of a functional group by atomic force microscopy. (a) Carboxylic acid groups are chemically attached to a gold-coated AFM tip. (b) Schematic views of the experiment. (Inset) Interaction between the gold tip coated with —COOH groups and the sample coated with both —CH_3 and —COOH groups. *(Adapted from C. D. Frisbie, L. F. Rozsnyai, A. Noy, M. S. Wrighton, and C. M. Lieber,* Science, **1994**, 265, *2072. With permission.)*

Figure 21-26 Chemical Force Microscopy (CFM) scan of regions coated with either methyl (—CH_3) or carboxylic acid (—COOH) groups. (a) An ordinary AFM scan of a surface with will-defined regions that terminate in either —CH_3 or —COOH groups. There is no evidence in the image of the presence of the functional groups becuse the regions are structurally similar. (b) When the AFM tip is coated with —COOH groups, there is greater frictional force between the tip and the —COOH groups on the surface than with the —CH_3 groups; hence, the surface appears raised and is of a lighter color. (c) The contrast is reversed when a tip with —CH_3 groups is used. *(Reprinted with permission from R. Baum,* Chem. Eng. News, ***1994***, *72, 6. Copyright 1994 American Chemical Society. Digital image courtesy of C. M. Lieber and A. Noy.*

image was produced from the surface of a sample of iodine atoms absorbed on platinum. The hexagonal pattern of absorbed iodine atoms is interrupted by a defect where an iodine atom is absent, which appears as a divot in the bottom center of the image. The scan that is shown represents a 2.5 nm × 2.5 nm area of the platinum surface. This example clearly shows how STM may be used to reveal structure of solid surfaces at the atomic level.

Figure 21-24 is an image of two intertwined DNA molecules on a mica surface obtained by a tapping mode AFM scan. Such measurements allow biochemists to study the structure of DNA and other biomolecules with relative ease.

An intriguing application of atomic force microscopy is illustrated in Figure 21-25.[23] In this experiment, a gold-coated Si_3N_4 tip is immersed for a time in a solution containing an organosulfur compound such as 11-thioundecanoic acid. The result is that the end of the acid molecules opposite the carboxylic acid group becomes covalently attached to the probe tip, so that the tip is then effectively coated with carboxylic acid groups as shown in Figure 21-25a. When the probe tip is then scanned across a surface having various organic functional groups attached as shown in Figure 21-25b, the differences in frictional forces between the acid groups on the tip and other functional groups on the surface of the sample result in an image that is a map of the positions of surface functional groups. This technique is termed *chemical force microscopy,* or CFM (Figure 21-26). CFM studies such as these illustrate that scanning probe microscopy provides quite specific qualitative analytical information as well as information on the spatial arrangement of analytes on surfaces.

[23] C. D. Frisbie, L. F. Rozsnyai, A. Noy, M. S. Wrighton, and C. M. Lieber, *Science,* **1994,** *265,* 2071.

21E QUESTIONS AND PROBLEMS

21-1 Describe the mechanism of the production of an MNN Auger electron.

21-2 Describe how it is possible to distinguish between XPS peaks and Auger electron peaks.

21-3 Explain why the information from an XPS chemical shift must also be contained in an X-ray absorption edge.

21-4 An XPS electron was found to have a kinetic energy of 1073.5 eV when a Mg K_α source was employed (λ = 9.8900 Å). The electron spectrometer had a work function of 14.7 eV.

(a) Calculate the binding energy for the emitted electron.

(b) If the signal was from a S(2*s*) electron, was the analyte S^{2-}, S^0, SO_3^{2-}, or SO_4^{2-}?

(c) What would the kinetic energy have been if an Al K_α source had been used (λ = 8.3393 Å)?

(d) If the ejected electron with the Mg K_α source had been an Auger electron, what would its kinetic energy be with the Al K_α source?

21-5 An XPS electron was found to have a kinetic energy of 1052.6 eV when ejected with an Al K_α source (λ = 8.3393 Å) and measured in a spectrometer with a work function of 27.8 eV. The electron is believed to be a N(1*s*) electron in $NaNO_3$.

(a) What was the binding energy for the electron?

(b) What would be the kinetic energy of the electron if a Mg K_α (λ = 9.8900 Å) source were used?

(c) How could one be sure that a peak was an XPS and not an Auger electron peak?

(d) At what binding and kinetic energies would a peak for $NaNO_2$ be expected when the Al K_α source was used with the same spectrometer?

21-6 Assume the CRT display of a scanning electron microscope is 20 cm wide. What scan width would be required to achieve a magnification of (a) 100 (b) 100,000?

21-7 Differentiate between elastic and inelastic and elastic scattering of electrons.

21-8 (a) Name the two types of scanning probe microscopes.
(b) How do they differ?
(c) What are the advantages of each?
(d) What are the disadvantages of each?

21-9 What is raster scanning?

Section IV

Electroanalytical Chemistry

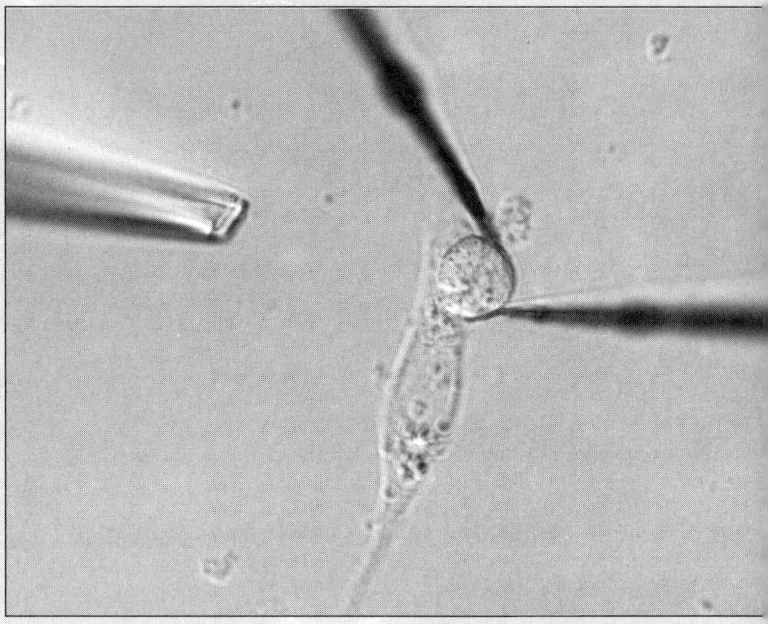

The photo shows two microelectrodes next to a bovine adrenal medullary cell along with a small pipette used to introduce reagents. The cell is from the medulla of the adrenal gland, which is the source of adrenaline, or epinephrine, a catecholamine that is released to induce the "fight or flight" mechanism. These cells also release norepinephrine, another catecholamine, and they have a diameter of about 15 μm. The catecholamines can be detected by voltammetry, which is discussed in Chapter 25. When one of the adrenal cells is stimulated appropriately, 300-nm vesicles containing 0.5 M catecholamine release spikes of the compound containing 10^6 to 10^7 molecules. The approximately 1-μm electrodes permit the determination of the catecholamines and the resolution of individual secretion sites. For details, see K. Pihel, T. J. Schroeder, and R. M. Wightman, *Anal. Chem.*, **1994**, *66*, 4532, and T. J. Schroeder, J. A. Jankowski, J. Senyshyn, R. W. Holtz, and R. M. Wightman, *J. Biol. Chem.*, **1994**, *269*, 17215. (Photo courtesy of R. M. Wightman.)

In this section the theory and methodology of electroanalytical chemistry are explored. Chapter 22 provides a general foundation for the study of subsequent chapters in this section. Terminology and conventions of electrochemistry as well as theoretical and practical aspects of the measurement of electrochemical potentials and currents are presented. Chapter 23 comprises the many methods and applications of potentiometry, while constant-potential coulometry, constant-current coulometry, and electrogravimetry are discussed in Chapter 24. This section concludes with a discussion of the many faces of the important and widely used technique of voltammetry.

Chapter 22

An Introduction to Electroanalytical Chemistry

Electroanalytical chemistry encompasses a group of quantitative analytical methods that are based upon the electrical properties of a solution of the analyte when it is made part of an electrochemical cell.[1] Electroanalytical techniques are capable of producing exceptionally low detection limits and a wealth of characterization information describing electrochemically addressable systems. Such information includes the stoichiometry and rate of interfacial charge transfer, rate of mass transfer, extent of adsorption or chemisorption, and rates and equilibrium constants for chemical reactions.

Electroanalytical methods have certain general advantages over other types of procedures discussed in this book. First, electrochemical measurements are often specific for a particular oxidation state of an element. For example, electrochemical methods make possible the determination of the concentration of each of the species in a mixture of cerium(III) and cerium(IV), whereas most other analytical methods are capable of revealing only the total cerium concentration. A second important advantage of electrochemical methods is that the instrumentation is relatively inexpensive. The most expensive electrochemical instrument costs perhaps \$20,000, and the price for a typical multipurpose commercial instrument will lie in the \$4000 to \$5000 range. In contrast, many spectroscopic instruments cost \$50,000 to \$250,000 or more. A third feature of certain electrochemical methods, which may be an advantage or a disadvantage, is that they provide information about activities rather than concentrations of chemical species. Ordinarily, in physiological studies, activities of ions such as calcium and potassium are of greater significance than are concentrations.

---

[1] Some reference works on electrochemistry and its applications include: A. J. Bard and L. R. Faulkner, *Electrochemical Methods.* New York: Wiley, 1980; D. T. Sawyer, A. Sobkowiak, and J. L. Roberts, *Electrochemistry for Chemists,* 2nd ed. New York: Wiley, 1995; J. Koryta and J. Dvorak, *Principles of Electrochemistry.* New York: Wiley, 1987; *Laboratory Techniques in Electroanalytical Chemistry,* 2nd ed., P. T. Kissinger and W. R. Heineman, Eds. New York: Marcel Dekker, 1996; Z. Galus, *Fundamentals of Electrochemical Analysis,* 2nd ed. New York: Ellis Horwood, 1994. The classical, and still useful, monograph dealing with electroanalytical chemistry is J. J. Lingane, *Electroanalytical Chemistry,* 2nd ed. New York: Interscience, 1958.

The intelligent application of the various electroanalytical methods that are described in the three chapters that follow this one requires an understanding of the basic theory and the practical aspects of the operation of direct current (dc) electrochemical cells. This chapter is devoted largely to these matters.[2]

22A ELECTROCHEMICAL CELLS

A dc electrochemical cell consists of two electrical conductors called *electrodes,* each immersed in a suitable electrolyte solution. For a current to develop in a cell, it is necessary (1) that the electrodes be connected externally by means of a metal conductor, (2) that the two electrolyte solutions be in contact to permit movement of ions from one to the other, and (3) that an electron transfer reaction can occur at each of the two electrodes. Figure 22-1 shows an example of a simple electrochemical cell. It consists of a zinc electrode immersed in a solution of zinc sulfate and a copper electrode in a solution of copper sulfate. The two solutions are joined by a *salt bridge,* which consists of a tube filled with a solution that is saturated with potassium chloride, or sometimes, some other electrolyte. The two ends of the tube are equipped with porous plugs that permit the movement of ions across them but prevent siphoning of liquid from one electrolyte solution to the other. The purpose of the bridge is to isolate the contents of the two halves of the cell while maintaining electrical contact between them. Isolation is necessary to prevent direct reaction between copper ions and the zinc electrode. The cell in Figure 22-1 contains two so-called *liquid junctions,* one being the interface between the zinc sulfate solution and the salt bridge; the second is at the other end of the salt bridge where the electrolyte solution of the bridge contacts the copper sulfate solution. As we shall show later, a small *junction potential* develops at each of these interfaces that may become important and influence the accuracy of the analysis.

22A-1 Conduction in a Cell

Charge is conducted by three distinct processes in various parts of the cell shown in Figure 22-1:

1. In the copper and zinc electrodes, as well as in the external conductor, electrons serve as carriers, mov-

[2] For brief reviews of developments in electrochemical instrumentation, see S. Borman, *Anal. Chem.,* **1987,** *59,* 347A; J. Osteryoung, *Science,* **1982,** *218,* 261.

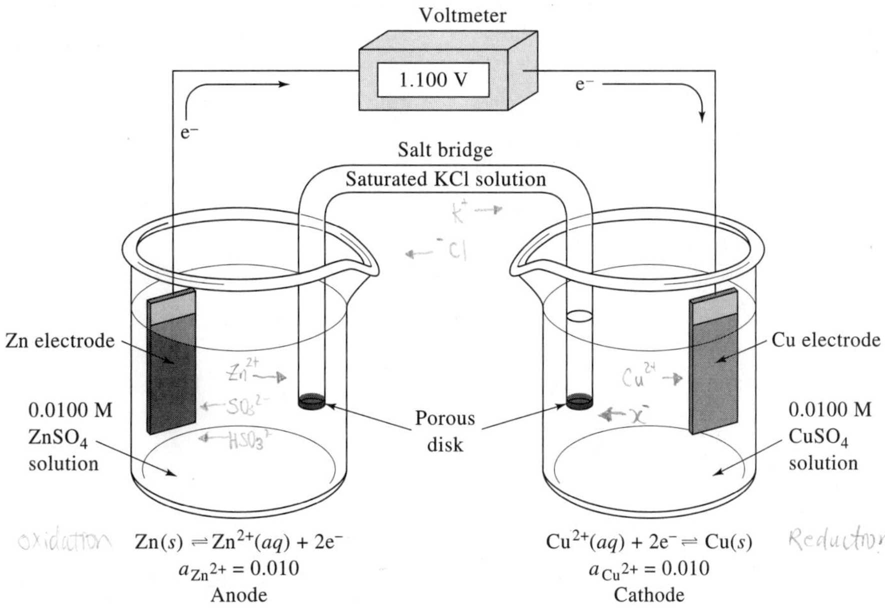

Figure 22-1 A galvanic electrochemical cell with a salt bridge.

ing from the zinc through the conductor to the copper.

2. Within the solutions the flow of electricity involves migration of both cations and anions. In the half cell on the left, zinc ions migrate away from the electrode, whereas sulfate and hydrogen sulfate ions move toward it; in the other compartment, copper ions move toward the electrode and anions away from it. Within the salt bridge, electricity is carried by migration of potassium ions to the right and chloride ions to the left. Thus, all of the ions in the three solutions participate in the flow of electricity.
3. A third process occurs at the two electrode surfaces. Here, an oxidation or a reduction reaction provides a mechanism whereby the ionic conduction of the solution is coupled with the electron conduction of the electrode to provide a complete circuit for the flow of charge. The two electrode processes are described by the equations

$$Zn(s) \rightleftharpoons Zn^{2+} + 2e^- \qquad (22\text{-}1)$$

$$Cu^{2+} + 2e^- \rightleftharpoons Cu(s) \qquad (22\text{-}2)$$

22A-2 Solution Structure; The Double Layer

It is important to realize that electrochemical measurements involve heterogeneous systems because an electrode can only donate or accept electrons from a species that is present in a layer of solution that is immediately adjacent to the electrode. Thus, this layer may have a composition that differs significantly from that of the bulk of the solution.

For example, let us consider the structure of the solution immediately adjacent to an electrode when a positive potential is first applied to that electrode. Immediately after impressing the potential, there will be a momentary surge of current, which rapidly decays to zero if no reactive species is present at the surface of the electrode. This current is a charging current that creates an excess (or a deficiency) of negative charge at the surface of the two electrodes. As a consequence of ionic mobility, however, the layers of solution immediately adjacent to the electrodes acquire an opposing charge. This effect is illustrated in Figure 22-2a. The surface of the metal electrode is shown as having an excess of positive charge as a consequence of an applied positive potential. The charged solution layer consists of two parts: (1) a compact inner layer (d_0 to d_1), in which the potential decreases linearly with distance from the electrode surface and (2) a diffuse layer (d_1 to d_2), in which the decrease is exponential (see Figure 22-2b). This assemblage of charge at the electrode surface and in the solution adjacent to the surface is termed an *electrical double layer.*

22A-3 Faradaic and Nonfaradaic Currents

Two types of processes can conduct currents across an electrode/solution interface. One kind involves a direct transfer of electrons via an oxidation reaction at one electrode and a reduction reaction at the other. Processes of this type are called *faradaic processes* because they are governed by Faraday's law, which states that the amount of chemical reaction at an electrode is proportional to the current; the resulting currents are called *faradaic currents.*

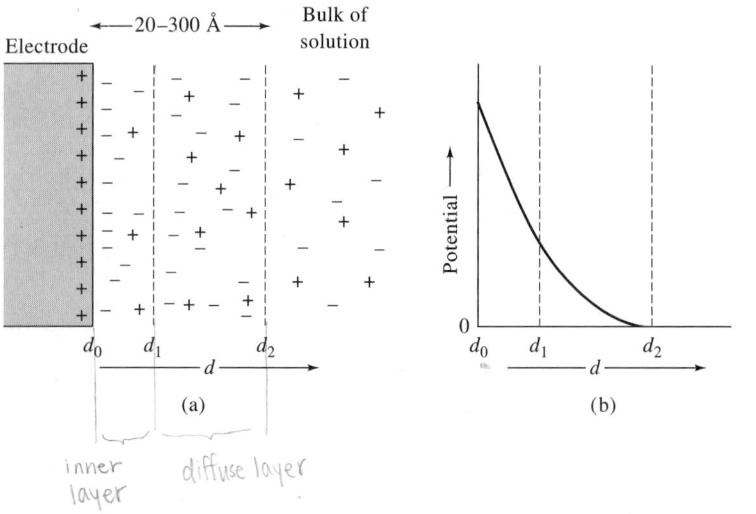

Figure 22-2 Electrical double layer formed at electrode surface as a result of an applied potential.

Under some conditions a cell will exhibit a range of potentials where faradaic processes are precluded at one or both of the electrodes for thermodynamic or kinetic reasons. Here, conduction of continuous alternating currents can still take place. With such currents, reversal of the charge relationship occurs with each half-cycle as first negative and then positive ions are attracted alternately to the electrode surface. Electrical energy is consumed and converted to heat by friction associated with this ionic movement. Thus, each electrode surface behaves as one plate of a capacitor, the capacitance of which may be large (several hundred to several thousand microfarads per cm^2). The capacitive current increases with frequency and with electrode area; by controlling these variables, it is possible to arrange conditions such that essentially all the alternating current in a cell is carried across the electrode interface by this nonfaradaic process.

To understand the basic difference between a faradaic and a nonfaradaic current, imagine an electron traveling down the external circuit to an electrode surface. When the electron reaches the solution interface, it can do one of only two things. It can remain at the electrode surface and increase the charge on the double layer, which constitutes a nonfaradaic current. Alternatively, it can leave the electrode surface and transfer to a species in the solution, thus becoming a part of a faradaic current.

22A-4 Mass Transfer in Cells with the Passage of Current

Because an electrode can probe only a very thin layer of solution at the electrode surface (d_0 to d_1 in Figure 22-2a), a faradaic current requires continuous mass transfer of reactive species from the bulk of the solution to the electrode surface. Three mechanisms bring about this mass transfer: *convection, migration,* and *diffusion.* Convection involves mechanical motion of the solution as a result of stirring or the flow of the solution past the surface of the electrode. Migration is the movement of ions through the solution brought about by electrostatic attraction between the ions and the charged electrode. Diffusion is the motion of species brought about by a concentration gradient. A detailed discussion of these mechanisms of mass transfer is found in Section 22E-3.

22A-5 Galvanic and Electrolytic Cells

The net cell reaction that occurs in the cell shown in Figure 22-1 is the sum of the two *half-cell* reactions shown as Equations 22-1 and 22-2. That is,

$$Zn(s) + Cu^{2+} \rightleftharpoons Zn^{2+} + Cu(s)$$

The potential that develops in this cell is a measure of the tendency for this reaction to proceed toward equilibrium. Thus, as shown in Figure 22-1, when the copper and zinc ion activities are 0.0100 M, a potential of 1.100 V develops, which shows that the reaction is far from equilibrium. As the reaction proceeds, the potential becomes smaller and smaller, ultimately reaching 0.000 V when the system achieves equilibrium.

Cells, such as the one shown in Figure 22-1, that are operated in a way that produces electrical energy are called *galvanic cells.* In contrast, *electrolytic cells* consume electrical energy. For example, the cell under discussion could be made electrolytic by connecting the negative terminal of a dc power supply to the zinc electrode and the positive terminal to the copper electrode. If the output of this supply was made somewhat greater than 1.1 V, the two electrode reactions would be reversed and the net cell reaction would become

$$Cu(s) + Zn^{2+} \rightleftharpoons Cu^{2+} + Zn(s)$$

A cell in which reversing the direction of the current simply reverses the reactions at the two electrodes is termed a *chemically reversible cell.*

22A-6 Anodes and Cathodes

By definition, the *cathode* of an electrochemical cell is the electrode at which reduction occurs, while the *anode* is the electrode where oxidation takes place. These definitions apply to both galvanic and electrolytic cells. For the galvanic cell shown in Figure 22-1, the copper electrode is the cathode and the zinc electrode is the anode. In contrast, where this same cell is operated as an electrolytic cell, the copper electrode would be the anode and the zinc electrode the cathode.

22A-7 Cells without Liquid Junctions

The interface between two solutions containing different electrolytes or different concentrations of the same electrolyte is called a *liquid junction.* Often, electrochemical cells contain one or more liquid junctions. For example, the cell shown in Figure 22-1 has two liquid junctions, one between the zinc sulfate solution and one end of the salt bridge, the other between the copper sulfate solution and the salt bridge. Liquid junctions are sometimes important in electrochemical measurements because a small *junction potential* develops at these interfaces that influences the magnitude of the overall measured cell potentials. The source and effect of junction potentials is discussed in some detail in Section 22B-2 and throughout Chapter 23.

Sometimes it is possible and advantageous to prepare cells in which the electrodes share a common electrolyte and thus eliminate the effect of junction potentials. An example of a cell of this type is shown in Figure 22-3. Here, the reaction at the silver cathode can be written as

$$AgCl(s) + e^- \rightleftharpoons Ag(s) + Cl^-(aq)$$

Hydrogen is consumed at the platinum anode:

$$H_2(g) \rightleftharpoons 2H^+(aq) + 2e^-$$

The overall cell reaction is then obtained by multiplying each term in the first equation by 2 and adding. That is,

$$2AgCl(s) + H_2(g) \rightleftharpoons 2Ag(s) + 2H^+(aq) + 2Cl^-(aq)$$

The direct reaction between hydrogen and solid silver chloride is so slow that the common electrolyte can be employed without significant loss of cell efficiency due to direct reaction between cell components.

The cathode reaction in this cell is of interest because it can be considered to be the result of a two-step process that is described by the equations

$$AgCl(s) \rightleftharpoons Ag^+ + Cl^-$$

$$Ag^+ + e^- \rightleftharpoons Ag(s)$$

Solution of the sparingly soluble silver chloride occurs in the first step to provide an essentially constant concentration of silver ions that are then reduced in the second step.

The anodic reaction in this cell is also a two step process that can be formulated as

$$H_2(g) \rightleftharpoons H_2(aq)$$

$$H_2(aq) \rightleftharpoons 2H^+ + 2e^-$$

Here hydrogen gas is bubbled across the surface of a platinum electrode so that the concentration of the gas at the surface is maintained at a constant level throughout. Note that in this case the inert platinum electrode plays no direct part in the reaction but serves only as a recipient of electrons.

The cell in Figure 22-3 behaves as a galvanic cell that develops a potential of about 0.46 V. This cell is also chemically reversible and can be operated as an electrolytic cell by applying an external potential of somewhat greater than 0.46 V.

22A-8 Schematic Representation of Cells

To simplify the description of cells, chemists often employ a shorthand notation. For example, the cells shown in Figures 22-1 and 22-3 can be described by

$$Zn\,|\,ZnSO_4(a_{Zn^{2+}} = 0.0100)\,\|\,CuSO_4(a_{Cu^{2+}} = 0.0100)\,|\,Cu$$

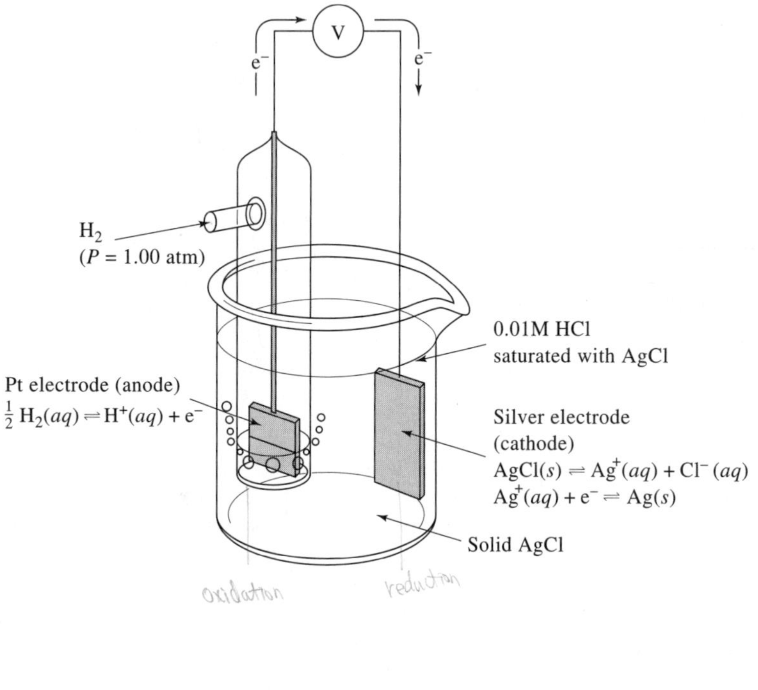

Figure 22-3 A galvanic cell without a liquid junction.

$$\mathrm{Pt,H_2}(p = 1\ \mathrm{atm})\,|\,\mathrm{H^+(0.01\ M),\ Cl^-(0.01\ M),AgCl(sat'd)}\,|\,\mathrm{Ag}$$

By convention, *the anode and information about the solution with which it is in contact is always listed on the left.* Single vertical lines represent phase boundaries across which potential differences may develop. Thus, in the first example, a part of the cell potential is the potential difference across the phase boundary between the zinc electrode and the zinc sulfate solution. Small potential differences also develop across liquid junctions; thus, two vertical lines are inserted between the zinc and copper sulfate solutions, which correspond to the two junctions at each end of the salt bridge. The cathode is then represented symbolically with another vertical line separating the electrolyte solution from the copper electrode. Because the potential of a cell is dependent upon activities of the cell components, it is common practice to provide activity or concentration data for the cell constituents in parentheses.

In the second cell, there are only two phase boundaries because the electrolyte is common to both electrodes. A shorthand representation of this cell would be

$$\mathrm{Pt}\,|\,\mathrm{H_2(sat'd),HCl(0.01\ M),Ag^+}(1.8 \times 10^{-8}\ \mathrm{M})\,|\,\mathrm{Ag}$$

Here, the molecular hydrogen concentration is that of a saturated solution; in the absence of partial pressure data, 1.00 atm is implied. The indicated molar silver ion concentration was computed from the solubility-product constant for silver chloride.

22B POTENTIALS IN ELECTROANALYTICAL CELLS

Electroanalytical methods may be based on the measurement of either (1) the current in an electrochemical cell at a fixed potential or (2) the potential of a cell while the current is fixed at some constant level. In general, however, in an electrochemical experiment the experimenter can only control the potential of the cell at a desired level and measure the current that results, or vice versa. Choosing to control one variable precludes any independent control of the other.

In this section we first consider the thermodynamics of electrochemical cells and the relationship between the activities of the participants in typical cell reactions and the observed potential of the cell. We then describe the source of junction potentials encountered in most electrochemical cells. In Section 22C that follows, we consider the potentials of individual cathodes and anodes making up cells.

22B-1 The Thermodynamics of Cell Potentials

It is important to understand that the potential of an electrochemical cell is related to the *activities* of the reactants and products of the cell reaction and indirectly to their molar concentrations. Often, we shall make the approximation that these activities are equal to molar concentrations, but it should always be borne in mind that this assumption may produce errors in calculated potentials. The relationship between the activity of a chemical species and its concentration is reviewed in Appendix 2.

Recall that the activity a_X of the species X is given by

$$a_X = \gamma_X[\mathrm{X}] \qquad (22\text{-}3)$$

Here, γ_X is the activity coefficient of solute X and the bracketed term is the molar concentration of X. In some of the examples, however, we will for convenience assume that the activity coefficient approaches unity so that the molar concentration and the activity of a species are identical.

This section deals with the way in which activities of reactants and products affect the potential of an electrochemical cell. We will use as an example the cell illustrated in Figure 22-3 for which the cell reaction is

$$2\mathrm{AgCl}(s) + \mathrm{H_2}(g) \rightleftharpoons 2\mathrm{Ag}(s) + 2\mathrm{Cl^-} + 2\mathrm{H^+} \qquad (22\text{-}4)$$

The equilibrium constant K for this reaction is given by

$$K = \frac{a_{\mathrm{H^+}}^2 \cdot a_{\mathrm{Cl^-}}^2 \cdot a_{\mathrm{Ag}}^2}{p_{\mathrm{H_2}} \cdot a_{\mathrm{AgCl}}^2}$$

where a's are the activities of the various species indicated by the subscripts and $p_{\mathrm{H_2}}$ is the partial pressure of hydrogen in atmospheres.

In Appendix 2, it is shown that the activity of a pure solid is unity when it is present in excess (that is, $a_{\mathrm{Ag}} = a_{\mathrm{AgCl}} = 1.00$). Therefore, the foregoing equation simplifies to

$$K = \frac{a_{\mathrm{H^+}}^2 \cdot a_{\mathrm{Cl^-}}^2}{p_{\mathrm{H_2}}} \qquad (22\text{-}5)$$

It is convenient to define a second quantity Q such that

$$Q = \frac{(a_{\mathrm{H^+}})_i^2 (a_{\mathrm{Cl^-}})_i^2}{(p_{\mathrm{H_2}})_i} \qquad (22\text{-}6)$$

Here, the subscript i indicates that the terms in the parentheses are instantaneous activities and *not equilibrium activities*. The quantity Q, therefore, is not a constant but changes continuously until equilibrium is reached; at that point, Q becomes equal to K, and the i subscripts are deleted.

From thermodynamics, it can be shown that the change in free energy ΔG for a cell reaction (that is, the maximum work obtainable at constant temperature and pressure) is given by

$$\Delta G = RT \ln Q - RT \ln K = RT \ln \frac{Q}{K} \qquad (22\text{-}7)$$

where R is the gas constant (8.316 J mol^{-1} K^{-1}) and T is the temperature in kelvins; ln refers to the logarithm to the base e. This relationship implies that the magnitude of the free energy for the system is dependent upon how far the system is from the equilibrium state. It can also be shown that the cell potential E_{cell} is related to the free energy of the reaction by the relationship

$$\Delta G = -nFE_{cell} \qquad (22\text{-}8)$$

where F is the faraday (96,485 coulombs per mole of electrons) and n is the number of moles of electrons associated with the oxidation/reduction process (in this example, $n = 2$).

Substitution of Equations 22-6 and 22-8 into 22-7 yields, upon rearrangement,

$$\begin{aligned} E_{cell} &= -\frac{RT}{nF} \ln Q + \frac{RT}{nF} \ln K \\ &= -\frac{RT}{nF} \ln \frac{(a_{H^+})_i^2 (a_{Cl^-})_i^2}{} + \frac{RT}{nF} \ln K \qquad (22\text{-}9) \end{aligned}$$

The last term in this equation is a constant, which is called the *standard electrode potential*, E^0_{cell} for the cell. That is,

$$E^0_{cell} = \frac{RT}{nF} \ln K \qquad (22\text{-}10)$$

Substitution of Equation 22-10 into Equation 22-9 yields

$$E_{cell} = E^0_{cell} - \frac{RT}{nF} \ln \frac{(a_{H^+})_i^2 (a_{Cl^-})_i^2}{} \qquad (22\text{-}11)$$

Note that the standard potential is equal to the *cell potential when the reactants and products are at unit activity and pressure.*

Equation 22-11 is a form of the *Nernst equation*, named in honor of the 1920 Nobel laureate and German physical chemist Walther Nernst (1864–1941). It finds wide application in electroanalytical chemistry.

22B-2 Liquid Junction Potentials

When two electrolyte solutions of different composition are brought into contact with one another, a potential develops across the interface. This *junction potential* arises from an unequal distribution of cations and anions across the boundary due to differences in the rates at which these species diffuse.

Consider the liquid junction in the system

$$HCl(1\ M)|HCl(0.01\ M)$$

Both hydrogen ions and chloride ions tend to diffuse across this boundary from the more concentrated to the more dilute solution, the driving force for this movement being proportional to the concentration difference. The rate at which various ions diffuse under the influence of a fixed force varies considerably (that is, the *mobilities* are different). In the present example, hydrogen ions are several times more mobile than chloride ions. As a consequence, there is a tendency for the hydrogen ions to diffuse faster than the chloride ions; a separation of charge is the end result (see Figure 22-4). The more dilute side of the boundary becomes positively charged owing to the more rapid diffusion of hydrogen ions; the concentrated side, therefore, acquires a negative charge from the excess of slower-moving chlo-

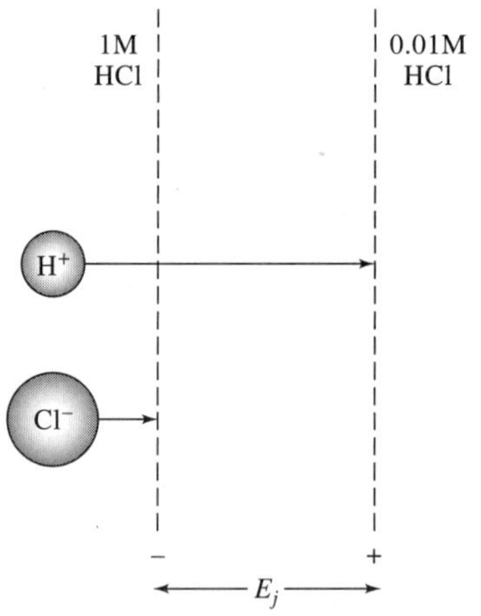

Figure 22-4 Schematic representation of a liquid junction showing the source of the junction potential E_j. The length of the arrows corresponds to the relative mobility of the two ions.

ride ions. The charge that develops tends to counteract the differences in mobilities of the two ions, and as a consequence, equilibrium is soon achieved. The junction potential difference resulting from this charge separation may amount to 30 mV or more.

In a simple system such as that shown in Figure 22-4, the magnitude of the junction potential can be calculated from a knowledge of the mobilities of the two ions involved. However, it is seldom that a cell of analytical importance has a sufficiently simple composition to permit such a computation.[3]

It has been found experimentally that the magnitude of the junction potential can be greatly reduced by introduction of a concentrated electrolyte solution (a salt bridge) between the two solutions. The effectiveness of a salt bridge improves not only as the concentration of the salt increases but also as the mobilities of the ions of the salt approach one another in magnitude. A saturated potassium chloride solution is particularly effective from both standpoints, its concentration being somewhat greater than 4 M at room temperature, and equally important, the mobility of its ions differing by only 4%. When chloride ion interferes, a concentrated solution of potassium nitrate can be substituted. With such bridges, the net junction potential typically amounts to a few millivolts or less, a negligible quantity in many, but not all, analytical measurements.

22C ELECTRODE POTENTIALS

It is useful to think of the cell reaction of an electrochemical cell as being made up of two *half-cell reactions,* each of which has a characteristic *electrode potential* associated with it. As will be shown later, these electrode potentials measure the driving force for the two half-reactions *when, by convention, they are both written as reductions.* Thus, the two half-cell or electrode reactions for the cell shown in Figure 22-3 are

$$2AgCl(s) + 2e^- \rightleftharpoons 2Ag(s) + 2Cl^-$$

$$2H^+ + 2e^- \rightleftharpoons H_2(g)$$

We now assume that electrode potentials E_{AgCl} and E_{H^+} are known for the two half-reactions. To obtain the cell reaction, the second half-reaction is subtracted from the first to give

$$2AgCl(s) + H_2 \rightleftharpoons 2Ag(s) + 2H^+ + 2Cl^-$$

Similarly, the cell potential E_{cell} is obtained by subtracting the electrode potential for the second reaction from the first. That is,

$$E_{cell} = E_{AgCl} - E_{H^+}$$

A more general statement of the last relationship is

$$E_{cell} = E_{cathode} - E_{anode} \qquad (22\text{-}12)$$

where $E_{cathode}$ and E_{anode} are the electrode potentials for the cathodic and anodic half-reactions.

22C-1 Nature of Electrode Potentials

The potential of an electrochemical cell is the *difference* between the potential of the cathode and that of the anode, and it is important to have a clear idea of what is meant by the potential of an electrode. This potential is a measure of an electrode's electron energy. For a metallic conductor immersed in a solution of an electrolyte, all excess charge resides *on its surface,* and it is possible to adjust the charge density on this surface by adjusting the output of an external power supply attached to the conductor. As this external source forces electrons onto the surface of an electrode, the electrons become more crowded, and their energy increases because of coulombic repulsion. A *negative* potential increase results. If the external circuitry withdraws enough electrons from the electrode, the surface will have a positive charge, and a *positive* potential develops. It is also possible to vary the electron energy, and thus the potential, of a metallic electrode by varying the composition of the solution in which it is immersed. For example, there is a potential at a platinum electrode immersed in a solution containing ferrocyanide and ferricyanide as result of the equilibrium

$$Fe(CN)_6^{3-} + e^- \rightleftharpoons Fe(CN)_6^{4-}$$

If the concentration of $Fe(CN)_6^{4-}$ is made much larger than that of $Fe(CN)_6^{3-}$, the former donates electrons to the metal, thus imparting a negative charge to the surface of the metal. Under this circumstance the potential of the electrode is negative. On the other hand, if $Fe(CN)_6^{3-}$ is present in large excess, electrons are removed from the electrode, causing surface layer ions to form in the solution and leaving a positive charge on the surface of the electrode. The platinum electrode then exhibits a positive potential.

[3] For methods for approximating junction potentials, see A. J. Bard and L. R. Faulkner, *Electrochemical Methods,* pp. 62–72. New York: Wiley, 1980.

It should be emphasized that *no method exists* for determining the absolute value of the potential of a single electrode, since all voltage-measuring devices determine only *differences* in potential. One conductor from such a device is connected to the electrode in question. In order to measure a potential difference, however, the second conductor must be brought in contact with the electrolyte solution of the half-cell in question. This latter contact inevitably involves a solid-solution interface and hence acts as a second half-cell at which a chemical reaction *must also take place* if charge is to flow. A potential will be associated with this second reaction. Thus, an absolute value for the desired half-cell potential is not measured; instead, what is measured is the difference between the potential of interest and the half-cell potential for the contact between the voltage-measuring device and the solution.

Our inability to measure absolute potentials for half-cell processes is not a serious handicap because *relative half-cell potentials,* measured against a common reference electrode, are just as useful. These relative potentials can be combined to give real cell potentials; in addition, they can be used to calculate equilibrium constants of oxidation/reduction processes.

In order to develop a useful list of relative half-cell or electrode potentials, it is necessary to have a carefully defined *reference electrode,* which is accepted by the entire chemical community. The *standard hydrogen electrode* (SHE), or the *normal hydrogen electrode* (NHE), is such a half-cell.

22C-2 The Standard Hydrogen Electrode (SHE)

Hydrogen gas electrodes were widely used in early electrochemical studies not only as reference electrodes but also as indicator electrodes for the determination of pH. The composition of this type of electrode can be represented as

$$\text{Pt,H}_2(p\ \text{atm})\,|\,\text{H}^+(a_{\text{H}^+} = x)$$

As suggested by the terms in parentheses, the potential at the platinum surface depends upon the hydrogen ion activity of the solution and upon the partial pressure of the hydrogen used to saturate the solution.

The half-cell shown on the left in Figure 22-5 illustrates the components of a typical hydrogen electrode. The conductor is fabricated from platinum foil, which has been *platinized,* a process in which the metal is coated with a finely divided layer of platinum by rapid chemical or electrochemical reduction of H_2PtCl_6. The finely divided platinum on the surface of the electrode does not reflect light as does polished platinum, so the electrode appears black. Thus, the finely divided platinum is called *platinum black.* The platinum black provides a large surface area to assure that the reaction

$$2\text{H}^+ + 2e^- \rightleftharpoons \text{H}_2(g)$$

proceeds rapidly at the electrode surface. As was pointed out earlier, the stream of hydrogen serves sim-

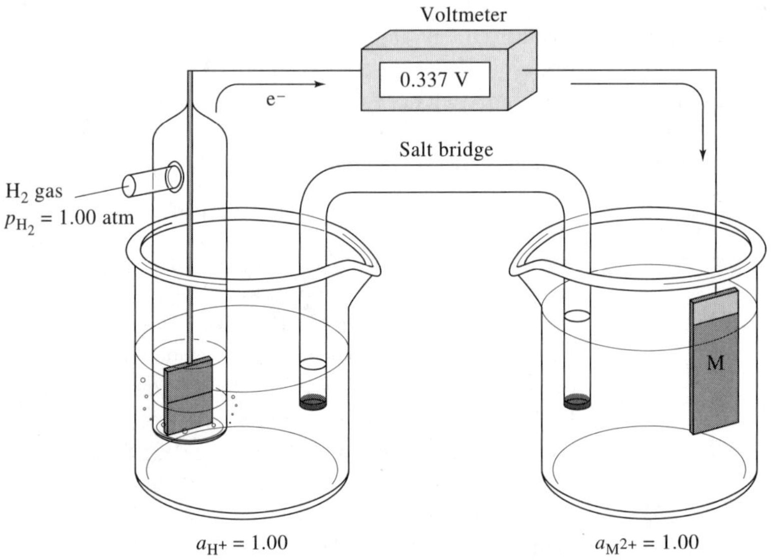

Figure 22-5 Definition of the standard electrode potential for $M^{2+}(aq) + 2e^- \rightleftharpoons M(s)$.

ply to keep the solution adjacent to the electrode saturated with respect to the gas.

The hydrogen electrode may act as an anode or a cathode, depending upon the half-cell with which it is coupled by means of the salt bridge shown in Figure 22-5. Hydrogen is oxidized to hydrogen ions when the electrode is an anode; the reverse reaction takes place as a cathode. Under proper conditions, then, the hydrogen electrode is electrochemically reversible.

The potential of a hydrogen electrode depends upon the temperature, the hydrogen ion activity in the solution, and the pressure of the hydrogen at the surface of the electrode. Values for these experimental variables must be carefully defined in order for the half-cell to serve as a reference. Specifications for the *standard hydrogen electrode* call for a hydrogen ion activity of unity and a partial pressure for hydrogen of exactly one atmosphere. *By convention, the potential of this electrode is assigned the value of exactly zero volt at all temperatures.*

22C-3 Practical Reference Electrodes

Although the standard hydrogen electrode is of great fundamental importance, the difficulty in preparing the electrode surface and controlling the activities of the reactants make it impractical enough so that it is seldom used for routine measurements by chemists. Instead, reference electrodes that are simple to prepare, more rugged, and easier to use are normally substituted for the hydrogen gas electrode. One of the most common of these is the silver/silver chloride electrode. This electrode can be prepared by applying an oxidizing potential to a silver wire immersed in a dilute solution of hydrochloric acid. A thin coating of silver chloride is formed that adheres tightly to the wire. The wire is then immersed in a saturated solution of potassium chloride. A salt bridge connects the potassium chloride solution to the electrode system being studied. The potential of this electrode is about 0.2 V positive with respect to the standard hydrogen electrode. The electrode half-reaction is

$$AgCl(s) + e^- \rightleftharpoons Cl^- + Ag(s)$$

A second widely used reference electrode is the saturated calomel electrode (SCE), which consists of a pool of mercury in contact with a solution that is saturated with mercury(I) chloride (calomel) as well as potassium chloride. Platinum wire dipping in the mercury provides electrical contact to the other conductor and a salt bridge to the second electrolyte completes the circuit. The potential of this reference is about 0.24 V positive. The electrode reaction is

$$Hg_2Cl_2(s) + 2e^- \rightleftharpoons 2Cl^- + 2Hg(l)$$

More detailed descriptions of the silver/silver chloride and the calomel reference electrode systems will be found in Section 23A.

22C-4 Definition of Electrode Potential

Electrode potentials are *defined* as *cell potentials* for a cell consisting of the electrode in question *acting as a cathode* and the standard hydrogen electrode *acting as an anode.* It should be emphasized that, despite its name, an electrode potential is in fact the potential of an electrochemical *cell* that contains a carefully defined reference electrode as the anode. It could more properly be called a "relative electrode potential" (but seldom is). Note that this cell may be galvanic or electrolytic depending upon the electron energy of the electrode in question. Thus, when this energy is greater than that of the standard hydrogen electrode, the electrode potential will be negative; when the electron energy of the electrode in question is less than that of the standard hydrogen electrode, the electrode potential will be positive.

The cell in Figure 22-5 illustrates the definition of the electrode potential for the half-reaction

$$M^{2+} + 2e^- \rightleftharpoons M(s)$$

Here, the half-cell on the right (the cathode) consists of a strip of the metal M in contact with a solution of M^{2+}. The half-cell on the left (the anode) is a standard hydrogen electrode. By definition, the potential E observed on the voltage-measuring device is the electrode potential for the M/M^{2+} couple. Here, we assume that the junction potentials across the salt bridge are zero. If we further assume that the activity of M^{2+} in the solution is exactly 1.00, the potential is called the *standard electrode potential* for the system and is given the symbol E^0. That is, the standard electrode potential for a half-reaction is the electrode potential when the reactants and products are all at unit activity.

If M in the figure is copper, and if the copper ion activity in the solution is 1.00, the compartment on the right behaves as a cathode of the galvanic cell, and the observed potential is +0.337 V as shown in the figure. The spontaneous cell reaction is

$$Cu^{2+} + H_2(g) \rightleftharpoons Cu(s) + 2H^+$$

Because the copper electrode is the cathode, the measured potential is, *by definition,* the electrode potential for the Cu/Cu^{2+} half-cell. Note that the copper electrode is positive with respect to the hydrogen electrode; that is, electrons flow in the external circuit from the negative hydrogen anode to the copper cathode. For this reason, the electrode potential is given a positive sign, and we write

$$Cu^{2+} + 2e^- \rightleftharpoons Cu(s) \qquad E^0 = +0.337\text{ V}$$

If M in Figure 22-5 is cadmium instead of copper, and the solution has a cadmium ion activity of 1.00, the potential is observed to be −0.403 V. In this case, the cadmium acts as the anode of the galvanic cell so that electrons flow from the cadmium to the hydrogen electrode. For this reason, the potential is given a negative sign. The spontaneous cell reaction is

$$Cd(s) + 2H^+ \rightleftharpoons Cd^{2+} + 2e^- + H_2(g)$$

and we may write

$$Cd^{2+} + 2\,e^- \rightleftharpoons Cd(s) \qquad E^0 = -0.403\text{ V}$$

A zinc electrode in a solution of zinc ion at unity activity exhibits a potential of −0.763 V when coupled with the standard hydrogen electrode. Because it is the anode in the galvanic cell, its electrode potential is also negative.

The standard electrode potentials for the four half-cells just described can be arranged in the order

$$Cu^{2+} + 2e^- \rightleftharpoons Cu(s) \qquad E^0 = +0.337\text{ V}$$

$$2H^+ + 2e^- \rightleftharpoons H_2(g) \qquad E^0 = 0.000\text{ V}$$

$$Cd^{2+} + 2e^- \rightleftharpoons Cd(s) \qquad E^0 = -0.403\text{ V}$$

$$Zn^{2+} + 2e^- \rightleftharpoons Zn(s) \qquad E^0 = -0.763\text{ V}$$

The magnitudes of these standard electrode potentials show the relative strengths of the four ionic species as electron acceptors (oxidizing agents); that is, in decreasing strengths as oxidizing agents: $Cu^{2+} > H^+ > Cd^{2+} > Zn^{2+}$.

22C-5 Sign Conventions for Electrode Potentials

It is perhaps not surprising that the choice of sign for electrode potentials has led to much controversy and confusion in the course of the development of electrochemistry. In 1953, the International Union of Pure and Applied Chemistry (IUPAC), meeting in Stockholm, attempted to resolve this controversy. The sign convention adopted at this meeting is sometimes called the IUPAC or Stockholm convention; it is the sign convention used by most present-day chemists.

Any sign convention must be based upon half-cell processes written in a single way—that is, entirely as oxidations or as reductions. According to the IUPAC convention, the term *electrode potential* (or more exactly, *relative electrode potential*) is reserved exclusively for half-reactions written as reductions. There is no objection to using the term *oxidation potential* to indicate an electrode process written in the opposite sense, but *an oxidation potential should never be called an electrode potential.*

The sign of the electrode potential is determined by the actual sign of the electrode of interest when it is coupled with a standard hydrogen electrode in a galvanic cell. Thus, a zinc or a cadmium electrode will behave as the anode from which electrons flow through the external circuit to the standard hydrogen electrode. These metal electrodes are thus the negative terminals of such galvanic cells, and their electrode potentials are *assigned* negative values. Thus,

$$Zn^{2+} + 2e^- \rightleftharpoons Zn(s) \qquad E^0 = -0.763\text{ V}$$

$$Cd^{2+} + 2e^- \rightleftharpoons Cd(s) \qquad E^0 = -0.403\text{ V}$$

The potential for the copper electrode, on the other hand, is given a positive sign because the copper behaves as a cathode in a galvanic cell constructed from this electrode and the hydrogen electrode; electrons flow toward the copper electrode through the external circuit. It is thus the positive terminal of the galvanic cell and for copper, we may write

$$Cu^{2+} + 2e^- \rightleftharpoons Cu(s) \qquad E^0 = +0.337\text{ V}$$

It is important to reemphasize that electrode potentials and their signs apply to half-reactions *written as reductions.* Both zinc and cadmium are oxidized by hydrogen ion; the spontaneous reactions are thus oxidations. It is evident, then, that the *sign of the electrode potential will indicate whether or not the reduction is spontaneous with respect to the standard hydrogen electrode.* Thus, the positive sign for the copper electrode potential means that the reaction

$$Cu^{2+} + H_2(g) \rightleftharpoons 2H^+ + Cu(s)$$

proceeds toward the right under ordinary conditions. The negative electrode potential for zinc, on the other hand, means that the analogous reaction

$$Zn^{2+} + H_2(g) \rightleftharpoons 2H^+ + Zn(s)$$

does not ordinarily occur unless forced by application of an external potential; indeed, the equilibrium favors the species on the left.

22C-6 Effect of Activity on Electrode Potential

Let us consider the half-reaction

$$pP + qQ + \cdots\cdots + ne^- \rightleftharpoons rR + sS \cdots\cdots$$

where the capital letters represent formulas of reacting species (whether charged or uncharged), e^- represents the electron, and the lower-case italic letters indicate the number of moles of each species (including electrons) participating in the half-cell reaction. Employing the same arguments that were used in the case of the silver/silver cell in Section 22B-1, we obtain

$$E = E^0 - \frac{RT}{nF} \ln \frac{(a_R)_i^r \cdot (a_S)_i^s \cdots\cdots}{(a_P)_i^p \cdot (a_Q)_i^q \cdots\cdots}$$

At room temperature (298 K), the collection of constants in front of the logarithm has units of joules per coulomb or volt. Therefore,

$$\frac{RT}{nF} = \frac{8.316 \text{ J mol}^{-1}\text{ K}^{-1} \times 298 \text{ K}}{n \times 96487 \text{ C mol}^{-1}}$$

$$= \frac{2.568 \times 10^{-2} \text{ J C}^{-1}}{n} = \frac{2.568 \times 10^{-2}}{n} \text{ V}$$

Upon converting from natural to base ten logarithms by multiplication by 2.303, the foregoing equation can be written as

$$E = E^0 - \frac{0.0592}{n} \log \frac{(a_R)^r \cdot (a_S)^s \cdots\cdots}{(a_P)^p \cdot (a_Q)^q \cdots\cdots} \quad (22\text{-}13)$$

For convenience, we have also deleted the *i* subscripts, which were inserted earlier as a reminder that the bracketed terms represented nonequilibrium concentrations. Hereafter, *the subscripts will not be used;* the student should, however, be alert to the fact that the quotients that appear in this type of equation are *not equilibrium constants,* despite their similarity in appearance.

Equation 22-13 is a general statement of the *Nernst equation,* which can be applied both to half-cell reactions or cell reactions.

22C-7 The Standard Electrode Potential, E^0

An examination of Equation 22-13 reveals that the constant E^0 is equal to the half-cell potential when the logarithmic term is zero. This condition occurs whenever the activity quotient is equal to unity, one such instance being when the activities of all reactants and products are unity. Thus, the standard potential is often defined as the electrode potential of a half-cell reaction (vs. SHE) when all reactants and products are present at unit activity.

The standard electrode potential is an important physical constant that gives a quantitative description of the relative driving force for a half-cell reaction. Four facts regarding this constant should be kept in mind. (1) The electrode potential is temperature dependent; if it is to have significance, the temperature at which it is determined must be specified. (2) The standard electrode potential is a relative quantity in the sense that it is really the potential of an electrochemical cell in which the anode is a carefully specified reference electrode—the standard hydrogen electrode—whose potential is *assigned* a value of zero volt. (3) The sign of a standard potential is identical with that of the conductor in contact with the half-cell of interest in a galvanic cell, the other half of which is the standard hydrogen electrode. (4) The standard potential is a measure of the driving force for a half-reaction. As such, it is independent of the notation employed to express the half-cell process. Thus, the potential for the process

$$Ag^+ + e^- \rightleftharpoons Ag(s) \qquad E^0 = +0.799 \text{ V}$$

although dependent upon the concentration of silver ions, is the same regardless of whether we write the half-reaction as above or as

$$100\, Ag^+ + 100\, e^- \rightleftharpoons 100\, Ag(s) \qquad E^0 = +0.799 \text{ V}$$

To be sure, the Nernst equation must be consistent with the half-reaction as it has been written. For the first of these, it will be

$$E = 0.799 - \frac{0.0592}{1} \log \frac{1}{a_{Ag^+}}$$

and for the second

$$E = 0.799 - \frac{0.0592}{100} \log \frac{1}{(a_{Ag^+})^{100}}$$

$$= 0.799 - \frac{0.0592}{\cancel{100}} \left[\cancel{100} \log \frac{1}{a_{Ag^+}} \right]$$

Standard electrode potentials are available for numerous half-reactions. Many have been determined directly from voltage measurements of cells in which a hydrogen or other reference electrode constituted the other half of the cell. It is possible, however, to calculate E^0 values from equilibrium studies of oxidation/reduction systems and from thermochemical data relating to such reactions. Many of the values found in the literature were so obtained.[4]

For illustrative purposes, a few standard electrode potentials are given in Table 22-1; a more comprehensive table is found in Appendix 3. The species in the upper left-hand part of the equations in Table 22-1 are most easily reduced, as indicated by the large positive E^0 values; they are therefore the most effective oxidizing agents. Proceeding down the left-hand side of the table, each succeeding species is a less effective acceptor of electrons than the one above it. The half-cell reactions at the bottom of the table have little tendency to take place as written. On the other hand, they do tend to occur in the opposite sense, as oxidations. The most effective reducing agents, then, are those species that appear in the lower right-hand side of the equations in the table.

[4] Comprehensive sources for standard electrode potentials include *Standard Electrode Potentials in Aqueous Solutions,* A. J. Bard, R. Parsons, and J. Jordan, Eds. New York: Marcel Dekker, 1985; G. Milazzo and S. Caroli, *Tables of Standard Electrode Potentials.* New York: Wiley-Interscience, 1977; M. S. Antelman and F. J. Harris, *Chemical Electrode Potentials.* New York: Plenum Press, 1982. Some compilations are arranged alphabetically by element; others are tabulated according to the value of E^0.

TABLE 22-1 Standard Electrode Potentials*

Reaction	E^0 at 25°C, V
$Cl_2(g) + 2e^- \rightleftharpoons 2Cl^-$	+1.359
$O_2(g) + 4H^+ + 4e^- \rightleftharpoons 2H_2O$	+1.229
$Br_2(aq) + 2e^- \rightleftharpoons 2Br^-$	+1.087
$Br_2(l) + 2e^- \rightleftharpoons 2Br^-$	+1.065
$Ag^+ + e^- \rightleftharpoons Ag(s)$	+0.799
$Fe^{3+} + e^- \rightleftharpoons Fe^{2+}$	+0.771
$I_3^- + 2e^- \rightleftharpoons 3I^-$	+0.536
$Cu^{2+} + 2e^- \rightleftharpoons Cu(s)$	+0.337
$Hg_2Cl_2(s) + 2e^- \rightleftharpoons 2Hg(l) + 2Cl^-$	+0.268
$AgCl(s) + e^- \rightleftharpoons Ag(s) + Cl^-$	+0.222
$Ag(S_2O_3)_2^{3-} + e^- \rightleftharpoons Ag(s) + 2S_2O_3^{2-}$	+0.010
$2H^+ + 2e^- \rightleftharpoons H_2(g)$	0.000
$AgI(s) + e^- \rightleftharpoons Ag(s) + I^-$	−0.151
$PbSO_4(s) + 2e^- \rightleftharpoons Pb(s) + SO_4^{2-}$	−0.350
$Cd^{2+} + 2e^- \rightleftharpoons Cd(s)$	−0.403
$Zn^{2+} + 2e^- \rightleftharpoons Zn(s)$	−0.763

*See Appendix 3 for a more extensive list.

A compilation of standard potentials provides the chemist with information regarding the extent and direction of electron-transfer reactions between the tabulated species. On the basis of Table 22-1, for example, we see that zinc is more easily oxidized than cadmium, and we conclude that a piece of zinc immersed in a solution of cadmium ions will cause the deposition of metallic cadmium; conversely, cadmium has little tendency to reduce zinc ions. Table 22-1 also shows that iron(III) is a better oxidizing agent than triiodide ion; therefore, in a solution containing an equilibrium mixture of iron(III), iodide, iron(II), and triiodide ions, we can predict that the latter pair will predominate.

22C-8 Measurement of Electrode Potentials

Although the standard hydrogen electrode is the universal standard of reference, it should be understood that the electrode, as described, can never be realized in the laboratory; it is a *hypothetical* electrode to which experimentally determined potentials can be referred only by suitable computation. The reason that the electrode, as defined, cannot be prepared is that chemists lack the knowledge to produce a solution with a hydrogen ion activity of exactly unity; no adequate theory exists to permit evaluation of the activity coefficient of hydrogen ions in a solution in which the ionic strength is as great as unity, as required by the definition (see Appendix 2). Thus, the *concentration* of HCl or other acid required to give a hydrogen ion activity of unity cannot be calculated. Notwithstanding, data for more dilute solutions of acid, where activity coefficients are known, can be used to compute *hypothetical* potentials at unit activity. The example that follows illustrates how standard potentials can be obtained in this way.

EXAMPLE 22-1

D. A. MacInnes[5] found that a cell similar to that shown in Figure 22-3 developed a potential of 0.52053 V. The cell is described by

$$\text{Pt},H_2(1.00 \text{ atm})|HCl(3.215 \times 10^{-3} \text{ M}),AgCl(\text{sat'd})|Ag$$

Calculate the standard electrode potential for the half-reaction

$$AgCl(s) + e^- \rightleftharpoons Ag(s) + Cl^-$$

Here, the electrode potential for the cathode is

$$E_{\text{cathode}} = E^0_{AgCl} - 0.0592 \log a_{Cl^-} = E^0_{AgCl} - 0.0592 \log \gamma_{Cl^-} c_{HCl}$$

where γ_{Cl^-} is the activity coefficient of Cl^-. The second half-cell reaction is

$$H^+ + e^- \rightleftharpoons \frac{1}{2} Hg_2(g)$$

and

$$E_{\text{anode}} = E^0_{H_2} = \frac{0.0592}{1} \log \frac{P^2_{H_2}}{a_{H^+}}$$
$$= E^0_{H_2} = \frac{0.0592}{1} \log \frac{P^2_{H_2}}{\gamma_{H^+} c_{HCl}}$$

The measured potential is the difference between these potentials (Equation 22-12):

$$E_{\text{cell}} = (E^0_{AgCl} - 0.0592 \log \gamma_{Cl^-} c_{HCl}) - \left(0.000 - 0.0592 \log \frac{P^{1/2}_{H_2}}{\gamma_{H^+} c_{HCl}}\right)$$

Combining the two logarithmic terms gives

$$E_{\text{cell}} = E^0_{AgCl} - 0.0592 \log \frac{\gamma_{H^+} \gamma_{Cl^-} c^2_{HCl}}{P^{1/2}_{H_2}}$$

The activity coefficients for H^+ and Cl^- can be calculated from Equation a2-3 (Appendix 2) taking 3.215×10^{-3} for the ionic strength μ; these values are 0.945 and 0.939, respectively. Substitution of these activity coefficients and the experimental data into the foregoing equation gives, upon rearrangement,

$$E^0_{AgCl} = 0.5203 + 0.0592 \log \frac{(3.215 \times 10^{-3})^2 (0.945)(0.939)}{1.00^{1/2}}$$
$$= 0.2223 \approx 0.222 \text{ V}$$

(The mean for this and similar measurements at other concentrations of HCl was 0.222 V.)

[5] D. A. MacInnes, *The Principles of Electrochemistry*, p. 187. New York: Reinhold, 1939.

22C-9 Calculation of Half-Cell Potentials from E^0 Values

Typical applications of the Nernst equation to the calculation of half-cell potentials are illustrated in the following examples.

EXAMPLE 22-2

What is the electrode potential for a half-cell consisting of a cadmium electrode immersed in a solution that is 0.0150 M in Cd^{2+}?

From Table 22-1, we find

$$Cd^{2+} + e^- \rightleftharpoons Cd(s) \qquad E^0 = -0.403 \text{ V}$$

We will assume that $a_{Cd^{2+}} \approx [Cd^{2+}]$ and write

$$E = E^0 - \frac{0.0592}{2} \log \frac{1}{[Cd^{2+}]}$$

Substituting the Cd^{2+} concentration into this equation gives

$$E = -0.403 - \frac{0.0592}{2} \log \frac{1}{0.0150}$$
$$= -0.457 \text{ V}$$

The sign for the potential from the foregoing calculation indicates the direction of the reaction when this half-cell is coupled with the standard hydrogen electrode. The fact that it is negative shows that the reaction

$$Cd(s) + 2H^+ \rightleftharpoons H_2(g) + Cd^{2+}$$

occurs spontaneously. Note that the calculated potential is a larger negative number than the standard electrode potential itself. This follows from mass-law considerations because the half-reaction, *as written*, has less tendency to occur with the lower cadmium ion concentration.

EXAMPLE 22-3

Calculate the potential for a platinum electrode immersed in a solution prepared by saturating a 0.0150 M solution of KBr with Br_2.

Here, the half-reaction is

$$Br_2(l) + 2e^- \rightleftharpoons 2Br^- \qquad E^0 = 1.065\ V$$

Note that the term (l) in the equation indicates that the aqueous solution is kept saturated by presence of an excess of *liquid* Br_2. Thus, the overall process is the sum of the two equilibria

$$Br_2(l) \rightleftharpoons Br_2(\text{sat'd } aq)$$

$$Br_2(\text{sat'd } aq) + 2e^- \rightleftharpoons 2Br^-$$

Assuming that $[Br^-] = a_{Br^-}$, the Nernst equation for the overall process is

$$E = 1.065 - \frac{0.0592}{2}\log\frac{[Br^-]^2}{1.00}$$

Here, the activity of Br_2 in the pure liquid is constant and equal to 1.00 by definition. Thus,

$$E = 1.065 - \frac{0.0592}{2}\log(0.0150)^2$$

$$= 1.173\ V$$

EXAMPLE 22-4

Calculate the potential for a platinum electrode immersed in a solution that is 0.0150 M in KBr and 1.00×10^{-3} M in Br_2.

Here, the half-reaction used in the preceding example *does not apply because the solution is no longer saturated in* Br_2. Table 22-1, however, contains the half-reaction

$$Br_2(aq) + 2e^- \rightleftharpoons 2Br^- \qquad E^0 = 1.087\ V$$

The term (aq) implies that all of the Br_2 present is in solution and that 1.087 V is the electrode potential for the half-reaction when the Br^- and Br_2 *solution* activities are 1.00 mol/L. It turns out, however, that the solubility of Br_2 in water at 25°C is only about 0.18 mol/L. Therefore, the recorded potential of 1.087 V is based on a *hypothetical system that cannot be realized experimentally.* Nevertheless, this potential is useful because it provides the means by which potentials for undersaturated systems can be calculated. Thus, if we assume that activities of solutes are equal to their molar concentrations, we obtain

$$E = 1.087 - \frac{0.0592}{2}\log\frac{[Br^-]^2}{[Br_2]}$$

$$= 1.087 - \frac{0.0592}{2}\log\frac{(1.50 \times 10^{-2})^2}{1.00 \times 10^{-3}}$$

$$= 1.106\ V$$

Here, the Br_2 activity is 1.00×10^{-3} rather than 1.00, as was the situation when the solution was saturated and excess $Br_2(l)$ was present.

22C-10 Electrode Potentials in the Presence of Precipitation and Complex-Forming Reagents

As shown by the following example, reagents that react with the participants of an electrode process have a marked effect on the potential for that process.

EXAMPLE 22-5

Calculate the potential of a silver electrode in a solution that is saturated with silver iodide and has an iodide ion activity of exactly 1.00 (K_{sp} for AgI = 8.3×10^{-17}).

$$Ag^+ + e^- \rightleftharpoons Ag(s) \qquad E^0 = +0.799\ V$$

$$E = +0.799 - 0.0592 \log\frac{1}{a_{Ag^+}}$$

We may calculate a_{Ag^+} from the solubility-product constant. Thus,

$$a_{Ag^+} = \frac{K_{sp}}{a_{I^-}}$$

Substituting into the Nernst equation gives

$$E = +0.799 - \frac{0.0592}{1}\log\frac{a_{I^-}}{K_{sp}}$$

This equation may be rewritten as

$$E = +0.799 + 0.0592 \log K_{sp} - 0.0592 \log a_{I^-} \qquad (22\text{-}14)$$

If we substitute 1.00 for a_{I^-} and use 8.3×10^{-17} for K_{sp}, the solubility product for AgI at 25.0°C, we obtain

$$E = +0.799 + 0.0592 \log 8.3 \times 10^{-17} - 0.0592 \log 1.00$$

$$= -0.151 \text{ V}$$

This example shows that the half-cell potential for the reduction of silver ion becomes smaller in the presence of iodide ions. Qualitatively this is the expected effect because decreases in the concentration of silver ions diminish the tendency for their reduction.

Equation 22-14 relates the potential of a silver electrode to the iodide ion activity of a solution that is also saturated with silver iodide. *When the iodide ion activity is unity,* the potential is the sum of two constants; it is thus the standard electrode potential for the half-reaction

$$AgI(s) + e^- \rightleftharpoons Ag(s) + I^- \qquad E^0 = -0.151 \text{ V}$$

where

$$E^0_{AgI} = +0.799 + 0.0592 \log K_{sp}$$

The Nernst relationship for the silver electrode in a solution saturated with silver iodide can then be written as

$$E = E^0 - 0.0592 \log a_{I^-} = -0.151 - 0.0592 \log a_{I^-}$$

Thus, when in contact with a solution *saturated with silver iodide,* the potential of a silver electrode can be described *either* in terms of the silver ion activity (with the standard electrode potential for the simple silver half-reaction) *or* in terms of the iodide ion activity (with the standard electrode potential for the silver/silver iodide half-reaction). The latter is usually more convenient.

The potential of a silver electrode in a solution containing an ion that forms a soluble complex with silver ion can be treated in a fashion analogous to the foregoing. For example, in a solution containing thiosulfate and silver ions, complex formation occurs

$$Ag^+ + 2S_2O_3^{2-} \rightleftharpoons Ag(S_2O_3)_2^{3-}$$

$$K_f = \frac{a_{Ag(S_2O_3)_2^{3-}}}{a_{Ag^+} \cdot a^2_{S_2O_3^{2-}}}$$

where K_f is the *formation constant* for the complex. The half-reaction for a silver electrode in such a solution can be written as

$$Ag(S_2O_3)_2^{3-} + e^- \rightleftharpoons Ag(s) + 2S_2O_3^{2-}$$

The standard electrode potential for this half-reaction will be the electrode potential when both the complex and the complexing anion are at unit activity. Using the same approach as in the previous example, we find that

$$E^0 = +0.799 + 0.0592 \log \frac{1}{K_f} \qquad (22\text{-}15)$$

Data for the potential of the silver electrode in the presence of selected ions are given in the tables of standard electrode potentials in Appendix 3 and in Table 22-1. Similar information is also provided for other electrode systems. Such data often simplify the calculation of half-cell potentials.

22C-11 Some Limitations to the Use of Standard Electrode Potentials

Standard electrode potentials are of great importance in understanding electroanalytical processes. There are, however, certain inherent limitations to the use of these data that should be clearly appreciated.

Substitution of Concentrations for Activities

As a matter of convenience, molar concentrations—rather than activities—of reactive species are generally employed in making computations with the Nernst equation. Unfortunately, the assumption that these two quantities are identical is valid only in dilute solutions; with increasing electrolyte concentrations, potentials calculated on the basis of molar concentrations can be expected to depart from those obtained by experiment.

To illustrate, the standard electrode potential for the half-reaction

$$Fe^{3+} + e^- \rightleftharpoons Fe^{2+}$$

is +0.771 V. Neglecting activities, we would predict that a platinum electrode immersed in a solution that contained 1 mol/L each of Fe^{2+} and Fe^{3+} ions in addition to perchloric acid would exhibit a potential numerically equal to this value relative to the standard hydrogen electrode. In fact, however, a potential of +0.732 V is observed experimentally when the perchloric acid concentration is 1 M. The reason for the discrepancy is seen if we write the Nernst equation in the form

$$E = E^0 - 0.0592 \log \frac{\gamma_{Fe^{2+}}[Fe^{2+}]}{\gamma_{Fe^{3+}}[Fe^{3+}]}$$

where $\gamma_{Fe^{2+}}$ and $\gamma_{Fe^{3+}}$ are the respective activity coefficients. The activity coefficients of the two species are less than one in this system because of the high ionic strength imparted by the perchloric acid and the iron salts. More important, however, the activity coefficient of the iron(III) ion is smaller than that of the iron(II) ion. As a consequence, the ratio of the activity coefficients as they appear in the Nernst equation would be larger than one, and the potential of the half-cell would be smaller than the standard potential.

Activity coefficient data for ions in solutions of the types commonly encountered in oxidation/reduction titrations and electrochemical work are fairly limited; consequently, molar concentrations rather than activities must be used in many calculations. Appreciable errors may result. Such calculations are still of value, however, because relative changes in values will often be close to correct, and direction of changes will be accurate.

Effect of Other Equilibria

The application of standard electrode potentials is further complicated by the occurrence of solvation, dissociation, association, and complex-formation reactions involving the species of interest. An example of this problem is encountered in the behavior of the potential of the iron(III)/iron(II) couple. As noted earlier, an equimolar mixture of these two ions in 1 M perchloric acid has an electrode potential of +0.73 V. Substitution of hydrochloric acid of the same concentration alters the observed potential to +0.70 V; a value of +0.6 V is observed in 1 M phosphoric acid. These differences arise because iron(III) forms more stable complexes with chloride and phosphate ions than does iron(II). As a result, the actual concentration of uncomplexed iron(III) in such solutions is less than that of uncomplexed iron(II), and the net effect is a shift in the observed potential.

Phenomena such as these can be taken into account only if the equilibria involved are known and constants for the processes are available. Often, however, such information is lacking; the chemist is then forced to neglect such effects and hope that serious errors do not flaw the calculated results.

Formal Potentials

In order to compensate partially for activity effects and errors resulting from side reactions, such as those described in the previous section, Swift proposed substituting a quantity called the *formal potential* $E^{0'}$ in place of the standard electrode potential in oxidation/reduction calculations.[6] The formal potential of a system is the potential of the half-cell with respect to the standard hydrogen electrode when the *concentrations* of reactants and products are 1 M and the concentrations of any other constituents of the solution are carefully specified. Thus, for example, the formal potential for the reduction of iron(III) is +0.732 V in 1 M perchloric acid and +0.700 V in 1 M hydrochloric acid. Use of these values in place of the standard electrode potential in the Nernst equation will yield better agreement between calculated and experimental potentials, provided the electrolyte concentration of the solution approximates that for which the formal potential was measured. Application of formal potentials to systems differing greatly as to kind and concentration of electrolyte can, however, lead to errors greater than those encountered with the use of standard potentials. The table in Appendix 3 contains selected formal potentials as well as standard potentials; in subsequent chapters, we shall use whichever is the more appropriate.

[6] E. H. Swift, *A System of Chemical Analysis,* p. 50. San Francisco: Freeman, 1939.

Reaction Rates

It should be realized that the existence of a half-reaction in a table of electrode potentials does not necessarily imply that a real electrode exists whose potential responds to the half-reaction. Many of the data in such tables have been obtained from calculations based upon equilibrium or thermal measurements rather than from the actual measurement of the potential for an electrode system. For some half-reactions, no suitable electrode is known; thus, the standard electrode potential for the process,

$$2CO_2 + 2H^+ + 2e^- \rightleftharpoons H_2C_2O_4 \qquad E^0 = -0.49\ V$$

has been arrived at indirectly. The electrode reaction is not reversible, and the rate at which carbon dioxide combines to give oxalic acid is negligibly slow. No electrode system is known whose potential varies in the expected way with the ratio of activities of the reactants and products. Nonetheless, the potential is useful for computational purposes.

22D CALCULATION OF CELL POTENTIALS FROM ELECTRODE POTENTIALS

An important use of standard electrode potentials is the calculation of the potential obtainable from a galvanic cell or the potential required to operate an electrolytic cell. These calculated potentials (sometimes called *thermodynamic potentials*) are theoretical in the sense that they refer to cells in which there is essentially no current; additional factors must be taken into account when there is a current in the cell. Furthermore, they do not take into account junction potentials within the cell. Normally, these are small enough to be neglected without serious error.

As shown earlier (Equation 22-12), the electromotive force of a cell is obtained by combining half-cell potentials as follows:

$$E_{cell} = E_{cathode} - E_{anode}$$

where E_{anode} and $E_{cathode}$ are the *electrode potentials* for the two half-reactions constituting the cell.

Consider the hypothetical cell

$$Zn\,|\,ZnSO_4(a_{Zn^{2+}} = 1.00)\|CuSO_4(a_{Cu^{2+}} = 1.00)\,|\,Cu$$

The overall cell process involves the oxidation of elemental zinc to zinc(II) and the reduction of copper(II) to the metallic state. Because the activities of the two ions are specified as unity, the standard potentials are also the electrode potentials. The cell diagram also specifies that the zinc electrode is the anode. Thus, using E^0 data from Table 22-1,

$$E_{cell} = +0.337 - (-0.763) = +1.100 \text{ V}$$

The positive sign for the cell potential indicates that the reaction

$$Zn(s) + Cu^{2+} \rightarrow Zn^{2+} + Cu(s)$$

occurs spontaneously and that this is a galvanic cell.

The foregoing cell, diagrammed as

$$Cu\,|\,Cu^{2+}(a_{Cu^{2+}} = 1.00)\|Zn^{2+}(a_{Zn^{2+}} = 1.00)\,|\,Zn$$

implies that the copper electrode is now the anode. Thus,

$$E_{cell} = -0.763 - (+0.337) = -1.100 \text{ V}$$

The negative sign indicates the nonspontaneity of the reaction

$$Cu(s) + Zn^{2+} \rightarrow Cu^{2+} + Zn(s)$$

The application of an external potential greater than 1.100 V is required to cause this reaction to occur.

EXAMPLE 22-6

Calculate potentials for the following cell employing **(a)** concentrations and **(b)** activities:

$$Zn\,|\,ZnSO_4(c_{ZnSO_4}), PbSO_4(\text{sat'd})\,|\,Pb$$

where $c_{ZnSO_4} = 5.00 \times 10^{-4}$, 2.00×10^{-3}, 1.00×10^{-2}, 2.00×10^{-2}, and 5.00×10^{-2}.

(a) In a neutral solution, little HSO_4^- will be formed; thus, we may assume that

$$[SO_4^{2-}] = c_{ZnSO_4} = 5.00 \times 10^{-4}$$

The half-reactions and standard potentials are

$$PbSO_4(s) + 2e^- \rightleftharpoons Pb(s) + SO_4^{2-} \qquad E^0 = -0.350 \text{ V}$$

$$Zn^{2+} + 2e^- \rightleftharpoons Zn \qquad E^0 = -0.763 \text{ V}$$

The potential of the lead electrode is given by

$$E_{Pb} = -0.350 - \frac{0.0592}{2}\log 5.00 \times 10^{-4}$$

$$= -0.252 \text{ V}$$

The zinc ion concentration is also 5.00×10^{-4} and

$$E_{Zn} = -0.763 - \frac{0.0592}{2}\log\frac{1}{5.00 \times 10^{-4}}$$

$$= -0.860 \text{ V}$$

Since the Pb electrode is specified as the cathode,

$$E_{cell} = -0.252 - (-0.860) = 0.608$$

Cell potentials at the other concentrations can be derived in the same way. Their values are given in column (a) of Table 22-2.

(b) To obtain activity coefficients for Zn^{2+} and SO_4^{2-} we must first calculate the ionic strength with the aid of Equation a2-2 (Appendix 2). Here, we assume that the concentrations of Pb^{2+}, H^+, and OH^- are negligible with respect to the concentrations of Zn^{2+} and SO_4^{2-}. Thus, the ionic strength is

$$\mu = \frac{1}{2}[5.00 \times 10^{-4} \times (2)^2 + 5.00 \times 10^{-4} \times (2)^2]$$

$$= 2.00 \times 10^{-3}$$

TABLE 22-2 Calculated Potentials for a Cell Based on (a) Concentrations and (b) Activities. (See Example 22-6.)

M $ZnSO_4$ x	Ionic Strength μ	(a) E(calc)	(b) E(calc)	E(exptl)*
5.00×10^{-4}	2.00×10^{-3}	0.608	0.613	0.611
2.00×10^{-3}	8.00×10^{-3}	0.572	0.582	0.583
1.00×10^{-2}	4.00×10^{-2}	0.531	0.549	0.553
2.00×10^{-2}	8.00×10^{-2}	0.513	0.537	0.542
5.00×10^{-2}	2.00×10^{-1}	0.490	0.521	0.529

*Experimental data from: I. A. Cowperthwaite and V. K. LaMer, *J. Amer. Chem. Soc.*, **1931,** *53,* 4333.

In Table a2-1, we find for SO_4^{2-}, $\alpha_A = 0.4$ and for Zn^{2+}, $\alpha_A = 0.6$. Substituting these values into Equation a2-3 gives for sulfate ion

$$-\log \gamma_{SO_4} = \frac{0.509 \times 2^2 \times \sqrt{2.00 \times 10^{-3}}}{1 + 3.28 \times 0.4\sqrt{2.00 \times 10^{-3}}}$$

$$= 8.59 \times 10^{-2}$$

$$\gamma_{SO_4} = 0.820$$

$$a_{SO_4} = 0.820 \times 5.00 \times 10^{-4}$$

$$= 4.10 \times 10^{-4}$$

Repeating the calculations employing $\alpha_A = 0.4$ for Zn^{2+} yields

$$\gamma_{Zn} = 0.825$$

$$a_{Zn} = 4.13 \times 10^{-4}$$

The Nernst equation for the Pb electrode now becomes

$$E_{Pb} = -0.350 - \frac{0.0592}{2} \times \log 4.10 \times 10^{-4}$$

$$= -0.250$$

For the zinc electrode

$$E_{Zn} = -0.763 - \frac{0.0592}{2} \times \log \frac{1}{4.13 \times 10^{-4}}$$

$$= -0.863$$

and

$$E_{cell} = -0.250 - (-0.863) = 0.613 \text{ V}$$

Values at other concentrations are listed in column (b) of Table 22-2.

It is of interest to compare the calculated cell potentials shown in the columns labeled (a) and (b) in Table 22-2 with the experimental results shown in the last column. Clearly, the use of activities provides a significant improvement at the higher ionic strengths.

22E CURRENTS IN ELECTROCHEMICAL CELLS

Only one general type of electroanalytical method is based upon measurements that are made in the absence of appreciable current, namely, potentiometric methods, which are treated in Chapter 23. The remaining several methods, considered in Chapters 24 and 25, all involve electrical currents and current measurements. Thus, we need to consider the behavior of cells when significant currents are present.

As noted earlier, electricity is carried within a cell by the movement of ions. With small currents, Ohm's law is usually obeyed, and we may write $E = IR$ where E is the potential difference in volts responsible for movement of the ions, I is the current in amperes, and R is the resistance in ohms of the electrolyte to the current. The resistance depends upon the kinds and concentrations of ions in the solution.

When dc electricity is carried through an electrochemical cell, the measured cell potential normally departs from that derived from thermodynamic calculations such as those shown in Section 22B. This departure can be traced to a number of phenomena, including ohmic resistance and several *polarization ef-*

fects, such as *charge-transfer overvoltage, reaction overvoltage, diffusion overvoltage,* and *crystallization overvoltage.* Generally, these phenomena have the effect of reducing the potential of a galvanic cell or increasing the potential needed to develop a current in an electrolytic cell.

22E-1 Ohmic Potential; *IR* Drop

To develop a current in either a galvanic or an electrolytic cell, a driving force in the form of a potential is required to overcome the resistance of the ions to movement toward the anode and the cathode. Just as in metallic conduction, this force follows Ohm's law and is equal to the product of the current in amperes and the resistance of the cell in ohms. The force is generally referred to as the *ohmic potential,* or the *IR drop.*

The net effect of *IR* drop is to increase the potential required to operate an electrolytic cell and to decrease the measured potential of a galvanic cell. Therefore, the *IR* drop is always *subtracted* from the theoretical cell potential. That is,[7]

$$E_{cell} = E_{cathode} - E_{anode} - IR \qquad (22\text{-}16)$$

EXAMPLE 22-7

The following cell has a resistance of 4.00 Ω. Calculate its potential when it is producing a current 0.100 A.

$$Cd\,|\,Cd^{2+}(0.0100\ M)\|Cu^{2+}(0.0100\ M)\,|\,Cu$$

Substitution into the Nernst equation reveals that the electrode potential for the Cu electrode is 0.278 V, while for the Cd electrode it is −0.462 V. Thus, the thermodynamic cell potential is

$$E = E_{Cu} - E_{Cd}$$

$$= 0.278 - (-0.462) = 0.740\ V$$

and the potential to yield the desired current is

$$E_{cell} = 0.740 - IR$$

$$= 0.740 - (0.100 \times 4.00) = 0.340\ V$$

Note that the potential of this cell drops dramatically when there is a current in the cell.

[7] Here and in the subsequent discussion we will assume that the junction potential is negligible relative to the other potentials.

EXAMPLE 22-8

Calculate the potential required to generate a current of 0.100 A in the reverse direction in the cell shown in Example 22-7.

$$E = E_{Cd} - E_{Cu}$$

$$= -0.462 - 0.278 = -0.740\ V$$

$$E_{cell} = -0.740 - (0.100 \times 4.00)$$

$$= -1.140\ V$$

Here, an external potential greater than 1.140 V is needed to cause Cd^{2+} to deposit and Cu to dissolve at a rate required for a current of 0.100 A.

22E-2 Polarization

Several important electroanalytical methods are based upon *current-voltage curves,* which are obtained by measuring the variation in current in a cell as a function of its potential. Equation 22-16 predicts that at constant electrode potentials, a linear relationship should exist between the cell voltage and the current. In fact, departures from linearity are often encountered; under these circumstances, the cell is said to be *polarized.* Polarization may arise at one or both electrodes.

As an introduction to this discussion, it is worthwhile considering current-voltage curves for an ideal polarized and an ideal nonpolarized electrode. Polarization at a single electrode can be studied by coupling it with an electrode that is not readily polarized. Such an electrode is characterized by being large in area and being based on a half-cell reaction that is rapid and reversible. Design details of nonpolarized electrodes will be found in subsequent chapters.

Ideal Polarized and Nonpolarized Electrodes and Cells

The ideal polarized electrode is one in which current remains constant and independent of potential over a considerable range. Figure 22-6a is a current-voltage curve for an electrode that behaves ideally in the region between *A* and *B.* Figure 22-6b depicts the current-voltage relationship for a depolarized electrode that behaves ideally in the region between *A* and *B.* Here the potential is independent of the current.

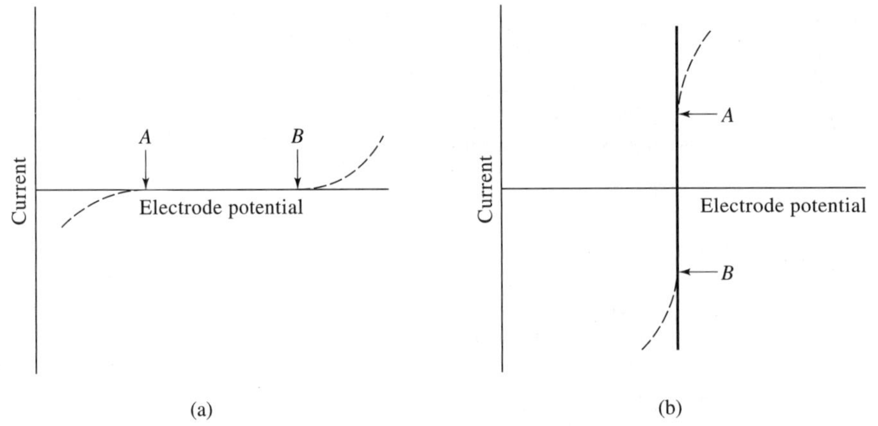

Figure 22-6 Current-voltage curves for an ideal (a) polarized and (b) nonpolarized electrode. Dashed lines show departure from ideal behavior by real electrodes.

Figure 22-7 is a current-voltage curve for a *cell* having electrodes that exhibit ideal nonpolarized behavior between points *A* and *B*. Because of the internal resistance of the cell, the current-voltage curve has a finite slope equal to *R* (Equation 22-16) rather than the infinite slope for the ideal nonpolarized electrode shown in Figures 22-6b. Beyond points *A* and *B*, polarization occurs at one or both electrodes, resulting in departures from the ideal straight line. The upper half of the curve gives the current-voltage relationship when the cell is operating as an electrolytic cell; the lower half describes its behavior as a galvanic cell.[8] Note that when polarization arises in an electrolytic cell, a higher potential is required to achieve a given current. Similarly, polarization of a galvanic cell produces a potential that is lower than expected.

Sources of Polarization

Figure 22-8 depicts three regions of a half-cell where polarization can occur. These regions include the electrode itself, a surface film of solution immediately adjacent to the electrode, and the bulk of the solution. For this half-cell, the overall electrode reaction is

$$\text{Ox} + ne^- \rightleftharpoons \text{Red}$$

Any one of the several intermediate steps shown in the figure may, however, limit the rate at which this overall reaction occurs and thus the magnitude of the current. One of these steps in the reaction, called *mass transfer,* involves movement of Ox from the bulk of the solution to the surface film. When this step (or the reverse mass transfer of Red to the bulk) limits the rate of the overall reaction and thus the current, *concentration polarization* is said to exist. Some half-cell reactions proceed by an intermediate chemical reaction in which the species such as Ox′ or Red′ form; this intermediate is then the actual participant in the electron transfer process. If the rate of formation or decomposition of such an intermediate limits the current, *reaction polarization* is said to

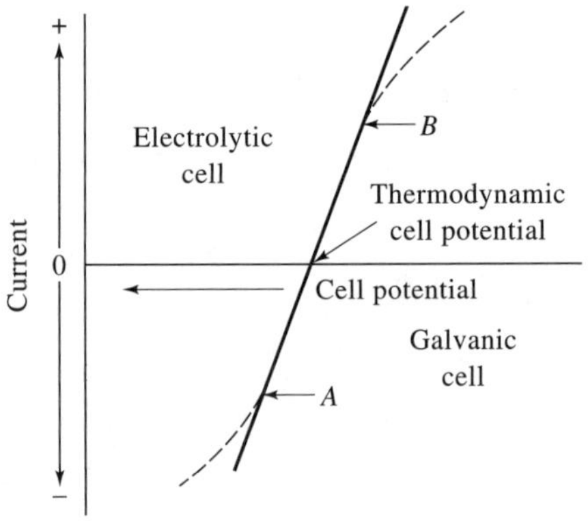

Figure 22-7 Current-voltage curve for a cell showing ideal nonpolarized behavior between *A* and *B* (solid line) and polarized behavior (dashed line).

[8] Here, the IUPAC current convention is being followed with cathodic currents being given a negative sign. Unfortunately, this convention was not followed in the early development of some techniques. Thus, for historical reasons we will not always follow the IUPAC current sign convention.

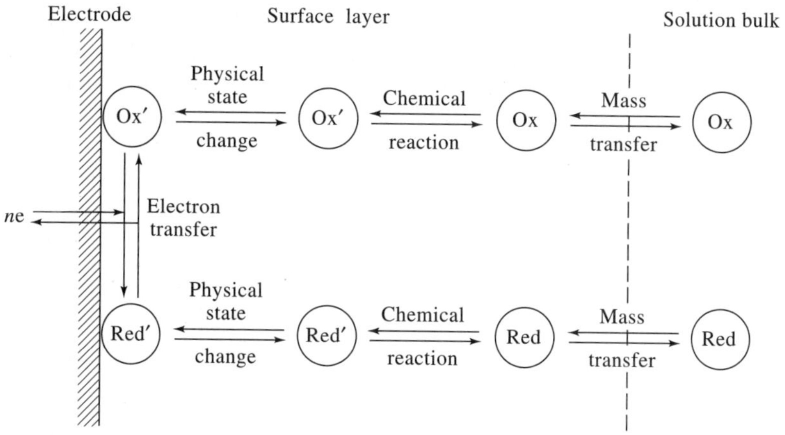

Figure 22-8 Steps in the reaction Ox + $ne^- \rightleftharpoons$ Red at an electrode. Note that the surface layer is only a few molecules thick. *(Adapted from A. J. Bard and L. R. Faulkner,* Electrochemical Methods, *p. 21. New York: Wiley, 1980. Reprinted by permission of John Wiley & Sons, Inc.)*

be present. In some instances, the rate of a physical process such as adsorption, desorption, or crystallization is current limiting. Here, *adsorption, desorption,* or *crystallization polarization* is occurring. Finally, *charge-transfer polarization* is encountered, where current limitation arises from the slow rate of electron transfer from the electrode to the oxidized species in the surface film or from the reduced species to the electrode. It is not unusual to encounter half-cells in which several types of polarization are occurring simultaneously.

Overvoltage

The degree of polarization of an electrode is measured by the *overvoltage* or *overpotential* η, which is the difference between the actual electrode potential E and the thermodynamic or equilibrium potential E_{eq}. That is,

$$\eta = E - E_{eq} \tag{22-17}$$

where $E < E_{eq}$. It is important to realize that polarization always reduces the electrode potential for a system. Thus, as we have indicated, E is always smaller than E_{eq}, and η is *always negative.*

Concentration Polarization

Concentration polarization arises when the rate of transport of reactive species to the electrode surface is insufficient to maintain the current demanded by Equation 22-16. With onset of concentration polarization, a *diffusion overvoltage* develops.

For example, consider a cell made up of an ideal nonpolarized anode and a polarizable cathode consisting of a small cadmium electrode immersed in a solution of cadmium ions. The reduction of cadmium ions is a rapid and reversible process so that when a potential is applied to this electrode, the *surface layer* of the solution comes to equilibrium with the electrode *essentially instantaneously.* That is, a brief current is generated that reduces the surface concentration of cadmium ions to the equilibrium concentration, c_0, given by

$$E = E^0_{Cd} - \frac{0.0592}{2} \log \frac{1}{c_0} \tag{22-18}$$

If no mechanism existed for transport of cadmium ions from the bulk of the solution to the surface film, the current would rapidly decrease to zero as the concentration of the film approached c_0. As we shall see, however, several mechanisms do indeed exist that bring cadmium ions from the bulk of the solution into the surface layer at a constant rate. As a consequence, the large initial current decreases rapidly to a constant level that is determined by the rate of ion transport.

It is important to appreciate that for a rapid and reversible electrode reaction, the concentration of the surface layer may always be considered to be the equilibrium concentration, which is determined by the instantaneous electrode potential (Equation 22-18). It is also important to realize that the surface concentration c_0 is often far different from that of the bulk of the solution because while the surface equilibrium is essentially instantaneously achieved, attainment of equilibrium between the electrode and the bulk of the solution often requires minutes or even hours.

For a current of the magnitude required by Equation 22-16 to be maintained, it is necessary that reactant be brought from the bulk of the solution to the surface layer at a rate dc/dt that is given by

$$I = dQ/dt = nFdc/dt$$

where dQ/dt is the rate of flow of electrons in the electrode (or the current I), n is the number of electrons appearing in the half-reaction, and F is the faraday. The rate of concentration change can be written as

$$\frac{dc}{dt} = AJ$$

where A is the surface area of the electrode in square meters (m^2) and J is the flux in mol s^{-1} m^{-2}. The two equations can then be combined to give

$$I = nFAJ \qquad (22\text{-}19)$$

When this demand for reactant cannot be met by the mass transport process, the IR drop in Equation 22-16 becomes smaller than theoretical, and a diffusion overvoltage appears that just offsets the decrease in IR. Thus, with the appearance of concentration polarization, Equation 22-16 becomes

$$E_{cell} = E_{cathode} - E_{anode} - IR + \eta_{cathode}$$

where $\eta_{cathode}$ represents the overpotential associated with the cathode. Note that in this example, it was assumed that the anode was an ideal nonpolarized one. A more general equation is

$$E_{cell} = E_{cathode} - E_{anode} + \eta_{cathode} + \eta_{anode} - IR \qquad (22\text{-}20)$$

where η_{anode} is the anodic overvoltage. Note that the overvoltage associated with each electrode always carries a negative sign and has the effect of reducing the overall potential of the cell.

22E-3 Mechanisms of Mass Transport

It is important now to investigate the mechanisms by which ions or molecules are transported from the bulk of the solution to a surface layer (or the reverse) because a knowledge of these mechanisms provides insights into how concentration polarization can be prevented or induced as required. As we noted in Section 22A-4, three mechanisms of mass transport can be recognized: (1) *diffusion,* (2) *migration,* (3) *convection.*

Whenever a concentration difference develops between two regions of a solution, as it does when a species is reduced at a cathode surface (or oxidized at an anode surface), ions or molecules move from the more concentrated region to the more dilute as a result of diffusion. The rate of diffusion dc/dt is given by

$$dc/dt = k(c - c_0) \qquad (22\text{-}21)$$

where c is the reactant concentration in the bulk of the solution, c_0 is its equilibrium concentration at the electrode surface, and k is a proportionality constant. As shown earlier, *the value of c_0 is fixed by the potential of the electrode and can be calculated from the Nernst equation.* As higher potentials are applied to the electrode, c_0 becomes smaller and smaller, and the diffusion rate becomes greater and greater. Ultimately, however, c_0 becomes negligible with respect to c; the rate then becomes constant. That is, when $c_0 \rightarrow 0$,

$$dc/dt = kc \qquad (22\text{-}22)$$

Under this circumstance, concentration polarization is said to be complete, and the electrode operates as an ideal polarized electrode.

The process by which ions move under the influence of an electrostatic field is called migration. It is often the primary process by which mass transfer occurs in the bulk of the solution in a cell. The electrostatic attraction (or repulsion) between a particular ionic species and the electrode becomes smaller as the total electrolyte concentration of the solution becomes greater. It may approach zero when the reactive species is but a small fraction, say 1/100, of the total concentration of ions with a given charge.

Reactants can also be transferred to or from an electrode by mechanical means. Thus, forced convection, such as stirring or agitation, tends to decrease concentration polarization. Natural convection resulting from temperature or density differences also contributes to material transport.

To summarize, concentration polarization is observed when diffusion, migration, and convection are insufficient to transport the reactant to or from an electrode surface at a rate demanded by the theoretical current. Concentration polarization causes the potential of a galvanic cell to be smaller than the value predicted on the basis of the thermodynamic potential and the IR drop. Similarly, in an electrolytic cell, a potential more negative than the theoretical is required to maintain a given current.

Concentration polarization is important in several electroanalytical methods. In some applications, steps are taken to eliminate it; in others, however, it is essential to the method, and every effort is made to promote it. The degree of concentration polarization is influenced experimentally by: (1) the reactant concentration, with polarization becoming more probable at low concentrations; (2) the total electrolyte concentration, with polarization becoming more probable at high concentrations; (3) mechanical agitation, with polarization decreasing in well-stirred solutions; and (4) electrode size, with polarization effects decreasing as the electrode surface area increases.

22E-4 Charge-Transfer Polarization

Charge-transfer polarization arises when the rate of the oxidation or reduction reaction at one or both electrodes is not sufficiently rapid to yield currents of the size demanded by theory. The overvoltage arising from charge-transfer polarization has the following characteristics:

1. Overvoltages increase with current density (current density is defined as the amperes per square centimeter of electrode surface).
2. Overvoltages usually decrease with increases in temperature.
3. Overvoltages vary with the chemical composition of the electrode, often being most pronounced with softer metals such as tin, lead, zinc, and particularly mercury.
4. Overvoltages are most marked for electrode processes that yield gaseous products such as hydrogen or oxygen; they are frequently negligible where a metal is being deposited or where an ion is undergoing a change of oxidation state.
5. The magnitude of overvoltage in any given situation cannot be predicted exactly because it is determined by a number of uncontrollable variables.[9]

The overvoltage associated with the evolution of hydrogen and oxygen is of particular interest to the chemist. The difference between the overvoltage of these gases on smooth and on platinized platinum surfaces is notable. This difference is primarily due to the much larger surface area associated with platinized electrodes, which results in a *real* current density that is significantly smaller than is apparent from the overall dimensions of the electrode. A platinized surface is always employed in construction of hydrogen reference electrodes in order to lower the current density to a point where the overvoltage is negligible.

The high overvoltage associated with the formation of hydrogen permits the electrolytic deposition of several metals that require potentials at which hydrogen would otherwise be expected to interfere. For example, it is readily shown from their standard potentials that rapid formation of hydrogen should occur well below the potential required for the deposition of zinc from a neutral solution. Nevertheless, quantitative deposition of zinc can be attained provided a mercury or copper electrode is used; because of the high overvoltage of hydrogen on these metals, little or no gas is evolved during the electrodeposition.

The magnitude of overvoltage can, at best, be only crudely approximated from empirical information available in the literature. Calculation of cell potentials in which overvoltage plays a part cannot, therefore, be very accurate. As with diffusion overvoltage, charge-transfer and physical overvoltages always carry a negative sign and have the effect of making the observed potential smaller than the thermodynamic electrode potential.

22F TYPES OF ELECTROANALYTICAL METHODS

A wide variety of electroanalytical methods have been proposed. Those that have found fairly general use, and are discussed in this book, are shown in Figure 22-9. These methods are divided into interfacial methods and bulk methods, with the former finding much wider usage. Interfacial methods are based upon phenomena that occur at the interface between electrode surfaces and the thin layer of solution just adjacent to these surfaces. Bulk methods, in contrast, are based upon phenomena that occur in the bulk of the solution; every effort is made to avoid interfacial effects.

Interfacial methods can be divided into two major categories, static and dynamic, based upon whether the electrochemical cells are operated in the absence or presence of current. The static methods, which involve potentiometric measurements, are of singular importance because of their speed and selectivity. Potentiometric methods are dealt with in Chapter 23.

[9] Overvoltage data for various gaseous species at different electrode surfaces are to be found in *Handbook of Analytical Chemistry,* L. Meites, Ed., p. 5-184. New York: McGraw-Hill, 1963.

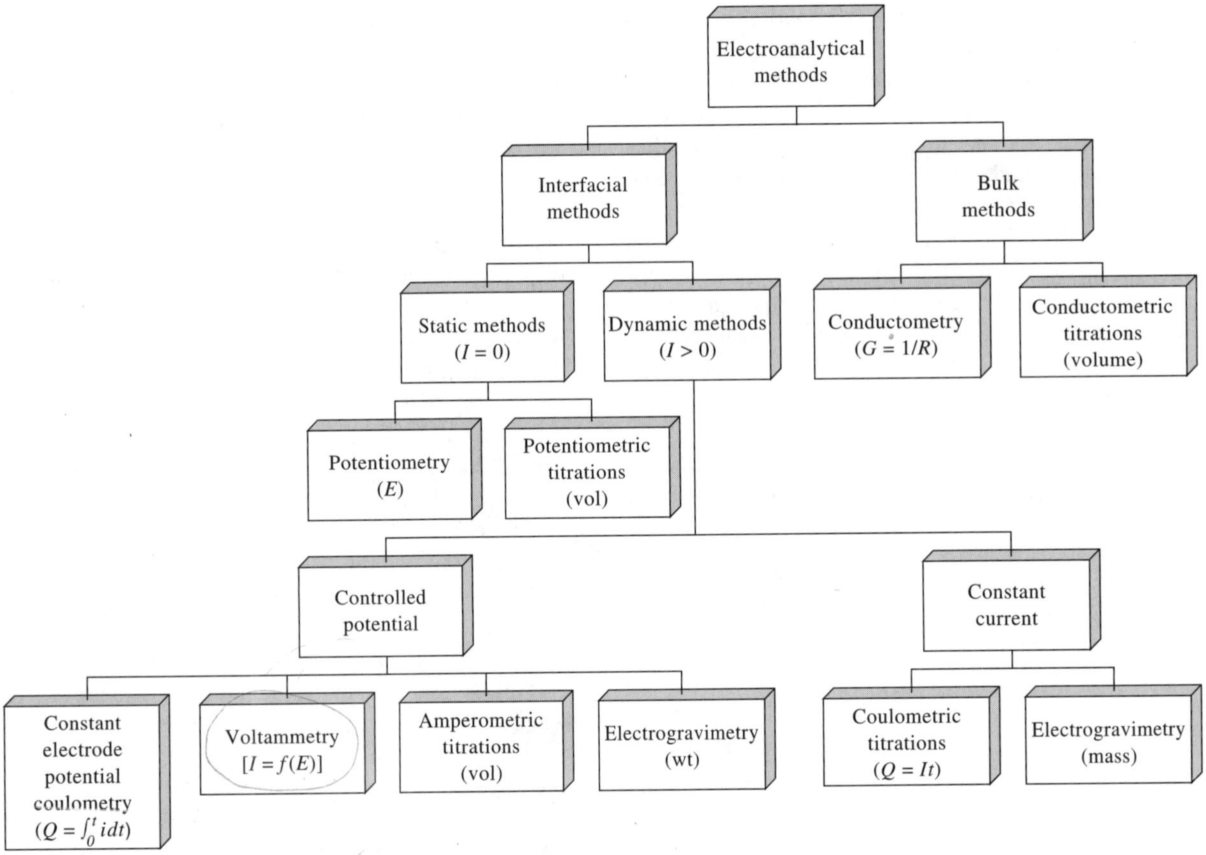

Figure 22-9 Summary of common electroanalytical methods. Quantity measured given in parentheses. (I = current, E = potential, R = resistance, G = conductance, Q = quantity of charge, t = time, vol = volume of a standard solution, wt = weight of an electrodeposited species.)

Dynamic interfacial methods, in which currents in electrochemical cells play a vital part, are of several types. In three of the methods shown on the left in Figure 22-9, the potential of the cell is controlled while measurements of other variables are carried out. Generally, these methods are sensitive and have relatively wide dynamic ranges (typically, 10^{-3} to 10^{-8} M). Furthermore, many of these procedures can be carried out with microliter or even nanoliter volumes of sample. Thus, detection limits in the picomole range may be realized.

In constant-current dynamic methods, the current in the cell is held constant while data are collected. Dynamic methods of both kinds are discussed in Chapters 24 and 25.

Most of the electroanalytical techniques shown in Figure 22-9 have been used as detectors in various chromatographic procedures.

22G QUESTIONS AND PROBLEMS

22-1 Calculate the electrode potentials of the following half-cells.

(a) Ag^+(0.0152 M) | Ag

(b) Fe^{3+}(2.35×10^{-4} M),Fe^{2+}(0.200 M) | Pt

(c) AgBr(sat'd),Br^-(0.100 M) | Ag

22-2 Calculate the electrode potentials of the following half-cells.

(a) $HCl(3.50\ M)|H_2(0.950\ atm),Pt$

(b) $IO_3^-(0.235\ M),I_2(5.00 \times 10^{-4}\ M),H^+(2.00 \times 10^{-3}\ M)|Pt$

(c) $Ag_2CrO_4(sat'd),CrO_4^{2-}(0.0250\ M)|Ag$

22-3 For each of the following half-cells, compare electrode potentials derived from (1) concentration and (2) activity data.

(a) $HCl(0.0100\ M),NaCl(0.0400\ M)|H_2(1.00\ atm),Pt$

(b) $Fe(ClO_4)_2(0.0111\ M), Fe(ClO_4)_3(0.0111\ M)|Pt$

22-4 For each of the following half-cells, compare electrode potentials derived from (1) concentration and (2) activity data.

(a) $Sn(ClO_4)_2(2.00 \times 10^{-5}\ M),Sn(ClO_4)_4(1.00 \times 10^{-5}\ M)|Pt$

(b) $Sn(ClO_4)_2(2.00 \times 10^{-5}\ M),Sn(ClO_4)_4(1.00 \times 10^{-5}\ M),NaClO_4(0.0500\ M)|Pt$

22-5 Calculate the potential of a silver electrode in contact with the following:

(a) a solution that is 0.0200 M in I^- and saturated with AgI.

(b) a solution that is 0.0060 M in CN^- and 0.0400 M in $Ag(CN)_2^-$.

(c) the solution that results from mixing 25.0 mL of 0.0400 M KBr with 20.0 mL of 0.200 M Ag^+.

(d) the solution that results from mixing 25.0 mL of 0.0400 M Ag^+ with 20.0 mL of 0.200 M KBr.

22-6 Calculate the electrode potentials for the following systems:

(a) $Cr_2O_7^{2-}(5.00 \times 10^{-3}\ M),Cr^{3+}(1.00 \times 10^{-2}\ M),H^+(0.200\ M)|Pt$

(b) $UO_2^{2+}(0.100\ M),U^{4+}(0.200\ M),H^+(0.600\ M)|Pt$

22-7 Calculate the theoretical potential of each of the following cells. Is the cell as written galvanic or electrolytic?

(a) $Pt|Cr^{3+}(2.00 \times 10^{-4}\ M),Cr^{2+}(1.00 \times 10^{-3}\ M)||$
$Pb^{2+}(6.50 \times 10^{-2}\ M)|Pb$

(b) $Hg|Hg_2^{2+}(4.00 \times 10^{-2}\ M)||H^+(3.00 \times 10^{-2}\ M),$
$V^{3+}(2.00 \times 10^{-2}\ M), VO^{2+}(6.00 \times 10^{-3}\ M)|Pt$

(c) $Pt|Fe^{3+}(2.00 \times 10^{-2}\ M),Fe^{2+}(6.00 \times 10^{-5}\ M)||$
$Sn^{2+}(3.50 \times 10^{-2}\ M),Sn^{4+}(1.50 \times 10^{-4}\ M)|Pt$

22-8 Calculate the theoretical potential of each of the following cells. Is the cell galvanic or electrolytic as written?

(a) $Bi|BiO^+(0.0400\ M),H^+(0.200\ M)||I^-(0.100\ M),AgI(sat'd)|Ag$

(b) $Zn|Zn^{2+}(7.50 \times 10^{-4}\ M)||Fe(CN)_6^{4-}(4.50 \times 10^{-2}\ M),Fe(CN)_6^{3-}(7.00 \times 10^{-2}\ M)|Pt$

(c) $Pt,H_2(0.100\ atm)|HCl(4.50 \times 10^{-4}\ M), AgCl(sat'd)|Ag$

22-9 Compute E^0 for the process

$$Ni(CN)_4^{2-} + 2e^- \rightleftharpoons Ni(s) + 4CN^-$$

given that the formation constant for the complex is 1.0×10^{22}.

22-10 The solubility product constant for PbI_2 is 7.1×10^{-9} at 25°C. Calculate E^0 for the process

22-11 Calculate the standard potential for the half-reaction

$$BiOCl(s) + 2H^+ + 3e^- \rightleftharpoons Bi(s) + Cl^- + H_20$$

given that K_{sp} for BiOCl has a value of 8.1×10^{-19}.

22-12 Calculate the standard potential for the half-reaction

$$Al(C_2O_4)_2^- + 3e^- \rightarrow Al(s) + 2C_2O_4^{2-}$$

if the formation constant for the complex is 1.3×10^{13}.

22-13 From the standard potentials

$$Tl^+ + e^- \rightleftharpoons Tl(s) \qquad E^0 = -0.336\ V$$

$$TlCl(s) + e^- \rightleftharpoons Tl(s) + Cl^- \qquad E^0 = -0.557\ V$$

calculate the solubility product constant for TlCl.

22-14 From the standard potentials

$$Ag_2SeO_4(s) + 2e^- \rightleftharpoons 2Ag(s) + SeO_4^{2-} \qquad E^0 = 0.355\ V$$

$$Ag^+ + e^- \rightleftharpoons Ag(s) \qquad E^0 = 0.799\ V$$

calculate the solubility product constant for Ag_2SeO_4.

22-15 A current of 0.0750 A is to be drawn from the cell

$$Pt\,|\,V^{3+}(1.0 \times 10^{-5}\ M), V^{2+}(1.65 \times 10^{-1}\ M)\|Br^-(0.150\ M), AgBr(sat'd)\,|\,Ag$$

As a consequence of its design, the cell has an internal resistance of 6.74 Ω. Calculate the potential to be expected initially.

22-16 The cell

$$Pt\,|\,V(OH)_4^+(1.04 \times 10^{-4}\ M), VO^{2+}(7.15 \times 10^{-2}\ M), H^+(2.75 \times 10^{-3}\ M)\|\ Cu^{2+}(5.00 \times 10^{-2}\ M)\,|\,Cu$$

has an internal resistance of 2.24 Ω. What will be the initial potential if 0.0300 A is drawn from this cell?

22-17 The resistance of the galvanic cell

$$Pt\,|\,Fe(CN)_6^{4-}(3.60 \times 10^{-2}\ M), Fe(CN)_6^{3-}(2.70 \times 10^{-3}\ M)\|\ Ag^+(1.65 \times 10^{-2}\ M)\,|\,Ag$$

is 4.10 Ω. Calculate the initial potential when 0.0106 A is drawn from this cell.

Chapter 23

Potentiometry

Potentiometric methods of analysis are based upon measurements of the potential of electrochemical cells in the absence of appreciable currents. Since the beginning of the twentieth century, potentiometric techniques have been used for the location of end points in titrimetric methods of analysis. Of more recent origin are methods in which ion concentrations are obtained directly from the potential of an ion-selective membrane electrode. Such electrodes are relatively free from interference and provide a rapid and convenient means for quantitative estimations of numerous important anions and cations.[1]

The equipment required for potentiometric methods is simple and inexpensive and includes a ***reference electrode****, an* ***indicator electrode****, and a* ***potential measuring device****. The design and properties of each of these components are described in the initial sections of this chapter. Following these discussions, consideration is given to the analytical applications of potentiometric measurements.*

[1] For further reading on potentiometric methods, see E. P. Serjeant, *Potentiometry and Potentiometric Titrations*. New York: Wiley, 1984.

23A REFERENCE ELECTRODES

In most electroanalytical applications, it is desirable that the half-cell potential of one electrode be known, constant, and completely insensitive to the composition of the solution under study. An electrode that fits this description is called a *reference*.[2] Employed in conjunction with the reference electrode is an *indicator* or *working electrode*, whose response depends upon the analyte concentration.

The ideal reference electrode: (1) is reversible and obeys the Nernst equation, (2) exhibits a potential that is constant with time, (3) returns to its original potential after being subjected to small currents, and (4) exhibits little hysteresis with temperature cycling. Although no reference electrode completely meets these ideals, several come surprisingly close.

[2] For a detailed discussion of reference electrodes, see D. J. G. Ives and G. J. Janz, *Reference Electrodes*. New York: Academic Press, 1961. For descriptions of typical commercially available reference electrodes, see R. D. Caton Jr., *J. Chem. Educ.*, **1973**, *50*, A571; **1974**, *51*, A7; *The Beckman Handbook of Applied Electrochemistry*, 2nd ed., Bulletin 7707A. Irvine, CA: Beckman Instruments, 1982.

23A-1 Calomel Electrodes

Calomel reference electrodes consist of mercury in contact with a solution that is saturated with mercury(I) chloride (calomel) and that also contains a known concentration of potassium chloride. Calomel half-cells can be represented as follows:

$$\text{Hg} \mid \text{Hg}_2\text{Cl}_2(\text{sat'd}), \text{KCl}(x\text{M}) \parallel$$

where x represents the molar concentration of potassium chloride in the solution.[3] The electrode potential for this half-cell is determined by the reaction

$$\text{Hg}_2\text{Cl}_2(s) + 2e^- \rightleftharpoons 2\text{Hg}(l) + 2\text{Cl}^-$$

and is dependent upon the chloride concentration x. Thus, this quantity must be specified in describing the electrode.

Table 23-1 lists the composition and the potentials for three commonly encountered calomel electrodes. Note that each solution is saturated with mercury(I) chloride (calomel) and that the cells differ only with respect to the potassium chloride concentration.

The saturated calomel electrode (SCE) is widely used by analytical chemists because of the ease with which it can be prepared.[4] Compared with the other calomel electrodes, however, its temperature coefficient is significantly larger (see Table 23-1). A further disadvantage is that when the temperature is changed, the potential comes to a new value only slowly because of the time required for solubility equilibrium for the potassium chloride and for the calomel to be reestablished. The potential of the saturated calomel electrode at 25°C is 0.2444 V.

Several convenient calomel electrodes are available commercially; typical are the two illustrated in Figure 23-1. The body of each electrode consists of an outer glass or plastic tube that is 5 to 15 cm in length and 0.5 to 1.0 cm in diameter. A mercury/mercury(I) chloride paste in saturated potassium chloride is contained in an inner tube, which is connected to the saturated potassium chloride solution in the outer tube through a small opening.

[3] By convention, a reference electrode is *always treated as an anode,* as in the diagram shown. This practice is consistent with the IUPAC convention for electrode potentials, which is discussed in Section 21C-2 and in which the reference is the standard hydrogen electrode as the anode or the electrode on the left in a cell diagram.

[4] Note that the term "saturated" in the name refers to the concentration of KCl (about 4.6 M) and not to the concentration of Hg_2Cl_2; the latter is saturated in all calomel electrodes.

TABLE 23-1 Potentials of Reference Electrodes in Aqueous Solutions

Temperature, °C	Electrode Potential (V), vs. SHE: 0.1 M[c] Calomel[a]	3.5 M[c] Calomel[b]	Saturated[c] Calomel[a]	3.5 M[b,c] Ag/AgCl	Saturated[b,c] Ag/AgCl
10		0.256		0.215	0.214
12	0.3362		0.2528		
15	0.3362	0.254	0.2511	0.212	0.209
20	0.3359	0.252	0.2479	0.208	0.204
25	0.3356	0.250	0.2444	0.205	0.199
30	0.3351	0.248	0.2411	0.201	0.194
35	0.3344	0.246	0.2376	0.197	0.189
38	0.3338		0.2355		
40		0.244		0.193	0.184

[a] Data from: R. G. Bates in *Treatise on Analytical Chemistry,* 2d ed., I. M. Kolthoff and P. J. Elving, Eds., Part I, Vol. 1, p. 793, Wiley: New York, 1978.
[b] Data from: D. T. Sawyer and J. L. Roberts Jr., *Experimental Electrochemistry for Chemists,* p. 42, Wiley: New York, 1974.
[c] "M" and "saturated" refer to the concentration of KCl and *not* Hg_2Cl_2.

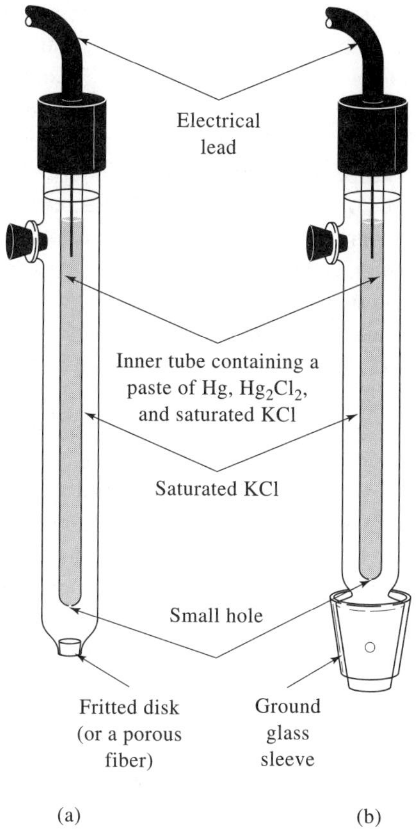

Figure 23-1 Typical commercial calomel reference electrodes.

For electrode (a), contact with the indicator electrode system is made by means of a fritted porcelain plug, a porous fiber, or a piece of porous Vycor ("thirsty glass") sealed in the end of the outer tubing. This type of junction has a relatively high resistance (2000 to 3000 Ω) and a limited current-carrying capacity; on the other hand, contamination of the analyte solution due to leakage of potassium chloride is minimal. The sleeve-type electrode shown in Figure 23-1b has a much lower resistance but tends to leak small amounts of saturated potassium chloride into the sample. Before it is used, the ground-glass collar of this electrode is loosened and turned so that a drop or two of the KCl solution flows from the hole and wets the entire inner ground surface. Better electrical contact to the analyte solution is thus established. The sleeve-type electrode is particularly useful for measurements of nonaqueous solutions and samples in the form of slurries, sludges, viscous solutions, and colloidal suspensions.

23A-2 Silver/Silver Chloride Electrodes

The most widely marketed reference electrode system consists of a silver electrode immersed in a solution of potassium chloride that has been saturated with silver chloride

$$\mathrm{Ag} \mid \mathrm{AgCl(sat'd), KCl(}x\mathrm{M)} \parallel$$

The electrode potential is determined by the half-reaction

$$\mathrm{AgCl}(s) + e^- \rightleftharpoons \mathrm{Ag}(s) + \mathrm{Cl}^-$$

Normally, this electrode is prepared with either a saturated or a 3.5 M potassium chloride solution; potentials for these electrodes are given in Table 23-1. Commercial models of this electrode are similar in external appearance and shape to the two calomel electrodes pictured in Figure 23-1. In silver/silver chloride electrodes, however, the internal tube is replaced by a silver wire that is coated with a layer of silver chloride; this wire is immersed in a potassium chloride solution that is saturated with silver chloride. Similar junctions are used for both types of electrodes.

Silver/silver chloride electrodes have the advantage that they can be used at temperatures greater than 60°C, whereas calomel electrodes cannot. On the other hand, mercury(II) ions react with fewer sample components than do silver ions (which can react with proteins for example); such reactions can lead to plugging of the junction between the electrode and the analyte solution.

23A-3 Precautions in the Use of Reference Electrodes

In using reference electrodes, such as those shown in Figure 23-1, the level of the internal liquid should always be kept above that of the sample solution to prevent contamination of the electrode solution and plugging of the junction due to reaction of the analyte solution with silver or mercury(I) ions from the internal solution. Junction plugging is probably the most common source of erratic cell behavior in potentiometric measurements.[5]

With the liquid level above the analyte solution, some contamination of the sample is inevitable. In most instances, the amount of contamination is so slight to be of no concern. In determining ions such as chloride, potassium, silver, and mercury, however, precaution must often

[5] For a useful discussion about the care and maintenance of reference electrodes, see J. E. Fisher, *Amer. Lab.*, **1984**, (6), 54.

be taken to avoid this source of error. A common way is to interpose a second salt bridge between the analyte and the reference electrode; this bridge should contain a noninterfering electrolyte, such as potassium nitrate or sodium sulfate. Double-junction electrodes based upon this design are offered by several instrument makers.

23B METALLIC INDICATOR ELECTRODES

An ideal indicator electrode responds rapidly and reproducibly to changes in activity of the analyte ion. Although no indicator electrode is absolutely specific in its response, a few are now available that are remarkably selective. There are two types of indicator electrodes: *metallic* and *membrane*. This section deals with metallic indicator electrodes.

Four types of metallic indicator electrodes can be recognized: *electrodes of the first kind, electrodes of the second kind, electrodes of the third kind,* and *redox electrodes.*

23B-1 Electrodes of the First Kind

Metallic *electrodes of the first kind* are in direct equilibrium with the cation derived from the electrode metal. Here, a single reaction is involved. For example, for a copper indicator electrode, we may write

$$Cu^{2+} + 2e^- \rightleftharpoons Cu(s)$$

The potential E_{ind} of this electrode is given by

$$\begin{aligned} E_{ind} &= E^0_{Cu} - \frac{0.0592}{2}\log\frac{1}{a_{Cu^{2+}}} \\ &= E^0_{Cu} - \frac{0.0592}{2}\,pCu \qquad (23\text{-}1) \end{aligned}$$

where pCu is the negative logarithm of the copper(II) ion activity $a_{Cu^{2+}}$. Thus, the copper electrode provides a direct measure of the pCu of the solution.[6]

[6] The results of potentiometric measurements are usually expressed in terms of a parameter, the *p-function,* which is directly proportional to the measured potential. The p-function then provides a measure of activity in terms of a convenient, small, and ordinarily positive, number. Thus, for a solution with a calcium ion activity of 2.00×10^{-6} M, we may write pCa = $-\log(2.00 \times 10^{-6}) = 5.699$. Note that as the activity of calcium increases, its p-function decreases. Note also that because the activity was given to three significant figures, we are entitled to keep this number of figures *to the right of the decimal point* in the computed pCa because these are the only numbers that carry information about the original 2.00. The 5 in the value for pCa provides information about the position of the decimal point in the original number only.

Electrode systems of the first kind are not widely used for potentiometric analyses for several reasons. For one, they are not very selective and respond not only to their own cations but also to other more easily reduced cations. For example, a copper electrode cannot be used for the determination of copper(II) ions in the presence of silver(I) ions, which are also reduced at the copper surface. In addition, many metal electrodes, such as zinc and cadmium, can only be used in neutral or basic solutions because they dissolve in the presence of acids. Third, some metals are so easily oxidized that their use is restricted to solutions that have been deaerated. Finally, certain harder metals—such as iron, chromium, cobalt, and nickel—do not provide reproducible potentials. Moreover, for these electrodes, plots of pX versus activity yield slopes that differ significantly and irregularly from the theoretical ($-0.0592/n$). For these reasons, the only electrode systems of the first kind that have been used are Ag/Ag^+ and Hg/Hg_2^{2+} in neutral solutions, and Cu/Cu^{2+}, Zn/Zn^{2+}, Cd/Cd^{2+}, Bi/Bi^{3+}, Tl/Tl^+, and Pb/Pb^{2+} in deaerated solutions.

23B-2 Electrodes of the Second Kind

A metal electrode can often be made responsive to the activity of an anion with which its ion forms a precipitate or a stable complex ion. For example, silver can serve as an *electrode of the second kind* for halide and halide-like anions. To prepare an electrode to determine chloride ion it is only necessary to saturate the layer of the analyte solution adjacent to a silver electrode with silver chloride. The electrode reaction can then be written as

$$AgCl(s) + e^- \rightleftharpoons Ag(s) + Cl^- \qquad E^0 = 0.222\ V$$

Application of the Nernst equation gives

$$\begin{aligned} E_{ind} &= 0.222 - 0.0592\log a_{Cl^-} \\ &= 0.222 + 0.0592\,pCl \qquad (23\text{-}2) \end{aligned}$$

A convenient way of preparing a chloride-sensitive electrode is to make a pure silver wire the anode in an electrolytic cell containing potassium chloride. The wire becomes coated with an adherent silver halide deposit, which will rapidly equilibrate with the surface layer of a solution in which it is immersed. Because the solubility of the silver chloride is low, an electrode formed in this way can be used for numerous measurements.

An important electrode of the second kind for measuring the activity of EDTA anion Y^{4-} is based upon the response of a mercury electrode in the presence of a small concentration of the stable EDTA complex of Hg(II).[7] The half-reaction for the electrode process can be written as

$$HgY^{2-} + 2e^- \rightleftharpoons Hg(l) + Y^{4-} \qquad E^0 = 0.21 \text{ V}$$

for which

$$E_{ind} = 0.21 - \frac{0.0592}{2} \log \frac{a_{Y^{4-}}}{a_{HgY^{2-}}}$$

To employ this electrode system, it is necessary to introduce a small concentration of HgY^{2-} into the analyte solution at the outset. The complex is so stable (for HgY^{2-}, $K_f = 6.3 \times 10^{21}$) that its activity remains essentially constant over a wide range of Y^{4-} activities. Therefore, the potential equation can be written in the form

$$E_{ind} = K - \frac{0.0592}{2} \log a_{Y^{4-}}$$

$$= K + \frac{0.0592}{2} \text{pY} \qquad (23\text{-}3)$$

where the constant K is equal to

$$K = 0.21 - \frac{0.0592}{2} \log \frac{1}{a_{HgY^{2-}}}$$

This electrode is useful for establishing end points for EDTA titrations.

23B-3 Electrodes of the Third Kind

A metal electrode can, under some circumstances, be made to respond to a different cation. It then becomes an *electrode of the third kind.* As an example, a mercury electrode has been used for the determination of the pCa of calcium-containing solutions. As in the previous example, a small concentration of the EDTA complex of Hg(II) is introduced into the solution. As before (Equation 23-3) the potential of a mercury electrode in this solution is given by

$$E_{ind} = K - \frac{0.0592}{2} \log a_{Y^{4-}}$$

If, in addition, a small volume of a solution containing the EDTA complex of calcium is introduced, a new equilibrium is established, namely,

$$CaY^{2-} \rightleftharpoons Ca^{2+} + Y^{4-} \qquad K_f = \frac{a_{Ca^{2+}} \cdot a_{Y^{4-}}}{a_{CaY^{2-}}}$$

Combining the formation-constant expression for CaY^{2-} with the potential expression yields

$$E_{ind} = K - \frac{0.0592}{2} \log \frac{K_f a_{CaY^{2-}}}{a_{Ca^{2+}}}$$

which can be written as

$$E_{ind} = K - \frac{0.0592}{2} \log K_f a_{CaY^{2-}} - \frac{0.0592}{2} \log \frac{1}{a_{Ca^{2+}}}$$

If a constant amount of CaY^{2-} is used in the analyte solution and in the solutions for standardization, we may write

$$E_{ind} = K' - \frac{0.0592}{2} \text{pCa} \qquad (23\text{-}4)$$

where

$$K' = K - \frac{0.0592}{2} \log K_f a_{CaY^{2-}}$$

Thus, the mercury electrode has become an electrode of the third kind for calcium ion.

23B-4 Metallic Redox Indicators

Electrodes fashioned from platinum, gold, palladium, or other inert metals often serve as indicator electrodes for oxidation/reduction systems. In these applications, the inert electrode acts as a source or sink for electrons transferred from a redox system in the solution. For example, the potential of a platinum electrode in a solution containing Ce(III) and Ce(IV) ions is given by

$$E_{ind} = E^0 - 0.0592 \log \frac{a_{Ce^{3+}}}{a_{Ce^{4+}}}$$

Thus, a platinum electrode can serve as the indicator electrode in a titration in which Ce(IV) serves as the standard reagent.

---

[7] Ethylenediaminetetraacetic acid, commonly called EDTA and formulated H_4Y, is a widely used complexing agent that is used as a titrant for the determination of many cations. The compound has the structure

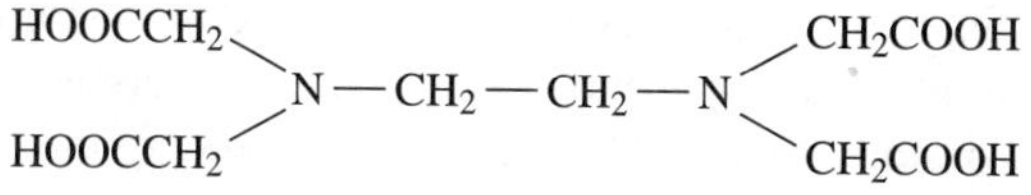

The four carboxylate groups, as well as the two amine groups form bonds with all cations. Regardless of the charge on the cation, the complex formation reaction can be formulated as

$$M^{n+} + H_4Y \rightleftharpoons MY^{(n-4)} + 4H^+$$

It should be noted, however, that electron-transfer processes at inert electrodes are frequently not reversible. As a consequence, inert electrodes do not respond in a predictable way to many of the half-reactions found in a table of electrode potentials. For example, a platinum electrode immersed in a solution of thiosulfate and tetrathionate ions does not develop reproducible potentials because the electron-transfer process

$$S_4O_6^{2-} + 2e^- \rightleftharpoons 2S_2O_3^{2-}$$

is slow and, therefore, not reversible at the electrode surface.

23C MEMBRANE INDICATOR ELECTRODES[8]

A wide variety of membrane electrodes are available from commercial sources that permit the rapid and selective determination of numerous cations and anions by direct potentiometric measurements. Often, membrane electrodes are called *ion-selective electrodes* because of the high selectivity of most of these devices. They are also referred to as *pIon electrodes* because their output is usually recorded as a p-function, such as pH, pCa, or pNO_3 (see footnote 6).

23C-1 Classification of Membranes

Table 23-2 lists the various types of ion-selective membrane electrodes that have been developed. These differ in the physical or chemical composition of the membrane. The general mechanism by which an ion-selective potential develops in these devices depends on the nature of the membrane and is entirely different from the source of potential in metallic indicator electrodes. We have seen that the potential of a metallic electrode arises from the tendency of an oxidation/reduction reaction to occur at an electrode surface. In membrane electrodes, in contrast, the observed potential is a kind of junction potential that develops across a membrane that separates the analyte solution from a reference solution.

TABLE 23-2 Types of Ion-Selective Membrane Electrodes

A. Crystalline Membrane Electrodes
   1. Single crystal
      Example: LaF_3 for F^-
   2. Polycrystalline or mixed crystal
      Example: Ag_2S for S^{2-} and Ag^+

B. Noncrystalline Membrane Electrodes
   1. Glass
      Examples: silicate glasses for Na^+ and H^+
   2. Liquid
      Examples: liquid ion exchangers for Ca^{2+} and neutral carriers for K^+
   3. Immobilized liquid in a rigid polymer
      Examples: polyvinyl chloride matrix for Ca^{2+} and NO_3^-

23C-2 Properties of Ion-Selective Membranes

All of the ion-selective membranes in the electrodes shown in Table 23-2 share common properties, which lead to the sensitivity and selectivity of membrane electrodes toward certain cations or anions. These properties include:

1. **Minimal solubility.** A necessary property of an ion-selective medium is that its solubility in analyte solutions (usually aqueous) approaches zero. Thus, many membranes are formed from large molecules or molecular aggregates such as silica glasses or polymeric resins. Ionic inorganic compounds of low solubility, such as the silver halides, can also be converted into membranes.
2. **Electrical conductivity.** A membrane must exhibit some electrical conductivity, albeit small. Generally, this conduction takes the form of migration of singly charged ions within the membrane.
3. **Selective reactivity with the analyte.** A membrane or some species contained within the membrane matrix must be capable of selectively binding the analyte ion. Three types of binding are encountered: ion-exchange, crystallization, and complexation. The former two are the more common, and the at-

[8] Some suggested sources for additional information on this topic are: A. Evans, *Potentiometry and Ion-Selective Electrodes*. New York: Wiley, 1987; J. Koryta and K. Stulik, *Ion Selective Electrodes*, 2nd ed. Cambridge: Cambridge University Press, 1983; *Ion-Selective Methodology*, A. K. Covington, Ed. Boca Raton, FL: CRC Press, 1979; R. P. Buck, in *Comprehensive Treatise of Electrochemistry*, J. D. M. Bockris, B. C. Conway, H. E. Yeager, Eds., Vol. 8, Chapter 3. New York: Plenum Press, 1984.

tention here will be largely focused on these types of bindings.

23C-3 The Glass Electrode for pH Measurements

We begin our discussion of how pIon membrane electrodes are constructed and how they function by examining in some detail the glass electrode for pH measurements. The choice of this type of membrane to begin our discussion is, perhaps, appropriate because the glass pH electrode predates all other membrane electrodes by several decades.

Since the early 1930s, the most convenient way for determining pH has been by measuring the potential difference across a glass membrane separating the analyte solution from a reference solution of fixed acidity. The phenomenon upon which this measurement is based was first recognized by Cremer[9] in 1906 and systematically explored by Haber[10] a few years later. Figure 23-2 pictures the glass-electrode system used in Haber's early studies. General use of the glass electrode for pH measurements was delayed for two decades, however, until the invention of the vacuum tube permitted the convenient measurement of potentials across glass membranes having resistances of 100 MΩ or more. Systematic studies of the pH sensitivity of glass membranes led ultimately in the late 1960s to the development and marketing of membrane electrodes for two dozen or more ions such as K^+, Na^+, Ca^{2+}, F^-, and NO_3^-.

[9] M. Cremer, *Z. Biol.*, **1906**, *47*, 562.
[10] F. Haber and Z. Klemensiewicz, *Z. Phys. Chem.*, **1909**, *67*, 385.

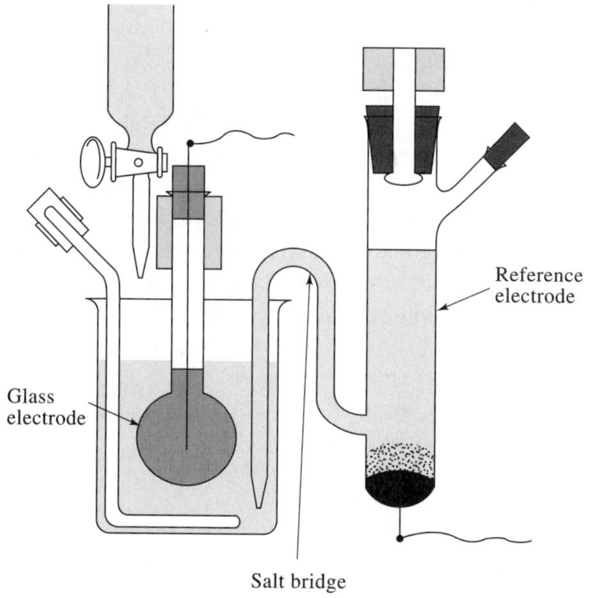

Figure 23-2 An early glass/calomel electrode system for measuring pH. *(From Haber and Klemensiewicz,* Z. Phys. Chem., ***1909,*** *67, 385.)*

Figure 23-3 shows a typical modern *cell* for measuring pH. The cell consists of a glass indicator electrode and a silver/silver chloride or a saturated calomel reference electrode immersed in the solution whose pH is to be determined. The indicator electrode consists of a thin, pH-sensitive glass membrane sealed onto one end of a heavy-walled glass or plastic tube. A small volume of dilute hydrochloric acid saturated with silver chloride is contained in the tube (the inner solution in some electrodes is a buffer containing chloride ion). A silver wire in this solution forms a silver/silver chloride reference electrode, which is connected to one of the terminals of a potential-measuring device. The reference electrode is connected to the other terminal.

Figure 23-4, which is a schematic representation of the cell in Figure 23-3, shows that this cell contains *two* reference electrodes: (1) the *external* silver/silver chloride electrode (ref 1) and (2) the *internal* silver/silver chloride electrode (ref 2). While the internal reference electrode is a part of the glass electrode, *it is not the pH-sensing element.* Instead, *it is the thin glass membrane at the tip of the electrode that responds to pH.*

The Composition and Structure of Glass Membranes

Much systematic investigation has been devoted to the effects of glass composition on the sensitivity of membranes to protons and other cations, and a number of formulations are now used for the manufacture of electrodes. Corning 015 glass, which has been widely used for membranes, consists of approximately 22% Na_2O, 6% CaO, and 72% SiO_2. This membrane is specific in its response toward hydrogen ions up to a pH of about 9. At higher pH values, however, the glass becomes somewhat responsive to sodium, as well as to other singly charged cations. Other glass formulations are now in use in which sodium and calcium ions are replaced to various degrees by barium and lithium ions. These membranes have superior selectivity at high pH.

Figure 23-5, is a two-dimensional view of the structure of a silicate glass membrane. Each silicon

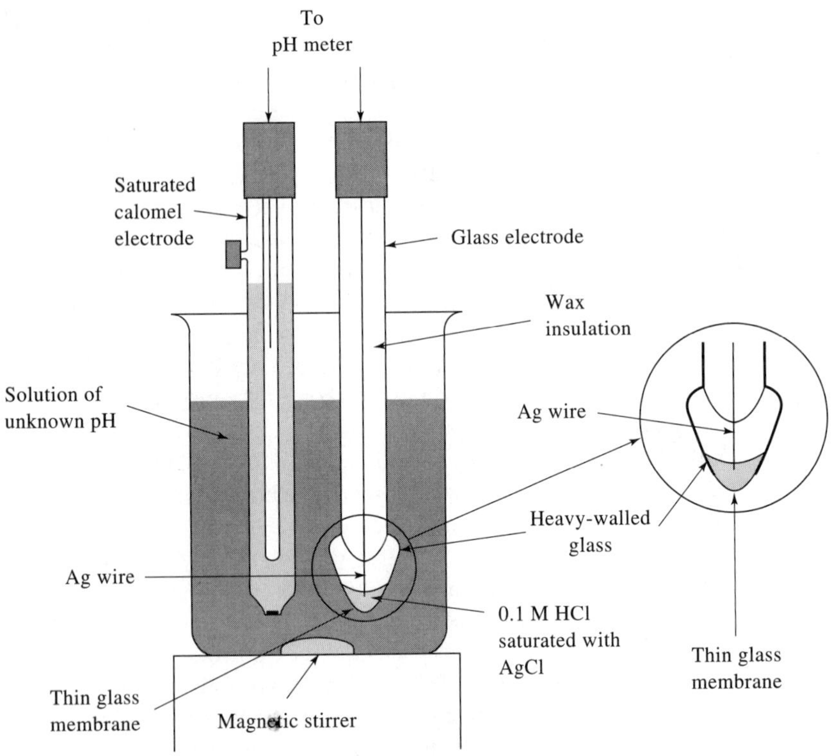

Figure 23-3 Typical electrode system for measuring pH.

atom is shown as being bonded to three oxygen atoms in the plane of the paper. In addition, each is bonded to another oxygen above or below the plane. Thus, the glass consists of an infinite three-dimensional network of SiO_4^{4-} groups in which each silicon is bonded to four oxygens and each oxygen is shared by two silicons. Within the interstices of this structure are sufficient cations to balance the negative charge of the silicate groups. Singly charged cations, such as sodium and lithium, are mobile in the lattice and are responsible for electrical conduction within the membrane.

The Hygroscopicity of Glass Membranes

The surface of a glass membrane must be hydrated before it will function as a pH electrode. The amount of water involved is approximately 50 mg per cubic centimeter of glass. Nonhygroscopic glasses show no pH function. Even hygroscopic glasses lose their pH sensitivity after dehydration by storage over a desiccant. The effect is reversible, however, and the response of a glass electrode is restored by soaking it in water.

The hydration of a pH-sensitive glass membrane involves an ion-exchange reaction between singly

Reference electrode 1 — Glass electrode

External analyte soln — Internal analyte soln

$$\mathrm{Ag}\,|\,\mathrm{AgCl\ (sat'd)},\ [\mathrm{Cl^-}] = 1.0\ \mathrm{M}\ ||\ [\mathrm{H_3O^+}] = a_1\ |\ \text{Glass membrane}\ |\ [\mathrm{H_3O^+}] = a_2,\ [\mathrm{Cl^-}] = 1.0\ \mathrm{M},\ \mathrm{AgCl\ (sat'd)}\ |\ \mathrm{Ag}$$

$E_{ref\,1}\ E_j$ — E_1 E_2: $E_b = E_1 - E_2$ — Reference electrode 2, $E_{ref\,2}$

Figure 23-4 Diagram of a cell for measurement of pH.

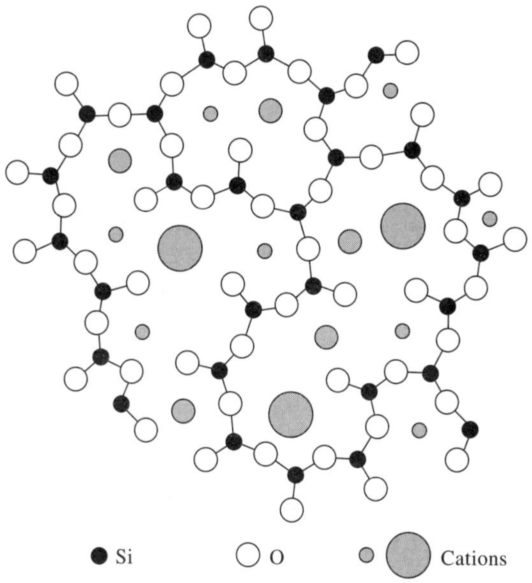

Figure 23-5 Cross-sectional view of a silicate glass structure. In addition to the three Si—O bonds shown, each silicon is bonded to an additional oxygen atom, either above or below the plane of the paper. *(Adapted with permission from G. A. Perley,* Anal. Chem., *1949, 21, 395. Copyright 1949 American Chemical Society.)*

charged cations in the glass lattice and protons from the solution. The process involves univalent cations exclusively because di- and trivalent cations are too strongly held within the silicate structure to exchange with ions in the solution. In general, then, the ion-exchange reaction can be written as

$$\underset{\text{soln}}{H^+} + \underset{\text{glass}}{Na^+Gl^-} \rightleftharpoons \underset{\text{soln}}{Na^+} + \underset{\text{glass}}{H^+Gl^-} \quad (23\text{-}5)$$

The equilibrium constant for this process is so large that the surface of a hydrated glass membrane ordinarily consists entirely of silicic acid (H^+Gl^-) groups. An exception to this situation exists in highly alkaline media, where the hydrogen ion concentration is extremely small and the sodium ion concentration is large; here, a significant fraction of the sites are occupied by sodium ions.

Electrical Conduction across Glass Membranes

To serve as an indicator for cations, a glass membrane must conduct electricity. Conduction within the hydrated gel layer involves the movement of hydrogen ions. Sodium ions are the charge carriers in the dry interior of the membrane. Conduction across the solution/gel interfaces occurs by the reactions

$$\underset{\text{soln}_1}{H^+} + \underset{\text{glass}_1}{Gl^-} \rightleftharpoons \underset{\text{glass}_1}{H^+Gl^-} \quad (23\text{-}6)$$

$$\underset{\text{glass}_2}{H^+Gl^-} \rightleftharpoons \underset{\text{soln}_2}{H^+} + \underset{\text{glass}_2}{Gl^-} \quad (23\text{-}7)$$

where subscript 1 refers to the interface between the glass and the analyte solution and subscript 2 refers to the interface between the internal solution and the glass. The positions of these two equilibria are determined by the hydrogen ion activities in the solutions on the two sides of the membrane. The surface at which the greater dissociation occurs becomes negative with respect to the other surface where less dissociation has taken place. A boundary potential E_b thus develops across the membrane. The magnitude of the boundary potential depends upon the ratio of the hydrogen ion activities of the two solutions. It is this potential difference that serves as the analytical parameter in potentiometric pH measurements with a membrane electrode.

Membrane Potentials

The lower part of Figure 23-4 shows four potentials that develop in a cell when pH is being determined with a glass electrode. Two of these, E_{ref1} and E_{ref2}, are reference electrode potentials. The third potential is the junction potential E_j across the salt bridge that separates the calomel electrode from the analyte solution. Junction potentials are found in all cells used for the potentiometric measurement of ion concentration. The fourth, and most important, potential shown in Figure 23-4 is the *boundary potential, E_b, which varies with the pH of the analyte solution.* The two reference electrodes simply provide electrical contacts with the solutions so that changes in the boundary potential can be measured.

Figure 23-4 reveals that the potential of a glass electrode has two components: the fixed potential of a silver/silver chloride electrode E_{ref2} and the pH-dependent boundary potential E_b. Not shown is a third potential, called the *asymmetry potential,* which is found in most membrane electrodes and which changes slowly with time. The source of the asymmetry potential is obscure.

The Boundary Potential

As shown in Figure 23-4, the boundary potential consists of two potentials, E_1 and E_2, each of which is associated with one of the two gel/solution interfaces. The boundary potential is simply the difference between these potentials:

$$E_b = E_1 - E_2 \quad (23\text{-}8)$$

It can be demonstrated from thermodynamic considerations[11] that E_1 and E_2 in Equation 23-8 are related to the hydrogen ion activities at each face by Nernst-like relationships:

$$E_1 = j_1 - \frac{0.0592}{n} \log \frac{a_1'}{a_1} \qquad (23\text{-}9)$$

$$E_2 = j_2 - \frac{0.0592}{n} \log \frac{a_2'}{a_2} \qquad (23\text{-}10)$$

where j_1 and j_2 are constants and a_1 and a_2 are activities of H^+ in the *solutions* on the external and internal sides of the membrane respectively. The terms a_1' and a_2' are the activities of H^+ at the external and internal *surfaces* of the glass making up the membrane.

If the two membrane surfaces have the same number of negatively charged sites (as they normally do) from which H^+ can dissociate, then j_1 and j_2 are identical; so also are a_1' and a_2'. By substituting Equations 23-9 and 23-10 into Equation 23-8 and employing the equalities $j_1 = j_2$ and $a_1' = a_2'$, we obtain upon rearrangement

$$E_b = E_1 - E_2 = 0.0592 \log \frac{a_1}{a_2} \qquad (23\text{-}11)$$

Thus, the boundary potential E_b depends only upon the hydrogen ion activities of the solutions on either side of the membrane. For a glass pH electrode, the hydrogen ion activity of the internal solution a_2 is held constant so that Equation 23-11 simplifies to

$$E_b = L' + 0.0592 \log a_1 = L' - 0.0592 \text{ pH} \qquad (23\text{-}12)$$

where

$$L' = -0.0592 \log a_2$$

The boundary potential is then a measure of the hydrogen ion activity of the external solution.

The significance of the potentials and the differences shown in Equation 23-11 is illustrated by the potential profiles shown in Figure 23-6. The profiles are plotted across the membrane from the analyte solution on the left through the membrane to the internal solution on the right.

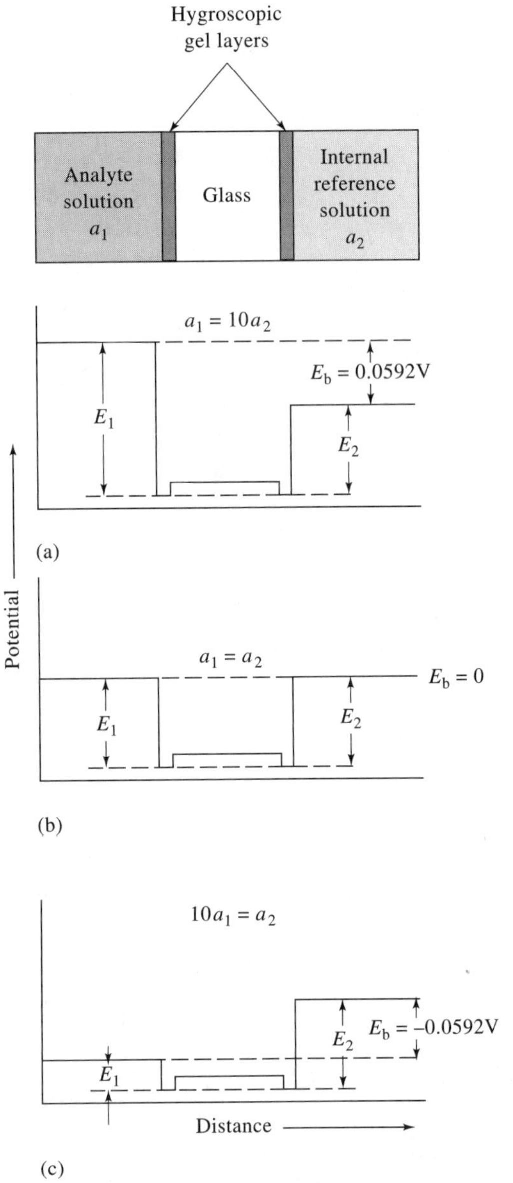

Figure 23-6 Potential profile across a glass membrane from the analyte solution to the internal reference solution. The reference electrode potentials are not shown.

The Potential of the Glass Electrode

As noted earlier, the potential of a glass indicator electrode E_{ind} has three components: (1) the boundary potential, given by Equation 23-12, (2) the potential of the internal Ag/AgCl reference electrode E_{ref2}, and (3) a small asymmetry potential, E_{asy}. In equation form,

$$E_{ind} = E_b + E_{ref2} + E_{asy}$$

Substitution of Equation 23-12 for E_b gives

$$E_{ind} = L' + 0.0592 \log a_1 + E_{ref2} + E_{asy}$$

[11] G. Eisenman, *Biophys. J.*, **1962**, *2* (part 2), 259.

or

$$E_{ind} = L + 0.0592 \log a_1 = L - 0.0592 \text{ pH} \quad (23\text{-}13)$$

where L is a combination of the three constant terms. That is,

$$L = L' + E_{ref2} + E_{asy} \quad (23\text{-}14)$$

Note the similarity between Equation 23-13 and Equations 23-1 and 23-4 for metallic cation indicator electrodes. It is important to emphasize that although the latter two equations are similar in form to Equation 23-13, the sources of the potential of the electrodes that they describe *are totally different*—one is a redox potential, whereas the other is a boundary potential at which no oxidation/reduction reaction occurs.

The Alkaline Error

Glass electrodes respond to the concentration of both hydrogen ion and alkali metal ions in basic solution. The magnitude of this *alkaline error* for four different glass membranes is shown in Figure 23-7 (curves *C* to *F*). These curves refer to solutions in which the sodium ion concentration was held constant at 1 M while the pH was varied. Note that the error is negative (that is, pH values that were measured were lower than the true values), which suggests that the electrode is responding to sodium ions as well as to protons. This observation is confirmed by data obtained for solutions containing different sodium ion concentrations. Thus at pH 12, the electrode with a Corning 015 membrane (curve *C* in Figure 23-7) registered a pH of 11.3 when immersed in a solution having a sodium ion concentration of 1 M but registered 11.7 in a solution that was 0.1 M in this ion. All singly charged cations induce an alkaline error whose magnitude depends upon both the cation in question and the composition of the glass membrane.

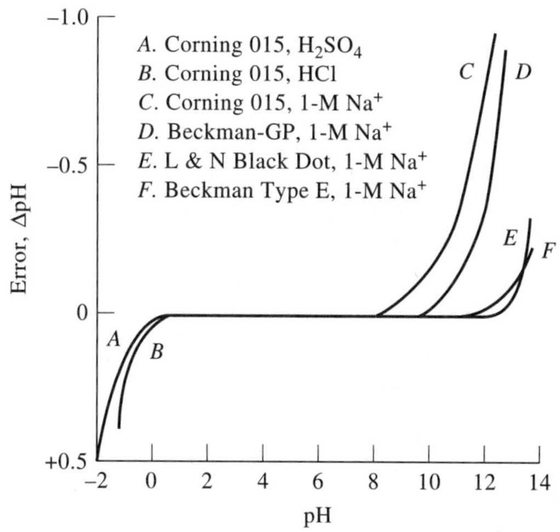

Figure 23-7 Acid and alkaline error of selected glass electrodes at 25°C. *(From R. G. Bates,* Determination of pH: Theory and Practice, *p. 316. New York: Wiley, 1964. Reprinted by permission of John Wiley & Sons, Inc.)*

The alkaline error can be satisfactorily explained by assuming an exchange equilibrium between the hydrogen ions on the glass surface and the cations in solution. This process is simply the reverse of that shown in Equation 23-5, or

$$\underset{\text{glass}}{H^+Gl^-} + \underset{\text{soln}}{B^+} \rightleftharpoons \underset{\text{glass}}{B^+Gl^-} + \underset{\text{soln}}{H^+}$$

where B^+ represents some singly charged cation, such as sodium ion. In this case, the activity of the sodium ions relative to that of the hydrogen ions becomes so large that the electrode responds to both species.

Selectivity Coefficients

The effect of an alkali metal ion on the potential across a membrane can be accounted for by inserting an additional term in Equation 23-12 to give

$$E_b = L' + 0.0592 \log (a_1 + k_{H,B}b_1) \quad (23\text{-}15)$$

where $k_{H,B}$ is the *selectivity coefficient* for the electrode and b_1 is the activity of the alkali metal ion. Equation 23-15 applies not only to glass indicator electrodes for hydrogen ion but also to all other types of membrane electrodes. Selectivity coefficients range from zero (no interference) to values greater than unity. A selectivity coefficient of unity means the electrode responds equally to the analyte ion and the interfering ion. If an electrode for ion A responds 20 times more strongly to ion B than to ion A, then $k_{A,B}$ has a value of 20. If the response of the electrode to ion C is 0.001 of its response to A (a much more desirable situation), $k_{A,C}$ is 0.001.[12]

The product $k_{H,B}b_1$ for a glass pH electrode is ordinarily small relative to a_1 provided the pH is less than 9; under these conditions, Equation 23-15 simplifies to Equation 23-13. At high pH values and at high concen-

[12] For a collection of selectivity coefficients for all types of ion-selective electrodes, see Y. Umezawa, *CRC Handbook of Ion Selective Electrodes: Selectivity of Coefficients.* Boca Raton, FL: CRC Press, 1990.

trations of a singly charged ion, however, the second term in Equation 23-15 assumes a more important role in determining E_b, and an alkaline error is encountered. For electrodes specifically designed for work in highly alkaline media (curve E in Figure 23-7), the magnitude of $k_{H,B}b_1$ is appreciably smaller than for ordinary glass electrodes.

The Acid Error

As shown in Figure 23-7, the typical glass electrode exhibits an error, opposite in sign to the alkaline error, in solutions of pH less than about 0.5; pH readings tend to be too high in this region. The magnitude of the error depends upon a variety of factors and is generally not very reproducible. The causes of the acid error are not well understood.

23C-4 Glass Electrodes for Other Cations

The alkaline error in early glass electrodes led to investigations concerning the effect of glass composition upon the magnitude of this error. One consequence has been the development of glasses for which the alkaline error is negligible below about pH 12. Other studies have discovered glass compositions that permit the determination of cations other than hydrogen. This application requires that the hydrogen ion activity a_1 in Equation 23-15 be negligible relative to $k_{H,B}b_1$; under these circumstances, the potential is independent of pH and is a function of pB instead. Incorporation of Al_2O_3 or B_2O_3 in the glass has the desired effect. Glass electrodes that permit the direct potentiometric measurement of such singly charged species as Na^+, K^+, NH_4^+, Rb^+, Cs^+, Li^+, and Ag^+ have been developed. Some of these glasses are reasonably selective toward particular singly charged cations. Glass electrodes for Na^+, Li^+, NH_4^+, and total concentration of univalent cations are now available from commercial sources.

23C-5 Crystalline Membrane Electrodes

The most important type of crystalline membranes is manufactured from an ionic compound or a homogeneous mixture of ionic compounds. In some instances the membrane is cut from a single crystal; in others, disks are formed from the finely ground crystalline solid by high pressures or by casting from a melt. The typical membrane has a diameter of about 10 mm and a thickness of 1 or 2 mm. To form an electrode, the membrane is sealed to the end of a tube made from a chemically inert plastic such as Teflon or polyvinyl chloride.

Conductivity of Crystalline Membranes

Most ionic crystals are insulators and do not have sufficient electrical conductivity at room temperature to serve as membrane electrodes. Those that are conductive are characterized by having a small singly charged ion that is mobile in the solid phase. Examples are fluoride ion in certain rare earth fluorides, silver ion in silver halides and sulfides, and copper(I) ion in copper(I) sulfide.

The Fluoride Electrode

Lanthanum fluoride, LaF_3, is a nearly ideal substance for the preparation of a crystalline membrane electrode for the determination of fluoride ion. Although this compound is a natural conductor, its conductivity can be enhanced by doping with europium fluoride, EuF_2. Membranes are prepared by cutting disks from a single crystal of the doped compound.

The mechanism of the development of a fluoride-sensitive potential across a lanthanum fluoride membrane is quite analogous to that described for glass, pH-sensitive membranes. That is, at the two interfaces, ionization creates a charge on the membrane surface as shown by the equation

$$\underset{\text{solid}}{LaF_3} \rightleftharpoons \underset{\text{solid}}{LaF_2^+} + \underset{\text{soln}}{F^-}$$

The magnitude of the charge is dependent upon the fluoride ion concentration of the solution. Thus, the side of the membrane encountering the lower fluoride ion concentration becomes positive with respect to the other surface; it is this charge difference that provides a measure of the difference in fluoride concentration of the two solutions. The potential of a cell containing a lanthanum fluoride electrode is given by an equation analogous to Equation 23-13. That is,

$$E_{ind} = L - 0.0592 \log a_{F^-} = L + 0.0592 \text{ pF} \quad (23\text{-}16)$$

Note that the signs of the second terms on the right are reversed because an anion is being determined (see also Equation 23-2).

Commercial lanthanum fluoride electrodes come in various shapes and sizes and are available from several sources. Most are rugged and can be used at temperatures between 0° and 80°C. The response of the fluoride electrode is linear down to 10^{-6} M (0.02 ppm), where the solubility of lanthanum fluoride begins to contribute to the concentration of fluoride ion in the analyte solution. The only ion that interferes directly with fluoride measurements is hydroxide ion; this interference becomes serious at a pH greater than eight. At a pH less than five, hydrogen ions also interfere in *total* fluoride determinations; here, undissociated hydrogen fluoride forms to which the electrode is not responsive. In most respects, the fluoride ion electrode approaches the ideal for selective electrodes.

Electrodes Based on Silver Salts

Membranes prepared from single crystals or pressed disks of the various silver halides act selectively toward silver and halide ions. Generally, their behavior is far from ideal, however, owing to low conductivity, low mechanical strength, and a tendency to develop high photoelectric potentials. It has been found, however, that these disadvantages are minimized if the silver salts are mixed with crystalline silver sulfide in an approximately 1:1 molar ratio. Homogeneous mixtures are formed from equimolar solutions of sulfide and halide ions by precipitation with silver nitrate. After washing and drying, the product is shaped into disks under a pressure of about 10^5 pounds per square inch. The resulting disk exhibits good electrical conductivity owing to the mobility of the silver ion in the sulfide matrix.

Membranes constructed either from silver sulfide or from a mixture of silver sulfide and another silver salt are useful for the determination of both sulfide and silver ions. Toward silver ions, the electrical response is similar to a metal electrode of the first kind (although the mechanism of activity is totally different). Toward sulfide ions, the electrical response of a silver sulfide membrane is similar to that of an electrode of the second kind (Section 23B-2). Here, upon immersion in the solution to be analyzed, dissolution of a tiny amount of silver sulfide quickly saturates the film of liquid adjacent to the electrode. The solubility, and thus the silver ion concentration, depends, however, upon the sulfide concentration of the analyte.

Crystalline membranes are also available that consist of a homogeneous mixture of silver sulfide with sulfides of copper(II), lead, or cadmium. Toward these divalent cations, electrodes from these materials have electrical responses similar to electrodes of the third kind (Section 23B-3). It should be noted that these divalent sulfides, by themselves, are not conductors and thus do not exhibit selective-ion activity.

Table 23-3 is a representative list of solid-state electrodes that are available from commercial sources.

23C-6 Liquid Membrane Electrodes

Liquid membranes are formed from immiscible liquids that selectively bond certain ions. Membranes of this type are particularly important because they permit the direct potentiometric determination of the activities of several polyvalent cations and of certain singly charged anions and cations as well.

Early liquid membranes were prepared from immiscible, liquid ion exchangers, which were retained in a porous inert solid support. As shown schematically in Figure 23-8, a porous, hydrophobic (that is, water-repelling) plastic disk (typical dimensions: 3 × 0.15 mm) served to hold the organic layer between the two aqueous solutions. Wick action caused the pores of the disk to stay filled with the organic liquid from the reservoir in the outer of the two concentric tubes. For divalent cation determinations, the inner tube contained an aqueous standard solution of MCl_2, where M^{2+} was the

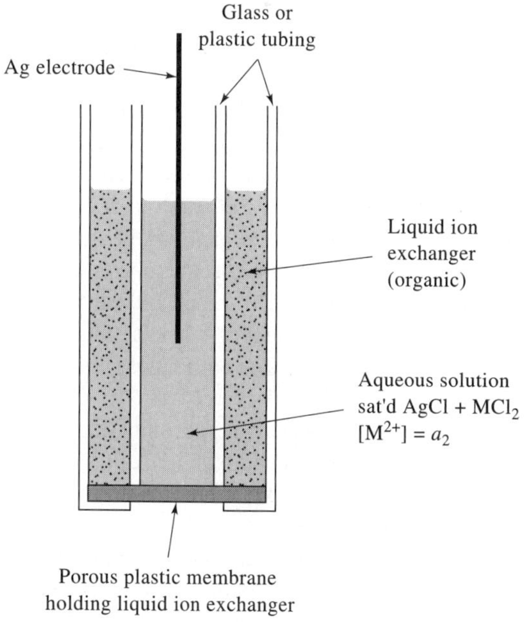

Figure 23-8 Liquid membrane electrode sensitive to M^{2+}.

TABLE 23-3 Commercial Solid-State Electrodes[a]

Analyte Ion	Concentration Range, M	Interferences[b]
Br^-	10^0 to 5×10^{-6}	mr: 8×10^{-5} CN^-; 2×10^{-4} I^-; 2 NH_3; 400 Cl^-; 3×10^4 OH^-. mba: S^{2-}
Cd^{2+}	10^{-1} to 10^{-7}	Fe^{2+} + Pb^{2+} may interfere. mba: Hg^{2+}, Ag^+, Cu^{2+}
Cl^-	10^0 to 5×10^{-5}	mr: 2×10^{-7} CN^-; 5×10^{-7} I^-; 3×10^{-3} Br^-; 10^{-2} $S_2O_3^{2-}$; 0.12 NH_3; 80 OH^-. mba: S^{2-}
Cu^{2+}	10^{-1} to 10^{-8}	high levels Fe^{2+}, Cd^{2+}, Br^-, Cl^-. mba: Hg^{2+}, Ag^+, Cu^+
CN^-	10^{-2} to 10^{-6}	mr: 10^{-1} I^-; 5×10^3 Br^-; 10^6 Cl^-. mba: S^{2-}
F^-	sat'd to 10^{-6}	0.1 M OH^- gives <10% interference when $[F^-] = 10^{-3}$ M
I^-	10^0 to 5×10^{-8}	mr: 0.4 CN^-; 5×10^3 Br^-; 10^5 $S_2O_3^{2-}$; 10^6 Cl^-
Pb^{2+}	10^{-1} to 10^{-6}	mba: Hg^{2+}, Ag^+, Cu^{2+}
Ag^+/S^{2-}	10^0 to 10^{-7} Ag^+ 10^0 to 10^{-7} S^{2-}	Hg^{2+} must be less than 10^{-7} M
SCN^-	10^0 to 5×10^{-6}	mr: 10^{-6} I^-; 3×10^{-3} Br^-; 7×10^{-3} CN^-; 0.13 $S_2O_3^{2-}$; 20 Cl^-; 100 OH^-. mba: S^{2-}

[a]From: *Handbook of Electrode Technology,* pp. 10–13, Appendix, Orion Research: Cambridge, MA, 1982. With permission.

[b]mr: maximum ratio $\left(\frac{c_{\text{interferent}}}{c_{\text{analyte}}}\right)$ for no interference.

mba: must be absent.

cation whose activity was to be determined. This solution was also saturated with AgCl to form a Ag/AgCl reference electrode with the silver lead wire.

As an alternative to the use of a porous disk as a rigid supporting medium, it has recently been found possible to immobilize liquid exchangers in tough polyvinyl chloride membranes. Here, the liquid ion exchanger and polyvinyl chloride are dissolved in a solvent such as tetrahydrofuran. Evaporation of the solvent leaves a flexible membrane, which can be cut and cemented to the end of a glass or plastic tube. Membranes formed in this way behave in much the same way as those in which the ion exchanger is held as an actual liquid in the pores of a disk. Currently, most liquid membrane electrodes are of this newer type.

The active substances in liquid membranes are of three kinds: (1) cation exchangers; (2) anion exchangers; and (3) neutral macrocyclic compounds, which selectively complex certain cations.

One of the most important liquid membrane electrodes is selective toward calcium ion in approximately neutral media. The active ingredient in the membrane is a cation exchanger consisting of an aliphatic diester of phosphoric acid dissolved in a polar solvent. The diester contains a single acidic proton; thus, two molecules react with the divalent calcium ion to form a dialkyl phosphate with the structure

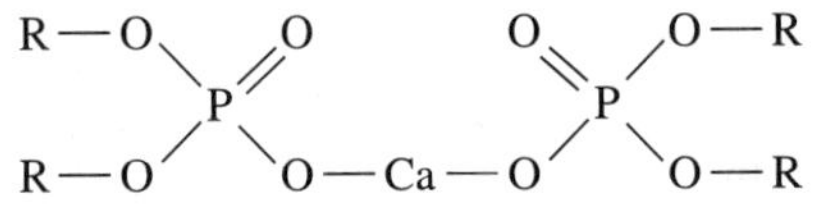

calcium dialkyl phosphate

Here, R is an aliphatic group containing from 8 to 16 carbon atoms. In commercial electrodes of this type, R is usually a C_9 group. The internal aqueous solution in contact with the exchanger (see Figure 23-8) contains a fixed concentration of calcium chloride and a silver/silver chloride reference electrode. The porous disk (or the polyvinyl chloride membrane) containing the ion-exchange liquid separates the analyte solution from the reference calcium chloride solution. The equilibrium established at each interface can be represented as

$$\underset{\text{organic}}{[(RO)_2POO]_2Ca} \rightleftharpoons \underset{\text{organic}}{2(RO)_2POO^-} + \underset{\text{aqueous}}{Ca^{2+}}$$

Note the similarity of this equation to Equation 23-7 for the glass electrode. The relationship between potential and pCa is also analogous to that for the glass electrode (Equation 23-12). Thus,

$$E_{ind} = L + \frac{0.0592}{2} \log a_1$$
$$= L' - \frac{0.0592}{2} \log pCa \qquad (23\text{-}17)$$

In this case, however, the second term on the right is divided by two because the cation is divalent.

The calcium membrane electrode has proved to be a valuable tool for physiological studies because this ion plays important roles in nerve conduction, bone formation, muscle contraction, cardiac conduction and contraction, and renal tubular function. At least some of these processes are influenced more by calcium ion *activity* than by calcium ion concentration; activity, of course, is the parameter measured by the electrode.

Another liquid specific-ion electrode of great value for physiological studies is that for potassium. The selectivity of a liquid membrane for potassium relative to sodium is especially important because both of these ions are present in all living systems and play important roles in neural transmission. A number of liquid membrane electrodes show promise of meeting these needs; one is based upon the antibiotic valinomycin, an uncharged macrocyclic ether that has a strong affinity for potassium ion. Of great importance is the observation that a liquid membrane consisting of valinomycin in diphenyl ether is about 10^4 times as responsive to potassium ion as to sodium ion.[13] Figure 23-9 is a photomicrograph of a valinomycin ultramicroelectrode for monitoring the potassium activity in the interior of a single cell. In this case, no physical membrane is needed to separate the internal solution from the analyte because of the small diameter of the opening at the tip (<1 μm) and the fact that the interior of the glass was made hydrophobic by a coating of silicone.

Table 23-4 lists some typical commercially available liquid membrane electrodes. The anion-sensitive electrodes employ a solution of an anion exchanger in an organic solvent. As mentioned earlier, many of the

[13] M. S. Frant and J. W. Ross Jr., *Science*, **1970**, *167*, 987.

TABLE 23-4 Liquid Membrane Electrodes[a,b]

Analyte Ion	Concentration Range, M	Interferences[c]
Ca^{2+}	10^0 to 5×10^{-7}	10^{-5} Pb^{2+}; 4×10^{-3} Hg^{2+}, H^+, 6×10^{-3} Sr^{2+}; 2×10^{-2} Fe^{2+}; 4×10^{-2} Cu^{2+}; 5×10^{-2} Ni^{2+}; 0.2 NH_3; 0.2 Na^+; 0.3 $Tris^+$; 0.3 Li^+; 0.4 K^+; 0.7 Ba^{2+}; 1.0 Zn^{2+}; 1.0 Mg^{2+}
BF_4^-	10^0 to 7×10^{-6}	5×10^{-7} ClO_4^-; 5×10^{-6} I^-; 5×10^{-5} ClO_3^-; 5×10^{-4} CN^-; 10^{-3} Br^-; 10^{-3} NO_2^-; 5×10^{-3} NO_3^-; 3×10^{-3} HCO_3^-; 5×10^{-2} Cl^-; 8×10^{-2} $H_2PO_4^-$, HPO_4^{2-}, PO_4^{3-}; 0.2 OAc^-; 0.6 F^-; 1.0 SO_4^{2-}
NO_3^-	10^0 to 7×10^{-6}	10^{-7} ClO_4^-; 5×10^{-6} I^-; 5×10^{-5} ClO_3^-; 10^{-4} CN^-; 7×10^{-4} Br^-; 10^{-3} HS^-; 10^{-2} HCO_3^-; 2×10^{-2} CO_3^{2-}; 3×10^{-2} Cl^-; 5×10^{-2} $H_2PO_4^-$, HPO_4^{2-}, PO_4^{3-}; 0.2 OAc^-; 0.6 F^-; 1.0 SO_4^{2-}
ClO_4^-	10^0 to 7×10^{-6}	2×10^{-3} I^-; 2×10^{-2} ClO_3^-; 4×10^{-2} CN^-, Br^-; 5×10^{-2} NO_2^-, NO_3^-; 2 HCO_3^-, CO_3^{2-}, Cl^-, $H_2PO_4^-$, HPO_4^{2-}, PO_4^{3-}, OAc^-, F^-, SO_4^{2-}
K^+	10^0 to 10^{-6}	3×10^{-4} Cs^+; 6×10^{-3} NH_4^+, Tl^+; 10^{-2} H^+; 1.0 Ag^+, $Tris^+$; 2.0 Li^+, Na^+
Water Hardness (Ca^{2+} + Mg^{2+})	10^{-3} to 6×10^{-6}	3×10^{-5} Cu^{2+}, Zn^{2+}; 10^{-4} Ni^{2+}; 4×10^{-4} Sr^{2+}; 6×10^{-5} Fe^{2+}; 6×10^{-4} Ba^{2+}; 3×10^{-2} Na^+; 0.1 K^+

[a]From: *Handbook of Electrode Technology*, pp. 10–13, Appendix, Orion Research: Cambridge, MA, 1982. With permission.
[b]All of the electrodes except the last are of the newer type in which the liquid ion exchanger or neutral carrier is supported in a plastic matrix.
[c]The numbers in front of each ion represent the molar concentration of that ion that gives a 10% error when the analyte concentration is 10^{-3} M.

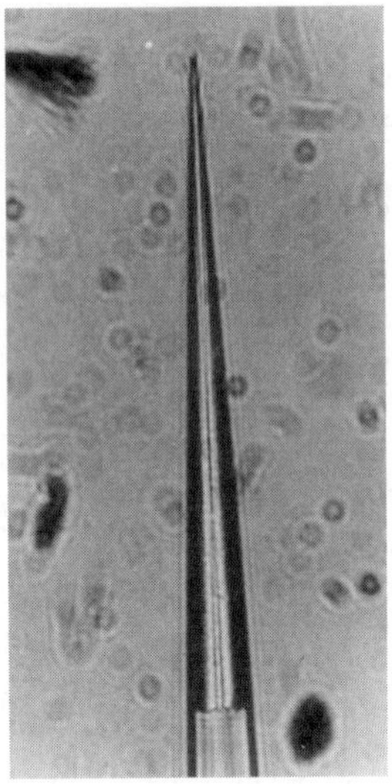

Figure 23-9 Photograph of a potassium liquid-ion exchanger microelectrode with 125 μm of ion exchanger inside the tip. The magnification of the original photo was 400×. *(From J. L. Walker,* Anal. Chem., ***1971****, 43(3)N, 91A. Reproduced by permission of the American Chemical Society.)*

so-called liquid membrane electrodes are in fact solids in which the liquid is held in a polymer matrix. These electrodes are somewhat more convenient to use than the older porous disk electrodes.

23D ION-SELECTIVE FIELD-EFFECT TRANSISTORS (ISFETs)

In Section 2C-3 (Figure 2-19), we described the metal oxide semiconductor field-effect transistor (MOSFET), which is widely used in computers and other electronic circuits as a switch to control current flow in circuits. One of the problems in employing this type of device in electronic circuits has been its pronounced sensitivity to ionic surface impurities, and a great deal of money and effort has been expended by the electronic industry in minimizing or eliminating this sensitivity in order to produce stable transistors.

Beginning in 1970, a number of scientists have been attempting to exploit the sensitivities of MOSFETs to surface ionic impurities for the selective potentiometric determination of various ions. These studies have led to the development of a number of different *ion-selective field-effect transistors* termed ISFETs. The theory of their selective-ion sensitivity is well understood and is described in the section that follows.[14]

23D-1 Mechanism of ISFET Ion-Selective Behavior

An ion-selective field-effect transistor is very similar in construction and function to an *n*-channel enhancement mode MOSFET (page 43). The ISFET differs only in that variation in the concentration of the ions of interest provides the variable gate voltage to control the conductivity of the channel. As shown in Figure 23-10, instead of the usual metallic contact, the gate of the ISFET is covered with an insulating layer of silicon nitride (Si_3N_4). The analyte solution, containing hydronium ions in this example, is in contact with this insulating layer and with a reference electrode. The surface of the gate insulator functions very much like the surface of a glass electrode. Protons from the hydronium ions in the test solution are adsorbed by available microscopic sites on the silicon nitride. Any change in the hydronium ion concentration of the solution results in a change in the concentration of adsorbed protons. The change in concentration of adsorbed protons then gives rise to a changing electrochemical potential between the gate and source, which in turn changes the conductivity of the channel of the ISFET. The conductivity of the

[14] For brief reviews of ISFETs, see J. N. Zermel, *Anal. Chem.*, **1975**, *47*, 255A; T. Matsuo and M. Esachi, *Sensors and Activation*, **1981**, *1*, 77; Y. G. Vlasov and Z. Fresnius, *Anal. Chem.*, **1989**, *335*, 96.

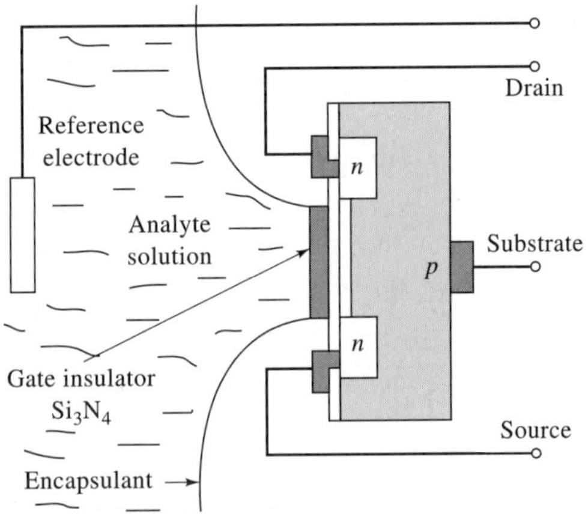

Figure 23-10 An ion-selective field-effect transistor (ISFET) for measuring pH.

channel can be monitored electronically to provide a signal that is proportional to the logarithm of the concentration of hydronium ion in the solution. Note that the entire ISFET except the gate insulator is coated with a polymeric encapsulant to insulate all electrical connections from the analyte solution.

23D-2 Application of ISFETs

The ion-sensitive surface of the ISFET is naturally sensitive to pH changes, but the device may be rendered sensitive to other species by coating the silicon nitride gate insulator with a polymer containing molecules that tend to form complexes with species other than hydronium ion. Furthermore, several ISFETs may be fabricated on the same substrate so that multiple measurements may be made simultaneously. All of the ISFETs may detect the same species to enhance accuracy and reliability, or each ISFET may be coated with a different polymer so that measurements of several different species may be made. Their small size (approximately 1 to 2 mm^2), rapid response time relative to glass electrodes, and ruggedness suggest that ISFETs may be the ion detectors of the future for many applications.

ISFETs offer a number of significant advantages over membrane electrodes, including ruggedness, small size, inertness toward harsh environments, rapid response, and low electrical impedance. In contrast to membrane electrodes, ISFETs do not require hydration before use and can be stored indefinitely in the dry states. Despite these many advantages, no ISFET specific ion electrode appeared upon the market until the early 1990s, over 20 years after their invention. The reason for this delay is that manufacturers were unable to develop the technology of encapsulating the devices in order to create a product that did not exhibit drift and instability.

In the early 1990s several companies came out with an ISFET for the determination of pH. At the time of writing, it has not been possible to learn details of the surface layer on these electrodes that is responsible for their selective behavior toward protons. Nor is information comparing the performance characteristics of these electrodes with the classical glass electrode. Certainly, they offer the several advantages listed in the previous paragraph. Just what their disadvantages are, if any, is not clear at this time. It seems likely, however, that ISFETs that are selective for ions other than hydrogen will be appearing on the market in the future.

23E MOLECULAR-SELECTIVE ELECTRODE SYSTEMS

Two types of membrane electrode systems have been developed that act selectively toward certain types of *molecules*. One of these is used for the determination of dissolved gases, such as carbon dioxide and ammonia. The other, which is based upon biocatalytic membranes, permits the determination of a variety of organic compounds, such as glucose and urea.

23E-1 Gas-Sensing Probes

During the past three decades, several gas-sensing electrochemical devices have become available from commercial sources. In the manufacturers' literature these devices are generally called gas-sensing "electrodes." As can be seen in Figure 23-11, these devices are not, in fact, electrodes but instead are electrochemical cells made up of a specific ion and a reference electrode immersed in an internal solution that is retained by a thin gas-permeable membrane. Thus, *gas-sensing probes* is a more suitable name for these gas sensors.

Gas-sensing probes are remarkably selective and sensitive devices for determining dissolved gases or

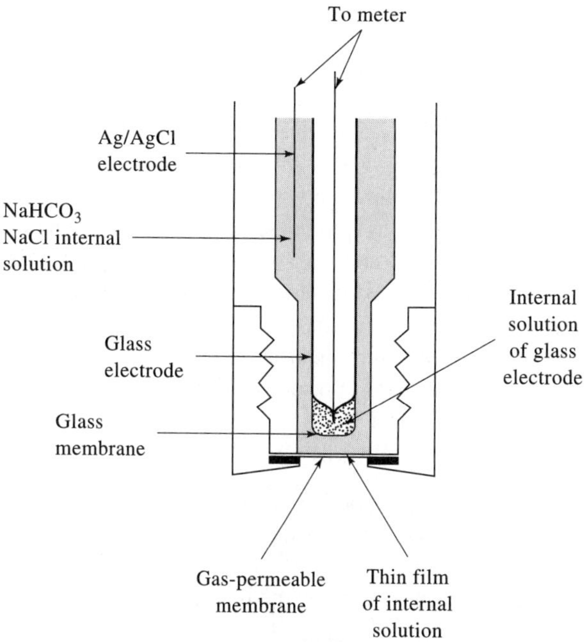

Figure 23-11 Schematic of a gas-sensing probe for carbon dioxide.

ions that can be converted to dissolved gases by pH adjustment.

Membrane Probe Design

Figure 23-11 is a schematic showing details of a gas-sensing probe for carbon dioxide. The heart of the probe is a thin, porous membrane, which is easily replaceable. This membrane separates the analyte solution from an internal solution containing sodium bicarbonate and sodium chloride. A pH-sensitive glass electrode having a flat membrane is fixed in position so that a very thin film of the internal solution is sandwiched between it and the gas-permeable membrane. A silver/silver chloride reference electrode is also located in the internal solution. It is the pH of the film of liquid adjacent to the glass electrode that provides a measure of the carbon dioxide content of the analyte solution on the other side of the membrane.

Gas-Permeable Membranes

Two types of membrane material are encountered, microporous and homogeneous. Microporous materials are manufactured from hydrophobic polymers such as polytetrafluoroethylene or polypropylene, which have a porosity (void volume) of about 70% and a pore size of less than 1 μm. Because of the water-repellent properties of the film, water molecules and electrolyte ions are excluded from the pores; gaseous molecules, on the other hand, are free to move in and out of the pores by *effusion* and thus across this barrier. Typically, microporous membranes are about 0.1 mm in thickness.

Homogeneous films, in contrast, are solid polymeric substances through which the analyte gas passes by dissolving in the membrane, diffusing, and then desolvating into the internal solution. Silicone rubber is the most widely used material for construction. Homogeneous films are generally thinner than microporous ones (0.01 to 0.03 mm) in order to hasten the transfer of gas and thus the rate of response of the system.

Mechanism of Response

When a solution containing dissolved carbon dioxide is brought into contact with the microporous membrane shown in Figure 23-11, the gas *effuses* through the membrane, as described by the reactions

$$\underset{\text{external solution}}{CO_2\,(aq)} \rightleftharpoons \underset{\text{membrane pores}}{CO_2(g)} \rightleftharpoons \underset{\text{internal solution}}{CO_2(aq)}$$

Because the pores are numerous and small, equilibrium is rapidly established (in a few seconds to a few minutes) between the external solution and the thin film of internal solution adjacent to the membrane. In the internal solution, another equilibrium is established that causes the pH of the internal surface film to change; that is,

$$CO_2(aq) + H_2O \rightleftharpoons HCO_3^- + H^+$$

A glass electrode immersed in the film of internal solution then detects the pH change. The overall reaction for the processes just described is given by

$$\underset{\text{external solution}}{CO_2(aq)} + H_2O \rightleftharpoons H^+ + \underset{\text{internal solution}}{HCO_3^-}$$

for which we may write

$$\frac{(a_{H^+})_{int}\,(a_{HCO_3^-})_{int}}{(a_{CO_2})_{ext}} = K$$

for a neutral species such as CO_2, $a_{CO_2} = [CO_2]$. Thus

$$\frac{(a_{H^+})_{int}\,(a_{HCO_3^-})_{int}}{[CO_2]_{ext}} = K$$

If the concentration of HCO_3^- in the internal solution is made relatively high so that its activity is not altered significantly by the carbon dioxide from the sample, then

$$\frac{(a_{H^+})_{int}}{[CO_2]_{ext}} = \frac{K}{(a_{HCO_3^-})_{int}} = K_g \qquad (23\text{-}18)$$

where K_g is a new constant. This equation can be rearranged to give

$$(a_{H^+})_{int} = a_1 = K_g[CO_2]_{ext} \qquad (23\text{-}19)$$

where a_1 is the internal hydrogen ion activity.

Substitution of this relationship into Equation 23-13 gives the potential of the glass electrode as a function of the concentration of CO_2 in the external solution. That is,

$$\begin{aligned} E_{ind} &= L'' + 0.0592 \log K_g[CO_2]_{ext} \\ &= L' + 0.0592 \log [CO_2]_{ext} \qquad (23\text{-}20) \end{aligned}$$

where $L'' = L' + 0.0592 \log K_g - E_{ref}$.

Thus, the potential of the cell consisting of the internal reference and indicator electrode is determined by the CO_2 concentration of the external solution. Note that *no electrode comes directly in contact* with the analyte. Note also that the only species that will interfere with the measurement are dissolved gases that can pass through the membrane and can additionally affect the pH of the internal solution.

The possibility exists for increasing the selectivity of the gas-sensing probe by employing an internal electrode that is sensitive to some species other than hydrogen ion; for example, a nitrate-sensing electrode can be used to provide a cell that would be sensitive to nitrogen dioxide. Here, the equilibrium would be

$$\underset{\text{external}}{2NO_2(aq)} + H_2O \rightleftharpoons \underset{\text{internal solution}}{NO_2^- + NO_3^- + 2H^+}$$

This electrode permits the determination of NO_2 in the presence of gases such as SO_2, CO_2, and NH_3, which would also alter the pH of the internal solution.

Table 23-5 lists representative gas-sensing probes that are commercially available. An oxygen-sensitive cell system is also on the market; it, however, is based on a voltammetric measurement and is discussed in Chapter 25.

23E-2 Biocatalytic Membrane Electrodes

Since about 1970, considerable effort has been devoted to combining the selectivity of enzyme-catalyzed reactions and electrochemical transducers to give highly selective *biosensors* for the determination of compounds of biological and biochemical interest.[15] In these devices the sample is brought into contact with an immobilized enzyme where the analyte undergoes a catalytic reaction to yield a species such as ammonia, carbon dioxide, hydrogen ions, or hydrogen peroxide. The concentration of this product, which is proportional to the analyte concentration, is then determined by the transducer. The most common transducers in these devices are membrane electrodes, gas-sensing probes, and voltammetric devices, which are discussed in Chapter 25.

Biosensors based upon membrane electrodes are attractive from several standpoints. First, in principle, complex organic molecules can be determined with the convenience, speed, and ease that characterize ion-selective measurements of inorganic species. Second, biocatalysts permit reactions to occur under mild conditions of temperature and pH and at minimal substrate concentrations. Third, combining the selectivities of the enzymatic reaction and the electrode response yields procedures that are free from most interferences.

The main limitation to enzymatic procedures is the high cost of enzymes, particularly when used for routine or continuous measurements. This disadvantage has led to the use of immobilized enzyme media in which a small amount of enzyme can be used for the repetitive analysis of hundreds of samples. Two general techniques are used. In one, the sample is passed through a fixed bed of immobilized enzyme and then to the detector. In the second, a porous layer of the immobilized enzyme is attached directly to the surface of the ion-selective electrode, thus forming an *enzyme electrode*. In such devices, the reaction product reaches the selective membrane surface by diffusion.

[15] For reviews of biosensors, see G. A. Rechnitz, *Science*, **1981**, *214*, 287; G. A. Rechnitz, *Anal. Chem.*, **1982**, *54*, 1194A; G. A. Rechnitz, *J. Chem. Educ.*, **1983**, *60*, 282. For an extensive treatment of the subject, see *Biosensors: Fundamentals and Applications*, A. P. F. Turner, I. Karube, and G. S. Wilson, Eds. New York: Oxford University Press, 1987.

TABLE 23-5 Commercial Gas-Sensing Probes

Gas	Equilibrium in Internal Solution	Sensing Electrode
NH_3	$NH_3 + H_2O \rightleftharpoons NH_4^+ + OH^-$	Glass, pH
CO_2	$CO_2 + H_2O \rightleftharpoons HCO_3^- + H^+$	Glass, pH
HCN	$HCN \rightleftharpoons H^+ + CN^-$	Ag_2S, pCN
HF	$HF \rightleftharpoons H^+ + F^-$	LaF_3, pF
H_2S	$H_2S \rightleftharpoons 2H^+ + S^{2-}$	Ag_2S, pS
SO_2	$SO_2 + H_2O \rightleftharpoons HSO_3^- + H^+$	Glass, pH
NO_2	$2NO_2 + H_2O \rightleftharpoons NO_2^- + NO_3^- + 2H^+$	Immobilized ion exchange, pNO_3

Immobilization of enzymes can be accomplished in several ways, including physical entrapment in a polymer gel, physical absorption on a porous inorganic support such as alumina, covalent bonding of the enzyme to a solid surface such as glass beads or a polymer, or copolymerization of the enzyme with a suitable monomer.

Figure 23-12 shows two types of enzyme electrodes that have been proposed for the determination of blood urea nitrogen (BUN), an important routine clinical test. In the presence of the enzyme urease, urea is hydrolyzed according to the reaction

$$(NH_2)_2CO + 2H_2O + H^+ \rightarrow 2NH_4^+ + HCO_3^- \quad \rightleftharpoons \quad 2NH_3 + 2H^+ \qquad (23\text{-}21)$$

The electrode in Figure 23-12a is a glass ion electrode that responds to the ammonium *ion* formed by the reaction shown in the upper part of Equation 23-21. The electrode in Figure 23-12b is an ammonia gas probe that responds to the *molecular* ammonia in equilibrium with the ammonium ion. Unfortunately, both electrodes have limitations. The glass electrode responds to all monovalent cations and its selectivity coefficients for NH_4^+ over Na^+ and K^+ are such that interference arises in most media of biological interest (such as blood). The ammonia gas probe suffers from a different handicap, namely a pH incompatibility between enzyme and sensor. Thus, the enzyme requires a pH of about 7 for maximum catalytic activity, but the sensor's maximum response occurs at a pH that is greater than 8 to 9 (where essentially all of the NH_4^+ has been converted to NH_3). Thus the sensitivity of the electrode is limited. Both limitations are overcome by use of a fixed-bed enzyme system where the sample at a pH of about 7 is pumped over the enzyme. The resulting solution is then made alkaline and the liberated ammonia determined with an ammonia gas probe. Automated instruments based on this technique have been on the market for several years.

To date, despite a considerable effort, no commercial enzyme electrodes based upon potential measurements are available, due at least in part to limitations such as those cited in the previous paragraph. Enzymatic electrodes based upon voltammetric measurements are, however, offered by a commercial source. These electrodes are discussed briefly in Chapter 25.

23E-3 Disposable Multilayer pIon Systems

Recently, disposable electrochemical cells, based on pIon electrodes, have become available. These systems, which have been designed for the routine determination of various ions in clinical samples, are described briefly in Section 33D-3 and depicted in Figure 33-18.

23F INSTRUMENTS FOR MEASURING CELL POTENTIALS

A prime consideration in the design of an instrument for measuring cell potentials is that its resistance must be large with respect to the cell. If it is not, significant error results as a consequence of the *IR* drop in the cell. This effect is demonstrated by the example that follows.

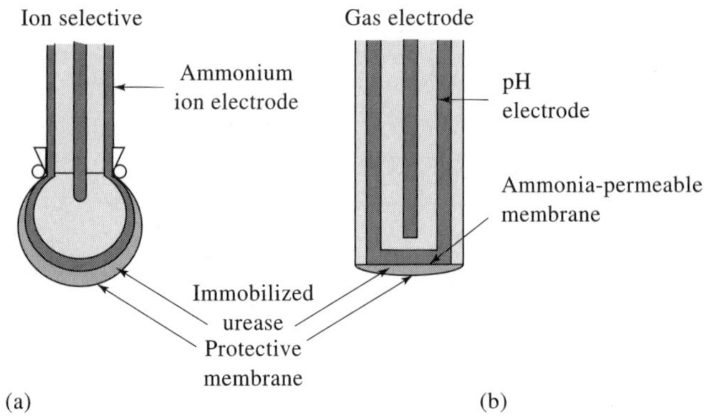

Figure 23-12 Enzyme electrodes for measuring urea. *(Reprinted with permission from D. N. Gray, M. H. Keyes, and B. Watson,* Anal. Chem., ***1977,*** 49, *1069A. Copyright 1977 American Chemical Society.)*

EXAMPLE 23-1

The true potential of a glass/calomel electrode system is 0.800 V; its internal resistance is 20 MΩ. What would be the relative error in the measured potential if the measuring device has a resistance of 100 MΩ?

Here, the circuit can be considered to consist of a potential source E_s and two resistors in series, R_s being that of the source and R_M that of the measuring device. That is,

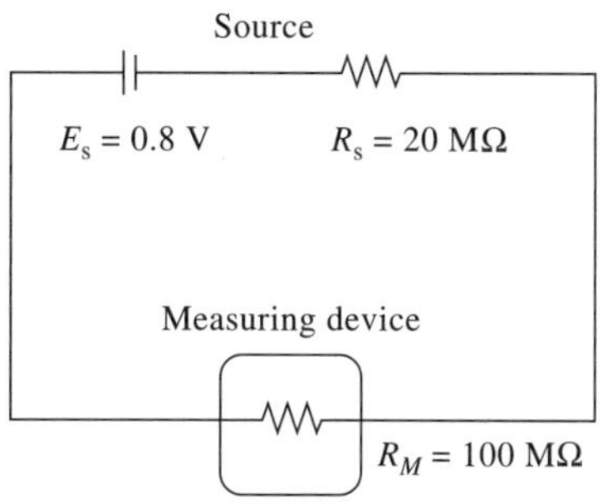

From Ohm's law, we may write

$$E_s = IR_s + IR_M$$

where I is the current in this circuit consisting of the cell and the measuring device. The current is then given by

$$I = \frac{0.800 \text{ V}}{(20 + 100) \times 10^6 \ \Omega} = 6.67 \times 10^{-9} \text{ A}$$

The potential drop across the measuring device (which is the potential indicated by the device, E_i) is IR_M. Thus,

$$E_i = 6.67 \times 10^{-9} \times 100\% \times 10^6 = 0.667 \text{ V}$$

and

$$\text{rel error} = \frac{0.667 - 0.800}{0.800} \times 100\%$$
$$= \frac{-0.133}{0.800} \times 100\% = -17\%$$

It is readily shown that to reduce the instrumental error due to IR drop to 1% relative, the resistance of the potential measuring device must be about 100 times greater than the cell resistance; for a relative error of 0.1%, the resistance must be 1000 times greater. Because the electrical resistance of cells containing selective-ion electrodes may be 100 MΩ or more, potential measuring devices to be used with these electrodes generally have internal resistance of 10^{12} Ω or more.

It is important to appreciate that an error in potential, such as that shown in Example 23-1 (−0.133 V), would have an enormous effect on the accuracy of a concentration measurement based upon that potential. Thus, as shown on page 614, a 0.001 V uncertainty in potential leads to a relative error of about 4% in the determination of the hydrogen ion concentration of a solution by potential measurement with a glass electrode. An error of the size found in Example 23-1 would result in a concentration uncertainty of two orders of magnitude or more.

Two types of instruments have been employed in potentiometry—the potentiometer and the direct-reading electronic voltmeter. Both instruments are referred to as *pH meters* when their internal resistances are sufficiently high to be used with glass and other membrane electrodes; with the advent of the many new specific ion electrodes, *pIon* or *ion meters* would perhaps be a more descriptive name. Modern ion meters are generally of the direct-reading type; thus, they are the only ones that will be described here.

23F-1 Direct-Reading Instruments

Numerous direct-reading pH meters are available commercially. Generally, these are solid-state devices employing a field-effect transistor or a voltage follower as the first amplifier stage in order to provide the needed high internal resistance. Figure 23-13 gives a schematic of a simple, battery-operated pIon meter that can be built for approximately $50 (exclusive of the electrodes). Here, the output of the ion electrode is connected to a high-resistance field-effect transistor. The output from the operational amplifier is displayed on a meter having a scale extending from −100 to 100 μA. These extremes correspond to a pH range of 1 to 14.

23F-2 Commercial Instruments

A wide variety of ion meters are available from several instrument manufacturers. For example, a 1975 publication[16] listed over one hundred models offered by

[16] J. A. Hauber and C. R. Dayton, *Amer. Lab.*, **1975**, *7* (4), 73.

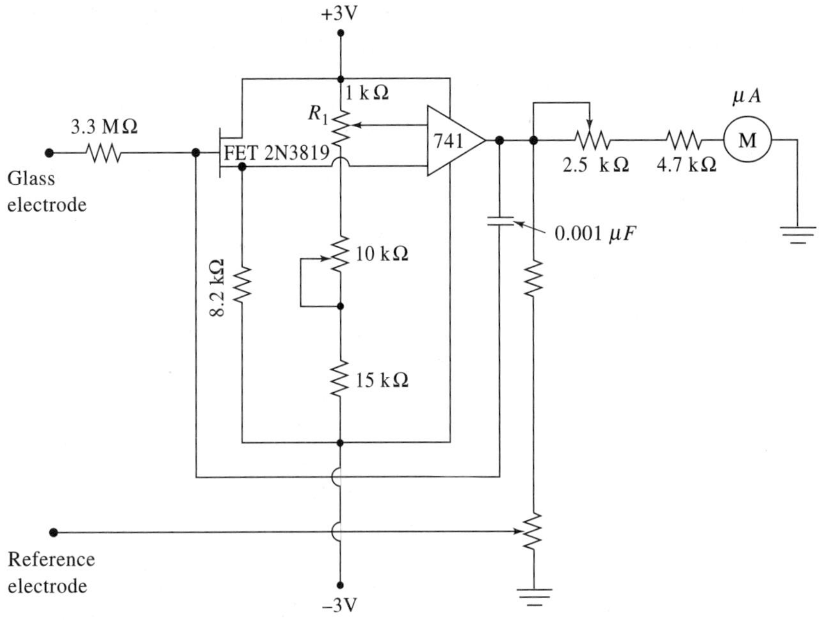

Figure 23-13 A simple pIon meter based upon a field-effect transistor and an operational amplifier. *(From D. Sievers,* J. Chem. Educ., ***1981,*** 58, *281. With permission.)*

more than 20 companies. Although the information in this paper is now dated, it does provide a realistic picture of the kinds and numbers of pIon meters that are currently available to the scientist. Four categories of meters based on price and readability are described. These include *utility* meters, which are portable, usually battery-operated instruments that currently range in price from less than \$100 to somewhat over \$500. Generally, utility meters are readable to 0.1 pH unit or better. *General-purpose* meters are line-operated instruments, which are readable to 0.05 pH unit or better. Some offer such features as digital readout, automatic temperature compensation, and scale expansions so that full-scale covers 1.4 units instead of 0 to 14 units, and a millivolt scale. Currently, prices for general-purpose meters range from \$300 to \$900. *Expanded-scale* instruments are generally readable to 0.01 pH unit or better and cost from about \$700 to \$1500. Most offer full-scale ranges of 0.5 to 2 pH units (as well as 0 to 7 and 0 to 14 ranges), four-digit readout, push-button control, millivolt scale, and automatic temperature compensation. *Research* meters are readable to 0.001 pH unit or better, generally have a five-digit display, and cost in the range of about \$1500 to \$2200. It should be pointed out that the readability of these instruments is usually significantly better than the sensitivity of most ion-selective electrodes.

23G DIRECT POTENTIOMETRIC MEASUREMENTS

The determination of an ion or molecule by direct potentiometric measurement is rapid and simple, requiring only a comparison of the potential developed by the indicator electrode in the test solution with its potential when immersed in one or more standard solutions of the analyte. Because most indicator electrodes are selective, preliminary separation steps are seldom required. In addition, direct potentiometric measurements are rapid and readily adapted to the continuous and automatic monitoring of ion activities.

23G-1 The Sign Convention and Equations for Direct Potentiometry

The sign convention for potentiometry is consistent with the convention described in Chapter 22 for standard electrode potentials.[17] In this convention, the indi-

[17] According to Bates, the convention being described here has been endorsed by standardizing groups in the United States and Great Britain as well as by IUPAC. See R. G. Bates, in *Treatise on Analytical Chemistry,* 2nd ed., I. M. Kolthoff and P. J. Elving, Eds., Part I, Vol. 1, pp. 831–832. New York: Wiley, 1978.

cator electrode is treated as the *cathode* and the reference electrode as the *anode.*[18] For direct potentiometric measurements, the potential of a cell can then be expressed as a sum of an indicator electrode potential, a reference electrode potential, and a junction potential:

$$E_{cell} = E_{ind} - E_{ref} + E_j \qquad (23\text{-}22)$$

In Sections 23B and 23C, we describe the response of various types of indicator electrodes to analyte activities. For the cation X^{n+} at 25°C, the electrode response takes the general *Nernstian* form

$$\begin{aligned} E_{ind} &= L - \frac{0.0592}{n} \log \frac{1}{[a_X]} \\ &= L - \frac{0.0592}{n} \text{pX} \qquad (23\text{-}23) \end{aligned}$$

where L is a constant and a_X is the activity of the cation. For metallic indicator electrodes, L is ordinarily the standard electrode potential (see Equation 23-1); for membrane electrodes, L is the summation of several constants, including the time-dependent asymmetry potential of uncertain magnitude (see Equation 23-14).

Substitution of Equation 23-23 into Equation 23-22 yields, with rearrangement,

$$\text{pX} = -\log a_X = -\frac{E_{cell} - (E_j - E_{ref} + L)}{0.0592/n}$$

The constant terms in parentheses can be combined to give a new constant K:

$$\text{pX} = -\log a_X = -\frac{E_{cell} - K}{0.0592/n} \qquad (23\text{-}24)$$

where

$$K = E_j - E_{ref} + L \qquad (23\text{-}25)$$

For an anion A^{n-}, the sign of Equation 23-24 is reversed:

$$\text{pA} = \frac{E_{cell} - K}{0.0592/n} \qquad (23\text{-}26)$$

All direct potentiometric methods are based upon Equation 23-24 or 23-26. The difference in sign in the two equations has a subtle but important consequence in the way that ion-selective electrodes are connected to pH meters and pIon meters. When the two equations are solved for E_{cell}, we find that for cations

$$E_{cell} = K - \frac{0.0592}{n} \text{pX} \qquad (23\text{-}27)$$

and for anions

$$E_{cell} = K + \frac{0.0592}{n} \text{pA} \qquad (23\text{-}28)$$

Equation 23-27 shows that an increase in pX results in a *decrease* in E_{cell} with a cation-selective electrode. Thus, when a high-resistance voltmeter is connected to the cell in the usual way, with the indicator electrode attached to the positive terminal, the meter reading decreases as pX increases. To eliminate this problem, instrument manufacturers generally reverse the leads so that cation-sensitive electrodes are connected to the *negative* terminal of the voltage-measuring device. Meter readings then increase with increases in pX. Anion-selective electrodes, on the other hand, are connected to the *positive* terminal of the meter so that increases in pA also yield larger readings.

23G-2 The Electrode Calibration Method

The constant K in Equations 23-27 and 23-28 is made up of several constants, at least one of which, the junction potential, cannot be computed from theory or measured directly. Therefore, before these equations can be used for the determination of pX or pA, K must be evaluated *experimentally* with one or more standard solutions of the analyte.

In the electrode-calibration method, K is determined by measuring E_{cell} for one or more standard solutions of known pX or pA. The assumption is then made that K is unchanged when the standard is replaced with analyte. The calibration is ordinarily performed at the time pX or pA for the unknown is determined. With membrane electrodes recalibration may be necessary if measurements extend over several hours because of the slowly changing asymmetry potential.

The direct electrode calibration method offers the advantages of simplicity, speed, and applicability to the continuous monitoring of pX or pA. Two important disadvantages attend its use, however. One of these is that the accuracy of a measurement obtained by this procedure is limited by the inherent uncertainty caused by the junction potential E_j; unfortunately, this uncertainty can never be totally eliminated. The second disadvantage of this procedure is that results of an analysis are in terms

[18] In effect, the sign convention for electrode potentials described in Section 22C-5 also designates the indicator electrode as the cathode by stipulating that half-reactions always be written as reductions; the standard hydrogen electrode, which is the reference electrode in this case, is then the anode.

of activities rather than concentrations (for some applications this is an advantage rather than a disadvantage).

Inherent Error in the Electrode Calibration Procedure

A serious disadvantage of the electrode calibration method is the existence of an inherent uncertainty that results from the assumption that K in Equation 23-24 or 23-26 remains constant between calibration and analyte determination. This assumption can seldom, if ever, be exactly true because the electrolyte composition of the unknown will almost inevitably differ from that of the solution employed for calibration. The junction potential E_j contained in K (Equation 23-25) will vary slightly as a consequence, even though a salt bridge is used. This uncertainty will frequently be on the order of 1 mV or more; unfortunately, because of the nature of the potential/activity relationship, such an uncertainty has an amplified effect on the inherent accuracy of the analysis. The magnitude of the uncertainty in analyte concentration can be estimated by differentiating Equation 23-24 with respect to K while holding E_{cell} constant:

$$-\log_{10} e \frac{da_X}{a_X} = -0.434 \frac{da_X}{a_X} = -\frac{dK}{0.0592/n}$$

$$\frac{da_X}{a_X} = \frac{ndK}{0.0256}$$

Upon replacing da_X and dK with finite increments and multiplying both sides of the equation by 100, we obtain

$$\% \text{ rel error} = \frac{\Delta a_X}{a_X} \times 100\% = 3.9 \times 10^3\, n\Delta K\% \quad (23\text{-}29)$$

The quantity $\Delta a_X/a_X$ is the relative error in a_X associated with an absolute uncertainty ΔK in K. If, for example, ΔK is ± 0.001 V, a relative error in activity of about $\pm 4n\%$ can be expected. *It is important to appreciate that this uncertainty is characteristic of all measurements involving cells that contain a salt bridge and that this uncertainty cannot be eliminated by even the most careful measurements of cell potentials or the most sensitive and precise measuring devices;* nor does it appear possible to devise a method for completely eliminating the uncertainty in K that is the source of this problem.

Activity Versus Concentration

Electrode response is related to activity rather than to analyte concentration. Ordinarily, however, the scientist is interested in concentration, and the determination of this quantity from a potentiometric measurement requires activity coefficient data. More often than not, activity coefficients will be unavailable because the ionic strength of the solution is either unknown or so high that the Debye-Hückel equation is not applicable. Unfortunately, the assumption that activity and concentration are identical may lead to serious errors, particularly when the analyte is polyvalent.

The difference between activity and concentration is illustrated by Figure 23-14, where the lower curve gives the change in potential of a calcium electrode as a function of calcium chloride concentration (note that the activity or concentration scale is logarithmic). The nonlinearity of the curve is due to the increase in ionic strength—and the consequent decrease in the activity coefficient of the calcium—as the electrolyte concentration becomes larger. When these concentrations are converted to activities, the upper curve is obtained; note that this straight line has the Nernstian slope of 0.0296 (0.0592/2).

Activity coefficients for singly charged ions are less affected by changes in ionic strength than are coefficients for species with multiple charges. Thus, the effect shown in Figure 23-14 will be less pronounced for electrodes that respond to H^+, Na^+, and other univalent ions.

In potentiometric pH measurements, the pH of the standard buffer employed for calibration is generally

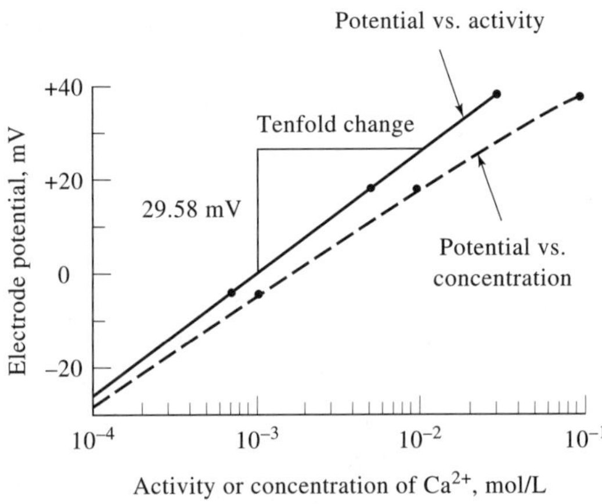

Figure 23-14 Response of a calcium ion electrode to variations in the calcium ion concentration and activity of solutions prepared from pure calcium chloride. *(Courtesy of Orion Research, In., Cambridge, MA.)*

based on the activity of hydrogen ions. Thus, the resulting hydrogen ion results are also on an activity scale. If the unknown sample has a high ionic strength, the hydrogen ion *concentration* will differ appreciably from the activity measured.

23G-3 Calibration Curves for Concentration Measurement

An obvious way of correcting potentiometric measurements to give results in terms of concentration is to make use of an empirical calibration curve such as the lower curve in Figure 23-14. For this approach to be successful, however, it is essential that the ionic composition of the standards closely approximate that of the analyte—a condition that is difficult to realize experimentally for complex samples.

Where electrolyte concentrations are not too great, it is often helpful to swamp both the samples and the calibration standards with a measured excess of an inert electrolyte. Under these circumstances, the added effect of the electrolyte in the sample becomes negligible, and the empirical calibration curve yields results in terms of concentration. This approach has been employed for the potentiometric determination of fluoride in public water supplies with a lanthanum fluoride electrode. Here, both samples and standards are diluted on a 1:1 basis with a solution containing sodium chloride, a citrate buffer, and an acetate buffer (this mixture, which fixes both ionic strength and pH, is sold under the name "Total Ionic Strength Adjusting Buffer" or TISAB); the diluent is sufficiently concentrated so that the samples and standards do not differ significantly in ionic strength. The procedure permits a rapid measurement of fluoride ion in the 1 ppm range, with a precision of about 5% relative.

Calibration curves are also useful for electrodes that do not respond linearly to pA.

23G-4 Standard Addition Method

The standard addition method, described in Section 1E-2, is equally applicable to potentiometric determinations. Here, the potential of the electrode system is measured before and after addition of a small volume (or volumes) of a standard to a known volume of the sample. The assumption is made that this addition does not alter the ionic strength and thus the activity coefficient γ_X of the analyte. It is further assumed that the added standard does not significantly alter the junction potential.

The standard addition method has been applied to the determination of chloride and fluoride in samples of commercial phosphors.[19] In this application, solid-state indicator electrodes for chloride and fluoride were used in conjunction with a reference electrode; the added standard contained known quantities of the two anions. The relative standard deviation for the measurement of replicate samples was found to be 0.7% for fluoride and 0.4% for chloride. When the standard addition method was not used, relative errors for the analyses appeared to range between 1% and 2%.

23G-5 Potentiometric pH Measurements with a Glass Electrode[20]

The glass electrode is unquestionably the most important indicator electrode for hydrogen ion. It is convenient to use and is subject to few of the interferences that affect other pH-sensing electrodes. Glass electrodes are available at relatively low cost and come in many shapes and sizes. A common variety is illustrated in Figure 23-3; the reference electrode is usually a silver/silver chloride electrode.

The glass electrode is a remarkably versatile tool for the measurement of pH under many conditions. The electrode can be used without interference in solutions containing strong oxidants, reductants, gases, and proteins (a calomel reference electrode rather than a silver/silver chloride reference electrode, is normally used in the presence of proteins because silver ions react with proteins); the pH of viscous or even semisolid fluids can be determined. Electrodes for special applications are available. Included among these are small electrodes for pH measurements in a drop (or less) of solution or in a cavity of a tooth, microelectrodes that permit the measurement of pH inside a living cell, systems for insertion in a flowing liquid stream to provide a continuous monitoring of pH, a small glass electrode that can be swallowed to indicate the acidity of the stomach contents (the reference electrode is kept in the mouth), and combination electrodes that contain both indicator and reference electrodes in a single probe.

---

[19] L. G. Bruton, *Anal. Chem.*, **1971**, *43*, 579.

[20] For a detailed discussion of potentiometric pH measurements, see R. G. Bates, *Determination of pH: Theory and Practice*, 2nd ed. New York: Wiley, 1973.

Summary of Errors Affecting pH Measurements with the Glass Electrode

The ubiquity of the pH meter and the general applicability of the glass electrode tend to lull the chemist into the attitude that any measurement obtained with such an instrument is surely correct. It is well to guard against this false sense of security because there are distinct limitations to the electrode system. These have been discussed in earlier sections and include the following:

1. **The alkaline error.** Modern glass electrodes become somewhat sensitive to alkali-metal ions at pH values greater than 11 to 12.
2. **The acid error.** At a pH less than 0.5, values obtained with a glass electrode tend to be somewhat high.
3. **Dehydration.** Dehydration of the electrode may cause unstable performance and errors.
4. **Errors in low ionic strength solutions.** It has been found that significant errors (as much as 1 or 2 pH units) may occur when the pH of low ionic strength samples, such as lake or stream samples, are measured with a glass/calomel electrode system.[21] The prime source of such errors has been shown to be nonreproducible junction potentials, which apparently result from partial clogging of the fritted plug or porous fiber that is used to restrict the flow of liquid from the salt bridge into the analyte solution. In order to overcome this problem, free diffusion junctions (FDJ) of various types have been designed, and one is produced commercially. In the latter, an electrolyte solution is dispensed from a syringe cartridge through a capillary tube, the tip of which is in contact with the sample solution (see Figure 23-15). Before each measurement, 6 μL of electrolyte is dispensed so that a fresh portion of electrolyte is in contact with the analyte solution.
5. **Variation in junction potential.** It should be reemphasized that variation in the junction potential between standard and sample leads to a fundamental uncertainty in the measurement of pH for which a correction cannot be applied. Absolute values more reliable than 0.01 pH unit are generally unobtainable. Even reliability to 0.03 pH unit requires considerable care. On the other hand, it is often possible to detect pH *differences* between similar solutions or pH *changes* in a single solution that are as small as 0.001 unit. For this reason, many pH meters are designed to permit readings to less than 0.01 pH unit.
6. **Error in the pH of the standard buffer.** Any inaccuracies in the preparation of the buffer used for calibration, or changes in its composition during storage, will be propagated as errors in pH measurements. A common cause of deterioration is the action of bacteria on organic components of buffers.

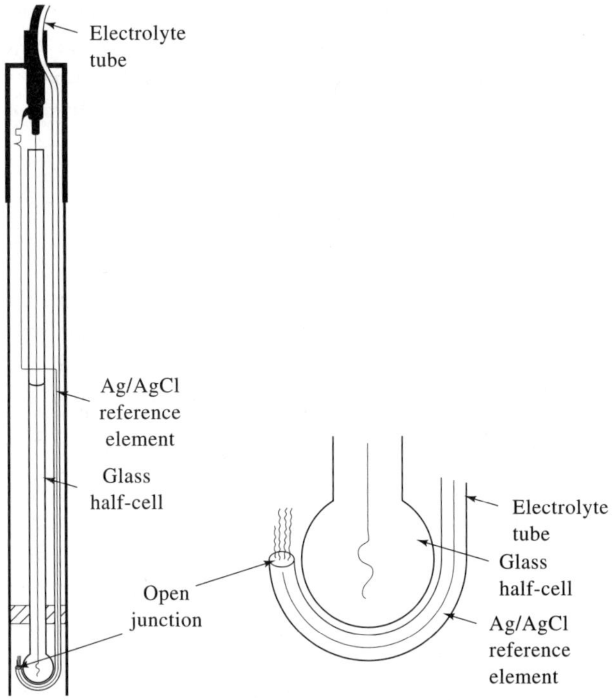

Figure 23-15 A combination pH electrode system with a free diffusion junction. *(Courtesy of Hach Company, Loveland, CO. With permission.)*

The Operational Definition of pH

The utility of pH as a measure of the acidity or alkalinity of aqueous media, the wide availability of commercial glass electrodes, and the relatively recent proliferation of inexpensive solid-state pH meters have made the potentiometric measurement of pH one of the most common analytical techniques in all of science. It is thus extremely important that pH be defined in a manner that is easily duplicated at various times and various

[21] See W. Davison and C. Woof, *Anal. Chem.*, **1985**, *57*, 2567; T. R. Harbinson and W. Davison, *Anal. Chem.*, **1987**, *59*, 2450; A. Kopelove, S. Franklin, and G. M. Miller, *Amer. Lab.*, **1989**, (6), 40.

laboratories throughout the world. To meet this requirement, it is necessary to define pH in operational terms—that is, by the way the measurement is made. Only then will the pH measured by one worker be the same as that measured by another.

The operational definition of pH endorsed by the National Institute of Standards and Technology (NIST), similar organizations in other countries, and the IUPAC is based upon the direct calibration of the meter with carefully prescribed standard buffers followed by potentiometric determination of the pH of unknown solutions.

Consider, for example, the glass/calomel system in Figure 23-3. When these electrodes are immersed in a standard buffer at 25°C, Equation 23-24 applies and we can write

$$pH_S = -\frac{E_S - K}{0.0592}$$

where E_S is the cell potential when the electrodes are immersed in the standard buffer. Similarly, if the cell potential is E_U when the electrodes are immersed in a solution of unknown pH also at 25°C, we have

$$pH_U = -\frac{E_U - K}{0.0592}$$

TABLE 23-6 Values of NIST Primary-Standard pH Solutions from 0° to 60°C[a]

Temperature, °C	Sat'd (25°C) KH Tartrate	0.05 *m* KH_2 Citrate[b]	0.05 *m* KHphthalate	0.025 *m* KH_2PO_4/ 0.025 *m* Na_2HPO_4	0.008695 *m* KH_2PO_4/ 0.03043 *m* Na_2HPO_4	0.01 *m* $Na_4B_4O_7$	0.025 *m* $NaHCO_3$/ 0.025 *m* Na_2CO_3
0	—	3.863	4.003	6.984	7.534	9.464	10.317
5	—	3.840	3.999	6.951	7.500	9.395	10.245
10	—	3.820	3.998	6.923	7.472	9.332	10.179
15	—	3.802	3.999	6.900	7.448	9.276	10.118
20	—	3.788	4.002	6.881	7.429	9.225	10.062
25	3.557	3.776	4.008	6.865	7.413	9.180	10.012
30	3.552	3.766	4.015	6.853	7.400	9.139	9.966
35	3.549	3.759	4.024	6.844	7.389	9.102	9.925
40	3.547	3.753	4.035	6.838	7.380	9.068	9.889
45	3.547	3.750	4.047	6.834	7.373	9.038	9.856
50	3.549	3.749	4.060	6.833	7.367	9.011	9.828
55	3.554	—	4.075	6.834	—	8.985	—
60	3.560	—	4.091	6.836	—	8.962	—
Buffer capacity[c] (mol/pH unit)	0.027	0.034	0.016	0.029	0.016	0.020	0.029
ΔpH1/2 for 1:1 dilution[d]	0.049	0.024	0.052	0.080	0.07	0.01	0.079

[a]Adapted from R. G. Bates, *Determination of pH,* 2nd ed., p. 73. New York: Wiley, 1973.
[b]m = molality (mol solute/kg H_2O).
[c]The buffer capacity is the number of moles of a monoprotic strong acid or base that causes 1.0 L to change pH by 1.0.
[d]Change in pH that occurs when one volume of buffer is diluted with one volume of H_2O.

By subtracting the first equation from the second and solving for pH_U, we find

$$pH_U = pH_S - \frac{(E_U - E_S)}{0.0592} \qquad (23\text{-}30)$$

Equation 23-30 has been adopted throughout the world as the *operational definition of pH.*[22]

Workers at the National Institute of Standards and Technology and elsewhere have used cells without liquid junctions to study primary-standard buffers extensively. Some of the properties of these buffers are presented in Table 23-6 and discussed in detail elsewhere.[23] For general use, the buffers can be prepared from relatively inexpensive laboratory reagents. For careful work, certified buffers can be purchased from the NIST.

23H POTENTIOMETRIC TITRATIONS

The potential of a suitable indicator electrode is conveniently employed to establish the equivalence point for a titration (a *potentiometric titration*).[24] A potentiometric titration provides different information than does a direct potentiometric measurement. For example, the direct measurement of 0.100 M acetic and 0.100 M hydrochloric acid solutions with a pH-sensitive electrode yield widely different pH values because the former is only partially dissociated. On the other hand, potentiometric titrations of equal volumes of the two acids require the same amount of standard base for neutralization.

The potentiometric end point is widely applicable and provides inherently more accurate data than the corresponding method employing indicators. It is particularly useful for titration of colored or turbid solutions and for detecting the presence of unsuspected species in a solution. Unfortunately, it is more time consuming than a titration that makes use of an indicator unless an automatic titrator is used. We will not discuss details of the potentiometric titration techniques here because most elementary analytical texts treat this subject thoroughly.[25]

---

22 Equation 23-30 applies only to solutions at 25°C. A more general equation is

$$pH_U = pH_S - \frac{(E_U - E_S)F}{2.303\,RT} = pH_S - \frac{(E_U - E_S)}{1.984 \times 10^{-4}\,T}$$

where T is the temperature of the sample and the standard buffer and pH_S is the pH of the standard at temperature T.

23 R. G. Bates, *Determination of pH,* 2nd ed., Chapter 4. New York: Wiley, 1973.

24 For a monograph on this method, see E. P. Sergeant, *Potentiometry and Potentiometric Titrations.* New York: Wiley, 1984.

25 For example, see D. A. Skoog, D. M. West, and F. J. Holler, *Fundamentals of Analytical Chemistry,* 7th ed., Chapter 19. Philadelphia: Saunders College Publishing, 1996.

23I QUESTIONS AND PROBLEMS

23-1 List the advantages and disadvantages of a potentiometric titration relative to a titration with visual indicators.

23-2 What is the source of the alkaline error in pH measurement with a glass electrode?

23-3 Compare the average temperature coefficients in the range of 15° to 35°C for the five reference electrodes listed in Table 23-1.

23-4 What occurs when the pH-sensitive tip of a newly manufactured glass electrode is immersed in water?

23-5 Differentiate between an electrode of the second kind and an electrode of the third kind.

23-6 A solution of methylamine is titrated with HCl using a glass/calomel electrode system. Show how K_b can be obtained from the pH at the point of half-neutralization.

23-7 What is the source of
(a) the boundary potential in a membrane electrode?
(b) a junction potential in a glass/calomel electrode system?
(c) the potential of a crystalline membrane electrode used to determine the concentration of F^-?

23-8 Calculate the theoretical potential of the following cells. (In each case assume that activities are approximately equal to molar concentrations and that the temperature is 25°C.)
(a) SCE‖Fe^{3+}(0.0150 M),Fe^{2+}(0.0222 M)|Pt
(b) SCE‖Zn^{2+}(0.00165 M)|Zn
(c) Saturated Ag/AgCl reference‖Ti^{3+}(0.0110 M),Ti^{2+}(0.0350 M)|Pt
(d) Saturated Ag/AgCl reference‖I_3^-(0.00562 M),I^-(0.00313 M)|Pt

23-9 (a) Calculate the standard potential for the reaction

$$CuBr(s) + e^- \rightleftharpoons Cu(s) + Br^-$$

For CuBr, $K_{sp} = 5.2 \times 10^{-9}$.
(b) Give a schematic representation of a cell with a copper indicator electrode as an anode and a saturated calomel electrode as a cathode that could be used for the determination of Br^-.
(c) Derive an equation that relates the measured potential of the cell in (b) to pBr (assume that the junction potential is zero).
(d) Calculate the pBr of a bromide-containing solution that is saturated with CuBr and contained in the cell described in (b) if the resulting potential is −0.071 V.

23-10 (a) Calculate the standard potential for the reaction

$$Ag_3AsO_4(s) + 3e^- \rightleftharpoons 3Ag(s) + AsO_4^{3-}$$

For Ag_3AsO_4, $K_{sp} = 1.2 \times 10^{-22}$.
(b) Give a schematic representation of a cell with a silver indicator electrode as the cathode and a saturated calomel electrode as an anode that could be used for determining AsO_4^{3-}.
(c) Derive an equation that relates the measured potential of the cell in (b) to $pAsO_4$ (assume that the junction potential is zero).
(d) Calculate the $pAsO_4$ of a solution that is saturated with Ag_3AsO_4 and contained in the cell described in (b) if the resulting potential is 0.312 V.

23-11 The following cell was employed for the determination of $pCrO_4$:

$$Ag|Ag_2CrO_4(\text{sat'd}),CrO_4^{2-}(x\text{M})\|SCE$$

Calculate $pCrO_4$ if the cell potential is −0.402 V.

23-12 The following cell was employed to determine the pSO_4 of a solution:

$$Hg|Hg_2SO_4(\text{sat'd}),SO_4^{2-}(x\text{M})\|SCE$$

Calculate the pSO_4 if the potential was −0.577 V.

23-13 The formation constant for the mercury(II) acetate complex is

$$Hg^{2+} + 2OAc^- \rightleftharpoons Hg(OAc)(aq) \qquad K_f = 2.7 \times 10^8$$

Calculate the standard potential for the half-reaction

$$Hg(OAc)_2(aq) + 2e^- \rightleftharpoons Hg(l) + 2OAc^-$$

23-14 The standard electrode potential for the reduction of the Cu(II) complex of EDTA is given by

$$CuY^{2-} + 2e^- \rightleftharpoons Cu(s) + Y^{4-} \qquad E^0 = -0.22 \text{ V}$$

Calculate the formation constant for the reaction

$$Cu^{2+} + Y^{4-} \rightleftharpoons CuY^{2-}$$

23-15 Calculate the potential of the cell (neglecting the junction potential)

$$Hg|HgY^{2-}(2.61 \times 10^{-4}), Y^{4-}(x\text{M})\|SCE$$

where Y^{4-} is the EDTA anion, and the concentration of Y^{4-} is (a) 1.21×10^{-1} M, (b) 1.21×10^{-3} M, and (c) 1.21×10^{-5} M.

$$HgY^{2-} + 2e^- \rightleftharpoons Hg(l) + Y^{4-} \qquad E^0 = 0.21 \text{ V}$$

23-16 The following cell was found to have a potential of 0.124 V:

$$SCE\|Cu^{2+}(3.25 \times 10^{-3}\text{ M})|\text{membrane electrode for } Cu^{2+}$$

When the solution of known copper activity was replaced with an unknown solution, the potential was found to be 0.105 V. What was the pCu of this unknown solution? Neglect the junction potential.

23-17 The following cell was found to have a potential of 1.020 V:

$$Cd|CdX_2(\text{sat'd}), X^-(0.0200\text{ M})\|SCE$$

Calculate the solubility product of CdX_2, neglecting the junction potential.

23-18 The following cell was found to have a potential of 0.515 V:

$$Pt, H_2(1.00\text{ atm})|HA(0.200\text{ M}), NaA(0.300\text{ M})\|SCE$$

Calculate the dissociation constant of HA, neglecting the junction potential.

23-19 A 40.00-mL aliquot of 0.0500 M HNO_2 is diluted to 75.0 mL and titrated with 0.0800 M Ce^{4+}. Assume the hydrogen ion concentration is at 1.00 M throughout the titration. (Use 1.44 V for the formal potential of the cerium system.)

(a) Calculate the potential of the indicator cathode with respect to a saturated calomel reference electrode after the addition of 10.0, 25.00, 40.00, 49.00, 49.9, 50.1, 51.00, 55.00, and 60.00 mL of cerium(IV).

(b) Draw a titration curve for these data.

23-20 A glass/calomel electrode system was found to develop a potential of −0.0412 V when used with a buffer of pH 6.00; with an unknown solution the potential was observed to be −0.2004 V.

(a) Calculate the pH and $[H^+]$ of the unknown.

(b) Assume that K in Equation 23-24 is uncertain by ±0.001 V as a consequence of a difference in the junction potential between standardization and measurement. What is the range of $[H^+]$ associated with this uncertainty?

(c) What is the relative error in $[H^+]$ associated with the uncertainty in E_j?

23-21 The following cell was found to have a potential of 0.2714 V:

$$\text{SCE} \| \text{Mg}^{2+}(a = 3.32 \times 10^{-3}\ \text{M}) | \text{membrane electrode for Mg}^{2+}$$

(a) When the solution of known magnesium activity was replaced with an unknown solution, the potential was found to be 0.1901 V. What was the pMg of this unknown solution?

(b) Assuming an uncertainty of ± 0.002 V in the junction potential, what is the range of Mg^{2+} activities within which the true value might be expected?

(c) What is the relative error in $[Mg^{2+}]$ associated with the uncertainty in E_j?

23-22 A 0.400-g sample of toothpaste was boiled with a 50-mL solution containing a citrate buffer and NaCl to extract the fluoride ion. After cooling, the solution was diluted to exactly 100 mL. The potential of a selective ion/calomel system in a 25.0-mL aliquot of the sample was found to be -0.1823 V. Addition of 5.0 mL of a solution containing 0.00107 mg F^-/mL caused the potential to change to -0.2446 V.

Calculate the weight-percent F^- in the sample.

Chapter 24

Coulometry

Three electroanalytical methods are based upon electrolytic oxidation or reduction of an analyte for a sufficient period to assure its quantitative conversion to a new oxidation state. These methods are: ***constant-potential coulometry, constant-current coulometry,*** *or* ***coulometric titrations,*** *and* ***electrogravimetry.*** *In electrogravimetric methods, the product of the electrolysis is weighed as a deposit on one of the electrodes. In the two coulometric procedures, on the other hand, the quantity of electricity needed to complete the electrolysis serves as a measure of the amount of analyte present.*

The three methods generally have moderate selectivity, sensitivity, and speed; in many instances, they are among the most accurate and precise methods available to the chemist, with uncertainties of a few tenths percent relative being common. Finally, in contrast to all of the other methods discussed in this text, these three require no calibration against standards; that is, the functional relationship between the quantity measured and the weight of analyte can be derived from theory.

A description of electrogravimetric methods is found in many elementary textbooks.[1] Thus, only the two coulometric methods will be considered in this chapter.

24A CURRENT-VOLTAGE RELATIONSHIPS DURING AN ELECTROLYSIS

An electrolysis can be performed in one of three ways: (1) with the applied cell potential held constant, (2) with the electrolysis current held constant, or (3) with the potential of the working electrode held constant. It is useful to consider the consequences of each of these modes of operation. For all three, the behavior of the cell is governed by an equation similar to Equation 22-20.

$$E_{appl} = E_c - E_a + (\eta_{cc} + \eta_{ck}) + (\eta_{ac} + \eta_{ak}) - IR \quad (24\text{-}1)$$

where E_{appl} is the applied potential from an external source and E_c and E_a are the reversible, or thermody-

[1] For example, see D. A. Skoog, D. M. West, and F. J. Holler, *Fundamentals of Analytical Chemistry,* 7th ed., Chapter 20. Philadelphia: Saunders College Publishing, 1995.

namic, potentials associated with the cathode and anode, respectively; their values can, of course, be calculated from standard potentials by means of the Nernst equation. The terms η_{cc} and η_{ck} are overvoltages due to concentration polarization and kinetic polarization at the cathode; η_{ac} and η_{ak} are the corresponding anodic overvoltages (see Section 22E-2). It is important to appreciate that the four overvoltages *always carry a negative sign* because they are potentials that must be overcome in order for charge to pass through the cell.[2]

Of the seven potential terms on the right side of Equation 24-1, only E_c and E_a are derivable from theoretical calculations; the others can only be evaluated empirically.

24A-1 Operation of a Cell at a Fixed Applied Potential

The simplest way of performing an analytical electrolysis is to maintain the applied cell potential at a more or less constant value. Ordinarily, however, this procedure has distinct limitations. These limitations can be understood by considering current-voltage relationships in a typical electrolytic cell. For example, consider a cell for the determination of copper(II), which consists of two platinum electrodes, each with a surface area of 150 cm^2, immersed in 200 mL of a solution that is 0.0220 M with respect to copper(II) ion and 1.00 M with respect to hydrogen ion. The cell resistance is 0.50 Ω. With application of a suitable potential difference between the two electrodes, copper is deposited upon the cathode, and oxygen is evolved at a partial pressure of 1.00 atm at the anode. The overall cell reaction is

$$Cu^{2+} + H_2O \rightarrow Cu(s) + \frac{1}{2}O_2(g) + 2H^+$$

Initial Thermodynamic Potential

Standard potential data for the two half-reactions in the cell under consideration are

$$Cu^{2+} + 2e^- \rightleftharpoons Cu(s) \qquad E^0 = 0.34\ V$$

$$\frac{1}{2}O_2(g) + 2H^+ + 2e^- \rightleftharpoons H_2O \qquad E^0 = 1.23\ V$$

Using the method shown in Example 22-6, the thermodynamic potential for this cell can be shown to be −0.94 V. Thus, no current would be expected at less negative applied potentials; at greater potentials, a linear increase in current should be observed in the absence of kinetic or concentration polarization.

Estimation of Required Potential

For the cell under consideration, kinetic polarization occurs only at the anode where oxygen is evolved, the cathodic reaction being rapid and reversible. That is, η_{ck} in Equation 24-1 is zero. Furthermore, concentration polarization at the anode is negligible at all times because the anodic reactant (water) is in large excess compared with concentration changes brought about by the electrolysis; therefore, the surface layer will never become depleted of reactant, and η_{ac} in Equation 24-1 is also negligible. Finally, it is readily shown by suitable substitutions into the Nernst equation that the increase in hydrogen ion concentration due to the anodic reaction results in a negligible change in electrode potential (<0.01 V); that is, the thermodynamic anode potential will remain constant throughout the electrolysis. Thus for this example, Equation 24-1 reduces to

$$E_{appl} = E_c - E_a + \eta_{cc} + \eta_{ak} - IR \qquad (24\text{-}2)$$

Let us now assume that we wish to operate the cell initially at a current of 1.5 A, which corresponds to a 0.010 A/cm^2 current density. In a table in the literature[3] it is found that η_{ak}, the oxygen overvoltage, will be about −0.85 V. Furthermore, concentration polarization will be negligible at the outset because the concentration of copper ions is initially high. Equation 24-2 then becomes

$$E_{appl} = 0.29 - 1.23 + 0 + (-0.85) - 0.050 \times 1.5$$

$$= -2.54\ V$$

Thus, a rough estimate of the potential required to produce an initial current of 1.5 A is −2.5 V.

Current Changes during an Electrolysis at Constant Applied Potential

It is useful to consider the changes in current in the cell under discussion when the potential is held constant at −2.5 V throughout the electrolysis. Here, the current

[2] Throughout this chapter, junction potentials are neglected because they are small enough to be of no pertinence to the discussion.

[3] *Handbook of Analytical Chemistry,* L. Meites, Ed., p. 5–184. New York: McGraw-Hill, 1963.

would be expected to decrease with time owing to depletion of copper ions in the solution as well as to an increase in cathodic concentration polarization. In fact, with the onset of concentration polarization, the current decrease becomes exponential in time. That is,

$$I_t = I_0 e^{-kt}$$

where I_t is the current t min after the onset of polarization and I_0 is the initial current. Lingane[4] showed that values for the constant k can be computed from the relationship

$$k = \frac{25.8\ DA}{V\delta}$$

where D is the diffusion coefficient in cm^2/s, or the rate at which the reactant diffuses under a unit concentration gradient. The quantity A is the electrode surface area in cm^2, V is the volume of the solution in cm^3, and δ is the thickness of the surface layer in which the concentration gradient exists. Typical values for D and δ are 10^{-5} cm^2/s and 2×10^{-3} cm. (The constant 25.8 includes the factor of 60 for converting D to cm^2/min, thus making k compatible with the units of t in the equation for I_t.)

When the initial applied potential is -2.5 V, it is found that concentration polarization, and thus an exponential decrease in current, occurs essentially immediately after application of the potential. Figure 24-1a depicts this behavior; the curve shown was derived for the cell under consideration with the aid of the foregoing two equations. After 30 min, the current has decreased from the initial 1.5 A to 0.08 A; by this time, approximately 96% of the copper has been deposited.

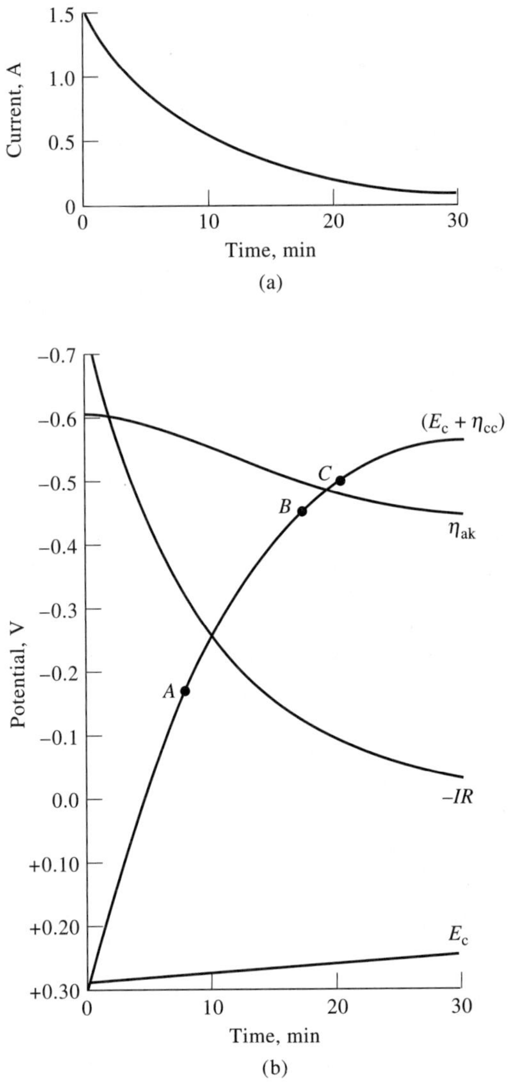

Figure 24-1 Changes in (a) current and (b) potentials during the electrolytic deposition of Cu^{2+}. Points A and B are potentials at which Pb and Cd would begin to codeposit if present. Point C is the potential at which H_2 might begin to form at the cathode. Any of these processes would distort curve ABC.

Potential Changes During an Electrolysis at Constant Applied Potential

It is instructive to consider the changes in the various potentials in Equation 24-2 as the electrolysis under consideration proceeds; Figure 24-1b depicts these changes.

As mentioned earlier, the thermodynamic anode potential remains substantially unchanged throughout the electrolysis because of the large excess of the reactant (water) and the small change in the concentration of reaction product (H^+). The reversible cathode potential E_c, on the other hand, becomes smaller (more negative) as the copper concentration decreases. The curve for E_c in Figure 24-1b was derived by substituting the calculated copper concentration after various electrolysis periods into the Nernst equation. Note that the negative shift in potential is approximately linear with time over a considerable period.

The IR drop shown in Figure 24-1b parallels the current changes shown in Figure 24-1a. The negative sign is employed for consistency with Equation 24-2.

[4] See J. J. Lingane, *Electroanalytical Chemistry,* 2nd ed., pp. 223–229. New York: Interscience, 1958.

As shown by the topmost curve of Figure 24-1b, the oxygen overvoltage η_{ak} also becomes less negative as the current, and thus the current density, falls. The data for this curve were obtained from more extensive compilations that are found in the reference in footnote 3.

The most significant feature of Figure 24-1b is the curve representing the change in *total cathode potential* ($E_c + \eta_{ck}$) as a function of time. It is evident that as *IR* and η_{ak} become less negative, one or more of the other potentials in Equation 24-2 *must become more negative.* Because of the large excess of reactant and product at the anode, its potential remains substantially constant. Thus, the *only* potentials that can change are those associated with the cathode; as seen from a comparison of the curves labeled E_c and ($E_c + \eta_{cc}$), it is evident that most of this negative drift is a consequence of an increase in concentration polarization (η_{cc} becomes more negative). That is, even with vigorous stirring, copper ions are not brought to the electrode surface at a sufficient rate to prevent polarization. The result is a rapid decrease in *IR* and a corresponding negative drift of the total cathode potential.

The shift in cathode potential that accompanies concentration polarization often leads to codeposition of other species and loss of selectivity. For example, points *A* and *B* on the cathode potential curve indicate the approximate potentials and times when lead and cadmium ions would begin to deposit if present in concentrations about equal to the original copper ion concentrations. Another event that would occur in the absence of lead, cadmium, or other easily reduced species would be the evolution of hydrogen at about point *C* (this process was not taken into account in deriving the curve in Figure 24-1a).

The interferences just described could be avoided by decreasing the applied potential by several tenths of a volt so that the negative drift of the cathode potential could never reach a level at which the interfering ions react. The consequence, however, is a diminution in current and, ordinarily, an enormous increase in the time required to complete the analysis.

At best, an electrolysis at constant cell potential can only be employed to separate easily reduced cations, such as Pb(II), Cd(II), Ag(I), Tl(I), Cu(II), and Zn(II) from those that are more difficult to reduce than hydrogen ion, such as Al(III). Evolution of hydrogen would be expected near the end of the electrolysis and would prevent interference by cations that are reduced at more negative potentials.

24A-2 Constant-Current Electrolysis

The analytical electrodeposition under consideration, as well as others, can be carried out by maintaining the current, rather than the applied potential, at a constant level. Here, periodic increases in the applied potential are required as the electrolysis proceeds.

In the preceding section, it was shown that concentration polarization at the cathode causes a decrease in current. Initially, this effect can be partially offset by increasing the applied potential. Electrostatic forces would then postpone the onset of concentration polarization by enhancing the rate at which copper ions are brought to the electrode surface. Soon, however, the solution becomes sufficiently depleted in copper ions so that diffusion, electrostatic attraction, and stirring cannot keep the electrode surface supplied with sufficient copper ions to maintain the desired current. When this occurs, further increases in E_{appl} cause rapid changes in η_{cc} and thus the cathode potential; codeposition of hydrogen (or other reducible species) then takes place. The cathode potential ultimately becomes stabilized at a level fixed by the standard potential and the overvoltage for the new electrode reaction; further large increases in the cell potential are no longer necessary to maintain a constant current. Copper continues to deposit as copper(II) ions reach the electrode surface; the contribution of this process to the total current, however, becomes smaller and smaller as the deposition becomes more and more nearly complete. The alternative process, such as reduction of hydrogen or nitrate ions, soon predominates. The changes in cathode potential under conditions of constant current are shown in Figure 24-2.

24A-3 Electrolysis at Constant Working Electrode Potentials

From the Nernst equation, it is seen that a tenfold decrease in the concentration of an ion being deposited requires a negative shift in potential of only $0.0592/n$ V. Electrolytic methods, therefore, are potentially reasonably selective. For example, as the copper concentration of a solution is decreased from 0.10 M to 10^{-6} M, the thermodynamic cathode potential E_c changes from an initial value of +0.31 to +0.16 V. In theory, then, it should be feasible to separate copper from any element that does not deposit within this 0.15-V potential range. Species that deposit quantitatively at potentials more positive than +0.31 V could be eliminated with a prereduction; ions that require potentials more negative than

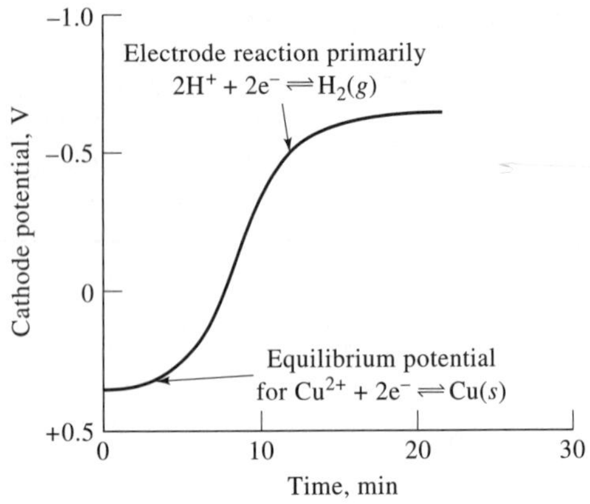

Figure 24-2 Changes in cathode potential during the deposition of copper with a constant current of 1.5 Å. Here, the cathode potential is equal to ($E_c + \eta_{cc}$).

+0.16 V would not interfere with the copper deposition. Thus, if we are willing to accept a reduction in analyte concentration to 10^{-6} M as a quantitative separation, it follows that divalent ions differing in standard potentials by about 0.15 V or greater can, theoretically, be separated quantitatively by electrodeposition, provided their initial concentrations are about the same. Correspondingly, about 0.30- to 0.10-V differences are required for univalent and trivalent ions, respectively.

An approach to these theoretical separations, within a reasonable electrolysis period, requires a more sophisticated technique than the ones thus far discussed because concentration polarization at the cathode, if unchecked, will prevent all but the crudest of separations. The change in cathode potential is governed by the decrease in *IR* drop (Figure 24-1b). Thus, where relatively large currents are employed at the outset, the change in cathode potential can ultimately be expected to be large. On the other hand, if the cell is operated at low current levels so that the variation in cathode potential is lessened, the time required for completion of the deposition may become prohibitively long. An obvious answer to this dilemma is to initiate the electrolysis with an applied cell potential that is sufficiently high to ensure a reasonable current; the applied potential is then continuously decreased to keep the cathode potential at the level necessary to accomplish the desired separation. Unfortunately, it is not feasible to predict the required changes in applied potential on a theoretical basis because of uncertainties in variables affecting the deposition, such as overvoltage effects and perhaps conductivity changes. Nor, indeed, does it help to measure the potential across the two electrodes, since such a measurement gives only the overall cell potential, E_{appl}. The alternative is to measure the potential of the working electrode against a third electrode whose potential in the solution is known and constant—that is, a reference electrode. The potential impressed across the working electrode and its counter electrode can then be adjusted to the level that will impart the desired potential to the cathode (or anode) with respect to the reference electrode. This technique is called *controlled cathode* (or *anode*) *potential electrolysis* or sometimes *potentiostatic electrolysis.*

Experimental details for performing a controlled cathode potential electrolysis are presented in a later section. For the present, it is sufficient to note that the potential difference between the reference electrode and the cathode is measured with an electronic voltmeter. The potential difference applied between the working electrode and its counter electrode is controlled with a voltage divider so that the cathode potential is maintained at a level suitable for the separation. Figure 24-3 is a schematic of a simple manual apparatus that would permit deposition at a constant cathode potential.

An apparatus of the type shown in Figure 24-3 can be operated at relatively high initial applied potentials to give high currents. As the electrolysis progresses, however, a lowering of the applied potential across *AC* is required. This decrease, in turn, diminishes the cur-

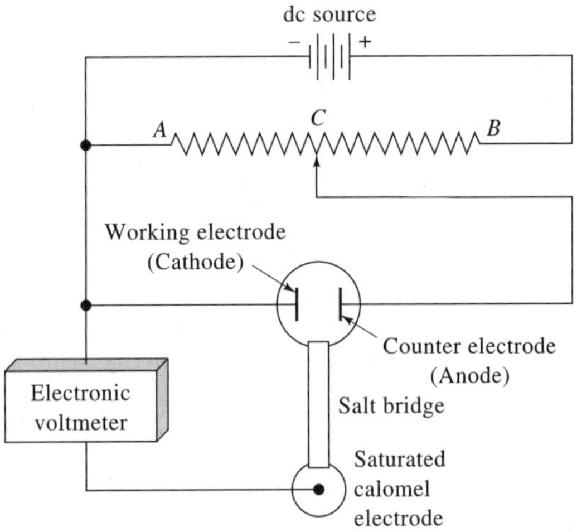

Figure 24-3 Apparatus for electrolysis at a controlled cathode potential. Contact *C* is continuously adjusted to maintain the cathode potential at the desired level.

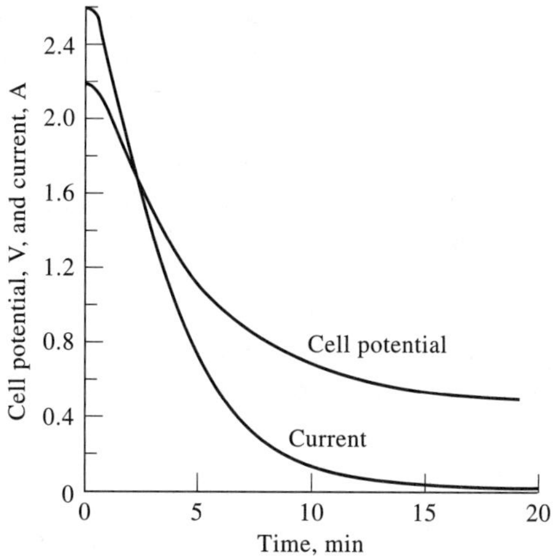

Figure 24-4 Changes in applied potential and current during a controlled-cathode-potential electrolysis. Deposition of copper upon a cathode maintained at −0.36 vs. SCE. *(Experimental data from J. J. Lingane,* Anal. Chem. Acta, ***1948,*** *2, 590. With permission.)*

rent. Completion of the electrolysis will be indicated by the approach of the current to zero. The changes that occur in a typical constant cathode potential electrolysis are depicted in Figure 24-4. In contrast to the electrolytic methods described earlier, this technique demands constant attention during operation. Usually, some provision is made for automatic control; otherwise, the operator time required represents a major disadvantage to the controlled cathode potential method. A simple potentiostat, such as that shown in Figure 24-6, is suitable for the control of the potential of a working electrode with respect to the reference electrode.

24B AN INTRODUCTION TO COULOMETRIC METHODS OF ANALYSIS

Coulometry encompasses a group of analytical methods that involve measuring the quantity of electricity (in coulombs) needed to convert the analyte quantitatively to a different oxidation state. In common with gravimetric methods, coulometry offers the advantage that the proportionality constant between the measured quantity (coulombs in this case) and the weight of analyte can be derived from known physical constants; thus, calibration or standardization is not ordinarily required. Coulometric methods are often as accurate as gravimetric or volumetric procedures; they are usually faster and more convenient than the former. Finally, coulometric procedures are readily adapted to automation.[5]

24B-1 Units for Quantity of Electricity

The quantity of electricity or charge can be stated in units of the *coulomb* (C) and the *faraday* (F). *The coulomb is the quantity of charge that is transported in one second by a constant current of one ampere.* Thus, for a constant current of I amperes for t seconds, the number of coulombs Q is given by the expression

$$Q = It \tag{24-3}$$

For a variable current, the number of coulombs is given by the integral

$$Q = \int_0^t I dt \tag{24-4}$$

The faraday is the charge in coulombs associated with one mole of electrons. The charge of the electron is 1.60218×10^{-19} C, and we may, therefore write

$$1\text{F} = 6.02218 \times 10^{23} \frac{e^-}{\text{mol } e^-} \times 1.60218 \times 10^{-19} \frac{\text{C}}{e^-} = 96{,}485 \frac{\text{C}}{\text{mol } e^-}$$

The relationship between the charge and the amount of a species electrolyzed is given by Faraday's law, which is

$$Q = nFN \tag{24-5}$$

EXAMPLE 24-1

A constant current of 0.800 A was used to deposit copper at the cathode and oxygen at the anode of an electrolytic cell. Calculate the grams of each product that was formed in 15.2 min, assuming no other redox reactions.

[5] For summaries of coulometric methods, see E. Bishop, in *Comprehensive Analytical Chemistry,* C. L. Wilson and D. W. Wilson, Eds., Vol. IID. New York: Elsevier, 1975; D. J. Curran, in *Laboratory Techniques in Electroanalytical Chemistry,* 2nd ed., P. T. Kissinger and W. R. Heineman, Eds., Chapter 25. New York: Marcel Dekker, 1996; A. J. Bard and L. B. Faulkner, *Electrochemical Methods,* Chapter 10. New York: Wiley, 1980; G. W. C. Milner and G. Phillips, *Coulometry in Analytical Chemistry.* New York: Pergamon Press, 1967.

The equivalent weights are determined from consideration of the two half-reactions

$$Cu^{2+} + 2e^- \rightarrow Cu(s)$$

$$2H_2O \rightarrow 4e^- + O_2(g) + 4H^+$$

From Equation 24-3, we find

$$Q = 0.800\text{ A} \times 15.2\text{ min} \times 60\text{ s/min}$$

$$= 729.6\text{ A}\cdot\text{s} = 729.6\text{ C}$$

$$\text{wt Cu} = \frac{729.6\text{ C}}{96{,}485\text{ C/mol e}^-} \times \frac{1\text{ mol Cu}}{2\text{ mol e}^-} \times \frac{63.5\text{ g Cu}}{\text{mol Cu}}$$

$$= 0.240\text{ g Cu}$$

and

$$\text{wt O}_2 = \frac{729.6\text{ C}}{96{,}485\text{ C/mol e}^-} \times \frac{1\text{ mol O}_2}{4\text{ mol e}^-} \times \frac{32.0\text{ g O}_2}{\text{mol O}_2}$$

$$= 0.0605\text{ g O}_2$$

24B-2 Types of Coulometric Methods

Two general techniques are used for coulometric analysis, namely *potentiostatic* and *amperostatic*. The first involves maintaining the potential of the *working electrode* (the electrode at which the analytical reaction occurs) at a constant level such that quantitative oxidation or reduction of the analyte occurs without involvement of less reactive species in the sample or solvent. Here, the current is initially high but decreases rapidly and approaches zero as the analyte is removed from the solution (see Figure 24-4). The quantity of electricity required is most commonly measured with an electronic integrator, although other charge-measuring devices and techniques have also been used.

The amperostatic method of coulometry makes use of a constant current, which is continued until an indicator signals completion of the analytical reaction. The quantity of electricity required to attain the end point is then calculated from the magnitude of the current and the time of its passage. The latter method has enjoyed wider application than the former; it is frequently called a *coulometric titration* for reasons that will become apparent later.

A fundamental requirement of all coulometric methods is that the species determined interacts with 100% current efficiency. This requirement means that each faraday of electricity must bring about a chemical change in the analyte that corresponds to one mole of electrons. Current efficiency of 100% does not, however, imply that the analyte must necessarily participate directly in the electron-transfer process at the electrode. Indeed, more often than not, the substance being determined is involved wholly or in part in a reaction that is secondary to the electrode reaction. For example, at the outset of the oxidation of iron(II) at a platinum anode, all of the current results from the reaction

$$Fe^{2+} \rightleftharpoons Fe^{3+} + e^-$$

As the concentration of iron(III) decreases, however, concentration polarization will cause the anode potential to rise until decomposition of water occurs as a competing process. That is,

$$2H_2O \rightleftharpoons O_2(g) + 4H^+ + 4e^-$$

The charge required to complete the oxidation of iron(II) would then exceed that demanded by theory. To avoid the consequent error, an unmeasured excess of cerium(III) can be introduced at the start of the electrolysis. This ion is oxidized at a lower anode potential than is water:

$$Ce^{3+} \rightleftharpoons Ce^{4+} + e^-$$

The cerium(IV) produced diffuses rapidly from the electrode surface, where it then oxidizes an equivalent amount of iron(II):

$$Ce^{4+} + Fe^{2+} \rightarrow Ce^{3+} + Fe^{3+}$$

The net effect is an electrochemical oxidation of iron(II) with 100% current efficiency even though only a fraction of the iron(II) ions are directly oxidized at the electrode surface.

The coulometric determination of chloride provides another example of an indirect process. Here, a silver electrode serves as the anode, and silver ions are produced by the current. These cations diffuse into the solution and precipitate the chloride. A current efficiency of 100% with respect to the chloride ion is achieved even though this ion is neither oxidized nor reduced in the cell.

24C POTENTIOSTATIC COULOMETRY

In potentiostatic coulometry, the potential of the working electrode is maintained at a constant level that will cause the analyte to react quantitatively with the current without involvement of other components in the sample. An analysis of this kind possesses all the advantages of an electrogravimetric method and is not subject to the limitation imposed by the need for a weighable product. The technique can therefore be applied to systems that yield deposits with poor physical properties as well as to reactions that yield no solid product at all. For example, arsenic may be determined coulometrically by the electrolytic oxidation of arsenous acid (H_3AsO_3) to arsenic acid (H_3AsO_4) at a platinum anode. Similarly, the analytical conversion of iron(II) to iron(III) can be accomplished with suitable control of the anode potential.

24C-1 Instrumentation

The instrumentation for potentiostatic coulometry consists of an electrolysis cell, a potentiostat, and an integrating device for determining the number of coulombs by means of Equation 24-4.

Cells

Figure 24-5 illustrates two types of cells that are used for potentiostatic coulometry. The first consists of a platinum-gauze working electrode and a platinum-wire counter electrode, which is separated from the test solution by a porous tube containing the same supporting electrolyte as the test solution (Figure 24-5a). Separating the counter electrode is sometimes necessary to prevent its reaction products from interfering in the analysis. A saturated calomel reference electrode is joined to the test solution by means of a salt bridge. Often this bridge also contains the same electrolyte as the test solution.

The second type of cell is a mercury pool type. A mercury cathode is particularly useful for separating easily reduced elements as a preliminary step in an analysis. For example, copper, nickel, cobalt, silver, and cadmium are readily separated from ions such as aluminum, titanium, the alkali metals, and phosphates. The precipitated elements dissolve in the mercury; little hy-

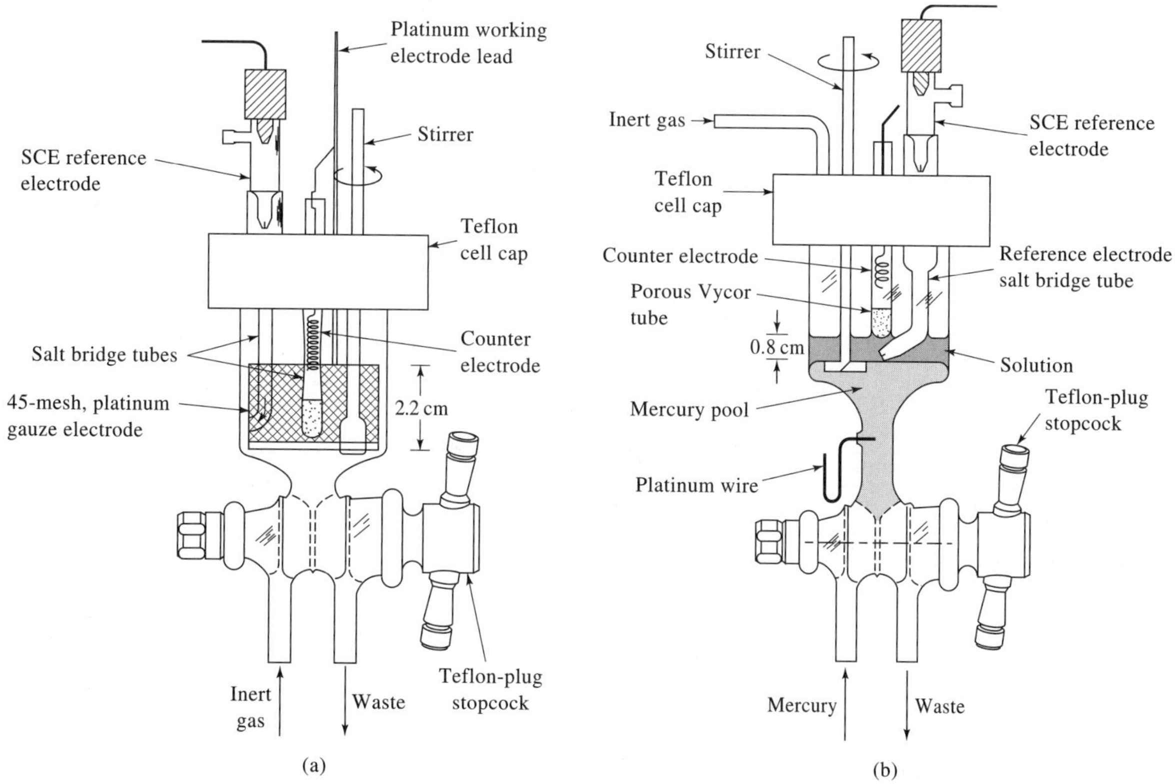

Figure 24-5 Electrolysis cells for potentiostatic coulometry. Working electrode: (a) platinum gauze; (b) mercury pool. *(Reprinted with permission from J. E. Harrar and C. L. Pomernacki,* Anal. Chem., ***1973,*** *45, 57. Copyright 1973 American Chemical Society.)*

drogen evolution occurs even at high applied potentials because of large overvoltage effects. A coulometric cell such as that shown in Figure 24-5b is also useful for coulometric determination of metal ions and certain types of organic compounds as well.

Potentiostats

A potentiostat is an electronic device that maintains the potential of a working electrode at a constant level relative to a reference electrode. Two such devices are shown in Figure 3-12 (page 64). Figure 24-6c is a schematic of an apparatus for potentiostatic coulometry, which contains a somewhat different type of potentiostat. In order to understand how this circuit works, consider the equivalent circuit shown in Figure 24-6a. The two resistances in this diagram correspond to resistances in two parts of the electrochemical cell shown in the Figure 24-6b. Here, R_s is the cell resistance between the counter electrode and the tip P of the reference electrode, and R_u is the so-called *uncompensated cell resistance,* which is the cell resistance between P and the working electrode. Because of the extremely high resistance of the inputs to the operational amplifier, there is no current in the feedback loop to the inverting input, but the potential difference between P and the inverting input of the operational amplifier is simply the reference electrode potential E_{SCE}.

Recall that in the noninverting configuration, the operational amplifier works to keep E_1 and E_2 equal and that the cell current I_c is supplied by the operational amplifier to maintain this condition. If we consider the path between the inverting input and the circuit common at the output, we see that

$$E_2 = E_1 = E_{SCE} + I_c R_u = E_{SCE} + E_c$$

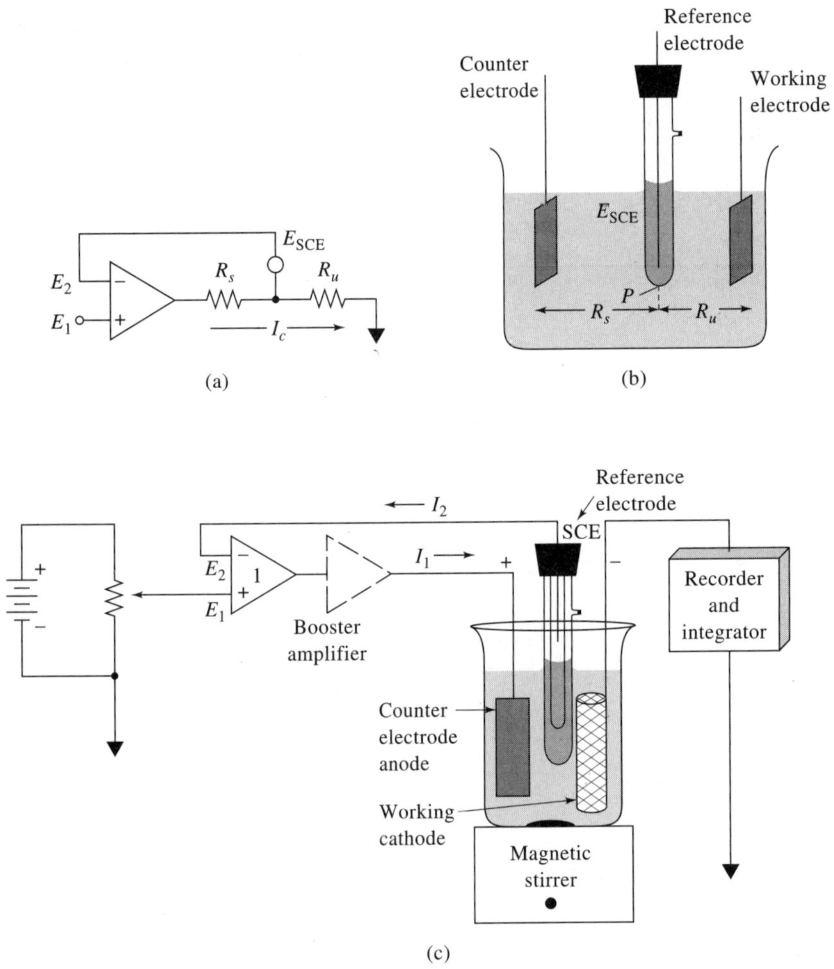

Figure 24-6 Schematic of a system for potentiostatic coulometry. (a) Equivalent circuit. (b) Resistance within the cell. (c) Practical circuit.

where E_c, the cathode potential, is essentially equal to the potential difference between P and the working cathode (see Figure 24-6b). Since E_1 and E_{SCE} are constant, $I_c R_u$ must also be constant. If R_u or R_s change in any way during the electrolysis, the operational amplifier output voltage changes in such a way as to maintain $E_c = I_c R_u$ at a constant level. If R_u increases as a result of an increase in the cell resistance or concentration polarization, the output voltage of the operational amplifier decreases, which leads to a decrease in I_c; if R_u decreases, the operational amplifier output voltage increases correspondingly to maintain E_c constant.

The practical circuit in Figure 24-6c shows other components that are necessary to carry out potentiostatic coulometry. This circuit includes a variable voltage source at the noninverting input of the operational amplifier so that the potentiostat control potential can be varied, a booster amplifier to supply the high currents that are often necessary, and a recorder and integrator. The presence of the booster amplifier has no effect on the potential control circuit.

Integrators

As shown in Section 3E-3, integrators can be constructed from operational amplifier circuits. Most modern apparatus for potentiostatic coulometry, however, employ digital integrators to determine the number of coulombs required to complete an electrolysis.

24C-2 Applications

Controlled potential coulometric methods have been applied to the determination of some 55 elements in inorganic compounds.[6] Mercury appears to be favored as the cathode, and methods for the deposition of two dozen or more metals at this electrode have been described. The method has found widespread use in the nuclear energy field for the relatively interference-free determination of uranium and plutonium.

The controlled potential coulometric procedure also offers possibilities for the electrolytic determination (and synthesis) of organic compounds. For example, Meites and Meites[7] have demonstrated that trichloroacetic acid and picric acid are quantitatively reduced at a mercury cathode whose potential is suitably controlled:

$$Cl_3CCOO^- + H^+ + 2e^- \rightarrow Cl_2HCCOO^- + Cl^-$$

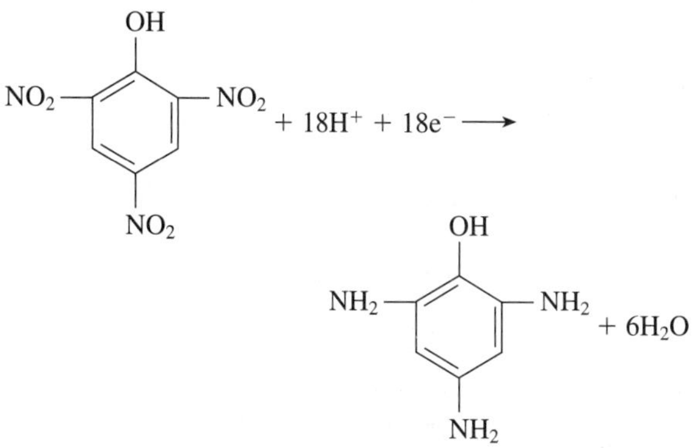

Coulometric measurements permit the analysis of these compounds with a relative error of a few tenths of a percent.

Variable current coulometric methods are frequently used to monitor continuously and automatically the concentration of constituents in gas or liquid streams. An important example is the determination of small concentrations of oxygen.[8] A schematic of the apparatus is shown in Figure 24-7. The porous silver cathode serves to break up the incoming gas into small bubbles; the reduction of oxygen takes place quantitatively within the pores. That is,

$$O_2(g) + 2H_2O + 4e^- \rightleftharpoons 4OH^-$$

The anode is a heavy cadmium sheet; here the half-cell reaction is

$$Cd(s) + 2OH^- \rightleftharpoons Cd(OH)_2(s) + 2e^-$$

Note that a galvanic cell is formed so that no external power supply is required. Nor is a potentiostat necessary because the potential of the working anode can never become great enough to cause oxidation of other species. The current produced is passed through a standard resistor and the potential drop is recorded. The oxygen concentration is proportional to this potential, and the chart paper can be made to display the instantaneous oxygen concentration directly. The instrument is

[6] For a summary of the applications, see J. E. Harrar, *Electroanalytical Chemistry*, A. J. Bard, Ed., Vol. 8. New York: Marcel Dekker, 1975; E. Bishop, in *Comprehensive Analytical Chemistry*, C. L. Wilson and D. W. Wilson, Eds., Vol. IID, Chapter XV. New York: Elsevier, 1975.

[7] T. Meites and L. Meites, *Anal. Chem.*, **1955**, *27*, 1531; **1956**, *28*, 103.

[8] For further details, see F. A. Keidel, *Ind. Eng. Chem.*, **1960**, *52*, 490.

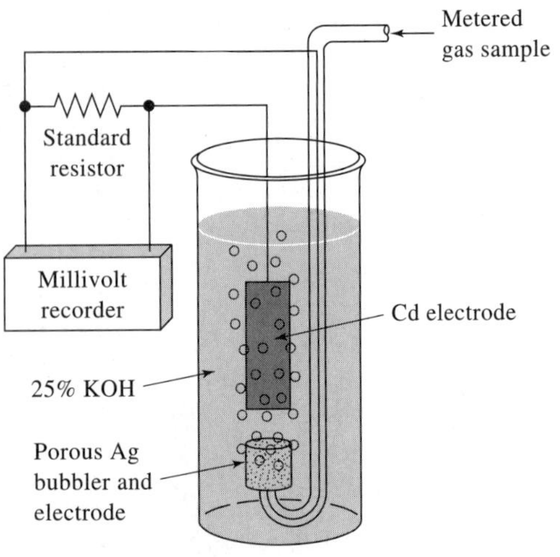

Figure 24-7 An instrument for continuously recording the O_2 content of a gas stream.

reported to provide oxygen concentration data in the range from 1 ppm to 1%.

24D COULOMETRIC TITRATIONS (AMPEROSTATIC COULOMETRY)

A coulometric titration employs a titrant that is electrolytically generated by a constant current. In some analyses, the active electrode process involves only generation of the reagent; an example is the titration of halides by silver ions produced at a silver anode. In other titrations, the analyte may also be directly involved at the generator electrode; an example of the latter is the coulometric oxidation of iron(II)—in part by electrolytically generated cerium(IV) and in part by direct electrode reaction (Section 24B-2). Under any circumstance, the net process must approach 100% current efficiency with respect to a single chemical change in the analyte.

The current in a coulometric titration is carefully maintained at a constant and accurately known level by means of an *amperostat;* the product of this current in amperes and the time in seconds required to reach an end point yields the number of coulombs, which is proportional to the quantity of analyte involved in the electrolysis. The constant current aspect of this operation precludes the quantitative oxidation or reduction of the unknown species entirely at the generator electrode because concentration polarization of the solution is inevitable before the electrolysis can be complete. The electrode potential must then rise if a constant current is to be maintained (Section 24A-2). Unless this potential rise produces a reagent that can react with the analyte, the current efficiency will be less than 100%. In a coulometric titration, then, at least part (and frequently all) of the reaction involving the analyte occurs away from the surface of the working electrode.

A coulometric titration, like a more conventional volumetric procedure, requires some means of detecting the point of chemical equivalence. Most of the end points applicable to volumetric analysis are equally satisfactory here; visual observations of color changes of indicators, as well as potentiometric, amperometric, and conductance measurements have all been successfully applied.

24D-1 Electrical Apparatus

Coulometric titrators are available from several laboratory supply houses. In addition, they can be readily assembled from components available in most analytical laboratories.

Figure 24-8 depicts the principal components of a typical coulometric titrator. Included are a source of constant current and a switch that simultaneously initiates the current and starts an electronic timer. Also required is the means for accurately measuring the current; in Figure 24-8, the potential drop across the standard resistor R_{std} is used for this measurement.

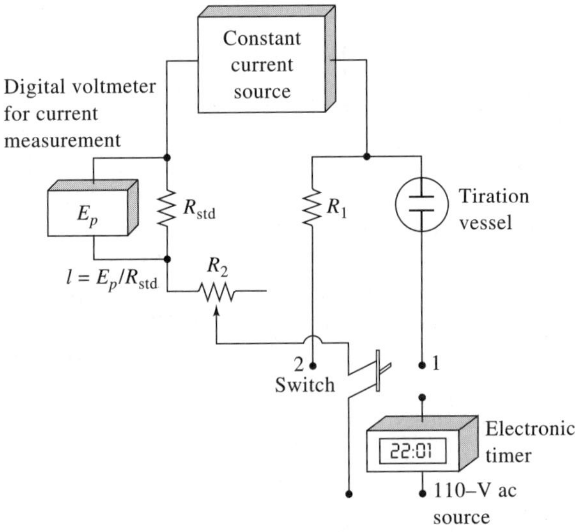

Figure 24-8 Schematic of a coulometric titration apparatus.

Many electronic or electromechanical amperostats are described in the literature. The ready availability of inexpensive operational amplifiers makes their construction a relatively simple matter (for example, see Figure 3-13, page 64).

Cells for Coulometric Titrations

A typical coulometric titration cell is shown in Figure 24-9. It consists of a generator electrode at which the reagent is formed and an auxiliary electrode to complete the circuit. The generator electrode, which should have a relatively large surface area, is often a rectangular strip or a wire coil of platinum; a gauze electrode such as the cathode shown in Figure 24-5a can also be employed.

The products formed at the second electrode frequently represent potential sources of interference. For example, the anodic generation of oxidizing agents is often accompanied by the evolution of hydrogen from the cathode; unless this gas is allowed to escape from the solution, reaction with the oxidizing agent becomes a likelihood. To eliminate this type of difficulty, the second electrode is isolated by a sintered disk or some other porous medium.

An alternative to isolation of the auxiliary electrode is a device such as that shown in Figure 24-10 in which the reagent is generated externally. The apparatus is so arranged that flow of the electrolyte continues briefly after the current is discontinued, thus flushing the residual reagent into the titration vessel. Note that the apparatus shown in Figure 24-10 provides either hydrogen or hydroxide ions depending upon which arm is used. The apparatus has also been used for generation of other reagents such as iodine produced by oxidation of iodide at the anode.

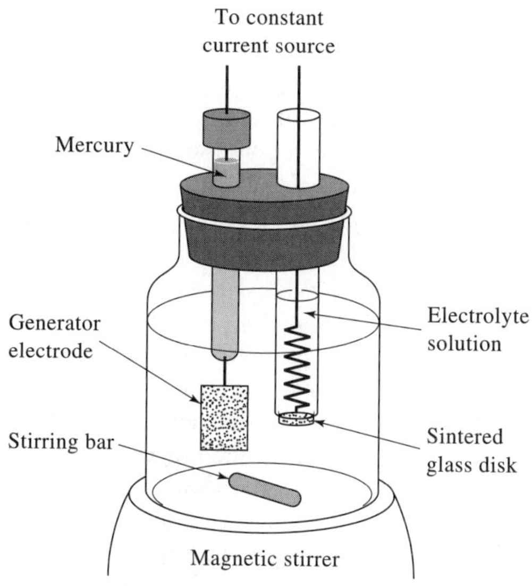

Figure 24-9 A typical coulometric titration cell.

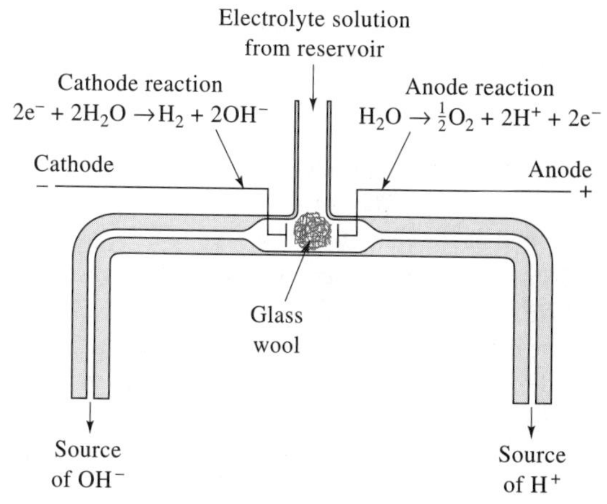

Figure 24-10 A cell for external generation of acid and base.

24D-2 Applications of Coulometric Titrations[9]

Coulometric titrations have been developed for all types of volumetric reactions. Selected applications are described in the following paragraphs.

Neutralization Titrations

Both weak and strong acids can be titrated with a high degree of accuracy using hydroxide ions generated at a cathode by the reaction

$$2H_2O + 2e^- \rightarrow 2OH^- + H_2(g)$$

The cells shown in Figures 24-9 and 24-10 can be employed. A convenient alternative involves substitution of a silver wire as the anode and the addition of chloride

[9] Applications of the coulometric procedure are summarized by E. Bishop, in *Comprehensive Analytical Chemistry,* C. L. Wilson and D. W. Wilson, Eds., Vol. IID, Chapters XVIII–XXIV. New York: Elsevier, 1975.

or bromide ions to the solution of the analyte. The anode reaction then becomes

$$Ag(s) + Br^- \rightleftharpoons AgBr(s) + e^-$$

Clearly, the silver bromide will not interfere with the neutralization reaction as would the hydrogen ions that are formed at most anodes.

Both potentiometric and indicator end points can be employed for these titrations. The problems associated with the estimation of the equivalence point are identical with those encountered in a conventional volumetric analysis. A real advantage to the coulometric method, however, is that interference by carbonate ion is far less troublesome; it is only necessary to eliminate carbon dioxide from the solution containing the analyte by aeration with a carbon dioxide free gas before beginning the analysis.

The coulometric titration of strong and weak bases can be performed with hydrogen ions generated at a platinum anode.

$$H_2O \rightleftharpoons \frac{1}{2}O_2(g) + 2H^+ + 2e^-$$

Here, the cathode must be isolated from the solution or external generation must be employed to prevent interference from the hydroxide ions produced at that electrode.

Precipitation and Complex-Formation Titrations

A variety of coulometric titrations involving anodically generated silver ions have been developed (see Table 24-1). A cell, such as that shown in Figure 24-9, can be employed with a generator electrode constructed from a length of heavy silver wire. End points are detected potentiometrically or with chemical indicators. Similar analyses, based upon the generation of mercury(I) ion at a mercury anode, have been described.

An interesting coulometric titration makes use of a solution of the amine mercury(II) complex of ethylenediaminetetraacetic acid (H_4Y).[10] The complexing agent is released to the solution as a result of the following reaction at a mercury cathode:

$$HgNH_3Y^{2-} + NH_4^+ + 2e^- \rightleftharpoons Hg + 2NH_3 + HY^{3-} \qquad (24\text{-}6)$$

Because the mercury chelate is more stable than the corresponding complexes with calcium, zinc, lead, or copper, complexation of these ions will not occur until the electrode process frees the ligand.

[10] C. N. Reilley and W. W. Porterfield, *Anal. Chem.*, **1956**, *28*, 443.

Oxidation/Reduction Titrations

Table 24-2 indicates the variety of reagents that can be generated coulometrically and the analyses to which they have been applied. Electrogenerated bromine has proved to be particularly useful among the oxidizing agents and forms the basis for a host of methods. Of interest also are some of the unusual reagents not ordinarily encountered in volumetric analysis because of the instability of their solutions; these include dipositive silver ion, tripositive manganese, and the chloride complex of unipositive copper.

Comparison of Coulometric and Volumetric Titrations

It is of interest to point out several analogies between coulometric and volumetric methods and apparatus. Both require an observable end point and are subject to a titration error as a consequence. Furthermore, in both techniques, the amount of analyte is determined through evaluation of its combining capacity—in the one case, for a standard solution and, in the other, for electrons. Also similar demands are made of the reactions; that is, they must be rapid, essentially complete, and free of side reactions. Finally, a close analogy exists between the various components of the apparatus shown in Figure 24-8 and the apparatus and solutions employed in a conventional volumetric analysis. The constant current source of known magnitude serves the same function as the standard solution in a volumetric method. The timer and switch correspond closely to the buret, the switch performing the same function as a stopcock. During the early phases of a coulometric titration, the switch is kept closed for extended periods; as the end point is approached, however, small additions of "reagent" are achieved by closing the switch for shorter and shorter intervals. The similarity to the operation of a buret is obvious.

Some real advantages can be claimed for a coulometric titration in comparison with the classical volumetric process. Principal among these is the elimination of problems associated with the preparation, standardization, and storage of standard solutions. This advantage is particularly important with labile reagents such as chlorine, bromine, or titantium(III) ion; owing to their instability, these species are inconvenient as volumetric reagents. Their utilization in coulometric analy-

TABLE 24-1 Summary of Applications of Coulometric Titrations Involving Neutralization, Precipitation, and Complex-Formation Reactions

Species Determined	Generator-Electrode Reaction	Secondary Analytical Reaction
Acids	$2H_2O + 2e^- \rightleftharpoons 2OH^- + H_2$	$OH^- + H^+ \rightleftharpoons H_2O$
Bases	$H_2O \rightleftharpoons 2H^+ + \frac{1}{2}O_2 + 2e^-$	$H^+ + OH^- \rightleftharpoons H_2O$
Cl^-, Br^-, I^-	$Ag \rightleftharpoons Ag^+ + e^-$	$Ag^+ + Cl^- \rightleftharpoons AgCl(s)$, etc.
Mercaptans	$Ag \rightleftharpoons Ag^+ + e^-$	$Ag^+ + RSH \rightleftharpoons AgSR(s) + H^+$
Cl^-, Br^-, I^-	$2\,Hg \rightleftharpoons Hg_2^{2+} + 2e^-$	$Hg_2^{2+} + 2Cl^- \rightleftharpoons Hg_2Cl_2(s)$, etc.
Zn^{2+}	$Fe(CN)_6^{3-} + e^- \rightleftharpoons Fe(CN)_6^{4-}$	$3Zn^{2+} + 2K^+ + 2Fe(CN)_6^{4-} \rightleftharpoons K_2Zn_3[Fe(CN)_6]_2(s)$
Ca^{2+}, Cu^{2+}, Zn^{2+}, and Pb^{2+}	See Equation 24-6	$HY^{3-} + Ca^{2+} \rightleftharpoons CaY^{2-} + H^+$, etc.

sis is straightforward, however, because they undergo reaction with the analyte immediately after being generated.

Where small quantities of reagent are required, a coulometric titration offers a considerable advantage. By proper choice of current, microquantities of a substance can be introduced with ease and accuracy; the equivalent volumetric process requires small volumes of very dilute solutions, a recourse that is always difficult.

A single constant current source can be employed to generate precipitation, complex formation, oxidation/reduction, or neutralization reagents. Furthermore, the coulometric method is readily adapted to automatic titrations, because current control is easily accomplished.

TABLE 24-2 Summary of Applications of Coulometric Titrations Involving Oxidation/Reduction Reactions

Reagent	Generator-Electrode Reaction	Substance Determined
Br_2	$2Br^- \rightleftharpoons Br_2 + 2e^-$	As(III), Sb(III), U(IV), Tl(I), I^-, SCN^-, NH_3, N_2H_4, NH_2OH, phenol, aniline, mustard gas; 8-hydroxyquinoline
Cl_2	$2Cl^- \rightleftharpoons Cl_2 + 2e^-$	As(III), I^-
I_2	$2I^- \rightleftharpoons I_2 + 2e^-$	As(III), Sb(III), $S_2O_3^{2-}$, H_2S
Ce^{4+}	$Ce^{3+} \rightleftharpoons Ce^{4+} + e^-$	Fe(II), Ti(III), U(IV), As(III), I^-, $Fe(CN)_6^{4-}$
Mn^{3+}	$Mn^{2+} \rightleftharpoons Mn^{3+} + e^-$	$H_2C_2O_4$, Fe(II), As(III)
Ag^{2+}	$Ag^+ \rightleftharpoons Ag^{2+} + e^-$	Ce(III), V(IV), $H_2C_2O_4$, As(III)
Fe^{2+}	$Fe^{3+} + e^- \rightleftharpoons Fe^{2+}$	Cr(VI), Mn(VII), V(V), Ce(IV)
Ti^{3+}	$TiO^{2+} + 2H^+ + e^- \rightleftharpoons Ti^{3+} + H_2O$	Fe(III), V(V), Ce(IV), U(VI)
$CuCl_3^{2-}$	$Cu^{2+} + 3Cl^- + e^- \rightleftharpoons CuCl_3^{2-}$	V(V), Cr(VI), IO_3^-
U^{4+}	$UO_2^{2+} + 4H^+ + 2e^- \rightleftharpoons U^{4+} + 2H_2O$	Cr(VI), Ce(IV)

Coulometric titrations are subject to five potential sources of error: (1) variation in the current during electrolysis, (2) departure of the process from 100% current efficiency, (3) error in the measurement of current, (4) error in the measurement of time, and (5) titration error due to the difference between the equivalence point and the end point. The last of these difficulties is common to volumetric methods as well; where the indicator error is the limiting factor, the two methods are likely to have comparable reliability.

With simple instrumentation, currents constant to 0.2% relative are easily achieved; with somewhat more sophisticated apparatus, control to 0.01% is obtainable. In general, then, errors due to current fluctuations are seldom of importance.

Although generalizations concerning the magnitude of uncertainty associated with the electrode process are difficult, current efficiencies of 99.5% to better than 99.9% are often reported in the literature. Currents are readily measured to ±0.1% relative.

To summarize, then, the current-time measurements required for a coulometric titration are inherently as accurate as or more accurate than the comparable volume-molarity measurements of a classical volumetric analysis, particularly where small quantities of reagent are involved. Often, however, the accuracy of a titration is not limited by these measurements but by the sensitivity of the end point; in this respect, the two procedures are equivalent.

24D-3 Automatic Coulometric Titrations

A number of instrument manufacturers offer automatic coulometric titrators. Most of these employ the potentiometric end point. Some of the commercial instruments are multipurpose and can be used for the determination of a variety of species. Others are designed for a single analysis. Examples of the latter include: chloride titrators in which silver ion is generated coulometrically; sulfur dioxide monitors, where anodically generated bromine oxidizes the analyte to sulfate ions; carbon dioxide monitors in which the gas, absorbed in monoethanolamine, is titrated with coulometrically generated base; and water titrators in which Karl Fischer reagent is generated electrolytically.

24E QUESTIONS AND PROBLEMS

24-1 Lead is to be deposited at a cathode from a solution that is 0.125 M in Pb^{2+} and 0.250 M in $HClO_4$. Oxygen is evolved at a pressure of 0.800 atm at a 30-cm^2 platinum anode. The cell has a resistance of 0.950 Ω.

(a) Calculate the thermodynamic potential of the cell.
(b) Calculate the *IR* drop if a current of 0.300 A is to be used.
(c) Estimate the total applied potential required to begin electrodeposition at the specified conditions.
(d) What potential will be required when the Pb^{2+} concentration is 0.00100 M, assuming that all other variables remain unchanged?

24-2 Calculate the minimum difference in standard electrode potentials needed to lower the concentration of the metal M_1 to 1.00×10^{-4} M in a solution that is 2.00×10^{-1} M in the less reducible metal M_2 where

(a) M_2 is univalent and M_1 is divalent.
(b) M_2 and M_1 are both divalent.
(c) M_2 is trivalent and M_1 is univalent.
(d) M_2 is divalent and M_1 is univalent.
(e) M_2 is divalent and M_1 is trivalent.

24-3 It is desired to separate and determine bismuth, copper, and silver in a solution that is 0.0650 M in BiO^+, 0.175 M in Cu^{2+}, 0.0962 M in Ag^+, and 0.500 M in $HClO_4$.

(a) Using 1.00×10^{-6} M as the criterion for quantitative removal, determine whether or not separation of the three species is feasible by controlled cathode potential electrolysis.
(b) If any separations are feasible, evaluate the range (vs. SCE) within which the cathode potential should be controlled for the deposition of each.
(c) Calculate the potential (vs. SCE) needed to deposit the third ion quantitatively.

24-4 Halide ions can be deposited at a silver anode, the reaction being

$$Ag(s) + X^- \rightarrow AgX(s) + e^-$$

Suppose that a cell was formed by immersing a silver anode in an analyte solution that was 0.0250 M Cl^-, Br^-, and I^- ions and connecting the half-cell to a saturated calomel cathode via a salt bridge.
(a) Which halide would form first and at what potential? Is the cell galvanic or electrolytic?
(b) Could I^- and Br^- be separated quantitatively? (Take 1.00×10^{-5} M as criteria for quantitative removal of an ion.) If a separation is feasible, what range of cell potential could be used?
(c) Repeat part (b) for I^- and Cl^-.
(d) Repeat part (b) for Br^- and Cl^-.

24-5 What cathode potential (vs. SCE) would be required to lower the total Hg(II) concentration of the following solutions to 1.00×10^{-6} M (assume reaction product in each case is elemental Hg):
(a) an aqueous solution of Hg^{2+}?
(b) a solution with an equilibrium SCN^- concentration of 0.100 M?

$$Hg^{2+} + 2SCN^- \rightleftharpoons Hg(SCN)_2(aq) \qquad K_f = 1.8 \times 10^7$$

(c) a solution with an equilibrium Br^- concentration of 0.250 M?

$$HgBr_4^{2-} + 2e^- \rightleftharpoons Hg(l) + 4Br^- \qquad E^0 = 0.223 \text{ V}$$

24-6 Calculate the time needed for a constant current of 0.800 A to deposit 0.100 g of
(a) Tl(III) as the element on a cathode.
(b) Tl(I) as the Tl_2O_3 on an anode.
(c) Tl(I) as the element on a cathode.

24-7 At a potential of −1.0 V (vs. SCE), carbon tetrachloride in methanol is reduced to chloroform at a Hg cathode:

$$2CCl_4 + 2H^+ + 2e^- + 2Hg(l) \rightarrow 2CHCl_3 + Hg_2Cl_2(s)$$

At −1.80 V, the chloroform further reacts to give methane:

$$2CHCl_3 + 6H^+ + 6e^- + 6Hg(l) \rightarrow 2CH_4 + 3Hg_2Cl_2(s)$$

A 0.750-g sample containing CCl_4, $CHCl_3$, and inert organic species was dissolved in methanol and electrolyzed at −1.0 V until the current approached zero. A coulometer indicated that 11.63 C had been used. The reduction was then continued at −1.80 V; an additional 68.6 C were required to complete the reaction. Calculate the percent CCl_4 and $CHCl_3$ in the mixture.

24-8 A 6.39-g sample of an ant-control preparation was decomposed by wet-ashing with H_2SO_4 and HNO_3. The arsenic in the residue was reduced to the trivalent state with hydrazine. After the excess reducing agent had been removed, the arsenic(III) was oxidized with electrolytically generated I_2 *in a faintly alkaline medium:*

$$HAsO_3^{2-} + I_2 + 2HCO_3^- \rightarrow HAsO_4^{2-} + 2I^- + 2CO + H_2O$$

The titration was complete after a constant current of 101.1 mA had been passed for 12 min and 36 s. Express the results of this analysis in terms of the percentage As_2O_3 in the original sample.

24-9 A 0.0809-g sample of a purified organic acid was dissolved in an alcohol/water mixture and titrated with coulometrically generated hydroxide ions. With a current of 0.0324 A, 251 s were required to reach a phenolphthalein end point. Calculate the equivalent weight of the acid.

24-10 Traces of aniline can be determined by reaction with an excess of electrolytically generated Br_2:

$$C_6H_5NH_2 + 3Br_2 \rightarrow C_6H_2Br_3NH_2 + 3H^+ + 3Br^-$$

The polarity of the working electrode is then reversed, and the excess bromine is determined by a coulometric titration involving the generation of Cu(I):

$$Br_2 + 2Cu^+ \rightarrow 2Br^- + 2Cu^{2+}$$

Suitable quantities of KBr and copper(II) sulfate were added to a 25.0-mL sample containing aniline. Calculate the micrograms of $C_6H_5NH_2$ in the sample from the accompanying data:

Working Electrode Functioning as	Generation Time (min) with a Constant Current of 1.00 mA
anode	3.76
cathode	0.270

Chapter 25

Voltammetry

Voltammetry *comprises a group of electroanalytical methods in which information about the analyte is derived from the measurement of current as a function of applied potential obtained under conditions that encourage polarization of an indicator, or working, electrode. Generally, in order to enhance polarization, the working electrodes in voltammetry are* ***microelectrodes*** *having surface areas of a few square millimeters at the most and in some applications, a few square micrometers or less.*

At the outset, it is worthwhile pointing out the basic differences between voltammetry and the two types of electrochemical methods that were discussed in earlier chapters. Voltammetry is based upon the measurement of a current that develops in an electrochemical cell under conditions of complete concentration polarization. In contrast, potentiometric measurements are made at currents that approach zero and where polarization is absent. Voltammetry differs from coulometry in that with the latter, measures are taken to minimize or compensate for the effects of concentration polarization. Furthermore in voltammetry a minimal consumption of analyte takes place, whereas in coulometry essentially all of the analyte is converted to another state.

Historically, the field of voltammetry developed from ***polarography****, which is a particular type of voltammetry that was discovered by the Czechoslovakian chemist Jaroslav Heyrovsky in the early 1920s.*[1] *Polarography, which is still an important branch of voltammetry, differs from other types of voltammetry in that the working microelectrode takes the form of a* ***dropping mercury electrode*** *(DME). The construction and unique properties of this electrode are discussed in a later section.*

Voltammetry is widely used by inorganic, physical, and biological chemists for nonanalytical purposes, including fundamental studies of oxidation and reduction processes in various media, adsorption processes on surfaces, and electron transfer mechanisms at chemically modified electrode surfaces. At one time, voltammetry (particularly classical polarography) was an important tool used by chemists for the determination of inorganic ions and certain organic species in aqueous solutions. In the late 1950s and the early 1960s, however, many of these analytical appli-

[1] J. Heyrovsky, *Chem. Listy,* **1922,** *16,* 256. Heyrovsky was awarded the 1959 Nobel Prize in chemistry for his discovery and development of polarography.

cations were supplanted by various spectroscopic methods, and voltammetry became a less important method of analysis except for certain special applications, such as the determination of molecular oxygen in solutions.

In the mid-1960s, several major modifications of classical voltammetric techniques were developed that enhance significantly the sensitivity and selectivity of the method. At about this same time, the advent of low-cost operational amplifiers made possible the commercial development of relatively inexpensive instruments that incorporated many of these modifications and made them available to all chemists. The result has been a recent resurgence of interest in applying voltammetric methods to the determination of a host of species, particularly those of pharmaceutical, environmental, and biological interest.[2] *Furthermore, voltammetry coupled with high-performance liquid chromatography has become a powerful tool for the analysis of complex mixtures of different kinds. Modern voltammetry also continues to be a potent tool used by various kinds of chemists interested in studying oxidation and reduction processes as well as adsorption processes.*[3]

25A EXCITATION SIGNALS IN VOLTAMMETRY

In voltammetry, a variable potential *excitation signal* is impressed upon an electrochemical cell containing a microelectrode. This excitation signal elicits a characteristic current response upon which the method is based. The waveforms of four of the most common excitation signals used in voltammetry are shown in Figure 25-1. The classical voltammetric excitation signal is the linear scan shown in Figure 25-1a, in which the dc potential applied to the cell increases linearly (usually over a 2- to 3-V range) as a function of time. The current that develops in the cell is then recorded as a function of time (and thus as a function of the applied potential).

Two pulse-type excitation signals are shown in Figures 25-1b and 1c. Currents are measured at various times during the lifetime of these pulses. With the triangular waveform shown in Figure 25-1d, the potential is cycled between two values, first increasing linearly to a maximum and then decreasing linearly with the same numerical slope to its original value. This process may be repeated numerous times with the current being recorded as a function of time. A complete cycle may take 100 or more seconds or be completed in less than one second.

The last column in Figure 25-1 lists the types of voltammetry that employ the various excitation signals. These techniques are discussed in the sections that follow.

25B VOLTAMMETRIC INSTRUMENTATION

Figure 25-2 is a schematic showing the components of a modern apparatus for carrying out linear-scan voltammetric measurements. The cell is made up of three electrodes immersed in a solution containing the analyte and also an excess of a nonreactive electrolyte called a *supporting electrolyte.* One of the three electrodes is the microelectrode, or *working electrode,* whose potential is varied linearly with time. Its dimensions are kept small in order to enhance its tendency to become polarized. The second electrode is a reference electrode (commonly a saturated calomel or a silver/silver chloride electrode) whose potential remains constant throughout the experiment. The third electrode is a *counter electrode,* which is often a coil of platinum wire or a pool of mercury that simply serves to conduct electricity from the signal source through the solution to the microelectrode.

The signal source is a linear-scan voltage generator similar to the integration circuit shown in Figure 3-14c (page 65). The output from this type of source is described by Equation 3-20. Thus for a constant dc input potential of E_i, the output potential E_0 is given by

$$E_o = -\frac{E_i}{R_i C_f}\int_0^t dt = -\frac{E_i}{R_i C_f}t \qquad (25\text{-}1)$$

The output signal from the source is fed into a potentiostatic circuit similar to that shown in Figure 24-6c. The electrical resistance of the control circuit containing the

[2] For a brief summary of several of these modified voltammetric techniques, see J. B. Flato, *Anal. Chem.,* **1972,** *44,* 75A.

[3] Some general references that deal with voltammetry include *Analytical Voltammetry,* M. R. Smyth and F. G. Vos, Eds. New York: Elsevier, 1992; A. J. Bard and L. R. Faulkner, *Electrochemical Methods.* New York: Wiley, 1980; *Laboratory Techniques in Electroanalytical Techniques,* 2nd ed. P. T. Kissinger and W. R. Heineman, Eds. New York: Marcel Dekker, 1996.

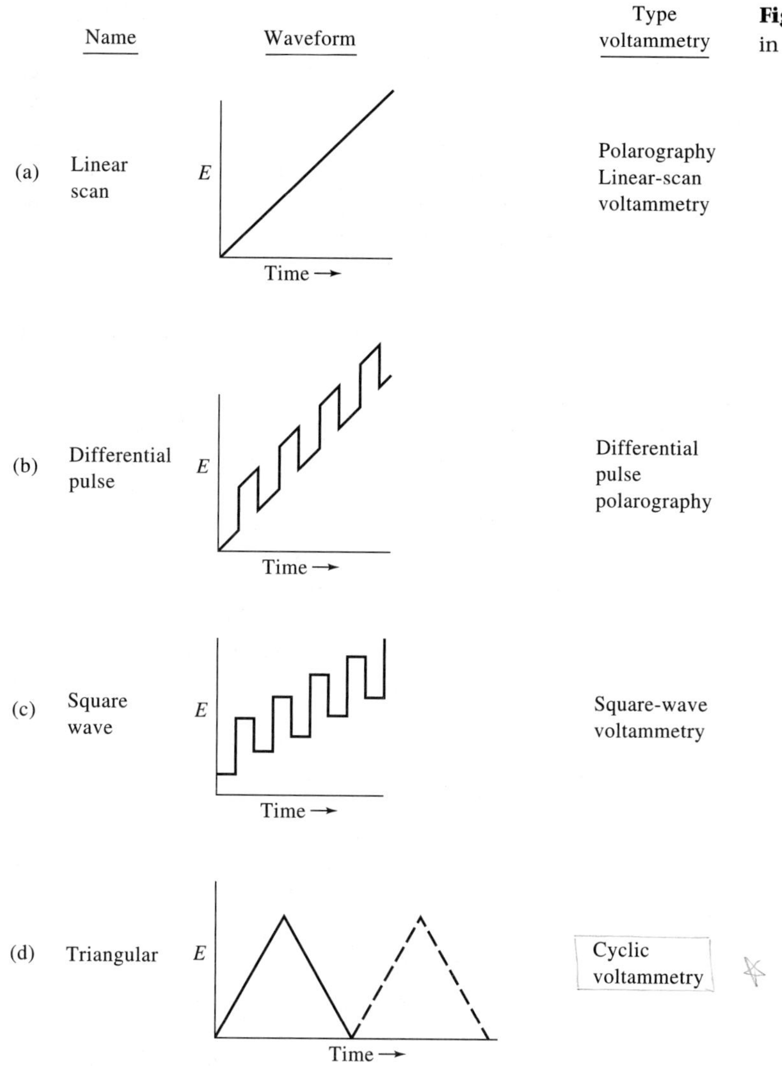

Figure 25-1 Potential excitation signals used in voltammetry.

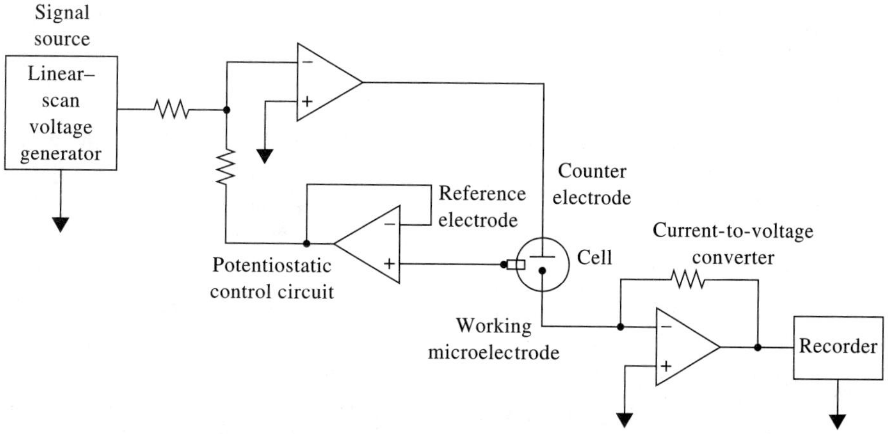

Figure 25-2 A system for potentiostatic three-electrode linear-scan voltammetry.

reference electrode is so large (>10^{11} Ω) that essentially no current is present in it. Thus, the entire current from the source is carried from the counter electrode to the microelectrode. Furthermore, the control circuit adjusts this current so that the potential between the microelectrode and the reference electrode is identical to the output potential from the linear voltage generator. The resulting current is then converted to a voltage and recorded as a function of time, which is directly proportional to the potential between the microelectrode/reference pair.[4] It is important to emphasize that the independent variable in this experiment is the potential of the *microelectrode* versus the reference electrode and not the potential between the microelectrode and the counter electrode. The working electrode is at virtual common potential throughout the course of the experiment.

25B-1 Microelectrodes[5]

The microelectrodes employed in voltammetry take a variety of shapes and forms. Often, they are small flat disks of a conductor that are press fitted into a rod of an inert material, such as Teflon or Kel-F that has imbedded in it a wire contact (see Figure 25-3a). The conductor may be an inert metal, such as platinum or gold; pyrolytic graphite or glassy carbon; a semiconductor, such as tin or indium oxide; or a metal coated with a film of mercury. As shown in Figure 25-4, the range of potentials that can be used with these electrodes in aqueous solutions varies and depends not only upon electrode material but also upon the composition of the solution in which it is immersed. Generally, the positive potential limitations are caused by the large currents that develop due to oxidation of the water to give molecular oxygen. The negative limits arise from the reduction of water giving hydrogen. Note that relatively large negative potentials can be tolerated with mercury electrodes owing to the high overvoltage of hydrogen on this metal.

Mercury microelectrodes have been widely employed in voltammetry for several reasons. One is the relatively large negative potential range just described. Furthermore, a fresh metallic surface is readily formed by simply producing a new drop. The ability to obtain a fresh surface readily is important because the currents measured in voltammetry are quite sensitive to cleanliness and freedom from irregularities. An additional advantage to mercury electrodes is that many metal ions are reversibly reduced to amalgams at the surface of a mercury electrode, which simplifies the chemistry. Mercury microelectrodes take several forms. The simplest of these is a mercury film electrode formed by electrodeposition of the metal onto a disk electrode, such as that shown in Figure 25-3a. Figure 25-3b illustrates a *hanging mercury drop electrode* (HMDE). The electrode, which is available from commercial sources, consists of a very fine capillary tube connected to a mercury-containing reservoir. The metal is forced out of the capillary by a piston arrangement driven by a micrometer screw. The micrometer permits formation of drops having surface areas that are reproducible to five percent or better.

Figure 25-3c shows a typical *dropping mercury electrode* (DME), which was used in nearly all early polarographic experiments. It consists of roughly 10 cm of a fine capillary tubing (i.d. ≈ 0.05 mm) through which mercury is forced by a mercury head of perhaps 50 cm. The diameter of the capillary is such that a new drop forms and breaks every 2 to 6 s. The diameter of the drop is 0.5 to 1 mm and is highly reproducible. In some applications, the drop time is controlled by a mechanical knocker that dislodges the drop at a fixed time after it begins to form.

Figure 25-3d shows a commercially available mercury electrode, which can be operated as a dropping mercury electrode or a hanging drop electrode. The mercury is contained in a plastic-lined reservoir about ten inches above the upper end of the capillary. A compression spring forces the polyurethane-tipped plunger

[4] Early voltammetry was performed with a two-electrode system rather than the three-electrode system shown in Figure 25-2. With a two-electrode system, the second electrode is either a large metal electrode, such as a pool of mercury, or a reference electrode large enough to prevent its polarization during an experiment. This second electrode combines the functions of the reference electrode and the counter electrode in Figure 25-2. Here, it is assumed that the potential of this second electrode remains constant throughout a scan so that the microelectrode potential is simply the difference between the applied potential and the potential of the second electrode. With solutions of high electrical resistance, however, this assumption is not valid because the *IR* drop became significant and increases as the current increases. Distorted voltammograms are the consequence. Almost all voltammetry is now performed with three-electrode systems.

[5] Most of this chapter will deal with voltammetric measurements with working microelectrodes having dimensions in the millimeter range. Since the late 1970s, however, there has been a growing interest in studies with electrodes having dimensions in the micrometer range. We will term such electrodes *ultramicroelectrodes* or *microscopic electrodes*. Such electrodes offer several advantages over classical microelectrodes. We shall describe some of the unique characteristics of ultramicroelectrodes in Section 25G.

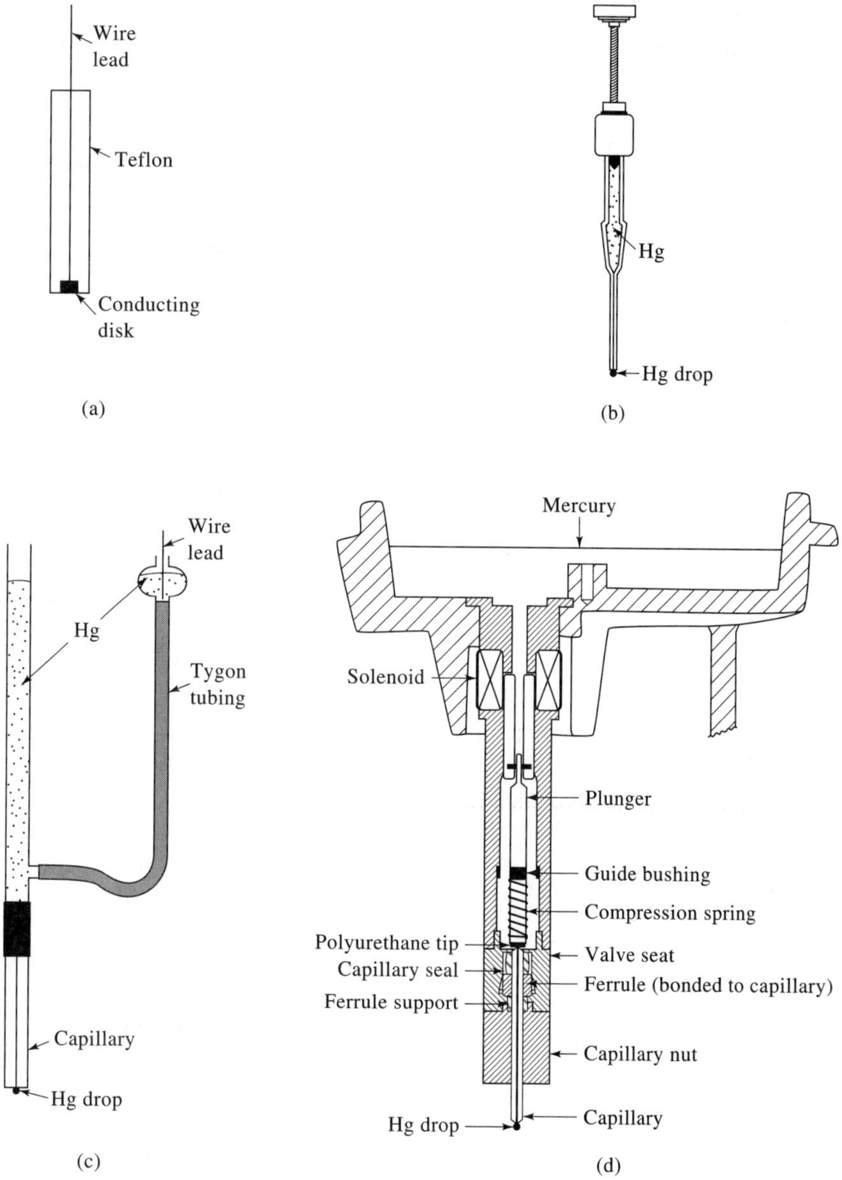

Figure 25-3 Some common types of microelectrodes: (a) a disk electrode, (b) a hanging mercury drop electrode, (c) a dropping mercury electrode, (d) a static mercury drop electrode.

against the head of the capillary, thus preventing a flow of mercury. This plunger is lifted upon activation of the solenoid by a signal from the control system. The capillary is much larger in diameter (0.15 mm) than the typical one. As a result the formation of the drop is extremely rapid. After 50, 100, or 200 ms, the valve is closed, leaving a full-sized drop in place until it is dislodged by a mechanical knocker that is built into the electrode support block. This system has the advantage that the full-sized drop forms quickly and current measurements can be delayed until the surface area is stable and constant. This procedure largely eliminates the large current fluctuations that are encountered with the classical dropping electrode.

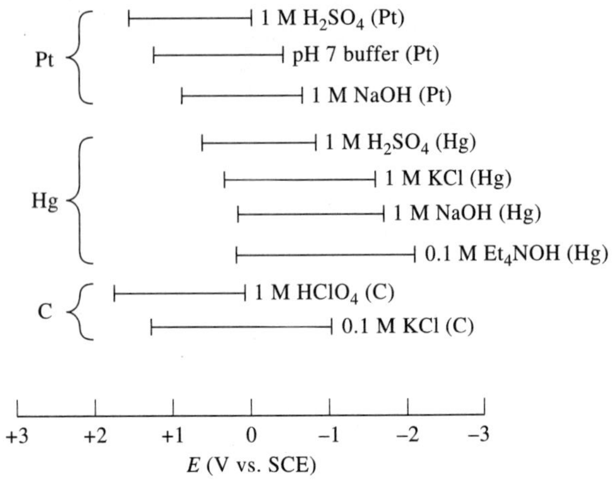

Figure 25-4 Potential ranges for three types of electrodes in various supporting electrolytes. *(Adapted from A. J. Bard and L. R. Faulkner,* Electrochemical Methods, *back cover. New York: Wiley, 1980. Reprinted by permission of John Wiley & Sons, Inc.)*

25B-2 Voltammograms

Figure 25-5 illustrates the appearance of a typical linear-scan voltammogram for an electrolysis involving the reduction of an analyte species A to give a product P at a mercury film microelectrode. Here, the microelectrode is assumed to be connected to the negative terminal of the linear-scan generator so that the applied potentials are given a negative sign as shown. By convention, cathodic currents are always treated as being positive, whereas anodic currents are given a negative sign. In this hypothetical experiment, the solution is assumed to be about 10^{-4} M in A, 0.0 M in P, and 0.1 M in KCl, which serves as the supporting electrolyte. The half-reaction at the microelectrode is the reversible reaction

$$A + ne^- \rightleftharpoons P \qquad E^0 = -0.26 \text{ V} \qquad (25\text{-}2)$$

For convenience, we have neglected the charges on A and P and also have assumed that the standard potential for the half-reaction is −0.26 V.

Linear-scan voltammograms generally assume a sigmoidal-shaped curve called a *voltammetric wave.* The constant current beyond the steep rise is called the diffusion-limited current, or simply the *limiting current* i_l because it arises from a limitation in the rate at which the reactant can be brought to the surface of the electrode by mass-transport processes. Limiting currents are generally directly proportional to reactant concentration. Thus, we may write

$$i_l = kc_A$$

where c_A is the analyte concentration and k is a constant. Quantitative linear-scan voltammetry is based upon this relationship.

The potential at which the current is equal to one half the limiting current is called the *half-wave potential* and given the symbol $E_{1/2}$. After correction for the reference electrode potential (0.242 V with a saturated calomel electrode), the half-wave potential is closely related to the standard potential for the half-reaction but is usually not identical to that constant. Half-wave potentials are sometimes useful for identification of the components of a solution.

In order to obtain reproducible limiting currents rapidly, it is necessary that either (1) the solution or the microelectrode be in continuous and reproducible motion or (2) that a dropping electrode, such as a dropping mercury electrode, be employed. Linear-scan voltammetry in which the solution or the electrode is kept in motion is called *hydrodynamic voltammetry.* Voltammetry that employs a dropping electrode is called *polarography.* We shall consider both.

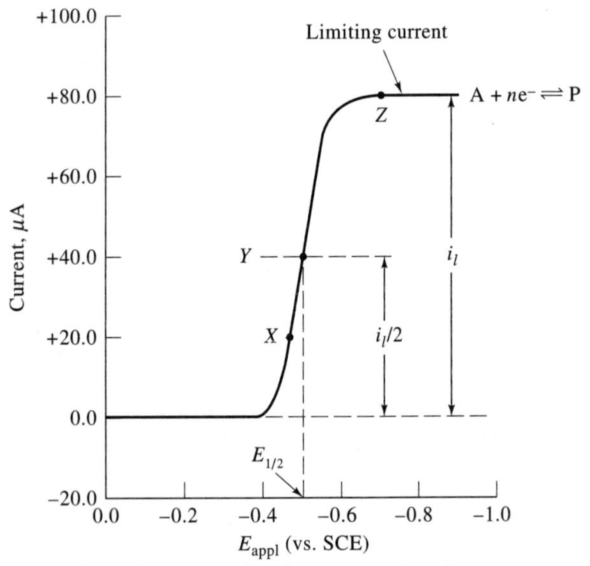

Figure 25-5 Linear-scan voltammogram for the reduction of a hypothetical species A to give a product P.

25C HYDRODYNAMIC VOLTAMMETRY

Hydrodynamic voltammetry is performed in several ways. One method involves stirring the solution vigorously while it is in contact with a fixed microelectrode.

A typical cell for hydrodynamic voltammetry is pictured in Figure 25-6. Here stirring is accomplished with an ordinary magnetic stirrer. Alternatively, the microelectrode is rotated at a constant high speed in the solution, thus providing the stirring action (see Figure 25-17a). Still another way of carrying out hydrodynamic voltammetry involves causing the analyte solution to flow through a tube in which the microelectrode is mounted (Figure 25-14). The last technique is becoming widely used for detecting oxidizable or reducible analytes as they exit from a liquid chromatographic column (Section 28C-6).

As described in Section 22E-3, during an electrolysis, reactant is carried to the surface of an electrode by three mechanisms: migration under the influence of an electric field, convection resulting from stirring or vibration, and diffusion due to concentration differences between the film of liquid at the electrode surface and the bulk of the solution. In voltammetry, every effort is made to minimize the effect of migration by introducing an excess of an inactive supporting electrolyte. When the concentration of supporting electrolyte exceeds that of the analyte by 50- to 100-fold, the fraction of the total current carried by the analyte approaches zero. As a result the rate of migration of the analyte toward the electrode of opposite charge becomes essentially independent of applied potential.

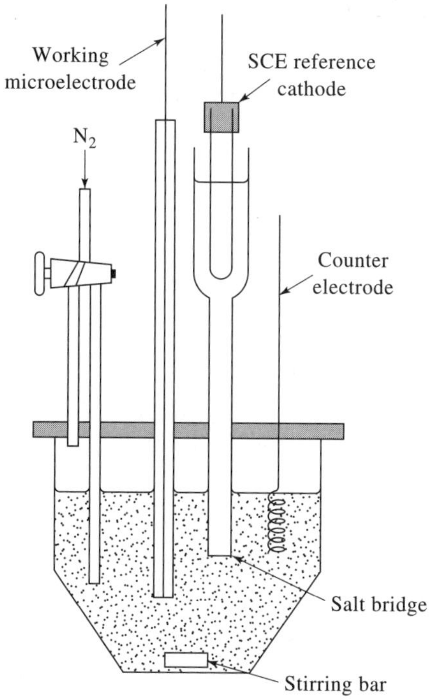

Figure 25-6 A three-electrode cell for hydrodynamic voltammetry.

25C-1 Concentration Profiles at Microelectrode Surfaces during Electrolyses

Throughout this discussion we will consider that the electrode reaction shown in Equation 25-2 takes place at a mercury-coated microelectrode in a solution of A that also contains an excess of a supporting electrolyte. We will assume that the initial concentration of A is c_A while that of P is zero and that P is not soluble in the mercury. We also assume that the reduction reaction is rapid and reversible so that the concentrations of A and P in the film of solution immediately adjacent to the electrode is given at any instant by the Nernst equation:

$$E_{\text{appl}} = E_A^0 - \frac{0.0592}{n} \log \frac{c_P^0}{c_A^0} - E_{\text{ref}} \quad (25\text{-}3)$$

where E_{appl} is the potential between the microelectrode and the reference electrode and c_P^0 and c_A^0 are the molar concentrations of P and A *in a thin layer of solution at the electrode surface only.* We also assume that because the electrode is so very small, the electrolysis, over short periods of time, does not alter the bulk concentration of the solution appreciably. As a consequence, the concentration of A in the bulk of the solution c_A is unchanged by the electrolysis and the concentration of P in the bulk of the solution c_P continues to be, for all practical purpose, zero ($c_P \approx 0$).

Profiles for Planar Electrodes in Unstirred Solutions

Before describing the behavior of a microelectrode in this solution under hydrodynamic conditions, it is instructive to consider what occurs when a potential is applied to a planar electrode, such as that shown in Figure 25-3a, in the absence of convection—that is, in an unstirred solution. Under these conditions mass transport of the analyte to the electrode surface occurs by diffusion alone.

Let us assume that a square-wave excitation potential E_{appl} is applied to the microelectrode for a period of τ s as shown in Figure 25-7a. Let us further assume that E_{appl} is large enough so that the ratio c_P^0/c_A^0 in Equation 25-3 is 1000 or greater. Under this condition, the concentration of A at the electrode surface is, for all practical purposes, immediately reduced to zero ($c_A^0 \rightarrow 0$).

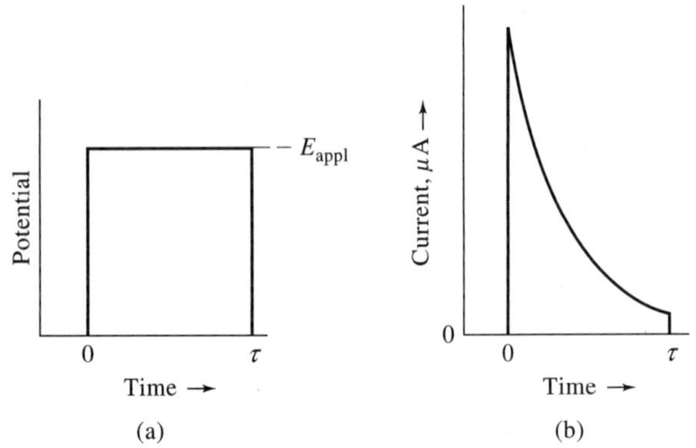

Figure 25-7 Current response to a stepped potential for a planar microelectrode in an unstirred solution. (a) Excitation potential. (b) Current response.

The current response to this step-excitation signal is shown in Figure 25-7b. Initially, the current rises to a peak value that is required to convert essentially all of A in the surface layer of solution to P. Diffusion from the bulk of the solution then brings more A into this surface layer where further reduction occurs. The current required to keep the concentration of A at the level required by Equation 25-2 decreases rapidly with time, however, because A must travel greater and greater distances to reach the surface layer where it can be reduced. Thus, as seen in Figure 25-7b, the current drops off rapidly after its initial surge.

Figure 25-8 shows concentration profiles for A and P after 0, 1, 5, and 10 ms of electrolysis in the system under discussion. Here, the concentration of A (solid lines) and P (dashed line) are plotted as a function of distance from the electrode surface. The graph on the left shows that the solution is homogeneous before application of the stepped potential with the concentration of A being c_A at the electrode surface and in the bulk of the solution as well; the concentration of P is zero in both of these regions. One millisecond after application of the potential, the profiles have changed dramatically. At the surface of the electrode, the concentration of A has been reduced to essentially zero while the concentration of P has increased and become equal to the original concentration of A; that is, $c_P^0 = c_A$. Moving away from the surface, the concentration of A increases linearly with distance and approaches c_A at about 0.01 mm from the surface. A linear decrease in the concentration of P occurs in this same region. As shown in the figure, with time, these concentration gradients extend farther and farther into the solution.

The current i required to produce these gradients is given by the slopes of the straight line portions of the solid lines in Figure 25-7b. That is,

$$i = \frac{\partial c_A}{\partial x} \tag{25-4}$$

As shown in the figure this slope becomes smaller with time; so also does the current.

It is not practical to obtain limiting currents with planar electrodes in unstirred solutions because the cur-

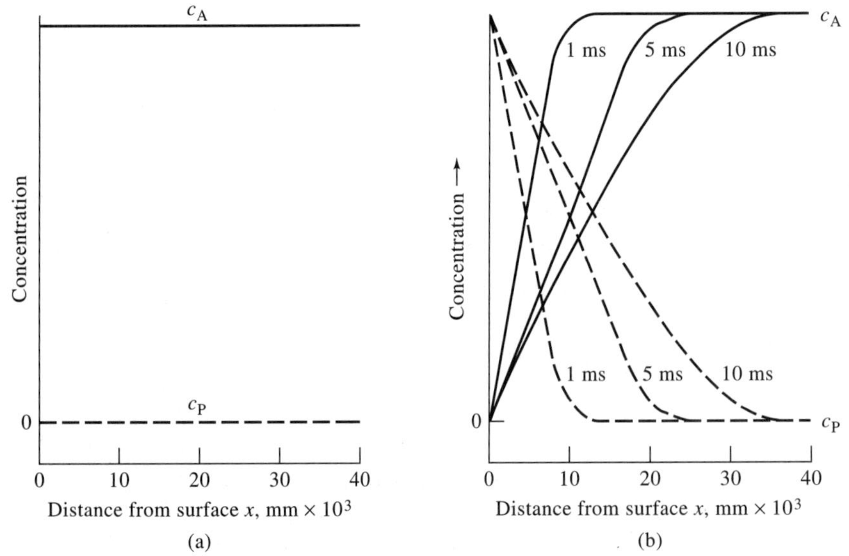

Figure 25-8 Concentration distance profiles during the diffusion-controlled reduction of A to give P at a planar microelectrode. (a) E_{appl} = 0 V. (b) E_{appl} = point z in Figure 25-5; elapsed time: 1, 5, and 10 ms.

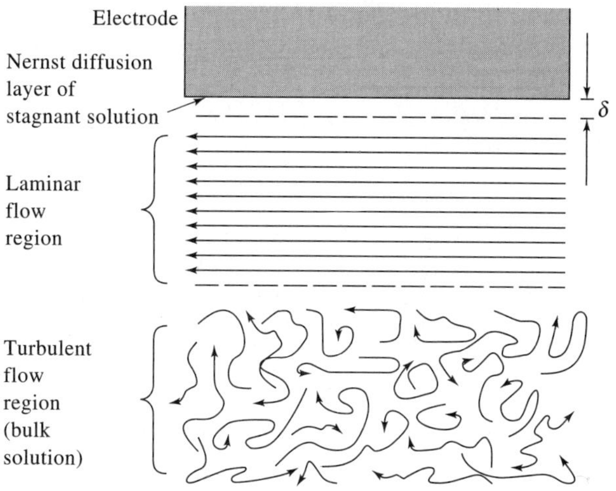

Figure 25-9 Flow patterns at the surface of a microelectrode in a stirred solution.

rents continually decrease with time as the slopes of the concentration profiles become smaller.

Profiles for Microelectrodes in Stirred Solutions

Let us now consider concentration/distance profiles when the reduction described in the previous section is performed at a microelectrode immersed in a solution that is stirred vigorously. In order to understand the effect of stirring, it is necessary to develop a picture of liquid flow patterns in a stirred solution containing a small planar electrode. As shown in Figure 25-9, three types of flow can be identified. (1) *Turbulent flow,* in which liquid motion has no regular pattern, occurs in the bulk of the solution away from the electrode. (2) As the surface is approached, a transition to *laminar flow* takes place. In laminar flow, layers of liquid slide by one another in a direction parallel to the electrode surface. (3) At δ cm from the surface of the electrode, the rate of laminar flow approaches zero as a result of friction between the liquid and the electrode, giving a thin layer of stagnant solution that is called the *Nernst diffusion layer.* It is only within the stagnant Nernst diffusion layer that the concentration of reactant and product vary as a function of distance from the electrode surface. That is, throughout the laminar flow and turbulent flow regions, convection maintains the concentration of A at its original value and the concentration of P at a level that is vanishingly small.

Figure 25-10 shows two sets of concentration profiles for A and P at three potentials shown as *X, Y,* and *Z* in Figure 25-5. In Figure 25-10a, the solution is divided into two regions. One makes up the bulk of the solution, and consists of both the turbulent and laminar flow regions shown in Figure 25-9, where mass transport takes place by mechanical convection brought about by the stirrer. The concentration of A throughout this region is c_A, whereas c_P is essentially zero. The second region is the Nernst diffusion layer, which is immediately adjacent to the electrode surface and has a thickness of δ cm. Typically, δ ranges from 10^{-2} to 10^{-3} cm, depend-

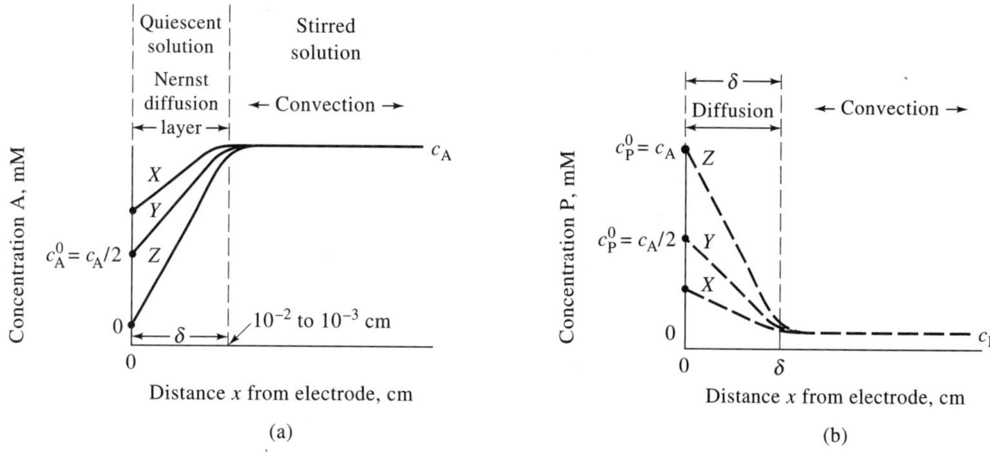

Figure 25-10 Concentration profiles at an electrode/solution interface during the electrolysis $A + ne^- \rightarrow P$ from a stirred solution of A. See Figure 25-5 for potentials corresponding to curves *X, Y,* and *Z.*

ing upon the efficiency of the stirring and the viscosity of the liquid. Within the static diffusion layer, mass transport takes place by diffusion alone, just as was the case with the unstirred solution. With the stirred solution, however, diffusion is limited to a narrow layer of liquid, which even with time, cannot extend out indefinitely into the solution. As a consequence, steady, diffusion-controlled currents are realized shortly after application of a voltage.

As is shown in Figure 25-10, at potential *X*, the equilibrium concentration of A at the electrode surface has been reduced to about 80% of its original value while the equilibrium concentration P has increased by an equivalent amount; that is, $c_P^0 = c_A - c_A^0$. At potential *Y*, which is the half-wave potential, the equilibrium concentrations of the two species at the surface are approximately the same and equal to $c_A/2$. Finally, at potential *Z* and beyond, the surface concentration of A approaches zero, while that of P approaches the original concentration of A, c_A. Thus, at potentials more negative than Z essentially all A ions entering the surface layer are instantaneously reduced to P. As is shown in Figure 25-10b, at potentials greater than Z the concentration of P in the surface layer remains constant at $c_P^0 = c_A$ because of diffusion of P back into the stirred region.

25C-2 Voltammetric Currents

The current at any point in the electrolysis we have just been considering is determined by the rate of transport of A from the outer edge of the diffusion layer to the electrode surface. Because the product of the electrolysis P diffuses away from the surface and is ultimately swept away by convection, a continuous current is required to maintain the surface concentrations demanded by the Nernst equation. Convection, however, maintains a constant supply of A at the outer edge of the diffusion layer. Thus, a steady-state current results that is determined by the applied potential. This current is a quantitative measure of how fast A is being brought to the surface of the electrode, and this rate is given by $\partial c_A/\partial x$ where x is the distance in centimeters from the electrode surface. For a planar electrode, it can be shown that the current is given by the expression

$$i = nFAD_A\left(\frac{\partial c_A}{\partial x}\right) \qquad (25\text{-}5)$$

where i is the current in A, n is the number of moles of electrons per mole of analyte, F is the faraday, A is the electrode surface area in cm^2, D_A is the diffusion coefficient for A in cm^2 s^{-1}, and c_A is the concentration of A in mol cm^{-3}. Note that $\partial c_A/\partial x$ is the slope of the initial part of the concentration profiles shown in Figure 25-10a, and these slopes can be approximated by $(c_A - c_A^0)/\delta$. Under this circumstance, Equation 25-5 reduces to

$$i = \frac{nFAD_A}{\delta}(c_A - c_A^0) = k_A(c_A - c_A^0) \qquad (25\text{-}6)$$

where the constant k_A is equal to $nFAD_A/\delta$.

Equation 25-6 shows that as c_A^0 becomes smaller as a result of a larger negative applied potential, the current increases until the surface concentration approaches zero, at which point the current becomes constant and independent of the applied potential. Thus, when $c_A^0 \to 0$, the current becomes the limiting current i_l, and Equation 25-6 reduces to Equation 25-7[6]

$$i_l = \frac{nFAD_A}{\delta}c_A = k_A c_A \qquad (25\text{-}7)$$

This derivation is based upon an oversimplified picture of the diffusion layer in that the interface between the moving and stationary layers is viewed as a sharply defined edge where transport by convection ceases and transport by diffusion begins. Nevertheless, this simplified model does provide a reasonable approximation of the relationship between current and the variables that affect the current.

Current/Voltage Relationships for Reversible Reactions

In order to develop an equation for the sigmoidal curve shown in Figure 25-5, we substitute Equation 25-7 into Equation 25-6 and rearrange, which gives

$$c_A^0 = \frac{i_l - i}{k_A} \qquad (25\text{-}8)$$

The surface concentration of P can also be expressed in terms of the current by employing a relationship similar to Equation 25-5. That is,

---

[6] Dimensional analysis of this equation leads to

$$\frac{n\left(\frac{\text{mol e}^-}{\text{mol analyte}}\right)F\left(\frac{\text{C}}{\text{mol e}^-}\right)A\ (\text{cm}^2)D_A\left(\frac{\text{cm}^2}{\text{s}}\right)}{\delta\ (\text{cm})} \times c_A\left(\frac{\text{mol analyte}}{\text{cm}^3}\right) = i_l\left(\frac{\text{C}}{\text{s}}\right)$$

By definition, one coulomb per second is one ampere.

$$i = -\frac{nFAD_P}{\delta}(c_P - c_P^0) \qquad (25\text{-}9)$$

where the minus sign results from the negative slope of the concentration profile for P. Note that D_P is now the diffusion coefficient of P. But we have said earlier that throughout the electrolysis the concentration of P approaches zero in the bulk of the solution and, therefore, when $c_P \approx 0.0$

$$i = \frac{-nFAD_P c_P^0}{\delta} = k_P c_P^0$$

where $k_P = -nAD_P/\delta$. Rearranging gives

$$c_P^0 = i/k_P \qquad (25\text{-}10)$$

Substituting Equations 25-8 and 25-10 into Equation 25-3 yields, after rearrangement,

$$E_{appl} = E_A^0 - \frac{0.0592}{n}\log\frac{k_A}{k_P} - \frac{0.0592}{n}\log\frac{i}{i_l - i} - E_{ref} \qquad (25\text{-}11)$$

When $i = i_l/2$, the third term on the right side of this equation becomes equal to zero, and, by definition, E_{appl} is the half-wave potential. That is,

$$E_{appl} = E_{1/2} = E_A^0 - \frac{0.0592}{n}\log\frac{k_A}{k_P} - E_{ref} \qquad (25\text{-}12)$$

Substituting this expression into Equation 25-11 gives an expression for the voltammogram in Figure 25-5. That is,

$$E_{appl} = E_{1/2} - \frac{0.0592}{n}\log\frac{i}{i_l - i} \qquad (25\text{-}13)$$

Often, the ratio k_A/k_P in Equation 25-11 and in Equation 25-12 is nearly unity, so that we may write for the species A

$$E_{1/2} \approx E_A^0 - E_{ref} \qquad (25\text{-}14)$$

Current/Voltage Relationships for Irreversible Reactions

Many voltammetric electrode processes, particularly those associated with organic systems, are irreversible, which leads to drawn-out and less well-defined waves. The quantitative description of such waves requires an additional term (involving the activation energy of the reaction) in Equation 25-11 to account for the kinetics of the electrode process. Although half-wave potentials for irreversible reactions ordinarily show some dependence upon concentration, diffusion currents remain linearly related to concentration; such processes are, therefore, readily adapted to quantitative analysis provided that suitable standards are available for calibration.

Voltammograms for Mixtures of Reactants

Ordinarily, the reactants of a mixture will behave independently of one another at a microelectrode; a voltammogram for a mixture is thus simply the sum of the waves for the individual components. Figure 25-11 shows the voltammograms for a pair of two-component mixtures. The half-wave potentials of the two reactants differ by about 0.1 V in curve *A* and by about 0.2 V in curve *B*. Note that a single voltammogram may permit the quantitative determination of two or more species provided there is sufficient difference between succeeding half-wave potentials to permit evaluation of individual diffusion currents. Generally, 0.1 to 0.2 V is required if the more easily reducible species undergoes a two-electron reduction; a minimum of about 0.3 V is needed if the first reduction is a one-electron process.

Anodic and Mixed Anodic/Cathodic Voltammograms

Anodic waves as well as cathodic waves are encountered in voltammetry. An example of an anodic wave is illustrated in curve *A* of Figure 25-12, where the electrode reaction involves the oxidation of iron(II) to iron(III) in the presence of citrate ion. A limiting current

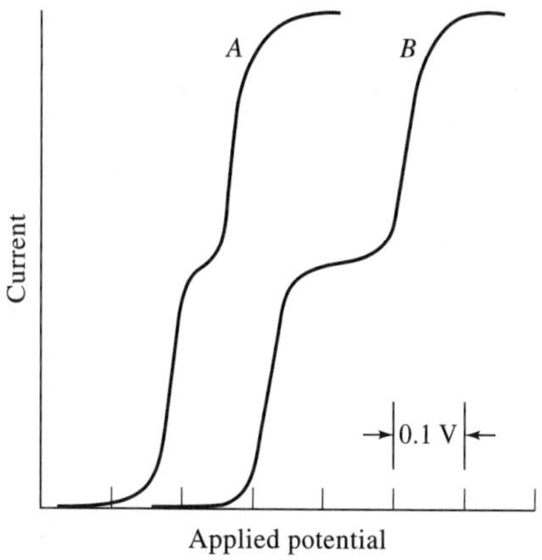

Figure 25-11 Voltammograms for two-component mixtures. Half-wave potentials differ by 0.1 V in curve *A*, and by 0.2 V in curve *B*.

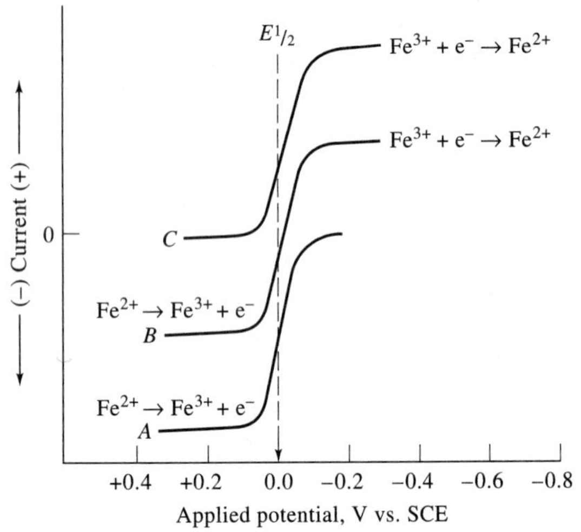

Figure 25-12 Voltammetric behavior of iron(II) and iron(III) in a citrate medium. Curve *A:* anodic wave for a solution in which $[Fe^{2+}] = 1 \times 10^{-4}$. Curve *B:* anodic/cathodic wave for a solution in which $[Fe^{2+}] = [Fe^{3+}] = 0.5 \times 10^{-4}$. Curve *C:* cathodic wave for a solution in which $[Fe^{3+}] = 1 \times 10^{-4}$.

is obtained at about +0.1 V (vs. SCE), which is due to the half-reaction

$$Fe^{2+} \rightleftharpoons Fe^{3+} + e^-$$

As the potential is made more negative, a decrease in the anodic current occurs; at about −0.02 V, the current becomes zero because the oxidation of iron(II) ion has ceased.

Curve *C* represents the voltammogram for a solution of iron(III) in the same medium. Here, a cathodic wave results from reduction of the iron(III) to the divalent state. The half-wave potential is identical with that for the anodic wave, indicating that the oxidation and reduction of the two iron species are perfectly reversible at the microelectrode.

Curve *B* is the voltammogram of an equimolar mixture of iron(II) and iron(III). The portion of the curve below the zero-current line corresponds to the oxidation of the iron(II); this reaction ceases at an applied potential equal to the half-wave potential. The upper portion of the curve is due to the reduction of iron(III).

25C-3 Oxygen Waves

Dissolved oxygen is easily reduced at a microelectrode. Thus, as shown in Figure 25-13, an aqueous solution saturated with air exhibits two distinct waves attributable to this element. The first results from the reduction of oxygen to hydrogen peroxide:

$$O_2(g) + 2H^+ + 2e^- \rightleftharpoons H_2O_2$$

The second corresponds to the further reduction of the hydrogen peroxide:

$$H_2O_2 + 2H^+ + 2e^- \rightleftharpoons 2H_2O$$

As would be expected from stoichiometric considerations, the two waves are of equal height.

Voltammetric measurements offer a convenient and widely used method for determining dissolved oxygen in solutions. However, the presence of oxygen often interferes with the accurate determination of other species. Thus, oxygen removal is ordinarily the first step in amperometric procedures. Deaeration of the solution for several minutes with an inert gas *(sparging)* accomplishes this end; a stream of the same gas, usually nitrogen, is passed over the surface during analysis to prevent reabsorption of oxygen.

25C-4 Applications of Hydrodynamic Voltammetry

Currently, the most important uses of hydrodynamic voltammetry include: (1) detection and determination of chemical species as they exit from chromatographic columns or flow-injection apparatus; (2) routine determination of oxygen and certain species of biochemical interest, such as glucose, lactose, and sucrose; (3) detec-

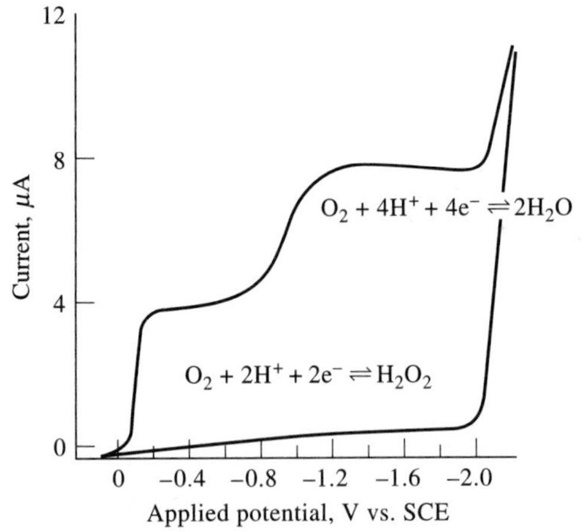

Figure 25-13 Voltammogram for the reduction of oxygen in an air-saturated 0.1-M-KCl solution. The lower curve is for oxygen-free 0.1 M KCl.

tion of end points in coulometric and volumetric titrations; (4) fundamental studies of electrochemical processes.

Voltammetric Detectors[7] in Chromatography and Flow-Injection Analysis

Hydrodynamic voltammetry is becoming widely used for detection and determination of oxidizable or reducible compounds or ions that have been separated by high-performance liquid chromatography or by flow-injection methods. In these applications a thin layer cell such as that shown in Figure 25-14 is used. In these cells the working electrode is typically imbedded in the wall of an insulating block that is separated from a counter electrode by a thin spacer as shown. The volume of such a cell is typically 0.1 to 1 μL. A potential corresponding to the limiting current region for analytes is applied between the metal or glassy carbon working electrode and a silver/silver chloride reference electrode that is located downstream from the detector. In this type of application, detection limits as low as 10^{-9} to 10^{-10} M of analyte are obtained. This application of hydrodynamic voltammetry is considered further in Section 28C-6.

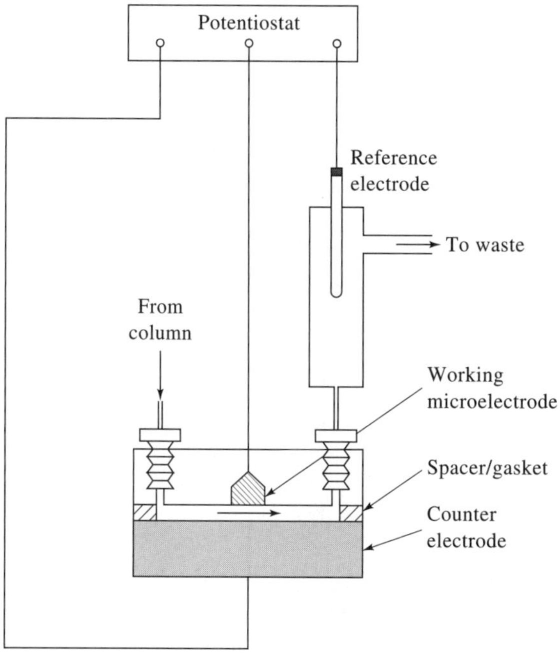

Figure 25-14 A voltammetric system for detecting electroactive species as they exit from a column.

Voltammetric Sensors

A number of voltammetric systems are produced commercially for the determination of specific species that are of interest in industry and research. These devices are sometimes called electrodes or detectors but are, in fact, complete voltammetric cells and are better referred to as sensors. Two of these devices are described here.

Oxygen Sensors. The determination of dissolved oxygen in a variety of aqueous environments, such as seawater, blood, sewage, effluents from chemical plants, and soils, is of tremendous importance to industry, biomedical and environmental research, and clinical medicine. One of the most common and convenient methods for making such measurements is with the *Clark oxygen sensor,* which was patented by L. C. Clark Jr., in 1956.[8] A schematic of the Clark oxygen sensor is shown in Figure 25-15. The cell consists of a platinum disk cathodic working electrode imbedded in a centrally located cylindrical insulator. Surrounding the lower end of this insulator is a ring-shaped silver anode. The tubular insulator and electrodes are mounted inside a second cylinder that contains a buffered solution of potassium chloride. A thin (≈20 μm) replaceable, oxygen permeable membrane of Teflon or polyethylene is held in place at the bottom end of the tube by an O-ring. The thickness of the electrolyte solution between the cathode and the membrane is approximately 10 μm.

When the oxygen sensor is immersed in a flowing or stirred solution of the analyte, oxygen diffuses through the membrane into the thin layer of electrolyte immediately adjacent to the disk cathode, where it diffuses to the electrode and is immediately reduced to water. In contrast to a normal hydrodynamic electrode, two

[7] Voltammetric detectors are a particular type of transducer called *limiting-current transducers.* In this discussion and subsequent discussions involving voltammetric transducers, we use the more common term, *voltammetric detector.* When a voltammetric transducer is inherently selective for a particular species by virtue of control of various experimental variables or when it is covered with a chemically selective layer of polymer or other membranous material, we refer to it as a *voltammetric sensor.* For a discussion of transducers, detectors, sensors, and their definitions, see Chapter 1.

[8] For a detailed discussion of the Clark oxygen sensor, see M. L. Hitchman, *Measurement of Dissolved Oxygen,* Chapters 3–5. New York: Wiley, 1978.

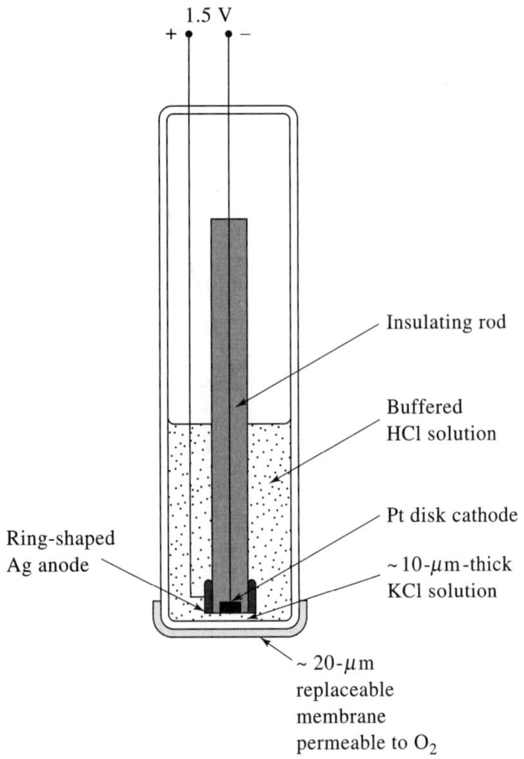

Figure 25-15 The Clark voltammetric oxygen sensor. Cathode reaction: $O_2 + 4H^+ + 4e^- \rightleftharpoons 2H_2O$. Anodic reaction: $Ag + Cl^- \rightarrow AgCl(s) + e^-$.

diffusion processes are involved—one through the membrane and the other through the solution between the membrane and the electrode surface. In order for a steady-state condition to be reached in a reasonable period (10 to 20 s), the thickness of the membrane and the electrolyte film must be 20 μm or less. Under these conditions, it is the rate of equilibration of the transfer of oxygen across the membrane that determines the rate at which steady-state currents are achieved.

Enzyme-Based Sensors. A number of enzyme-based voltammetric sensors are offered commercially. An example is a glucose sensor that is widely used in clinical laboratories for the routine determination of glucose in blood serums. This device is similar in construction to the oxygen sensor shown in Figure 25-15. The membrane in this case is more complex and consists of three layers. The outer layer is a polycarbonate film that is permeable to glucose but impermeable to proteins and other constituents of blood. The middle layer is an immobilized enzyme (see Section 23E-2)—glucose oxidase in this instance. The inner layer is a cellulose acetate membrane, which is permeable to small molecules, such as hydrogen peroxide. When this device is immersed in a glucose-containing solution, glucose diffuses through the outer membrane into the immobilized enzyme, where the following catalytic reaction occurs:

$$\text{glucose} + O_2 \rightarrow H_2O_2 + \text{gluconic acid}$$

The hydrogen peroxide then diffuses through the inner layer of membrane and to the electrode surface, where it is oxidized to give oxygen. That is,

$$H_2O_2 + OH^- \rightleftharpoons O_2 + H_2O + 2e^-$$

The resulting current is directly proportional to the glucose concentration of the analyte solution.

Several other sensors are available that are based upon the voltammetric measurement of hydrogen peroxide produced by enzymatic oxidations of other species of clinical interest. These analytes include sucrose, lactose, ethanol and L-lactate. A different enzyme is, of course, required for each species.

Amperometric Titrations

Hydrodynamic voltammetry can be used to estimate the equivalence point of titrations, provided at least one of the participants or products of the reaction involved is oxidized or reduced at a microelectrode. Here, the current at some fixed potential in the limiting current region is measured as a function of the reagent volume (or of time if the reagent is generated by a constant-current coulometric process). Plots of the data on either side of the equivalence point are straight lines with different slopes; the end point is established by extrapolation to their intersection.

Amperometric titration curves typically take one of the forms shown in Figure 25-16. Figure 25-16a represents a titration in which the analyte reacts at the electrode while the reagent does not. Figure 25-16b is typical of a titration in which the reagent reacts at the microelectrode and the analyte does not. Figure 25-16c corresponds to a titration in which both the analyte and the titrant react at the microelectrode.

Two types of amperometric electrode systems are encountered. One employs a single polarizable microelectrode coupled to a reference; the other uses a pair of identical solid-state microelectrodes immersed in a stirred solution. For the first, the microelectrode is often a rotating platinum electrode constructed by sealing a

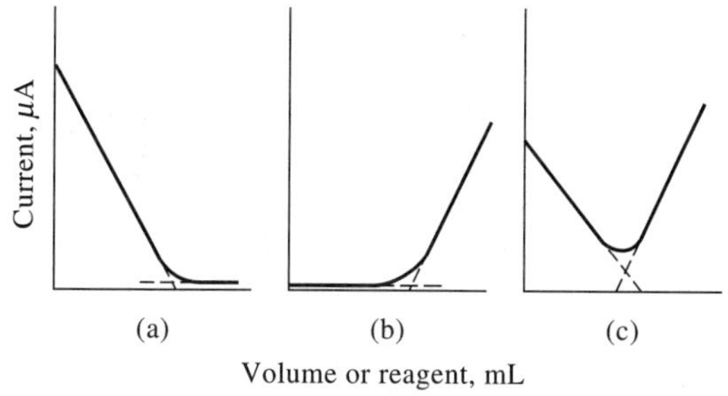

Figure 25-16 Typical amperometric titration curves: (a) analyte is reduced, reagent is not; (b) reagent is reduced, analyte is not; (c) both reagent and analyte are reduced.

platinum wire into the side of a glass tube that is connected to a stirring motor. A dropping mercury electrode may also be used, in which case the solution is not stirred.

Amperometric titrations with one indicator electrode have, with one notable exception, been confined to titrations in which a precipitate or a stable complex is the product. Precipitating reagents include silver nitrate for halide ions, lead nitrate for sulfate ion, and several organic reagents, such as 8-hydroxyquinoline, dimethylglyoxime, and cupferron, for various metallic ions that are reducible at microelectrodes. Several metal ions have also been determined by titration with standard solutions of EDTA. The exception just noted involves titrations of organic compounds, such as certain phenols, aromatic amines, olefins; hydrazine; and arsenic(III) and antimony(III) with bromine. The bromine is often generated coulometrically; it has also been formed by adding a standard solution of potassium bromate to an acidic solution of the analyte that also contains an excess of potassium bromide. Bromine is formed in the acidic medium by the reaction

$$BrO_3^- + 5Br^- + 6H^+ \rightarrow 3Br_2 + 3H_2O$$

This type of titration has been carried out with a rotating platinum electrode or twin platinum microelectrodes. No current is observed prior to the equivalence point; after chemical equivalence, a rapid increase in current takes place due to electrochemical reduction of the excess bromine.

The use of a pair of identical metallic microelectrodes to establish the equivalence point in amperometric titrations offers the advantage of simplicity of equipment and the avoidance of having to prepare and maintain a reference electrode. This type of system has been incorporated into equipment designed for the routine automatic determination of a single species, usually with a coulometric generated reagent. An example of this type of system is an instrument for the automatic determination of chloride in samples of serum, sweat, tissue extracts, pesticides, and food products. Here, the reagent is silver ion coulometrically generated from a silver anode. The indicator system consists of a pair of twin silver microelectrodes between which is imposed a potential difference of perhaps 0.1 V. Short of the equivalence point in the titration of chloride ion, there is essentially no current because no easily reduced species is present in the solution. Consequently, electron transfer at the cathode is precluded, and that electrode is completely polarized. Note that the anode is not polarized because the reaction

$$Ag \rightleftharpoons Ag^+ + e^-$$

occurs in the presence of a suitable cathodic reactant or depolarizer.

After the equivalence point has been passed, the cathode becomes depolarized owing to the presence of a significant amount of silver ions, which can react to give silver. That is,

$$Ag^+ + e^- \rightleftharpoons Ag$$

A current develops as a result of this half-reaction and the corresponding oxidation of silver at the anode. The magnitude of the current is, as in other amperometric methods, directly proportional to the concentration of the excess reagent. Thus, the titration curve is similar to that shown in Figure 25-16b. In the automatic titrator just mentioned, the amperometric detection current signal causes the coulometric generator current to cease; the chloride concentration is then computed from the magnitude of the titration current and the generation time. The instrument is said to have a range of 1 to 999.9 mM Cl^- per liter, a precision of 0.1% relative, and an accuracy of 0.5% relative. Typical titration times are 20 s.

Fundamental Studies with Rotating Electrodes

In order to carry out theoretical studies of oxidation/reduction reactions, it is often of interest to know how k_A in Equation 25-7 is affected by the hydrodynamics of the system. The most common method for obtaining a rigorous description of the hydrodynamic flow of stirred solution is based upon a *rotating disk electrode,* such as that illustrated in Figures 25-17a and 17b. When the disk electrode is rotated rapidly, the flow pattern

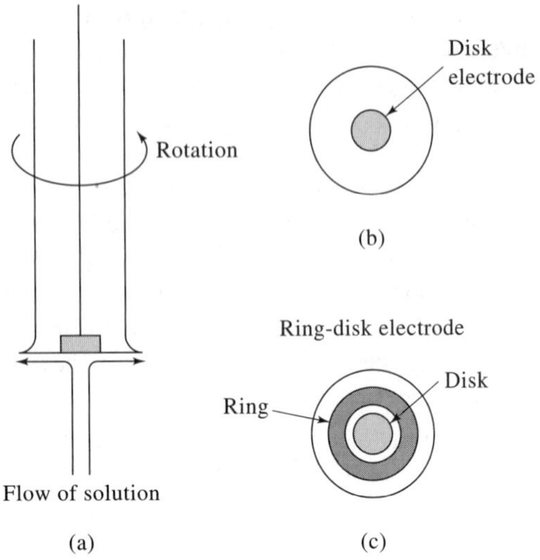

Figure 25-17 (a) Side view of a rotating disk electrode showing solution flow pattern. (b) Bottom view of a disk electrode. (c) Bottom view of a ring-disk electrode.

shown by the arrows in the figure is set up. Here, liquid at the surface of the disk moves out horizontally from the center of the device, which results in an upward axial flow to replenish the displaced liquid. A rigorous treatment of the hydrodynamics is possible in this case and leads to the equation

$$i_l = 0.620nFAc_A \qquad (25\text{-}15)$$

The terms in this equation have the same meaning as those in Equation 25-5. Voltammograms for reversible systems generally have the ideal shape shown in Figure 25-5. Numerous studies of the kinetics and the mechanisms of electrochemical reactions have been performed with rotating disk electrodes. This type of electrode has found little or no use in analysis, however, because it is not very convenient to set up and use.

The *rotating ring-disk electrode* is a modified rotating disk electrode that is a useful device for studying electrode reactions; it has little use in analysis. Figure 25-17c shows that a ring-disk electrode contains a second ring-shaped electrode that is electrically isolated from the center disk. In using this device, a reactive species is generated electrochemically at the disk. This species is then swept passed the ring where it is detected voltammetrically. Figure 25-18 shows voltammograms from a typical ring-disk experiment. The curve on the left is the voltammogram for the reduction of oxygen to hydrogen peroxide at the disk electrode. The curve on the right is the *anodic* voltammogram for the oxidation of the hydrogen peroxide as it flows past the ring electrode. Note that when the potential of the disk electrode becomes sufficiently negative so that the reduction product is hydroxide rather than hydrogen peroxide, the current in the ring electrode decreases to zero. Studies of this type provide much useful information about mechanisms and intermediates in electrochemical reactions.

25D CYCLIC VOLTAMMETRY[9]

In cyclic voltammetry (CV), the current response of a small stationary electrode in an unstirred solution is excited by a triangular potential wave form, such as that shown in Figure 25-19. In this example, the potential is

[9] For brief reviews, see P. T. Kissinger and W. R. Heineman, *J. Chem. Educ.*, **1983**, *60*, 702; D. H. Evans, K. M. O'Connell, T. A. Petersen, and M. J. Kelly, *J. Chem. Educ.*, **1983**, *60*, 290.

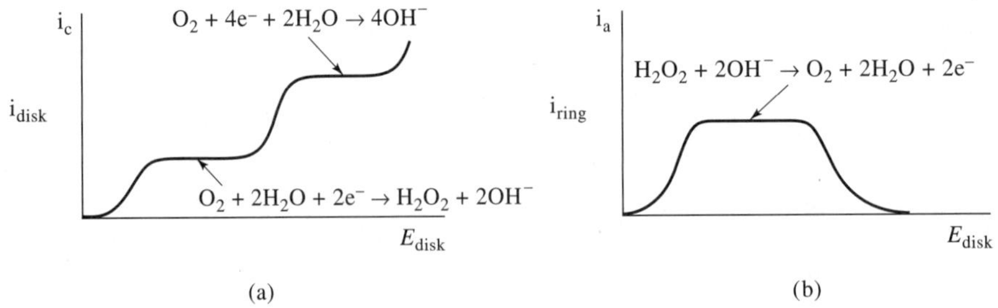

Figure 25-18 Disk (a) and ring (b) current for reduction of oxygen at the rotating ring-disk electrode. *(From* Laboratory Techniques in Electroanalytical Chemistry, *P. T. Kissinger and W. R. Heineman, Eds., p. 117. New York: Marcel Dekker, 1984. With permission.)*

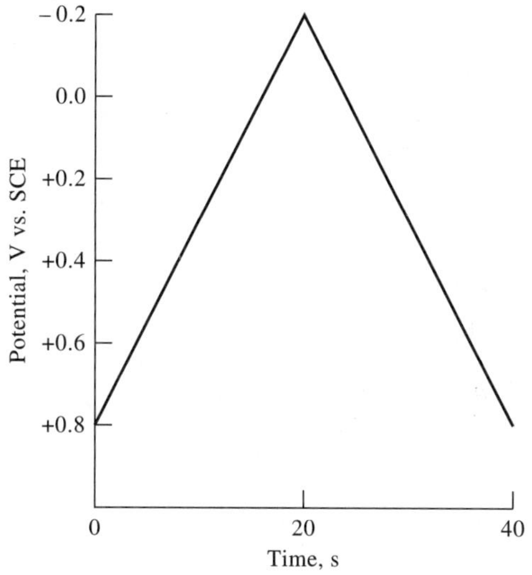

Figure 25-19 Cyclic voltammetric excitation signal used to obtain voltammogram in Figure 25-20.

first varied linearly from +0.8 V to −0.15 V versus a saturated calomel electrode whereupon the scan direction is reversed, and the potential is returned to its original value of +0.8 V. The scan rate in either direction is 50 mV/s. This excitation cycle is often repeated several times. The potentials at which reversal takes place (in this case, −0.15 and +0.8 V) are called *switching potentials.* The range of switching potentials chosen for a given experiment is one in which a diffusion-controlled oxidation or reduction of one or more analytes occurs. The direction of the initial scan may be either negative, as shown, or positive, depending upon the composition of the sample (a scan in the direction of more negative potentials is termed a *forward scan,* while one in the opposite direction is called a *reverse scan*). Generally, cycle times range from 1 ms or less to 100 s or more. In this example, the cycle time is 40 s.

Figure 25-20 shows the current response when a solution that is 6 mM in $K_3Fe(CN)_6$ and 1 M in KNO_3 is subjected to the cyclic excitation signal shown in Figure 25-19. The working electrode was a stationary platinum microelectrode and the reference electrode was a saturated calomel electrode. At the initial potential of +0.8 V, a tiny anodic current is observed, which immediately decreases to zero as the scan is continued. This initial negative current arises from the oxidation of water to give oxygen (at more positive potentials, this current rapidly increases and becomes quite large at about +0.9 V). No current is observed between a potential of +0.7 and +0.4 V because no reducible or oxidizable species is present in this potential range. When the potential becomes somewhat less positive, +0.4 V, a cathodic current develops (point *B*) due to the reduction of the hexacyanoferrate(III) ion to hexacyanoferrate(II) ion. The reaction at the cathode is then

$$Fe(CN)_6^{3-} + e^- \rightleftharpoons Fe(CN)_6^{4-}$$

A rapid increase in the current occurs in the region of *B* to *D* as the surface concentration of $Fe(CN)_6^{3-}$ becomes smaller and smaller. The current at the peak is made up of two components. One is the initial current surge required to adjust the surface concentration of the reactant to its equilibrium concentration as given by the Nernst equation. The second is the normal diffusion-controlled current. The first current then decays rapidly (points *D* to *F*) as the diffusion layer is extended farther and farther away from the electrode surface (see also Figure

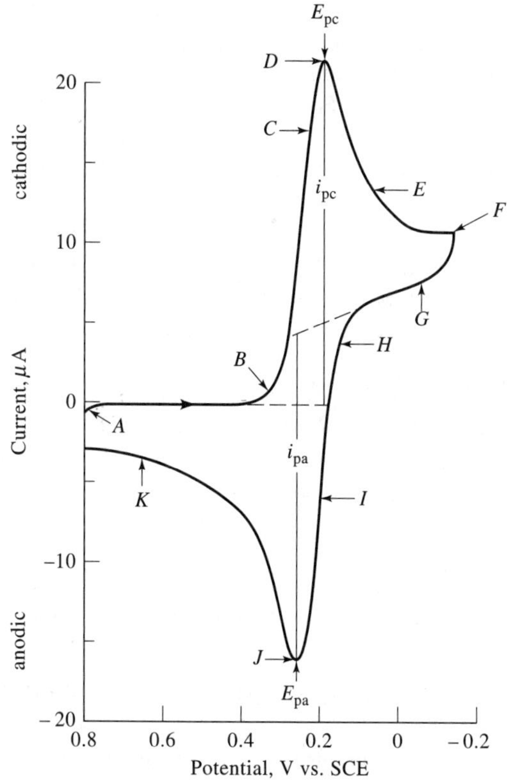

Figure 25-20 Cyclic voltammogram for a solution that is 6.0 mM in $K_3Fe(CN)_6$ and 1.0 M in KNO_3. *(From P. T. Kissinger and W. H. Heineman,* J. Chem. Educ., *1983,* 60, *702. With permission.)*

25-7a). At point −0.15 V (*F*), the scan direction is switched. The current, however, continues to be cathodic even though the scan is toward more positive potentials because the potentials are still negative enough to cause reduction of $Fe(CN)_6^{3-}$. Once the potential becomes positive enough so that reduction of $Fe(CN)_6^{3-}$ can no longer occur, the current goes to zero and then becomes anodic. The anodic current results from the reoxidation of $Fe(CN)_6^{4-}$ that has accumulated near the surface during the forward scan. This anodic current peaks and then decreases as the accumulated $Fe(CN)_6^{4-}$ is used up by the anodic reaction.

Important parameters in a cyclic voltammogram are the cathodic peak potential E_{pc}, the anodic peak potential E_{pa}, the cathodic peak current i_{pc}, and the anodic peak current i_{pa}. How these parameters are established is illustrated in Figure 25-20. For a reversible electrode reaction, anodic and cathodic peak currents are approximately equal in absolute value, but opposite in sign, and the difference in peak potentials is $0.0592/n$, where n is the number of electrons involved in the half-reaction.

The primary use of cyclic voltammetry is as a tool for fundamental and diagnostic studies that provides qualitative information about electrochemical processes under various conditions. As an example, consider the cyclic voltammogram for the agricultural insecticide parathion that is shown in Figure 25-21.[10] Here, the switching potentials were about −1.2 V and +0.3 V. The initial forward scan was, however, started at 0.0 V and not +0.3 V. Three peaks are observed. The first cathodic peak (*A*) results from a four-electron reduction of the parathion to give a hydroxylamine derivative

$$\phi NO_2 + 4e^- + 4H^+ \rightarrow \phi NHOH + H_2O \qquad (25\text{-}16)$$

The anodic peak at *B* arises from the oxidation of the hydroxylamine to a nitroso derivative during the reverse scan. The electrode reaction is

$$\phi NHOH \rightarrow \phi NO + 2H^+ + 2e^- \qquad (25\text{-}17)$$

The cathodic peak at *C* results from the oxidation of the nitroso compound to the hydroxylamine as shown by the equation

$$\phi NO + 2e^- + 2H^+ \rightarrow \phi NHOH \qquad (25\text{-}18)$$

Cyclic voltammograms for authentic samples of the two intermediates confirmed the identities of the compounds responsible for peaks *B* and *C*.

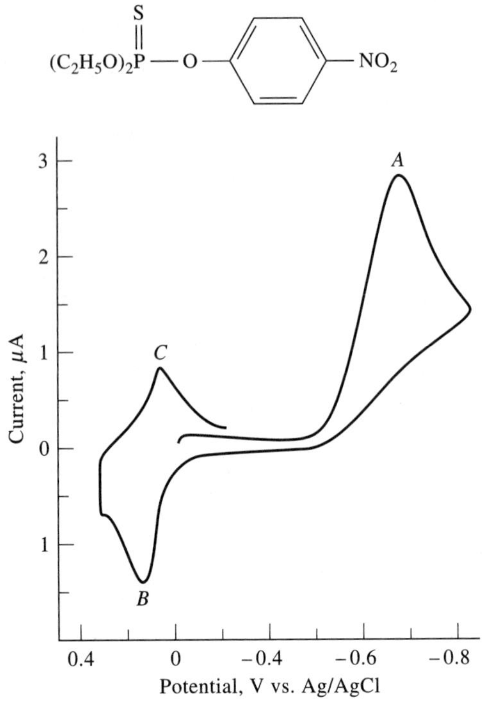

Figure 25-21 Cyclic voltammogram of the insecticide parathion in 0.5 M pH 5 sodium acetate buffer in 50% ethanol. Hanging mercury drop electrode. Scan rate: 200 m V/s. *(From W. R. Heineman and P. T. Kissinger,* Amer. Lab., ***1982*** *(11), 34. Copyright 1982 by International Scientific Communications, Inc.)*

Cyclic voltammetry, while not used for routine quantitative analyses, has become an important tool for the study of mechanisms and rates of oxidation/reduction processes, particularly in organic and metalorganic systems. This method is normally the first technique selected for investigation of an electrochemically addressable system. Often cyclic voltammograms will reveal the presence of intermediates in oxidation/reduction reactions (for example, see Figure 25-21). Usually, platinum is used for fabrication of microelectrodes used with this technique.

25E POLAROGRAPHY

Fri May 13, 05

Polarography was the first type of voltammetry to be invented and used. It differs from hydrodynamic voltammetry in two regards. First, convection is avoided, and second, a dropping mercury electrode, such as that shown in Figure 25-3c is used as the working electrode. A consequence of the first difference is that polaro-

[10] This discussion and the voltammogram are from W. R. Heineman and P. T. Kissinger, *Amer. Lab.*, **1982** (11), 29.

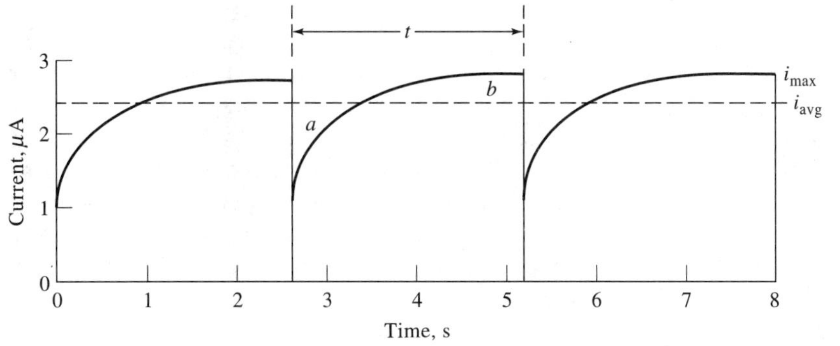

Figure 25-22 Effect of drop growth on polarographic current. For i_{avg}, area a = area b and $i_{avg} \approx 6/7\ i_{max}$.

graphic limiting currents are controlled by diffusion alone rather than by both diffusion and convection. Because convection is absent, polarographic limiting currents are generally one or more orders of magnitude smaller than hydrodynamic limiting currents.

25E-1 Linear-Scan Polarography

Early polarographic methods were based upon measuring current as a function of the potential applied to a dropping mercury electrode. In most experiments, the dropping electrode was attached to the negative terminal of a dc source, and the potential of this source was increased linearly until the current increased exponentially as a consequence of reduction of the solvent or the supporting electrolyte.

Polarographic Currents

The current in a cell containing a dropping electrode undergoes periodic fluctuations corresponding in frequency to the drop rate. As a drop dislodges from the capillary, the current falls to zero (see Figure 25-22); it then increases rapidly as the electrode area grows because of the greater surface to which diffusion can occur. The *average current* is the hypothetical *constant* current that in the drop time t would produce the same quantity of charge as the fluctuating current does during this same period. In order to determine the average current, it is necessary to reduce the large fluctuations in the current by a low-pass filter (Section 2B-5) or by sampling the current near the end of each drop, where the change in current with time is relatively small. As shown in Figure 25-23 curve *A*, low-pass filtering limits

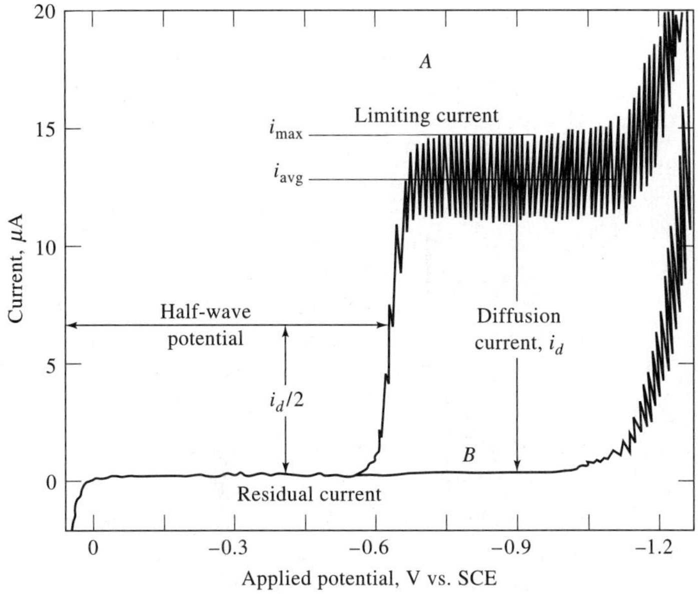

Figure 25-23 Polarograms for *(A)* a 1 M solution of HCl that is 5×10^{-4} M in Cd^{2+} and *(B)* a 1 M solution of HCl. *(From D. T. Sawyer, A. Sobkowiak, and J. L. Roberts Jr.,* Experimental Electrochemistry for Chemists, *2nd ed., p. 59. New York: Wiley, 1995. Reprinted by permission of John Wiley & Sons, Inc.)*

the oscillations to a reasonable magnitude; the average current (or, alternatively, the maximum current) is then readily determined, provided the drop time t is reproducible. Note the effect of irregular drops in the upper part of curve *A,* probably caused by vibration of the apparatus.

Polarograms

Figure 25-23 shows two polarograms, one for a solution that is 1.0 M in hydrochloric acid and 5×10^{-4} M in cadmium ion (curve *A*), while the second is for the acid in the absence of cadmium ion (curve *B*). The polarographic wave in curve *A* arises from the reaction

$$Cd^{2+} + 2e^- + Hg \rightleftharpoons Cd(Hg) \qquad (25\text{-}19)$$

where Cd(Hg) represents elemental cadmium dissolved in mercury, giving an amalgam. The sharp increase in current at about -1 V in both polarograms is caused by the reduction of hydrogen ions to give hydrogen. Examination of the polarogram for the supporting electrolyte alone reveals that a small current, called the *residual current,* is present in the cell even in the absence of cadmium ions.

As in hydrodynamic voltammetry, limiting currents are observed when the magnitude of the current is limited by the rate at which analyte can be brought up to the electrode surface. In polarography, however, the only mechanism of mass transport is diffusion, and for this reason, polarographic limiting currents are usually termed *diffusion currents* and given the symbol i_d. As shown in Figure 25-23, the diffusion current is the difference between the limiting and the residual current. The diffusion current is directly proportional to analyte concentration.

Diffusion Current at Dropping Electrodes

In deriving an equation for polarographic diffusion currents, it is necessary to take into account the rate of growth of the spherical electrode, which is related to the drop time in seconds t and the rate of flow of mercury through the capillary m in mg/s and the diffusion coefficient of the analyte D in cm^2/s. These variables are taken into account in the *Ilkovic equation*

$$(i_d)_{max} = 706\ nD^{1/2}m^{2/3}t^{1/6}c$$

where $(i_d)_{max}$ is the maximum current in microamperes, and c is the analyte concentration in *millimoles per liter.* To obtain an expression for the average current rather than the maximum, the constant in the foregoing equation becomes 607 rather than 706. That is,

$$(i_d)_{avg} = 607\ nD^{1/2}m^{2/3}t^{1/6}c \qquad (25\text{-}20)$$

Note that either the average or the maximum current can be used in quantitative polarography.

The product $m^{2/3}t^{1/6}$ in the Ilkovic equation, called the *capillary constant,* describes the influence of dropping electrode characteristics upon the diffusion current; both m and t are readily evaluated experimentally. Comparison of diffusion currents from different capillaries is thus possible.

Residual Currents

Figure 25-24 shows a residual current curve (obtained at high sensitivity) for a 0.1 M solution of hydrogen chloride. This current has two sources. The first is the reduction of trace impurities that are almost inevitably present in the blank solution; contributors here include small amounts of dissolved oxygen, heavy metal ions from the distilled water, and impurities present in the salt used as the supporting electrolyte.

A second component of the residual current is the so-called *charging* or *condenser current* resulting from a flow of electrons that charge the mercury droplets with respect to the solution; this current may be either negative or positive. At potentials more negative than about -0.4 V, an excess of electrons from the dc source provides the surface of each droplet with a negative charge. These excess electrons are carried down with the drop as it breaks; since each new drop is charged as it forms, a small but continuous current results. At applied potentials less negative than about -0.4 V, the mercury tends to be positive with respect to the solution; thus, as each drop is formed, electrons are repelled from the surface toward the bulk of mercury, and a negative current is the result. At about -0.4 V, the mercury surface is uncharged, and the charging current is zero. The charging current is a type of *nonfaradaic current* (see Section 22A-3) in the sense that charge is carried across an electrode/solution interface without an accompanying oxidation/reduction process.

Ultimately, the accuracy and sensitivity of the polarographic method depend upon the magnitude of the nonfaradaic residual current and the accuracy with which a correction for its effect can be determined.

Comparison of Currents from Dropping and Stationary Planar Electrodes

As indicated in Section 25B-3, constant currents are not obtained with a planar electrode in reasonable periods of time because concentration gradients out from the

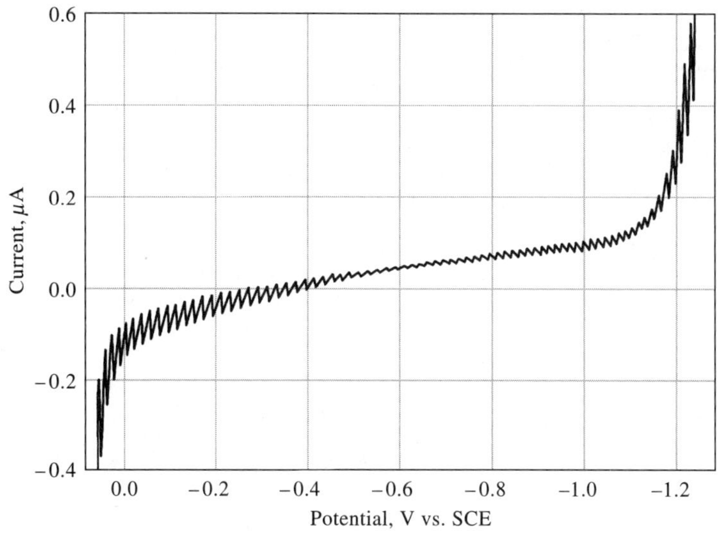

Figure 25-24 Residual current curve for a 0.1 M solution of HCl.

electrode surface are constantly changing with time. In contrast, the dropping electrode exhibits constant reproducible currents essentially instantaneously after an applied voltage adjustment. This behavior represents an advantage of the dropping mercury electrode that accounts for its widespread use in the early years of voltammetry.

The rapid achievement of constant currents arises from the highly reproducible nature of the drop formation process and, equally important, the fact that the solution in the electrode area becomes homogenized each time a drop breaks from the capillary. Thus, a concentration gradient is developed only during the brief lifetime of the drop. As we have noted, current changes due to an increase in surface area occur during each lifetime. Changes in the diffusion gradient *dc/dx* also occur during this period. But these changes are entirely reproducible, leading to currents that are also highly reproducible.

Effect of Complex Formation on Polarographic Waves

We have already seen (Section 22C-10) that the potential for the oxidation or reduction of a metallic ion is greatly affected by the presence of species that form complexes with that ion. It is not surprising, therefore, that similar effects are observed with polarographic half-wave potentials. The data in Table 25-1 show clearly that the half-wave potential for the reduction of a metal complex is generally more negative than that for reduction of the corresponding simple metal ion. In fact, this negative shift in potential permits the elucidation of the composition of the complex ion and the determination of its formation constant *provided that the electrode reaction is reversible.* Thus, for the reactions

$$M^{n+} + Hg + ne^- \rightleftharpoons M(Hg)$$

and

$$M^{n+} + xA^- \rightleftharpoons MA_x^{(n-x)+}$$

Lingane[11] derived the following relationship between the molar concentrations of the ligand c_L and the shift in half-wave potential brought about by its presence:

$$(E_{1/2})_c - E_{1/2} \rightleftharpoons -\frac{0.0592}{n}\log K_f - \frac{0.0592x}{n}\log c_L \qquad (25\text{-}21)$$

where $(E_{1/2})_c$ and $E_{1/2}$ are the half-wave potentials for the complexed and uncomplexed cations, respectively, K_f is the formation constant for the complex, and x is the molar combining ratio of complexing agent to cation.

Equation 25-21 makes it possible to evaluate the formula for the complex. Thus, a plot of the half-wave potential against log c_L for several ligand concentrations gives a straight line, the slope of which is

[11] J. J. Lingane, *Chem. Rev.*, **1941**, *29*, 1.

TABLE 25-1 Effect of Complexing Agents on Polarographic Half-Wave Potentials at the Dropping Mercury Electrode

Ion	Noncomplexing Media	1 M KCN	1 M KCl	1 M NH_3, 1 M NH_4Cl
Cd^{2+}	−0.59	−1.18	−0.64	−0.81
Zn^{2+}	−1.00	NR*	−1.00	−1.35
Pb^{2+}	−0.40	−0.72	−0.44	−0.67
Ni^{2+}	—	−1.36	−1.20	−1.10
Co^{2+}	—	−1.45	−1.20	−1.29
Cu^{2+}	+0.02	NR*	+0.04	−0.24
			−0.22	−0.51

*No reduction occurs before involvement of the supporting electrolyte.

$0.0592x/n$. If n is known, the combining ratio of ligand to metal ion x is readily calculated. Equation 25-21 can then be used to calculate K_f.

Advantages and Disadvantages of the Dropping Mercury Electrode

In the past, the dropping mercury electrode was the most widely used microelectrode for voltammetry because of its several unique features. The first is the unusually high overvoltage associated with the reduction of hydrogen ions. As a consequence, metal ions such as zinc and cadmium can be deposited from acidic solution even though their thermodynamic potentials suggest that deposition of these metals without hydrogen formation is impossible. A second advantage is that a new metal surface is generated continuously; thus the behavior of the electrode is independent of its past history. In contrast, solid metal electrodes are notorious for their irregular behavior, which is related to adsorbed or deposited impurities. A third unusual feature of the dropping electrode, which has already been described, is that reproducible average currents are *immediately* realized at any given potential regardless of whether this potential is approached from lower or higher settings.

One serious limitation of the dropping electrode is the ease with which mercury is oxidized; this property severely limits the use of the electrode as an anode. At potentials greater than about +0.4 V, formation of mercury(I) occurs, giving a wave that masks the curves of other oxidizable species. In the presence of ions that form precipitates or complexes with mercury(I), this behavior occurs at even lower potentials. For example, in Figure 25-23, the beginning of an anodic wave can be seen at 0 V due to the reaction

$$2Hg + 2Cl^- \rightleftharpoons Hg_2Cl_2(s) + 2e^-$$

Incidentally, this anodic wave can be used for the determination of chloride ion.

Another important disadvantage of the dropping mercury electrode is the nonfaradaic residual or charging current, which limits the sensitivity of the classical method to concentrations of about 10^{-5} M. At lower concentrations, the residual current is likely to be greater than the diffusion current, a situation that prohibits accurate measurement of the latter. As will be shown later, methods are now available for enhancing detection limits by one to two orders of magnitude.

Finally, the dropping mercury electrode is cumbersome to use and tends to malfunction as a result of clogging. An annoying aspect of the dropping mercury electrode is that it often produces current maxima such as those shown in Figure 25-25. Although the cause or causes of maxima are not fully understood, there is considerable empirical knowledge of methods for eliminating them. Generally, the addition of traces of such high-molecular-weight substances as gelatin, Triton X-100 (a commercial surface-active agent, or surfactant), methyl red, or other dyes will cause a maximum to disappear. Care must be taken to avoid large amounts of these reagents, however, because the excess may reduce the magnitude of the diffusion current. The proper amount

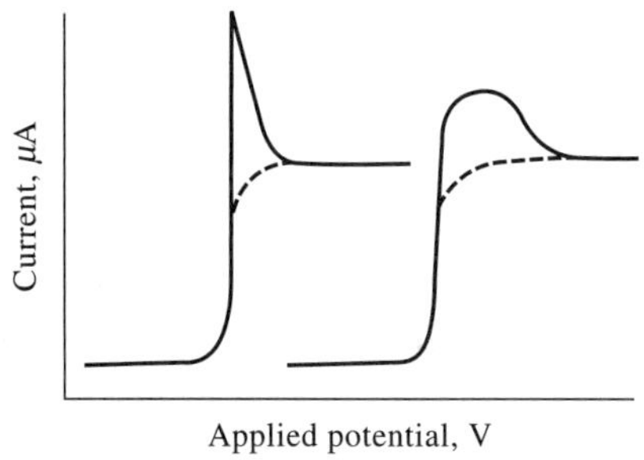

Figure 25-25 Typical current maxima with a dropping mercury electrode.

of suppressor must be determined by trial and error; the amount required varies widely from analyte to analyte.

Current-Sampled (TAST) Polarography

A simple modification of the classical polarographic technique, and one that is incorporated into modern voltammetric instruments, involves measurement of current only for a period near the end of the lifetime of each drop. Here, a mechanical knocker is generally used to detach the drop after a highly reproducible time interval (usually 0.5 to 5 s). Because of this feature, the method is sometimes called *tast* polarography (from German *tasten,* to touch). *Current-sampled polarography* is more descriptive than tast polarography and thus to be preferred inasmuch as devices other than the mechanical knocker are now used for drop detachment (for example, see Figure 25-3d).

Figure 25-26 contrasts the appearance of a classical and current-sampled polarograms. In obtaining the latter, the potential was scanned linearly at 5 mV/s. Rather than recording the current continuously, however, it was instead sampled for a 5-ms period just before termination of each drop. Between sampling periods, the recorder was maintained at its last current level by means of a sample-and-hold circuit. As is apparent from the figure, a major advantage of current sampling is that it substantially reduces the large current fluctuations due to the continuous growth and fall of drops that occur with the dropping electrode. Note in Figure 25-22, that the current near the end of the life of a drop is nearly constant, and it is this current only that is recorded in the current-sampled technique. The result is a smoothed curve consisting of a series of steps, which are significantly smaller than the current fluctuations encountered in normal polarography. The improvements in precision and detection limit by current sampling alone are not, unfortunately, as great as might be hoped. For example, Bond and Canterford[12] showed that the detection limits for copper could be lowered from about 3×10^{-6} M for conventional polarography to 1×10^{-6} M with the current-sampled method, a marginal change at best.

25E-2 Pulse Polarographic and Voltammetric Methods

By the 1960s, linear-scan polarography ceased to be an important analytical tool in most laboratories. The reason for the decline in use of this once popular technique was not only the appearance of several more convenient spectroscopic methods but also the inherent disadvantages of the method including slowness, inconvenient apparatus, and, particularly, poor detection limits. These limitations were largely overcome by pulse methods and the development of electrodes such as those shown in Figure 25-3d. We shall discuss the two most important pulse techniques, *differential-pulse po-*

[12] A. M. Bond and D. R. Canterford, *Anal. Chem.,* **1972,** *44,* 721.

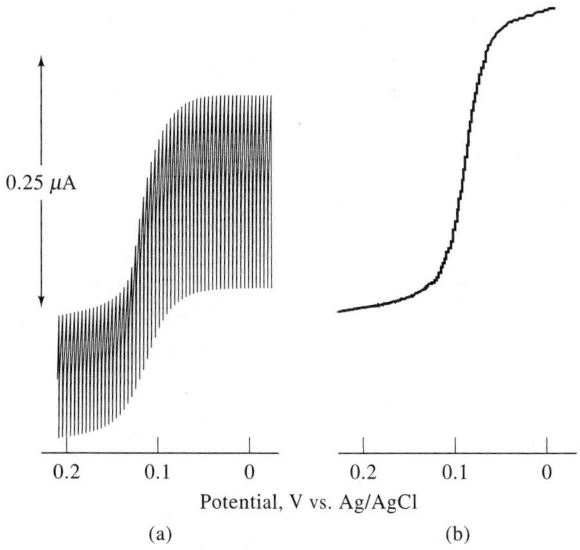

Figure 25-26 Comparison of classical (a) and current-sampled (b) polarographic waves for a 1×10^{-4} M solution of Cu^{2+} in 1 M $NaNO_3$. *(Reprinted with permission from A. M. Bond and D. R. Canterford,* Anal. Chem., ***1972,*** 44, *721. Copyright 1972 American Chemical Society.)*

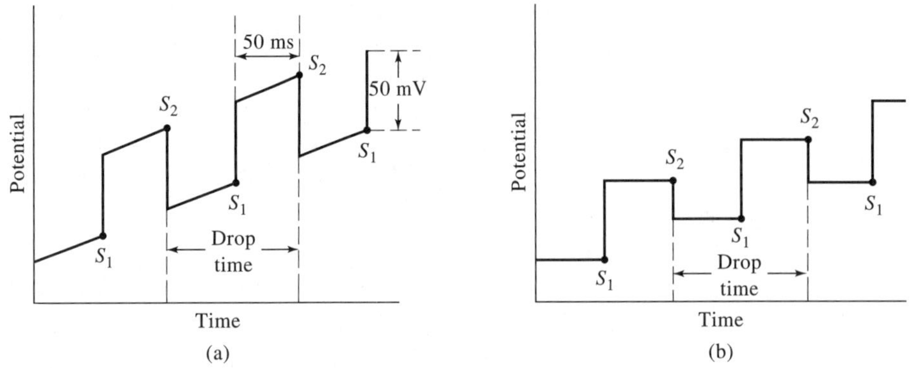

Figure 25-27 Excitation signals for differential pulse polarography.

larography and *square-wave polarography.* Both methods have also been applied with electrodes other than the dropping mercury electrode, in which case the procedures are termed *differential* and *square-wave voltammetry.* The idea behind all pulse-polarographic methods is to measure the current at a time when the difference between the desired faradaic curve and the interfering charging current is large.

Differential-Pulse Polarography

Figure 25-27 shows the two most common excitation signals that are employed in commercial instruments for differential pulse polarography. The first (27a), which is used in analog instruments, is obtained by superimposing a periodic pulse on a linear scan. The second (27b), which is ordinarily used in digital instruments, involves combination of a pulse output with a staircase signal. In either case, a 50-mV pulse is applied during the last 50 ms of the lifetime of the mercury drop. Here again, to synchronize the pulse with the drop, the latter is detached at an appropriate time by a mechanical means.

As shown in Figure 25-27, two current measurements are made alternately—one (at S_1), which is 16.7 ms prior to the dc pulse and one for 16.7 ms (at S_2) at the end of the pulse. The *difference in current per pulse* (Δi) is recorded as a function of the linearly increasing voltage. A differential curve results, consisting of a peak (see Figure 25-28) the height of which is directly proportional to concentration. For a reversible reaction, the peak potential is approximately equal to the standard potential for the half-reaction.

One advantage of the derivative-type polarogram is that individual peak maxima can be observed for substances with half-wave potentials differing by as little as 0.04 to 0.05 V; in contrast, classical and normal-pulse polarography require a potential difference of about 0.2 V for resolution of waves. More important, however, differential-pulse polarography increases the sensitivity of the polarographic method. This enhancement is illustrated in Figure 25-29. Note that a classical polarogram for a solution containing 180 ppm of the antibiotic tetracycline gives two barely discernible waves; differential-pulse polarography, in contrast, provides well-defined peaks at a concentration level that is 2×10^{-3} that for the classic wave, or 0.36 ppm. Note also that the current scale for Δi is in nanoamperes. Generally, detection limits with differential-pulse polarography are two to three orders of magnitude lower than those for classical polarography and lie in the range of 10^{-7} to 10^{-8} M.

The greater sensitivity of differential-pulse polarography can be attributed to two sources. The first is an enhancement of the faradaic current, while the second is a decrease in the nonfaradaic charging current.

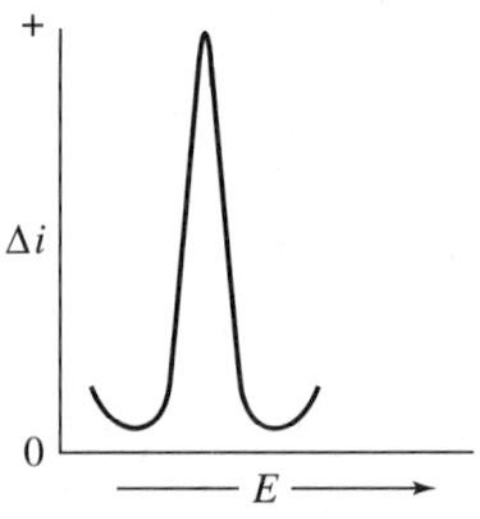

Figure 25-28 Voltammogram for a differential pulse polarography experiment. Here, $\Delta i = i_{S_2} - i_{S_1}$. (See Figure 25-27.)

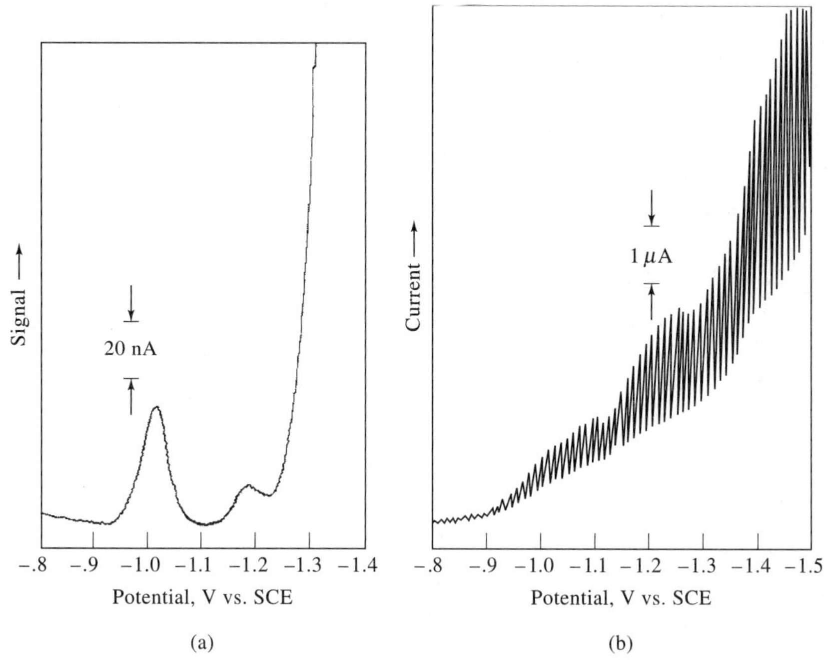

Figure 25-29 (a) Differential pulse polarogram: 0.36 ppm tetracycline-HCl in 0.1 M acetate buffer, pH 4, PAR Model 174 polarographic analyzer, dropping mercury electrode, 50-mV pulse amplitude, 1-s drop. (b) DC polarogram: 180 ppm tetracycline-HCl in 0.1 M acetate buffer, pH 4, similar conditions. *(Reprinted with permission from J. B. Flato,* Anal. Chem., ***1972,*** *44(11), 75A. Copyright 1972 American Chemical Society.)*

To account for the former, let us consider the events that must occur in the surface layer around an electrode as the potential is suddenly increased by 50 mV. If a reactive species is present in this layer, there will be a surge of current that lowers the reactant concentration to that demanded by the new potential (see Figure 25-6b). As the equilibrium concentration for that potential is approached, however, the current decays to a level just sufficient to counteract diffusion; that is, to the diffusion-controlled current. In classical polarography, the initial surge of current is not observed because the time scale of the measurement is long relative to the lifetime of the momentary current. On the other hand, in pulse polarography, the current measurement is made before the surge has completely decayed. Thus, the current measured contains both a diffusion-controlled component and a component that has to do with reducing the surface layer to the concentration demanded by the Nernst expression; the total current is typically several times larger than the diffusion current. It should be noted that when the drop is detached, the solution again becomes homogeneous with respect to the analyte. Thus, at any given voltage, an identical current surge accompanies each voltage pulse.

When the potential pulse is first applied to the electrode, a surge in the nonfaradaic current also occurs as the charge on the drop increases. This current, however, decays exponentially with time and approaches zero near the end of the life of a drop when its surface area is changing only slightly (see Figure 25-22). Thus, by measuring currents at this time only, the nonfaradaic residual current is greatly reduced, and the signal-to-noise ratio is larger. Enhanced sensitivity results.

Reliable instruments for differential-pulse polarography are now available commercially at reasonable cost. The method has thus become the most widely used analytical polarographic procedure and is especially useful for determining trace concentrations of heavy metal ions.

Square-Wave Polarography[13]

Square-wave polarography is a type of pulse polarography that offers the advantage of great speed and high sensitivity. An entire voltammogram is obtained in less than 10 ms. With a dropping mercury electrode, the scan is performed during the last few milliseconds of the life of a single drop when the charging current is essentially constant. Square-wave voltammetry has also been used with hanging drop electrodes and with chromatographic detectors.

[13] For further information on square-wave voltammetry, see J. G. Osteryoung and R. A. Osteryoung, *Anal. Chem.*, **1985,** *57,* 101A.

Figure 25-30c shows the excitation signal in square-wave voltammetry, which is obtained by superimposing the pulse train shown in 25-30b onto the staircase signal in 25-30a. The length of each step of the staircase and the period of the pulses (τ) are identical and usually about 5 ms. The potential step of the staircase ΔE_s is typically 10 mV. The magnitude of the pulse $2E_{sw}$ is often 50 mV. Operating under these conditions, which corresponds to a pulse frequency of 200 Hz, a 1-V scan requires 0.5 s. For a reversible reduction reaction, the size of a pulse is great enough so that oxidation of the product formed on the forward pulse occurs during the reverse pulse. Thus, as shown in Figure 25-31, the forward pulse produces a cathodic current i_1, whereas the reverse pulse gives an anodic current i_2. Usually the difference in these currents Δi is plotted to give voltammograms. This difference is directly proportional to concentration; the potential of the peak corresponds to the polarographic half-wave potential. Because of the speed of the measurement, it is possible and practical to increase the precision of analyses by signal-averaging data from several voltammetric scans. Detection limits for square-wave voltammetry are reported to be 10^{-7} to 10^{-8} M.

Commercial instruments for square-wave voltammetry have recently become available from several manufacturers, and as a consequence, it seems likely that this technique will gain considerable use for analysis of inorganic and organic species. It has also been suggested that square-wave voltammetry can be used in detectors for high-performance liquid chromatography.

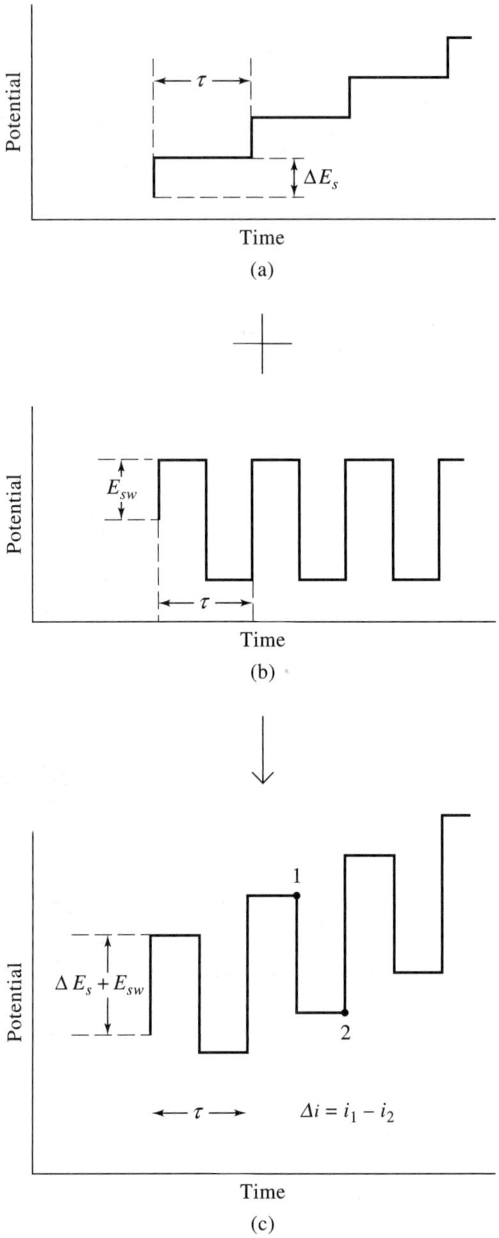

Figure 25-30 Generation of a square-wave voltammetry excitation signal. The staircase signal in (a) is added to the pulse train in (b) to give the square-wave excitation signal in (c). The current response Δi is equal to the current at potential 1 minus the current at potential 2.

25E-3 Applications of Polarography

In the past, linear-scan polarography was used for the quantitative determination of a wide variety of inorganic and organic species, including molecules of biological and biochemical interest. Currently, pulse methods have supplanted the classical method almost completely because of their greater sensitivity, convenience, and selectivity. Generally, quantitative applications are based upon calibration curves in which peak heights are plotted as a function of analyte concentration. In some instances the standard addition method is employed in lieu of calibration curves. In either case, it is essential that the composition of standards resemble as closely as possible the composition of the sample, both as to electrolyte concentrations and pH. When this is done, relative precisions and accuracies in the 1% to 3% range can often be realized.

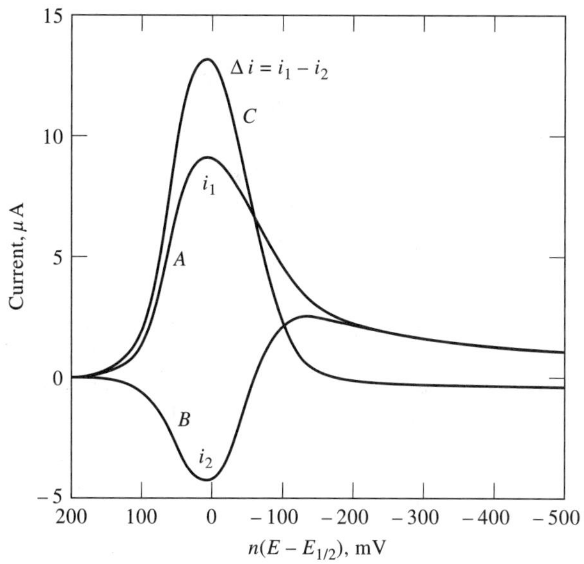

Figure 25-31 Current response for a reversible reaction to excitation signal in Figure 25-30. *A:* forward current i_1. *B:* reverse current i_2. *C:* current difference $i_1 - i_2$. *(Reprinted with permission from J. J. O'Dea, J. Osteryoung, and R. A. Osteryoung,* Anal. Chem., ***1981***, 53, *695. Copyright 1981 American Chemical Society.)*

Inorganic Applications

The polarographic method is widely applicable to the analysis of inorganic substances. Most metallic cations, for example, are reduced at the dropping electrode. Even the alkali and alkaline-earth metals are reducible, provided the supporting electrolyte does not react at the high potentials required; here, the tetraalkyl ammonium halides are useful electrolytes because of their high reduction potentials.

The successful polarographic determination of cations frequently depends upon the supporting electrolyte that is used. To aid in this selection, tabular compilations of half-wave potential data are available.[14] The judicious choice of anion often enhances the selectivity of the method. For example, with potassium chloride as a supporting electrolyte, the waves for iron(III) and copper(II) interfere with one another; in a fluoride medium, however, the half-wave potential of the former is shifted by about −0.5 V, while that for the latter is altered by only a few hundredths of a volt. The presence of fluoride thus results in the appearance of well-separated waves for the two ions.

The polarographic method is also applicable to the analysis of such inorganic anions as bromate, iodate, dichromate, vanadate, selenite, and nitrite. In general, polarograms for these substances are affected by the pH of the solution because the hydrogen ion is a participant in their reduction. As a consequence, strong buffering to some fixed pH is necessary to obtain reproducible data (see next section).

Organic Polarographic Analysis

Almost from its inception, the polarographic method has been used for the study and determination of organic compounds with many papers being devoted to this subject. Several common functional groups are reduced at the dropping electrode, thus making possible the determination of wide variety of organic compounds.[15] The number of functional groups that can be oxidized at a dropping mercury electrode is relatively limited, however, because anodic potentials greater than +0.4 V (vs. SCE) cannot be employed because of oxidation of mercury. Oxidizable organic functional groups can, however, be studied voltammetrically with platinum, gold, or carbon microelectrodes.

Effect of pH on Polarograms. Organic electrode processes ordinarily involve hydrogen ions, the typical reaction being represented as

$$R + nH^+ + ne^- \rightleftharpoons RH_n$$

where R and RH_n are the oxidized and reduced forms of the organic molecule. Half-wave potentials for organic compounds are therefore markedly pH dependent. Furthermore, alteration of the pH may result in a change in the reaction product. For example, when benzaldehyde is reduced in a basic solution, a wave is obtained at about −1.4 V, attributable to the formation of benzyl alcohol:

$$C_6H_5CHO + 2H^+ + 2e^- \rightarrow C_6H_5CH_2OH$$

---

[14] For example, see *Handbook of Analytical Chemistry,* L. Meites, Ed. New York: McGraw-Hill, 1963; D. T. Sawyer, A. Sobkowiak and J. L. Roberts, *Experimental Electrochemistry for Chemists,* 2nd ed., pp. 102–130. New York: Wiley, 1995.

[15] For a detailed discussion of organic polarographic analysis, see P. Zuman, *Organic Polarographic Analysis.* Oxford: Pergamon Press, 1964 and *Topics in Organic Polarography,* P. Zuman, Ed. New York: Plenum Press, 1970.

If the pH is less than 2, however, a wave occurs at about −1.0 V that is just half the size of the foregoing one; here, the reaction involves the production of hydrobenzoin:

$$2C_6H_5CHO + 2H^+ + 2e^- \rightarrow C_6H_5CHOHCHOHC_6H_5$$

At intermediate pH values, two waves are observed, indicating the occurrence of both reactions.

It should be emphasized that an electrode process that consumes or produces hydrogen ions will alter the pH of the solution *at the electrode surface,* often drastically, unless the solution is well buffered. These changes affect the reduction potential of the reaction and cause drawn-out, poorly defined waves. Moreover, where the electrode process is altered by pH, as in the case of benzaldehyde, nonlinearity in the diffusion current/concentration relationship will also be encountered. Thus, in organic polarography good buffering is generally vital for the generation of reproducible half-wave potentials and diffusion currents.

Solvents for Organic Polarography. Solubility considerations frequently dictate the use of solvents other than pure water for organic polarography; aqueous mixtures containing varying amounts of such miscible solvents as glycols, dioxane, acetonitrile, alcohols, Cellosolve, or acetic acid have been employed. Anhydrous media such as acetic acid, formamide, diethylamine, and ethylene glycol have also been investigated. Supporting electrolytes are often lithium or tetraalkyl ammonium salts.

Reactive Functional Groups. Organic compounds containing any of the following functional groups can be expected to produce one or more polarographic waves.

1. **The carbonyl group,** including aldehydes, ketones, and quinones, produce polarographic waves. In general, aldehydes are reduced at lower potentials than ketones; conjugation of the carbonyl double bond also results in lower half-wave potentials.
2. **Certain carboxylic acids** are reduced polarographically, although simple aliphatic and aromatic monocarboxylic acids are not. Dicarboxylic acids such as fumaric, maleic, or phthalic acid, in which the carboxyl groups are conjugated with one another, give characteristic polarograms; the same is true of certain keto and aldehydo acids.
3. **Most peroxides and epoxides** yield polarograms.
4. **Nitro, nitroso, amine oxide, and azo groups** are generally reduced at the dropping electrode.
5. **Most organic halogen groups** produce a polarographic wave, which results from replacement of the halogen group with an atom of hydrogen.
6. **The carbon/carbon double bond** is reduced when it is conjugated with another double bond, an aromatic ring, or an unsaturated group.
7. **Hydroquinones and mercaptans** produce anodic waves.

In addition, a number of other organic groups cause catalytic hydrogen waves that can be used for analysis. These include amines, mercaptans, acids, and heterocyclic nitrogen compounds. Numerous applications to biological systems have been reported.[16]

25F STRIPPING METHODS

Stripping methods encompass a variety of electrochemical procedures having a common, characteristic initial step.[17] In all of these procedures, the analyte is first deposited on a microelectrode, usually from a stirred solution. After an accurately measured period, the electrolysis is discontinued, the stirring is stopped, and the deposited analyte is determined by one of the voltammetric procedures that have been described in the previous section. During this second step in the analysis, the analyte is redissolved or stripped from the microelectrode; hence the name attached to these methods. In *anodic stripping methods,* the microelectrode behaves as a cathode during the deposition step and as an anode during the stripping step, with the analyte being oxidized back to its original form. In a *cathodic stripping method,* the microelectrode behaves as an anode during the deposition step and as a cathode during stripping. The deposition step amounts to an electrochemical preconcentration of the analyte; that is, the concentration of the analyte in the surface of the microelectrode is far greater than it is in the bulk solution. As a result of the preconcentration step, stripping methods yield the lowest detection limits of all voltammetric procedures. For example, anodic stripping with pulse polarography can

[16] M. Brezina and P. Zuman, *Polarography in Medicine, Biochemistry and Pharmacy.* New York: Interscience, 1958; *Polarography of Molecules of Biological Significance,* W. F. Smyth, Ed. New York: Academic Press, 1979.

[17] For detailed discussions of stripping methods, see J. Wang, *Stripping Analysis.* Deerfield Beach, FL: VCH Publishers, 1985; A. M. Bond, *Modern Polarographic Methods in Analytical Chemistry,* Chapter 9. New York: Marcel Dekker, 1980.

reach nanomolar detection limits for environmentally important species, such as Pb^{2+}, Ca^{2+}, and Tl^{+}.

Figure 25-32a illustrates the voltage excitation program that is followed in an anodic stripping method for determining cadmium and copper in an aqueous solution of these ions. A linear scan voltammetric method is used to complete the analysis. Initially, a constant cathodic potential of about −1 V is applied to the microelectrode, which causes both cadmium and copper ions to be reduced and deposited as metals. The electrode is maintained at this potential for several minutes until a significant amount of the two metals has accumulated at the electrode. The stirring is then stopped for perhaps 30 s while the electrode is maintained at −1 V. The potential of the electrode is then decreased linearly to less negative values while the current in the cell is recorded as a function of time, or potential. Figure 25-32b shows the resulting voltammogram. At a potential somewhat more negative than −0.6 V, cadmium starts to be oxidized, causing a sharp increase in the current. As the deposited cadmium is consumed, the current peaks and then decreases to its original level. A second peak for oxidation of the copper is then observed when the potential has decreased to approximately −0.1 V. The heights of the two peaks are proportional to the weights of deposited metal.

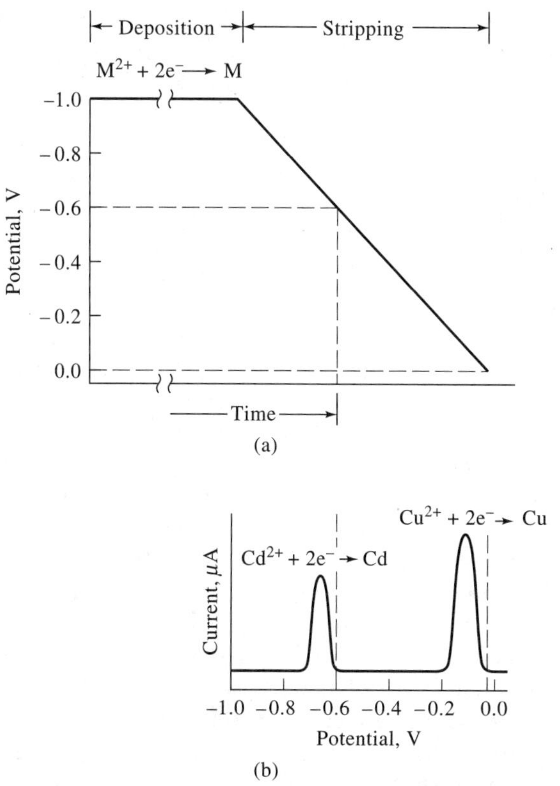

Figure 25-32 (a) Excitation signal for stripping determination of Cd^{2+} and Cu^{2+}. (b) Stripping voltammogram.

Stripping methods are of prime importance in trace work because the concentrating aspects of the electrolysis permit the determination of minute amounts of an analyte with reasonable accuracy. Thus, the analysis of solutions in the 10^{-6} to 10^{-9} M range becomes feasible by methods that are both simple and rapid.

25F-1 Electrodeposition Step

Ordinarily, only a fraction of the analyte is deposited during the electrodeposition step; hence, quantitative results depend not only upon control of electrode potential but also upon such factors as electrode size, length of deposition, and stirring rate for both the sample and standard solutions employed for calibration.

Microelectrodes for stripping methods have been formed from a variety of materials including mercury, gold, silver, platinum, and carbon in various forms. The most popular electrode is the *hanging mercury drop electrode* (HMDE), which consists of a single drop of mercury in contact with a platinum wire. Hanging drop electrodes are available from several commercial sources. These electrodes often consist of a microsyringe with a micrometer for exact control of drop size. The drop is then formed at the tip of a capillary by displacement of the mercury in the syringe-controlled delivery system (see Figure 25-3b). The system shown in Figure 25-3d is also capable of producing a hanging drop electrode.

To carry out the determination of a metal ion by anodic stripping, a fresh hanging drop is formed, stirring is begun, and a potential is applied that is a few tenths of a volt more negative than the half-wave potential for the ion of interest. Deposition is allowed to occur for a carefully measured period that can vary from a minute or less for 10^{-7} M solutions to 30 min or longer for 10^{-9} M solutions. It should be emphasized that these times seldom result in complete removal of the ion. The electrolysis period is determined by the sensitivity of the method ultimately employed for completion of the analysis.

25F-2 Voltammetric Completion of the Analysis

The analyte collected in the hanging drop electrode can be determined by any of several voltammetric procedures. For example, in a linear anodic scan procedure, as described at the beginning of this section, stirring is discontinued for perhaps 30 s after termination of the deposition. The voltage is then decreased at a linear fixed rate from its original cathodic value, and the resulting anodic current is recorded as a function of the applied voltage. This linear scan produces a curve of the type shown in Figure 25-32b. Analyses of this type are generally based on calibration with standard solutions of the cations of interest. With reasonable care, analytical precisions of about 2% relative can be obtained.

Most of the other voltammetric procedures described in the previous section have also been applied to the stripping step. The most widely used of these appears to be an anodic differential pulse technique. Often, narrower peaks are produced by this procedure, which is desirable when mixtures are to be analyzed. Another method of obtaining narrower peaks is to use a mercury film electrode. Here, a thin mercury film is electrodeposited on an inert microelectrode such as glassy carbon. Usually, the mercury deposition is carried out simultaneously with the analyte deposition. Because the average diffusion path length from the film to the solution interface is much shorter than that in a drop of mercury, escape of the analyte is hastened; the consequence is narrower and larger voltammetric peaks, which leads to greater sensitivity and better resolution of mixtures. On the other hand, the hanging drop electrode appears to give more reproducible results, especially at higher analyte concentrations. Thus, for most applications the hanging drop electrode is employed. Figure 25-33 is a differential-pulse anodic stripping polarogram for a mixture of cations present at concentrations of 25 ppb showing good resolution and adequate sensitivity for many purposes.

Many other variations of the stripping technique have been developed. For example, a number of cations have been determined by electrodeposition on a platinum cathode. The quantity of electricity required to remove the deposit is then measured coulometrically. Here again, the method is particularly advantageous for trace analyses. Cathodic stripping methods for the halides have also been developed. Here, the halide ions are first deposited as mercury(I) salts on a mercury anode. Stripping is then performed by a cathodic current.

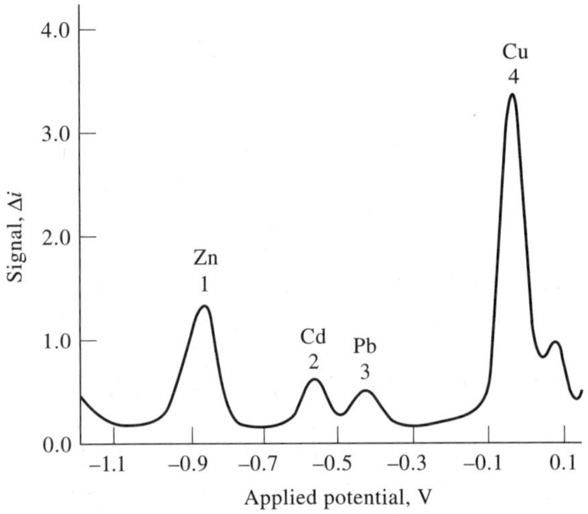

Figure 25-33 Differential pulse anodic stripping voltammogram of 25 ppm zinc, cadmium, lead, and copper. *(Reprinted with permission from W. M. Peterson and R. V. Wong,* Amer. Lab., ***1981,*** 13*(11), 116. Copyright 1981 by International Scientific Communications, Inc.)*

25F-3 Adsorptive Stripping Methods

Adsorptive stripping methods are quite similar to the anodic and cathodic stripping methods we have just considered. Here, a microelectrode, most commonly a hanging mercury drop electrode, is immersed in a stirred solution of the analyte for several minutes. Deposition of the analyte then occurs by physical adsorption on the electrode surface rather than by electrolytic deposition. After sufficient analyte has accumulated, the stirring is discontinued and the deposited material determined by linear scan or pulsed voltammetric measurements. Quantitative information is based upon calibration with standard solutions that are treated in the same way as samples.

Many organic molecules of clinical and pharmaceutical interest have a strong tendency to be adsorbed from aqueous solutions onto a mercury surface, particularly if the surface is maintained at about -0.4 V (vs. SCE) where the charge on the mercury is zero (see page 658). With good stirring, adsorption is rapid, and only 1 to 5 min are required to accumulate sufficient analyte for analysis from 10^{-7} M solutions and 10 to 20 min for 10^{-9} M solutions. Figure 25-34 illustrates the sensitivity of differential pulse adsorptive stripping voltammetry when it is applied to the determination of riboflavin in a 5×10^{-10} M solution. Many other examples of this type can be found in the recent literature.

Adsorptive stripping voltammetry has also been applied to the determination of a variety of inorganic cations at very low concentrations. In these applications the cations are generally complexed with surface active

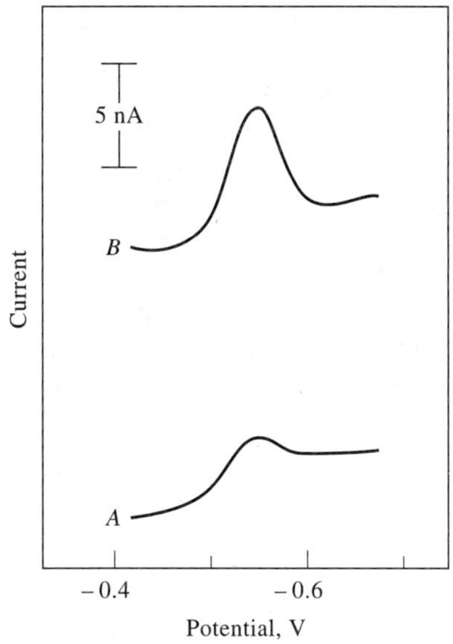

Figure 25-34 Differential pulse voltammogram for 5×10^{-10} M riboflavin. Adsorptive preconcentrations for 5 *(A)* and 30 *(B)* min at −0.2 V. *(From J. Wang,* Amer. Lab., *1985 (5), 43. Copyright 1985 by International Scientific Communications, Inc.)*

complexing agents, such as dimethylglyoxime, catechol, and bipyridine. Detection limits in the 10^{-10} to 10^{-11} M range have been reported.

25G VOLTAMMETRY WITH ULTRAMICROELECTRODES

During the last decade, a number of voltammetric studies have been carried out with microelectrodes that have dimensions that are smaller by an order of magnitude or more than the microelectrodes we have described so far. The electrochemical behavior of these tiny electrodes is significantly different from classical microelectrodes and appears to offer advantages in certain analytical applications.[18] Such electrodes are often called *microscopic electrodes,* or *ultramicroelectrodes* to distinguish them from classical microelectrodes. The dimensions of such electrodes are generally smaller than about 20 μm and may be as small as a few tenths of a micrometer. As shown in Figure 25-35, these miniature microelectrodes take several forms. The most common is a planar electrode formed by sealing a carbon fiber with a radius of 5 μm or a gold or platinum wire having dimensions from 0.3 to 20 μm into a fine capillary tube; the fiber or wires are then cut flush with the ends of the tubes. Cylindrical electrodes are also used in which a small portion of the wire extends from the end of the tube. This geometry has the advantage of larger currents but the disadvantages of being fragile and difficult to clean and polish. Band electrodes are attractive because they can be fabricated on a nanometer scale in one dimension and their behavior is determined by this dimension except that the magnitude of their currents increases with length. Electrodes of this type with 20 Å widths have been constructed by sandwiching metal films between glass or epoxy insulators. Ultramicro mercury electrodes are formed by electrodeposition of the metal onto carbon or metal disk electrodes.

Generally, with ultramicroelectrodes there is no need to employ the three-electrode system used in con-

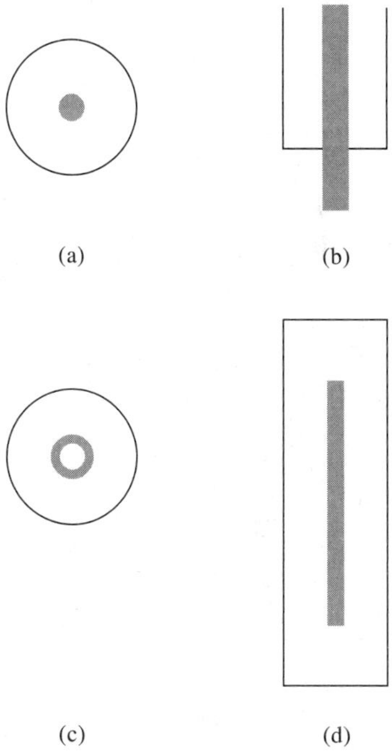

Figure 25-35 Some typical configurations for ultramicroelectrodes. (a) Disk electrode (end view); (b) cylindrical electrode (side view); (c) ring electrode (end view); (d) band electrode (end view). In each case the dark area represents a metal or carbon conductor. The light areas are insulators.

[18] See R. M. Wightman, *Science,* **1988,** *240,* 415; *Anal. Chem.,* **1981,** *53,* 1325A; J. Heinze, *Agnew. Chem., Int. Ed.,* **1993,** *32,* 1268; R. M. Wightman and D. O. Wipf, in *Electroanalytical Chemistry,* Volume XV, A. J. Bard, Ed. New York: Marcel Dekker, 1989.

ventional voltammetry because the currents are so very small (in the picoampere to nanoampere range) that the *IR* drop does not distort the voltammetric waves the way microampere currents do. The low currents also make it possible to perform voltammetric measurements in high resistance, nonaqueous solvents, as encountered in normal phase high-performance liquid chromatography.

One of the reasons for the early interest in microscopic microelectrodes was the desire to study the chemical process going on inside organs of living species, as in mammalian brains. One approach to this problem was to use electrodes that are small enough to not cause significant alteration in the function of the organ. One of the outcomes from these studies was the realization that ultramicroelectrodes have certain advantages that justify their application to other kinds of analytical problems. Among these advantages is the very small *IR* losses, which make them applicable to solvents having low dielectric constants, such as toluene. Second, capacitive charging currents, which often limit detection with ordinary microelectrodes, are reduced to insignificant proportions as the electrode size is diminished because the electrode capacitance decreases as the electrode area becomes smaller. Third, the rate of mass transport to and from an electrode increases as the size of an electrode diminishes; as a consequence, steady-state currents are established in unstirred solutions in less than a microsecond rather than in a millisecond or more as is the case with classical microelectrodes. Such high-speed measurements permit the study of intermediates in rapid electrochemical reactions. Undoubtedly, the future will see many more applications of these ultramicroelectrodes.

25H QUESTIONS AND PROBLEMS

25-1 Distinguish between
(a) voltammetry and polarography.
(b) linear-scan polarography and pulse polarography.
(c) differential-pulse polarography and square-wave polarography.
(d) a hanging drop mercury electrode and a dropping mercury electrode.
(e) a limiting current and a residual current.
(f) a limiting current and a diffusion current.
(g) laminar flow and turbulent flow.
(h) the standard electrode potential and the half-wave potential for a reversible reaction at a microelectrode.
(i) normal stripping methods and adsorptive stripping methods.

25-2 Define
(a) voltammograms.
(b) hydrodynamic voltammetry.
(c) Nernst diffusion layer.
(d) a mercury film electrode.
(e) half-wave potential.

25-3 Why is it necessary to buffer solutions in organic voltammetry?

25-4 List the advantages and disadvantages of the dropping mercury electrode compared with platinum or carbon microelectrodes.

25-5 Suggest how Equation 25-13 could be used to determine the number of electrons *n* involved in a reversible reaction at a microelectrode.

25-6 Quinone undergoes a reversible reduction at a dropping mercury electrode. The reaction is

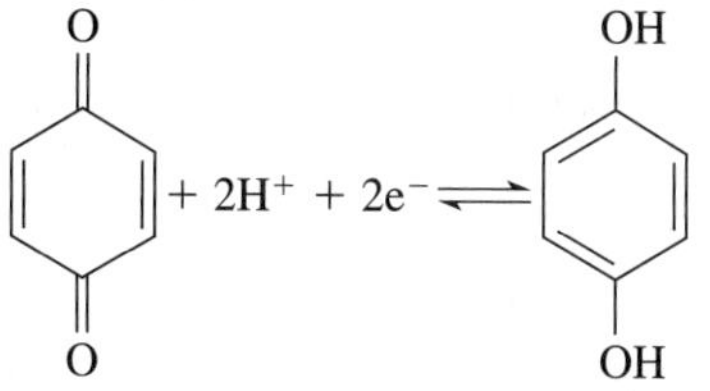

$E^0 = 0.599$ V

(a) Assume that the diffusion coefficient for quinone and hydroquinone are approximately the same and calculate the approximate half-wave potential (vs. SCE) for the reduction of hydroquinone at a rotating disk electrode from a solution buffered to a pH of 7.0.

(b) Repeat the calculation in (a) for a solution buffered to a pH of 5.0.

25-7 What are the sources of the residual current in linear-scan polarography? Why are residual currents smaller with current sampled polarography?

25-8 The polarogram for 20.0 mL of solution that was 3.65×10^{-3} M in Cd^{2+} gave a wave for that ion with a diffusion current of 31.3 μA. Calculate the percentage change in concentration of the solution if the current in the limiting current region were allowed to continue for (a) 5 min, (b) 10 min, (c) 30 min.

25-9 Calculate the mass in milligrams of cadmium per milliliter of sample, based on the following data (corrected for residual current):

	Volumes Used, mL				
Solution	**Sample**	**0.400 M KCl**	**2.00×10^{-3} M Cd^{2+}**	**H_2O**	**Current, μA**
(a)	15.0	20.0	0.00	15.0	79.7
	15.0	20.0	5.00	10.0	95.9
(b)	10.0	20.0	0.00	20.0	49.9
	10.0	20.0	10.0	10.0	82.3
(c)	20.0	20.0	0.00	10.0	41.4
	20.0	20.0	5.00	5.00	57.6
(d)	15.0	20.0	0.00	15.0	67.9
	15.0	20.0	10.0	5.00	100.3

25-10 The following polarographic data were obtained for the reduction of Pb^{2+} to its amalgam from solutions that were 2.00×10^{-3} M in Pb^{2+}, 0.100 M in KNO_3, and that also had the following concentrations of the anion A^-. From the half-wave potentials, find the formula of the complex as well as its formation constant.

Concn A^-, M	$E_{1/2}$ vs. SCE, V
0.0000	−0.405
0.0200	−0.473
0.0600	−0.507
0.1007	−0.516
0.300	−0.547
0.500	−0.558

25-11 Shown below is the polarogram for a solution that was 1.0×10^{-4} M in KBr and 0.1 M in KNO_3. Offer an explanation of the wave that occurs at +0.12 V and of the rapid change in current that starts at about +0.48 V. Would the wave at 0.12 V have any analytical applications? Explain.

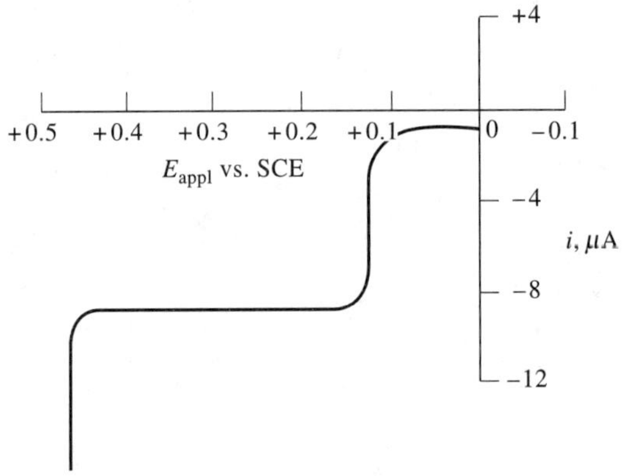

25-12 The following reaction is reversible and has a half-wave potential of −0.349 V when carried out at a dropping mercury electrode from a solution buffered to pH 2.5.

$$Ox + 4H^+ + 4e^- \rightleftharpoons R$$

Predict the half-wave potential at pH: (a) 1.0, (b) 3.5, (c) 7.0.

25-13 Why are stripping methods more sensitive than other voltammetric procedures?

25-14 What are the advantages of performing voltammetry with ultramicroelectrodes?

Section V

Separation Methods

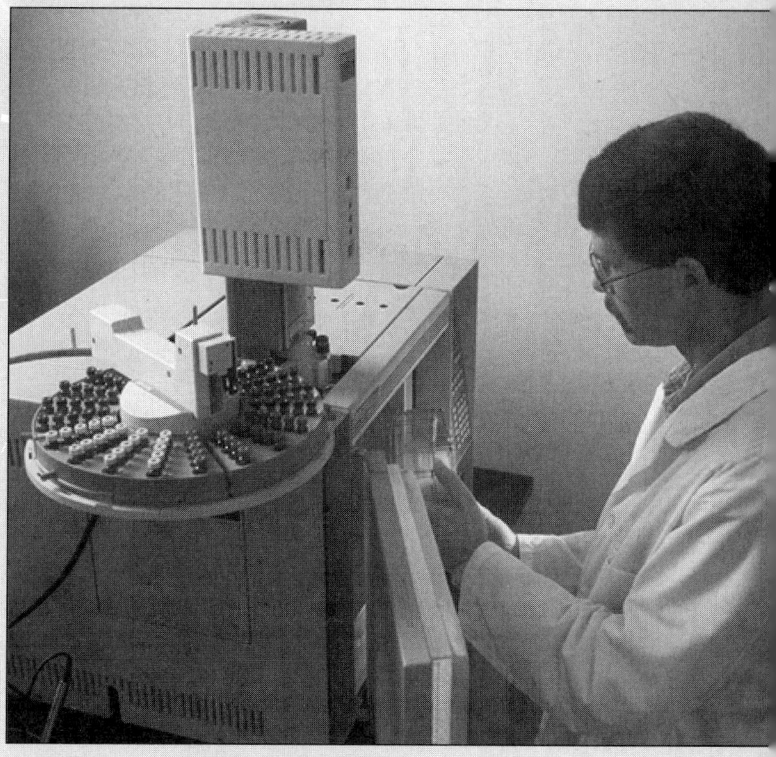

The photo shows a technician installing a capillary column in a state-of-the-art gas chromatograph. Highly automated instruments of this type are used in many fields of science and technology for the separation of components of complex mixtures. (Photo courtesy of Bioanalytical Systems, Inc., West Lafayette, IN.)

The methods described in this section are used to separate various components of analytical samples prior to their determination by instrumental means. Chapter 26 begins this section with an introduction to the terminology and theory of separations with emphasis on the optimization of experimental variables for efficient qualitative and quantitative analysis. This treatment is followed by a thorough discussion of the theory and practice of gas chromatography in Chapter 27. During recent years, the use of high-performance liquid chromatography in many areas of science has experienced enormous growth; the details of this mainstay of the analytical repertoire are discussed in Chapter 28. Chapter 29 describes the widely used technique of supercritical fluid chromatography and many of its applications. This section concludes with a discussion of the emerging field of capillary electrophoresis and some of its many applications.

Chapter 26

An Introduction to Chromatographic Separations

Generally, methods for chemical analysis are, at best, selective; few, if any, are truly specific. Consequently, the separation of the analyte from potential interferences is, more often than not, a vital step in analytical procedures. Until the middle of the twentieth century, analytical separations were largely carried out by such classical methods as precipitation, distillation, and extraction. By now, however, analytical separations are most commonly carried out by chromatography and electrophoresis, particularly with samples that are multicomponent and complex. This chapter and Chapters 27 through 29 are devoted to the several types of chromatography that are in current use. Chapter 30 describes separations by electrophoresis.

***Chromatography** is a powerful separation method that finds application to all branches of science. Chromatography was invented and named by the Russian botanist Mikhail Tswett at the beginning of the twentieth century. He employed the technique to separate various plant pigments such as chlorophylls and xanthophylls by passing solutions of these compounds through a glass column packed with finely divided calcium carbonate. The separated species appeared as colored bands on the column, which accounts for the name he chose for the method (Greek **chroma** meaning "color" and **graphein** meaning to "write").*

The applications of chromatography have grown explosively in the last fifty years, owing not only to the development of several new types of chromatographic techniques but also to the growing need by scientists for better methods for characterizing complex mixtures. The tremendous impact of these methods on science is attested by the 1952 Nobel Prize that was awarded to A. J. P. Martin and R. L. M. Synge for their discoveries in the field. Perhaps more impressive is a list of the 12 Nobel Prizes between 1937 and 1972 that were based upon work in which chromatography played a vital role.[1] By now, this list has undoubtedly increased.

[1] See L. S. Ettre, in *High-Performance Liquid Chromatography,* C. Horvath, Ed., Vol. 1, p. 4. New York: Academic Press, 1980.

26A A GENERAL DESCRIPTION OF CHROMATOGRAPHY

Chromatography encompasses a diverse and important group of methods that permit the scientist to separate closely related components of complex mixtures; many of these separations are impossible by other means.[2] In all chromatographic separations the sample is transported in a *mobile phase,* which may be a gas, a liquid, or a supercritical fluid. This mobile phase is then forced through an immiscible *stationary phase,* which is fixed in place in a column or on a solid surface. The two phases are chosen so that the components of the sample distribute themselves between the mobile and stationary phase to varying degrees. Those components that are strongly *retained* by the stationary phase move only slowly with the flow of mobile phase. In contrast, components that are weakly held by the stationary phase travel rapidly. As a consequence of these differences in mobility, sample components separate into discrete *bands,* or *zones,* that can be analyzed qualitatively and/or quantitatively.

26A-1 Classification of Chromatographic Methods

Chromatographic methods can be categorized in two ways. The first classification is based upon the physical means by which the stationary and mobile phases are brought into contact. In *column chromatography,* the stationary phase is held in a narrow tube through which the mobile phase is forced under pressure. In *planar* chromatography, the stationary phase is supported on a flat plate or in the interstices of a paper; here, the mobile phase moves through the stationary phase by capillary action or under the influence of gravity. The discussion in this and the next three chapters focuses on column chromatography. Section 28H is devoted to planar methods. It is important to point out here, however, that the equilibria upon which the two types of chromatography are based are identical and that the theory developed for column chromatography is readily adapted to planar chromatography as well.

A more fundamental classification of chromatographic methods is one based upon the types of mobile and stationary phases and the kinds of equilibria involved in the transfer of solutes between phases. Table 26-1 lists three general categories of chromatography: *liquid chromatography, gas chromatography,* and *supercritical-fluid chromatography.* As the names imply, the mobile phases in the three techniques are liquids, gases, and supercritical fluids, respectively. As shown in column 2 of the table, several specific chromatographic methods fall into each of the first two general categories.

It is noteworthy that only liquid chromatography can be performed either in columns or on plane surfaces; gas chromatography and supercritical-fluid chromatography, on the other hand, are restricted to column procedures so that the column walls contain the mobile phase.

26A-2 Elution Chromatography on Columns

Figure 26-1 shows schematically how two substances A and B are separated on a column by *elution chromatography* with a liquid mobile phase. Elution involves washing a species through a column by continuous addition of fresh solvent. As shown in the figure, a single portion of the sample is introduced at the head of a column (time t_0 in Figure 26-1), whereupon the components of the sample distribute themselves between the two phases. Introduction of additional mobile phase (the *eluent*) forces the solvent containing a part of the sample down the column, where further partition between the mobile phase and fresh portions of the stationary phase occurs (time t_1). Simultaneously, partitioning between the fresh solvent and the stationary phase takes place at the site of the original sample.

Continued additions of solvent carry solute molecules down the column in a continuous series of transfers between the mobile and the stationary phases. Because solute movement can only occur in the mobile phase, however, the average *rate* at which a solute zone migrates down the column *depends upon the fraction of time it spends in that phase.* This fraction is small for

[2] General references on chromatography include *Chromatography: Fundamentals and Applications of Chromatography and Electrophoretic Methods, Part A: Fundamentals, Part B: Applications,* E. Heftmann, Ed. New York: Elsevier, 1983; P. Sewell and B. Clarke, *Chromatographic Separations.* New York: Wiley, 1988; *Chromatographic Theory and Basic Principles,* J. A. Jonsson, Ed. New York: Marcel Dekker, 1987; R. M. Smith, *Gas and Liquid Chromatography in Analytical Chemistry.* New York: Wiley, 1988; K. Robards, P. R. Haddad, and P. E. Jackson, *Principles and Practice of Modern Chromatographic Methods.* New York: Academic Press, 1994; R. P. W. Scott, *Techniques and Practice of Chromatography.* New York: Marcel Dekker, 1995; J. C. Giddings, *Unified Separation Science.* New York: Wiley, 1991.

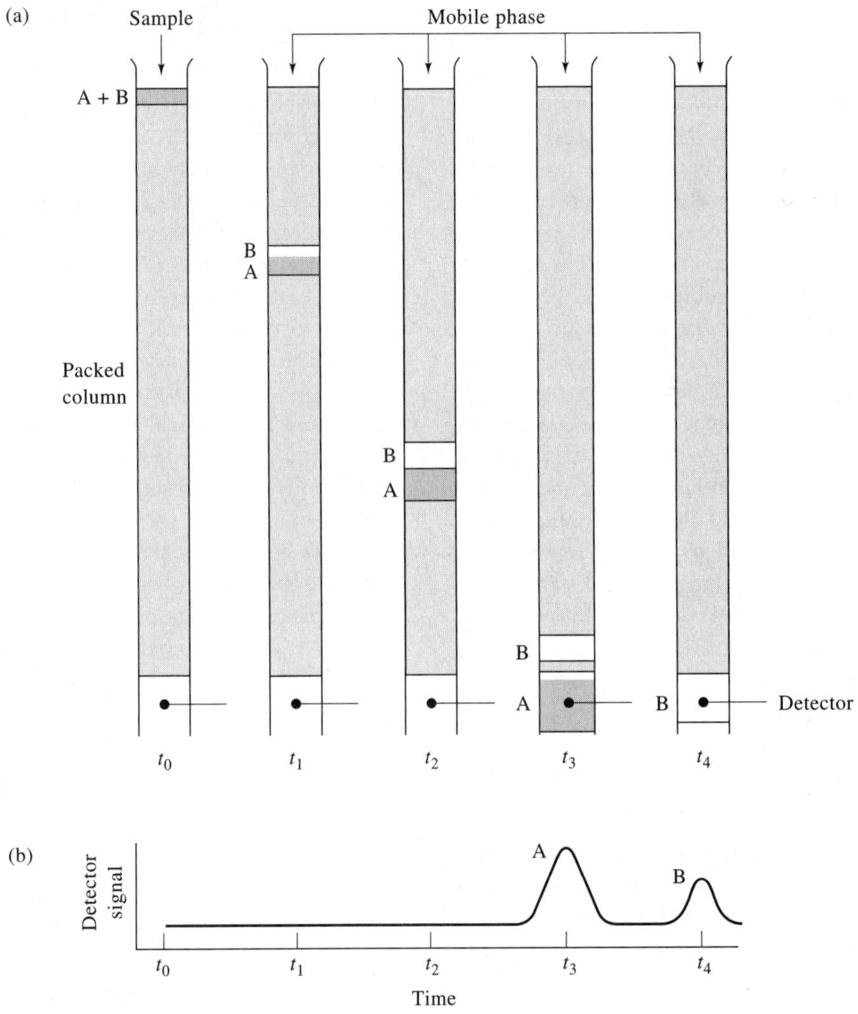

Figure 26-1 (a) Diagram showing the separation of a mixture of components A and B by column elution chromatography. (b) The output of the signal detector at the various stages of elution shown in (a).

solutes that are strongly retained by the stationary phase (for example, compound B in Figure 26-1), and it is large where retention in the mobile phase is more likely (component A). Ideally, the resulting differences in rates cause the components in a mixture to separate into bands, or zones, located along the length of the column (see time t_2 in Figure 26-1). Isolation of the separated species is then accomplished by passing a sufficient quantity of mobile phase through the column to cause the individual zones to pass out the end, where they can be detected or collected (times t_3 and t_4 in Figure 26-1).

Analyte Dilution

Figure 26-1 illustrates an important general characteristic of the separation process—namely, dilution of analytes nearly always accompanies separations. Thus, the size of the original zone containing the analytes in the figure is noticeably smaller than either of the two zones that reach the detector, meaning that significant dilution of the analytes has occurred while they were being separated. As a result, the detectors employed for separated analytes must often be more sensitive than would be required if the separation process was unnecessary.

TABLE 26-1 Classification of Column Chromatographic Methods

General Classification	Specific Method	Stationary Phase	Type of Equilibrium
Liquid chromatography (LC) (mobile phase: liquid)	Liquid-liquid, or partition	Liquid adsorbed on a solid	Partition between immiscible liquids
	Liquid-bonded phase	Organic species bonded to a solid surface	Partition between liquid and bonded surface
	Liquid-solid, or adsorption	Solid	Adsorption
	Ion exchange	Ion-exchange resin	Ion exchange
	Size exclusion	Liquid in interstices of a polymeric solid	Partition/sieving
Gas chromatography (GC) (mobile phase: gas)	Gas-liquid	Liquid adsorbed on a solid	Partition between gas and liquid
	Gas-bonded phase	Organic species bonded to a solid surface	Partition between liquid and bonded surface
	Gas-solid	Solid	Adsorption
Supercritical-fluid chromatography (SFC) (mobile phase: supercritical fluid)		Organic species bonded to a solid surface	Partition between supercritical fluid and bonded surface

Chromatograms

If a detector that responds to solute concentration is placed at the end of the column and its signal is plotted as function of time (or of volume of the added mobile phase), a series of peaks is obtained, as shown in the lower part of Figure 26-1. Such a plot, called a *chromatogram,* is useful for both qualitative and quantitative analysis. The positions of peaks on the time axis may serve to identify the components of the sample; the areas under the peaks provide a quantitative measure of the amount of each component.

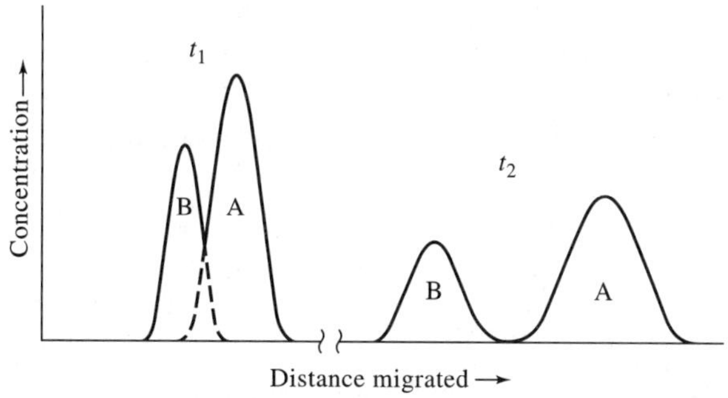

Figure 26-2 Concentration profiles of analyte bands A and B at two different times in their migration down the column in Figure 26-1. The times t_1 and t_2 are indicated in Figure 26-1.

The Effects of Migration Rates and Zone Broadening on Resolution

Figure 26-2 shows concentration profiles for solutes A and B at an early (t_1) and late (t_2) stage of elution from the chromatographic column shown in Figure 26-1.[3] Species B is more strongly retained; thus, B lags during the migration. It is apparent that movement down

[3] Note that the relative positions of bands for A and B in the concentration profile shown in Figure 26-2 appear to be the reverse of the peaks in the chromatogram shown in the lower part of Figure 26-1. The difference, of course, is that the abscissa in the former is distance along the column while in the latter it is time. Thus in the chromatogram in Figure 26-1, the *front* of a peak lies to the left and the *tail* to the right; for the concentration profile or zone the reverse obtains.

the column increases the distance between the two zones. At the same time, however, broadening of both zones takes place, which lowers the efficiency of the column as a separating device. While zone broadening is inevitable, conditions can ordinarily be found where it occurs more slowly than band separation. Thus, as shown in Figure 26-1, a clean resolution of species is often possible provided the column is sufficiently long.

Figure 26-3 shows two methods for improving the separation of a hypothetical two-component mixture. In Figure 26-3b conditions have been altered so that the first component moves down the column at a faster pace and the second at a slower. In Figure 26-3c the rates of zone broadening for the two species have been decreased. Both of these measures clearly lead to a cleaner separation. Variables that influence the relative migration rate of solute zones through a stationary phase are described in the next section. In Section 26C, consideration is given to factors that influence the rate of zone broadening.

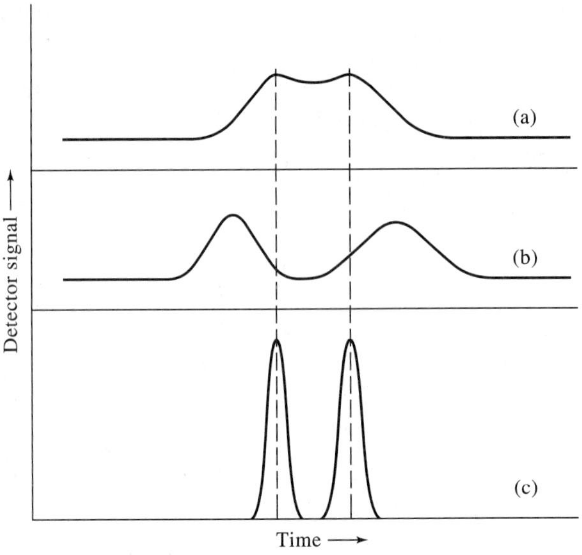

Figure 26-3 Two-component chromatograms illustrating two methods of improving separators: (a) original chromatogram with overlapping peaks; improvements brought about by (b) an increase in band separation, and (c) a decrease in band spread.

26B MIGRATION RATES OF SOLUTES[4]

The effectiveness of a chromatographic column in separating two solutes depends in part upon the relative rates at which the two species are eluted. These rates are determined by the magnitude of the equilibrium constants for the reactions by which the solutes distribute themselves between the mobile and stationary phases.

26B-1 Distribution Constants

Often, the distribution equilibria involved in chromatography are described by relatively straightforward equations that involve the transfer of an analyte between the mobile and stationary phases. Thus, for the solute species A, we may write

$$A_{mobile} \rightleftharpoons A_{stationary}$$

The equilibrium constant K for this reaction is called the *distribution constant*, the *partition ratio*, or the *partition coefficient*,[5] and is defined as

$$K = \frac{c_S}{c_M} \qquad (26\text{-}1)$$

where c_S is the molar concentration of the solute in the stationary phase and c_M is its molar concentration in the mobile phase. Ideally, K is constant over a wide range of solute concentrations; that is, c_S is directly proportional to c_M. Chromatography in which Equation 26-1 applies is termed *linear chromatography* and results in such characteristics as symmetric Gaussian type peaks and retention times that are independent of the amount

[4] The common practice of using the word "migration" as a generic description of molecular motion in chromatography should not be confused with *migration* as a specific type of molecular motion wherein ions move under the influence of an electric field. See Section 22A-4.

[5] For recent recommendations of the IUPAC Commission on Analytical Nomenclature for chromatography, see L. S. Ettre, *Pure Appl. Chem.*, **1993**, *65*, 819. The committee recommends the use of the term *distribution constant* K_c rather than the older term *partition coefficient*, or *partition ratio*, *K*. In our discussion, we use the recommended term *distribution constant* but retain the symbol *K*. Consult Section 3.9.01, page 846 of the aforementioned reference for the definition of K_c and the rationale for its use. Be aware that both terms will be found in the chromatographic literature.

of analyte injected. We will limit our theoretical discussions to linear chromatography exclusively.

26B-2 Retention Time

Figure 26-4 is a typical chromatogram for a sample containing a single analyte. The time it takes after sample injection for the analyte peak to reach the detector is called the *retention time* and is given the symbol t_R. The small peak on the left is for a species that is *not* retained by the column. Often the sample or the mobile phase will contain an unretained species. When they do not, such a species may be added to aid in peak identification. The time t_M for the unretained species to reach the detector is sometimes called the *dead time.* The rate of migration of the unretained species is the same as the average rate of motion of the mobile phase molecules.

The average linear rate of solute migration $\bar{v}$ is

$$\bar{v} = \frac{L}{t_R} \tag{26-2}$$

where L is the length of the column packing. Similarly, the average linear rate of movement u of the molecules of the mobile phase is

$$u = \frac{L}{t_M} \tag{26-3}$$

where t_M, the *dead time,* is the time required for an average molecule of the mobile phase to pass through the column.

26B-3 The Relationship Between Retention Time and Distribution Constant

In order to relate the retention time of a solute to its distribution constant, we express its migration rate as a fraction of the velocity of the mobile phase:

$$\bar{v} = u \times \text{fraction of time solute spends in mobile phase}$$

This fraction, however, also equals the average number of moles of a solute in the mobile phase at any instant divided by the total number of moles of that solute in the column:

$$\bar{v} = u \times \frac{\text{moles of solute in mobile phase}}{\text{total moles of solute}}$$

The total number of moles of a solute in the mobile phase is equal to the molar concentration c_M of the solute in that phase multiplied by the phase volume V_M. Similarly, the number of moles of a solute in the stationary phase is given by the product of the concentration c_S of the solute in the stationary phase and its volume V_S. Therefore,

$$\bar{v} = u \times \frac{c_M V_M}{c_M V_M + c_S V_S} = u \times \frac{1}{1 + c_S V_S / c_M V_M}$$

Substitution of Equation 26-1 into this equation gives an expression for the rate of solute migration as a function of its distribution constant and as

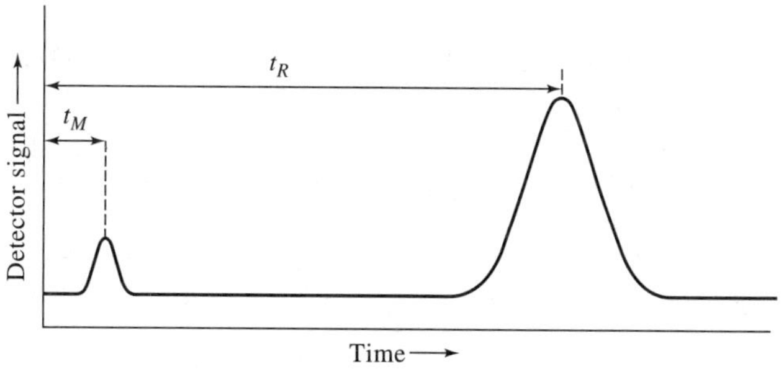

Figure 26-4 A typical chromatogram for a two-component mixture. The small peak on the left represents a species that is not retained on the column and so reaches the detector almost immediately after elution is started. Thus its retention time t_M is approximately equal to the time required for a molecule of the mobile phase to pass through the column.

a function of the volumes of the stationary and mobile phases:

$$\bar{\nu} = u \times \frac{1}{1 + KV_S/V_M} \quad (26\text{-}4)$$

The two volumes can be estimated from the method by which the column is prepared.

26B-4 The Rate of Solute Migration: The Retention Factor

The *retention factor*, or *capacity factor*, is an important parameter that is widely used to describe the migration rates of solutes on columns.[6] For a solute A, the retention factor k'_A is defined as

$$k'_A = \frac{K_A V_S}{V_M} \quad (26\text{-}5)$$

where K_A is the distribution constant for the species A. Substitution of Equation 26-5 into 26-4 yields

$$\bar{\nu} = u \times \frac{1}{1 + k'_A} \quad (26\text{-}6)$$

In order to show how k'_A can be derived from a chromatogram, we substitute Equations 26-2 and 26-3 into Equation 26-6:

$$\frac{L}{t_R} = \frac{L}{t_M} \times \frac{1}{1 + k'_A} \quad (26\text{-}7)$$

This equation rearranges to

$$k'_A = \frac{t_R - t_M}{t_M} \quad (26\text{-}8)$$

As shown in Figure 26-4, t_R and t_M are readily obtained from a chromatogram. When the retention factor for a solute is much less than unity, elution occurs so rapidly that accurate determination of the retention times is difficult. When the retention factor is larger than perhaps 20 to 30, elution times become inordinately long. Ideally, separations are performed under conditions in which the retention factors for the solutes in a mixture lie in the range between 2 and 10.

26B-5 Relative Migration Rates: The Selectivity Factor

The *selectivity factor* α of a column for the two species A and B is defined as

$$\alpha = \frac{K_B}{K_A} \quad (26\text{-}9)$$

where K_B is the distribution constant for the more strongly retained species B and K_A is the distribution constant for the less strongly held, or more rapidly eluted, species A. By this definition α *is always greater than unity.*

Substitution of Equation 26-5 and the analogous equation for solute B into Equation 26-9 provides, after rearrangement, a relationship between the selectivity factor for two solutes and their retention factors:

$$\alpha = \frac{k'_B}{k'_A} \quad (26\text{-}10)$$

where k'_B and k'_A are the retention factors for B and A, respectively. Substitution of Equation 26-8 for the two solutes in Equation 26-10 gives an expression that permits the determination of α from an experimental chromatogram:

$$\alpha = \frac{(t_R)_B - t_M}{(t_R)_A - t_M} \quad (26\text{-}11)$$

In Section 26D-1 we show how to use the selectivity factor and retention factors to compute the resolving power of a column.

26C ZONE BROADENING AND COLUMN EFFICIENCY

The *rate*, or *kinetic theory of chromatography* successfully explains in quantitative terms the shapes of chromatographic peaks and the effects of several variables on the breadth of these peaks.[7] A detailed discussion of

[6] The IUPAC Commission on Analytical Nomenclature now recommends that k' be called the *retention factor* rather than the older name *capacity factor* and that it be symbolized k. We use the recommended term but retain the symbol k' because it is still widely used in the literature. For details regarding the rationale for this change as well as for other changes in nomenclature, see L. S. Ettre, *Pure Appl. Chem.*, **1993**, *65*, 843 (Section 3.7.12).

[7] For a detailed presentation of the rate theory, see J. C. Giddings, *Dynamics of Chromatography*, Part I. New York: Marcel Dekker, 1965; J. C. Giddings, *Unified Separation Science*, New York: Wiley, 1990; R. P. W. Scott, *Contemporary Liquid Chromatography* (Vol. XI of *Techniques of Chemistry*, A. Weissberger, Ed.), Chapter 2. New York: Wiley, 1976. For a shorter presentation, see J. C. Giddings, *J. Chem. Educ.*, **1958**, *35*, 588; **1967**, *44*, 704.

this theory, which is based upon a random-walk mechanism, is beyond the scope of this text. We can, however, give a qualitative picture of chromatographic zones and why these zones broaden as they move down a column. This discussion then leads to a consideration of variables that improve column efficiency by reducing broadening.

26C-1 The Shapes of Chromatographic Peaks

Examination of the peaks in a chromatogram (Figures 26-1 and 26-4) or the bands on columns (Figure 26-2) reveals a similarity to normal error or Gaussian curves (Figure a1-3, Appendix 1), which are obtained when replicate values of a measurement are plotted as a function of the frequency of their occurrence.[8] As was shown in Section a1B-1, normal error curves can be rationalized by assuming that the uncertainty associated with any single measurement is the summation of a much larger number of small, individually undetectable and random uncertainties, each of which has an equal probability of being positive or negative. The most common occurrence is for these uncertainties to cancel one another, thus leading to the mean value. With less likelihood, the summation may cause results that are greater or smaller than the mean. The consequence is a symmetric distribution of replicate data around the mean value. In a similar way, the Gaussian shape of an ideal chromatographic zone can be attributed to the additive combination of the random motions of the myriad solute molecules in the chromatographic zone.

It is instructive to first consider the behavior of an individual solute molecule, which, during migration, undergoes many thousands of transfers between the stationary and the mobile phase. The time it spends in either phase after a transfer is highly irregular and depends upon its accidentally gaining sufficient thermal energy from its environment to accomplish a reverse transfer. Thus, in some instances, the residence time in a given phase may be transitory; in others, the period may be relatively long. Recall that the molecule is eluted *only during residence in the mobile phase;* as a result, its migration down the column is also highly irregular. Because of variability in the residence time, the average rate at which individual molecules move relative to the mobile phase varies considerably. Certain individual particles travel rapidly by virtue of their accidental inclusion in the mobile phase for a majority of the time. Others, in contrast, may lag because they happen to have been incorporated in the stationary phase for a greater-than-average time. The consequence of these random individual processes is a symmetric spread of velocities around the mean value, which represents the behavior of the average molecule.

The breadth of a band increases as it moves down the column because more time is allowed for spreading to occur. Thus, the zone breadth is directly related to residence time in the column and inversely related to the velocity at which the mobile phase flows.

[8] Some chromatographic peaks are nonideal and exhibit *tailing* or *fronting*. In the former case the tail of the peak, appearing to the right on the chromatogram, is drawn out while the front is steepened. With fronting, the reverse is the case. A common cause of tailing and fronting is a nonlinear distribution constant. Fronting also arises when too large a sample is introduced onto a column. Distortions of this kind are undesirable because they lead to poorer separations and less reproducible elution times. In the discussion that follows, tailing and fronting are assumed to be absent or minimal.

26C-2 Methods for Describing Column Efficiency

Two related terms are widely used as quantitative measures of chromatographic column efficiency: (1) *plate height H* and (2) *plate count plates N.* The two are related by the equation

$$N = L/H \tag{26-12}$$

where L is the length (usually in centimeters) of the column packing. The efficiency of chromatographic columns increases as the plate count becomes greater and as the plate height becomes smaller. Enormous differences in efficiencies are encountered in columns as a result of differences in column type and in mobile and stationary phases. Efficiencies in terms of plate numbers can vary from a few hundred to several hundred thousand; plate heights ranging from a few tenths to one thousandth of a centimeter or smaller are not uncommon.

The genesis of the terms "plate height" and "number of theoretical plates" is a pioneering theoretical study of Martin and Synge in which they treated a chromatographic column as if it were similar to a distillation column that is made up of numerous discrete but contiguous narrow layers called *theoretical plates.*[9] At each

[9] A. J. P. Martin and R. L. M. Synge, *Biochem. J.*, **1941**, *35*, 1358.

plate, equilibration of the solute between the mobile and stationary phase was assumed to take place. Movement of the solute down the column was then treated as a stepwise transfer of equilibrated mobile phase from one plate to the next.

The *plate theory* successfully accounts for the Gaussian shape of chromatographic peaks and their rate of movement down a column. It was ultimately abandoned in favor of the *rate theory,* however, because it fails to account for peak broadening in a mechanistic way. Nevertheless, the original terms for efficiency have been carried over to the rate theory. This nomenclature is perhaps unfortunate because it tends to perpetuate the myth that a column contains plates where equilibrium conditions exist. In fact, the equilibrium state can never be realized with the mobile phase in constant motion.

The Definition of Plate Height

As shown in Section a1B-1 of Appendix 1, the breadth of a Gaussian curve is directly related to the variance σ^2 or the standard deviation σ of a measurement. Because chromatographic bands are generally assumed to be Gaussian in shape, it is convenient to define the efficiency of a column in terms of variance per unit length of column. That is, the plate height H is given by

$$H = \frac{\sigma^2}{L} \tag{26-13}$$

This definition of plate height is illustrated in Figure 26-5, which shows a column having a packing L cm in length. Above this schematic is a plot showing the distribution of molecules along the length of the column at the moment the analyte peak reaches the end of the packing (that is, at the retention time t_R). The curve is Gaussian, and the locations of $L - 1\sigma$ and $L + 1\sigma$ are indicated as broken vertical lines. Note that L carries units of centimeters and σ^2 units of centimeters squared; thus H represents a linear distance in centimeters (Equation 26-13). In fact, the plate height can be thought of as the length of column (at the end of the column) that contains a fraction of analyte that lies between $L - \sigma$ and L. Because the area under a normal error curve bounded by $\pm\sigma$ is about 68% of the total area, the plate height, as defined, contains approximately 34% of the analyte.

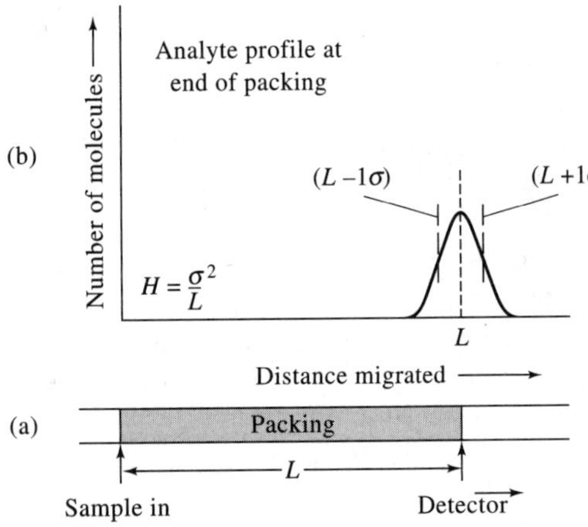

Figure 26-5 Definition of plate height $H = \sigma^2/L$.

The Experimental Evaluation of H and N

Figure 26-6 is a typical chromatogram with time as the abscissa. The variance of the solute peak, which can be obtained by a simple graphical procedure, has units of seconds squared and is usually designed at τ^2 to distinguish it from σ^2, which has units of centimeters squared. The two standard deviations τ and σ are related by

$$\tau = \frac{\sigma}{L/t_R} \tag{26-14}$$

where L/t_R is the average linear velocity of the solute in centimeters per second.

Figure 26-6 illustrates a simple means for approximating τ and σ from an experimental chromatogram. Tangents at the inflection points on the two sides of the chromatographic peak are extended to form a triangle with the base line of the chromatogram. The area of this triangle can be shown to be approximately 96% of the total area under the peak. In Section a1B-1 it was shown that about 96% of the area under a Gaussian peak is included within plus or minus two standard deviations ($\pm 2\sigma$) of its maximum. Thus, the intercepts shown in Figure 26-6 occur at approximately $\pm 2\tau$ from the maximum, and $W = 4\tau$, where W is the magnitude of the base of the triangle. Substituting this relationship into Equation 26-14 and rearranging yield

$$\sigma = \frac{LW}{4t_R} \tag{26-15}$$

Substitution of this equation for σ into Equation 26-13 gives

$$H = \frac{LW^2}{16t_R^2} \tag{26-16}$$

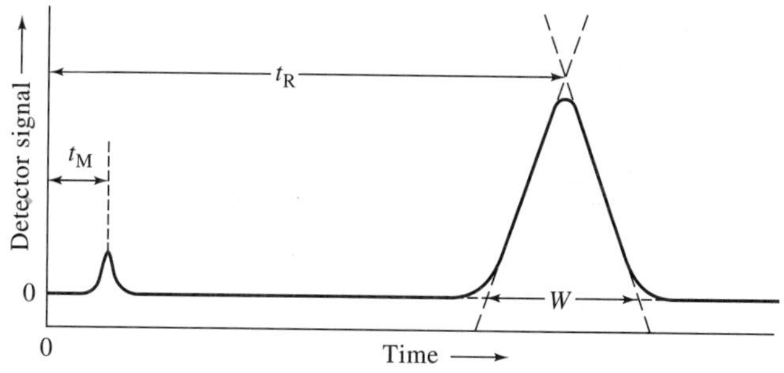

Figure 26-6 Determination of the standard deviation τ from a chromatographic peak: $W = 4\tau$.

To obtain N, we substitute into Equation 26-12 and rearrange to get

$$N = 16\left(\frac{t_R}{W}\right)^2 \tag{26-17}$$

Thus, N can be calculated from two time measurements, t_R and W; to obtain H, the length of the column packing L must also be known.

Another method for approximating N, which some workers believe to be more reliable, is to determine $W_{1/2}$, the width of the peak at half its maximum height. The plate count is then given by

$$N = 5.54\left(\frac{t_R}{W_{1/2}}\right)^2 \tag{26-18}$$

The plate count N and the plate height H are widely used in the literature and by instrument manufacturers as measures of column performance. For these numbers to be meaningful in comparing two columns, it is essential that they be determined with the *same compound.*

March 29. 05 Tuesday

26C-3 Kinetic Variables Affecting Zone Broadening

Zone broadening has been shown to be the consequence of the finite rate at which several mass-transfer processes occur during migration of a solute down a column. Some of these rates are controllable by adjustment of experimental variables, thus permitting improvement in separations. Table 26-2 lists the most important of these variables. Their effects on column efficiency, as measured by plate height H, are described in the paragraphs that follow.

smaller plate height = higher efficiency

The Effect of Mobile-Phase Flow Rate

The magnitude of kinetic effects on column efficiency clearly depends upon the length of time the mobile phase is in contact with the stationary phase, which in turn depends upon the flow rate of the mobile phase. For this reason, efficiency studies have generally been carried out by determining H (by means of Equation 26-17 or 26-18 and Equation 26-12) as a function of mobile-phase velocity u. The data obtained from such studies are typified by the two plots shown in Figure 26-7, one for liquid chromatography and the other for gas chromatography. Both show a minimum in H (or a maximum in efficiency) at low flow rates. The minima for liquid chromatography generally occur at flow rates that are well below those for gas chromatography and often so low that they are not observed under normal operating conditions. The rate theory of chromatographic zone broadening, which is described in this section, accurately predicts the shape of a plot of H versus u, which is often called a *van Deemter plot* after the originator of the theory.

TABLE 26-2 Variables That Affect Column Efficiency

Variable	Symbol	Usual Units
Linear velocity of mobile phase	u	$cm \cdot s^{-1}$
Diffusion coefficient in mobile phase*	D_M	$cm^2 \cdot s^{-1}$
Diffusion coefficient in stationary phase*	D_S	$cm^2 \cdot s^{-1}$
Retention factor (Equation 26-8)	k'	unitless
Diameter of packing particle	d_p	cm
Thickness of liquid coating on stationary phase	d_f	cm

*Increases as temperature increases and viscosity decreases.

smaller plate height = higher efficiency

As suggested by Figure 26-7, flow rates for liquid chromatography are significantly smaller than those used in gas chromatography. This means that gas chromatographic separations are typically completed in shorter times than are liquid chromatographic separations. Furthermore, as shown in the figure, plate heights for liquid chromatographic columns are an order of magnitude or more smaller than those encountered with gas chromatographic columns. Offsetting this advantage, however, is the fact that it is impractical to employ liquid columns that are longer than about 25 to 50 cm (because of high pressure drops), whereas gas chromatographic columns may be 50 m or more in length. Consequently, the total number of plates, and thus overall column efficiency, are usually superior with gas chromatographic columns. Thus, one comparison between gas chromatography and liquid chromatography is that gas chromatography is capable of faster separations and higher-efficiency separations, although not necessarily both simultaneously.

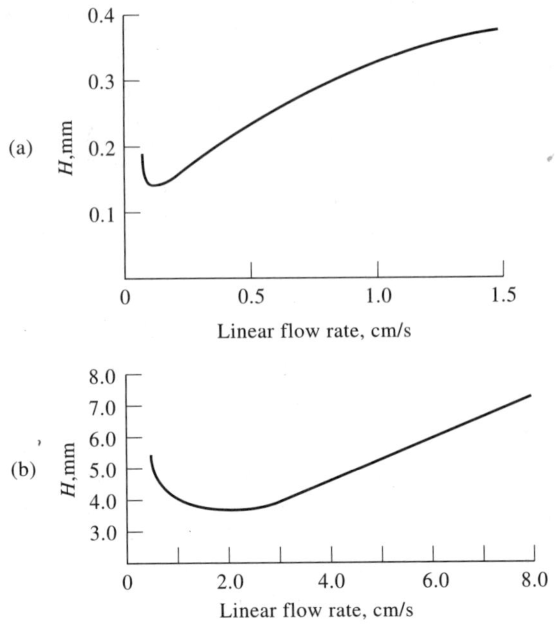

Figure 26-7 Effect of mobile-phase flow rate on plate height for (a) liquid chromatography and (b) gas chromatography.

Relationship between Plate Height and Column Variables

Over the last forty years, an enormous amount of theoretical and experimental effort has been devoted to developing quantitative relationships describing the effects of the variables in Table 26-2 on plate heights for various types of columns. Perhaps a dozen or more mathematical expressions relating plate heights to column variables have been put forward and applied with varying degrees of success. It is apparent that none of these equations is entirely adequate to explain the complex physical interactions and effects that lead to zone broadening. Some, though imperfect, have been of considerable use, however, in pointing the way toward improved column performance. One of these is presented here.

Mathematical approximation of the behavior of chromatographic columns began with the studies in the 1950s by Dutch chemical engineers that led to the *van Deemter equation,* which can be written in the form

$$
\begin{aligned}
H &= A + B/u + Cu \\
&= A + B/u + (C_S + C_M)u \qquad (26\text{-}19)
\end{aligned}
$$

where H is the plate height in centimeters, u is the linear velocity of the mobile phase in centimeters per second, and the quantities A, B, and C are coefficients related to the phenomena of *multiple flow paths, longitudinal diffusion,* and *mass transfer between phases,* respectively. As indicated on the right, the C coefficient can be broken into two coefficients, one related to the stationary phase (C_S) and one related to the mobile phase (C_M).

More recent theoretical studies have led to elaboration of this expression, but it has been shown experimentally that the basic van Deemter equation is quite satisfactory to understand much of column efficiency.[10] Note that the van Deemter equation contains terms linearly and inversely proportional to, as well as independent of, the mobile phase velocity.

Let us now examine in some detail the variables that affect the four terms in Equation 26-19, A, B/u, $C_S u$, $C_M u$. These terms are identified in Table 26-3 and described in the paragraphs that follow.

The Multipath Term (A). Zone broadening arises in part from the multitude of pathways by which a molecule (or ion) can find its way through a packed column. As shown in Figure 26-8, the length of these pathways

[10] E. Katz, K. L. Ogan, and R. P. W. Scott, *J. Chromatogr.,* **1983,** *270,* 51.

TABLE 26-3 Kinetic Processes That Contribute to Peak Broadening

Process	Term in Equation 26-19	Relationship to Column* and Analyte Properties
Multiple flow paths	A	$A = 2\lambda d_P$
Longitudinal diffusion	B/u	$\frac{B}{u} = \frac{2\gamma D_M}{u}$
Mass transfer to and from liquid stationary phase	$C_S u$	$C_S u = \frac{f_S(k')d_f^2}{D_S} u$
Mass transfer in mobile phase	$C_M u$	$C_M u = \frac{f_M(k')d_p^2}{D_M} u$

*u, D_S, D_M, d_f, d_p, k' are as defined in Table 26-2.
$f(x)$ = function of x.
λ, γ: constants that depend on the quality of the packing.
B: coefficient of longitudinal diffusion.
C_S, C_M: coefficients of mass transfer in stationary and mobile phases, respectively.

may differ significantly; thus, the residence time in the column for molecules of the same species is also variable. Solute molecules then reach the end of the column over a time interval, which leads to a broadened band. As shown in the third column of Table 26-3, this effect, which is sometimes called *eddy diffusion,* is directly proportional to the diameter of the particles making up the column packing.

Multipath broadening may be partially offset by ordinary diffusion, which results in molecules' being transferred from a stream following one pathway to a stream following another. If the velocity of flow is very low, a large number of these transfers will occur, and each molecule in its movement down the column samples numerous flow paths, spending a brief time in each. As a consequence, the rate at which each molecule moves down the column tends to approach that of the average. Thus, at low mobile-phase velocities, the molecules are not significantly dispersed by the multiple-path nature of the packing. At moderate or high velocities, however, sufficient time is not available for diffusion averaging to occur, and band broadening due to the different path lengths is observed.

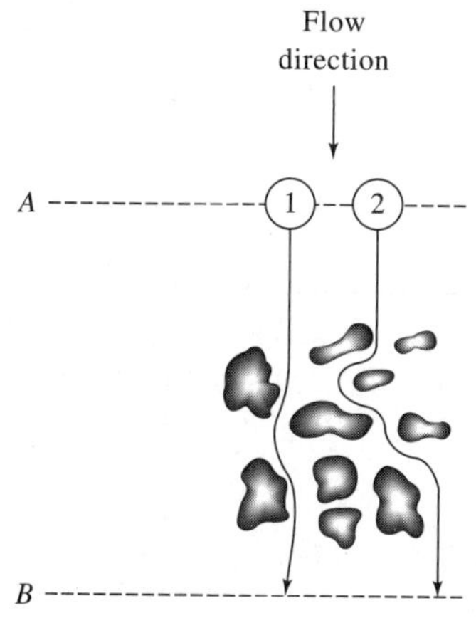

Figure 26-8 Typical pathways of two molecules during elution. Note that distance traveled by molecule 2 is greater than that traveled by molecule 1. Thus, molecule 2 would arrive at B later than molecule 1.

The Longitudinal Diffusion Term (B/u). Longitudinal diffusion in column chromatography is a band-broadening process in which solutes diffuse from the concentrated center of a zone to the more dilute regions ahead of and behind the zone center—that is, toward and opposed to the direction of flow of the mobile phase. As shown by the second equation in Table 26-3, the longitudinal diffusion term is directly proportional to the mobile-phase diffusion coefficient D_M, which is a constant equal to the rate of migration under a unit concentration gradient. The constant γ is called the obstructive factor, which recognizes that longitudinal diffusion is hindered by the packing. With packed columns, this constant typically has a value of about 0.6; for unpacked capillary columns, its value is unity.

The contribution of longitudinal diffusion is seen to be inversely proportional to the mobile phase velocity. Such a relationship is not surprising when we consider that the analyte is in the column for a briefer period when the flow rate is high. Thus, diffusion from the center of the band to the two edges has less time to occur. The negative slopes at low flow rates in both graphs of Figure 26-7 are a consequence of longitudinal diffusion. Note that the effect is much less pronounced in liquid chromatography because diffusion coefficients in liquids are orders of magnitude smaller than those in gases. In fact, for most liquids B/u approaches zero relative to the other terms in Equation 26-19. Thus, the minimum shown Figure 26-7a is sometimes not observed.

Mass-Transfer Coefficients (C_S and C_M). The need for the two mass-transfer coefficients C_S and C_M in Equation 26-19 arises because the equilibrium between the mobile and the stationary phase is established so slowly that a chromatographic column always operates under nonequilibrium conditions. Consequently, analyte molecules at the front of a band are swept ahead before they have time to equilibrate with the stationary phase and thus be retained. Similarly, equilibrium is not reached at the trailing edge of a band, and molecules are left behind in the stationary phase by the fast-moving mobile phase.

Band broadening from mass-transfer effects arises because the many flowing streams of a mobile phase within a column and the layer making up the stationary phase both have finite widths. Consequently, time is required for solute molecules to diffuse from the interior of these phases to their interface where transfer occurs. This time lag results in the persistence of nonequilibrium conditions along the length of the column. If the rates of mass transfer within the two phases were infinite, broadening of this type would not occur.

Note that the extent of both longitudinal broadening and mass-transfer broadening depend upon the rate of diffusion of analyte molecules but that the direction of diffusion in the two cases is different. Longitudinal broadening arises from the tendency of molecules to move in directions that tend to parallel the flow, whereas mass-transfer broadening occurs from diffusion that tends to be at right angles to the flow. As a consequence, the extent of longitudinal broadening is *inversely* related to flow rate. For mass-transfer broadening, in contrast, the faster the mobile phase moves, the less time there is for equilibrium to be approached. Thus, as shown by the last two terms in Equation 26-19, the mass-transfer effect on plate height is directly proportional to the rate u of the movement of the mobile phase.

The Stationary Phase Mass-Transfer Term ($C_S u$). The third equation in Table 26-3 reveals that when the stationary phase is an immobilized liquid, the mass-transfer coefficient is proportional to a complex function $f_S(k')$ of the retention factor k', is directly proportional to the square of the thickness of the film on the support particles d_f^2 and is inversely proportional to the diffusion coefficient D_S of the solute in the film. The effects of this relatonship can be understood by realizing that these factors influence the average frequency at which analyte moleules reach the interface where transfer to the mobile phase can occur. For example, with thick films, molecules must on the average travel farther to reach the surface; with smaller diffusion coefficients, they travel more slowly. The consequence of both factors is a lower rate of mass transfer and an increase in plate height.

The Mobile Phase Mass-Transfer Term ($C_M u$). As shown by the fourth equation in Table 26-3, the mobile phase mass-transfer coefficient C_M is a complex function $f(k')$ of the retention factor k', is directly proportional to the square of the particle diameter of the packing d_p^2 and is inversely proportional to the mobile phase diffusion coefficient D_M.

Effect of Mobile-Phase Velocity on Terms in Equation 26-19. Figure 26-9 provides an idea of the quality of the fit of the van Deemter equation to typical data. The eluted analyte was benzyl acetate in a solution of *n*-hexane containing a few percent of ethyl acetate. The points on the upper curve are experimental points, and the solid curve was obtained from a numerical fit of the van Deemter equation to the data. The lower plots in the figure also show the contribution of the multipath effect, longitudinal diffusion, and the combined mass-transfer effects.

Summary of Methods for Reducing Zone Broadening. Two important controllable variables that affect column efficiency are the diameter of the particles making up the packing and the diameter of the column. The effect of particle diameter is demonstrated by the data shown in Figure 26-10. To take advantage of the effect

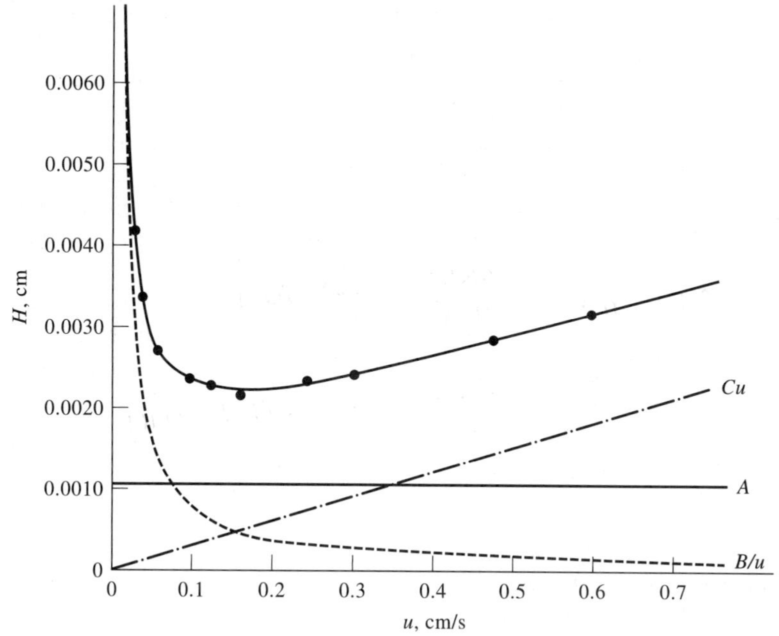

Figure 26-9 A van Deemter plot for a packed liquid chromatographic column. The points on the upper curve are experimental. The contributions of the various rate terms are shown by the lower curves: *A*, multipath effect; *B/u*, longitudinal diffusion; *Cu*, mass transfer for both phases. *(From E. Katz, K. L. Ogan, and R. P. W. Scott,* J. Chromatogr., *1983*, 270, *51. With permission.)*

of column diameter, narrower and narrower columns have been used in recent years.

With gaseous mobile phases, the rate of longitudinal diffusion can be reduced appreciably by lowering the temperature and thus the diffusion coefficient D_M. The consequence is significantly smaller plate heights at low temperatures. This effect is usually not noticeable in liquid chromatography because diffusion is slow enough so that the longitudinal diffusion term has little effect on overall plate height.

With liquid stationary phases, the thickness of the layer of adsorbed liquid should be minimized since C_S in Equation 26-19 is proportional to the square of this variable d_f (see third equation in Table 26-3).

26D OPTIMIZATION OF COLUMN PERFORMANCE

A chromatographic separation is optimized by varying experimental conditions until the components of a mixture are separated cleanly with a minimum expenditure of time. Optimization experiments are aimed at either (1) reducing zone broadening or (2) altering relative migration rates of the components. As we have shown in Section 26C, zone broadening is increased by those kinetic variables that increase the plate height of a column. Migration rates, on the other hand, are varied by changing those variables that affect retention and selectivity factors of the solution (Section 26B).

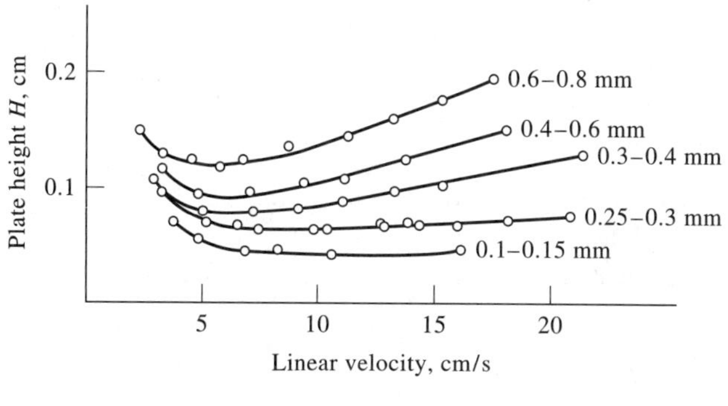

Figure 26-10 Effect of particle size on plate height. The numbers to the right are particle diameters. *(From J. Boheman and J. H. Purnell, in* Gas Chromatography, 1958, *D. H. Desty, Ed. New York: Academic Press, 1958. With permission of Butterworths, Stoneham, MA.)*

26D-1 Column Resolution

The *resolution* R_s of a column provides a quantitative measure of its ability to separate two analytes. The significance of this term is illustrated in Figure 26-11, which consists of chromatograms for species A and B on three columns having different resolutions. Column resolution is defined as

$$R_s = \frac{\Delta Z}{W_A/2 + W_B/2} = \frac{2\Delta Z}{W_A + W_B} = \frac{2[(t_R)_B - (t_R)_A]}{W_A + W_B} \qquad (26\text{-}20)$$

where all of terms on the right side of the equation are apparent in the figure.

It is evident from Figure 26-11 that a resolution of 1.5 gives an essentially complete separation of the two components, whereas a resolution of 0.75 does not. At a resolution of 1.0, zone A contains about 4% B and zone B contains a similar amount of A. At a resolution of 1.5, the overlap is about 0.3%. The resolution for a given stationary phase can be improved by lengthening the column, thus increasing the number of plates. An adverse consequence of the added plates, however, is an increase in the time required for the separation.

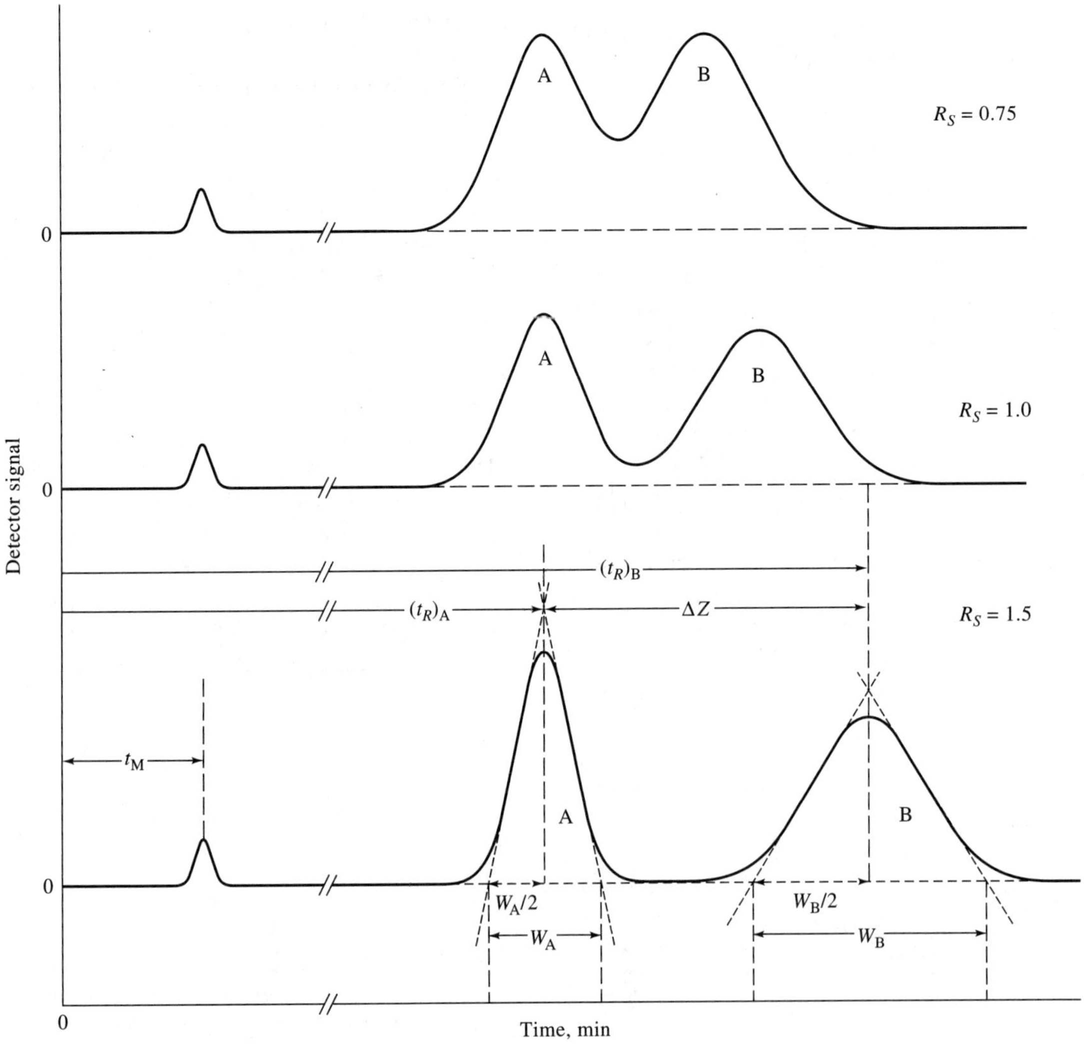

Figure 26-11 Separations at three resolutions. Here, $R_s = 2\Delta Z/(W_A + W_B)$.

26D-2 The Effect of Retention and Selectivity Factors on Resolution

It is useful to develop a mathematical relationship between the resolution of a column and the retention factors k'_A and k'_B for two solutes, the selectivity factor α, and the number of plates N making up the column. To this end, we will assume that we are dealing with two solutes A and B having retention times that are close enough to one another that we can assume

$$W_A = W_B \approx W$$

Equation 26-20 then takes the form

$$R_s = \frac{(t_R)_B - (t_R)_A}{W}$$

Equation 26-17 permits the expression of W in terms of $(t_R)_B$ and N, which can then be substituted into the foregoing equation to give

$$R_s = \frac{(t_R)_B - (t_R)_A}{(t_R)_B} \times \frac{\sqrt{N}}{4}$$

Substituting Equation 26-8 and rearranging leads to an expression for R_S in terms of the retention factors for A and B. That is,

$$R_s = \frac{k'_B - k'_A}{1 + k'_B} \times \frac{\sqrt{N}}{4}$$

Let us eliminate k'_A from this expression by substituting Equation 26-10 and rearranging. Thus,

$$R_s = \frac{\sqrt{N}}{4}\left(\frac{\alpha - 1}{\alpha}\right)\left(\frac{k'_B}{1 + k'_B}\right) \qquad (26\text{-}21)$$

Often it is desirable to calculate the number of theoretical plates required to achieve a desired resolution. An expression for this quantity is obtained by rearranging Equation 26-21 to give

$$N = 16R_s^2\left(\frac{\alpha}{\alpha - 1}\right)^2\left(\frac{1 + k'_B}{k'_B}\right)^2 \qquad (26\text{-}22)$$

Simplified forms of Equations 26-21 and 26-22 are sometimes encountered where these equations are applied to a pair of solutes whose distribution constants are similar enough to make their separation difficult. Thus, when $K_A \approx K_B$, it follows from Equation 26-5 that $k'_A \approx k'_B = k'$ and from Equation 26-9, $\alpha \rightarrow 1$. With these approximations, Equations 26-21 and 26-22 reduce to

$$R_s = \frac{\sqrt{N}}{4}(\alpha - 1)\left(\frac{k'}{1 + k'}\right) \qquad (26\text{-}23)$$

$$N = 16R_s^2\left(\frac{1}{\alpha - 1}\right)^2\left(\frac{1 + k'}{k'}\right)^2 \qquad (26\text{-}24)$$

where k' is the average of k'_A and k'_B.

26D-3 The Effect of Resolution on Retention Time

Before considering in detail the significance of the four equations just derived, it is worthwhile developing an equation for a related performance characteristic for a column, namely, the time required to complete the separation of solutes A and B. Clearly, what is desired in chromatography is the highest possible resolution in the shortest possible elapsed time. Unfortunately, these two properties cannot both be maximized under the same conditions, and a compromise must always be struck.

The time for completion of a separation is determined by the velocity $\bar{v}_B$ of the slower-moving solute, as given in Equation 26-2. That is,

$$\bar{v}_B = \frac{L}{(t_R)_B}$$

Combining this expression with 26-6 and 26-12 yields, after rearranging,

$$(t_R)_B = \frac{NH(1 + k'_B)}{u}$$

where $(t_R)_B$ is the time required to bring the peak for B to the end of the column when the velocity of the mobile phase is u. When this equation is combined with Equation 26-22 and rearranged, we find that

$$(t_R)_B = \frac{16R_s^2 H}{u}\left(\frac{\alpha}{\alpha - 1}\right)^2 \frac{(1 + k'_B)^3}{(k'_B)^2} \qquad (26\text{-}25)$$

26D-4 Variables That Affect Column Performance

Equations 26-21 and 26-25 are significant because they serve as guides to the choice of conditions that are likely to allow the user of chromatography to achieve the sometimes elusive goal of a clean separation in a minimum of time. An examination of these equations reveals that each is made up of three parts. The first, which is related to the kinetic effects that lead to band broadening, consists of $\sqrt{N}$ or H/u. The second and third terms are related to the thermodynamics of the constituents being separated—that is, to the relative magnitude of their distribution constants and the volumes of the mobile and stationary phases. The second term in Equations 26-21 and 26-25, which is the quotient containing α, is a selectivity term that depends solely upon the properties of the two solutes. The third term, which is the quotient containing k'_B, depends upon the properties of both the solute and the column.

In seeking optimum conditions for achieving a desired separation, it must be kept in mind that the fundamental parameters, α, k', and N (or H) can be adjusted more or less independently. Thus, α and k' can be varied most easily by varying temperature or the composition of the mobile phase. Less conveniently, a different type of column packing can be employed. As we have seen, it is possible to change N by changing the length of the column, and to change H by altering the flow rate of the mobile phase, the particle size of the packing, the viscosity of the mobile phase (and thus D_M or D_S), and the thickness of the film of adsorbed liquid constituting the stationary phase (see Table 26-3). van deemter plot

Variation in N

An obvious way to improve resolution is to increase the number of plates in the column (Equation 26-21). As shown by Example 26-1, which follows, this expedient is usually expensive in terms of time required to complete the separation unless the increase in N is realized by a reduction in H rather than by lengthening the column.

EXAMPLE 26-1

Substances A and B were found to have retention times of 16.40 and 17.63 min, respectively, on a 30.0-cm column. An unretained species passed through the column in 1.30 min. The peak widths (at base) for A and B were 1.11 and 1.21 min, respectively. Calculate: **(a)** the column resolution, **(b)** the average number of plates in the column, **(c)** the plate height, **(d)** the length of column required to achieve a resolution of 1.5, **(e)** the time required to elute substance B on the longer column, and **(f)** the plate height required for a resolution of 1.5 on the original 30-cm column and in the original time.

$(t_R)_A = 16.4$ min $L = 30$ cm $W_A = 1.11$ min
$(t_R)_B = 17.63$ min $t_M = 1.30$ min $W_B = 1.21$ min

Substituting into Equation 26-20 gives

(a) $$R_S = \frac{2(17.63 \text{ min} - 16.40 \text{ min})}{(1.11 \text{ min} + 1.21 \text{ min})} = 1.06$$

(b) Equation 26-17 permits computation of N. Thus,

$$N = 16\left(\frac{16.40 \text{ min}}{1.11 \text{ min}}\right)^2 = 3.49 \times 10^3$$

and

$$N = 16\left(\frac{17.63 \text{ min}}{1.21 \text{ min}}\right)^2 = 3.40 \times 10^3$$

$$N_{av} = \left(\frac{3.49 \times 10^3 + 3.40 \times 10^3}{2}\right) = 3.44 \times 10^3 = 3.4 \times 10^3$$

(c) $$H = L/N = \frac{30.0 \text{ cm}}{3.44 \times 10^3} = 8.7 \times 10^{-3} \text{ cm}$$

(d) k' and α do not change with increasing N and L. Thus, substituting N_1 and N_2 into Equation 26-21 and dividing one of the resulting equations by the other yields

$$\frac{(R_s)_1}{(R_s)_2} = \frac{\sqrt{N_1}}{\sqrt{N_2}}$$

where subscripts 1 and 2 refer to the original and the longer columns, respectively. Substituting the appropriate values for N_1, $(R_s)_1$, and $(R_s)_2$ gives

$$\frac{1.06}{1.5} = \frac{\sqrt{3.44 \times 10^3}}{\sqrt{N_2}}$$

$$N_2 = 3.44 \times 10^3\left(\frac{1.5}{1.06}\right)^2 = 6.9 \times 10^3$$

Substitution into Equation 26-12 yields

$$L = N \times H = 6.9 \times 10^3 \times 8.7 \times 10^{-3} \text{ cm} = 60 \text{ cm}$$

(e) Substituting $(R_s)_1$ and $(R_s)_2$ into Equation 26-25 and dividing yields

$$\frac{(t_R)_1}{(t_R)_2} = \frac{(R_s)_1^2}{(R_s)_2^2} = \frac{17.63}{(t_R)_2} = \frac{(1.06)^2}{(1.5)^2}$$

and

$$(t_R)_2 = 35 \text{ min}$$

Thus, to obtain the improved resolution of 1.5 requires that the time of separation be approximately doubled.

(f) Substituting H_1 and H_2 into Equation 26-25 and dividing one of the resulting equations by the second gives

$$\frac{(\cancel{t_R})_B}{(\cancel{t_R})_B} = \frac{(R_s)_1^2}{(R_s)_2^2} \times \frac{H_1}{H_2}$$

where the subscripts 1 and 2 refer to the original and new plate heights, respectively. Rearranging gives

$$H_2 = H_1\frac{(R_s)_1^2}{(R_s)_2^2} = 8.7 \times 10^{-3}\ \text{cm}\ \frac{(1.06)^2}{(1.5)^2}$$

$$= 4.3 \times 10^{-3}\ \text{cm}$$

Thus to achieve a resolution of 1.5 in 17.63 min on a 30-cm column, the plate height would need to be halved.

Variation in H

In Example 26-1f, it is shown that a significant improvement in resolution can be achieved at no cost in time if the plate height can be reduced. Table 26-3 reveals the variables that may be manipulated to accomplish this end. Note that decreases in particle size of the packing lead to marked improvements in H. For liquid mobile phases, where B/u is ordinarily negligible, reduced plate heights can also be achieved by reducing the solvent viscosity, thus increasing the diffusion coefficient in the mobile phase.

Variation in the Retention Factor

Often, a separation can be improved significantly by manipulation of the retention factor k'_B. Increases in k'_B generally enhance resolution (but at the expense of elution time). To determine the optimum range of values for k'_B, it is convenient to write Equation 26-21 in the form

$$R_s = \frac{Qk'_B}{1 + k'_B}$$

and Equation 26-25 as

$$(t_R)_B = Q'\frac{(1 + k'_B)^3}{(k'_B)^2}$$

where Q and Q' contain the rest of the terms in the two equations. Figure 26-12 is a plot of R_s/Q and $(t_R)_B/Q'$ as a function of k'_B, assuming that Q and Q' remain approximately constant. It is clear that values of k'_B greater than about 10 are to be avoided because they provide little increase in resolution but markedly increase the time required for separations. The minimum in the elution time curve occurs at a value of k'_B of about 2. Often, then, the optimal value of k'_B, taking into account both resolution and expended time, lies in the range of 1 to 5.

Usually, the easiest way to improve resolution is by optimizing k'. For gaseous mobile phases, k' can often be improved by temperature increase. For liquid mobile phases, change in the solvent composition often permits manipulation of k' in such a way as to yield better separations. An example of the dramatic effect that relatively simple solvent changes can bring about is demonstrated in Figure 26-13. Here, modest variations in the methanol/water ratios convert unsatisfactory chromatograms (a and b) to ones with well-separated peaks for each component (c and d). For most purposes, the chromatogram shown in (c) would be the best since it shows adequate resolution in a minimum time.

Variation in the Selectivity Factor

When α approaches unity, optimizing k' and increasing N are not sufficient to give a satisfactory separation of two solutes in a reasonable time. Under this circumstance, a means must be sought to increase α while maintaining k' in the optimum range of 1 to 10. Several options are available; in decreasing order of their

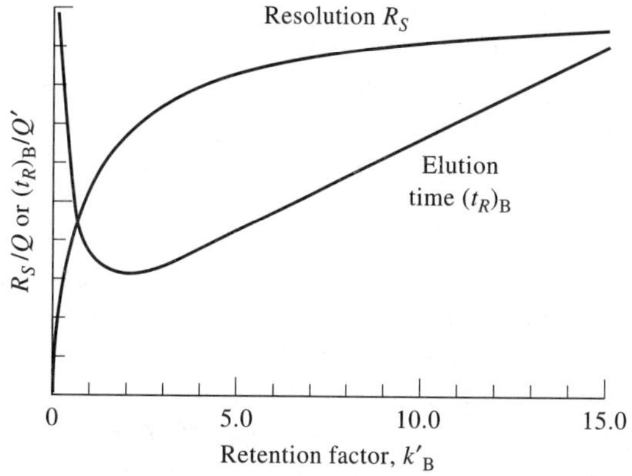

Figure 26-12 Effect of retention factor k'_B on resolution R_s and elution time $(t_R)_B$. It is assumed that Q and Q' remain constant with variation in k'_B.

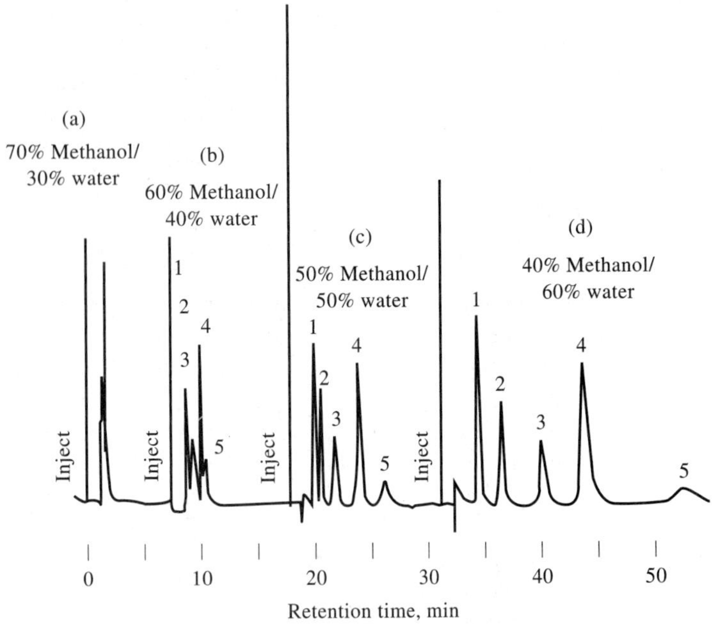

Figure 26-13 Effect of solvent variation on chromatograms. Analytes: (1) 9,10-anthraquinone; (2) 2-methyl-9,10-anthraquinone; (3) 2-ethyl-9,10-anthraquinone; (4) 1,4-dimethy-9,10-anthraquinone; (5) 2-*t*-butyl-9,10-anthraquinone. *(Courtesy of DuPont Instrument Systems, Wilmington, DE.)*

desirability as determined by promise and convenience, the options include: (1) changing the composition of the mobile phase, including changes in pH; (2) changing the column temperature; (3) changing the composition of the stationary phase; (4) using special chemical effects.

An example of the use of option (1) has been reported for the separation of anisole ($C_6H_5OCH_3$) and benzene.[11] With a mobile phase that was a 50% mixture of water and methanol, k' for the two solutes was 4.5 and 4.7, respectively, while α was only 1.04. Substitution of an aqueous mobile phase containing 37% tetrahydrofuran gave k' values of 3.9 and 4.7 and an α value of 1.20. Peak overlap was significant with the first solvent system and negligible with the second.

For separations involving ionizable acids or bases, alteration in pH of the mobile phase often allows manipulation of α values without major changes in k'; enhanced separation efficiencies result.

A less convenient, but often highly effective, method of improving α while still maintaining values for k' in their optimal range, is to alter the chemical composition of the stationary phase. To take advantage of this option, most laboratories that carry out chromatographic separations frequently, maintain several columns, which can be interchanged with a minimum of effort.

Increases in temperature usually cause increases in k' but have little effect on α values in liquid-liquid and liquid-solid chromatography. In contrast, with ion-exchange chromatography, the effect of temperature can have large enough effects to make exploration of this option worthwhile before resorting to a change in column packing.

A final method for enhancing resolution is to incorporate into the stationary phase a species that complexes or otherwise interacts with one or more components of the sample. A well-known example of the use of this option arises where an adsorbent impregnated with a silver salt improves the separation of olefins as a consequence of the formation of complexes between silver ions and unsaturated organic compounds.

26D-5 The General Elution Problem

Figure 26-14 illustrates hypothetical chromatograms for a six-component mixture made up of three pairs of components having widely differing distribution constants and thus retention factors. In chromatogram (a), conditions have been adjusted so that retention factors for components 1 and 2 (k'_1 and k'_2) are in the optimal range of 2 to 5. The corresponding factors for the other components are, however, far larger than the optimum. Thus, the peaks for 5 and 6 appear only after an inordinate time; furthermore, these peaks are so broadened that they may be difficult to identify unambiguously.

[11] L. R. Snyder and J. J. Kirkland, *Introduction to Modern Liquid Chromatography,* 2nd ed., p. 75. New York: Wiley, 1979.

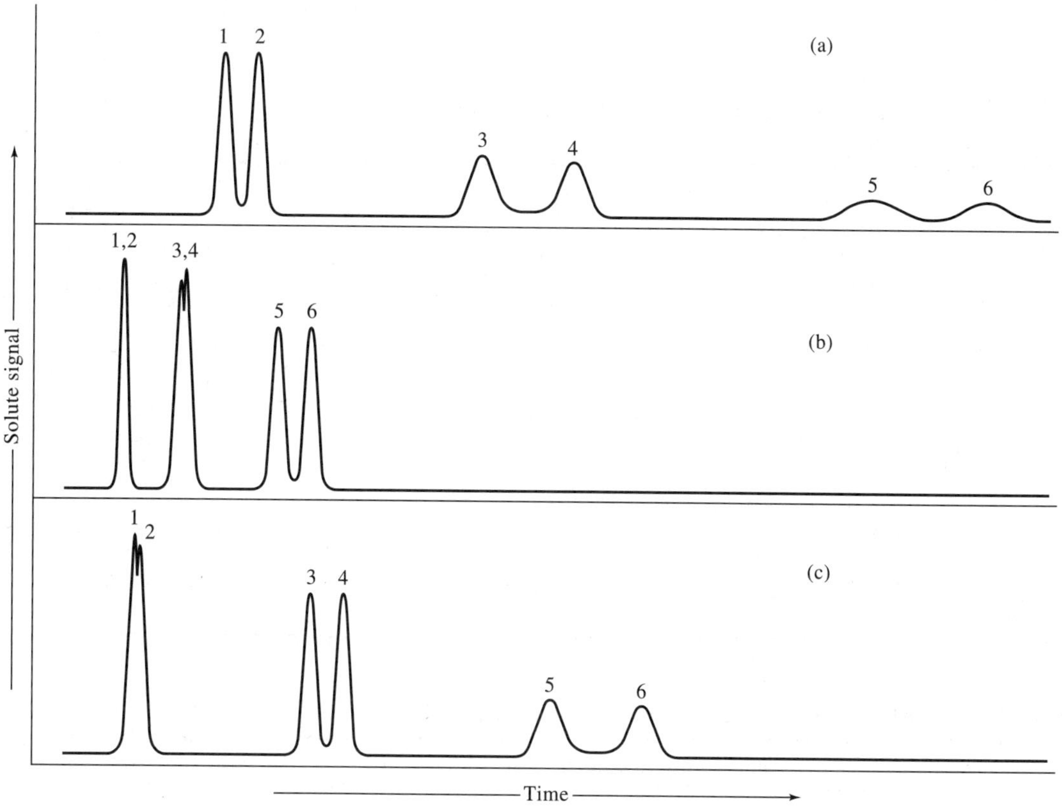

Figure 26-14 Illustration of the general elution problem in chromatography.

As shown in chromatogram (b), changing conditions to optimize the separation of components 5 and 6, bunches the peaks for the first four components to the point where their resolution is unsatisfactory. Here, however, the total elution time is ideal.

A third set of conditions, in which k' values for components 3 and 4 are optimal, results in chromatogram (c). Again, separation of the other two pairs is not entirely satisfactory.

The phenomenon illustrated in Figure 26-14 is encountered often enough to be given a name—the *general elution problem.* A common solution to this problem is to change conditions that determine the values of k' as the separation proceeds. These changes may be performed in a stepwise manner or continuously. Thus, for the mixture shown in Figure 26-14, conditions at the outset could be those producing chromatogram (a). Immediately after elution of components 1 and 2, however, conditions could be changed to those that were optimal for separating 3 and 4 (as in chromatogram c). With the appearance of peaks for these components, the elution could be completed under conditions used for producing chromatogram (b). Often such a procedure leads to satisfactory peaks for all of the components in a mixture in minimal time.

For liquid chromatography, variations in k' are brought about by variations in the composition of the mobile phase during elution *(gradient elution* or *solvent programming).* For gas chromatography, temperature increases *(temperature programming)* serve to achieve optimal conditions for separations.

26E SUMMARY OF IMPORTANT RELATIONSHIPS FOR CHROMATOGRAPHY

The number of quantities, terms, and relationships employed in chromatography is large and often confusing. Tables 26-4 and 26-5 serve to summarize the most important definitions and equations that will be used in the next three chapters.

TABLE 26-4 Important Chromatographic Experimental Quantities and Relationships

Name	Symbol of Experimental Quantity	Determined From
Migration time, nonretained species	t_M	Chromatogram (Figure 26-6)
Retention times, species A and B	$(t_R)_A$, $(t_R)_B$	Chromatogram (Figure 26-6)
Adjusted retention time, species A	$(t'_R)_A$	$(t'_R)_A = (t'_R)_A - t_M$
Peak widths, species A and B	W_A, W_B	Chromatogram (Figure 26-6)
Length of column packing	L	Direct measurement
Flow rate	F	Direct measurement
Volume of stationary phase	V_S	Packing preparation data
Concentration of analyte in mobile and stationary phases	c_M, c_S	Analysis and preparation data

TABLE 26-5 Important Derived Quantities and Relationships

Name	Calculation of Derived Quantities	Relationship to Other Quantities
Linear mobile-phase velocity	$u = L/t_M$	
Volume of mobile phase	$V_M = t_M F$	
Retention factor	$k' = (t_R - t_M)/t_M$	$k' = \dfrac{KV_S}{V_M}$
Distribution constant	$K = \dfrac{k'V_M}{V_S}$	$K = \dfrac{c_S}{c_M}$
Selectivity factor	$\alpha = \dfrac{(t_R)_B - t_M}{(t_R)_A - t_M}$	$\alpha = \dfrac{k'_B}{k'_A} = \dfrac{K_B}{K_A}$
Resolution	$R_s = \dfrac{2[(t_R)_B - (t_R)_A]}{W_A + W_B}$	$R_s = \dfrac{\sqrt{N}}{4}\left(\dfrac{\alpha - 1}{\alpha}\right)\left(\dfrac{k'_B}{1 + k'_B}\right)$
Number of plates	$N = 16\left(\dfrac{t_R}{W}\right)^2$	$N = 16R_s^2\left(\dfrac{\alpha}{\alpha - 1}\right)^2\left(\dfrac{1 + k'_B}{k'_B}\right)^2$
Plate height	$H = L/N$	
Retention time	$(t_R)_B = \dfrac{16R_s^2 H}{u}\left(\dfrac{\alpha}{\alpha - 1}\right)^2 \dfrac{(1 + k'_B)^3}{(k'_B)^2}$	

26F APPLICATIONS OF CHROMATOGRAPHY

Chromatography has grown to be the premiere method for separating closely related chemical species. In addition, it can be employed for qualitative identification and quantitative determination of separated species.[12] This section considers some of the general characteristics of chromatography as a tool for completion of an analysis.

26F-1 Qualitative Analysis

A chromatogram provides only a single piece of qualitative information about each species in a sample, namely, its retention time or its position on the stationary phase after a certain elution period. Additional data can, of course, be derived from chromatograms involving different mobile and stationary phases and various elution temperatures. Still, the amount of information obtainable by chromatography is small compared with the amount provided by a single IR, NMR, or mass spectrum. Furthermore, spectral wavelength or frequency data can be determined with much higher precision than can their chromatographic counterpart (t_R).

The foregoing should not be interpreted to mean that chromatography lacks important qualitative applications. Indeed, it is a widely used tool for recognizing the presence or absence of components of mixtures containing a limited number of possible species whose identities are known. For example, 30 or more amino acids in a protein hydrolysate can be detected with a relatively high degree of certainty by means of a chromatogram. Even here, however, confirmation of identity requires spectral or chemical investigation of the isolated components. Note, however, that positive spectroscopic identification would ordinarily be impossible on as complex a sample as the foregoing without a preliminary chromatographic separation. Thus, chromatography is often a vital precursor to qualitative spectroscopic analyses.

It is important to note that while chromatograms may not lead to positive identification of species present in a sample, they often provide sure evidence of the *absence* of certain compounds. Thus, if the sample does not produce a peak at the same retention time as a standard run under identical conditions, it can be assumed that the compound in question is absent (or is present at a concentration level below the detection limit of the procedure).

[12] For a recent review article on recent development and applications of analytical chromatography, see M. V. Novotny, *Science,* **1989,** *246,* 51; *LC-GC,* **1992,** *10,* 211.

26F-2 Quantitative Analysis

Chromatography owes its precipitous growth during the past four decades in part to its speed, simplicity, relatively low cost, and wide applicability as a separating tool. It is doubtful, however, if its use would have become as widespread had it not been for the fact that it can also provide useful quantitative information about the separated species. It is important, therefore, to discuss some of the quantitative aspects that apply to all types of chromatography.

Quantitative column chromatography is based upon a comparison of either the height or the area of the analyte peak with that of one or more standards. For planar chromatography, the area covered by the separated species serves as the analytical parameter. If conditions are properly controlled, these parameters vary linearly with concentration.

Analyses Based on Peak Height

The height of a chromatographic peak is obtained by connecting the base lines on either side of the peak by a straight line and measuring the perpendicular distance from this line to the peak. This measurement can ordinarily be made with reasonably high precision. It is important to note, however, that peak heights are inversely related to peak widths. Thus, accurate results are obtained with peak heights only if variations in column conditions do not alter the peak widths during the period required to obtain chromatograms for sample and standards. The variables that must be controlled closely are column temperature, eluent flow rate, and rate of sample injection. In addition, care must be taken to avoid overloading the column. The effect of sample injection rate is particularly critical for the early peaks of a chromatogram. Relative errors of 5% to 10% due to this cause are not unusual with syringe injection.

Analyses Based on Peak Areas

Peak areas are independent of broadening effects due to the variables mentioned in the previous paragraph. From this standpoint, therefore, areas are a more satisfactory analytical variable than peak heights. On the other hand, peak heights are more easily measured and, for narrow peaks, more accurately determined.

Most modern chromatographic instruments are equipped with digital electronic integrators that permit precise estimation of peak areas. If such equipment is not available, a manual estimate must be made. A simple method, which works well for symmetric peaks of reasonable widths, is to multiply the height of the peak by its width at one half the peak height. Other methods involve the use of a planimeter or cutting out the peak and determining its weight relative to the weight of a known area of recorder paper. In general, manual integration techniques provide areas that are reproducible at the 2% to 5% level; digital integrators are at least an order of magnitude more precise.[13]

Calibration and Standards

The most straightforward method for quantitative chromatographic analyses involves the preparation of a series of standard solutions that approximate the composition of the unknown. Chromatograms for the standards are then obtained and peak heights or areas are plotted as a function of concentration. A plot of the data should yield a straight line passing through the origin; analyses are based upon this plot. Frequent restandardization is necessary for highest accuracy.

The most important source of error in analyses by the method just described is usually the uncertainty in the volume of sample; occasionally the rate of injection is also a factor. Ordinarily, samples are small (~1 μL), and the uncertainties associated with injection of a reproducible volume of this size with a microsyringe may amount to several percent relative. The situation is exacerbated in gas-liquid chromatography, where the sample must be injected into a heated sample port; here, evaporation from the needle tip may lead to large variations in the volume injected. Errors in sample volume can be reduced to perhaps 1% to 2% relative by means of a rotary sample valve such as that shown in Figure 27-4.

The Internal Standard Method

The highest precision for quantitative chromatography is obtained by use of internal standards because the uncertainties introduced by sample injection are avoided. In this procedure, a carefully measured quantity of an internal standard substance is introduced into each standard and sample, and the ratio of analyte to internal standard peak areas (or heights) serves as the analytical parameter. For this method to be successful, it is necessary that the internal standard peak be well separated from the peaks of all other components of the sample ($R_s > 1.25$); the standard peak should, on the other hand, appear close to the anaylte peak. With a suitable internal standard, precisions of better than 1% relative can usually be achieved.

The Area Normalization Method

Another approach that avoids the uncertainties associated with sample injection is the area normalization method. Complete elution of all components of the sample is required. In the normalization method, the areas of all eluted peaks are computed; after correcting these areas for differences in the detector response to different compound types, the concentration of the analyte is found from the ratio of its area to the total area of all peaks. The following example illustrates the procedure.

Unfortunately, it is often not practical to arrange conditions so that all of the components of a mixture are eluted from al column in a reasonalbe period. As a consequence, the area normalization method has limited applications.

---

[13] See *Chromatographic Integration Methods,* R. M. Smith, Ed.: Boca Raton, FL, 1990.

EXAMPLE 26-2

The area normalization method was applied to the determination of normal-, secondary-, iso-, and tertiary butyl alcohol. In order to determine the relative response factor for the alcohols, a standard solution of the alcohols was prepared and its gas chromatogram observed. The results were as follows:

Alcohol	Weight Taken, g	Weight % Alcohol	Peak area, A, cm2	Weight % Area	Relative Response Factor, *F*
n-Butyl	0.1731	24.61	3.023	8.141	1.000
i-Butyl	0.1964	27.92	3.074	9.083	1.116
s-Butyl	0.1514	21.52	3.112	6.915	0.849
t-Butyl	0.1826	25.96	3.004	8.642	1.062
	Σ wt = 0.7035	Σ % = 100.00	Σ A = 12.213		

The relative response factors were obtained by dividing the data in column 5 by 8.141, the first entry in column 5.

A sample containing only the four alcohols yielded the following data. Calculate the weight percent of each alcohol present.

Alcohol	Peak Area, cm^2	Area × *F*	Weight % Alcohol
n-Butyl	1.731	1.731	18.18
i-Butyl	3.753	4.188	43.99
s-Butyl	2.845	2.415	25.36
t-Butyl	1.117	1.186	12.46
		Σ = 9.521	99.99

26G QUESTIONS AND PROBLEMS

26-1 Define:
- (a) elution
- (b) mobile phase
- (c) stationary phase
- (d) distribution constant
- (e) retention time
- (f) retention factor
- (g) selectivity factor
- (h) plate height
- (i) longitudinal diffusion
- (j) eddy diffusion
- (k) column resolution
- (l) eluent

26-2 Describe the general elution problem.

26-3 List the variables that lead to zone broadening.

26-4 What is the difference between gas-liquid and liquid-liquid chromatography?

26-5 What is the difference between liquid-liquid and liquid-solid chromatography?

26-6 What variables are likely to affect the α value for a pair of analytes?

26-7 How can the retention factor for a solute be manipulated?

26-8 Describe a method for determining the number of plates in a column.

26-9 What are the effects of temperature variation on chromatograms?

26-10 Why does the minimum in a plot of plate height versus flow rate occur at lower flow rates with liquid chromatography than with gas chromatography?

26-11 What is gradient elution?

26-12 The following data apply to a column for liquid chromatography:

length of packing	24.7 cm
flow rate	0.313 mL/min
V_M	1.37 mL
V_S	0.164 mL

A chromatogram of a mixture of species A, B, C, and D provided the following data:

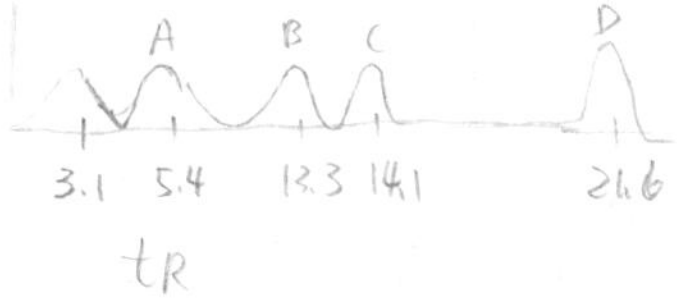

	Retention Time, min	Width of Peak Base (W), min
nonretained	3.1	—
A	5.4	0.41
B	13.3	1.07
C	14.1	1.16
D	21.6	1.72

Calculate
(a) the number of plates from each peak.
(b) the mean and the standard deviation for N.
(c) the plate height for the column.

26-13 From the data in Problem 26-12, calculate for each of A, B, C, and D
(a) the retention factor.
(b) the distribution constant.

26-14 From the data in Problem 26-12, for species B and C, calculate
(a) the resolution.
(b) the selectivity factor, α.
(c) the length of column necessary to give a resolution of 1.5.
(d) the time required to separate B and C with a resolution of 1.5.

26-15 From the data in Problem 26-12 for species C and D, calculate
(a) the resolution.
(b) the length of column required to give a resolution of 1.5.

26-16 The following data were obtained by gas-liquid chromatography on a 40-cm packed column:

Compound	t_R, min	W, min
air	1.9	—
methylcyclohexane	10.0	0.76
methylcyclohexene	10.9	0.82
toluene	13.4	1.06

Calculate
(a) an average number of plates from the data.
(b) the standard deviation for the average in (a).
(c) an average plate height for the column.

26-17 Referring to Problem 26-16, calculate the resolution for
(a) methylcyclohexene and methylcyclohexane.
(b) methylcyclohexene and toluene.
(c) methylcyclohexane and toluene.

26-18 If a resolution of 1.5 was desired in resolving methylcyclohexane and methylcyclohexene in Problem 26-16,
(a) how many plates would be required?
(b) how long would the column have to be if the same packing were employed?
(c) what would be the retention times for methylcyclohexene on the column in Problem 26-18b?

26-19 If V_S and V_M for the column in Problem 26-16 were 19.6 and 62.6 mL, respectively, and a nonretained air peak appeared after 1.9 min, calculate the
(a) retention factor for each of the three compounds.
(b) distribution constant for each of the three compounds.
(c) selectivity factor for methylcyclohexane and methylcyclohexene.
(d) selectivity factor for methylcyclohexene and toluene.

26-20 List variables that lead to (a) zone broadening and (b) zone separation.

26-21 What would be the effect on a chromatographic peak of introducing the sample at too slow a rate?

26-22 From distribution studies, species M and N are known to have distribution constants between water and hexane of 6.01 and 6.20 ($K = [M]_{H_2O}/[H]_{hex}$). The two species are to be separated by elution with hexane in a column packed with silica gel containing adsorbed water. The ratio V_S/V_M for the packing is known to be 0.422.
(a) Calculate the retention factor for each of the solutes.
(b) Calculate the selectivity factor.
(c) How many plates will be needed to provide a resolution of 1.5?
(d) How long a column is needed if the plate height of the packing is 2.2×10^{-3} cm?
(e) If a flow rate of 7.10 cm/min is employed, what time will be required to elute the two species?

26-23 Repeat the calculations in Problem 26-22 assuming $K_M = 5.81$ and $K_N = 6.20$.

26-24 The relative peak areas obtained from a gas chromatogram of a mixture of methyl acetate, methyl propionate, and methyl *n*-butyrate were 17.6, 44.7, and 31.1, respectively. Calculate the percentage of each compound if the respective relative detection responses were 0.65, 0.83, and 0.92.

26-25 The relative areas for the five gas chromatographic peaks obtained in the separation of five steroids are given below. Also shown are the relative responses of the detector to the five compounds. Calculate the percentage of each component in the mixture.

Compound	Peak Area, Relative	Detection Response, Relative
Dehydroepiandrosterone	27.6	0.70
Estradiol	32.4	0.72
Estrone	47.1	0.75
Testosterone	40.6	0.73
Estriol	27.3	0.78

Chapter 27

Gas Chromatography

In gas chromatography (GC), the sample is vaporized and injected onto the head of a chromatographic column. Elution is brought about by the flow of an inert gaseous mobile phase. In contrast to most other types of chromatography, the mobile phase does not interact with molecules of the analyte; its only function is to transport the analyte through the column. Two types of gas chromatography are encountered: ***gas-solid chromatography*** *(GSC) and* ***gas-liquid chromatography*** *(GLC). Gas-liquid chromatography finds widespread use in all fields of science, where its name is usually shortened to* ***gas chromatography*** *(GC), despite the fact that this usage neglects gas-solid chromatography as a legitimate kind of chromatography.*

Gas-solid chromatography *is based upon a solid stationary phase in which retention of analytes is the consequence of physical adsorption. Gas-solid chromatography has limited application owing to semipermanent retention of active or polar molecules and severe tailing of elution peaks (a consequence of the nonlinear character of adsorption process). Thus, this technique has not found wide application except for the separation of certain low-molecular-weight gaseous species; it is, therefore, discussed only briefly in Section 27E.*

Gas-liquid chromatography is based upon the partition of the analyte between a gaseous mobile phase and a liquid phase immobilized on the surface of an inert solid. The concept of gas-liquid chromatography was first enunciated in 1941 by Martin and Synge, who were also responsible for the development of liquid-liquid partition chromatography. More than a decade was to elapse, however, before the value of gas-liquid chromatography was demonstrated experimentally.[1] Three years later, in 1955, the first commercial apparatus for gas-liquid chromatography appeared on the market. Since that time, the growth in applications of this technique has been phenomenal.[2] By 1985, it was estimated that as many as 200,000 gas chromatographs were in use throughout the world.[3]

[1] A. J. Jones and A. J. P. Martin, *Analyst,* **1952,** *77,* 915.

[2] For monographs on GLC, see J. Willet, *Gas Chromatography.* New York: Wiley, 1987; W. Jennings, *Analytical Gas Chromatography.* Orlando, FL: Academic Press, 1987; *Modern Practice of Gas Chromatography,* 3rd ed., R. L. Grob, Ed. New York: Wiley-Interscience, 1995; M. L. Lee, F. Yang, and K. Bartle, *Open Tubular Gas Chromatography: Theory and Practice.* New York: Wiley, 1984.

[3] R. Schill and R. R. Freeman, in *Modern Practice of Gas Chromatography,* 2nd ed., R. L. Grob, Ed., p. 294. New York: Wiley, 1985.

27A PRINCIPLES OF GAS-LIQUID CHROMATOGRAPHY

The general principles of chromatography, which were developed in Chapter 26, and the mathematical relationships summarized in Section 26E are applicable to gas chromatography with only minor modifications that arise from the compressibility of gaseous mobile phases.

27A-1 Retention Volumes

To take into account the effects of pressure and temperature in gas chromatography, it is sometimes useful to use *retention volumes* rather than the retention times that were employed in Section 26B. The relationship between the two is

$$V_R = t_R F \qquad (27\text{-}1)$$

and

$$V_M = t_M F \qquad (27\text{-}2)$$

where F is the average volumetric flow rate within the column; V and t are retention volumes and times, respectively, and the subscripts R and M refer to species that are retained and not retained on the column. The average flow rate is not directly measurable, however; instead, only the rate of gas flow as it exits the column is conveniently determined experimentally. Normally, this rate is measured by means of a soap-bubble meter, which is illustrated in Figure 27-2. The average flow rate F is then

$$F = F_m \times \frac{T_C}{T} \times \frac{(P - P_{H_2O})}{P} \qquad (27\text{-}3)$$

where T_C is the column temperature in kelvin, T is the temperature at the meter, F_m is the measured flow rate, and P is the gas pressure at the end of the column. Usually P and T are the ambient pressure and temperature. In the soap-bubble meter, the gas becomes saturated with water. Thus, the pressure must be corrected for the vapor pressure of water, P_{H_2O}.

Both V_R and V_M depend upon the average pressure *within the column*—a quantity that lies intermediate between the inlet pressure P_i and the outlet pressure P (atmospheric pressure). The *pressure drop correction factor* j is used to account for the fact that the pressure within the column is a nonlinear function of the P_i/P ratio. *Corrected retention volumes* V_R^0 and V_M^0, which correspond to volumes at the average column pressure are obtained from the relationships

$$V_R^0 = jt_R F \text{ and } V_M^0 = jt_M F \qquad (27\text{-}4)$$

where j can be calculated from the relationship

$$j = \frac{3[(P_i/P)^2 - 1]}{2[(P_i/P)^3 - 1]} \qquad (27\text{-}5)$$

The *specific retention volume* V_g is then defined as

$$V_g = \frac{V_R^0 - V_M^0}{W} \times \frac{273}{T_C} \qquad (27\text{-}6)$$

$$= \frac{jF(t_R - t_M)}{W} \times \frac{273}{T_C}$$

where W is the mass of the stationary phase, a quantity determined at the time of column preparation, and T_C is the column temperature in kelvin.

27A-2 Relationship Between V_g and K

It is of interest to relate V_g to the distribution constant K. To do so, we substitute the expression relating t_R and t_M to k' (Equation 26-8) into Equation 27-6, which gives

$$V_g = \frac{jFt_M k'}{W} \times \frac{273}{T_C}$$

Combining this expression with Equation 27-4 yields

$$V_g = \frac{V_M^0 k'}{W} \times \frac{273}{T_C}$$

Substituting Equation 26-5 for k' gives (here, V_M^0 and V_M are identical)

$$V_g = \frac{KV_S}{W} \times \frac{273}{T_C}$$

The density of ρ_S of the liquid on the stationary phase is given by

$$\rho_S = \frac{W}{V_S}$$

Thus,

$$V_g = \frac{K}{\rho_S} \times \frac{273}{T_C} \qquad (27\text{-}7)$$

Note that V_g at a given temperature depends only upon the distribution constant of the solute and the density of the liquid making up the stationary phase. As such, it should in principle be a useful parameter for identifying species. The literature contains large numbers of specific retention volumes; unfortunately these data are widely scattered and often unreliable.

27A-3 Effect of Mobile-Phase Flow Rate

Equation 26-19 and the relationships shown in Table 26-3 are fully applicable to gas chromatography. The longitudinal diffusion term (B/u) is more important in gas-liquid chromatography, however, than in other chromatographic processes because of the much larger diffusion rates in gases (10^4 times greater than liquids). As a consequence, the minima in curves relating plate height H to flow rate are usually considerably broadened in gas chromatography (see Figures 26-7 and 26-10).

27B INSTRUMENTS FOR GAS-LIQUID CHROMATOGRAPHY

Today, well over 30 instrument manufacturers offer some 130 different models of gas-chromatographic equipment at costs that vary from perhaps \$1500 to \$40,000. In the last two decades, many changes and improvements in gas chromatographic instruments have appeared in the marketplace. In the 1970s, electronic integrators and computer-based data-processing equipment became common. The 1980s saw computers being used for automatic control of most instrument parameters, such as column temperature, flow rates, and sample injection; development of very high-performance instruments at moderate costs; and perhaps most important, the development of open tubular columns that are capable of separating a multitude of analytes in relatively short times.

The basic components of an instrument for gas chromatography are illustrated in Figure 27-1. A description of each component follows. Note that in the illustration, the gas flow is split before it enters the column. This type of arrangement is employed when a type of detector is used that measures a change in property of the gas stream in the presence of analyte molecules. With other types of detectors, flow splitters are normally not used.

27B-1 Carrier Gas Supply

Carrier gases, which must be chemically inert, include helium, nitrogen, and hydrogen. As will be shown later, the choice of gases is often dictated by the type of de-

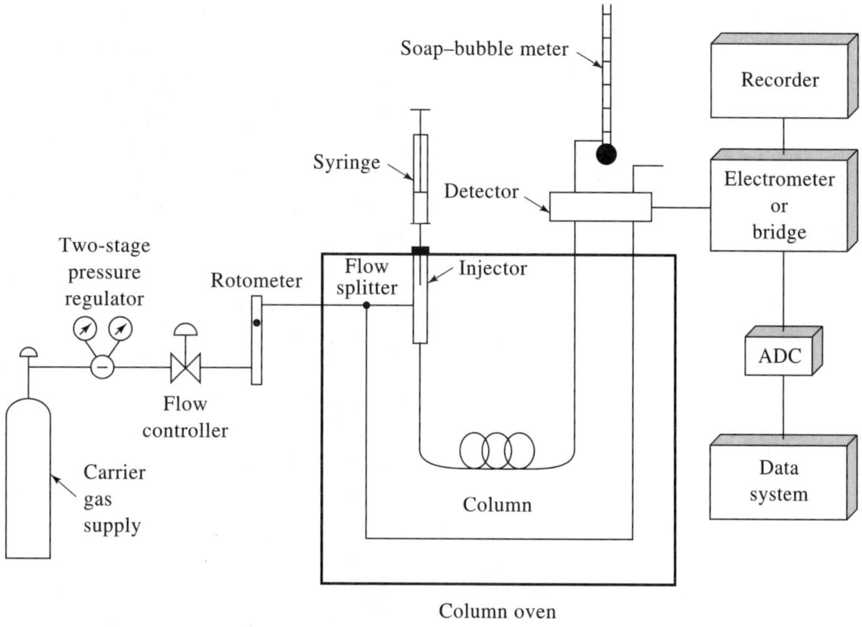

Figure 27-1 Schematic of a gas chromatograph.

tector used. Associated with the gas supply are pressure regulators, gauges, and flow meters. In addition, the carrier gas system often contains a molecular sieve to remove water or other impurities.

Flow rates are normally controlled by a two-stage pressure regulator at the gas cylinder and some sort of pressure regulator or flow regulator mounted in the chromatograph. Inlet pressures usually range from 10 to 50 psi (above room pressure), which lead to flow rates of 25 to 150 mL/min with packed columns and 1 to 25 mL/min for open-tubular capillary columns. Generally, it is assumed that flow rates will be constant if the inlet pressure remains constant. Flow rates can be established by a rotameter at the column head; this device, however, is not as accurate as a simple soap-bubble meter, which, as shown in Figure 27-1, is located at the end of the column. A soap film is formed in the path of the gas when a rubber bulb containing an aqueous solution of soap or detergent is squeezed; the time required for this film to move between two graduations on the buret is measured and converted to volumetric flow rate (see Figure 27-2). Many modern commercial gas chromatographs are equipped with electronic flow meters that are computer controlled to maintain the flow rate at any desired level.

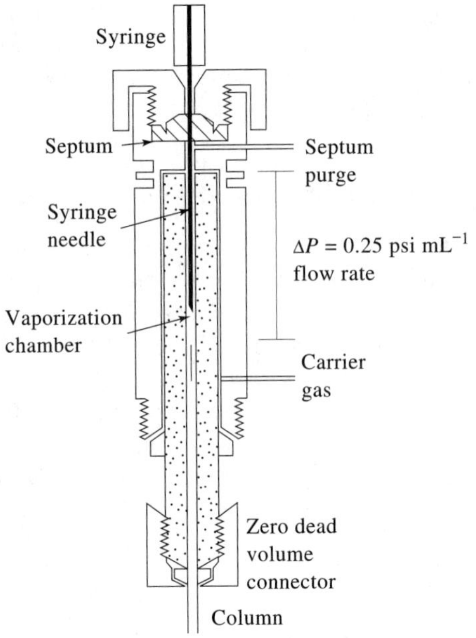

Figure 27-3 Cross-sectional view of a microflash vaporizer direct injector.

27B-2 Sample Injection System

Column efficiency requires that the sample be of suitable size and be introduced as a "plug" of vapor; slow injection of oversized samples causes band spreading and poor resolution. The most common method of sample injection involves the use of a microsyringe to inject a liquid or gaseous sample through a self-sealing, silicone-rubber diaphragm or septum into a flash vaporizer port located at the head of the column (the sample port is ordinarily about 50°C above the boiling point of the least volatile component of the sample). Figure 27-3 is a schematic of a typical injection port. For ordinary analytical columns, sample sizes vary from a few tenths of a microliter to 20 μL. Capillary columns require much smaller samples ($\approx 10^{-3}$ μL); here, a sample splitter system is employed to deliver only a small fraction of the injected sample to the column head, with the remainder going to waste.

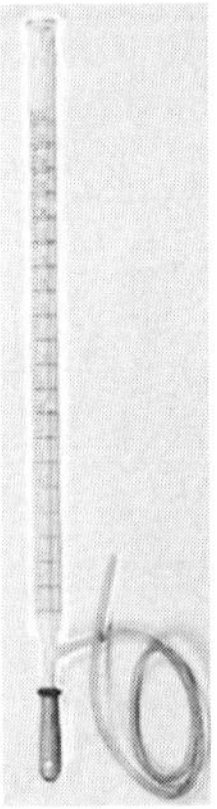

Figure 27-2 A soap-bubble flow meter. *(Courtesy of Chrompack Inc., Raritan, NJ.)*

For quantitative work, more reproducible sample sizes for both liquids and gases are obtained by means of a rotary sample valve such as that shown in Figure 27-4. With such a device, errors due to sample size can be reduced to 0.5% to 2% relative. The sampling loop in Figure 27-4a is filled by injection of an excess of sample. Rotation of the valve by 45 deg then introduces the reproducible volume *ACB* into the mobile phase. Solid samples are introduced as solutions or, alternatively, are sealed into thin-walled vials that can be inserted at the

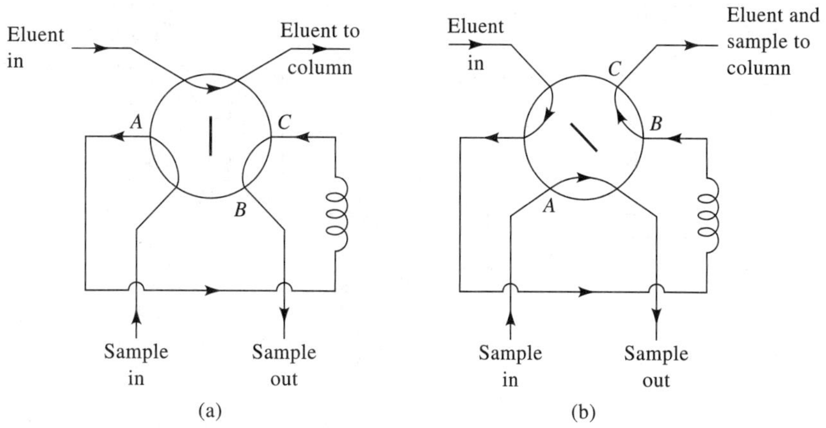

Figure 27-4 A rotary sample valve: valve position (a) for filling sample loop *ACB* and (b) for introduction of sample into column.

head of the column and punctured or crushed from the outside.

27B-3 Column Configurations and Column Ovens

Two general types of columns are encountered in gas chromatography, *packed* and *open tubular,* or *capillary.* To date, the vast majority of gas chromatography has been carried out on packed columns. Currently, however, this situation is changing rapidly, and it seems probable that in the near future, packed columns will be replaced by the more efficient and faster open tubular columns except for certain special applications.

Chromatographic columns vary in length from less than 2 m to 50 m or more. They are constructed of stainless steel, glass, fused silica, or Teflon. In order to fit into an oven for thermostating, they are usually formed as coils having diameters of 10 to 30 cm. A detailed discussion of columns, column packings, and stationary phases is found in Section 27C.

Column temperature is an important variable that must be controlled to a few tenths of a degree for precise work. Thus, the column is ordinarily housed in a thermostated oven. The optimum column temperature depends upon the boiling point of the sample and the degree of separation required. Roughly, a temperature equal to or slightly above the average boiling point of a sample results in a reasonable elution time (2 to 30 min). For samples with a broad boiling range, it is often desirable to employ *temperature programming,* whereby the column temperature is increased either continuously or in steps as the separation proceeds. Figure 27-5c shows the improvement in a chromatogram brought about by temperature programming.

In general, optimum resolution is associated with minimal temperature; the cost of lowered temperature, however, is an increase in elution time and therefore the time required to complete an analysis. Figures 27-5a and 27-5b illustrate this principle.

27B-4 Detection Systems

Dozens of detectors have been investigated and used during the development of gas chromatography. In the sections that follow immediately, we describe the most widely used of these. In Section 27D-3, we consider instruments in which gas chromatographs are coupled to mass spectrometers and infrared spectrophotometers. Here, the spectral device serves not only to detect the appearance of analytes at the end of the column but also to identify them.

Characteristics of the Ideal Detector

The ideal detector for gas chromatography has the following characteristics:

1. Adequate sensitivity. Just what constitutes adequate sensitivity cannot be described in quantitative terms. For example, the sensitivities of the detectors we are

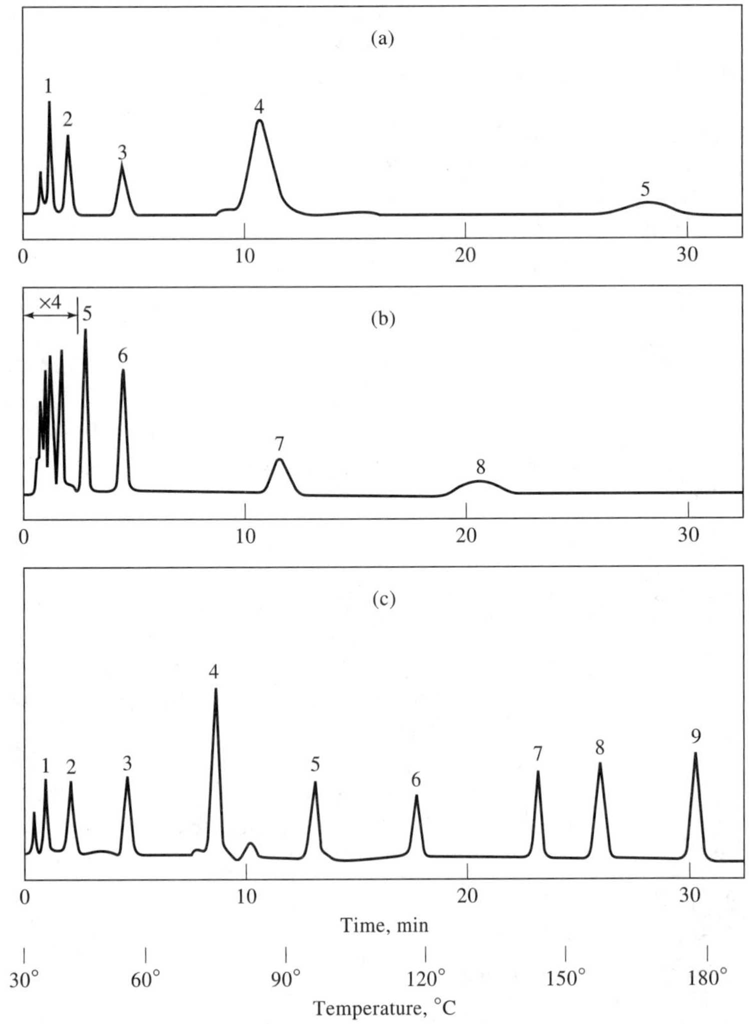

Figure 27-5 Effect of temperature on gas chromatograms: (a) isothermal at 45°C; (b) isothermal at 145°C; (c) programmed at 30° to 180°C. *(From W. E. Harris and H. W. Habgood,* Programmed Temperature Gas Chromatography, *p. 10. New York: Wiley, 1966. Reprinted by permission of John Wiley & Sons, Inc.)*

about to describe differ by a factor of 10^7. Yet all are widely used and clearly adequate for certain tasks; the least sensitive are not, however, satisfactory for certain applications. In general, the sensitivities of present-day detectors lie in the range of 10^{-8} to 10^{-15} g solute/s.

2. Good stability and reproducibility.
3. A linear response to solutes that extends over several orders of magnitude.
4. A temperature range from room temperature to at least 400°C.
5. A short response time that is independent of flow rate.
6. High reliability and ease of use. The detector should, to the extent possible, be foolproof in the hands of inexperienced operators.
7. Similarity in response toward all solutes or alternatively a highly predictable and selective response toward one or more classes of solutes.
8. Nondestructive of sample.

Needless to say, no detector exhibits all of these characteristics, and it seems unlikely that such a detector will ever be designed.

Flame Ionization Detectors (FID)

The flame ionization detector is the most widely used and generally applicable detector for gas chromatography. With a burner such as that shown in Figure 27-6, the effluent from the column is mixed with hydrogen and air and then ignited electrically. Most organic compounds, when pyrolyzed at the temperature of a hydro-

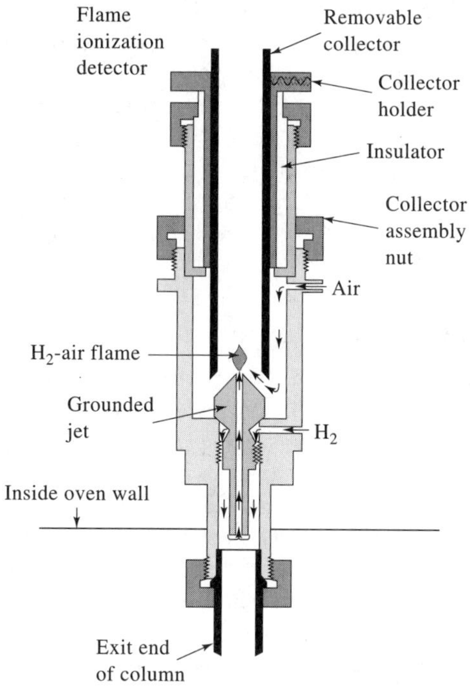

Figure 27-6 A typical flame ionization detector. *(Courtesy of Hewlett-Packard Company.)*

gen/air flame, produce ions and electrons that can conduct electricity through the flame. A potential of a few hundred volts is applied across the burner tip and a collector electrode located above the flame. The resulting current ($\approx 10^{-12}$ A) is then directed into a high-impedance operational amplifier for measurement.

The ionization of carbon compounds in a flame is a poorly understood process, although it is observed that the number of ions produced is roughly proportional to the number of *reduced* carbon atoms in the flame. Because the flame ionization detector responds to the number of carbon atoms entering the detector per unit of time, it is a *mass-sensitive,* rather than a concentration-sensitive, device.

Functional groups, such as carbonyl, alcohol, halogen, and amine, yield fewer ions or none at all in a flame. In addition, the detector is insensitive toward noncombustible gases such as H_2O, CO_2, SO_2, and NO_x. These properties make the flame ionization detector a most useful general detector for the analysis of most organic samples, including those that are contaminated with water and the oxides of nitrogen and sulfur. The insensitivity of the flame ionization detector to water makes it particularly useful for the detection of pollutants in natural water samples.

The flame ionization detector exhibits a high sensitivity (~10^{-13} g/s), large linear response range (~10^7), and low noise. It is generally rugged and easy to use. A disadvantage of the flame ionization detector is that it is destructive of sample.

Thermal Conductivity Detectors (TCD)

A very early detector for gas chromatography, and one that still finds wide application, is based upon changes in the thermal conductivity of the gas stream brought about by the presence of analyte molecules. This device is sometimes called a *katharometer.* The sensing element of a katharometer is an electrically heated element whose temperature at constant electrical power depends upon the thermal conductivity of the surrounding gas. The heated element may be a fine platinum, gold, or tungsten wire or, alternatively, a semiconducting thermistor. The resistance of the wire or thermistor gives a measure of the thermal conductivity of the gas; in contrast to the wire detector, the thermistor has a negative temperature coefficient. Figure 27-7a is a cross-sectional view of one of the temperature-sensitive elements in a thermal conductivity detection system.

Figure 27-7b shows the arrangement of detector elements in a typical detector unit. Two pair of elements are employed, one pair being located in the flow of the effluent from the column and the other in the gas stream *ahead* of the sample injection chamber. (These elements are labeled "sample" and "reference" in Figure 27-7b.) Alternatively, the gas stream may be split as shown in Figure 27-1. In either case, the effect of thermal conductivity of the carrier gas is canceled, and the effects of variation in flow rate, pressure, and electrical power are minimized. The resistances of the twin-detector pairs are usually compared by incorporating them into two arms of a simple Wheatstone bridge circuit such as that shown in Figure 27-7b.

A modulated single-filament thermal conductivity detector was introduced in 1979; this device offers higher sensitivity, freedom from baseline drift, and reduced equilibration time. Here, the analytical and reference gases are passed alternately over a tiny filament held in a ceramic detector cell, which has a volume of only 5 μL. The gas-switching device operates at a frequency of 10 Hz. The output from the filament is thus a 10-Hz electrical signal whose amplitude is proportional to the difference in thermal conductivity of the analyti-

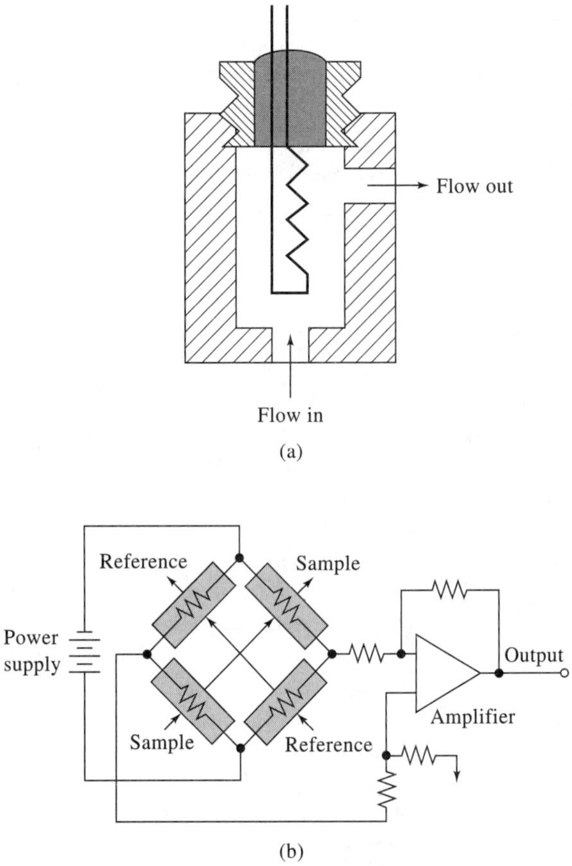

Figure 27-7 Schematic of (a) a thermal conductivity detector cell, and (b) an arrangement of two sample detector cells and two reference detector cells. *(From J. V. Hinshaw,* LC-GC, ***1990****, 8, 298. With permission.)*

cal and reference gases. Because the amplifier circuit responds only to a 10-Hz signal, thermal noise in the system is largely eliminated.

The thermal conductivities of helium and hydrogen are roughly six to ten times greater than those of most organic compounds. Thus, in the presence of even small amounts of organic materials, a relatively large decrease in the thermal conductivity of the column effluent takes place; consequently, the detector undergoes a marked rise in temperature. The conductivities of other carrier gases more closely resemble those of organic constituents; therefore, a thermal conductivity detector dictates the use of hydrogen or helium in order to achieve good sensitivity.

The advantage of the thermal conductivity detector is its simplicity, its large linear dynamic range ($\sim 10^5$), its general response to both organic and inorganic species, and its nondestructive character, which permits collection of solutes after detection. A limitation of the katharometer is its relatively low sensitivity ($\sim 10^{-8}$ g solute/mL carrier gas). Other detectors exceed this sensitivity by factors as large as 10^4 to 10^7. It should be noted that the low sensitivity of thermal detectors often preclude their use with capillary columns because of the very small samples that can be accommodated by such columns.

Sulfur Chemiluminescence Detector (SCD)

A recent addition to the variety of available detectors for gas chromatography is the sulfur chemiluminescence detector, which is based upon the reaction between certain sulfur compounds and ozone. The resulting luminescence intensity is proportional to the concentration of sulfur. This detector has proven particularly useful for the determination of pollutants such as mercaptans. In the sulfur chemiluminescence detector, the eluent is mixed with hydrogen and air and is burned as in the flame ionization detector. The resulting gases are then mixed with ozone, and the intensity of the resulting emission is measured. The sulfur chemiluminescence detector has also been adapted to supercritical fluid chromatography.

Electron-Capture Detectors (ECD)[4]

The electron-capture detector has become one of the most widely used detectors for environmental samples because this detector selectivity detects halogen containing compounds, such as pesticides and polychlorinated biphenyls. This type of detector operates in much the same way as a proportional counter for measurement of X-radiation (Section 12B-4). As shown in Figure 27-8, the effluent from the column is passed over a β emitter, usually nickel-63. An electron from the emitter causes ionization of the carrier gas (often nitrogen) and the production of a burst of electrons. In the absence of organic species, a constant standing current between a pair of electrodes results from this ionization process. The current decreases markedly, however, in the presence of those organic molecules that tend to capture electrons. The response is nonlinear unless the potential across the detector is pulsed.

The electron-capture detector is selective in its response, being highly sensitive to molecules containing

[4] For a recent description of commercial electron-capture detectors, see D. Noble, *Anal. Chem.*, **1995**, *67*, 442A.

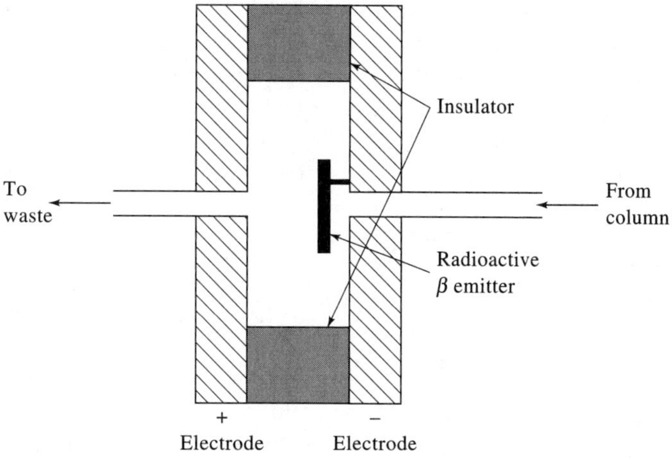

Figure 27-8 A schematic of an electron-capture detector.

electronegative functional groups such as halogens, peroxides, quinones, and nitro groups. It is insensitive to functional groups such as amines, alcohols, and hydrocarbons. An important application of the electron-capture detector has been for the detection and determination of chlorinated insecticides.

Electron-capture detectors are highly sensitive and possess the advantage of not altering the sample significantly (in contrast to the flame detector). On the other hand, their linear response range is usually limited to about two orders of magnitude.

Atomic Emission Detectors (AED)

The atomic emission detector is available commercially.[5] In this device (see Figure 27-9), the eluent is introduced into a microwave-energized helium plasma that is coupled to a diode array optical emission spectrometer. The plasma is sufficiently energetic to atomize all of the elements in a sample and to excite their char-

[5] See B. D. Quimby and J. J. Sullivan, *Anal. Chem.*, **1990**, *62*, 1027, 1034.

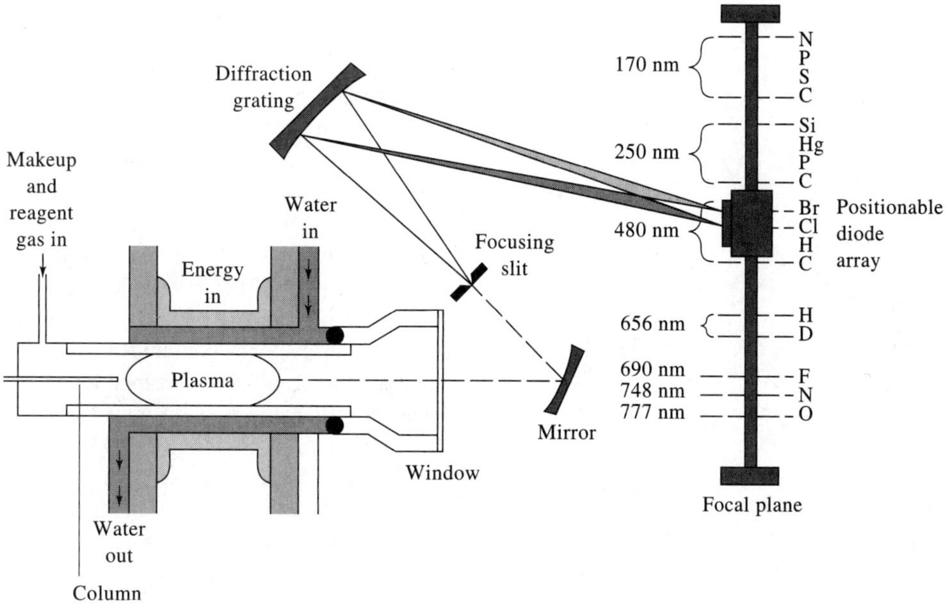

Figure 27-9 An atomic emission detector. *(Courtesy of Hewlett-Packard Company.)*

acteristic atomic emission spectra. These spectra are then observed with a spectrometer that employs a movable, flat diode array capable of detecting emitted radiation from about 170 to 780 nm. As shown on the right of the figure, the positionable diode array is capable of monitoring simultaneously two to four elements at any given setting. At the present time, the software supplied with the detector allows measurement of the concentration of 15 elements. Presumably, future software will permit detection of other elements.

Figure 27-10 illustrates the power of this type of detector. The sample in this case consisted of a gasoline containing a small concentration of methyl tertiary butyl ether (MTBE), an antiknock agent, as well as several aliphatic alcohols in low concentrations. The upper spectrum, obtained by monitoring the carbon emission line at 198 nm, consists of a myriad of peaks that would be impossible to sort out and identify. In contrast, when the oxygen line at 777 nm is used to obtain the chromatogram (Figure 27-10b), peaks for the various alcohols and for MTBE are clearly evident and readily identifiable.

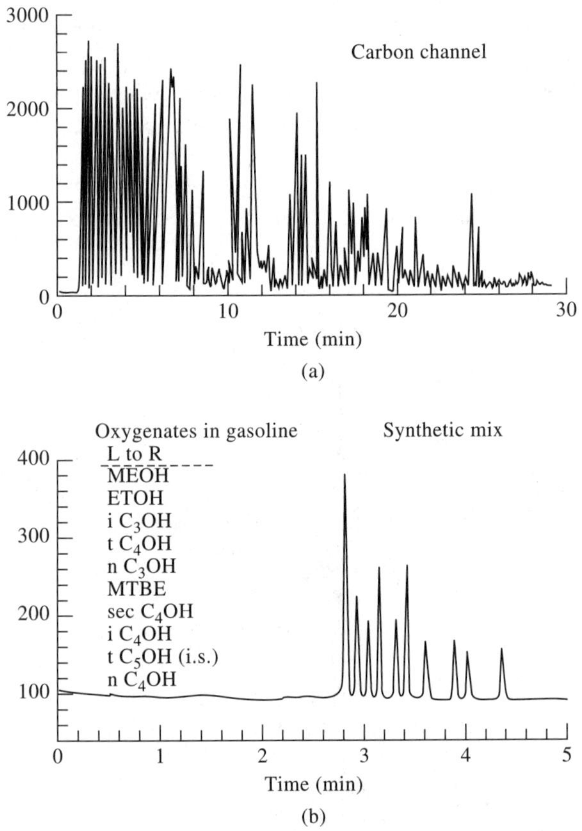

Figure 27-10 Chromatograms for a gasoline sample containing a small amount of MTBE and several aliphatic alcohols: (a) monitoring the line for carbon; (b) monitoring the line for oxygen. *(Courtesy of Hewlett-Packard Company.)*

Thermionic Detectors (TID)

The thermionic detector is selective toward organic compounds containing phosphorus and nitrogen. Its response to a phosphorus atom is approximately ten times greater than to a nitrogen atom and 10^4 to 10^6 larger than a carbon atom. Compared with the flame ionization detector, the thermionic detector is approximately 500 times more sensitive to phosphorus-containing compounds and 50 times more sensitive to nitrogen-bearing species. These properties make thermionic detection particularly useful for detecting and determining the many phosphorus-containing pesticides.

A thermionic detector is similar in structure to the flame detector shown in Figure 27-6. The column effluent is mixed with hydrogen, passes through the flame tip assembly, and is ignited. The hot gas then flows around an electrically heated rubidium silicate bead, which is maintained at about 180 V with respect to the collector. The heated bead forms a plasma having a temperature of 600° to 800°C. Exactly what occurs in the plasma to produce unusually large numbers of ions from phosphorus- or nitrogen-containing molecules is not understood; but large ion currents result, which are useful for determining compounds containing these two elements.

Other Types of Detectors

The *flame photometric detector* (FPD) has been widely applied to the analysis of air and water pollutants, pesticides, and coal hydrogenation products. It is a selective detector that is primarily responsive to compounds containing sulfur and phosphorus. In this detector, the eluent is passed into a low-temperature hydrogen/air flame, which converts part of the phosphorus to an HPO species that emits bands of radiation centered about 510 and 526 nm. Sulfur in the sample is simultaneously converted to S_2, which emits a band centered at 394 nm. However, the sulfur luminescence detector mentioned earlier provides lower detection limits and wider linear working range than does the flame photometric detector. Suitable filters are employed to isolate these bands, and their intensity is recorded photometrically. Other elements that have been detected by flame photometry include the halogens, nitrogen, and several metals, such as tin, chromium, selenium, and germanium. It should be noted that the sulfur luminescence detector mentioned earlier is more sensitive and provides a greater working range for sulfur than does the flame photometric detector.

In the *photoionization detector,* the column eluent is irradiated with an intense beam of ultraviolet radiation varying in energy from 8.3 to 11.7 eV (λ = 149 to 106 nm), which causes ionization of the molecules. Application of a potential across a cell containing the ions leads to an ion current, which is amplified and recorded.

Two other detectors that have gained importance in gas chromatography are the mass spectrometer and the Fourier transform infrared spectrometer. The application of these instruments are discussed in some detail in later sections of this chapter.

27C GAS CHROMATOGRAPHIC COLUMNS AND STATIONARY PHASES

Historically, all of the pioneering gas-liquid chromatographic studies in the early 1950s were carried out on packed columns in which the stationary phase was a thin film of liquid adsorbed on the surface of a finely divided, inert solid support. From theoretical studies made during this early period, it became apparent, however, that unpacked columns having inside diameters of a few tenths of a millimeter should provide separations that were much superior to packed columns in both speed and column efficiency. In such *capillary columns,* the stationary phase was a uniform film of liquid a few tenths of a micrometer thick that coated the interior of a capillary tubing uniformly. In the late 1950s such *open tubular columns* were constructed, and the predicted performance characteristics were experimentally confirmed in several laboratories with open tubular columns having 300,000 plates or more described.[6] Despite such spectacular performance characteristics, capillary columns did not gain widespread use until more than two decades after their invention. The reasons for the delay were several, including: small sample capacities; fragileness of columns; mechanical problems associated with sample introduction and connection of the column to the detector; difficulties in coating the column reproducibly; short lifetimes of poorly prepared columns; tendencies of columns to clog; and patents, which limited commercial development to a single manufacturer (the original patent expired in 1977). By the late 1970s these problems had become manageable, and several instrument companies began to offer open tubular columns at a reasonable cost. As a consequence, a major growth in the use of open tubular columns has occurred in the last few years.

27C-1 Open Tubular Columns[7]

Open tubular, or capillary, columns are of two basic types, namely, *wall-coated open tubular* (WCOT) and *support-coated open tubular* (SCOT). Wall-coated columns are simply capillary tubes coated with a thin layer of the stationary phase. In support-coated open tubular columns, the inner surface of the capillary is lined with a thin film ($\approx$30 μm) of a support material, such as diatomaceous earth. This type of column holds several times as much stationary phase as does a wall-coated column and thus has a greater sample capacity. Generally, the efficiency of a SCOT column is less than that of a WCOT column but significantly greater than that of a packed column.

Early WCOT columns were constructed of stainless steel, aluminum, copper, or plastic. Subsequently glass was used. Often the glass was etched with gaseous hydrochloric acid, strong aqueous hydrochloric acid, or potassium hydrogen fluoride to give a rough surface, which bonded the stationary phase more tightly. The newest WCOT columns, which first appeared in 1979, are *fused-silica open tubular columns* (FSOT columns). Fused-silica capillaries are drawn from specially purified silica that contain minimal amounts of metal oxides. These capillaries have much thinner walls than their glass counterparts. The tubes are given added strength by an outside protective polyimide coating, which is applied as the capillary tubing is being drawn. The resulting columns are quite flexible and can be bent into coils having diameters of a few inches. Silica open tubular columns are available commercially and offer several important advantages such as physical strength, much lower reactivity toward sample components, and flexibility. For most applications, they have replaced the older type WCOT glass columns.

[6] In 1987, a world record for length of an open tubular column and number of theoretical plates was set, as attested in the *Guinness Book of Records,* by Chrompack International Corporation of the Netherlands. The column was a fused silica column drawn in one piece and having an internal diameter of 0.32 mm and a length of 2100 m or 1.3 miles. The column was coated with a 0.1 μm film of polydimethyl siloxane. A 1300-m section of this column contained over 2 million plates.

[7] For a monograph on open tubular gas chromatography, see M. Lee, F. Yang, and K. Bartle, *Open Tubular Column Gas Chromatography: Theory and Practice.* New York: Wiley, 1984.

The most widely used silica open tubular columns have inside diameters of 320 and 260 μm. Higher-resolution columns are also sold with diameters of 200 and 150 μm. Such columns are more troublesome to use and are more demanding of the injection and detection systems. Thus, a sample splitter must be used to reduce the size of the sample injected onto the column and a more sensitive detector system with a rapid response time is required. Recently, 530-μm capillaries, sometimes called *megabore* columns, have appeared on the market and will tolerate sample sizes that are similar to those for packed columns. The performance characteristics of megabore open tubular columns are not as good as those of smaller diameter columns but significantly better than those of packed columns.

Table 27-1 compares the performance characteristics of fused-silica capillary columns with other types of wall-coated columns as well as with support-coated and packed columns.

27C-2 Packed Columns

Present-day packed columns are fabricated from glass, metal (stainless steel, copper, aluminum), or Teflon tubes that typically have lengths of 2 to 3 m and inside diameters of 2 to 4 mm. These tubes are densely packed with a uniform, finely divided packing material, or solid support, that is coated with a thin layer (0.05 to 1 μm) of the stationary liquid phase. In order to fit in a thermostating oven, the tubes are formed as coils having diameters of roughly 15 cm.

Solid Support Materials

The solid support in a packed column serves to hold the liquid stationary phase in place so that as large a surface area as possible is exposed to the mobile phase. The ideal support consists of small, uniform, spherical particles with good mechanical strength and a specific surface area of at least 1 m^2/g. In addition, the material should be inert at elevated temperatures and be uniformly wetted by the liquid phase. No substance that meets all of these criteria perfectly is yet available.

Today, the most widely used support material is prepared from naturally occurring diatomaceous earth, which is made up of the skeletons of thousands of species of single-celled plants (diatoms) that inhabited ancient lakes and seas (see Figure 21-17a). Such plants received their nutrients and disposed of their wastes via molecular diffusion through their pores. As a consequence, their remains are well-suited as support materials because gas chromatography is also based upon the same kind of molecular diffusion.

Particle Size of Supports

As shown in Figure 26-10 (page 687), the efficiency of a gas-chromatographic column increases rapidly with decreasing particle diameter of the packing. The pressure difference required to maintain a given flow rate of carrier gas, however, varies inversely as the square of the particle diameter; the latter relationship has placed lower limits on the size of particles employed in gas chromatography because it is not convenient to use pressure differences that are greater than about 50 psi. As a result, the usual support particles are 60 to 80 mesh (260 to 170 μm) or 80 to 100 mesh (170 to 149 μm).

27C-3 Adsorption on Column Packings or Capillary Walls

A problem that has plagued gas chromatography from its inception has been the physical adsorption of polar or polarizable analyte species, such as alcohols or aromatic hydrocarbons, on the silicate surfaces of column packings or capillary walls. Adsorption results in distorted peaks, which are broadened and often exhibit a tail. It has been established that adsorption is the consequence of silanol groups that form on the surface of silicates by reaction with moisture. Thus, a fully hydrolyzed silicate surface has the structure

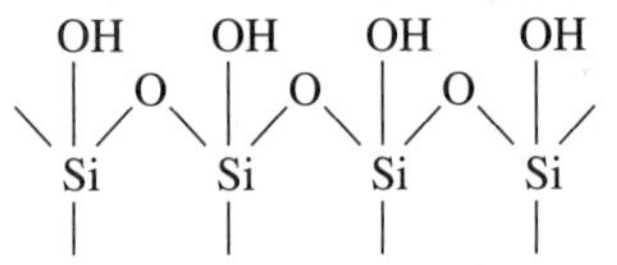

The SiOH groups on the support surface have a strong affinity for polar organic molecules and tend to retain them by adsorption.

Support materials can be deactivated by silanization with dimethylchlorosilane (DMCS). The reaction is

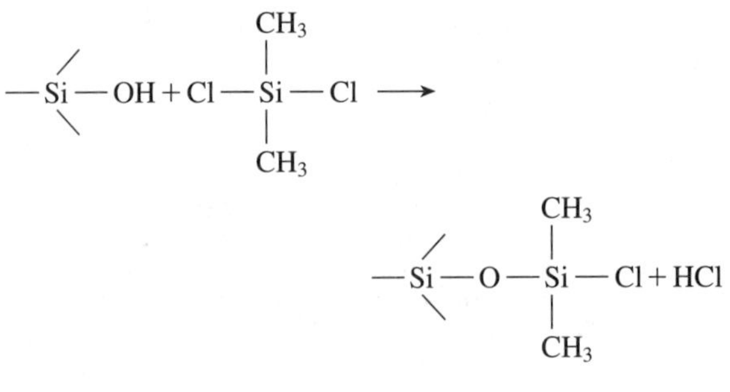

TABLE 27-1 Properties and Characteristics of Typical Gas-Chromatographic Columns

	Type of Column*			
	FSOT	WCOT	SCOT	Packed
Length, m	10–100	10–100	10–100	1–6
Inside diameter, mm	0.1–0.53	0.25–0.75	0.5	2–4
Efficiency, plates/m	2000–4000	1000–4000	600–1200	500–1000
Total plates	$(20–400) \times 10^3$	$(10–400) \times 10^3$	$(6–120) \times 10^3$	$(1–10) \times 10^3$
Sample size, ng	10–75	10–1000	10–1000	$10–10^6$
Relative back pressure	Low	Low	Low	High
Relative speed	Fast	Fast	Fast	Slow
Chemical inertness	Best →			Poorest
Flexible?	Yes	No	No	No

*FSOT: Fused-silica, open tubular column.
WCOT: Wall-coated, open tubular column.
SCOT: Support-coated open tubular column.

Upon washing with methanol, the second chloride is replaced by a methoxy group. That is,

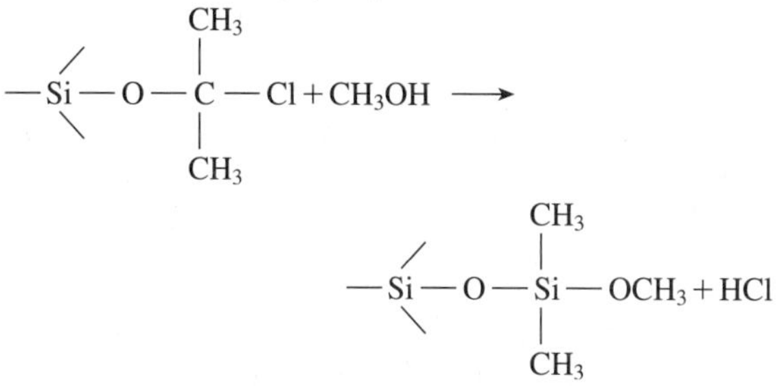

Silanized surfaces of column packings may still show a residual adsorption, which apparently arises from metal oxide impurities in the diatomaceous earth. Acid washing prior to silanization removes these impurities. Fused silica that is used for manufacturing open tubular columns is largely free of this type of impurity. As a consequence, fewer problems with adsorption are encountered with fused-silica columns.

27C-4 The Stationary Phase

Desirable properties for the immobilized liquid phase in a gas-liquid chromatographic column include: (1) *low volatility* (ideally, the boiling point of the liquid should be at 100°C higher than the maximum operating temperature for the column); (2) *thermal stability;* (3) *chemical inertness;* (4) *solvent characteristics* such that k' and α (Sections 26B-4 and 26B-5) values for the solutes to be resolved fall within a suitable range.

A myriad of solvents have been proposed as stationary phases in the course of the development of gas-liquid chromatography. By now, only a handful—perhaps a dozen or less—suffice for most applications. The proper choice among these solvents is often critical to the success of a separation. Qualitative guidelines exist for making this choice, but in the end, the best stationary phase can only be determined in the laboratory.

The retention time for a solute on a column depends upon its distribution constant (Equation 26-1), which in turn is related to the chemical nature of the stationary phase. Clearly, to be useful in gas-liquid chromatography, the immobilized liquid must generate different distribution constants for different solutes. In addition, however, these constants must not be extremely large or extremely small because the former lead to prohibitively long retention times and the latter result in such short retention times that separations are incomplete.

To have a reasonable residence time in the column, a species must show some degree of compatibility (solubility) with the stationary phase. Here, the principle of "like dissolves like" applies, where "like" refers to the polarities of the solute and the immobilized liquid. Polarity is the electrical field effect in the immediate vicinity of a molecule and is measured by the dipole moment of the species. Polar stationary phases contain functional groups such as —CN, —CO, and —OH. Hydrocarbon-type stationary phases and dialkyl siloxanes are nonpolar, whereas polyester phases are highly polar. Polar solutes include alcohols, acids, and amines; solutes of medium polarity include ethers, ketones, and aldehydes. Saturated hydrocarbons are nonpolar. Generally, the polarity of the stationary phase should match that of the sample components. When the match is good, the order of elution is determined by the boiling point of the eluents.

Some Widely Used Stationary Phases

Table 27-2 lists the most widely used stationary phases for both packed and open tubular column gas chromatography in order of increasing polarity. These six liquids can probably provide satisfactory separations for 90% or more of the samples encountered by the scientist.

Five of the liquids listed in Table 27-2 are polydimethyl siloxanes that have the general structure

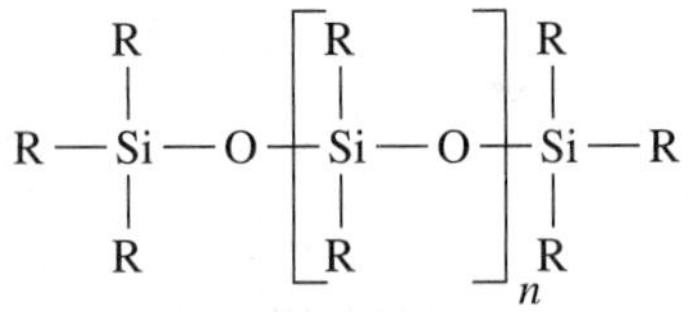

In the first of these, polydimethyl siloxane, the —R groups are all $—CH_3$, giving a liquid that is relatively nonpolar. In the other polysiloxanes shown in the table, a fraction of the methyl groups are replaced by functional groups such as phenyl ($—C_6H_5$), cyanopropyl ($—C_3H_6CN$), and trifluoropropyl ($—C_3H_6CF_3$). The percentage descriptions in each case gives the amount of substitution of the named group for methyl groups on the polysiloxane backbone. Thus, for example, 5% phenyl polydimethyl siloxane has a phenyl ring bonded to 5% by number of the silicon atoms in the polymer. These substitutions increase the polarity of the liquids to various degrees.

The fifth entry in Table 27-2 is a polyethylene glycol having the structure

$$HO—CH_2—CH_2—(O—CH_2—CH_2)_n—OH$$

It finds widespread use for separating polar species. Figure 27-11 illustrates applications of the phases listed in Table 27-2 for open tubular columns.

TABLE 27-2 Some Common Stationary Phases for Gas-Liquid Chromatography

Stationary Phase	Common Trade Name	Maximum Temperature, °C	Common Applications
Polydimethyl siloxane	OV-1, SE-30	350	General-purpose nonpolar phase; hydrocarbons; polynuclear aromatics; drugs; steroids; PCBs
Poly(phenylmethyldimethyl) siloxane (10% phenyl)	OV-3, SE-52	350	Fatty acid methyl esters; alkaloids; drugs; halogenated compounds
Poly(phenylmethyl) siloxane (50% phenyl)	OV-17	250	Drugs; steroids; pesticides; glycols
Poly(trifluoropropyldimethyl) siloxane	OV-210	200	Chlorinated aromatics; nitroaromatics; alkyl-substituted benzenes
Polyethylene glycol	Carbowax 20M	250	Free acids; alcohols; ethers; essential oils; glycols
Poly(dicyanoallyldimethyl) siloxane	OV-275	240	Polyunsaturated fatty acids; rosin acids; free acids; alcohols

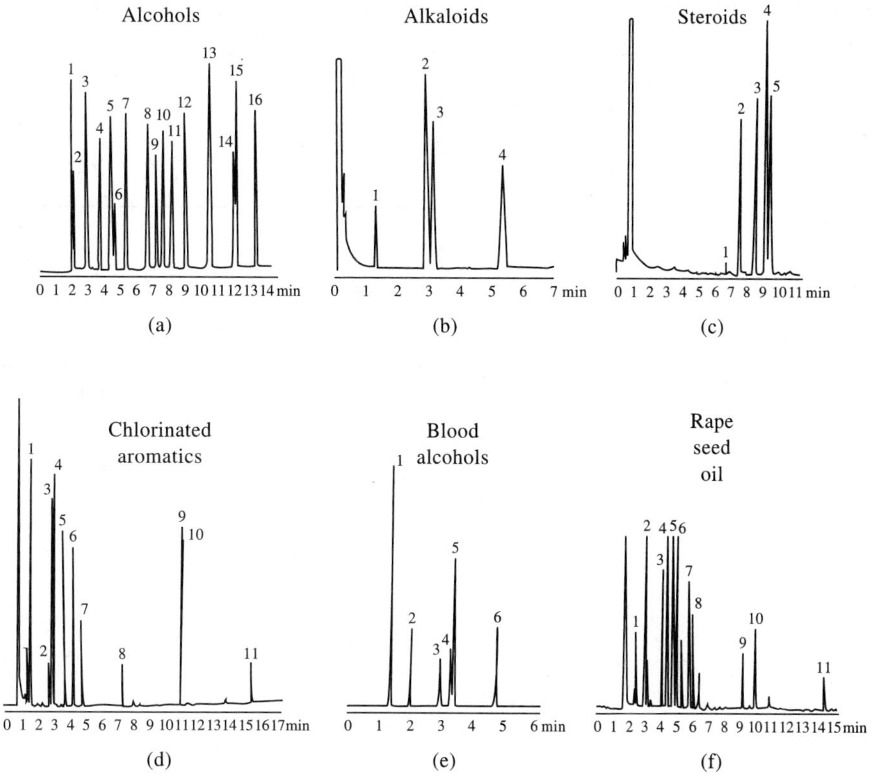

Figure 27-11 Typical chromatograms from open tubular columns coated with (a) polydimethyl siloxane; (b) 5(phenylmethyldimethyl) siloxane; (c) 50(phenylmethyldimethyl) siloxane; (d) 50% poly(trifluoropropyl-dimethyl) siloxane; (e) polyethylene glycol; (f) 50% poly(cyanopropyl-dimethyl) siloxane. *(Courtesy of J & W Scientific.)*

Bonded and Cross-Linked Stationary Phases

Commercial columns are advertised as having bonded and/or cross-linked stationary phases. The purpose of bonding and cross-linking is to provide a longer-lasting stationary phase that can be rinsed with a solvent when the film becomes contaminated. With use, untreated columns slowly loose their stationary phase due to "bleeding," in which a small amount of immobilized liquid is carried out of the column during the elution process. Bleeding is exacerbated when a column must be rinsed with a solvent to remove contaminants. Chemical bonding and cross-linking inhibit bleeding.

Bonding involves attaching a monomolecular layer of the stationary phase to the silica surface of the column by a chemical reaction. For commercial columns, the nature of the reactions is ordinarily proprietary.

Cross-linking is carried out *in situ* after a column is coated with one of the polymers listed in Table 27-2. One way of cross-linking is to incorporate a peroxide into the original liquid. When the film is heated, reaction between the methyl groups in the polymer chains is initiated by a free radical mechanism. The polymer molecules are then cross-linked through carbon-to-carbon bonds. The resulting films are less extractable and have considerably greater thermal stability than do untreated films. Cross-linking has also been initiated by exposing the coated columns to gamma radiation.

Film Thickness

Commercial columns are available having stationary phases that vary in thickness from 0.1 to 5 μm. Film thickness primarily affects the retentive character and the capacity of a column. Thick films are used with highly volatile analytes because such films retain solutes for a longer time, thus providing a greater time for separation to take place. Thin films are useful for separating species of low volatility in a reasonable length of time. For most applications with 0.26- or 0.32-mm columns, a film thickness of 0.26 μm is recommended. With megabore columns, 1- to 1.5-μm films are often used. Today, columns with 8-μm films are marketed.

Chiral Stationary Phases

During the past decade, there has been a good deal of effort devoted to developing methods for the separation of enantiomers by gas or liquid chromatography. Two approaches have been employed. One is based upon forming derivatives of the analyte with an optically active reagent that forms a pair of diastereomers that can be separated on an achiral column. The alternative method is to use a chiral liquid as the stationary phase. A number of amino acid–derived chiral phases have been developed for this purpose, and it is likely that more liquid phases of this type will appear in the near future.[8] The structure of one of these liquids that has been used for separation of optically active amino acids follows:

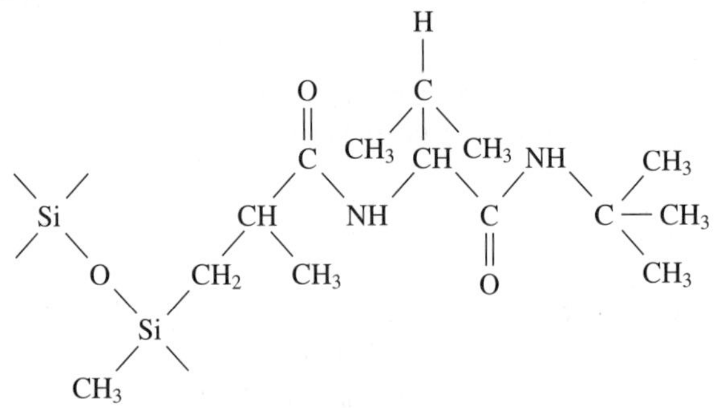

L-valine-1-butylamide on carboxymethylsiloxane

27D APPLICATIONS OF GAS-LIQUID CHROMATOGRAPHY (GLC)

In evaluating the importance of GLC, it is necessary to distinguish between the two roles the method plays. The first is as a tool for performing separations; in this capacity, it is unsurpassed when applied to complex organic, metal-organic, and biochemical systems that are made up of volatile species or species that can be derivitized to yield volatile substances. The second, and distinctly different, function is that of providing the means for completion of an analysis. Here, retention times or volumes are employed for qualitative identification, while peak heights or peak areas provide quantitative information. For qualitative purposes, GLC is much more limited than most of the spectroscopic methods considered in earlier chapters. As a consequence, an important trend in the field has been in the direction of combining the remarkable fractionation qualities of GLC with the superior identification properties of such instruments as mass, infrared, and NMR spectrometers (see Section 27D-3).

27D-1 Qualitative Analysis

Gas chromatograms are widely used as criteria of purity for organic compounds. Contaminants, if present, are revealed by the appearance of additional peaks; the areas under these peaks provide rough estimates of the extent of contamination. The technique is also useful for evaluating the effectiveness of purification procedures.

In theory, retention times should be useful for the identification of components in mixtures. In fact, however, the applicability of such data is limited by the number of variables that must be controlled in order to obtain reproducible results. Nevertheless, gas chromatography provides an excellent means of confirming the presence or absence of a suspected compound in a mixture, provided an authentic sample of the substance is available. No new peaks in the chromatogram of the mixture should appear upon addition of the known compound, and enhancement of an existing peak should be observed. The evidence is particularly convincing if the effect can be duplicated on different columns and at different temperatures.

Selectivity Factors

We have seen (Section 26B-3) that the selectivity factor α for compounds A and B is given by the relationship

$$\alpha = \frac{K_B}{K_A} = \frac{(t_R)_B - t_M}{(t_R)_A - t_M} = \frac{(t_R')_B}{(t_R')_A}$$

If a standard substance is chosen as compound B, then α can provide an index for identification of compound A, which is largely independent of column variables other than temperature; that is, numerical tabulations of selectivity factors for pure compounds relative to a

---

[8] For a review of chiral stationary separations by gas chromatography, see J. V. Hinshaw, *LC-GC*, **1993**, *11*, 644.

common standard can be prepared and then used for the characterization of solutes. Unfortunately, finding a universal standard that yields selectivity factors of reasonable magnitude for all types of analytes is not possible. Thus, the amount of selectivity factor data available in the literature is presently limited.

The Retention Index

The retention index I was first proposed by Kovats in 1958 as a parameter for identifying solutes from chromatograms.[9] The retention index for any given solute can be derived from a chromatogram of a mixture of that solute with at least two normal alkanes having retention times that bracket that of the solute. That is, normal alkanes are the standards upon which the retention index scale is based. By definition, the retention index for a normal alkane is equal to 100 times the number of carbons in the compound *regardless of the column packing, the temperature, or other chromatographic conditions.* The retention index for all compounds other than normal alkanes vary, often by several hundred retention index units, with column variables.

It has long been known that within a homologous series, a plot of the logarithm of adjusted retention time t'_R ($t'_R = t_R - t_M$) versus the number of carbon atoms is linear, provided the lowest member of the series is excluded. Such a plot for C_4 to C_9 normal alkane standards is shown in Figure 27-12. Also indicated on the ordinate are log retention times for three compounds on the same column and at the same temperature. Their retention indexes are then obtained by multiplying the corresponding abscissa values by 100. Thus, the retention index for toluene is 749, while for benzene it is 644.

Normally a graphical procedure is not required in determining retention indexes. Instead adjusted retention data are derived by interpolation from a chromatogram of a mixture of the solute of interest and two or more alkane standards.

It is important to reiterate that the retention index *for a normal alkane* is independent of temperature and column packing. Thus, I for heptane, by definition, is always 700. In contrast, retention indexes of all other solutes may, and often do, vary widely from one column to another. For example, the retention index for acenapthene on a cross-linked polydimethyl siloxane stationary phase at 140°C is 1460. With 5% phenylpolydimethyl siloxane as the stationary phase, it is 1500 at the same temperature, while with polyethylene glycol as the stationary phase, the retention index is 2084.

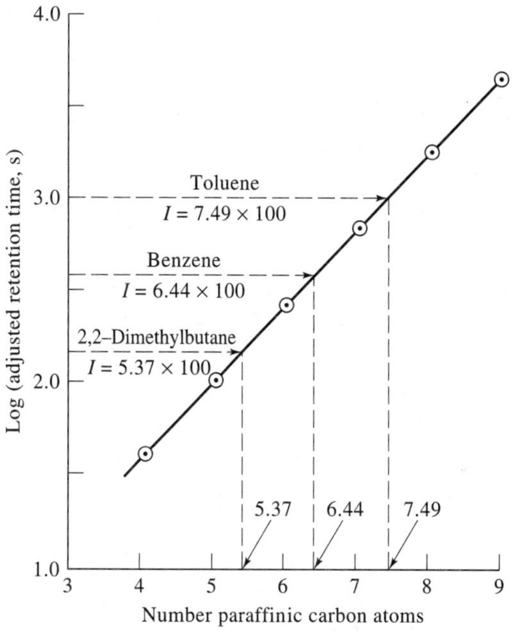

Figure 27-12 Graphical illustration of the method for determining retention indexes for three compounds. Stationary phase: squalane. Temperature: 60°C.

The retention index system has the advantage of being based upon readily available reference materials that cover a wide boiling range. In addition, the temperature dependence of retention indexes is relatively small. In 1984, Sadtler Research Laboratories introduced a library of retention indexes measured on four types of fused-silica open tubular columns. The computerized format of the database allows retention index searching and possible identity recall with a desktop computer.[10]

27D-2 Quantitative Analysis

The detector signal from a gas-liquid chromatographic column has had wide use for quantitative and semiquantitative analyses. An accuracy of 1% relative is attainable under carefully controlled conditions. As with

[9] E. Kovats, *Helv. Chim. Acta,* **1958,** *41,* 1915.

[10] See J. F. Sprouse and A. Varano, *Amer. Lab.,* **1984** (9), 54.

most analytical tools, reliability is directly related to the control of variables; the nature of the sample also plays a part in determining the potential accuracy. The general discussion of quantitative chromatographic analysis given in Section 26F-2 applies to gas chromatography as well as to other types; therefore, no further consideration of this topic is given here.

27D-3 Interfacing Gas Chromatography with Spectroscopic Methods

Gas chromatography is often coupled with the selective techniques of spectroscopy and electrochemistry, thus giving so-called *hyphenated methods* that provide the chemist with powerful tools for identifying the components of complex mixtures.[11]

In early hyphenated methods, the eluates from the chromatographic column were collected as separate fractions in a cold trap after being detected by a nondestructive and nonselective detector. The composition of each fraction was then investigated by nuclear magnetic resonance, infrared, or mass spectroscopy, or by electroanalytical measurements. A serious limitation to this approach was the very small (usually micromolar) quantities of solute contained in a fraction; nonetheless, the general procedure proved useful for the qualitative analysis of many multicomponent mixtures.

A second general method, which now finds widespread use, involves the application of a selective detector to monitor the column effluent continuously. Generally, these procedures require computer control of instruments and computer memory for storage of spectral data for subsequent display as spectra and chromatograms.

Gas Chromatography/Mass Spectrometry (GC/MS)

Several instrument manufacturers offer gas-chromatographic equipment that can be directly interfaced with rapid-scan mass spectrometers of various types.[12] The flow rate from capillary columns is generally low enough that the column output can be fed directly into the ionization chamber of the mass spectrometer. A schematic of a typical system is shown in Figure 27-13. For packed columns and megabore capillary columns, however, a jet separator such as that shown in Figure 27-14 must be employed to remove most of the carrier gas from the analyte. In this device, the exit gases flow through a nozzle of an all-glass jet separator, which increases the momentum of the heavier analyte molecules so that 50% or more of them travel in a more or less straight path to the skimmer. The light helium atoms, in contrast, are deflected by the vacuum and are thus pumped away.

Most quadrupole and magnetic sector mass spectrometers are offered with accessories that permit interfacing with gas-chromatographic equipment. In addition, the Fourier transform mass spectrometer described in Section 20C-4 has been coupled to gas-liquid columns.[13] Its speed and high sensitivity are particularly advantageous for this application.

Beginning in the late 1970s, several mass spectrometers designed specifically as gas chromatographic detectors appeared on the market. Generally, these are compact quadrupole instruments, which are less expensive ($25,000 to $50,000) and easier to use and maintain than the multipurpose mass spectrometers described in Chapter 20.[14]

The simplest mass detector for use in gas chromatography is the *ion trap detector* (ITD)[15] (see Section 20C-3 and Figure 20-15). In this instrument, ions are created from the eluted sample by electron impact or chemical ionization and stored in a radio-frequency field (see Figure 27-15). The trapped ions are then ejected from the storage area to an electron multiplier detector. The ejection is controlled so that scanning on the basis of mass-to-charge ratio is possible. The ion trap detector is remarkably compact and less expensive than quadrupole instruments.

Mass spectrometric detectors ordinarily have several display modes, which fall into two categories: real time and computer reconstructed. Within each of these categories are a choice of total ion current chromatograms (a plot of the sum of all ion currents as a function of time), selected ion current chromatograms (a plot of ion currents for one or a few ions as a function

---

[11] For a review on hyphenated methods, see T. Hirschfeld, *Anal. Chem.*, **1980**, *52*, 297A; C. L. Wilkins, *Science*, **1983**, *222*, 291; *Anal. Chem.*, **1987**, *59*, 571A; P. R. Griffiths et al., *Anal. Chem.*, **1986**, *58*, 1349A.

[12] For additional information, see G. M. Message, *Practical Aspects of Gas Chromatography/Mass Spectrometry*. New York: Wiley, 1984; J. Masucci and G. W. Caldwell, in *Modern Practice of Gas Chromatography*, 3rd ed., R. L. Grob, Ed., Chapter 6. New York: Wiley-Interscience, 1995; C. L. Wilkins, *Anal. Chem.*, **1987**, *59*, 571A; J. Fjeldsted and J. Truche, *Amer. Lab.*, **1989** (10), 33.

[13] E. B. Ledford et al., *Anal. Chem.*, **1980**, *52*, 2550.

[14] For descriptions of several benchtop GC/MS instruments, see F. Wach, *Anal. Chem.*, **1994**, *66*, 927A.

[15] G. C. Stafford Jr., P. E. Kelley, and D. C. Bradford, *Amer. Lab.*, **1983**, *15* (6), 51; S. A. Borman, *Anal. Chem.*, **1983**, *55*, 726A.

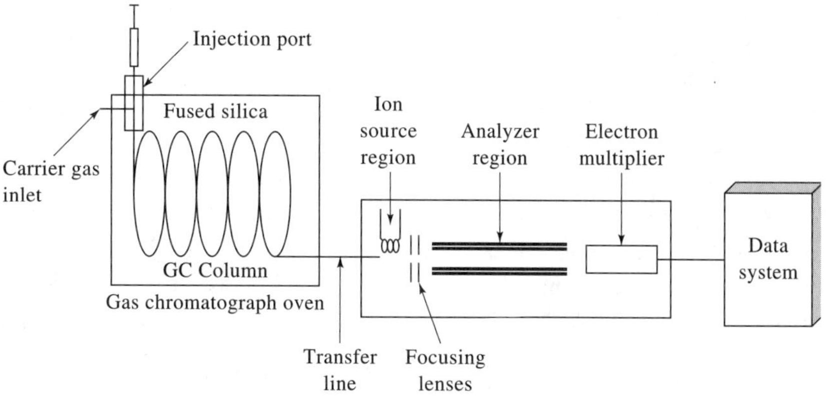

Figure 27-13 Schematic of a typical capillary gas chromatography/mass spectrometer.

of time), and mass spectra of various peaks. Real-time mass spectra appear on an oscilloscope screen equipped with mass markers; the mass chromatogram may appear on the oscilloscope screen or as a real-time plot. After a separation is complete, computer-reconstructed chromatograms can be displayed on the screen or can be printed out. Reconstructed mass spectra for each peak can also be displayed or printed. Some instruments are further equipped with spectral libraries for compound identification.

Gas chromatography/mass spectrometry instruments have been used for the identification of hundreds of components that are present in natural and biological systems. For example, these procedures have permitted characterization of the odor and flavor components of foods, identification of water pollutants, medical diagnosis based on breath components, and studies of drug metabolites.

An example of one application of GC/MS is shown in Figure 27-16. The upper figure is a computer-reconstructed mass chromatogram of a sample trapped from an environmental chamber during the combustion of cloth treated with a fire-retarding chemical. The ordinate here is the total ion current while the abscissa

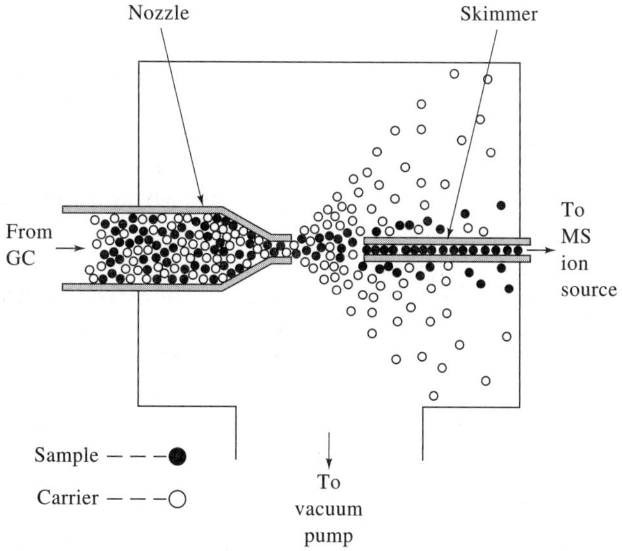

Figure 27-14 Schematic of a jet separator. *(Courtesy of DuPont Instrument Systems, Wilmington, DE.)*

is retention time. The lower figure is the computer-reconstructed mass spectrum of peak 12 on the chromatogram. Here, relative ion currents are plotted as a function of mass number.

Gas Chromatography/Fourier Transform Infrared Spectroscopy (GC/FTIR)

Coupling capillary column gas chromatographs with Fourier transform infrared spectrometers provides a potent means for separating and identifying the components of difficult mixtures. Several instruments of this type are now offered commercially.[16]

[16] See C. Fujimoto and K. Jinno, *Anal. Chem.*, **1992**, *64*, 476A; P. R. Griffiths, S. L. Pentoney, A. Giorgetti, and K. H. Shafter, *Anal. Chem.*, **1986**, *58*, 1349A; C. L. Wilkins, *Science*, **1983**, *222*, 291; P. R. Griffiths, J. A. de Haseth, and L. V. Azarraga, *Anal. Chem.*, **1983**, *55*, 1361A; S. A. Borman, *Anal. Chem.*, **1982**, *54*, 901A.

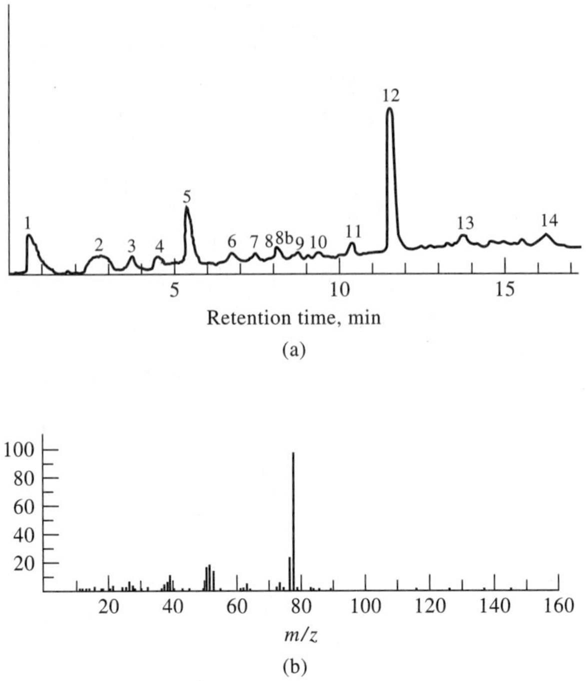

Figure 27-16 Typical output from a GC/MS instrument. The upper curve is a computer-reconstructed chromatogram. The peaks correspond to: (1) air, (2) water, (3) hydrogen cyanide, (4) unknown, (5) acetaldehyde, (6) ethanol, (7) acetonitrile, (8) acetone, (8b) unknown, (9) carbon disulfide, (10) unknown, (11) unknown, (12) benzene, (13) toluene, (14) xylene. The lower plot is the computer-reconstructed mass spectrum for peak 12 (benzene). *(Reprinted with permission from T. W. Sickels and D. T. Stafford,* Amer. Lab., ***1977***, 12*(5), 17. Copyright 1977 by International Scientific Communications, Inc.)*

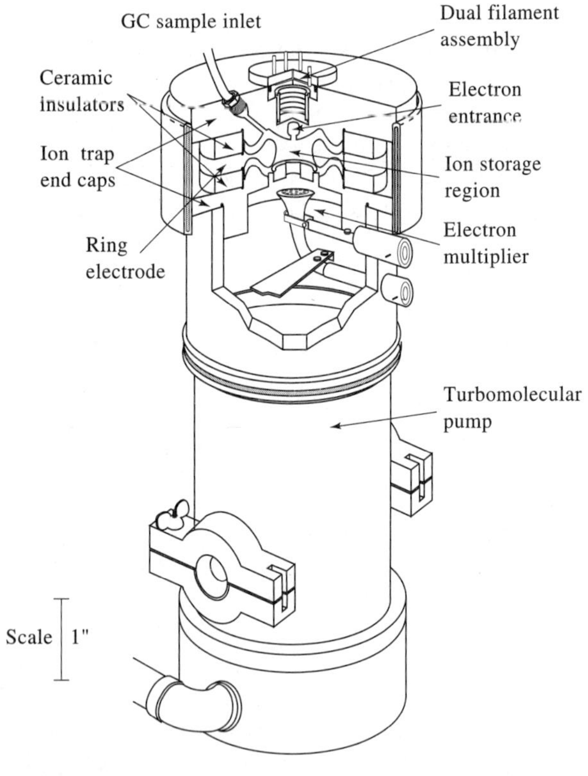

Figure 27-15 Schematic of the ion trap detector. *(Reprinted with permission from G. C. Stafford Jr., P. E. Kelley, and D. C. Bradford,* Amer. Lab., ***1983***, 15*(6), 51. Copyright 1983 by International Scientific Communications, Inc.)*

As with GC/MS, the interface between the column and the detector is critical. In this instance, a narrow light pipe having a length of 10 to 40 cm and an inside diameter of 1 to 3 mm is connected to the column by means of narrow tubing. The light pipe, a version of which is shown schematically in Figure 27-17, consists of a Pyrex tube that is internally coated with gold. Transmission of radiation occurs by multiple reflections off the wall. Often the light pipe is heated in order to avoid condensation of the sample components. Light pipes of this type are designed to maximize the path length for enhanced sensitivity while minimizing the dead volume to lessen band broadening. The radiation detectors are generally highly sensitive, liquid nitrogen–cooled mercury/cadmium telluride devices.

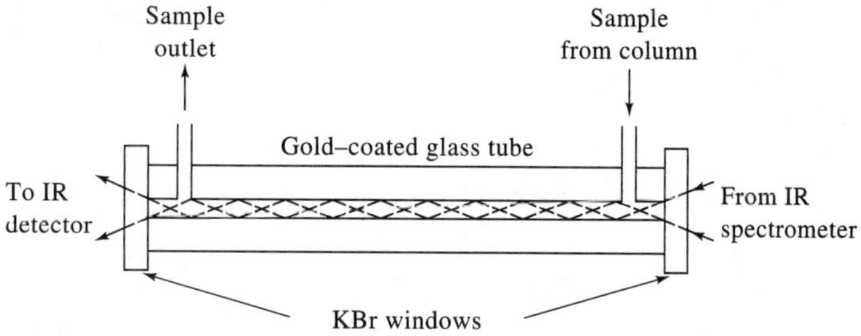

Figure 27-17 A typical light pipe for GC/IR instruments.

Scanning is triggered by the output from a nondestructive chromatographic peak detector and begins after a brief delay to allow the component to travel from the detector region to the infrared cell. The spectral data are digitized and stored in a computer from which printed spectra are ultimately derived.

Difficulty is sometimes encountered when attempts are made to compare spectra for the gaseous effluents from a column with library spectra that have been obtained with liquid or solid samples. Gaseous spectra contain rotational fine structure, which is absent in liquid or solid spectra; significant differences in appearance are the result. Another difference between gas-phase and liquid-phase spectra is the absence of bands in the former due to intermolecular interactions, such as hydrogen bonding in alcohols and acids.

As with GC/MS, digitized spectral libraries and search systems are being developed to handle the enormous amount of data that are produced by MS/FTIR instruments, even with relatively simple samples.[17]

27E GAS-SOLID CHROMATOGRAPHY

Gas-solid chromatography is based upon adsorption of gaseous substances on solid surfaces. Distribution coefficients are generally much larger than those for gas-liquid chromatography. Consequently, gas-solid chromatography is useful for the separation of species that are not retained by gas-liquid columns, such as the components of air, hydrogen sulfide, carbon disulfide, nitrogen oxides, carbon monoxide, carbon dioxide, and the rare gases.

Gas-solid chromatography is performed with both packed and open tubular columns. For the latter, a thin layer of the adsorbent is affixed to the inner walls of the capillary. Such columns are sometimes called *porous layer open tubular columns,* or PLOT columns. Two types of adsorbents are encountered, molecular sieves and porous polymers.

27E-1 Molecular Sieves

Molecular sieves are aluminum silicate ion exchangers, whose pore size depends upon the kind of cation present. Commercial preparation of these materials are available in particle sizes of 40/60 mesh to 100/120 mesh. The sieves are classified according to the maximum diameter of molecules that can enter the pores. Commercial molecular sieves come in pore sizes of 4, 5, 10, and 13 Å. Molecules smaller than these dimensions penetrate into the interior of the particles where adsorption takes place. For such molecules, the surface area is enormous when compared with the area available to larger molecules. Thus molecular sieves can be used to separate small molecules from large. For example, a 6-ft, 5-Å packing at room temperature will easily separate a mixture of helium, oxygen, nitrogen, methane, and carbon monoxide in the order given. Figure 27-18a shows a typical molecular sieve chromatogram. In this application two packed columns were used, one an ordinary gas-liquid column and the other a molecular sieve column. The former retains only the carbon dioxide and passes the remaining gases at rates corresponding to the carrier rate. When the carbon dioxide is eluted from the first column, a switch directs the flow around the second column briefly to avoid permanent adsorption of the carbon dioxide on the molecular sieve. After the carbon dioxide signal has returned to zero, the flow is switched back through the second col-

[17] For a description of one of these systems, see S. R. Lowry and D. A. Huppler, *Anal. Chem.*, **1981**, *53*, 889.

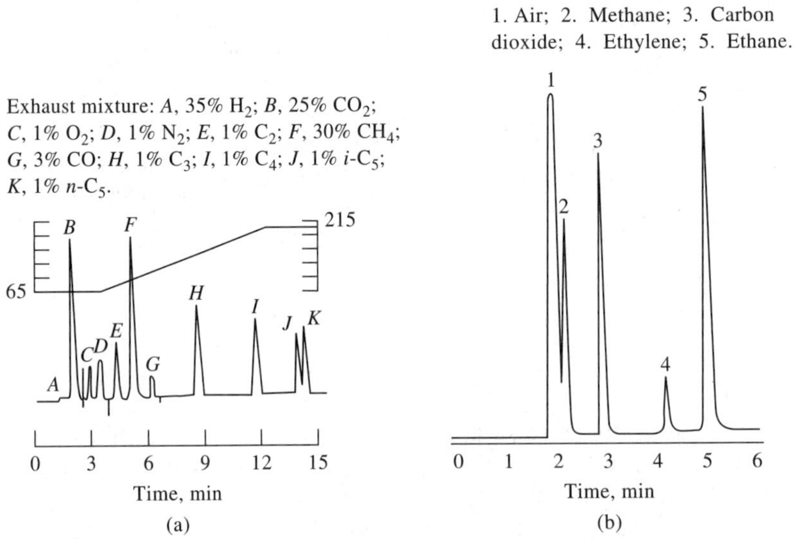

Figure 27-18 Typical gas-solid chromatographic separations: (a) $5' \times \frac{1}{8}''$ molecular sieve column; (b) 30 m × 0.53 mm PLOT column.

umn, thereby permitting separation and elution of the remainder of the sample components.

27E-2 Porous Polymers

Porous polymer beads of uniform size are manufactured from styrene cross-linked with divinylbenzene (Section 27G-2). The pore size of these beads is uniform and is controlled by the amount of cross-linking. Porous polymers have found considerable use in the separation of gaseous polar species such as hydrogen sulfide, oxides of nitrogen, water, carbon dioxide, methanol, and vinyl chloride. A typical application of an open tubular column lined with a porous polymer (PLOT column) is shown in Figure 27-18.

27F QUESTIONS AND PROBLEMS

27-1 How do gas-liquid and gas-solid chromatography differ?

27-2 What kind of mixtures are separated by gas-solid chromatography?

27-3 Why is gas-solid chromatography not used nearly as extensively as gas-liquid chromatography?

27-4 Define (a) retention volume, (b) corrected retention volume, (c) specific retention volume.

27-5 How does a soap-bubble flow meter work?

27-6 What is temperature programming as used in gas chromatography?

27-7 What is the difference between a concentration-sensitive and a mass flow–sensitive detector? Indicate which type of device the following detectors are: (a) thermal conductivity, (b) atomic emission, (c) thermionic, (d) electron capture, (e) flame photometry, (f) photoionization.

27-8 Describe the principle upon which each of the detectors listed in Problem 27-7 is based.

27-9 What are the principal advantages and the principal limitations of each of the detectors listed in Problem 27-7?

27-10 What is the packing material used in most packed gas-chromatographic columns?

27-11 How do the following open tubular columns differ?
(a) PLOT columns (b) WCOT columns (c) SCOT columns

27-12 What are megabore open tubular columns? Why are they used?

27-13 What are the advantages of fused-silica capillary columns compared with glass or metal columns?

27-14 What properties should the stationary-phase liquid for gas chromatography possess?

27-15 Why are gas-chromatographic stationary phases often bonded and cross-linked? What do these terms mean?

27-16 What is the effect of stationary-phase film thickness on gas chromatograms?

27-17 What are retention indexes? Describe how they are determined.

27-18 How is a chromatogram usually obtained in GC/MS? In GC/IR?

27-19 The same polar compound is gas chromatographed on an SE-30 (very nonpolar) column and then on a carbowax 20M (very polar column). How will $K = c_S/c_M$ vary between the two columns?

27-20 Use the retention data given below to calculate the retention index of 1-hexene.

Sample	Retention Time (min)
air	0.571
n-pentane	2.16
n-hexane	4.23
1-hexene	3.15

27-21 A GLC column was operated under the following conditions:

- **column:** 1.10 m × 2.0 mm, packed with Chromosorb P; weight of stationary liquid added, 1.40; density of liquid, 1.02 g/mL
- **pressures:** inlet, 26.1 psi above room; room, 748 torr
- **measured outlet flow rate:** 25.3 mL/min
- **temperature:** room, 21.2°C; column, 102.0°C
- **retention times:** air, 18.0 s; methyl acetate, 1.98 min; methyl propionate, 4.16 min; methyl *n*-butyrate, 7.93 min
- **peak widths at base:** 0.19, 0.39, and 0.79, respectively

Calculate

(a) the average flow rate in the column.
(b) the corrected retention volumes for air and the three esters.
(c) the specific retention volumes for the three components.
(d) distribution constants for each of the esters.
(e) a corrected retention volume and a retention time for methyl *n*-hexanoate.

27-22 From the data in Problem 27-21, calculate
(a) k' for each compound.
(b) α values for each adjacent pair of compounds.
(c) the average number of theoretical plates and plate height for the column.
(d) the resolution for each adjacent pair of compounds.

27-23 The stationary liquid in the column described in Problem 27-21 was didecylphthalate, a solvent of intermediate polarity. If a nonpolar solvent such as a silicone oil had been used instead, would the retention times for the three compounds be larger or smaller? Why?

27-24 Corrected retention times for ethyl, *n*-propyl, and *n*-butyl alcohols on a column employing a packing coated with silicone oil are 0.69, 1.51, and 3.57, respectively. Predict retention times for the next two members of the homologous series.

27-25 What would be the effect of the following on the plate height of a column? Explain.
(a) Increasing the weight of the stationary phase relative to the packing weight.
(b) Decreasing the rate of sample injection.
(c) Increasing the injection port temperature.
(d) Increasing the flow rate.
(e) Reducing the particle size of the packing.
(f) Decreasing the column temperature.

27-26 Calculate the retention index for each of the following compounds:

Compound	$(t_R - t_M)$
(a) Propane	1.29
(b) *n*-Butane	2.21
(c) *n*-Pentane	4.10
(d) *n*-Hexane	7.61
(e) *n*-Heptane	14.08
(f) *n*-Octane	26.11
(g) Toluene	16.32
(h) Butane-2	2.67
(i) *n*-Propanol	7.60
(j) Methylethyl ketone	8.40
(k) Cyclohexane	6.94
(l) *n*-Butanol	9.83

Chapter 28

High-Performance Liquid Chromatography

This chapter deals with the four basic types of chromatography in which the mobile phase is a liquid. The four include: ***partition chromatography; adsorption,*** or ***liquid-solid chromatography; ion exchange chromatography;*** and ***size exclusion,*** *or* ***gel, chromatography.*** *Most of this chapter deals with column applications of these four important types of chromatography. The final section, however, presents a brief description of planar liquid chromatography because this technique provides a simple and inexpensive way of deriving information on what is likely to be optimal conditions for carrying out separations on a column.*

Early liquid chromatography, including the original work by Tswett, was carried out in glass columns with diameters of 1 to 5 cm and lengths of 50 to 500 cm. To assure reasonable flow rates, the diameter of the particles of the solid stationary phase was usually in the 150- to 200-μm range. Even then, flow rates were low, amounting to a few tenths of a milliliter per minute. Thus separation times were long—often several hours. Attempts to speed up the classic procedure by application of vacuum or by pumping were not effective, however, because increases in flow rates acted to increase plate heights beyond the minimum in the typical plate height versus flow rate curve (see Figure 26-7a); decreased efficiencies were the result.

Early in the development of liquid chromatography, scientists realized that major increases in column efficiency could be brought about by decreasing the particle size of packings. It was not until the late 1960s, however, that the technology for producing and using packings with particle diameters as small as 3 to 10 μm was developed. This technology required sophisticated instruments operating at high pressures, which contrasted markedly with the simple glass columns of classic gravity-flow liquid chromatography. The name ***high-performance liquid chromatography*** *(HPLC) is employed to distinguish these newer procedures from the basic methods, which are still used for preparative purposes. This chapter deals exclusively with HPLC.*[1]

[1] A large number of books on liquid chromatography are available. Among these are R. W. Scott, *Liquid Chromatography for the Analyst.* New York: Marcel Dekker, 1994; M. T. Gilbert, *High Performance Liquid Chromatography.* Bristol, UK: Wright, 1987; P. R. Brown and R. A. Hartwick, *High Performance Liquid Chromatography.* New York: Wiley, 1989; V. R. Meyer, *Practical High Performance Liquid Chromatography.* New York: Wiley, 1988. For a recent review dealing with past, current, and future developments in HPLC, see P. R. Brown, *Anal. Chem.*, **1990**, *62*, 995A.

28A SCOPE OF HPLC

High-performance liquid chromatography is the most widely used of all of the analytical separation techniques, with annual sales of HPLC equipment approaching the billion dollar mark. The reasons for the popularity of the method is its sensitivity, its ready adaptability to accurate quantitative determinations, its suitability for separating nonvolatile species or thermally fragile ones, and above all, its widespread applicability to substances that are of prime interest to industry, to many fields of science, and to the public. Examples of such materials include: amino acids, proteins, nucleic acids, hydrocarbons, carbohydrates, drugs, terpenoids, pesticides, antibiotics, steroids, metal-organic species, and a variety of inorganic substances.

Figure 28-1 reveals that the various liquid chromatographic procedures tend to be complementary insofar as their areas of applications are concerned. Thus, for solutes having molecular weights greater than 10,000, exclusion chromatography is often used, although it is now becoming possible to handle such compounds by reversed-phase partition chromatography as well. For lower-molecular-weight ionic species, ion-exchange chromatography is widely used. Small polar but nonionic species are best handled by partition methods. In addition, this procedure is frequently useful for separating members of a homologous series. Adsorption chromatography is often chosen for separating nonpolar species, structural isomers, and compound classes such as aliphatic hydrocarbons from aliphatic alcohols.

28B COLUMN EFFICIENCY IN LIQUID CHROMATOGRAPHY

The discussion on band broadening in Section 26C-3 is generally applicable to liquid chromatography. The present section illustrates the important effect of station-

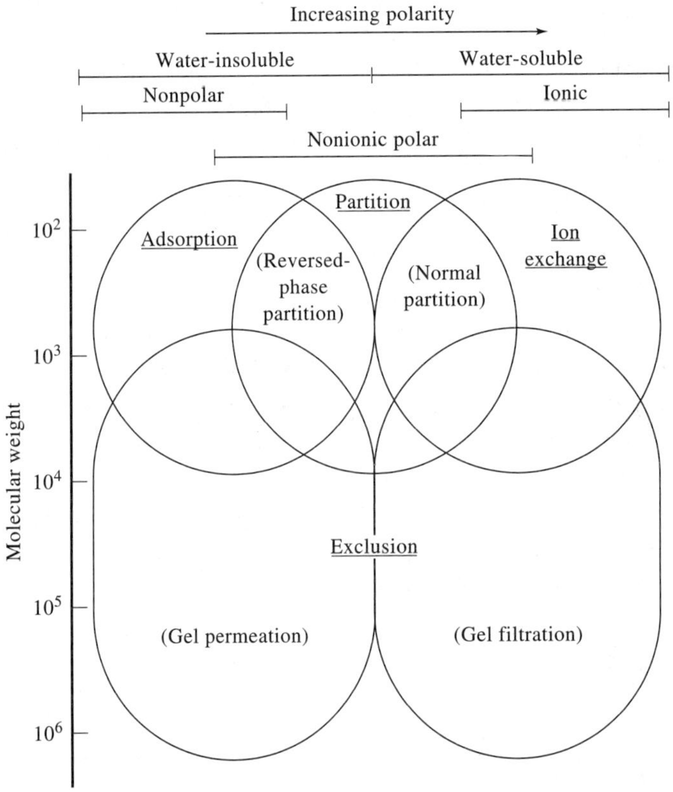

Figure 28-1 Applications of liquid chromatography. *(From D. L. Saunders, in* Chromatography, *3rd ed., E. Heftmann, Ed., p. 81. New York: Van Nostrand Reinhold, 1975. With permission.)*

ary-phase particle size and describes two additional sources of zone spreading that are sometimes of considerable importance in liquid chromatography.

28B-1 Effects of Particle Size of Packings

An examination of the mobile-phase mass transfer coefficient in Table 26-3 (page 685) reveals that C_M in Equation 26-19 is directly related to the square of the diameter d_p of the particles making up a packing. As a consequence, the efficiency of an HPLC column should improve dramatically as the particle size is decreased. Figure 28-2 is an experimental demonstration of this effect, where it is seen that a reduction of particle size from 45 to 6 μm results in a ten-fold or more decrease in plate height. Note that none of the plots in this figure exhibits the minimum that is predicted by Equation 26-19. Such minima are, in fact, observable in liquid chromatography (see Figure 26-7a) but usually at flow rates that are too low for most practical applications.

28B-2 Extra-Column Band Broadening in Liquid Chromatography

In liquid chromatography, significant band broadening sometimes occurs outside the column packing itself. This so-called *extra-column band broadening* occurs as the solute is carried through open tubes such as those found in the injection system, the detector region, and the piping connecting the various components of the system. Here, broadening arises from differences in flow rates between layers of liquid adjacent to the wall and the center of the tube. As a consequence, the center part of a solute band moves more rapidly than the peripheral part. In gas chromatography, extra-column spreading is largely offset by diffusion. Diffusion in liquids, however, is significantly slower, and band broadening of this type often becomes noticeable.

It has been shown that the contribution of extra-column effects H_{ex} to the total plate height is given by[2]

$$H_{ex} = \frac{\pi r^2 u}{24 D_M} \tag{28-1}$$

where u is the linear flow velocity in cm/s, r is the radius of the tube in cm, and D_M is the diffusion coefficient of the solute in the mobile phase in cm^2/s.

Extra-column broadening can become quite serious when small bore columns are used. Here, every effort must be made to reduce the radius of the extra-column components to 0.10 inch or less.

28B-3 Effect of Sample Size on Column Efficiency

Figure 28-3 shows the effect of sample mass (μg sample/g packing) on column efficiency for various types of liquid chromatography. Note the superior performance

[2] R. P. W. Scott and P. Kucera, *J. Chromatogr. Sci.*, **1971**, *9*, 641.

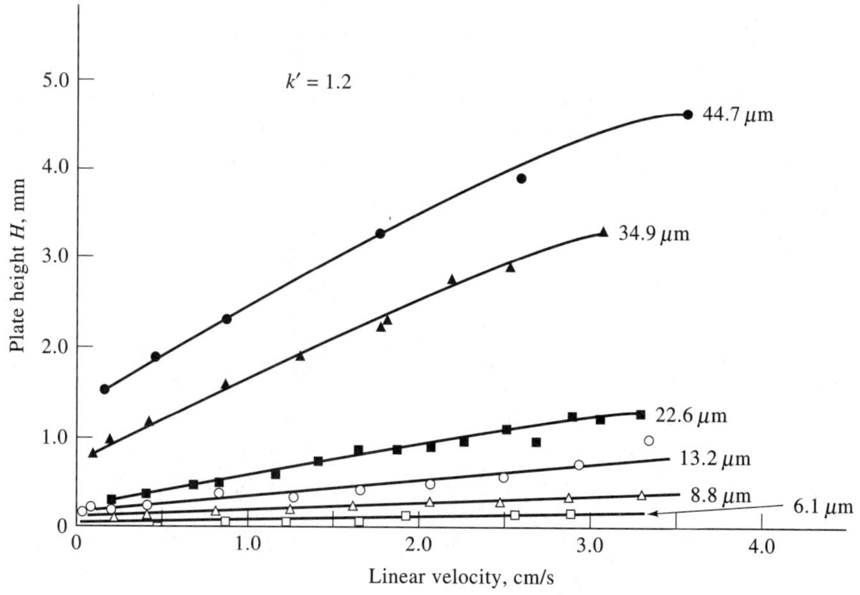

Figure 28-2 Effect of particle size of packing and flow rate upon plate height H in liquid chromatography. Column dimensions: 30 cm × 2.4 mm. Solute: N,N-diethyl-*n*-aminoazobenzene. Mobile phase: mixture of hexane, methylene chloride, isopropyl alcohol. *(From R. E. Majors, J. Chromatogr. Sci.,* ***1973****, 11, 92. With permission.)*

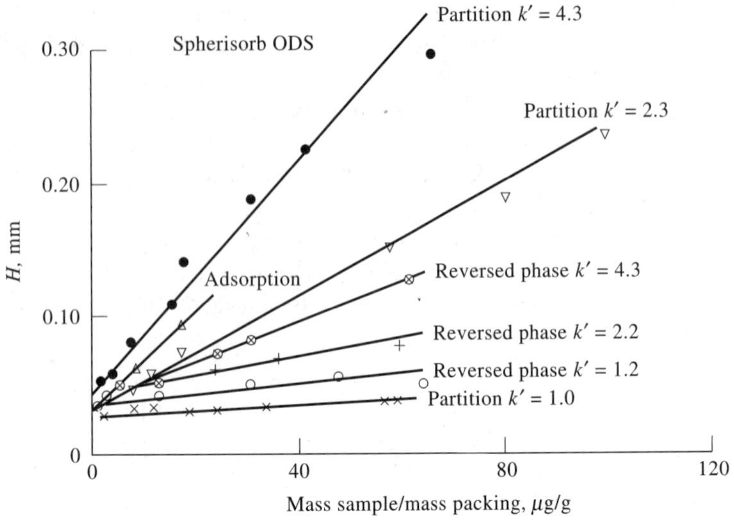

Figure 28-3 Effect of relative sample mass on plate height. *(From J. N. Done,* J. Chromatogr. Sci., ***1976,*** 125, *54. With permission.)*

of reversed-phase, bonded packings (Section 28D-1) compared with other types of packings.

28C INSTRUMENTS FOR LIQUID CHROMATOGRAPHY

In order to realize reasonable eluent flow rates with packings in the 2- to 10-μm particle sizes, which are common in modern liquid chromatography, pumping pressures of up to several thousand pounds per square inch are required. As a consequence of these high pressures, the equipment required for HPLC tends to be more elaborate and expensive than that encountered in other types of chromatography. Figure 28-4 is a schematic showing the important components of a typical high-performance liquid chromatograph; each is discussed in the paragraphs that follow.[3]

28C-1 Mobile-Phase Reservoirs and Solvent Treatment Systems

A modern HPLC apparatus is equipped with one or more glass or stainless steel reservoirs, each of which contains 200 to 1000 mL of a solvent. The reservoirs are often equipped with a means of removing dissolved gases—usually oxygen and nitrogen—that interfere by forming bubbles in the column and the detector systems. These bubbles cause band spreading; in addition, they often interfere with the performance of the detector. Degassers may consist of a vacuum pumping system, a distillation system, devices for heating and stirring the solvents or, as shown in Figure 28-4, systems for *sparging* in which the dissolved gases are swept out of the solution by fine bubbles of an inert gas of low solubility. Often the systems also contain a means of filtering dust and particulate matter from the solvents to prevent these particles from damaging the pumping or injection systems or clogging the column. It is not necessary that the degassers and filters be integral parts of the HPLC system as shown in Figure 28-4. For example, a convenient way of treating solvents before introduction into the reservoir is to filter them through a millipore filter under vacuum. This treatment removes gases as well as suspended matter.

A separation that employs a single solvent of constant composition is termed an *isocratic elution.* Frequently, separation efficiency is greatly enhanced by *gradient elution.* Here two or three solvent systems that differ significantly in polarity are employed. After elution is begun, the ratio of the solvents is varied in a programmed way, sometimes continuously and sometimes in a series of steps. Modern HPLC equipment is often equipped with devices that introduce solvents from two or more reservoirs into a mixing chamber at rates that vary continuously; the volume ratio of the solvents may then be altered linearly or exponentially with time.

[3] For a detailed discussion of HPLC systems, see N. A. Parris, *Instrumental Liquid Chromatography,* 2nd ed. New York: Elsevier, 1984; *Practice of High Performance Liquid Chromatography: Applications, Equipment, and Quantitative Analysis,* H. Engelhardt, Ed., Chapter 1. New York: Springer-Verlag, 1986; A. Katritzky and R. J. Offerman, *Crit. Rev. Anal. Chem.,* **1989,** *21*(4), 83.

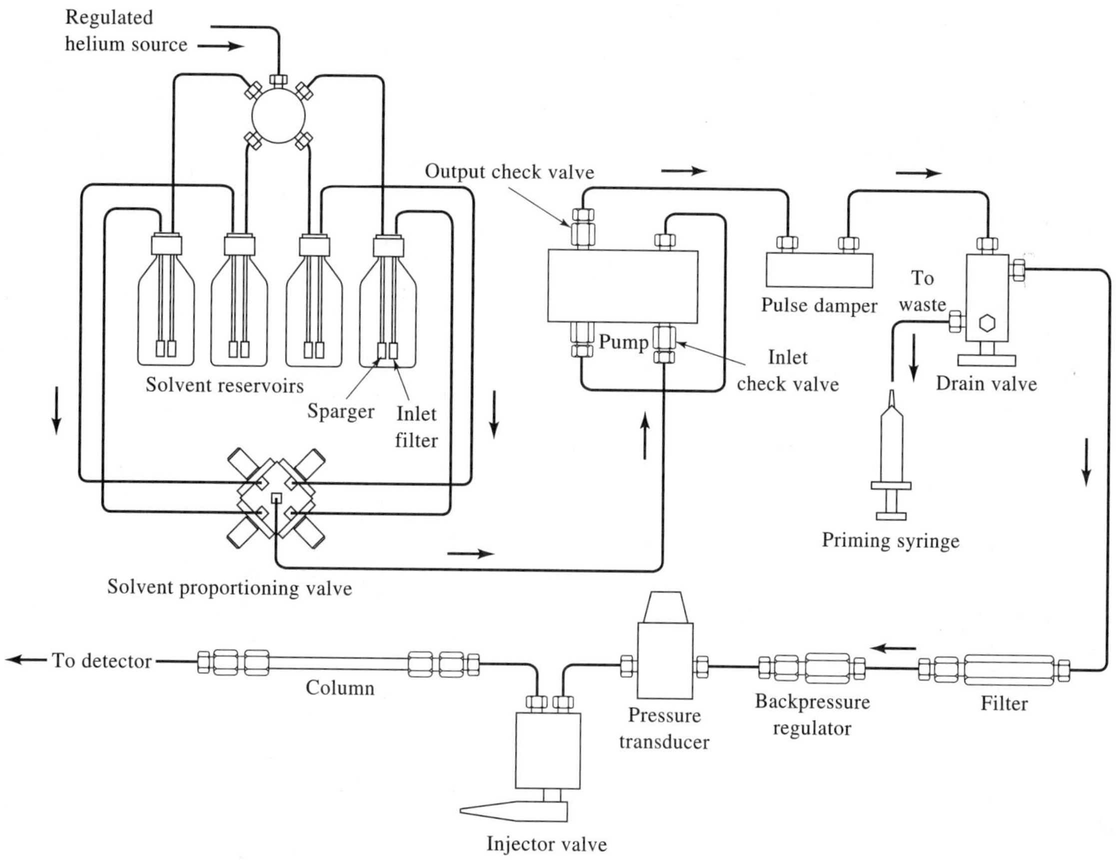

Figure 28-4 Schematic of an apparatus for HPLC. *(Courtesy of Perkin-Elmer Corporation, Norwalk, CT.)*

Figure 28-5 illustrates the advantage of a gradient eluent in the separation of a mixture of chlorobenzenes. Isocratic elution with a 50:50 (v/v) methanol/water solution yielded the curve in Figure 28-5b. The curve in Figure 28-5a is for gradient elution, which was initiated with a 40:60 mixture of the two solvents; the methanol concentration was then increased at the rate of 8%/min. Note that gradient elution shortened the time of separation significantly without sacrifice in resolution of the early peaks. Note also that gradient elution produces effects that are similar to those produced by temperature programming in gas chromatography (see Figure 27-5).

28C-2 Pumping Systems

The requirements for an HPLC pumping system are severe and include: (1) the generation of pressures of up to 6000 psi (lbs/in^2), (2) pulse-free output, (3) flow rates ranging from 0.1 to 10 mL/min, (4) flow control and flow reproducibility of 0.5% relative or better, (5) corrosion-resistant components (seals of stainless steel or Teflon). It should be noted that the high pressures generated by HPLC pumps do not constitute an explosion hazard because liquids are not very compressible. Thus, rupture of a component of the system results only in solvent leakage. Of course, such leakage may constitute a fire hazard.

Three types of pumps, each with its own set of advantages and disadvantages, are encountered, namely, reciprocating pumps, syringe or displacement-type pumps, and pneumatic or constant pressure pumps.

Reciprocating Pumps

Reciprocating pumps, which are currently used in about 90% of the commercially available HPLC systems, usually consist of a small chamber in which the solvent is pumped by the back and forth motion of a motor-driven piston (see Figure 28-6). Two ball check valves, which open and close alternately, control the flow of solvent into and out of a cylinder. The solvent is in direct contact with the piston. As an alternative, pressure may be transmitted to the solvent via a flexible diaphragm, which in turn is hydraulically pumped by a reciprocating piston. Reciprocating pumps have the disadvantage of producing a pulsed flow, which must be damped because its presence is manifested as baseline noise on the chromatogram. The advantages of reciprocating pumps

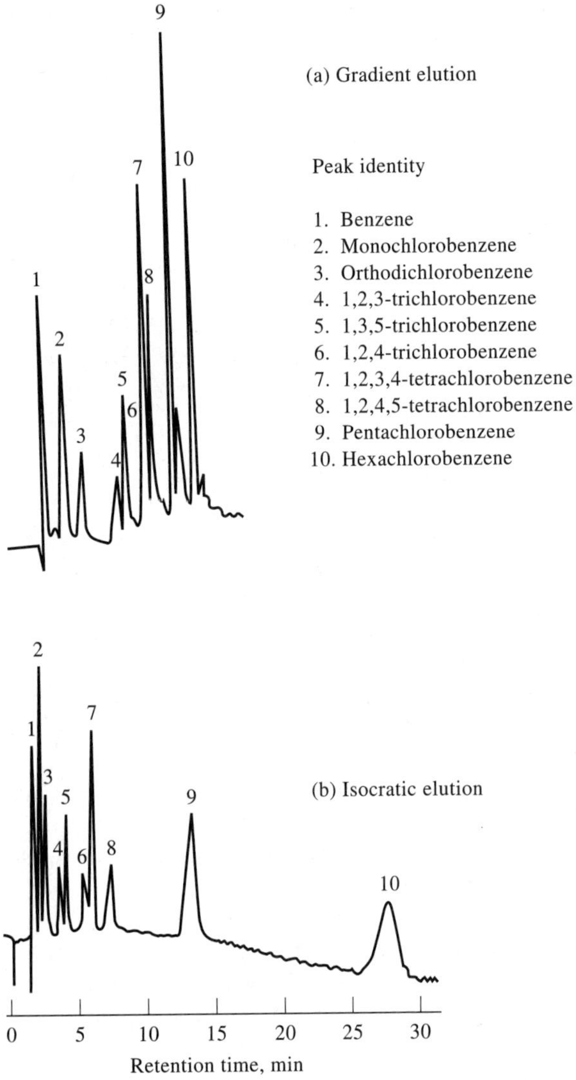

Figure 28-5 Improvement in separation efficiency by gradient elution. Column: 1 m × 2.1 mm i.d., precision-bore stainless; packing: 1% Permaphase® ODS. Sample: 5 μL of chlorinated benzenes in isopropanol. Detector: UV photometer (254 nm). Conditions: temperature, 60°C, pressure, 1200 psi. *(From J. J. Kirkland,* Modern Practice of Liquid Chromatography, *p. 88. New York: Interscience, 1971. Reprinted by permission of John Wiley & Sons, Inc.)*

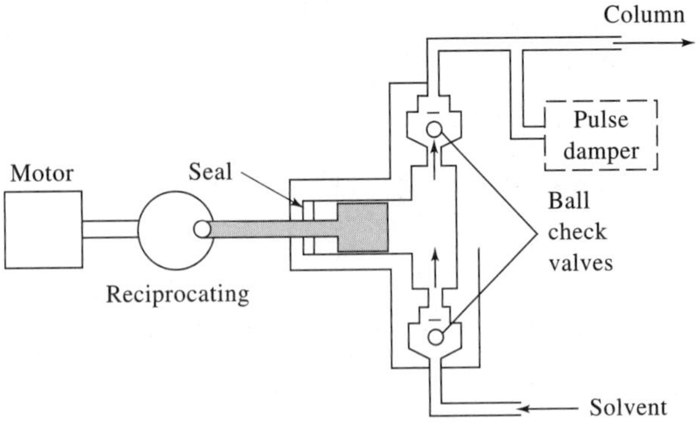

Figure 28-6 A reciprocating pump for HPLC.

include their small internal volume (35 to 400 μL), their high output pressures (up to 10,000 psi), their ready adaptability to gradient elution, and their constant flow rates, which are largely independent of column back-pressure and solvent viscosity.

Displacement Pumps

Displacement pumps usually consist of large, syringe-like chambers equipped with a plunger that is activated by a screw-driven mechanism powered by a stepping motor. Displacement pumps also produce a flow that tends to be independent of viscosity and back pressure. In addition the output is pulse free. Disadvantages include limited solvent capacity (≈250 mL) and considerable inconvenience when solvents must be changed.

Pneumatic Pumps

In the simplest pneumatic pumps, the mobile phase is contained in a collapsible container housed in a vessel that can be pressurized by a compressed gas. Pumps of this type are inexpensive and pulse free; they suffer from limited capacity and pressure output as well as a dependence of flow rate on solvent viscosity and column back-pressure. In addition, they are not amenable to gradient elution and are limited to pressures less than about 2000 psi.

Flow Control and Programming Systems

As part of their pumping systems, many commercial instruments are equipped with computer-controlled devices for measuring the flow rate by determining the pressure drop across a restrictor located at the pump outlet. Any difference in signal from a preset value is then used to increase or decrease the speed of the pump motor. Most instruments also have a means for varying the composition of the solvent either continuously or in a stepwise fashion. For example, the instrument shown in Figure 28-4 contains a proportioning valve that permits mixing of up to four solvents in a preprogrammed and continuously variable way.

28C-3 Sample Injection Systems

Often, the limiting factor in the precision of liquid chromatographic measurement lies in the reproducibility with which samples can be introduced onto the column packing. The problem is exacerbated by band broadening, which accompanies overloading columns. Thus, the volumes used must be minuscule—a few tenths of a microliter to perhaps 500 μL. Furthermore, it is convenient to be able to introduce the sample without depressurizing the system.

The earliest and simplest means of sample introduction was syringe injection through a self-sealing elastomeric septum. For this purpose, microsyringes designed to withstand pressures up to 1500 psi are used. In *stop-flow* injections, the flow of solvent is stopped momentarily, a fitting at the column head is removed, and the sample is injected directly onto the head of the column packing. After replacing the fitting, the system is again pressurized. The advantage of this technique is its simplicity. Unfortunately, the reproducibility of syringe injection is rarely better than 2% to 3% and is often considerably worse.

The most widely used method of sample introduction in liquid chromatography is based upon sampling loops, such as that shown in Figures 28-7 and 27-4. These devices are often an integral part of liquid-chromatographic equipment and have interchangeable loops providing a choice of sample sizes from 5 to 500 μL. Sampling loops of this type permit the introduction of samples at pressures up to 7000 psi with precisions of a few tenths percent relative. Micro sample injection valves, with sampling loops having volumes of 0.5 to 5 μL, are also available.

28C-4 Liquid-Chromatographic Columns

Liquid-chromatographic columns are ordinarily constructed from smooth-bore stainless steel tubing, although heavy-walled glass tubing is occasionally encountered. The latter is restricted to pressures that are lower than about 600 psi. Hundreds of packed columns differing in size and packing are available from several manufacturers. Their costs range from \$200 to \$500 generally.[4]

Analytical Columns

The majority of liquid-chromatographic columns range in length from 10 to 30 cm. Normally, the columns are straight, with added length, where needed, being gained

[4] For descriptions of recent commercially available HPLC columns, see R. E. Majors, *LC-GC*, 1995, *13*(8), 202; 1994, *12*(12), 890.

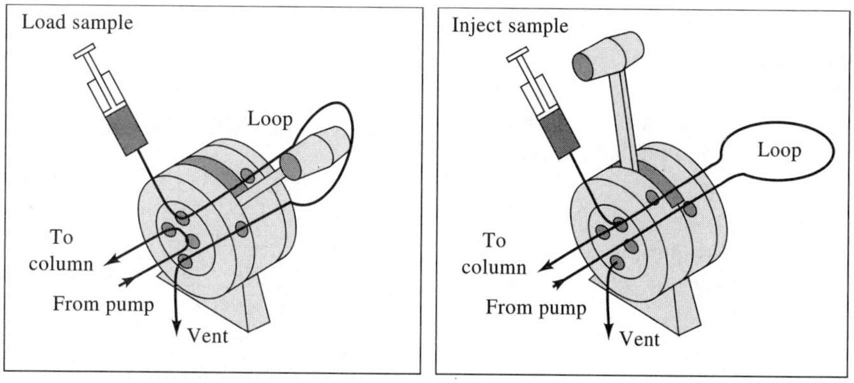

Figure 28-7 A sampling loop for liquid chromatography. *(Courtesy of Beckman Instruments, Fullerton, CA.)* With the valve handle as shown on the left, the loop is filled from the syringe, and the mobile phase flows from pump to column. When the valve is placed in the position on the right, the loop is inserted between the pump and the column so that the mobile phase sweeps the sample onto the column.

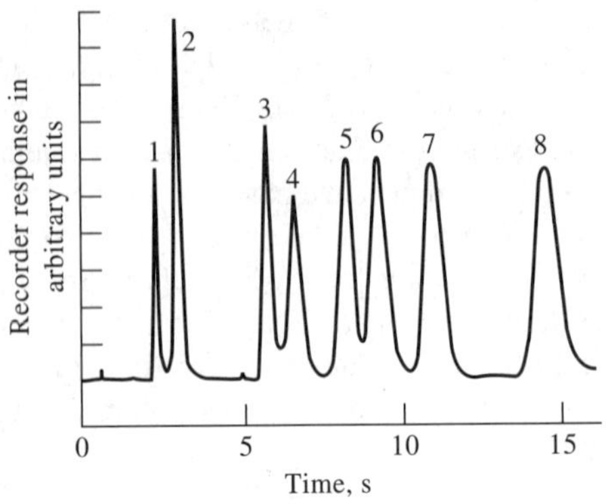

Figure 28-8 High-speed isocratic separation. Column dimensions: 4 cm length; 0.4 cm i.d. Packing: 3-μm sperisorb. Mobile phase: 4.1% ethyl acetate in *n*-hexane. Compounds: (1) *p*-xylene, (2) anisole, (3) benzyl acetate, (4) dioctyl phthalate, (5) dipentyl phthalate, (6) dibutyl phthalate, (7) dipropyl phthalate, (8) diethyl phthalate. *(From R. P. W. Scott,* Small Bore Liquid Chromatography Columns: Their Properties and Uses, *p. 156. New York: Wiley, 1984. Reprinted with permission of John Wiley & Sons, Inc.)*

by coupling two or more columns together. Occasionally coiled columns are encountered, although some loss in efficiency results from this configuration. The inside diameter of liquid columns is often 4 to 10 mm; the most common particle size of packings is 5 or 10 μm. Perhaps the most common column currently in use is one that is 25 cm in length, 4.6 mm in inside diameter, and packed with 5 μm particles. Columns of this type contain 40,000 to 60,000 plates/meter.

Recently, manufacturers have been producing high-speed, high-performance columns, which have smaller dimensions than those just described.[5] Such columns may have inside diameters that range from 1 to 4.6 mm and may be packed with 3- or 5-μm particles. Often, their lengths are as short as 3 to 7.5 cm. Such columns contain as many as 100,000 plates/meter and have the advantage of speed and minimal solvent consumption. The latter property is of considerable importance because the high-purity solvents required for liquid chromatography are expensive to purchase and to dispose of. Figure 28-8 illustrates the speed with which a separation can be performed on this type of column. Here, eight components of diverse type are separated in about 15 s. The column was 4 cm in length and had an inside diameter of 4 mm; it was packed with 3-μm particles.

Guard Columns

Usually, a short guard column is introduced before the analytical column to increase the life of the analytical column by removing not only particulate matter and contaminants from the solvents but also sample components that bind irreversibly to the stationary phase. In addition, in liquid-liquid chromatography, the guard column serves to saturate the mobile phase with the stationary phase so that losses of this solvent from the analytical column are minimized. The composition of the guard-column packing should be closely similar to that of the analytical column; the particle size is usually larger, however, to minimize pressure drop. When the guard column has become contaminated, it is repacked or discarded and replaced with a new one of the same type. Thus, the guard column is sacrificed to protect the more expensive analytical column.

Column Thermostats

For many applications, close control of column temperature is not necessary, and columns are operated at ambient temperature. Often, however, better chromatograms are obtained by maintaining column temperatures constant to a few tenths degree centigrade. Most modern commercial instruments are now equipped with column heaters that control column temperatures to a few tenths of a degree from near ambient to 100° to 150°C. Columns may also be fitted with water jackets fed from a constant temperature bath to give precise temperature control.

28C-5 Types of Column Packings

Two basic types of packings have been used in liquid chromatography, *pellicular* and *porous particle.* The former consist of spherical, nonporous, glass or polymer beads with typical diameters of 30 to 40 μm. A thin, porous layer of silica, alumina, a polystyrene-divinyl benzene synthetic resin, or an ion-exchange resin is deposited on the surface of these beads. For some applications, an additional coating is applied, which consists of a liquid stationary phase that is held in place by adsorption. Alternatively, the beads may be treated chemically to give an organic surface layer. Currently,

[5] See *Microcolumn High-Performance Liquid Chromatography,* P. Kucera, Ed. New York: Elsevier, 1984; *Small Bore Liquid Chromatography Columns: Their Properties and Uses,* R. P. W. Scott, Ed. New York: Wiley, 1984; M. Novotny, *Anal. Chem.,* **1988,** *60,* 500A.

pellicular packings are used largely for guard columns and not for analytical columns.

The typical porous particle packing for liquid chromatography consists of porous microparticles having diameters ranging from 3 to 10 μm; for a given size particle, every effort is made to minimize the range of particle sizes. The particles are composed of silica, alumina, the synthetic resin polystyrene-divinyl benzene, or an ion-exchange resin. Silica is by far the most common packing in liquid chromatography. Silica particles are prepared by agglomerating submicron silica particles under conditions that lead to larger particles having highly uniform diameters. The resulting particles are often coated with thin organic films, which are chemically or physically bonded to the surface.

28C-6 Detectors[6]

Unlike gas chromatography, liquid chromatography has no detectors that are as universally applicable and as reliable as the flame ionization and thermal conductivity detectors described in Section 27B-4. A major challenge in the development of liquid chromatography has been in detector improvement.

Characteristics of the Ideal Detector

The ideal detector for liquid chromatography should have all of the properties listed on page 706 for gas chromatography with the exception that the liquid chromatography detector need not be responsive over as great a temperature range. In addition, an HPLC detector should have minimal internal volume in order to reduce zone broadening (Section 28B-2).

Types of Detectors

Liquid chromatographic detectors are of two basic types. *Bulk property detectors* respond to a mobile-phase bulk property, such as refractive index, dielectric constant, or density, which is modulated by the presence of solutes. In contrast, *solute property detectors* respond to some property of solutes, such as UV absorbance, fluorescence, or diffusion current, that is not possessed by the mobile phase.

[6] For more detailed discussions of liquid chromatographic detectors, see R. P. W. Scott, *Liquid Chromatographic Detectors,* 2nd ed. Amsterdam: Elsevier, 1986; E. S. Yeung and R. E. Synovec, *Anal. Chem.,* **1986,** *58,* 1237A; C. A. Dorschel et al., *Anal. Chem.,* **1989,** *61,* 951A.

Table 28-1 lists the most common detectors for HPLC and some of their more important properties. A 1982 survey of 365 published papers in which liquid chromatography played an important role revealed that 71% were based upon detection by UV absorption, 15% by fluorescence, 5.4% by refractive index, 4.3% by electrochemical measurements, and another 4.3% by other measurements.[7] Of the UV absorption detectors, 39% were based upon one of the emission lines of mercury, 13% upon filtered radiation from a deuterium source, and 48% upon radiation emitted from a grating monochromator.

Absorbance Detectors

Figure 28-9 is a schematic of a typical, Z-shaped, flow-through cell for absorbance measurements on eluents from a chromatographic column. In order to minimize extra-column band broadening, the volume of such a cell is kept as small as possible. Thus, typically, volumes are limited to 1 to 10 μL and cell lengths to 2 to 10 mm. Most cells of this kind are restricted to pressures no greater than about 600 psi. Consequently, a pressure reduction device is often required.

Many absorbance detectors are double-beam devices in which one beam passes through the eluent cell and the other through a filter to reduce its intensity.

[7] See *Anal. Chem.,* **1982,** *54,* 327A.

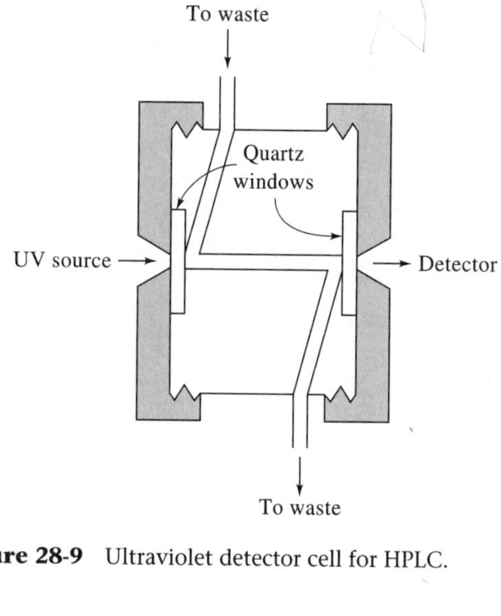

Figure 28-9 Ultraviolet detector cell for HPLC.

TABLE 28-1 Performances of LC Detectors

LC Detector	Commercially Available	Mass LOD (commercial detectors)[a]	Mass LOD (state of the art)[b]
Absorbance	Yes[c]	100 pg–1 ng	1 pg
Fluorescence	Yes[c]	1–10 pg	10 fg
Electrochemical	Yes[c]	10 pg–1 ng	100 fg
Refractive index	Yes	100 ng–1 μg	10 ng
Conductivity	Yes	500 pg–1 ng	500 pg
Mass spectrometry	Yes[d]	100 pg–1 ng	1 pg
FT–IR	Yes[d]	1 μg	100 ng
Light scattering[e]	Yes	10 μg	500 ng
Optical activity	No	—	1 ng
Element selective	No	—	10 ng
Photoionization	No	—	1 pg–1 ng

[a]Mass LOD is calculated for injected mass that yields a signal equal to five times the σ noise, using a mol wt of 200 g/mol, 10 μL injected for conventional or 1 μL injected for microbore LC.
[b]Same definition as *a*, above, but the injected volume is generally smaller.
[c]Commercially available for microbore LC also.
[d]Commercially available, yet costly.
[e]Including low-angle light scattering and nephelometry.
(From E. S. Yeung and R. E. Synovec, *Anal. Chem.*, **1986,** *58,* 1238. With permission.)

Matched photoelectric detectors are then used to compare the intensities of the two beams. Alternatively, a chopped beam system similar to that shown in Figure 9-13b is used in conjunction with a single phototube. In either case, the chromatogram consists of a plot of the log of the ratio of the two transduced signals as a function of time. Single-beam instruments are also encountered. Here, intensity measurements of the solvent system are stored in a computer memory and ultimately recalled for the calculation of absorbance.

Ultraviolet Absorbance Detectors with Filters. The simplest UV absorption detectors are filter photometers with a mercury lamp as the source. Most commonly the intense line at 254 nm is isolated by filters; with some instruments, lines at 250, 313, 334, and 365 can also be employed by substitution of filters. Obviously this type of detector is restricted to solutes that absorb at one of these wavelengths. As shown in Section 14B, several organic functional groups and a number of inorganic species exhibit broad absorption bands that encompass one or more of these wavelengths.

Deuterium or tungsten filament sources with interference filters also provide a simple means of detecting absorbing species as they are eluted from a column. Some modern instruments are equipped with filter wheels containing several filters that can be rapidly switched to detect various species as they are eluted. Such devices are particularly useful for repetitive, quantitative analyses where the qualitative composition of the sample is known so that a sequence of appropriate filters can be chosen. Often, the filter changes are computer controlled.

Ultraviolet Absorbance Detectors with Monochromators. Most HPLC manufacturers offer detectors that consist of a scanning spectrophotometer with grating optics. Some are limited to ultraviolet radiation; others encompass both ultraviolet and visible radiation. Several operational modes can be chosen. For example, the entire chromatogram can be obtained at a single wavelength; alternatively, when eluent peaks are sufficiently separated in time, different wavelengths can be chosen for each peak. Here again, computer control is often used to select the best wavelength for each eluent.

Where entire spectra are desired for identification purpose, the flow of eluent can be stopped for a sufficient period to permit scanning the wavelength region of interest.

The most powerful ultraviolet spectrophotometric detectors are diode-array instruments as described in Section 7E-3 and Figure 13-22.[8] Several manufacturers offer such instruments, which permit collection of data for an entire spectrum in approximately one second. Thus, spectral data for each chromatographic peak can be collected and stored as it appears at the end of the column. One form of presentation of the spectral data, which is helpful in identification of species and for choosing conditions for quantitative determination, is a three-dimensional plot such as that shown in Figure 28-10. Here, spectra were obtained at successive five-second intervals. The appearance and disappearance of each of the three steroids in the eluent is clearly evident.

Infrared Absorbance Detectors. Two types of infrared detectors are offered commercially. The first is similar in design to the instrument shown in Figure 16-14, with wavelength scanning being provided by three semicircular filter wedges. The range of this instrument is from 2.5 to 14.5 μm or 4000 to 690 cm^{-1}.

[8] See J. C. Miller, S. O. George, and B. G. Willis, *Science*, **1982**, *218*, 241; S. A. Borman, *Anal. Chem.*, **1983**, *55*, 836A; *Diode Array Detection in HPLC*, L. Huber and S. A. George, Eds. New York: Marcel Dekker, 1993.

The second, and much more sophisticated, type of infrared detector is based upon Fourier transform instruments similar to those discussed in Section 16C-1. Several of the manufacturers of Fourier transform instruments offer accessories that permit their use as HPLC detectors.

Infrared detector cells are similar in construction to those used with ultraviolet radiation except that windows are constructed of sodium chloride or calcium fluoride. Cell lengths range from 0.2 to 1.0 mm and volumes from 1.5 to 10 μL.

The simpler infrared instruments can be operated at one or more single wavelength settings; alternatively, the spectra for peaks can be scanned by stopping the flow at the time of elution. The Fourier transform instruments are used in an analogous way to the diode-array instrument for ultraviolet absorbance measurement described in the previous section.

A major limitation to the use of infrared detectors lies in the low transparency of many useful solvents. For example, the broad infrared absorption bands for water and the alcohols largely preclude the use of this detector for many applications.

Fluorescence Detectors

Fluorescence detectors for HPLC are similar in design to the fluorometers and spectrofluorometers described in Section 15B-2. In most, fluorescence is observed by a photoelectric detector located at 90 deg to the excitation

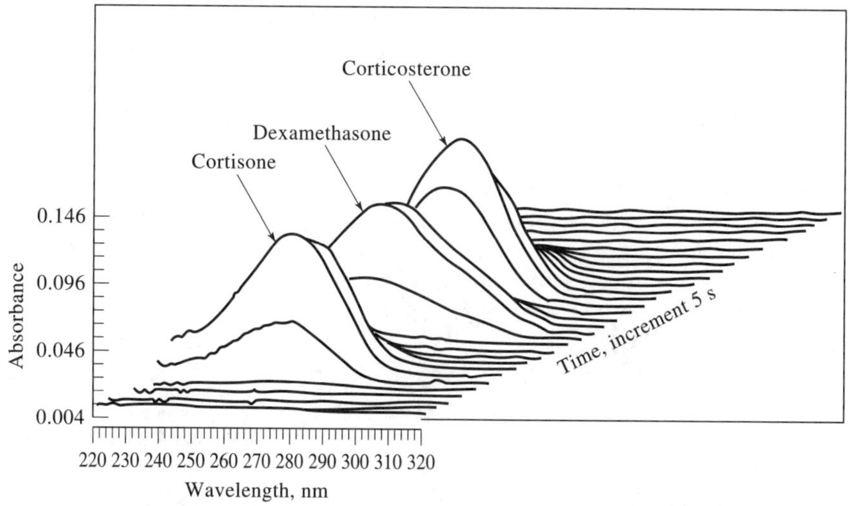

Figure 28-10 Absorption spectra of the eluent from a mixture of three steroids taken at 5-second intervals. *(Courtesy of Hewlett-Packard Company, Palo Alto, CA.)*

beam. The simplest detectors employ a mercury excitation source and one or more filters to isolate a band of emitted radiation. More sophisticated instruments are based upon a Xenon source and employ a grating monochromator to isolate the fluorescent radiation. Future developments in fluorescence detectors will probably be based upon tunable laser sources, which should lead to enhanced sensitivity and selectivity.[9]

As was pointed out in Chapter 15, an inherent advantage of fluorescence methods is their high sensitivity, which is typically greater by more than an order of magnitude than most absorbance procedures. This advantage has been exploited in liquid chromatography for the separation and determination of the components of samples that fluoresce. As was pointed out in Section 15C-2, fluorescent compounds are frequently encountered in the analysis of such materials as pharmaceuticals, natural products, clinical samples, and petroleum products. Often, the number of fluorescing species can be enlarged by preliminary treatment of samples with reagents that form fluorescent derivatives. For example, dansylchloride (5-dimethylaminonapthalene-1-sulphonyl chloride), which reacts with primary and secondary amines, amino acids, and phenols to give fluorescent compounds, has been widely used for the detection of amino acids in protein hydrolyzates.

Refractive-Index Detectors

Figure 28-11 is a schematic of a differential refractive-index detector in which the solvent passes through one half of the cell on its way to the column; the eluate then flows through the other chamber. The two compartments are separated by a glass plate mounted at an angle such that bending of the incident beam occurs if the two solutions differ in refractive index. The resulting displacement of the beam with respect to the photosensitive surface of a detector causes variation in the output signal, which, when amplified and recorded, provides the chromatogram.

Refractive-index detectors have the significant advantage of responding to nearly all solutes. That is, they are general detectors analogous to flame or thermal conductivity detectors in gas chromatography. In addition, they are reliable and unaffected by flow rate. They are, however, highly temperature sensitive and must be maintained at a constant temperature to a few thousandths of a degree centigrade. Furthermore, they are not as sensitive as most other types of detectors.

Evaporative Light Scattering Detector

Recently, a new type of general detector has become available commercially for HPLC, the *evaporative light scattering detector* (ELSD). In this detector, the column effluent is passed into a nebulizer where it is converted into a fine mist by a flow of nitrogen or air. The fine droplets are then carried through a controlled temperature drift tube where evaporation of the mobile phase occurs, leading to formation of fine particles of the analyte. The cloud of analyte particles then passes through a laser beam. The scattered radiation is detected at right angles to the flow by a silicon photodiode.

A major advantage of this type of detector is that its response is reportedly approximately the same for all nonvolatile solutes. In addition it is significantly more sensitive than the refractive index detector, with detection limits stated to be 5 ng/25 μL.

[9] See R. B. Green, *Anal. Chem.*, **1983**, *55*, 20A; E. S. Yeung and M. J. Sepaniak, *Anal. Chem.*, **1980**, *52*, 1465A.

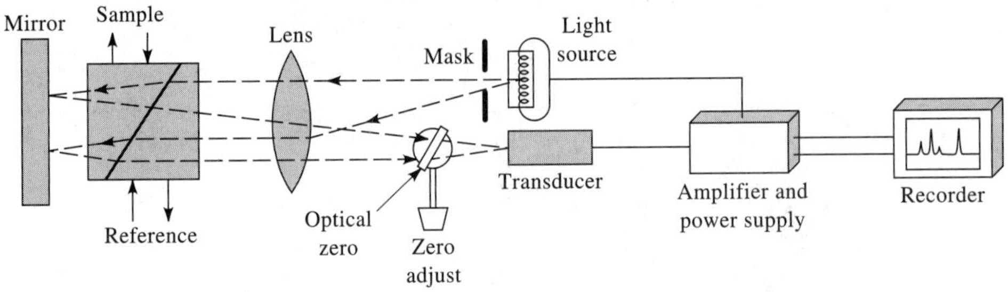

Figure 28-11 Schematic of a differential refractive-index detector. *(Courtesy of Waters Associates, Inc., Milford, MA 91757.)*

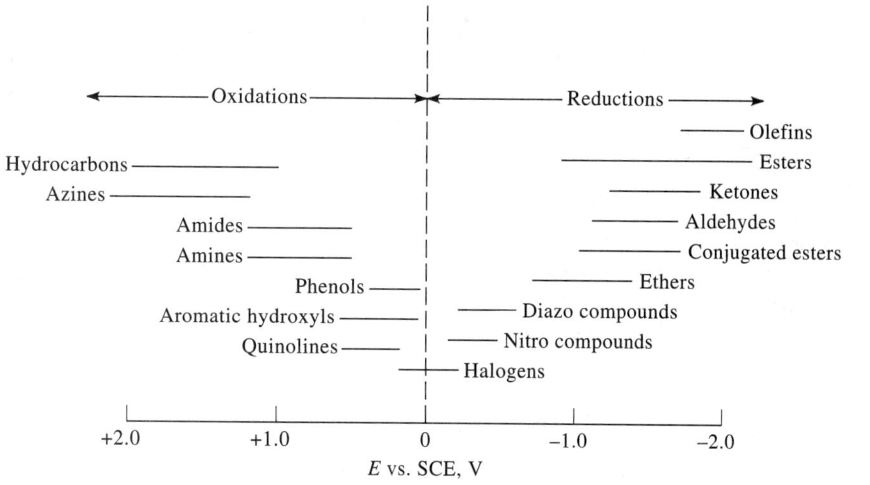

Figure 28-12 Potentially detectable organic functional groups by amperometric measurements. The horizontal lines show the range of oxidation or reduction potentials wherein compounds containing the indicated functional groups are electroactive.

Electrochemical Detectors[10]

Electrochemical detectors of several types are currently available from instrument manufacturers.[11] These devices are based upon amperometry, polarography, coulometry, and conductometry. The first three of these methods are discussed in Chapters 24 and 25.

Although electroanalytical procedures have as yet not been exploited to the extent of optical detectors, they appear to offer advantages, in many instances, of high sensitivity, simplicity, convenience, and widespread applicability. This last property is illustrated in Figure 28-12, which depicts the potential ranges at which oxidation or reduction of 16 organic functional groups occurs. In principle, then, species containing any of these groups could be detected by amperometric, polarographic, or coulometric procedures. Thus, electrochemical detection would appear to have the potential for fulfilling a longtime need of HPLC, namely, a sensitive general or universal detector.[12]

[10] For short reviews of electrochemical detectors, see P. T. Kissinger, *J. Chem. Educ.*, **1983**, *60*, 308; R. D. Rocklin, A. Henshall, and R. B. Rubin, *Amer. Lab.*, **1990** (3), 34; G. Horvai and E. Pungor, *Crit. Rev. Anal. Chem.*, **1989**, *21*(1), 1; G. Horvai and E. Pungor, *Chromatography*, **1987**, *2* (3), 15; D. C. Johnson, *Anal. Chem.*, **1990**, *62*, 589A.

[11] For a description of nine commercially available electrochemical detectors, see M. Warner, *Anal. Chem.*, **1994**, *66*, 601A.

[12] For a recent description of the use of voltammetric detection of a variety of organic functional groups with platinum or gold microelectrodes, see D. C. Johnson and W. R. LaCourse, *Anal. Chem.*, **1990**, *62*, 589A.

A variety of HPLC/electrochemical detector cells have been described in the literature, and several are available from commercial sources. Figure 28-13 is an example of a simple thin-layer type of flow-through cell for amperometric detection. Here, the electrode surface is part of a channel wall formed by sandwiching a 50-μm Teflon gasket between two machined blocks of Kel-F plastic. The indicator electrode is platinum, gold, glassy carbon, or a carbon paste. A reference electrode, and often a counter electrode, are located downstream

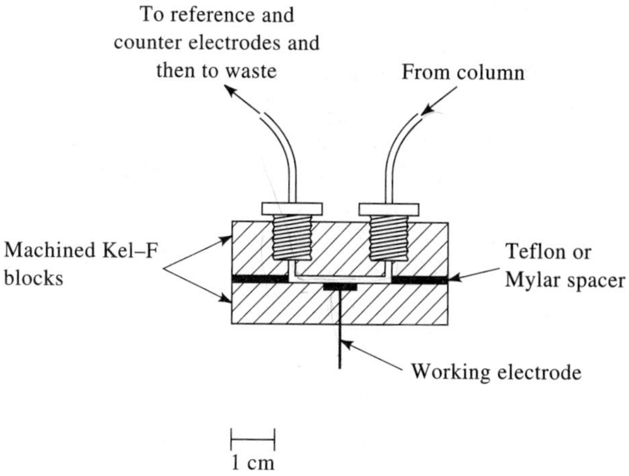

Figure 28-13 Amperometric thin-layer detector cell for HPLC.

from the indicator electrode block. The cell volume is 1 to 5 μL. A useful modification of this cell, which is available commercially, includes two working electrodes, which can be operated in series or in parallel.[13] The former configuration, in which the eluent flows first over one electrode and then over the second, requires that the analyte undergo a reversible oxidation (or reduction) at the upstream electrode. The second electrode then operates as a cathode (or an anode) to determine the oxidation (or reduction) product. This arrangement enhances the selectivity of the detection system. An interesting application of this system is for the detection and determination of the components in mixtures containing both thiols and disulfides. Here, the upstream mercury electrode reduces the disulfides at about −1.0 V. That is,

$$RSSR + 2H^+ + 2e^- \rightarrow 2RSH$$

A downstream mercury electrode is then oxidized in the presence of the thiols from the original sample as well as those formed at the upstream electrode. That is,

$$2RSH + Hg(l) \rightarrow Hg(SR)_2(s) + 2H^+ + 2e^-$$

In the parallel configuration, the two electrodes are rotated so that the axis between them is at 90 deg to the stream flow. The two can then be operated at different potentials (relative to a downstream reference electrode), which often gives an indication of peak purity. Alternatively, one electrode can be operated as a cathode and the other as an anode, thus making possible simultaneous detection of both oxidants and reductants.

Polarographic detectors have also been described in the literature. For example, an accessory is available for the mechanically controlled dropping electrode shown in Figure 25-3d; this device directs the eluent flow around the mercury droplet.[14] The potential of the electrode is then maintained at a suitable level during elution. Plots of current versus time provide elution patterns for species that are reduced at the chosen potential.

Conductometric and coulometric[15] detectors are also available from instrument manufacturers. The former are discussed in Section 28F-3.

[13] D. A. Roston, R. E. Shoup, and P. T. Kissinger, *Anal. Chem.*, **1982**, *54*, 1417A.

[14] For details, see S. K. Vohra, *Amer. Lab.*, **1981**, *13*(5), 66.

[15] For a description of series and parallel coulometric detectors, see R. W. Andrews et al., *Amer. Lab.*, **1982**, *14*(10), 140.

Mass Spectrometric Detectors

A fundamental problem in coupling liquid chromatography with mass spectrometry is the enormous mismatch between the relatively large solvent volumes from the former and the vacuum requirements of the latter. Several interfaces have been developed for solving this problem.[16] In one, which is available commercially, the eluent from the column is split, with only a tiny fraction being introduced directly into the mass spectrometer. Direct liquid introduction systems appear to hold considerable promise when used in conjunction with the new microbore columns, which typically have flow rates of 10 to 50 μL/min. In a second type of interface, which is also sold commercially, the effluent is deposited on a continuous, moving-belt or moving-wire that transports the solvent and analyte to a heated chamber for removal of the former by volatilization. Following solvent evaporation, the analyte residues on the belt or wire pass into the ion source area, where desorption-ionization occurs.

A new and promising interface, which is now available commercially, is called a *thermospray*.[17] A thermospray interface permits direct introduction of the total effluent from a column at flow rates as high as 2 mL/min. With this interface, the liquid is vaporized as it passes through a stainless steel, heated capillary tube to form an aerosol jet of solvent and analyte molecules. In the spray, the analyte is ionized through a charge exchange mechanism with a salt, such as ammonium acetate, which is incorporated in the eluent. Thus, the thermospray is not only an interface but also an ionization source. The resulting spectra are generally simple and provide molecular weight data but lack the detail that make electron impact spectra so useful for identification purposes. Furthermore, the thermospray interface is applicable only to polar analyte molecules and polar mobile phases that will dissolve a salt such as ammonium acetate. With these limitations, the thermospray interface provides spectra for a wide range of nonvolatile and thermally stable compounds such as peptides and nucleotides. Detection limits down to 1 to 10 picograms have been reported.

[16] See A. L. Yergey, C. G. Edmonds, I. A. S. Lewis, and M. L. Vestal, *Liquid Chromatography/Mass Spectrometry Techniques and Applications.* New York: Plenum Press, 1990; *Liquid Chromatography/Mass Spectrometry Applications in Agricultural, Pharmaceutical, and Environmental Chemistry.* M. A. Brown, Ed. Washington, DC: American Chemical Society, 1990; T. R. Covey, E. D. Lee, A. P. Bruins, and J. D. Henion, *Anal. Chem.*, **1986**, *58*, 1451A; M. L. Vestal, *Science*, **1984**, *226*, 275. See also *Spectra*, **1983**, *9*(1). The entire issue of this Finnigan MAT publication is devoted to LC/MS interfaces.

[17] M. L. Vestal, *Anal. Chem.*, **1983**, *55*, 750, 1741.

Recently a new interface has been introduced commercially that makes it possible to obtain either electron impact or chemical ionization spectra. In this device, thermal nebulization and desolvation occur simultaneously to produce a mixture of particulate solute molecules and gaseous mobile-phase molecules. This aerosol is accelerated through a nozzle into a vacuum region where the mobile phase molecules are pumped away. Here, separation is accomplished by a momentum separator similar to that shown in Figure 27-14. The particulate analyte molecules are then ionized in an electron beam or chemically.[18]

Computer control and data storage is generally used with mass spectrometric detectors. Both real-time and computer-reconstructed chromatograms and spectra of the eluted peaks can be obtained. At this date instruments for HPLC/MS are not as fully developed as instruments for GC/MS; it seems likely that this situation will change in the next few years.[19]

March 28 05 Monday 9:30 pm

28D PARTITION CHROMATOGRAPHY

Partition chromatography has become the most widely used of all of the four types of liquid chromatographic procedures. In the past, most of the applications have been to nonionic, polar compounds of low to moderate molecular weight (usually <3000). Recently, however, methods have been developed (derivatization and ion pairing) that have extended partition separations to ionic compounds.

Partition chromatography can be subdivided into *liquid-liquid* and *bonded-phase* chromatography. The difference in these techniques lies in the method by which the stationary phase is held on the support particles of the packing. With liquid-liquid, a liquid stationary phase is retained on the surface of the packing by physical adsorption. With bonded-phase, the stationary phase is bonded chemically to the support surfaces. Early partition chromatography was exclusively of the liquid-liquid type; by now, however, the bonded-phase method has become predominate because of certain disadvantages of liquid-liquid systems. One of these disadvantages is the loss of stationary phase by dissolution in the mobile phase, which requires periodic recoating of the support particles. Furthermore, stationary-phase solubility problems prohibit the use of liquid-phase packings for gradient elution. Our discussions will focus exclusively on bonded-phase partition chromatography.[20]

28D-1 Columns for Bonded-Phase Chromatography

The supports for the majority of bonded-phase packings for partition chromatography are prepared from rigid silica, or silica-based, compositions. These solids are formed as uniform, porous, mechanically sturdy particles commonly having diameters of 3, 5, or 10 μm. The surface of fully hydrolyzed silica (hydrolyzed by heating with 0.1 M HCl for a day or two) is made up of chemically reactive silanol groups. That is,

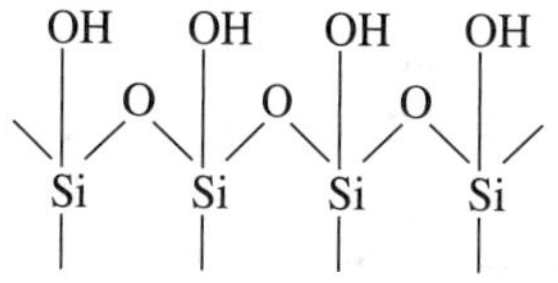

Typical silica surfaces contain about 8 μmol/m^2 of OH groups.

The most useful bonded-phase coatings are siloxanes formed by reaction of the hydrolyzed surface with an organochlorosilane. For example,

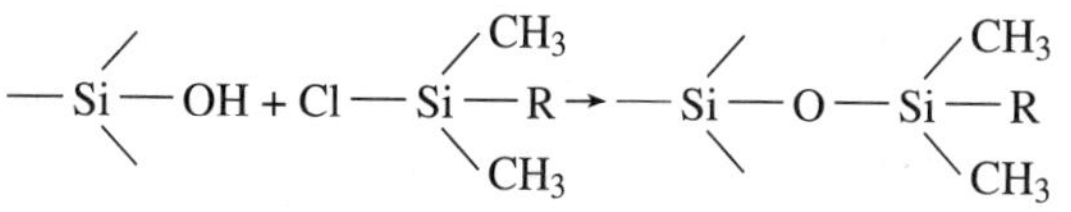

where R is an alkyl group or a substituted alkyl group.

Surface coverage by silanization is limited to 4 μmol/m^2 or less because of steric effects. The unreacted SiOH groups, unfortunately, impart an undesirable polarity to the surface, which may lead to tailing of chromatographic peaks, particularly for basic solutes. To lessen this effect, siloxane packings are frequently *capped* by further reaction with chlorotrimethylsilane that, because of its smaller size, can bond many of the unreacted silanol groups.

Reversed-Phase and Normal-Phase Packings

Two types of partition chromatography are distinguishable based upon the relative polarities of the mobile and stationary phases. Early work in liquid chromatography was based upon highly polar stationary phases such as water or triethyleneglycol supported on silica or alu-

[18] See R. C. Willoughby and R. F. Browner, *Anal. Chem.*, **1984**, *56*, 2626.

[19] See S. Borman, *Anal. Chem.*, **1987**, *59*, 769A.

[20] For a report on retention mechanism of bonded-phase chromatography, see J. G. Dorsey and W. T. Cooper, *Anal. Chem.*, **1994**, *66*, 857A.

mina particles; a relatively nonpolar solvent such as hexane or *i*-propylether then served as the mobile phase. For historic reasons, this type of chromatography is now referred to as *normal-phase chromatography.* In *reversed-phase chromatography,* the stationary phase is nonpolar, often a hydrocarbon, and the mobile phase is relatively polar (such as water, methanol, or acetonitrile).[21] In normal-phase chromatography, the *least* polar component is eluted first because in a relative sense, it is the most soluble in the mobile phase; *increasing* the polarity of the mobile phase has the effect of *decreasing* the elution time. In contrast, in the reversed-phase method, the *most* polar component appears first, and *increasing* the mobile phase polarity *increases* the elution time. These relationships are illustrated in Figure 28-14.

Bonded-phase packings are classified as reversed-phase when the bonded coating is nonpolar in character and as normal-phase when the coating contains polar functional groups. Perhaps three quarters of all high-performance liquid chromatography is currently being carried out in columns with reversed-phase packings. Most commonly, the R group of the siloxane in these coatings is a C_8 chain (*n*-octyl) or a C_{18} chain (*n*-octyldecyl). With such preparations, the long-chain hydrocarbon groups are aligned parallel to one another and perpendicular to the particle surface, giving a brush- or bristle-like structure. The mechanism by which these surfaces retain solute molecules is at present not entirely clear. Some scientists believe that from the standpoint of the solute molecules, the brush behaves as a liquid hydrocarbon medium similar in nature to an ordinary liquid-liquid stationary phase. Others prefer to view the brush coating as a modified surface at which physical adsorption occurs. The molecules of the mobile phase then compete with the analyte molecules for position on the organic surface. Regardless of the detailed mechanism of retention, a bonded coating can be treated as if it were a conventional, physically retained liquid.

Figure 28-15 illustrates the effect of chain length of the alkyl group upon performance. As expected, longer chains produce packings that are more retentive. In addition, longer chain lengths permit the use of larger samples. For example, the maximum sample size for a C_{18} packing is roughly double that for a C_4 preparation under similar conditions.

In most applications of reversed-phase chromatography, elution is carried out with a highly polar mobile phase such as an aqueous solution containing various concentrations of such solvents as methanol, acetonitrile, or tetrahydrofuran. In this mode, care must be

[21] For a monograph on reversed-phase chromatography, see A. M. Krstulovic and P. R. Brown, *Reversed-Phase High Performance Liquid Chromatography.* New York: Wiley, 1982.

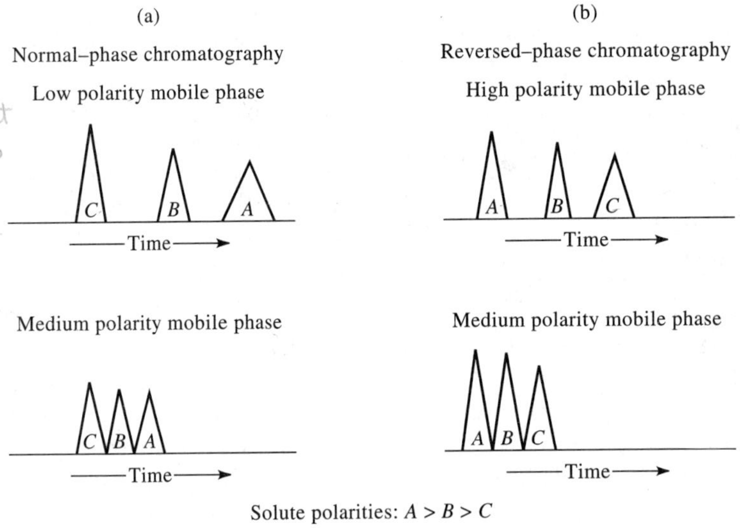

Figure 28-14 The relationship between polarity and elution times for normal-phase and reversed-phase chromatography.

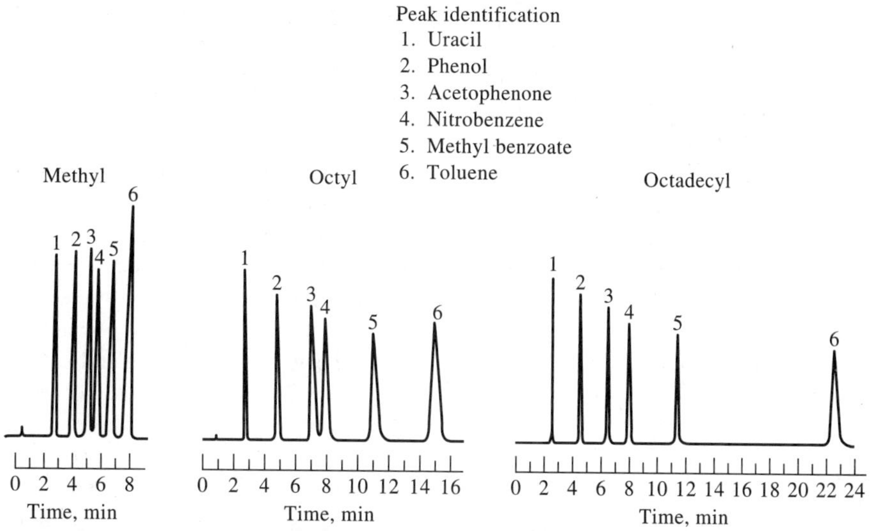

Figure 28-15 Effect of chain length on performance of reversed-phase siloxane columns packed with 5-μm particles. Mobile phase: 50/50 methanol/water. Flow rate: 1.0 mL/min.

taken to avoid pH values greater than about 7.5 because hydrolysis of the siloxane takes place, which leads to degradation or destruction of the packing.

In commercial normal-phase bonded packings, the R in the siloxane structure is a polar functional group such as cyano ($—C_2H_4CN$), diol ($—C_3H_6OCH_2CHOHCH_2OH$), amino ($—C_3H_6NH_2$), and dimethylamino ($C_3H_6N(CH_3)2$). The polarities of these packing materials vary over a considerable range with the cyano type being the least polar and the amino types the most. Diol packings are intermediate in polarity. With normal-phase packings, elution is carried with relatively nonpolar solvents, such as ethyl ether, chloroform, and *n*-hexane.

28D-2 Method Development in Partition Chromatography

Method development tends to be more complex in liquid chromatography than in gas chromatography because in a liquid mobile phase the sample components interact with *both* the stationary phase and the mobile phase. In contrast, in gas chromatography, the mobile phase behaves as an ideal gas and makes no contribution to the separation process; it serves simply to carry the sample components through the stationary phase. That is, in gas chromatography, separations are not significantly affected by whether the mobile phase is helium, nitrogen, or hydrogen. In marked contrast, the success of a partition chromatographic separation is often critically dependent upon whether the mobile phase is, say, acetonitrile, hexane, or dioxane.

Column Selection in Partition Chromatographic Separations

Successful chromatography with interactive mobile phases requires a proper balance of intermolecular forces among the *three* active participants in the separation process—the solute, the mobile phase, and the stationary phase. These intermolecular forces are described qualitatively in terms of the relative polarity of each of the three reactants. The polarities of various analyte functional groups in increasing order are: hydrocarbons < ethers < esters < ketones < aldehydes < amides < amines < alcohols. Water is more polar than compounds containing any of the preceding functional groups.

Often in choosing a column for a partition chromatographic separation, the polarity of the stationary phase is matched roughly with that of the analytes; a mobile phase of considerably different polarity is then used for elution. This procedure is generally more successful than one in which the polarities of the solute and mobile phase are matched but different from that of the stationary phase. Here, the stationary phase often cannot compete successfully for the sample components; retention times then become too short for practical application. At the other extreme, of course, is the situation where the polarity of the solute and stationary phases are too much alike and totally different from that of the mobile phase. Here, retention times become inordinately long.

In summary, then, polarities for solute, mobile phase, and stationary phase must be carefully blended if good partition chromatographic separations are to be realized in a reasonable time. Unfortunately, theories of mobile phase and stationary phase interactions with any given set of sample components are imperfect, and at best, a scientist can only narrow the choice of stationary phase to a general type. Having made this choice, the scientist must then perform a series of trial-and-error experiments in which chromatograms are obtained with various mobile phases until a satisfactory separation is realized. If resolution of all of the components of a mixture proves to be impossible, a different type of column may have to be chosen.

Mobile-Phase Selection in Partition Chromatography[22]

In Section 26D-4, three methods were described for improving the resolution of a chromatographic column; each is based upon varying one of the three parameters (N, k', and α) contained in Equation 26-21. In liquid chromatography, the retention factor k' is experimentally the most easily manipulated of the three because of the strong dependence of this constant upon the composition of the mobile phase. As noted earlier, for optimal performance, k' should be in the ideal range between 2 and 5; for complex mixtures, however, this range must often be expanded to perhaps 0.5 to 20 in order to provide time for peaks for all of the components to appear.

Sometimes, adjustment of k' alone does not suffice to produce individual peaks with no overlap; variation in selectivity factors α must then be resorted to. Here again, the simplest way of bringing about changes in α is by altering the mobile-phase composition, taking care, however, to keep k' within a reasonable range. Alternatively, α can be changed by choosing a different column packing.

[22] For an excellent monograph dealing with systematic methods for selecting stationary and mobile-phase combinations to accomplish various kinds of separations, see L. R. Snyder, J. L. Glajch, and J. J. Kirkland, *Practical HPLC Method Development.* New York: Wiley, 1988.

Effect of Solvent Strength on Retention Factors. Solvents that interact strongly with solutes are often termed "strong" solvents or polar solvents. Several indexes have been developed for quantitatively describing the polarity of solvents. The most useful of these for partition chromatography appears to be the *polarity index* P', which was developed by Snyder.[23] This parameter is based upon solubility measurements for the substance in question in three solvents: dioxane (a low dipole proton acceptor), nitromethane (a high dipole proton acceptor), and ethyl alcohol (a high dipole proton donor). The polarity index is a numerical measure of relative polarity of various solvents. Table 28-2 lists polarity indexes (and other properties) for a number of solvents that are employed in partition chromatography. Note that the polarity index varies from 10.2 for the highly polar water to −2 for the highly nonpolar fluoroalkanes. Any desired polarity index between these limits can be achieved by mixing two appropriate solvents. Thus, the polarity index P'_{AB} of a mixture of solvents A and B is given by

$$P'_{AB} = \phi_A P'_A + \phi_B P'_B \tag{28-2}$$

where P'_A and P'_B are polarity indexes of the two solvents and ϕ_A and ϕ_B are the volume fractions of each.

In Section 26D-4, it was pointed out that the easiest ways of improving the chromatographic resolution of two species is by manipulation of the retention factor k', which can in turn be varied by changing the polarity index of the solvent. Here, adjustment of P' is easily accomplished by the use of mobile phases that consist of a mixture of two solvents. Typically, a 2-unit change in P' results (very roughly) in a ten-fold change in k'. That is, for a normal-phase separation

$$\frac{k'_2}{k'_1} = 10^{(P'_1 - P'_2)/2} \tag{28-3}$$

where k'_1 and k'_2 are initial and final values of k' for a solute and P'_1 and P'_2 are the corresponding values for P'. For a reversed-phase column,

$$\frac{k'_2}{k'_1} = 10^{(P'_2 - P'_1)/2} \tag{28-4}$$

It should be emphasized that these equations apply only approximately. Nonetheless, as shown in Example 28-1, they can be useful.

[23] L. R. Snyder, *J. Chromatogr. Sci.*, **1978**, *16*, 223.

TABLE 28-2 Properties of Common Chromatographic Mobile Phases

Solvent	Refractive Index[a]	Viscosity, cP[b]	Boiling Point, °C	Polarity Index, P'	Eluent Strength,[c] ϵ^0
Fluoroalkanes[d]	1.27–1.29	0.4–2.6	50–174	<-2	−0.25
Cyclohexane	1.423	0.90	81	0.04	−0.2
n-Hexane	1.372	0.30	69	0.1	0.01
1-Chlorobutane	1.400	0.42	78	1.0	0.26
Carbon tetrachloride	1.457	0.90	77	1.6	0.18
i-Propyl ether	1.365	0.38	68	2.4	0.28
Toluene	1.494	0.55	110	2.4	0.29
Diethyl ether	1.350	0.24	35	2.8	0.38
Tetrahydrofuran	1.405	0.46	66	4.0	0.57
Chloroform	1.443	0.53	61	4.1	0.40
Ethanol	1.359	1.08	78	4.3	0.88
Ethyl acetate	1.370	0.43	77	4.4	0.58
Dioxane	1.420	1.2	101	4.8	0.56
Methanol	1.326	0.54	65	5.1	0.95
Acetonitrile	1.341	0.34	82	5.8	0.65
Nitromethane	1.380	0.61	101	6.0	0.64
Ethylene glycol	1.431	16.5	182	6.9	1.11
Water	1.333	0.89	100	10.2	Large

[a]At 25°C.
[b]The centipoise is a common unit of viscosity; in SI units, 1 cP = 1 mN · s · m^{-2}.
[c]On Al_2O_3. Multiplication by 0.8 gives ϵ^0 on SiO_2.
[d]Properties depend upon molecular weight. Range of data given.

EXAMPLE 28-1

In a reversed-phase column, a solute was found to have a retention time of 31.3 min, while an unretained species required 0.48 min for elution when the mobile phase was 30% (by volume) methanol and 70% water. Calculate **(a)** k' and **(b)** a water/methanol composition that should bring k' to a value of about 5.

(a) Application of Equation 26-8 yields

$$k' = (31.3 - 0.48)/0.48 = 64$$

(b) To obtain P' for the mobile phase we substitute polarity indexes for methanol and water from Table 28-2 into Equation 28-2 to give

$$P' = 0.30 \times 5.1 + 0.70 \times 10.2 = 8.7$$

Substitution of this result into Equation 28-4 gives

$$\frac{5}{64} = 10^{(P_2' - 8.7)/2}$$

Taking the log of both sides of this equation gives

$$-1.11 = \frac{P_2' - 8.7}{2} = 0.5\,P_2' - 4.35$$

$$P_2' = 6.5$$

Letting x be the volume fraction methanol in the new solvent mixture and substituting again into Equation 28-2, we find

$$6.5 = x \times 5.1 + (1 - x)10.2$$

$$x = 0.73 \text{ or } 73\%$$

Thus, a 73% methanol/27% water mixture should provide the desired value of k'.

Often, in reversed-phase separations, a solvent mixture of water and a polar organic solvent are employed. The retention factor is then readily manipulated by varying the water concentration as shown by Example 28-1. The effect of such manipulations is shown by the chromatograms in Figures 28-16a and 28-16b where the sample was a mixture of six steroids. With a 41% acetonitrile/59% water mixture, k' had a value of 5, and all of the analytes were eluted in such a short time (≈2 min) that the separation was quite incomplete. By increasing the percent water to 70, elution took place over 7 min, which doubled the value of k'. Now the total elution time was sufficient to achieve a separation, but the alpha value for compounds 1 and 3 was not great enough to resolve them.

Effect of Mobile Phase on Selectivities. In many cases, adjusting k' to a suitable level is all that is needed to give a satisfactory separation. When two bands still overlap, however, as in Figure 28-16, the selectivity factor α for the two species must be made larger. Such a change can be brought about most conveniently by changing the chemical nature of the mobile phase while holding the predetermined value of k' *more or less the same.*

For reversed-phase chromatography, a four-solvent optimization procedure has been developed for finding a solvent system that will in theory resolve a given mixture in a minimum of time.[24] Here, three compatible solvents are used for adjustment of α values. These include methanol, acetonitrile, and tetrahydrofuran. Water is then used to adjust the strength of the mixture in such a way as to yield a suitable value for k'.

Figure 28-16 illustrates the systematic, four-solvent approach to the development of a separation of six steroids by reversed-phase chromatography. The first two chromatograms show the results from initial experiments to determine the minimum value of k' required. With k' equal to 10, room exists on the time scale for discrete peaks; here, however, the α value for components 1 and 3 (and to a lesser extent 5 and 6) is not great enough for satisfactory resolution. Further experiments were then performed with the goal of finding better α values; in each case, the water concentration was adjusted to a level that yielded $k' = 10$. The results of experiments with methanol/water and tetrahydrofuran/water mixtures are shown in Figures 28-16c and 28-16d. Several additional experiments involving the systematic variation of pairs of the organic solvents were performed (in each case adjusting k' to 10 with water). Finally, the mixture shown in Figure 28-16e was chosen as the best mobile phase for the separation of the particular group of compounds.

For normal-phase operations a similar four-solvent system is used in which the selectivity solvents are ethyl ether, methylene chloride, and chloroform; the solvent strength adjustment is then made with *n*-hexane. With these four-solvent systems, optimization is said to be possible with a minimal number of experiments.

28D-3 Applications of Partition Chromatography

Reversed-phase bonded packings, when used in conjunction with highly polar solvents (often aqueous), approach the ideal, universal system for liquid chromatography. Because of their wide range of applicability, their convenience, and the ease with which k' and α can be altered by manipulation of aqueous mobile phases, these packings are frequently applied before all others for exploratory separations with new types of samples.

Table 28-3 lists a few typical examples of the multitude of uses of partition chromatography in various fields. A more complete picture of the ubiquity of this technique can be obtained by consulting recent reviews in *Analytical Chemistry.*[25]

Figure 28-17 illustrates two of many thousands of applications of bonded-phase partition chromatography to the analysis of consumer and industrial materials.

---

[24] See R. Lehrer, *Amer. Lab.*, **1981**, *13*(10), 113; J. L. Glajch et al., *J. Chromatogr. Sci.*, **1980**, *199*, 57.

[25] J. G. Dorsey et al., *Anal. Chem.*, **1996**, *68*, 515R; **1994**, *66*, 500R; **1992**, *66*, 353A; **1990**, *62*, 324R.

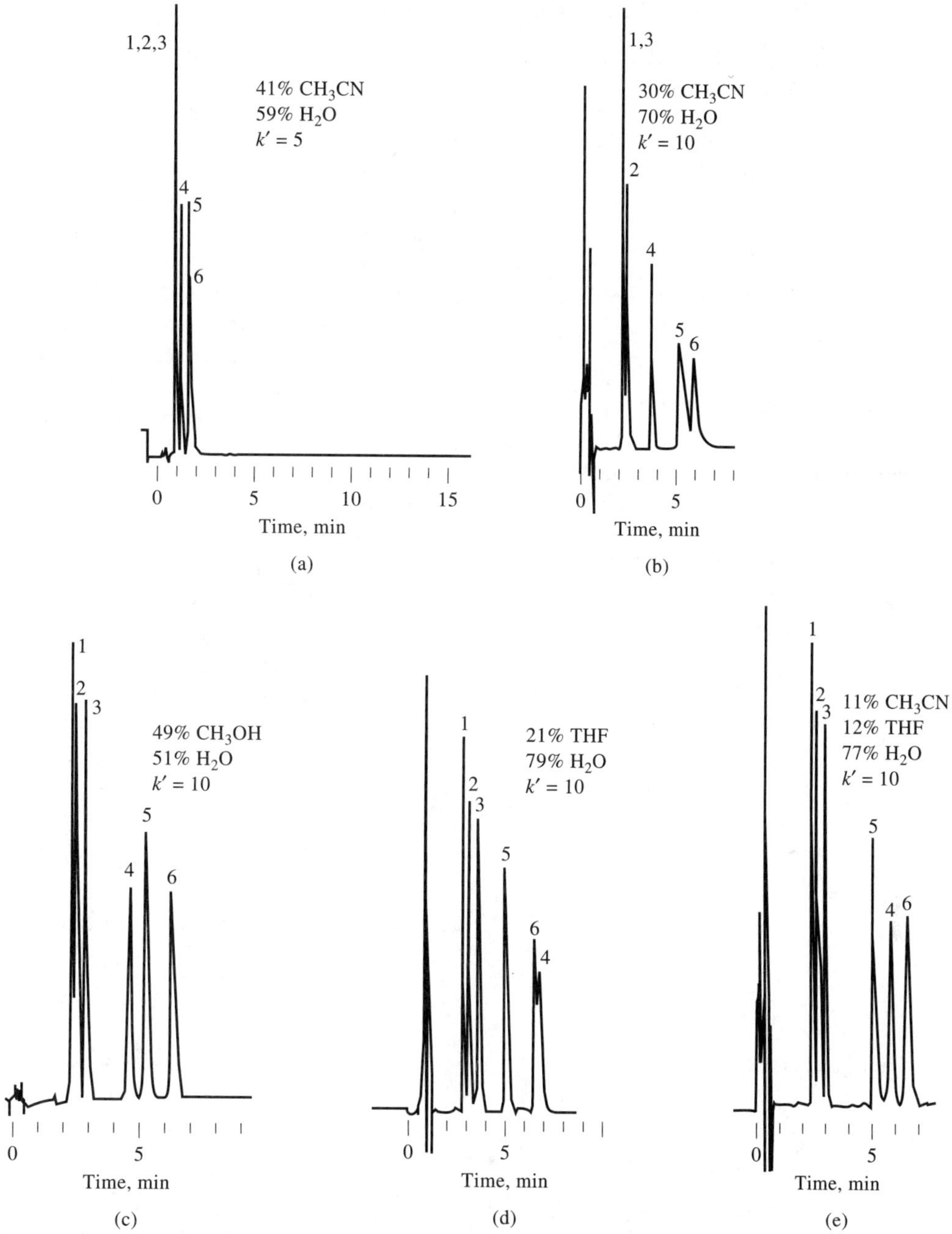

Figure 28-16 Systematic approach to the separation of six steroids. The use of water to adjust k' is shown in (a) and (b). The effects of varying α at constant k' are shown in (b), (c), (d), and (e). Column: 0.4 × 150 mm packed with 5 μm C_8 bonded, reversed-phase particles. Temperature: 50°C. Flow rate: 3.0 cm^3/min. Detector: UV 254 nm. THF = tetrahydrofuran. CH_3CN = acetonitrile. Compounds: (1) prednisone, (2) cortisone, (3) hydrocortisone, (4) dexamethasone, (5) corticosterone; (6) cortoexolone. *(Courtesy of DuPont Instrument Systems, Wilmington, DE.)*

TABLE 28-3 Typical Applications of Partition Chromatography

Field	Typical Mixtures
Pharmaceuticals	Antibiotics, Sedatives, Steroids, Analgesics
Biochemical	Amino acids, Proteins, Carbohydrates, Lipids
Food products	Artificial sweeteners, Antioxidants, Aflatoxins, Additives
Industrial chemicals	Condensed aromatics, Surfactants, Propellants, Dyes
Pollutants	Pesticides, Herbicides, Phenols, PCBs
Forensic chemistry	Drugs, Poisons, Blood alcohol, Narcotics
Clinical medicine	Bile acids, Drug metabolites, Urine extracts, Estrogens

Derivative Formation

In some instances, it is useful to convert the components of a sample to a derivative before, or sometimes after, chromatographic separation is undertaken. Such treatment may be desirable (1) to reduce the polarity of the species so that partition rather than adsorption or ion-exchange columns can be used, (2) to increase the detector response and thus sensitivity, for all of the sample components, and (3) to selectively enhance the detector response to certain components of the sample.

Figure 28-18 illustrates the use of derivatives to reduce polarity and enhance sensitivity. The sample was made up of 30 amino acids of physiological importance. Heretofore such a separation would be performed on an ion-exchange column with photometric detection based upon post-column reaction of the amino acids with a colorimetric reagent such as ninhydrin. The chromatogram shown in Figure 28-18 was obtained by automatic, *pre*-column derivative formation with orthophthalaldehyde. The substituted isoindoles[26] formed by

[26] For the structure of these compounds, see S. S. Simons Jr. and D. F. Johnson, *J. Amer. Chem. Soc.*, **1976**, *98*, 7098.

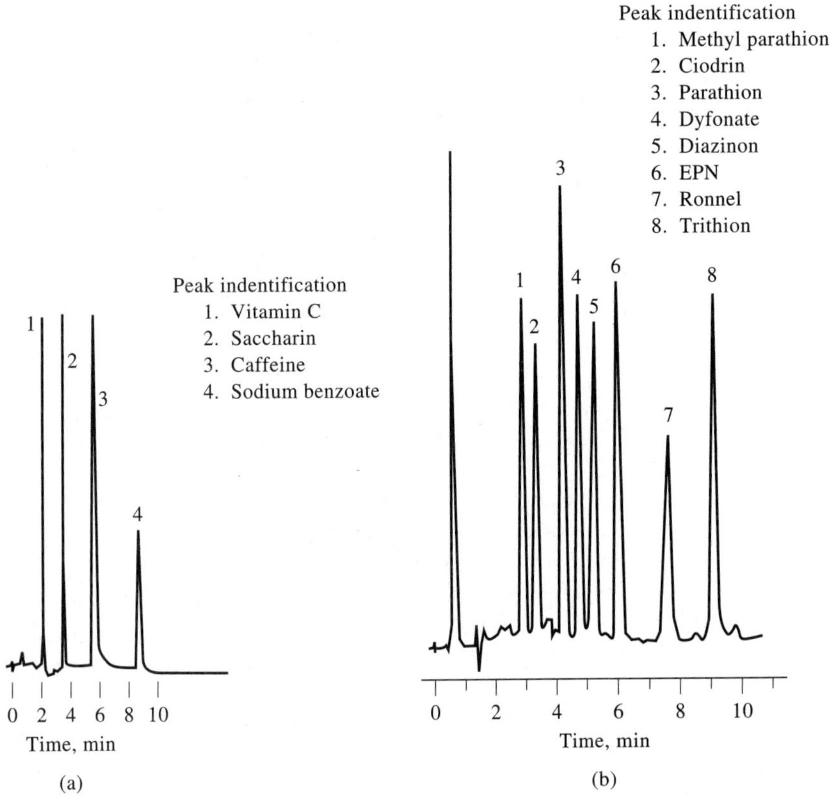

Figure 28-17 Typical applications of bonded-phase chromatography. (a) Soft-drink additives. Column: 4.6 × 250 mm paced with polar (nitrile) bonded-phase packings. Isocratic solvent: 6% HOAC/94% H_2O. Flow rate: 1.0 cm^3/min. *(Courtesy of DuPont Instrument Systems, Wilmington, DE.)* (b) Organophosphate insecticides. Column: 4.5 × 250 mm paced with 5-μm, C_8, bonded-phase particles. Gradient: 67% CH_3OH/33% H_2O to 80% CH_3/20% H_2O. Flow rate 2 mL/min. *(Courtesy of IBM Instruments Inc., Danbury, CT.)* Both used 254-nm UV detectors.

1. Phosphoserine
2. Aspartic acid
3. Glutamic acid
4. α-Amino adipic acid
5. Asparagine
6. Serine
7. Glutamine
8. Histidine
9. Glycine
10. Threonine
11. Citrulline
12. 1-Methylhistidine
13. 3-Methylhistidine
14. Arginine
15. β-Alanine
16. Alanine
17. Taurine
18. Anserine
19. β-Aminobutyric acid
20. β-Aminoisobutyric acid
21. Tyrosine
22. α-Aminobutyric acid
23. Methionine
24. Valine
25. Tryptophan
26. Phenylalanine
27. Isoleucine
28. Leucine
29. δ-Hydroxylysine
30. Lysine

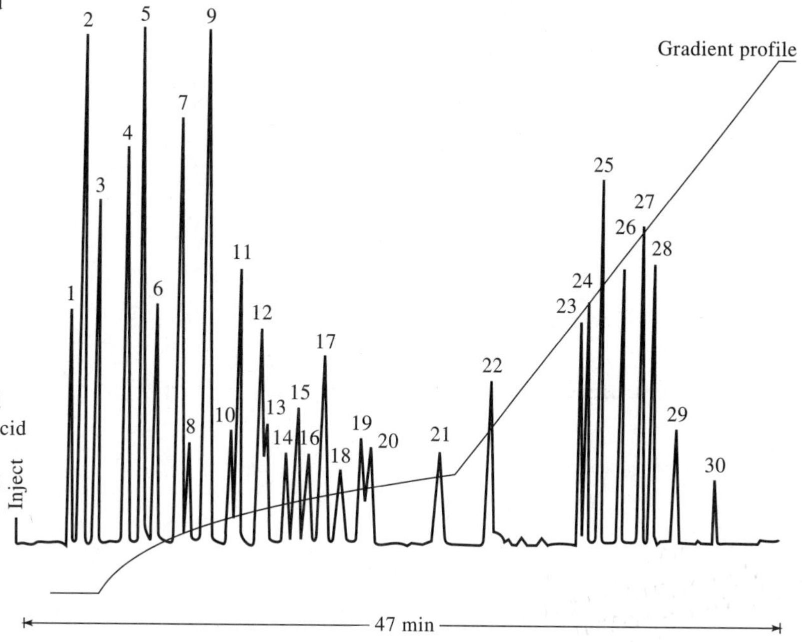

Figure 28-18 Chromatogram of orthophthalaldehyde derivatives of 30 amino acids of physiological importance. Column: 5 μm C_{18}, reversed-phase. Solvent A: 0.05 M Na_2HPO_4, pH 7.4, 96:2:2 $CH_3OH/THF/H_2O$. Fluorescence detector: excitation 334 nm; emission 425 nm. *(Reprinted with permission from R. Pfiefer et al.,* Amer. Lab., ***1983,*** *15(3), 86. Copyright 1983 by International Scientific Communications, Inc.)*

the reaction exhibit intense fluorescence at 425 nm, which permits detection down to a few picomoles (10^{-12} mol). Furthermore, the polarity of the derivatives are such that separation on a C_{18} reversed-phase packing becomes feasible. The advantages of this newer procedure are speed and smaller sample size.

Ion-Pair Chromatography[27]

Ion-pair (or *paired-ion*) *chromatography* is a type of reversed-phase partition chromatography that is used for the separation and determination of ionic species. The mobile phase in ion-pair chromatography consists of an aqueous buffer containing an organic solvent such as methanol or acetonitrile and an ionic compound containing a *counter ion* of opposite charge to the analyte. A counter ion is an ion that combines with the analyte ion to form an *ion pair,* which is a neutral species that is retained by a reversed-phase packing. Note that most of the counter ions listed in Table 28-4 contain alkyl groups to enhance retention of the resulting ion pair on the nonpolar stationary phase. Elution of the ion pairs is then accomplished with an aqueous solution of methanol or other water soluble organic solvent.

Applications of ion-pair chromatography frequently overlap those of ion-exchange chromatography, which are discussed in Section 28F. For the separation of small inorganic and organic ions, ion exchange is usually preferred unless selectivity is a problem. An example of where the ion-pair method provides better separations is for analyzing mixtures of chlorate and nitrate ion. For this pair of solutes, selectivity with an ion-exchange packing is poor, whereas resolution by ion-pair forma-

[27] See *Ion Pair Chromatography,* M. T. W. Hearn, Ed. New York: Marcel Dekker, 1985.

TABLE 28-4 Systems for Reversed-Phase Ion-Pair Chromatograms

Sample	Mobile Phase	Counter-Ion	Type Stationary Phase
Amines	0.1 M $HClO_4/H_2O$/acetonitrile	ClO_4^-	BP[a]
	$H_2O/CH_3OH/H_2SO_4$	$C_{12}H_{25}SO_3^-$	BP
Carboxylic acids	pH 7.4	$(C_4H_9)_4N^+$	BP
	pH 7.4	$(C_4H_9)_4N^+$	L[b]
Sulfonic acids	H_2O/C_3H_7OH	$(C_{16}H_{33})(CH_3)_3N^+$	BP
	pH 7.4	$(C_4H_9)_4N^+$	L[b]
	pH 3.8	Bis-(2-ethylhexyl)phosphate	L[c]
Dyes	pH 2–4; H_2O/CH_3OH	$(C_4H_9)_4N^+$	BP

[a]Bonded-phase.
[b]Adsorbed 1-pentanol.
[c]Adsorbed bis-(2-ethylhexyl)phosphoric acid/$CHCl_3$.

tion is excellent. With large ions, the resolution of ion-exchange separations is often impaired because ion-exchange resins have tight internal networks that slow the mass-transfer process with a consequent loss of efficiency. With surfactants, ion-pair chromatography is the method of choice not only because of the size of the ions involved but also because these species have a high affinity for ion exchangers, which makes their elution difficult.

Chromatography with Chiral Stationary Phase[28]

In the mid-1960s, chemists began to use chiral stationary phases (CSPs) in both gas and liquid chromatography to separate optically active isomers (enantiomers). This technique, while useful in gas chromatography, has proven to be considerably better adapted to column and planar high-performance liquid chromatography because differences in distribution constants for diastereomers becomes smaller at the elevated temperatures in gas chromatographic columns. Furthermore, high column temperatures often cause racemization of the chiral stationary phase.

More than a dozen chiral stationary phases are currently available from various manufacturers. All of these are coated on silica gel supports. The coating itself is generally a polymeric material to which is bonded an optically active isomer. For example, the *l* form of the amino acid proline

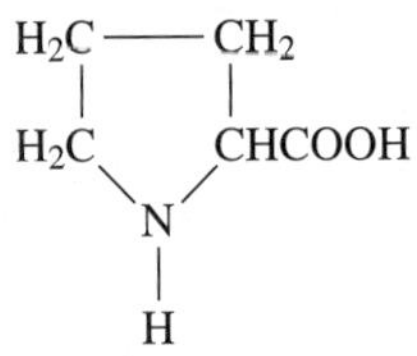

proline

has been bonded to a polystyrene-*p*-divinylbenzene cross-linked copolymer to give an optically active stationary phase for the separation of racemic mixtures of amino acids. In this application, copper ions are introduced into the solution of the analyte enantiomers to be separated. As shown in Figure 28-19, a ternary complex is formed between the stationary phase, the amino acid anions, and the copper cations. The formation constant for this complex differs for the *d* and *l* forms of the analyte amino acid, thus making their separation possible.

28E ADSORPTION CHROMATOGRAPHY

Adsorption, or liquid-solid, chromatography is the classic form of liquid chromatography first introduced by Tswett at the beginning of the twentieth century. In more recent times, it has been adapted

[28] For further information, see D. W. Armstrong, *Anal. Chem.*, **1987**, *59*, 84A; W. J. Lough, *Chiral Liquid Chromatography.* Glascow: Blackie, 1989; M. Zief and L. H. Crane. New York: Marcel Dekker, 1988.

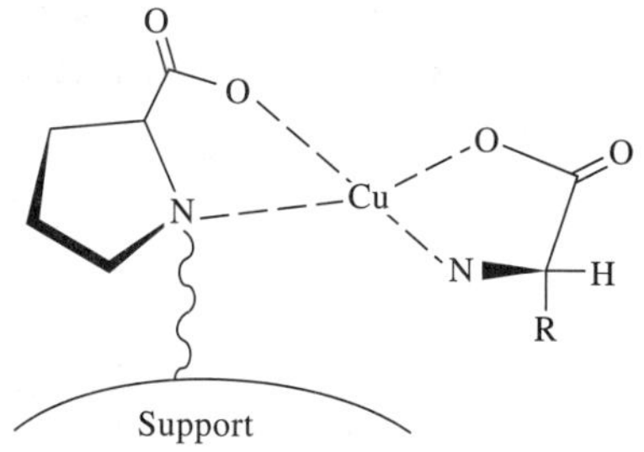

Figure 28-19 Schematic of the ternary complex formed between an L-proline bonded-phase, an analyte amino acid, and a copper(II) ion. *(Reprinted with permission from D. W. Armstrong,* Anal. Chem., *1987,* 59, *84A. Copyright 1981 American Chemical Society.)*

to and has become an important member of HPLC methods.[29]

The only stationary phases that are used for liquid-solid HPLC are silica and alumina, with the former being preferred for most, but not all, applications because of its higher sample capacity and its wider range of useful forms. With a few exceptions, the adsorption characteristics of the two substances parallel one another. With both, the order of retention times is: olefins < aromatic hydrocarbons < halides, sulfides < ethers < nitro compounds < esters ≈ aldehydes ≈ ketones < alcohols ≈ amines < sulfones < sulfoxides < amides < carboxylic acids.[30]

28E-1 Solvent Selection for Adsorption Chromatography

In liquid-solid chromatography, the only variable available to optimize k' and α is the composition of the mobile phase (in contrast to partition chromatography, where the column packing has a pronounced effect on α). Fortunately, in adsorption chromatography, enormous variations in resolution and retention time accompany variations in the solvent system, and only rarely can a suitable mobile phase not be found.

Solvent Strength

It has been found that the polarity index P', which was described in Section 28D-2, can also serve as a rough guide to the strengths of solvents for adsorption chromatography. A much better index, however, is the *eluent strength* ϵ^0, which is the adsorption energy per unit area of solvent.[31] This parameter depends upon the adsorbent, with ϵ^0 values for silica being about 0.8 of those on alumina. The values for ϵ^0 in the last column of Table 28-2 are for alumina. Note that solvent-to-solvent differences in ϵ^0 roughly parallel those for P'.

Choice of Solvent Systems

A useful technique for choosing a solvent system for adsorption chromatography is similar to that described for partition separations. That is, two compatible solvents are chosen, one of which is too strong (ϵ^0 too large) and the other of which is too weak. A suitable value for k' is then obtained by varying the volume ratio of the two. It has been found that an increase in ϵ^0 value by 0.05 unit usually decreases all k' values by a factor of 3 to 4. Thus, enormous variations in k' are possible, and some binary system involving the solvents in Table 28-2 can be found that will give adequate retention times for nearly any sample. Unfortunately ϵ^0 does not vary linearly with volume ratios, as was the case with P' in partition chromatography. Thus, calculating an optimal mixture is more difficult. Graphs have been developed, however, that relate ϵ^0 to composition for a number of common binary solvent mixtures.[32]

When overlapping peaks are encountered, exchanging one strong solvent for another while holding k' more or less constant, will change α values and often provide the desired resolution. This trial-and-error approach is tedious and sometimes unsuccessful. Methods for systematizing and shortening the search have been developed.[33] One is to carry out preliminary scouting by means of thin-layer chromatography, the open-bed version of adsorption chromatography (see Section 28H-3).

---

[29] See D. L. Saunders, *J. Chromatogr. Sci.*, **1977**, *15*, 372.

[30] Silica and alumina surfaces are highly polar, and elutions are usually performed with some of the less polar mobile phases. Thus, some chromatographers treat adsorption chromatography as a type of normal-phase partition chromatography. (For example, see J. G. Dorsey et al., *Anal. Chem.*, **1990**, *62*, 326R.)

[31] See L. R. Snyder, *Principles of Adsorption Chromatography*, Chapter 8. New York: Marcel Dekker, 1968.

[32] For this and other practical aspects of adsorption chromatography, see D. L. Saunders, *J. Chromatogr. Sci.*, **1977**, *15*, 372.

[33] See L. R. Snyder and J. J. Kirkland, *Introduction to Modern Liquid Chromatography*, 2nd ed., pp. 365–389. New York: Wiley, 1979.

28E-2 Applications of Adsorption Chromatography

Figure 28-1 shows that adsorption chromatography is best suited for nonpolar compounds having molecular weights less than perhaps 5000. Although some overlap exists between adsorption and partition chromatography, the methods tend to be complementary.

Generally, liquid-solid chromatography is best suited to samples that are soluble in nonpolar solvents and correspondingly have limited solubility in aqueous solvents such as those used in the reversed-phase partition procedure. As with partition chromatography, compounds with differing kinds or numbers of functional groups are usually separable. A particular strength of adsorption chromatography, which is not shared by other methods, is its ability to differentiate among the components of isomeric mixtures. Table 28-5 compares selectivities of partition and adsorption chromatography for several types of analytes. Note that resolution of homologs and benzologs is generally better with reversed-phase partition chromatography. Separation of isomers, however, is usually better with the adsorption procedure.

Figure 28-20 illustrates a typical application of adsorption chromatography.

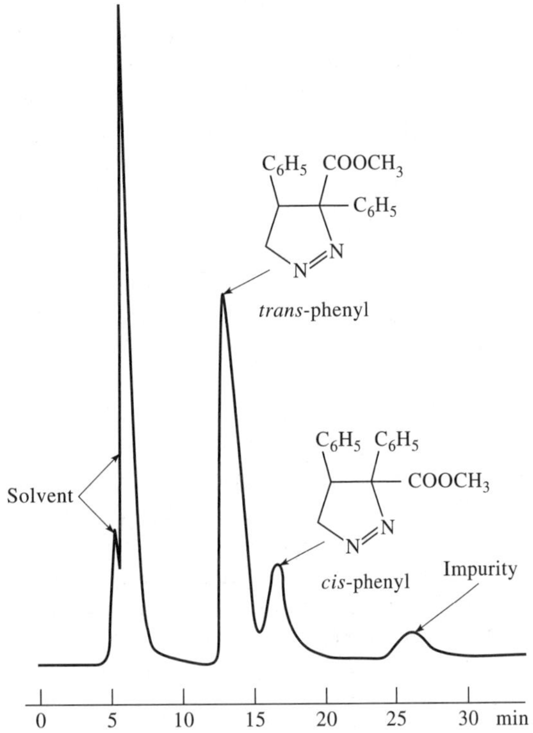

Figure 28-20 A typical application of adsorption chromatography: separations of *cis*- and *trans*-pyrazoline. Column: 100 × 0.3 cm pellicular silica. Mobile phase: 50% methylene chloride/isooctane. Temperature: ambient. Flow rate: 0.225 mL/min. Detector: UV, 254 nm. *(Courtesy of Hewlett-Packard Company, Palo Alto, CA.)*

28F ION-EXCHANGE CHROMATOGRAPHY

Ion-exchange chromatography (IC), which is often shortened to *ion chromatography* refers to modern and efficient methods of separating and determining ions based upon ion-exchange resins. Ion chromatography was first developed in the mid-1970s when it was shown that anion or cation mixtures can be readily resolved on HPLC columns packed with anion-exchange or cation-exchange resins. At that time, detection was generally performed with conductivity measurements. Currently, other detectors are also available for ion chromatography.[34]

Ion chromatography was an outgrowth of ion-exchange chromatography, which during the Manhattan project was developed for the separation of closely related rare earth cations with cation-exchange resins. This monumental work, which laid the theoretical groundwork for ion-exchange separations, was extended after World War II to many other types of materials; ultimately it led to automated methods for the separation and detection of amino acids and other ionic species in complex mixtures. The development of modern HPLC began in the late 1960s, but its application to ion-exchange separation of ionic species was delayed by the lack of a sensitive general method of detecting such eluted ionic species as alkali and alkaline earth cations and halide, acetate, and nitrate anions. This situation was remedied in 1975 by the development by workers at Dow Chemical Company of an eluent sup-

[34] For a brief review of modern ion chromatography, see J. F. Fritz, *Anal. Chem.*, **1987**, *59*, 335A; P. K. Dasgupta, *Anal. Chem.*, **1992**, *64*, 775A. For monographs on the subject, see H. Small, *Ion Chromatography.* New York: Plenum Press, 1989; D. T. Gjerde and J. S. Fritz, *Ion Chromatography,* 2nd ed. Mamaroneck, NY: Huethig, 1987; R. E. Smith, *Ion Chromatography Applications,* Boca Raton, FL: CRC Press, 1987; J. Weiss, *Ion Chromatography,* 2nd ed. New York: VCH, 1995; P. R. Haddad and P. E. Jackson, *Ion Chromatography: Principles and Applications.* New York: Elsevier, 1990. For a description of several commercially available IC systems, see D. Noble, *Anal. Chem.*, **1995**, *67*, 205A.

TABLE 28-5 Comparison of Selectivities of Adsorption and Reversed-Phase Chromatography*

Separation of	Compound	Adsorption α	Reversed-Phase α
Homologs	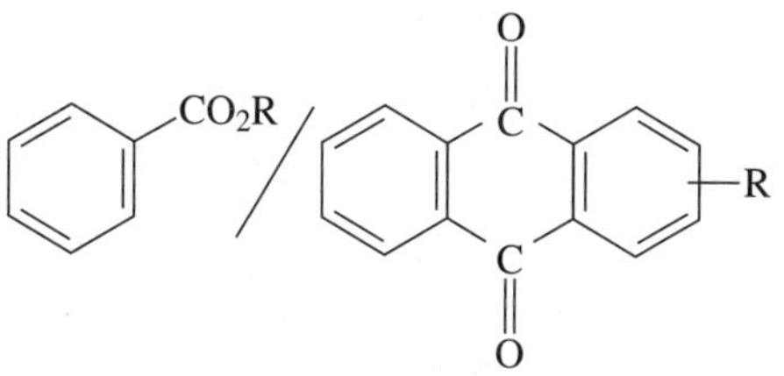		
	$R = C_1$	4.8	3.3
	C_2		6.5
	C_4	4.1	17
	C_{10}	3.6	
Benzologs	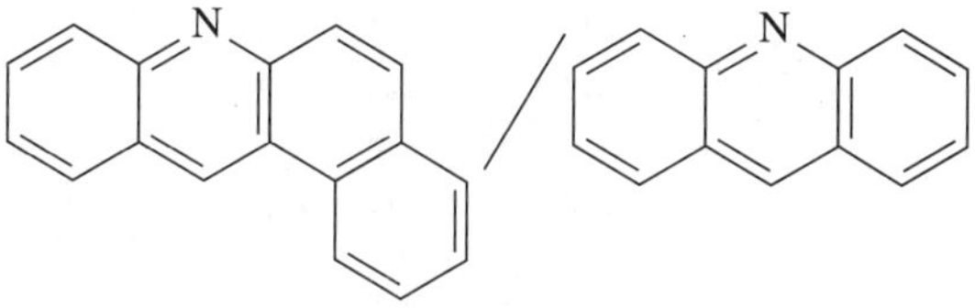	1.2	1.4
	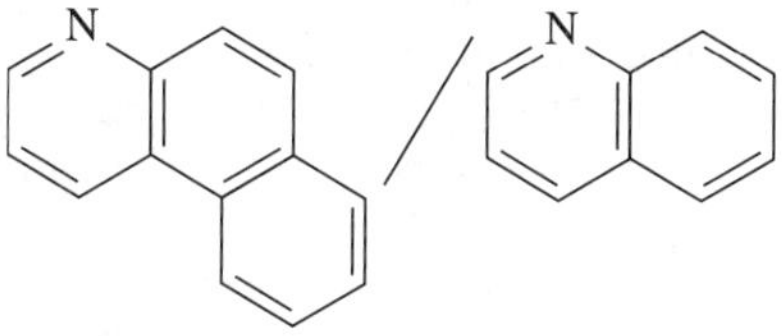	1.1	1.8
Isomers	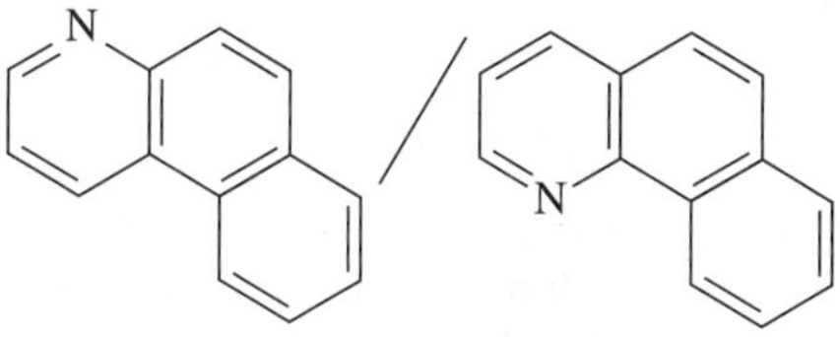	12.5	1.06
	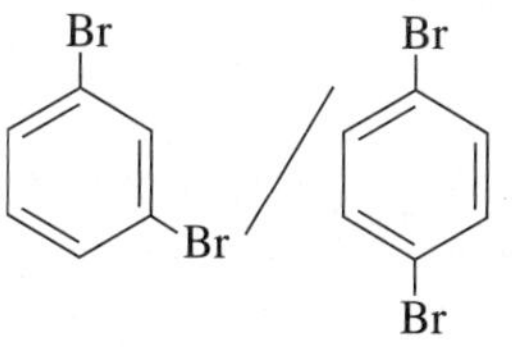	1.8	
	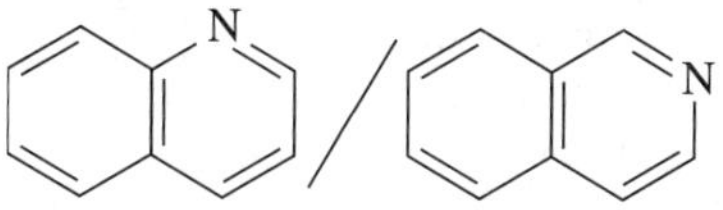	3.4	
	1,2,3,4, dibenzanthrocene, $C_{22}H_{14}$/Picene $C_{22}H_{14}$	20	

*Data from: L. R. Snyder and J. J. Kirkland, *Introduction to Modern Liquid Chromatography,* 2nd ed., pp. 357–358. New York: Wiley, 1979.

pressor technique, which made possible the conductometric detection of eluted ions.[35] This technique is described in Section 28F-3.

28F-1 Ion-Exchange Equilibria

Ion-exchange processes are based upon exchange equilibria between ions in solution and ions of like sign on the surface of an essentially insoluble, high-molecular-weight solid. Natural ion-exchangers, such as clays and zeolites, have been recognized and used for several decades. Synthetic ion-exchange resins were first produced in the mid-1930s for water softening, water deionization, and solution purification. The most common active sites for cation-exchange resins are the sulfonic acid group $—SO_3^-H^+$, a strong acid, and the carboxylic acid group $—COO^-H^+$, a weak acid. Anionic exchangers contain tertiary amine groups $—N(CH_3)_3^+OH^-$ or primary amine groups $—NH_3^+OH^-$; the former is a strong base and the latter a weak one.

When a sulfonic acid ion-exchanger is brought in contact with an aqueous solvent containing a cation M^{x+}, an exchange equilibria is set up that can be described by

$$\underset{\text{solid}}{xRSO_3^-H^+} + \underset{\text{solution}}{M^{x+}} \rightleftharpoons \underset{\text{solid}}{(RSO_3^-)_x M^{x+}} + \underset{\text{solution}}{xH^+}$$

where $RSO_3^-H^+$ represents *one* of many sulfonic acid groups attached to a large polymer molecule. Similarly a strong base exchanger interacts with the anion A^{x-} as shown by the reaction

$$\underset{\text{solid}}{xRN(CH_3)_3OH^-} + \underset{\text{solution}}{A^{x-}} \rightleftharpoons \underset{\text{solid}}{[RH(CH_3)_3^+]_xA^{x-}} + \underset{\text{solution}}{xOH^-}$$

As an example of the application of the mass-action law to ion-exchange equilibria, we will consider the reaction between a singly charged ion B^+ with a sulfonic acid resin held in a chromatographic column. From a neutral solution, initial retention of B^+ ions at the head of the column occurs because of the reaction

$$RSO_3^-H^+(s) + B^+(aq) \rightleftharpoons RSO_3^-B^+(s) + H^+(aq) \quad (28\text{-}5)$$

Here, the (s) and (aq) emphasize that the system contains a solid and an aqueous phase. Elution with a dilute solution of hydrochloric acid shifts the equilibrium in Equation 28-5 to the left, causing part of the B^+ ions in the stationary phase to be transferred to the mobile phase. These ions then move down the column in a series of transfers between the stationary and mobile phases.

The equilibrium constant K_{ex} for the exchange reaction shown in Equation 28-5 takes the form

$$\frac{[RSO_3^-B^+]_s[H^+]_{aq}}{[RSO_3^-H^+]_s[B^+]_{aq}} = K_{ex} \quad (28\text{-}6)$$

Here, $[RSO_3^-B^+]_s$ and $[RSO_3^-H^+]_s$ are concentrations (strictly activities) of B^+ and H^+ *in the solid phase.* Rearranging yields

$$\frac{[RSO_3^-B^+]_s}{[B^+]_{aq}} = K_{ex}\frac{[RSO_3^-H^+]_s}{[H^+]_{aq}} \quad (28\text{-}7)$$

During the elution, the aqueous concentration of hydrogen ions is much larger than the concentration of the singly charged B^+ ions in the mobile phase. Furthermore, the exchanger has an enormous number of exchange sites relative to the number of B^+ ions being retained. Thus, the overall concentrations $[H^+]_{aq}$ and $[RSO_3^-H^+]_s$ are not affected significantly by shifts in the equilibrium 28-5. Therefore, when $[RSO_3^-H^+]_s \gg [RSO_3^-B^+]_s$ and $[H^+]_{aq} \gg [B^+]_{aq}$ the right-hand side of Equation 28-7 is substantially constant, and we can write

$$\frac{[RSO_3^-B^+]_s}{[B^+]_{aq}} = K = \frac{c_S}{c_M} \quad (28\text{-}8)$$

where K is a constant that corresponds to the distribution constant as defined by Equation 26-1. All of the equations in Table 26-4 (Section 26E) can then be applied to ion-exchange chromatography in the same way as to the other types, which have already been considered.

Note that K_{ex} in Equation 28-6 represents the affinity of the resin for the ion B^+ *relative* to another ion (here, H^+). Where K_{ex} is large, a strong tendency exists for the solid phase to retain B^+; where K_{ex} is small, the reverse obtains. By selecting a common reference ion such as H^+, distribution ratios for different ions on a given type of resin can be experimentally compared. Such experiments reveal that polyvalent ions are much more strongly held than singly charged species. Within a given charge group, however, differences appear that are related to the size of the hydrated ion as well as to other properties. Thus, for a typical sulfonated cation-exchange resin, values for K_{ex} decrease in the order $Tl^+ > Ag^+ > Cs^+ > Rb^+ > K^+ > NH_4^+ > Na^+ >$

[35] H. Small, T. S. Stevens, and W. C. Bauman, *Anal. Chem.*, **1975**, *47*, 1801.

$H^+ > Li^+$. For divalent cations, the order is $Ba^{2+} > Pb^{2+} > Sr^{2+} > Ca^{2+} > Ni^{2+} > Cd^{2+} > Cu^{2+} > Co^{2+} > Zn^{2+} > Mg^{2+} > UO_2^{2+}$.

For anions, K_{ex} for a strong base resin decreases in the order $SO_4^{2-} > C_2O_4^{2-} > I^- > NO_3^- > Br^- > Cl^- > HCO_2^- > CH_3CO_2^- > OH^- > F^-$. This sequence is somewhat dependent upon type of resin and reaction conditions and should thus be considered only approximate.

28F-2 Ion-Exchange Packings

Historically, ion-exchange chromatography was performed on small, porous beads formed during emulsion copolymerization of styrene and divinylbenzene. The presence of divinylbenzene (usually ≈ 8%) results in cross-linking, which imparts mechanical stability to the beads. In order to make the polymer active toward ions, acidic or basic functional groups are then bonded chemically to the structure. The most common groups are sulfonic acid and quaternary amines.

Figure 28-21 shows the structure of a strong acid resin. Note the cross-linking that holds the linear polystyrene molecules together. The other types of resins have similar structures except for the active functional group.

Porous polymeric particles are not entirely satisfactory for chromatographic packings because of the slow rate of diffusion of analyte molecules through the micropores of the polymer matrix and because of the compressibility of the matrix. To overcome this problem, two newer types of packings have been developed and are in more general use than the porous polymer type. One is a pellicular bead packing in which the surface of a relatively large, (30 to 40 μm) nonporous, spherical, glass, or polymer bead is coated with a synthetic ion-exchange resin. A second type of packing is prepared by coating porous microparticles of silica, such as those used in adsorption chromatography, with a thin film of the exchanger. With either type, faster diffusion in the polymer film leads to enhanced efficiency. On the other hand, the sample capacity of these particles is less, particularly for the pellicular type.

28F-3 Inorganic Applications of Ion-Exchange Chromatography

The mobile phase in ion-exchange chromatography must have the same general properties that are required for other types of chromatography. That is, it must dissolve the sample, have a solvent strength that leads to reasonable retention times (correct k' values), and interact with solutes in such a way as to lead to selectivity (suitable α values). The mobile phases in ion-exchange chromatography are aqueous solutions that may contain moderate amounts of methanol or other water-miscible organic solvents; these mobile phases also contain ionic species, often in the form of a buffer. Solvent strength

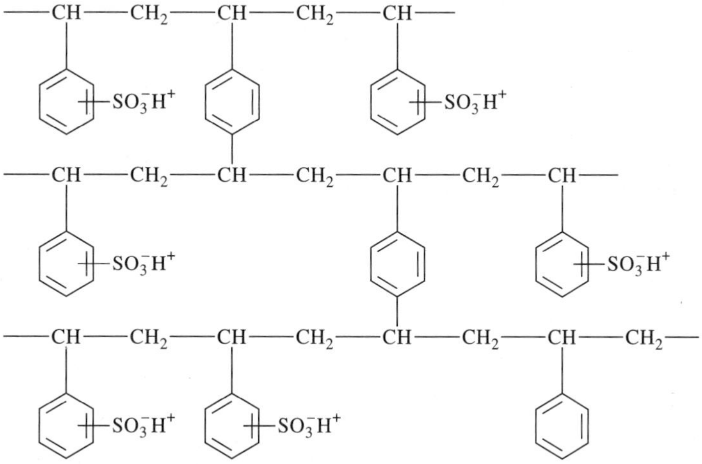

Figure 28-21 Structure of a cross-linked polystyrene ion-exchange resin. Similar resins are used in which the $—SO_3^-H^+$ group is replaced by $—COO^-H^+$, $—NH_3^+OH^-$, and $—N(CH_3)_3^+OH^-$ groups.

and selectivity are determined by the kind and concentration of these added ingredients. In general, the ions of the mobile phase compete with analyte ions for the active sites on the ion-exchange packing.

Ion-Exchange Chromatography with Eluent Suppressor Columns

As noted earlier, the widespread application of ion chromatography for the determination of inorganic species was inhibited by the lack of a good general detector, which would permit quantitative determination of ions on the basis of chromatographic peak areas. Conductivity detectors are an obvious choice for this task. They can be highly sensitive, they are universal for charged species, and, as a general rule, they respond in a predictable way to concentration changes. Furthermore, such detectors are simple, inexpensive to construct and maintain, easy to miniaturize, and ordinarily give prolonged, trouble-free service. The only limitation to conductivity detectors proved to be a serious one, which delayed their general use. This limitation arises from the high electrolyte concentration required to elute most analyte ions in a reasonable time. As a consequence, the conductivity from the mobile-phase components tends to swamp that from analyte ions, thus greatly reducing the detector sensitivity.

In 1975, the problem of high eluent conductance was solved by the introduction of a so-called eluent suppressor column immediately following the analytical ion-exchange column. The suppressor column is packed with a second ion-exchange resin that effectively converts the ions of the solvent to a molecular species of limited ionization without affecting the analyte ions. For example, when cations are being separated and determined, hydrochloric acid is often chosen as the eluting reagent, and the suppressor column is an anion-exchange resin in the hydroxide form. The product of the reaction in the suppressor is water. That is,

$$H^+(aq) + Cl^-(aq) + Resin^+OH^-(s) \rightarrow Resin^+Cl^-(s) + H_2O$$

The analyte cations are, of course, not retained by this second column.

For anion separations, the suppressor packing is the acid form of a cation-exchange resin. Here, sodium bicarbonate or carbonate may serve as the eluting agent. The reaction in the suppressor is then

$$Na^+(aq) + HCO_3^-(aq) + Resin^-H^+(s) \rightarrow Resin^-Na^+(s) + H_2CO_3(aq)$$

Here, the largely undissociated carbonic acid does not contribute significantly to the conductivity.

An inconvenience associated with the original suppressor columns was the need to regenerate them periodically (typically, every 8 to 10 hr) in order to convert their packings back to the original acid or base form. Recently, however, fiber membrane suppressors have become available that operate continuously, as shown in Figure 28-22. Here, the eluent and the suppressor solution flow in opposite directions on either side of the permeable ion-exchange membranes labeled *M*. For the analysis of anions, the membranes are cation-exchange resins; for cations, they are anion exchangers. When, for example, sodium ions are to be removed from the eluent, acid for regeneration flows continuously in the suppressor stream. Sodium ions from the eluent exchange with hydrogen ions of the membrane and then migrate across the membrane where they exchange with hydrogen ions of the regeneration reagent. The device has a remarkably high exchange rate; for example, it is capable of removing essentially all of the sodium ions from a 0.1 M

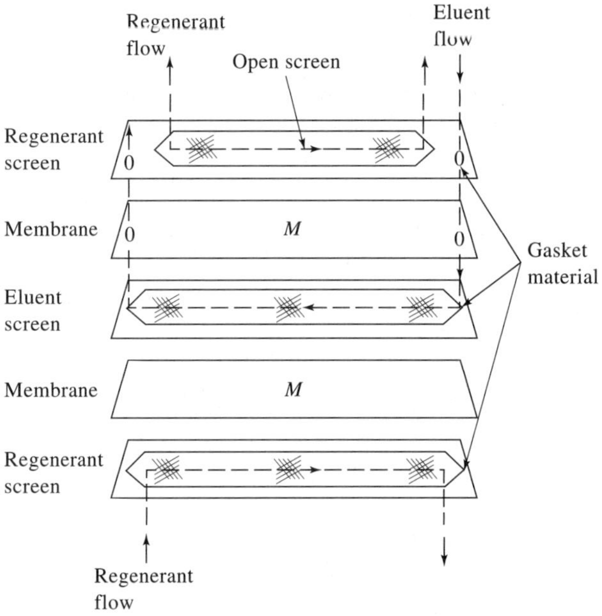

Figure 28-22 A micromembrane suppressor. Eluent flows through a narrow channel that contains a plastic screen that reduces the void volume and appears to increase mass-transfer rates. The eluent is separated from the suppressor solution by 50-μm exchange resins. Regenerant flow is in the direction opposite to eluent flow. *(Courtesy of Dionex Corporation, Sunnyvale, CA.)*

solution of sodium hydroxide when the eluent flow rate is 2 mL/min.

In recently designed commercial instruments, regeneration of suppressor solutions is performed automatically with electrogenerated hydrogen or hydroxyl ions so that interruptions in the use of the instruments for regeneration are not required.[36]

Figure 28-23 shows two applications of ion chromatography based upon a suppressor column and conductometric detection. In each case, the ions were present in the parts per million range, and the sample sizes were 50 μL in one case and 100 μL in the other. The method is particularly important for anion analysis because no other rapid and convenient method for handling mixtures of this type now exists.

[36] See A. Henshall et al., *Amer. Lab.*, **1992** (11), 20R; R. Saari-Nordhaus and J. M. Anderson, *Amer. Lab.*, **1996** (2), 33N.

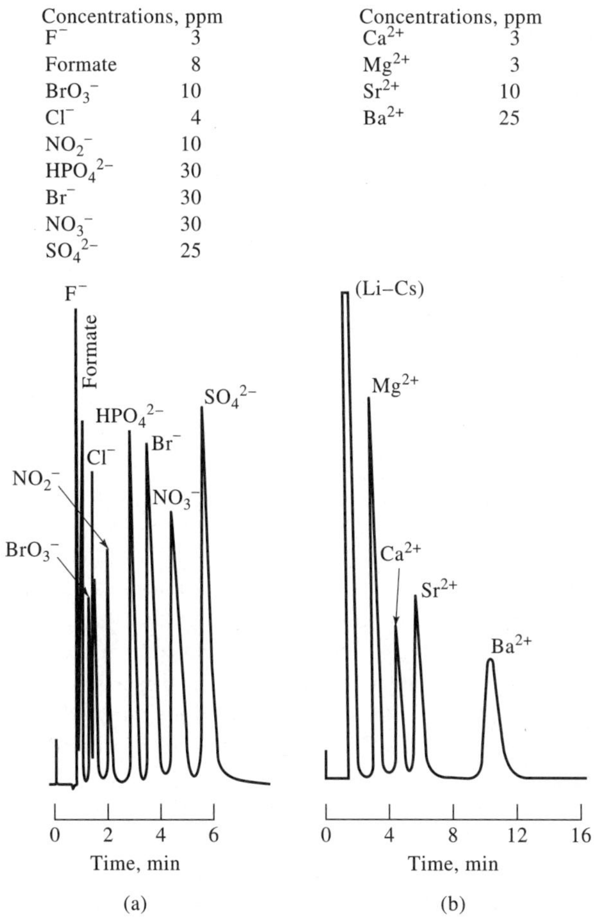

Figure 28-23 Typical applications of ion chromatography. (a) Separation of anions on an anion-exchange column. Eluent: 0.0028 M $NaHCO_3$/0.0023 M Na_2CO_3. Sample size: 50 μL. (b) Separation of alkaline earth ions on a cation-exchange column. Eluent: 0.025 M phenylenediamine dihydrochloride/0.0025 M HCl. Sample size: 100 μL. *(Courtesy of Dionex Corporation, Sunnyvale, CA.)*

Single-Column Ion Chromatography[37]

Equipment has also become available commercially for ion chromatography in which no suppressor column is used. This approach depends upon the small differences in conductivity between the eluted sample ions and the prevailing eluent ions. To amplify these differences, low-capacity exchangers are used, which makes possible elution with species having low equivalent conductances.[38] Single-column chromatography tends to be somewhat less sensitive and to have a more limited range than ion chromatography with a suppressor column.

An indirect photometric method that permits the separation and detection of nonabsorbing anions and cations without a suppressor column has recently been described.[39] Here also, no suppressor column is used, but instead, anions or cations that do absorb are used to displace the analyte ions from the column. When the analyte ions are displaced from the exchanger, their place is taken by an equal number of eluent ions (provided, of course, that the charge on the analyte and eluent ions is the same). Thus, the absorbance of the eluate decreases as analyte ions exit from the column. Figure 28-24 shows a chromatogram obtained by this procedure. Here, the eluent was a dilute solution of disodium phthalate, phthalate ion being the ultraviolet-absorbing displacing ion. Indirect detection has a more restricted working range than does direct detection.

28F-4 Organic and Biochemical Applications of Ion-Exchange Chromatography

Ion-exchange chromatography has been applied to a variety of organic and biochemical systems, including drugs and their metabolites, serums, food preservatives, vitamin mixtures, sugars, and pharmaceutical prepara-

[37] See D. T. Gjerde and J. S. Fritz, *Ion Chromatography,* 2nd ed. Mamaroneck, NY: Huethig Verlag, 1987.

[38] See J. S. Fritz, D. J. Gjerde, and R. M. Becker, *Anal. Chem.*, **1980**, *52*, 1519.

[39] H. Small and T. E. Miller, *Anal. Chem.*, **1982**, *54*, 462.

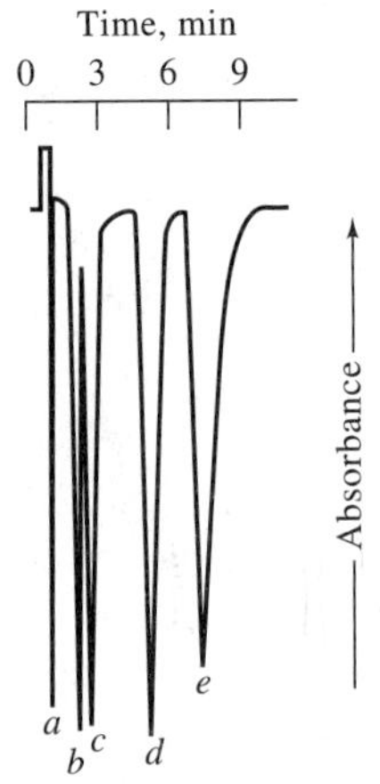

Figure 28-24 Indirect photometric detection of several anions by elution. Eluent: 10^{-3} M disodium phthalate, 10^{-3} M boric acid, pH 10. Flow rate: 5 mL/min. Sample volume: 0.02 mL. UV detector. Sample ions: (a) 18-μg carbonate; (b) 1.4-μg chloride, (c) 3.8-μg phosphate; (d) 5-μg azide; (e) 10-μg nitrate. *(Reprinted with permission from H. Small, Anal. Chem., **1985,** 55, 240A. Copyright 1983 American Chemical Society.)*

tions. An example of one of these applications is shown in Figure 28-25 in which 1×10^{-8} mol each of 17 amino acids was separated on a cation-exchange column.

28F-5 Ion-Exclusion Chromatography

Ion-exclusion chromatography is not a form of ion chromatography because neutral species rather than ions are being separated. Nevertheless, it is convenient to discuss it here because ion-exclusion chromatography, like ion chromatography, employs ion-exchange columns to achieve separations.

The theory and applications of ion-exclusion chromatography are conveniently illustrated by the separation of simple carboxylic acids shown by the chromatogram in Figure 28-26. Here, the stationary phase was a cation-exchange resin in its acidic form, and elution was accomplished with a dilute solution of hydrochloric acid. The analytical column was followed by a suppressor column that was packed with a cation-exchange resin in the silver form. The hydrogen ions of the eluent were exchanged for silver ions, which then precipitated the chloride ions, thus removing the ions contributed by the eluent. The undissociated form of the analyte acids were distributed between the mobile phase in the column and the immobilized liquid held in the pores of the packing. The distribution constants for the various acids are primarily related to the inverse of their dissociation constants, although other factors also play a part in the extent to which various species are distributed between the two phases.

Ion-exclusion chromatography finds numerous applications for identification and determination of acidic species in such materials as milk, coffee, wine, and many other products of commerce. Salts of weak acids can also be analyzed because they are converted to the corresponding acid by the hydrogen ions in the exchanger. Weak bases and their salts can also be determined by ion-exclusion chromatography. In these applications an anion-exchange column in the hydroxide form is employed.

28G SIZE-EXCLUSION CHROMATOGRAPHY

Size-exclusion chromatography, which has also been called *gel permeation,* or *gel filtration chromatography,* is a powerful technique that is particularly applicable to high-molecular-weight species.[40] Packings for size-exclusion chromatography consist of small (≈10-μm) silica or polymer particles containing a network of uniform pores into which solute and solvent molecules can diffuse. While in the pores, molecules are effectively trapped and removed from the flow of the mobile phase. The average residence time in the pores depends upon the effective size of the analyte molecules. Molecules that are larger than the average pore size of the packing are excluded and thus suffer essentially no retention; such species are the first to be eluted. Molecules having diameters that are significantly smaller than the pores can penetrate or permeate throughout the pore maze and are thus entrapped for the greatest time; these are last to be eluted. Between these two extremes are intermediate-size molecules whose average penetration into the pores of the packing depends upon their diameters. Within this group, fractionation occurs, which is directly related to molecular size and to some extent molecular shape. Note that size-exclusion separations differ from the other procedures we have been considering in that no chemical or

[40] For monographs on this subject, see *Size Exclusion Chromatography,* B. J. Hunt and S. R. Holding, Eds. New York: Chapman and Hall, 1988; *Aqueous Size-Exclusion Chromatography,* P. L. Dubin, Ed. Amsterdam, Neth.: Elsevier, 1988; *Handbook of Size Exclusion Chromatography,* C. S. Wu, Ed. New York: Marcel Dekker, 1995.

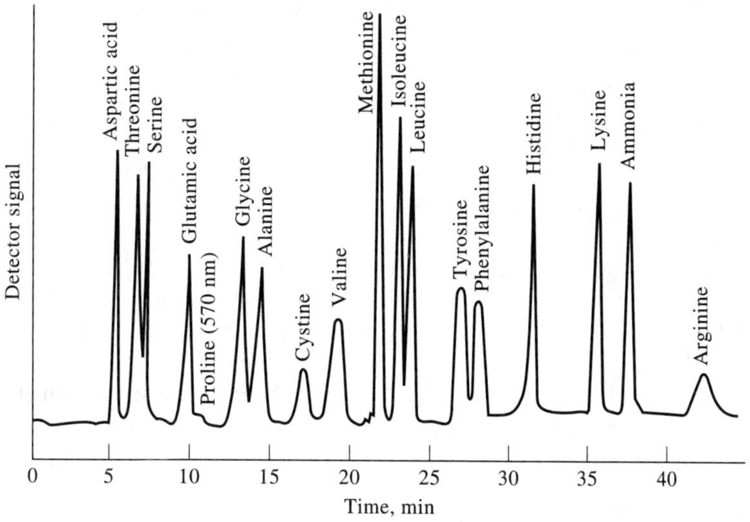

Figure 28-25 Separation of amino acids on an ion-exchange column. Packing: cation exchange with particle size of 8 μm. Pressure: 2700 psi. *(Reprinted with permission from J. R. Benson,* Amer. Lab., *1972,* 4*(10), 60. Copyright 1972 by International Scientific Communications, Inc.)*

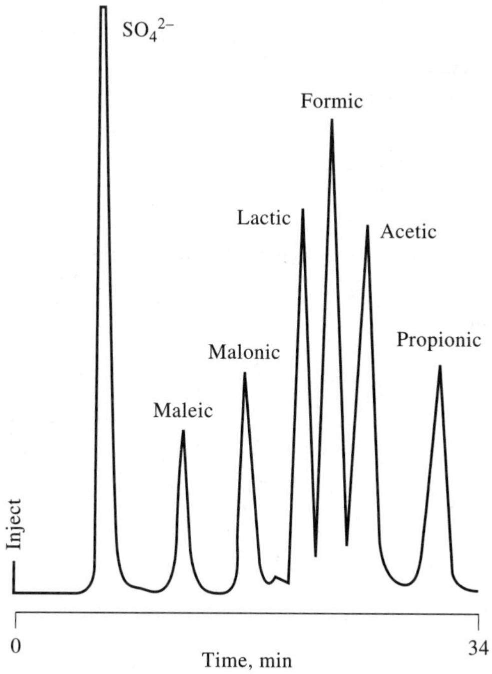

Figure 28-26 An ion-exclusion chromatogram for a mixture of six weak acids.

physical interaction between analytes and the stationary phase are involved. Indeed, every effort is made to avoid such interactions because they lead to impaired column efficiencies. Also note that unlike other forms of chromatography, there is an upper limit to retention time because no analyte species is retained longer than those that totally permeate the stationary phase.

28G-1 Column Packing

Two types of packing for size-exclusion chromatography are encountered: polymer beads and silica-based particles, both of which have diameters of 5 to 10 μm. The latter have the advantages of greater rigidity, which leads to easier packing and permits the use of higher pressures; greater stability, which permits the use of a wider range of solvents, including water; more rapid equilibration with new solvents; and stability at higher temperatures. The disadvantages of silica-based particles include their tendency to retain solutes by adsorption and their potential for catalyzing the degradation of solute molecules.

Most early size-exclusion chromatography was carried out on cross-linked styrene-divinylbenzene copolymers similar in structure (except that the sulfonic acid groups are absent) to that shown in Figure 28-21. The pore size of these polymers is controlled by the extent of cross-linking and hence the relative amount of divinylbenzene present during manufacture. As a consequence, polymeric packings having several different average pore sizes are marketed. Originally, styrene-divinylbenzene gels were hydrophobic and thus could only be used with nonaqueous mobile phases. Now, however, hydrophilic gels are available, making possible the use of aqueous solvents for the separation of large, water-soluble molecules such as sugars. These hydrophilic gels are sulfonated divinylbenzenes or polyacrylamides.

Porous glasses and silica particles, having average pore sizes ranging from 40 Å to 2500 Å, are now available commercially. In order to reduce adsorption, the surfaces of these particles are often modified by reaction with organic substituents. For example, the surface of one hydrophilic packing has the structure

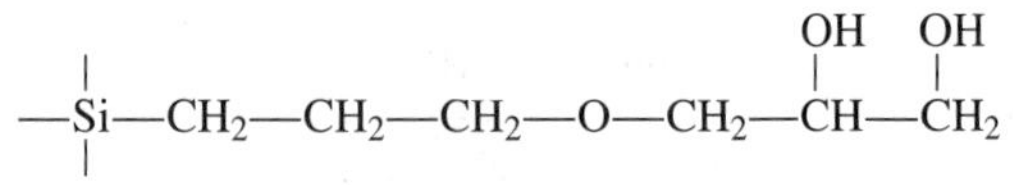

Table 28-6 lists the properties of some typical commercial size-exclusion packings.

28G-2 Theory of Size-Exclusion Chromatography

The total volume V_t of a column packed with a porous polymer or silica gel is given by

$$V_t = V_g + V_i + V_o \qquad (28\text{-}9)$$

where V_g is the volume occupied by the solid matrix of the gel, V_i is the volume of solvent held in its pores, and V_o is the free volume outside the gel particles. Assuming no mixing or diffusion, V_o also represents the theoretical volume of solvent required to transport through the column those components too large to enter the pores of the gel. In fact, however, some mixing and diffusion will occur, and as a consequence the nonretained components will appear in a Gaussian-shaped band with a concentration maximum at V_o. For components small enough to enter freely into the pores of the gel, band maxima will appear at the end of the column at an eluent volume corresponding to $(V_i + V_o)$. Generally, V_i, V_o, and V_g are of the same order of magnitude; thus, a gel column permits separation of the large molecules of a sample from the small molecules with a minimal volume of eluate.

Molecules of intermediate size are able to transfer to some fraction K of the solvent held in the pores; the elution volume V_e for these retained molecules is

$$V_e = V_o + KV_i \qquad (28\text{-}10)$$

Equation 28-10 applies to all of the solutes on the column. For molecules too large to enter the gel pores, $K = 0$ and $V_e = V_o$; for molecules that can enter the pores unhindered, $K = 1$ and $V_e = (V_o + V_i)$. In deriving Equation 28-10, the assumption was made that no interaction, such as adsorption, occurs between the solute molecules and the gel surfaces. With adsorption, the amount of interstitially held solute will in-

TABLE 28-6 Properties of Typical Commercial Packings for Size-Exclusion Chromatography

Type	Particle Size, μm	Average Pore Size, Å	Molecular Weight Exclusion Limit*
Polystyrene-divinylbenzene	10	10^2	700
		10^3	$(0.1 \text{ to } 20) \times 10^4$
		10^4	$(1 \text{ to } 20) \times 10^4$
		10^5	$(1 \text{ to } 20) \times 10^5$
		10^6	$(5 \text{ to } > 10) \times 10^6$
Silica	10	125	$(0.2 \text{ to } 5) \times 10^4$
		300	$(0.03 \text{ to } 1) \times 10^5$
		500	$(0.05 \text{ to } 5) \times 10^5$
		1000	$(5 \text{ to } 20) \times 10^5$

*Molecular weight above which no retention occurs.

crease; with small molecules, K will then be greater than unity.

Equation 28-10 rearranges to

$$K = (V_e - V_o)/V_i = c_S/c_M \qquad (28\text{-}11)$$

where K is the distribution constant for the solute (see Equation 26-1). Values of K range from zero for totally excluded large molecules to unity for small molecules. The distribution constant is a valuable parameter for comparing data from different packings. In addition, it makes possible the application of all of the equations in Table 26-5 to exclusion chromatography.

The useful molecular weight range for a size-exclusion packing is conveniently illustrated by means of a calibration curve such as that shown in the upper part of Figure 28-27. Here, molecular weight, which is directly related to the size of solute molecules, is plotted against retention volume V_R, where V_R is the product of the retention time and the volumetric flow rate. Note that the ordinate scale is logarithmic. The *exclusion limit* defines the molecular weight of a species beyond which no retention occurs. All species having greater molecular weight than the exclusion limit are so large that they are not retained and elute together to give peak A in the chromatogram shown in Figure 28-27b. The *permeation limit* is the molecular weight below which the solute molecules can penetrate into the pores completely. All molecules below this molecular

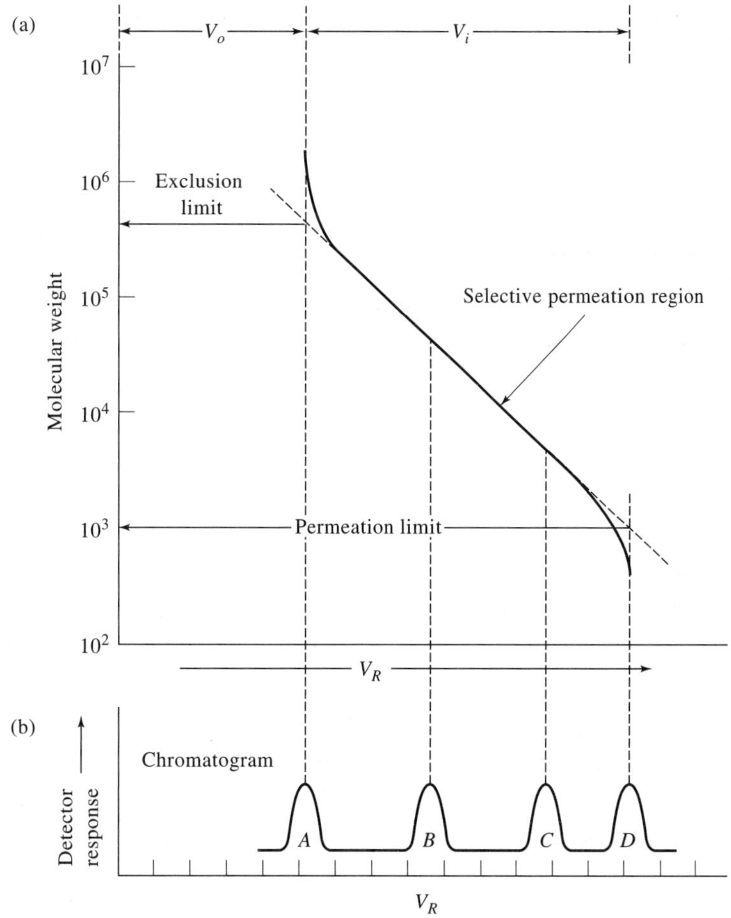

Figure 28-27 (a) Calibration curve for a size-exclusion column. (b) Chromatogram showing peak A containing all compounds with molecular weights greater than the exclusion limit, peaks B and C consisting of compounds within the selective permeation region, and peak D containing all compounds smaller than the permeation limit.

weight are so small that they elute as the single band labeled *D*. As molecular weights decrease from the exclusion limit, solute molecules spend more and more time, on the average, in the particle pores and thus move progressively more slowly. It is in the selective permeation region that fractionation occurs, yielding individual solute peaks such as *B* and *C* in the chromatogram.

Experimental calibration curves, similar in appearance to the hypothetical one in Figure 28-27a, are readily obtained by means of standards. Often, such curves are supplied by manufacturers of packing materials.

28G-3 Application of Size-Exclusion Chromatography

Size-exclusion methods are subdivided into *gel filtration* and *gel permeation* chromatography. The former use aqueous solvents and hydrophilic packings. The latter are based upon nonpolar organic solvents and hydrophobic packings. The methods are complementary in the sense that the one is applied to water-soluble samples and the other to substances soluble in less polar organic solvents.

One useful application of the size-exclusion procedure is to the separation of high-molecular-weight, natural-product molecules from low-molecular-weight species and from salts. For example, a gel with an exclusion limit of several thousand can clearly separate proteins from amino acids and low-molecular-weight peptides.

Another useful application of gel permeation chromatography is to the separation of homologs and oligomers. These applications are illustrated by the two examples shown in Figure 28-28. The first shows the separation of a series of fatty acids ranging in molecular weight from 116 to 344 on a polystyrene-based packing with an exclusion limit of 1000. The second is a chromatogram of a commercial epoxy resin, again on a polystyrene packing. Here, *n* refers to the number of monomeric units (mw = 264) in the molecules.

Figure 28-29 illustrates an application of size-exclusion chromatography to the determination of glucose, sucrose, and fructose in four types of fruit juices. The packing, which had an exclusion limit of 1000, was a cross-linked polystyrene polymer made hydrophilic by sulfonation. A 25-cm column with this packing contained 7600 plates at 80°C, the temperature used.

Another major application of size-exclusion chromatography is to the rapid determination of molecular weight or molecular-weight distribution of larger polymers or natural products. Here, the elution volumes of the sample are compared with elution volumes for a series of standard compounds that possess the same chemical characteristics.

The most important advantages of size-exclusion procedures include: (1) short and well-defined separation times [all solutes leave the column between V_o and $(V_o + V_i)$ in Equation 28-9 and Figure 28-27]; (2) narrow bands, which lead to good sensitivity; (3) freedom from sample loss because solutes do not interact with the stationary phase; and (4) absence of column deactivation brought about by interaction of solute with the packing.

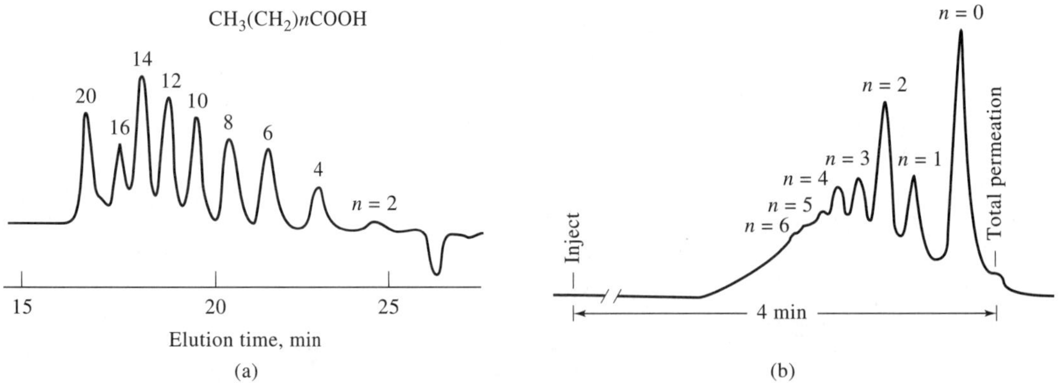

Figure 28-28 Applications of size-exclusion chromatography. (a) Separation of fatty acids. Column: polystyrene based, 7.5 × 600 nm, with exclusion limit of 1×10^3. Mobile phase: tetrahydrofuran. Flow rate: 1.2 mL/min. Detector: refractive index. (b) Analysis of a commercial epoxy resin (*n* = number monomeric units in the polymer). Column: porous silica 6.2 × 250 mm. Mobile phase: tetrahydrofuran. Flow rate: 1.3 mL/min. Detector: UV absorption. *(Courtesy of DuPont Instrument Systems, Wilmington, DE.)*

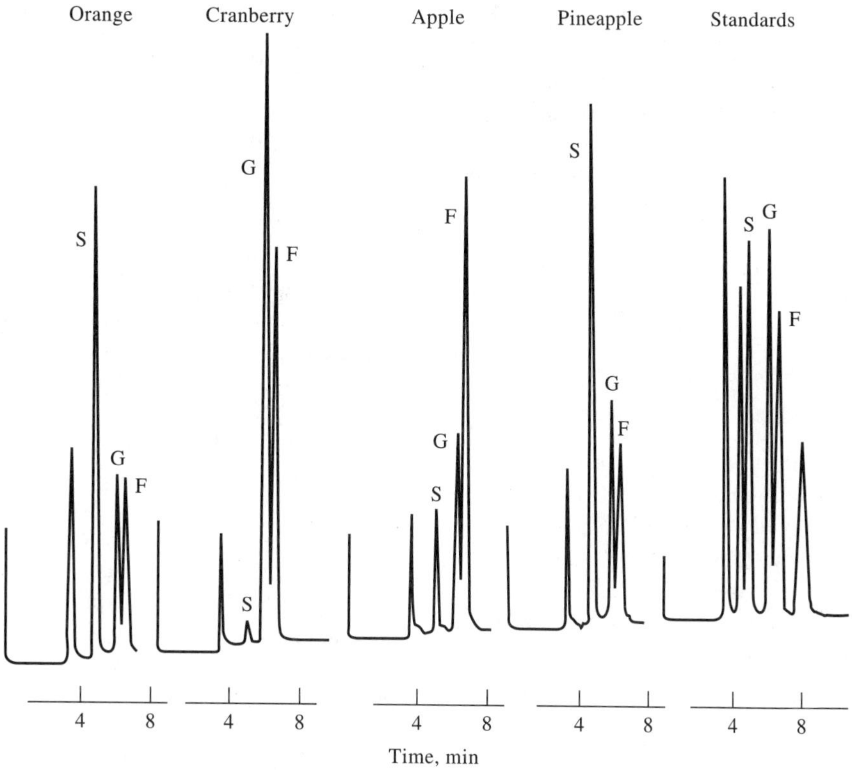

Figure 28-29 Determination of glucose (G), fructose (F), and sucrose (S) in canned juices. *(Courtesy of Perkin-Elmer Corporation, Norwalk, CT.)*

The disadvantages are: (1) only a limited number of bands can be accommodated because the time scale of the chromatogram is short; and (2) inapplicability to samples of similar size, such as isomers. Generally, a 10% difference in molecular weight is required for reasonable resolution.

28H THIN-LAYER CHROMATOGRAPHY

Planar chromatographic methods include *thin-layer chromatography* (TLC) and *paper chromatography* (PC), and *electrochromatography.* Each makes use of a flat, relatively thin layer of material that is either self-supporting or is coated on a glass, plastic, or metal surface. The mobile phase moves through the stationary phase by capillary action, sometimes assisted by gravity or an electrical potential. Planar chromatography is sometimes called two-dimensional chromatography, although this description is not strictly correct inasmuch as the stationary phase does have a finite thickness.

Currently, most planar chromatography is based upon the thin-layer technique, which is faster, has better resolution, and is more sensitive than its paper counterpart. This section is devoted to thin-layer methods only.

28H-1 The Scope of Thin-Layer Chromatography

In terms of theory, the types of stationary and mobile phases, and applications, thin-layer and column liquid chromatography are remarkably similar. In fact, an important use of thin-layer chromatography, and the reason this topic is included here, is to serve as a guide to the development of optimal conditions for performing separations by column liquid chromatography. The advantages of following this procedure are the speed and low cost of the exploratory thin-layer experiments. Some chromatographers, in fact, take the position that thin-layer experiments should always precede column experiments.

In addition to its use in developing column chromatographic methods, thin-layer chromatography has become the workhorse of the drug industry for the all-important determination of product purity. It has also found widespread use in clinical laboratories and is the backbone of many biochemical and biological studies. Finally, it finds widespread use in the industrial laboratories.[41] As a consequence of these many areas of application, it has been estimated that at least as many analyses are performed by thin-layer chromatography as by high-performance column liquid chromatography.[42]

28H-2 How Thin-Layer Separations Are Performed

Typical thin-layer separations are performed on flat glass or plastic plates that are coated with a thin and adherent layer of finely divided particles; this layer constitutes the stationary phase. The particles are similar to those described in the discussion of adsorption, normal- and reversed-phase partition, ion-exchange, and size-exclusion column chromatography. Mobile phases are also similar to those employed in high-performance column liquid chromatography.

Thin-Layer Plates

Thin-layer plates are available from several commercial sources at costs that range from \$1 to \$10 per plate. The common plate sizes in centimeters are 5×20, 10×20, and 20×20. Commercial plates come in two categories, conventional and high-performance. The former have thicker layers (200 to 250 μm) of particles having nominal particle sizes of 20 μm or greater. High-performance plates usually have film thicknesses of 100 μm and particle diameters of 5 μm or less. High-performance plates, as their name implies, provide sharper separations in shorter times. Thus, a conventional plate typically will exhibit 2000 theoretical plates in 12 cm with a development time of 25 min. The corresponding figures for a high-performance plate are 4000 theoretical plates in 3 cm requiring 10 min for development. High-performance plates suffer from the disadvantage of having a significantly smaller sample capacity.

Sample Application

Sample application is perhaps the most critical aspect of thin-layer chromatography, particularly for quantitative measurements. Usually the sample, as a 0.01% to 0.1% solution, is applied as a spot 1 to 2 cm from the edge of the plate. For best separation efficiency, the spot should have a minimal diameter—about 5 mm for qualitative work and smaller for quantitative analysis. For dilute solutions, three or four repetitive applications are used, with drying in between.

Manual application of samples is performed by touching a capillary tube containing the sample to the plate or by use of a hypodermic syringe. A number of mechanical dispensers, which increase the precision and accuracy of sample application, are now offered commercially.

Plate Development

Plate development is the process in which a sample is carried through the stationary phase by a mobile phase; it is analogous to elution in liquid chromatography. The most common way of developing a plate is to place a drop of the sample near one edge of the plate and mark its position with a pencil. After the sample solvent has evaporated, the plate is placed in a closed container saturated with vapors of the developing solvent. One end of the plate is immersed in the developing solvent, with care being taken to avoid direct contact between the sample and the developer (Figure 28-30). The developing solvent travels up the plate, being drawn by capillary action between the fine particles. As the developing solvent travels past the point of sample application, it dissolves

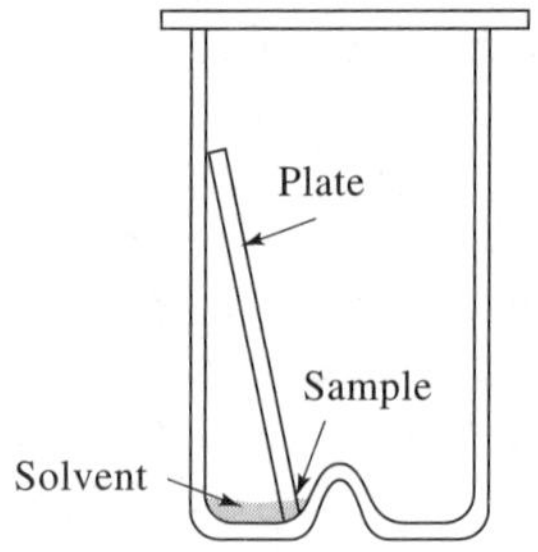

Figure 28-30 A typical development chamber.

[41] Monographs devoted to the principles and applications of thin-layer chromatography include: R. Hamilton and S. Hamilton, *Thin Layer Chromatography.* New York: Wiley, 1987; B. Fried and J. Sherma, *Thin-Layer Chromatography,* 3rd ed. New York: Marcel Dekker, 1994; J. C. Touchstone, *Practice of Thin-Layer Chromatography,* 2nd ed. New York: Wiley, 1983. For briefer reviews, see D. C. Fenimore and C. M. Davis, *Anal. Chem.,* **1981,** *53,* 253A; S. J. Costanzo, *J. Chem. Educ.,* **1984,** *61,* 1015; C. F. Poole and S. K. Poole, *Anal. Chem.,* **1989,** *61,* 1257A.

[42] T. H. Mauch II, *Science,* **1982,** *216,* 161; J. C. Touchstone, *LC-GC,* **1993,** *11*(6), 404.

the sample and carries it up the plate, with the sample distributing itself between the moving solvent and the stationary phase. After the developer has traversed one half or two thirds of the length of the plate, the plate is removed from the container and dried. The positions of the components are then determined in any of several ways.

Locating Analytes on the Plate

Several methods are employed to locate sample components after separation. Two common methods, which can be applied to most organic mixtures, involve spraying with a solution of iodine or sulfuric acid, both of which react with organic compounds to yield dark products. Several specific reagents (such as ninhydrin) are also useful for locating separated species.

Another method of detection is based upon incorporating a fluorescent material into the stationary phase. After development, the plate is examined under ultraviolet light. The sample components quench the fluorescence of the material so that all of the plate fluoresces except where the nonfluorescing sample components are located.

Figure 28-31 is an idealized drawing showing the appearance of a plate after development. Sample 1 contained two components, whereas sample 2 contained one. Frequently, the spots on a real plate exhibit tailing,

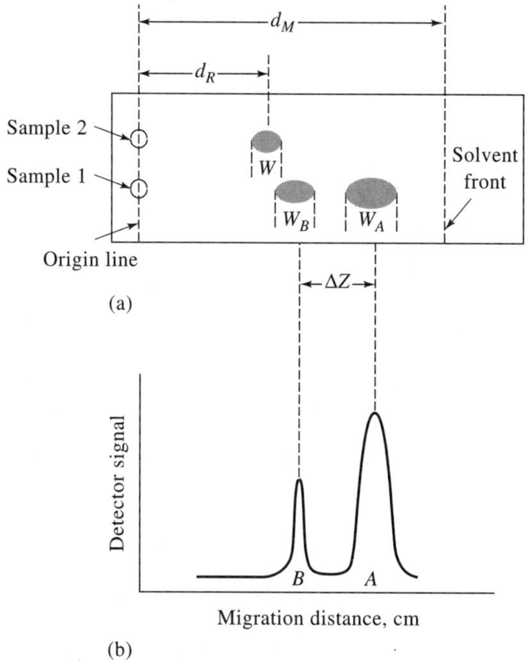

Figure 28-31 Thin-layer chromatograms.

giving spots that are not symmetrical as are those in the figure.

28H-3 Performance Characteristics of Thin-Layer Plates

Most of the terms and relationships developed for column chromatography in Section 26B can, with slight modification, be applied to thin-layer chromatography as well. One new term, the *retardation factor* or R_F factor, is required.

The Retardation Factor

The thin-layer chromatogram for a single solute is shown as chromatogram 2 in Figure 28-31. The retardation factor for this solute is given by

$$R_F = \frac{d_R}{d_M} \qquad (28\text{-}12)$$

where d_R and d_M are linear distances measured from the origin line. Values for R_F can vary from 1 for solutes that are not retarded to a value that approaches 0. It should be noted that if the spots are not symmetric, as they are in Figure 28-31, the measurement of d_R is based on the position of maximum intensity.

The Retention Factor

All of the equations in Table 26-5 are readily adapted to thin-layer chromatography. In order to apply these equations, it is only necessary to relate d_R and d_M as defined in Figure 28-31 to t_R and t_M, which are defined in Figure 26-4 (page 679). To arrive at these relationships, consider the single solute that appears in chromatogram 2 in Figure 28-31. Here, t_M and t_R correspond to times required for the mobile phase and the solute to travel a fixed distance—in this case, d_R. For the mobile phase, this time is equal to the distance divided by its linear velocity u, or

$$t_M = d_R/u \qquad (28\text{-}13)$$

The solute does not reach this same point, however, until the mobile phase has traveled the distance d_M. Therefore,

$$t_R = d_M/u \qquad (28\text{-}14)$$

Substitution of Equations 28-13 and 28-14 into Equation 26-8 yields

$$k' = \frac{d_M - d_R}{d_R} \qquad (28\text{-}15)$$

The retention factor k' can also be expressed in terms of the retardation factor by rewriting Equation 28-15 in the form

$$k' = \frac{1 - d_R/d_M}{d_R/d_M} = \frac{1 - R_F}{R_F} \qquad (28\text{-}16)$$

Retention factors derived in this way can be used for method development in column chromatography as described in Section 28D-2. Obtaining retention factors by thin-layer chromatography is, however, usually simpler and more rapid than obtaining the data from experiments on a column.

Plate Heights

Approximate plate heights can also be derived for a given type of packing by thin-layer chromatographic measurements. Thus, for sample 2 in Figure 28-31, the plate count is given by the equation

$$N = 16\left(\frac{d_R}{W}\right)^2 \qquad (28\text{-}17)$$

where d_R and W are defined in the figure. The plate height is then given by

$$H = d_R/N \qquad (28\text{-}18)$$

28H-4 Applications of Thin-Layer Chromatography

Qualitative Thin-Layer Chromatography

The data from a single chromatogram usually do not provide sufficient information to permit identification of the various species present in a mixture because of the variability of R_F values with sample size, the thin-layer plate, and the conditions extant during development. In addition, the possibility always exists that two quite different solutes may exhibit identical or nearly identical R_F values under a given set of conditions.

Variables That Influence R_F. At best, R_F values can be reproduced to but two significant figures; among several plates, one significant figure may be a more valid statement of precision. The most important factors that determine the magnitude of R_F include thickness of the stationary phase, moisture content of the mobile and stationary phases, temperature, degree of saturation of the developing chamber with mobile phase vapor, and sample size. Complete control of these variables is generally not practical. Partial amelioration of their effects can often be realized, however,

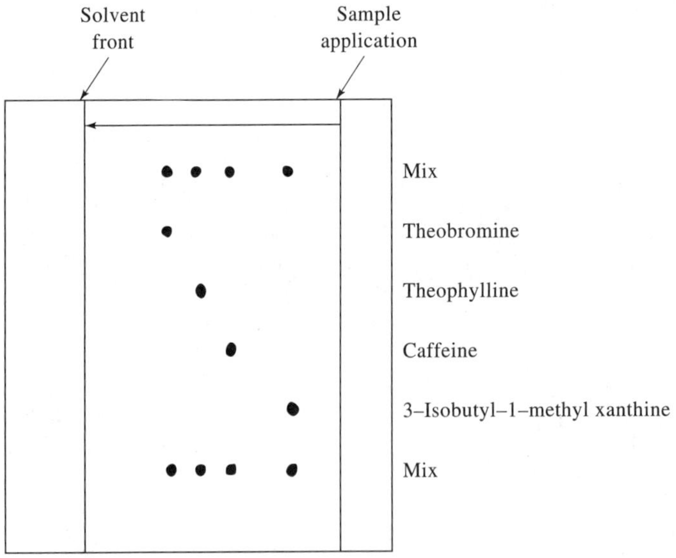

Figure 28-32 Separation of xanthine derivatives on a C-18 reversed-phase plate. Mobile phase: methanol/0.1 M K_2HPO_4 (55:45 v/v). Detection: iodine vapor. Development time: 1 hr, R_F values: theobromine 0.68, theophylline 0.56, caffeine 0.44, 3-isobutyl-methyl xanthine 0.21.

by substituting a relative retention factor R_X for R_F, where

$$R_X = \frac{\text{travel distance of analyte}}{\text{travel distance of a standard substance}}$$

Use of Authentic Substances. A method that often provides tentative identification of the components of a sample is to apply to the plate the unknown and solutions of purified samples of species likely to be present in that unknown. A match in R_F values between a spot for the unknown and that for a standard provides strong evidence as to the identity of one of the components of the sample (see Figure 28-32). Confirmation is always necessary, however. A convenient confirmatory test is to repeat the experiment with different stationary and mobile phases as well as with different visualization reagents.

Elution Methods. The identity of separated analyte species can also be confirmed or determined by a scraping and dissolution technique. Here, the area containing the analyte is scraped from the plate with a razor or a spatula and the contents collected on a piece of glazed paper. After transfer to a test tube or other container, the analyte is dissolved with a suitable solvent and separated from the stationary phase by centrifugation or filtration. Identification is then carried out by such techniques as mass spectrometry, nuclear magnetic resonance, or infrared spectroscopy.

Two-Dimensional Planar Chromatography. Figure 28-33 illustrates the separation of amino acids in a mixture by development in two dimensions. The sample was placed in one corner of a square plate and development was performed in the ascending direction with solvent A. This solvent was then removed by evaporation, and the plate was rotated 90 deg, following which an ascending development with solvent B was performed. After solvent removal, the positions of the amino acids were determined by spraying with ninhydrin, a reagent that forms a pink to purple product with amino acids. The spots were identified by comparison of their positions with those of standards.

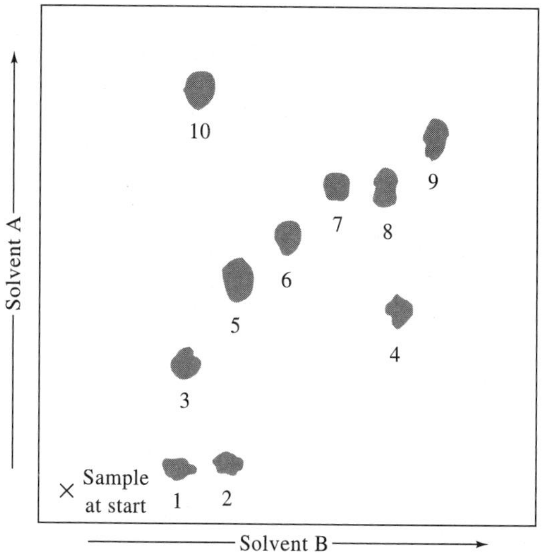

Figure 28-33 Two-dimensional thin-layer chromatogram (silica gel) of some amino acids. Solvent A: toluene/2-chloroethanol/pyridine. Solvent B: chloroform/benzyl alcohol/acetic acid. Amino acids: (1) aspartic acid, (2) glutamic acid, (3) serine, (4) β-alanine, (5) glycine, (6) alanine, (7) methionine, (8) valine, (9) isoleucine, and (10) cysteine.

Quantitative Analysis

A semiquantitative estimate of the amount of a component present can be obtained by comparing the area of a spot with that of a standard. Better data can be obtained by scraping the spot from the plate, extracting the analyte from the stationary-phase solid, and measuring the analyte by a suitable physical or chemical method. In a third method, a scanning densitometer can be employed to measure the radiation emitted from the spot by fluorescence or reflection.

28I QUESTIONS AND PROBLEMS

28-1 List the kinds of substances to which each of the following kinds of chromatography is the most applicable:
(a) gas-liquid
(b) liquid-partition
(c) reversed-phase partition
(d) ion-exchange
(e) gel-permeation
(f) gel-filtration
(g) gas- solid
(h) liquid-adsorption
(i) ion-pair

28-2 Describe three general methods for improving resolution in partition chromatography.

28-3 Describe a way to manipulate the retention factor of a solute in partition chromatography.

28-4 How can the selectivity factor be manipulated in (a) gas chromatography and (b) liquid chromatography?

28-5 In preparing a benzene/acetone gradient for an alumina HPLC column, is it desirable to increase or decrease the proportion of benzene as the column is eluted?

28-6 Define the following terms:
(a) sparging
(b) isocratic elution
(c) gradient elution
(d) stop-flow injection
(e) pellicular packing
(f) extra-column broadening
(g) reversed-phase packing
(h) normal-phase packing
(i) bulk property detector
(j) solute property detector
(k) electrophoresis micelles

28-7 What is meant by the linear response range of a detector?

28-8 What is a guard column in partition chromatography?

28-9 In what way are normal-phase partition chromatography and adsorption chromatography similar?

28-10 List the desirable characteristics of HPLC detectors.

28-11 Describe some of the techniques that are used to interface a liquid chromatograph with a mass spectrometer.

28-12 List the differences in properties and roles of the mobile phases in gas and liquid chromatography. How do these differences influence the characteristics of the two methods?

28-13 In a normal-phase partition column, a solute was found to have a retention time of 29.1 min, while an unretained sample had a retention time of 1.05 min when the mobile phase was 50% by volume chloroform and 50% n-hexane. Calculate (a) k' for the solute and (b) a solvent composition that would bring k' down to a value of about 10.

28-14 The mixture of solvents in Problem 28-13 did not provide a satisfactory separation of two solutes. Suggest how the mobile phase might be altered to improve the resolution.

28-15 Suggest a type of liquid chromatography that would be suitable for the separation of
(a) 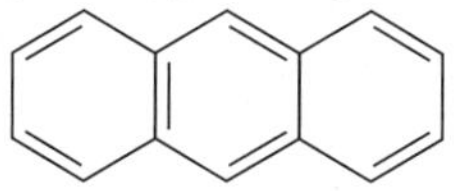 and 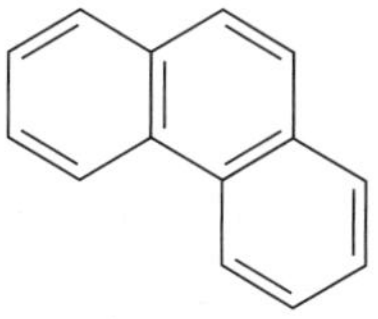

(b) CH_3CH_2OH and $CH_3CH_2CH_2OH$,
(c) Ba^{2+} and Sr^{2+}.
(d) C_4H_9COOH and $C_5H_{11}COOH$.
(e) high-molecular-weight glucosides.

28-16 What is a suppressor column and why is it employed?

28-17 On a silica gel column, a compound was found to have a retention time of 28 min when the mobile phase was toluene. Which solvent, carbon tetrachloride or chloroform, would be more likely to shorten the retention time? Explain.

28-18 For a normal-phase separation, predict the order of elution of
(a) *n*-hexane, *n*-hexanol, benzene.
(b) ethyl acetate, diethyl ether, nitrobutane.

28-19 For a reversed-phase separation, predict the order of elution of the solutes in Problem 28-18.

28-20 Estimate the distribution constant for compounds B and C in Figure 28-27 if the retention volume for compound A was 5.1 mL and for compound D was 14.2 mL.

28-21 What is a mass chromatogram and how is it obtained?

28-22 Two chromatographic peaks were found to have average peak widths of 9 s with a flow rate of 2.5 mL/min. What is the maximum volume of sample that can be taken if the loss of resolution is to be less than (a) 5% and (b) 20%?

Chapter

29

Supercritical Fluid Chromatography and Extraction

***D**uring the past two decades, two new techniques that are based upon the use of supercritical fluids have been developed and promise to play an important role in the analysis of environmental, biomedical, and food samples. These new methods are **supercritical fluid chromatography** (SFC) and **supercritical fluid extraction** (SFE). Instruments for both of these techniques have been commercialized during the last decade and their use appears to be growing rapidly in the analytical community. In this chapter we describe the theory, instrumentation, and applications of both of these methods.*

29A PROPERTIES OF SUPERCRITICAL FLUIDS

The *critical temperature* of a substance is the temperature above which a distinct liquid phase cannot exist, regardless of pressure. The vapor pressure of a substance at its critical temperature is its *critical pressure.* At temperatures and pressures above its critical temperature and pressure (its *critical point*), a substance is called a *supercritical fluid.* Supercritical fluids have densities, viscosities, and other properties that are intermediate between those of the substance in its gaseous and in its liquid state. Table 29-1 compares certain properties of supercritical fluids to those of typical gases and liquids. The properties chosen are those that are of importance in gas, liquid, and supercritical fluid chromatography and extractions.

Table 29-2 lists properties of four of perhaps two dozen compounds that have been employed as mobile phases in supercritical fluid chromatography. Note that their critical temperatures, and the pressures at these temperatures, are well within the operating conditions of ordinary HPLC.

An important property of supercritical fluids, which is related to their high densities (0.2 to 0.5 g/cm^3), is their remarkable ability to dissolve large, nonvolatile molecules. For example, supercritical carbon dioxide readily dissolves *n*-alkanes containing from 5 to over 30 carbon atoms, di-*n*-alkylphthalates in which the alkyl groups contain 4 to 16 carbon atoms, and various polycyclic aromatic hydrocarbons made up of several rings. It is perhaps noteworthy that certain important industrial processes are based upon the high solubility of organic species in supercritical carbon

TABLE 29-1 Comparison of Properties of Supercritical Fluids with Liquids and Gases (all of the data are order-of-magnitude only)

	Gas (STP)	Supercritical Fluid	Liquid
Density (g/cm^3)	$(0.6–2) \times 10^{-3}$	0.2–0.5	0.6–2
Diffusion coefficient (cm^2/s)	$(1–4) \times 10^{-1}$	10^{-3}–10^{-4}	$(0.2–2) \times 10^{-5}$
Viscosity ($g\ cm^{-1}\ s^{-1}$)	$(1–3) \times 10^{-4}$	$(1–3) \times 10^{-4}$	$(0.2–3) \times 10^{-2}$

dioxide. For example, this medium has been employed for extracting caffeine from coffee beans to give decaffeinated coffee and for extracting nicotine from cigarette tobacco.

A second important property of many supercritical fluids is that analytes dissolved in them can be easily recovered by simply allowing the solutions to equilibrate with the atmosphere at relatively low temperatures. Thus an analyte dissolved in supercritical carbon dioxide, the most commonly used solvent, can be recovered by simply reducing the pressure and allowing the fluid to evaporate under ambient laboratory conditions. This property is particularly important with thermally unstable analytes. Another advantage of many supercritical fluids is that they are inexpensive, innocuous, and nontoxic substances that can be allowed to evaporate into the atmosphere with no harmful environmental effects. Supercritical fluid carbon dioxide is especially attractive for extractions and chromatography. Finally, supercritical fluids have the advantage of solute diffusivities that are an order of magnitude higher and viscosities that are an order of magnitude lower than liquid solvents. As we shall show, these last two advantages are important in both chromatography and extractions with supercritical fluids.

29B SUPERCRITICAL FLUID CHROMATOGRAPHY

Supercritical fluid chromatography (SFC) is a hybrid of gas and liquid chromatography that combines some of the best features of each.[1] This technique is an important third kind of column chromatography that is beginning to find use in many industrial, regulatory, and academic laboratories. In 1985, several instrument manufacturers began to offer equipment specifically designed for supercritical fluid chromatography, and since that time, the use of such equipment has expanded at a rapid pace.

Supercritical fluid chromatography is of importance because it permits the separation and determination of a group of compounds that are not conveniently

[1] For reviews of this technique, see M. D. Palmieri, *J. Chem. Educ.*, **1988**, *65*, A254; **1989**, *66*, A141; P. R. Griffiths, *Anal. Chem.*, **1988**, *60*, 593A; R. D. Smith, B. W. Wright, and C. R. Yonker, *Anal. Chem.*, **1988**, *60*, 1323A; M. L. Lee and K. E. Markides, *Science*, **1987**, *235*, 1342. For a monograph on the subject, see *Supercritical Fluid Chromatography*, R. M. Smith, Ed. London: The Royal Society of Chemistry, 1988; C. M. White, *Modern Supercritical Fluid Chromatography*. Heidelberg, FRG: Alfred Huethig Verlag, 1988.

TABLE 29-2 Properties of Some Supercritical Fluids*

Fluid	Critical Temperature, °C	Critical Pressure, atm	Critical Point Density, g/mL	Density at 400 atm, g/mL
CO_2	31.3	72.9	0.47	0.96
N_2O	36.5	71.7	0.45	0.94
NH_3	132.5	112.5	0.24	0.40
n-Butane	152.0	37.5	0.23	0.50

*From M. L. Lee and K. E. Markides, *Science*, **1987**, *235*, 1345. With permission.

handled by either gas or liquid chromatography. These compounds (1) are either nonvolatile or thermally labile so that gas chromatographic procedures are inapplicable and (2) contain no functional groups that make possible detection by the spectroscopic or electrochemical techniques employed in liquid chromatography. Chester has estimated that up to 25% of all separation problems faced by chemists today involve mixtures containing such intractable species.[2]

29B-1 Instrumentation and Operating Variables

As mentioned earlier, the pressures and temperatures required for creating supercritical fluids derived from several common gases and liquids lie well within the operating limits of ordinary HPLC equipment. Thus, as shown in Figure 29-1, instruments for supercritical fluid chromatography are similar in most regards to the instruments for HPLC described in Section 28C. There are two important differences between the two, however. First, a thermostated column oven, similar to that used in gas chromatography (Section 27B-3), is required to provide precise temperature control of the mobile phase; second, a restrictor, or back-pressure device is used to maintain the pressure in the column at a desired level and to convert the eluent from a supercritical fluid to a gas for transfer to the detector. A typical restrictor for a 50- or 100-μm open-tubular column consists of a 2- to 10-cm length of 5- to 10-μm capillary tubing attached directly to the end of the column. Alternatively, the restrictor may be an integral part of the column formed by drawing down the end of the column in a flame. The former permits the use of interchangeable restrictors having different inside diameters, thus providing a range of flow rates at any given pumping pressure.

As shown in Figure 29-1, a commercial instrument for SFC is ordinarily equipped with one or more microprocessors to control such instrument variables as pumping pressure, oven temperature, and detector performance.

Effects of Pressure

Pressure changes in supercritical chromatography have a pronounced effect on the retention, or capacity, factor k'. This effect is a consequence of the increase in density of the mobile phase with increases in pressure. Such density increases cause a rise in solvent power of the mobile phase, which in turn shortens the elution time for eluents. As an example of the effect of pressure increases, it is found that when the carbon dioxide pressure in a packed column is increased from 70 to 90 atm, the elution time for hexadecane decreases from about 25 to 5 min. This effect is general and produces results that are analogous to those obtained with gradient elution in gas and liquid chromatography. Figure 29-2 illustrates the improvement in chromatograms realized by pressure programming. The most common pressure profiles used in supercritical fluid chromatography are often constant (isobaric) for a given length of time followed by a linear or asymptotic increase to a final pressure.

[2] T. L. Chester, *J. Chromatogr. Sci.*, **1986**, *24*, 226.

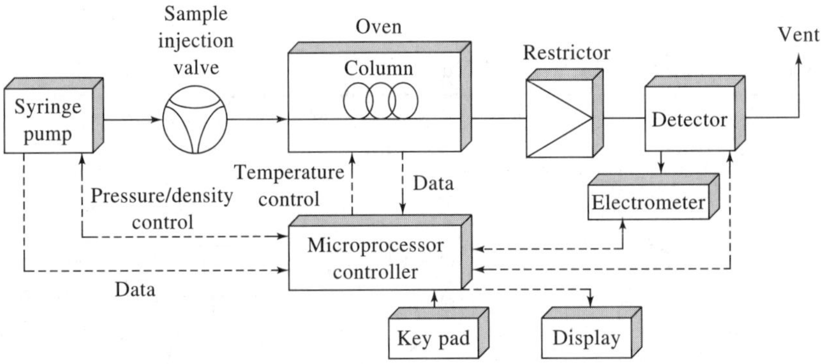

Figure 29-1 Schematic of an instrument for supercritical fluid chromatography.

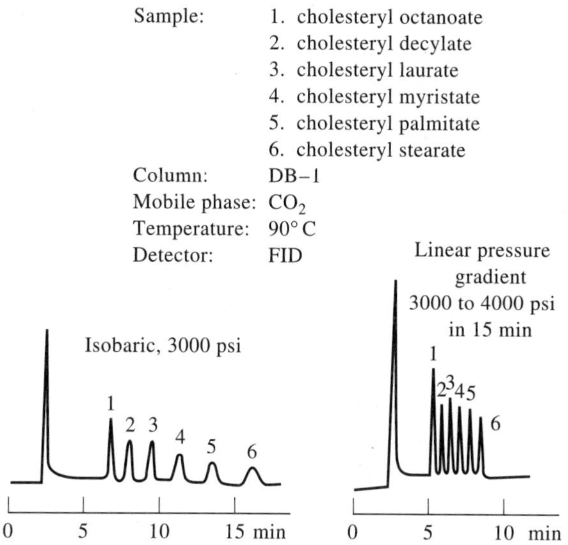

Figure 29-2 Effect of pressure programming in supercritical fluid chromatography. *(Courtesy of Brownlee Labs., Santa Clara, CA.)*

Stationary Phases

Both open-tubular and packed columns are used for SFC, although currently the former are favored. Open-tubular columns are similar to the fused-silica columns described in Section 27C-1 with internal coatings of bonded and cross-linked siloxanes of various types. Column lengths are often 10 or 20 m and inside diameters are 0.05 or 0.10 mm. Film thicknesses vary from 0.05 to 1 μm. Packed columns, similar to those used in partition liquid chromatography, are also employed in SFC. These columns vary in diameter from 0.5 mm or less to 4.6 mm with particle diameters ranging from 3 to 10 μm. The coatings are similar to those used in partition HPLC.

Mobile Phases

The most widely used mobile phase for supercritical fluid chromatography is carbon dioxide. It is an excellent solvent for a variety of nonpolar organic molecules. In addition, it transmits in the ultraviolet, is odorless, nontoxic, readily available, and remarkably inexpensive when compared with other chromatographic mobile phases. Carbon dioxide's critical temperature of 31°C and its critical pressure of 72.9 atm permit a wide selection of temperatures and pressures without exceeding the operating limits of modern HPLC equipment. In some applications, particularly with polar analytes, polar organic modifiers such as methanol are introduced in small concentrations (≈1%) to modify α values for analytes.

A number of other substances have served as mobile phases for supercritical chromatography, including ethane, butane, nitrous oxide, dichlorodifluoromethane, diethyl ether, ammonia, and tetrahydrofuran.

Detectors

A major advantage of SFC over HPLC is that the flame ionization detector of gas chromatography can be employed. As indicated in Section 27B-4, this detector exhibits a *general response* to organic compounds, is highly sensitive, and is largely trouble free. Mass spectrometers are also more easily adapted as detectors for SFC than HPLC. Several of the detectors used in liquid chromatography find use in SFC as well, including ultraviolet and infrared absorption, fluorescence emission, thermionic, and flame photometric detectors.

29B-2 Comparison of Supercritical to Other Types of Chromatography

The data in Tables 29-1 and 29-2 reveal that several physical properties of supercritical fluids are intermediate between gases and liquids. As a consequence, this new type of chromatography combines some of the characteristics of both gas and liquid chromatography. For example, like gas chromatography, supercritical fluid chromatography is inherently faster than liquid chromatography because the lower viscosity makes possible the use of higher flow rates. Diffusion rates in supercritical fluids are intermediate between those in gases and in liquids. As a consequence, band broadening is greater in supercritical fluids than in liquids but less than in gases. Thus, the intermediate diffusivities and viscosities of supercritical fluids should in theory result in faster separations than are achieved with liquid chromatography accompanied by lower zone spreading than is encountered in gas chromatography.

Figures 29-3 and 29-4 compare the performance characteristics of a packed column when elution is performed with supercritical carbon dioxide and a conventional liquid mobile phase. In Figure 29-3 it is seen that at a linear mobile-phase velocity of 0.6 cm/s, the supercritical column yields a plate height of 0.013 mm while the plate height with a liquid eluent is three times as large or 0.039 mm. Thus, a reduction in peak width by a

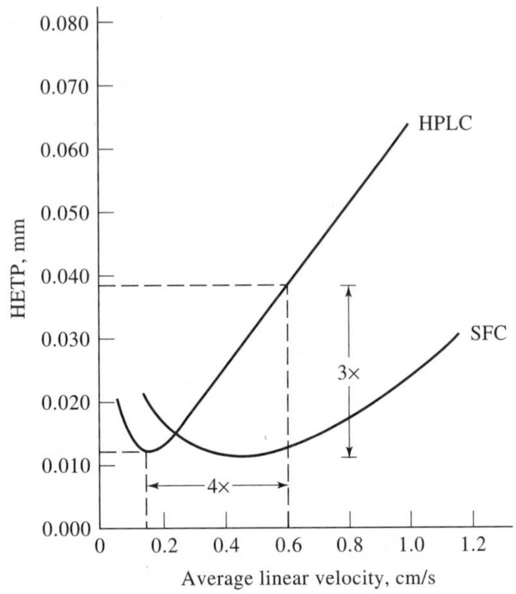

Figure 29-3 Performance characteristics of a 5-μm ODS column when elution is carried out with a conventional mobile phase (HPLC) and supercritical carbon dioxide SFC. *(From D. R. Gere,* Application Note 800-3. *Hewlett-Packard Company, 1983. With permission.)*

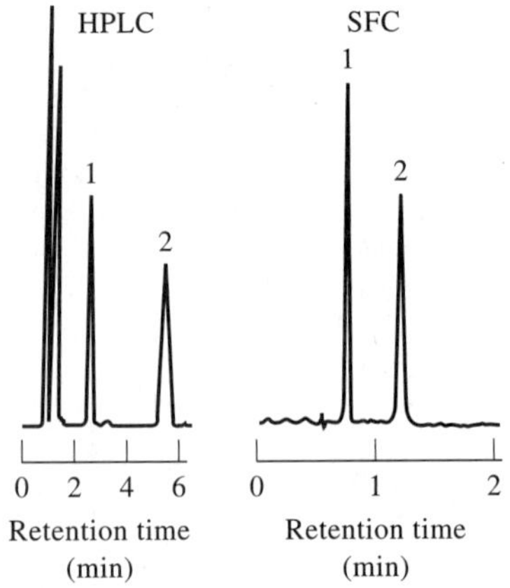

Figure 29-4 Comparison of chromatograms obtained by conventional partition chromatography (HPLC) and supercritical fluid chromatography (SFC). Column: 20 cm × 4.6 mm packed with 10 μm reversed-phase bonded. Analytes: (1) biphenyl; (2) terphenyl. For HPLC: mobile phase, 65/35% CH_3OH/H_2O; flow rate, 4 mL/min; linear velocity, 0.55 cm/s; sample size, 10 μL. For SFC: mobile phase, CO_2; flow rate, 5.4 mL/min; linear velocity, 0.76/s; sample size, 3 μL. *(From D. R. Gere, T. J. Stark, and T. N. Tweeten,* Application Note 800-4. *Hewlett-Packard Company, 1983. With permission.)*

factor of $\sqrt{3}$ should be realized. Alternatively, there is a gain of 4 in linear velocity at a plate height corresponding to the minima in the HPLC curve; this gain would result in a reduction of analysis time by a factor of 4. These advantages are reflected in the two chromatograms shown in Figure 29-4.

It is worthwhile comparing the role of the mobile phase in gas, liquid, and supercritical fluid chromatography. Ordinarily in gas chromatography, the mobile phase serves but one purpose—zone movement. As we have seen in Chapter 28, in liquid chromatography, the mobile phase provides not only transport of solute molecules but also interactions with solutes that influence selectivity factors (α). When a molecule dissolves in a supercritical medium, the process resembles volatilization but at a much lower temperature than would normally be used in gas chromatography. Thus, at a given temperature, the vapor pressure for a large molecule in a supercritical fluid may be 10^{10} times greater than in the absence of that fluid. As a consequence, high-molecular-weight compounds, thermally unstable species, polymers, and large biological molecules can be eluted from a column at reasonably low temperatures. Interactions between solute molecules and the molecules of a supercritical fluid must occur to account for their solubility in these media. The solvent power is thus a function of the chemical composition and the density of the fluid. Therefore, in contrast to gas chromatography, the possibility exists for varying α by changing the mobile phase.

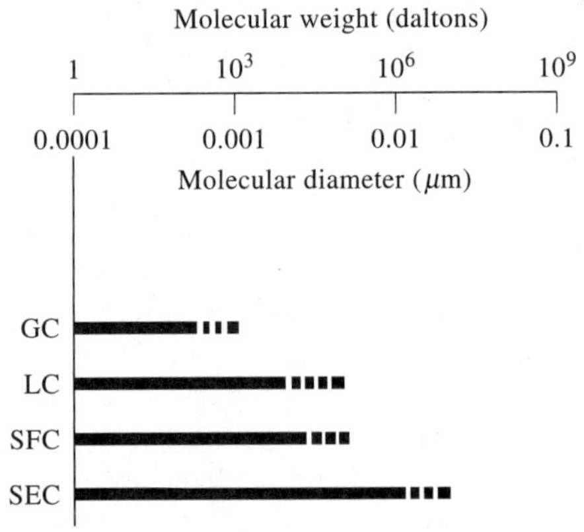

Figure 29-5 Range of molecular weights and sizes over which various column chromatographic techniques can be applied. *(From M. L. Lee and K. E. Markide,* Science, ***1987,*** 235. *With permission.)*

Figure 29-5 compares application range of supercritical fluid chromatography with gas chromatography, liquid partition chromatography, and size-exclusion chromatography (SEC). Note that liquid and supercritical fluid chromatography are applicable over molecular weight ranges that are several orders of magnitude greater than gas chromatography. As we have noted earlier, size-exclusion chromatography can be applied to even larger molecules.

29B-3 Applications

By now, supercritical fluid chromatography has been applied to a wide variety of materials, including natural products, drugs, foods, pesticides and herbicides, surfactants, polymers and polymer additives, fossil fuels, and explosives and propellants. Figures 29-6, 29-7, and 29-8 illustrate three typical and diverse applications of SFC. Figure 29-6 shows the separation of a series of dimethyl polysiloxane oligomers ranging in molecular weight from 400 to 700 Da. This chromatogram was obtained by using a 10 m × 100-μm inside diameter fused-silica capillary coated with a 0.25-μm film of 5% phenyl polysiloxane. The mobile phase was CO_2 at 140°C, and the following pressure program was used: 80 atm for 20 min, then a linear gradient from 80 to 280 atm at 5 atm/min. A flame ionization detector was used.

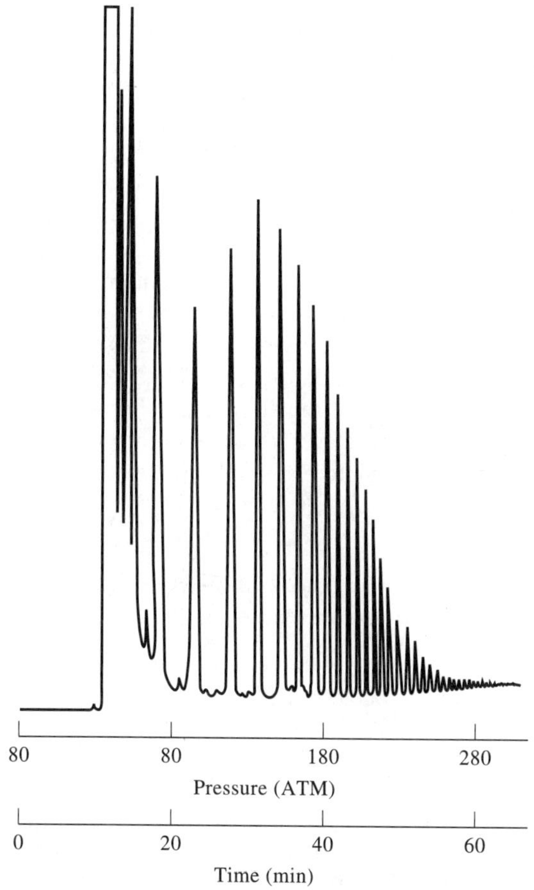

Figure 29-6 Separation of oligomers of dimethylpolysiloxane by supercritical fluid chromatography. *(From C. M. White and R. K. Houck,* HRC & CC, ***1986****, 9, 4. With permission.)*

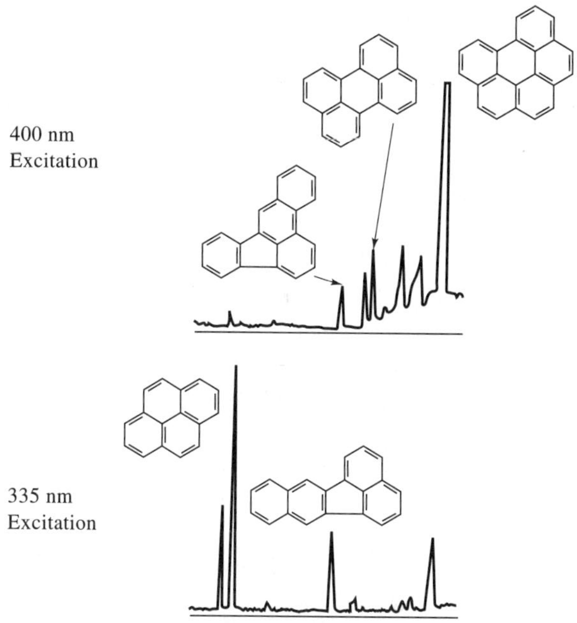

Figure 29-7 Portions of the supercritical fluid chromatograms of polycyclic aromatics in a carbon-black extract, illustrating the selectivity achieved by fluorescence excitation at two wavelengths. *(From C. M. White and R. K. Houck,* HRC & CC, ***1986****, 9, 4. With permission.)*

Figure 29-7 illustrates the separation of polycyclic aromatic hydrocarbons extracted from a carbon black. Detection was by fluorescence excited by two different wavelengths. Note the selectivity provided by this technique. The chromatogram was obtained by using a 40-m × 50-μm inside diameter capillary coated with a 0.25-μm film of 50% phenyl polysiloxane. The mobile phase was pentane at 210°C, and the following program was used: initial mobile phase density held at 0.07 g/mL for 24 min, then an asymptotic density program to 0.197 g/mL.

Figure 29-8 illustrates a separation of the oligomers in a sample of the nonionic surfactant Triton X-100. Detection involved measuring the total ion current produced by chemical ionization mass spectrometry. The mobile phase was carbon dioxide containing 1% by volume of methanol. The column was a 30-m capillary column that was coated with a 1-μm film of 5% phenyl polysiloxane. The column pressure was increased linearly at a rate of 2.5 bar/min.

Several reviews on applications of the supercritical fluid chromatography are now available in the recent literature.[3]

[3] T. L. Chester and J. D. Pinkston, *Anal. Chem.*, **1990**, *62*, 394R; M. D. Palmieri, *J. Chem. Educ.*, **1989**, *66*, A141; C. M. White and R. K. Houk, *HRC & CC*, **1986**, *9*(1), 4.

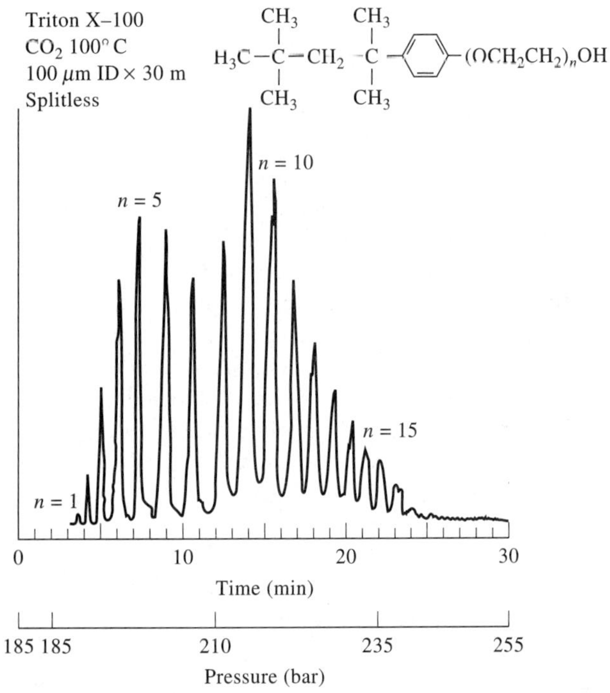

Figure 29-8 Chromatograms for the nonionic surfactant Triton X-100 with total current mass spectrometric detection. *(Reprinted with permission from R. D. Smith and H. R. Udseth,* Anal. Chem., *1987, 59, 17. Copyright 1987 American Chemical Society.)*

29C SUPERCRITICAL FLUID EXTRACTION

Often, the analysis of complex materials requires as a preliminary step separation of the analyte or analytes from a sample matrix. Ideally, an analytical separation method should be rapid, simple, and inexpensive; should give quantitative recovery of analytes without loss or degradation; should yield a solution of the analyte that is sufficiently concentrated to permit the final measurement to be made without the need for concentration; and should generate little or no laboratory wastes that have to be disposed of.

For many years, one of the most common methods for performing analytical separations on complex environmental, pharmaceutical, food, and petroleum samples was based upon extraction of bulk samples with hydrocarbon or chlorinated organic solvents using a Soxhlet extractor. Unfortunately, liquid extractions frequently fail to meet several of the ideal criteria listed in the previous paragraph. They usually require several hours or more to achieve satisfactory recoveries of analytes and sometimes never do. The solvent costs are often high. The solution of the recovered analytes are often so dilute that a concentration step must follow the extraction. Analyte degradation or loss as well as atmospheric pollution may accompany this concentration step.

Beginning in the mid-1980s, chemists began to explore the use of supercritical fluids for the separation of analytes from the matrix of many samples of interest to industry and governmental agencies because use of reagents of this type avoided many of the problems described in the previous paragraphs for organic liquid extractants.[4]

29C-1 Advantages of Supercritical Fluid Extraction

In this section we list the several advantages of supercritical fluid extractions (SFC).

[4] For more detailed information about SFE, see S. B. Hawthorne, *Anal. Chem.*, **1990**, *62*, 633A; J. L. Hedrick, L. J. Mulcahey, and L. T. Taylor, *Mikrochim. Acta*, **1992**, *108*, 115; L. T. Taylor, *Anal. Chem.*, **1995**, *67*, 364A; *Supercritical Fluid Extraction and Its Use in Chromatographic Sample Preparation*, S. A. Westwood. Ed. Boca Raton: CRC Press, 1993; M. D. Luque de Castro, M. Valcarcel, and M. T. Tena, *Analytical Supercritical Fluid Extraction*. New York: Springer Verlag, 1994; L. T. Taylor, *Supercritical Fluid Extraction*. New York: Wiley, 1996.

1. SFC is generally fast. The rate of mass transfer between a sample matrix and an extraction fluid is determined by the rate of diffusion of a species in the fluid and the viscosity of the fluid—the greater the diffusion rate and the lower the viscosity, the greater will be the rate of mass transfer. As we have noted earlier, both of these variables are more favorable for supercritical fluids than for typical liquid solvents. As a consequence, supercritical fluid extractions can generally be completed in 10 to 60 minutes, whereas extractions with an organic liquid may require several hours or even days.
2. The solvent strength of a supercritical fluid can be varied by changes in the pressure and to a lesser extent in the temperature. In contrast, the solvent strength of an organic liquid is essentially constant regardless of conditions. This property allows the conditions for extraction with a supercritical fluid to be optimized for a given class of analytes.
3. Many supercritical fluids are gases at ambient conditions. Thus recovery of analytes becomes simple compared to organic liquids, which must be vaporized by heating, which leads to the possibility of decompositions of thermally unstable analytes or loss of volatile analytes. In contrast, a supercritical fluid can be separated from the analyte by simply releasing pressure. Alternatively, the analyte stream can be bubbled through a small vial containing a good solvent for the analyte, which will dissolve in the small volume of solvent.
4. Some supercritical fluids are cheap, inert, and nontoxic. Thus they are readily disposed of after an extraction is completed by allowing them to evaporate into the atmosphere.

29C-2 Instrumentation

The instrumentation for SFE can be relatively simple as shown in Figure 29-9. Instrument components include a fluid source, most commonly a tank of carbon dioxide; a syringe pump having a pressure rating of at least 400 atm and a flow rate for the pressurized fluid of at least 2 mL/min; a valve to control the flow of the critical fluid into a heated extraction cell having a capacity of a few mL; an exit valve leading to a flow restrictor that depressurizes the fluid and transfers it into a collection device. In the simplest instruments, the flow restrictor is simply a 10- to 50-cm piece of capillary tubing. In modern sophisticated commercial instruments, the restrictors are variable and controlled manually or automatically. Several instrument manufacturers offer various types of SFE apparatus.[5]

A SFE system can be operated in one of two ways. In the *dynamic extraction* mode, the valve between the extraction cell and the restrictor remains open so that the sample is continually supplied with fresh supercritical fluid and the extracted material flows into the collection vessel where depressurization occurs. In the *static extraction* mode, the valve between the extraction cell and the restrictor is closed and the extraction cell is pressurized under static conditions. After a suitable period, the exit valve is opened and the cell contents are transferred through the restrictor by a dynamic flow of fluid from the pump. The dynamic mode is more widely used than the static mode.

[5] See F. Wach, *Anal. Chem.*, **1994**, *66*, 369A.

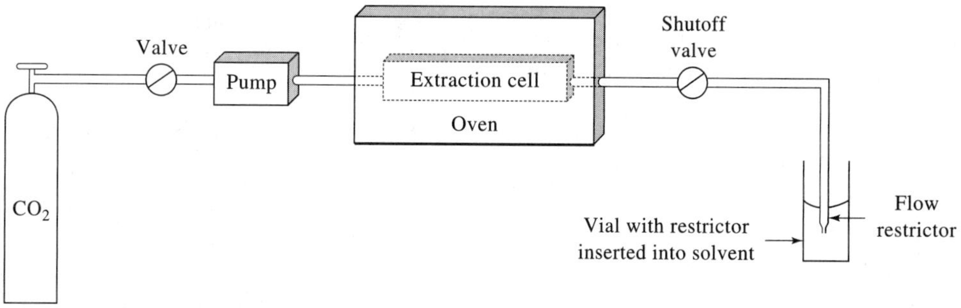

Figure 29-9 A typical arrangement for off-line SFC. The shutoff valve is required for static SFC but not dynamic SFC.

29C-3 Supercritical Fluid Choice

Two dozen or more supercritical fluids have been described as extraction media, but as in SFC by far, the most widely used substance is carbon dioxide or carbon dioxide containing an organic modifier. The best choice of fluid is determined by a number of variables, including polarity and solubility of the analytes and the matrix components, physical nature of the matrix, concentration of the analytes, moisture content of the sample, and kinetic considerations.[6] Unfortunately, the theory of supercritical fluid extractions is currently imperfect, and final conditions in most cases must be determined empirically. Carbon dioxide has been the fluid of choice in most studies. It is an excellent solvent for nonpolar species, such as alkanes and terpenes, and a moderately good extraction medium for moderately polar species, such as polycyclic aromatic hydrocarbons, polychlorinated biphenyls, aldehydes, esters, alcohols, organic chloro pesticides, and fats. It is generally not a good extraction medium for highly polar compounds unless modified by the addition of strongly polar modifiers such as methanol. Modifiers can be introduced into the extraction system either by means of a second pump, or by injecting the modifier into the sample prior to extraction.

A variety of modifiers have been used to enhance the polarity of supercritical fluid carbon dioxide, including several of the lower-molecular-weight alcohols, propylene carbonate, 2-methoxyethanol, methylene chloride, and certain organic acids. The most common is a few percent of methanol. Figure 29-10 demonstrates the improved efficiency in the presence of a small amount of methanol in the extraction of various materials from soil samples.

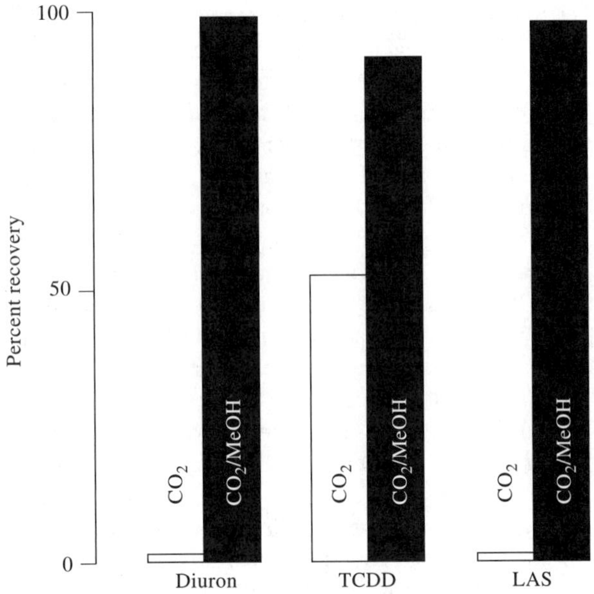

Figure 29-10 Comparison of extraction efficiencies obtained by using CO_2 and CO_2 modified with methanol. Samples were a soil. All extractions were for 30 min. Diuron is a common herbicide that is an aromatic substituted derivative of urea. TCDD is 2,3,7,8-tetrachlorodibenzo-*p*-dioxin. LAS is linear alkylbenzenesulfonate detergent.

29C-4 Off-Line and On-Line Extractions

Two types of methods have been used to collect analytes after extraction: *off-line* and *on-line.* In off-line collection, which is the simpler of the two, the analytes are collected by immersing the restrictor in a few milliliters of solvent and allowing the gaseous supercritical fluid to escape into the atmosphere (see Figure 29-9). Analytes have also been collected on adsorbents, such as silica. The adsorbed analytes are then eluted with a small volume of a liquid solvent. In either case the separated analytes are then identified by any of several optical, electrochemical, or chromatographic methods.

In the on-line method, the effluent from the restrictor, after depressurization, is transferred directly to a chromatographic system. In most cases the latter is a gas chromatograph or a supercritical fluid chromatograph, although occasionally a liquid chromatographic instrument has also been used. The principal advantages of an on-line system is the elimination of sample handling between the extraction and the measurement and the potential for enhanced sensitivity because no dilution of the analyte occurs.

29C-5 Typical Applications of SFE

The last reference in footnote 4 lists over 100 analytical applications of supercritical fluid extractions that have appeared in the literature between 1988 and 1993. A majority of these have been for the analysis of environmental samples. Others have been for the analysis of foods, biomedical, and industrial samples of various types. Table 29-3 provides a few typical applications of off-line and on-line supercritical fluid extractions.

[6] For a review of the effects of many of these variables, see M. E. P. McNally, *Anal. Chem.,* **1995,** *67,* 308A.

TABLE 29-3 Some Typical Applications of Supercritical Fluid Extraction

Material	Analyte*	SF	Extraction Time, min	Off-line (1), On-line (2)
Soils	Pesticides	CO_2	20	1
River sediments	PAHs	CO_2/5% MeOH	120	1
Smoke, urban dust	PAHs	CO_2	15	2
Railroad bed soil	PCBs, PAHs	CO_2/MeOH	45	1
Foods	Fat	CO_2/MeOH	12	1
Spices, bubble gum	Aromas and fragrances	CO_2	10	2
Serum	Cholesterol	CO_2	30	1
Coal, fly ash	PCBs, dioxins	CO_2	15	2
Polymers	Additives and oligomers	CO_2	15	2
Animal tissue	Drug residues	CO_2	9	1

*PAHs = polycyclicaromatic hydrocarbons; PCBs = polychlorinated biphenyls.

29D QUESTIONS AND PROBLEMS

29-1 Define
(a) critical temperature and critical pressure of a gas.
(b) supercritical fluid.

29-2 What properties of a supercritical fluid are important in chromatography?

29-3 How do instruments for supercritical fluid chromatography differ from those for (a) HPLC and (b) GC?

29-4 Describe the effect of pressure on supercritical fluid chromatograms.

29-5 List some of the advantageous properties of supercritical CO_2 as a mobile phase for chromatographic separations.

29-6 Compare supercritical fluid chromatography with other column chromatographic methods. GC, LC

29-7 For supercritical carbon dioxide, predict the effect that the following changes will have upon the elution time in an SFC experiment.
(a) Increase the flow rate (at constant temperature and pressure)
(b) Increase the pressure (at constant temperature and flow rate)
(c) Increase the temperature (at constant pressure and flow rate)

29-8 For supercritical fluid extractions, differentiate between
(a) on-line and off-line processes.
(b) static and dynamic extractions.

29-9 List the advantages of supercritical fluid extractions over liquid extractions.

29-10 How are analytes generally recovered after a supercritical fluid extraction?

Chapter

30

Capillary Electrophoresis and Capillary Electrochromatography

*This chapter deals with two relatively new separation methods, **capillary electrophoresis** and **capillary electrochromatography**, that employ a common type of instrumentation. Both are carried out in a buffer-filled capillary tube that is typically 10 to 100 μm in internal diameter and 40 to 100 cm in length. The tube extends between two buffer reservoirs that also hold platinum electrodes. The sample is introduced into one end of the tubing, and a dc potential in the 20- to 30-kV range is applied between the two electrodes throughout the separation. The separated analytes are observed by a detector that is located at the end of the tube that is opposite to that where the sample was introduced. Figure 30-1 is a schematic of a typical instrument that can be used for either capillary electrophoresis or capillary electrochromatography.*

*Despite the similarity in equipment, the principles upon which the two separation methods are based are fundamentally different. With capillary electrophoresis, separations depend upon the difference in rates at which analyte **ions** migrate under the influence of an electric field. Thus, this technique is only applicable to analytes that are ionic in the buffered medium contained in the capillary. In contrast, capillary electrochromatography is applicable to both ionic and neutral analyte species. Here electroosmotic flow of the buffer solution acts as a pump to move analyte molecules past a second phase that is capable of retaining analyte molecules to various extents. Thus, separations are based upon differences in distribution equilibria as in HPLC. The source of the pumping action by electroosmotic flow is described in detail in Section 30B-3.*

*In this chapter, we first consider the principles of electrophoretic separations, with particular emphasis on capillary electrophoresis, and the applications of this technique to various types of analytical problems. We then turn to two types of electrochromatography, the most widely used of which to date is **micellar electrokinetic chromatography.***

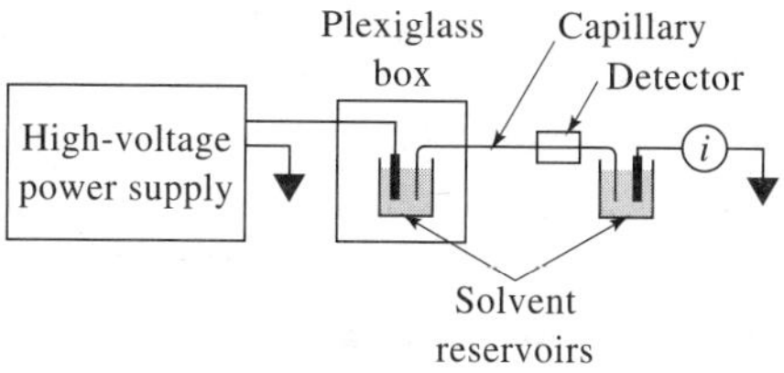

Figure 30-1 Schematic of a capillary zone electrophoresis system.

30A AN OVERVIEW OF ELECTROPHORESIS[1]

Electrophoresis is a separation method based on the differential rate of migration of charged species in a buffer solution across which has been applied a dc electric field. This separation technique was first developed by the Swedish chemist Arne Tiselius in the 1930s for the study of serum proteins; he was awarded the 1948 Nobel Prize for this work.

Electrophoresis has been applied to a variety of difficult analytical separation problems: inorganic anions and cations, amino acids, catecholamines, drugs, vitamins, carbohydrates, peptides, proteins, nucleic acids, nucleotides, polynucleotides, and numerous other species.

A particular strength of electrophoresis is its unique ability to separate charged macromolecules of interest in the biotechnology industry and in biochemical and biological research. For many years, electrophoresis has been the powerhouse method of separation of proteins (enzymes, hormones, antibodies) and nucleic acids (DNA, RNA), for which it offers unparalleled resolution. For example, in order to sequence DNA it is necessary to distinguish between long-chain polynucleotides having molecular weights ranging from perhaps 200 to 500 and differing by only a single nucleotide. Only electrophoresis has sufficient resolving power to handle this problem. The Human Genome Initiative is a large-scale project with the goal of determining the complete sequence for human DNA. The separation requirements for this project are staggering. Human DNA is estimated to contain three billion nucleotides. However, a modern DNA-sequencing laboratory determines the sequence of fewer than a thousand nucleotides in an entire day.

An electrophoretic separation is performed by injecting a small band of the sample into an aqueous buffer solution that is contained in a narrow tube or on a flat porous support medium such as paper or a semisolid gel. A high dc potential is applied across the length of the buffer by means of a pair of electrodes located at either end of the buffer. This potential causes ions of the sample to migrate toward one or the other of the electrodes. The rate of migration of a given species depends upon its charge and also upon its size. Separations are then based upon differences in charge-to-size ratios for the various analytes in a sample. The larger this ratio, the faster an ion migrates in the electrical field.

30A-1 Types of Electrophoresis

Electrophoretic separations are currently performed in two quite different formats: one is called *slab electrophoresis* and the other *capillary electrophoresis.* The first is the classic method that has been used for many years to separate complex, high-molecular-weight species of biological and biochemical interest. Slab separations are carried out on a thin flat layer or *slab* of a porous semisolid gel containing an aqueous buffer solution within its pores. Ordinarily this slab has dimensions of a few centimeters on a side and, like a chromatographic thin layer plate, is capable of separating several samples simultaneously. Samples are introduced as spots or bands on the slab, and a dc potential is applied across the slab for a fixed period. When the separations are judged complete, the current is discontinued and the separated species are visualized by staining in much the same way as was described for thin-layer chromatography in Section 28H-2.

Slab electrophoresis is currently the most widely used separation tool of the biochemist and the biologist. A perusal of monographs, textbooks, or journals in the life science fields usually reveals hundreds of photographs of developed electrophoretic slabs. Capillary electrophoresis, which is an instrumental version of electrophoresis, has been developed and used only in the last decade and one half and has become an important separation tool used by chemists and life scientists. In many cases, this new method of performing electrophoretic separations appears to be a satisfactory substitute for slab electrophoresis with several important advantages, which will be described subsequently.

[1] For a more extensive discussion of electrophoresis, see J. W. Jorgenson, *Anal. Chem.,* **1986,** *58,* 743A; A. T. Andrews, *Electrophoresis: Theory, Techniques, and Clinical Applications.* Cambridge: Oxford University Press, 1986; M. Melvin, *Electrophoresis.* New York: Wiley, 1987.

In this chapter, we shall say very little about classic slab electrophoresis because the literature on this technique is so enormous that it is beyond the scope of this book. In addition, the slab electrophoresis is not really an instrumental method, whereas capillary electrophoresis is. We shall also discuss *electrochromatography,* which is a hybrid technique of electrophoresis and liquid chromatography.

30A-2 The Basis for Electrophoretic Separations

The *migration velocity* v of an ion in centimeters per second in an electric field is equal to the product of the field strength E (V cm^{-1}) and the *electrophoretic mobility* μ_e ($cm^2V^{-1}s^{-1}$). That is,

$$v = \mu_e E \tag{30-1}$$

The electrophoretic mobility is in turn proportional to the ionic charge on the analyte and inversely proportional to frictional retarding factors. The electric field acts only on ions. If two species differ either in charge or in the frictional forces they experience in moving through the buffer, they will be separated from each other. Neutral species are not separated. The frictional retarding force on an analyte ion is determined by the size and shape of the ion and the viscosity of the medium in which it migrates. For ions of the same size, the greater the charge, the greater the driving force and the faster the rate of migration. For ions of the same charge, the smaller the ion, the smaller the frictional forces and the faster the rate of migration. The ion's *charge-to-size ratio* combines these two effects. Note that in contrast to chromatography, only one phase is involved in an electrophoretic separation.

30B CAPILLARY ELECTROPHORESIS

As useful as conventional slab electrophoresis has been, and continues to be, this type of electrophoretic separation is typically slow, labor intensive, difficult to automate, and does not yield very precise quantitative information. During the mid- and late 1980s, there was explosive growth in research and application of electrophoresis performed in capillary tubes as well as the appearance of several commercially available instruments. *Capillary electrophoresis* yields high-speed, high-resolution separations on exceptionally small sample volumes (0.1 to 10 nL in contrast to slab electrophoresis, which requires samples in the μL range). Additionally, the separated species are eluted from one end of the capillary, so quantitative detectors, like those of HPLC, can be used instead of the cumbersome staining techniques of slab electrophoresis.[2]

30B-1 Migration Rates and Plate Heights in Capillary Electrophoresis

As we have already seen in Equation 30-1, an ion's migration velocity v depends upon the electric field strength. The electric field in turn is determined by the magnitude of the applied potential (V, in volts) and the length L over which it is applied. Thus

$$v = \mu_e \cdot \frac{V}{L} \tag{30-2}$$

This relationship indicates that high applied potentials are desirable to achieve rapid ionic migration and a rapid separation. It is desirable to have rapid separations, but it is even more important to achieve high-resolution separations. So we need to examine the factors that determine resolution in electrophoresis.

30B-2 Plate Heights in Capillary Electrophoresis

In chromatography, both longitudinal diffusion and mass-transfer resistance contribute to band broadening. However, since only a single phase is involved for electrophoresis, only longitudinal diffusion need be considered. It has been shown for electrophoresis, that the plate count (N) is given by

$$N = \frac{\mu_e V}{2D} \tag{30-3}$$

where D is the diffusion coefficient of the solute in $cm^2\ s^{-1}$. Because resolution increases as the plate count increases, it is desirable to use high applied potentials in order to achieve high-resolution separations. Note that for electrophoresis, contrary to the situation in chro-

---

[2] For review articles on capillary electrophoresis, see J. W. Jorgenson, *Anal. Chem.,* **1986,** *58,* 743A; A. G. Ewing, R. A. Wallingford, and T. M. Olefirowicz, *Anal. Chem.,* **1989,** *61,* 292A; H. H. Lauer and J. B. Ooms, *Anal. Chim. Acta,* **1991,** *250,* 45. For recent books, see D. R. Baker, *Capillary Electrophoresis.* New York: Wiley, 1995; R. Weinberger, *Practical Capillary Electrophoresis.* New York: Academic Press, 1993; S. Y. Li, *Capillary Electrophoresis.* New York: Elsevier, 1992; *Capillary Electrophoresis,* P. Camilleri, Ed. Boca Raton: CRC Press, 1993; F. Foret, L. Krivankova, and P. Bocek, *Capillary Zone Electrophoresis.* New York: VCH, 1993.

matography, the plate count does not increase with the length of the column.

With gel slab electrophoresis, joule heating limits the magnitude of the applied potential to about 500 V. Here is where one of the strengths of the capillary format compared with the slab format is realized. The long length and small cross-sectional area of the capillary means that the solution resistance through the capillary is exceptionally high. Because power dissipation is inversely proportional to resistance, much higher potentials can be applied. Additionally, the high surface-to-volume ratio of the capillary provides efficient cooling. As a consequence of these two factors, band broadening due to thermally driven convective mixing does not occur to any significant extent. Potentials of 20,000 to 60,000 V are commonly used with capillary electrophoresis, and these high potentials lead to corresponding improvements in speed and resolution over that seen in slab format. Capillary electrophoresis peak widths often approach the theoretical limit set by longitudinal diffusion. Capillary electrophoresis normally yields plate counts in the range of 100,000 to 200,000 as compared to 5,000 to 20,000 plates typical for HPLC. Plate counts of 3,000,000 have been reported for capillary zone electrophoresis of dansylated amino acids,[3] and plate counts of 10,000,000 have been reported for capillary gel electrophoresis of polynucleotides.[4]

[3] R. D. Smith, J. A. Olivares, N. T. Nguyen, and H. R. Udseth, *Anal. Chem.*, **1988**, *60*, 436.

[4] A. Guttman, A. S. Cohen, D. N. Heiger, and B. L. Karger, *Anal. Chem.*, **1990**, *62*, 137.

30B-3 Electroosmotic Flow

When a high potential is applied across a capillary tube containing a buffer solution, *electroosmotic flow* usually occurs, in which the solvent migrates toward the cathode or the anode. The rate of migration can be appreciable. For example, a 50-mM pH 8 buffer has been found to flow through a 50-cm capillary toward the cathode at approximately 5 cm/min with an applied potential of 25 kV.[5]

As shown in Figure 30-2, the cause of electroosmotic flow is the electric double layer that develops at the silica/solution interface. Above pH 3, the inside wall of a silica capillary is negatively charged due to ionization of the surface silanol groups (Si—OH). Buffer cations congregate in an electrical double layer adjacent to the negative surface of the silica capillary. The properties of an electrical double layer were described in Section 22A-2. The cations in the diffuse outer layer of the double layer are attracted toward the cathode, or negative electrode; and since the cations are solvated, they drag the bulk solvent along with them. As shown in Figure 30-3, electroosmosis leads to bulk solution flow that has a flat profile across the tube rather than a parabolic profile as with the pressure-driven flow encountered in liquid chromatography. Because the profile is essentially flat, electroosmotic flow does not contribute significantly to band broadening in the way pressure-driven flow does in chromatography.

[5] J. D. Olechno, J. M. Y. Tso, J. Thayer, and A. Wainright, *Amer. Lab.*, **1990**, *22*(17), 51.

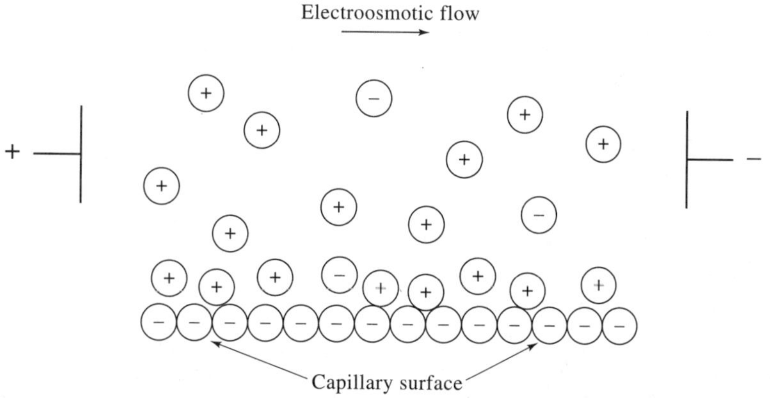

Figure 30-2 Charge distribution at a silica/capillary interface and resulting electroosmotic flow. *(From A. G. Ewing, R. A. Wallingford, and T. M. Olefirowicz,* Anal. Chem., ***1989,*** 61, *294A. With permission.)*

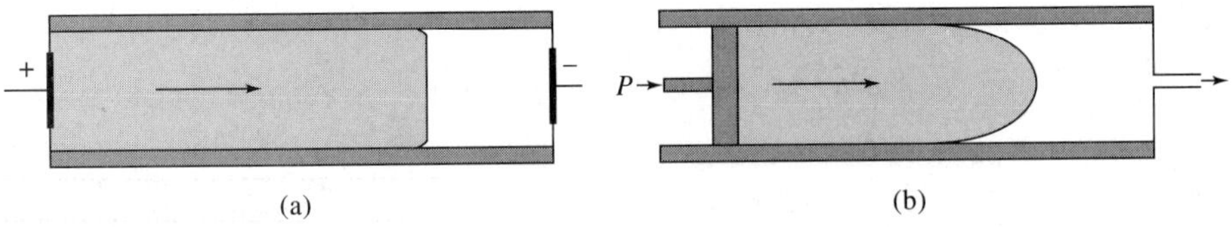

Figure 30-3 Flow profiles for liquids under (a) electroosmotic pressure and (b) hydrodynamic pressure.

The rate of electroosmotic flow is generally greater than the electrophoretic migration velocities of the individual ions and effectively becomes the mobile-phase pump of capillary zone electrophoresis. Even though analytes migrate according to their charges within the capillary, the electroosmotic flow rate is usually sufficient to sweep all positive, neutral, and even negative species toward the same end of the capillary, so that all can be detected as they pass by a common point. The resulting *electropherogram* looks just like a chromatogram.

The electroosmotic flow velocity v is given by an equation similar to Equation 30-1. That is,

$$v = \mu_{eo}E \qquad (30\text{-}4)$$

In the presence of electroosmosis, an ion's velocity is the sum of its migration velocity and the velocity of electroosmotic flow. Thus,

$$v = (\mu_e + \mu_{eo})E \qquad (30\text{-}5)$$

Note that μ_e for an anion will carry a negative sign.

As a result of electroosmosis, order of elution in a typical capillary electrophoretic separation is first the fastest cation followed by successively slower cations, then all the neutrals in a single zone, and finally the slowest anion followed by successively faster anions (see Figure 30-4). In some instances, the rate of electroosmotic flow may not be great enough to surpass the rate at which some of the anions move toward the anode, in which case these species move toward the anode.

It is possible to reverse the direction of the normal electroosmotic flow by adding a cationic surfactant to the buffer. The surfactant adsorbs on the capillary wall and makes the wall positively charged. Now buffer anions congregate near the wall and are swept toward the cathode, or positive electrode. This ploy is often used to speed up the separation of anions.

Electroosmosis is often desirable in certain types of capillary electrophoresis, but in other types it is not. Electroosmotic flow can be eliminated by coating the inside capillary wall with a reagent like trimethylchlorosilane to eliminate the surface silanol groups.

30B-4 Instrumentation for Capillary Electrophoresis

As shown in Figure 30-1, the instrumentation for capillary electrophoresis is simple.[6] A buffer-filled fused-silica capillary, that is typically 10 to 100 μm in internal diameter and 40 to 100 cm long, extends between two buffer reservoirs that also hold platinum electrodes. Sample introduction is performed at one end, and detec-

[6] For a review of current commercially available capillary electrophoresis instruments, see C. Henry, *Anal. Chem.*, **1996**, *64*, 747A.

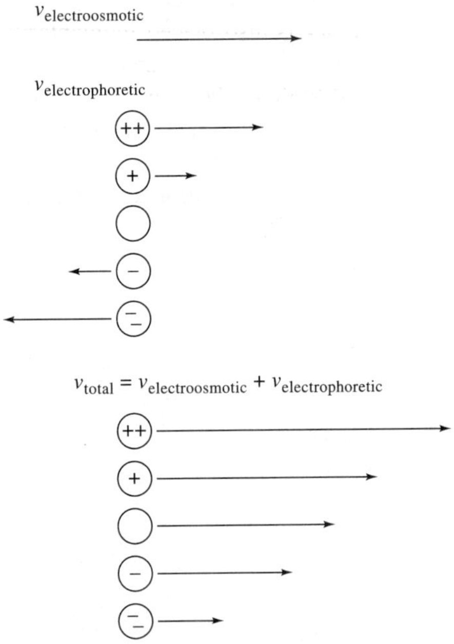

Figure 30-4 Velocities in the presence of electroosmotic flow. The length of the arrow next to an ion indicates the magnitude of its velocity; the direction of the arrow indicates the direction of motion. The negative electrode would be to the right, and the positive electrode to the left of this section of solution.

tion at the other. The polarity of the high-voltage power supply can be as indicated, or can be reversed to allow rapid separation of anions.

Although the instrumentation is conceptually simple, there are significant experimental difficulties in sample introduction and detection due to the very small volumes involved. Since the volume of a normal capillary is 4 to 5 μL, injection and detection volumes must be on the order of a few nanoliters or less.

Sample Introduction

The most common sample introduction methods are electrokinetic injection and pressure injection. With electrokinetic injection, one end of the capillary and its electrode are removed from their buffer compartment and placed in a small cup containing the sample. A potential is then applied for a measured time, causing the sample to enter the capillary by a combination of ionic migration and electroosmotic flow. The capillary end and electrode are then placed back into the regular buffer solution for the duration of the separation. This injection technique discriminates by injecting larger amounts of the more mobile ions relative to the slower-moving ions.

With pressure injection, the sample introduction end of the capillary is also placed momentarily into a small cup containing the sample, and a pressure difference is then used to drive the sample solution into the capillary. The pressure difference can come from applying a vacuum at the detector end, by pressurizing the sample, or by elevating the sample end. Pressure injection does not discriminate due to ion mobility, but cannot be used in gel-filled capillaries.

For both electrokinetic injection and pressure injection, the volume injected is controlled by the duration of the injection. Injections of 5 to 50 nL are common, and volumes below 100 pL have been reported. For a buffer with density and viscosity near the values for water, a height differential of 5 cm for 10 s injects about 6 nL with a 75-μm inside diameter capillary.[7]

Microinjection tips constructed from capillaries pulled down to very small diameters allow sampling from picoliter environments such as single cells or substructures within single cells. This technique has been employed to study amino acids and neurotransmitters from single cells.

[7] J. D. Olechno, J. M. Y. Tso, J. Thayer, and A. Wainright, *Amer. Lab.*, **1990**, *22*(18), 30.

Detection

Because the separated analytes move past a common point in most types of capillary electrophoresis, detectors are similar in design and function to those described for HPLC. One difference in behavior of detectors is encountered, however, because in capillary electrophoresis each ion migrates at a rate determined by its electrophoretic mobility. Thus, analyte bands pass through the detector at different rates, which results in peak areas that are somewhat dependent upon retention times. In contrast, in HPLC all species pass through the detector at the velocity of the mobile phase, and peak areas are independent of retention times. Ordinarily this time dependence is of no concern.

Table 30-1 lists the detection methods that have been reported through 1988 for capillary electrophoresis. The second column of the table lists representative detection limits for these detectors.

Absorbance Methods. Both fluorescence and absorbance detectors are widely used in capillary electrophoresis, although the latter are more common because they are more generally applicable. In order to keep the detection volume on the nL scale or smaller, detection is performed *on-column.* In this case a small section of the protective polyimide coating is removed from the exterior of the capillary by burning, dissolution, or scraping. That section of the capillary then serves as the detector cell. Unfortunately, the path length for such measurements is no more than 50 to 100 μm, which restricts detection limits in concentration terms; because such small volumes are involved, however, mass detection limits are equal to or better than those for HPLC.

In order to improve the sensitivity of absorbance measurements, several techniques have been suggested for increasing the path length of the measurements. Three of these are shown in Figure 30-5. One of these, which is commercially available, involves bending the end of the capillary into a "z" shape as shown in Figure 30-5a, giving a path length of 3 mm. Unfortunately, the improvement in sensitivity with this technique is much smaller than would be expected, probably because of inadequate focusing of the light.

Figure 30-5b shows a second stratagem for increasing path lengths. Here a bubble is formed near the end of the capillary. In the commercial version of this technique, the bubble for a 50-μm capillary has an inside diameter of 150 μm, thus giving a threefold increase in path length.

TABLE 30-1 Detection Modes Developed for Capillary Electrophoresis[a]

Detection Principle	Representative Detection Limit[b] (moles detected)
Spectrometry	
Absorption[c]	10^{-15}–10^{-13}
Fluorescence	
Precolumn derivatization	10^{-17}–10^{-20}
On-column derivatization	8×10^{-16}
Postcolumn derivatization	2×10^{-17}
Indirect fluorescence	5×10^{-17}
Thermal lens[c]	4×10^{-17}
Raman[c]	2×10^{-15}
Mass spectrometry	1×10^{-17}
Electrochemical	
Conductivity[c]	1×10^{-16}
Potentiometry	Not reported
Amperometry	7×10^{-19}
Radiometry[c]	1×10^{-19}

[a]From A. G. Ewing, R. A. Wallingford, and T. M. Olefirowicz, *Anal. Chem.*, **1989,** *61,* 298A. With permission.
[b]Detection limits quoted have been determined with a wide variety of injection volumes that range from 18 pl to 10 nL.
[c]Mass detection limit converted from concentration detection limit using a 1-nL injection volume.

A third method for increasing the path length of radiation by reflection is shown in Figure 30-5c. In this technique, a reflective coating of silver is deposited on the end of the capillary. In this case the radiation beam undergoes numerous reflections until it exits the capillary.

Indirect Detection. Indirect absorbance detection has been used for detection of species that are difficult to detect because of low molar absorptivities without derivatization. An ionic chromophore is placed in the electrophoresis buffer. The detector then receives a constant signal due to the presence of this substance. The analyte displaces some of these ions, just as in ion-exchange chromatography, so that the detector signal decreases during the passage of an analyte band through the detector. The analyte is then determined from the decrease in absorbance. The electropherogram in Figure 30-6 was generated by using indirect absorbance detection with 4-mM chromate ion as the chromophore; this ion absorbs strongly at 254 nm in the buffer.

Fluorescence Detection. Just as in HPLC, fluorescence detection yields increased sensitivity and selectivity for fluorescent analytes or fluorescent derivatives. Laser-based instrumentation is preferred in order to focus the excitation radiation on the small capillary and to achieve the low detection limits available from intense sources. Laser fluorescence detection has allowed detection of only 10 zeptomoles or 6000 molecules.[8]

Electrochemical Detection. Two types of electrochemical detection have been used with capillary electrophoresis: conductivity and amperometry. One of the problems with electrochemical detection has been that

[8] S. Wu and N. Dovichi, *J. Chromatogr.,* **1989,** *480,* 141.

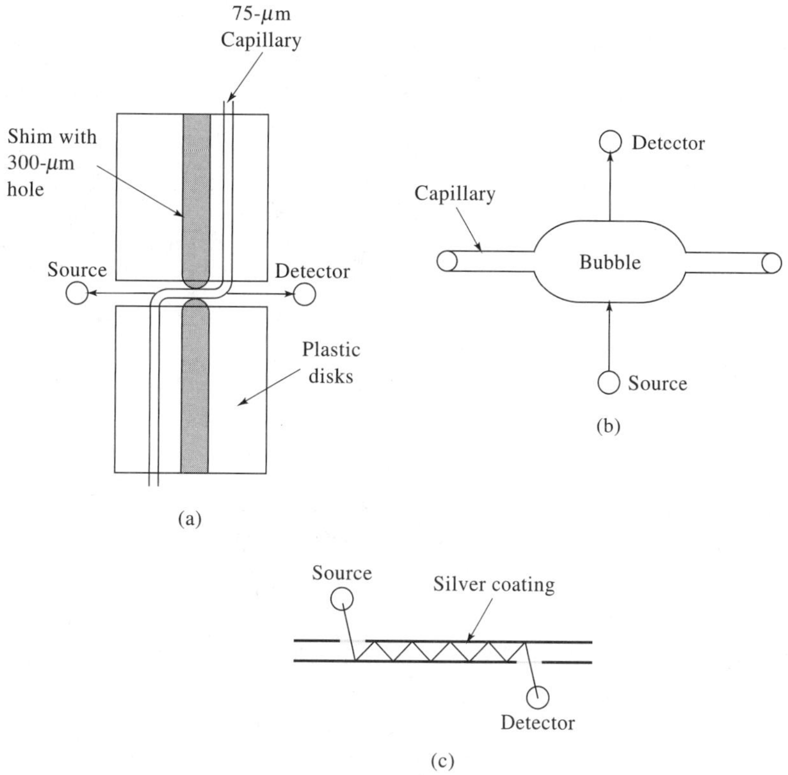

Figure 30-5 Three types of cells for improving the sensitivity of detection by absorbance measurements: (a) the 3-mm z cell, (b) the 150-μm bubble cell, (c) the multireflection cell.

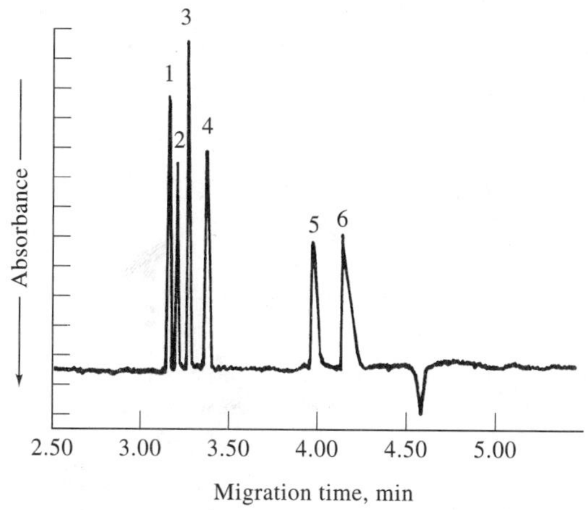

Figure 30-6 Electropherogram of a six-anion mixture by indirect detection with 4-nM chromate ion at 254 nm. Peak: (1) bromide (4 ppm), (2) chloride (2 ppm), (3) sulfate (4 ppm), (4) nitrate (4 ppm), (5) fluoride (1 ppm), (6) phosphate (6 ppm).

of isolating the detector electrodes from the high potential required for the separation. One method for isolation involves inserting of a porous glass or graphite joint between the end of the capillary and a second capillary containing the detector electrodes.

Mass Spectrometric Detection. The very small volumetric flow rates of under 1 μL/min from electrophoresis capillaries makes it feasible to couple the effluent from the capillary of an electrophoretic device directly to the ionization source of a mass spectrometer. Currently the most common sample introduction/ionization interface used for this purpose is electrospray (Section 20B-4), although fast atom bombardment has also been applied. Capillary electrophoresis with mass-spectrometric detection is only a decade and a half old but has become of major interest to biologists and biochemists for the determination of large molecules that occur in nature, such as proteins, DNA fragments, and peptides.

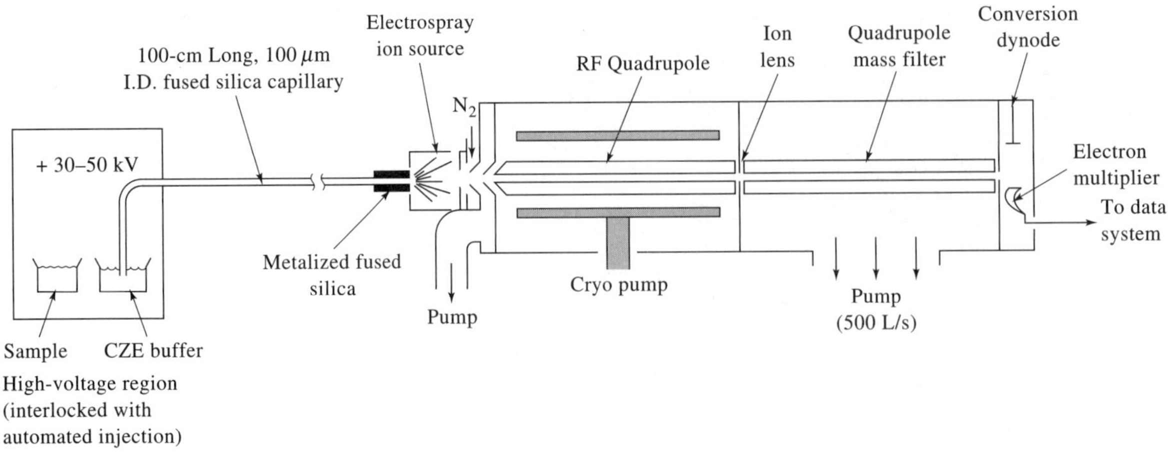

Figure 30-7 An instrument for capillary electrophoresis/mass spectrometry. The voltage between the buffer solution on the left and the metalized silica capillary is 30 to 50 kV. The flow of nitrogen is 3 to 5 kV. The flow of nitrogen at ≈70°C for desolvation is 3 to 6 L/min. *(From R. D. Smith, J. A. Olivares, N. T. Nguyen, and H. R. Udseth,* Anal. Chem., ***1988****,* 60, *437. With permission.)*

Figure 30-7 is a schematic of a typical electrospray interface coupled to a quadrupole mass spectrometer. Note that the second buffer compartment has been dispensed with, and the 30- to 50-kV potential for the electrophoretic separation is applied between the remaining buffer compartments and the opposite end of the capillary, which is metalized. A 5-kV potential is also applied between the capillary tip and ground to charge the electrospray. Figure 30-8 shows the electrospray mass spectrum obtained for vasotocin, a polypeptide having a molecular weight of 1050. Note the presence of doubly and triply charged species. With higher-molecular-weight species, ions bearing a dozen or more charges are often encountered. This fact makes it possible to detect high-molecular-weight analytes with a quadrupole instrument with a relatively modest mass range.

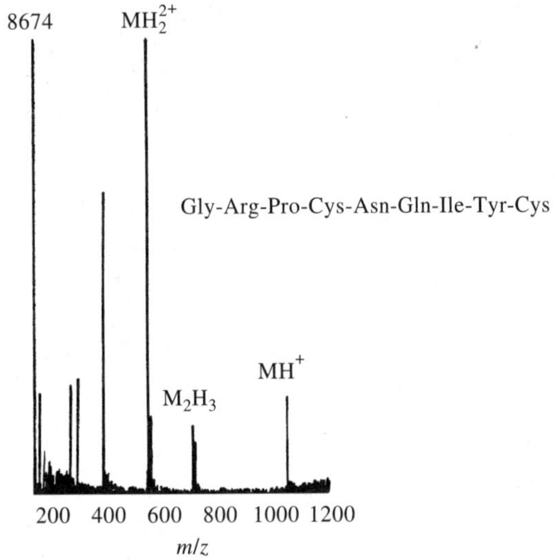

Figure 30-8 ESI mass spectrum for vasotocin. *(From R. D. Smith, J. A. Olivares, N. T. Nguyen, and H. R. Udseth,* Anal. Chem., ***1988****,* 60, *438. With permission.)*

Typical detection limits for capillary electrophoresis/mass spectrometry (CE/MS) are on the order of a few tens of femtomoles for molecules having a molecular weight of 100,000 or more.

30C APPLICATIONS OF CAPILLARY ELECTROPHORESIS[9]

Capillary electrophoretic separations are performed in several ways called modes. It is noteworthy that these modes were first employed in slab electrophoresis and were subsequently adapted for capillary electrophoretic separations. These modes include *capillary zone electrophoresis (CZE), capillary gel electrophoresis (CGE), capillary isoelectric focusing (CIEF),* and *capillary iso-*

[9] For an excellent review of applications of capillary electrophoresis and electrochromatography to the biological sciences, see B. L. Karger, A. S. Cohen, and A. Gutman, *J. Chromatogr.,* **1989,** *492,* 585.

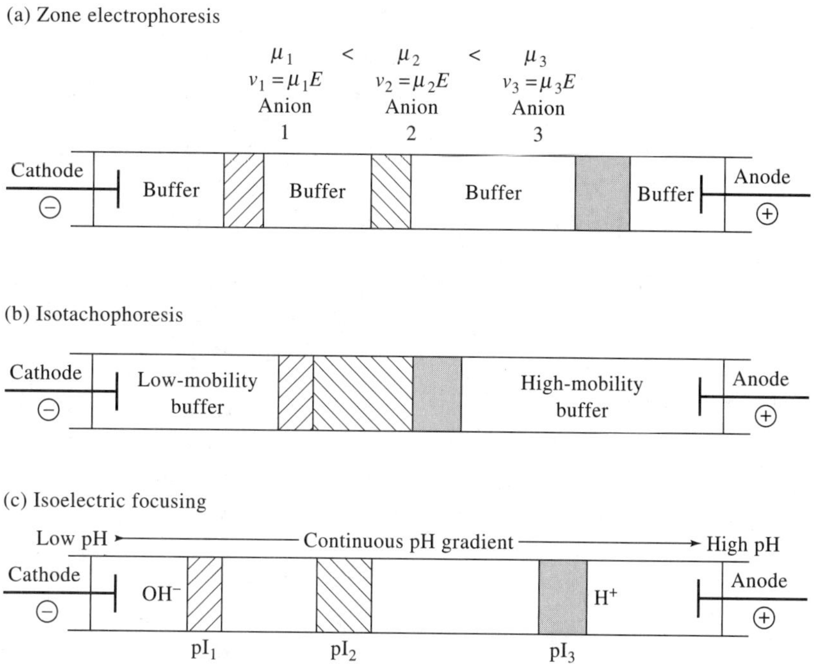

Figure 30-9 Three modes of separation by electrophoresis.

tachophoresis (CITP). The sections that follow illustrate typical applications of each of these techniques.

30C-1 Capillary Zone Electrophoresis

In capillary zone electrophoresis (CZE), the buffer composition is constant throughout the region of the separation. The applied potential causes the different ionic components of the mixture to each migrate according to its own mobility and to separate into zones that may be completely resolved or may be partially overlapped. Completely resolved zones have regions of buffer between them (see Figure 30-9a). The situation is analogous to elution column chromatography, where regions of mobile phase are located between zones containing separated analytes.

Separation of Small Ions

For most electrophoretic separations of small ions, it has been found best from the standpoint of decreased time to have the analyte ions move in the same direction as the electroosmotic flow. Thus, for cation separations the walls of the capillary are untreated, and the electroosmotic flow and the cation movement is toward the cathode. For the separation of anions, on the other hand, the electroosmotic flow is usually reversed by treating the walls of the capillary with an alkyl ammonium salt, such as cetyl trimethylammonium bromide. The positively charged ammonium ions become attached to the negatively charged silica surface and, in turn, create a negatively charged double layer of solution, which is attracted toward the anode, thus reversing the electroosmotic flow.

In the past, the most common method for analysis of small anions has been ion-exchange chromatography. For cations, the preferred techniques have been atomic absorption spectroscopy and inductively coupled plasma emission spectroscopy. Recently, however, capillary electrophoretic methods have begun to compete with these traditional methods for small-ion analysis. Several major reasons for adoption of electrophoretic methods have been recognized: lower equipment costs, smaller sample size requirements, much greater speed, and better resolution.

The initial cost of equipment and the expense of maintenance for electrophoresis is generally significantly lower than those for ion chromatographic and atomic spectroscopic instruments. Thus, commercial

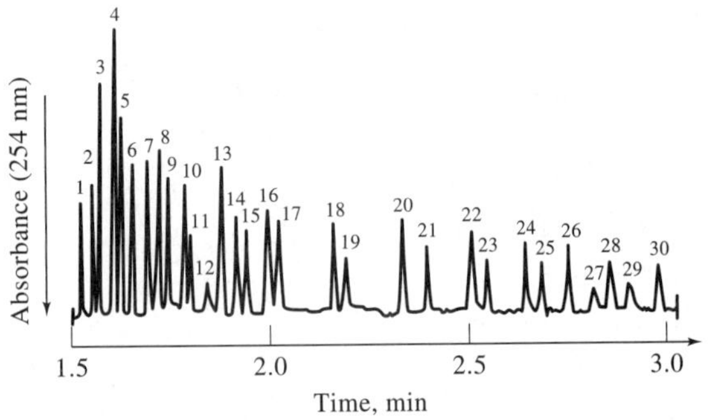

Figure 30-10 Electropherogram showing the separation of 30 anions. Capillary internal diameter: 50 μm (fused silica). Detection: indirect UV, 254 nm. Peaks: 1 = thiosulfate (4 ppm), 2 = bromide (4 ppm), 3 = chloride (2 ppm), 4 = sulfate (4 ppm), 5 = nitrite (4 ppm), 6 = nitrate (4 ppm), 7 = molybdate (10 ppm), 8 = azide (4 ppm), 9 = tungstate (10 ppm), 10 = monofluorophosphate (4 ppm), 11 = chlorate (4 ppm), 12 = citrate (2 ppm), 13 = fluoride (1 ppm), 14 = formate (2 ppm), 15 = phosphate (4 ppm), 16 = phosphite (4 ppm), 17 = chlorite (4 ppm), 18 = galactarate (5 ppm), 19 = carbonate (4 ppm), 20 = acetate (4 ppm), 21 = ethanesulfonate (4 ppm), 22 = propionate (5 ppm), 23 = propanesulfonate (4 ppm), 24 = butyrate (5 ppm), 25 = butanesulfonate (4 ppm), 26 = valerate (5 ppm), 27 = benzoate (4 ppm), 28 = *l*-glutamate (5 ppm), 29 = pentanesulfonate (4 ppm), 30 = *d*-gluconate (5 ppm). *(From W. A. Jones and P. Jandik,* J. Chromatogr., ***1991,*** *546, 445. With permission.)*

electrophoretic instruments are marketed in the price range of \$15,000 to \$40,000.[10]

Sample sizes for electrophoresis are in the nanoliter range, whereas microliter or larger samples are usually needed for other types of small-ion analysis. Thus, electrophoretic methods are more sensitive than the other methods on a mass basis (but usually not on a concentration basis).

Figure 30-10 illustrates the unsurpassed quickness and resolution of electrophoretic separations of small anions. Here, 30 anions were separated cleanly in just over three minutes. Typically an ion-exchange separation of only three or four anions could be accomplished in this brief time period. Figure 30-11 further illustrates the speed at which separations can be carried out. Here, 19 cations were separated in less than two minutes.

Separation of Molecular Species

A variety of small synthetic herbicides, pesticides, and pharmaceuticals that are ionic or can be derivatized to yield ions have been separated and analyzed by CZE. Figure 30-12 is illustrative of this type of application in which three anti-inflammatory drugs, which are carboxylic derivatives, are separated in less than 15 minutes.

Proteins, amino acids, and carbohydrates have all been separated in minimum times by CZE. In the case of neutral carbohydrates, the separations are preceded by formation of negatively charged borate complexes. These complexes are readily formed if a borate buffer is used as the separation medium. The separation of protein mixtures is illustrated by Figure 30-13.

30C-2 Capillary Gel Electrophoresis

Capillary gel electrophoresis (CGE) is generally performed in a porous gel polymer matrix, the pores of which contain a buffer mixture in which the separation is carried out. In early slab electrophoresis studies, the primary purpose of the polymeric medium was to reduce analyte dispersion by convection and diffusion and to provide a medium that could be conveniently handled for detection and scanning. It subsequently developed that this type of medium could provide a molecular sieving action that retarded the migration of analyte species to various extents depending upon the pore size of the polymer and the size of the analyte ions. This type of sieving action is particularly helpful in separating macromolecules such as proteins, DNA fragments, and oligonucleotides that have substantially the same charge but differ in size. Currently, most macroscale electrophoresis separations are carried on a gel slab. Some capillary electrophoretic separations of species that differ in size are also performed in gels contained in capillary tubes.

The most common type of gel used in electrophoresis is a polyacrylamide polymer formed by polymerizing acrylamide ($CH_2{=}CH{-}CO{-}NH_2$) in the presence of a cross-linking agent. The pore size of the polymer depends upon the ratio of monomer to cross-linking agent. Increases in the amount of cross-linking agent results in smaller pore size. Other gels that have been used for capillary gel electrophoresis include agarose, a polysaccharide extracted from a marine alga; methyl cellulose; and polyethylene glycol. A typical separation of a standard protein mixture

[10] See M. Warner, *Anal. Chem.*, **1994**, *66*, 1137A.

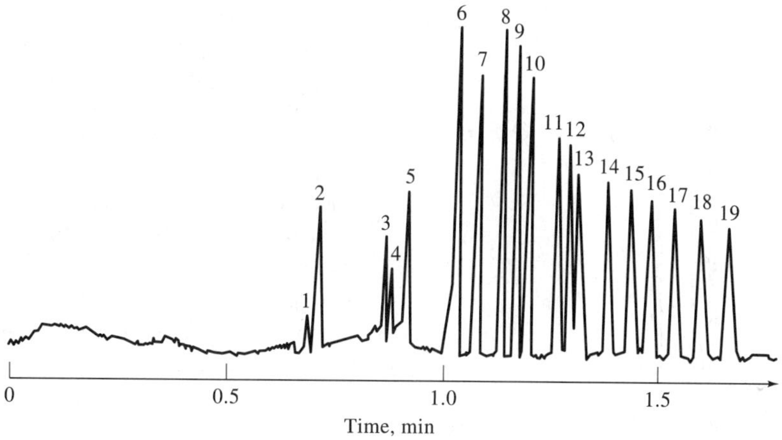

Figure 30-11 Separation of alkali, alkaline earths, and lanthanides. Capillary: 36.5 cm × 75-μm fused silica, +30 kV. Injection: hydrostatic, 20 s at 10 cm. Detection: indirect UV, 214 nm. Peaks: 1 = rubidium (2 ppm), 2 = potassium (5 ppm), 3 = calcium (2 ppm), 4 = sodium (1 ppm), 5 = magnesium (1 ppm), 6 = lithium (1 ppm), 7 = lanthanum (5 ppm), 8 = cerium (5 ppm), 9 = praseodymium (5 ppm), 10 = neodymium (5 ppm), 11 = samarium (5 ppm), 12 = europium (5 ppm), 13 = gadolinium (5 ppm), 14 = terbium (5 ppm), 15 = dysprosium (5 ppm), 16 = holmium (5 ppm), 17 = erbium (5 ppm), 18 = thulium (5 ppm), 19 = ytterbium (5 ppm). *(From P. Jandik, W. R. Jones, O. Weston, and P. R. Brown,* LC-GC, ***1991,*** *9, 634. With permission.)*

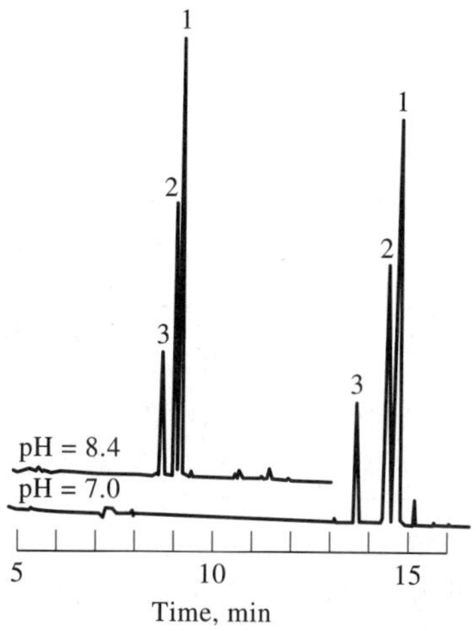

Figure 30-12 Separation of antiinflammatory drugs by CZE. Detection: absorption at 215 nm. Analytes: (1) naproxen, (2) ibuprofen, (3) tolmetin. *(Reprinted with permission from A. Wainright,* J. Microcolumn. Sep., ***1990,*** *2, 166.)*

in a polyethylene glycol gel is illustrated in Figure 30-14.

30C-3 Capillary Isotachophoresis

In capillary isotachophoresis (CITP) all analyte bands ultimately migrate at the same velocity; hence, the name from *iso* for same and *tach* for speed. In any particular application, either cations or anions can be separated, but not both at the same time. In a separation, the sample is injected between two buffers: a leading one containing ions of a higher mobility than any of the analyte ions and a terminating one with ions of a lower mobility than the sample ions. For example, in separating anions, fast-moving chloride ions might be contained in the leading buffer, and slow-moving heptanoate ions in the terminating buffer. For a separation of anions, the leading electrolyte solution is connected to the anode, and the terminating one to the cathode.

When the potential field is first applied in an isotachophoretic separation, analyte ions migrate as in zone electrophoresis, each ion with its unique velocity given by the product $\mu_e E$. This difference in migration rates re-

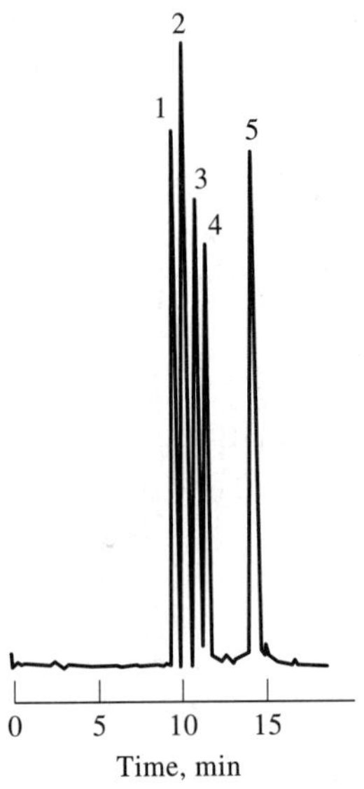

Figure 30-13 CZE separation of a model protein mixture. Conditions: pH 2.7 buffer; absorbance detection at 214 nm; 22 kV, 10 μA. Peaks identified in following table.

Model Proteins Separated at pH 2.7

Peak No.	Proteins	Mol. wt.	Isoelectric Point, pI
1	Cytochrome c	12,400	10.7
2	Lysozyme	14,100	11.1
3	Trypsin	24,000	10.1
4	Trypsinogen	23,700	8.7
5	Trypsin inhibitor	20,100	4.5

sults in the separation of the various analyte species into adjacent bands, with the fastest species located in a band directly adjacent to the leading buffer and the slowest just ahead of the terminal buffer. After the bands have formed, they then move at the same velocity. The reason that the bands have the same velocity is that the potential field becomes smaller for the more mobile bands and greater for the slower bands, so that the current is the same, as it must be, in all parts of the buffer. The ionic current that results from the flow of ions in the buffer is analogous to the dc current in a circuit consisting of several resistors connected in series to a battery. Here, the current must be identical in all of the resistors. Hence the potential across each of the resistors must vary in such a way that Ohm's law is obeyed.

When equilibrium is reached in an isotachophoretic experiment, a condition is reached in which each sample component is migrating in a band that is sandwiched between a band that contains the next-slower-moving ions and the next-faster-moving band as shown in Figure 30-9b. The boundary between bands is sharp. If a solute species starts to diffuse into the next-faster band, it encounters a lower field, which reduces its velocity until it drops back into its original band. Note in Figure 30-9b that, in contrast to zone electrophoresis or elution chromatography, the analyte bands are immediately adjacent to one another and not separated by bands of the buffer.

30C-4 Capillary Isoelectric Focusing

Capillary isoelectric focusing (CIEF), is used to separate amphiprotic species, such as amino acids and proteins that contain a weak carboxylic acid group and a weak base amine group.

Properties of Amphiprotic Compounds

An *amphiprotic* compound is a species that in solution is capable of both donating and accepting a proton. A typical amino acid, such as glycine, is an amphiprotic compound. When glycine is dissolved in water, three important equilibria operate:

$$NH_2CH_2COOH \rightleftharpoons NH_3^+CH_2COO^- \quad (30\text{-}6)$$

$$NH_3^+CH_2COO^- + H_2O \rightleftharpoons NH_2CH_2COO^- + H_3O^+ \quad K_a = 2 \times 10^{-10} \quad (30\text{-}7)$$

$$NH_3^+CH_2COO^- + H_2O \rightleftharpoons NH_3^+CH_2COOH + OH^- \quad K_b = 2 \times 10^{-12} \quad (30\text{-}8)$$

The first reaction constitutes a kind of internal acid/base reaction and is analogous to the reaction one would observe between a simple carboxylic acid and an amine. The typical aliphatic amine has a base dissociation constant of 10^{-4} to 10^{-5}, while many carboxylic acids have acid dissociation constants of about the same magnitude. The consequence is that the first reaction proceeds far to the right, with the product or products being the predominant species in the solution.

The amino acid product in Equation 30-6, bearing both a positive and a negative charge, is called a *zwitterion.* As shown by the equilibrium constants for the reactions shown as Equations 30-7 and 30-8, the zwitterion of glycine is stronger as an acid than as a base. Thus, an aqueous solution of glycine is somewhat acidic.

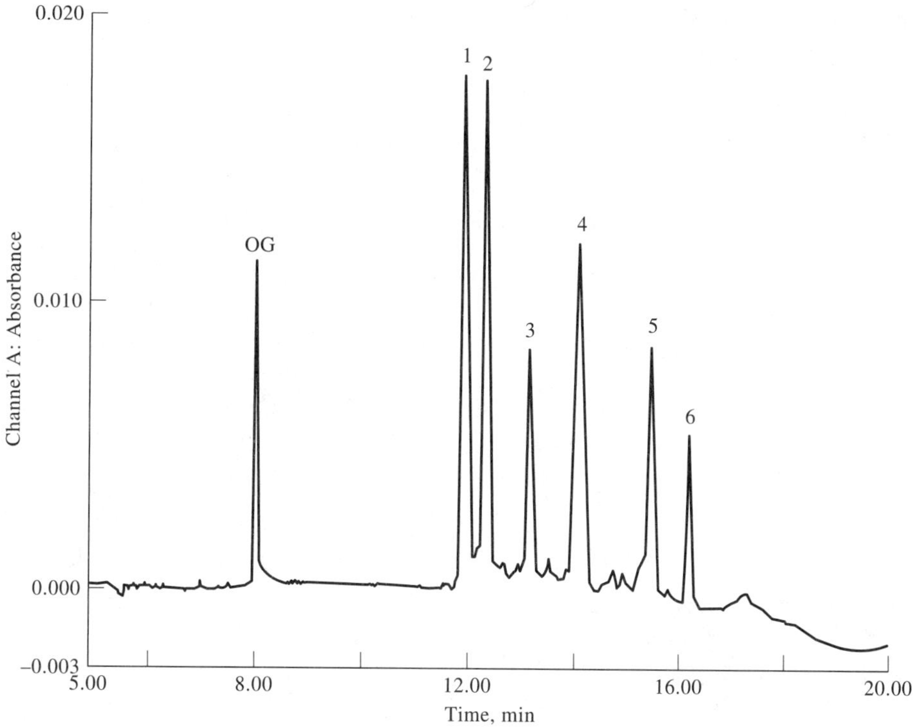

Figure 30-14 Capillary gel separation of SDS denatured proteins in a polyethylene glycol column. UV absorption detection at 214 nm. Proteins: (1) α-lactalbumin, (2) soybean trypsin inhibitor, (3) carbonic anhydrase, (4) ovalbumin, (5) bovine serum albumin, (6) phosphorylase B. *(Reprinted with permission from* Anal. Chem., ***1992****,* 64*, 2665. With permission.)*

The zwitterion of an amino acid, containing as it does a positive and a negative charge, has no tendency to migrate to an electric field, whereas the singly charged anionic and cationic species are attracted to electrodes of opposite charge. No *net* migration of the amino acid occurs in an electric field when the pH of the solvent is such that the concentrations of the anionic and cationic forms are identical. The pH at which no net migration occurs is called the *isoelectric point* (pI), and is an important physical constant for characterizing amino acids. The isoelectric point is readily related to the ionization constants for the species. Thus, for glycine

$$K_a = \frac{[H_3O^+][NH_2CH_2COO^-]}{[N\overset{+}{H_3}CH_3COO^-]}$$

$$K_b = \frac{[OH^-][N\overset{+}{H_3}CH_2COOH]}{[N\overset{+}{H_3}CH_2COO^-]}$$

At the isoelectric point,

$$[NH_2CH_2COO^-] = [N\overset{+}{H_3}CH_2COOH]$$

Thus, if we divide K_a by K_b and substitute this relationship, we obtain for the isoelectric point

$$\frac{K_a}{K_b} = \frac{[H_3O^+][NH_2CH_2COO^-]}{[OH^-][N\overset{+}{H_3}CH_2COOH]} = \frac{[H_3O^+]}{[OH^-]}$$

Substitution of $K_w/[H_3O^+]$ for $[OH^-]$ and rearrangement yields

$$[H_3O^+] = \sqrt{\frac{K_aK_w}{K_b}}$$

The isoelectric point pI for glycine occurs at a pH of 6.0 as shown by the following:

$$[H_3O^+] = \left(\frac{2 \times 10^{-10}}{2 \times 10^{-12}} \times 1 \times 10^{-14}\right)^{1/2} = 1 \times 10^{-6}$$

$$pI = -\log 1 \times 10^{-6} = 6$$

Separation of Amphiprotic Species

In isoelectric separations of amphiprotic species, the separation is performed in a buffer mixture that varies in pH continuously along its length. This pH gradient is prepared from a mixture of several different *ampholytes* in an aqueous solution. Ampholytes are amphoteric compounds usually containing carboxylic and amino groups. Ampholyte mixtures having different pH ranges can be prepared or are available from several commercial sources.

To perform an isoelectric focusing experiment in a capillary tube, the analyte mixture is dissolved in a dilute solution of the ampholytes, which is then transferred to the tube. One end of the capillary is then inserted in a solution of strong base, such as sodium hydroxide, that also holds the cathode. The other end of the tube is immersed in a solution of a strong acid, such as phosphoric, that also holds the anode. When the potential is applied, hydrogen ions begin to migrate from the anode compartment toward the cathode. Hydroxide ions from the cathode begin to move in the opposite direction. If a component of the ampholyte or the analyte has a net negative charge, it migrates toward the positive anode. As it migrates it passes into continuously lower pH regions, where progressive protonation of the species occurs, which lowers its negative charge. Ultimately it reaches the pH where its net charge is zero (its isoelectric point). Migration of the species then ceases. This process goes on for each ampholyte species and ultimately provides a continuous pH gradient throughout the tube. Analyte ions also migrate until they reach their isoelectric points. These processes then result in the separation of each analyte into a narrow band that is located at the pH of its isoelectric point (see Figure 30-9c). Very sharp focusing is realized in such systems. Note that isoelectric focusing separations are based upon differences in equilibrium properties of the analytes (K_a,K_b) rather than on differences in rates of migration. Once each analyte has migrated to a region where it is neutral, the positions of bands become constant and no longer change with time.

Mobilization of Focused Bands

In order to detect the focused bands in a capillary isoelectric focusing separation, it is necessary to move, or *mobilize,* the contents of the capillary so that the bands pass the detector located at one end. This mobilization can be accomplished through application of a pressure difference, just as for sample loading, or can be brought about by simply changing the solution in the electrode compartment. During the focusing step, equal numbers of H^+ and OH^- ions would enter opposite ends of the capillary, so the pH gradient would remain stable. Suppose that sodium chloride is added to the sodium hydroxide solution after focusing is finished. Now both Cl^- and OH^- migrate into that end of the column, and the sum of these two concentrations is balanced by H^+ entering the opposite end. That means that there is now less OH^- than H^+ flowing into the capillary. The pH decreases at the cathode end. The pH gradient is no longer stable. It moves toward the cathode end, and along with it go the focused bands. The bands that pass the detector first are the ones corresponding to proteins with the most alkaline isoelectric points. Figure 30-15 shows an electropherogram for the separation of several proteins by capillary isoelectric focusing. The pI for each protein is shown above the peak. Mobilization was accomplished by adding sodium chloride to the anode compartment.

30D CAPILLARY ELECTROCHROMATOGRAPHY[11]

Electrochromatography is a hybrid of capillary electrophoresis and HPLC that offers some of the best features of each of the two techniques. Two types of capillary electrochromatography have been developed since the early 1980s: *packed column* and *micellar electrokinetic capillary.* To date, the latter has found more widespread applications.

Capillary electrochromatography (CEC) appears to offer several advantages over either of the parent techniques. First, like HPLC, it is applicable to the separation of uncharged species. Second, like capillary electrophoresis, it provides highly efficient separations on microvolumes of sample solution without the need for a high-pressure pumping system. In electrochromatography, a mobile phase is transported through a stationary phase by electroosmotic-flow pumping rather than by mechanical pumping, thus simplifying the system significantly. In addition, electroosmotic pumping leads to the plug flow profile shown in Figure 30-3a rather than the hydrodynamic profile shown in Figure 30-3b. The flat face of plug flow leads to less band broadening than the hydrodynamic profile. Greater separation efficiencies are the consequence.

---

[11] See R. Weinberger, *Practical Capillary Electrophoresis,* Chapter 8. New York: Academic Press, 1993.

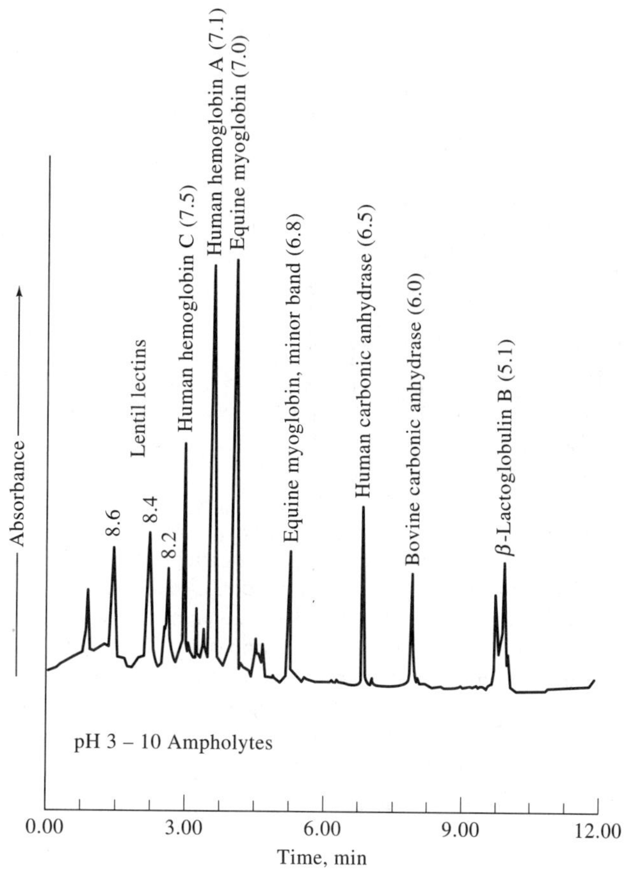

Figure 30-15 Capillary isoelectric focusing of proteins. Isoelectric pI listed above the peaks. Detection was by UV absorption. *(From T. Wehr, M. Zhu, R. Rodriquez, D. Burke, and K. Duncan,* Amer. Biotech. Lab., *1990, 8(11), 22. With permission.)*

30D-1 Packed Column Electrochromatography

Electrochromatography based upon packed columns is the least mature of the various electroseparation techniques. In this method a polar solvent is usually driven by electroosmotic flow through a capillary that is packed with a reversed-phase HPLC packing. Separations depend upon the distribution of the analyte species between the mobile phase and the liquid stationary phase held on the packing. Figure 30-16 shows a typical electrochromatogram for the separation of 16 polyaromatic hydrocarbons in a 33-cm-long capillary having an inside diameter of 75 μm. The mobile phase was made up of acetonitrile in a 4-mM sodium borate solution. The stationary phase consisted of 3-μm octadecylsilica particles.

30D-2 Micellar Electrokinetic Capillary Chromatography[12]

The capillary electrophoretic methods we have described thus far are not applicable to the separation of uncharged solutes. In 1984, however, Terabe and collaborators[13] described a modification of the method that permitted the separation of low-molecular-weight aro-

---

[12] For a more extensive discussion of electrokinetic capillary chromatography, see S. Terabe, "Electrokinetic Chromatography: An Interface between Electrophoresis and Chromatography," *Trends Anal. Chem.*, **1989**, *8*, 129.

[13] S. Terabe, K. Otsuka, K. Ichikawa, A. Tsuchiya, and T. Ando, *Anal. Chem.*, **1984**, *56*, 111; S. Terabe, K. Otsuka, and T. Ando, *Anal. Chem.*, **1985**, *57*, 841. For a discussion of optimization of separations by this technique, see J. P. Foley, *Anal. Chem.*, **1990**, *62*, 1302.

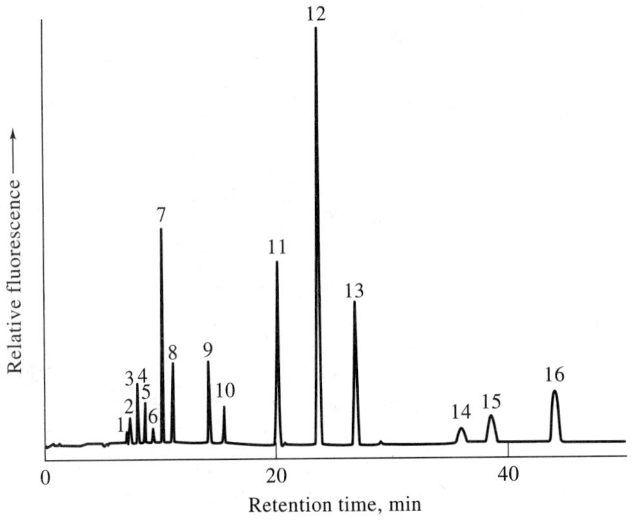

Figure 30-16 Electrochromatogram showing the electrochromatographic separation of the 16 PAHs. The peaks are identified as follows: (~10^{-6} to 10^{-8} M of each compound): (1) naphthalene, (2) acenaphthylene, (3) acenaphthene, (4) fluorene, (5) phenanthrene, (6) anthracene, (7) fluoranthene, (8) pyrene, (9) benz[*a*]anthracene, (10) chrysene, (11) benzo[*b*]fluoranthene, (12) benzo[*k*]fluoranthene, (13) benzo[*a*]pyrene, (14) dibenz[*a,h*]anthracene, (15) benzo[*ghi*]perylene, and (16) indeo[1,2,3-*cd*]pyrene. *(From C. Yan, R. Dadoo, H. Zhao, D. J. Rakestroaw, and R. N. Zare,* Anal. Chem., ***1995****, 67, 2026. With permission.)*

matic phenols and nitro compounds with equipment such as that shown in Figure 30-1. This technique involved introduction of surfactant, such as sodium dodecyl sulfate (SDS), at a concentration level at which *micelles* form. Micelles form in aqueous solutions when the concentration of an ionic species having a long-chain hydrocarbon tail is increased above a certain level called the *critical micelle concentration* (CMC). At this point the ions begin to form spherical aggregates made up to 40 to 100 ions whose hydrocarbon tails are in the interior of the aggregate and whose charged ends are exposed to water on the outside. Micelles constitute a stable second phase that is capable of absorbing nonpolar compounds into the hydrocarbon interior of the particles, thus *solubilizing* the nonpolar species. Solubilization is commonly encountered when a greasy material or surface is washed with a detergent solution.

Capillary electrophoresis carried out in the presence of micelles is termed *micellar electrokinetic capillary chromatography* and given the acronym MECC or MEKC. In this technique, surfactants are added to the operating buffer in amounts that exceed the critical micelle concentration. For most applications to date, the surfactant has been sodium dodecyl sulfate. The surface of an anionic micelle of this type has a large negative charge, which gives it a large electrophoretic mobility toward the positive electrode. Most buffers, however, exhibit such a high electroosmotic flow rate toward the negative electrode that the anionic micelles are carried toward that electrode also, but at a much reduced rate. Thus, during an experiment, the buffer mixture consists of a faster-moving aqueous phase and a slower-moving micellar phase. When a sample is introduced into this system, the components distribute themselves between the aqueous phase and the hydrocarbon phase in the interior of the micelles. The positions of the resulting equilibria depend upon the polarity of the solutes. With polar solutes the aqueous solution is favored; with nonpolar compounds, the hydrocarbon environment is preferred.

The system just described is quite similar to what exists in a liquid partition chromatographic column except that the "stationary phase" is moving along the length of the column but at a much slower rate than the mobile phase. The mechanism of separations is identical in the two cases and depends upon differences in

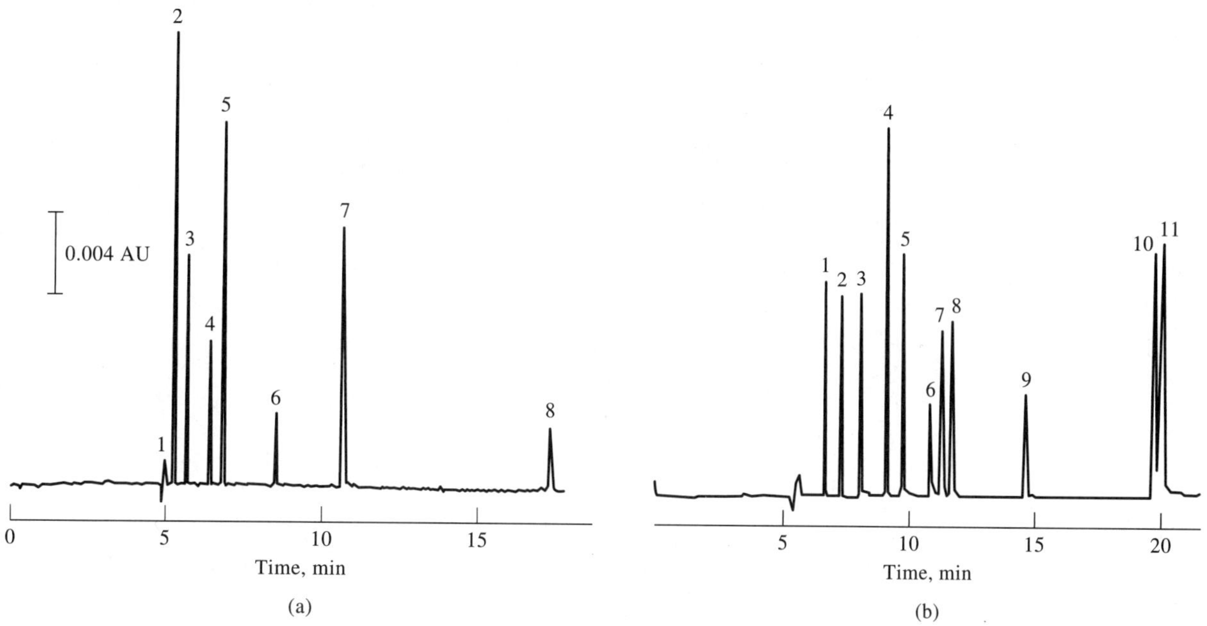

Figure 30-17 Typical separation by MECC. (a) Some test compounds: 1 = methanol, 2 = resorcinol, 3 = phenol, 4 = *p*-nitroaniline, 5 = nitrobenzene, 6 = toluene, 7 = 2-naphthol, 8 = Sudan III; capillary, 0.05-mm I.D., 500 mm to the detector; applied voltage, ca. 15 kV; current detection absorbance at 210 nm. (b) Analysis of a cold medicine: 1 = acetaminophen, 2 = caffeine, 3 = sulpyrine, 4 = naproxen, 5 = guaiphenesin, 10 = noscapine, 11 = chloropheniramine and tipepidine; applied voltage, 20 kV; capillary, as in (a); detection absorption at 220 nm. *(From S. Terabe,* Trends Anal. Chem., *1989,* 8, *129. With permission.)*

distribution coefficients for analytes between the mobile aqueous phase and the hydrocarbon *pseudostationary phase.* The process is thus a true chromatographic one; hence, the name micellar electrokinetic capillary *chromatography.* Figure 30-17 illustrates two typical separations by MECC.

Capillary chromatography in the presence of micelles appears to have a promising future. One advantage that this hybrid technique has over HPLC is much higher column efficiencies (100,000 plates or more). In addition, changing the second phase in MECC is simple, involving only the changing of the micellar composition of the buffer. In contrast, in HPLC, the second phase can only be altered by changing the type of column packing.

30E QUESTIONS AND PROBLEMS

- sample injection (2 way)
- How CE separate neutral analyte

30-1 What is electroosmotic flow? Why does it occur?

30-2 Suggest a way in which electroosmotic flow might be repressed.

30-3 Why does pH affect separation of amino acids by electrophoresis?

30-4 What is the principle of separation by capillary zone electrophoresis?

30-5 What is the principle of micellar electrokinetic capillary chromatography? How does it differ from capillary zone electrophoresis?

30-6 Describe a major advantage of micellar electrokinetic capillary chromatography over conventional liquid chromatography.

Section VI

Miscellaneous Methods

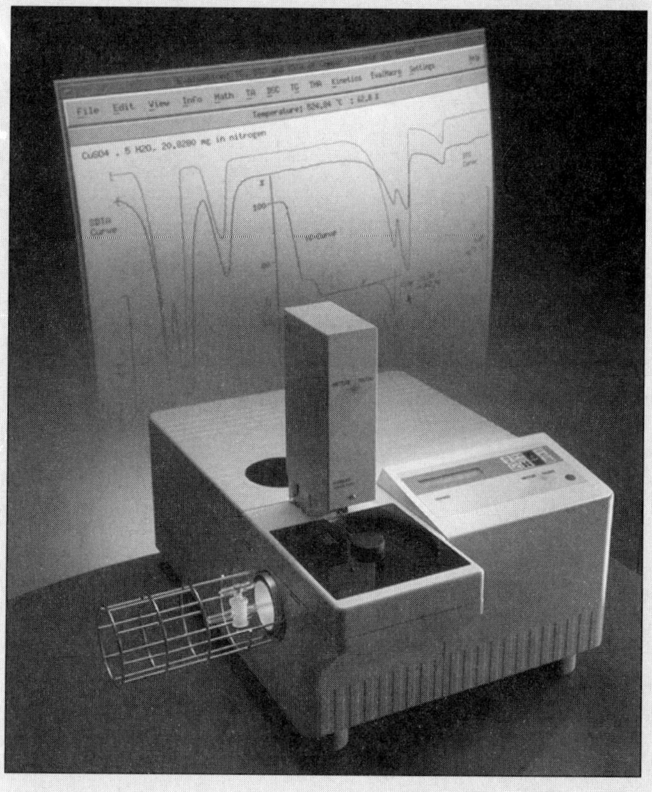

The instrument in the photo is a thermogravimetric analyzer capable of both TGA and DTA, which are detailed in Chapter 31. In TGA mode, the instrument accepts samples of up to 5 g mass and performs its analysis with a dynamic range of 5 g and resolution of 0.1 mg. Temperature accuracy is 0.25°C with 0.005°C resolution over a range of 1600°C. The instrument can handle up to 34 samples in series. As is suggested by the photo, the instrument is under the control of a computer for experimental setup, data collection, data analysis, and data presentation. This instrument is typical of today's electronically controlled, highly automated instruments for chemical analysis that find wide use in characterizing materials in industrial, academic, and governmental settngs. (Photo courtesy of Mettler-Toledo, Inc., Hightstown, NH.)

Section VI consists of three chapters devoted to miscellaneous instrumental methods of chemical analysis. Thermogravimetry (TG), differential thermal analysis (DTA), and differential scanning calorimetry (DSC) are discussed in Chapter 31. The theory and practice of radiochemical methods, including neutron activation analysis and isotope dilution techniques, are described in Chapter 32. This section and the book close with Chapter 33, which examines the principles, instrumentation, and methods of automated analysis.

Chapter 31

Thermal Methods

A generally accepted definition of **thermal analysis** *is: "A group of techniques in which a physical property of a substance and/or its reaction products is measured as a function of temperature whilst the substance is subjected to a controlled temperature program."*[1] *Well over a dozen thermal methods can be recognized, which differ in the properties measured and the temperature programs.*[2] *These methods find widespread use for both quality control and research applications on industrial products, such as polymers, pharmaceuticals, clays and minerals, metals, and alloys. We shall confine our discussion to three of the methods, which provide primarily chemical rather than physical information about samples of matter. These methods include* **thermogravimetry** *(TG),* **differential thermal analysis** *(DTA), and* **differential scanning calorimetry** *(DSC).*

[1] R. C. Mackenzie, *Thermochim. Acta,* **1979**, *28,* 1.

[2] For a detailed description of most of these techniques, see W. W. Wendlandt, *Thermal Analysis,* 3rd ed. New York: Wiley, 1985; M. E. Brown, *Introduction to Thermal Analysis: Techniques and Applications.* New York: Chapman and Hall, 1988; B. Wunderlich, *Thermal Analysis.* Boston: Academic Press, 1990; P. J. Haines, *Thermal Methods of Analysis.* London: Blackie, 1995.

31A THERMOGRAVIMETRIC METHODS (TG)

In a thermogravimetric analysis the mass of a sample in a controlled atmosphere is recorded continuously as a function of temperature or time as the temperature of the sample is increased (usually linearly with time). A plot of mass or mass percent as a function of time is called a *thermogram,* or a *thermal decomposition curve.*[3]

31A-1 Instrumentation

Modern commercial instruments for thermogravimetry consists of: (1) a sensitive analytical balance, (2) a furnace, (3) a purge gas system for providing an inert (or sometimes reactive) atmosphere, and (4) a microcomputer/microprocessor for instrument control and data acquisition and display. In addition, a purge gas switching system is a common option for applications in which the purge gas must be changed during an experiment.

[3] For a brief review of thermogravimetry, see C. M. Earnest, *Anal. Chem.,* **1984**, *56,* 1471A.

The Balance

A number of different thermobalance designs are available commercially that are capable of providing quantitative information about samples ranging in mass from 1 mg to 100 g. The most common type of balance, however, has a range of 5 to 20 mg. Although the sample holder must be housed in the furnace, the rest of the balance must be thermally isolated from the furnace. Figure 31-1 is a schematic of one thermobalance design. A change in sample mass causes a deflection of the beam, which interposes a light shutter between a lamp and one of two photodiodes. The resulting imbalance in the photodiode current is amplified and fed into coil *E*, which is situated between the poles of a permanent magnet *F*. The magnetic field generated by the current in the coil restores the beam to its original position. The amplified photodiode current is monitored and transformed into mass or mass loss information by the data acquisition system. In most cases mass versus temperature data can either be plotted in real time or stored for further manipulation or display at a later time.

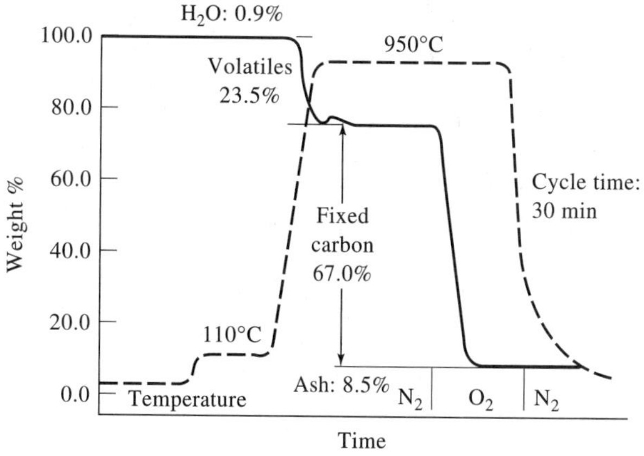

Figure 31-2 A controlled atmospheric thermogram for a bituminous coal sample. A nitrogen atmosphere was employed for approximately 18 min followed by an oxygen atmosphere for 4 to 5 min. The analysis was then completed in nitrogen. *(Reprinted with permission from C. M. Earnest, Anal. Chem., **1984**, 56, 1478A. Copyright 1984 American Chemical Society.)*

The Furnace

The temperature range for most furnaces for thermogravimetry is from ambient to 1500°C. Often the heating and cooling rate of the furnace can be selected from somewhat greater than zero to as high as 200°C/min. Insulation and cooling of the exterior of the furnace is required to avoid heat transfer to the balance. Nitrogen or argon are usually used to purge the furnace and prevent oxidation of the sample. For some analyses, it is desirable to switch purge gases as the analysis proceeds. Figure 31-2 provides an example in which the purge gas was automatically switched from nitrogen to oxygen and then back to nitrogen. The sample in this case was a bituminous coal. Nitrogen was employed during the first 18 min while the moisture content and the percent volatiles were recorded. The gas was then switched to oxygen for 4 to 5 min, which caused oxidation of carbon to carbon dioxide. Finally, the analysis was finished with a nitrogen purge to give a measure of the ash content.

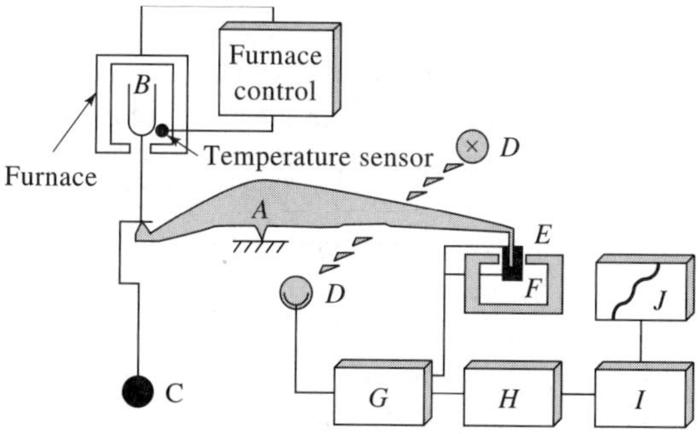

Figure 31-1 Components of a thermal balance: *A*, beam; *B*, sample cup and holder; *C*, counterweight; *D*, lamp and photodiodes; *E*, coil; *F*, magnet; *G*, control amplifier; *H*, tare calculator; *I*, amplifier; *J*, recorder. *(Courtesy of Mettler Instrument Corp., Hightstown, NJ.)*

Instrument Control/Data Handling

The temperature recorded in a thermogram is ideally the actual temperature of the sample. This temperature can, in principle, be obtained by immersing a small thermocouple directly in the sample. Such a procedure is seldom followed, however, because of possible catalytic decomposition of samples, potential contamination of samples, and weighing errors resulting from the thermocouple leads. As a consequence of these problems, recorded temperatures are generally measured with a small thermocouple located as close as possible to the sample container. The recorded temperatures then generally lag or lead the actual sample temperature.

Modern thermobalances usually use a computerized temperature control routine that automatically compares the voltage output of the thermocouple with a voltage versus temperature table that is stored in read-only memory (ROM). The microcomputer uses the difference between the temperature of the thermocouple and the temperature specified in ROM to adjust the voltage to the heater. Using this method it is possible to achieve excellent agreement between the specified temperature program and the temperature of the sample. Typical run-to-run reproducibility for a particular program falls within ±2°C throughout an instrument's entire operating range.

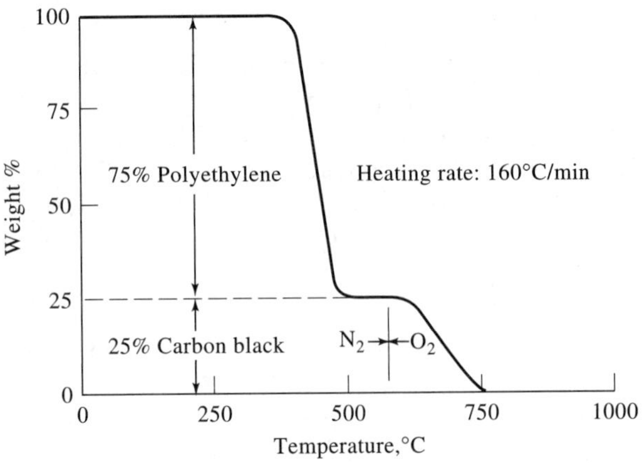

Figure 31-4 Thermogravimetric determination of carbon black in polyethylene. *(From J. Gibbons,* Amer. Lab., ***1981,*** *13, (1), 33. Copyright 1981 by International Scientific Communications, Inc.)*

31A-2 Applications

The information provided by thermogravimetric methods is more limited than that obtained with the other two thermal methods described in this chapter because here a temperature variation must bring about a change in mass of the analyte. Thus, thermogravimetric methods are largely limited to decomposition and oxidation reactions and to such physical processes as vaporization, sublimation, and desorption.

Perhaps the most important applications of thermogravimetric methods are found in the study of polymers. Thermograms provide information about decomposition mechanisms for various polymeric preparations. In addition, the decomposition patterns are characteristic for each kind of polymer and in some cases can be used for identification purposes. Figure 31-3 shows decomposition patterns for five polymers obtained by thermogravimetry.

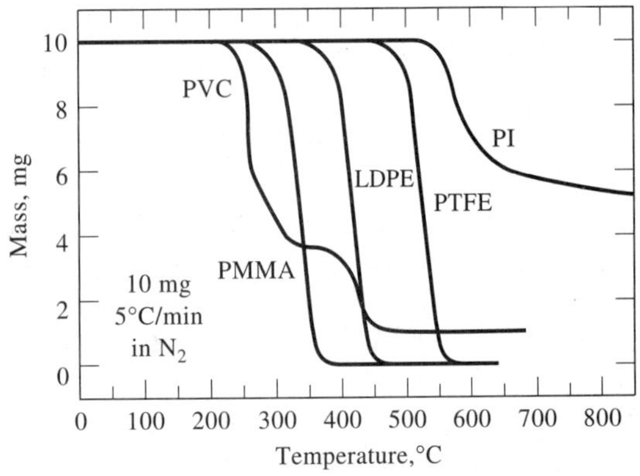

Figure 31-3 Thermograms for some common polymeric materials. PVC = polyvinyl chloride; PMMA = polymethyl methacrylate; LDPE = low density polyethylene; PTFE = polytetrafluoroethylene; PI = aromatic polypyromellitimide. *(From J. Chiu, in* Thermoanalysis of Fiber-Forming Polymers. *R. F. Schwenker, Ed., p. 26. New York: Interscience, 1966. Reprinted by permission of John Wiley & Sons, Inc.)*

Figure 31-4 illustrates how a thermogram is used for quantitative analysis of a polymeric material. The sample is polyethylene that has been formulated with fine carbon-black particles to inhibit degradation from exposure to sunlight. This analysis would be difficult by most other analytical methods.

Figure 31-5 is a recorded thermogram obtained by increasing the temperature of pure $CaC_2O_4 \cdot H_2O$ at a rate of 5°C/min. The clearly defined horizontal regions correspond to temperature ranges in which the indicated calcium compounds are stable. This figure illustrates one of the important applications of thermogravimetry, namely, that of defining thermal conditions necessary to produce a pure form for the gravimetric determination of a species.

Figure 31-6a illustrates an application of thermogravimetry to the quantitative analysis of a mixture of calcium, strontium, and barium ions. The three are first precipitated as the monohydrated oxalates. The mass in the temperature range between 320° and 400°C is that of the three anhydrous compounds, CaC_2O_4, SrC_2O_4, and BaC_2O_4, while the mass between about 580° and 620°C corresponds to the weight of the three carbonates. The weight change in the next two steps results from the loss of carbon dioxide, as first CaO and then SrO are formed. Clearly, sufficient data are available in the thermogram to calculate the weight of each of the three elements present in the sample.

Figure 31-6b is the derivative of the thermogram shown in Figure 31-6a. The data acquisition systems of

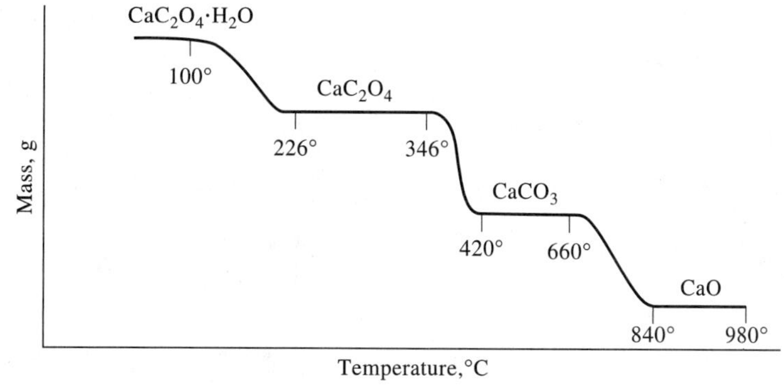

Figure 31-5 A thermogram for decomposition of $CaC_2O_4 \cdot H_2O$ in an inert atmosphere. *(From S. Peltier and C. Duval,* Anal. Chim. Acta, ***1947,*** 1, *345. With permission.)*

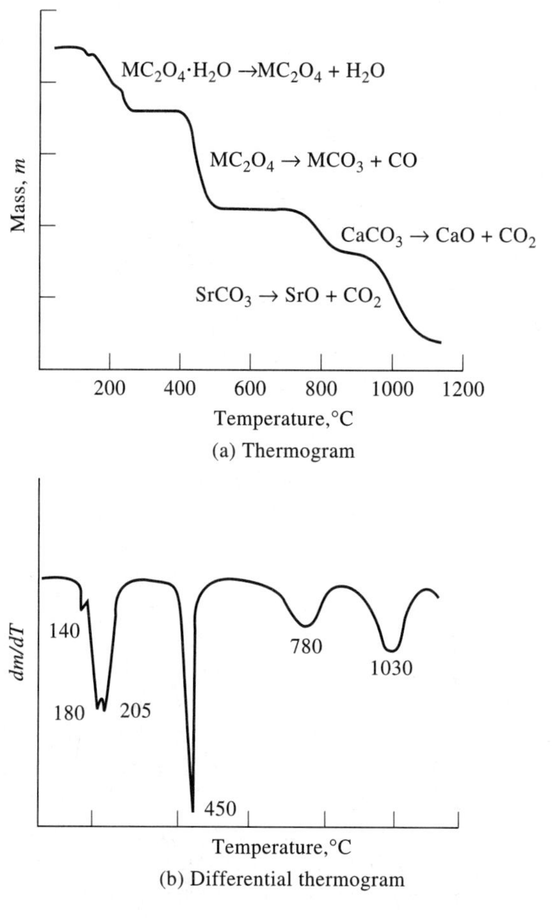

Figure 31-6 Decomposition of $CaC_4O_4 \cdot H_2O$, $SrC_2O_4 \cdot H_2O$, and $BaC_2O_4 \cdot H_2O$. *(From L. Erdey, G. Liptay, G. Svehla, and F. Paulik,* Talanta, ***1962,*** 9, *490. With permission.)*

most modern instruments are capable of providing such a curve as well as the thermogram itself. The derivative curve may reveal information that is not detectable in the ordinary thermogram. For example, the three peaks at 140°, 180°, and 205°C suggest that the three hydrates lose moisture at different temperatures. However, all appear to lose carbon monoxide simultaneously and thus yield a single sharp peak at 450°C.

31B DIFFERENTIAL THERMAL ANALYSIS (DTA)

Differential thermal analysis is a technique in which the difference in temperature between a substance and a reference material is measured as a function of temperature while the substance and reference material are subjected to a controlled temperature program. Usually, the temperature program involves heating the sample and reference material in such a way that the temperature of the sample T_s increases linearly with time. The difference in temperature ΔT between the sample temperature and the reference temperature $T_r (\Delta T = T_r - T_s)$ is then monitored and plotted against sample temperature to give a differential thermogram such as that shown in Figure 31-7. The significance of the various parts of this curve are described in Section 31B-2.

31B-1 Instrumentation

Figure 31-8 is a schematic of the furnace compartment of a differential thermal analyzer. A few milligrams of the sample *(S)* and an inert reference substance *(R)* are

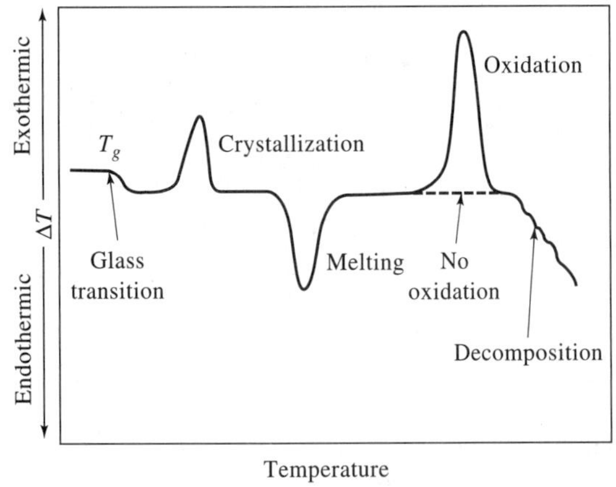

Figure 31-7 Schematic differential thermogram showing types of changes encountered with polymeric materials. *(From R. M. Schulken Jr., R. E. Roy Jr., and R. H. Cox,* J. Polymer Sci., *Part C,* ***1964,*** *6, 18. Reprinted by permission of John Wiley & Sons, Inc.)*

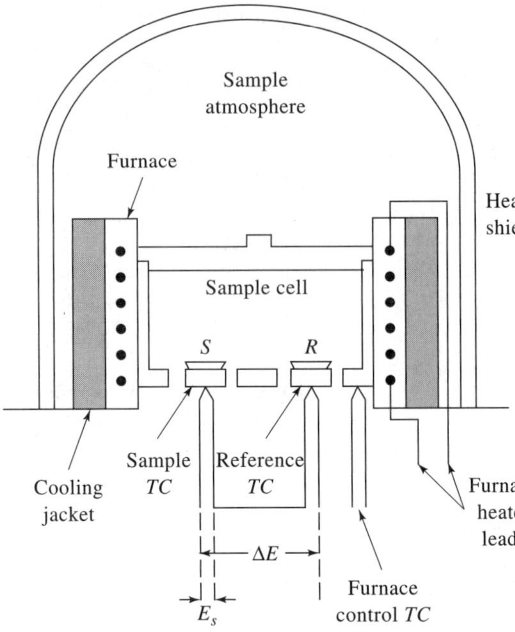

Figure 31-8 Schematic of a typical instrument for differential thermal analysis (TC = thermocouple).

contained in small aluminum dishes that are located above sample and reference thermocouples in an electrically heated furnace.[4] The reference material is an inert substance such as alumina, silicon carbide, or glass beads.

The output potential E_s from the sample thermocouple passes into a microcomputer where it is made to control the current input to the furnace in such a way that the sample temperature increases linearly and at a predetermined rate. The sample thermocouple signal is also converted to temperature T_s and is then recorded as the abscissa of the differential thermogram. The output across the sample and reference thermocouples ΔE is amplified and converted to a temperature difference ΔT, which serves as the ordinate of the thermogram.

Generally, the sample and reference chamber in differential thermal apparatus are designed to permit the circulation of an inert gas, such as nitrogen, or a reactive gas, such as oxygen or air. Some systems also have the capability of operating at high and low pressures.

31B-2 General Principles

Figure 31-7 is an idealized differential thermogram obtained by heating a polymer over a sufficient temperature range to cause its ultimate decomposition. The initial decrease in ΔT is due to the *glass transition,* a phenomenon observed initially when many polymers are heated. The glass transition temperature T_g is the characteristic temperature at which glassy amorphous polymers become flexible or rubber-like because of the onset of the concerted motion of large segments of the polymer molecules. Upon being heated to a certain temperature T_g, the polymer changes from a glass to a rubber. Such a transition involves no absorption or evolution of heat so that no change in enthalpy results—that is, $\Delta H = 0$. The heat capacity of the rubber is, however, different from that of the glass, which results in the lowering of the baseline, as shown in the figure. No peak results during this transition, however, because of the zero enthalpy change.

Two maxima and a minimum are observed in the thermogram in Figure 31-7, all of which are called peaks. The two maxima are the result of exothermic processes in which heat is evolved from the sample, thus causing its temperature to rise; the minimum labeled "melting" is the consequence of an endothermic process in which heat is absorbed by the analyte. Upon being heated to a characteristic temperature, many amorphous polymers begin to crystallize as microcrystals, giving off heat in the process. Crystal formation is

[4] For a description of several DTA instruments from commercial sources, see L. Voress, *Anal. Chem.,* **1994,** *66,* 1035A.

responsible for the first exothermic peak shown in Figure 31-7. The area under such a peak becomes larger the slower the heating rate because more and more crystals have time to form and grow under this circumstance.

The second peak in the figure is endothermic and involves melting of the microcrystals formed in the initial exothermic process. The third peak is exothermic and is encountered only if the heating is performed in the presence of air or oxygen. This peak is the result of the exothermic oxidation of the polymer. The final negative change in ΔT results from the endothermic decomposition of the polymer to produce a variety of products.

As suggested in Figure 31-7, differential thermal analysis peaks result from both physical changes and chemical reactions induced by temperature changes in the sample. Physical processes that are endothermic include fusion, vaporization, sublimation, absorption, and desorption. Adsorption and crystallization are generally exothermic. Chemical reactions may also be exothermic or endothermic. Endothermic reactions include dehydration, reduction in a gaseous atmosphere, and decomposition. Exothermic reactions include oxidation in air or oxygen, polymerization, and catalytic reactions.

Peak areas in differential thermograms depend upon the mass of the sample, m, the enthalpy, ΔH, of the chemical or physical process, and certain geometric and heat conductivity factors. These variables are related by the equation

$$A = -kGm\Delta H = -k'm\Delta H \qquad (31\text{-}1)$$

where A is the peak area, G is a calibration factor that depends upon the sample geometry, and k is a constant related to the thermal conductivity of the sample. The convention of assigning a negative sign to an exothermic enthalpy change accounts for the negative sign in the equation. For a given species, k' remains constant provided that a number of variables such as heating rate, particle size, and placement of the sample relative to the sample thermocouple are carefully controlled. Under these circumstances Equation 31-1 can be used to determine: (1) the mass of a particular analyte if k' and ΔH can be determined by calibration; (2) the enthalpy change if k' and m are known.

31B-3 Applications

Differential thermal analysis finds widespread use in determining the thermal behavior and composition of naturally occurring and manufactured products. The number of applications is impressive and can be appreciated by examination of a two-volume monograph and recent reviews in *Analytical Chemistry*.[5] A few illustrative applications follow.

Differential thermal analysis is a powerful and widely used tool for studying and characterizing polymers. Figure 31-7 illustrates the type of physical and chemical changes in polymeric materials that can be studied by differential thermal methods. Note that thermal transitions for a polymer often take place over an extended temperature range because even a pure polymer is a mixture of homologs, and not a single chemical species.

[5] *Differential Thermal Analysis,* R. C. Mackenzie, Ed. New York: Academic Press, 1970; D. Dollimore, *Anal. Chem.,* **1996,** *68,* 63R; **1994,** *66,* 17R; **1992,** *64,* 147R; **1990,** *62,* 44R; **1988,** *60,* 274R.

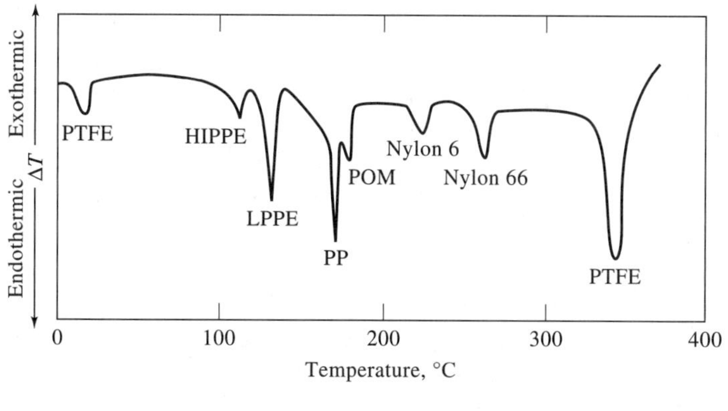

Figure 31-9 Differential thermogram for a mixture of seven polymers. PTFE = polytetrafluoroethylene; HIPPE = high-pressure (low-density) polyethylene; LPPE = low-pressure (high-density) polyethylene; PP = polypropylene; POM = polyoxymethylene. *(From J. Chiu,* DuPont Thermogram, ***1965,*** *2(3), 9. With permission.)*

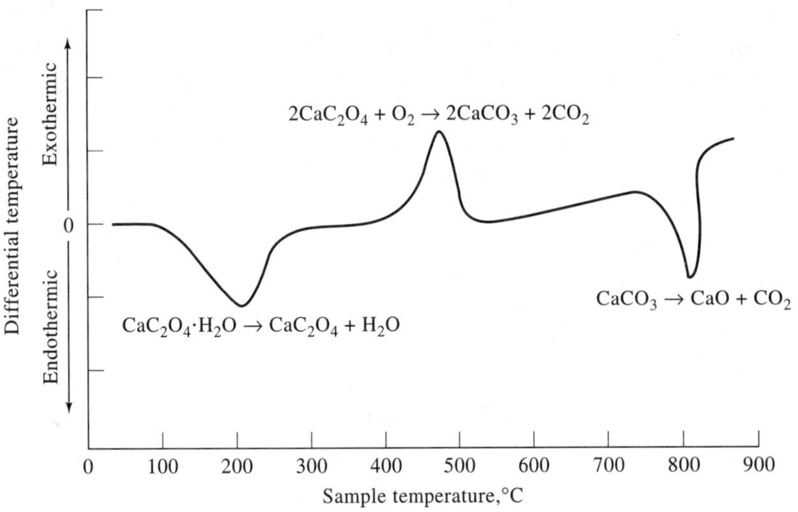

Figure 31-10 Differential thermogram of $CaC_2O_4 \cdot H_2O$ in the presence of O_2; the rate of temperature increase was 8°C/min. *(From* Handbook of Analytical Chemistry, *L. Meites, Ed., pp. 8–14. New York: McGraw-Hill, 1963. With permission.)*

Figure 31-9 is a differential thermogram of a physical mixture of seven commercial polymers. Each peak corresponds to the melting point of one of the components. Polytetrafluoroethylene (PTFE) has an additional low-temperature peak that arises from a crystalline transition. Clearly, differential thermal analysis has the potential use of identifying polymers.

Differential thermal measurements have been used for studies of the thermal behavior of pure inorganic compounds as well as such inorganic substances as silicates, ferrites, clays, oxides, ceramics, catalysts, and glasses. Information is provided about such processes as fusion desolvation, dehydration, oxidation, reduction, adsorption, and solid-state reactions.

Figure 31-10 demonstrates the use of differential thermal analysis for studying the thermal behavior of a simple inorganic species. The differential thermogram was obtained by heating calcium oxalate monohydrate in a flowing stream of air. The two minima indicate that the sample became cooler than the reference material as a consequence of the two endothermic reactions that are shown by the equations below the minima. The single maximum indicates that the oxidation of calcium oxalate to give calcium carbonate and carbon dioxide is exothermic. When an inert gas, such as nitrogen, is substituted for air as the purge gas, three minima are encountered because decomposition of calcium oxalate is now endothermic, with the products being calcium carbonate and carbon monoxide.

An important use of differential thermal analysis is for the generation of phase diagrams and the study of phase transitions. An example is shown in Figure 31-11, which is a differential thermogram of sulfur, in which the peak at 113°C corresponds to the solid-phase change from the rhombic to the monoclinic form, whereas the peak at 124°C corresponds to the melting point of the element. Liquid sulfur is known to exist in at least three forms, and the peak at 179°C apparently involves these transitions, while the peak at 446°C corresponds to the boiling point of sulfur.

The differential thermal method provides a simple and accurate way of determining the melting, boiling,

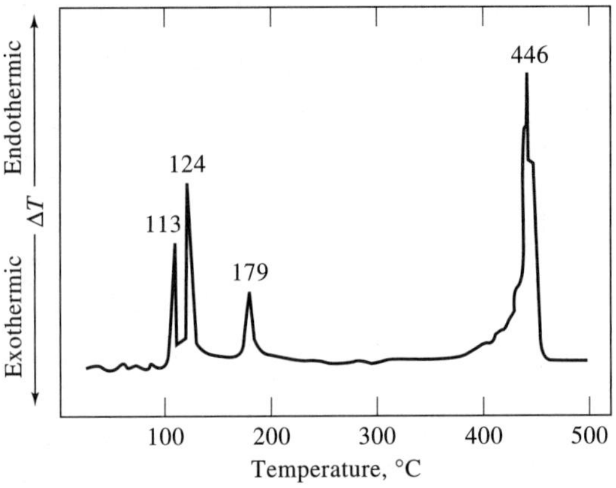

Figure 31-11 Differential thermogram for sulfur. *(Reprinted with permission from J. Chiu,* Anal. Chem., ***1963,*** *35, 933. Copyright 1963 American Chemical Society.)*

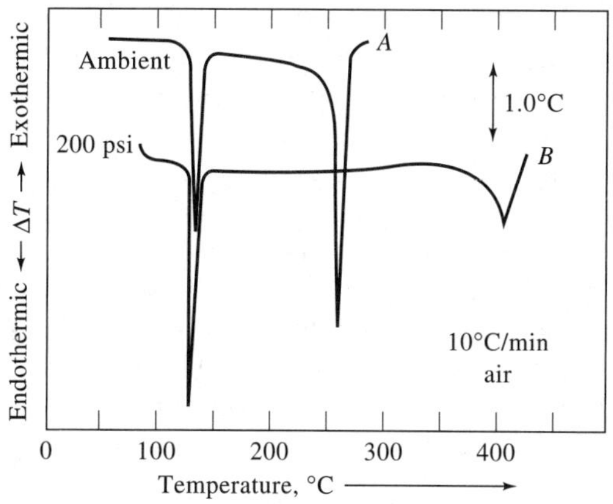

Figure 31-12 Differential thermogram for benzoic acid. Curve *A:* at atmospheric pressure; curve *B:* at 200 lbs/in^2. *(From P. F. Levy, G. Nieuweboer, and L. C. Semanski,* Thermochim. Acta, *1970,* 1, *433. With permission.)*

and decomposition points of organic compounds. Generally, the data appear to be more consistent and reproducible than those obtained with a hot stage or a capillary tube. Figure 31-12 shows thermograms for benzoic acid at atmospheric pressure *(A)* and at 200 psi *(B)*. The first peak corresponds to the melting point and the second to the boiling point of the acid.

31C DIFFERENTIAL SCANNING CALORIMETRY (DSC)

Differential scanning calorimetry is a thermal technique in which differences in heat flow into a substance and a reference are measured as a function of sample temperature while the two are subjected to a controlled-temperature program. The basic difference between differential scanning calorimetry and differential thermal analysis is that the former is a calorimetric method in which differences in *energy* are measured. In contrast, in differential thermal analysis, differences in temperature are recorded. The temperature programs for the two methods are similar. Differential scanning calorimetry has by now become the most widely used of all thermal methods.

31C-1 Instrumentation

Two types of methods are used to obtain differential scanning calorimetry data. In *power compensated DSC* the sample and reference material are heated by separate heaters in such a way that their temperatures are kept equal while these temperatures are increased (or decreased) linearly. In *heat flux DSC,* the difference in heat flow into the sample and reference is measured as the sample temperature is increased—or decreased—linearly. Although the two methods provide the same information, the instrumentation for the two is fundamentally different.[6]

Power Compensated DSC

Figure 31-13 is a schematic showing the design of a power compensated calorimeter for performing DSC measurements. The instrument has two independent furnaces, one for heating the sample and the other for heating the reference. In a commercial model based upon this design, the furnaces are small, weighing about a gram each, a feature that leads to rapid rates of heating, cooling, and equilibration. The furnaces are imbedded in a large temperature-controlled heat sink. Above the furnaces are the sample and reference holders, which have platinum resistance thermometers imbedded in them to monitor the temperatures of the two materials continuously.

Two control circuits are employed in obtaining differential thermograms with the instrument shown in Figure 31-13, one for average-temperature control and one for differential-temperature control. In the average-temperature control circuit, a programmer provides an electrical signal that is proportional to the desired average temperature of the sample and reference holders as a function of time. This signal is compared in a com-

[6] For a description of several commercially available instruments of both types, see D. Noble, *Anal. Chem.*, **1995**, *67*, 323A.

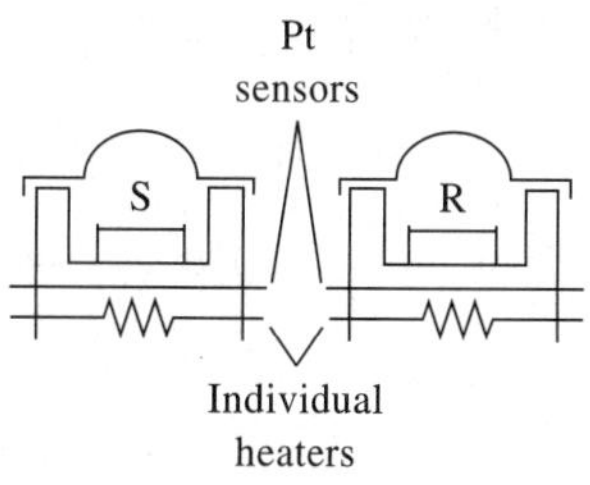

Figure 31-13 Schematic of DSC sample holder and furnaces. *(Courtesy of W. W. Wendlandt,* Thermal Analysis, *3rd ed., p. 346. New York: Wiley, 1986. Reprinted by permission of John Wiley & Sons, Inc.)*

puter with the average of the signals from the sample and reference detectors imbedded in the sample and reference holders. Any difference between the programmer signal and the average platinum sensor signal is used to adjust the average temperature of the sample and reference. The average temperature then serves as the abscissa for the thermogram.

In the differential-temperature circuit, sample and reference signals from the platinum resistance sensors are fed into a differential amplifier via a comparator circuit that determines which is greater. The amplifier output then adjusts the power input to the two furnaces in such a way that their temperatures are kept identical. That is, throughout the experiment, the sample and reference are isothermal. A signal that is proportional to the difference in power input to the two furnaces is also transmitted to the data acquisition system. This difference in power, usually in milliwatts, is the information most frequently plotted as a function of sample temperature.

Heat Flux DSC

Figure 31-14 is a schematic of a commercial heat flux DSC cell. Heat flows into both the sample and reference material via an electrically heated constantan thermoelectric disk.[7] Small aluminum sample and reference pans sit on raised platforms formed on the constantan disk. Heat is transferred through the disks and up into the sample and reference via the two pans. The differential heat flow to the sample and reference is monitored by Chromel/constantan area thermocouples formed by the junction between the constantan platform and Chromel® disks attached to the underside of the platforms. It can be shown that the differential heat flow into the two pans is directly proportional to the difference in output of the two thermocouple junctions.[8] The sample temperature is estimated by means of the Chromel/alumel junction under the sample disk.

31C-2 Applications

The DSC curve for an amorphous sample of polyethylene terphthalate is shown in Figure 31-15. The ordinate of this plot is energy input in milliwatts. Note the similarity in appearance of this curve and the DTA plot shown in Figure 31-7. The two initial peaks in both figures arise from microcrystal formation and melting. A glass transition is also evident in both cases, but no oxidation peak is found in the DSC curve because the experiment was carried out in an atmosphere of nitrogen.

[7] Constantan is an alloy with 60% copper and 40% nickel. Chromel is a trademark for a series of alloys containing chromium, nickel, and at times iron.

[8] R. A. Baxter, in *Thermal Analysis,* R. F. Schwenker and P. D. Garn, Eds., Vol. 1, pp. 68–70. New York: Academic Press, 1969.

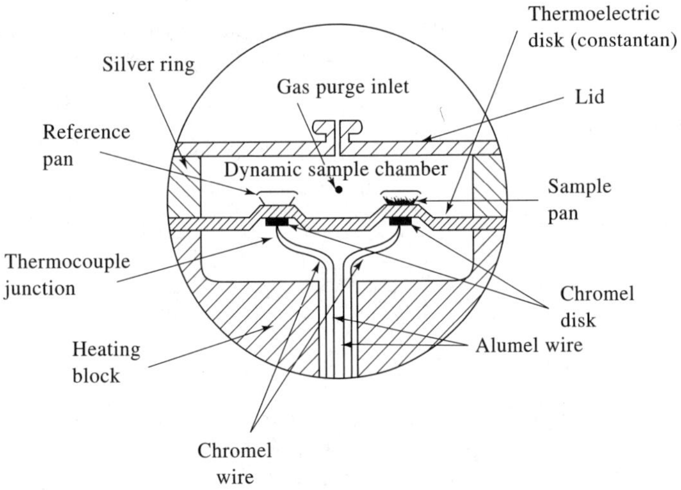

Figure 31-14 Schematic of a heat flux DSC cell. *(Courtesy of DuPont Instrument Systems, Wilmington, DE.)*

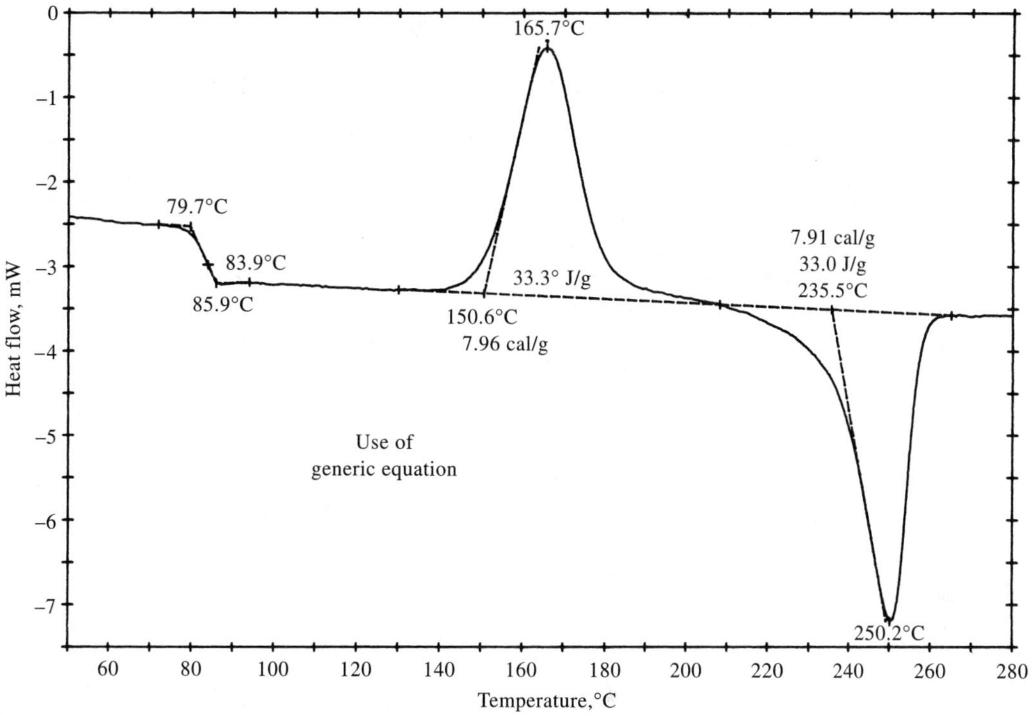

Figure 31-15 Differential scanning calorimetry output from a thermal instrument showing the thermal transition for polyethylene terphthalate. *(Courtesy of DuPont Instrument Systems, Wilmington, DE.)*

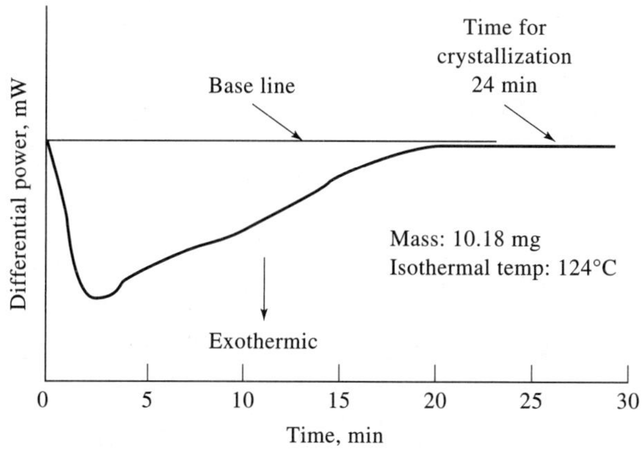

Figure 31-16 DSC curve for the isothermal crystallization of polyethylene.

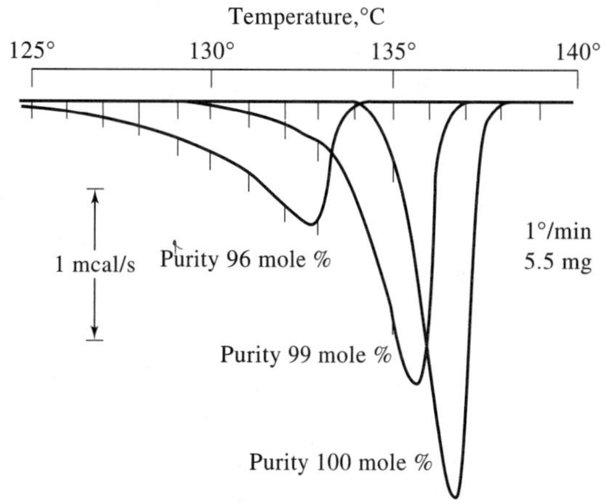

Figure 31-17 Differential scanning calorimetry study of samples of the drug phenacetin. *(Reprinted with permission from H. P. Vaughan and J. P. Elder,* Amer. Lab., ***1974,*** *6(1), 58. Copyright 1974 by International Scientific Communications, Inc.*

Differential scanning calorimetric experiments are usually performed in the temperature scan mode, but isothermal experiments are occasionally encountered. Figure 31-16 is an illustration of the use of DSC to monitor the isothermal crystallization of polyethylene. The area under the exothermic peak in this experiment can be used to estimate the degree of crystallization that has occurred at this temperature. Note that at 124°C, 24 min are required to develop maximum crystallinity. By performing similar experiments at a series of temperatures it is possible to completely characterize the crystallization behavior of this material.

Differential thermal methods have found widespread use in the pharmaceutical industry for testing the purity of drug samples. An example is shown in Figure 31-17 in which DSC curves are used to determine the purity of phenacetin preparations. Generally, curves of this type provide purity data with relative uncertainties of ±10%.

31D QUESTIONS AND PROBLEMS

31-1 Describe what quantity is measured and how the measurement is performed for each of the following techniques: (a) thermogravimetric analysis; (b) differential thermal analysis; (c) differential scanning calorimetry; (d) isothermal differential scanning calorimetry.

31-2 A 0.6025-g sample was dissolved, and the Ca^{2+} and Ba^{2+} ions present were precipitated as $BaC_2O_4 \cdot H_2O$ and $CaC_2O_4 \cdot H_2O$. The oxalates were then heated in a thermogravimetric apparatus leaving a residue that weighed 0.5713 g in the 320° to 400°C range and 0.4673 g in the 580° to 620°C range. Calculate the % Ca and % Ba in the sample.

31-3 The following table summarizes some data about three iron(III) chlorides.

Compound	Molecular Weight	Melting Point (°C)
$FeCl_3 \cdot 6H_2O$	270	37
$FeCl_3 \cdot 5/2\ H_2O$	207	56
$FeCl_3$	162	306

Sketch the thermogravimetric curve anticipated if a 25.0-mg sample of $FeCl_3 \cdot 6H_2O$ is heated from 0° to 400°C.

31-4 Why are the two low-temperature endotherms in Figure 31-12 coincident whereas the high-temperature peaks are displaced from each other?

31-5 It should be possible to at least partially characterize an oil shale sample using techniques discussed in this chapter. Briefly discuss two techniques that it would be appropriate to use for this purpose. Sketch typical thermal curves and discuss the information that might be obtained and problems that might be anticipated.

31-6 In thermal analysis methods, why is the thermocouple for measuring sample temperature seldom immersed directly into the sample?

31-7 List the types of physical changes that can yield exothermic and endothermic peaks in DTA and DSC.

31-8 List the types of chemical changes that can yield exothermic and endothermic peaks in DTA and DSC.

31-9 Why are the applications of thermal gravimetry more limited than those for DSC and DTA?

31-10 Why does the glass transition for a polymer yield no exothermic or endothermic peak?

31-11 Describe the difference between power compensators and heat flux DSC instruments.

Chapter 32

Radiochemical Methods

The availability of both natural and artificial radioactive isotopes has made possible the development of analytical methods (radiochemical methods) that are both sensitive and specific.[1] *These procedures are often characterized by good accuracy and widespread applicability; in addition, some minimize or eliminate chemical separations that are required in other analytical methods.*

Radiochemical methods are of three types based upon the origin of the radioactivity. In ***activation analysis****, activity is induced in one or more elements of the sample by irradiation with suitable radiation or particles (most commonly thermal neutrons from a nuclear reactor); the resulting radioactivity is then measured. In the second category are methods in which the radioactivity is physically introduced into the sample by adding a measured amount of a radioactive species called a* ***tracer****. The most important class of quantitative methods based upon introduction of radioactive species are* ***isotope dilution*** *methods in which a weighed quantity of radioactively tagged analyte having a known activity is added to a measured amount of the sample. After thorough mixing to assure homogeneity, a fraction of the component of interest is isolated and purified; the analysis is then based upon the activity of this isolated fraction. In addition, organic chemists often utilize reagents that have been labeled with radioactive tracers in order to elucidate reaction mechanisms. The third class of methods involves measurement of radioactivity that is naturally occurring in a sample. Examples of this type of method are the measurement of radon in household air and uranium in pottery and ceramic materials.*

[1] For a detailed treatment of radiochemical methods, see G. Friedlander, J. W. Kennedy, E. S. Macias, and J. M. Miller, *Nuclear and Radiochemistry,* 3rd ed. New York: Wiley, 1981; *Treatise on Analytical Chemistry,* P. J. Elving, V. Krivan, and I. M. Kolthoff, Eds., Part I, Vol. 14. New York: Wiley, 1986; H. J. Arnikar, *Essentials of Nuclear Chemistry,* 2nd ed. New York: Wiley, 1987; W. D. Ehmann and D. E. Vance, *Radiochemistry and Nuclear Methods of Analysis.* New York: Wiley, 1991.

32A RADIOACTIVE ISOTOPES

With one exception, all atomic nuclei are made up of a collection of protons and neutrons; the exception, of course, is the hydrogen nucleus, which consists of a proton only. The chemical properties of an atom are determined by its atomic number Z, which is the number of protons contained in its nucleus. The sum of the number of neutrons and protons in a nucleus is the mass number A.[2] Isotopes of elements are atoms having the same atomic number but different mass numbers. That is, the nuclei of isotopes of an element contain the same number of protons but different numbers of neutrons.

Stable isotopes are those that have never been observed to decay spontaneously. *Radioactive isotopes (radionuclides),* in contrast, undergo spontaneous disintegration, which ultimately leads to stable isotopes. The disintegration or *radioactive decay* of isotopes occurs with the emission of electromagnetic radiation in the form of X-rays or gamma rays (γ rays); with the formation of electrons, positrons, and the helium nucleus; or by *fission* in which a nucleus breaks up into smaller nuclei.

32A-1 Radioactive Decay Products

Table 32-1 lists the most important (from a chemist's viewpoint) types of radiation from radioactive decay. Four of these types—alpha particles, beta particles, gamma-ray photons, and X-ray photons—can be detected and counted by the various detection systems described in Section 12B-4. Thus, most radiochemical methods of analyses are based upon counting of pulses of electricity produced when these decay particles or photons strike a radiation detector.

32A-2 Decay Processes

Several types of radioactive decay processes yield the products listed in Table 32-1.

Alpha Decay

Alpha decay is a common radioactive process encountered with heavier isotopes. Isotopes with mass numbers less than about 150 ($Z \approx 60$) seldom yield alpha particles. The alpha particle is a helium nucleus having

TABLE 32-1 Characteristics of Common Radioactive Decay Products

Product	Symbol	Charge	Mass Number
Alpha particle	α	+2	4
Beta particles			
Negatron	β^-	−1	1/1840 (~0)
Positron	β^+	+1	1/1840 (~0)
Gamma ray	γ	0	0
X-Ray	χ	0	0
Neutron	n	0	1
Neutrino	ν	0	0

a mass of 4 and a charge of +2. An example of alpha decay is shown by the equation

$$ {}^{238}_{92}\text{U} \rightarrow {}^{234}_{90}\text{Th} + {}^{4}_{2}\text{He} \qquad (32\text{-}1) $$

Here, uranium-238 (^{234}U) is converted to thorium-234 (^{234}Th), a *daughter* nuclide having an atomic number that is two less than the *parent.*

Alpha particles from a particular decay process are either monoenergetic or are distributed among relatively few discrete energies. For example, the decay process shown as Equation 32-1 proceeds by two distinct pathways. The first, which accounts for 77% of the decays, produces an alpha particle with an energy of 4.196 MeV.[3] The second pathway (23% of decays) produces an alpha particle having an energy of 4.149 MeV; this reaction is accompanied by the release of a 0.047-MeV gamma ray.

Alpha particles progressively lose their energy as a result of collisions as they pass through matter, and are ultimately converted into helium atoms through capture of two electrons from their surroundings. Their relatively large mass and charge render alpha particles highly effective in producing ion pairs within the matter through which they pass; this property makes their detection and measurement easy. Because of their high mass and charge, alpha particles have a low penetrating

---

[2] The nuclear composition of the isotopes of the element X is indicated by the symbol ${}^{A}_{Z}$X or sometimes more simply as ^{A}X, where A is the number of protons plus the number of neutrons (the mass number), and Z is the atomic number or the number of protons.

[3] Energies associated with nuclear reactions are usually reported in millions of electron volts (MeV) or thousands of electron volts (keV).

power in matter. The identity of an isotope that is an alpha emitter can often be established by measuring the length (or range) over which the emitted alpha particles produce ion pairs within a particular medium (often air).

Alpha particles are relatively ineffective for producing artificial isotopes because of their low penetrating power.

Beta Decay

Any nuclear reaction in which the atomic number Z changes but the mass number A does not is classified as β decay. Three types of β decay are encountered: *negatron formation, positron formation,* and *electron capture.* Examples of the three processes are:

$$^{14}_{6}\text{C} \rightarrow {}^{14}_{7}\text{N} + \beta^- + \bar{\nu}$$

$$^{65}_{30}\text{Zn} \rightarrow {}^{65}_{29}\text{Cu} + \beta^+ + \nu$$

$$^{48}_{24}\text{Cr} + {}^{0}_{1}\text{e} \rightarrow {}^{48}_{23}\text{V} + \text{X-rays}$$

Here, $\bar{\nu}$ and ν in the first two equations represent an antineutrino and a neutrino particle of no significance in analytical chemistry. The third equation illustrates a beta-decay process called *electron capture.* Here, the capture of an electron by the $^{48}_{24}$Cr nucleus produces $^{48}_{23}$V, but this process leaves one of the atomic orbitals (usually the 1*s*, or K orbital, in which case the process is called *K capture*) of the vanadium deficient by one electron. X-Ray emission results when an electron from one of the outer orbitals fills the void left by the capture process. Note that the emission of an X-ray photon is *not a nuclear process* but the capture of an electron by the nucleus is.

Two types of β particles are created by radioactive decay. Negatrons (β^-) are electrons that form when one of the neutrons in the nucleus is converted to a proton. In contrast, the positron (β^+), with the mass of the electron, forms when the number of protons in the nucleus is decreased by one. The positron has a transitory existence, its ultimate fate being annihilation by reaction with an electron to yield two 0.511-MeV gamma-ray photons.

In contrast to alpha emission, beta emission is characterized by production of particles with a continuous spectrum of energies ranging from nearly zero to some maximum that is characteristic for each decay process. The beta particle is not nearly as effective as the alpha particle in producing ion pairs in matter because of its small mass (about 1/7000 that of an alpha particle); at the same time, its penetrating power is substantially greater. Beta energies are frequently expressed in terms of the thickness of an absorber, ordinarily aluminum, required to stop the particle.

Gamma-Ray Emission

Many alpha and beta emission processes leave a nucleus in an excited state, which then returns to the ground state in one or more quantized steps with the release of monoenergetic gamma rays. It is important to note that gamma rays, except for their source, are indistinguishable from X-rays of the same energy. Thus, gamma rays are produced by nuclear relaxations, while X-rays derive from electronic relaxations. The gamma-ray emission spectrum is characteristic for each nucleus and is thus useful for identifying radioisotopes.

Not surprisingly, gamma radiation is highly penetrating. Upon interaction with matter, gamma rays lose energy by three mechanisms; the one that predominates depends upon the energy of the gamma-ray photon. With low-energy gamma radiation, the *photoelectric effect* predominates. Here, the gamma-ray photon disappears after ejecting an electron from an atomic orbital (usually a K orbital) of the target atom. The photon energy is totally consumed in overcoming the binding energy of the electron and in imparting kinetic energy to the ejected electron. With relatively energetic gamma rays, the *Compton effect* is encountered. In this instance, an electron is also ejected from an atom but acquires only a part of the photon energy. The photon, now with diminished energy, recoils from the electron and then goes on to further Compton or photoelectric interactions. If the gamma-ray photon possesses sufficiently high energy (at least 1.02 MeV), *pair production* can occur. Here, the photon is totally absorbed in creating a positron and an electron in the field surrounding a nucleus.

X-Ray Emission

Many processes result in loss of inner-shell electrons from an atom. X-Rays are then formed from *electronic* transitions in which outer electrons fill the vacancies created by the nuclear process. One of the processes is electron capture, which was discussed earlier. A second process, which may lead to X-rays, is *internal conversion,* a type of nuclear process that is an alternative gamma-ray emission. In this instance, an electromagnetic interaction between the excited nucleus and an extranuclear electron results in the ejection of an orbital electron with a kinetic energy equal to the difference between the energy of the nuclear transition and the binding energy of the electron (see Section 12A-3). The emission of this so-called *internal conversion electron* leaves a vacancy in the K, L, or higher orbital; X-rays are emitted as the orbital is filled by an electronic transition.

32A-3 Radioactive Decay Rates

Radioactive decay is a completely random process. Thus, while no prediction can be made concerning the lifetime of an individual nucleus, the behavior of a large ensemble of like nuclei can be described by the first-order rate expression

$$-\frac{dN}{dt} = \lambda N \qquad (32\text{-}2)$$

where N represents the number of radioactive nuclei of a particular kind in the sample at time t and λ is the characteristic *decay constant* for the radioisotope. Upon rearranging this equation and integrating over the interval between $t = 0$ and $t = t$ (during which the number of radioactive nuclei in the sample decreases from N_0 to N), we obtain

$$\ln\frac{N}{N_0} = -\lambda t \qquad (32\text{-}3)$$

or

$$N = N_0 e^{-\lambda t} \qquad (32\text{-}4)$$

The *half-life* $t_{1/2}$ of a radioactive isotope is defined as the time required for one half the number of radioactive atoms in a sample to undergo decay; that is, for N to become equal to $N_0/2$. Substitution of $N_0/2$ for N in Equation 32-3 leads to

$$t_{1/2} = \frac{\ln 2}{\lambda} \approx \frac{0.693}{\lambda} \qquad (32\text{-}5)$$

Half-lives of radioactive species range from small fractions of a second to many billions of years.

The *activity* A of a radionuclide is defined as its disintegration rate. Thus, from Equation 32-2, we may write

$$A = -\frac{dN}{dt} = \lambda N \qquad (32\text{-}6)$$

Activity is given in units of s^{-1}. The *becquerel* (Bq) corresponds to 1 decay per second. That is, 1 Bq = 1 s^{-1}. An older, but still widely used, unit of activity is the Curie (Ci), which was originally defined as the activity of 1 g of radium-226. One curie is equal to 3.70×10^{10} Bq. Usually in analytical radiochemistry, activities of analytes range from a nanocurie or less to a few microcuries.

In the laboratory, absolute activities are seldom measured because detection efficiencies are seldom 100%. Instead, the *counting rate* R is employed, where $R = cA$. Substituting this relationship into Equation 32-6 yields

$$R = cA = c\lambda N \qquad (32\text{-}7)$$

Here, c is a constant called the *detection coefficient,* which depends upon the nature of the detector, the efficiency of counting disintegrations, and the geometric arrangement of sample and detector. The decay law given by Equation 32-4 can then be written in the form

$$R = R_0 e^{-\lambda t} \qquad (32\text{-}8)$$

EXAMPLE 32-1

In an initial counting period, the counting rate for a particular sample was found to be 453 cpm (counts per minute). In a second experiment performed 420 min later, the same sample exhibited a counting rate of 285 cpm. If it can be assumed that all counts are the result of the decay of a single isotope, what is the half-life of that isotope?

Equation 32-8 can be used to calculate the decay constant, λ,

$$285\text{ cpm} = 453\text{ cpm } e^{-\lambda(420\text{ min})}$$

$$\ln\left(\frac{285\text{ cpm}}{453\text{ cpm}}\right) = -\lambda(420\text{ min})$$

$$\lambda = 1.10 \times 10^{-3}\text{ min}^{-1}$$

Equation 32-5 can be used to calculate the half-life

$$t_{1/2} = \frac{\ln 2}{\lambda} = \frac{0.693}{1.10 \times 10^{-3}\text{ min}^{-1}} = 630\text{ min}$$

32A-4 Counting Statistics[4]

As will be shown in Section 32B, radioactivity is measured by means of a detector that produces a pulse of electricity for each atom undergoing decay. Quantitative information about decay rates is obtained by counting these pulses for a specified period. Table 32-2 shows typical decay data obtained by successive one-minute counts of a radioactive source. Because the decay process is random, considerable variation among the data is observed. Thus, the counts per minute range from a low of 132 to a high of 187.

Although radioactive decay is random, the data, particularly for low counts, are not distributed according to Equation a1-13 (Appendix 1), because the decay process does not follow Gaussian behavior. The reason that decay data are not normally distributed lies in the fact that radioactivity consists of a series of discrete events that cannot vary continuously as can the indeterminate errors for which the Gaussian distribution applies. Furthermore, negative counts are not possible. Therefore, the data cannot be distributed symmetrically about the mean.

In order to describe accurately radioactive behavior, it is necessary to assume a *Poisson distribution,* which is given by the equation

$$y = \frac{\mu^{x_i}}{x_i!}e^{-\mu} \tag{32-9}$$

Here, y is the frequency of occurrence of a given count x_i and μ is the mean for a large set of counting data.[5]

The data plotted in Figure 32-1 were obtained with the aid of Equation 32-9. These curves show the deviation $(x_i - \mu)$ from the true average count that would be expected if 1000 replicate observations were made on the same sample. Curve *A* gives the distribution for a substance for which the true average count μ for a selected period is 5; curves *B* and *C* correspond to samples having the true means of 15 and 35. Note that the *absolute* deviations become greater with increases in μ, but the *relative* deviations become smaller. Note also that for the two smaller number of counts the distribution is distinctly not symmetric around the average; this lack of symmetry is a consequence of the fact that a negative count is impossible, while a finite likelihood always exists that a given count can exceed the average by severalfold.

TABLE 32-2 Variations in One-Minute Counts from a Radioactive Source

Minutes	Counts	Minutes	Counts
1	180	7	168
2	187	8	170
3	166	9	173
4	173	10	132
5	170	11	154
6	164	12	167

Total counts = 2004
Average counts/min = $\bar{x}$ = 167

Standard Deviation of Counting Data

In contrast to Equation a1-13 (Appendix 1) for a Gaussian distribution, Equation 32-9 for a Poisson distribution contains no corresponding standard deviation term, and indeed it can be shown that the breadth of curves such as those in Figure 32-1 are dependent only upon the total number of counts for any given period.[6] That is,

$$\sigma_M = \sqrt{M} \tag{32-10}$$

where M is the number of counts for any given period and σ_M is the standard deviation for a Poisson distribution.

The relative standard deviation σ_M/M, is given by

$$\frac{\sigma_M}{M} = \frac{\sqrt{M}}{M} = \frac{1}{\sqrt{M}} \tag{32-11}$$

Thus, although the absolute standard deviation increases with the number of counts, the relative standard deviation decreases.

The counting rate R is equal to M/t. To obtain the standard deviation in R, we apply Equation a1-28 (Appendix 1), which gives

$$\sigma_R^2 = \left(\frac{\partial R}{\partial M}\right)^2 \sigma_M^2 + \left(\frac{\partial R}{\partial t}\right)^2 \sigma_T^2$$

[4] For a more complete discussion, see G. Friedlander, J. W. Kennedy, E. S. Macias, and J. M. Miller, *Nuclear and Radiochemistry,* 3rd ed., Chapter 9. New York: Wiley, 1981.

[5] In the derivation of this equation it is assumed that the counting period is short with respect to the half-life so that no significant change in the number of radioactive atoms occurs. Further restrictions include a detector that responds to the decay of a single isotope only and an invariant counting geometry so that the detector responds to a constant fraction of the decay events that occur.

[6] See footnote 4.

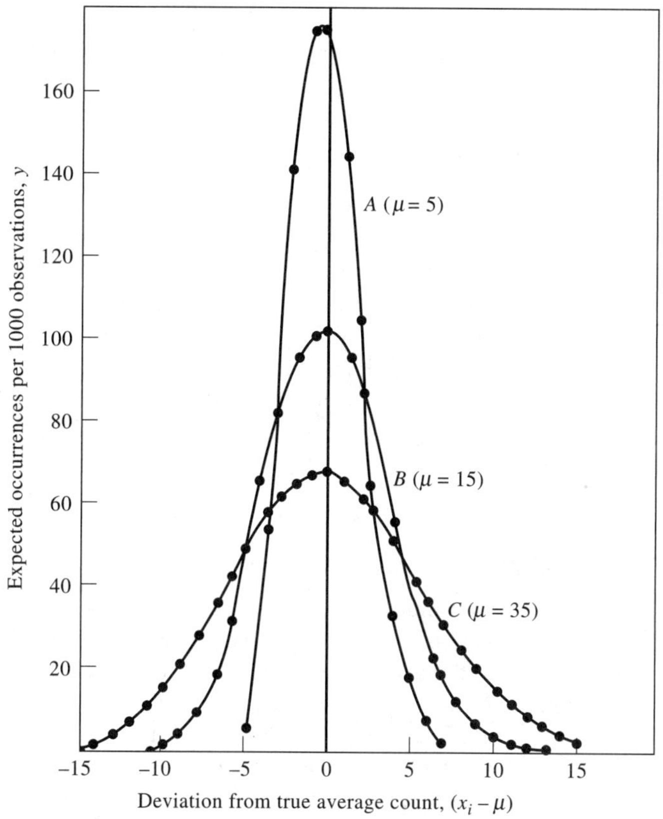

Figure 32-1 Deviation from true average count, $(x_i - \mu)$.

Generally, time can be measured with such high precision that $\sigma_T^2 \sim 0$. The partial derivative of R with respect to M is $1/t$. Thus,

$$\sigma_R^2 = \frac{\sigma_M^2}{t^2}$$

Taking the square root of this equation and substituting Equation 32-10 gives

$$\sigma_R = \frac{\sqrt{M}}{t} = \frac{\sqrt{Rt}}{t} = \sqrt{\frac{R}{t}} \qquad (32\text{-}12)$$

$$\frac{\sigma_R}{R} = \frac{\sqrt{R/t}}{R} = \sqrt{\frac{1}{Rt}} \qquad (32\text{-}13)$$

EXAMPLE 32-2

Calculate the absolute and relative standard deviations in the counting rate for **(a)** the first entry in Table 32-2 and **(b)** the mean of all of the data in the table.

(a) Applying Equation 32-12 gives

$$\sigma_R = \frac{\sqrt{M}}{t} = \frac{\sqrt{180}}{1\text{ min}} = 13.4\text{ cpm}$$

$$\frac{\sigma_R}{R} = \frac{13.4\text{ cpm}}{180\text{ cpm}} \times 100\% = 7.4\%$$

(b) For the entire set,

$$\sigma_R = \frac{\sqrt{2004}}{12} = 3.73\text{ cpm} = 3.7\text{ cpm}$$

$$\frac{\sigma_R}{R} = \frac{3.73}{167} \times 100\% = 2.2\%$$

Confidence Intervals for Counts

In Section a1B-2 (Appendix 1), the confidence interval for a measurement was defined as the limits around a measured quantity within which the true mean can be expected to fall with a stated probability. When the measured standard deviation is believed to be a good approximation of the true standard deviation ($s \rightarrow \sigma$), the confidence limit CL is given by Equation a1-18:

$$\text{CL for } \mu = \bar{x} \pm z\sigma$$

For counting rates, this equation takes the form

$$\text{CL for } R = R \pm z\sigma_R \qquad (32\text{-}14)$$

where z is dependent upon the desired level of confidence. Some values for z are given in Table a1-3.

EXAMPLE 32-3

Calculate the 95% confidence limits for **(a)** the first entry in Table 32-2 and **(b)** the mean of all of the data in the table.

(a) In Example 32-2, we found that $\sigma_R = 13.4$ cpm. Table a1-3 (Appendix 1) reveals that $z = 1.96$ at the 95% confidence level. Thus, for R

$$95\% \text{ CL} = 180 \text{ cpm} \pm 1.96 \times 13.4 \text{ cpm}$$
$$= 180 \ (\pm 26) \text{ cpm}$$

(b) In this instance, σ_R was found to be 3.73 cpm, and

$$95\% \text{ CL for } R = 167 \text{ cpm} \pm 1.96 \times 3.73 \text{ cpm}$$
$$= 167 \ (\pm 7) \text{ cpm}$$

Thus, there are 95 chances in 100 that the true rate for R (for the average of 12 min of counting) lies between 160 and 174 counts/min. For the single count in part **(a)**, 95 out of 100 times the true rate will lie between 154 and 206 counts/min.

Figure 32-2 illustrates the relationship between total counts and tolerable levels of uncertainty as calculated from Equation 32-14. Note that the horizontal axis is logarithmic; thus, a decrease in the relative uncertainty of a factor of ten requires that the number of counts be increased by a factor of one hundred.

Background Corrections

The count recorded in a radiochemical analysis includes a contribution from sources other than the sample. Background activity can be traced to the existence of minute quantities of radon isotopes in the atmosphere, to the materials used in construction of the laboratory, to accidental contamination within the laboratory, to cosmic radiation, and to the release of radioactive materials into the Earth's atmosphere. In order to obtain a true assay, then, it is necessary to correct the total count for background. The counting period required to establish the background correction frequently differs from

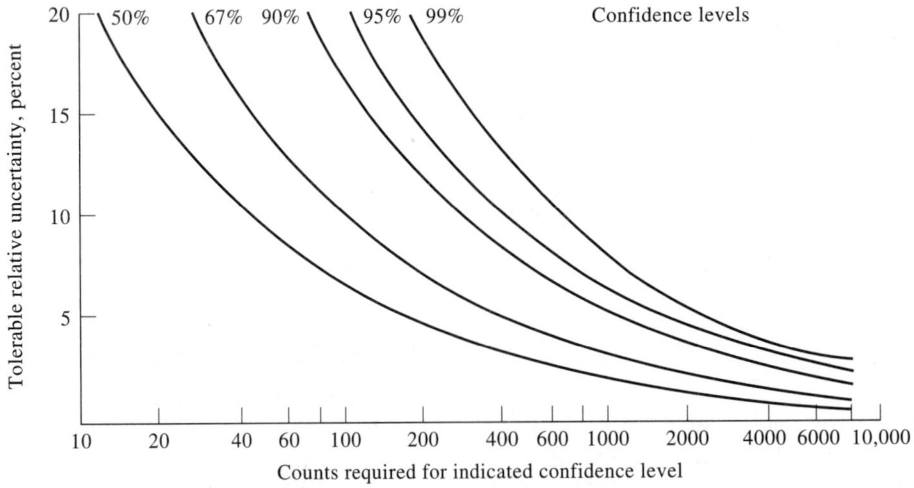

Figure 32-2 Relative uncertainty in counting.

that for the sample; as a result, it is more convenient to employ counting rates. Then,

$$R_c = R_x - R_b \tag{32-15}$$

where R_c is the corrected counting rate and R_x and R_b are the rates for the sample and the background, respectively. The standard deviation of the corrected counting rate can be obtained by application of Equation (1) in Table a1-5 (Appendix 1). Thus,

$$\sigma_{R_c} = \sqrt{\sigma_{R_x}^2 + \sigma_{R_b}^2}$$

Substituting Equation 32-12 into this equation leads to

$$\sigma_{R_c} = \sqrt{\frac{R_x}{t_x} + \frac{R_b}{t_b}} \tag{32-16}$$

EXAMPLE 32-4

A sample yielded 1800 counts in a 10-min period. Background was found to be 80 counts in 4 min. Calculate the absolute uncertainty in the corrected counting rate at the 95% confidence level.

$$R_x = \frac{1800}{10} = 180 \text{ cpm}$$

$$R_b = \frac{80}{4} = 20 \text{ cpm}$$

Substituting into Equation 32-16 yields

$$\sigma_{R_c} = \sqrt{\frac{180}{10} + \frac{80}{4}} = 6.2 \text{ cpm}$$

At the 95% confidence level,

$$\text{CL for } R_c = (180 - 20) \pm 1.96 \times 6.2$$
$$= 160 \pm 12 \text{ cpm}$$

Here, the chances are 95 in 100 that the true count lies between 148 and 172 cpm.

32B INSTRUMENTATION

Radiation from radioactive sources can be detected and measured in essentially the same way as X-radiation (Sections 12B-4 and 12B-5). Gas-filled transducers, scintillation counters, and semiconductor detectors are all sensitive to alpha and beta particles, and to gamma rays because absorption of these particles produces ionization or photoelectrons, which can in turn produce thousands of ion pairs. A detectable electrical pulse is thus produced for each particle reaching the transducer.

32B-1 Measurement of Alpha Particles

In order to minimize self-absorption, alpha-emitting samples are generally counted as thin deposits prepared by electrodeposition or by distillation and condensation. Often these deposits are then sealed with thin windows and counted in windowless gas-flow proportional counters or ionization chambers. Alternatively, they are placed immediately adjacent to a solid-state detector, often in a vacuum for counting. Liquid scintillation counting (see next section) is becoming increasingly important for counting alpha emitters because of the ease of sample preparation and the higher sensitivity for alpha-particle detection.

As mentioned earlier, alpha-particle spectra consist of characteristic, discrete energy peaks, that are useful for identification. Pulse-height analyzers (Section 12B-5) permit the recording of alpha-particle spectra.

32B-2 Measurement of Beta Particles

For beta sources having energies greater than about 0.2 MeV, a uniform layer of the sample is ordinarily counted with a thin-windowed Geiger or proportional tube counter. For low-energy beta emitters, such as carbon-14, sulfur-35, and tritium, liquid scintillation counters (page 285) are preferable. Here, the sample is dissolved in a solution of the scintillating compound. A vial containing the solution is then placed between two photomultiplier tubes housed in a light-tight container. The output from the two tubes is fed into a *coincidence counter,* an electronic device that records a count only when pulses from the two transducers arrive at the same time. The coincidence counter reduces background noise from the detectors and amplifiers because of the low probability that such noise will affect both systems simultaneously. Liquid scintillation counting of beta radiation probably accounts for the majority of radioisotopic determinations because of the widespread use of this technique in clinical laboratories. Because beta spectra are ordinarily continuous, pulse-height analyzers are less useful.

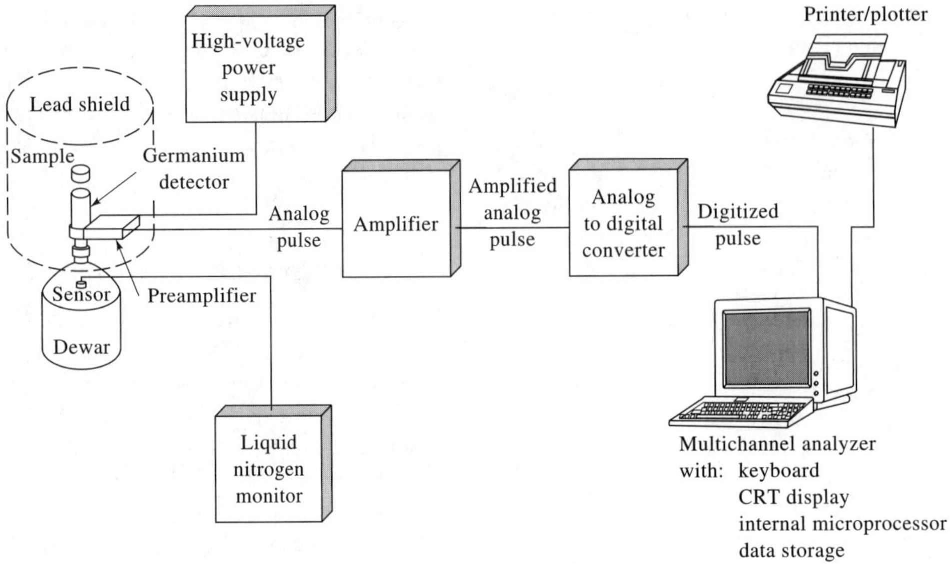

Figure 32-3 Schematic of a gamma spectrometer equipped with a high purity germanium detector. *(Courtesy of Nuclear Data, Inc., Schamburg, Illinois.)*

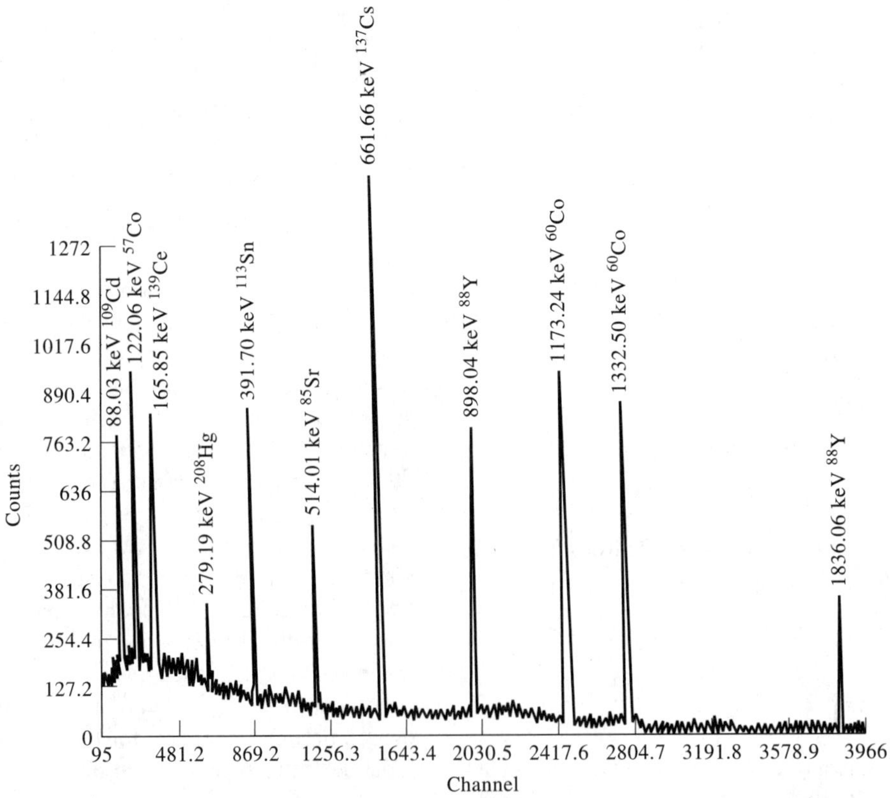

Figure 32-4 Gamma ray spectrum of a calibrated reference source.

32B-3 Measurement of Gamma Radiation

Gamma radiation is detected and measured by the methods described in Sections 12B-4 and 12B-5 for X-radiation. Interference from alpha- and beta-radiation is readily avoided by filtering the radiation with a thin piece of aluminum or Mylar.[7]

Gamma-ray spectrometers are similar to the pulse-height analyzers described in Section 12B-5. A common arrangement is shown in Figure 32-3. Figure 32-4 shows a typical gamma-ray reference spectrum obtained with a 4000-channel analyzer. Here, the characteristic peaks for the various elements are superimposed upon a continuum that arises primarily from the Compton effect.

Figure 32-5 is a schematic of a *well-type* scintillation counter that is used for gamma-ray counting. Here, the sample is contained in a small vial and placed in a cylindrical hole or well in the scintillating crystal of the counter.

[7] For a textbook on the application of gamma spectrometers in radiochemistry, see G. Gilmore and J. D. Hemingway, *Practical Gamma-Ray Spectrometry.* New York: Wiley, 1995.

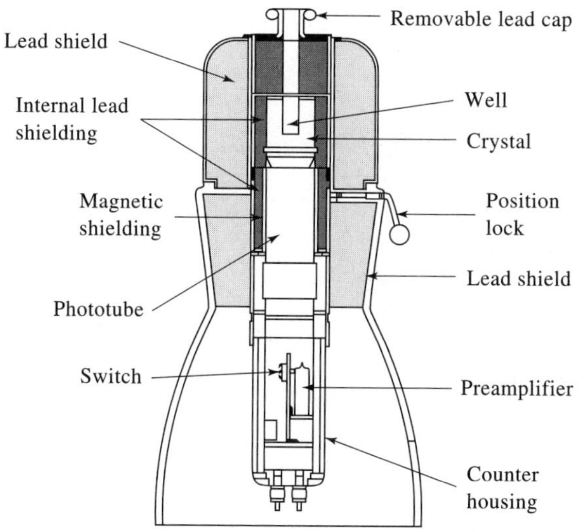

Figure 32-5 A well-type scintillation counter. *(Courtesy of Texas Nuclear Division, Ramsey Engineering Co. Austin, TX. Formerly Nuclear-Chicago Corporation.)*

32C NEUTRON ACTIVATION METHODS

Activation methods are based upon the measurement of radioactivity that has been induced in samples by irradiation with neutrons or charged particles, such as hydrogen, deuterium, or helium ions.[8]

32C-1 Neutrons and Neutron Sources

Three sources of neutrons are employed in neutron activation methods: *reactors, radionuclides,* and *accelerators.* All three produce highly energetic neutrons (in the MeV range), which are usually passed through a moderating material that reduces their energies to a few hundredths of an electron volt. Energy loss to the moderator occurs by *elastic scattering* in which neutrons bounce off nuclei in the moderator material, transferring part of their kinetic energy to each nucleus they strike. Ultimately, the nuclei come to temperature equilibrium with their surroundings. Neutrons having this energy (about 0.04 eV) are called *thermal neutrons,* and the process of moderating high-energy neutrons to thermal conditions is called *thermalization.* The most efficient moderators are low-molecular-weight substances, such as water, deuterium oxide, and paraffin.

Most activation methods are based upon thermal neutrons, which react efficiently with most elements of analytical interest. For some of the lighter elements, however, such as nitrogen, oxygen, fluorine, and silicon, *fast neutrons* having energies of about 14 MeV are more efficient for inducing radioactivity. Such high-energy neutrons are commonly produced by accelerators.

Reactors

Nuclear reactors are sources of copious thermal neutrons and are, therefore, widely used for activation analyses. A typical research reactor will have a neutron flux of 10^{11} to 10^{14} $n\ cm^{-2}s^{-1}$. These high neutron densities lead to detection limits that for many elements range from 10^{-3} to 10 μg.

[8] Monographs on neutron activation methods include: *Nondestructive Activation Analysis,* S. Amiel, Ed. New York: Elsevier, 1981; S. J. Parry, *Activation Spectrometry in Chemical Analysis.* New York: Wiley, 1991; *Activation Analysis,* Vol. I and II, Z. B. Alfassi, Ed. Boca Raton, FL: CRC Press, 1989; W. D. Ehmann and D. E. Vance, *Crit. Rev. Anal. Chem.,* **1989,** *20*(6), 405.

Radioactive Neutron Sources

Radioactive isotopes are convenient and relatively inexpensive sources of neutrons for activation analyses. Their neutron flux densities range from perhaps 10^5 to 10^{10} n cm^{-2}s^{-1}. As a consequence, detection limits are generally not as good as those in which a reactor serves as a source.

One common radioactive source of neutrons is a transuranium element that undergoes spontaneous fission to yield neutrons. The most common example of this type of source consists of ^{238}Cf (californium), which has a half-life of 2.6 years. About 3% of its decay occurs by spontaneous fission, which yields 3.8 neutrons per fission. Thermal flux densities of about 3×10^7 n cm^{-2}s^{-1} or higher are obtainable from this type of source.

Neutrons can also be produced by preparing an intimate mixture of an alpha emitter such as plutonium, americium, or curium with a light element such as beryllium. A commonly used source of this kind is based upon the reaction

$$^{9}_{4}\text{Be} + ^{4}_{2}\text{He} \rightarrow ^{12}_{6}\text{C} + ^{1}_{0}n + 5.7\text{ MeV}$$

To produce thermal neutrons, a paraffin container is typically employed as a moderator.

Accelerators

Benchtop charged particle accelerators are commercially available for the generation of neutrons. A typical generator consists of an ion source that delivers deuterium ions to an area where they are accelerated through a potential of about 150 kV to a target containing tritium absorbed on titanium or zirconium. The reaction is

$$^{2}_{1}\text{H} + ^{3}_{1}\text{H} \rightarrow ^{4}_{2}\text{He} + ^{1}_{0}n$$

Neutrons produced in this way have energies of about 14 MeV and are useful for activating the lighter elements.

32C-2 Interactions of Neutrons with Matter

The basic characteristics of neutrons are given in Table 32-1. Free neutrons are not stable, and they decay with a half-life of about 10.3 min to give protons and electrons. Free neutrons do not, however, generally exist long enough to disintegrate in this way because of their great tendency to react with ambient material. The high reactivity of neutrons arises from their zero charge that permits them to approach charged nuclei without interference from coulombic forces.

Neutron capture is the most important reaction for activation methods. Here, a neutron is captured by the analyte nucleus to give an isotope with the same atomic number, but with a mass number that is greater by one. The new nuclide is in a highly excited state because it has acquired about 8 MeV of energy by binding the neutron. This excess energy is released by *prompt gamma ray emission* or emission of one or more nuclear particles, such as neutrons, protons, or alpha particles. An example of a reaction that produces prompt gamma rays is

$$^{23}_{11}\text{Na} + ^{1}_{0}n \rightarrow ^{24}_{11}\text{Na} + \gamma$$

Usually, equations of this type are written in the abbreviated form

$$^{23}_{11}\text{Na}(n, \gamma)^{24}_{11}\text{Na}$$

The prompt gamma rays formed by capture reactions are of analytical interest in special cases, but the radionuclide produced (^{24}Na) is generally of far greater utility.

32C-3 Theory of Activation Methods

When exposed to a flux of neutrons, the rate of formation of radioactive nuclei from a single isotope can be shown to be

$$\frac{dN^*}{dt} = N\phi\sigma$$

where dN^*/dt is the formation rate of active particles in neutrons per second (n/s), N is the number of stable target atoms, ϕ is the average flux in cm^{-2}s^{-1}, and σ is the capture cross section in cm^2/target atom.[9] The last is a measure of the probability of the nuclei reacting with a neutron at the particle energy employed. Tables of reaction cross section for thermal neutrons list values for σ in *barns* b where 1b = 10^{-24} cm^2/target atom.

Once formed, the radioactive nuclei decay at a rate $-dN^*/dt$ given by Equation 32-2. That is,

$$\frac{-dN^*}{dt} = \lambda N^*$$

[9] Here we use the symbol N^* to distinguish the number of radioactive nuclei from the number of stable nuclei N.

Thus during irradiation with a uniform flux of neutrons, the net rate of formation of active particles is

$$\frac{dN^*}{dt} = N\phi\sigma - \lambda N^*$$

When this equation is integrated from time 0 to *t*, one obtains

$$N^* = \frac{N\phi\sigma}{\lambda}[1 - \exp(-\lambda t)]$$

Substitution of Equation 32-5 into the exponential term yields

$$N^* = \frac{N\phi\sigma}{\lambda}\left[1 - \exp\left(-\frac{0.693t}{t_{1/2}}\right)\right]$$

The last equation can be rearranged to give the product λN^*, which is the activity *A* (see Equation 32-6). Thus,

$$A = \lambda N^* = N\phi\sigma\left[1 - \exp\left(-\frac{0.693t}{t_{1/2}}\right)\right]$$

$$= N\phi\sigma S \qquad (32\text{-}17)$$

where *S* is the *saturation factor,* which is equal to one minus the exponential term.

The foregoing equation can be written in terms of experimental rate measurements by substituting Equation 32-7 to give

$$R = N\phi\sigma c\left[1 - \exp\left(-\frac{0.693t}{t_{1/2}}\right)\right] = N\phi\sigma cS \qquad (32\text{-}18)$$

Figure 32-6 is a plot of this relationship at three levels of neutron flux. The abscissa is the ratio of the irradiation time to the half-life of the isotope ($t/t_{1/2}$). In each case, the counting rate approaches a constant value where the rates of formation and disintegration of the isotope approach one another. Clearly, irradiation for periods beyond four or five half-lives for an isotope will result in little improvement in sensitivity.

In many analyses, irradiation of the sample and standards is carried out for a long enough period to reach saturation. Under this circumstance, all of the terms except *N* on the right side of Equation 32-18 are constant, and the number of analyte radionuclides is directly proportional to the counting rate. If the parent, or target, nuclide is naturally occurring, the weight of the analyte *w* can be obtained from *N* by multiplying it by Avogadro's number, the natural abundance of the analyte isotope, and the chemical atomic weight. Since all of these are constants, the weight of analyte is directly proportional to the counting rate. Thus, if we use the

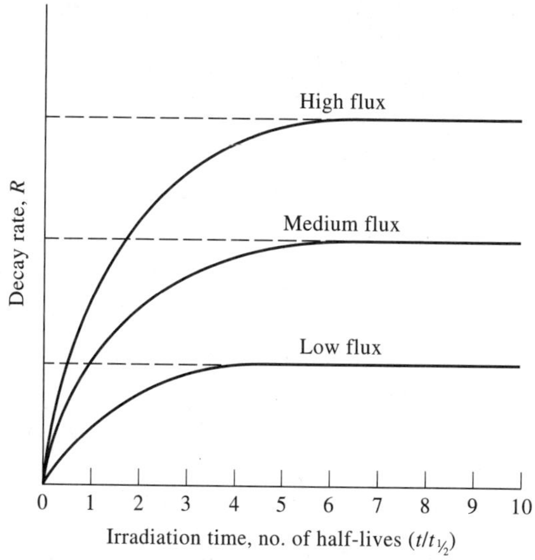

Figure 32-6 The effect of neutron flux and time upon the activity induced in a sample.

subscripts *x* and *s* to represent sample and standard, respectively, we may write

$$R_x = kw_x \qquad (32\text{-}19)$$

$$R_s = kw_s \qquad (32\text{-}20)$$

where *k* is a proportionality constant. Dividing one equation by the other and rearranging leads to the basic equation for computing the weight of analyte in an unknown:

$$w_x = \frac{R_x}{R_s}w_s \qquad (32\text{-}21)$$

32C-4 Experimental Considerations in Activation Methods

Figure 32-7 is a block diagram showing the flow of sample and standards in the two most common types of activation methods, *destructive* and *nondestructive.* In both procedures the sample and one or more standards are irradiated simultaneously with neutrons (or other types of radiation). The samples may be solids, liquids, or gases, although the first two are more common. The standards should approximate the sample as closely as possible both physically and chemically. Generally, the samples and standards are contained in small polyethylene vials; heat-sealed quartz vials are also used on occasion. Care is taken to be sure the samples and standards

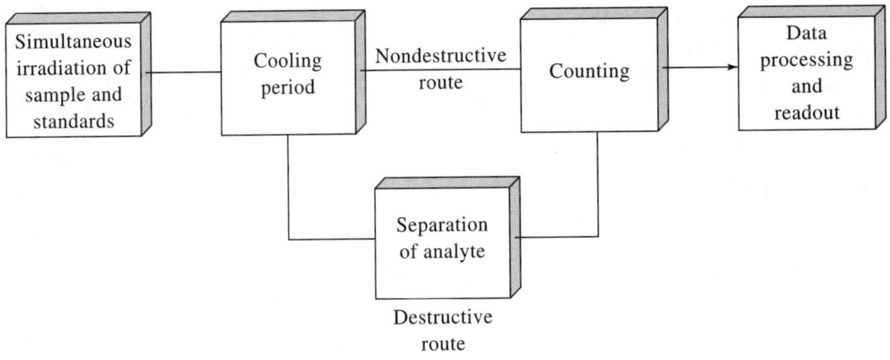

Figure 32-7 Flow diagram for two types of neutron activation methods.

are exposed to the same neutron flux. The time of irradiation is dependent upon a variety of factors and often is determined empirically. Frequently, an exposure time of roughly three to five times the half-life of the analyte product is employed (see Figure 32-6). Irradiation times generally vary from several minutes to several hours.

After irradiation is terminated, the sample and standards are often allowed to decay (or "cool") for a period that again varies from a few minutes to several hours or more. During cooling, short-lived interferences decay away so that they do not affect the outcome of the analysis. Another reason for allowing an irradiated sample to cool is to reduce the health hazard associated with counting the material.

Nondestructive Methods

As shown in Figure 32-7, in the nondestructive method, the sample and standards are counted directly after cooling. Here, the ability of a gamma-ray spectrometer to discriminate among the radiation of different energies provides selectivity. Equation 32-21 is then used to compute the amount of analyte in the unknown.

EXAMPLE 32-5

Two 5.00-mL aliquots of river water were taken for neutron activation analysis. Exactly 1.00 mL of a standard solution containing 1.00 μg of Al^{3+} was added to one aliquot, and 1.00 mL of deionized water was introduced into the other. The two samples were then irradiated simultaneously in a homogeneous neutron flux. After a brief cooling period, the gamma radiation from decay of ^{28}Al was counted. The solution that was diluted with water gave a counting rate of 2315 cpm, whereas the solution containing the added Al^{3+} gave a reading of 4197 cpm. Calculate the weight of Al in the 5.00-mL sample.

Here, we are dealing with a simple standard addition problem that can be solved by substituting into Equations 32-19 and 32-20. Thus,

$$2315 = kw_x$$

$$4197 = k(w_x + w_s) = k(w_x + 1.00)$$

Solving these two equations leads to

$$w_x = 1.23\ \mu g$$

Clearly, success of the nondestructive method requires that the spectrometer be able to isolate the gamma-ray signal produced by the analyte from signals arising from the other components. Whether or not an adequate resolution is possible depends upon the complexity of the sample, the presence or absence of elements that produce gamma rays of about the same energy as that of the element of interest, and the resolving power of the spectrometer. Improvements in resolving power, which have been made in the past three decades as a consequence of the development of high-purity germanium detectors (Section 12B-4), have greatly broadened the scope of the nondestructive method. The great advantage of the nondestructive approach is its simplicity in terms of sample handling and the minimal operator time required to complete an analysis. In fact, modern activation equipment is largely automated.

Destructive Methods

As shown in the lower pathway in Figure 32-7, a destructive method requires that the analyte be separated from the other components of the sample prior to counting. In this case a known amount of the irradiated sample is dissolved and the analyte separated by precipitation, extraction, ion exchange, or chromatography. The isolated material or a known fraction thereof is then counted for its gamma—or beta—activity. As in the nondestructive method, standards may be irradiated simultaneously and treated in an identical way. Equation 32-21 is then used to compute the results of the analysis.

32C-5 Application of Neutron Activation

Neutron activation methods offer several advantages including high sensitivity, minimal sample preparation, and ease of calibration. Often these procedures are nondestructive and for this reason are applied to the analysis of art objects, coins, forensic samples, and archaeological specimens. The major disadvantages of activation methods are their need for large and expensive equipment and special facilities for handling and disposing of radioactive materials. Another handicap is the long time required to complete analyses when long-lived radionuclides are used.

Scope

Figure 32-8 illustrates that neutron activation is potentially applicable to the determination of 69 elements. In addition, four of the inert gases form active isotopes with thermal neutrons and thus can also be determined. Finally, three additional elements—oxygen, nitrogen, and yttrium—can be activated with fast neutrons from an accelerator. A list of types of materials to which the method has been applied is impressive and includes metals, alloys, archeological objects, semiconductors, biological specimens, rocks, minerals, and water. Acceptance of evidence developed from activation analysis by courts of law has led to its widespread use in forensic chemistry. Here, the high sensitivity and nondestructive aspect of the method are particularly useful. Most applications have involved the determination of traces of various elements.

Accuracy

The principal errors that arise in activation analyses are due to self-shielding, unequal neutron flux at sample and standard, counting uncertainties, and errors in counting due to scattering, absorption, and differences in geometry between sample and standard. The errors from these causes can usually be reduced to less than 10% relative; uncertainties in the range of 1% to 3% are frequently obtainable.

Sensitivity

The most important characteristic of the neutron activation method is its remarkable sensitivity for many elements. Note in Figure 32-8, for example, that as little as 10^{-5} μg of several elements can be detected. Note also the wide variations in sensitivities among the elements; thus, about 50 μg of iron are required for detection in contrast to 10^{-6} μg for europium.

The efficiency of chemical recovery, if required prior to radioassay, may limit the sensitivity of an activation analysis. Other factors include the sensitivity of the detection equipment for the emitted radiation, the extent to which activity in the sample decays between irradiation and assay, the time available for counting, and the magnitude of the background count with respect to that for the sample. A high rate of decay is desirable from the standpoint of minimizing the duration of the counting period. Concomitant with high decay rates, however, is the need to establish with accuracy the time lapse between the cessation of irradiation and the commencement of counting. A further potential complication is associated with counting rates that exceed the resolving time of the detecting system; under these circumstances, a correction must be introduced to account for the difference between elapsed (clock) and live (real) counting times.

32D ISOTOPE DILUTION METHODS

Isotope dilution methods, which predate activation procedures, have been and still are extensively applied to problems in all branches of chemistry. These methods are among the most selective available to chemists. Both stable and radioactive isotopes are employed in the isotope dilution technique. The latter are the more convenient, however, because of the ease with which the concentration of the isotope can be determined. We

																	F
																	–
																	1
	Na	Mg											Al	Si	P	S	Cl
β	5×10^{-3}	5×10^{-1}											1×10^{-1}	5×10^{-2}	5×10^{-1}	5	1×10^{-2}
γ	5×10^{-3}	5×10^{-1}											1×10^{-2}	500	–	200	1×10^{-1}
	K	Ca	Sc	Ti	V	Cr	Mn	Fe	Co	Ni	Cu	Zn	Ga	Ge	As	Se	Br
β	5×10^{-2}	1.0	1×10^{-2}	5×10^{-1}	5×10^{-3}	–	5×10^{-5}	50	5×10^{-3}	5×10^{-2}	1×10^{-3}	1×10^{-1}	5×10^{-3}	5×10^{-3}	1×10^{-3}	–	5×10^{-3}
γ	5×10^{-2}	5	5×10^{-2}	5×10^{-2}	1×10^{-3}	1	5×10^{-5}	200	1×10^{-1}	5×10^{-1}	1×10^{-3}	1×10^{-1}	5×10^{-3}	5×10^{-2}	5×10^{-3}	5	5×10^{-3}
	Rb	Sr		Zr	Nb	Mo		Ru	Rh	Pd	Ag	Cd	In	Sn	Sb	Te	I
β	5×10^{-2}	5×10^{-3}		1	5×10^{-3}	5×10^{-1}		1×10^{-2}	1×10^{-3}	5×10^{-4}	5×10^{-3}	5×10^{-2}	5×10^{-5}	5×10^{-1}	5×10^{-3}	5×10^{-2}	5×10^{-3}
γ	5	5×10^{-3}		1	1	1×10^{-1}		5×10^{-2}	5×10^{-4}	5	5×10^{-3}	5×10^{-1}	1×10^{-4}	5×10^{-1}	1×10^{-2}	5×10^{-2}	5×10^{-2}
	Cs	Ba	La	Hf	Ta	W	Re	Os	Ir	Pt	Au	Hg		Pb	Bi		
β	5×10^{-1}	5×10^{-2}	1×10^{-3}	–	5×10^{-2}	1×10^{-3}	5×10^{-4}	5×10^{-2}	1×10^{-4}	5×10^{-2}	5×10^{-4}	–		10	5×10^{-1}		
γ	5×10^{-1}	1×10^{-1}	5×10^{-3}	1	5×10^{-1}	5×10^{-3}	1×10^{-3}	–	1×10^{-3}	5×10^{-1}	5×10^{-4}	1×10^{-2}		–	–		

	Ce	Pr	Nd		Sm	Eu	Gd	Tb	Dy	Ho	Er	Tm	Yb	Lu
β	1×10^{-1}	5×10^{-4}	1×10^{-1}		5×10^{-4}	5×10^{-6}	1×10^{-2}	5×10^{-2}	1×10^{-6}	1×10^{-4}	1×10^{-3}	1×10^{-2}	1×10^{-3}	5×10^{-5}
γ	1×10^{-1}	5×10^{-2}	1×10^{-1}		5×10^{-3}	5×10^{-4}	5×10^{-2}	1×10^{-1}	5×10^{-6}	1×10^{-4}	1×10^{-3}	1×10^{-1}	1×10^{-3}	5×10^{-5}
	Th		U											
β	5×10^{-2}		5×10^{-3}											
γ	5×10^{-2}		5×10^{-3}											

Figure 32-8 Estimated sensitivities of neutron activation methods. Upper numbers correspond to β sensitivities in micrograms, lower numbers to γ sensitivities in micrograms. In each case samples were irradiated for 1 hr or less in a thermal flux of 1.8×10^{-12} neutrons/cm^2/s. *(From V. P. Guinn and H. R. Lukens Jr., in* Trace Analysis: Physical Methods, *G. H. Morrison, Ed., p. 345. New York: Wiley, 1965.)*

shall limit this discussion to methods employing radioactive species.

32D-1 Principles of the Isotope Dilution Procedure

Isotope dilution methods require the preparation of a quantity of the analyte in a radioactive form. A known weight of this isotopically labeled species is then mixed with the sample to be analyzed. After treatment to assure homogeneity between the active and nonactive species, a part of the analyte is isolated chemically in the form of a purified compound. By counting a weighed portion of this product, the extent of dilution of the active material can be calculated and related to the amount of nonactive substance in the original sample. Note that quantitative recovery of the species is not required. Thus, in contrast to the typical analytical separation, steps can be employed to assure a highly pure product on which to base the analysis. It is this independence from the need for quantitative isolation that leads to the high selectivity of the isotope dilution method.

In developing an equation that relates the activity of the isolated and purified mixture of analyte and tracer to the original amount of the analyte, let us assume that W_t grams of the tracer having a counting rate of R_t cpm is added to a sample containing W_x grams of inactive analyte. The count for the resulting $(W_x + W_t)$ grams of the mixture will be the same as that of the W_t grams of the tracer, or R_t. If now W_m grams of the isolated and purified mixture of the active and inactive species is counted, its counting rate R_m will be $W_m/(W_x + W_t)$ of R_t because of dilution. Therefore we may write

$$R_m = R_t\left(\frac{W_m}{W_x + W_t}\right) \qquad (32\text{-}22)$$

which rearranges to

$$W_x = \frac{R_t}{R_m}W_m - W_t \qquad (32\text{-}23)$$

Thus, the weight of the species originally present is obtained from the four measured quantities on the right-hand side of Equation 32-23.

EXAMPLE 32-6

To a sample of a protein hydrolysate, a chemist added 1.00 mg of tryptophan, which was labeled with ^{14}C and exhibited a counting rate of 584 cpm above background. After this labeled compound was thoroughly mixed with the sample, the mixture was passed through an ion-exchange column. The fraction of effluent containing only tryptophan was collected, and from it an 18.0 mg sample of pure tryptophan was isolated. The isolated sample had a count of 204 cpm in the same counter. What was the weight of tryptophan in the original sample?

Substituting into Equation 32-23 gives

$$W_x = \frac{584 \text{ cpm}}{204 \text{ cpm}} \times 18.0 \text{ mg} - 1.00 \text{ mg} = 50.5 \text{ mg}$$

32D-2 Application of the Isotope Dilution Method

The isotopic dilution technique has been employed for the determination of about 30 elements in a variety of matrix materials.[10] Isotopic dilution procedures have also been most widely used for the determination of compounds that are of interest in organic chemistry and biochemistry. Thus, methods have been developed for the determination of such diverse substances as vitamin D, vitamin B_{12}, sucrose, insulin, penicillin, various amino acids, corticosterone, various alcohols, and thyroxine. Isotopic dilution analysis has had less widespread application since the advent of activation methods. Continued use of the procedure can be expected, however, because of the relative simplicity of the equipment required. In addition, the procedure is often applicable where the activation method fails.

---

[10] It is of interest that the dilution principle has also had other applications. One example is its use in the estimation of the size of salmon spawning runs in Alaskan coastal streams. Here, a small fraction of the salmon are trapped, mechanically tagged, and returned to the river. A second trapping then takes place perhaps 10 miles upstream, and the fraction of tagged salmon is determined. The total salmon population is readily calculated from this information and from the number originally tagged. The assumption must, of course, be made that the fish population becomes homogenized during its travel between stations.

32E QUESTIONS AND PROBLEMS

Throughout these problems unity detection coefficients should be assumed. It should also be assumed that all counts and counting rates have been corrected for background unless specific background values are given.

32-1 Identify X in each of the following nuclear reactions:

(a) ${}^{68}_{30}\text{Zn} + {}^{1}_{0}n \rightarrow {}^{65}_{28}\text{Ni} + \text{X}$
(b) ${}^{30}_{15}\text{P} \rightarrow {}^{30}_{14}\text{Si} + \text{X}$
(c) ${}^{214}_{82}\text{Pb} \rightarrow {}^{214}_{83}\text{Bi} + \text{X}$
(d) ${}^{235}_{92}\text{U} + {}^{1}_{0}n \rightarrow 4({}^{1}_{0}n) + {}^{72}_{30}\text{Zn} + \text{X}$
(e) ${}^{130}_{52}\text{Te} + {}^{2}_{1}\text{H} \rightarrow {}^{131}_{53}\text{I} + \text{X}$
(f) ${}^{64}_{29}\text{Cu} + \text{X} \rightarrow {}^{64}_{28}\text{Ni}$

32-2 Potassium-42 is a β emitter with a half-life of 12.36 hr. Calculate the fraction of this isotope remaining in a sample after (a) 1 hr, (b) 10 hr, (c) 20 hr, (d) 75 hr.

32-3 Calculate the fraction of the following isotopes that remains after 27 hr (half-lives are given in parentheses):

(a) iron-59 (44.51 days).
(b) titanium-45 (3.078 hr).
(c) calcium-47 (4.536 days).
(d) phosphorus-33 (25.3 days).

32-4 A $PbSO_4$ sample contains 1 microcurie of Pb-200 ($t_{1/2}$ – 21.5 hr). What storage period is needed to assure that its activity is less than 0.01 microcurie?

32-5 Estimate the standard deviation and the relative standard deviation associated with counts of (a) 100, (b) 750, (c) 7.00×10^3, (d) 2.00×10^4.

32-6 Estimate the absolute and relative uncertainty associated with a measurement involving 800 counts at the

(a) 50% confidence level.
(b) 90% confidence level.
(c) 99% confidence level.

32-7 For a particular sample the total counting rate (sample plus background) was 300 cpm, and this value was obtained over a 14.0-min counting period. The background was counted for 2.0 min and gave 9 cpm. Estimate

(a) the corrected counting rate R_c.
(b) the standard deviation associated with the corrected counting rate σ_{R_c}.
(c) the 90% confidence level associated with the corrected counting rate.

32-8 The background counting rate of a laboratory was found to be approximately 9 cpm when measured over a 3-min period. The goal is to keep the relative standard deviation of the corrected counting rate at less than 5%; ($\sigma_{R_c}/R_c = 0.05$). What total number of counts should be collected if the total counting rate is (a) 90 cpm and (b) 300 cpm?

32-9 A sample of ^{64}Cu exhibits 3250 cpm. After 10.0 hr, the same sample gave 2230 cpm. Calculate the half-life of ^{64}Cu.

32-10 One half of the total activity in a particular sample is due to ^{38}Cl ($t_{1/2}$ = 87.2 min). The other half of the activity is due to ^{35}S ($t_{1/2}$ = 37.5 days). The beta emission of ^{35}S must be measured because this nuclide emits no gamma photons. Therefore, it is desirable to wait until the activity of the ^{38}Cl has decreased to a negligible level. How much time must elapse before the activity of the ^{38}Cl has decreased to only 0.1% of the remaining activity due to ^{35}S?

32-11 Prove that the relative standard deviation of the counting rate σ_R/R is simply $M^{-1/2}$, where M is the number of counts.

32-12 A crystal of potassium fluoride is to be studied via NAA. The following table summarizes the behavior of all naturally occurring isotopes.

Natural Abundance			
100%	$^{19}_{9}F\ (n, \gamma)^{20}_{9}F$	$\xrightarrow{t_{1/2} = 11\ s}$	$^{20}_{10}Ne + \beta^- + \bar{\nu}$
93%	$^{39}_{19}K\ (n, \gamma)^{40}_{19}K$		
0.01%	$^{40}_{19}K\ (n, \gamma)^{41}_{19}K$		
7%	$^{41}_{19}K\ (n, \gamma)^{42}_{19}K$	$\xrightarrow{t_{1/2} = 12.4\ hr}$	$^{42}_{20}Ca + \beta^- + \bar{\nu}$

^{19}F, ^{20}Ne, ^{39}K, ^{41}K, and ^{42}Ca are stable, and we will assume that ^{40}K is also stable because it has a half-life of 1.3×10^9 yr. What sort of irradiation and detection sequence would you use if you wanted to base your analysis on (a) fluorine and (b) potassium?

32-13 Refer to Problem 32-12 and calculate the activity due to ^{20}F and ^{42}K in a 58-mg (1.0 millimole) sample of pure potassium fluoride that has been irradiated for 60 s. The thermal neutron cross sections for ^{19}F and ^{41}K are 0.0090×10^{-24} cm^2/(target atom) and 1.1×10^{-24} cm^2/(target atom), respectively. Assume a flux of 1.0×10^{13} neutrons/(cm^2 s).

32-14 Under what conditions can the second term in Equation 32-23 be ignored?

32-15 A 2.00-mL solution containing 0.120 microcurie per milliliter of tritium was injected into the bloodstream of a dog. After allowing time for homogenization, a 1.00-mL sample of the blood was found to have a counting rate of 15.8 counts per second. Calculate the blood volume of the animal.

32-16 The penicillin in a mixture was determined by adding 0.981 mg of the ^{14}C-labeled compound having a specific activity of 5.42×10^3 cpm/mg. After equilibration, 0.406 mg of pure crystalline penicillin was isolated. This material had a net activity of 343 cpm. Calculate the mass in milligrams of penicillin in the sample.

32-17 In an isotope dilution experiment, chloride was determined by adding 5.0 mg of sodium chloride containing ^{38}Cl ($t_{1/2}$ = 37.3 min) to a sample. The specific activity of the added NaCl was 4.0×10^4 cps/mg. What was the total amount of chloride present in the original sample if 400 mg of pure AgCl was isolated and if this material had a counting rate of 35 cps above background 148 min after the addition of the radiotracer?

32-18 A 10.0-g sample of protein was hydrolyzed. A 30-mg portion of ^{14}C-labeled threonine, with specific activity of 1000 cpm/mg, was added to the hydrolysate. After mixing, 60.0 mg of pure threonine was isolated and found to have a specific activity of 20 cpm/mg.

(a) What percent of this protein is threonine?

(b) Had ^{11}C ($t_{1/2}$ = 20.3 min) been used instead of ^{14}C, how would the answer change?

Assume all numerical values are the same, and that the elapsed time between the two specific activity measurements was 32 min.

32-19 The streptomycin in 500 g of a broth was determined by addition of 1.34 mg of the pure antibiotic containing ^{14}C; the specific activity of this preparation was found to be 223 cpm/mg for a 30-min count. From the mixture, 0.112 mg of purified streptomycin was isolated, which produced a count of 654 counts in 60.0 min. Calculate the concentration in parts per million streptomycin in the sample.

32-20 Show, via a calculation, that the average kinetic energy of a population of thermal neutrons is approximately 0.04 eV.

32-21 Naturally occurring manganese is 100% ^{55}Mn. This nuclide has a thermal neutron capture cross section of 13.3×10^{-24} cm^{-1}. The product of neutron capture is ^{56}Mn, which is a beta and gamma emitter with a 2.50-hr half-life. When Figure 32-8 was produced, it was assumed that the sample was irradiated for 1 hr at a neutron flux of 1.8×10^{12} neutrons/(cm^2 s) and that 10 cpm above background could be detected.

(a) Calculate the minimum mass of manganese that could be detected.

(b) Suggest why the calculated value is lower than the tabulated value.

Chapter 33

Automated Methods of Analysis

One of the major developments in analytical chemistry during the last four decades has been the appearance of commercial automatic analytical systems, which provide analytical data with a minimum of operator intervention. Initially, these systems were designed to fulfill the needs of clinical laboratories, where perhaps 30 or more species are routinely determined for diagnostic and screening purposes. Domestically, hundreds of millions of clinical analyses are performed annually; the need to keep their cost at a reasonable level is obvious. These two considerations motivated the development of early automatic analytical systems. Now, such instruments find application in fields as diverse as the control of industrial processes and the routine determination of a wide spectrum of species in air, water, soils, and pharmaceutical and agricultural products.[1]

[1] For monographs on automatic methods, see J. K. Foreman and P. B. Stockwell, *Automatic Chemical Analysis*. New York: Wiley, 1975; M. Valcarcel and M. D. Luque de Castro, *Automatic Methods of Analysis*. New York: Elsevier, 1988; V. Cerda and G. Ramis, *An Introduction to Laboratory Automation*. New York: Wiley, 1990.

33A AN OVERVIEW OF AUTOMATIC INSTRUMENTS AND INSTRUMENTATION

At the outset, it should be noted that the International Union of Pure and Applied Chemists recommends that a distinction be made between *automatic* and *automated systems*. By IUPAC terminology, automatic devices do not modify their operation as a result of feedback from an analytical transducer. For example, an automatic acid/base titrator adds reagent to a solution and simultaneously records pH as a function of volume of reagent. In contrast, an automated instrument contains one or more feedback systems that control the course of the analysis. Thus, some automated titrators compare the potential of a glass electrode to its theoretical potential at the equivalence point and use the difference to control the rate of addition of acid or base. While this distinction between automatic and automated may be useful, it is not one that is followed by the majority of authors, nor is it followed in this presentation.

33A-1 Advantages and Disadvantages of Automatic Analyses

In the proper context, automated instruments offer a major economic advantage because of their savings in labor costs. For this advantage to be realized, however, it is necessary that the volume of work for the instrument be large enough to offset the original capital investment, which is often large, and the extensive effort that is usually required to put the automatic system in full operation. For laboratories in which large numbers of routine analyses are performed daily, the savings realized by automation can be enormous. With respect to savings in labor costs, it is worthwhile noting that most automated instruments require less skilled, and thus less expensive, operating personnel; on the other hand, more skilled supervisors may be necessary.

A second major advantage of automated instruments is their speed, which is frequently significantly greater than that of manual devices. Indeed, this speed often makes possible the continuous monitoring of the composition of products as they are being manufactured. This information in turn permits alteration of conditions to improve quality or yield. Continuous monitoring is also useful in medicine where analytical results can be used to determine patients' current condition and their response to therapy.

A third advantage of automation is that a well-designed analyzer can usually produce more reproducible results over a long period of time than can an operator employing a manual instrument. Two reasons can be cited for the higher precision of an automated device. First, machines do not suffer from fatigue, which has been demonstrated to adversely affect the results obtained manually, particularly near the end of a working day. A more important contributing factor to precision is the high reproducibility of the timing sequences of automated instruments—a reproducibility that can seldom be matched in human manual operations. For example, automatic analyzers permit the use of colorimetric reactions that are incomplete or that yield products whose stabilities are inadequate for manual measurement. Similarly, separation techniques, such as solvent extraction or dialysis, where analyte recoveries are incomplete, are still applicable when automated systems are used. In both instances, the high reproducibility of the timing of the operational sequences assures that samples and standards are processed in exactly the same way and for exactly the same length of time.

33A-2 Unit Operations in Chemical Analysis

All analytical methods can be broken down into a series of eight steps, or unit operations, any one of which can be automated. Table 33-1 lists these steps in the order in which they occur in a typical analysis. In some cases, it is possible to dispense with one or more of these operations, but several are required in every analysis.

In a totally automatic method, all of the unit operations just listed are performed without human intervention. That is, with a totally automatic instrument, an unmeasured quantity of the untreated sample is introduced into the device, and ultimately an analytical result is produced as a printout or graph. Such instruments are common in clinical laboratories but are less frequently encountered in industrial and university settings because no totally automatic instrument exists that can accommodate the wide variety of sample matrices and compositions that are encountered in such laboratories.

33A-3 Types of Automatic Analytical Systems

Automatic analytical systems are of two general types: *discrete analyzers* and *continuous-flow analyzers;* occasionally, a combination of the two is encountered. In a discrete instrument, individual samples are maintained as separate entities and kept in separate vessels throughout each unit operation listed in Table 33-1. In continuous-flow systems, in contrast, the sample becomes a part of a flowing stream where the several unit operations take place as the sample is carried from the injection or sample introduction point to a flow-through measuring unit, and thence to waste. Both discrete and continuous instruments are generally computer controlled.

Because discrete instruments are based upon the use of individual containers, cross-contamination among samples is totally eliminated. On the other hand, interactions among samples is always a concern in continuous systems, particularly as the rate of sample throughput is increased. Here, special precautions are required to minimize sample contamination.

Modern continuous-flow analyzers are generally mechanically simpler and less expensive than their discrete counterparts. Indeed, in many continuous systems, the only moving parts are peristaltic pumps and switching valves. Both of these components are inexpensive

TABLE 33-1 Unit Operations in a Chemical Analysis

Operation	Typical Examples	Typical Type Automation*
1. Sample preparation	Grinding, homogenizing, drying	D
2. Sample definition	Determining sample weight or volume	D
3. Sample dissolution	Treating with solvent and diluting	C, D
	Heating, igniting, fusing	D
4. Separation	Precipitating and filtering	D
	Extracting, dialyzing, and chromatographing	C, D
5. Measurement	Determining absorbance, emission intensity, potential, current, and conductivity	C, D
	Titrating and weighing	D
6. Calibration	Running standards	C, D
7. Data reduction	Calculating result, analyzing data for accuracy and precision	C, D
8. Data presentation	Printing out numerical results, plotting data	C, D

*D = discrete; C = continuous flow

and reliable. In contrast, discrete systems often have a number of moving parts such as syringes, valves, and mechanical devices for transporting samples or packets of reagents from one part of the system to another. In the most sophisticated discrete systems, the various unit operations are performed by versatile computerized robots, which are capable of performing operations in much the same way as a human operator can.

As indicated in the third column in Table 33-1, some unit operations are not possible with continuous-flow systems because such systems are only capable of handling fluid samples. Thus, when solid materials are to be analyzed or when grinding, weighing, ignition, fusion, or filtration is required in an analysis, automation is only possible with a discrete system.

In this chapter, we first discuss *flow-injection analysis* (FIA), a recent and important type of continuous-flow method. We then describe several types of discrete automatic systems, some of which are based upon robotics.

33B FLOW-INJECTION ANALYSIS

Flow-injection methods, in their present form, were first described by Ruzicka and Hansen in Denmark and Stewart and coworkers in the United States in the mid 1970s.[2]

Flow-injection methods are an outgrowth of segmented-flow procedures, which were widely used in clinical laboratories in the 1960s and 1970s for automatic routine determination of a variety of species in blood and urine samples for medical diagnostic purposes. In segmented-flow systems, which were manufactured by a single company in this country, samples were carried through the system to a detector by a flowing aqueous solution that contained closely spaced air bubbles. The purpose of the air bubbles was to prevent excess sample dispersion, to promote turbulent mixing of samples and reagents, and to scrub the walls of the conduit thus preventing cross-contamination between successive samples. The discoverers of flow-injection analysis found, however, that excess dispersion and cross-contamination are nearly completely avoided in a properly designed system without air bubbles and that mixing of samples and reagent could be easily realized.[3]

[2] K. K. Stewart, G. R. Beecher, and P. E. Hare, *Anal. Biochem.*, **1976**, *70*, 167; J. Ruzicka and E. H. Hansen, *Anal. Chim. Acta*, **1975**, *78*, 145.

[3] For monographs on flow-injection analysis, see J. Ruzicka and E. H. Hansen, *Flow Injection Analysis*, 2nd ed. New York: Wiley, 1988; M. Valcarcel and M. D. Luque de Castro, *Flow Injection Analysis. Principles and Applications*. Chichester, England: Ellis Horwood, 1987; B. Karlberg and G. E. Pacey, *Flow Injection Analysis. A Practical Guide*, New York: Elsevier, 1989.

The absence of air bubbles imparts several important advantages to flow-injection measurements including: (1) higher analysis rates (typically 100 to 300 samples/hr), (2) enhanced response times (often less than 1 min between sample injection and detector response), (3) much more rapid start-up and shut-down times (less than 5 min for each), and (4) except for the injection system, simpler and more flexible equipment. The last two advantages are of particular importance because they make it feasible and economic to apply automated measurements to a relatively few samples of a nonroutine kind. That is, no longer are continuous-flow methods restricted to situations where the number of samples is large and the analytical method highly routine. As a consequence of these advantages, segmented-flow systems have been largely replaced by flow-injection methods (and also by discrete systems based upon robotics).

33B-1 Instrumentation

Figure 33-1a is a flow diagram of the simplest type of flow-injection systems. Here, a colorimetric reagent for chloride ion is moved by a peristaltic pump directly into a valve that permits injection of samples into the flowing stream. The sample and reagent then pass through a 50-cm reactor coil where the reagent diffuses into the sample plug and produces a colored product by the sequence of reactions

$$Hg(SCN)_2(aq) + 2Cl^- \rightleftharpoons HgCl_2(aq) + 2SCN^-$$

$$Fe^{3+} + SCN^- \rightleftharpoons \underset{\text{red}}{Fe(SCN)^{2+}}$$

From the reactor coil, the solution passes into a flow-through photometer equipped with a 480-nm interference filter.

The recorder output from this system for a series of standards containing from 5 to 75 ppm of chloride are shown on the left of Figure 33-1b. Note that four injections of each standard were made to demonstrate the reproducibility of the system. The two curves to the right in the figure are high-speed recorder scans of one of the samples containing 30 (R_{30}) and another containing 75 (R_{75}) ppm chloride. These curves demonstrate that cross-contamination is minimal in an unsegmented stream. Thus, less than 1% of the first analyte is present in the flow cell after 28 s, the time of the next injection (S_2). This system has been successfully used for the routine determination of chloride ion in brackish and waste waters as well as in serum samples.

Sample and Reagent Transport System

Ordinarily, the solution in a flow-injection analysis is moved through the system by a peristaltic pump, a device in which a fluid (liquid or gas) is squeezed through plastic tubing by rollers. Figure 33-2 illustrates the operating principle of the peristaltic pump. Here, the

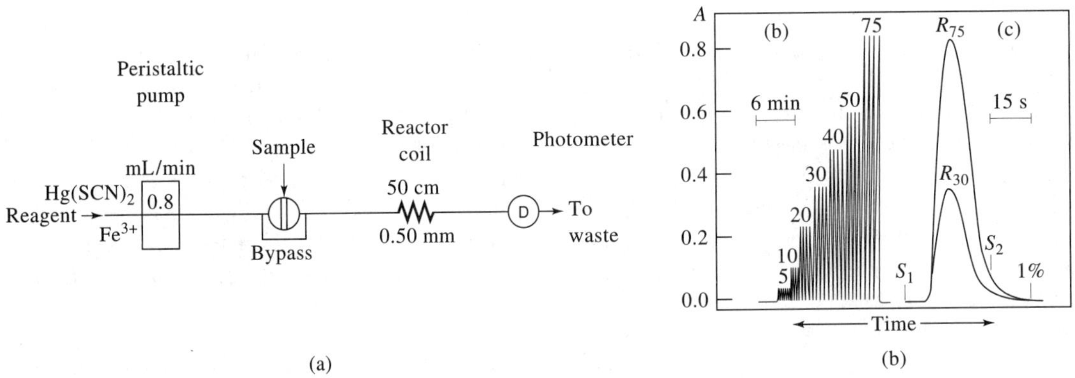

Figure 33-1 Flow-injection determination of chloride: (a) flow diagram, (b) recorder readout for quadruplicate runs on standards containing 5 to 75 ppm of chloride ion, and (c) fast scan of two of the standards to demonstrate the low analyte carryover (less than 1%) from run to run. Note that the point marked 1% corresponds to where the response would just begin for a sample injected at time S_2. *(From J. Ruzicka and E. H. Hansen,* Flow Injection Methods, *2nd ed., p. 16. New York: Wiley, 1988. Reprinted by permission of John Wiley & Sons, Inc.)*

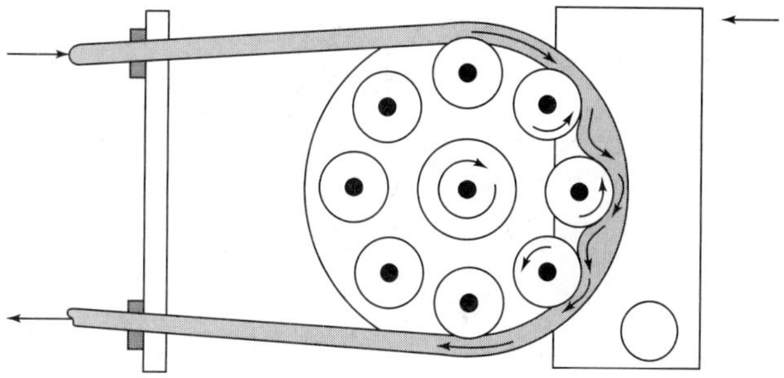

Figure 33-2 Diagram showing one channel of a peristaltic pump. Usually, several additional tubes may be located under the one shown (below the plane of the paper). *(From B. Karlberg and G. E. Pacey,* Flow Injection Analysis. A Practical Guide, *p. 34. New York: Elsevier, 1989. With permission of Elsevier Science Publishers.)*

spring-loaded cam, or band, pinches the tubing against two or more of the rollers at all times, thus forcing a continuous flow of fluid through the tubing. Modern pumps generally have 8 to 10 rollers, arranged in a circular configuration, so that half are squeezing the tube at any instant. This design leads to a flow that is relatively pulse free. The flow rate is controlled by the speed of the motor, which should be greater than 30 rpm, and the inside diameter of the tubing. A wide variety of tube sizes (i.d. = 0.25 to 4 mm) are available commercially and permit flow rates as small as 0.0005 mL/min and as great as 40 mL/min. The rollers of typical commercial peristaltic pumps are long enough so that several tubes can be handled simultaneously.

As shown in Figure 33-1a, flow-injection systems often contain a coiled section of tubing (typical coil diameters are about 1 cm or less) whose purpose it is to enhance axial dispersion and to increase radial mixing of the sample and reagent, both of which lead to more symmetric peaks.

Sample Injectors and Detectors

The injectors and detectors employed in flow-injection analysis are similar in kind and performance requirements to those used in HPLC. Sample sizes for flow-injection procedures range from 5 to 200 μL, with 10 to 30 μL being typical for most applications. For a successful analysis, it is vital that the sample solution be injected rapidly as a pulse or plug of liquid; in addition, the injections must not disturb the flow of the carrier stream. A typical method for introducing sample is shown in Figure 33-1a. With the sampling valve in the position shown, the flow of reagents continues through the bypass while the sample flows through the valve. When the valve is turned 90 deg, the sample enters the flow as a single, well-defined zone. For all practical purposes, flow through the bypass ceases with the valve in this position because the diameter of the sample loop is significantly greater than that of the bypass tubing. Although injection with a syringe is sometimes used, the most satisfactory way of sample introduction is based upon sampling loops similar to those encountered in chromatography (Figures 27-4 and 28-7).

Detection in flow-injection procedures has been carried out by atomic absorption and emission instruments, fluorometers, electrochemical systems, refractometers, spectrophotometers, and photometers. The last is perhaps the most common.

Separations in FIA

Separations by dialysis, by liquid/liquid extraction, and by gaseous diffusion are readily carried out automatically with flow-injection systems.[4]

Dialysis and Gas Diffusion. Dialysis is often used in continuous-flow methods to separate inorganic ions, such as chloride or sodium, or small organic molecules, such as glucose, from high-molecular-weight species such as proteins. Small ions and molecules diffuse relatively rapidly through thin hydrophilic membranes of cellulose acetate or nitrate, whereas large molecules do

[4] For a review of applications of FIA for sample preparation and separations, see G. D. Clark, D. A. Whitman, G. D. Christian, and J. Ruzicka, *Crit. Rev. Anal. Chem.*, **1990**, *21*(5), 357.

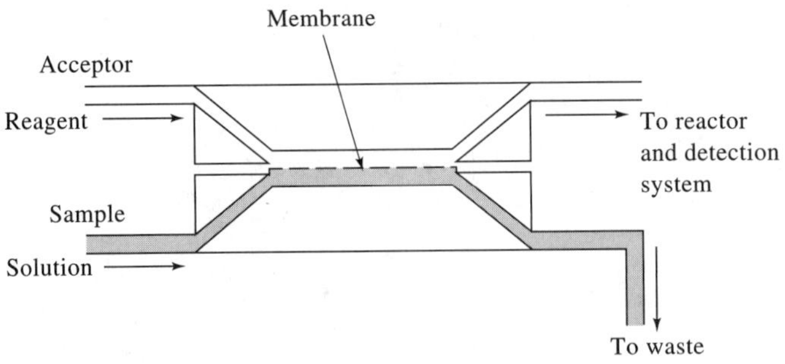

Figure 33-3 A dialysis module. *(From M. Valcarcel and M. D. Luque de Castro,* Automatic Methods of Analysis, *p. 105. New York: Elsevier, 1988. With permission Elsevier Science Publishers.)*

not. Dialysis usually precedes the determination of ions and small molecules in whole blood or serum.

Figure 33-3 is a diagram of a dialysis module in which analyte ions or small molecules diffuse from the sample solution through a membrane into an acceptor stream that often contains a reagent that reacts with the analyte to form a colored species, which can then be determined colorimetrically. Large molecules, which interfere in the determination, remain in the original stream and are carried to waste. The membrane is supported between two plastic plates in which congruent channels have been cut to accommodate the two stream flows. The transfer of smaller species through this membrane is usually incomplete (often less than 50%). Thus, successful quantitative analysis requires close control of temperature and flow rates for both samples and standards. Such control is readily achieved in flow-injection systems.

Gas diffusion from a donor stream containing a gaseous analyte to an acceptor stream containing a reagent that permits its determination is a highly selective technique that has found considerable use in flow-injection analysis. The separations are carried out in a module similar to that shown in Figure 33-3. In this application, however, the membrane is usually a hydrophobic microporous material, such as Teflon or isotactic polypropylene. An example of use of this type of separation technique is found in a method for determining total carbonate in an aqueous solution. Here the sample is injected into a carrier stream of dilute sulfuric acid, which is then directed into a gas-diffusion module, where the liberated carbon dioxide diffuses into an acceptor stream containing an acid/base indicator. This stream then passes through a photometric detector, which yields a signal that is proportional to the carbonate content of the sample.

Extraction. Another common separation technique readily adapted to continuous-flow methods is extraction. Figure 33-4a is a flow diagram for a system for the colorimetric determination of an inorganic cation by extracting an aqueous solution of the sample with chloroform containing a complexing agent, such as 8-hydroxyquinoline. At point *A*, the organic solution is injected into the sample-containing carrier stream. Figure 33-4b shows that the stream becomes segmented at this point and is made up of successive bubbles of the aqueous solution and the organic solvent. Extraction of the metal complex occurs in the reactor coil. Separation of the immiscible liquids takes place in the simple T-shaped separator shown in Figure 33-4c. The separator contains a Teflon strip or fiber that guides the heavier organic layer out of the lower arm of the T, where it then flows through the detector labeled FC in Figure 33-4a. This type of separator can be employed for lighter liquids by inverting the T.

It is important to reiterate that none of the separation procedures in FIA methods is ever complete. The lack of completeness is of no consequence, however, because unknowns and standards are treated in an identical way. As was pointed out earlier, the timing sequences in automatic instruments are sufficiently reproducible so that loss of precision and accuracy does not accompany incomplete separations, as is the case with manual operations.

33B-2 Principles of Flow-Injection Analysis

Immediately after injection with a sampling valve, the sample zone in a flow-injection apparatus has the rectangular concentration profile shown in Figure 33-5a.

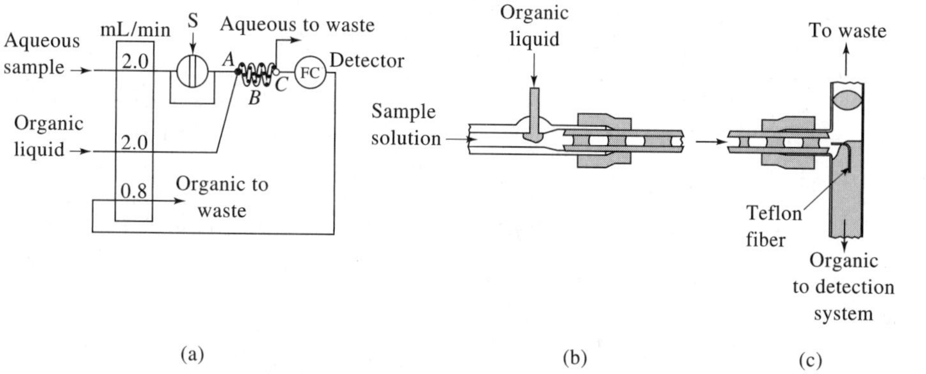

Figure 33-4 (a) Flow diagram of a flow-injection system containing an extraction module (*ABC*). (b) Details of *A*, the organic injector system. (c) Details of *C*, the separator. *(Adapted from J. Ruzicka and E. H. Hansen,* Flow Injection Analysis, *2nd ed. New York: Wiley, 1988. With permission of John Wiley & Sons.)*

As it moves through the tubing, band broadening or *dispersion* takes place. The shape of the resulting zone is determined by two phenomena. The first is convection arising from laminar flow in which the center of the fluid moves more rapidly than the liquid adjacent to the walls, thus creating the parabolic-shaped front and the skewed zone profile shown in Figure 33-5b. Broadening also occurs as a consequence of diffusion. Two types of diffusion can, in principle, occur: radial, or perpendicular to the flow direction and longitudinal, or parallel to the flow. It has been shown that the latter is of no significance in narrow tubing, whereas radial diffusion is always important under this circumstance. In fact, at low flow rates it may be the major source of dispersion. When such conditions exist, the symmetrical distribution shown in Figure 33-5d is approached. In fact, flow-injection analyses are usually performed under conditions in which dispersion by both convection and radial diffusion occurs; peaks like that in Figure 33-5c are then obtained. Here, the radial dispersion from

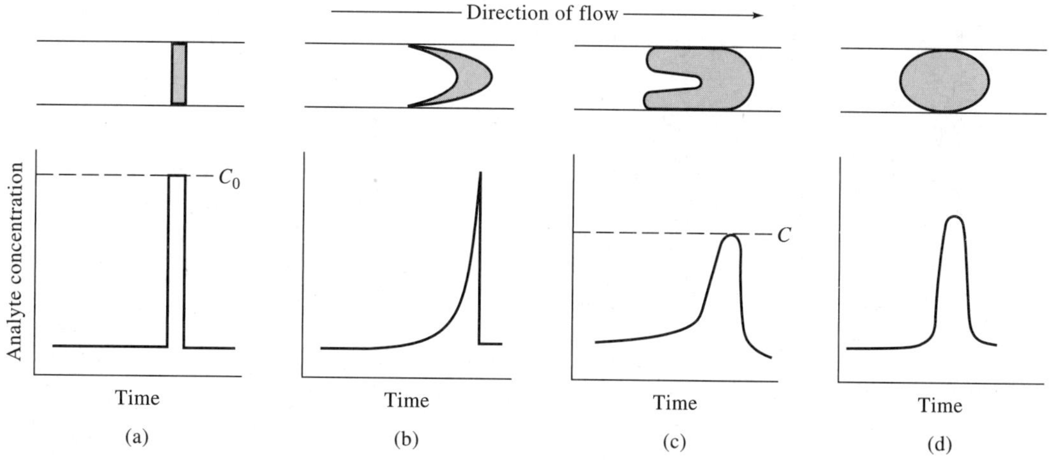

Figure 33-5 Effect of convection and diffusion on concentration profiles of analytes at the detector: (a) no dispersion, (b) dispersion by convection, (c) dispersion by convection and radial diffusion, and (d) dispersion by diffusion. *(Reprinted with permission from D. Betteridge,* Anal. Chem., ***1978,*** 50, *836A. Copyright 1978 American Chemical Society.)*

the walls toward the center serves the important function of essentially freeing the walls of analyte and thus eliminating cross-contamination between samples.

Dispersion

Dispersion D is defined by the equation

$$D = c_0/c \tag{33-1}$$

where c_0 is the analyte concentration of the injected sample and c is the peak concentration at the detector (see Figure 33-5c). Dispersion is readily measured by injecting a dye solution of known concentration c_0 and then measuring the absorbance in the flow-through cell. After calibration, c is calculated from Beer's law.

Dispersion is influenced by three interrelated and controllable variables: sample volume, tube length, and pumping rate. The effect of sample volume on dispersion is shown in Figure 33-6a; here, the other two variables were held constant. Note that at large sample volumes, the dispersion becomes unity. Under these circumstances, no appreciable mixing of sample and carrier takes place, and thus no sample dilution has occurred. Most flow-injection analyses, however, involve interaction of the sample with the carrier or an injected reagent. Here, dispersion greater than unity is necessary. For example, a dispersion of 2 would be required if sample and carrier are to be mixed in a 1:1 ratio.

The dramatic effect of sample volume on peak height shown in Figure 33-6a emphasizes the need for highly reproducible injection volumes when dispersions of 2 and greater are used. Other conditions also must be closely controlled if good precision is to be obtained.

Figure 33-6b demonstrates the effect of tube length on dispersion when sample size and pumping rate are constant. Here, the number above each peak gives the length of sample travel in centimeters.

33B-3 Applications of Flow-Injection Analysis

In the flow-injection literature, the terms *limited dispersion, medium dispersion,* and *large dispersion* are frequently encountered where they refer to dispersions of 1 to 3, 3 to 10, and greater than 10, respectively. Methods based on all three types of dispersion have been developed.

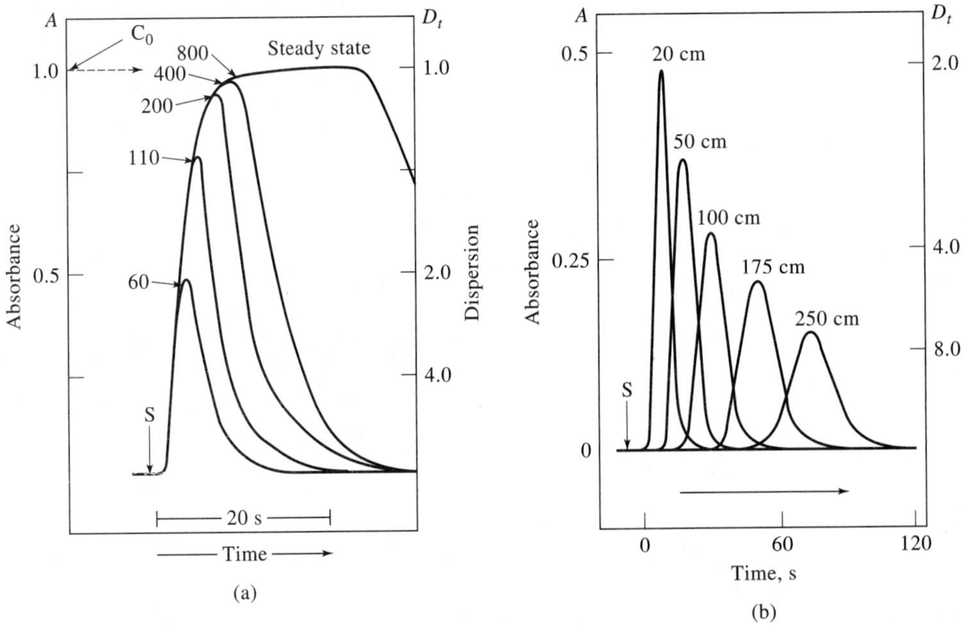

Figure 33-6 Effect of sample volume and length of tubing on dispersion. (a) Tube length: 20 cm; flow rate: 1.5 mL/min; indicted volumes are in μL. (b) Sample volume: 60 μL; flow rate: 1.5 mL/min. *(From J. Ruzicka and E. H. Hansen,* Anal. Chim. Acta, ***1980,*** 114, *21. With permission.)*

Limited-Dispersion Applications

Limited-dispersion flow-injection techniques have found considerable application for high-speed feeding of such detector systems as flame atomic absorption and emission as well as inductively coupled plasma. As these detectors are normally used, sample aliquots are aspirated directly into the flame or plasma, and a steady-state signal is measured. With the flow-injection procedure, in contrast, a blank reagent is pumped through the system to the detector continuously to give a baseline output; samples are then injected periodically, and the resulting transient analyte signals are recorded. Sampling rates of up to 300 samples/hr have been reported.

Limited-dispersion injection has also been used with electrochemical detectors such as specific-ion electrodes and voltammetric microelectrodes. The justification for using flow-injection methods for obtaining such data as pH, pCa, or pNO_3 is the small sample size required ($\approx$25 μL) and the short measurement time ($\approx$10 s). That is, measurements are made well before steady-state equilibria are established, which for many specific-ion electrodes may require a minute or more. With flow-injection measurements, transient signals for sample and standards provide equally accurate analytical data. For example, it has been reported that pH measurements on blood serum can be accomplished at a rate of 240/hr with a precision of ± 0.002 pH. Results are displayed within 5 s after sample injection.

In general, limited-dispersion conditions are realized by reducing as much as possible the distance between injector and detector, slowing the pumping

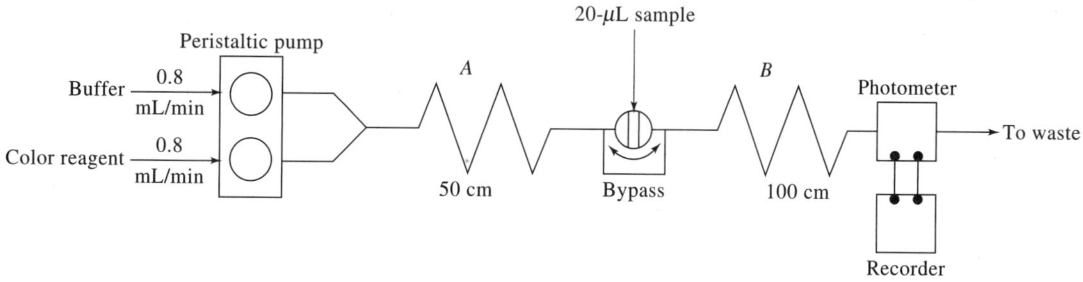

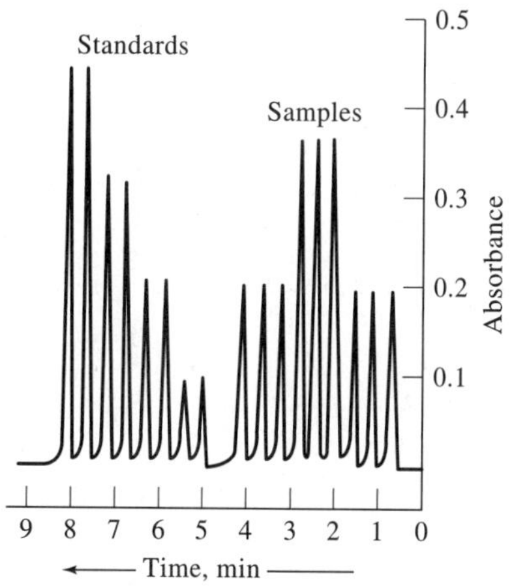

Figure 33-7 (a) Flow-injection apparatus for determining calcium in water by formation of a colored complex with *o*-cresolphthalein complexone at pH 10. All tubing had an inside diameter of 0.5 mm. *A* and *B* are reaction coils having the indicated lengths. (b) Recorder output. Three sets of curves at right are for triplicate injections of three samples. Four sets of peaks on the left are for duplicate injections of standards containing 5, 10, 15, and 20 ppm calcium. *(From E. H. Hansen, J. Ruzicka, and A. K. Ghose,* Anal. Chim. Acta, ***1978,*** 100, *151. With permission.)*

speed, and increasing the sample volume. Thus, for the pH measurements just described, the length of 0.5-mm tubing was only 10 cm, and the sample size was 30 μL.

Medium-Dispersion Applications

Figure 33-7a illustrates a medium-dispersion system for the colorimetric determination of calcium in serum, milk, and drinking water. Here, a borax buffer and a color reagent are combined in a 50-cm mixing coil *A* prior to sample injection. The recorder output for three samples in triplicate and four standards in duplicate are shown in Figure 33-7b.

Figure 33-8 illustrates a yet more complicated system designed for the spectrophotometric determination of caffeine in drug preparations after extraction of the caffeine into chloroform. The chloroform solvent, after cooling in an ice bath to minimize evaporation, is mixed with the alkaline sample stream in a T-tube. After passing through the 2-m extraction coil, the mixture enters a T-tube separator, which is differentially pumped so that about 35% of the organic phase passes into the flow cell, the other 65% accompanying the aqueous solution to waste. In order to avoid contaminating the flow cell with water, Teflon fibers, which are not wetted by water, were twisted into a thread and inserted in the inlet to the T-tube in such a way as to form a smooth downward bend. The chloroform flow then follows this bend to the photometer cell.

Stopped-Flow Methods

It was noted earlier that dispersion in small-diameter tubing decreases with flow rate. In fact, it has been found that dispersion ceases almost entirely when flow is stopped. This fact has been exploited to increase the sensitivity of measurements by allowing time for reactions to go further toward completion without dilution of the sample zone by dispersion. In this type of appli-

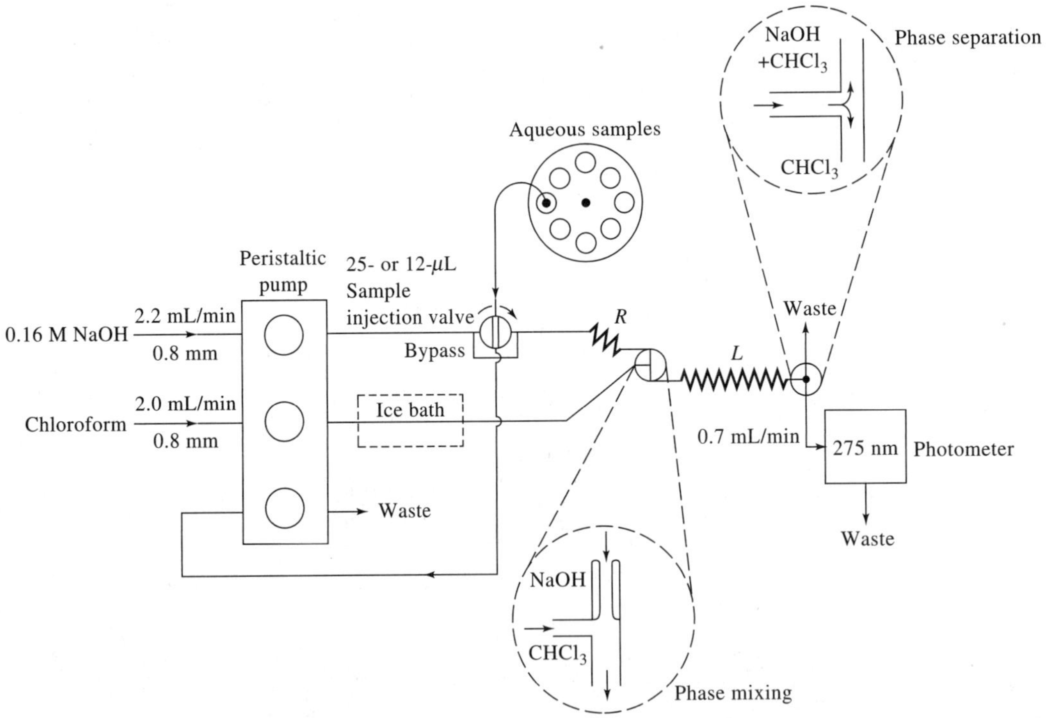

Figure 33-8 Flow-injection apparatus for the determination of caffeine in acetylsalicylic acid preparations. With valve rotated at 90 deg, the flow in the bypass is essentially zero because of its small diameter. *R* and *L* are Teflon coils with 0.8-mm inside diameters; *L* has a length of 2 m, while the distance from the injection point through R to the mixing point is 0.15 m. *(Adapted from B. Karlberg and S. Thelander,* Anal. Chim. Acta, ***1978,*** 98, 2. *With permission.)*

cation, a timing device is required to turn the pump off at precisely timed and regular intervals.

A second application of the stop-flow technique is for kinetic measurements. In this application, the flow is stopped with the reaction mixture in the flow cell where the changes in the concentration of reactants or products can be followed as a function of time. This technique was first applied to the enzymatic determination of glucose based upon the use of the enzyme glucose dehydrogenase.[5] The reaction is carried out in the presence of the coenzyme nicotinamide adenine dinucleotide, which serves as the chromophoric agent ($\lambda_{max} = 340$ nm). As many as 120 samples/hr can be analyzed in this way. The procedure has the considerable virtue of consuming less than one unit of the expensive enzyme per sample.

Flow-Injection Titrations

Titrations can also be performed continuously in a flow-injection apparatus. Here, the injected sample is combined with a carrier in a mixing chamber that promotes large dispersion. The mixture is then transported to a confluence fitting, where it is mixed with the reagent, which contains an indicator. If the detection is set to respond to the color of the indicator in the presence of excess analyte, peaks such as those shown in Figure 33-9 are obtained. Here, an acid is being titrated with a standard solution of sodium hydroxide, which contains bromothymol blue indicator. With injection of samples, the solution changes from blue to yellow and remains yellow until the acid is consumed and the solution again becomes blue. As shown in the figure, the concentration of analyte is determined from the widths of the peaks at half height. Titrations of this kind can be performed at a rate of 60 samples/hr.

[5] J. Ruzicka and E. H. Hansen, *Anal. Chim. Acta,* **1979**, *106*, 207.

33B-4 An Automated System for Monitoring Mercury Concentrations

Figure 33-10 is a schematic of a fully automated continuous-flow apparatus for monitoring the mercury content of wastewater from a laboratory building. The system contains two peristaltic pumps labeled 1 and 2. Pump 1—which handles the sample, a refrigerated water supply to the condenser, and air—produces flow rates of about 3 mL/min. Pump 2, which introduces three reagents, delivers flows of about 0.1 mL/min. The sample is first mixed with sulfuric acid and a 4% solution of potassium persulfate, a potent oxidizing agent. This mixture passes into a reaction tube maintained at a temperature of 80°C by a water bath. The reaction tube is a coiled Tygon tubing that is 10-m long and with a bore of

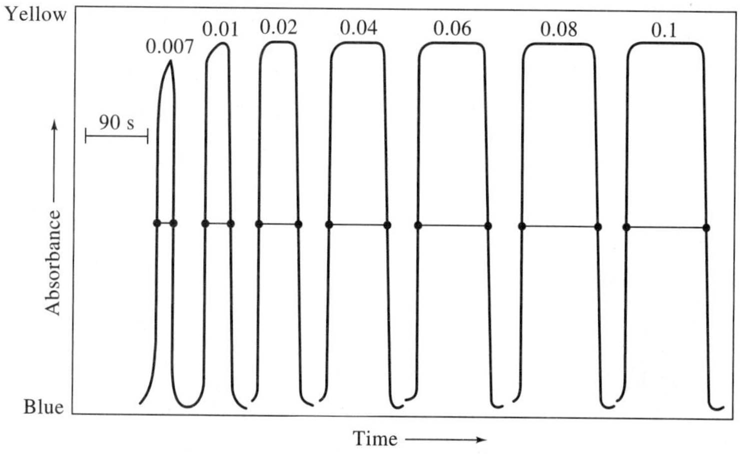

Figure 33-9 Flow-injection titration of HCl with 0.001 M NaOH. The molarities of the HCl solutions are shown at the top of the figure. The indicator was bromothymol blue. The time interval between the points is a measure of the acid concentration. *(From J. Ruzicka, E. H. Hansen, and M. Mosback,* Anal. Chim. Acta, ***1980****, 114, 29. With permission.)*

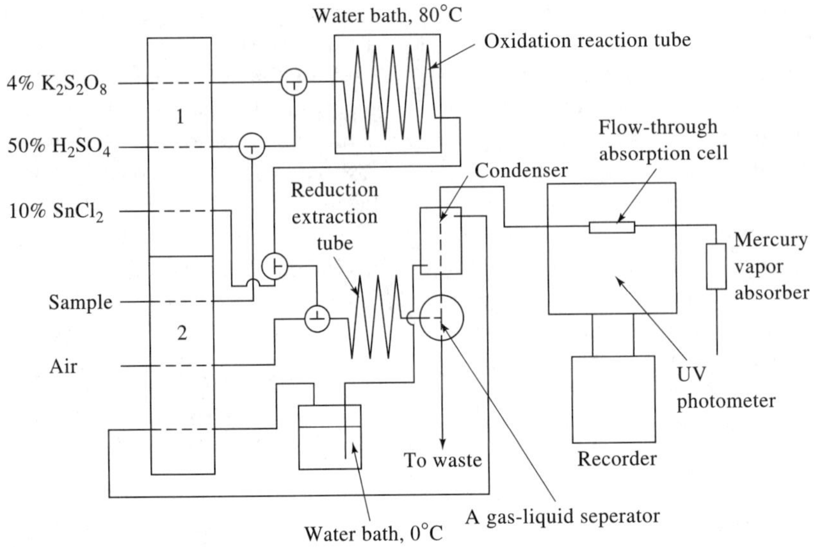

Figure 33-10 An automatic system for determining mercury in wastewater. (1) Peristaltic pump producing a flow rate of 3 mL/min. (2) Peristaltic pump producing a flow rate of 0.1 mL/min. *(Adapted from M. Gota,* TrAc, ***1983****, 2(4), 92. With permission of the Royal Society of Chemistry.)*

1 mm. Oxidation of organic matrix components takes place here, and mercury is converted to mercury(II) ions in the warm environment. The oxidized solution then passes into a second reaction tube after being mixed with a strong solution of tin(II) chloride and a stream of air bubbles. In this reactor tube mercury(II) is reduced to the elemental mercury and vaporized by the flow of air. The mixture then passes into a gas-liquid separator, with the vapor being carried into the condenser that is cooled to 0°C. The resulting dry mercury vapor is then carried into a photometer where its absorbance at 253.7 nm is obtained. These absorbance data are converted to mercury concentrations continuously and recorded. Figure 33-11 shows a typical readout from the apparatus over an hour and one-half period during an afternoon. For a 10-min period beginning at about 15:30, a significant increase in mercury concentration can be seen.

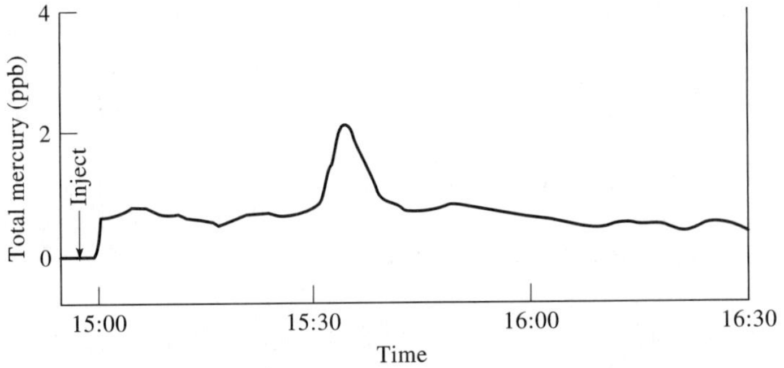

Figure 33-11 Continuous monitoring result from total mercury in sewage from chemical laboratories. *(Adapted from M. Gota,* TrAC, ***1983****, 2(4), 92. With permission of the Royal Society of Chemistry.)*

33C DISCRETE AUTOMATIC SYSTEMS

A wide variety of discrete automatic systems are offered by numerous instrument manufacturers. Some of these devices are designed to perform one or more of the unit operations listed in Table 33-1; others are capable of carrying out an entire analysis automatically. Some discrete systems have been designed for a specific analysis only—for example, the determination of nitrogen in organic compounds. Others can perform a variety of analyses of a given general type. For example, several automatic titrators are available that can perform neutralization, precipitation, complex formation, and oxidation/reduction titrations as directed by a user-programmed computer. In this section, we describe a few typical discrete systems.

33C-1 Automatic Sampling and Sample Definition of Liquids and Gases

Several dozen automatic devices for sampling liquids and gases are currently available from instrument manufacturers. Figure 33-12a illustrates the principle of reversible pump samplers. This device consists of a movable probe, which is a syringe needle or a piece of fine plastic tubing supported by an arm that periodically lifts the tip of the needle or tube from the sample container and positions it over a second container in which the analysis is performed. This motion is synchronized with the action of a reversible peristaltic pump. As shown in Figure 33-12a, with the probe in the sample container, the pump moves the liquid from left to right for a brief period. The probe is then lifted and positioned over the container on the right, and the direction of pumping is reversed. Pumping is continued in this direction until the sample and the desired volume of diluent have been delivered. The probe then returns to its original position to sample the next container. Needless to say, the sample volume is always kept small enough so that none of the sample ever reaches the pump or diluent container. Reversible pump samplers are often used in conjunction with a circular rotating sample table. Tables of this kind generally accommodate 40 or more samples in plastic or glass cups or tubes. The rotation of the table is synchronized with the moving arm of the sampler so that samples are withdrawn sequentially.

Figure 33-12b illustrates a typical syringe-based sampler and diluter. Here again a movable sample probe is used. With the probe in the sample cup, the screw-driven syringe on the left withdraws a fixed volume of sample. Simultaneously, the syringe on the right withdraws a fixed volume of diluent (usually larger). The valve shown in the figure permits these two processes to go on independently. When the probe moves over the container on the right, both syringes empty, dispensing the two liquids into the analytical vessel.

Generally, syringe-type injectors are driven by computer-controlled stepping motors, which force the liquid out of the syringe body in series of identically sized pulses. Thus, for example, a 1-mL syringe pow-

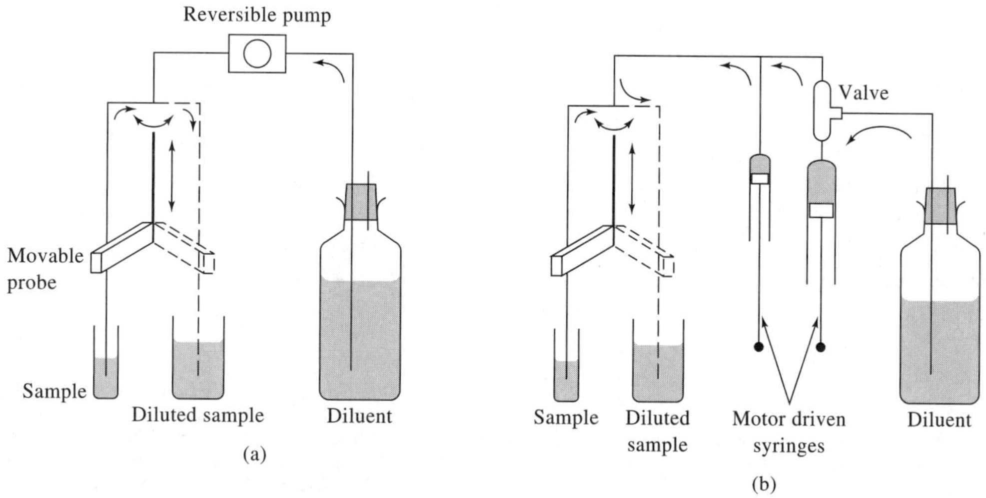

Figure 33-12 Automatic samples: (a) reversible pump type; (b) syringe type.

ered by a motor that requires 1000 steps to empty the syringe is controllable to 1×10^{-3} mL or 1 μL. For a 5000-step motor, the precision would be 0.2 μL.

33C-2 Robotics

With solids, sample preparation, definition, and dissolution involves such unit operations as grinding, homogenizing, drying, weighing, igniting, fusing, and treating with solvents. Each of these individual procedures has been automated. Only recently, however, have instruments appeared that can be programmed to perform several of these unit operations sequentially and without operator intervention. Generally, such instruments are based upon small laboratory robots, which first appeared on the market in the mid-1980s.[6] Figure 33-13 is a schematic of an automated laboratory system based on one of these robots. Central to the system is a horizontal arm, which is mounted on two vertical pillars and has four degrees of freedom in its movement. Its 360-deg rotational motion covers a circumference of about 150 in; its maximum reach is then 24 in. The device is equipped with a tonged hand, which has a 360-deg wrist-like motion that permits manipulating vials or tubes, pouring of liquids or solids, and shaking and swirling liquids in tubes. Consequently, the arm and hand are capable of carrying out many of the manual operations that are performed by the laboratory chemist. One important feature of the device is its ability to change hands. Thus, the robot can leave its tonged hand on the table and attach a syringe in its place for pipetting liquids.

The robotic system is controlled by a microprocessor that can be user programmed. Thus, the instrument can be instructed to bring samples to the master laboratory station where they can be diluted, filtered, partitioned, ground, centrifuged, homogenized, extracted, and treated with reagents. The device can also be instructed to heat and shake samples, dispense measured volumes of liquids, inject samples into a chromatographic column, and collect fractions from a column. In addition, the robot can be interfaced with an automatic electric balance for weighing samples.

A typical application of this device is for the routine screening of newly synthesized compounds for electrochemical activity.[7] Here, the robot cleans, fills, and deaerates the electrochemical cell, introduces samples, performs standard additions, starts and stops the voltammetric analyzer, and records data. Analyses are performed with a relative standard deviation of 2%.

Another robotic system has also been described that totally automates the acquisition of pH titration curves for solid samples.[8] This system performs automatically and sequentially weighing out of sample, dissolution, dilution, pH meter calibration, acquisition of titration data, selection of end point, and finally, report generation. Undoubtedly, more and more instruments of this kind can be expected in the near future.

33C-3 The Centrifugal Fast Scan Analyzer

A type of batch analyzer that is capable of analyzing as many as 16 samples simultaneously for a single constituent is based upon the use of a centrifuge to mix the samples with a reagent and to transfer the mixtures to cells for photometric or spectrophotometric measurement.[9] The system is such that conversion from one type of reagent to another is usually easy.

The principle of the instrument is seen in Figure 33-14, which is a cross-sectional view of the circular, plastic rotor of a centrifuge. The rotor has 17 dual compartments arranged radially around the axis of rotation. Samples and reagents are pipetted automatically into 16 of the compartments as shown; solvent and reagent are measured into the seventeenth as a blank. When the rotor reaches a rotation rate of about 350 rpm, reagent and liquids in the 17 compartments are mixed simultaneously and carried into individual cells located at the outer edge of the rotor; these cells are equipped with horizontal quartz windows. Mixing is hastened by drawing air through the mixtures. Radiation from an interference-filter photometer or a spectrophotometer passes through the cells and falls upon a photomultiplier tube. For each rotation, a series of electrical pulses

[6] For a description of laboratory robots, which are now available commercially, and their performance characteristics, see J. R. Strimaltis, *J. Chem. Educ.*, **1989**, *66*, A8; **1990**, *67*, A20; V. Berry, *Anal. Chem.*, **1990**, *62*, 337A; R. Dessy, *Anal. Chem.*, **1983**, *55*, 1100A, 1232A; W. J. Hurst and J. W. Mortimer, *Laboratory Robotics*. New York: VCH Publishers, 1987.

[7] M. L. Dittenhafer and J. D. McClean, *Anal. Chem.*, **1983**, *55*, 1242A.

[8] G. D. Owens and R. J. Eckstein, *Anal. Chem.*, **1982**, *54*, 2347.

[9] For a brief description of a typical system, see C. S. Scott and C. A. Burtis, *Anal. Chem.*, **1973**, *45*(3), 327A; C. D. Scott and C. A. Burtis, *Centrifugal Analysis in Clinical Chemistry*. New York: Praeger, 1980; B. B. Lentrichia, M. F. Turanchik, and K. K. Yeung, *Am. Biotechnol. Lab.*, **1987**, *5*(3), 17.

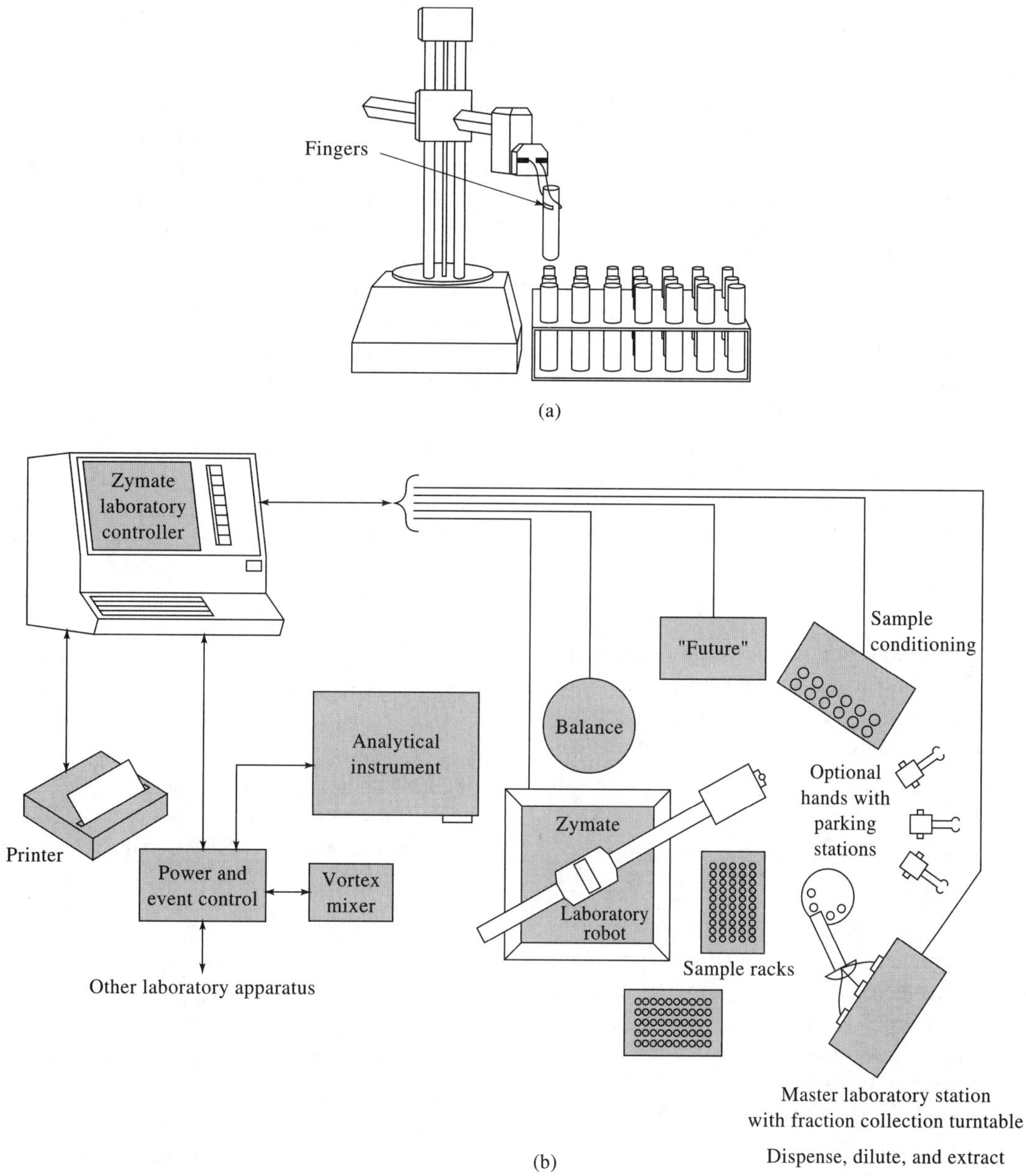

Figure 33-13 A robotic laboratory system: (a) robot arm and hand; (b) total system. *(Courtesy of Zymark Corporation, Hopkinton, MA.)*

are produced, 16 for the samples and 1 for the blank. Between each of these pulses is a signal corresponding to the dark current. The successive signals from the instrument are collected in the memory of a dedicated microcomputer for subsequent manipulation. Signal averaging can be employed to improve the signal-to-noise ratio.

One of the most important applications of the centrifugal fast analyzer is for the determination of enzymes. Ordinarily, enzyme analyses are based on the catalytic effect of the analyte upon a reaction that involves formation or consumption of an absorbing species. Here, a calibration curve relating the rate of appearance or disappearance of the absorbing species as a

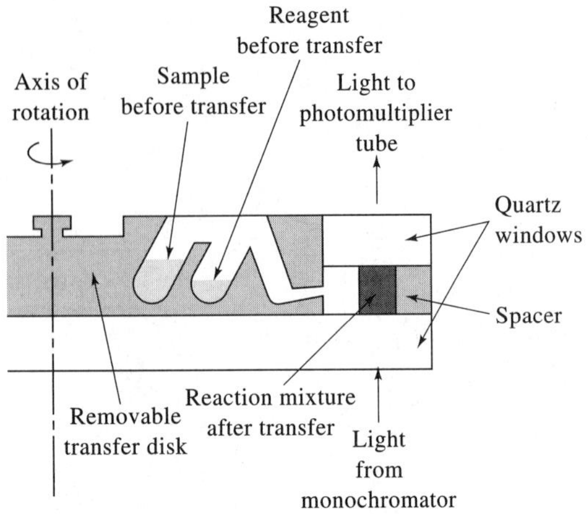

Figure 33-14 Rotor for a centrifugal fast analyzer. *(Reprinted with permission from R. L. Coleman, W. D. Shults, M. T. Kelley, and J. A. Dean,* Amer. Lab., ***1971,*** *3(7), 26. Copyright by International Scientific Communications, Inc.)*

function of enzyme concentration serves as the basis for the analysis. The centrifugal analyzer permits the simultaneous determination of the rates of 16 reactions under exactly the same conditions; thus, 16 simultaneous enzyme analyses are feasible.

33C-4 Automatic Organic Elemental Analyzers

Several manufacturers produce automatic instruments for analyzing organic compounds for one or more of the common elements including carbon, hydrogen, oxygen, sulfur, and nitrogen.[10] All of these instruments are based upon high-temperature oxidation of the organic compounds, which converts the elements of interest to gaseous molecules. In some instruments, the gases are separated on a chromatographic column, while in others separations are based upon specific absorbents. In most instruments, thermal conductivity detection serves to complete the determinations. Often these instruments are equipped with devices that automatically load the weighed samples into the combustion area.

Figure 33-15 is a schematic of a commercial automatic instrument for the determination of carbon, hydrogen, and nitrogen. In this instrument, samples are oxidized at 900°C under static conditions in a pure oxygen environment that produces a gaseous mixture of carbon dioxide, carbon monoxide, water, elemental nitrogen, and oxides of nitrogen. After 2 to 6 min in the oxygen environment, the products are swept with a stream of helium through a 750°C tube furnace where hot copper reduces the oxides of nitrogen to the element and also removes the oxygen as copper oxide. Additional copper oxide is also present to convert carbon monoxide to the dioxide. Halogens are removed by a silver wool packing.

The products from the reaction furnace pass into a mixing chamber where they are brought to a constant temperature. The resulting homogeneous mixture is then analyzed by passing it through a series of three precision thermal conductivity detectors, each detector consisting of a pair of sensing cells.

Between the first pair of cells is a magnesium perchlorate absorption trap that removes water. The differential signal then serves as a measure of the hydrogen in the sample. Carbon dioxide is removed in a second absorption trap. Again, the differential signal between the second pair of cells is a measure of carbon in the sample. The remaining gas, consisting of helium and nitrogen, passes through the third detector cell. The output of this cell is compared to that of a reference cell through which pure helium flows. The voltage differential across this pair of cells is related to the amount of nitrogen in the sample.

For oxygen analysis, the reaction tube is replaced by a quartz tube filled with platinized carbon. When the sample is pyrolyzed in helium and swept through this tube, all of the oxygen is converted to carbon monoxide, which is then converted to carbon dioxide by passage over hot copper oxide. The remainder of the procedure is the same as just described, with the oxygen concentration being related to the differential signal before and after absorption of the carbon dioxide.

For a sulfur analysis, the sample is combusted in an oxygen atmosphere in a tube packed with tungsten(VI) oxide or copper oxide. Water is removed by a dehydrating reagent located in the cool zone of the same tube. The dry sulfur dioxide is then separated and determined by the differential signal at what is normally the hydrogen detection bridge. In this instance, however, the sulfur dioxide is absorbed by a silver oxide reagent.

The instrument shown in Figure 33-15 can be fully automated, whereby up to 60 weighed samples contained in small capsules are loaded into a carousel sample tray for automatic sampling.

[10] For a description of these instruments, see T. S. Ma and R. C. Rittner, *Modern Organic Elemental Analysis,* Chapters 2, 3, and 4. New York: Marcel Dekker, 1979.

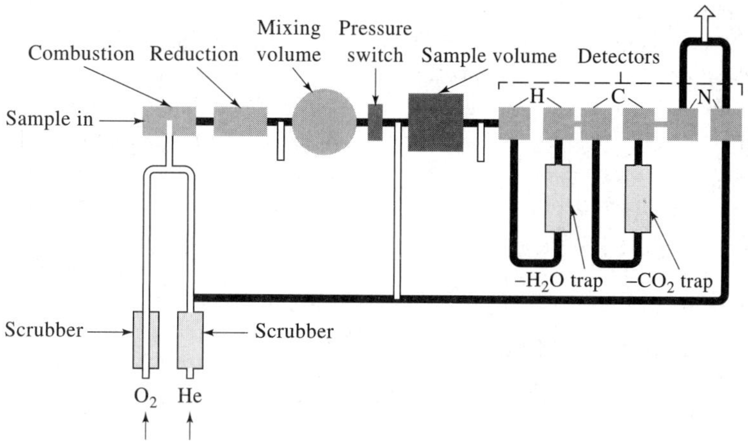

Figure 33-15 An automatic C, H, and N analyzer. *(Courtesy of Perkin-Elmer, Norwalk, CT.)*

33D ANALYSES BASED UPON MULTILAYER FILMS

During the past two decades, a technology has been developed for performing the various steps in a quantitative analysis automatically in discrete films arranged in multilayers supported on transparent, disposable plates having the size of a postage stamp. Here, a small drop of the sample (10 to 50 μL) is placed on the top layer of the multilayer film element where it spreads rapidly and uniformly. Water and low-molecular-weight components diffuse from the spreading layer through one or more reagent layers where the analyte interacts to produce a colored (or occasionally a fluorescent) product, which is then determined by reflectance photometry. To date, this technology has been applied to the routine determination of blood metabolites, such as glucose; serum enzymes, such as lactate dehydrogenase; and therapeutic drugs in samples of blood. In addition, a modification of the technique has been developed that permits the potentiometric determination of the electrolyte components, such as potassium, in blood serums. As noted earlier, hundreds of millions of such determinations are performed each year in clinical laboratories throughout the world, generally by means of automatic continuous-flow or discrete instruments. The cost of such instruments is large, which has required their location in centralized laboratories, where their expense can be amortized over a large volume of samples. The advent of multilayer film elements means that it is now economic to perform some of these routine clinical analyses automatically in decentralized locations—even in the individual physician's office and perhaps ultimately in the home.[11] In fact, glucose monitoring kits are now available for use by insulin-dependent diabetic patients.

33D-1 General Principles

The development of multilayer film elements for chemical analysis is an outgrowth of the technology developed by the color photographic industry for producing multilayer films wherein complex chemistries take place. For example, a typical instant color film is made up of as many as 15 separate layers having thicknesses of 1.5 to 5 μm. One of these layers contains a developing fluid, which, when released, diffuses through the remaining layers, wherein a series of chemical reactions occur that ultimately lead to the deposition of blue, green, and red dyes in those areas that have been photosensitized by exposure in a camera.

This same technology makes it feasible to perform automatic chemical analyses based upon a sequence of physical or chemical reactions. For example, the product of a chemical reaction produced in the first layer may be separated from interferences by selective diffusion through a second layer to a third layer where further reaction can occur. Each layer of a multilayer film thus offers a separate domain in which a chemical reac-

[11] For review articles describing this technology, see B. Walter, *Anal. Chem.*, **1983**, *55*, 499A; H. G. Curme et al., *Clin. Chem.*, **1978**, *24*, 1335; R. W. Spayd et al., *Clin. Chem.*, **1978**, *24*, 1343.

tion or a physical separation can be carried out. For each type of analysis, a complete set of reagents is miniaturized in a disposable dry form. Reconstitution of reagents and many other manual manipulations are usually replaced by the single step of applying the sample to the film element. Instruments based upon multilayer films are frequently called *dry film analyzers.*

33D-2 Film Structures

Most multilayer film elements consist of a transparent support, one or more reaction layers, a reflective layer, and a spreading or metering layer. Figure 33-16a is a cross-sectional diagram of a film element, which is commercially available, for the determination of glucose in serum. The chemistry that occurs in the reagent layer is described by the reactions

$$\text{glucose} + O_2 + 2H_2O \xrightarrow[\text{oxidase}]{\text{glucose}} \text{gluconic acid} + 2H_2O_2$$

$$2H_2O_2 + \text{indicator}_{(\text{reduced})} \xrightarrow{\text{peroxidase}} \text{indicator}_{(\text{oxid})} + 4H_2O$$

The reaction product is an oxidized dye, which absorbs strongly at 495 nm.

The reagent layer contains, in addition to the oxidation/reduction indicator, the enzymes glucose oxidase and peroxidase, and a pH-5.0 buffer. All are immobilized in a gelatin binder, which is approximately 10-μm thick. This film is supported on a rigid, transparent plastic film.

The spreading or metering layer upon which a drop of sample is placed, is typically 100-μm thick and is made up of cellulose acetate in which titanium dioxide is dispersed as a reflectant. The spreading layer serves three purposes. The first is to reflect the radiation from a source back through the reagent and support layers to the detector of a photometer. Its second function is to cause the sample to spread into a uniform layer. During the spreading, which takes 5 to 10 s, most of the fluid movement is lateral until all of the liquid is contained in the pore structure. As a result of this rapid lateral spreading with little penetration, the amount of fluid and analyte per unit area is relatively independent of drop volume. Thus, a 10% variation in sample size produces only a 1% change in concentration of analyte per unit of area. Furthermore, the concentration of analyte is found to be uniform throughout the width of the spot. A third function of the spreading layer is to retain cells, crystals, and particulate matter as well as large molecules such as proteins.

The determination of blood glucose with the plate just described is remarkably simple and can be fully au-

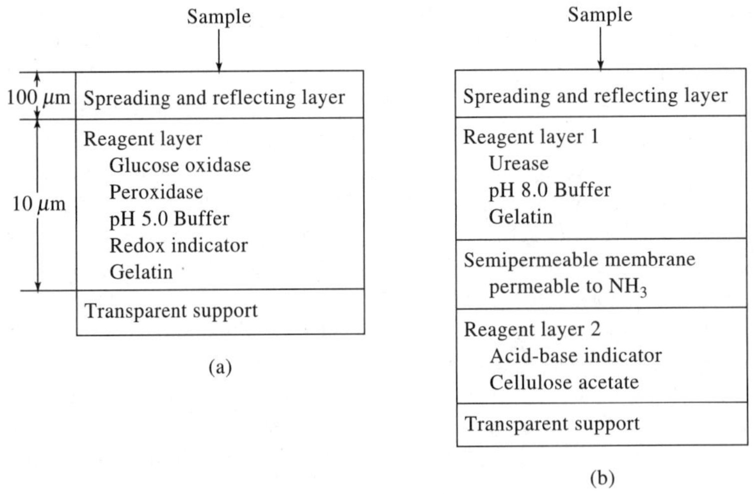

Figure 33-16 Cross section of two commercial film elements for the determination of (a) glucose and (b) blood urea nitrogen, BUN. Note that film thickness is not to scale.

tomated. A 10-μL sample is introduced onto the plate, which is then held in an incubator at 37(±0.05)°C for 7 min. The reflectance at 495 nm is measured with the instrument described in the next section.

Figure 33-16b depicts a somewhat more complicated film element for the determination of blood urea nitrogen (BUN). Here, two reagent layers are separated by a semipermeable layer of cellulose acetate butyrate that passes ammonia but excludes carbon dioxide and hydroxide ions from the acid/base indicator layer. The reactions in this instance are

$$(NH_2)_2CO + H_2O \xrightarrow[\text{pH 8.0}]{\text{urease}} 2NH_3 + CO_2$$

$$\underset{\text{colorless}}{NH_3 + HIn} \longrightarrow NH_4^+ + \underset{\text{colored}}{In^-}$$

33D-3 Instrumentation

Quantitative measurements of the products of multilayer film separations and reactions have been based on reflectance photometry, specific-ion potentiometry, and fluorescence.

Reflective Photometer

Figure 33-17 is a schematic of a manual instrument for measuring the diffuse reflectance from a multilayer film. Here, the sample while still in an incubator is illuminated by filtered radiation of a wavelength absorbed by the analyte. The source beam, which is at 45 deg to the film element, is viewed normal to that element. This geometry minimizes front surface reflection. The diameter of the illuminated spot is about 2.5 mm. Detection is by means of a photomultiplier tube.

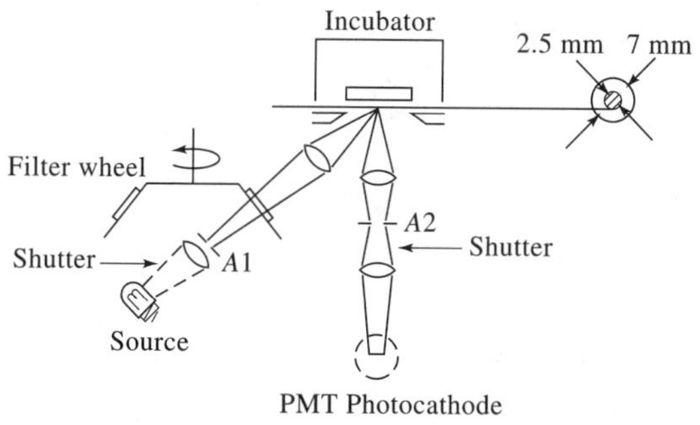

Figure 33-17 A manual reflectance photometer for use with thin-film elements. *(From H. G. Cume et al.,* Clin. Chem., *1978,* 24, *1336. With permission.)*

In reflectance spectroscopy, two types of reflection are encountered: specular or mirror-like, in which the angles of incidence and reflection are identical, and diffuse, which is reflection from a matte structure. The latter is the one that serves as the basis for reflectance spectroscopy. Diffuse reflection is not a surface phenomenon but results from scattering, transmission, and absorption interactions of the radiation in the volume of the illuminated film. Absorption within the volume reduces the reflected intensity.

Reflectance data are usually expressed in terms of the *percent reflectance* (%*R*), which is analogous to percent transmittance in absorption spectroscopy. Thus,

$$\%R = \frac{I_s}{I_r} \times 100\%$$

where I_s is the intensity of the beam reflected from the sample and I_r is the intensity from a reference standard, usually barium sulfate. As with transmittance, reflectance decreases nonlinearly with increases in the concentration of absorbing species. Several algorithms have been developed for linearizing this relationship. The specific algorithm used depends upon the reflection characteristic of the particular multilayer film, the nature of the illumination, and the geometry of the instrument.

Both manual and fully automated reflectance instruments are now available. With the latter, the operator needs only provide the sample and specify the tests to be performed. The instrument then chooses the appropriate film element from a cassette, selects the appropriate radiation filter, calibrates the instrument, applies the sample, and prints out the result. This instrument is capable of performing more than 500 determinations per hour involving any of 16 available tests.

Potentiometry

Multilayer film technology has been extended to the manufacture of single-test, disposable, ion-selective membrane systems for the determination of potassium, sodium, and chloride ions as well as the determination of several other analytes. A schematic of the cell used for potassium ion determination in serum samples is shown in Figure 33-18. The dimensions of the device are 2.8 by 2.4 cm. Here, two identical film elements are

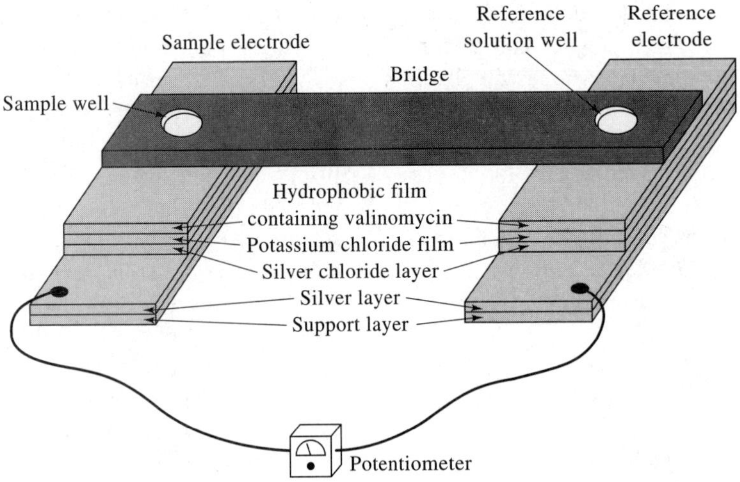

Figure 33-18 Multilayer selection-ion cell for potassium. The dimensions of the device are 2.8 × 2.4 cm × 150 μm thickness. *(Reprinted with permission from B. Walter,* Anal. Chem., ***1983,*** 55, *508A. Copyright 1983 American Chemical Society.)*

TABLE 33-2 Performance Data for Some Thin-Film Elements*

Analyte	Dynamic Range	Precision, % Relative Standard Deviation
Albumin	72.5–869 μmol/L	4.9
Ammonia	0.01–12 mmol/L	5
Bilirubin	3.4–445 μmol/L	2.5
Calcium	0.25–4.0 mmol/L	1.5
Carbon dioxide	5–55 mmol/L	5–7
Chloride	50–175 mmol/L	1.5
Cholesterol	0.39–14.3 mmol/L	5.2
Creatinine	4.4–1459 μmol/L	4
Glucose	1.1–34.7 mmol/L	2.1
Potassium	1–14 mmol/L	2.0
Sodium	75–250 mmol/L	1.3
Triglycerides	0–6.5 mmol/L	2.7
Urea	0.7–42.8 mmol/L	3.3
Uric acid	29.7–1010 μmol/L	2.3

*From: T. L. Shirey, *Clin. Biochem.*, **1983,** *16,* 147. With permission.

coupled by a paper salt bridge. Approximately 10 μL of sample is placed in one well and 10 μL of standard potassium solution in the other. These solutions rapidly diffuse laterally to activate the salt bridge. The top layer of the film element, which is the ion-selective membrane, is made up of valinomycin in a hydrophobic plastic. As was described earlier (Section 23C-6), valinomycin selectively binds potassium ions, which causes a potential to develop across the film interface. The remaining three layers in each element make up a silver/silver chloride reference electrode.

Unlike conventional specific-ion measurements, the potential of the sample cell is referred directly to that of a cell containing the standard. The potassium concentration in the unknown is then obtained from the potential difference between these two cells. The performance of this and the other ion-selective cells appears to be comparable with that of conventional specific-ion electrode systems.

33D-4 Performance and Applications

Thin-film elements and detectors are now offered by several manufacturers for a growing number of clinical tests. Currently, the number of different types of assays is about three dozen. Table 33-2 lists some representative determinations that can be performed with these devices. Generally, the volume of sample is small (10 to 50 μL), and the time required to complete an analysis is short—usually 1 to 10 min. The dynamic ranges shown in the table are sufficiently large to accommodate approximately 98% of the samples encountered in a clinical laboratory. Samples outside this range can usually be handled by suitable dilutions.

Extensive performance testing of these new devices generally reveals a good correlation between the data they produce and the results by standard procedures. Precision of 1% to 10% relative are reported depending upon the type of test, which again is comparable with the data from automated standard methods.

33E QUESTIONS AND PROBLEMS

33-1 List sequentially a set of laboratory unit operations that might be used to
(a) ascertain the presence or absence of lead in flakes of dry paint.
(b) determine the iron content of multiple vitamin/mineral tablets.

33-2 Sketch a flow-injection system that could be used for the determination of K^+ and Na^+ in blood based upon flame photometric measurements.

33-3 Sketch a flow-injection system that might be employed for determining lead in the aqueous effluent from an industrial plant based upon the extraction of lead ions with a carbon tetrachloride solution of dithizone, which reacts with lead ion to form an intensely colored product.

33-4 Sketch a flow-injection apparatus for the determination of sodium sulfite in aqueous samples.

33-5 Sketch a top view of a robot finger arrangement, equipped with photoelectric "sight," that would permit the pickup of 2-mL capped vials.

33-6 Sketch a flow diagram (indicating columns, detectors, and switching valves) that is designed to satisfy the following requirements. First, the solvent peak must be rapidly separated from two analytes that are of much lower volatility than the solvent. Second, the solvent peak should not pass into the analytical column. Third, the two analytes, which differ considerably in polarity, must be separated and ultimately determined quantitatively.

Appendix 1

Evaluation of Analytical Data

This appendix describes the types of errors that are encountered in analytical chemistry and how their magnitudes are estimated and reported. Estimation of the probable accuracy of results is a vital part of any analysis because data of unknown reliability are essentially worthless.

a1A PRECISION

Two terms are widely used in discussions of the reliability of data, *precision* and *accuracy.*

a1A-1 Precision

Precision describes the reproducibility of results; that is, the agreement between numerical values for two or more replicate measurements, or measurements that have been made *in exactly the same way.* Generally, the precision of an analytical method is readily obtained by simply repeating the measurement.

Three terms are widely used to describe the precision of a set of replicate data including *standard deviation, variance,* and *coefficient of variation.* These terms have statistical significance and are defined in Section a1B-1.

a1A-2 Accuracy

Accuracy describes the correctness of an experimental result. Strictly speaking, the only type of measurement that can be completely accurate is one that involves counting objects. All other measurements contain errors and give only an approximation of the truth.

Accuracy is a relative term in the sense that what is an accurate or inaccurate method very much depends on the needs of the scientist and the difficulty of the analytical problem. For example, an analytical method that yields results that are within ±10%, or one part per billion, of the correct amount of mercury in a sample of fish tissue that contains 10 parts per billion of the metal would usually be considered to be reasonably accurate. In contrast, a procedure that yields results that are with ±10% of the correct amount of mercury in an ore that contains 20% of the metal would usually be deemed unacceptably inaccurate.

Accuracy is expressed in terms of either absolute error or relative error. The *absolute error* E_a of the mean (or average) $\bar{x}$ of a small set of replicate analyses is given by the relationship

$$E_a = \bar{x} - x_t \qquad \text{(a1-1)}$$

where x_t is an accepted value of the quantity being measured. Often, it is useful to express the accuracy in terms of *relative error,* where

$$\text{relative error} = \frac{\bar{x} - x_t}{x_t} \times 100\% \qquad \text{(a1-2)}$$

Frequently, the relative error is expressed as a percentage as shown; in other cases the quotient is multiplied by 1000 to give the error in parts per thousand (ppt).

Note that both absolute and relative errors bear a sign, a positive sign indicating that the measured result is greater than its true value and a negative sign the reverse.

We will be concerned with two types of errors, *random*, or *indeterminate, errors* and *systematic*, or *determinate, errors*.[1] The error in the mean of a set of replicate measurements is then the sum of these two types of errors:

$$E_a = E_r + E_s \qquad \text{(a1-3)}$$

where E_r is the random error associated with a measurement, and E_s is the systematic error.

Random Errors

Whenever analytical measurements are repeated on the same sample, a scatter of data such as that shown in Table a1-1 are obtained because of the presence of random, or indeterminate errors—that is, the presence of random errors is reflected in the imprecision of the data. The data in columns 2, 4, and 6 of the table are absorbances (Section 12A) obtained with a spectrophotometer on 50 replicate red solutions produced by treating identical aqueous samples containing 10 ppm of Fe(III) with an excess of thiocyanate ion. The measured absorbances are directly proportional to iron concentration.

The distribution of random errors in these data is more easily comprehended if they are organized into equal-size, contiguous data groups, or *cells*, as shown in Table a1-2. The relative frequency of occurrence of data in each cell is then plotted as in Figure a1-1*A* to give a bar graph called a *histogram*.

It is reasonable to suppose that if the number of analyses were much larger than that shown in Table a1-2 and that if the size of the cells were made much smaller then, ultimately, a smooth curve such as that shown in Figure a1-1*B* would be obtained. A smooth curve of this type is called a *Gaussian curve*, or a *normal error curve*. It is found empirically that the results of replicate chemical analyses are frequently distributed in an approximately Gaussian, or normal, form.

The frequency distribution exhibited by a Gaussian curve has the following characteristics:

1. The most frequently observed result is the mean μ of the set of data.
2. The results cluster symmetrically around this mean value.
3. Small divergences from the central mean value are found more frequently than are large divergences.
4. In the absence of systematic errors, the mean of a large set of data approaches the true value.

[1] A third type of error, which is encountered occasionally, is *gross error*, which arises in most instances from the carelessness, ineptitude, laziness, or bad luck of the experimenter. Typical sources include transposition of numbers in recording data, spilling of a sample, using the wrong scale on a meter, accidental introduction of contaminants, and reversing the sign on a meter reading. A gross error in a set of replicate measurements appears as an *outlier*—a datum that is noticeably different from the other data in the set. We will not consider gross errors in this discussion.

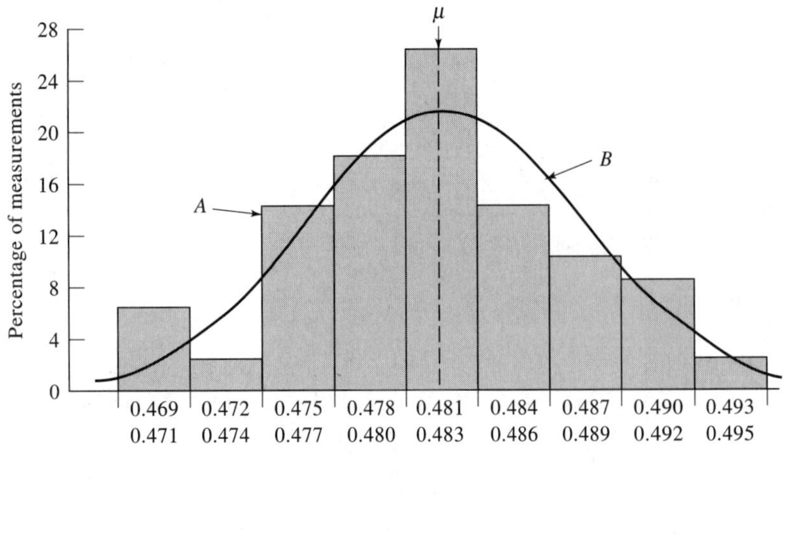

Figure a1-1 *A*, Histogram showing distribution of the 50 results in Table a1-1. *B*, Gaussian curve for data with the same mean and same standard deviation as the data in *A*.

TABLE a1-1 Replicate Absorbance Measurements[a]

Trial	Absorbance, A	Trial	Absorbance, A	Trial	Absorbance, A
1	0.488	18	0.475	35	0.476
2	0.480	19	0.480	36	0.490
3	0.486	20	0.494[b]	37	0.488
4	0.473	21	0.492	38	0.471
5	0.475	22	0.484	39	0.486
6	0.482	23	0.481	40	0.478
7	0.486	24	0.487	41	0.486
8	0.482	25	0.478	42	0.482
9	0.481	26	0.483	43	0.477
10	0.490	27	0.482	44	0.477
11	0.480	28	0.491	45	0.486
12	0.489	29	0.481	46	0.478
13	0.478	30	0.469[c]	47	0.483
14	0.471	31	0.485	48	0.480
15	0.482	32	0.477	49	0.483
16	0.483	33	0.476	50	0.479
17	0.488	34	0.483		
Mean absorbance = 0.482					
Standard deviation = 0.0056					

[a]Data are listed in the order obtained.
[b]Maximum value.
[c]Minimum value.

Characteristic 4 means that in principle it is always possible to reduce the random error of an analysis to something that approaches zero. Unfortunately, it is seldom practical to achieve this goal, however, because to do so requires performing 20 or more replicate analyses. Ordinarily, a scientist can only afford the time for two or three replicated measurements, and a significant random error is to be expected for the mean of such a small number of replicate measurements.

Statisticians usually use μ to represent the mean of an infinite collection of data (see Figure a1-1*B*) and $\bar{x}$ for the mean of a small set of replicate data. The random error E_r for the mean of the small set is then given by

$$E_r = \bar{x} - \mu \qquad \text{(a1-4)}$$

The mean for a finite set of data rapidly approaches the true mean when the number of measurements N increases beyond 20 or 30. Thus, as shown in the following example we can sometimes determine the random error in an individual datum or in the mean of a small set of data.

EXAMPLE a1-1

Calculate the random error for **(a)** the second datum in Table a1-1 and **(b)** the mean for the first three entries in the table.

TABLE a1-2 Frequency Distribution of Data from Table a1-1

Absorbance Range A	Number in Range y	Relative Frequency, y/N^a
0.469 to 0.471	3	0.06
0.472 to 0.474	1	0.02
0.475 to 0.477	7	0.14
0.478 to 0.480	9	0.18
0.481 to 0.483	13	0.26
0.484 to 0.486	7	0.14
0.487 to 0.489	5	0.10
0.490 to 0.492	4	0.08
0.493 to 0.495	1	0.02

[a]N = total number of measurements = 50.

The mean for the entire set of data is 0.482 and because this mean is for 50 measurements, we may assume that the random error in this mean is approximately zero. Thus, the limiting mean μ can be taken as 0.482.

(a) Here, the random error for a single measurement x_2 is

$$E_r = x_2 - \mu = 0.480 - 0.482 = -0.002$$

(b) The mean $\bar{x}$ for the first three entries in the table is

$$\bar{x} = \frac{0.488 + 0.480 + 0.486}{3} = 0.485$$

Substituting into Equation a1-4 gives

$$E_r = \bar{x} - \mu = 0.485 - 0.482 = +0.003$$

The random nature of indeterminate errors makes it possible to treat these effects by the methods of statistics. Statistical techniques are considered in Section a1B.

Systematic Errors—Bias

Systematic errors have a definite value, have an assignable cause, and are of the same sign and magnitude for replicate measurements made in exactly the same way. Systematic errors lead to *bias* in a measurement technique. Bias is illustrated by the two curves in Figure a1-2, which show the frequency distribution of replicate results in the analysis of identical samples by two methods that have random errors of identical size. Method A has no bias so that the limiting mean is the true value x_t. Method B has a bias that is given by

$$\text{bias} = \mu_B - x_t = \mu_B - \mu_A \qquad \text{(a1-5)}$$

Note that bias affects all of the data in a set and that it bears a sign.

Systematic errors are of three types: *instrumental, personal,* and *method.*

Instrumental Errors. Typical sources of instrumental errors include drift in electronic circuits, leakage in vacuum systems, temperature effects on detectors, currents induced in circuits from 110-V power lines, decreases in voltages of batteries with use, and calibration errors in meters, weights, and volumetric equipment.

Systematic instrument errors are commonly detected and corrected by calibration with suitable standards. Periodic calibration of instruments is always desirable because the response of most instruments change with time as a consequence of wear, corrosion, or mistreatment.

Personal Errors. Personal errors are those introduced into a measurement by judgments that the experimentalist must make. Examples include estimating the position of a pointer between two scale divisions, the color of a solution at the end point in a titration, the level of a

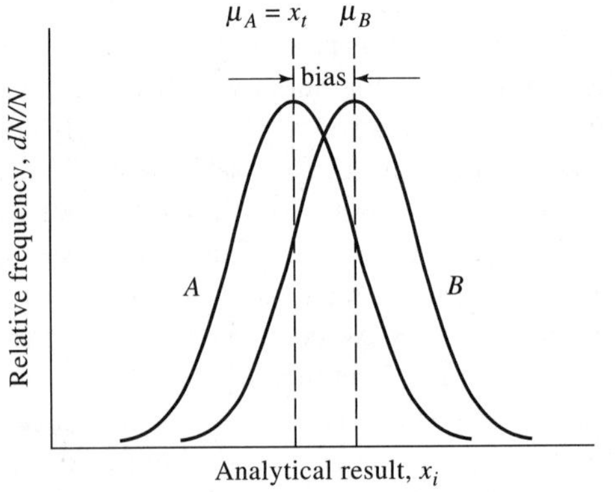

Figure a1-2 Illustration of bias = $\mu_B - \mu_A = \mu_B - x_i$

liquid with respect to a graduation in a pipet, or the relative intensity of two light beams. Judgments of this type are often subject to systematic, unidirectional uncertainties. For example, one person may read a pointer consistently high, another may be slightly slow in activating a timer, and a third may be less sensitive to color. Color blindness or other physical handicaps often exacerbate determinate personal errors.

Number bias is another source of personal systematic error that is widely encountered and varies considerably from person to person. The most common bias encountered in estimating the position of a needle on a scale is a preference for the digits 0 and 5. Also prevalent is a preference for small digits over large and even ones over odd.

A near-universal source of personal error is prejudice. Most of us, no matter how honest, have a natural tendency to estimate scale readings in a direction that improves the precision in a set of results or causes the results to fall closer to a preconceived notion of the true value for the measurement.

Most personal errors can be minimized by care and self-discipline. Thus, most scientists develop the habit of systematically double-checking instrument readings, notebook entries, and calculations. Robots, automated systems, computerized data collection, and computerized instrument control have the potential of minimizing or eliminating personal type systematic errors.

Method Errors. Method-based errors are often introduced from nonideal chemical and physical behavior of the reagents and reactions upon which an analysis is based. Possible sources include slowness or incompleteness of chemical reactions, losses by volatility, adsorption of the analyte on solids, instability of reagents, contaminants, and chemical interferences.

Systematic method errors are usually more difficult to detect and correct than are instrument and personal errors. The best and surest way involves *validation* of the method by using it for the analysis of standard materials that resemble the samples to be analyzed both in physical state and in chemical composition. The analyte concentrations of these standards must, of course, be known with a high degree of certainty. For simple materials, standards can sometimes be prepared by blending carefully measured amounts of pure compounds. Unfortunately, more often than not, materials to be analyzed are sufficiently complex to preclude this simple approach.

The National Institute of Standards and Technology[2] offers for sale a variety of *standard reference materials* (SRMs) that have been specifically prepared for the validation of analytical methods.[3] The concentration of one or more constituents in these materials has been determined by (1) a previously validated reference method, (2) two or more independent, reliable measurement methods, or (3) analyses from a network of cooperating laboratories, technically competent and thoroughly familiar with material being tested. Most standard reference materials are substances that are commonly encountered in commerce or in environmental, pollution, clinical, biological, and forensic studies. A few examples include trace elements in coal, fuel oil, urban particulate matter, sediments from estuaries, and water; lead in blood samples; cholesterol in human serum; drugs of abuse in urine; and a wide variety of elements in rocks, minerals, and glasses. In addition, several commercial supply houses now offer a variety of analyzed materials for method testing.[4]

a1B STATISTICAL TREATMENT OF RANDOM ERRORS

Randomly distributed data of the kind described in the section labeled Random Errors are conveniently analyzed by the techniques of statistics, which are considered in the next several sections.[5]

[2] In 1989, the name of the National Bureau of Standards (NBS) was changed to the National Institute of Standards and Technology (NIST). At this time, several of the NIST publications still bear the NBS label. Information on the NIST, its products, and its services is available on the World Wide Web at http://nvl.nist.gov/.

[3] See U.S. Department of Commerce, *NIST Standard Reference Materials Catalog* 1995–96, NIST Special Publication 260. Washington, DC: Government Printing Office, 1995. The current version of the catalog is available at the World Wide Web site mentioned in footnote 2. For a description of the NIST reference material program, see R. A. Alvarez, S. D. Rasberry, and G. A Uriano, *Anal. Chem.*, **1982,** *54,* 1226A; G. A. Uriano, *ASTM Standardization News,* **1979,** *7,* 8.

[4] See C. Veillon, *Anal. Chem.*, **1986,** *58,* 851A.

[5] For a more detailed treatment of statistics, see R. Calcutt and R. Boddy, *Statistics for Analytical Chemistry.* New York: Chapman and Hall, 1983; J. Mandel, in *Treatise on Analytical Chemistry,* 2nd ed., I. M. Kolthoff and P. J. Elving, Eds., Part I, Vol. 1, Chapter 5. New York: Wiley, 1978; R. L. Anderson, *Practical Statistics for Analytical Chemistry.* New York: Van Nostrand Reinhold, 1987.

a1B-1 Populations and Samples

In the statistical treatment of data, it is assumed that the handful of replicate experimental results obtained in the laboratory is a minute fraction of the infinite number of results that could in principle be obtained given infinite time and an infinite amount of sample. Statisticians call the handful of data a *sample* and view it as a subset of an infinite *population,* or *universe,* of data that in principle exists. The laws of statistics apply strictly to populations only; when applying these laws to a sample of laboratory data, we must assume that the sample is truly representative of the population. Because there is no assurance that this assumption is valid, statements about random errors are necessarily uncertain and must be couched in terms of probabilities.

Definition of Some Terms Used in Statistics

Population Mean (μ). The *population mean,* or *limiting mean,* of a set of replicate data is defined by the equation

$$\mu = \lim_{N \to \infty} \frac{\sum_{i=0}^{N} x_i}{N} \tag{a1-6}$$

where x_i represents the value of the ith measurement. As indicated by this equation, the mean of a set of measurements approaches the population mean as N, the number of measurements, approaches infinity. It is important to add that in the absence of bias, μ *is the true value for the quantity being measured.*

Population Standard Deviation (σ) and the Population Variance (σ^2). The population standard deviation and the population variance provide statistically significant measures of the precision of a population of data. The population standard deviation is given by the equation

$$\sigma = \sqrt{\lim_{N \to \infty} \frac{\sum_{i=0}^{N} (x_i - \mu)^2}{N}} \tag{a1-7}$$

where x_i is again the value for the ith measurement. Note that the population standard deviation is the root-mean-square of the individual *deviations from the mean* for the population.

Statisticians prefer to express the precision of data in terms of *variance,* which is simply the square of the standard deviation (σ^2), because variances are additive. That is, if there are n independent sources of random error in a system, then the total variance σ_t^2 is given by the relationship

$$\sigma_t^2 = \sigma_1^2 + \sigma_2^2 + \cdots \sigma_n^2 \tag{a1-8}$$

where $\sigma_1^2, \sigma_2^2, \cdots \sigma_n^2$ are the individual variances of the sources.

Chemists generally prefer to describe the precision of measurements in terms of standard deviation rather than variance because the standard deviation carries the same units as the measurement itself.

Sample Mean ($\bar{x}$). The *sample mean* is the mean, or average, of a finite set of data. Because N in this case is a finite number, $\bar{x}$ often differs somewhat from the population mean μ, and thus the true value, of the quantity being measured. The use of a different symbol for the sample mean emphasizes this important distinction.

Sample Standard Deviation (*s*) and Sample Variance (s^2). The *standard deviation (s)* for a sample of data that is of limited size is given by the equation

$$s = \sqrt{\frac{\sum_{i=1}^{N} (x_i - \bar{x})^2}{N - 1}} \tag{a1-9}$$

Note that the sample standard deviation differs in three ways from the population standard deviation as defined by Equation a1-7. First, σ is replaced by s in order to emphasize the difference between the two terms. Second, the true mean μ is replaced by $\bar{x}$, the sample mean. Finally, $(N - 1)$, which is defined as the *number of degrees of freedom,* appears in the denominator rather than N.[6]

Relative Standard Deviation (RSD) and Coefficient of Variation (CV). Relative standard deviations are often more informative than are absolute standard devi-

[6] By definition, the number of degrees of freedom is the number of data that remain independent when s is computed. The standard deviation of a set of experimental data is calculated based on $(N - 1)$ degrees of freedom because the mean $\bar{x}$ is used in the calculation. By substituting the mean and any subset of $(N - 1)$ data into the algebraic equation for the mean, the numerical value of the one excluded datum can be calculated. The fact that the numerical value of any data point can be calculated from the remaining data and the mean illustrates that one degree of freedom is lost any time a mean is used in the calculation of any subsequent statistic.

ations. The relative standard deviation of a data sample is given by

$$\text{RSD} = \frac{s}{\bar{x}} \times 10^z \qquad \text{(a1-10)}$$

When $z = 2$, the relative standard deviation is given as a percent; when it is 3, the deviation is reported in parts per thousand. The relative standard deviation expressed as a percent is also known as the *coefficient of variation* (CV) for the data. That is,

$$\text{CV} = \frac{s}{\bar{x}} \times 100\% \qquad \text{(a1-11)}$$

In dealing with a population of data, σ and μ are used in place of s and $\bar{x}$ in Equations a1-10 and a1-11.

An Alternate Way of Calculating Sample Standard Deviations. In calculating s with a handheld calculator that does not have a standard deviation function, the following algebraic identity to Equation a1-9 is somewhat more convenient to use:

$$s = \sqrt{\frac{\Sigma x_i^2 - (\Sigma x_i)^2/N}{N - 1}} \qquad \text{(a1-12)}$$

EXAMPLE a1-2

The following replicate data were obtained for the concentration of SO_2 in the air near a paper mill: 1.96, 1.91, 1.88, and 1.94 parts per million (ppm). Calculate **(a)** the mean, **(b)** the absolute standard deviation, and **(c)** the coefficient of variation for the data.

	ppm SO_2	
	x_i	x_i^2
x_1	1.96	3.8416
x_2	1.91	3.6481
x_3	1.88	3.5344
x_4	1.94	3.7636
	$\Sigma x_1 = 7.69$	$\Sigma x_i^2 = 14.7877$

(a) $\bar{x} = 7.69/4 = 1.9225 = 1.92$ ppm SO_2

(b) Applying Equation a1-12, we obtain

$$s = \sqrt{\frac{14.7877 - (7.69)^2/4}{4 - 1}}$$

$$= \sqrt{\frac{14.7877 - 14.784025}{3}}$$

$$= \sqrt{\frac{0.003675}{3}} = 0.035 \text{ ppm } SO_2$$

(c) $\text{CV} = \dfrac{0.035 \text{ ppm}}{1.92 \text{ ppm}} \times 100\% = 1.8\%$

Note that the difference between Σx_i^2 and $(\Sigma x_i)^2/N$ in Example a1-2 is so small that premature rounding would have led to a serious error in the computed value of s. Because of this source of error, Equation a1-12 should never be used to calculate the standard deviation for numbers containing five or more digits; instead, Equation a1-9 should be used. It is also important to note that handheld calculators and small computers with a standard deviation function usually use a version of Equation a1-12. Consequently, large errors in s are to be expected when these devices are applied to data having five or more significant figures.[7]

The Normal Error Law

In Gaussian statistics, the results of replicate measurements arising from indeterminate errors are assumed to be distributed according to the *normal error law,* which states that the fraction of a population of observations, dN/N, whose values lie in the region x to $(x + dx)$ is given by

$$\frac{dN}{N} = \frac{1}{\sigma\sqrt{2\pi}} e^{-(x-\mu)^2/2\sigma^2} dx \qquad \text{(a1-13)}$$

Here, μ and σ are the population mean and the standard deviation, and N is the number of observations. The two curves shown in Figure a1-3a are plots of Equation a1-13. The standard deviation for the data in curve B is twice that for the data in curve A.

Note that $(x - \mu)$ in Equation a1-13 is the *absolute deviation of the individual values of x from the mean* in

[7] See H. E. Solberg, *Anal. Chem.,* **1983,** *55,* 1661; P. M. Wanek et al., *Anal. Chem.,* **1982,** *54,* 1877.

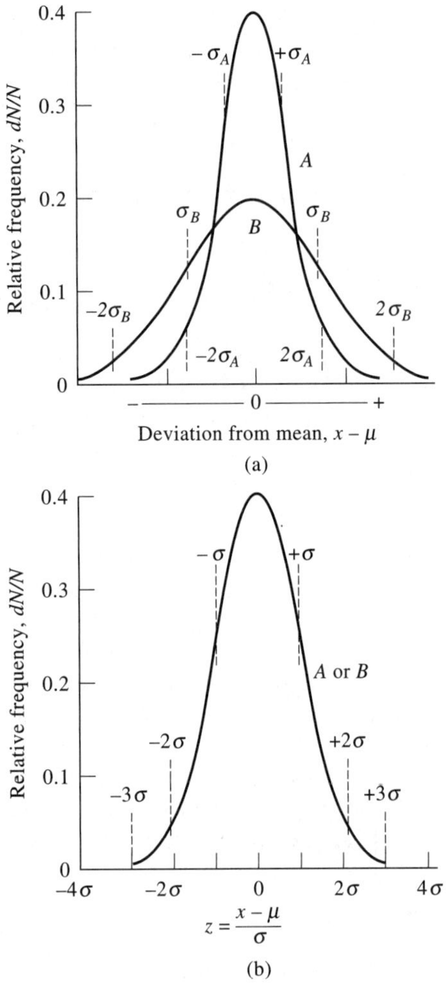

Figure a1-3 Normal error curves. The standard deviation for B is twice that for A; that is, $\sigma_B = 2\sigma_A$. (a) The abscissa is the deviation from the mean in the units of the measurement. (b) The abscissa is the deviation from the mean relative to σ. Thus, A and B produce identical curves.

whatever units are used in the measurement. It is, however, more convenient to express the deviations from the mean in units of standard deviation z where

$$z = (x - \mu)/\sigma \qquad \text{(a1-14)}$$

Taking the derivative of this equation with respect to x gives

$$dz = dx/\sigma \qquad \text{(a1-15)}$$

Substitution of these two relationships into Equation a1-13 leads to an equation that expresses the distribution in terms of the single variable z. That is,

$$\frac{dN}{N} = \frac{1}{\sqrt{2\pi}} e^{-z^2/2}\, dz \qquad \text{(a1-16)}$$

We may convert this expression to finite intervals by integrating between N_1 and N_2 as shown in the following section.

The Normal Error Curve

Figure a1-3b shows another way of plotting the data for the two curves in Figure a1-3a. The abscissa is now z, the deviations from the mean of the data in units of standard deviation (Equation a1-14). This type of plot yields a single curve regardless of the magnitude of the standard deviation of the data. The general properties of this curve include: (1) zero deviation from the mean occurring with maximum frequency, (2) symmetrical distribution of positive and negative deviations about this maximum, and (3) exponential decrease in frequency as the magnitude of the deviations increases. Thus, small random errors are much more common than large errors.

Areas Under Regions of the Normal Error Curve. The area under the curve in Figure a1-3 is the integral of Equation a1-16 and is determined as follows:

$$\frac{\Delta N}{N} = \int_{-a}^{a} \frac{1}{\sqrt{2\pi}} e^{-z^2/2}\, dz = \operatorname{erf}\left(\frac{a\sqrt{2}}{2}\right)$$

where erf(b) is the so-called *error function,* which is given by

$$\operatorname{erf}(b) = \int_0^b \frac{2}{\sqrt{\pi}} e^{-x^2}\, dx$$

The fraction of the population between any specified limits is given by the area under the curve between these limits. For example, the area under the curve between $z = -1$ and $z = +1$ is given by the definite integral

$$\frac{\Delta N}{N} = \int_{-1}^{1} \frac{1}{\sqrt{2\pi}} e^{-z^2/2}\, dx = \operatorname{erf}\left(\frac{\sqrt{2}}{2}\right) = 0.683$$

Thus, 0.683 or 68.3% of the total area under the curve lies within $\pm 1\sigma$ of the mean. That is, $\Delta N/N = 0.683$, which means that 68.3% of a population of data lie within $\pm 1\sigma$ of its mean value. For similar calculations with $z = 2$ and $z = 3$, we find that 95.5% lie within $\pm 2\sigma$ and 99.7% within $\pm 3\sigma$. Values for $(x - \mu)$ corresponding to $\pm 1\sigma$, $\pm 2\sigma$, and $\pm 3\sigma$ are indicated by broken vertical lines in Figure a1-3.

The properties of the normal error curve are useful because they permit statements to be made about the probable magnitude of the net random error in a given measurement or set of measurements *provided the standard deviation is known.* Thus, one can say that the chances are 68.3 out of 100 that the random error associated with any single measurement is smaller than $\pm 1\sigma$, that the chances are 95.5 out of 100 that the error is less than $\pm 2\sigma$, and so forth. Clearly, the standard deviation is a useful parameter for estimating and reporting the probable net random error for an analytical method.

Standard Error of a Mean. The figures on percentage distribution just cited refer to the probable error of a *single* measurement. If a set of samples, each containing N data, are taken randomly from a population of data, the means of samples will show less and less scatter as N increases. The standard deviation of the means of the samples is known as the *standard error* of the mean and is given the symbol s_m. It can be shown that the standard error is inversely proportional to the square root of the number of data used to calculate the means. That is,

$$s_m = s/\sqrt{N} \qquad \text{(a1-17)}$$

The mean and the standard deviation for a set of data are statistics of primary importance in all types of science and engineering. The mean is important because it usually provides the best estimate of the variable of interest. The standard deviation of the mean is equally important because it provides information about the precision and thus the random error associated with the measurement.

Method for Obtaining a Good Approximation of σ

In order to apply a statistical relationship directly to finite samples of data, it is necessary to know that the sample standard deviation s for the data is a good approximation of the population standard deviation σ. Otherwise, statistical inferences must be modified to take into account the uncertainty in s. In this section, we consider methods for obtaining reliable values for s from small samples of experimental data.

Performing Preliminary Experiments. Uncertainty in the calculated value for s decreases as the number of measurements N in Equation a1-9 increases. Figure a1-4 shows the relative error in s as a function of N. Note that when N is greater than about 20, s and σ can be assumed, for most purposes, to be identical. Thus, when a method of measurement is not excessively time-consuming and when an adequate supply of sample is available, it is sometimes feasible and economical to carry out preliminary experiments whose sole purpose is that of obtaining a reliable standard deviation for the method.

Pooling Data. For analyses that are time-consuming, the foregoing procedure is seldom practical. In such cases, however, precision data from a series of similar samples accumulated in the course of time can be pooled to provide an estimate of s that is superior to the

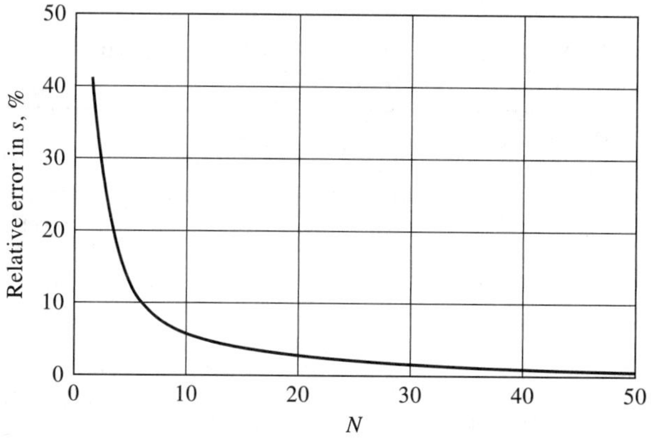

Figure a1-4 Relative error in s as a function of N.

$$s_{\text{pooled}} = \sqrt{\frac{\sum_{i=1}^{N_1}(x_i - \bar{x}_1)^2 + \sum_{j=1}^{N_2}(x_j - \bar{x}_2)^2 + \sum_{k=1}^{N_3}(x_k - \bar{x}_3)^2 + \ldots + \sum_{p=1}^{N_{n_t}}(x_p - \bar{x}_{n_t})^2}{N_1 + N_2 + N_3 + \ldots - n_t}}$$

value for any individual subset. Again, one must assume the same sources of random error are present in all the samples. This assumption is usually valid provided the samples have similar compositions and each has been analyzed identically. To obtain a pooled estimate of s, deviations from the mean for each subset are squared; the squares for all of the subsets are then summed and divided by an appropriate number of degrees of freedom. The pooled standard deviation s_{pooled} is obtained by extracting the square root of the quotient. One degree of freedom is lost in computing the mean for each of the n_t subsets. Thus, the number of degrees of freedom for the pooled s is equal to the total number of measurements minus n_t as shown in the equation at the top of the page, where i, j, k, and p are the indices of each data subset, $N_1, N_2, N_3 \ldots N_{n_t}$ are the numbers of data in each subset, and n_t is the number of subsets. Example a1-3 shows of this type of calculation.

Note that in Example a1-3 one degree of freedom was lost for each of the seven samples. Because the remaining number of degrees of freedom is greater than 20, however, the computed s can be considered to be a good approximation of σ; that is, we may assume that $s \rightarrow \sigma$.

a1B-2 Confidence Limits (CL)

The true or population mean (μ) of a measurement is a constant that must always remain unknown. However, in the absence of systematic errors, limits can be set within which the population mean can be expected to lie with a given degree of probability. The limits obtained in this manner are called *confidence limits.*

The confidence limit, which is calculated from the sample standard deviation, depends on the certainty with which s is known. If there is reason to believe that s is a good approximation of σ, then the confidence limits can be significantly narrower than if the estimate of s is based on only two or three measurements.

Confidence Limit When s is a Good Approximation of σ

Figure a1-5 is a set of normal error curves in which the abscissa is the quantity z, which represents the deviation from the mean in units of the population standard deviation (Equation a1-14). The number in the center of each curve gives the percent of the total area under the curve that is encompassed by the indicated values of $-z$ and $+z$. For example, 50% of the area under any Gaussian curve lies between -0.67σ and $+0.67\sigma$; 80% of the area lies between -1.29σ and $+1.29\sigma$. From the latter limits then, we may assert, with 80% probability of being correct, that the population mean lies within $\pm 1.29\sigma$ of any single measurement we make. Here, the *confidence level* is 80% and the *confidence interval* is $\pm z\sigma = \pm 1.29\sigma$. A general statement for the confidence limit (CL) of a single measurement is obtained by rearranging Equation a1-14 remembering that z can take positive or negative values. Thus,

$$\text{CL for } \mu = x \pm z\sigma \qquad \text{(a1-18)}$$

Equation a1-18 applies to the result of a single measurement. Application of Equation a1-17 shows the confidence interval is decreased by $\sqrt{N}$ for the average of N replicate measurements. Thus, a more general form of Equation a1-18 is

$$\text{CL for } \mu = \bar{x} \pm \frac{z\sigma}{\sqrt{N}} \qquad \text{(a1-19)}$$

Values for z at various confidence levels are found in Table a1-3.

TABLE a1-3 Confidence Levels for Various Values of z

Confidence Level, %	z	Confidence Level, %	z
50	0.67	96	2.00
68	1.00	99	2.58
80	1.29	99.7	3.00
90	1.64	99.9	3.29
95	1.96		

EXAMPLE a1-3

The mercury in samples of seven fish taken from the Mississippi River was determined with a method based on the absorption of radiation by gaseous elemental mercury. Calculate a pooled estimate of the standard deviation for the method, based on the first three columns of data in the table that follows:

Specimen Number	Number of Samples Measured	Hg Content, ppm	Mean, ppm Hg	Sum of Squares of Deviations from Means
1	3	1.80, 1.58, 1.64	1.673	0.0259
2	4	0.96, 0.98, 1.02, 1.10	1.015	0.0115
3	2	3.13, 3.35	3.240	0.0242
4	6	2.06, 1.93, 2.12, 2.16, 1.89, 1.95	2.018	0.0611
5	4	0.57, 0.58, 0.64, 0.49	0.570	0.0114
6	5	2.35, 2.44, 2.70, 2.48, 2.44	2.482	0.0685
7	4	1.11, 1.15, 1.22, 1.04	1.130	0.0170
	Number of measurements 28			Sum of sum of squares = 0.2196

The values in columns 4 and 5 for specimen 1 were computed as follows

x_i	$\lvert(x_i - \bar{x})\rvert$	$(x_i - \bar{x})^2$
1.80	0.127	0.0161
1.58	0.093	0.0087
1.64	0.033	0.0011
5.02		Sum of squares = 0.0259

$$\bar{x} = \frac{5.02}{3} = 1.673$$

The other data in columns 4 and 5 were obtained similarly. Then

$$s_{\text{pooled}} = \sqrt{\frac{0.0259 + 0.0115 + 0.0242 + 0.0611 + 0.0114 + 0.0685 + 0.0170}{28 - 7}}$$

$$= 0.100 \text{ ppm Hg}$$

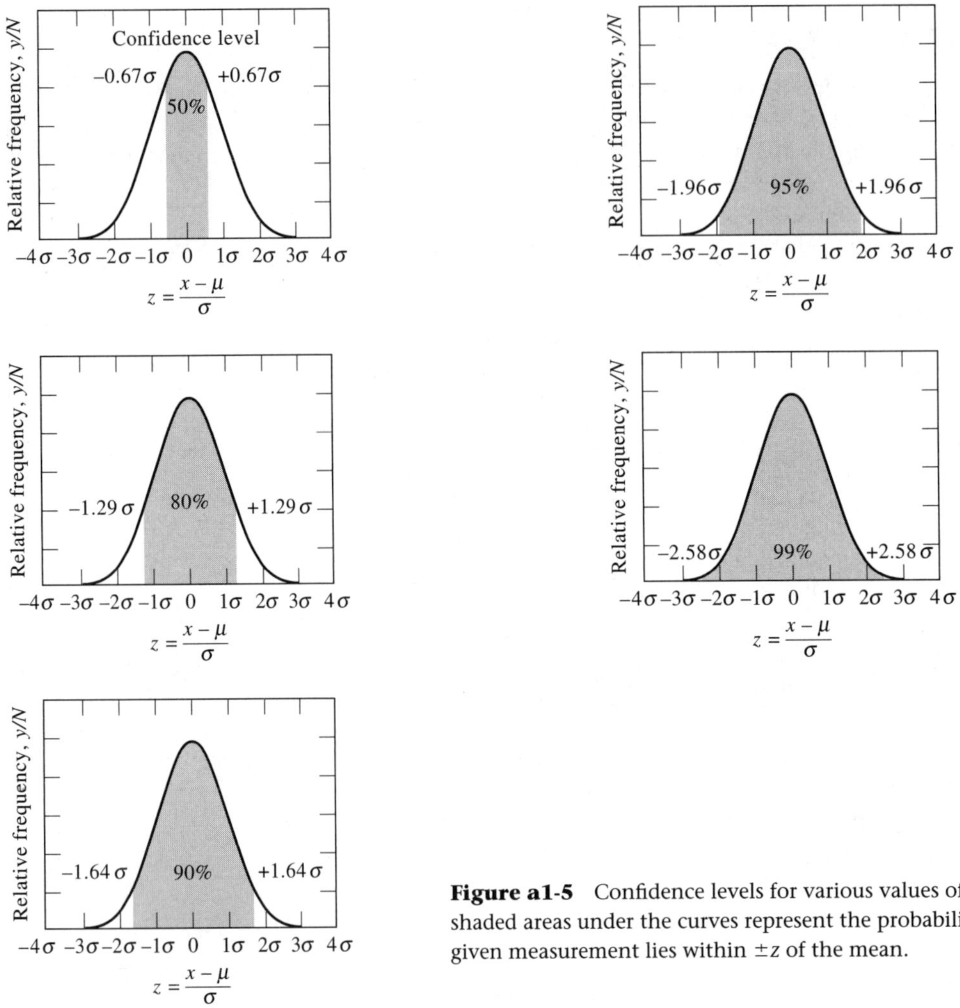

Figure a1-5 Confidence levels for various values of z. The shaded areas under the curves represent the probability that a given measurement lies within $\pm z$ of the mean.

EXAMPLE a1-4

Calculate the 50% and the 95% confidence limits for the mean value (1.67 ppm Hg) for sample 1 in Example a1-3. Again, $s \approx \sigma = 0.10$.

Applying Equation a1-19 to the three measurements yields

$$50\%\ \text{CL} = 1.67 \pm \frac{0.67 \times 0.10}{\sqrt{3}}$$

$$= 1.67 \pm 0.04$$

$$95\%\ \text{CL} = 1.67 \pm \frac{1.96 \times 0.10}{\sqrt{3}}$$

$$= 1.67 \pm 0.11$$

From Example a1-4, we conclude that probability is 50% that μ, the population mean (and *in the absence of systematic error* the true value), will lie between the limits of 1.63 and 1.71 ppm Hg. Furthermore, there is a 95% chance that μ will be found between 1.56 and 1.78 ppm Hg.

EXAMPLE a1-5

How many replicate measurements of specimen 1 in Example a1-3 would be needed to decrease the 95% confidence interval to ±0.07 ppm Hg?

The pooled value for s is a good estimate for σ. For a confidence interval of ±0.07 ppm Hg, substitution into Equation a1-19 leads to

$$0.07 = \pm\frac{zs}{\sqrt{N}} = \pm\frac{1.96 \times 0.10}{\sqrt{N}}$$

$$\sqrt{N} = \pm\frac{1.96 \times 0.10}{0.07} = 2.80$$

$$N = 7.8$$

We conclude, then, that 8 measurements would provide a slightly better than 95% chance of the population mean lying within ±0.07 ppm of the experimental mean.

A consideration of Equation a1-19 indicates that the confidence interval for an analysis can be halved by using the mean of four measurements in our calculations. Sixteen measurements are required to narrow the limit by another factor of two. It is apparent that a point of diminishing return is rapidly reached in acquiring additional data. Thus, we ordinarily take advantage of the relatively large gain afforded by averaging two to four measurements but can seldom afford the time required for further increases in confidence.

In data analysis, it is essential to always keep in mind that confidence limits based on Equation a1-19 apply only *in the absence of systematic errors.*

Confidence Limits When σ Is Unknown

Frequently, a chemist must make use of an unfamiliar method wherein limitations in time or amount of available sample preclude an accurate estimation of σ. Here, a single set of replicate measurements must provide not only a mean but also a precision estimate. As indicated earlier, s calculated from a small set of data may be subject to considerable uncertainty; thus, confidence limits must be broadened when a good estimate of σ is unavailable.

To account for the potential variability of s, use is made of the statistical parameter t, which is defined as

$$t = (x - \mu)/s \tag{a1-20}$$

Note the similarity between Equations a1-20 and a1-14. In contrast to z in Equation a1-14, t is dependent not only on the desired confidence level but also on the number of degrees of freedom available in the calculation of s. Table a1-4 provides values for t for a few degrees of freedom; more extensive tables are found in most mathematical handbooks. Note that the values for t become equal to those for z (Table a1-3) as the number of degrees of freedom becomes infinite. The variable t is often called *student's t* after the pseudonym of the British mathematician W. S. Gossett who first described the variable and demonstrated its use.

The confidence limit for the mean $\bar{x}$ of N replicate measurements can be computed from t by an equation analogous to Equation a1-19; that is,

$$\text{CL for } \mu = \bar{x} \pm \frac{ts}{\sqrt{N}} \tag{a1-21}$$

TABLE a1-4 Values of *t* for Various Levels of Probability

Degrees of Freedom	Factor for Confidence Interval, %				
	80	90	95	99	99.9
1	3.08	6.31	12.7	63.7	637
2	1.89	2.92	4.30	9.92	31.6
3	1.64	2.35	3.18	5.84	12.9
4	1.53	2.13	2.78	4.60	8.60
5	1.48	2.02	2.57	4.03	6.86
6	1.44	1.94	2.45	3.71	5.96
7	1.42	1.90	2.36	3.50	5.40
8	1.40	1.86	2.31	3.36	5.04
9	1.38	1.83	2.26	3.25	4.78
10	1.37	1.81	2.23	3.17	4.59
11	1.36	1.80	2.20	3.11	4.44
12	1.36	1.78	2.18	3.06	4.32
13	1.35	1.77	2.16	3.01	4.22
14	1.34	1.76	2.14	2.98	4.14
∞	1.29	1.64	1.96	2.58	3.29

EXAMPLE a1-6

A chemist obtained the following data for the alcohol content of a sample of blood; percent ethanol = 0.084, 0.089, and 0.079. Calculate the 95% confidence limit for the mean assuming **(a)** no additional knowledge about the precision of the method and **(b)** that on the basis of previous experiences it is known that $s \rightarrow \sigma = 0.006\%$ ethanol.

(a) $\Sigma x_i = 0.084 + 0.089 + 0.079 = 0.252$

$$\Sigma x_i^2 = 0.0070566 + 0.007921 + 0.006241 = 0.021218$$

$$s = \sqrt{\frac{0.021218 - (0.252)^2/3}{3 - 1}} = 0.0050$$

Here, $\bar{x} = 0.252/3 = 0.084$. Table a1-4 indicates that $t = \pm 4.30$ for two degrees of freedom and 95% confidence. Thus,

$$95\%\ \text{CL} = \bar{x} \pm \frac{ts}{\sqrt{N}}$$

$$= 0.084 \pm \frac{4.3 \times 0.0050}{\sqrt{3}}$$

$$= 0.084 \pm 0.012$$

(b) Because a good value of σ is available

$$95\%\ \text{CL} = \bar{x} \pm \frac{z\sigma}{\sqrt{N}}$$

$$= 0.084 \pm \frac{1.96 \times 0.006}{\sqrt{3}}$$

$$= 0.084 \pm 0.007$$

Note from Example a1-6 that a good estimate of σ nearly halves the confidence interval.

a1B-3 Test for Bias

As noted in Section a1A-2, bias in an analytical method is generally detected by the analysis of one or more standard reference materials whose composition is known. In all likelihood, the experimental mean of such an analysis $\bar{x}$ will differ from the true value μ supplied for the standard. In this case, a judgment must be made whether this difference is the consequence of random error in the analysis of the reference material or is a result of bias in the method used.

A common way of treating this problem statistically is to compare the experimental difference $\bar{x} - \mu$ with the difference that could be expected at a certain probability level if there were no bias. If the experimental $\bar{x} - \mu$ is larger than the calculated difference, bias is likely. If, on the other hand, the experimental value is equal to or smaller than the computed difference, the presence of bias has not been demonstrated.

This test for bias makes use of the t statistics discussed earlier. Here, we rearrange Equation a1-21 to give

$$\bar{x} - \mu = \pm\frac{ts}{\sqrt{N}} \qquad \text{(a1-22)}$$

where N is the number of replicate measurements used in the test. If a good estimate of σ is available, then the equation can be modified by replacing t with z and s with σ. If the experimental value of $\bar{x} - \mu$ is larger than the value of $\bar{x} - \mu$ calculated from Equation a1-22, the presence of bias in the method is suggested. If, on the other hand, the value calculated using Equation a1-22 is larger, no bias has been demonstrated.

EXAMPLE a1-7

A new procedure for the rapid determination of sulfur in kerosenes was tested on a sample known from its method of preparation to contain 0.123% S. The results were % S = 0.112, 0.118, 0.115, and 0.119. Do the data indicate the presence of bias in the method?

$$\Sigma x_i = 0.112 + 0.118 + 0.115 + 0.119 = 0.464$$

$$\bar{x} = 0.464/4 = 0.116\%\ \text{S}$$

$$\bar{x} - \mu = 0.116 - 0.123 = -0.007\%\ \text{S}$$

$$\Sigma x_i^2 = 0.012544 + 0.013924 + 0.013225 + 0.014161 = 0.053854$$

$$s = \sqrt{\frac{0.053854 - (0.464)^2/4}{4 - 1}} = \sqrt{\frac{0.000030}{3}}$$

$$= 0.0032$$

From Table a1-4, we find that at the 95% confidence level, t has a value of 3.18 for three degrees of freedom. Thus,

$$\frac{ts}{\sqrt{4}} = \frac{3.18 \times 0.0032}{\sqrt{4}} = \pm 0.0051$$

An experimental mean can be expected to deviate by ± 0.0051 or greater no more frequently than 5 times in 100. Thus, if we conclude that $\bar{x} - \mu = -0.007$ is a significant difference and that bias is present, we will, on the average, be wrong fewer than 5 times in 100.

If we make a similar calculation using the value for t at the 99% confidence level, $ts/\sqrt{N}$ assumes a value of 0.0093. Thus, if we insist on being wrong no more often than 1 time in 100, we must conclude that no bias has been *demonstrated.* Note that this statement is different from saying that there is no bias.

EXAMPLE a1-8

Suppose we know from past experience that the method described in Example a1-7 had a population standard deviation of 0.0032% S. That is, $s \rightarrow \sigma = 0.0032$. Is the presence of bias suggested at the 99% confidence level?

Here we write that

$$\bar{x} - \mu = \pm\frac{z\sigma}{\sqrt{N}} = \pm\frac{2.58 \times 0.0032}{\sqrt{4}} = \pm 0.00413$$

The experimental difference of -0.007 is significantly larger than this number. Thus, bias is strongly suggested.

a1B-4 Propagation of Measurement Uncertainties

A typical instrumental method of analysis involves several experimental measurements each of which is subject to an indeterminate uncertainty and each of which contributes to the net indeterminate error of the final result. For the purpose of showing how such indeterminate uncertainties affect the outcome of an analysis, let us assume that a result x is dependent on the experimental variables, $p, q, r, \cdots$, each of which fluctuates in a random and independent way. That is, x is a function of $p, q, r, \cdots$, so that we may write

$$x = f(p, q, r, \cdots) \qquad \text{(a1-23)}$$

The uncertainty dx_i (that is, the deviation from the mean) in the ith measurement of x will depend on the size and sign of the corresponding uncertainties $dp_i, dq_i, dr_i, \cdots$. That is,

$$dx_i = f(dp_i, dq_i, dr_i, \cdots)$$

The total variation in dx as a function of the uncertainties in $p, q, r, \cdots$ can be derived by taking the partial derivative of x (Equation a1-23) with respect to each of the variables $p, q, r, \cdots$ That is,

$$dx = \left(\frac{\partial x}{\partial p}\right)_{q,r,\cdots} dp + \left(\frac{\partial x}{\partial q}\right)_{p,r,\cdots} dq + \left(\frac{\partial x}{\partial r}\right)_{p,q,\cdots} dr + \cdots \qquad \text{(a1-24)}$$

In order to develop a relationship between the standard deviation of x and the standard deviations of p, q, and r, we must square the foregoing equation. Thus,

$$(dx)^2 = \left[\left(\frac{\partial x}{\partial p}\right)_{q,r,\cdots} dp + \left(\frac{\partial x}{\partial q}\right)_{p,r,\cdots} dq + \left(\frac{\partial x}{\partial r}\right)_{p,q,\cdots} dr + \cdots\right]^2 \qquad \text{(a1-25)}$$

This equation must then be summed between the limits of $i = 1$ to $i = N$, where N again is the total number of replicate measurements.

In squaring Equation a1-24, two types of terms from the right-hand side of the equation emerge: (1) square terms and (2) cross terms. Square terms take the form

$$\left(\frac{\partial x}{\partial p}\right)^2 dp^2, \left(\frac{\partial x}{\partial q}\right)^2 dq^2, \left(\frac{\partial x}{\partial r}\right)^2 dr^2, \cdots$$

Square terms are always positive and can, therefore, *never cancel.* In contrast, cross terms may be either positive or negative in sign. Examples are

$$\left(\frac{\partial x}{\partial p}\right)\left(\frac{\partial x}{\partial q}\right) dpdq, \left(\frac{\partial x}{\partial p}\right)\left(\frac{\partial x}{\partial r}\right) dpdr, \cdots$$

If dp, dq, and dr represent *independent* and *random uncertainties,* some of the cross terms will be negative and others positive. Thus, the *sum of all such terms should approach zero,* particularly when N is large.[8]

As a consequence of the tendency of cross terms to cancel, the sum of Equation a1-25 from $i = 1$ to $i = N$ can be assumed to be made up exclusively of squared terms. This sum then takes the form

$$\Sigma(dx_i)^2 = \left(\frac{\partial x}{\partial p}\right)^2 \Sigma(dp_i)^2 + \left(\frac{\partial x}{\partial q}\right)^2 \Sigma(dq_i)^2 + \left(\frac{\partial x}{\partial r}\right)^2 \Sigma(dr_i)^2 + \cdots \qquad \text{(a1-26)}$$

[8] If the variables are not independent, the cross terms must be kept regardless of the size of N. See S. L. Meyer, *Data Analysis for Scientists and Engineers.* New York: Wiley, 1975.

Dividing through by N gives

$$\frac{\Sigma(dx_i)^2}{N} = \left(\frac{\partial x}{\partial p}\right)^2 \frac{\Sigma(dp_i)^2}{N} + \left(\frac{\partial x}{\partial q}\right)^2 \frac{\Sigma(dq_i)^2}{N} + \left(\frac{\partial x}{\partial r}\right)^2 \frac{\Sigma(dr_i)^2}{N} + \cdots \quad \text{(a1-27)}$$

From Equation a1-7, however, we see that

$$\frac{\Sigma(dx_i)^2}{N} = \frac{\Sigma(x_i - \mu)^2}{N} = \sigma_x^2$$

where σ_x^2 is the variance of x. Similarly,

$$\frac{\Sigma(dp_i)^2}{N} = \sigma_p^2$$

and so forth. Thus, Equation a1-27 can be written in terms of the variances of the quantities; that is,

$$\sigma_x^2 = \left(\frac{\partial x}{\partial p}\right)^2 \sigma_p^2 + \left(\frac{\partial x}{\partial q}\right)^2 \sigma_q^2 + \left(\frac{\partial x}{\partial r}\right)^2 \sigma_r^2 \cdots \quad \text{(a1-28)}$$

The example that follows illustrates how Equation a1-28 is used to give the variance of a quantity calculated from several experimental data.

EXAMPLE a1-9

The number of plates N in a chromatographic column can be computed with Equation 26-17 (Chapter 26):

$$N = 16\left(\frac{t_R}{W}\right)^2$$

where t_R is the retention time and W is the width of the chromatographic peak in the same units as t_R. The significance of these terms is explained in Figure 26-6.

Hexachlorobenzene exhibited a high-performance liquid chromatographic peak at a retention time of 13.36 min. The width of the peak at its base was 2.18 min. The standard deviation s for the two time measurements was 0.043 and 0.061 min, respectively. Calculate **(a)** the number of plates in the column and **(b)** the standard deviation for the computed result.

(a) $N = 16\left(\frac{13.36}{2.18}\right)^2 = 601$ plates

(b) Substituting s for σ in Equation a1-28 gives

$$s_N^2 = \left(\frac{\partial N}{\partial t_R}\right)_W^2 s^2 t_R + \left(\frac{\partial N}{\partial W}\right)_{tr}^2 s_W^2$$

Taking partial derivatives of the original equation

$$\left(\frac{\partial N}{\partial t_R}\right)_W = \frac{32 t_R}{W^2} \text{ and } \left(\frac{\partial N}{\partial W}\right)_{t_R} = \frac{-32 t_R^2}{W^3}$$

Substituting these relationships into the previous equation gives

$$s_N^2 = \left(\frac{32 t_R}{W^2}\right)^2 s_{t_R}^2 + \left(\frac{-32 t_R^2}{W^3}\right)^2 s_W^2$$

$$= \left(\frac{32 \times 13.36 \text{ min}}{(2.18 \text{ min})^2}\right)^2 (0.061 \text{ min})^2 + \left(\frac{-32(13.36 \text{ min})^2}{(2.18 \text{ min})^3}\right)^2 (0.043 \text{ min})^2 = 592.1$$

$$s_N = \sqrt{592.1} = 24.3 = 24 \text{ plates}$$

Thus, $N = 6.0(\pm 0.2) \times 10^2$ plates.

a1B-5 Rounding Results from Arithmetic Calculations

Equation a1-28 is helpful in deciding how the results of arithmetic calculations should be rounded. For example, consider the case where the result x is computed by the relationship.

$$x = p + q - r$$

where p, q, and r are experimental quantities having sample standard deviations of s_p, s_q, and s_r, respectively. Applying Equation a1-18 (using sample rather than population standard deviations) gives

$$s_x^2 = \left(\frac{\partial x}{\partial p}\right)_{q,r}^2 s_p^2 + \left(\frac{\partial x}{\partial q}\right)_{q,r}^2 s_q^2 + \left(\frac{\partial x}{\partial r}\right)_{p,q}^2 s_r^2$$

But,

$$\left(\frac{\partial x}{\partial p}\right)_{q,r} = \left(\frac{\partial x}{\partial q}\right)_{p,r} = 1 \text{ and } \left(\frac{\partial x}{\partial r}\right)_{p,q} = -1$$

Therefore, the variance of x is given by

$$s_x^2 = s_p^2 + s_q^2 + s_r^2$$

or the standard deviation of the result is given by

$$s_x = \sqrt{s_p^2 + s_q^2 + s_r^2}$$

Thus, the *absolute* standard deviation of a sum or difference is equal to the square root of the sum of the squares of the *absolute* standard deviation of the numbers making up the sum or difference.

Proceeding in this same way yields the relationships shown in column 3 of Table a1-5 for other types of arithmetic operations. Note that in several calculations, relative variances such as $(s_x/x)^2$ and $(s_p/p)^2$ are combined rather than absolute standard deviations.

EXAMPLE a1-10

Calculate the standard deviation of the result of

$$\frac{[14.3(\pm 0.2) - 11.6(\pm 0.2)] \times 0.050(\pm 0.001)}{[820(\pm 10) \pm 1030(\pm 5)] \times 42.3(\pm 0.4)} = 1.725(\pm ?) \times 10^{-6}$$

First we must calculate the standard deviation of the sum and the difference. The standard deviation s_p for the difference in the numerator is given by

$$s_p = \sqrt{(\pm 0.2)^2 + (\pm 0.2)^2} = \pm 0.283$$

For the sum in the denominator, the standard deviation s_q is

$$s_q = \sqrt{(\pm 10)^2 + (\pm 5)^2} = \pm 11.2$$

We may then rewrite the equation as

$$\frac{2.7(\pm 0.283) \times 0.050(\pm 0.001)}{1850(\pm 11.2) \times 42.3(\pm 0.4)} = 1.725 \times 10^{-6}$$

The equation now contains only products and quotients, and Equation (2) in Table a1-5 applies:

$$\frac{s_x}{x} = \sqrt{\left(\frac{\pm 0.283}{2.7}\right)^2 + \left(\frac{\pm 0.001}{0.050}\right)^2 + \left(\frac{\pm 11.2}{1850}\right)^2 + \left(\frac{\pm 0.4}{42.3}\right)^2} = \pm 0.107$$

To obtain the absolute standard deviation, we write

$$s_x = \pm 0.107x = \pm 0.0107 \times 1.725 \times 10^{-6} = \pm 0.185 \times 10^{-6}$$

and the answer is rounded to $1.7(\pm 0.2) \times 10^{-6}$.

EXAMPLE a1-11

Calculate the absolute deviations of the results of the following computations. The absolute standard deviation for each quantity is given in parentheses.

(a) $x = \log[2.00(\pm 0.02) \times 10^{-4}] = -3.6990 \pm ?$
(b) $x = \text{antilog}[1.200(\pm 0.003)] = 15.849 \pm ?$
(c) $x = \text{antilog}[45.4(\pm 0.3)] = 2.5119 \times 10^{45} \pm ?$

TABLE a1-5 Error Propagation in Arithmetic Calculations

Type of Calculation	Example*	Standard Deviation of x	
Addition or Subtraction	$x = p + q - r$	$s_x = \sqrt{s_p^2 + s_q^2 + s_r^2}$	(1)
Multiplication or Division	$x = p \cdot q/r$	$\frac{s_x}{x} = \sqrt{\left(\frac{s_p}{p}\right)^2 + \left(\frac{s_q}{q}\right)^2 + \left(\frac{s_r}{r}\right)^2}$	(2)
Exponentiation	$x = p^y$	$\frac{s_x}{x} = y\frac{s_p}{p}$	(3)
Logarithm	$x = \log_{10} p$	$s_x = 0.434\frac{s_p}{p}$	(4)
Antilogarithm	$x = \text{antilog}_{10} p$	$\frac{s_x}{x} = 2.303 s_p$	(5)

*p, q, and r are experimental variables whose standard deviations are s_p, s_q, and s_r, respectively.

(a) Referring to Equation (4) in Table a1-5, we see

$$s_x = \pm 0.434 \times \frac{0.02 \times 10^{-4}}{2.00 \times 10^{-4}} = \pm 0.004$$

Thus,

$$\log [2.00\ (\pm 0.02) \times 10^{-4}] = -3.699\ (\pm 0.004)$$

(b) Using Equation (5) in Table a1-5, we obtain

$$\frac{s_x}{x} = 2.303 \times (\pm 0.003) = \pm 0.0069$$

$$s_x = \pm 0.0069x = \pm 0.0069 \times 15.849 = 0.109$$

Therefore, antilog [1.200 (±0.003)]
= 15.8 ± 0.1

(c) $$\frac{s_x}{x} = 2.303\ (\pm 0.3) = \pm 0.691$$

$$s_x = \pm 0.691 \times 2.511 \times 10^{45} = \pm 1.7 \times 10^{45}$$

Therefore, antilog [45.4 (±0.3)]
= 2.5 (±1.7) × 10^{45}

Example a1-12 demonstrates that a large absolute error is associated with the antilogarithm of a number with few digits beyond the decimal point. This large uncertainty arises from the fact that the numbers to the left of the decimal (the characteristic) serve only to locate the decimal point. The large error in the antilogarithm results from the relatively large uncertainty in the *mantissa* of the number (that is, 0.4 ± 0.3).

a1C METHOD OF LEAST SQUARES

Most analytical methods are based on a calibration curve in which a measured quantity y is plotted as a function of the known concentration x of a series of standards. Figure a1-6 shows a typical calibration curve, which was computed for the chromatographic determination of isooctane in hydrocarbon samples. The ordinate (the dependent variable) is the area under the chromatographic peak for isooctane, and the abscissa (the independent variable) is the mole percent of isooctane. As is typical (and desirable), the plot approximates a straight line. Note, however, that not all the data fall exactly on the line because of the random errors in the measuring process. Thus, we must try to fit a "best" straight line through the points. A common

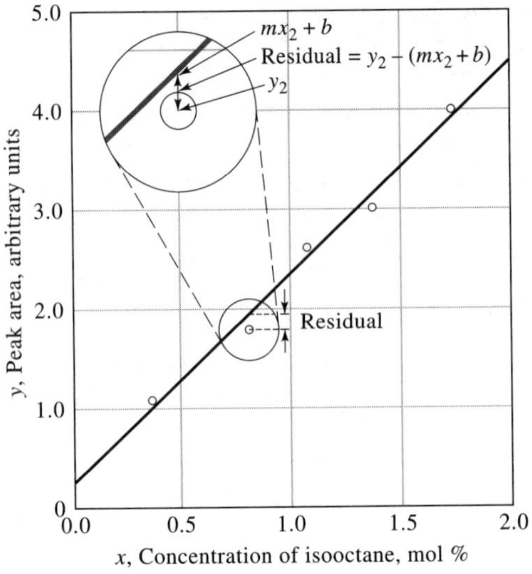

Figure a1-6 Calibration curve for determining isooctane in hydrocarbon mixtures.

way of finding such a line is the *method of least squares.*

In applying the method of least squares, we assume that a straight line is a good model for the relationship between the area of a peak (y) and its analyte concentration (x) as given by the equation

$$y = mx + b$$

where m is the slope of the straight line and b is the intercept. The slope and intercept are referred to as the *parameters* of the model, which in this case is a straight line. We also assume that any deviation of individual points from the straight line results from error in the area measurement and that there is no error in the values of x—that is, the concentrations of the standard solutions are known exactly.

As illustrated in Figure a1-5, the vertical deviation of each point from the straight line is called a *residual.* The line generated by the least-squares method is the one that minimizes the sum of the squares of the residuals for all of the points.

For convenience, we define three quantities S_{xx}, S_{yy}, and S_{xy} as follows:

$$S_{xx} = \Sigma(x_i - \bar{x})^2 = \Sigma x_i^2 - \frac{(\Sigma x_i)^2}{N} \qquad \text{(a1-29)}$$

$$S_{yy} = \Sigma(y_i - \bar{y})^2 = \Sigma y_i^2 - \frac{(\Sigma y_i)^2}{N} \qquad \text{(a1-30)}$$

$$S_{xy} = \Sigma(x_i - \bar{x})(y_i - \bar{y}) = \Sigma x_i y_i - \frac{\Sigma x_i \Sigma y_i}{N} \quad \text{(a1-31)}$$

Here, x_i and y_i are the coordinates of the individual data points, N is the number of pairs of data used in preparation of the calibration curve, and $\bar{x}$ and $\bar{y}$ are the average values for the variables, or

$$\bar{x} = \Sigma x_i/N \text{ and } \bar{y} = \Sigma y_i/N$$

Note that S_{xx} and S_{yy} are the sum of the squares of the deviations from the mean for the individual values of x and y. The equivalent expressions shown to the far right in Equations a1-29, a1-30, and a1-31 are more convenient when a handheld calculator is being used.

Six useful quantities can be computed from S_{xx}, S_{yy}, and S_{xy}.

1. The slope of the line m:

$$m = S_{xy}/S_{xx} \quad \text{(a1-32)}$$

2. The intercept b:

$$b = \bar{y} - m\bar{x} \quad \text{(a1-33)}$$

3. The standard deviation about regression, or standard error of the estimate, s_c of the residuals ($y_c - mx - b$):

$$s_r = \sqrt{\frac{S_{yy} - m^2 S_{xx}}{N - 2}} \quad \text{(a1-34)}$$

4. The standard deviation of the slope s_m:

$$s_m = s_r/\sqrt{S_{xx}} \quad \text{(a1-35)}$$

5. The standard deviation s_b of the intercept:

$$s_b = s_r \sqrt{\frac{\Sigma x_i^2}{N\Sigma x_i^2 - (\Sigma x_i)^2}} \quad \text{(a1-36)}$$

$$= s_r \sqrt{\frac{1}{N - (\Sigma x_i)^2/\Sigma x_i^2}}$$

6. The standard deviation s_c for analytical results obtained from the parameters of the least-squares fit of the data:

$$s_c = \frac{s_r}{m}\sqrt{\frac{1}{M} + \frac{1}{N} + \frac{(\bar{y}_c - \bar{y})^2}{m^2 S_{xx}}} \quad \text{(a1-37)}$$

Equation a1-37 permits the calculation of the standard deviation from the mean $\bar{y}_c$, of a set of M replicate analyses when a calibration curve that contains N points is used; recall that $\bar{y}$ is the mean value of y for the N calibration data.

EXAMPLE a1-12

Carry out a least-squares analysis of the experimental data provided in the first two columns in Table a1-6 and plotted in Figure a1-6.

Columns 3, 4, and 5 of the table contain computed values for x_i^2, y_i^2, and $x_i y_i$; their sums appear as the last entry of each column. Note that the number of digits carried in the computed values should be the *maximum allowed by the calculator or computer,* and *rounding should not be performed until the calculation is complete.*

We now substitute into Equations a1-29, a1-30, and a1-31 to obtain

$$S_{xx} = \Sigma x_i^2 - (\Sigma x_i)^2/N = 6.90201 - (5.365)^2/5 = 1.14537$$

$$S_{yy} = \Sigma y_i^2 - (\Sigma y_i)^2/N = 36.3775 - (12.51)^2/5 = 5.07748$$

$$S_{xy} = \Sigma x_i y_i - \Sigma x_i \Sigma y_i/N = 15.81992 - 5.365 \times 12.51/5 = 2.39669$$

Substitution of these quantities into Equations a1-32 and a1-33 yields

$$m = 2.39669/1.14537 = 2.0925 = 2.09$$

$$b = \frac{12.51}{5} - 2.0925 \times \frac{5.365}{5} = 0.2567 = 0.26$$

Thus, the equation for the least-squares line is

$$y = 2.09x + 0.26$$

Substitution into Equation a1-34 yields the standard deviation for the residuals:

$$s_r = \sqrt{\frac{S_{yy} - m^2 S_{xx}}{N - 2}}$$

$$= \sqrt{\frac{5.07748 - (2.0925)^2 \times 1.14537}{5 - 2}}$$

$$= 0.144 = 0.14$$

and substitution into Equation a1-35 gives the standard deviation of the slope:

$$s_m = s_r\sqrt{S_{xx}} = 0.144/\sqrt{1.14537} = 0.13$$

The standard deviation of the intercept is obtained from Equation a1-36. Thus

$$s_b = 0.144\sqrt{\frac{1}{5 - (5.365)^2/6.90201}} = 0.16$$

EXAMPLE a1-13

The calibration curve computed in Example a1-12 was used for the chromatographic determination of isooctane in a hydrocarbon mixture. A peak area of 2.65 was obtained. Calculate the mole percent of isooctane and the standard deviation for the result if the area was **(a)** the result of a single measurement and **(b)** the mean of four measurements.

In either case,

$$x = \frac{y - 0.26}{2.09} = \frac{2.65 - 0.26}{2.09} = 1.14 \text{ mol } \%$$

(a) Substituting into Equation a1-37, we obtain

$$s_c = \frac{0.14}{2.09}\sqrt{\frac{1}{1} + \frac{1}{5} + \frac{(2.65 - 12.51/5)^2}{(2.09)^2 \times 1.145}}$$
$$= 0.074 \text{ mol } \%$$

(b) For the mean of four measurements,

$$s_c = \frac{0.14}{2.09}\sqrt{\frac{1}{4} + \frac{1}{5} + \frac{(2.65 - 12.51/5)^2}{(2.09)^2 \times 1.145}}$$
$$= 0.046 \text{ mol } \%$$

TABLE a1-6 Calibration Data for a Chromatographic Method for the Determination of Isooctane in a Hydrocarbon Mixture

Mole Percent Isooctane, x_i	Peak Area, y_i	x_i^2	y_i^2	$x_i y_i$
0.352	1.09	0.12390	1.1881	0.38368
0.803	1.78	0.64481	3.1684	1.42934
1.08	2.60	1.16640	6.7600	2.80800
1.38	3.03	1.90440	9.1809	4.18140
1.75	4.01	3.06250	16.0801	7.01750
5.365	12.51	6.90201	36.3775	15.81992

a1D QUESTIONS AND PROBLEMS

a1-1 Consider the following sets of data:

A	B	C	D
61.45	3.27	12.06	2.7
61.53	3.26	12.14	2.4
61.32	3.24		2.6
	3.24		2.9
	3.28		
	3.23		

Calculate: (a) the mean for each data set, and decide how many degrees of freedom are associated with the calculation of $\bar{x}$; (b) the absolute standard deviation of each set, and decide how many degrees of freedom are associated with the calculation of s; (c) the standard error of the mean of each set; (d) the coefficient of variation for the individual data points from the mean.

a1-2 The accepted value for the quantity that provided each of the sets of data in Problem a1-1 is: A 61.71, B 3.28, C 12.23, D 2.75. Calculate: (a) the absolute error for the mean of each set; (b) the percent relative error for each mean.

a1-3 A particular method for the analysis of copper yields results that are low by 0.5 mg. What will be the percent relative error due to this source if the weight of copper in a sample is (a) 25 mg? (b) 100 mg? (c) 250 mg? (d) 500 mg?

a1-4 The method described in Problem a1-3 is to be used to analyze an ore that contains about 4.8% copper. What minimum sample weight should be taken if the relative error due to a 0.5-mg loss is to be smaller than (a) 0.1%? (b) 0.5%? (c) 0.8%? (d) 1.2%?

a1-5 A certain instrumental technique has a standard deviation of 1.0%. How many replicate measurements are necessary if the standard error of the mean is to be 0.01%?

a1-6 A certain technique is known to have a mean of 0.500 and standard deviation of 1.84×10^{-3}. It is also known that Gaussian statistics apply. How many replicate determinations are necessary if the standard error of the mean is not to exceed 0.100%?

a1-7 A constant solubility loss of approximately 1.8 mg is associated with a particular method for the determination of chromium in geological samples. A sample containing approximately 18% Cr was analyzed by this method. Predict the relative error (in parts per thousand) in the results due to this systematic error if the sample taken for analysis weighed 0.400 g.

a1-8 Following are data from a continuing study of calcium ion in the blood plasma of several individuals:

Subject	Mean Calcium Content, mg/100 mL	Number of Observations	Derivation of Individual Results from Mean Values
1	3.16	5	0.14, 0.09, 0.06, 0.00, 0.11
2	4.08	4	0.07, 0.12, 0.10, 0.01
3	3.75	5	0.13, 0.05, 0.08, 0.14, 0.07
4	3.49	3	0.10, 0.13, 0.07
5	3.32	6	0.07, 0.10, 0.11, 0.03, 0.14, 0.05

(a) Calculate s for each set of values.
(b) Pool the data and calculate s for the analytical method.

a1-9 A method for determining the particulate lead content of air samples is based on drawing a measured quantity of air through a filter and performing the analysis on circles cut from the filter. Calculate the individual values for s as well as a pooled value for the accompanying data.

Sample	μg Pb/m³ Air
1	1.5, 1.2, 1.3
2	2.0, 2.3, 2.3, 2.2
3	1.8, 1.7, 1.4, 1.6
4	1.6, 1.3, 1.2, 1.5, 1.6

a1-10 Estimate the absolute standard deviation and the coefficient of variation for the results of the following calculations. Round each result so that it contains only significant digits. The numbers in parentheses are absolute standard deviations.
(a) $y = 6.75(\pm 0.03) + 0.843(\pm 0.001) - 7.021(\pm 0.001) = 0.572$
(b) $y = 67.1(\pm 0.3) \times 1.03(\pm 0.02) \times 10^{-17} = 6.9113 \times 10^{-16}$
(c) $y = 243(\pm 1) \times \dfrac{760(\pm 2)}{1.006(\pm 0.006)} = 183578.5$

(d) $y = \dfrac{143(\pm 6) - 64(\pm 3)}{1249(\pm 1) + 77(\pm 8)} = 5.9578 \times 10^{-2}$

(e) $y = \dfrac{1.97(\pm 0.01)}{243(\pm 3)} = 8.106996 \times 10^{-3}$

a1-11 Estimate the absolute standard deviation and the coefficient of variation for the results of the following calculations. Round each result to include only significant figures. The numbers in parentheses are absolute standard deviations.

(a) $y = -1.02(\pm 0.02) \times 10^{-7} - 3.54(\pm 0.2) \times 10^{-8} = -1.374 \times 10^{-7}$

(b) $y = 100.20(\pm 0.08) - 99.62(\pm 0.06) + 0.200(\pm 0.004) = 0.780$

(c) $y = 0.0010(\pm 0.0005) \times 18.10(\pm 0.02) \times 200(\pm 1) = 3.62$

(d) $y = [33.33(\pm 0.03)]^3 = 37025.927$

(e) $y = \dfrac{1.73(\pm 0.03) \times 10^{-14}}{1.63(\pm 0.04) \times 10^{-16}} = 106.1349693$

a1-12 Based on extensive past experience, it is known that the standard deviation for an analytical method for gold in sea water is 0.025 ppb. Calculate the 99% confidence limits for an analysis using this method, based on

(a) a single measurement.

(b) three measurements.

(c) five measurements.

a1-13 An established method of analysis for chlorinated hydrocarbons in air samples has a standard deviation of 0.030 ppm.

(a) Calculate the 95% confidence limit for the mean of four measurements obtained by this method.

(b) How many measurements should be made if the 95% confidence limit is to be ±0.017?

a1-14 The standard deviation in a method for the analysis of carbon monoxide in automotive exhaust gases has been found, on the basis of extensive past experience, to be 0.80 ppm.

(a) Estimate the 90% confidence limit for a triplicate analysis.

(b) How many measurements would be needed for the 90% confidence limit for the set to be 0.50 ppm?

a1-15 The certified percentage of nickel in a particular NIST reference steel sample is 1.12%, and the standard deviation associated with this number is 0.03%. A new spectrometric method for the determination of nickel produced the following percentages: 1.10, 1.08, 1.09, 1.12, 1.09. Is there an indication of bias in the method at the 95% level?

a1-16 A titrimetric method for the determination of calcium in limestone was tested by analysis of an NIST limestone containing 30.15% CaO. The mean result of four analyses was 30.26% CaO, with a standard deviation of 0.085%. By pooling data from several analyses, it was established that $s \rightarrow \sigma = 0.094\%$ CaO.

(a) Do the data indicate the presence of a determinate error at the 95% confidence level?

(b) Do the data indicate the presence of a determinate error at the 95% confidence level if no pooled value for σ was available?

a1-17 In order to test the quality of the work of a commercial laboratory, duplicate analyses of a purified benzoic acid (68.8% C, 4.953% H) sample was requested. It is assumed that the relative standard deviation of the method is $s_r \rightarrow \sigma_r = 4$ ppt for carbon and 6 ppt for hydrogen. The means of the reported results are 68.5% C and 4.882% H.

At the 95% confidence level, is there any indication of determinate error in either analysis?

a1-18 The diameter of a sphere has been found to be 2.15 cm, and the standard deviation associated with the mean is 0.02 cm. What is the best estimate of the volume of the sphere, and what is the standard deviation associated with the volume?

a1-19 A given pH meter can be read with a standard deviation of ±0.01 pH units throughout the range 2 to 12. Calculate the standard deviation of $[H_3O^+]$ at each end of this range.

a1-20 A solution is prepared by weighing 5.0000 g of compound X into a 100-mL volumetric flask. The balance could be used with a precision of 0.2 mg reported as a standard deviation, and the volumetric flask could be filled with a precision of 0.15 mL also reported as a standard deviation. What is the estimated standard deviation of concentration in g/mL?

a1-21 Estimate the absolute standard deviation in the result computed from the following operations (the numbers in parentheses are absolute standard deviations for the numbers they follow); report the result to the appropriate number of significant figures.
(a) $x = \log 878(\pm 4) = 2.94349$
(b) $x = \log 0.4957(\pm 0.0004) = -0.30478$
(c) $p = \text{antilogarithm } 3.64(\pm 0.01) = 4365.16$
(d) $p = \text{antilogarithm } -7.191(\pm 0.002) = 6.44169 \times 10^{-8}$

a1-22 The sulfate ion concentration in natural water can be determined by measuring the turbidity that results when an excess of $BaCl_2$ is added to a measured quantity of the sample. A turbidimeter, the instrument used for this analysis, was calibrated with a series of standard Na_2SO_4 solutions. The following data were obtained in the calibration:

mg SO_4^{2-}/L, c_x	Turbidimeter Reading, R
0.00	0.06
5.00	1.48
10.00	2.28
15.0	3.98
20.0	4.61

Assume that a linear relationship exists between the instrument reading and concentration.
(a) Plot the data and draw a straight line through the points by eye.
(b) Determine the parameters of a least-squares equation for the relationship between the variables.
(c) Compare the straight line from the relationship computed in (b) with that in (a).
(d) Calculate the standard deviation for the slope and the intercept for the least-squares line.
(e) Calculate the concentration of sulfate in a sample yielding a turbidimeter reading of 3.67. Calculate the absolute standard deviation of the result and the coefficient of variation.
(f) Repeat the calculations in (e), assuming that the 3.67 was a mean of six turbidimeter readings.

a1-23 The following data were obtained in calibrating a calcium ISE for the determination of pCa. Assume that there is a linear relationship between the potential *E* and pCa.

pCa	*E*, mV
5.00	−53.8
4.00	−27.7
3.00	+2.7
2.00	+31.9
1.00	+65.1

(a) Plot the data and draw a line through the points by eye.
(b) Calculate the parameters of a least-squares expression for the best straight line through the points. Plot this line.
(c) Calculate the standard deviation for the slope and the intercept of the least-squares line.
(d) Calculate the pCa of a serum solution in which the electrode potential was 20.3 mV. Calculate the absolute and relative standard eviations for pCa if the result was from a single voltage measurement.
(e) Calculate the absolute and relative standard deviations for pCa if the millivolt reading in (d) was the mean of two replicate measurements. Repeat the calculation based on the mean of eight measurements.
(f) Calculate the molar calcium ion concentration for the sample described in (d).
(g) Calculate the absolute and relative standard deviations in the calcium ion concentration if the measurement was performed as described in (e).

a1-24 The following are relative peak areas for chromatograms of standard solutions of methyl vinyl ketone (MVK).

Concn MVK, mmol/L	Relative Peak Area
0.500	3.76
1.50	9.16
2.50	15.03
3.50	20.42
4.50	25.33
5.50	31.97

(a) Calculate the parameters of a least-squares expression, assuming the variables bear a linear relationship to one another.
(b) Plot the least-squares line as well as the experimental points.
(c) Calculate the standard deviation of the slope and intercept of the line.
(d) Two samples containing MVK yielded relative peak areas of 6.3 and 27.5. Calculate the concentration of MVK in each solution.
(e) Assume that the results in (d) represent a single measurement as well as the mean of four measurements. Calculate the respective absolute and relative standard deviations.

Appendix 2

Activity Coefficients

The relationship between the *activity* a_X of a species and its molar concentration [X] is given by the expression

$$a_X = \gamma_X[X] \quad \text{(a2-1)}$$

where γ_X is a dimensionless quantity called the *activity coefficient.* The activity coefficient, and thus the activity of X, varies with the *ionic strength* of a solution such that the use of a_X instead of [X] in an electrode potential calculation, or in the other equilibrium calculations, renders the numerical value obtained independent of the ionic strength. Here, the ionic strength μ is defined by the equation

$$\mu = \frac{1}{2}(c_1Z_1^2 + c_2Z_2^2 + c_3Z_3^2 + \cdots) \quad \text{(a2-2)}$$

where $c_1, c_2, c_3, \cdots$ represent the molar concentration of the various ions in the solution and $Z_1, Z_2, Z_3, \cdots$ are their respective charges. Note that an ionic strength calculation requires taking account of *all* ionic species in a solution, not just the reactive ones.

EXAMPLE a2-1

Calculate the ionic strength of a solution that is 0.0100 M in $NaNO_3$ and 0.0200 M in $Mg(NO_3)_2$.

Here, we will neglect the contribution of H^+ and OH^- to the ionic strength since their concentrations are so low compared with those of the two salts. The molarities of Na^+, NO_3^-, and Mg^{2+} are 0.0100, 0.0500, and 0.0200, respectively. Then

$$c_{Na^+} \times (1)^2 = 0.0100 \times 1 = 0.0100$$

$$c_{NO_3^-} \times (1)^2 = 0.0500 \times 1 = 0.0500$$

$$c_{Mg^{2+}} \times (2)^2 = 0.0200 \times 2^2 = \underline{0.0800}$$

$$\text{Sum} = 0.1400$$

$$\mu = \frac{1}{2} \times 0.1400 = 0.0700$$

a2A PROPERTIES OF ACTIVITY COEFFICIENTS

Activity coefficients have the following properties:

1. The activity coefficient of a species can be thought of as a measure of the effectiveness with which that species influences an equilibrium in which it is a participant. In very dilute solutions, where the ionic strength is minimal, ions are sufficiently far apart that they do not influence one another's behavior. Here, the effectiveness of a common ion on the position of equilibrium becomes dependent only on its molar concentration and independent of other ions. Under these circumstances, the activity coefficient becomes equal to unity and [X] and a in Equation a2-1 are identical. As the ionic strength becomes greater, the behavior of an individual ion is influenced by its nearby neighbors. The result is a decrease in effectiveness of the ion in altering the position of chemical equilibria. Its activity coefficient then becomes less than unity. We may summarize this behavior in terms of Equation a2-1. At moderate ionic strengths, $\gamma_X < 1$; as the solution approaches infinite dilution ($\mu \rightarrow 0$), $\gamma_X \rightarrow 1$ and thus $a_X \rightarrow [X]$.

At high ionic strengths, the activity coefficients for some species increase and may even become greater than one. The behavior of such solutions is difficult to interpret; we shall confine our discussion to regions of low to moderate ionic strengths (that is, where $\mu < 0.1$). The variation of typical activity coefficients as a function of ionic strength is shown in Figure a2-1.

2. In dilute solutions, the activity coefficient for a given species is independent of the specific nature of the electrolyte and depends only on the ionic strength.
3. For a given ionic strength, the activity coefficient of an ion departs further from unity as the charge carried by the species increases. This effect is shown in Figure a2-1. The activity coefficient of an uncharged molecule is approximately one, regardless of ionic strength.
4. Activity coefficients for ions of the same charge are approximately the same at any given ionic strength. The small variations that do exist can be correlated with the effective diameter of the hydrated ions.

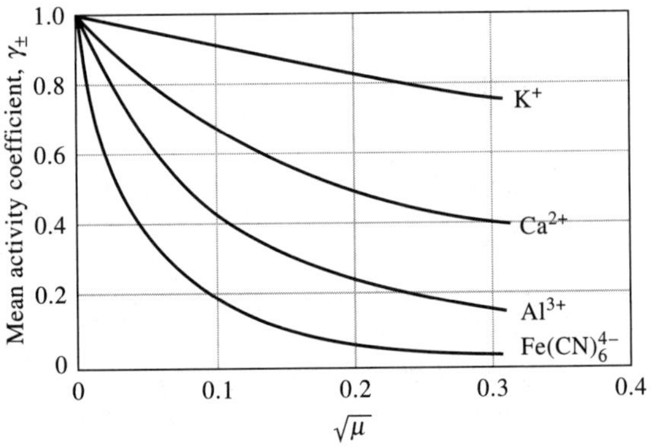

Figure a2-1 Effect of ionic strength on activity coefficients.

5. The product of the activity coefficient and molar concentration of a given ion describes its effective behavior in all equilibria in which it participates.

a2B EXPERIMENTAL EVALUATION OF ACTIVITY COEFFICIENTS

Although activity coefficients for individual ions can be calculated from theoretical considerations, their experimental measurement is, unfortunately, impossible. Instead, only a mean activity coefficient for the positively and negatively charged species in a solution can be derived.

For the electrolyte A_mB_n, the mean activity coefficient $\gamma_\pm$ is defined by the equation

$$\gamma_\pm = (\gamma_A^m \gamma_B^n)^{1/(m+n)}$$

The mean activity coefficient can be measured in any of several ways, but it is impossible experimentally to resolve this term into the individual activity coefficients γ_A and γ_B. For example, if A_mB_n is a precipitate, we can write

$$K_{sp} = [A]^m[B]^n\gamma_A^m\gamma_B^n$$
$$= [A]^m[B]^n\gamma_\pm^{(m+n)}$$

By measuring the solubility of A_mB_n in a solution in which the electrolyte concentration approaches zero (that is, where γ_A and $\gamma_B \rightarrow 1$), we could obtain K_{sp}. A second solubility measurement at some ionic strength, μ_1 would give values for [A] and [B]. These data would then permit the calculation of $\gamma_A^m\gamma_B^n = \gamma_\pm^{(m+n)}$ for ionic strength μ_1. It is important to understand that there are insufficient experimental data to permit the calculation of the *individual* quantities γ_A and γ_B, however, and that there appears to be no additional experimental information that would permit evaluation of these quantities. This situation is general; the *experimental* determination of individual activity coefficients appears to be impossible.

a2C THE DEBYE-HÜCKEL EQUATION

In 1923, P. Debye and E. Hückel derived the following theoretical expression, which permits the calculation of activity coefficients of ions:[1]

$$-\log \gamma_A = \frac{0.509\, Z_A^2 \sqrt{\mu}}{1 + 3.28\, \alpha_A \sqrt{\mu}} \qquad \text{(a2-3)}$$

[1] P. Debye and E. Hückel, *Physik. Z.*, **1923**, *24*, 185.

where

γ_A = activity coefficient of the species A
Z_A = charge on the species A
μ = ionic strength of the solution
α_A = the effective diameter of the hydrated ion in angstrom units

The constants 0.509 and 0.328 are applicable to solutions at 25°C; other values must be used at different temperatures.

Unfortunately, there is considerable uncertainty in the magnitude of α_A in Equation a2-3. Its value appears to be approximately 3 Å for most singly charged ions so that for these species, the denominator of the Debye-Hückel equation reduces to approximately $(1 + \sqrt{\mu})$. For ions with higher charge, α_A may be larger than 10 Å. It should be noted that the second term of the denominator becomes small with respect to the first when the ionic strength is less than 0.01; under these circumstances, uncertainties in α_A are of little significance in calculating activity coefficients.

Kielland[2] has calculated values of α_A for numerous ions from a variety of experimental data. His "best values" for effective diameters are given in Table a2-1. Also presented are activity coefficients calculated from Equation a2-3 using these values for the size parameter.

For ionic strengths up to about 0.01, activity coefficients from the Debye-Hückel equation lead to results from equilibrium calculations that agree closely with experiment; even at ionic strengths of 0.1, major discrepancies are generally not encountered. At higher ionic strengths, however, the equation fails, and experimentally determined mean activity coefficients must be used. Unfortunately, many electrochemical calculations involve solutions of high ionic strength for which no experimental activity coefficients are available. Concentrations must thus be used instead of activities; uncertainties that vary from a few percent relative to an order of magnitude may be expected.

[2] J. Kielland, *J. Amer. Chem. Soc.*, **1937**, *59*, 1675.

TABLE a2-1 Activity Coefficients for Ions at 25°C[a]

Ion	Activity Coefficients at Indicated Ionic Strength					
	α_X, nm	0.001	0.005	0.01	0.05	0.1
H_3O^+	0.9	0.967	0.933	0.914	0.86	0.83
Li^+, $C_6H_5COO^-$	0.6	0.965	0.929	0.907	0.84	0.80
Na^+, IO_3^-, HSO_3^-, HCO_3^-, $H_2PO_4^-$, $H_2AsO_4^-$, OAc^-	0.4–0.45	0.964	0.928	0.902	0.82	0.78
OH^-, F^-, SCN^-, HS^-, ClO_3^-, ClO_4^-, BrO_3^-, IO_4^-, MnO_4^-	0.35	0.964	0.926	0.900	0.81	0.76
K^+, Cl^-, Br^-, I^-, CN^-, NO_2^-, NO_3^-, $HCOO^-$	0.3	0.964	0.925	0.899	0.80	0.76
Rb^+, Cs^+, Tl^+, Ag^+, NH_4^+	0.25	0.964	0.924	0.898	0.80	0.75
Mg^{2+}, Be^{2+}	0.8	0.872	0.755	0.69	0.52	0.45
Ca^{2+}, Cu^{2+}, Zn^{2+}, Sn^{2+}, Mn^{2+}, Fe^{2+}, Ni^{2+}, Co^{2+}, Phthalate^{2-}	0.6	0.870	0.749	0.675	0.48	0.40
Sr^{2+}, Ba^{2+}, Cd^{2+}, Hg^{2+}, S^{2-}	0.5	0.868	0.744	0.67	0.46	0.38
Pb^{2+}, CO_3^{2-}, SO_3^{2-}, $C_2O_4^{2-}$	0.45	0.868	0.742	0.665	0.46	0.37
Hg_2^{2+}, SO_4^{2-}, $S_2O_3^{2-}$, CrO_4^{2-}, HPO_4^{2-}	0.40	0.867	0.740	0.660	0.44	0.36
Al^{3+}, Fe^{3+}, Cr^{3+}, La^{3+}, Ce^{3+}	0.9	0.738	0.54	0.44	0.24	0.18
PO_4^{3-}, $Fe(CN)_6^{3-}$	0.4	0.725	0.50	0.40	0.16	0.095
Th^{4+}, Zr^{4+}, Ce^{4+}, Sn^{4+}	1.1	0.588	0.35	0.255	0.10	0.065
$Fe(CN)_6^{4-}$	0.5	0.57	0.31	0.20	0.048	0.021

[a]From J. Kielland, *J. Am. Chem. Soc.*, **1937,** *59,* 1675. By courtesy of the American Chemical Society.

Appendix 3

Some Standard and Formal Electrode Potentials

Half-Reaction	E^0, V[a]	Formal Potential, V[b]
Aluminum		
$Al^{3+} + 3e^- \rightleftharpoons Al(s)$	−1.662	
Antimony		
$Sb_2O_5(s) + 6H^+ + 4e^- \rightleftharpoons 2SbO^+ + 3H_2O$	+0.581	
Arsenic		
$H_3AsO_4 + 2H^+ + 2e^- \rightleftharpoons H_3AsO_3 + H_2O$	+0.559	0.577 in 1 M HCl, $HClO_4$
Barium		
$Ba^{2+} + 2e^- \rightleftharpoons Ba(s)$	−2.906	
Bismuth		
$BiO^+ + 2H^+ + 3e^- \rightleftharpoons Bi(s) + H_2O$	+0.320	
$BiCl_4^- + 3e^- \rightleftharpoons Bi(s) + 4Cl^-$	+0.16	
Bromine		
$Br_2(l) + 2e^- \rightleftharpoons 2Br^-$	+1.065	1.05 in 4 M HCl
$Br_2(aq) + 2e^- \rightleftharpoons 2Br^-$	+1.087[c]	
$BrO_3^- + 6H^+ + 5e^- \rightleftharpoons \frac{1}{2}Br_2(l) + 3H_2O$	+1.52	
$BrO_3^- + 6H^+ + 6e^- \rightleftharpoons Br^- + 3H_2O$	+1.44	
Cadmium		
$Cd^{2+} + 2e^- \rightleftharpoons Cd(s)$	−0.403	
Calcium		
$Ca^{2+} + 2e^- \rightleftharpoons Ca(s)$	−2.866	

Half-Reaction	E^0, V[a]	Formal Potential, V[b]
Carbon		
$C_6H_4O_2$ (quinone) + $2H^+ + 2e^- \rightleftharpoons C_6H_4(OH)_2$	+0.699	0.696 in 1 M HCl, $HClO_4$, H_2SO_4
$2CO_2(g) + 2H^+ + 2e^- \rightleftharpoons H_2C_2O_4$	−0.49	
Cerium		
$Ce^{4+} + e^- \rightleftharpoons Ce^{3+}$		+1.70 in 1 M $HClO_4$; +1.61 in 1 M HNO_3; +1.44 in 1 M H_2SO_4
Chlorine		
$Cl_2(g) + 2e^- \rightleftharpoons 2Cl^-$	+1.359	
$HClO + H^+ + e^- \rightleftharpoons \frac{1}{2}Cl_2(g) + H_2O$	+1.63	
$ClO_3^- + 6H^+ + 5e^- \rightleftharpoons \frac{1}{2}Cl_2(g) + 3H_2O$	+1.47	
Chromium		
$Cr^{3+} + e^- \rightleftharpoons Cr^{2+}$	−0.408	
$Cr^{3+} + 3e^- \rightleftharpoons Cr(s)$	−0.744	
$Cr_2O_7^{2-} + 14H^+ + 6e^- \rightleftharpoons 2Cr^{3+} + 7H_2O$	+1.33	
Cobalt		
$Co^{2+} + 2e^- \rightleftharpoons Co(s)$	−0.277	
$Co^{3+} + e^- \rightleftharpoons Co^{2+}$	+1.808	
Copper		
$Cu^{2+} + 2e^- \rightleftharpoons Cu(s)$	+0.337	
$Cu^{2+} + e^- \rightleftharpoons Cu^+$	+0.153	
$Cu^+ + e^- \rightleftharpoons Cu(s)$	+0.521	
$Cu^{2+} + I^- + e^- \rightleftharpoons CuI(s)$	+0.86	
$CuI(s) + e^- \rightleftharpoons Cu(s) + I^-$	−0.185	
Fluorine		
$F_2(g) + 2H^+ + 2e^- \rightleftharpoons 2HF(aq)$	+3.06	
Hydrogen		
$2H^+ + 2e^- \rightleftharpoons H_2(g)$	0.000	−0.005 in 1 M HCl, $HClO_4$
Iodine		
$I_2(s) + 2e^- \rightleftharpoons 2I^-$	+0.5355	
$I_2(aq) + 2e^- \rightleftharpoons 2I^-$	+0.615[c]	
$I_3^- + 2e^- \rightleftharpoons 3I^-$	+0.536	
$ICl_2^- + e^- \rightleftharpoons \frac{1}{2}I_2(s) + 2Cl^-$	+1.056	
$IO_3^- + 6H^+ + 5e^- \rightleftharpoons \frac{1}{2}I_2(s) + 3H_2O$	+1.196	
$IO_3^- + 6H^+ + 5e^- \rightleftharpoons \frac{1}{2}I_2(aq) + 3H_2O$	+1.178[c]	
$IO_3^- + 2Cl^- + 6H^+ + 4e^- \rightleftharpoons ICl_2^- + 3H_2O$	+1.24	
$H_5IO_6 + H^+ + 2e^- \rightleftharpoons IO_3^- + 3H_2O$	+1.601	
Iron		
$Fe^{2+} + 2e^- \rightleftharpoons Fe(s)$	−0.440	
$Fe^{3+} + e^- \rightleftharpoons Fe^{2+}$	+0.771	0.700 in 1 M HCl; 0.732 in 1 M $HClO_4$; 0.68 in 1 M H_2SO_4
$Fe(CN)_6^{3-} + e^- \rightleftharpoons Fe(CN)_6^{4-}$	+0.36	0.71 in 1 M HCl; 0.72 in 1 M $HClO_4$, H_2SO_4

Half-Reaction	E^0, V[a]	Formal Potential, V[b]
Lead		
$Pb^{2+} + 2e^- \rightleftharpoons Pb(s)$	−0.126	−0.14 in 1 M $HClO_4$; −0.29 in 1 M H_2SO_4
$PbO_2(s) + 4H^+ + 2e^- \rightleftharpoons Pb^{2+} + 2H_2O$	+1.455	
$PbSO_4(s) + 2e^- \rightleftharpoons Pb(s) + SO_4^{2-}$	−0.350	
Lithium		
$Li^+ + e^- \rightleftharpoons Li(s)$	−3.045	
Magnesium		
$Mg^{2+} + 2e^- \rightleftharpoons Mg(s)$	−2.363	
Manganese		
$Mn^{2+} + 2e^- \rightleftharpoons Mn(s)$	−1.180	
$Mn^{3+} + e^- \rightleftharpoons Mn^{2+}$		1.51 in 7.5 M H_2SO_4
$MnO_2(s) + 4H^+ + 2e^- \rightleftharpoons Mn^{2+} + 2H_2O$	+1.23	
$MnO_4^- + 8H^+ + 5e^- \rightleftharpoons Mn^{2+} + 4H_2O$	+1.51	
$MnO_4^- + 4H^+ + 3e^- \rightleftharpoons MnO_2(s) + 2H_2O$	+1.695	
$MnO_4^- + e^- \rightleftharpoons MnO_4^{2-}$	+0.564	
Mercury		
$Hg_2^{2+} + 2e^- \rightleftharpoons 2Hg(l)$	+0.788	0.274 in 1 M HCl; 0.776 in 1 M $HClO_4$; 0.674 in 1 M H_2SO_4
$2Hg^{2+} + 2e^- \rightleftharpoons Hg_2^{2+}$	+0.920	0.907 in 1 M $HClO_4$
$Hg^{2+} + 2e^- \rightleftharpoons Hg(l)$	+0.854	
$Hg_2Cl_2(s) + 2e^- \rightleftharpoons 2Hg(l) + 2Cl^-$	+0.268	0.244 in sat'd KCl; 0.282 in 1 M KCl; 0.334 in 0.1 M KCl
$Hg_2SO_4(s) + 2e^- \rightleftharpoons 2Hg(l) + SO_4^{2-}$	+0.615	
Nickel		
$Ni^{2+} + 2e^- \rightleftharpoons Ni(s)$	−0.250	
Nitrogen		
$N_2(g) + 5H^+ + 4e^- \rightleftharpoons N_2H_5^+$	−0.23	
$HNO_2 + H^+ + e^- \rightleftharpoons NO(g) + H_2O$	+1.00	
$NO_3^- + 3H^+ + 2e^- \rightleftharpoons HNO_2 + H_2O$	+0.94	0.92 in 1 M HNO_3
Oxygen		
$H_2O_2 + 2H^+ + 2e^- \rightleftharpoons 2H_2O$	+1.776	
$HO_2^- + H_2O + 2e^- \rightleftharpoons 3OH^-$	+0.88	
$O_2(g) + 4H^+ + 4e^- \rightleftharpoons 2H_2O$	+1.229	
$O_2(g) + 2H^+ + 2e^- \rightleftharpoons H_2O_2$	+0.682	
$O_3(g) + 2H^+ + 2e^- \rightleftharpoons O_2(g) + H_2O$	+2.07	
Palladium		
$Pd^{2+} + 2e^- \rightleftharpoons Pd(s)$	+0.987	
Platinum		
$PtCl_4^{2-} + 2e^- \rightleftharpoons Pt(s) + 4Cl^-$	+0.73	
$PtCl_6^{2-} + 2e^- \rightleftharpoons PtCl_4^{2-} + 2Cl^-$	+0.68	
Potassium		
$K^+ + e^- \rightleftharpoons K(s)$	−2.925	

Half-Reaction	E^0, V[a]	Formal Potential, V[b]
Selenium		
$H_2SeO_3 + 4H^+ + 4e^- \rightleftharpoons Se(s) + 3H_2O$	+0.740	
$SeO_4^{2-} + 4H^+ + 2e^- \rightleftharpoons H_2SeO_3 + H_2O$	+1.15	
Silver		
$Ag^+ + e^- \rightleftharpoons Ag(s)$	+0.799	0.228 in 1 M HCl; 0.792 in 1 M $HClO_4$; 0.77 in 1 M H_2SO_4
$AgBr(s) + e^- \rightleftharpoons Ag(s) + Br^-$	+0.073	
$AgCl(s) + e^- \rightleftharpoons Ag(s) + Cl^-$	+0.222	0.228 in 1 M KCl
$Ag(CN)_2^- + e^- \rightleftharpoons Ag(s) + 2CN^-$	−0.31	
$Ag_2CrO_4(s) + 2e^- \rightleftharpoons 2Ag(s) + CrO_4^{2-}$	+0.446	
$AgI(s) + e^- \rightleftharpoons Ag(s) + I^-$	−0.151	
$Ag(S_2O_3)_2^{3-} + e^- \rightleftharpoons Ag(s) + 2S_2O_3^{2-}$	+0.017	
Sodium		
$Na^+ + e^- \rightleftharpoons Na(s)$	−2.714	
Sulfur		
$S(s) + 2H^+ + 2e^- \rightleftharpoons H_2S(g)$	+0.141	
$H_2SO_3 + 4H^+ + 4e^- \rightleftharpoons S(s) + 3H_2O$	+0.450	
$SO_4^{2-} + 4H^+ + 2e^- \rightleftharpoons H_2SO_3 + H_2O$	+0.172	
$S_4O_6^{2-} + 2e^- \rightleftharpoons 2S_2O_3^{2-}$	+0.08	
$S_2O_8^{2-} + 2e^- \rightleftharpoons 2SO_4^{2-}$	+2.01	
Thallium		
$Tl^+ + e^- \rightleftharpoons Tl(s)$	−0.336	−0.551 in 1 M HCl; −0.33 in 1 M $HClO_4$, H_2SO_4
$Tl^{3+} + 2e^- \rightleftharpoons Tl^+$	+1.25	0.77 in 1 M HCl
Tin		
$Sn^{2+} + 2e^- \rightleftharpoons Sn(s)$	−0.136	−0.16 in 1 M $HClO_4$
$Sn^{4+} + 2e^- \rightleftharpoons Sn^{2+}$	+0.154	0.14 in 1 M HCl
Titanium		
$Ti^{3+} + e^- \rightleftharpoons Ti^{2+}$	−0.369	
$TiO^{2+} + 2H^+ + e^- \rightleftharpoons Ti^{3+} + H_2O$	+0.099	0.04 in 1 M H_2SO_4
Uranium		
$UO_2^{2+} + 4H^+ + 2e^- \rightleftharpoons U^{4+} + 2H_2O$	+0.334	
Vanadium		
$V^{3+} + e^- \rightleftharpoons V^{2+}$	−0.256	0.21 in 1 M $HClO_4$
$VO^{2+} + 2H^+ + e^- \rightleftharpoons V^{3+} + H_2O$	+0.359	
$V(OH)_4^+ + 2H^+ + e^- \rightleftharpoons VO^{2+} + 3H_2O$	+1.00	1.02 in 1 M HCl, $HClO_4$
Zinc		
$Zn^{2+} + 2e^- \rightleftharpoons Zn(s)$	−0.763	

[a]G. Milazzo, S. Caroli, and V. K. Sharma, *Tables of Standard Electrode Potentials.* London: Wiley, 1978.

[b]E. H. Swift and E. A. Butler, *Quantitative Measurements and Chemical Equilibria.* New York: Freeman, 1972.

[c]These potentials are hypothetical because they correspond to solutions that are 1.00 M in Br_2 or I_2. The solubilities of these two compounds at 25°C are 0.18 M and 0.0020 M, respectively. In saturated solutions containing an excess of $Br_2(l)$ or $I_2(s)$, the standard potentials for the half-reaction $Br_2(l) + 2e^- \rightleftharpoons 2Br^-$ or $I_2(s) + 2e^- \rightleftharpoons 2I^-$ should be used. In contrast, at Br_2 and I_2 concentrations less than saturation, these hypothetical electrode potentials should be employed.

Appendix

4

Compounds Recommended for the Preparation of Standard Solutions of Some Common Elements[a]

Element	Compound	FW	Solvent[b]	Notes
Aluminum	Al metal	26.98	Hot dil HCl	a
Antimony	$KSbOC_4H_4O_6 \cdot \frac{1}{2}H_2O$	333.93	H_2O	c
Arsenic	As_2O_3	197.84	dil HCl	i,b,d
Barium	$BaCO_3$	197.35	dil HCl	
Bismuth	Bi_2O_3	465.96	HNO_3	
Boron	H_3BO_3	61.83	H_2O	d,e
Bromine	KBr	119.01	H_2O	a
Cadmium	CdO	128.40	HNO_3	
Calcium	$CaCO_3$	100.09	dil HCl	i
Cerium	$(NH_4)_2Ce(NO_3)_6$	548.23	H_2SO_4	
Chromium	$K_2Cr_2O_7$	294.19	H_2O	i,d
Cobalt	Co metal	58.93	HNO_3	a
Copper	Cu metal	63.55	dil HNO_3	a
Fluorine	NaF	41.99	H_2O	b
Iodine	KIO_3	214.00	H_2O	i
Iron	Fe metal	55.85	HCl, hot	a

Element	Compound	FW	Solvent[b]	Notes
Lanthanum	La_2O_3	325.82	HCl, hot	f
Lead	$Pb(NO_3)_2$	331.20	H_2O	a
Lithium	Li_2CO_3	73.89	HCl	a
Magnesium	MgO	40.31	HCl	
Manganese	$MnSO_4 \cdot H_2O$	169.01	H_2O	g
Mercury	$HgCl_2$	271.50	H_2O	b
Molybdenum	MoO_3	143.94	1 M NaOH	
Nickel	Ni metal	58.70	HNO_3, hot	a
Phosphorus	KH_2PO_4	136.09	H_2O	
Potassium	KCl	74.56	H_2O	a
	$KHC_8H_4O_4$	204.23	H_2O	i,d
	$K_2Cr_2O_7$	294.19	H_2O	i,d
Silicon	Si metal	28.09	NaOH, concd	
	SiO_2	60.08	HF	j
Silver	$AgNO_3$	169.87	H_2O	a
Sodium	NaCl	58.44	H_2O	i
	$Na_2C_2O_4$	134.00	H_2O	i,d
Strontium	$SrCO_3$	147.63	HCl	a
Sulfur	K_2SO_4	174.27	H_2O	
Tin	Sn metal	118.69	HCl	
Titanium	Ti metal	47.90	H_2SO_4, 1:1	a
Tungsten	$Na_2WO_4 \cdot 2H_2O$	329.86	H_2O	h
Uranium	U_3O_8	842.09	HNO_3	d
Vanadium	V_2O_5	181.88	HCl, hot	
Zinc	ZnO	81.37	HCl	a

[a]The data in this table were taken from a more complete list assembled by B. W. Smith and M. L. Parsons. *J. Chem. Educ.*, **1973,** *50,* 679. Unless otherwise specified, compounds should be dried to constant weight at 110°C.

[b]Unless otherwise specified, acids are concentrated analytical grade.

a Approaches primary standard quality.

b Highly toxic.

c Loses ½ H_2O at 110°C. After drying, fw = 324.92. The dried compound should be weighed quickly after removal from the desiccator.

d Available as a primary standard from the National Institute of Standards and Technology.

e H_3BO_3 should be weighed directly from the bottle. It loses 1 H_2O at 100°C and is difficult to dry to constant weight.

f Absorbs CO_2 and H_2O. Should be ignited just before use.

g May be dried at 110°C without loss of water.

h Loses both waters at 110°C, fw = 293.82. Keep in desiccator after drying.

i Primary standard.

j HF highly toxic and dissolves glass.

Appendix

5

Acronyms and Abbreviations of Significance in Analytical Chemistry

Abbreviations and acronyms are ubiquitous in analytical chemistry. This appendix is a compilation of abbreviations and acronyms that are used routinely in analytical chemistry, including many that are encountered in this text. Notice that acronyms for basic techniques are often combined to represent a more sophisticated technique. Thus, FT (Fourier transform) is combined with MS (mass spectrometry), IR (infrared), or NMR (nuclear magnetic resonance) to give FTMS, FTIR, and FTNMR, respectively. Some of these combined terms have separate listings here. You may notice some of the following trends among the acronyms and abbreviations in this collection.

- Those ending in C usually indicate a type of chromatography; for example, GC means gas chromatography.
- Those ending in D often stand for a type of chromatographic detector; for example, FID indicates flame ionization detector.
- Those ending in E usually denote a type of electrode; for example, SCE signifies saturated calomel electrode.
- If you know one acronym, you often know parts of several other acronyms because a particular analytical technique often appears in several forms with only slightly different names. Hence, the acronyms for the different forms may be similar. For example, gas chromatographic open tubular (OT) columns have been used in several forms, wall coated (WCOT), support coated (SCOT), and porous layer (PLOT). In the case of signal-to-noise ratio, two different acronyms have the same meaning: SNR and S/N. The opposite situation occurs with the acronym AES, which can stand for either atomic emission spectroscopy or Auger electron spectrometry. These techniques are very different in theory, instrumentation, and application, and so, within the context of a particular field or type of measurement, there should be little confusion between the abbreviations.

Acronym or Abbreviation	Name
AAS	**A**tomic **A**bsorption **S**pectroscopy
ACS	**A**merican **C**hemical **S**ociety
ADC	**A**nalog-to-**D**igital **C**onverter

Acronym or Abbreviation	Name
AED	**A**tomic **E**mission **D**etector
AES	**A**tomic **E**mission **S**pectroscopy
AES	**A**uger **E**lectron **S**pectroscopy
AFM	**A**tomic **F**orce **M**icroscopy
AFS	**A**tomic **F**luorescence **S**pectroscopy
ALU	**A**rithmetic **L**ogic **U**nit
amu	**A**tomic **M**ass **U**nit
AOAC	**A**ssociation of **O**fficial **A**nalytical **C**hemists
AOTF	**A**cousto-**O**ptic **T**unable **F**ilter
ASCII	**A**merican **S**tandard **C**ode for **I**nformation **I**nterchange (computers)
ASTM	**A**merican **S**ociety for **T**esting **M**aterials
ASV	**A**nodic **S**tripping **V**oltammetry
ATR	**A**ttenuated **T**otal **R**eflectance
BCD	**B**inary-**C**oded **D**ecimal
BJT	**B**ipolar **J**unction **T**ransistor
BUN	**B**lood **U**rea **N**itrogen
CARS	**C**oherent **A**nti-**S**tokes **R**aman **S**pectroscopy
CCD	**C**harge-**C**oupled **D**evice
CE	**C**apillary **E**lectrophoresis
CGE	**C**apillary **G**el **E**lectrophoresis
CHC	**C**hlorinated **H**ydro**C**arbons
CI	**C**hemical **I**onization source
CID	**C**harge-**I**njection **D**evice
CIEF	**C**apillary **I**so**E**lectric **F**ocusing electrophoresis
CITP	**C**apillary **I**so**T**acho**P**horesis
CL	**C**hemi**L**uminescence
CL	**C**onfidence **L**evel
CL	**C**onfidence **L**imit
CMC	**C**ritical **M**icelle **C**oncentration
CMRR	**C**ommon **M**ode **R**ejection **R**atio
CPU	**C**entral **P**rocessing **U**nit
CRT	**C**athode **R**ay **T**ube

Continued

Acronym or Abbreviation	Name
CSP	**C**hiral **S**tationary **P**hase partition chromatography
CT	**C**omputerized **T**omography
CTD	**C**harge **T**ransfer **D**etector
CV	**C**yclic **V**oltammetry
CVD	**C**hemical **V**apor **D**eposition
CW	**C**ontinuous **W**ave (laser)
CZE	**C**apillary **Z**one **E**lectrophoresis
DAC	**D**igital-to-**A**nalog **C**onverter
DBR	**D**istributed **B**ragg **R**eflector
DCP	**D**irect **C**urrent **P**lasma
DCPMS	**D**irect **C**urrent **P**lasma **M**ass **S**pectrometry
DCU	**D**ecade **C**ounting **U**nit
DMCS	**D**i**M**ethyl**C**hloro**S**ilane
DME	**D**ropping **M**ercury **E**lectrode
DMM	**D**igital **M**ulti**M**eter
DNA	**D**eoxyribo**N**ucleic **A**cid
DPV	**D**ifferential **P**ulse **V**oltammetry
DRIFTS	**D**iffuse **R**eflectance **I**nfrared **F**ourier **T**ransform **S**pectroscopy
DSC	**D**ifferential **S**canning **C**alorimetry
DTA	**D**ifferential **T**hermal **A**nalysis
DVM	**D**igital **V**olt**M**eter
EAAS	**E**lectrothermal **A**tomic **A**bsorption **S**pectrometry
EAROM	**E**lectrically **A**lternable **R**ead **O**nly **M**emory
EC	**E**lectro**C**hromatography
ECD	**E**lectron **C**apture **D**etector
EDL	**E**lectrodeless **D**ischarge **L**amp
EDTA	**E**thylene**D**iamine**T**etraacetic **A**cid
EDXFS	**E**nergy **D**ispersive **X**-ray **F**luorescence **S**pectroscopy
EI	**E**lectron **I**mpact, as in **EI MS**
ELSD	**E**vaporative **L**ight **S**cattering **D**etector
EMI	**E**lectro**M**agnetic **I**nduction
EMR	**E**lectro**M**agnetic **R**adiation

Acronym or Abbreviation	Name
EOF	Electro**O**smotic **F**low
EPA	**E**nvironmental **P**rotection **A**gency
EPROM	**E**lectronically **P**rogrammable **R**ead **O**nly **M**emory
ERT	**E**lectrical **R**esistance **T**omography
ESCA	**E**lectron **S**pectroscopy for **C**hemical **A**nalysis; same as **XPS**
ESI	**E**lectro**S**pray **I**onization mass spectrometry
ESR	**E**lectron **S**pin **R**esonance spectroscopy
ETAAS	**E**lectro**T**hermal **A**tomic **A**bsorption **S**pectroscopy
ETAFS	**E**lectro**T**hermal **A**tomic **F**luorescence **S**pectroscopy
ETV	**E**lectro**T**hermal **V**aporization
FAAS	**F**lame **A**tomic **A**bsorption **S**pectroscopy
FAB	**F**ast **A**tom **B**ombardment **(MS)**
FD	**F**ield **D**esorption
FDA	**F**ood and **D**rug **A**dministration
FDJ	**F**ree **D**iffusion **J**unction
FES	**F**lame **E**mission **S**pectroscopy
FET	**F**ield **E**ffect **T**ransistor
FFP	**F**iberoptic **F**abry-**P**erot interferometer
FI	**F**ield **I**onization **(MS)**
FIA	**F**low **I**njection **A**nalysis
FID	**F**lame **I**onization **D**etector
FID	**F**ree **I**nduction **D**ecay
FL	**FL**uorescence
FPD	**F**lame **P**hotometric **D**etector
FSOT	**F**used **S**ilica **O**pen **T**ubular column
FT	**F**ourier **T**ransform
FTIR	**F**ourier **T**ransform **I**nfrared **S**pectroscopy
FT/NMR	**F**ourier **T**ransform **N**uclear **M**agnetic **R**esonance
FTIR-PAS	**F**ourier **T**ransform **I**nfra**R**ed **P**hoto**A**coustic **S**pectroscopy
FTMS	**F**ourier **T**ransform **M**ass **S**pectrometry
FTNMR	**F**ourier **T**ransform **N**uclear **M**agnetic **R**esonance

Continued

Acronym or Abbreviation	Name
FWHM	**F**ull **W**idth at **H**alf **M**aximum
FY	**F**iscal **Y**ear
GC	**G**as **C**hromatography
GC/FTIR	**G**as **C**hromatography/**F**ourier **T**ransform **I**nfra**R**ed spectrometry
GC/IR	**G**as **C**hromatography/**I**nfra**R**ed spectroscopy
GC/MS	**G**as **C**hromatography/**M**ass **S**pectroscopy
GD	**G**low **D**ischarge
GDAAS	**G**low **D**ischarge **A**tomic **A**bsorption **S**pectroscopy
GDAES	**G**low **D**ischarge **A**tomic **E**mission **S**pectroscopy
GDMS	**G**low **D**ischarge **M**ass **S**pectrometry
GLC	**G**as-**L**iquid **C**hromatography
GSC	**G**as-**S**olid **C**hromatography
HCL	**H**ollow **C**athode **L**amp
HDME	**H**anging **D**rop **M**ercury **E**lectrode
HETP	**H**eight **E**quivalent to a **T**heoretical **P**late
HPLC	**H**igh-**P**erformance **L**iquid **C**hromatography
IC	**I**on **C**hromatography
ICP	**I**nductively **C**oupled **P**lasma
ICP-AES	**I**nductively **C**oupled **P**lasma-**A**tomic **E**mission **S**pectroscopy
ICPMS	**I**nductively **C**oupled **P**lasma-**M**ass **S**pectrometry
IEC	**I**on **E**xchange **C**hromatography
IEF	**I**so**E**lectric **F**ocusing
IMA	**IM**muno**A**ssay
IR	**I**nfra**R**ed
ISE	**I**on **S**elective **E**lectrode
ISFET	**I**on-**S**elective **F**ield **E**ffect **T**ransistor
ITD	**I**on **T**rap **D**etector
ITMS	**I**on **T**rap **M**ass **S**pectrometry
IUPAC	**I**nternational **U**nion of **P**ure and **A**pplied **C**hemistry
LA	**L**aser **A**blation
LAN	**L**ocal **A**rea **N**etwork
LASS	**LA**ser **S**park **S**pectrometry

Acronym or Abbreviation	Name
LC	Liquid Chromatography
LC/MS	Liquid Chromatography/Mass Spectroscopy
LCD	Liquid Crystal Display
LCEC	Liquid Chromatography with Electrochemical Detection
LD	Laser Desorption source
LED	Light Emitting Diode
LIBS	Laser-Induced Breakdown Spectroscopy
LIF	Laser-Induced Fluorescence
LIFI	Laser-Induced Fluorescence Imaging
LIMS	Laboratory Information Management Systems
LIMS/DM	Laboratory Information Management System/Data Management
LIMS/SM	Laboratory Information Management System/Sample Management
LLL	Lawrence Livermore Lab
LMMS	Laser Microprobe Mass Spectrometry
LOD	Limit Of Detection
LOGS	Laser OptoGalvanic Spectroscopy
LOL	Limit Of Linearity
LOQ	Limit Of Quantitation
LOW	Liquid Observation Well
LSB	Least Significant Bit
LSER	Linear Solvation Energy Relationships
MALDI	Matrix-Assisted Laser Desorption/Ionization
MDMI	Monodisperse Dried Microparticulate Injector
MECC	Micellar Electrokinetic Capillary Chromatography
MIP	Microwave-Induced Plasma Source
MIPMS	Microwave-Induced Plasma source, Mass Spectrometry
MOSFET	Metal Oxide Semiconductor Field Effect Transistor
MRI	Magnetic Resonance Imaging
MS	Mass Spectrometry
MS/MS	Mass Spectrometry/Mass Spectrometry
MSB	Most Significant Bit
MSD	Mass Selective Detector (now called simply MS)
NAA	Neutron Activation Analysis

Continued

Acronym or Abbreviation	Name
NAPL	Non-Aqueous Phase Liquid
NBS	National Bureau of Standards (the predecessor to NIST)
NIH	National Institutes of Health
NIR	Near InfraRed spectroscopy
NIST	National Institute of Standards and Technology
NMR	Nuclear Magnetic Resonance
NRL	Naval Research Lab
NSF	National Science Foundation
NSOM	Near-field Scanning Optical Microscopy
OA	Operational Amplifier
ODS	OctoDecylSilane
ONR	Office of Naval Research
op amp	OPerational AMPlifier
OSHA	Occupational Safety and Health Administration
OT	Open Tubular
PAGE	PolyAcrylamide Gel Elcctrophoresis
PAH	Polynuclear Aromatic Hydrocarbons
PAS	PhotoAcoustic Spectrometry
PAWS	Portable Acoustic Wave Sensor
PC	Paired-ion Chromatography
PC	Paper Chromatography
PCE	PerChloroEthylene
PCF	Plasma Centrifugal Furnace
PD	Plasma Desorption
PDA	PhotoDiode Array
PI	Principal Investigator
PID	PhotoIonization Detector
PITTCON	PITTsburgh CONference on Analytical Chemistry and Applied Spectroscopy
PLOT	Porous Layer Open Tubular column
PLS	Partial Least Squares
PM	PhotoMultiplier tube
PMT	PhotoMultiplier Tube
PNA	PolyNuclear Aromatics

Acronym or Abbreviation	Name
ppm	**P**arts **P**er **M**illion
PROM	**P**rogram **R**ead **O**nly **M**emory
QCM	**Q**uartz **C**rystal **M**icrobalance
R&D	**R**esearch and **D**evelopment
RAM	**R**andom **A**ccess **M**emory
RID	**R**efractive **I**ndex **D**etector
ROM	**R**ead **O**nly **M**emory
RPC	**R**everse **P**hase **C**hromatography
RRS	**R**esonance **R**aman **S**pectroscopy
RSD	**R**elative **S**tandard **D**eviation
RTR	**R**eal-**T**ime **R**adiography
SAW	**S**urface **A**coustic **W**ave
SCD	**S**ulfur **C**hemiluminescence **D**etector
SCE	**S**aturated **C**alomel **E**lectrode
SCF	**S**uper**C**ritical **F**luid
SCOT	**S**upport **C**oated **O**pen **T**ubular
SCP	**SC**anning **P**robe microscopy
SDS	**S**odium **D**odecyl **S**ulfate
SEAC	**S**ociety for **E**lectro**A**nalytical **C**hemistry
SEC	**S**ize **E**xclusion **C**hromatography
SEM	**S**canning **E**lectron **M**icroscope
SERS	**S**urface **E**nhanced **R**aman **S**pectroscopy
SFC	**S**upercritical **F**luid **C**hromatography
SFE	**S**upercritical **F**luid **E**xtraction
SGE	**S**lab **G**el **E**lectrophoresis
SHE	**S**tandard **H**ydrogen **E**lectrode
SIMS	**S**econdary **I**on **M**ass **S**pectrometry
SNR (or S/N)	**S**ignal-to-**N**oise **R**atio
SPE	**S**olid **P**hase **E**xtraction
SPM	**S**canning **P**robe **M**icroscopy
SSMS	**S**park **S**ource **M**ass **S**pectrometry
STM	**S**canning **T**unneling **M**icroscope (or **M**icroscopy)
STNG	**S**ealed-**T**ube **N**eutron **G**enerator

Continued

Acronym or Abbreviation	Name
STP	Standard Temperature and Pressure
TA	Thermal Analysis
TCD	Thermal Conductivity Detector
TCE	TriCholoroEthylene
TDGC/MS	Time Domain Gas Chromatography/Mass Spectrometry
TDR	Time Domain Reflectometry
TFR	Thin Film Resonators
TG	ThermoGravimetry,
TGA	ThermoGravimetric Analysis
TID	ThermIonic Detector
TIMS	Thermal Ionization Mass Spectrometry
TIRS	Transient InfraRed Spectroscopy
TISAB	Total Ionic Strength Adjusting Buffer
TLC	Thin-Layer Chromatography
TMA	ThermoMechanical Analysis
TMS	TetraMethylSilane
TOF	Time-Of-Flight mass spectrometry
TS	Thermospray Source
UPS	Ultraviolet Photoelectron Spectroscopy
USDA	United States Department of Agriculture
USGS	United States Geological Survey
UV	UltraViolet
VOC	Volatile Organic Compound
VOST	Volatile Organic Sampling Train
VTVM	Vacuum Tube VoltMeter
WCOT	Wall Coated Open Tubular column
XES	X-Ray Emission Spectrometry
XFS	X-Ray Fluorescence Spectroscopy
XPS	X-Ray Photoelectron Spectroscopy
XRF	X-Ray Fluorescence
YAG	Yttrium Aluminum Garnet
ZDV	Zero Dead Volume

Answers to Selected Problems

Chapter 1

1-9 (a) $m = 0.0670\ \text{ppm}^{-1}$

(b), (c)

c_x, ppm	$\gamma = m/s_x$	CV, %
2.00	7.1	5.4
6.00	8.0	2.0
10.00	8.0	1.2
14.00	7.9	0.89
18.00	6.1	0.88

(d) $c_m = 0.35$ ppm X

1-10 $c_x = 9.00 \times 10^{-4}$ M

1-11 (c) $S_F = 3.16(\pm 0.02)V_s + 3.25(\pm 0.02)$
(d) 0.410 μg/mL
(e) 0.003 μg/mL

Chapter 2

2-1 (a) $R_1 = 50\ \Omega$; $R_2 = 200\ \Omega$; $R_3 = 250\ \Omega$
(b) 5.0 V
(c) 0.020 A
(d) 0.20 W

2-2 (a) 4.4 V
(b) 0.039 W
(c) 30%

2-3 (a) −15%
(b) −1.7%
(c) −0.17%

2-4 (a) 74 kΩ
(b) 740 kΩ

2-5
(a) $V_1 = 1.21$ V; $V_2 = V_3 = 1.73$ V; $V_4 = 12.1$ V
(b) $I_1 = I_5 = 1.21 \times 10^{-2}$ A; $I_2 = 3.5 \times 10^{-3}$ A; $I_3 = I_4 = 8.6 \times 10^{-3}$ A
(c) $P = 1.5 \times 10^{-2}$ W
(d) 2.9 V

2-6
(a) 0.085 W
(b) 8.0×10^{-3}
(c) 8.00 V
(d) 5.3 V
(e) 16 V

2-7 0.535 V

2-10
(a) 20 kΩ
(b) 1.98 MΩ

2-11 −0.4%

2-12 −3.7%

2-13

t, s	0.00	0.01	0.10	1.0	10
I, μA	2.40	2.39	2.28	1.46	1.62×10^{-2}

2-14
(a) 0.69 s
(b) 0.069 s
(c) 6.9×10^{-5} s

2-15
(a) 0.15 s
(b) 0.015 s
(c) 1.5×10^{-5} s

2-16
(a) 1.75 ms
(b)

t, ms	*i*, μA	v_R, V	v_C, V
0	500	25.0	0.0
1	282	14.1	11
2	159	8.0	17
3	90	4.5	20
4	51	2.5	22
5	29	1.4	24
10	1.6	0.08	24.9

2-17 (a) 1.00 s

(b)

t, ms	i, μA	v_R, V	v_C, V
0	0.75	15.0	0.0
1	0.28	5.5	9.5
2	0.10	2.0	13.0
3	3.7×10^{-2}	0.75	14.2
4	1.4×10^{-2}	0.27	14.7
5	5.0×10^{-3}	0.10	14.9
10	3.0×10^{-5}	0.00	15.0

2-18

	X_C, Ω	Z, Ω	φ, V
(a)	4.8×10^6	4.8×10^6	−90
(b)	4.8×10^3	2.1×10^4	−13.6
(c)	4.8	2.0×10^4	0.0
(d)	4.8×10^6	4.8×10^6	−90
(e)	4.8×10^3	4.8×10^3	−87.6
(f)	4.8	2.0×10^2	−1.4
(g)	4.8×10^5	4.8×10^5	−90
(h)	4.8×10^2	2.1×10^3	−13.6
(i)	0.48	2.0×10^3	0.0

2-19

$(V_p)_o/(V_p)_i$	f	$(V_p)_o/(V_p)_i$	f
0.01	4.2×10^3	0.60	57
0.10	4.2×10^2	0.80	32
0.20	2.1×10^2	0.90	21
0.40	97	0.99	6.1

2-20

$(V_p)_o/(V_p)_i$	f	$(V_p)_o/(V_p)_i$	f
0.01	1.6×10^1	0.60	1.2×10^3
0.10	1.6×10^2	0.80	2.1×10^3
0.20	3.2×10^2	0.90	3.3×10^3
0.40	7.0×10^2	0.99	1.1×10^4

Chapter 3

3-1

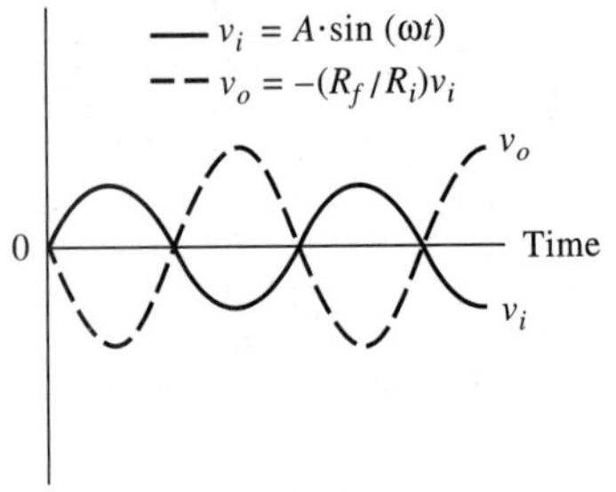

(a)

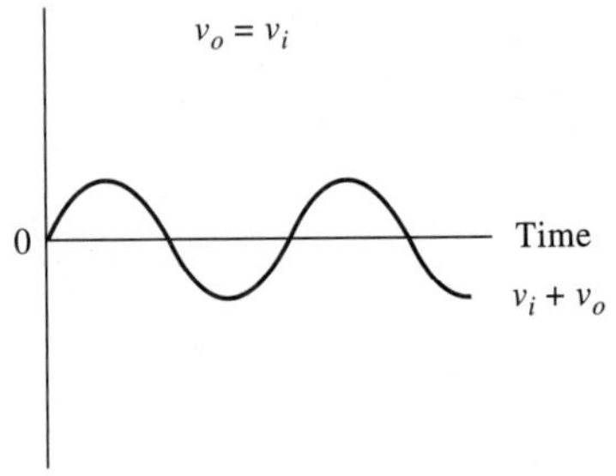

(b)

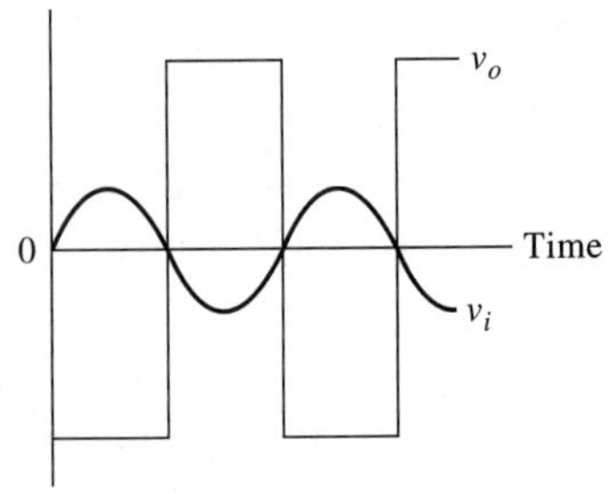

(c)

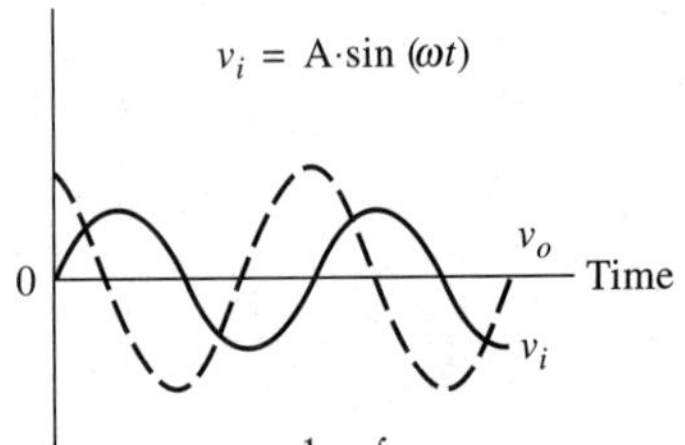

$$v_o = \frac{-1}{R_iC}\int v_i\,dt$$

$$= \frac{-A}{R_iC}\ -\cos(\omega \text{t})$$

$$= \frac{A}{R_iC}\ \cos(\omega \text{t})$$

(d)

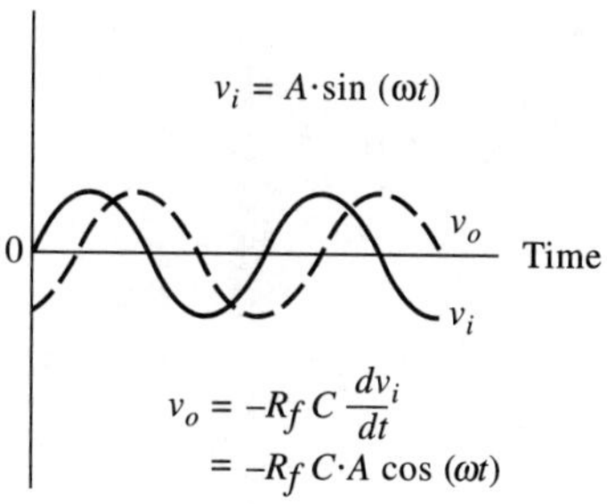

(e)

3-2 (a) $v_o \cong -27.3$ mV or -23.6 mV when the relatively small value of A is taken into account.
(b) 0.910 μA
(c) 0.910 μA

3-3 $t_r = 6.7$ ns; $dV/dt = 1500$ V/μs (assuming a 10-V change in 6.7 ns)

3-4 +0.04%

3-5

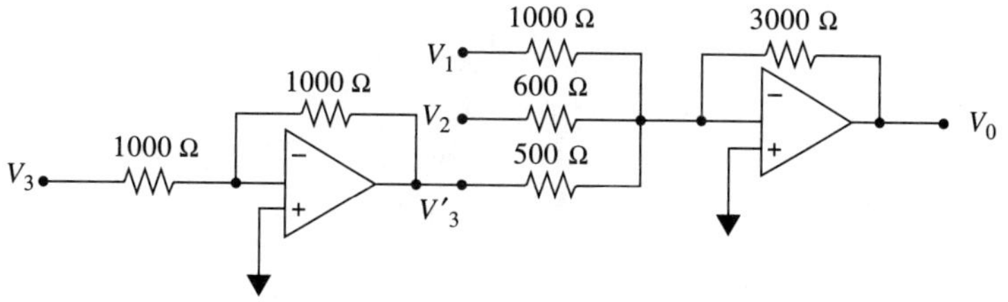

3-6

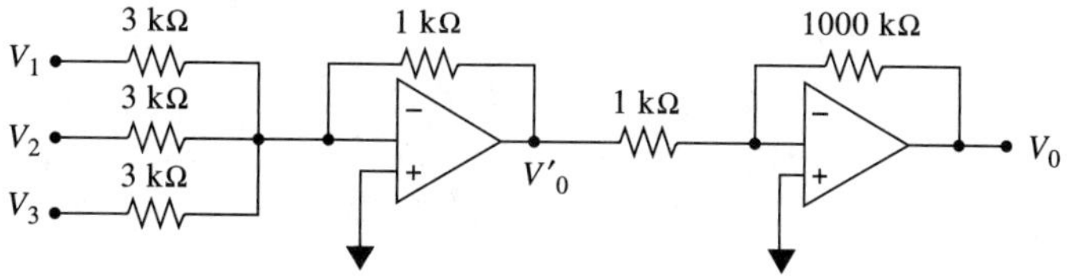

3-7

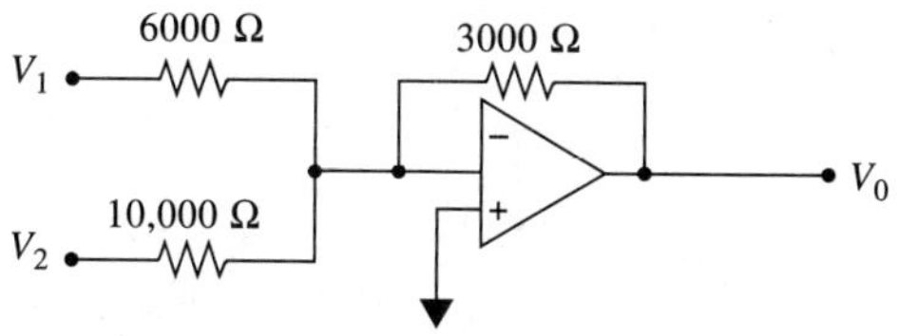

3-8 $R_f = 1.00$ kΩ and $R_1 = 250$ Ω

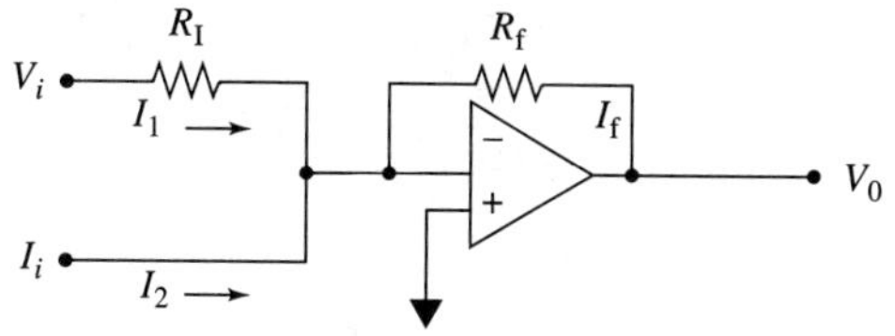

3-9 (a) $$V_o = \frac{V_1R_{f1}R_{f2}}{R_1R_4} + \frac{V_2R_{f1}R_{f2}}{R_2R_4} - \frac{V_3R_{f2}}{R_3}$$

(b) $$V_o = V_1 + 4V_2 - 40V_3$$

3-10 $$V_o = 10V_1 + 6V_2 - 3V_3 - 2V_4$$

3-11

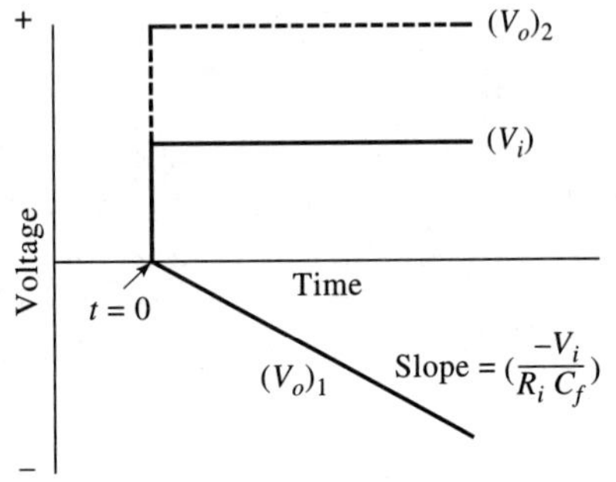

3-12 $$V_o = \left(\frac{R_1 + R_2}{R_1}\right)V_i$$

3-13 $$V_o = -\left(\frac{X}{R - X}\right)V_i$$

3-14 $$V_o = -(5V_1 + 20V_2)\int_o^t dt$$

3-15 $$V_o = V_2 - V_1$$

3-16 34 cm above ground.

3-17

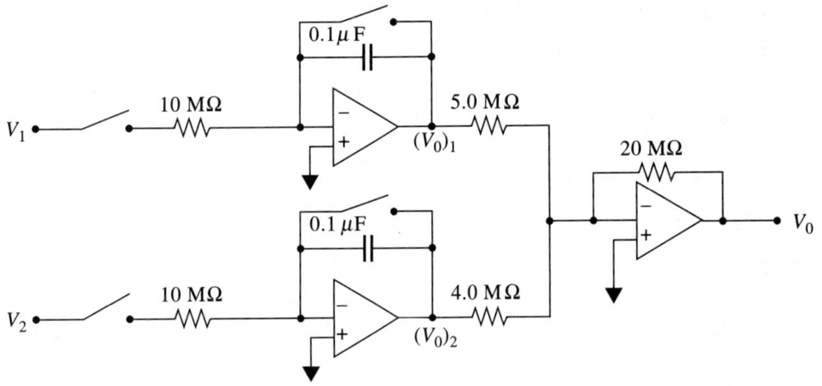

3-18

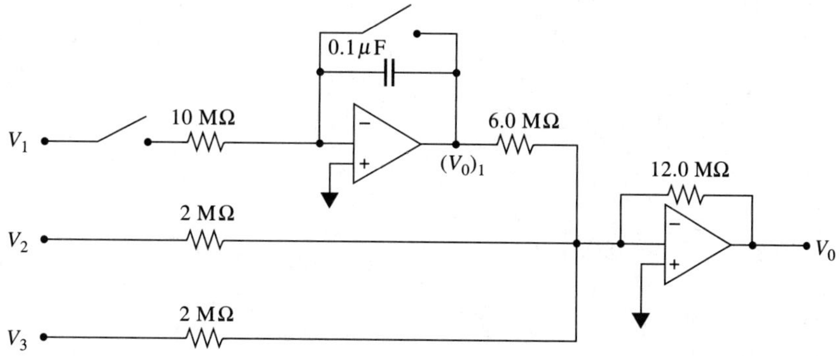

3-19 V_o after 1, 3, 5, and 7 s is −8, −24, −40, and −56 mV, respectively.

Chapter 4

4-1
(a) 11000
(b) 1001111
(c) 10001000
(d) 1001000101

4-2
(a) 5
(b) 21
(c) 226
(d) 841

4-3
(a) 0.039 V
(b) 0.0024 V
(c) 0.00015 V

4-6
(a) 1 Hz
(b) 20 Hz

4-7 62.5 kHz

4-8
(a) 68,719,476,736

(b) 1.84×10^{19}

Chapter 5

5-7
(a) $S/N = 358$
(b) $n = 18$

5-8 (a) $S/N = 5.3$
(b) $n = 28$

5-9 (a) 1.28×10^{-4} V; noise reduced by a factor of 100.

5-10 $n = 100$

5-11 $(S/N)_{50} = 7.1(S/N)_1$; $(S/N)_{200} = 14.1(S/N)_1$

5-12 $(S/N)_D = 30(S/N)_A$

5-13 (a) 100 points
(b) 200 points

Chapter 6

6-2 1.11×10^{18} Hz; 7.36×10^{-16} J; 4.59×10^3 eV

6-3 $\nu = 5.246 \times 10^{13}$ Hz
$\overline{\nu} = 1750$ cm^{-1}
$E = 3.476 \times 10^{-20}$ J
$E = 20.93$ kJ/mol

6-4 $\lambda = 136$ cm
$E = 1.46 \times 10^{-25}$ J

6-5 $\nu_{cell} = 2.10 \times 10^{10}$ cm s^{-1}
$\nu = 5.09 \times 10^{14}$ Hz
$\lambda = 413$ nm

6-6 2.42

6-7 167 nm

6-8 469 nm

6-9 (a) 9.63×10^{-20} J
(b) 4.60×10^5 m/s

6-10 2.4 μm

6-11 (a) 436 nm
(b) 448 nm

6-12 17.3%

6-14	(a)	42.2%
	(b)	4.73%
	(c)	97.3%
6-15	(a)	0.474
	(b)	0.0357
	(c)	1.76
6-16	(a)	64.9%
	(b)	21.8%
	(c)	99%
6-17	(a)	0.775
	(b)	0.337
	(c)	2.06

6-18 3.89×10^{-3} M

6-19 1.98×10^{-4} M

Chapter 7

7-3	(a)	725 nm
	(b)	1.45 μm
	(c)	2.90 μm
7-4	(a)	1.46×10^7 W/m^2
	(b)	9.10×10^5 W/m^2
	(c)	5.69×10^4 W/m^2
7-5	(a)	1010 and 967 nm
	(b)	3.86×10^6 W/m^2 and 4.61×10^6 W/m^2

7-9 (a) 1.69 μm (b) 2.27 μm, 0.151 μm, etc.

7-10 For first-order interference, the thickness should decrease linearly from 0.265 μm to 1.52 μm.

7-13 $\lambda/\Delta\lambda = 720$; 1.39 cm^{-1}

7-14 (a) 15.4 and 7.7 μm (b) 10.6 and 5.3 μm

7-16 $f = 1.95$

7-17 The second lens has 2.5 times the light-gathering power of the first lens.

7-18 (a) $R = 2.50 \times 10^4$
(b) $D_1^{-1} = 0.50$ nm/mm; $D_2^{-1} = 0.25$ nm/mm

7-19 (a) $D^{-1} = 0.77$ nm/mm
(b) $R = 6.0 \times 10^4$
(c) $\Delta\lambda = 0.0093$ nm

7-22 (a) 8.33×10^4 Hz (c) 3.33×10^3 Hz
(b) 3.57×10^4 Hz (d) 1.25×10^3 Hz

7-23 (a) 2.1 cm
(b) 0.313 cm

Chapter 8

8-8 (a) 0.028 Å
(b) 0.033 Å

8-9 For Na and Mg^+, respectively, N_j/N_0
(a) 2.7×10^{-5}; 7.1×10^{-11}
(b) 6.6×10^{-4}; 6.1×10^{-8}
(c) 5.1×10^{-2}; 5.7×10^{-4}

8-10 (a) $N_{4s}/N_{3s} = 1.2 \times 10^{-5}$
(b) $N_{4s}/N_{3s} = 1.8 \times 10^{-2}$

Chapter 9

9-11 104 mm

9-12

	Height	T, K	$I_{(x\ cm)}/I_{(2\ cm)}$
(a)	2.0	1973	1.00
(b)	3.0	2136	2.07
(c)	4.0	2092	1.72
(d)	5.0	1998	1.13

9-14

	Na	Mg^+
(a)	2.7×10^{-5}	7.1×10^{-11}
(b)	6.6×10^{-4}	6.1×10^{-8}
(c)	5.1×10^{-2}	5.7×10^{-4}

9-15 (a) 1.2×10^{-5}
(b) 1.8×10^{-2}

9-16 0.099 (±0.008) μg Pb/mL

9-20 0.297 ppm

9-21 (b) $R = 0.920\, c_{Na_2O} + 3.18$
(c) $s_m = 0.015$; $s_r = 0.96$; $s_b = 0.74$
(d)

	A	B	C
$\%Na_2O$	0.257	0.389	0.736
s	0.013	0.012	0.015
$(S_c/c) \times 100\%$	4.9%	3.2%	2.0%

9-22 (b) $A = 8.81 \times 10^{-3}$ mL + 0.202
(c) $s_m = 4.1 \times 10^{-5}$; $s_r = 1.3 \times 10^{-3}$
(d) 28.0 ppm Cr
(e) 0.22 ppm Cr

Chapter 12

12-1 $\lambda_0 = 0.155$ Å

12-2

	K_β, kV	L_β, kV
(a)	112	17.2
(b)	3.59	No line
(c)	15.0	1.75
(d)	67.4	9.67

12-3 (a) 2.5 Å
(b) 1.6 Å
(c) 1.1 Å
(d) 1.0 Å
(e) 0.54 Å
(f) 0.47 Å

12-4 (a) 24 Å
(b) 14 Å
(c) 8.9 Å
(d) 8.3 Å
(e) 4.0 Å
(f) 3.5 Å

12-5 2.3×10^{-3} cm

12-6 (a) 3.94 cm^2g
(b) 0.126

12-7 2.73×10^{-3} cm

12-8 (a) 1.39%
(b) 0.261%

12-9 For Fe, Se, and Ag respectively,
(a) 80.9, 42.9, and 21.1 deg
(b) 51.8, 28.5, and 14.2 deg
(c) 36.4, 20.3, and 10.1 deg

12-10 (a) 135 deg
(b) 99.5 deg

12-11 (a) 4.05 kV
(b) 1.32 kV
(c) 20.9 kV
(d) 25.0 kV

12-12 0.300 ($\pm$0.002)% Mn

Chapter 13

13-1 (a) 88.9%
(b) 12.1%
(c) 41.8%
(d) 54.8%
(e) 32.7%
(f) 19.9%

13-2 (a) 0.593
(b 0.246
(c) 0.484
(d) 1.45
(e) 1.07
(f) 0.269

13-3 (a) 79.1%
(b) 1.46%
(c) 17.5%
(d) 30.1%
(e) 10.7%
(f) 39.4%

13-4
(a) 0.894
(b) 0.547
(c) 0.785
(d) 1.75
(e) 1.37
(f) 0.570

13-5 1.80×10^4 L cm^{-1} mol^{-1}

13-6 1.57×10^3

13-7
(a) 0.582
(b) 26.2%
(c) 1.25×10^{-5} M

13-8
(a) 0.175
(b) 0.350
(c) 66.8% and 44.7%
(d) 0.476

13-9 0.059

13-10
(a) 34.4%
(b) 6.92%
(c) 2.90×10^3

13-11 (a)

$c_{K_2Cr_2O_7}$, M	A_{430}	A_{600}
3.00×10^{-4}	1.54	1.06
2.00×10^{-4}	0.935	0.777
1.00×10^{-4}	0.383	0.455
0.500×10^{-4}	0.149	0.261
0.250×10^{-4}	0.056	0.145

13-12

$c_{K_2Cr_2O_7}$, M	A_{345}	A_{370}	A_{400}
4.00×10^{-4}	0.826	1.649	0.621
3.00×10^{-4}	0.654	1.352	0.512
2.00×10^{-4}	0.470	1.018	0.388
1.00×10^{-4}	0.266	0.613	0.236

13-25
(a) 1.69 μm
(b) 2.27, 1.51, 1.13 μm, etc.

Chapter 14

14-1
(b) $A = 0.0781\, c_{Fe} + 0.0148$
(c) 1.23×10^{-2}
(d) 8.1×10^{-4}

14-2

	c_{Fe}, ppm	s_c	s_c (mean of 3)
(a)	1.18	0.20	0.15
(b)	9.04	0.17	0.11
(c)	19.5	0.20	0.15

14-4 20.1 ppm

14-5 1.84×10^{-3}%

14-6
(a) 8.82×10^{-5} M Ni and 8.88×10^{-5} M Co
(b) 9.8×10^{-5} M Ni and 1.66×10^{-4} M Co

14-7
(a) $[A] = 7.65 \times 10^{-5}$ $[B] = 2.34 \times 10^{-5}$
(b) $[A] = 5.45 \times 10^{-5}$ $[B] = 3.75 \times 10^{-5}$

14-8
(a) At 485 nm, $\epsilon_{In} = 104$ and $\epsilon_{HIn} = 908$
At 625 nm, $\epsilon_{HIn} = 1646$ and $\epsilon_{HIn} = 352$
(b) $K_a = 2.07 \times 10^{-6}$
(c) pH = 3.735
(d) $K_a = 2.33 \times 10^{-6}$
(e) $A_{475} = 0.091$ and $A_{625} = 0.306$

14-9
(a) 1.4%
(b) 1.4%
(c) 7.6%
(d) 3.0%
(e) 67%
(f) 40%

14-12 1.80×10^{8}

14-3 1.53×10^{6}

Chapter 15

15-6
(b) $I = 22.43\, c + 0.03107$
(c) $s_m = 0.45$; $s_r = 0.29$; $s_b = 0.23$
(d) 0.541
(e) 0.026
(f) 0.018

15-7 (b) $I = 0.970\ c_x + 6.13$
(c) $s_m = 0.030$; $s_b = 0.28$; $s_y = 0.336$
(d) 1.39 ppm Zn^{2+}
(e) 0.077 ppm Zn^{2+}

15-8 (c) $R = -8.96\ c_F + 66.5$
(d) $s_m = 0.198$; $s_r = 0.664$; $s_b = 0.556$
(e) 14.8 ppb F^-
(f) 0.35 ppb F^-

15-9 2.83×10^{-5} M

Chapter 16

16-1 (a) $1.91 \times 10^3\ N/m$
(b) $2.08 \times 10^3\ cm^{-1}$

16-2 (a) $4.81 \times 10^2\ N/m$
(b) $2.07 \times 10^3\ cm^{-1}$

16-3 (a) inactive
(b) active
(c) active
(d) active
(e) inactive
(f) active
(g) inactive

16-4 $3.03 \times 10^3\ cm^{-1}$ and 2.85×10^3 to $3.33 \times 10^3\ cm^{-1}$

16-5 $1.4 \times 10^4\ cm^{-1}$ or 0.70 μm

16-6 $1.3 \times 10^4\ cm^{-1}$ or 0.75 μm

16-7 3 peaks

16-9 (a) 25 cm
(b) 0.25 cm

16-10 (a) $N_1/N_0 = 8.9 \times 10^{-7}$
(b) $N_2/N_0 = 7.9 \times 10^{-13}$

16-12 1600

16-13 (a) 3400 Hz
(b) 3420 Hz
(c) 3430 Hz

Chapter 17

17-1 (a) C=O stretch
(b) CCl_4, $CHCl_3$, C_6H_{12}, $C_2H_4Cl_4$
(c) 0.015 mg/mL

17-2 vinyl alcohol, $CH_2{=}CH{-}CH_2O$

17-3

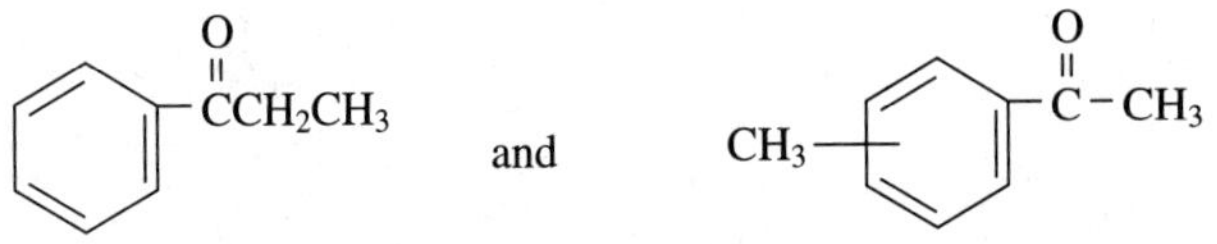

17-4 acrolein $CH_2{=}CH{-}CHO$ with H_2O contaminant

17-5 propanonitrile $CH_3CH_2{\equiv}N$

17-9 0.0071 cm

17-10 0.021 cm

17-11 7.9×10^{-3} cm

Chapter 18

18-3

	(a)		(b)	
$\Delta\bar{\nu}$, cm^{-1}	**Stokes, nm**	**Anti-Stokes, nm**	**Stokes, nm**	**Anti-Stokes, nm**
218	641.7	624.2	493.2	482.9
314	645.6	620.5	495.6	480.6
459	651.7	614.9	499.2	477.3
762	664.9	603.7	506.8	470.5
790	666.1	602.7	507.6	469.9

18-4 (a) 2.83
(b) The efficiency of the detector system is a function of λ.

18-6 (a) 0.342 at 20°C and 0.367 at 40°C
(b) 0.105 and 0.121
(c) 0.0206 and 0.0264

18-7 (a) $I_\perp/I_\parallel = 0.77$
(b) $I_\perp/I_\parallel = 0.012$ polarized
(c) $I_\perp/I_\parallel = 0.076$ polarized
(d) $I_\perp/I_\parallel = 0.76$

Chapter 19

19-6 6 spin states; $m = 5/2, 3/2, 1/2; -1/2, -3/2, -5/2$

19-7
(a) 1.0×10^8 Hz
(b) 2.6×10^7 Hz
(c) 9.6×10^7 Hz
(d) 4.1×10^7 Hz

19-9 $N_j/N_0 = 0.9999959$

19-13 $\Delta E(^{13}C)/\Delta E(^{1}H) = 0.251$

19-14
(a) $\nu_0(^{19}F) = 200$ MHz
(b) $\nu_0(^{31}P) = 86$ MHz

19-15 $N_j/N_0 = 0.99999109$

19-16 $\nu_0(^{31}P) = 24$ MHz and $\nu_0(^{1}H) = 59$ MHz
The ^{1}H signal will be much more intense than the ^{31}P signal and will be split into a doublet. The ^{31}P signal will be made up of 10 peaks with areas proportional to the coefficients in the expansion of $(1 \times x)^9$.

19-17

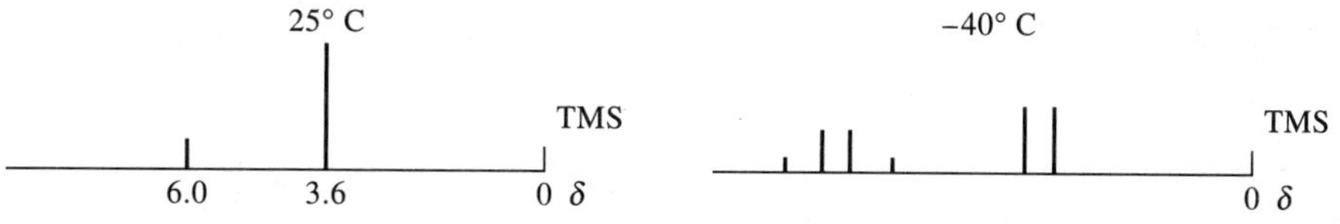

19-18

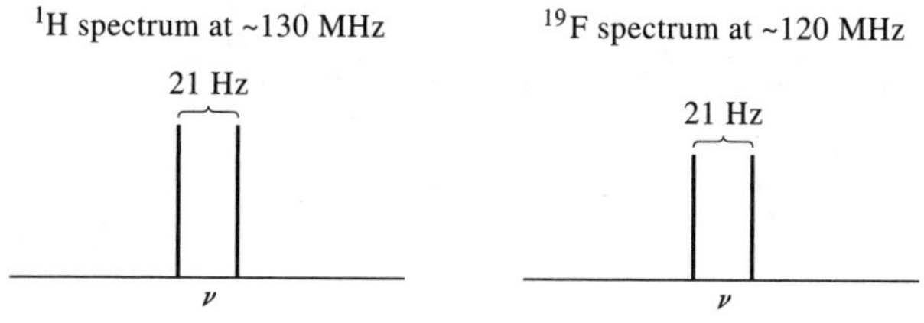

(a)

(b)

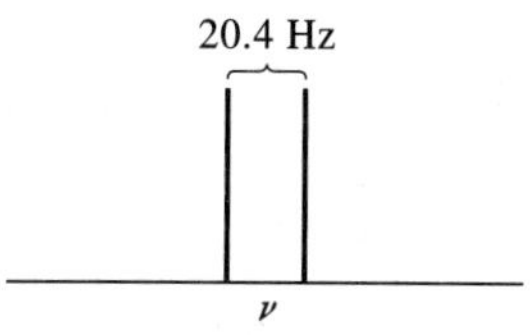

10 peaks with areas proportional to the coefficients in the expansion of $(1 + x)^9$

(c)

19-22

(a) Doublet (areas ∝ 3) at δ = 2.2 and a quartet (area ∝ 1) at δ = 9.7 to 9.8 (area ratio 1:3:3:1).
(b) Singlet (area ∝ 3) at δ = 2.2 and a singlet (area ∝ 1) at δ = 11 to 12.
(c) Triplet (area ∝ 3) at δ = 1.6 (area ratio 1:2:1) and a quartet (area ∝ 2) at δ = 4.4 (area ratio 1:3:3:1).

19-23

(a) Singlet at δ = 2.1.
(b) Singlet (area ∝ 3) at δ = 2.1, a quartet (area ∝ 2) at δ = 2.4 (area ratio 1:3:3:1), and a triplet (area ∝ 3) at δ = 1.1 (area ratio 1:2:1).
(c) Singlet (area ∝ 3) at δ = 2.1, a doublet (area ∝ 6) at δ = 1.1, and 7 peaks (area ∝ 1) centered at δ = 2.6 (area ratio 1:7:21:35:35:21:7:1).

19-24

(a) Singlet at δ = 1.2 to 1.4.
(b) Singlets at δ = 3.2 and 3.4 with peak areas in the ratio of 3:2, respectively.
(c) Triplet (area ∝ 3) at δ = 1.2 and a quartet (area ∝ 2) at δ = 3.4.

19-25

(a) Singlets at δ = 2.2 and 6.5 to 8 with peak areas in the ratio of 3:5 respectively.
(b) Triplet (area ∝ 3) at δ = 1.1, quartet (area ∝ 2) at δ = 2.6, and a singlet (area ∝ 5) at δ = 6.5 to 8.
(c) Doublet (area ∝ 9) at δ 0.9 to 1 and 10 peaks (area ∝ 1) at δ = 1.5.

19-27

$$\overset{\displaystyle Br}{\overset{|}{CH_3CH_2CHCOOH}}$$

19-28

$$\overset{\displaystyle O}{\overset{\|}{CH_3CH_2CCH_3}}$$

19-29

$$\overset{\displaystyle O}{\overset{\|}{CH_3COC_2H_5}}$$

19-30

(a)

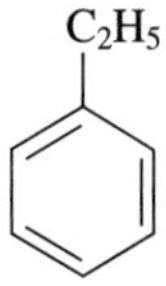

(b)

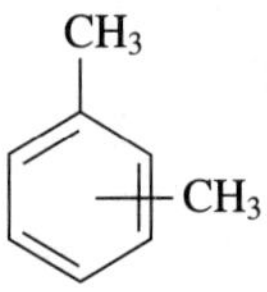

19-31

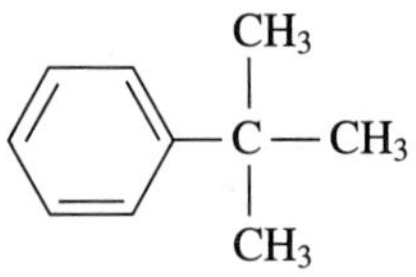

19-32

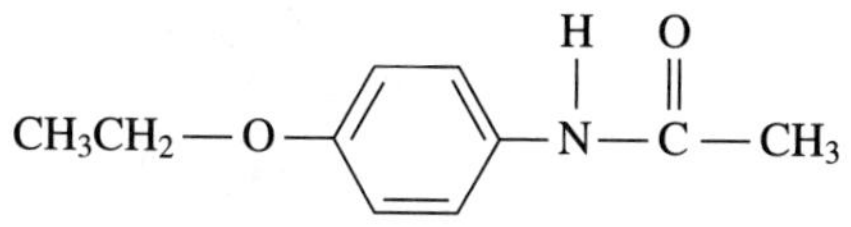

Chapter 20

20-4	(a)	140 V
20-5	(a)	0.126 T to 0.498 T
	(b)	3000 V to 192 V
20-6	4.48 V	
20-7	42.7 μs	
20-10	(a)	2.22×10^3
	(b)	769
	(c)	7.09×10^4
	(d)	992
20-11	(a)	1.96
	(b)	1.30
20-14	(b)	80.92

20-15 $m = 131$ due to $^{35}Cl_3CCH_2^+$, $m = 133$ due to $^{37}Cl^{35}Cl_2CCH_2^+$, $m = 135$ due to $^{37}Cl_2{}^{35}ClCCH_2^+$, $m = 117$ due to $^{35}Cl_3C^+$, $m = 119$ due to $^{37}Cl^{35}Cl_2C^+$, $m = 121$ due to $^{37}Cl_2{}^{35}ClC^+$

20-16 $m = 84$ due to $^{35}Cl_2{}^{12}CH_2^+$, $m = 85$ due to $^{35}Cl_2{}^{13}CH_2^+$, $m = 86$ due to $^{37}Cl^{35}Cl^{12}CH_2^+$, $m = 87$ due to $^{37}Cl^{35}Cl^{13}CH_2^+$, $m = 88$ due to $^{37}Cl_2{}^{12}CH_2^+$

Chapter 21

21-4	(a)	165.4 eV
	(b)	SO_3^{2-}
	(c)	1306.6 eV
	(d)	1073.5 eV
21-5	(a)	406.3 eV
	(b)	8.19.5 eV
	(c)	403.4 eV

Chapter 22

22-1	(a)	0.691 V
	(b)	0.598 V
	(c)	0.132 V
22-2	(a)	0.033 V
	(b)	0.998 V
	(c)	0.493 V
22-3	(a1)	−0.118 V
	(a2)	−0.122V
	(b1)	0.771 V
	(b2)	0.751 V
22-4	(a1)	0.145 V
	(a2)	0.143 V
	(b1)	0.145 V
	(b2)	0.125 V
22-5	(a)	−0.050 V
	(b)	−0.130 V
	(c)	0.729 V
	(d)	0.143 V
22-6	(a)	1.25 V
	(b)	0.299 V
22-7	(a)	0.288 V, galvanic
	(b)	−0.599 V, electrolytic
	(c)	−0.836 V, electrolytic
22-8	(a)	−0.357 V, electrolytic
	(b)	1.227 V, galvanic
	(c)	0.589 V, galvanic
22-9	−0.90 V	
22-10	−0.367 V	
22-11	−0.037 V	
22-12	−1.92 V	
22-13	1.8×10^{-4}	
22-14	1.0×10^{-15}	

22-15 0.122 V

22-16 −0.298 V

22-17 0.357 V

Chapter 23

23-8
(a) 0.517 V
(b) −0.925
(c) −0.598
(d) 0.493

23-9
(a) 0.031 V
(b) SCE‖CuBr(sat'd),Br^-(xM) | Cu
(c) $pBr = \dfrac{E_{cell} - E^0_{CuBr} + E_{SCE}}{0.0592}$
(d) 2.40

23-10
(a) 0.366 V
(b) SCE‖Ag_3AsO_4(sat'd),AsO_4^{3-}(xM) | Ag
(c) $pAsO_4 = \dfrac{\left(E_{cell} - E^O_{Ag_3AsO_4} + E_{SCE}\right)3}{} = \dfrac{(E_{cell} - 0.122)3}{0.0592}$
(d) 9.62

23-11 6.76

23-12 6.96

23-13 0.604 V

23-14 6.6×10^{18}

23-15
(a) 0.113 V
(b) 0.054 V
(c) −0.005 V

23-16 10.22

23-17 1.0×10^{-16}

23-18 3.97×10^{-5}

23-19

Vol Ce^{4+}	*E* vs. SCE	Vol Ce^{4+}	*E* vs SCE
10.00	0.68 V	50.10	1.04 V
25.00	0.70	51.00	1.10
40.00	0.71	55.00	1.14
49.00	0.75	60.00	1.16
49.90	0.78		

23-20
(a) 8.69 and 2.05×10^{-9}
(b) 1.97×10^{-9} and 2.13×10^{-9}
(c) ±3.9%

23-21
(a) 5.226
(b) 5.004×10^{-6} and 6.831×10^{-6}
(c) −15.9% and −14.8%

23-22 4.28×10^{-4}%

Chapter 24

24-1
(a) −1.346 V
(b) 0.285 V
(c) −1.630 V
(d) −1.692 V

24-2
(a) 0.077 V
(b) 0.097 V
(c) 0.223 V
(d) 0.216 V
(e) 0.0582 V

24-3
(a) Ag^+ can be separated from both Cu^{2+} and BiO^+.
(b) To separate Ag^+, keep the cell potential between 0.495 V and 0.200 V.
(c) To precipitate Bi quantitatively, the cell potential must be −0.540 V.

24-4
(a) AgI forms first at a cell potential of 0.300 V. The cell is galvanic.
(b) Separation feasible if cell potential kept between 0.099 V and 0.076 V.
(c) Separation feasible if cell potential kept between 0.099 V and −0.073 V.
(d) Separation of Br^- from Cl^- not feasible.

24-5
(a) 0.432 V
(b) 0.247 V
(c) −0.127 V

24-6 (a) 2.951 min
(b) 1.967 min
(c) 0.984 min

24-7 2.471% CCl_4 and 1.854% $CHCl_3$

24-8 0.613% As_2O_3

24-9 9.60×10^2 g/equiv

24-10 33.7 μg $C_6H_5NH_2$

Chapter 25

25-6 (a) −0.059 V vs. SCE
(b) +0.059 V vs. SCE

25-8 (a) −0.067%
(b) −0.13%
(c) −0.40%

25-9 (a) 0.369
(b) 0.346
(c) 0.144
(d) 0.314 mg Cd/mL

25-10 For PbA_2, $K_f = 7.0 \times 10^5$

25-12 (a) −0.260 V
(b) −0.408 V
(c) −0.615 V

Chapter 26

26-12 (a) $N_A = 2775$ $N_B = 2472$ $N_C = 2364$ $N_D = 2523$
(b) $\overline{N} = 2.5 \times 10^3$ $s = 0.2 \times 10^3$
(c) $H = 0.0097$ cm

26-13 (a) $k'_A = 0.74$ $k'_B = 3.3$ $k'_C = 3.5$ $k'_D = 6.0$
(b) $K_A = 6.2$ $K_B = 27$ $K_C = 30$ $K_D = 50$

26-14 (a) $R_s = 0.72$
(b) $\alpha_{C,B} = 1.08$
(c) $L = 108$ cm
(d) $t_R = 62$ min

26-15 (a) $R_S = 5.2$
(b) $L = 2.1$ cm

26-16 (a) $\overline{N} = 2.7 \times 10^3$
(b) $s = 0.1 \times 10^3$
(c) $H = 0.015$ cm

26-17 (a) $R_S = 1.1$
(b) $R_S = 2.7$
(c) $R_S = 3.7$

26-18 (a) $N = 4.7 \times 10^3$ plates
(b) $L = 69$ cm
(c) $t_R = 19$ min

26-19 (a) $k'_1 = 4.3$; $k'_2 = 4.7$; $k'_3 = 6.1$
(b) $K_1 = 14$; $K_2 = 15$; $K_3 = 19$
(c) $\alpha_{2,1} = 1.11$ (d) $\alpha_{3,2} = 1.28$

26-22 (a) $k'_M = 2.54$; $k'_N = 2.62$ (d) $L = 1.8 \times 10^2$ cm
(b) $\alpha = 1.03$ (e) $t_R = 91$ min
(c) $N = 8.1 \times 10^4$ plates

26-23 (a) $k'_M = 2.45$; $k'_N = 2.62$ (d) $L = 35$ cm
(b) $\alpha = 1.07$ (e) $t_R = 18$ min
(c) $N = 1.6 \times 10^4$ plates

26-24 % methyl acetate = 14.8
% methyl propionate = 8.1
% methyl *n*-butyrate = 37.1

26-25 % 1 = 15.0; % 2 = 18.1; % 3 = 27.4; % 4 = 23; % 5 = 16.5

Chapter 27

27-20 558

27-21 (a) 30.2 mL/min
(b) 8.4 mL, 55.3 mL, 116.2 mL, 221.5 mL
(c) 24.4 mL, 56.1 mL, 111 mL
(d) $K_1 = 34.2$, $K_2 = 78.6$, $K_3 = 156$
(e) $t_R = 36.0$ min; $V_R^0 = 1.01 \times 10^3$ mL

27-22 (a) $k'_1 = 5.6$; $k'_2 = 13$; $k'_3 = 25$
(b) $\alpha_{2,1} = 2.3$; $\alpha_{3,2} = 2.0$
(c) $\overline{N} = 1.7 \times 10^3$ plates; $H = 0.064$ cm
(d) $(R_S)_{2,1} = 7.5$; $(R_S)_{3,2} = 6.4$

27-24 $t_R = 8.00$ and 18.2 min

27-26
(a) 300
(b) 400
(c) 500
(d) 600
(e) 700
(f) 800
(g) 726
(h) 431
(i) 629
(j) 616
(k) 585
(l) 642

Chapter 28

28-13
(a) $k'_1 = 26.7$
(b) 71% $CHCl_3$ and 29% *n*-hexane

28-18
(a) *n*-hexane, benzene, hexanol
(b) diethyl ether, ethyl acetate, nitromethane

28-19
(a) hexanol, benzene, *n*-hexane
(b) nitromethane, ethyl acetate, diethyl ether

28-20 $K_B = 0.38$; $K_C = 0.73$

Appendix 1

a1-1

		A	B	C	D
(a)	$\bar{x}$	61.43	3.25	12.10	2.65
	df	3	6	2	4
(b)	s	0.11	0.020	0.057	0.21
	df	2	5	1	3
(c)	s_m	0.064	0.0082	0.040	0.10
(d)	CV	0.18%	0.60%	0.47%	7.9%

a1-2

	A	B	C	D
E_a	−0.28	−0.030	−0.13%	−0.10
% rel error	−0.45%	−0.91%	−1.1%	−3.6%

a1-3
(a) −2.0%
(b) −0.5%
(c) −0.2%
(d) −0.1%

a1-4
(a) 10 g
(b) 2.1 g
(c) 1.3 g
(d) 0.87 g

a1-5 $N = 1 \times 10^4$

a1-6 13.5 = 14

a1-7 −25 ppt

a1-8
(a) 0.10, 0.099, 0.11, 0.13, 0.10 mg Ca/100 mL
(b) 0.11 mg Ca/100 mL

a1-9
$s = 0.15, 0.14, 0.17, 0.18$ μg Pb/m^3
$s_{pooled} = 0.16$ μg Pb/m^3

a1-10

	s_y	CV	y
(a)	0.03	5.2%	0.57 (±0.03)
(b)	0.14×10^{-16}	2.0%	$6.9 (\pm 0.1) \times 10^{-16}$
(c)	1.4×10^3	0.77%	$1.84 (\pm 0.01) \times 10^5$
(d)	5.07×10^{-3}	8.5%	$6.0 (\pm 0.5) \times 10^{-2}$
(e)	1.08×10^{-4}	1.3%	$8.1 (\pm 0.1) \times 10^{-3}$

a1-11

	s_y	CV	y
(a)	0.03×10^{-7}	2.0%	$-1.37 (\pm 0.03) \times 10^{-7}$
(b)	0.10	13%	0.8 (±0.1)
(c)	1.8	50%	4. (±2.)
(d)	100	0.27%	$3.70 (\pm 0.01) \times 10^4$
(e)	3.2	3%	106. (±3.)

a1-12
(a) $\bar{x} \pm 0.064$ ppb
(b) $\bar{x} \pm 0.037$ ppb
(c) $\bar{x} \pm 0.029$ ppb

a1-13 (a) $\bar{x} \pm 0.029$ ppm
(b) 12 measurements

a1-14 (a) $\bar{x} \pm 0.76$ ppm
(b) 7 measurements

a1-15 Bias is suggested.

a1-16 (a) Determinate error is suggested.
(b) No determinate error demonstrated.

a1-17 (a) For C, no determinate error indicated.
(b) For H, determinate error is suggested.

a1-18 $V = 5.20\ \text{cm}^3$ $s_V = 0.15\ \text{cm}^3$

a1-19 At pH = 2.0, s for $[H_3O^+] = 2.3 \times 10^{-4}$
At pH = 12.0, s for $[H_3O^+] = 2.3 \times 10^{-14}$

a1-20 $s = 7.5 \times 10^{-5}$ g/mL

a1-21 (a) 2.943 (±0.002)
(b) −0.3048 (±0.0004)
(c) $4.4\ (\pm 0.1) \times 10^3$
(d) $6.44\ (\pm 0.03) \times 10^{-8}$

a1-22 (b) $R = 0.232c_x + 0.162$
(d) $s_m = 0.0174$; $s_b = 0.21$
(e) 15.1 mg SO_4^{2-}/L; $s_c = 1.4$ mg SO_4^{2-}/L; CV = 9.3%
(f) $s_c = 0.81$ mg SO_4^{2-}/L; CV = 5.4%

a1-23 (b) $E = -29.7\ \text{pCa} + 92.9$
(c) $s_m = 0.68$; $s_b = 2.24$
(d) pCa = 2.44; $s_c = 0.080$; $(s_c)c = 3.3\%$
(e) For mean of 2, $s_c = 0.061$; $(s_c)c = 2.5\%$
For mean of 8, $s_c = 0.043$; $(s_c)c = 1.8\%$
(f) $[Ca^{2+}] = 3.6 \times 10^{-3}$
(g) For mean of 2, $s = 5.1 \times 10^{-4}$; $(s_c)c = 14\%$
For mean of 8, $s = 3.6 \times 10^{-4}$; $(s_c)c = 9.9\%$

a1-24 (a) $A = 5.57c_{MVK} + 0.902$
(c) $s_m = 0.096$; $s_b = 0.33$
(d)

A	c_{MVK}, mmol/L	N	s_cc mmol/L	$(s_c)_c$, %
6.3	0.969	1	0.086	8.9
6.3	0.969	4	0.058	6.0
27.5	4.78	1	0.084	1.8
27.5	4.78	4	0.056	1.2

Index

Page numbers followed by *t* refer to tabular entries; page numbers preceded by A refer to appendix references.

USEFUL CONVERSION FACTORS AND RELATIONSHIPS

Length	Mass
SI unit: Meter (m) 1 kilometer = 1000. meters = 0.62137 mile 1 meter = 100. centimeters 1 centimeter = 10. millimeters 1 nanometer = 1.00×10^{-9} meter 1 picometer = 1.00×10^{-12} meter 1 inch = 2.54 centimeter (exactly) 1 Ångstrom = 1.00×10^{-10} meter	**SI unit: Kilogram (kg)** 1 kilogram = 1000. grams 1 gram = 1000. milligrams 1 pound = 453.59237 grams = 16 ounces 1 ton = 2000. pounds

Volume	Pressure
SI unit: Cubic meter (m^3) 1 liter (L) = 1.00×10^{-3} m^3 = 1000. cm^3 = 1.056688 quarts 1 gallon = 4.00 quarts	**SI unit: Pascal (Pa)** 1 pascal = 1 N/m^2 = 1 $kg/m \cdot s^2$ 1 atmosphere = 101.325 kilopascals = 760. mmHg = 760 torr = 14.70 lb/in^2

Energy	Temperature
SI unit: Joule (J) 1 joule = 1 kg m^2/s^2 = 0.23901 calorie = 1 C × 1 V 1 calorie = 4.184 joules	**SI unit: kelvin (K)** T (K) = T(°C)(1 K/°C) + 273.15 K T (°C) = (5°C/9°F)[T(°F) − 32°F] T (°F) = (9°F/5°C)T(°C) + 32°F

SYMBOLS FOR COMMON PHYSICAL AND CHEMICAL QUANTITIES

A	absorbance, area	S/N	signal to noise ratio
a	absorptivity, activity	T	transmittance, temperature
B	magnetic field	t	time
C	capacitance	V	dc voltage, volume
D	diffusion coefficient	v	ac voltage, velocity
d	diameter, spacing	X	reactance
E	electrical potential, energy	$\overline{x}$	sample mean
e^-	electron	Z	impedance
F	faraday		
f	frequency		
G	conductance, free energy		
H	enthalpy		
I	dc current	γ	(gamma) activity coefficient
i	ac current	ϵ	(epsilon) molar absorptivity
K	equilibrium constant	λ	(lambda) wavelength
L	inductance	μ	(mu) population mean
n	number of moles	ν	(nu) frequency
n	spectral order	$\overline{\nu}$	wavenumber
P	radiant or electrical power	ρ	(rho) density
Q	quantity of dc charge	σ	(sigma) population standard deviation
q	quantity of ac charge	τ	(tau) period
R	electrical resistance, gas constant	ϕ	(phi) phase angle
s	sample standard deviation	ω	(omega) angular velocity